DYNAMICS OF CHOLINERGIC FUNCTION

ADVANCES IN BEHAVIORAL BIOLOGY

Recent Volumes in this Series

A Continuation Order Plan is available for this series. A continuation order will bring delivery of each new volume immediately upon publication. Volumes are billed only upon actual shipment. For further information please contact the publisher.

DYNAMICS OF CHOLINERGIC FUNCTION

Edited by
ISRAEL HANIN
Loyola University Stritch School of Medicine
Chicago, Illinois

PLENUM PRESS • NEW YORK AND LONDON

Library of Congress Cataloging in Publication Data

Conference on Dynamics of Cholinergic Function (1983: Oglebay Park, W. Va.)
  Dynamics of cholinergic function.
  (Advances in behavioral biology; v. 30)

  "Based on the Conference on Dynamics of Cholinergic Function, held October 31–
November 4, 1983, at Wilson Lodge at Oglebay Park, West Virginia"—T.p. verso.
  Includes bibliographies and index.
  1. Cholinergic mechanisms—Congresses. 2. Cholinergic receptors—Congresses. I.
Hanin, Israel. II. Title. III. Series. [DNLM: 1. Cholinergic Fibers—physiology—con-
gresses. 2. Parasympatholytics—metabolism—congresses. WL610 C748d 1983]
QP364.7.C66 1983   599'.01'88   86-18732
ISBN 0-306-42384-7

Based on the Conference on Dynamics of Cholinergic Function,
held October 31–November 4, 1983,
at Wilson Lodge at Oglebay Park, West Virginia

© 1986 Plenum Press, New York
A Division of Plenum Publishing Corporation
233 Spring Street, New York, N.Y. 10013

PREFACE

This book incorporates the proceedings of the Fifth International Cholinergic Conference, which took place in Oglebay Park, West Virginia, USA, on October 30th to November 4th, 1983. A scenic forty-five minute ride from the City of Pittsburgh, surrounded by championship golf courses, luxurious woods and a picturesque lake, Oglebay provided relaxed and beautiful surroundings, conducive to contemplation, stimulating discussions and, thought-provoking scientific sessions.

Over 160 individuals from all over the world participated in the sessions. The meeting was sub-divided into oral presentations, round table discussions and poster sessions, and centered upon ten key topics of cholinergic relevance (see Table of Contents).

Following in the tradition of the four International Conferences which had preceded this one, the Conference featured the most up-to-date developments in the area of cholinergic mechanisms, and provided for ample and productive discussion of new fields and directions in this area. Moreover, both senior investigators in the field as well as recent newcomers to this sphere of investigation participated in the proceedings.

This book, touches on a wide array of mechanisms and applications - from the preclinical to the clinical level. It should thus be useful as a comprehensive resource, with the cholinergic system as a focal hub.

The Conference could not have been as successful as it turned out to be, without the support of a number of important contributors. These include the U.S. Army Medical Research and Development Command, (Grant #DAMD 17-83-G-9534), the National Institute on Aging, (Grant #1 R13 AG04301-01), the Fogarty International Center, American Cyanamid Company (Lederle Laboratories), Astra Lakemedel AB, Ciba-Geigy Corporation, Fidia Research Laboratories, Hoechst-Roussel Pharmaceuticals, Inc., KabiVitrum AB, Thomas J. Lipton, Inc., Merck Sharp & Dohme Research Laboratories, UCB Secteur Pharmaceutique, Unilever Research Laboratorium and The Upjohn Company.

I am particularly grateful to members of the Scientific Committee, which consisted of: John P. Blass, M.D., Ph.D. (USA), Alan M. Goldberg, Ph.D. (USA), Edith Heilbronn, Ph.D. (Sweden), Bo Holmstedt, M.D. (Sweden), Donald J. Jenden, M.D. (USA), Alexander G. Karczmar, M.D., Ph.D. (USA), Giancarlo Pepeu, M.D. (Italy), Earl Usdin, Ph.D. (USA), and Victor P. Whittaker, M.D. (Germany). They provided valuable suggestions pertaining to the scientific program of the Conference, and assisted in the planning for, and execution of the Conference. My extra special thanks in this regard go to Giancarlo Pepeu and his charming wife Ileana, as well as to my good friend Alan Goldberg.

All the chapters in this book went through an extensive process of review, revision, and retyping, in an attempt to provide uniformity in style, format and quality to the book. This explains the reason for the lag time between the Conference itself, and the eventual publication of its proceedings. We decided, however, to sacrifice haste in the interest of quality.

Special recognition goes to Jacqueline Molinaro, Diana Donnelly, Carol Kristofic, and especially to Joyce O'Leary for carrying out their monumental task of retyping and formatting these entire proceedings.

Last but not least, the success of a conference is dependent upon the participants themselves. The diverse international cholinergic community was well represented at this meeting. It is hoped that this Conference served the purpose of bringing together old friends, helped to create new acquaintanceships, and resulted in a number of new, fruitful, collaborative ventures.

Israel Hanin, Ph.D.
Chicago, IL   USA

CONTENTS

I. CHOLINERGIC PATHWAYS:
ANATOMY OF THE CENTRAL NERVOUS SYSTEM

II. AGING, SDAT AND OTHER CLINICAL CONDITIONS

III.   CHOLINERGIC PRE- AND POST-SYNAPTIC RECEPTORS

IV. ACETYLCHOLINE RELEASE

V. CHOLINESTERASES, ANTICHOLINESTERASES AND REACTIVATORS

VI. ACETYLCHOLINE SYNTHESIS, METABOLISM AND PRECURSORS

VII.   SECOND MESSENGER MECHANISMS

VIII.   INTERACTION OF ACETYLCHOLINE
WITH OTHER NEUROTRANSMITTER SYSTEMS

IX. CHOLINERGIC MECHANISMS IN PHYSIOLOGICAL FUNCTION, INCLUDING CARDIOVASCULAR EVENTS

CHOLINERGIC SYSTEMS IN THE CENTRAL NERVOUS SYSTEM:

RETROSPECTION, ANATOMIC DISTRIBUTION, AND FUNCTIONS

L.L. Butcher and N.J. Woolf

Department of Psychology
and Brain Research Institute
Los Angeles, California 90024 U.S.A.

INTRODUCTION

In 1906, Reid Hunt (Fig. 1), in a paper with Taveau (23), made the commentary, remarkable for its prescience, that:

> "I frequently obtained extracts of the suprarenal (and also of the brain) which caused a fall of blood pressure...and which were also more powerful than cholin...I also got results...which led me to think that at least some of these results were to be attributed to a precursor of cholin or to some compound of cholin...From these observations it seemed not impossible that...cholin compounds...might arise in the body; and, further, that such compounds may have some importance in certain pathological conditions...Acetylcholin, the first of this series, is a substance of extraordinary physiological activity. In fact, I think it safe to state that, as regards its effect upon the circulation, it is the most powerful substance known...We have not determined the cause of the fall of blood pressure from acetyl-cholin, but from the fact that it can be prevented entirely by atropine, I am inclined to think that it is due to an effect upon the terminations of the vagus in the heart." (p. 1789)

The important contributions that Hunt made to early work on acetylcholine were recognized by Dale in 1914 (11):

1

Figure 1. Reid Hunt, cholinergic scientist nonpareil. Hunt was born on April 20, 1870, in Martinsville, Ohio, and he died on March 10, 1948, in Belmont, Massachusetts.

> *"Reid Hunt found evidence of the existence of a substance in the supra-renal gland, which was not choline itself, but easily yielded that base in the process of extraction. If acetylcholine, however, or any substance of comparable activity, existed in the supra-renal gland in quantities sufficient for chemical detection, its action would inevitably overpower that of adrenine in a gland extract."* (pp. 188-189)

Dale took a less equivocal point of view in 1938, however, when more data became available and when armed with the wisdom of hindsight:

> *"If Reid Hunt...had examined more in detail the action of acetylcholine, the intense depressor activity of which he described...I think it must have been realized that acetylcholine would be a more suitable and likely parasympathetic transmitter than muscarine."* (p. 614)

Since the time of Hunt's obversations and conjectures and of the immediately succeeding era, considerable experimental evidence has accumulated indicating that acetylcholine is involved importantly in intercellular communication processes in both the central and peripheral nervous systems (for recent reviews, see refs. 4, 6,

14, 21, 25, 33, 35, 40, 43).   Indeed, considerable progress has been made toward understanding the biochemistry, physiology, and pharmacology of this messenger.

Results have been less compelling, however, for the anatomy of central cholinergic systems, the chief reason being, until recently, an absence of histochemical methods for identifying unequivocally cholinergic neurons and their projections (4, 6, 7). Although sensitive and specific chemical and biochemical procedures exist for most of the known cholinergic indices -
including acetylcholine (26), choline (26) and high affinity uptake of choline (e.g., see ref. 29), choline-O-acetyltransferase (ChAT; e.g., see ref. 17), acetylcholinesterase (AChE; e.g., see ref. 40), nicotinic (e.g., see ref. 37) and muscarinic (38, 49) receptors-histochemical procedures for these same entities, with the exception of AChE and cholinergic receptors are either non-existent, as is the case for acetylcholine and choline, or are in their infancy, as is the current situation for ChAT and methods based on high affinity uptake (4, 6).

Those histochemical procedures based on the use of radiolabelled compounds (cholinergic receptors, high affinity uptake of choline), although useful for identifying neuronal sites where cholinergic mechanisms may operate, typically evince poor cellular morphology (e.g., see ref. 10), and, accordingly, are not preferred for detailed anatomic investigations.   This leaves histochemical procedures for two cholinergic indices, those for ChAT and AChE, that possess the dual advantages of demonstrating acceptable morphologic detail while being useful for identifying cholinergic neurons at the same time (4, 6, 7).   Both of these histochemical approaches will be emphasized in the present chapter.

ANATOMY OF BRAIN CHOLINERGIC SYSTEMS; METHODS AND RESULTS

Using a combined procedure that permist the visualization on the same tissue section of (a) retrogradely and anterogradely transported neuronal labels; (b) immunohistochemically demonstrated ChAT (polyclonal and monoclonal antibodies); (c) AChE (pharmacohistochemical regimen); and (d) Nissl substance and myelin (5, 7, 8, 48), we have mapped the distribution and projection patterns of putative cholinergic neurons in the central nervous system of the rat. These neurons evince two organizational schemata: those that are intrinsically organized, local circuit cells and those that are projection neurons. Local circuit neurons are found in the caudate-putamen complex, nucleus accumbens, olfactory tubercule, and possibly also the cerebral cortex and hippocampus (7, 46). Five separable, but not necessarily mutually exclusive, systems of presumed cholinergic projection neurons have been ascertained: (a) the basal forebrain cholinergic system (components: medial septal nucleus, nuclei of the vertical and horizontal limbs of the

diagonal band, ventral pallidum/lateral preoptic area, nucleus preopticus magnocellularis, substantia innominata, nucleus basalis,
and nucleus of the ansa lenticularis) projecting widely to the cerebral
cortex, hippocampus, olfactory bulbs, amygdala, and brainstem, particularly the habenular nuclei and interpeduncular nucleus (for
detailed discussion of the anatomy and projection patterns of this
system, see refs. 3, 7, 47, 48); (b) the pedunculopontine cholinergic
system and dorsolateral tegmental nucleus projecting to the thalamus,
habenula, interpeduncular nucleus, pretectal nuclei, subthalamic
nucleus, lateral hypothalamus, entopeduncular nucleus, globus pallidus,
and basal forebrain cholinergic complex (7); (c) the brainstem reticular formation that putatively provides, in part, cholinergic afferents
to the spinal cord (7); (d) the various somatic and parasympathetic
cell bodies of cranial nerves III-VII and IX-XII (6, 7); and (e)
various somata in the spinal cord, including prominently the α-
and γ-motor neurons of the ventral horn and the preganglionic sympathetic neurons of the intermediolateral cell column at thoracic
and lumbar levels (6, 7). Some of these systems are schematically
represented in Fig. 2.

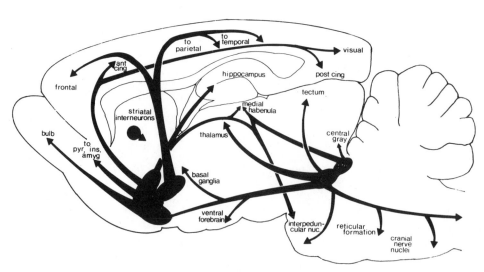

Figure 2. Schematic representation of some of the major cholinergic
cellular aggregates and projections, including their trajectories,
in the rat central nervous system. Abbreviations: ms, medial septal
nucleus; td, nuclei of the diagonal band; poma, nucleus preopticus
magnocellularis; si, substantia innominata; bas, nucleus basalis;
cing, cingulate cortex; pyr, pyriform cortex; ins, insular cortex;

amyg, amygdala; tpp, penduculopontine cholinergic complex; dltn, dorsolateral tegmental nucleus.

FUNCTIONS OF CENTRAL CHOLINERGIC SYSTEMS

Despite the fact that the neuroanatomy of central cholinergic systems is becoming increasingly understood, there is presently a relative paucity of information concerning their functions. Nonetheless, various experimental findings suggest that some of these systems are involved in neuronal processes, both normal and pathologic, operating at the highest levels of brain physiology and integration, including, among other functions, learning and memory, sleep and wakefulness, and motor behavior (Table 1).

Table 1

Some Normal and Pathologic Behaviors and Physiologic
Processes in which Acetylcholine has been Implicated

Function/Dysfunction	Selected Experimental Evidence
Motor Activity	
(a) Posture, reflexes, and gait	Spinal neurons are cholinergic [36]; intrathecal injections of acetylcholine or neostigmine depress spinal reflexes [28]; oxotremorine produces ataxia and spasticity [20].
(b) Catalepsy	Systemic administration of arecoline produces catalepsy [9].
(c) Tremor	Intracaudate infusions of carbachol, eserine, and DFP and systemic injection of oxotremorine elicit tremors [20, 30].
(d) Circling	Intracarotid injections of DFP produce contralateral circline [18].
Temperature Regulation	Infusions of carbachol or acetylcholine and eserine into the hypothalamus produce hyperthermia in monkeys [34].
Nociception	Oxotremorine, DFP, and pilocarpine produce analgesia [20, 27].

Table 1 (continued):

Ingestive Behavior

   (a)   Drinking   Injections of carbachol into the hypothalamus, preoptic area, septum, and hippocampus elicit drinking in rats [31].

   (b)   Feeding   Intrahypothalamic infusions of carbachol elicit feeding in rabbits [41].

Aggression   Neostigmine injections into the lateral hypothalamus increase mouse-killing by rats [1]; "rage" is produced in cats by infusions of carbachol into the anteromedial hypothalamus [2]; ChAT inhibition reduces isolation-induced aggression in mice (see ref. [39]); oxotremorine produces a rage-like state in cats and monkeys [20].

Sleep and Wakefulness   Depletion of acetylcholine with hemicholinium-3 decreases amounts of REM sleep [15]; acetylcholine release from cortex is greatest during EEG desynchrony of wakefulness and REM sleep [24].

Learning and Memory   Cerebroventricular injections of hemicholinium-3 increase number of trials to criterion in a conditioned avoidance task [39]; physostigmine produces retrograde amnesia [22]; scopolamine impairs recall in humans [16].

Disease States

   (a)   Parkinsonism   Atropine ameliorates tremor [19].

   (b)   Huntington's Chorea   Physostigmine and choline decrease involuntary movements [13]; benztropine exacerbates those movements [44].

   (c)   Schizophrenia   Physostigmine and arecoline ameliorate symptoms [32].

   (d)   Mania   Predominantly euphoric manics become less manic after physostigmine [13].

Table 1 (continued):

(e) Alzheimer's Disease   ChAT and AChE are reduced in
     and Senile Dementia   the cerebral cortex and hippocampus
     of the Alzheimer Type   [42]; cholinergic neurons in the
                                     basal forebrain project to the
                                     cortex, hippocampus and amygdala
                                     [3, 47, 48]; neurons in the basal
                                     forebrain cholinergic system undergo
                                     degeneration [45].

CONSPECTUS

      Until recently, the anatomy of central cholinergic systems,
with the exception of the septo-hippocampal projection, was largely
unknown (e.g., see ref. 7).   Fortunately, the situation has now
changed dramatically. Much of the impetus for this renewed interest
in cholinergic neuroanatomy can be traced, we believe, to three
major breakthroughs: (a) the development and use of the pharmacohistochemical regimen for AChE; (b) the development and use of morphologically acceptable and specific methods for the immunohistochemical demonstration of ChAT; and (c) the development and use of
methods based on the combination of AChE and ChAT histochemistry
with anterograde and retrograde tracing procedures, other histologic
and histochemical thechniques, and chemical and biochemical methods
for ChAT and acetylcholine (for review, see ref. 7). Use of these
procedures has permitted confirmation of a cholinergic component
to the septo-hippocampal pathway and has allowed the discovery,
description, and delineation of major cholinergic projection systems
deriving from the basal forebrain, pedunculopontine complex, and
dorsolateral tegmental nucleus, among others, as well as detailed
analyses of the intrinsically organized constellation of cholinergic
cells in the caudate-putamen complex (e.g., see refs. 6, 7, 46,
47, 48). No doubt additional cholinergic systems and refinements
in the descriptions of existing cellular aggregates and pathways
will be ascertained and delineated as more data become available.
The linking of cholinergic mechanisms with human neuropathologic
disorders such as Alzheimer's disease further suggests that the
investigation of cholinergic neuroanatomy will remain an exciting
and fruitful area of study for many years to come, both in its own
right and in relation to analyses of the functions of central cholinergic systems.

Acknowledgements.   This research was supported by USPHS grant NS
10928.   Dr. Felix Eckenstein is thanked for pertinent discussions
and valuable collaboration.

REFERENCES

1. Bandler, R.J. (1969): Nature (Lond.) 224:1035-1036.
2. Baxter, B.L. (1966): Pharmacologist 8:205.
3. Bigl, V., Woolf, N.J. and Butcher, L.L. (1982):   Brain Res.
    Bull. 8:727-749.
4. Butcher, L.L. (1978):   In Cholinergic Mechanisms and Psycho-
    pharmacology, (ed) D.J. Jenden, Plenum Press, New York, pp.
    93-124.
5. Butcher, L.L. and Bilezikjian, L. (1975): Eur. J. Pharmacol.
6. Butcher, L.L. and Woolf, N.J. (1982): In chemical Transmission
    in the Brain, Progr. Brain Res., Vol. 55, (eds) R.M. Buijs, P.
    Pevet, D.F. Swaab, Elsevier Biomedical Press, Amsterdam, pp.
    3-40.
7. Butcher, L.L. and Woolf, N.J. (1984):   In Handbook of chemical
    Neuroanatomy, Vol. 2, (eds) A. Bjorklund, T. Hokfelt and M.J.
    Kuhar, Elseview Biomedical Press, Amsterdam, pp. -
8. Butcher, L.L., Talbot, K. and Bilezikjian, L. (1975): J. Neural
    Transmission 37:127-153.
9. Costall, B. and Olley, J.E. (1971): Neuropharmacology 10:297-
    306.
10. Csillik, B., Haarstad, V.B. and Knyihar, E. (1970): J. Histo-
    chem. Cytochem. 18:58-60.
11. Dale, H.H. (1914): J. Pharmacol. Exp. Ther. 6:147-190.
12. Dale, H.H. (1938): J. Mt. Sinai Hosp. 4:401-429.
13. Davis, K.L., Berger, P.A., Hollister, L.E., DoAmaral, J.R. and
    Barchas, J.D. (1978):   In Cholinergic Mechanisms and Psycho-
    pharmacology, (ed) D.J. Jenden, Plenum Press, New York, pp.
    755-779.
14. DeFeudis, F.V. (1974): Central Cholinergic Systems and Behaviour,
    Academic Press, New York.
15. Domino, E.F., Yamamoto, K. and Dren, A.T. (1968): In Anticholi-
    nergic Drugs and Behavior Functions in Animals and Man, Progr.
    Brain Res., Vol. 28, (eds) P.B. Bradley and M. Fink, Elsevier,
    Amsterdam, pp. 113-133.
16. Drachman, D.A. and Leavitt, J.L. (1974): Arch. Neurol. 30:113-
    121.
17. Fonnum, F. (1975): J. Neurochem. 24:407-409.
18. Freedman, A.M. and Himwich, H.E. (1949): Am. J. Physiol. 156:125-
    128.
19. Friedman, A.H. and Everett, G.M. (1964):   In Advances in Phar-
    macology, Vol. 3, (eds) S. Garattini and P.A. Shore, Academic
    Press, New York, pp. 83-127.
20. George, R., Haslett, W.L. and Jenden, D.J. (1962):   Life Sci.
    1:361-363.
21. Goldberg, A.M. and Hanin, I. (eds) (1976):   Biology of Chol-
    inergic Function, Raven Press, New York.
22. Hamburg, M.D. (1967): Science 156:973-974.
23. Hunt, R. and Taveau, R. de M. (1906): Brit. Med. J. 2:1788-
    1791.

24. Jasper, H.H. and Tessier, J. (1971): Science 172:601-602.
25. Jenden, D.J. (ed) (1978): Cholinergic Mechanisms and Psychopharmacology, Plenum Press, New York.
26. Jenden, D.J. and Hanin, I. (1974): In Choline and Acetylcholine: Handbook of Chemical Assay Methods, (ed) I. Hanin, Raven Press, New York, pp. 135-150.
27. Karczmar, A.G. (1978): In Cholinergic Mechanisms and Psychopharmacology, (ed) D.J. Jenden, Raven Press, New York, pp. 679-708.
28. Kremer, M. (1942): Quart. J. Exp. Physiol. 31:337-357.
29. Kuhar, M.J. (1973): Life Sci. 13:1623-1634.
30. Lalley, P.M., Rossi, G.V. and Baker, W.W. (1970): Exp. Neurol. 27:258-275.
31 Levitt, R.A. and Boley, R.P. (1970): Physiol. Behav. 5:693-695.
32. Lloyd, K.G. (1978): In Cholinergic-Monoaminergic Interactions in the Brain, (ed) L.L. Butcher, Academic Press, New York, pp. 363-392.
33. Morgane, P.J. and Stern, W.C. (1974): In Advances in Sleep Research, Vol. I, (ed) E. Weitzman, Spectrum, New York, pp. 1-31.
34. Myers, R.D. and Yaksh, T.L. (1969): J. Physiol. (Lond.) 202:483-500.
35. Pepeu, G. and Ladinsky, H. (eds) (1981): Cholinergic Mechanisms: Phylogenetic Aspects, Central and Peripheral Synapses, and Clinical Significance, Plenum Press, New York.
36. Phillis, J.W. (1970): The Pharmacology of Synapses, Pergamon Press, Oxford.
37. Polz-Tejera, G., Schmidt, J. and Karten, H.J. (1975): Nature (Lond.) 258:349-351.
38. Rotter, A., Birdsall, N.J.M., Burgen, A.S.V., Field, P.M., Hulme, E.C. and Raisman, G. (1979): Brain Res. Rev. 1:141-165.
39. Russell, R.W. (1978): In Cholinergic Mechanisms and Psychopharmacology, (ed) D.J. Jenden, Raven Press, New York, pp. 709-731.
40. Silver, A. (1974): The Biology of Cholinesterases, American Elsevier, New York.
41. Sommer, S.R., Novin, D. and Levine M. (1967): Science 156:983-984.
42. Terry, R.D. and Daview, P. (1980): Ann. Rev. Neurosci. 3:77-95.
43. Waser, P.G. (ed) (1975): Cholinergic Mechanisms, Raven Press, New York.
44. Weiner, W.J. and Klawans, H.L. (1978): In Cholinergic-Monoaminergic Interactions in the Brain, (ed) L.L. Butcher, Academic Press, New York, pp. 335-362.
45. Whitehouse, P.J., Price, D.L., Clark, A.W., Coyle, J.T. and DeLong, M.R. (1981): Ann. Neurol. 10:122-126.
46. Woolf, N.J. and Butcher, L.L. (1981): Brain Res. Bull. 7:487-507.

47. Woolf, N.J. and Butcher, L.L. (1982): Brain Res. Bull. 8:751-
     763.
48. Woolf, N.J., Eckenstein, F. and Butcher, L.L. (1983): Neurosci.
     Lett. 40:93-98.
49. Yamamura, H.I. and Snyder, S.H. (1974): Proc. Nat. Acad. Sci.
     U.S.A. 71:1725-1729.

CHOLINERGIC NEURONS AND CHOLINERGIC PROJECTIONS

IN THE MAMMALIAN CNS

P.L. McGeer, E.G. McGeer, H. Kimura and J.-F. Peng
Kinsmen Laboratory of Neurological Research
University of British Columbia
Vancouver, B.C., Canada

INTRODUCTION

Considerable information regarding the distribution of cholinergic neurons in the mammalian central nervous system has now been accumulated. This is due to the development of reliable markers for these cells. The most significant one is immunohistochemical staining for choline acetyltransferase (ChAT), the enzyme for acetylcholine synthesis. Many laboratories have now produced antibodies to ChAT from various sources (13, 14, 17, 18, 19, 37, 47, 58, 66, 69, 70). They are of two general kinds: polyclonal, from the serum of mammals immunized against enzyme purified from various ChAT-rich sources, and monoclonal, raised by the hybridoma technique using mouse myeloma cell lines. Two important ancillary techniques are histochemistry for acetylcholinesterase (AChE), particularly after diisopropylfluorophosphate (DFP) administration, and retrograde choline transport. The former method is based on the observation that many known cholinergic cells have higher levels of AChE than cholinoceptive cells and this difference can be revealed by the re-synthesis rate of AChE following administration of an irreversible inhibitor such as DFP. The retrograde choline transport technique is based upon the high affinity uptake system for choline characteristic of cholinergic nerve terminals.

Each technique has its assets and limitations. Immunohistochemistry, despite its elegance, is laden with potential artifactual problems as well as difficulties in sensitivity. Background staining, introduced by the many proteins that are involved in the multiple sandwich techniques that are commonly used, is always present. The antibodies themselves may be of marginal titer to reveal the specific structures that are sought, and the large molecular weights of the proteins create penetration problems. Appropriate fixation is necessary for any good histochemical method, but that fixation may destroy or impair the very protein recognition sites that form the basis

11

of the method. Monoclonal antibodies, because they are directed against a single recognition site, are particularly vulnerable to these weaknesses.

A major difficulty in ChAT immunohistochemistry is that there seem to be many axosomatic terminals and, if the fixation is not good, the staining of the terminals on a cholinoceptive neuron cannot readily be distinguished from the cell body staining of a cholinergic neuron (32, 33). This may be a major factor underlying some of the confusion now existing in the literature.

The cholinesterase histochemical technique, while much simpler to carry out, suffers from the enormous disadvantage of not being specific for cholinergic cells. The high concentration of cholines-terase in some non-cholinergic groups, such as the catecholamine neurons of the substantia nigra and locus coeruleus (1, 7, 49), raises questions about cholinergic specificity of every high AChE-containing cell group. Retrograde choline transport is, as yet, of unproven specificity. Only a small amount of the choline entering brain is directed at the production of acetylcholine, and the diverse uses of this molecule make it likely that it will find its way into other neurons as well as cholinergic ones. Nonetheless, the technique does show promise in that it has been shown to reveal a number of known cholinergic pathways.

The technique of lesioning is a classical one for establishing specific biochemical neuronal pathways in brain. It is the use of this technique which helped establish the presence of the first known cholinergic pathway in brain, from the septum to the hippo-campus (20, 39, 67). However, a problem with this technique is the possibility of interrupting fibers of passage. Thus, the early suggestion that a cholinergic tract extended from the habenula to the interpeduncular nucleus has needed to be revised on the basis that cholinergic fibers emanating from the medial basal forebrain are apparently interrupted by habenular lesions (23). The use of kainic or ibotenic acid as a lesioning tool has helped to alleviate some potential problems with this technique because these excitotoxins spare axons of passage, and has allowed resolution of some doubtful data on pathways in cholinergic as well as many other systems (12, 41).

At this stage, a combination of techniques is required to es-tablish the credibility that any given cell group is truly cholinergic.

CHOLINERGIC STRUCTURES

The most comprehensive maps of cholinergic structures in brain so far proposed are those of Kimura et al. (32, 33) for the cat and rat. These were prepared using Fab fragments of high titer monospecific antibodies to ChAT isolated from the serum of rabbits

immunized against human neostriatal ChAT (60). Many of these same
structures have also been revealed in the rat and human through
the use of monoclonal antibodies to human neostriatal ChAT prepared
by the standard hybridoma technique in BALB/c mice (45, 55).

Based on the available data to date, at least four major cholinergic systems, with a high probability of a fifth, are known to
exist in brain. In addition, a number of other, minor systems should
be included (see Figure 1).

The major, confirmed cholinergic systems in brain are the following:

1) The medial forebrain complex. This is a more or less continuous
sheet of giant cholinergic cells which starts anteriorly on the

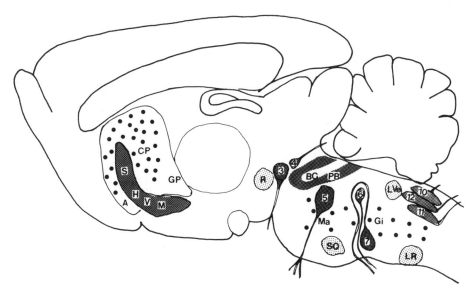

Figure 1. Sagittal view of rat brain illustrating ChAT-containing
neurons. Major systems are indicated by black dots or heavy stippling, minor ones by light stippling.
    Figure Captions: A-Nucleus accumbens; Am-Amygdala; BC-Brachium
conjunctivum; CP-Caudate-putamen; Gi-Gigantocellular division of
the reticular formation; GP-Globus pallidus; H-Horizontal limb of
the diagonal band; Ha-Habenula; IC-Inferior colliculus; Ip-Inter-
peduncular nucleus; LR-Lateral reticular nucleus; LVe-Lateral vestibular nucleus; M-Nucleus basalis of Meynert; Ma-Magnocellular
division of the reticular formation; PB-Parabrachial complex; R-
Red nucleus; S-Medial septum; SN-Substantia nigra; SO-Superior ol-
ive; V-Vertical limb of the diagonal band

medial surface of the cortex and extends in a caudo-lateral direction, always maintaining its position close to the medial and ventral
surfaces of the brain, and terminating towards the caudal aspect
of the lentiform nucleus. The names usually given to the various
sub-regions of this complex from rostral to caudal order are the
following: medial septal nucleus, nucleus of the vertical limb of
the diagonal band of Broca, nucleus of the horizontal limb of the
diagonal band of Broca, and the nucleus basalis of Meynert (S, V,
H and M in Figure 1). This cell complex, originally confirmed to
be cholinergic in the rat by immunohistochemistry using polyclonal,
monospecific rabbit antisera (34), has now been confirmed by at
least four groups using polyclonal and monoclonal antibodies in
various species (3, 25, 46, 71). We have additionally studied and
reported on this cell group in cats and humans using both Fab fragments
from rabbit polyclonal antibodies and monoclonal antibodies to human
striatal ChAT (45, 54, 55). In addition, this is a very prominent
cell complex revealed by the AChE-DFP histochemical technique (6,
27, 36, 61).

2) **Striatal interneurons.** It has long been known from lesions studies
that an internal system of cholinergic neurons exists in the striatum
(43). A group of giant cells within the striatum has now been identified as being cholinergic both by ChAT immunohistochemistry (32,
33) and by AChE-DFP histochemistry (8, 35, 73). These giant cells
have also been stained by other groups using ChAT immunohistochemistry
(2, 24, 71). The ChAT and AChE levels of the caudate, putamen,
and accumbens are extremely high but these giant neurons represent
no more that 1% of the neuronal population (29). It cannot be said
with certainty that there are not other types of cholinergic cells
in the striatum although these cells are certainly the most prominent
ones.

3) **Motor nuclei for peripheral nerves.** Cholinergic cells, equivalent
to anterior horn cells, exist in cranial nerve nuclei 3-7 and 9-12
as the source for efferent fibers to skeletal muscle and autonomic
ganglia (3, 32, 33, 52). The counterparts in the spinal cord, in
the anterior and lateral horns, also stain positively (44).

4) **Parabrachial (pedunculopontine) system (PB in Fig. 1).**   This
system of cholinergic cells is the most intense and concentrated
one in the brain stem. It surrounds the brachium conjunctivum com-
mencing in the most rostral aspects of the pons and following the
direction of the brachium conjunctivum (superior cerebellar peduncules)
in a caudodorsal direction. Various subnuclei are separately identified in this particular region. The most commonly described nucleus
is the pedunculopontine tegmental nucleus in the lateral aspect at
the most rostral portion of the complex. Caudal to this nucleus
are three adjoining nuclei, the dorsal parabrachial, the Kolliker-Fuse,
and the ventral parabrachial nuclei. They surround the cerebellar
peduncles in the dorsal, lateral, and ventral regions, respectively.

Slightly, more dorsally placed are the cuneiform nucleus and the dorsolateral tegmental nucleus. Armstrong et al. (3) and Mufson et al. (52) have also reported cells in this area to be positive for ChAT. These neurons also stained intensely by the AChE-DFP method (H. Kimura, personal communication).

5) Reticular formation cells. In addition to these four, extensively confirmed systems, a fifth, which should be classified as major, exists as a scattered collection of very large cells extending throughout the gigantocellular and magnocellular tegmental fields of the reticular formation (Gi and Ma in Fig. 1). Caudally, the giganto and magnocellular ChAT containing neurons are gradually aggregated medially towards the grannular layer of the raphe and ventrally to the area near the inferior olivary nucleus. Thus, these cells extend continuously from the pons into the medulla as a longitudinally oriented cluster. Using antibodies to ChAT, the only group beyond our own reporting these to stain positively is that of Cozzari and Hartman (13) using an antibody to bovine caudate nucleus. However, these cells do stain intensely by the AChE-DFP method (65, 74). They have also been shown to concentrate labeled choline following spinal cord injections (57).

The minor systems in the brain which are probably also cholinergic are the following:

1). **Magnocellular neurons of the red nucleus.** These neurons, in addition to staining in our laboratory with polyclonal and monoclonal anti-ChAT antibodies in the cat and rat (32, 33), have also been reported to stain positively by Sofroniew et al. (72) using monoclonal antibodies to the rat. These magnocellular neurons have also been shown to concentrate labeled choline by retrograde flow (57). This would suggest that the reticulospinal tract is at least partially cholinergic.

2). **Lateral reticular nucleus.** These neurons stained positively in our hands (32, 33) and also concentrated labeled choline by retrograde flow (57). They are also AChE-DFP positive (LR in Fig. 1).

3). **Olivary neurons.** The nuclei of the superior olivary complex of both cat and rat stain positively by our methods (32, 33) (SO in Fig. 1).

4). **Vestibular nuclei.** We found positive staining of Deiters nucleus in the cat (32) and the lateral vestibular nucleus in the rat (33) (LVe in Fig. 1).

5). Small ChAT positive cells were noted in the **intermedial-medial region**, near the central canal in the rat both by us (33) and by Houser et al. (24).

The only structures reported as positive but not included in Figure 1 are the **ventromedial cells of the medial habenula** (24), the small **interneurons of the sensory cortex** (16, 24, 25) and the hippocampal interneurons mentioned by Matthews et al. (48). McGeer et al. (44) and Hattori et al. (23) had noted positive **cortical and medial habenula neurons** but uncertainty about distinguishing between cholinergic and intensely cholinoceptive neurons was raised by the results obtained following improvements in the methods of fixation (32, 33). It should be anticipated that some differences in immunohistochemical results will occur between laboratories since particular antibodies may recognize different sites on ChAT, or because the conformation of ChAT may vary according to the method of fixation or its location within the CNS. Thus, extensive confirmation, using multiple techniques, will be required before structures can definitively be accepted as cholinergic. In these particular cases, the evidence from other histochemical (65) and lesion studies is controversial. This evidence, as well as literature on cholinergic pathways, has been recently reviewed (40, 42) and only very pertinent or later references will be cited in the following paragraphs.

CHOLINERGIC PATHWAYS

At this stage less is known about cholinergic pathways than about cholinergic cell groups. Acetylcholine and ChAT are ubiquitous in brain and therefore every area contains some cholinergic terminals from one of the main cell groups (9, 32, 33). The known or suspected pathways which serve these various areas are depicted in Fig. 2. Solid arrows indicate the well documented pathways while the dotted arrows indicate those pathways where more work remains to be done.

The most extensively studied cholinergic cell mass is of the mediobasal forebrain which projects to the neo-, paleo- and archicortex. The projection of the most rostromedial part of the complex to the hippocampus was the first to be identified (39). Projections from the caudolateral areas, including the nuclei of the diagonal band and the substantia innominata, were revealed later (15, 27, 30). The relationship of these pathways to the cholinergic system was originally suggested by Shute and Lewis (78), based on AChE staining. Their concept was reinforced by work showing that lesions of the area cause reductions in ChAT (26, 35). Since then, the distribution has been extensively mapped using retrograde tracing methods in the monkey (27, 59), retrograde tracing combined with ChAT immunohistochemistry and with AChE-DFP histochemistry in the monkey (50) or with only AChE-DFP histochemistry in the rat (6, 27, 35, 51). Heavy projections also go to the amygdala (53). The projections are more or less topographically arranged but with considerable overlap of the various cortical fields. The ChAT-positive terminals and cholinoceptive neurons seen in various regions of the cat cortex have been mapped in detail (31).

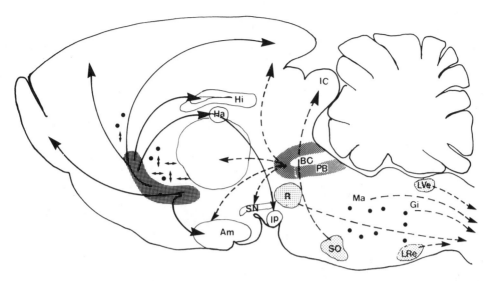

Figure 2. Sagittal view of rat brain indicating cholinergic pathways. Well established routes are indicated by solid arrows; suspected routes by dotted arrows. **Figure Captions:** See Figure 1.

One of the most intense cholinergic projections in the rat is to the interpeduncular nucleus. One source for this projection comes from the above mentioned forebrain complex (11, 21, 22). Lesions of this area, however, do not totally eliminate ChAT from the interpeduncular nucleus, raising the probability that a second source exists. This second source must also traverse the fasciculus retroflexus because lesions of that tract almost totally eliminate ChAT in the interpeduncular nucleus (28). A possible source for the remainder may be the medial habenula where ChAT positive cells have been reported (23, 24) although other staining and lesion evidence is controversial (40, 42).

The extrapyramidal systems are known to be intrinsic to the basal ganglia as previously described.

The parabrachial-pedunculopontine system is the major cholinergic cell group requiring intensive elucidation. The rostral aspect, the pedunculopontine nucleus, projects to the substantia nigra (5) and this pathway appears to be partly cholinergic (38). Projections from this area have been described to the cortex (2, 63) which is also innervated by some fibers from the Kolliker-Fuse nucleus (63). Saper and Lowey (64) have also described efferent projections from these nuclei into the midline, medial and ventral basal thalamic nuclei, the hypothalamus and amygdala. Therefore, it would appear as if this region is a major supplier of cholinergic afferents to all parts of the diencephalon, particularly the intralaminar and midline thalamic nuclei. There may also be descending cholinergic projections (64).

The magnocellular division of the red nucleus is the principle source of the descending rubrospinal tract. It crosses in the ventro-tegmental decussation and intermingles with the corticospinal tract. Fibers terminate on the somata and dendrites of large and small cells within lamina 5, 6, and 7 of the dorsal and central parts of the spinal cord. Collateral fibers enter into the cerebellum, and the lateral reticular nucleus (10). Magnocellular and gigantocellular cells of the reticular formation also descend in the spinal cord, terminating in lamina 7, 8, and 9 (4, 10, 56). Ascending reticular fibers from the nucleus reticularis gigantocellularis also send ascending fibers primarily to the thalamus (10).

The lateral reticular nucleus projects to the cerebellum and to the spinal cord. The superior olivary nucleus projects to the inferior colliculus (10).

These then are the main known projection areas of nuclei that contain cholinergic cells. Other populations than the cholinergic ones are undoubtedly contained in these cellular masses and, therefore, the projection fields can, at this stage, only be considered tentatively cholinergic. However, retrograde choline transport from the spinal cord to the lateral reticular nucleus and to the large cells of the reticular formation strongly support the notion that at least some of these fibers are cholinergic (57).

REFERENCES

1. Albanese, A. and Butcher, L.L. (1979): Neurosci. Lett. 14:101-104.
2. Armstrong, D.M., Saper, C.B., Levey, A.I., Wainer, B.H. and Terry, R.D. (1982): Soc. Neurosci. Abstrs. 8:662.
3. Armstrong, D.M., Saper, C.B., Levey, A.I., Wainer, B.H. and Terry, R.D. (1983): J. Comp. Neurol. 216:53-68.
4. Basbaum, A.I., Clanton, C.H. and Fields, H.L. (1978): J. Comp. Neurol. 178:209-224.
5. Beckstead, R.M., Domesick, V.B. and Nauta, W.J.H. (1979): Brain Res. 175:191-217.
6. Bigl, V., Woolf, N.J. and Butcher, L.L. (1982): Brain Res. Bull. 8:727-729.
7. Butcher, L.L. and Marchand, R. (1978): Eur. J. Pharmacol. 52:415-417.
8. Butcher, L.L., Talbot, K. and Bilezikjian, L. (1975): J. Neural. Transm. 37:127-153.
9. Chao, L.P., Kan, K.J.K. and Hung, F.-M. (1982): Brain Res. 235:65-82.

10. Carpenter, M.B. and Sutin, J. (1983): Human Neuroanatomy, 8th edition, Williams and Wilkins, Baltimore, pp. 291, 296, 334, 372.
11. Contestible, A. and Fonnum, F. (1983): Brain Res. 275:287-297.
12. Coyle, J.T., Molliver, M.E. and Kuhar, M.J. (1978): J. Comp. Neurol. 180:301-324.
13. Cozzari, C. and Hartman, B.K. (1977): Proc. Intl. Soc. Neurochem. 6:140.
14. Dietz, G.W. Jr. and Salvaterra, P.M. (1980): J. Biol. Chem. 255:10612-10617.
15. Divac, I. (1975): Brain Res. 9:385-398.
16. Eckenstein, F. and Thoenen, H. (1983): Neurosci. Lett. 36:211-215.
17. Eckenstein, F. and Thoenen, H. (1982): The EMBR Journal 1:363-368.
18. Eng, L.F., Uyeda, C.T., Chao, L.P. and Wolfgram, F. (1974): Nature 250:243-245.
19. Fex, J., Altschuler, R.A., Parakkal, M.H. and Eckenstein, F. (1982): Soc. Neurosci. Abstrs. 8:41.
20. Fonnum, F. (1970): J. Neurochem. 17:1029-1037.
21. Gottesfeld, Z. and Jacobowitz, D.M. (1978): Brain Res. 156:329-332.
22. Gottesfeld, Z. and Jacobowitz, D.M. (1979): Brain Res. 176:391-394.
23. Hattori, T., McGeer, E.G., Singh, V.K. and McGeer, P.L. (1977): Exp. Neurol. 55:666-679.
24. Houser, C.R., Crawford, G.D., Barber, R.P., Salvaterra, P.M. and Vaughn, J.E. (1983): Brain Res. 266:97-119.
25. Houser, C.R., Crawford, G.D., Salvaterra, P.M. and Vaughn, J.E. (1983): Soc. Neurosci. Abstrs. 9:576.
26. Johnston, M.V. and Coyle, J.T. (1979): Brain Res. 170:135-155.
27. Jones, E.G., Burton, H., Saper, C.B. and Swanson, L.W. (1976): J. Comp. Neurol. 167:385-420.
28. Katoka, K., Nakamura, Y. and Hassler, R. (1973): Brain Res. 62:264-267.
29. Kemp, J.M. and Powell, T.P.S. (1971): Phil. Trans. B. 262:383-401.
30. Kievet, J. and Kuypers, H.G.J. (1975): Science 187:660-662.
31. Kimura, H., McGeer, E.G., Peng, F. and McGeer, P.L. (1983): In: Structure and Function of peptidergic and Aminergic Neurons, (eds) Y. Sano, Y. Ibata and E.Z. Zimmerman. Japan Scientific Societies Press, Tokyo, pp. 26, 274.
32. Kimura, H., McGeer, P.L., Peng, J.H. and McGeer, E.G. (1981): J. Comp. Neurol. 200:151-201.
33. Kimura, H., McGeer, P.L. and Peng, J.H. (1984): In: Handbook of Chemical Neuroanatomy, (eds) A. Bjorklund and T. Hokfelt. (eds) Elsevier/North Holland Biomedical Press B.V.
34. Kimura, H., McGeer, P.L., Peng, J.H. and McGeer, E.G. (1980): Science 208:1057-1059.
35. Lehmann, J. and Fibiger, H.C. (1978): J. Neurochem. 30:615-624.
36. Lehmann, J.C., Nagy, J.I., Atmadja, S. and Fibiger, H.C. (1980): Neuroscience 5:1161-1174.

37. Levey, A.I., Aoki, M., Fitch, F.W. and Wainer, B.H. (1981): Brain Res. 218:383-387.
38. McGeer, E.G., McGeer, P.L. and Staines, W.A. (1984): Can. J. Neurol. Sci. 11:89-99.
39. McGeer, E.G., Wada, J.A., Terao, A. and Jung, E. (1969): Exp. Neurol. 24:277-284.
40. McGeer, P.L., Kimura, H., McGeer, E.G. and Peng, J.H. (1982): In: Compartmentation of Cholinergic Systems in the Central Nervous System, (ed) H. Bradford. Plenum Press, N.Y. and Lond., pp. 255-289.
41. McGeer, P.L., and McGeer, E.G. (1982): In: Critical Reviews in Toxicology, Vol. 10, (ed) L. Goldberg, CRC Press, Boca Raton, Florida, pp. 1-26.
42. McGeer, P.L. and McGeer, E.G. (1984): In: Handbook of Neurochemistry, Vol. 6, (ed) A. Lajtha, Plenum Press, New York, pp. 379-410.
43. McGeer, P.L., McGeer, E.G., Fibiger, H.C. and Wickson, V. (1971): Brain Res. 35:308-314.
44. McGeer, P.L., McGeer, E.G., Singh, V.K. and Chase, W.H. (1974): Brain Res. 81:373-379.
45. McGeer, P.L., McGeer, E.G., Suzuki, J., Dolman, C.E. and Nagai, T. (1984): Neurology 34:741-745.
46. Maley, B., Elde, R., Wainer, B. and Levey, A. (1982): Soc. Neurosci. Abstrs. 8:518.
47. Malthe-Sorenssen, D., Lea, T., Fonnum, F. and Eskeland, T. (1978): J. Neurochem. 30:35-46.
48. Matthews, D.A., Salvaterra, P.M., Crawford, G.D., Houser, C.R. and Vaughn, J.E. (1983): Soc. Neurosci. Abstrs. 9:79.
49. Meibach, R.C. and Weaver, L.M. (1979): J. Neural Transm. 44:87-96.
50. Mesulam, M.M., Mufson, E.J., Levey, A.I. and Wainer, B.H. (1983): J. Comp. Neurol. 214:170-197.
51. Mesulam, M.M. and Van Hoesen, G.W. (1976): Brain Res. 109:152-157.
52. Mufson, E.J., Levey, A., Wainer, B. and Mesulam, M.M. (1982): Soc. Neurosci. Abstrs. 8:135.
53. Nagai, T., McGeer, P.L. and McGeer, E.G. (1982): J. Neurosci. 2:513-520.
54. Nagai, T. McGeer, P.L., Peng, J.H., McGeer, E.G. and Dolman, C.E. (1983): Neurosci. Lett. 36:195-199.
55. Nagai, T., Pearson, T., Peng, F., McGeer, E.G. and McGeer, P.L. (1983): Brain Res. 265:300-306.
56. Nyberg-Hansen, R. (1964): J. Comp. Neurol. 122:355-368.
57. Pare, M.F., Jones, B.E. and Beaudet, A. (1982): Soc. Neurosci. Abstrs. 8:517.
58. Park, D.H., Ross, M.E., Pickel, V.M., Reis, D.J. and Joh, T.H. (1982): Neurosci. Lett. 34:129-135.
59. Pearson, R.C.A., Gatter, K.C., Brodal, P. and Powell, T.P.S. (1983): Brain Res. 259:132-136.

60. Peng, J.H., Kimura, H., McGeer, P.L. and McGeer, E.G. (1981): Neurosci. Lett. 21:281-285.

61. Peng, J.H., McGeer, P.L., Kimura, H., Sung, S.C. and McGeer, E.G. (1980): Neurochem. Res. 5:943-961.

62. Peng, J.H., McGeer, P.L. and McGeer, E.G. (1982): J. Neuro-immunol. 3:113-121.

63. Saper, C.B. (1982): Brain Res. 242:33-40.

64. Saper, C.B. and Loewy, A.D. (1980): Brain Res. 197:297-317.

65. Satoh, K., Armstrong, D.M. and Fibiger, H.C. (1983): Soc. Neurosci. Abstrs. 9:80.

66. Shuster, L. and O'Toole, C. (1974): Life Sciences 15:645-656.

67. Shute, C.C.D. and Lewis, P.R. (1963): Nature 199:1160-1164.

68. Shute, C.C.D. and Lewis, P.R. (1967): Brain 90:497-522.

69. Singh, V.K. and McGeer, P.L. (1974): Life Sciences 15:901-913.

70. Slemmon, J.R., Salvaterra, P.M., Crawford, G.D. and Roberts, E. (1982): J. Biol. Chem. 257:3847-3852.

71. Sofroniew, M.V., Edkenstein, F., Thoenen, H. and Cuello, A.C. (1982): Neurosci. Lett. 33:7-12.

72. Sofroniew, M.V., Eckenstein, F., Thoenen, H. and Cuello, A.C. (1982): Soc. Neurosci. Abstrs. 8:516.

73. Vincent, S.R., Staines, W.A. and Fibiger, H.C. (1983): Neurosci. Lett. 35:111-114.

74. Vincent, S.R., Satoh, K. and Fibiger, H.C. (1983): Soc. Neurosci. Abstrs. 9:576.

CORTICAL CHOLINERGIC INNERVATION: DISTRIBUTION AND SOURCE IN MONKEYS

R.G. Struble[*], J.Lehmann[+], S.J. Mitchell[#], L.C. Cork[**], J.T. Coyle[++], D.L. Price[##], M.R. DeLong[***] and P.G. Antuono[+++]

[*]Neuropathology Laboratory, Department of Pathology; [+]Department of Psychiatry and Behavioral Sciences; [#]Department of Neurology; [**]Division of Comparative Medicine, Neuropathology Laboratory, and Department of Pathology; [++]Departments of Neuroscience, Pediatrics, Pharmacology and Experimental Therapeutics, and Psychiatry and Behavioral Sciences; [##]Neuropathology Laboratory, Departments of Pathology, Neurology and Neuroscience; [***]Departments of Neurology and Neuroscience; [+++]Department of Psychiatry and Behavioral Sciences

The Johns Hopkins University School of Medicine
600 North Wolfe Street
Baltimore, Maryland 21205 U.S.A.

INTRODUCTION

Cholinergic neurons of the primate basal forebrain, termed the Ch1-4 system (18), can be identified by their chromophilia, intensity with which they stain for acetylcholinesterase (AChE) activity, and content of choline acetyltransferase (ChAT)-like immunoreactivity (10, 13, 18, 20). These cholinergic nerve cells are located in the medial septal nucleus (Ch1), nucleus of the diagonal band of Broca (dbB) (Ch2), nucleus basalis of Meynert (nbM) (Ch4), and in laminae around the globus pallidus. Anterograde and retrograde transport studies in primates (e.g., 12, 18, 21, 27) and combined tracing and AChE histochemistry (17) show that these neurons project directly to cortex, and suggest that neurons in Ch1, Ch2, and Ch4 provide the major source of cholinergic innervation of the amygdala, hippocampus, and neocortex.

In Alzheimer's disease (AD) and its late-life variant, senile dementia of the Alzheimer's type (SDAT), the predominant neurochemical abnormalities are marked decrements in the activities of ChAT and

23

AChE, high affinity uptake of [^{3}H]choline, and synthesis of acetylcholine (3). Postmortem studies of patients with AD/SDAT have shown that these neurons of the Ch system clearly are at risk, although the extent of cell loss appears to vary with the age of the patient at death, with younger patients showing a greater cell loss than older patients (22, 26, 29, 31). The evidence for dysfunction and death of neurons of the Ch1-4 system in AD/SDAT points out the need for better understanding of the patterns and sources of cortical cholinergic innervation.

Consequently, we undertook two studies to delineate more clearly the variability of cortical cholinergic innervation and the contribution of the Ch system, particularly the Ch4, to this cholinergic innervation. In the first study, ChAT activity was assessed in multiple samples of neocortex from seven normal cynomolgus monkeys. In the second study, the nbM was lesioned in order to determine the contribution of the Ch system to cortical cholinergic innervation.

REGIONAL ChAT ACTIVITY IN PRIMATE NEOCORTEX

Seven female cynomolgus monkeys were anesthetized with pentobarbital. After the brain was removed from the calvarium, cuts (approximately 2 mm per side) perpendicular to the pia were made through the full extent of cortex from 16 separate neocortical sites, and fine forceps were used to separate the cortical tissue from the underlying white matter. After freezing with dry ice and storing at -80°C, tissue samples were homogenized and assayed for both ChAT and glutamic acid decarboxylase (GAD) activity (4, 9, 32).

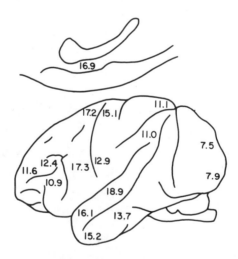

Figure 1. Distribution of ChAT activity in monkey neocortex. Overlying the site of the sample, the values of enzyme activity (nmol/mg/hr) represent the mean of seven animals.

There were consistent regional variations in ChAT activity (Fig. 1); temporal and motor cortices showed the highest levels of enzyme activity, with prefrontal cortex having intermediate levels, and the values in occipital cortex being lowest. These differences were statistically significant (f = 12.6; df = 15, 90; p < 0.005). In addition, there was considerable interanimal variability in average ChAT activity, ranging from 0.8- to 1.2-fold of the group mean (f = 19,1; df = 6, 90; p < 0.005). GAD activity did not show a regional variation, although there were significant interanimal differences (p < 0.005).

CONTRIBUTIONS OF Ch4 TO ChAT ACTIVITY IN PRIMATE NEOCORTEX

Before lesions in the basal forebrain system were made, it was first necessary to locate neurons of the nbM. The location of neurons in the Ch system were first mapped electrophysiologically (19) and then multiple injections of ibotenic acid were placed in the main body of the nbM in two rhesus and one cynomolgus monkey. Ibotenic acid was used to make the lesion, because this toxin destroys neuronal perikarya at the injection site while sparing axons of passage (25). Since cortical projections from the Ch system are almost totally ipsilateral (21), we compared reductions in cortical ChAT activity ipsilateral to the lesion with the enzyme activity in the homologous cortical region contralateral to the lesion. Seven to ten days after injection of the excitotoxin, animals were anesthetized and prepared as above; samples were obtained bilaterally as in the previous mapping study from comparable cortical areas. These samples were assayed for ChAT and, in some cases, GAD. This latter enzyme was included to control for the possibility of direct damage to cortex or a nonspecific decrease in cortical enzymatic markers. After cortical samples were obtained, the remaining brain was immersed in buffered formalin. To localize the lesion and assess both basal forebrain and cortical AChE activities, sections were stained with cresyl violet or reacted to demonstrate AChE activity.

The intended target of the excitoxic lesions was the nbM and, in particular, that portion of the nbM lying subjacent to the globus pallidus and the anterior commissure. Lesions varied from a relatively small lesion in a subcomponent of Ch4 in one animal to substantial destruction of the major portion of Ch4 in another animal. Figure 2 shows maps of ChAT activity following lesions of Ch4, with values expressed as a percent of ChAT activity in the homologous contralateral site of two lesioned animals.

With the smaller lesion (Fig. 2A) located within the anterior nbM, there was a relatively modest reduction in cortical ChAT activity ipsilateral to the lesion (maximal levels of 32% in the ventral motor cortex). With the larger lesion involving substantial parts of the nbM, cortical ChAT activity was decreased up to 69%in the dorsal motor cortex and 50-60% in the prefrontal cortex.

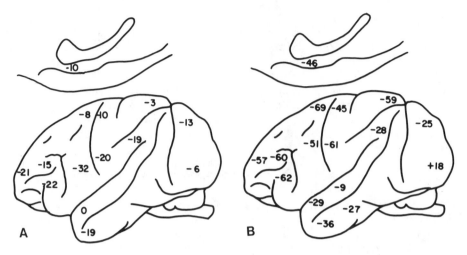

Figure 2. Patterns of cortical ChAT activity decrease following ibotenic acid lesions of the basal forebrain. Numerical values are percent reductions on the side ipsilateral to the lesion as compared to the normal contralateral side. A: This drawing shows changes in ChAT activity following a small lesion in the subpallidal region of the nbM. Note that there are modest decrements of ChAT activity, primarily in the ventral motor cortex. B: A more extensive lesion, involving a large portion of the nbM, causes substantial decreases in cortical ChAT activity ipsilateral to the lesion, particularly in the dorsal motor and prefrontal cortices.

CHOLINERGIC INNERVATION OF NORMAL NEOCORTEX

ChAT activity is not homogeneously distributed throughout the neocortex of monkeys. A greater than two-fold variation is present between regions of highest and lowest activities. This finding contrasts with results reported for rodents (11), in which ChAT activity appears to be homogeneously distributed throughout the neocortex. These data do not provide information on the variability within a particular gyrus; studies are presently in progress to address this question. This finding does indicate that there are significant species differences in the distribution of cholinergic innervation. The observation that there is considerable heterogeneity in cholinergic innervation in primate cortex emphasizes the importance for careful sampling of cortical regions in primates. This heterogeneity is particularly important in assessing changes in ChAT activity in pathological material.

CONTRIBUTION OF THE Ch SYSTEM TO CORTICAL CHOLINERGIC INNERVATION

Since cortical ChAT activity is reduced up to 70% in some cortical regions following incomplete lesions of Ch4, we conclude that the majority of cortical cholinergic ChAT activity represents enzymes derived from axons of the Ch system. This maximum reduction of 70% is similar to that occurring in rats following large subcortical lesions (11, 15) or cortical undercuttings (8).

Changes and patterns of ChAT activity can be interpreted to show a topography of cholinergic innervation from the nbM (Ch4). With one lesion involving a limited part of Ch4, there was a reduction in ChAT activity, most severe in the ventral part of the motor cortex (Fig. 2A). When the more caudal and lateral parts of Ch4, including neurons in the laminae of the globus pallidus were lesioned (not shown), there was a significant decrement in ChAT activity in the temporal and inferior visual cortices. This finding is consistent with retrograde transport studies which have shown that, following injections into visual cortex, border cells in the posterior part of Ch4 are labeled (27). Finally, when the lesion (Fig. 2B) involved the dbB and anterior portion of the nbM, there were major reductions in ChAT activity in motor and prefrontal cortices. Observations derived from anterograde and retrograde transport studies are in agreement with these data (14, 18, 21). Based on our studies and those in the literature, we have schematically depicted the general topography of Ch projections to cortex (Fig. 3).

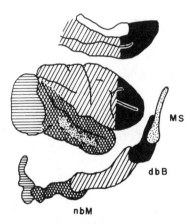

Figure 3. This diagram shows the lateral surface of the cortex of a nonhuman primate and subcomponents of the Ch system, labeled medial septum (Ch1), diagonal band of Broca (Ch2), and nucleus basalis of Meynert (Ch4). On the basis of our lesion studies and anatomical investigations, a highly simplified schematic topography of Ch1-4 projections is depicted.

Although we have presented arguments for a topographical distribution of Ch projections, a highly collateralized system cannot
be excluded rigorously on the basis of the present data. The finding
that >60% reductions in some regions of cortex are not accompanied
by reductions in other cortical regions of the same animal suggests
that it is unlkely that the majority of Ch4 neurons give rise to
diffuse neocortical projections. This conclusion is in agreement
with studies using double-retrograde labeling which have not found
major amounts of double labeling (16, 24, 33).

To date, the function of the Ch1-4 system (particularly Ch4)
in behavior is unclear. In 1975, Divac (7) suggested that this
system might play a role in the electrical activity of the brain,
in behavior, and perhaps in reward reinforcement or attention processes. Studies by DeLong et al. (5, 6) suggest that these large
neurons change firing frequencies during reward and feeding. Presently, it is unclear whether alterations in firing are related to
the reward itself or to the motor activity associated with the reward.
Variability in cortical cholinergic innervation does suggest that
cholinergic modulation of cortical functioning is not uniform, e.g.,
cholinergic modulation of cortical functions may be more important
in tasks involving temporal or motor cortices (both of which have
generally high levels of ChAT activity) and less important for visual
functions.

CONCLUSIONS

In senile dementia of the Alzheimer type (AD/SDAT), the presence
of cognitive deficits correlates with the magnitude of cortical
cholinergic abnormalities (23, 28), suggesting that the cholinergic
deficit in AD/SDAT is a significant factor in the expression of
dementia. Quantitative studies of numbers of neurons in the Ch1-4
system are consistent with this hypothesis (2, 26, 30, 31). The
data presented herein confirm that the Ch4 is the major source for
cortical cholinergic innervation in primates, and behavioral studies
of lesioned animals should allow better appreciation of the role
of cholinergic dysfunctions in cognitive abnormalities (1). As we
begin to understand the role of the Ch system in normal and abnormal
behaviors, the contributions of pathologies involving other neurotransmitter systems should also become clearer. With this information,
we may be able to develop therapeutic strategies for cholinergic
dysfunctions and for treatments of other transmitter-related abnormalities occurring in individuals with AD/SDAT.

Acknowledgements. The authors acknowledge helpful discussions with
Drs. Cheryl A. Kitt, John C. Hedreen, William C. Mobley, Lary C.
Walker, Bruce H. Wainer, and Peter J. Whitehouse. The authors thank
Mrs. Carla Jordon and Ms. Nancy Cook for their assistance with the
preparation of the manuscript. Ms. Barbara Holden provided excellent
technical assistance. This work was supported by grants from the

U.S. Public Health Service (NIH AG 03359, NS 07179, NS 15721, MH 00868, MH 26654, NS 18414, NS 15417, and ES 07894) and a grant from The Johns Hopkins University. Drs. Cork and Coyle are recipients of Research Career Development Awards (NS 00488 and MH 00125, respectively).

REFERENCES

1. Aigner, T., Aggleton, J., Mitchell, S.J., Price, D., DeLong, M. and Mishkin, M. (1983): Soc. Neurosci. Abstr. 9:826.
2. Arendt, T., Bigl, V., Arendt, A. and Tennstedt, A. (1983): Acta Neuropathol. (Berlin) 61:101-108.
3. Bowen, D.M. (1983): In Banbury Report 15: Biological Aspects of Alzheimer's Disease (ed) R. Katzman, CSH, pp. 219-231.
4. Bull, G. and Oderfeld-Nowak, B. (1971): J. Neurochem. 18: 935-941.
5. DeLong, M.R. (1971): J. Neurophysiol. 34:414-427.
6. DeLong, M.R. (1972): Brain Res. 40:127-135.
7. Divac, I. (1975): Brain Res. 93:385-398.
8. Fibiger, H.C. and Lehmann, J. (1981): In Cholinergic Mechanisms: Phylogenetic Aspects, Central and Peripheral Synapses, and Clinical Significance. Advances in Behavioral Biology, Vol. 25 (eds) G. Pepeu and H. Ladinsky, Plenum Press, New York, pp. 663-672.
9. Fonnum, F. (1975): J. Neurochem. 24:407-409.
10. Hedreen, J.C., Bacon, S.J., Cork, L.C., Kitt, C.A., Crawford, G.D., Salvaterra, P.M. and Price, D.L. (1983): Neurosci. Lett. 43:173-177.
11. Johnston, M.V., McKinney, M. and Coyle, J.T. (1981): Exp. Brain Res. 43:159-172.
12. Kievit, J. and Kuypers, H.G.J.M. (1975): Brain Res. 85:261-266.
13. Kitt, C.A., Hedreen, J.C., Bacon, S.J., Becher, M.W., Salvaterra, P.M., Levy, A.I., Wainer, B.H. and Price, D.L. (1983): Soc. Neurosci.Abstr. 9:966.
14. Kitt, C.A., Price, D.L., DeLong, M.R., Struble, R.G., Mitchell, S.J. and Hedreen, J.C. (1982): Soc. Neurosci. Abstr. 8:212.
15. Lehmann, J., Nagy, J.I., Atmadja, S. and Fibiger, H.C. (1980): Neuroscience 5:1161-1174.
16. McKinney, M., Coyle, J.T. and Hedreen, J.C. (1983): J. Comp. Neurol. 217:103-121.
17. Mesulam, M.M. and Van Hoesen, G.W. (1976): Brain Res. 109: 152-157.
18. Mesulam, M.M., Mufson, E.J., Levey, A.I. and Wainer, B.H. (1983): J. Comp. Neurol. 214;170-197.
19. Mitchell, S.J., Richardson, R.T., Baker, F.H. and DeLong, M.R. (1982): Soc. Neurosci. Abstr. 8:212.
20. Parent, A., Gravel, S. and Olivier, A. (1979): In Advances in Neurology, Vol. 24 (eds) L.J. Poirier, T.L. Sourkes and P.J. Bedard, Raven Press, New York, pp. 1-11.
21. Pearson, R.C.A., Gatter, K.C., Brodal, P. and Powell, T.P.S.

(1983):   Brain Res. 259:132-136.

22.   Perry, R.H., Candy, J.M. and Perry, E.K. (1983):   Banbury
      Report 15: Biological Aspects of Alzheimer's Disease (ed) R.
      Katzman, CSH, pp. 351-361.
23.   Perry, E.K., Tomlinson, B.E., Blessed, G., Bergmann, K.
      Gibson, P.H. and Perry, R.H. (1978):   Br. Med. J. 2:1457-1459.
24.   Price, J.L. and Stern, R. (1983):   Brain Res. 269:352-356.
25.   Schwarcz, R., Hokfelt, T., Fuxe, K., Johsson, G., Goldstein,
      M. and Terenius, L. (1979):   Exp. Brain Res. 37:199-216.
26.   Tagliavini, F. and Pilleri, G. (1983):   Lancet 1:469-470.
27.   Tigges, J., Tigges, M., Cross, N.A., McBride, R.L., Letbetter,
      W.D. and Anschel, A. (1982):   J. Comp. Neurol. 209:29-40.
28.   Tomlinson, B.E., Blessed, G. and Roth, M. (1970):   J. Neurol.
      Sci. 11:205-242.
29.   Whitehouse, P.J., Hedreen, J.C., Jones, B.E. and Price, D.L.
      (1983):   Ann. Neurol. 14:149-150.
30.   Whitehouse, P.J., Hedreen, J.C., White, C.L. III and Price,
      D.L. (1983):   Ann. Neurol. 13:243-248.
31.   Whitehouse, P.J., Price, D.L., Struble, R.G., Clark, A.W.,
      Coyle, J.T. and DeLong, M.R. (1982):   Science 215:1237-1239.
32.   Wilson, S.H., Schrier, R.K., Farber, J.L., Thompson, E.J.,
      Rosenberg, R.N., Blume, A.J. and Nirenberg, M.W. (1972):   J.
      Biol. Chem. 147:3159-3169.
33.   Woolf, N.J. and Butcher, L.L. (1982):   Brain Res. Bull. 8:
      751-763.

IDENTIFICATION OF PUTATIVE M_1 MUSCARINIC RECEPTORS USING [^{3}H]PIRENZEPINE: CHARACTERIZATION OF BINDING AND AUTORADIOGRAPHIC LOCALIZATION IN HUMAN STELLATE GANGLIA

M. Watson, W.R. Roeske, T.W. Vickroy, K. Akiyama,
J.K. Wamsley[*], P.C. Johnson and H.I. Yamamura

Departments of Pharmacology, Biochemistry, Psychiatry, Pathology,
Internal Medicine, and the Arizona Research Laboratories
University of Arizona Health Sciences Center
Tucson, Arizona 85724 U.S.A.
and
[*]Departments of Psychiatry, Anatomy, and Pharmacology
University of Utah School of Medicine
Salt Lake City, Utah 84132 U.S.A.

INTRODUCTION

The advent of selective antimuscarinic drugs such as pirenzepine has led to increased acceptance of the concept of distinct subclasses of muscarinic receptors. This chapter focuses upon our studies of [^{3}H]pirenzepine ([^{3}H]PZ), which we suggested is an effective ligand for the investigation of putative M_1 receptor- effector mechanisms (26-31). Data indicating that [^{3}H]PZ labels putative M_1 muscarinic receptors with high affinity is discussed in relation to the growing body of evidence which supports the M_1/M_2 hypothesis. Emphasis is placed upon our studies of human stellate ganglia. Several relevant reviews have recently appeared in the literature (3, 22, 23, 29).

A unique regional distribution of high affinity [^{3}H]PZ binding provides strong evidence for muscarinic receptor subtypes (31). Moreover, while ions exert potent effects upon [^{3}H]PZ binding, the selectivity exhibited by [^{3}H]PZ is insensitive to guanine nucleotide regulatory effects (27, 31). M_1 and M_2 sites are not readily interconvertible and it appears that the putative M_1 site may be linked to another effector such as phosphatidylinositol turnover while the putative M_2 site may be coupled to a guanine regulatory protein (23, 29, 31).

Agonist binding to the muscarinic acetylcholine receptor (AChR) is best described by a multiple affinity state model in most tissues (3, 22). Regulators such as guanine nucleotides, sulfhydryl reagents, and ions exert effects at closely related sites, thus altering the relative proportions of various affinity states (2, 3, 6, 14, 17, 20). However, tissue-specific effects suggest a differential regulatory mechanism of high affinity agonist binding properties of putative M_1/M_2 subtypes (22-24). Magnesium ions enhance, while N-ethylmaleimide (NEM) and guanyl-5'-yl imidodiphosphate [Gpp(NH)p] inhibit high affinity agonist binding in myocardial and cerebellar tissue. In contrast, magnesium ions and Gpp(NH)p demonstrate markedly weaker effects in the striatum and hippocampus and are virtually without effect on the high affinity muscarinic agonist binding site of the cerebral cortex. Interestingly, NEM enhances high affinity agonist binding in the hippocampus, striatum and cortex. The latter three tissues, unlike the former, also demonstrate significantly higher levels of high affinity [^{3}H]PZ binding (27-33). In concert with data obtained from kinetic studies (24, 27), these results further substantiate the M_1/M_2 concept, suggesting that distinct muscarinic AChR subtypes show differential agonist binding and regulation by interactions with separate allosteric sites, which still remain to be elucidated (22-24).

We have recently reviewed the binding of agonists to multiple muscarinic receptors (22) and the implications for different receptor-effector coupling mechanisms (23). Our present hypothesis is that there are multiple agonist affinity states for each of the muscarinic receptor subtypes (22).

Pirenzepine (Gastrozepin, LS 519), is a tricyclic compound which possesses selective antimuscarinic properties (4, 7, 9-12, 17, 26-33). It is clinically useful in the treatment of gastric ulcer disease and produces a significantly lowered incidence of typical antimuscarinic side effects (13). Although the high affinity site with which pirenzepine interacts to inhibit gastric acid secretion is not known, its hydrophilic nature allows only minimal penetration through the blood-brain-barrier, thus suggesting a peripheral site of action.

Interestingly, functional heterogeneity of muscarinic receptors was first demonstrated in neurons of sympathetic ganglia (5, 12). McNeil A-343 [4-(m-chlorophenylcarbamoyloxy)-2-butynyltrimethyl-ammonium chloride], which exhibited little effect upon muscarinic AChR's in the isolated heart of jejunum, was shown to evoke a rise in blood pressure by selectively stimulating muscarinic AChR's in sympathetic ganglia (16, 19). Other studies in ganglia, including various in vitro studies such as the report of tissue-specific muscarinic effects of the antagonist 4-dephenylacetoxy-N-methylpiperidine methiodide (4-DAMP) (1), and in vivo studies showing discriminating effects in the lower esophageal sphincter of the opossum (8) also provided early evidence of muscarinic receptor heterogeneity. However, results obtained from indirect binding studies of the non-classical antagonist pirenzepine (PZ), which suggested that PZ may be a selective muscarinic antagonist (9, 10), attracted more widespread interest. Studies of PZ produced many observations of complexities in the nature of antagonist binding (3, 9, 10, 26-33). Pharmacologic studies indicate that PZ shows over 20-fold greater affinity for muscarinic AChR's mediating depolarization of rat superior cervical ganglia than for muscarinic AChR's mediating smooth muscle contraction in rat ileal tissue (4). Similarly, while atropine and PZ were equi-effective inhibitors at the ganglionic site which mediates the rise in arterial pressure of the rat, PZ was nearly two orders of magnitude less potent in blocking the vagally-mediated bradycardia in the atrium (11). PZ was also shown to produce differential effects upon gastric and cardiac function in the dog (12).

We have now extended our previous findings regarding the selectivity of [³H]PZ binding in the central nervous system (28, 33) and various other peripheral tissues (30, 31) by characterizing the muscarinic receptor binding sites of human stellate ganglia (26). As illustrated in this chapter, we find that there are two populations of muscarinic sites within this tissue based upon the differential high affinity labeling of a sub-population of sites by [³H]PZ (26-33) as compared to the total number of sites labelled by the classical antagonist [³H](-)quinuclidinyl benzilate ([³H](-)QNB) (31) under similar conditions.

METHODS

Human stellate ganglia were obtained as previously described (26) and tissue samples were then stored at -20°C. Homogenates were prepared using a Brinkmann polytron (2x, 15sec with 30sec pause; setting 8.0) and filtered with cheesecloth. Protein content was determined according to the method of Lowry et al. (15) using bovine serum albumin as the standard.

The specific binding of [³H]PZ (75 Ci/mmole, New England Nuclear) and [³H](-)QNB (33.1 Ci/mmole, New England Nuclear) was determined using our previously described rapid filtration assay (31) in modified Krebs phosphate buffer (120mM NaCl, 4.8mM KCl, 1.2mM MgSO₄, 1.3 mM CaCl₂, 20.3 mM NaH₂PO₄, 3.2 mM HCl 10mM Dglucose, pH 7.4). For autoradiography, freshly excised ganglia were coated with plastic embedding medium (OCT Compound, Lab-Tek Products, Naperville, Illinois) prior to microtome sectioning. The ganglia were then frozen onto microtome chucks and slices were prepared essentially as described by Wamsley and Palacios (25). Each section (10 microns) was mounted onto chrome-alum/ gelatin-coated slides for preincubation in modified Krebs phosphate buffer for 30 mins at 25°C. Sections were labeled with either 20nM [³H]PZ or 1nM [³H](-)QNB for 60 or 120 mins respectively at 25°C, using 1 μM atropine for non-specific binding.

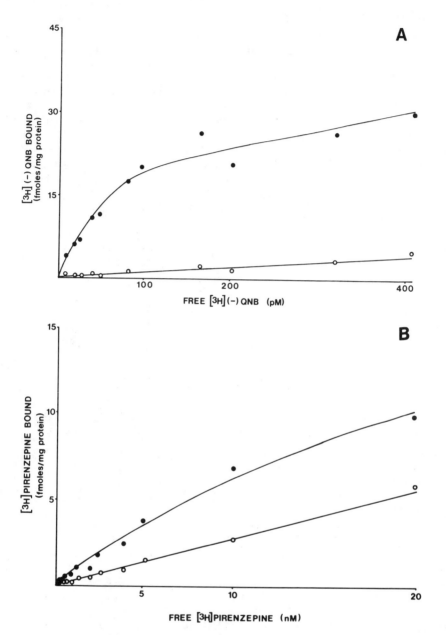

Figure 1. Homogenates of human stellate ganglia were incubated in modified Krebs phosphate buffer with various concentrations of either (A) [³H](-)QNB at 25°C for 2hr or (B) [³H]PZ at 25°C for 1hr. Values for specific (●) and nonspecific tissue binding (○), as distinguished by differences in the presence and absence of atropine (1 µM), represent the mean of six separate determinations obtained from six

Incubations were then terminated by rinsing each section once (PZ) or twice (QNB) for 5 mins (0-4°C) in buffer and another rinse for 1 min (0-4°C) in distilled water. After drying, autoradiographs were prepared by exposure of the sections to tritium-sensitive film (LKB-ultrofilm, LKB Products) for 3-4 months at 4°C.

RESULTS

The results of these experiments illustrate that there are muscarinic receptor binding sites in human stellate ganglia. [³H](-)QNB binding is depicted in Fig. 1A, and the parallel data for [³H]PZ is shown in Fig. 1B. Analysis of binding isotherms for [³H](-)QNB yields a K_d (dissociation constant) of 59pM and a B_{max} value (receptor density) of 4.0 fmol/mg tissue. [³H]PZ yields a high affinity K_d of 14nM and a B_{max} value of 2.1 fmol/mg tissue. Hill slopes for both ligands were approximately one, and no significant improvement in fit was obtained using a two- site binding model for either ligand.

The K_d values are in good agreement with our previous studies in various other tissues and different buffer conditions (26-33). Furthermore, over 50 percent of the number of sites labeled by [³H](-)QNB demonstrate high affinity [³H]PZ binding, providing further support for the concept that pirenzepine is a selective antagonist. Also, this is the highest proportion of [³H]PZ sites showing high affinity in any peripheral tissue we have examined to date. These sites also demonstrate pharmacologic specificity as indicated by our inhibition studies (Table 1).

These data are similar to our previous results in rat cerebral cortex using a centrifugation assay at 0°C in 50mM sodium-potassium phosphate buffer (28). Dexetimide is 500 times more potent than levetimide, thus demonstrating stereospecificity. Both atropine and pirenzepine strongly inhibit [³H]PZ binding. Carbamylcholine and McNeil-A-343, classical and non-classical agonists, respectively, are weaker.

Under the conditions used for our autoradiographic studies, specifically bound ligand accounted for nearly 90 percent of total binding. The results of these studies are illustrated in Figure 2. Both [³H]PZ and [³H](-)QNB binding appeared to be localized primarily within the principal ganglion cells and nerve bundles.

---

distinct ganglia. Curves indicate the least squares best fit line for these saturation isotherms. While equilibrium binding of [³H]-(-)QNB yielded a K_d=59 pM and a B_{max}=33.0 fmoles/mg protein, parallel assays of [³H]PZ yielded a K_d=14 nM and a B_{max}=16.7 fmoles/mg protein. (Reprinted from Brain Research with permission.)

Table 1

Inhibitory Effects of Muscarinic Drugs on [^{3}H]PZ Binding
in Human Stellate Ganglia[a]

Drug	IC$_{50}$(nM)
Atropine	0.5
Dexetimide	2.0
Pirenzepine	2.2
Levetimide	1000
McNeil A-343	5000
Carbamylcholine	11000

[a]Homogenates were incubated with 10nM [^{3}H]PZ in modified Krebs phosphate buffer with multiple concentrations of each drug at 25°C for 1hr in a 2ml final volume. Each point represents the mean specific [^{3}H]PZ binding from at least 3 separate determinations from 3 different ganglia (26).

Following earlier demonstrations of functional heterogeneity in neurons of sympathetic ganglia (5, 21) and the selective stimulant action of McN-A-343 (16, 19), Goyal and Rattan proposed the M$_1$ (for intramural ganglia) and M$_2$ (for smooth muscle) subclassification of muscarinic receptors to explain the discriminating effects they observed in the lower esophageal sphincter of the opossum (8). Indirect binding (9, 10) and pharmacologic studies of PZ (4, 11), along with our direct [^{3}H]PZ binding studies (26-31), are in good agreement with this concept. Recently, we have further confirmed the selectivity of PZ in autoradiographic studies of the rat brain (32) and in homogenates of cultured cells of the neuroblastoma glioma hybrid (NG 108-15).

Table 2 summarizes the results of data from NG 108-15 cells (submitted for publication) indicating [^{3}H]PZ labels approximately half the number of [^{3}H](-)QNB sites. In addition to providing further support for the concept of M$_1$ and M$_2$ muscarinic receptor subtypes and for the selectivity of PZ, these results from autoradiographic studies and binding in homogenates of ganglia suggest that PZ may influence gastrointestinal function indirectly via ganglia of the peripheral nervous system (26). Recent data indicating that PZ is only a weak inhibitor of acid secretion in isolated rat parietal cells (18) also support the contention that PZ's antisecretory effects are attributable to an indirect mechanism such as interaction with a high affinity site within intramural ganglia in the stomach or other peripheral ganglia (9).

Table 2

A Comparison of $[^3H](-)QNB$ and $[^3H]PZ$ Binding in
NG 108-15 Cells[a]

Ligand	N	K_d(nM)	B_{max} (fmol/mg prot)
$[^3H](-)QNB$	4	0.017 ± 0.0009	53.2 ± 1.4
$[^3H]PZ$	4	4.02 ± 1.50	27.8 ± 5.6

[a]Assays were conducted using 10 mM sodium-potassium phosphate buffer at 25°C for 90 mins in a final assay volume of 1ml using the previously described method of Watson et al. (31). Once washed homogenates were diluted to a final protein concentration of 0.3-0.4 mg per assay. Values ($\pm$SEM) for the K_d are the geometric mean, whereas B_{max} values are the arithmetic mean.

DISCUSSION AND SUMMARY

In support of the hypothesis that there is a subset of muscarinic receptors which are not critically linked to guanine nucleotides (6, 23, 27, 29-31), PZ binds with high affinity to a large proportion of sites in the hippocampus, corpus striatum, and cerebral cortex (31). A low density of high affinity PZ binding is found in tissues such as the cerebellum, heart and ileum, where low affinity PZ sites predominate and strong guanine nucleotide effects upon agonist binding have previously been noted (6, 22, 27-31). Thus, we have predicted that $[^3H]PZ$ will bind more favorably to tissues where little guanine nucleotide effect can be demonstrated (31).

The putative M_1 subtype, as defined by high affinity $[^3H]PZ$ binding, appears to be very weakly, if at all, linked to any guanine nucleotide regulatory protein. Guanine nucleotides do not modulate $[^3H]PZ$ binding and they produce only modest shifts (≤ 2 fold) of agonists to lower affinity in $[^3H]PZ$ labeled cerebral cortical membranes. M_1 effects may be more strongly linked to some other effector such as phosphatidylinositol turnover, while the M_2 subtype may be more closely associated with a guanine regulatory protein (29). Nonetheless, PZ discriminates between either different antagonist affinity states which are modulated by molecules associated with the receptor or detects small intrinsic differences in the primary structure of the receptor molecule itself (26, 33).

Until the determination is made whether structural dissimilarity, differential coupling, or some measure of both account for the selectivity exhibited by PZ and other compounds, an adequately precise definition of these muscarinic receptor subtypes will remain

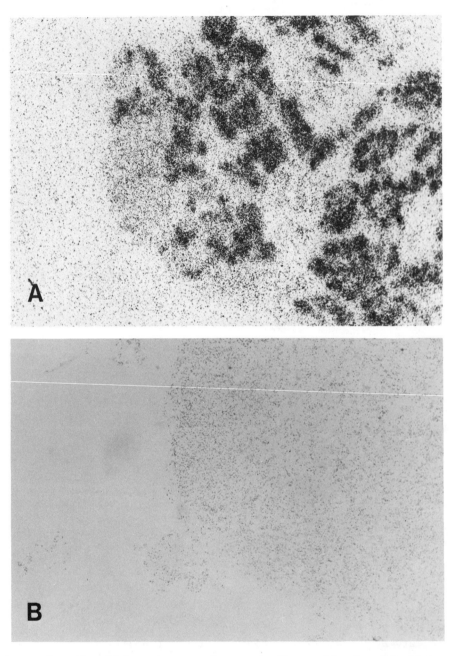

Figure 2. The light microscopic autoradiographic localization of [3H](-)QNB binding sites (A) and corresponding cresyl violet stained tissue section (B) is compared with specific [3H]PZ binding (C) and corresponding stained tissue section (D) in the human stellate ganglion

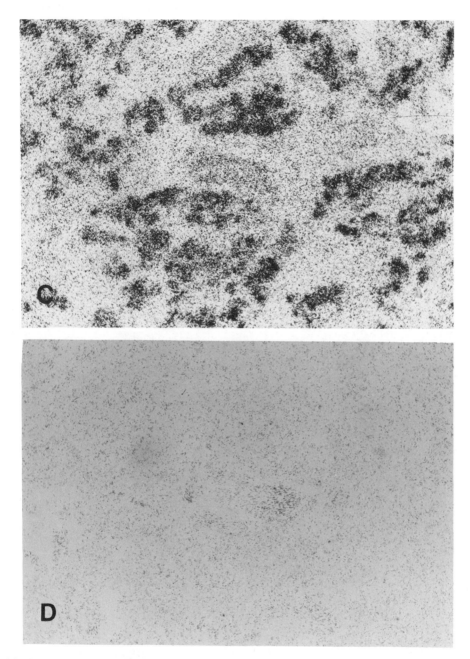

10 micron sections). Black grains indicate exposure to radioactivity from the muscarinic ligands associated primarily with the principal ganglionic neurons and nerve bundles (see text for methods).

elusive. However, the demonstration of heterogeneity by antagonists in the functionally relevant ganglion provides strong evidence for the growing literature in support of the concept of muscarinic AChR subtypes.

Acknowledgements. M. W. is a recipient of a Postdoctoral Fellowship Award and a Grant-in-Aid from the American Heart Association, Arizona Affiliate and an awardee of an Individual National Research Service Award from the NIMH (MH-09091). Portions of these studies were supported by USPHS grants MH-30626, MH-27257, HL-29565, Program Project Grant HL 20984, a RSDA Type II (MH-00095) from NIMH to H.I.Y., and an RSDA (HL-00776) from NHLBI to W.R.R. We thank Pat Gonzalez for fine secretarial assistance and Martha Ackerman for excellent technical help.

REFERENCES

1. Barlow, R.B., Berry, K.J., Glenton, P.A.M., Nikolaou, N.M. and Soh, K.S. (1976): Brit. J. Pharmacol. 58:613-620.
2. Berrie, C.P., Birdsall, N.J.M., Burgen, A.S.V. and Hulme, E.C. (1979): Biochem. Biophys. Res. Commun. 87:1000-1006.
3. Birdsall, N.J.M., Hulme, E.C., Hammer, R., and Stockton, J.S. (1980): In Psychopharmacology and Biochemistry of Neurotransmitter Receptors (eds) H.I. Yamamura, R. Olsen and E. Usdin, Elsevier/North-Holland, New York, pp. 97-100.
4. Brown, D.A., Forward, A. and March, J. (1980): Br. J. Pharmacol. 71:362-364.
5. Eccles, R.M. and Libet, B. (1961): J. Physiol. (London) 157:484-503.
6. Ehlert, F.J., Roeske, W.R. and Yamamura, H.I. (1980): J. Supramol. Struct. 14:149-162.
7. Evans, R.A., Watson, M., Yamamura, H.I. and Roeske, W.R. (1984): Clin. Res. 32:35A.
8. Goyal, R.K. and Rattan, S. (1978): Gastroenterol. 74:548-619.
9. Hammer, R. (1980): Scand. J. Gastroenterol. Suppl. 66:5-11.
10. Hammer, R., Berrie, C.P., Birdsall, N.J.M., Burgen, A.S.V. and Hulme, E.C. (1980): Nature (London) 223:90-92.
11. Hammer, R. and Giachetti, A. (1982): Life Sci. 31:2991-2998.
12. Hirschowitz, B.I., Fong, J. and Molina, E. (1983): J. Pharmacol. Exp. Ther. 255:263-268.
13. Jaup, B.H. and Dotevall, G. (1981): Scand. J. Gastroenterol. 16:769-773.
14. Katada, T. and Ui, M. (1982): Proc. Nat'l. Acad. Sci. USA 79:3129-3134.
15. Lowry, O.H., Rosebrough, N.J., Farr, A.L. and Randall, R.J. (1951): J. Biol. Chem. 193:265-275.
16. Murayama, S. and Unna, K.R. (1963): J. Pharmacol. Exp. Ther. 140:183-192.
17. Rosenberger, L.B., Roeske, W.R. and Yamamura, H.I. (1979): Eur. J. Pharmacol. 56:179-180.

18. Rosenfeld, G.C. (1983): Eur. J. Pharmacol. 86:99-101.
19. Roszkowski, A.P. (1961): J. Pharmacol. Exp. Ther. 132: 156-170.
20. Sokolovsky, M., Gerwitz, D., and Galron, R. (1980): Biochem. Biophys. Res. Commun. 94:487-492.
21. Takeshige, C. and Volle, R.J. (1964): Brit. J. Pharmacol. 23:80-89.
22. Vickroy, T.W., Watson, M., Yamamura, H.I. and Roeske, W.R. (1984): Fed. Proc. 43:2785-2790.
23. Vickroy, T.W., Watson, M., Yamamura, H.I. and Roeske, W.R. (1984): In Neurotransmitter Receptor Regulation, Interactions, and Coupling (eds) S. Kito, T. Segawa, K. Kuriyama, H.I. Yamamura and R.W. Olsen, Plenum Press, New York, pp. 99-114.
24. Vickroy, T.W., Yamamura, H.I. and Roeske, W.R. (1983): Biochem. Biophys. Res. Commun. 116:284-291.
25. Wamsley, J.K. and Palacios, J. (1983): In Handbook of Neurochemistry Vol. II (ed) A. Lajtha, Plenum Press, New York, pp. 27-52.
26. Watson, M., Roeske, W.R., Johnson, P.C. and Yamamura, H.I. (1984): Brain Res. 290:179-182.
27. Watson, M., Roeske, W.R. and Yamamura, H.I. (1983): Abstr. Soc. Neurosci. 9:962.
28. Watson, M., Roeske, W.R. and Yamamura, H.I. (1982): Life Sci. 31:2019-2023.
29. Watson, M., Vickroy, T.W., Roeske, W.R. and Yamamura, H.I. (1984): Trends in Pharmacological Sciences, Suppl. 1:9-11.
30. Watson, M., Vickroy, T.W., Yamamura, H.I. and Roeske, W.R. (1983): Circulation 68:1535.
31. Watson, M., Yamamura, H.I. and Roeske, W.R. (1983): Life Sci. 32:3001-3011.
32. Yamamura, H.I., Wamsley, J.K., Deshmukh, P. and Roeske, W.R. (1983): Eur. J. Pharmacol. 91:147-149.
33. Yamamura, H.I., Watson, M. and Roeske, W.R. (1983): In Central Nervous System Receptors: From Molecular Pharmacology to Behavior (eds) P. Mandel and F.V. Defeudis, Raven Press, New York, pp. 331-336.

NEUROTRANSMITTERS THAT ACT ON CHOLINERGIC MAGNOCELLULAR FOREBRAIN NUCLEI INFLUENCE CORTICAL ACETYLCHOLINE OUTPUT

G. Pepeu, F. Casamenti, P. Mantovani
and M. Magnani

Department of Pharmacology
University of Florence
Viale Morgagni 65
50134 Florence, Italy

INTRODUCTION

The amount of acetylcholine (ACh) released from the nerve endings depends upon the impulse flow within the cortical cholinergic network. The collection of ACh diffusing from the cortical surface makes it possible therefore to measure indirectly the activity of cholinergic neurons (24).

In recent studies it has been shown that in the rat (9, 27), cat (14), monkey (19) and man (6) the main, if not exclusive origin of the cortical cholinergic network, lies in the basal forebrain. By a detailed study of the organization of the projections from the cholinergic neurons of the basal forebrain to the neocortex and associated structures, four groups of cholinergic neurons in the basal forebrain, designated as Ch1 - Ch4 have been recognized (19).

Stimulation of Ch4 neurons, which correspond to the nucleus basalis magnocellularis of Meynert, is followed by an increase in cortical ACh output (24). Conversely a lesion of the nucleus basalis is associated with a 40% decrease in ACh output from the cerebral cortex (18).

The question arises as to which neuronal pathways impinge upon the magnocellular forebrain nuclei and modulate their activity. The activation or inhibition of these pathways would influence ACh release from the cortical endings of the ascending cholinergic fibers and would, in turn, affect cortical activity.

43

Wood and Richards (28) demonstrated that the local injection of muscimol, a GABA agonist, into the nucleus basalis decreases ACh turnover in the cerebral cortex. This finding suggests a GABAergic modulation of the cholinergic neurons.

In the present study we investigated the effect of the dopaminergic agonists amphetamine and apomorphine and of cholecystokinin (CCK) on ACh output from the cerebral cortex in the rat. It has been shown that CCK is present in cortical nerve terminals (26) and coexists with dopamine in mesolimbic neurons (11).

MATERIALS AND METHODS

Animals. Male Wistar rats, Nossan strain, weighing 200-250 g, were used in all experiments.

ACh output _in vivo_. ACh output from the cerebral cortex was investigated using the cortical cup technique, either in unanaesthetized freely moving rats (4), or in urethane anaesthetized rats (22). The composition of the Ringer solution filling the Perspex cylinder placed on the exposed cortex, was as follows (in mM concentrations): NaCl 150, KCl 5.6, $CaCl_2$ 1.6, $NaHCO_3$ 5.9, glucose 5.5 and eserine sulphate 0.15. The solution in the collecting cup (0.3 ml) was removed every 15 min, diluted with 0.12 ml of distilled water and bioassayed for ACh on the dorsal muscle of the leech (23). All results are expressed in terms of acetylcholine chloride.

In some of the anaesthetized rats systemic arterial blood pressure was simultaneously monitored from the right carotid artery by means of a pressure transducer (MARB, Italy) connected with a pen recorder.

Lesions. Unilateral electrolytic lesions were placed under ketamine anaesthesia (100 mg/kg i.p.) in the nucleus basalis and in the substantia nigra. A current of 1.0 mA was passed through a unipolar electrode for 30 sec. The following coordinates (Koenig and Klippel, 17) were used: for the nucleus basalis 0.2 anterior to bregma, 2.5 lateral, 7 mm below dura; for the substantia nigra 4.1 posterior to bregma, 1.4 and 2.2 lateral, 7 mm below dura.

In the sham-operated rats the electrode was lowered into the cortex without passing any current. The studies on ACh output were always carried out 20 days after the operation. At the end of the experiments a histological examination was carried out on each animal in order to assess the electrode placement, and the size of the lesion.

Drugs. Freshly prepared solutions of the following drugs were used: acetylcholine chloride (Sigma); physostigmine sulphate (BDH); 6-hydroxydopamine, urethane (Merck); ketamine (Ketalar Parke Davies); apomorphine, sulphated cholecystokinin octapeptide (Sigma); racemic naloxone hydrochloride (Endo Laboratories); ceruletide (caerulein diethylammonium hydrate, Farmitalia, kindly supplied by Dr. Roberto de Castiglione); amphetamine (Recordati); and proglumide (Rotta).

RESULTS

Dopaminergic agonists. As shown in Fig. 1, amphetamine, an indirect dopaminergic agonist, administered i.p. to unanaesthetized rats, brought about a dose dependent increase in ACh output from the cerebral cortex. The increase in ACh output was much smaller in rats with a unilateral lesion of the nucleus basalis, and it was completely abolished by a lesion of the substantia nigra.

Given these results, it seemed possible that amphetamine acted on the cholinergic neurons by releasing dopamine from fibers originating in the substantia nigra and directly impinging upon the cholinergic neurons of the nucleus basalis. This possibility appeared also to be supported by the finding that apomorphine, a direct dopaminergic agonist (2), stimulated ACh output from the cerebral cortex,

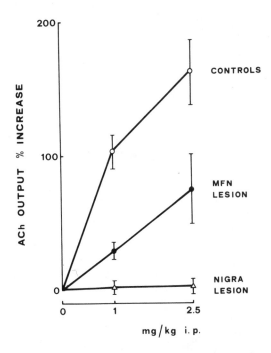

Figure 1.   Dose-effect relationship of amphetamine on ACh output from the cerebral cortex of unanaesthetized rats. Note the decrease of the effect in rats with a unilateral lesion of the magnocellular forebrain nuclei (MFN) and its suppression by a lesion of the nigra.

as shown in Fig. 2. The same figure also shows that apomorphine
lost its stimulatory effect in rats with a lesion of the magnocellular
forebrain nuclei or of the nigra.

The first lesion destroyed the cholinergic neurons; it also
destroyed the postsynaptic dopaminergic receptors on their membrane,
through which they are presumably activated by apomorphine. On
the other hand, after a lesion of the nigra the postsynaptic receptors were still present and supersensitivity developed (10).
Therefore the lack of effect on ACh output of apomorphine, a direct
dopaminergic agonist, was an unexpected finding in the rats with a
nigra lesion. Similarly apomorphine had no effect when the dopaminergic fibers were destroyed by i.c.v. injection of 4 μg of 6-
hydroxydopamine (6-OHDA), a selective toxin for catecholaminergic
neurons (13). These results could be explained by an indirect stimulatory effect of apomorphine on the cholinergic neurons.

In all groups of rats sniffing and stereotypes induced by apomorphine were present. This indicates that a cholinergic link is
not involved in these behavioral effects.

The possibility that the dopaminergic agonists might influence
the cholinergic neurons of the magnocellular forebrain nuclei indi-

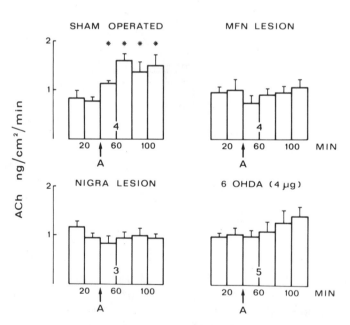

Figure 2. Effect of apomorphine (10 mg/kg i.p.) on ACh output from
the cerebral cortex in unanaesthetized rats. *p < 0.01.

rectly by acting on interneurons therefore was considered. An attempt
was made to identify the nature of such interneurons by pretreating
some of the rats with p-chlorophenylalanine (pCPA) 400 mg/kg i.p. 3
days before amphetamine administration. Neither the inhibition of
serotonin synthesis brought about by pCPA (16) nor the blockade of
GABA receptors by picrotoxin (15) appeared to modify the stimulatory
effect of amphetamine on ACh output.

Cholecystokinin and ceruletide. As shown in Fig. 3 the administration of CCK, 1.5 μg/kg i.p., to urethane anaesthetized rats brought
about a two-fold increase in ACh output. Conversely, the administration of 10 μg/kg of CCK was followed by a significant decrease
in ACh output, which reached a maximum after 45 min.

The results reported in Table 1 show that ceruletide, an analogue
of CCK, extracted from the skin of Hyla coerulea (1), produced the
same biphasic effect as CCK. Ceruletide was, however, slightly
less effective than CCK, since 5 μg/kg i.p. were needed to produce
the same stimulatory effect as 1.5 μg/kg of CCK.

Table 1 also summarizes the results of preliminary experiments
carried out in order to characterize the biphasic effect of the
two peptides on ACh output. First, it can be seen that neither
stimulation nor depression are influenced by urethane anaesthesia

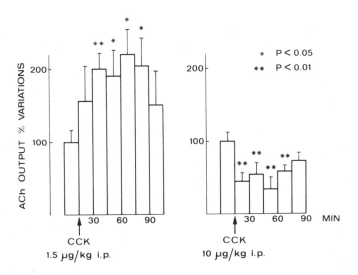

Figure 3. Effect of CCK-8 on ACh release from the cerebral cortex
of urethane anaesthetized rats. Statistically significant difference
from basal values: *p < 0.05; **p < 0.01.

Table 1

Maximal % Changes in ACh Output from Rat Cerebral Cortex
After CCK or Ceruletide Administration

Condition	Pretreatment	Rat #	Drug	Dose µg/kg i.p.	ACh % change ± SE	P
Anaesthesia	--	6	Ceruletide	5	+ 94 ± 38	0.05
"	--	8	"	10	- 37 ± 9	0.01
"	--	7	"	20	- 45 ± 13	0.05
"	Naloxone 1 mg/Kg i.p.	6	"	10	+ 91 ± 40	0.01
Freely moving	--	2	"	10	- 60	
"	Naloxone 1 mg/Kg i.p.	1	"	10	+ 84	
Anaesthesia	--	7	CCK	1.5	+119 ± 33	0.05
"	Proglumide 160 mg/Kg i.p.	4	CCK	1.5	- 49 ± 20	0.05
"	MFN lesion	2	"	1.5	+ 24	

since they both also occur in unanaesthetized rats. Pretreatment
of rats with naloxone 1 mg/kg i.p. antagonized the decrease in ACh
output, which was replaced by a short-lasting increase. On the
other hand, by pretreating the rats with proglumide 60 min earlier,
the stimulatory effect of CCK was prevented and a decrease in ACh
output occurred. Proglumide neither affected basal release nor
modified the increase in ACh output induced by 1 mg/kg amphetamine.
Finally the stimulatory effect of CCK was much smaller in rats with
a lesion of the magnocellular forebrain nuclei than in controls.

It may be noted that the administration of ceruletide to the
freely moving rats was followed by sedation, which was prevented
by naloxone.

DISCUSSION

Our experiments indicate that dopamine and the putative neurotransmitter CCK modulate the activity of the cholinergic neurons
of the nucleus basalis projecting to the cerebral cortex.

It is well known (3) that the dopaminergic system inhibits
directly striatal cholinergic interneurons. Conversely, our ex-

periments with apomorphine suggest that the nigro-striatal dopaminergic fibers which project to the pallidum (21) stimulate indirectly the cholinergic neurons of the nucleus basalis. Specific dopaminergic receptors are involved, since it has been shown that the stimulatory effect of amphetamine on ACh release is blocked by dopamine receptor antagonists (25).

CCK octapeptide and ceruletide are structurally similar peptides (7) which cause qualitatively identical effects under several experimental conditions (29). Their modulation of the cholinergic neurons projecting to the cortex is complex and appears to involve two types of receptors. Their stimulatory effect on ACh output was suppressed by proglumide, a specific CCK receptor antagonist (5). On the other hand the depressant effect was blocked by naloxone. This finding suggests that CCK and ceruletide also act upon opiate receptors either directly or through the release of endogenous opiates. In this connection it has been reported that CCK antagonizes the analgesic effect of morphine and endorphine (8, 12).

It is difficult at present to define the location of the receptors acted upon by CCK and ceruletide. The reduction of their stimulatory effect in rats with a lesion of the nucleus basalis would imply a location on the cholinergic neurons originating from this nucleus, either at a subcortical or cortical level.

The functional meaning of CCK modulation needs also to be clarified. However the decrease in ACh output associated with sedation induced by CCK and ceruletide can be considered part of the widespread CNS depression which occurs after the administration of large doses of these peptides (20).

Acknowledgement. This work was supported by CNR grants 81.00137 and 82.02043.

REFERENCES

1. Anastasi, A., Erspamer, V. and Endean, R. (1967): Experientia 23:699-700.
2. Anden, N.E., Rubenson, A. and Hokfelt, T. (1967): J. Pharm. Pharmac. 19:627-629.
3. Butcher, L.L. and Woolf, N.J. (1982): Brain Res. Bull. 9: 475-492.
4. Casamenti, F., Pedata, F., Corradetti, R. and Pepeu, G. (1980): Neuropharmacology 19:597-605.
5. Chiodo, L.A. and Bunney, B.S. (1983): Science 219:1449-1451.
6. Davies, P. and Feisullin, S. (1982): J. Neurochem. 39:1743-1747.
7. Erspamer, V. and Melchiorri, P. (1980): TIPS 1:391-395.
8. Faris, P.L., Komisararuk, B.R., Watkins, L.R. and Mayer, D.J.-(1983): Science 219:310-312.

9. Fibiger, H.C. (1982): Brain Res. Rev. 4:327-388.
10. Fuxe, K., Agnati, L.F., Kohler, C., Kuonen, D., Ogren, S.O., Andersson, K. and Hokfelt, T. (1981): J. Neural. Transm. 51: 3-37.
11. Hokfelt, T., Rehfeld, J.F., Skirboll, L., Ivemark, B., Goldstein, M. and Markey, K. (1980): Nature 285:476-478.
12. Itoh, S., Katsuura, G. and Maeda, Y. (1982): Eur. J. Pharmacol. 80:421-425.
13. Jonsson, G. (1980): Ann. Rev. Neurosci. 3:169-187.
14. Kimura, H., McGeer, P.L., Peng, J.H. and McGeer, E.G. (1981): J. Comp. Neurol. 200:151-201.
15. Kelly, J.S. and Renaud, L.P. (1973): Brit. J. Pharmacol. 48:369-386.
16. Koe, B.K. and Weissman, A. (1966): J. Pharmacol. Exp. Ther. 154:499-516.
17. Koenig, J.F.R. and Klippel, R.A. (1963): The rat brain, Krieger, New York.
18. Lo Conte, G., Casamenti, F., Bigl, V., Milaneschi, E. and Pepeu, G. (1982): Archs. Ital. Biol. 120: 176-188.
19. Mesulam, M.M., Mufson, E.J., Levey, A.I. and Wainer, B.H. (1983): J. Comp. Neurol. 214:170-197.
20. Morley, J.E. (1982): Life Sci. 30:479-493.
21. Moore, R.Y. and Bloom, F.E. (1978): Ann. Rev. Neurosci. 1: 129-164.
22. Mulas, A., Mulas, M.L. and Pepeu, G. (1974): Psychopharmacologia (Berlin) 39:223-230.
23. Murnaghan, M.F. (1958): Nature 182:317.
24. Pepeu, G. (1983): TIPS 4:416-418.
25. Pepeu, G. and Bartolini, A. (1968): Eur. J. Pharmacol. 4:254-263.
26. Pinget, M., Strauss, E. and Yalow, R.S. (1978): Proc. Natl. Acad. Sci. U.S.A. 75:6324-6326.
27. Wenk, H., Bigl, V. and Mayer, V. (1980): Brain Res. Rev. 2: 295-316.
28. Wood, P.L. and Richard, J. (1982): Neuropharmacology 21:969-972.
29. Zetler, G. (1980): Neuropharmacology 19:415-422.

BIOCHEMISTRY AND IMMUNOCYTOCHEMISTRY

OF CHOLINE ACETYLTRANSFERASE

P.M. Salvaterra,, G.D. Crawford, C.R. Houser,
D.A. Matthews, R.P. Barber and J.E. Vaughn

Division of Neurosciences
Beckman Research Institute
of the City of Hope
Duarte, California 91010   U.S.A.

INTRODUCTION

Choline acetyltransferase (ChAT, acetyl CoA-choline-0-acetyl-transferase, EC 2.3.1.6) is the enzyme responsible for catalyzing the synthesis of the important central and peripheral nervous system neurotransmitter, acetylcholine (ACh), and therefore the regulation of ChAT activity is likely to be a controlling factor in availability of ACh for neural transmission.   Thus, many studies have been done to establish the relationship of ChAT activity to ACh levels as well as the factors regulating enzyme activity (for reviews see 9, 19, 23).   In addition, ChAT may serve as a "phenotypic" marker protein for cells that make and use ACh as a neurotransmitter.   Stimulated in large part by this latter possibility, many laboratories have attempted to produce specific antibodies for ChAT that would be useful in immunocytochemical procedures to study "cholinergic" pathways (1, 2, 8, 11, 14, 17, 26).

Early attempts at specific antibody production were hampered by difficulties in obtaining rigorously characterized, pure ChAT for use as an antigen in conventional polyclonal antisera production (24).   These problems have been overcome in the last few years by appying monoclonal antibody production techniques to obtain ChAT specific antibodies.   Several groups have now reported the production of monoclonal antibodies to ChAT from a number of species (2, 3, 6, 7, 13, 16).   These monoclonal antibodies are being used to further characterize ChAT as a protein and determine its cellular and subcellar distribution.

We have recently produced monoclonal antibodies to ChAT prepared from either Drosophila (3) or rat brain (2), and have used these antibodies in a number of studies to characterize the enzyme protein, the interaction of antigen with the antibodies, and immunocytochemical studies of Drosophila and rat nervous system (11). This paper summarizes some of our recent results.

METHODS

ChAT was purified from Drosophila heads or rat brains according to published procedures (4, 5, 25). We have estimated that ChAT represented less than 10% of the proteins present in the original immunogen. BALB/c mice were immunized with partially purified ChAT and their spleen cells were used in conventional myeloma fusions (15), as described in detail elsewhere (2, 3). Anti-ChAT antibodies were detected in the growing hybridoma culture supernatants by an enzyme depletion screening assay. All antibodies obtained after cloning are thus anti-ChAT specific. Antibody producing cell lines were grown in large scale tissue culture or carried as ascites tumors. All antibodies were of the IgG_1 subclass and were purified before use by employing Protein A affinity chromatography (2, 3).

Antibodies were characterized as directly inhibiting by assaying ChAT activity (10) following a 3 to 24 hr incubation of enzyme and antibody. Noninhibiting antibodies were detected by including a goat anti-mouse Ig precipitation step in the assay and monitoring the disappearance of ChAT activity from solution and/or the appearance of ChAT activity in immune precipitates. The size of the immune-complex formed from the interaction of non-inhibiting antibodies (either alone or in pairs) with ChAT was determined by HPLC gel filtration (as described in 2).

Immunoblots of partially purified Drosophila ChAT were performed by slight modification of Towbin's procedure (27). Anti-rat antibodies have been coupled to cyanogen bromide activated Sepharose by standard procedures and used as an immunoaffinity column for purification of rat brain ChAT.

Immunocytochemistry was performed on material obtained from Sprague-Dawley rats perfused through the heart with a fixative solution containing 4% paraformaldehyde and 0.1% glutaraldehyde. Vibratome or cryostat sections were incubated in immunocytochemical reagents and processed for either light or electron microscopy (as described in detail in 11). ChAT was detected by the unlabeled peroxidase-antiperoxidase (PAP) method. In some cases, it was necessary to include a "double-bridging" step (21) to intensify the PAP reaction product. Spurious staining was reduced considerably by using a species-specific second antibody (goat anti-mouse Ig) that was depleted of antibodies that cross-react with rat Ig (12).

RESULTS AND DISCUSSION

In two separate fusions of spleen cells from animals injected with _Drosophila_ ChAT, we have obtained 12 different monoclonal antibodies (3, and unpublished results). All but one or two of these antibodies are directly inhibiting and they may interact with ChAT at or near the enzyme active site. Alternatively, they may interfere with enzyme activity by some unknown mechanism. The former possibility seems more likely since antibody mediated inhibition of enzyme activity can be partially prevented by including a large excess of acetyl CoA or CoA, but not choline, in the antibody-enzyme incubation mixture (3).

In contrast to the results obtained with anti-_Drosophila_ ChAT antibodies, we have obtained five non-inhibiting anti-rat brain ChAT antibodies (2). By determining the size of the immune complex formed when incubating pairs of anti-rat ChAT antibodies, we have been able to determine the relative positions of the antibody binding epitopes on the enzyme surface. All five epitopes appear to be close together since most antibodies interfere partially or completely with each other when binding to the enzyme (2). A summary of the relative positions of the antibody binding epitopes on the enzyme surface is presented in Figure 1. From the likely size of an antibody binding domain (2 nm^2) and the total area (35 nm^2) likely to be sterically blocked by the non-bound portions of an immunoglobulin (20), we conclude that all five antibody binding domains are within a <100 nm^2 area, compared to a total available surface area of 6400 nm^2 for a typical globular protein of the molecular weight observed for ChAT (67K daltons).

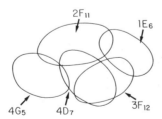

Figure 1. Venn diagram showing the relative relationship between antibody binding domains inferred from HPLC gel filtration of immune complexes (2). Each domain is represented as an oval. Independent domains do not intersect, interacting domains are shown with a slight overlap, and mutually exclusive domains are represented by major overlap.

On the basis of the above findings, mammalian and <u>Drosophila</u> ChAT would seem to be proteins with a so-called "main immunogenic region" responsible for initiating an immune response. It is interesting that this "main immunogenic region" seems to be different for ChAT from the two species (i.e., near the active site for <u>Drosophila</u> and not the active site for rat enzyme).

The immunological relationship of ChAT from different species has been studied by determining the cross-reactivity of the monoclonal antibodies with enzyme from several species. The two best characterized anti-<u>Drosophila</u> ChAT antibodies appear to be species specific. No reaction of these antibodies has been observed with ChAT prepared from other species (including other insects). In contrast, some of the anti-rat ChAT antibodies react with enzyme from a variety of species, including insects. The results are summarized in Table 1. At least part of the "main immunogenic region" on mammalian ChAT thus appears to be highly conserved and may have some, as yet unknown, functional significance.

Table 1

Summary of anti-rat ChAT monoclonal antibody properties

Ab[a]	$K_D(M^{-1})$[b]	Cross-Reactions[c]
1E6	5.46×10^{-9}	rodents (hamster, mouse, etc.), avian, (chicken, quail), <u>Drosphila</u>, primate (monkey, human)
2F11	2.96×10^{-8}	--
3F12	1.66×10^{-9}	--
4D7	1.40×10^{-7}	rodent, avian, primate, <u>Drosphila</u>, <u>Aplysia</u>
4G5	2.11×10^{-7}	rodent, <u>Drosphila</u>, Aplysia

[a]All antibodies are I_gG_1 and are non-inhibiting.   [b]Determined from Scatchard analysis using rat brain ChAT as the test antigen.   [c]Ability of antibodies to remove ChAT activity (usually ammonium sulfate precipitated ChAT) from solution in a double precipitation assay.

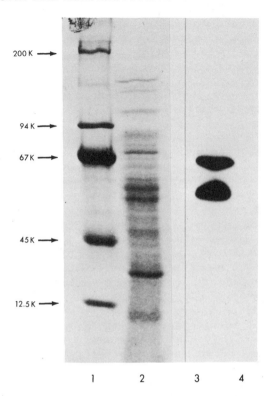

Figure 2. Immunoblot staining of <u>Drosophila</u> ChAT using a monoclonal antibody. The DEAE fraction (30 μl) from a <u>Drosophila</u> ChAT puri- fication was run on a 10% polyacrylamide gel. Lanes 1 and 2 are a Coomassie blue protein stain of the molecular weight standards and the DEAE fraction. Lanes 3 and 4 are a radioautograph of the nitro- cellulose replicate of the DEAE fraction and molecular weight markers stained with anti-<u>Drosophila</u> ChAT monoclonal antibody 1G4 plus [125I]- Protein A, as described in the text.

The nature of the antibody binding epitopes on ChAT can also be studied by observing antibody recognition when the protein is denatured and run on sodium dodecyl sulfate (SDS) gels or fixed to tissue. Both anti-<u>Drosophila</u> ChAT antibodies have been used to stain nitrocellulose replicas made from SDS gels of partially purified ChAT. Figure 2 shows an immunoblot of a DEAE fraction, from a Droso- phila ChAT purification (25), estimated to contain <0.1% ChAT protein. Two proteins corresponding in molecular weight (67K and 54K daltons) to the major structurally related proteins in completely purified <u>Drosphila</u> ChAT are stained. We have not been able to successfully stain immunoblots of rat brain ChAT under a variety of conditions using any of our anti-rat ChAT antibodies. Perhaps the determinants recognized by the anti-Drosophila ChAT antibody are represented by the primary sequence of amino acids, while those for the anti-rat ChAT antibodies are formed from higher order protein structure which is not recovered after SDS denaturation.

Very little is known about the molecular structure(s) of mammalian ChAT in the brain since it is generally present in relatively low concentrations and has proven to be very difficult to purify. We have reported the observation of three structurally related proteins following SDS gel electrophoresis of conventionally purified rat brain ChAT (4). Recently we have been able to use one of our monoclonal anti-rat brain ChAT antibodies to construct an immunoaffinity column for purification of the rat enzyme. Figure 3 shows a silver-stained SDS gel of rat ChAT eluted with urea from an antibody affinity column. Two major protein bands are observed in this preparation at 67K and 55K daltons. These bands have been peptide mapped and show structural homology (data not shown). The molecular weights of the proteins purified by immunoaffinity chromatography agree with the results of other investigators who have used immunoblotting procedures to identify ChAT peptides following SDS gel electrophoresis of partially purified enzyme (18). We are confident that one or, more likely, both proteins are ChAT since approximately 20% of the ChAT activity bound to the column can be recovered in the eluate under the elution conditions used after removal of the urea. In fact, the pattern of proteins observed in conventional or immuno-affinity purified rat ChAT is very similar to the pattern we obtain with completely purified Drosophila ChAT (25).

The specificity of one of our anti-rat brain ChAT antibodies for immunocytochemical studies has been confirmed by demonstrating ChAT-positive neurons in a number of well defined "cholinergic" systems (11). Figure 4 shows ChAT-positive reaction product in the cell bodies and dendrites of· spinal cord motor neurons as well as their axon terminations, including those at neuromuscular junctions. One surprising result of our immunocytochemical studies using anti-rat ChAT antibody is the relative ease of demonstrating immunoreactive protein in the cell bodies and dendrites as compared to axon terminals. Double-bridging procedures (21) are usually necessary for visualization of terminals, but not somata and dendrites. This result is in contrast to the observations with glutamic acid decarboxylase immunocytochemistry, where special procedures, such as colchicine injection, are often necessary to observe reaction product in cell bodies but terminals are routinely demonstrated (22). The significance of these observations is unknown and may possibly involve several reasons. For instance, there may be accessibility problems for the anti-ChAT antibodies in the terminal regions. This seems unlikely in view of the glutamic acid decarboxylase results. Perhaps the ChAT epitope recognized by antibody is subject to differential fixation in the proximal and distal regions of the cell, thus altering the ability of antibody to recognize it. There

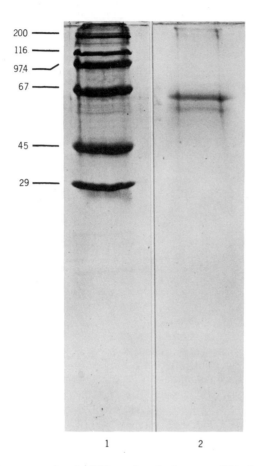

Figure 3.   Silver stained SDS gel of immunoaffinity purified rat
brain ChAT.   Lane 1:   molecular weight standards (in K daltons).
Lane 2:   Proteins released from a 4G5 monoclonal anti-rat ChAT affinity
column by urea.   The sample, a CM-Sephadex fraction (5 units of
activity, 0.029 μmoles/min/mg), was absorbed to the column by incubating overnight at 4°C.   The column was washed extensively with
buffer (75 mM PO$_4$, pH 7.4, 1 M NaCl, 0.5% Triton, 0.1% SDS, 0.1 mM
EDTA) before elution of specifically bound proteins with 8 M urea.

may even be different immunoreactive forms of ChAT in these different
cellular compartments.   This latter reason seems unlikely because
in vitro immunoassays have shown that all five anti-rat ChAT antibodies
can precipitate 100% of the ChAT activity from solution.   The differential ability of ChAT staining may be due to different concentrations

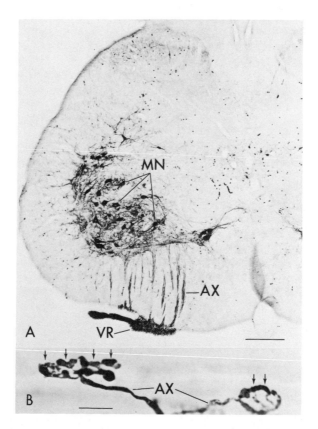

Figure 4. Photomicrographs of sections of the spinal cord and diaphragm. **A.** In the lumbar spinal cord, large ChAT positive motorneurons (MN) are present in the medial and lateral motor nuclei,
and fascicles of their axons (AX can be observed traversing the
ventral funiculus to gain access to the ventral root (VR). Scale
line, 300 μm. **B.** In the diaphragm, two ChAT-positive axons (AX)
can be followed to their terminations at motor end-plates, where
the axons expand into several smaller processes with terminal en-
largements (arrows). Scale line, 20 μm.

of ChAT in the terminal versus cell body regions. This quantitative
difference, as opposed to a qualitative difference, may explain
why the double-bridging procedures are successful for visualizing
ChAT in terminals. Interestingly, one other antibody (1E6) we have
used for immunocytochemical procedures stains cell bodies and proximal
dendrites, but it does not appear to detect axon terminals, even
with doublebridging, however, we have not yet optimized the conditions
for using 1E6 as an immunocytochemical reagent.

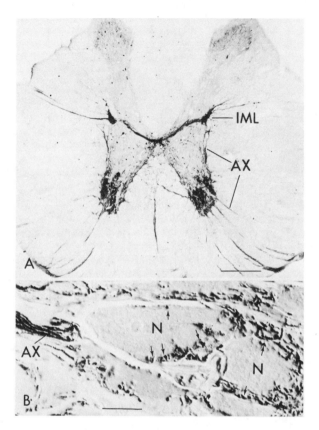

Figure 5. Photomicrographs of sections of the spinal cord and superior cervical ganglion.   A. In the thoracic spinal cord, ChATpositive preganglionic sympathetic neurons are present in the intermediolateral nucleus (IML), as well as in the motor nuclei.   Fascicles of somatic motor and pre-ganglionic axons are designated (AX).   Scale line, 500 μm. B. In the superior cervical ganglion, groups of ChATpositive pre-ganglionic axons (AX) can be observed, and their punctate terminations (arrows) are seen surrounding the unstained cell bodies of ganglionic neurons (N). Section photographed with Nomarski optics. Scale line, 20 μm.

Figure 5 shows intense ChAT-positive reaction product in the preganglionic sympathetic neuron cell bodies in the rat thoracic spinal cord, as well as in their axon terminations in the superior cervical ganglion.

One system holding special relevance for investigators interested in studying "cholinergic" pathways in mammalian brain is the

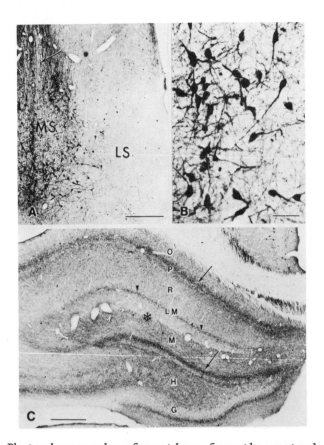

Figure 6. Photomicrographs of sections from the rostral forebrain.
A. ChAT-positive cell bodies are present in the medial septal nucleus
(MS), and vertically-oriented, stained fibers course through and
are especially prominent in the dorsal part of the necleus (arrow).
Very little staining for ChAT is present in the lateral septum (LS).
Scale line, 300 μm. **B.** Higher magnification from Figure 6A showing
ChAT-positive neurons of the medial septal nucleus. Scale line,
50 μm. **C.** Different densities of ChAT-positive terminals form a
laminated pattern in the hippocampus and dentate gyrus. In hippocampus, the highest density of terminals (upper arrow) is located
in the part of the stratum oriens (0) that borders stratum pyramidale
(P). A relatively low density of puncta is manifest throughout
stratum radiatum (R). The density of puncta increases in the deeper
part of stratum lacunosum-moleculare (L-M), forming a narrow, dense
band (arrowheads), and then decreases to the lowest hippocampal
level in the superficial part of stratum lacunosummoleculare adjacent
to the hippocampal fissure (*). In the dentate gyrus, highest densities of terminals are present in the supragranular region (lower
arrow), while moderate densities are located in stratum moleculare
(M) and the hilus (H). Relatively low densities of terminals are
observed in the stratum granulosum (G). Scale line, 30 μm.

septo-hippocampal pathway. Figure 6 shows the presence of many intensely staining cell bodies in the medial septal nucleus, and relatively sparce positive staining somata in the lateral septal nucleus. In the anterior hippocampal formation (Figure 6), ChAT-positive punctate structures are found in varying numbers in the different layers, creating a laminar pattern. Pronounced bands of staining are found: 1) immediately superficial to the granule cell bodies of the dentate gyrus; 2) just deep to the layer of pyramidal neuron cell bodies in the hippocampus; and 3) at the junction of stratum radiatum and stratum lacunosum-moleculare of the hippocampus. The laminar distribution of ChAT-positive terminals in the hippocampal formation corresponds closely with the pattern of acetylcholinesterase staining in this brain region. Lesions of the medial septal nuclei and the diagonal band virtually eliminate the laminar pattern of ChAT, with only small clusters of puncta remaining near the granule and pyramidal cell bodies (data not shown). In addition, we have detected a population of ChAT-positive cell bodies within the hippo-campal formation. This heterogenous group of cells was found through-out the septotemporal extent of the hippocampus, but their numbers and locations varied greatly in a given section. These intrinsic, ChAT-positive, neurons remained following septal lesions.

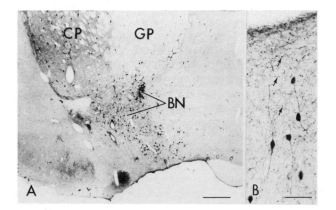

Figure 7. Photomicrographs of coronal sections of forebrain.
A. Numerous ChAT-positive cell bodies are present in the basal nucleus (BN) of the forebrain. At this rostral level, many stained neurons are located adjacent to, but do not invade, the relatively unstained globus pallidus (GP). ChAT-positive staining is present in both the cell bodies and neuropil of the caudate-putamen (CP). Scale line, 500 μm. B. Small fibers with ChAT-positive varicosities (arrows) form a diffuse and intricate network in layers I and II of the cerebral cortex. Small ChAT-positive cell bodies are also present in layer II and frequently exhibit a bipolar dendritic pattern. Scale line, 50 μm.

In the forebrain, numerous ChAT-positive neurons with multipolar dendrites are detected in the basal nucleus and are demonstrated in Figure 7. Figure 7 also shows the ChAT staining pattern in cerebral cortex. ChAT-positive fibers, oriented in all directions, are observed in all layers of the cortex, creating the impression of a meshwork. The fibers exhibit numerous small punctate varicosities and can often be traced through several cortical layers. In addition, small ChAT-positive cell bodies, many of which exhibit bipolar dendrites, are observed in layers II-VI.

Much exciting work remains to be done on ChAT and "cholinergic" systems. We are now pursuing many more detailed immunocytochemical studies using the ChAT monoclonal antibodies. In addition, we have begun studies aimed at cloning the gene for ChAT in hopes of understanding the molecular details of this important enzyme as well as the genetic regulation of its expression.

Acknowledgements. This work was supported by grants from NIH/ - NINCDS. We thank Lucia Correa for excellent technical assistance and Eve Hardy for preparing the manuscript.

REFERENCES

1. Cozzari, C. and Hartman, B.K. (1980): Proc. Natl. Acad. Sci. USA 77:7453-7457.
2. Crawford, G.D., Correa, L. and Salvaterra, P.M. (1982): Proc. Natl. Acad. Sci. USA 79:7031-7035.
3. Crawford, G.D., Slemmon, J.R. and Salvaterra, P.M. (1982): J. Biol. Chem. 257:3853-3856.
4. Dietz, G.W., Jr. and Salvaterra, P.M. (1980): J. Biol. Chem. 255:10612-10617.
5. Driskell, W.J., Weber, B.H. and Roberts, E. (1978): J. Neurochem. 30:1135-1141.
6. Eckenstein, F., Barde, Y.-A. and Thoenen, H. (1981): Neuroscience 6:993-1000.
7. Eckenstein, F. and Thoenen, H. (1982): EMBO Journal 1:363-368.
8. Eng., L.F., Uzeda, C.T., Chao, L.-P. and Wolfgram, F. (1974): Nature (Lond.) 250:243-245.
9. Fonnum, F. (1975): In Cholinergic Mechanisms (ed) R.G. Waser, Raven Press, New York, pp. 145-160.
10. Fonnum, F. (1975): J. Neurochem. 24:407-409.
11. Houser, C.R., Crawford, G.D., Barber, R.P., Salvaterra, P.M. and Vaughn, J.E. (1983): Brain Res. 266:97-119.
12. Houser, C.R., Barber, R.P., Crawford, G.D., Matthewes, D.A., Phelps, P.E., Salvaterra, P.M. and Vaughn, J.E. (1984): J. Histochem. Cytochem. 32:395-402.
13. Ichikawa, T., Ishida, I. and Deguchi, T. (1983): FEBS Lett. 155:306-310.
14. Kimura, H., McGeer, P.L., Peng, F. and McGeer, E.G. (1980): Science 208:1057-1059.

15. Kohler, G. and Milstein, C. (1976): Eur. J. Immunol. 6:511-519.
16. Levey, A.I., Aoki, M., Fitch, F.W. and Wainer, B.H. (1981): Brain Res. 218:383-387.
17. Levey, A.I., Armstrong, D.M., Atweh, S.F., Terry, R.D., and Wainer, B.H. (1983): Neuroscience 3:1-9.
18. Levey, A.I., Rye, D.B. and Wainer, B.H. (1982): J. Neurochem. 39:1652-1659.
19. Mautner, H.G., (1977): CRC Crit. Rev. Biochem. 4:341-370.
20. Medgyesi, G.A., Fust, G., Gergely, T. and Bazin, H. (1978): Immunochem. 15:125-129.
21. Ordronneau, P., Lindstrom, P.B.M. and Petrusz, P. (1981): J. Histochem. Cytochem. 29:1397-1404.
22. Ribak, C.E., Vaughn, J.E. and Saito, K. (1978): Brain Res. 140: 315-332.
23. Rossier, J. (1977): Int. Rev. Neurobiol. 20:283-337.
24. Rossier, J. (1981): Neuroscience 6:989-991.
25. Slemmon, J.R., Salvaterra, P.M., Crawford, G.D. and Roberts, E. (1982): J. Biol. Chem. 257:3847-3852.
26. Sofroniew, M.V., Eckenstein, F., Thoenen, H. and Cuello, A.C. (1982): Neurosci. Lett. 33:7-12.
27. Towbin, H., Staehlin, T. and Gordon, J. (1979): Proc. Natl. Acad. Sci. USA 76:4350-4354.

AUTORADIOGRAPHIC LOCALIZATION OF SUBTYPES OF MUSCARINIC AGONIST AND ANTAGONIST BINDING SITES: ALTERATIONS FOLLOWING CNS LESIONS

D.R. Gehlert, H.I. Yamamura[*], W.R. Roeske[*] and J.K. Wamsley

Departments of Psychiatry and Pharmacology
University of Utah School of Medicine
Salt Lake City, Utah 84132   U.S.A.

[*]Departments of Pharmacology, Biochemistry, Psychiatry,
Internal Medicine and the Arizona Research Laboratories
University of Arizona Health Sciences Center
Tucson, Arizona 85724   U.S.A.

INTRODUCTION

The use of radiolabelled ligands, specific for muscarinic receptors, has allowed the demonstration of several subtypes of muscarinic receptors. Agonists have been postulated to bind to three distinct affinity states which are believed to be different conformational states of the same receptor (4, 6, 7), while classical muscarinic antagonists are believed to bind to all of these agonist states with equal high affinity. However, the non-classical muscarinic antagonist pirenzepine has been shown to have selective antagonism in both the clinic and the laboratory. Low doses of pirenzepine will reduce gastric secretion with little effect on salivation, gut motility, ciliary muscle function in the eye, or vagal tone in the heart (20, 21, 25). Experimentally, pirenzepine has been found to selectively bind with high affinity to a dense receptor population in membrane preparations from several organs including parts of the brain and peripheral ganglia (42, 47; Yamamura et al., this book), while binding to a very low density of receptors in other peripheral organs, such as smooth muscle and heart (3, 42). The selective nature of pirenzepine binding has led to the postulation of two separate muscarinic receptors designated as M_1 and M_2 (17, 23, 43).

In this communication, we present autoradiographic evidence for both agonist and antagonist binding heterogeneity by detecting discretely localized areas of agonist and antagonist binding in the rat CNS. We also present applications of these techniques to evaluate alterations in muscarinic receptor populations after surgical or chemical lesioning.

METHODS

Before autoradiograms were generated, initial biochemical experiments were conducted for each ligand to determine optimal conditions for binding. These initial experiments required 10 micron tissue sections, cut in a cryostat from an area known to contain a large population of the desired receptor. These were then thaw-mounted onto cold, chrome-alum/gelatin coated slides with two sections per slide. Development of procedures for in vitro labelling for auto-radiography involved four basic steps: first, a rinse time needed to be generated; second, an incubation time selected; third, the appropriate concentration of the ligand needed to be determined; and finally, it was necessary to establish the specificity of the ligand for the receptor. In all these initial experiments, the point with the best specific to non-specific (signal to noise) ratio was selected.

Once these parameters had been determined, tissue sections were labelled for autoradiography. The parameters for labelling with muscarinic ligands are listed in Table 1. Slides were incubated and rinsed as before. After a dip in distilled water to remove buffer salts, the tissue sections were dried on cold metal pans with a stream of cool dry air. The slides were then desiccated in a refrigerator overnight.

Autoradiograms were generated either by apposition to tritium sensitive film (27) or coverslips coated with photographic emulsion (49). To appose the tritium sensitive film, the slides were affixed to photographic mounting board with double stick tape or spray mounting adhesive. This was then placed in an X-ray cassette, a piece of tritium sensitive film (LKB Ultrofilm; Rockville, MD) was placed over the slides and the X-ray cassette clamped shut. After an appropriate exposure period, the film was removed and developed. The autoradiographic image was viewed with a microscope under brightfield illumination and quantitated using a DADS Model 560 computer interfaced with an MPV Compact microphotometry system attached to a Leitz Ortho-plan microscope. Autoradiographic standards were prepared according to Unnerstall et al. (31) and included on the exposure of each film.

The coverslip technique involved the apposition of long, thin emulsion (Kodak NTB-3; Rochester, NY) coated coverslips (25x77 mm, Corning #0, Corning Glass Works; Corning, NY) over the slide mounted tissue sections. The emulsion coated coverslips were glued to the frosted end of the slides and clamped over the labelled tissues with a binder clip and a piece of Teflon. After an appropriate exposure period, the coverslips were bent back with spacers, developed and the tissue sections subsequently stained. After drying, the spacers were removed and the coverslips affixed with permanent mounting fluid. Autoradiographic grains were visualized directly over the tissue sections using darkfield (incident light) illumination. A very detailed consideration of all these procedures is available elsewhere (38).

Several different lesions were used in the research outlined in this chapter. Electrothermal lesions of the fimbria were made with a stereotaxically placed microelectrode attached to a lesion generator (35). Ligation of the sciatic and vagus nerves were performed by tying silk suture around the nerve trunk (40, 50). Chronic spinal cats were developed in collaboration with John Lane from the Department of Pharmacology at the Texas College of Osteopathic Medicine, and with James Smith from the Department of Psychiatry at the Louisiana State University in Shreveport. The spinal cords of cats were completely transected at a cervical level and the muscarinic receptors were examined at lumbar spinal levels (Lane et al., in preparation). Lesions of the brain as a result of chronic thiamine deficiency were developed by maintaining rats on a thiamine deficient diet or by treating them with pyrithiamine, a thiamine antagonist (13). Electrothermal lesions of the septum were performed in collaboration with Keith Crutcher in the Department of Anatomy at the University of Utah in order to examine muscarinic receptor alterations in the hippocampus (Crutcher et al., in preparation).

Muscarinic receptor subtypes were examined using the tritium sensitive film method of receptor autoradiography, and the alterations in density were quantitated using computer assisted microdensitometry.

RESULTS AND DISCUSSION

Muscarinic receptors exist in several agonist affinity states (6), while classical antagonists are all believed to bind with equal high affinity (19, 45). Multiple agonist affinity states can be demonstrated autoradiographically by labelling with [3H]-Nmethyl-scopolamine ([3H]-NMS) and displacing the high affinity sites with 10^{-4}M carbachol (Table 1). This concentration of carbachol displaces 98.4% of [3H]-NMS binding to the high affinity site, while only 27% of the binding to the low affinity sites is displaced (39). This technique was used to produce extensive maps of the two states in the rat brain (41). High and low affinity binding sites exist in discrete areas of the brain, a single subtype sometimes apparently being the sole representative in a single nucleus or area (Figure 1). High affinity agonist sites predominate in such regions as the

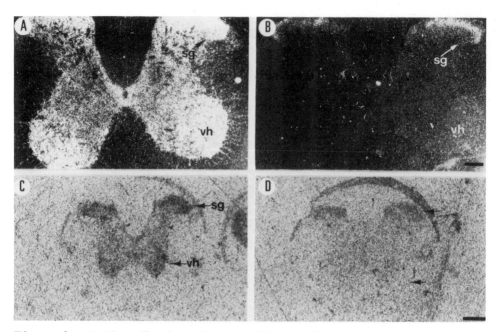

Figure 1. A.Distribution of autoradiographic grains over a section
of spinal cord labelled with [^{3}H]-NMS. The autoradiographic grains,
associated with regions of muscarinic receptor binding, reflect
the density and distribution of muscarinic receptors in the gray
matter of the spinal cord. The grains appear as white dots against
a dark background using darkfield optics and demonstrate the high
density of receptors in the substantia gelatinosa (sg) and lateral
portions of the ventral horn (vh). B. The section shown in this
photomicrograph was immediately adjacent to the one shown in A.
This particular section was incubated using exactly the same conditions, except for the added presence of 10^{-4}M carbachol in the
incubation medium. Thus, the binding of [^{3}H]-NMS to the high affinity
agonist sites has been displaced. Comparison of the areas labelled
in A and not in B show the distribution of the high affinity sites.
Since the nonspecific binding of [^{3}H]-NMS is very low and uniformly
distributed, the grains appearing in B reflect the distribution of
the low affinity sites. Low affinity sites thus predominate in
the sg while the rest of the spinal cord gray matter contains the
high affinity agonist sites. Bar = 250 microns. C. An autoradiogram
as it appeared on a region of tritium sensitive film overlying a
section of spinal cord labelled with [^{3}H]-QNB. In this case, the
autoradiographic grains appear as black dots against a light background. This antagonist labels regions of the cord similar to those
labelled with [^{3}H]-NMS. D. A section adjacent to the one shown
in C was used to generate this autoradiogram. This particular section
is labelled with [^{3}H]-PZ to depict the density and distribution of
the M_1 sites. Thus, the sg contains the M_1 muscarinic antagonist
receptor subtype while the rest of the spinal cord gray matter contains
the M_2 subtype. Bar = 400 microns.

nucleus of the diagonal band of Broca, lamina IV of the parietal cortex (34), medial septum, lateral geniculate body, ventral rostral nucleus of the lateral lemniscus, periaqueductal gray matter, superficial lamina of the superior colliculus, facial nerve nucleus, hypoglossal nerve nucleus, and ventral horn of the spinal cord (Table 2). Low affinity agonist sites predominate in the caudate-putamen, hippocampus, amygdala, and dorsal horn of the spinal cord. Direct labelling with the tritiated muscarinic agonist [^{3}H]-cis methyldioxolane ([^{3}H]-CD) has also been used to localize binding in discrete areas of the brain (Figure 2). Areas of specific binding for [^{3}H]-CD correlate well with areas that are carbachol displaceable (46). There are some discrepancies to this observation, however. For instance, [^{3}H]-CD labelled some of the sites present in the cortex, striatum, and in the dorsal horn of the spinal cord. These sites are not carbachol displaceable (using indirect labelling) and thus would presumably represent regions of low affinity binding. Thus, carbachol could be peculiar in its inability to bind to receptors in these areas. Alternatively, these could be regions where [^{3}H]-CD is labelling low affinity agonist sites since these particular areas are where the highest densities of low affinity sites are found, and the number of low affinity sites labelled with [^{3}H]-CD has yet to be determined. In most regions, it would appear that indirect and direct labelling techniques provide supporting information on the normal localization of the high affinity muscarinic agonist sites.

Muscarinic antagonist heterogeneity has been demonstrated using the non-classical muscarinic antagonist pirenzepine (3, 16, 43). To evaluate this heterogeneity, we have labelled a series of serial sections with [^{3}H]-pirenzepine ([^{3}H]-PZ) and [^{3}H]-quinuclidinyl benzilate ([^{3}H]-QNB), to localize areas of binding for non-classical and classical muscarinic antagonists, respectively (Table 1). [^{3}H]-QNB has previously been demonstrated, in an autoradiographic study, to bind to a single population of muscarinic receptors in the manner of a classical antagonist (37). Comparison of the areas directly labelled by these two antagonists revealed regions that were highly labelled with [^{3}H]-QNB, but were devoid of labelling by [^{3}H]-PZ (36, 48; Yamamura et al., this book). These regions included the nucleus of the diagonal band of Broca, lamina IV of the parietal cortex, superior colliculus, periaqueductal gray matter, facial nerve nucleus, nucleus tractus solitarius, hypoglossal nerve nucleus and the ventral horn of the spinal cord. Interestingly, all areas labelling with [^{3}H]-QNB, but not with [^{3}H]-PZ, are also those areas previously demonstrated to be carbachol displaceable (Table 2). Therefore, [^{3}H]-PZ appears to bind only to areas containing predominantly the low affinity state of the muscarinic receptor (Figures 1 and 2).

Could the subtype of muscarinic receptor labelled with [^{3}H]-PZ (the M_1 site) be the same as the low affinity agonist site? High affinity agonist receptors can be converted to a lower affinity

Table 1

Parameters for *in vitro* labelling of muscarinic receptors in tissue sections

Ligands	Concentration	Buffer	Incubation Time & Temp.	Rinse Time	References
M_1 and M_2 antagonists:					
[3H]-N methylscopolamine	1nM	A	60', rm. temp.	2 x 5'	39, 41
[3H]-Quinuclidinyl benzilate	1nM	B	120', rm. temp.	2 x 5'	36, 37, 48
M_1 selective antagonist:					
[3H]-Pirenzepine	20nM	B	60', rm. temp.	5'	36, 48
Indirect agonist site labelling:					
High and low affinity agonist sites were labelled with the antagonist [3H]-N methylscopolamine. High affinity agonist sites were selectively displaced on adjacent sections by including a 10^{-4} M concentration of carbachol in the incubation medium.					
Direct agonist site labelling:					
[3H]-cis methyl dioxolane	5nM	C	120', 0–4°C	3 sec.	46

Buffers: A. 0.05M phosphate buffered saline, pH 7.4; B. Modified Krebs phosphate buffer, pH 7.4; C. 10mM sodium–potassium phosphate buffer with 200mM sucrose, pH 7.4. In all cases the displacer used to determine nonspecific binding was 1 micromolar atropine.

Table 2

Regions of overlap between M_1 vs. M_2 antagonist sites and high vs. low affinity agonist sites

Area	Carbachol displaceable[A] [3H]-NMS binding	[3H]-CD[B]	[3H]-QNB[C]	[3H]-PZ[D]
Nucleus of the diagonal band of Broca	+	+	+	-
Lamina IV of the parietal cortex	+	+	+	-
Caudate-putamen	-	+	+	+
Hippocampus (CA1)				
Stratum oriens	-	-	+	+
Stratum radiatum	-	-	+	+
Lateral geniculate body	+	+	+	-
Superior colliculus	+	+	+	-
Periaqueductal gray	+	+	+	-
Facial nerve nucleus	+	+	+	-
Nucleus tractus solitarius	+	+	+	-
Hypoglossal nerve nucleus	+	+	+	-
Spinal cord				
dorsal horn	-	+	+	+
ventral horn	+	+	+	-

A. This column refers to the [3H]-N methylscopolamine binding displaceable in the presence of 10^{-4} M concentrations of carbachol added in the incubation medium. Data from references 39, 41. B. [3H]-cis methyl dioxolane binding. Data from reference 46. C. [3H]-Quinuclindinyl benzilate binding. Data from references 36, 37, 48. D. [3H]-Pirenzepine binding. Data from references 36, 48.

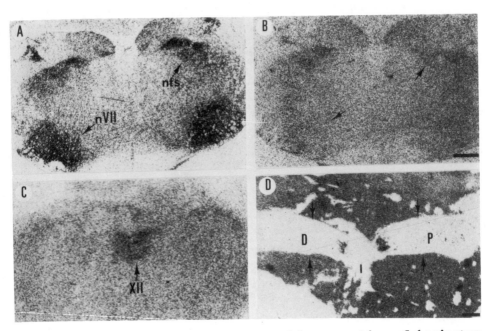

Figure 2. A. Autoradiogram generated by a section of brainstem incubated in [^{3}H]-QNB. The presence of muscarinic receptors in the nucleus tractus solitarius (nts) and the facial nerve nucleus (nVII) can be seen. B. This autoradiogram is from a section, adjacent to the one shown in A, which was labelled with [^{3}H]-PZ. Note the absolute lack of labelling in the facial nerve nucleus and the very low binding in the nts. The muscarinic receptor population in both of these nuclear areas thus represents predominantly the M_2 antagonist receptor subtype. Bar = 500 microns. C. Autoradiographic grain densities associated with a section of brainstem labelled with the muscarinic agonist [^{3}H]-CD. Specific labelling of the muscarinic receptors in the hypoglossal nerve nucleus (XII) can be appreciated. D. A section of ligated sciatic nerve, labelled with [^{3}H]-PZ. The nerve was ligated 24 hours before the animal was sacrificed, the nerve removed and cryostat sections of sciatic nerve prepared for autoradiography. The site of the ligature is indicated (1) along with the proximal (P) and distal (D) portions of the accompanying nerve. The arrows demarcate the borders of the nerve trunk. Note the accumulation of [^{3}H]-PZ labelled sites just proximal (toward the spinal cord) to the ligature. A smaller build-up can be seen distal to the ligature. Since these sites are labelled with [^{3}H]-PZ, they demonstrate that at least a portion of the receptors transported in the sciatic nerve are of the M_1 receptor subtype. Bar = 250 microns.

state in the presence of guanine nucleotides in the peripheral tissues (2, 9, 12). This same phenomenon has been recently demonstrated in the spinal cord (Gehlert et al., in preparation). High affinity agonist sites can be converted to a lower affinity state in the presence of guanosine triphosphate. If the muscarinic receptors in the ventral horn are shifted to a lower affinity state then the [^{3}H]-PZ binding in the ventral horn should increase if the low affinity agonist site and the [^{3}H]-PZ site are the same. The results of this experiment are shown in Figure 3. The low affinity agonist site produced by a GTP shift is not the same as the [^{3}H]-PZ site. The ventral horn of the spinal cord is an M_2 tissue, however, which may have multiple receptor agonist states which are GTP sensitive.

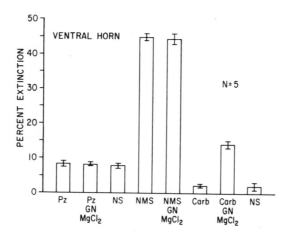

Figure 3. Effect of a guanine nucleotide (GN) on [^{3}H]-PZ (PZ) and [^{3}H]-NMS (NMS) binding in the ventral horn of the spinal cord. The density measurements were made on the tritium sensitive film overlying serial tissue sections incubated using the indicated conditions. The measurements are recorded as percent extinction and represent the average of readings from several sections plus or minus the standard error of the mean. PZ does not label muscarinic receptors in the ventral horn of the spinal cord regardless of whether or not GN (10 micromolar GppNHp) is present in the incubation media. NMS, however, binds to a dense population of sites in the ventral horn, and this binding is not affected by the presence of the GN. The sites labelled with NMS are displaceable with 10^{-4}M concentrations of carbachol (Carb), indicating they are high affinity agonist sites. The latter experiment, performed in the presence of the GN, shifts the high affinity sites to a lower affinity and retards their recognition by Carb. Thus, not as many sites are carbachol displaceable when the experiment is carried out in the presence of GN. This experiment shows that the carbachol sensitive high affinity agonist sites do not recognize PZ even when they are shifted to a low affinity conformation.

The M_1 and M_2 receptors thus appear to be distinct. This data supports other findings which indicate both M_1 and M_2 receptors may have multiple agonist affinity states associated with them.

The use of carbachol displacement of [^{3}H]-NMS to autoradiographically detect alterations in high and low affinity muscarinic receptor populations after lesioning, yields some intriguing insight into the dynamics of muscarinic receptors. Axonal transport of muscarinic receptors has been demonstrated in the dog splenic nerve (22), rat sciatic nerve (40), rat vagus nerve (50) and in the brain (35). In the vagus nerve, the orthograde transported receptor is in the high affinity agonist state which is moved by fast transport while low affinity receptor accumulates on the distal side of a ligature (40, 50). The presence of 10^{-5}M guanyl5'-yl imidodiphosphate (GppNHp), a non-hydrolyzable GTP analogue, and 10mM $MgCl_2$ produces a conversion of the high affinity (carbachol displaceable) sites on the proximal side of the ligature to a lower affinity state. This conversion gives evidence that the high and low affinity agonist sites are different conformations of the same receptor (probably M_2 in this case, although M_1's may also be undergoing transport).

Transport of muscarinic receptors can be demonstrated in the fimbria (septal-hippocampal pathway) by lesioning with a microelectrode attached to a radiofrequency lesion generator. This transport is similar to what is seen in the PNS, but also differs in several aspects. Low doses of colchicine, which may block fast transport relatively selectively (26), have little effect on the accumulation of receptors on the proximal side of the fimbria lesion when the injections are placed in the medial septum. High doses of colchicine which inhibit both fast and slow transport, effectively inhibit this transport when injected into the septum (Gehlert et al., in preparation). High affinity sites accumulate initially at the proximal side of the fimbria lesion followed later by a large build-up of low affinity agonist sites. Thus, either the high affinity sites are transported faster or they soon convert to low affinity sites as a result of the lesion. Also, these affinity states do not interconvert in the presence of guanine nucleotides, indicating that these central receptors are regulated differently. Interestingly, at 24 hours when a large proportion of sites at the proximal side of the lesion are of low affinity, [^{3}H]-PZ binds proximal to the lesion (Wamsley et al., in preparation). Thus, M_1 receptors are being transported in the fimbria. This muscarinic receptor subtype may have high and low affinity agonist states of the receptor which are not sensitive to guanine nucleotides.

The use of lesioning techniques can also be used to demonstrate postsynaptic muscarinic receptor up-regulation which may reflect supersensitivity. Irreversible spinal cord injury results in an initial flacid paralysis followed by the progressive development of hyperreflexia. To investigate the receptors potentially involved in this phenomenon, adult cat spinal cords were transected at the T5-T7 levels and were sacrificed at various time intervals up to three months. Lumbar spinal cord levels were removed at sacrifice, sectioned and labelled with [^{3}H]-NMS in the presence or absence of 100 micromolar concentrations of carbachol to detect changes in high and low affinity binding sites (Lane et al., in preparation). After two days, the substantia gelatinosa of the dorsal horn showed a significant increase in muscarinic receptor density due entirely to an increase in low affinity state of the receptor when compared to sham operated controls. The ventral horn displayed supersensitivity at two days characterized by an increase in the low affinity state with a corresponding decrease in the high affinity state. Density readings taken at 14 days demonstrated no change in total binding. However, an increase in low affinity sites in the ventral horn and a decrease in low affinity sites in the dorsal horn was observed. Studies have implicated the low affinity conformation as the functional state of the receptor in neurotransmission (5, 7, 30). Therefore, the shift in receptor populations to the low affinity state may represent an exaggerated functional capability due to a loss of presynaptic input (denervation supersensitivity). Increases in high affinity sites may represent newly synthesized receptors, transported to the sites but not yet incorporated into the active conformation.

Septal lesions have also recently been shown to increase the density of low affinity agonist sites in CA1 of the hippocampus (Crutcher et al., in preparation). The opposite situation is true in the human disease amyotrophic lateral sclerosis. This disease is characterized by a progressive weakness and atrophy of denervated muscles. There is a marked degeneration of motor neurons in the ventral horn of the spinal cord. In post-mortem spinal cords labelled for high and low affinity sites, there is a marked reduction in muscarinic receptors in the ventral horn of the spinal cord, characterized principally by a reduction in the high affinity agonist state (44).

The cholinergic system has also been implicated in the memory impairment seen in many psychiatric disorders (1). The memory deficits observed in the Wernicke-Korsakoff syndrome are thought to result from thiamine deficiency secondary to chronic alcoholism, but can be caused by other factors as well (10, 18, 24, 32). To investigate the muscarinic receptor role in this disease animals were made thiamine deficient by the method of Gubler (15) or by a thiamine deficient diet. Animals were sacrificed at the onset of severe thiamine deficiency marked by polyneuritic convulsions, piloerection, severe weight loss and anorexia. Brains were removed, sectioned and labelled to detect changes in high and low affinity binding sites. Thiamine deficient animals exhibited a marked alteration in muscarinic receptors when compared to their respective controls. Marked elevations in receptor densities were seen in areas such as the stratum oriens and the stratum lacunosum moleculare of the hippocampus, and the striatum (13). The elevations were characterized primarily by an increase in the low affinity agonist state of the receptor. Since thiamine deficiency has been reported to reduce acetylcholine levels in the midbrain and diencephalic areas, including the hippocampus (29, 33), the increase in receptor populations may represent an "up-regulation" (supersensitivity). This may explain the clinical usefulness, although temporary, of anticholinesterases in the treatment of memory disorders (28). Thiamine, physostigmine (an anticholines- terase), or arecoline (a muscarinic agonist) can reverse the alter- ations seen in behavioral paradigms induced by thiamine deficiency (14). Two other areas, the subiculum and the ventro-medial hypo- thalamus showed marked decreases in receptor density. The most prominent decrease was seen in the low affinity state of the receptor. The ventro-medial hypothalamus is thought to play a role in the control of appetite as the so-called "satiety center" (8, 11). Since one of the signs of severe thiamine deficiency is anorexia, the reduction in muscarinic receptor densities seen in the ventro- medial hypothalamus may represent a decreased functionality of the cholinergic system in controlling appetite, though the role of the cholinergic function in appetite is unknown.

In conclusion, quantitative in vitro receptor autoradiography is a valuable tool in assessing the dynamics of cholinergic func- tion. Changes in receptor populations can be detected with a high degree of precision in areas traditionally unavailable to dissection techniques (i.e., laminated structures, small brain nuclei). Using agonist displacement of a classical antagonist, alterations in both the high affinity and low affinity agonist state of the receptor can be determined. Oftentimes, this method can detect alterations which would be missed if receptors were measured in homogenate prepa- rations using a radiolabelled antagonist alone. Detection of changes in the high and low affinity state of the receptor can also add additional information about the etiology of the alteration.

Acknowledgements. These studies were supported by the following grants from the Public Health Service: MH-36365 (JKW), MH-30626, MH-27257 (HIY) and Program Project Grant HL-20984 (WRR). These studies were also supported by a grant from the Department of Defense to JKW (DAMD17-83-C-3023). HIY is a recipient of an RSDA type II award (MH-00095) and WRR is a recipient of a RSDA (HL-00776). We thank Jane Stout for her secretarial assistance.

REFERENCES

1. Bartus, R.J., Dean, R.L., Bear, B. and Lippa, A.S. (1982): Science 47:408-417.
2. Berrie, C.P., Birdsall, N.J.M., Burgen, A.S.V. and Hulme, E.C. (1979): Biochem. Biophys. Res. Commun. 87:1000-1005.

3. Birdsall, N.J.M., Burgen, A.S.V., Hammer, R., Hulme, E.C. and Stockton, J. (1980): Scand. J. Gastroenterol. 15,Suppl. 66:1-5.
4. Birdsall, N.J.M., Burgen, A.S.V., Hiley, C.R. and Hulme, E.C. (1976): J. Supramolec. Struct. 4:367-371.
5. Birdsall, N.J.M., Burgen, A.S.V. and Hulme, E.C. (1977): In: Cholinergic Mechanisms and Psychopharmacology, (ed) D.J. Jenden, Plenum Press, New York, pp. 25-33.
6. Birdsall, N.J.M., Burgen, A.S.V. and Hulme, E.C. (1978): Mol. Pharmacol. 14:723-736.
7. Birdsall, N.J.M. and Hulme, E.C. (1976): J. Neurochem. 27:7-16.
8. Bray, G.A. and York, D.A. (1979): Physiol. Rev. 59:719-809.
9. Burgisser, E., DeLeon, A. and Lefkowitz, R.J. (1982): Proc. Natl. Acad. Sci. 79:1732-1736.
10. Butters, N. (1980): In: Currents in Alcoholism, Vol. VIII, "Recent Advances in Research and Treatment," (ed) M. Galanter, New York, Grune and Stratton, pp. 205-232.
11. Colpaert, F.C. (1975): Behav. Biol. 15:27-44.
12. Ehlert, F.J., Roeske, W.R. and Yamamura, H.I. (1980): J. Supramol. Struct. 14:149-162.
13. Gehlert, D.R., Morey, W.A. and Wamsley, J.K.: J. Neurosci. Res., in press.
14. Gibson, G., Barclay, L. and Blass, J. (1982): Ann. N.Y. Acad. Sci. 378:382-403.
15. Gubler, C. (1961): J. Biol. Chem. 236:3112-3120.
16. Hammer, R., Berrie, C.P., Birdsall, N.J.M., Burgen, A.S.V. and Hulme, E.C. (1980): Nature 283:90-92.
17. Hammer, R. and Giachetti, A. (1982): Life Sci. 31:2991-2998.
18. Hoyumpa, A.M. (1980): Am. J. Clin. Nutr. 33:2750-2761.
19. Hulme, E.C., Birdsall, N.J.M., Burgen, A.S.V. and Metha, P. (1978): Mol. Pharmacol. 14:737-750.
20. Jannewein, H.M. (1979): In: Die Behandlund des Vleus Pepticum mit Pirenzepin, (eds) A.L. Blum and R. Hammer, Karl Demeter, Munich, pp. 41-48.
21. Jaup, B.H., Stockbrugger, R.W. and Dotevall, G. (1980): Scand. J. Gastroenterol., Suppl. 6, 89-102.
22. Laduron, P. (1980): Nature 286:287-288.
23. Laduron, P.M., Leysen, J.E. and Gorissen, H. (1981): Arch. Int. Pharmacodyn. 249:319-321.
24. Leevy, C.M. (1982): Ann. N.Y. Acad. Sci. 378:316-326.
25. Matsuo, Y. and Seki, A. (1979): Arzeneimittelforschung 29: 1028-1035.
26. McClure, W.O. (1972): Adv. Pharmacol. Chemother. 10:185-220.
27. Palacios, J.M., Niehoff, D.L. and Kuhar, M.J. (1981): Neurosci. Lett. 25:101-105.
28. Souhupova, B., Vojtechovsky, M. and Safratova, V. (1970): Act. Nerv. Super. (Praha) 12:91-93.
29. Speeg, K.V., Chen, D., McCandless, D.W. and Scheuker, S. (1970): Proc. Soc. Exp. Biol. Med. 135:1005-1009.
30. Strange, P.G., Birdsall, N.J.M. and Burgen, A.S.V. (1977): Biochem. Soc. Trans. 5:189-191.

31. Unnerstall, J.R., Niehoff, D.L., Kuhar, M.J. and Palacios, J. M. (1982): J. Neurosci. Meth. 6:59-73.
32. Victor, M., Adams, R.D. and Collins, G.H. (1971): The Wernicke-Korsakoff Syndrome. Oxford, Blackwell.
33. Vorhees, C.V., Schmidt, D.E. and Barrett, R.J. (1978): Brain Res. Bull. 3:493-496.
34. Wamsley, J.K. (1984): In Cerebral Cortex, Vol. 2, (eds) E.G. Jones and A. Peters, Plenum Press, New York, pp. 173-202.
35. Wamsley, J.K. (1983): Eur. J. Pharmacol. 86:309-310.
36. Wamsley, J.K., Gehlert, D.R., Roeske, W.R. and Yamamura, H.I. (1984): Life Sci. 34:1395-1402.
37. Wamsley, J.K., Lewis, M.S., Young, W.S. and Kuhar, M.J. (1981): J. Neurosci. 1:176-191.
38. Wamsley, J.K. and Palacios, J.M. (1983): In: Current Methods in Cellular Neurobiology, (eds) J. Barker and J. McKelvy, John Wiley and Sons, New York, pp. 241-268.
39. Wamsley, J.K., Zarbin, M.A., Birdsall, N.J.M. and Kuhar, M.J. (1980): Brain Res. 200:1-12.
40. Wamsley, J.K., Zarbin, M.A. and Kuhar, M.J. (1981): Brain Res. 217:155-162.
41. Wamsley, J.K., Zarbin, M.A. and Kuhar, M.J. (1983): Brain Res. Bull., 12:233-243.
42. Watson, M., Roeske, W.R. and Yamamura, H.I. (1982): Life Sci. 31:2019-2023.
43. Watson, M., Yamamura, H.I. and Roeske, W.R. (1983): Life Sci. 32:3001-3011.
44. Whitehouse, P.J., Wamsley, J.K., Zarbin, M.A., Price, D.L., Tourtellotte, W.W. and Kuhar, M.J. (1983): Ann. Neurol. 14:8-16.
45. Yamamura, H.I. and Snyder, S.H. (1974): Mol. Pharmacol. 10: 861-867.
46. Yamamura, H.I., Vickroy, T.W., Wamsley, J.K. and Roeske, W.R.: Brain Res., in press.
47. Yamamura, H.I., Watson, M. and Roeske, W.R. (1983): In: CNS Receptors: From Molecular Pharmacology to Behavior, (ed) F.V. DeFeudis, Raven Press, New York.
48. Yamamura, H.I., Wamsley, J.K., Deshmukh, P. and Roeske, W.R. (1983): Eur. J. Pharmacol. 91:147-149.
49. Young, W.S. III and Kuhar, M.J. (1979): Brain Res. 179:255-270.
50. Zarbin, M.A., Wamsley, J.K. and Kuhar, M.J. (1982): J. Neurosci. 2:934-941.

IMMUNOCYTOCHEMICAL EVIDENCE FOR THE INTRA-AXONAL

TRANSPORT OF nAChR-LIKE MATERIAL IN MOTOR NEURONS

A.B. Dahlström[1], S. Bööj[1], A.-G. Dahllöf[1],
J. Häggblad[2] and E. Heilbronn[2]

Institute of Neurobiology
University of Göteborg[1], Sweden
Unit of Neurochemistry and Neurotoxicology
University of Stockholm[2], Sweden

INTRODUCTION

The nicotinic acetylcholine receptor (nAChR) at the motor endplate of mammals, and in the electric organ of e.g. Torpedo, is an integral membrane protein (5). In the normal mammalian motor endplate nAChR is present in the postsynaptic membrane, i.e., in the sarcolemma of the innervated skeletal muscle. Upon denervation the amount of nAChR increases and the receptors will eventually spread over the whole surface of the muscle cell (3, 23). This indicates that the muscle cell is capable of manufacturing its own nAChR protein.

However, certain results from pharmacological and morphological investigations on motor endplates suggest the existence also of presynaptic nAChR sites. Earlier pharmacological studies are reviewed by e.g., Miyamoto (24) and recent pharmacological evidence is discussed by Häggblad and Heilbronn (this book; 11). The morphological evidence consists mainly of the demonstration of α-bungarotoxin (α-bgt) binding sites at the presynaptic membrane, in addition to the considerable toxin binding at the postsynaptic membrane. Bender et al. (4) located the bound α-bgt by the immunoperoxidase procedure, and found in human skeletal muscle a weak but significant staining of the presynaptic membrane in the synaptic region. Similar observations were made in frog and mouse muscle (8). A more direct way of studying the α-bgt binding at the EM level is to use α-bgt directly conjugated with horseradish peroxidase (HRP). This approach was used by Lentz et al. (20) who found a staining of the presynaptic membrane, particularly of the axolemma overlying the active zones in rat and frog. The significance of these observations has been debated, and a diffusion artefact has been suggested to cause the presynaptic labelling (16). However, using the HRP-α-bgt Lentz et al. (20) found no staining over Schwann cell fingers in the synapse, and in terminals removed from the endplates by enzyme digestion the axolemma was still labelled, supporting the true localization of α-bgt-binding receptor protein in the presynaptic membrane.

Presynaptic receptor sites may originate from the muscle by transsynaptic passage of receptor protein molecules. Another and more obvious source would, however, be the motor neuron, whose distal ends carry these presumptive receptor sites. Since the nAChR is such a complex molecule it would have to be manufactured in the cell soma and carried distally by axonal transport. To test this hypothesis the present investigation was undertaken. A short report on this work has been published earlier (6).[1]

MATERIAL AND METHODS

Male Sprague-Dawley rats (200 g) were used. Under ether anaesthesia the sciatic nerves were crushed with a silk suture or spinal roots were crushed with watchmakers forceps at the level of L4-L5 vertebrae. At various times postoperatively the animals were reanaesthetized with Nembutal, for intracardiac perfusion. After a preperfusion with warm phosphate buffered saline (PBS, with heparin) the animals were perfused with ice cold 4% formaldehyde in PBS (pH 7.4) for 3-4 min. The spinal cord, the crushed roots and the sciatic nerves were dissected and postfixed in the same fixative for 4 h at +4°C. Following thorough rinsing in PBS with 5% sucrose the specimens were frozen and cryostat sectioned at 10μ. The sections were placed on gelatin coated glass slides and incubated for demonstration of nAChR-like material. In all experiments the middle endplate carrying section of the diaphragm was dissected and included in the incubation series as a control for incubation efficiency.

In a separate set of experiments one group of rats was subjected to an axotomy operation (a cut) at the knee level of the sciatic nerve, 1-7 d before a 6 h crush 15-17 mm proximal to the initial crush. For controls, animals with the 6 h crush only were used. In these animals the sciatic nerve proximal and distal to the crush as well as the spinal cord (L3-L4) were dissected after the perfusion fixation.

Incubations. Four different antibody preparations against nAChR complex were used:

---

[1]Preliminary observations were communicated at the European Symposium on Cholinergic Transmission in Strasburg, May 1982, and presented at the meeting on "Molecular Aspects of Neurological Disorder" in Australia, September 1982 (6).

1). Rabbit anti Torpedo nAChR (13), IgG-fraction, was used in dilution 1:200, followed by fluorescein-isothiocyanate (FITC)conjugated swine anti-rabbit (SWAR) IgG, 1:30; 2 and 3). Two monoclonal antibodies (mcAb) against Torpedo nAChR, numbers 5.14 and 5.5 (prepared as described in 25, 30) were a gift from Dr. Sara Fuchs. They were used in a dilution of 1:50, followed by incubation in FITC-conjugated rabbit-anti-mouse (RAM) IgG, 1:20; 4). The IgG fraction from a patient with Myasthenia gravis, demonstrated to contain antibodies against nAChR-protein, was used in dilution 1:200, followed by FITC-conjugated rabbit antihomo IgG, 1:20. Incubations with primary antisera were performed at +4°C overnight, while secondary antisera (all purchased from Dakopatts, Copenhagen) were applied for 1 h at 37°C. All antisera contained 0.1% Triton X-100.

To demonstrate α-bgt-binding capacity in the tissue sections FITC- or rhodamine-conjugated α-bgt (12.5 nM) was used. Since the molecular ratio of fluorophor/α-bgt is only 1:1, giving a very low fluorescence yield, a second method to localize α-bgt- binding sites was used. After incubation with unlabelled α-bgt (10 nM) for 1 h at 37°C and rinsing in PBS x 3 for 30-60 min. at room temperature, the bound toxin was localized by incubation with rabbit-anti-α-bgt-serum (1:200) overnight at +4°C, and the sections were incubated in SWAR-FITC (vide supra).

The sections were examined in a Zeiss fluorescence photomicroscope equipped for incident activation. Tri-X-Pan (Kodak, 400 ASA) was used for photography.

For demonstration of AChE after immunoincubation the method of Koelle (17) as modified by Gruber and Zenker (9) was used.

RESULTS

Diaphragm. Motor endplates were regularly visualized using rabbit-- anti-Torpedo-IgG and mcAb 5.14 (Figs. 1a, b). With the myasthenic homo-IgG the background fluorescence was generally too high to enable the endplates to stand out, but a few endplates could usually be identified. With mcAb 5.5 endplates could not be observed. With labelled α-bgt endplates were strongly fluorescent.

Sciatic nerve. In intact nerves no immunofluorescence was present intraaxonally. Using the homo myasthenic IgG, however, the connective tissue component had an obvious background fluorescence. In operated nerves, accumulations of intraaxonal immunofluorescence were seen with all nAChR antibody preparations used. The fluorescence was strongest with the rabbit-anti-Torpedo-nAChR-IgG, and weakest with mcAb 5.5 (Figs. 2 a-d). Fluorescence was present on both sides of the crush area but always much stronger on the proximal than on the distal side.

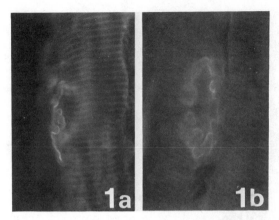

Figure 1. Motor endplates from rat diaphragm incubated for indirect
immunofluorescence (see text) with: a) rabbit-anti-Torpedo-nAChR-
IgG; and b) monoclonal antibody 5.14. The postsynaptic foldings
are seen. Fluorescence microphotographs x 360.

      About one half of the thick myelinated axons was found to accumu-
late the immunofluorescent material. Most of these axons were AChE
positive, and thus probably motor axons (Figs.
2a, b).

      The accumulations could not be seen clearly earlier than 3-6
h after crushing, in contrast to other axonally transported substances
and organelles, which can be seen to accumulate within one hour
after crushing (cf. 7). After 6 h, however, the accumulations built
up rapidly to reach maximum around 24 h postoperatively. One reason
for this late appearance may be that the nerve crush, interrupting
the axonal connection between soma and periphery, induced an increased
synthesis of the nAChR-like material, and thereby increased the
amounts of axonally transported material. In order to test this

Figure 2. Sections of rat sciatic nerve, crushoperated (arrows)
13 h before perfusion fixation. The sections were incubated for
indirect immunofluorescence (see text) with: (a) rabbit-antiTorpedo-
nAChR-IgG; (c) monoclonal antibody 5.14; and (d) monoclonal antibody
5.5. In (b), the same section as in (a) was after photography stained
for AChE (see text). Most of the axons with accumulations of immunoreactive material in (a) also have AChE accumulations in (b). Microphotographs x 180.

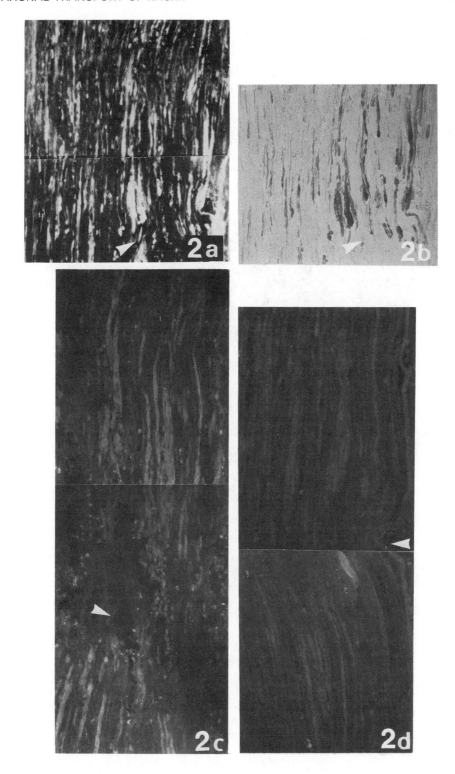

some rats were axotomized 1-7 d prior to a 6 h crush. All previously axotomized sciatic nerves contained considerably more immunofluorescent material proximal to the 6 h crush than the control rats, in which the accumulation were barely visible. Also, as judged by eye, the motor somata of the spinal cord of the axotomized animals appeared to have a stronger fluorescence intensity.

With fluorophor labelled α-bgt a very weak fluorescence was observed in 24 h crushed nerves. When α-bgt binding was investigated using rabbit-anti-α-bgt-antiserum, however, a clear fluorescence was observed in accumulated axons on either side of the lesion.

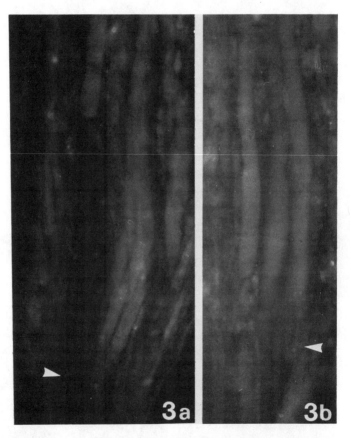

Figure 3. Sections of rat ventral root L3, crushoperated with forceps (arrows) 18 h before perfusion fixation, incubated with: (a) anti-Torpedo-nAChR-IgG; and (b) α-bgt followed by rabbit anti-α-bgt serum (see text). In both sections thick axons proximal to the crush contain accumulated immunofluorescent material, which in (a) cross-reacts with the antiTorpedo-nAChR-IgG, while in (b) it has bound α-bgt. Fluorescence microphotographs x 680.

Roots. In ventral roots crushed 12 or 18 h earlier, accumulations of immunoreactive material were present in most axons on the proximal side (Fig. 3). The order of immune reactivity (based on fluorescence intensity) of the antisera used was the same as for the sciatic nerve. Some immunofluorescence was also present on the distal side.

When α-bgt-incubation was followed by incubation with rabbit-anti-α-bgt-antiserum a medium fluorescence was noted in dilated axons proximal but also distal to the crush in ventral roots (Fig. 3b).

Spinal cord. In cross-sections of the lumbar intumescence the large motor cell bodies in the ventral horn contained immunoreactive material. The fluorescence intensity varied from very weak (with mcAb 5.5) to weak (mcAb 5.14) to medium with the rabbit-antiTorpedo-nAChR-IgG. The material was located in the cytoplasm with a patchy pattern, often concentrated near the margin of the cells (Fig. 4). Also the nerve cells in other areas of the gray matter were fluorescent, as well as glial cell bodies in the white matter (Fig. 4).

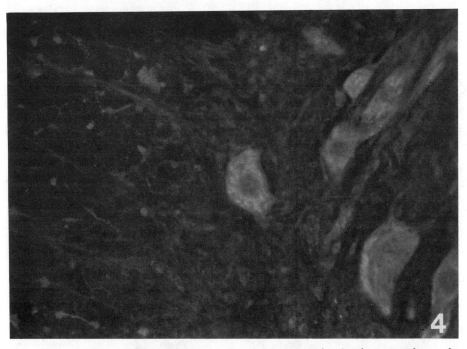

Figure 4. Cross-section from rat spinal cord, lumbar region, incubated for indirect immunofluorescence with anti-Torpedo-nAChR-IgG. Large motorneurons in the ventral horn contain immunoreactive material with a patchy appearance in the cytoplasm. Also glial cell bodies in the white matter contain immunoreactivity. Fluorescence microphotograph x 520.

The α-bgt-binding capacity appeared to be lower in the nerve
cell bodies than in the crushed axons; only weak to very weak fluorescence was observed after immunohistochemical detection of the
bound α-bgt.

Controls. a) When the primary antisera were excluded, no immunofluorescence was seen in ligated axons, perikarya or glia cells;
b) Two preparations of nAChR-protein were used to preadsorb two of
the four antibody preparations (rabbit-anti- Torpedo-nAChR-IgG and
mcAb 5.14), i.e., a microsac preparation from Torpedo electric organ
containing nAChR and a Triton-X solubilized preparation of nAChR
from denervated rat skeletal muscle. Both nAChR-preparations could,
when used for preadsorbtion experiments, prevent the occurrence of
immunofluorescence in axons, perikarya and glial cell bodies (Figs.
5a,b).

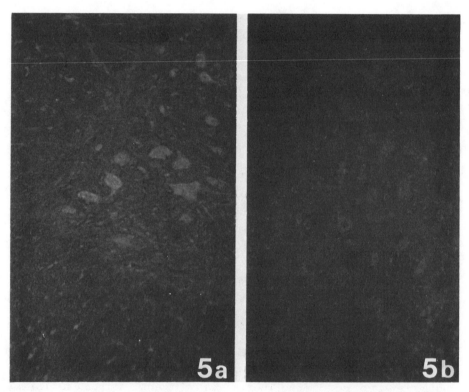

Figure 5.   Sections from rat spinal cord, lumbar region incubated
with: (a) anti-Torpedo-nAChR-IgG; and (b) with the same antibody
preadsorbed with a rat nAChR-preparation.   Fluorescence microphotographs x 130.

DISCUSSION

All four antibody preparations used have been demonstrated to contain antibodies against nAChR-protein. In controls, where the primary specific antibodies were excluded, or where preimmune serum, serum from normal human controls, or culture medium were applied instead of the anti-nAChR antibodies, no immunoreactive material was observed in axons or cell bodies. The immunoreactive material discovered with the antibodies is thus immunologically related to nAChR. The results after preadsorbtion with preparations of nAChR membrane bound, which abolished the immunostaining, support the view that the material observed is related to nAChR. However, the membrane-preparations used for preadsorbtion also contain other membrane components, and therefore the results do not suffice to definitely identify the material as nAChR.

Demonstration of α-bgt binding would provide further evidence. Such binding could not be convincingly demonstrated directly using FITC- or TRITC-labelled α-bgt, but a very low fluorescence could be detected upon careful examination (6). The molar ratio of fluorophor/α-bgt is about 1:1 and since the amount of nAChR-like material in the motor neurons is probably very low, detection of α-bgt binding requires an amplification step, e.g. indirect immunohistochemistry, to detect the bound α-bgt. When using this method a clear immunofluorescence was noted in ventral root axons (Fig. 3b) and a weak but obvious fluorescence in cell bodies, supporting the presence of α-bgt-binding sites, probably nAChR or subunits thereof, in motor nerves. Thus, all evidence so far indicate that a nAChR-like material is present in motor axons of the rat sciatic nerve, and in the neuronal perikarya of the spinal cord.

Our immunocytochemical evidence for the presence and axonal transport of nAChR protein in rat motor neurons is further supported by the observations by Aizenman et al. (1, 2). These authors, by using [125I] labelled α-bgt, have demonstrated the bidirectional axonal transport of α-bgt-binding sites in rat sciatic nerve, using an autoradiographic procedure. They also investigated the pharmacological nature of the α-bgt-binding sites in the nerve and found that the axonally transported α-bgt-binding sites behave more like toxin binding sites in the brain as opposed to the muscle nAChR. It is probable that the immunologically reactive material we see corresponds to the α-bgt-binding sites observed by Aizenman et al. (2).

Nincovic and Hunt (29) also found that α-bgt-binding sites accumulated bidirectionally in the sciatic nerve, but suggested that this occurred in sensory afferents. The reason for this suggestion was based on the physiological evidence on the action of ACh on sensory nerves and secondly on the lack of significant α-bgt-binding in the ventral horn of the spinal cord as compared with sensory ganglia. We have not so far investigated the sensory system in detail, but preliminary observations indicate that the amount of immunoreactive material in crushed dorsal roots is less than in ventral roots. Since we were able to demonstrate immunoreactive material in motor somata and ventral roots and a significant α-bgt binding in ventral roots this clearly shows that motor neurons do contain nAChR-like material. This does not, however, rule out the possibility that sensory nerves also contain similar material. In addition, the α-bgt-binding site may not be identical to immuno-reactive nAChR-like material and vice versa (see 25, 30).

The observation that glial cell bodies also contain immuno-reactive nAChR-like material is interesting, since Nachmansohn (cf. 28) suggested ACh-receptor to be present in the axon-Schwann cell membrane and since Villegas (34) has discussed ACh mechanisms with α-bgt-binding capacity to exist in Schwann cells, the peripheral counterpart of glial cells. In cultured glioma cells Hamprecht et al. (12) found sensitivity to ACh which could be blocked by atropine and α-bgt but not by d-tubocurarine. In this context it should be mentioned that, using mcAb 5.14, minute fluorescent dots are frequently observed over the myelin sheath of large myelinated axons in areas without intraaxonally accumulated material. With the fluorescence microscope the exact location (in axolemma· or myelin sheath) cannot be estab-lished.

The observation that the amount of nAChR-like material increases in previously axotomized nerves may indicate that the production of this material is controlled by retrograde information from the periphery. The material is probably formed by the granular ER, as suggested by the patchy immunofluorescence picture in the ventral horn cells (Fig. 4).

If the immunoreactive substance is indeed an nAChR, the material in anterograde transport could be destined to be incorporated into the presynaptic membrane, acting as an autoreceptor for the ACh release at the endplate (cf. 10, 11, 19)[2]. Since in the motor endplate the concentration of HRP-α-bgt-binding sites in the presynaptic membrane was higher near the active sites (17), where the ACh release is suggested to occur, this seems a plausible hypothesis. Known autoreceptor molecules, e.g. muscarinic AChR, have been demonstrated to be transported to the presynaptic site by axonal transport, in e.g. adrenergic axons of the cat hypogastric (18) and rat sciatic and vagus nerves (32, 35). A similar axonal transport of β_1-adreno-

---

[2]This suggested presynaptic autoreceptor is not identical with the pre-junctional receptors, the activation of which by ACh and related drugs causes antidromic repetitive firing in ventral roots. These pre-junctional receptors are probably located at the first node of Ranvier, and probably are not of physiologic significance (cf. 33).

ceptors (21) and of mAChR (31) has been reported to occur also in axons in the CNS.

It is noteworthy that accumulations of nAChR-like material occurred also on the distal side of the nerve crushes. This strongly indicated that a retrograde axonal transport of the receptor like material occurs, probably as a sign of membrane retrieval with return of used membrane constituents to the cell soma (cf. 16).

Ito et al. (15) could demonstrate that the total amount of ACh was about twice normal in myasthenic muscle and the release of ACh appeared to be higher than normal in muscle biopsies incubated without tetrodotoxin. When the toxin was added to the incubation medium (to abolish the spontaneous fasciculations occurring due to the presence of AChE-inhibitors in the medium) the resting release of ACh was, however, in the control range. In a later study on myasthenic muscle preparations Molenaar et al. (28) observed that the initial KCl induced release of ACh was more than twice that in control muscle, but after 20 min. the ACh release decreased markedly. In rats immunized with nAChR (experimental autoimmune myasthenia) the ACh release from diaphragm, induced by electrical stimulation or 50 mM KCl, was about twice that in normal diaphragms (10, 26). Further, after α-bgt pretreatment the ACh release from rat diaphragm, evoked by KCl or nerve stimulation, was markedly increased (11, 22). This may suggest that impulse or KCl triggered ACh release is increased when nAChR-sites are blocked. Such an increase may well be due to inactivation of presynaptic autoreceptors, although postsynaptic events have been discussed (14). Thus, the presynaptic nAChR would be involved in a negative feedback of ACh release (a possibility discussed by Molenaar and Polak, 27, and Häggblad and Heilbronn, 11).

Acknowledgements. Supported by the Swedish MRC (2207), by Greta and Einar Asker's Foundation, Kung Gustaf V:s 80-årsfond, Åke Wiberg's Foundation and Riksföreningen for Trafik-och Polioskadade.

We are grateful to Drs. Sara Fuchs and Daria Mochly-Rosen, The Weizmann Institute of Science, Rehovot, Israel, for a generous gift of the monoclonal antibodies. The skillful technical assistance of Mrs. Kerstin Lundmark is acknowledged.

REFERENCES

1. Aizenman, E., Millington, W.R., Zarbin, M.A., Bierkamper, G.G. and Kuhar, M.J. (1983): Fed. Proc. 42:1147.
2. Aizenman, E., Millington, W.R., Zarbin, M.A., Bierkamper, G.G. and Kuhar, M.J. (1984): This book.
3. Axelsson, J. and Thesleff, S. (1959): J. Physiol. (Lond.) 149: 178-193.

4.   Bender, A.N., Ringel, S.P. and Engel, W.K. (1976):   Neurology
     26:477-483.
5.   Briley, M.S. and Changeux, J.-P. (1977):   Int. Rev. Neurobiol.
     20:31-63.
6.   Dahlström, A. (1983):   In Molecular Aspects of Neurological
     Disorders (ed.) L. Austin, Academic Press Australia, pp. 139-146.
7.   Dahlström, A. (1983):   In Handbook of Neurochemistry ,Vol. 5
     (ed.) A. Lajtha, Plenum Press, New York, pp. 405-441.
8.   Daniels, M.P. and Vogel, Z. (1975):   Nature (Lond.) 253:339-341.
9.   Gruber, H. and Zenker, W. (1973):   Brain Res. 51:207-214.
10. Häggblad, J. and Heilbronn, E. (1983):   Br. J. Pharmacol. 80:
     471-476.
11. Häggblad, J. and Heilbronn, E. (1984):   This book.
12. Hamprecht, B., Kemper, W. and Amano, T. (1976):   Brain Res.
     101:129-135.
13. Heilbronn, E. and Mattsson, E. (1974):   J. Neurochem. 22:315-317.
14. Hohlfeld, R., Sterz, R. and Peper, K. (1981):   Pflügers Arch.
     391:213-218.
15. Ito, Y., Miledi, R., Molenaar, P.C., Vincent, A., Polak, R.L.,
     van Gelder, M. and Newsom-Davis, J. (1976):   Proc. R.   Soc.
     Lond. 192:475-480.
16. Jones, S.W. and Salpeter, M.M. (1983):   J. Neurosci. 3:326-331.
17. Koelle, G.B. (1955):   J. Pharm. Exp. Therap. 114:167-184.
18. Laduron, P. (1980):   Nature (Lond.) 286:287-288.
19. Lentz, T.L. and Chester, J. (1982):   Neurosci. 7:9-20.
20. Lentz, T.L., Mazurkiewicz, J.E. and Rosenthal, J. (1977):   Brain
     Res. 132:423-442.
21. Levin, B.E. (1982):   Science 217:555-557.
22. Miledi, R., Molenaar, P.C. and Polak, R.L. (1978):   Nature (Lond.)
     272:641-643.
23. Miledi, R. and Potter, L.T. (1971):   Nature (Lond.) 233:599-603.
24. Miyamoto, M.D. (1978):   Pharmacol. Reviews 29:221-247.
25. Mochly-Rosen, D. and Fuchs, S. (1981):   Biochemistry 20:5920-
     5924.
26. Molenaar, P.C. and Polak, R.L. (1980):   Progr. Pharmacol. 3/4:
     39-44.
27. Molenaar, P.C., Polak, R.L., Miledi, R., Alema, S., Vincent,
     A. and Newsom-Davis, J. (1979):   Progr. Brain Res. 49:449-458.
28. Nachmansohn, D. (1959):   In Chemical and Molecular Basis of
     Nerve Activity, Academic Press, New York.
29. Ninkovic, M. and Hunt, S. (1983):   Brain Res. 272:57-69.
30. Souroujon, M.C., Mochly-Rosen, D., Gordon, A.S. and Fuchs, S.
     (1983):   Muscle Nerve 6:303-311.
31. Wamsley, J.K. (1983):   Eur. J. Pharmacol. 86:309-310.
32. Wamsley, J.K., Zarbin, M.A. and Kuhar, M.J. (1981):   Brain Res.
     217:155-162.
33. Webb, S.N. and Bowman, W.C. (1974):   Clin. Exp. Pharmacol.
     Physiol. 1:123-134.
34. Villegas, J. (1978):   TINS, pp. 66-68.

35. Zarbin, M.A., Wamsley, J.K. and Kuhar, M.J. (1982): J. Neuro-
     sci. $\underline{2}$:934-941.

LATERALIZATION OF CHOLINERGIC AND ENERGY METABOLISM

RELATED ENZYMES IN HUMAN TEMPORAL CORTEX

S. Sorbi, L. Bracco, S. Piacentini,
A. Morandi and L. Amaducci

Department of Neurology
University of Florence
Firenze, Italy

INTRODUCTION

Neuroanatomical, neurochemical and functional studies have recently provided evidence of lateral asymmetry in mammalian brains, including human brain. For example, asymmetries in concentration, activity and distribution of several neuroconstituents (9), in the effect of drugs (10, 11) and in thickness of cerebral cortex (5) have been clearly observed in the rat brain. Several studies have also demonstrated lateral asymmetry in human brain. Probably the first asymmetry was described by Heschel (14), who observed right-left differences in cortical folding. Recently it has been shown that the planum temporale is significantly larger and longer on the left in the majority of infant and adult human brains (8).

However, little is known about neurochemical correlates of anatomical and functional human brain asymmetry. Oke and coinvestigators (15) provided evidence of a naturally occurring significant lateralization of norepinephrine in human thalamus. We have observed a significant lateralization of acetylcholine (ACh) synthesizing enzyme, choline acetyltransferase (ChAT), in the first temporal gyrus of human brain (1, 2, 17, 20). This finding has been recently confirmed, in other brain regions, by Glick et al. (12) who, reanalyzing data from Rossor et al. (19), found a left-bias of ChAT activity in cortical and subcortical areas. Glick et al. (12) also reported that some of these asymmetries significantly correlate with age. Moreover, they observed that other neurotransmitter related substances, namely glutamic acid decarboxylase (GAD), gamma aminobutyric acid (GABA), and dopamine (DA) which are asymmetrically distributed in human brain. Recent studies in vivo have also shown asymmetries in cerebral

93

blood flow and metabolism in stimulated normal and pathological human brains (4, 6, 13, 16, 17).

In this paper we report some of our more recent data on chemical asymmetries in human brain. We have extended our study of lateralization of the cholinergic system analyzing ChAT activity, and the density and the affinity of quinuclidinyl benzilate (QNB) reactive muscarinic receptors in the temporal lobes. Moreover, to investigate the relationship between the cholinergic lateralization in the human temporal cortex and the energy metabolism of the brain, we have analyzed the activities of some brain energy metabolism related enzymes, namely phosphofructokinase (PFK), hexokinase (HK), lactate dehydrogenase (LDH) and the pyruvate dehydrogenase complex (PDHC).

MATERIALS AND METHODS

Cortical samples were collected from seven caucasian right-handed males aged 61.3 $\pm$ 11 years (mean $\pm$ SD) who died suddenly with no clinical or pathological signs of neurological diseases. Autopsy was performed within 25 $\pm$ 3 hr (mean $\pm$ SD) after death and the brains immediately frozen and stored at -75°C until biochemical analysis. Dissection of the brains and collection of the temporal cortex samples was performed exactly as previously described (1, 2, 17, 20). Briefly, from both hemispheres of each brain four punches were collected on the first temporal gyrus: the first 1 cm caudally to the intersection between Sylvius's and Rolando's fissures, and the others at a distance of 1 cm from each other in a rostro-caudal direction. ChAT activity was assayed as previously described (1, 2, 17, 20); PDHC by the method of Sorbi and Blass (21); HK, PFK, and LDH by a modification of the method of Raker (18) and B_{max} and K_D of muscarinic receptors according to Burgen et al. (3).

Asymmetries are expressed, according to Glick et al. (12), using the formula

$$[(high\ side-low\ side)-1)] \times 100$$

Positive values indicate left-biased asymmetries while negative values indicate right-biased asymmetries.

RESULTS

Confirming our previous studies (1, 2, 17, 20) ChAT activity was always lateralized (Table 1).

ChAT in the first temporal gyrus was significantly greater (p < 0.05) on the left side in the majority of the samples in all seven brains studied (Student's t-test). B_{max} and K_D of QNB binding to muscarinic receptors were studied in three brains at the level of punch four and they were both significantly lateralized (Table 2).

Table 1

Choline Acetyltransferase Asymmetry in
Human First Temporal Gyrus

Punch	
1	+ 24.4 $\pm$ 23.6
2	+ 37.3 $\pm$ 27.2
3	+ 72.2 $\pm$ 34.3
4	+ 20.4 $\pm$ 15.0

Values are mean $\pm$ SE of asymmetries in 7 brains. Positive and negative values indicate left and right bias, respectively.

However, both B_{max} and K_D of QNB binding to muscarinic receptors were always right biased.

PDHC, PFK, LDH and HK activities also were always lateralized in all four punches collected in the first temporal gyrus. The activities of these energy metabolism related enzymes were greater on the right side in the anterior and central portion of the first temporal gyrus, but they were left biased in the most posterior portion of the first temporal gyrus (Table 3).

DISCUSSION

The results of this study provide further evidence that chemical lateralization is, indeed, present in human brain and that it is perhaps functionally significant. We have confirmed our previous results (1, 2, 17, 20) and those of Glick et al. (12) indicating that ChAT activity is greater on the left side of the brain.

However, the B_{max} and the K_D of QNB binding to muscarinic receptors exhibited a significant right bias in the most posterior portion of the first temporal gyrus. This finding indicates that on the left side, where ChAT activity is greater, the number of receptors may be lower but they may have a greater affinity for ACh. Alternatively, on the right side, ChAT activity may be lower while the number of receptors is greater but with lower affinity for ACh.

This negative correlation between ChAT activity and B_{max} and K_D in the most posterior portion of the first temporal gyrus seems worthy of consideration. In fact, the data from this anatomical area, the most interesting because of the anatomical asymmetries and the lateralization of language function, seem to indicate that

Table 2

B_{max} and K_D of QNB Binding to Muscarinic Receptor
Asymmetry in Human First Temporal Gyrus

Case		Left	Right
1	B_{max}	208	270
	K_D	0.09	0.11
2	B_{max}	161	333
	K_D	0.05	0.14
3	B_{max}	259	389
	K_D	0.04	0.11
Mean $\pm$ SD			
	B_{max}	209 $\pm$ 28	330 $\pm$ 34*
	K_D	0.06 $\pm$ 0.01	0.12 $\pm$ 0.001*

B_{max} is expressed as pmol/g protein. K_D is expressed as μM.
*$p < 0.05$ by Student's t-test.

Table 3

PDHC, PFK, HK and LDH Asymmetries in Human
First Temporal Gyrus

Punch	PDHC	PFK	HK	LDH
1	- 40 $\pm$ 16	- 52 $\pm$ 22	- 25 $\pm$ 23	- 16 $\pm$ 4
2	+ 5 $\pm$ 16	- 37 $\pm$ 26	- 26 $\pm$ 7	- 13 $\pm$ 5
3	- 22 $\pm$ 48	- 25 $\pm$ 14	- 9 $\pm$ 15	- 8 $\pm$ 1
4	+ 44 $\pm$ 43	+ 9 $\pm$ 22	+ 7 $\pm$ 16	+ 14 $\pm$ 6

Values are mean $\pm$ SE of asymmetries in four cases. Positive and
negative values indicate left and right bias, respectively.

negative and/or positive feedback may play an important role in determining and/or controlling left-right asymmetries.

This consideration agrees with that of Glick et al. (12) who found significant correlations between most of the substances they analyzed in human brain. We observed also that the activity of all four energy metabolism related enzymes had a very similar pattern of lateral asymmetry in the first temporal gyrus. PDHC, LDH, HK and PFK left-right asymmetries were in fact significant by the chi square test ($P < 0.05$). This positive correlation between all four enzymes is concordant with the intimate relationship existing among them, and supports the large body of evidence (4, 6, 16, 17) indicating that energy metabolism differs between the two sides of the human brain.

Acknowledgement: This paper has been supported in part by C.N.R. (Consiglio Nazionale delle Richerche, Rome, Italy) grants No. 82. 02268.56; 82.02072.56; and 82.02071.04.

REFERENCES

1. Amaducci, L., Sorbi, S., Albanese, A., Gainotti, G. (1981): Neurology 31:799-805.
2. Amaducci, L., Bracco, L., Sorbi, S., Albanese, A., Gainotti, G. (1982): In: Katsuki, International Congress Series No. 568; Neurology, pp. 289-297; Excerpta Medica.
3. Burgen, A.S.V., Hiley, C.R., Young, J.M. (1974): Brit. J. Pharmacol. 51:279-285.
4. Chase, T.N., Fedio, P., Foster, N., Pollard, S. (1983): Eur. Neurol. 22(Suppl 2):32.
5. Diamond, M.C., Johnson, R.E., Ehlert, J. (1979): Behav. Neur. Biol. 26:485-491.
6. Dupui, P., Guell, A., Bessols, G., Geraud, G., Bes, A. (1983): Eur. Neurol. 22(Suppl 2):27.
7. Gainotti, G., Sorbi, S., Albanese, A., Amaducci, L. (1981): Psychiat. Behav. 7:7-20.
8. Geschwind, N., Levitsky, W. (1968): Science 161:186-187.
9. Glick, S.D., Ross, D.A. (1981): Trends Neurosci. 4:196-199.
10. Glick, S., Meibach, R.C., Cox, R.D., Maayani, S.D. (1979): Life Sci. 25:395-400.
11. Glick, S.D., Weaver, L.M., Meibach, R.C. (1981): Psychopharmacol. 73:323-327.
12. Glick, S.D., Ross, D.A., Hough, L.B. (1982): Brain Res. 234: 53-63.
13. Gur, R.C., Gur, R.E., Obrist, W.D. et al. (1982): Science 217: 659-661.
14. Heschel, R.L. (1878): Uber die vordere quere schlafenwindung des menschlinchen grosshirns. Braumuller, Vienna.
15. Oke, A., Keller, R., Mefford, I., Adams, R.N. (1978): Science 200:1411-1413.

16. Phelps, M.E., Mazziotta, J.C., Kuhl, D.E., Nuwer, M., Packwood, J., Metter, J., Engel, J. (1981): Neurology 31:517-529.
17. Phelps, M.E., Kuhl, D.E., Mazziotta, J.C. (1981): Science 211: 1445-1448.
18. Racker, E. (1947): J. Biol. Chem. 167:843-854.
19. Rossor, M., Garett, N., Iversen, L. (1980): J. Neurochem. 35: 743-745.
20. Sorbi, S., Amaducci, L., Albanese, A., Gainotti, G. (1980): Boll. Soc. It. Biol. Sper. 56:2226-2270.
21. Sorbi, S., Blass, J.P. (1981): J. Biochem. Biophys. Meth. 5: 169-176.

IMMUNOLOGICAL APPROACH TO CHOLINERGIC TRANSMISSION: PRODUCTION OF MONOCLONAL ANTIBODIES AGAINST PRESYNAPTIC MEMBRANES ISOLATED FROM THE ELECTRIC ORGAN OF TORPEDO MARMORATA

L. Eder-Colli and S. Amato
Departement de Pharmacologie, C.M.U.
1211 Geneve 4 Switzerland

INTRODUCTION

The release of the neurotransmitter acetylcholine (ACh) has been thoroughly analyzed with the techniques of electrophysiology. The events taking place at the membrane of the cholinergic nerve endings during ACh release were more recently visualized by morphological analysis of freeze fractured membranes. However, the biochemical basis of the neurotransmitter release is still poorly understood. A major question is which components of the membrane of the cholinergic nerve endings are involved in the release mechanism? One means of identifying some of these components would be to apply the monoclonal antibody strategy (13); in this way, it should be possible to isolate a monoclonal antibody with an exquisite specificity towards a given component. This antibody would then be used as a tool to purify and further characterize the antigen.

One of the best sources of material to study cholinergic transmission is the electric organ of the fish Torpedo because of (i) the profuse and purely cholinergic innervation of this tissue and of (ii) the possibility of isolating from this tissue cholinergic nerve endings (synaptosomes) that are practically devoid of mitochondrial and postsynaptic contamination (10). We have prepared synaptosomal plasma membranes (SPM) from the electric organ of Torpedo marmorata and subsequently used these as immunogen in Balb/c mice. In the present chapter, we will first briefly describe some characteristics of the isolated SPM and then describe some properties of the monoclonal antibodies we have developed against these membranes.

METHODS

Isolation of SPM. Synaptosomes were obtained as described by Israël et al. (10). The purification of SPM was mainly based on the procedure described by Morel et al. (18), with some modifications (see Figure

1). Choline-acetyltransferase (ChAT), acetylcholinesterase (AChE), and lactate dehydrogenase (LDH) activities were determined as previously described (6, 12, 19). The nicotinic receptor to ACh (nRACh) was assayed according to Schmidt and Raftery (21) and proteins were assayed by the Amido Black method (20).

Immunization and cell fusions. Female Balb/c mice were immunized as previously described (5). The hybridization experiments were performed according to Galfré et al. (7). Immune responses were analyzed by the dot immunobinding assay (8), immunoblotting techniques (2, 3), and indirect immunofluorescence using a fluorescein-conjugated second antibody.

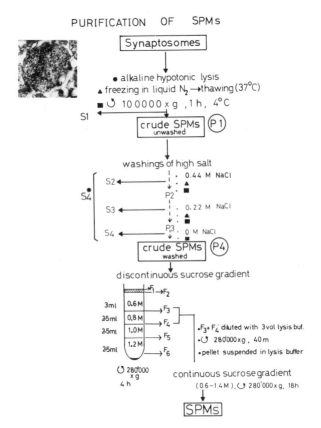

Figure 1. Flow chart of the preparation of SPM. The method of Morel et al. (18) was used with the following modification: washings with high salt solutions of the crude SPM were introduced in order to remove all the ChAT activity which binds ionically to membranes. We could improve our purification by using the last step shown, which is a continuous sucrose gradient. The symbols ▲ and ■ are used to indicate repetition of the freezing and thawing steps and of the high speed centrifugation.

Screening of the hybrids. To permit an early selection of hybridomas of interest, antibodies were tested for their reaction against purified SPM and membranes from the electric nerve which innervates the electric organ. Only those hybridomas secreting antibodies that reacted very positively against the purified SPM and not or only very faintly against the electric nerve membranes were chosen. This criterion was used because these nerve membranes are those most closely related to the membranes of the nerve endings, but are not specialized for ACh release.

RESULTS

Purification of the SPM. Due to the lack of a biochemical or morphological marker for measuring an enrichment in SPM, we assayed AChE activity which is present in high amounts in Torpedo synaptosomes (16, 17). Moreover we were also interested in following the distribution of ChAT activity during the SPM purification. This enzyme which is widely accepted as being cytoplasmic, was very recently described as actually existing in several forms: cytoplasmic, high salt-solubilized and membrane-bound (1).

The distribution of mean specific activities (SA) of different markers is shown in Table 1. The activity of LDH, a cytoplasmic marker, was taken as an index of occluded and non-specifically bound activity. The nRACh was used as a marker for contamination by postsynaptic membranes. After high salt treatment of the crude synaptosomal membranes (fractions P1 to P4) practically no LDH activity was left bound to the membranes and about 50% of the protein present in the intact synaptosomes was removed. In the P4 fraction, there was a rather important enrichment of the SA of nRACh, but the purified SPM had a two times lower SA for the nRACh than did the intact synaptosomes.

The SA of AChE decreased slightly during the preparation of synaptosomes to the P1 stage, but increased again in the P4 fraction; in the latter case, 42% of the AChE initially present was recovered. About 11% of the synaptosomal AChE could be measured in the supernatant obtained after lysis of the synaptosomes (S1, see Figure 1) and an additional 3% of the AChE could be washed off by high salt treatment (S4*). The purified SPM had a 6-fold higher SA for AChE than did the intact synaptosomes. There was a 2-fold decrease in the SA of ChAT from the intact synaptosomes to P1 and about a 4-fold decrease to P4. This could be explained by the loss of the cytoplasmic ChAT in P1 and the removal of the high salt soluble activity in P4, leaving 10.3% of the synaptosomal ChAT bound non-ionically to the membranes. The distribution of the SA of ChAT appeared therefore to be significantly different from that of the cytoplasmic marker LDH. The purified SPM had a higher SA for ChAT than did the P4 preparation. In Table 1, each of the two values indicated for the SA of ChAT were provided by a set of three different

Table 1

Distribution of particulate specific activities during the SPM purification and percent of activities recovered as membrane-bound activities.

Intact synaptosomes	P_1 (Syn)	P_4 (unwashed)	SPM (washed)		% of activity bound		
					P_1/Syn	P_4/Syn	SPM/P_1
ChAT nmol/h/mg	(n = 6) 664 ± 100	(n = 3) 357 ± 91	(n = 3) 133 ± 91	320-550	47 ± 12	10.3 ± 2.8	3.5
LDH ΔOD/min/mg	(n = 6) 18.9 ± 2.0	(n = 3) 1.9 ± 0.4	(n = 3) 1.3 ± 0.2	—	11 ± 3	1.1 ± 0.3	—
AChE mmol/h/mg	(n = 6) 3.3 ± 1.3	(n = 3) 2.2 ± 0.2	(n = 3) 2.7 ± 0.3	18	74 ± 18	42 ± 13	30
nRACh pmol/mg	(n = 2) 886	(n = 2) 981	(n = 2) 2950	450	96	97	2
Proteins mg/g electric organ	(n = 6) 0.113 ± 0.012	(n = 4) 0.098 ± 0.03	(n = 3) 0.058 ± 0.018	0.0035	86 ± 9	54 ± 11	3.6

experiments that were pooled together in the last step of the purification (i.e. the continuous sucrose gradient).

In our hands, the SA for ChAT in purified SPM varied widely from one set of experiments to another. That was not the case for the other markers. It should also be emphasized that the ChAT activity was generally more labile than AChE activity and therefore, at each step of the purification, a loss of ChAT probably by denaturation was encountered.

SDS-PAGE analysis. The polypeptide composition of the SPM was analyzed by one dimensional gel electrophoresis in SDS according to Laemmli (14). The patterns displayed (Fig. 2) show clearly an enrichment in SPM of a polypeptide migrating at about 66 kdaltons (kd) molecular weight (MW). This result is in good agreement with the findings of other authors (18). This protein band could not be detected in homogenates of the electric nerve, nerve membranes isolated from the electric nerve nor in a fraction highly enriched in nRACh (prepared according to Sobel et al. (22).

In partially purified synaptic vesicles, prepared according to Israël et al. (9), we could detect a polypeptide having a slightly slower migration than 66 kd. In SPM the 66 kd polypeptide only appeared after reduction of disulfide bridges; on the other hand

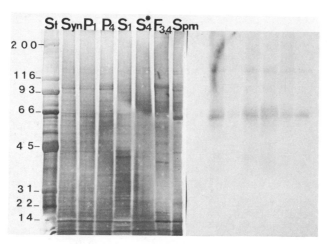

Figure 2. SDS-gel immunoblot identifying the SPM polypeptide antigen recognized by the monoclonal antibody 8/43 in all the steps of the purification of SPM. The gel was a linear gradient (6-15%) of polyacrylamide; 20 μg of each sample was loaded per lane. St indicates standards of MW; MW scale to the left is in kdaltons. The immunoblot is shown to the right and the corresponding gel to the left; silver nitrate staining. In SPM, the band around 66 kd (see arrow) appears to be highly enriched. For symbols see flow chart in Fig. 1.

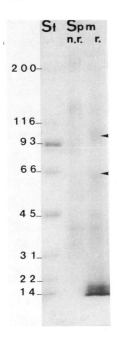

Figure 3. SDS-PAGE analysis of non-reduced and reduced SPM. Ten μg of the solubilized SPM in 1% SDS were subjected to electrophoresis on a linear gradient of polyacrylamide (6-15%). The stacking gel was 3% polyacrylamide. Reduction of the disulfide bridges was performed by adding to the solubilized SPM, 5 mM final concentration of dithiothreitol and then, by heating the sample in boiling water for 3 minutes. Iodoacetamide (40 mM final concentration) was then added for 15 minutes at room temperature and the samples centrifuged for 2 minutes in a Beckman microfuge. St indicates MW standards. Amido-Black staining was used (0.1% in 45% methanol, 10% acetic acid); only the major bands are stained. n.r., non reduced; r. reduced.

two high molecular weight polypeptides (130 and above 200 kd) were stained only in the unreduced SPM (Fig. 3). Therefore, the polypeptide at 66 kd might be a subunit of a larger polypeptide, for example, one of the two larger proteins.

Polyclonal and monoclonal antibodies. When injected into Balb/c mice, the purified SPM yielded antisera that, at a 1/5,000,000 dilution were still reacting positively against the antigen. Two weeks after the hybridization experiments, vigorous hybrid growth was observed in 100% of the cultures. Hybrid cells secreting anti-SPM antibodies were screened and selected according to the criterion described in "METHODS". From a total of two fusions we isolated, by limiting dilution, 16 stable hybridomas which reacted very positively against purified SPM and not or very weakly against electric nerve membranes (Fig. 4). Moreover all these monoclonal antibodies gave a positive, but fainter immune response against partially purified synaptic vesicles; some of them also recognized antigens present in a subfraction highly contaminated by postsynaptic membranes.

The monoclonal antibodies obtained were then screened by (i) indirect immunofluorescence and by (ii) immunoblotting of SDS-gels to identify the polypeptide antigens.

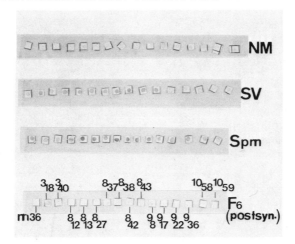

Figure 4. Cross-reactivity analysis of the monoclonal antibodies to SPM with other membrane fractions isolated from the <u>Torpedo</u> electric organ. The dot immunobinding assay was used. The binding of the monoclonal antibodies (code numbers indicated in F6 line) was detected by using a peroxidase conjugated rabbit antimouse I_gG. The membrane fractions are (i) electric nerve membranes (NM), i.e. axonal membranes partially isolated from electric nerves according to Itokawa and Cooper (1970); (ii) partially purified synaptic vesicles (SV); (iii) SPM (purified); and (iv) F6, a fraction containing more post-synaptic than presynaptic membranes. All these fractions were used at 0.5 mg/ml of protein and 0.5 μl dotted onto the nitrocellulose filter. The intensity of the staining of each spot is approximately proportional to the intensity of the immune reaction.

By indirect immunofluorescence using sections of the electric organ, all sixteen monoclonal antibodies labelled specifically, although at varying intensities, a thin and continuous band running all along the ventral side of the layered electrocytes (Fig. 5). This localization corresponds to the synaptic region of the electric organ. However, it is difficult to infer a presynaptic localization since at the light microscopy level one cannot resolve the presynaptic from the postsynaptic side of the synapse.

Using immunoblotting, we found that only seven of the sixteen monoclonal antibodies were able to label polypeptide bands. These were the monoclonal antibodies 8/12, 8/13, 8/27, 8/37, 8/38, 8/42 and 8/43. On the autoradiogram (Fig. 6) obtained by labelling the immune complex formed on the blot with [^{125}I]-Protein A, the antibodies seemed to recognize the same polypeptides: one, migrating around 66 kd was heavily labelled and another one, slightly labelled, migrating at 130 kd. This last band was not so clearly labelled when an immunoperoxidase reaction was used. Immunoblottings of

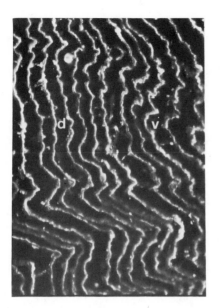

Figure 5. Indirect immunofluorescence analysis shows specific binding
of the monoclonal antibody 8/37 to synaptic regions of the electric
organ of <u>Torpedo</u> <u>marmorata</u>. Twenty μm sections were cut from blocks
of tissue which had first been fixed in 3% paraformaldehyde and
embedded in paraplast as previously described (4). After dewaxing,
the binding of the monoclonal antibody was revealed using fluorescein-conjugated rabbit antimouse IgG (left hand-side panel); antibody
adsorbed with purified SPM shows no fluorescence (right hand-side
panel). v, ventral; d, dorsal. Magnification: 100 x.

synaptic vesicles showed the same labelling pattern for all seven
antibodies of the series 8, but the stain was much fainter.

No staining at all could be detected by immunoblotting of electric
nerve membranes or with the fraction highly enriched in nRACh. The
66 kd polypeptide might well be the one which was shown to become
highly enriched in the purified SPM. This was investigated by using
a plate binding assay. The monoclonal antibody 8/43 was assayed
against the several fractions obtained at each successive step of
the purification (see Fig. 1): crude synaptosomal membranes (P1),
high salt treated membranes (P4), the supernatant S1, the pool of
supernatants of the high salt washings (S4*), the fractions collected
from the discontinuous sucrose gradient (F1 to F6) and finally the
purified SPM. The results (Fig. 7A) are expressed in cpm of $[^{125}I]$-
antimouse I_gG per μg of protein loaded in each well of the plastic
plate. It is evident that the specific activity of the antigen

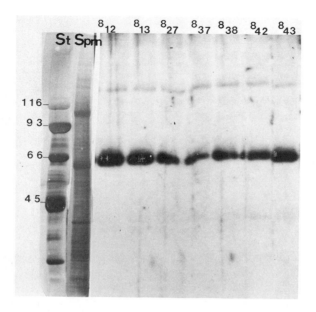

Figure 6. SDS-gel immunoblots identifying SPM polypeptide antigens recognized by the monoclonal antibodies of the series 8. The gel was a linear gradient of polyacrylamide (6-15%). Two polypeptides are detected: one is heavily marked at 66 kd and one is weakly marked at 130 kd. The binding of the antibodies is revealed using a purified rabbit IgG fraction against mouse IgG (0.5 mg/ml) followed by $[^{125}I]$- Protein A. Autoradiography was done on KODAX X-1R films using Ilford intensifying screens.

increased from the P1 fraction to the purified SPM. In all the steps of the purification, it was the same antigen that was labelled, which means that the antigenicity was preserved throughout the puri- fication (see immunoblot of Fig. 2). A surprising finding was that the supernatants (S1 and S4*) had both a rather high specific activity for these antigens. It was interesting to compare these results with the distribution of the SA of ChAT and AChE in the fractions. This is illustrated in Figure 7B. In fact, there was a close similarity between the distributions of the SA of the antigen and those of the two enzymatic "markers". Both ChAT and AChE appeared as particulate as well as "soluble" or "solubilized" forms.

DISCUSSION AND SUMMARY

Synaptosomal plasma membranes were isolated from the purely cholinergic nerve endings of the electric organ of <u>Torpedo</u> <u>marmo</u>- <u>rata</u>. These membranes were characterized by the rather high SA

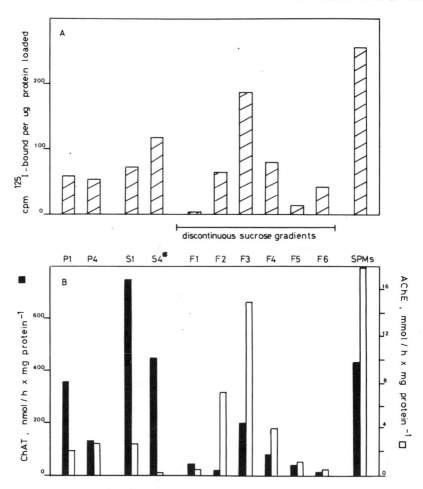

Figure 7. Comparison between the distribution of specific activities of antigen, ChAT and AChE during the purification of SPM.
A. Plate binding radioimmunoassay of the monoclonal antibody 8/43. Binding of the antibody was revealed by using [^{125}I]-Rabbit anti-mouse-IgG. The fractions were used at 100 μg/ml of protein and 100 μl of each loaded into the wells of the plastic plate.
B. Specific activities of ChAT and AChE. For the symbols from P1 to SPM see the flow chart of Fig. 1.

they had for AChE and for non-ionically bound ChAT. Our preparation had a 2 to 3 times higher SA for nRACh than did the SPM purified by Morel et al. (18). A band migrating at about 66 kd MW was highly enriched in purified SPM and might therefore be a marker for these membranes. This polypeptide appeared to be a subunit of a larger protein (possibly the 130 kd one). It is interesting that on SDS-PAGE performed under reducing conditions, a 5.6 S form of AChE isolated from Torpedo, was found to have a MW of 65 kd (15). Moreover, the MW of ChAT is accepted to be around 66 kd. Therefore it might well be possible than in pure SPM the band at 66 kd is AChE and/or ChAT.

We were able to isolate sixteen monoclonal antibodies which recognized components of the purified SPM but not of the electric nerve membranes. Among these, a panel of seven antibodies were further characterized. All of them labelled mainly a polypeptide of about 66 kd MW. This antigen was preserved all along the purification procedure; this is a prerequisite for assaying the effects of these antibodies on the functional properties of synaptosomes.

The antigen appeared to become more and more concentrated as the purification improved. Therefore, we suggest that the 66 kd protein is a marker for SPM, and the antigen recognized by the antibodies of the series 8, might well be the same protein. The similarities between the MW of the antigen, ChAT and AChE, as well as the fact that the three of them were found to become enriched in the same fractions (membrane and soluble), led us to suggest that the antigen might be AChE or ChAT, or both.

More work is now required to determine whether or not the monoclonal antibodies described here: (i) recognize the same protein; (ii) recognize the same antigenic site on the molecule; and (iii) interfere with the release of ACh.

Acknowledgements. This work was supported from the Swiss FNRS. We are indebted to S. Bonnet, N. Collet, F. Pillonel for technical and secretarial assistance, and to Drs. A. Kato and G.J. Jones for helpful discussion.

REFERENCES

1.   Benishin, C.G. and Caroll, P.T. (1983): J. Neurochem. 41: 1030-1039.
2.   Burke, B. Griffiths, G., Reggio, H. Louvard, D. and Warren, G. (1982): Embo J., 1:1621-1628.
3.   Burnette, W.N. (1981): Anal. Biochem. 112:195-203.
4.   Eder, L., Dunant, Y. and Baumann, M. (1978): J. Neurocytol. 7:637-647.
5.   Eder-Colli, L., Powell, J.F., Cuello, A.C. and Smith, A.D. (1982): Neurochem. Int. 4:383-388.
6.   Ellman, G.L., Courtney, K.D., Andres, V. Jr. and Featherstone, R.M. (1961): Biochem. Pharmacol. 7:88-95.
7.   Galfrè, G., Howe, S.C., Milstein, C., Butcher, G.W. and Howard, J.C. (1977): Nature (Lond.) 266:550-552.
8.   Hawkes, R., Niday, E. and Gordon, J. (1982): Anal. Biochem. 119:142-147.

9. Israël, M., Gautron, J. and Lesbats, B. (1970): J. Neurochem. 17:1441-1450.
10. Israël, M., Manaranche, R., Mastour-Frachon, P. and Morel, N. (1976): Biochem. J. 160:113-115.
11. Itokawa, Y. and Cooper, J.R. (1970): Biochim. Biophys. Acta 75:274-284.
12. Johnson, M.K. (1960): Biochem. J. 77:610-618.
13. Köhler, G. and Milstein, C. (1975): Eur. J. Immunol. 61: 511-519.
14. Laemmli, U.K. (1970): Nature (Lond.) 227:680-685.
15. Lee, S.L., Camp, S.J. and Taylor, P. (1982): J. Biol. Chem. 257:12302-12309.
16. Li, Z.Y. and Bon. C. (1983): J. Neurochem. 40:338-349.
17. Morel, N. and Dreyfus, P. (1982): Neurochem. Int. 4:283-288.
18. Morel, N., Manaranche, R., Israël, M. and Gulik-Krzywicki, T. (1982): J. Cell Biol. 93:349-356.
19. Rossier, J., Baumann, A. and Benda, P. (1973): FEBS Lett. 32:231-234.
20. Schaffner, W. and Weissmann, C. (1973): Anal. Biochem. 56:502-514.
21. Schmidt, J. and Raftery, M.A. (1973): Anal. Biochem. 52: 349-354.
22. Sobel, A., Weber, M. and Changeux, J.P. (1977): Eur. J. Biochem. 80:215-224.

REDUCED ACETYLCHOLINE SYNTHESIS IN ALZHEIMER'S DISEASE

IS A CLINICALLY RELEVANT CHANGE

D.M. Bowen*, D. Neary**, N.R. Sims* and J.S. Snowden**

*Department of Neurochemistry
Institute of Neurology
London, U.K.

**Department of Neurology
Royal Infirmary
Manchester, U.K.

(N.R. Sims is now at Burke Rehabilitation Center, White Plains, New York, U.S.A.)

INTRODUCTION

There are many causes of the clinical state of dementia. Excluding those patients with symptoms associated with tumors, infections (including Creutzfeldt-Jacob disease caused by a slow virus), vascular disease and the rare dementias (Huntington's chorea and Pick's disease), there remains a large group whose brains are atrophied and show an excess of senile degeneration (senile plaques and neurofibrillary degeneration) in the neocortex and hippocampus. When the condition occurs before the age of 65 years it is known as "presenile dementia" or Alzheimer's disease; after this age it has been called "senile dementia" or senile dementia of Alzheimer's type. There is no good reason on neuropathological grounds (10, 29) to maintain this distinction. The disease is generally accepted to be the commonest organic cause of intellectual deterioration (8) and is increasing in prevalence (16). There is still considerable uncertainty about its pathogenesis and little is known about its etiology. The first systematic examination of the incidence of senile degeneration in a "population" is described by Corsellis (9), who studied patients that died in Runwell Psychiatric Hospital (Essex, U.K.).

The present study of the biochemistry of Alzheimer's disease commenced in 1972, utilizing brains from Runwell Hospital, kindly provided by Professor Corsellis. Analyses were made of neocortex, caudate nucleus, and the entire temporal lobe for biochemical components to give indications of possible cellular and subcellular change (3-7, 24, 25). Although the caudate nucleus was scarcely affected, a 20% shrinkage of the temporal lobe was associated with a loss of about 33% of nerve cell components common to all nerve cells. There was a 52% loss of acetylcholine esterase and a 65% loss of choline acetyltransferase (ChAT). Indeed ChAT activity was the most depleted component of some 40 constituents measured. Loss of ganglioside NeuAc, a myelin enzyme, and RNA exceeded loss of putative markers of cell number (brain specific protein NS5 or 14.3.2, DNA and β-galactosidase activity) suggesting that shrinkage of neurones and of cell processes occurs in the neocortex, as well as an actual reduction in the nerve cell population.

Reduced activities of glutamate and aromatic amino acid decarboxylases also occurred; these are thought to be related to terminal agonal state (5). Only slight or no change occurred in four other enzymes, markers of capillaries and glial cells.

REASONS FOR INVESTIGATING BRAIN BIOPSY SAMPLES

Post-synaptic muscarinic receptors, as measured by QNB binding, are apparently not greatly affected in Alzheimer's disease, thus raising the possibility of therapy by the use of agents enhancing the activity of the cholinergic system. The recent implication of a cholinergic specific projection system to the cerebral cortex in the pathogenesis of the disease (18, 26) reinforces such a possibility. However, it cannot necessarily be inferred that deficiencies in cortical ChAT activity reflect a reduced ability of the brain to synthesize acetylcholine; indeed the results of animal studies suggest that ChAT may not be the rate limiting enzyme of acetylcholine synthesis. Nevertheless, the concept that the cortical abnormalities of the disease might be secondary to a primary failure of subcortical projection systems to the cortex is attractive because of its theoretical and possible therapeutic implications.

If it were the case that a cholinergic specific subcortical system were the sole or contributory cause of the psychological changes of Alzheimer's disease then consistent relationships ought to exist between measures of cholinergic activity, the nature and degree of pathological change in the cerebral cortex, and the severity of the dementia. In autopsy series of Alzheimer's disease in elderly subjects, the relationship between ChAT activity, the frequency of senile plaques and neurofibrillary tangles, and the degree of dementia have been studied (2, 15, 27, 28), but the results have not been in uniform agreement.

The lack of consistency in reported correlations between the neurochemical, pathological and psychological measures in Alzheimer's disease may be partly attributable to methodological differences, the tendency for autopsy studies to include the very old and more severely advanced forms of the disease and for psychological measures to be made some time before pathological and biochemical examination of the brain. Moreover, the effects of terminal illness and postmortem change on the brain may introduce errors into neurochemical assays.

In the present study of patients in the presenium, psychological assessment was carried out to provide measures of relative severity of dementia, with which pathological (data not presented) and chemical indices of impairment could be compared. Fresh cortical biopsy tissue permitted the assay of ChAT activity, and also the determination of acetylcholine synthesis in a preparation enriched in cortical synaptosomes. Additionally, choline uptake has been measured for comparison.

SELECTION AND BRIEF DESCRIPTION OF PATIENTS

The patients were drawn from a larger series of young individuals with progressive dementia and cortical atrophy in whom the precise clinical diagnosis had been uncertain. The general aim of this collaborative investigation of these patients has been to provide a diagnostic classification of this fatal group of disorders and to unearth potentially reversible pathological and neurochemical changes which might be drug responsive. The study group consisted of patients, in whom progressive deterioration of mental function had begun before the age of 65. In no patient was there evidence of cardiovascular or other systematic disease, or previous history of cerebral trauma or alcohol abuse.

METHODS

Each patient had a psychological and neurological assessment, and computed tomography of the brain before cerebral biopsy. Tissue samples obtained from biopsy were subjected to pathological and chemical analyses. All measurements were "blind", each being undertaken by independent assessors, without access to other information about the patient.

Cerebral Cortical Biopsy. The procedure of cerebral biopsy was approved by the appropriate ethical committees and consent was obtained from patients' relatives, and in some cases from the patients themselves. Biopsy was undertaken under general anaesthesia. Tissue was routinely dissected from the non-dominant (right) middle temporal gyrus. The temporal lobe was selected as the site for most cortical biopsies since it is known the   Alzheimer-like pathological changes are especially profuse in that region. Animal studies (20) indicate that neither the anaesthetic nor acute tissue ischaemia (during sample excision) will influence the biochemical data.

Pathological. One third of the fresh biopsy tissue was immediately placed in 10% neutral formalin, fixed overnight and routinely processed. Paraffin blocks were cut in 5 m sections and stained by conventional techniques, including Gros-Bielchowsky (GB) silver staining. Measurements of frequency of senile plaques were made on the GB stained sections using the method of Aherne and Diggle (1) as detailed elsewhere (14).

Biochemical. Neurosurgical samples from all patients were handled and processed for measuring acetylcholine synthesis by incorporation of $[U-^{14}C]$glucose into $[^{14}C]$acetylcholine in neocortical tissue prisms, according to Sims et al. (20). Incubations were carried out in the presence of high concentration of choline (2 mM), to minimize the possibility of an influence from endogenous choline. An indication of overall utilization of the radioactive glucose by the prisms was obtained by trapping the $[^{14}C]O_2$ produced (data not shown, see 23, also for incorporation into ninhydrin positive material and relationship between acetylcholine synthesis and adenine nucleotide content). The ChAT activity of cell free homogenates of neocortex was determined according to Fonnum (11). Uptake of choline was based on the technique of Polak et al. (17) for its estimation in slices of neocortex. Incubations were carried out in medium containing 0.82 uM $[^{3}H]$choline and 31 mM potassium ion (22). Prisms were prepared from 373 $\pm$ 134 mg of neocortex while much less tissue (47 $\pm$ 31 mg) was homogenized for the determination of ChAT activity. Control samples from patients aged 33-68 (mean 52 years) were taken from apparently normal neocortex of the temporal lobe during procedures for deep seated tumor removal. It has been shown that such samples provide a good indication of values for acetylcholine synthesis, $[^{14}C]O_2$ production, ChAT activity and choline uptake of normal tissue (8, 20, 22).

Psychological. A rating of magnitude of dementia from 0-9 was obtained on the basis of patients' performance on a variety of mental tests. Disorders of language, perceptuo-spatial function and memory were each rated from 0-3, in terms of perceived extent of impairment (absent, mild, moderate or severe), and the overall psychological rating represented a cumulation of these three assessments determined as follows:

Language:
1. Spontaneous speech
2. Series speech
3. Comprehension: object names, body parts, spatial and relational terms, metaphor and proverb
4. Repetition: digit span, phrases, word lists
5. Word finding: naming to confrontation and from

description, category production
6. Reading: aloud, for comprehension
7. Writing
8. Calculation
9. Use of gesture and pantomine

Visuo-Spatial Functions:
1. Visual identification: recognition of shapes, objects, famous faces
2. Spatial location: body parts, external objects
3. Visual tracking: maze tracing
4. Integration: picture interpretation
5. Replication: hand configurations, drawings, block constructions

Memory:
1. Name and address, immediate and delayed, free recall and forced-recognition
2. Sentence learning
3. Memory for short story, immediate and delayed, free and cued recall
4. Memory for drawings, free and cued recall

RESULTS

Cortical sections from most of the patients were found on light microscopy to contain the characteristic changes of Alzheimer's disease (senile plaques and neurones bearing neurofibrillary - tangles).

In comparison with control values, synthesis of $[^{14}C]$acetylcholine was markedly reduced in samples from demented patients which exhibited the histological features of Alzheimer's disease (Fig. 1). The alteration of $[^{14}C]$acetylcholine synthesis did not seem to be related to the metabolic activity of the tissue, as a similar mean reduction was observed for measurements made in the presence of either 5 mM K^+ (58.0% of control) or 31 mM K^+ (49.0%). The deficit was a feature of samples removed from both the frontal and temporal lobes.

ChAT activity was similarly related to the presence of histological features, and in samples from patients with confirmed Alzheimer's disease was reduced to a slightly greater extent (39.0% of mean control) than was $[^{14}C]$acetylcholine synthesis. Choline uptake was also found to reflect the changes in the other markers, the mean being significantly reduced for the six samples from Alzheimer's disease cases in which it was measured and within the control range for samples from demented patients which failed to show diagnostic histological features (Table 1).

Table 1

Choline uptake of samples from demented patients and neurosurgical controls compared with values for other presynaptic cholinergic markers in the same demented patients

	Choline uptake (in 31 mM K⁺) pmol/min/mg protein	[¹⁴C]acetylcholine synthesis (31 mM K⁺) dpm/min/mg protein	ChAT activity pmol/min/mg protein
Controls	3.04 ± 0.33 (6)a	6.6 ± 1.2 (22)	86 ± 21 (11)
Alzheimer's Disease	1.72 ± 0.58 (6)**b	3.9 ± 1.2 (6)**	37 ± 25 (6)
Other demented patients	2.71, 2.73	4.9, 8.5	70, 74

a. Values are expressed as mean ± S.D. with number of samples in parentheses where appropriate.
b. **p < 0.01 c.f. control (Wilcoxon rank test).

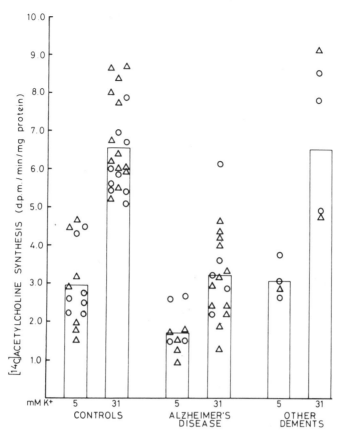

Figure 1. [^{14}C]acetylcholine synthesis measured in the presence of either 5 mM K$^+$ or 31 mM K$^+$ in samples of temporal neocortex ($\triangle$) and frontal neocortex (o) from neurosurgical controls, patients with histologically confirmed Alzheimer's disease and other demented patients. Samples from patients with Alzheimer's disease were significantly different (p < 0.01; Wilcoxon rank test) from control samples measured under the same conditions.

The relationship between individual values was examined for the three cholinergic parameters for both the demented group as a whole and for the subgroup of confirmed Alzheimer's disease patients alone. There was a correlation between ChAT activity and [^{14}C]acetylcholine synthesis for both analyses. Choline uptake correlated with each of the other measures for the total population of demented patients, but there was not a significant relationship for the Alzheimer's disease patients alone (which may have been a reflection of the small numbers involved). Synthesis progressively declined as psychological impairment (i.e., degree of dementia) increased (Fig. 2).

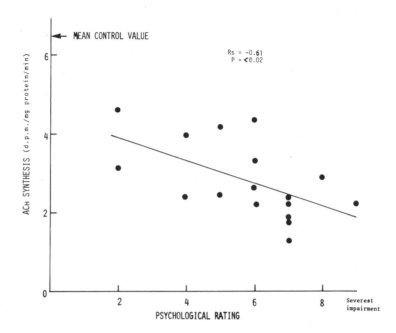

Figure 2. Relationship between [^{14}C]acetylcholine synthesis and psychological rating (Spearman rank correlation test) with correction for ties. (Line of best fit by linear regression analysis). Synthesis determined in 31 mM K$^+$ with temporal lobe cortical samples from Alzheimer's patients.

DISCUSSION

Although changes in ChAT activity have frequently been interpreted as indicating changes in acetylcholine synthesis in Alzheimer's disease, the present study (19, 20, 22) provides the first direct evidence of a reduced capacity for synthesis in this disease. A close relationship was revealed between [^{14}C]acetylcholine synthesis and ChAT activity, indicating that the enzyme may be used to provide a fairly reliable estimate of residual synthetic capacity in the neocortex in Alzheimer's disease.

As seen in animal studies, choline uptake and acetylcholine synthesis are generally closely related but there is not obligatory coupling (13), and under adverse conditions the two activities may diverge (12). ChAT appears not to be rate-limiting in the synthesis process. Assuming that in human tissue the three markers also are not closely associated, the most plausible explanation of the similar reductions of these in Alzheimer's disease is that there is a loss of functionally active cholinergic nerve endings with parallel reductions of all associated parameters. An alternative explanation might be postulated if there was a reduction of ChAT activity per terminal so that it became rate-limiting in the disease human tissue. However, under such circumstances it is likely that synthesis would be more affected at the higher rate produced in response to K^+ depolarization (i.e., in 31 mM K^+) whereas the proportional response to K^+ was found to be largely preserved in the diseased tissue.

A radiochemical assay for acetylcholine synthesis was chosen for sensitivity and convenience in dealing with the limited amounts of human tissue available. The low concentrations of acetylcholine produced and the high content of choline in the media have precluded the direct estimation of acetylcholine specific activity in human samples. Thus the possibility must be considered that there is dilution of the radiolabel in the metabolic process from glucose to acetylcholine, leading to apparent changes in acetylcholine synthesis. Studies with this prism system using rat neocortex (20) as well as animal studies in a number of similar preparations have consistently shown that there is little or no dilution of radiolabel in the production of acetylcholine. Furthermore, changes in the ChAT activity and choline uptake are unlikely to reflect changes in pool sizes of glucose metabolites, and yet these related closely to the decrease in [^{14}C]acetylcholine synthesis in Alzheimer's disease. Moreover, if dilution in intermediate metabolic pools were producing an apparent decrease in synthesis, this would be expected to have an apparently greater effect when there were lower rates of glucose metabolism, or at shorter incubation times, such that any dilution of the label wuld be proportionately greater. However, the percentage reduction was similar at both K^+ concentrations; and in addition in two samples from patients with Alzheimer's disease on which incubations were also done for 30 min (results not shown), values corrected for time were almost identical (96% and 106%) to those obtained from 60 min incubations. Hence the reduction in [^{14}C]acetylcholine synthesis would seem to reflect a genuine reduced capacity for synthesis of the neurotransmitter.

Of the two measures of cholinergic function determined on most Alzheimer samples ChAT activity correlated significantly with senile plaque frequency (data not shown) whereas the values for acetylcholine synthesis did not. However, the psychological rating was significantly related to acetylcholine synthesis only. If loss of cholinergic synapses results in a proportional loss of ChAT apoenzyme, then a significant correlation between an index of structural synaptic change, i.e., senile plaque frequency, might be expected to occur. If, however, remaining neurones are capable of compensatory increases in the rate of acetylcholine synthesis, then a similar proportional relationship between plaques might not obtain. It may be of relevance that the percentage reduction in ChAT activity was greater than that for acetylcholine synthesis. Since acetylcholine synthesis did correlate significantly with the severity of dementia then it would appear to be a more sensitive index of the physiologically active pool of acetylcholine in the cortex than is ChAT activity. This may be because assays of acetylcholine synthesis at high potassium ion concentrations actually reflect the release of neurotransmitter from synaptic nerve endings (21).

Acknowledgements. We gratefully acknowledge the cooperation of the large number of people involved in the collection and classification of specimens, including Professor L.W. Duchen, Drs. D.M.A. Mann and F. Scaravilli. The research was supported by the Medical Research Council, Brain Research Trust and the Miriam Marks Charitable Trust.

REFERENCES

1. Aherne, W.A. and Diggle, P.J. (1978): J. Microscopy 114: 285-293.
2. Blessed, G., Tomlinson, B.E. and Roth, M. (1968): Brit. J. Psychiat. 114:797-811.
3. Bowen, D.M., Smith, C.B. and Davison, A.N. (1973): Brain 96:849-856.
4. Bowen, D.M., White, P., Flack, R.H.A., Smith, C.B. and Davison, A.N. (1974): Lancet i:1247-1250.
5. Bowen, D.M., Smith, C.B., White, P. and Davison, A.N. (1976): Brain 99:459-496.
6. Bowen, D.M., Smith, C.B., White, P., Flack, R.H.A., Carrasco, L.H., Gedye, J.L. and Davison, A.N. (1977): Brain 100:427-453.
7. Bowen, D.M., White, P., Spillane, J.A., Goodhardt, M.J., Curzon, G., Iwangoff, P., Meier-Ruge, W. and Davison, A.N. (1979): Lancet i:11-14.
8. Bowen, D.M., Benton, J.S., Spillane, J.A., Smith, C.C.T. and Allen, S.J. (1982): J. Neurol. Sci. 57:191-202.
9. Corsellis, J.A.N. (1962): Mental Illness and the Ageing Brain, Oxford University Press, Oxford.
10. Corsellis, J.A.N. (1976): In Greenfield's Neuropathology, (eds) W. Blackwood and J.A.N. Corsellis, Edward Arnold, London, pp. 796-849.
11. Fonnum, F. (1975): J. Neurochem. 24:407-409.
12. Jope, R.S., Weiler, M.H. and Jenden, D.J. (1978): J. Neurochem. 30: 949-954.
13. Kessler, P.D. and Marchbanks, R.M. (1979): Nature 279:542-544.
14. Mann, D.M.A., Neary, D., Yates, P.O., Lincoln, J., Snowden, J.S. and Stanworth, P. (1981): Neuropathol. Appl. Neurobiol. 7:37-47.
15. Perry, E.K., Tomlinson, B.E., Bergmann, K., Gibson, P.H. and Perry, R.H. (1978): Br. Med. J. 2:1457-1459.
16. Plum, F. (1979): Nature 279:372-373.
17. Polak, R.L., Molenaar, P.C. and Van Gelder, M. (1977): J. Neurochem. 29:477-485.
18. Price, D.L., Whitehouse, P.J., Struble, R.G., Clarke, A.W.,

Coyle, J.T., DeLong, M.R. and Hedreen, J.C. (1982): Neuroscience Commentaries 1:84-92.

19. Sims, N.R., Bowen, D.M., Smith, C.C.T., Flack, R.H.A., Davison, A.N., Snowden, J.S. and Neary, D. (1980): Lancet i:333-335.

20. Sims, N.R., Bowen, D.M. and Davison, A.N. (1981): Biochem. J. 196:867-876.

21. Sims, N.R., Marek, K.L., Bowen, D.M. and Davison, A.N. (1982): J. Neurochem. 38:488-492.

22. Sims, N.R., Bowen, D.M., Allen, S.J., Smith, C.C.T., Neary, D., Thomas, D.J. and Davison, A.N. (1983): J. Neurochem. 40: 503-509.

23. Sims, N.R., Bowen, D.M., Neary, D. and Davison, A.N. J. Neurochem. 41:1329-1334.

24. Smith, C.B. and Bowen, D.M. (1976): J. Neurochem. 27:1521-1526.

25. White, P., Hiley, C.R., Goodhardt, M.J., Carrasco, L.H., Keet, J.P., Williams, I.E.I. and Bowen, D.M. (1977): Lancet i:668-670.

26. Whitehouse, P.J., Price, D.L., Struble, R.G., Clark, A.W., Coyle, J.T. and DeLong, M.R. (1982): Science 215:1237-1239.

27. Wilcock, G.K. and Esiri, M.M. (1982): J. Neurol. Sci. 56:343-356.

28. Wilcock, G.K., Esiri, M.M., Bowen, D.M. and Smith, C.C.T. (1982): J. Neurol. Sci. 57:407-417.

29. Wilcock, G.K. and Esiri, M.M. (1983): Lancet ii:346-348.

CORTICAL CHOLINERGIC HYPOFUNCTION AND BEHAVIORAL

IMPAIRMENT PRODUCED BY BASAL FOREBRAIN LESIONS IN THE RAT

B.E. Lerer[1], E. Gamzu[2] and E. Friedman[3]

[1]CNS Diseases Research
DuPont Pharmaceuticals
Wilmington, Delaware 19898 U.S.A.

[2]Pharmacology I
Hoffman-LaRoche Inc.
Nutley, New Jersey 07110
U.S.A.

[3]Departments of Psychiatry
and Pharmacology
New York University Medical Center
New York, New York 10016 U.S.A.

INTRODUCTION

The magnocellular nuclei of the basal forebrain (MNBF) provide extensive cholinergic innervation to fronto-parietal cortex (5, 13, 27, 28). In the rat, these large cells comprise the homologue of the human nucleus basalis of Meynert, which has been implicated in the cholinergic hypothesis of cognitive dysfunction in Alzheimer's disease (AD, for a review see 3). AD patients suffer severe and progressive declines in cognitive and behavioral functions that are correlated with pronounced deficits in neocortical choline acetyltransferase (ChAT) and acetylcholinesterase activities (23). This suggests that cortical cholinergic dysfunction may underlie certain learning and memory deficits in AD. The present communication is part of an ongoing effort aimed at investigating the role of the MNBF-cortical pathway in memory and developing a useful animal model of AD (8, 18).

In previous studies, we used the excitotoxin kainic acid (KA) to produce bilateral MNBF lesions that depleted cortical ChAT and resulted in behavioral deficits in passive avoidance indicative of memory disruption, and in perseverative behavior in a spontaneous alternation task (8, 17, 18). Others have independently confirmed that MNBF lesions disrupt learning and memory (7, 19).

In the present study, we have confirmed the cortical ChAT and passive avoidance deficits resulting from bilateral KA lesions of the MNBF. We further report that the deficit in perseverative behavior resulting from such a lesion interferes with instrumental learning in a serial reversal of a spatial discrimination task (9). This task is sensitive to the memory deficits produced by hippocampal damage (22); moreover, it is appetitively motivated and so broadens the scope of testing of behavioral effects of MNBF lesions beyond shock-induced behavior.

Because of the reported passive avoidance deficits, we were interested in whether bilateral MNBF lesions would interfere with learning in an active avoidance paradigm. Septal and hippocampal damage facilitated, and caudate damage impaired, shuttlebox avoidance acquisition (1, 10). Rats with unilateral electrolytic MNBF lesions were impaired in shuttlebox avoidance (19), and rats bilaterally lesioned with ibotenic acid were impaired in acquisition of shelf-jump avoidance (7). We have consequently examined acquisition of a lever-press response in a signalled avoidance procedure.

Finally, lesioned and control rats were compared for spontaneous motor activity, because differences in activity may sometimes lead to performance changes that are independent of learning and memory processes (2, 11).

MATERIALS AND METHODS

Male Sprague-Dawley rats, 230-270 g at the time of surgery, were subjects in these experiments. Bilateral MNBF lesions were induced by injection into each hemisphere of 1 μg KA in 1 μl 0.2 M sodium-phosphate buffer (pH 7.2), at the following coordinates: 0.7 mm posterior to bregma; 2.7 mm lateral to the midsagittal suture; and 7.0 mm ventral to the brain surface (with the bregma and lambda skull sutures horizontal). There were 3 control groups: (i) SHAM - The sham-operated controls ($\underline{n}$ = 10) were anesthetized and had burr holes drilled in their skulls. (ii) CORTICAL KA - These control rats ($\underline{n}$ = 6) were injected with KA in the cortical area directly dorsal to the MNBF (0.5 mm ventral to the brain surface). (iii) UNOPERATED - These control animals ($\underline{n}$ = 6) were not exposed to any drugs or surgery.

Behavioral testing began three weeks after surgery when MNBF lesioned rats had recovered from the lesion sequellae (8, 17), and only if they weighed 20 g more than their preoperative baseline weight. Because of health and equipment limitations, not all rats were tested in all behavioral procedures.

Passive Avoidance. Dark avoidance training took place in a step-through shuttlebox. Scrambled 1.5 mA footshock was applied until the rat remained in the lighted compartment for 60 sec (for details see 8). Retention testing occurred 24 hr later (maximum latency 600 sec).

Active Avoidance Acquisition. Lever-press avoidance acquisition was measured in four daily 1 hr sessions of 60 trials. A single lever-press during the 15 sec warning tone or during the subsequent 15 sec of 1.0 mA footshock terminated all stimuli for that trial and was recorded as an avoidance or an escape response, respectively. The inter-trial interval was 30 sec.

Serial Discrimination Reversal. Rats were food deprived and trained in a 1 hr session to respond for 45 mg food pellets. The required response was poking the snout into a recessed food trough and breaking a photo beam; each time the rat inserted its head into the trough a response was recorded. Rats were trained to respond on a fixed ratio (FR) reinforcement schedule. Discrimination testing began the following day and lasted for 8 days. The test chamber was modified to include a second food trough. At the start of the testing session, both troughs were activated. The first trough to register a response was designated "correct" for the first component. In this first component, a rat was required to break the photo beam of the "correct" trough on a FR4 schedule until five reinforcements were obtained or 10 min elapsed. Subsequently, the second component began, with the discrimination reversed. That is, the rat was now required to respond at the second trough to obtain reinforcement. Thereafter, the components alternated, each one lasting until five reinforcements were obtained or 10 min elapsed. The session ended after the fourteenth consecutive component was completed or after 60 min elapsed. Responses at the "incorrect" trough were designated errors, and the total number of errors per session were analyzed.

When behavioral testing was completed, subjects were sacrificed by decapitation and the brains rapidly removed and dissected over ice. The cortex was subdivided into frontal, parietal and occipital regions, as described by Kelly and Moore (14); the striatum and hippocampus were also removed. Samples were stored at -80°C until assayed. ChAT was assayed by a modification of the method of Kuhar et al. (15), under saturating conditions of 20 mM choline and 2 mM [^{14}C]-Acetylcoenzyme A (specific activity = 53.0 mCi/mmol, NEN, combined with unlabelled Acetylcoenzyme A to give 1000 d.p.m./ - nmol). The product, [^{14}C]-acetylcholine, was counted in a liquid scintillation counter. The protein content of aliquots of tissue samples was measured by the method of Lowry et al. (20), using bovine serum albumin as a standard. Amines and metabolites were separated by HPLC on a 5 μM ODS reverse phase C_{18} column and quantified by electrochemical detection using a glassy carbon electrode set at +0.75 V (21). Epinine was added as an internal standard.

In most cases, the data from the three control groups were pooled as there were minimal and nonsignificant behavioral and biochemical differences between these groups.

RESULTS

MNBF-lesioned rats displayed motor deficits for 2-5 days, and aphagia and adipsia for up to 2 wk after surgery. Control rats injected with KA in the cortical area directly above MNBF (group CORTICAL KA) exhibited only a brief (6-10 hr) period of post-operative hyperactivity.

In the passive avoidance task, there was no significant difference between the MNBF-lesioned and control groups in the initial latency to enter the dark compartment. There was also no difference between groups in the number of shocks required on the training day. During the retention test, however, the mean latency to enter the dark compartment was significantly shorter for the MNBFlesioned group (255.0 $\pm$ 79.6, mean $\pm$ S.E.M.) than for the control group (531.3 $\pm$ 51.8; $p < 0.05$, df = 18, two-tailed Student's t-test).

A repeated-measures analysis of variance (ANOVA) of the mean number of daily avoidances in the lever-press active avoidance test showed a significant improvement in learning over days (F = 6.20, $p < 0.001$), but there were no significant group or interaction effects ($p > 0.50$). The mean number of avoidances for the combined control groups on Days 1-4 for the MNBF-lesioned group were: 8.6, 7.9, 10.5 and 10.9, respectively.

In the serial discrimination reversal task, all groups showed evidence of acquiring the task, as seen in the overall decrease in the mean number of daily errors (Figure 1). However, the performance of the MNBF-lesioned rats was significantly poorer than that of the controls from Day 3 through Day 8 ($p < 0.01$, df = 28, Student's t-test).

Motor activity was measured in a circular chamber with photosensor circuits. Subjects were individually tested for 30 min during the middle of the diurnal light cycle. In the first 5 min of the session, mean activity rates were 119.7 $\pm$ 11.8 (S.E.M.) photosensor counts/min for the MNBF-lesioned group; 108.7 $\pm$ 7.6 for the UNOPERATED controls; 108.2 $\pm$ 10.4 for the SHAM group. These rates were not significantly different ($p > 0.20$, df = 3,33, 1-Way ANOVA).

ChAT activity was significantly decreased in cortical areas of MNBF-lesioned rats as compared to controls (Table 1). In frontal and parietal areas, ChAT was decreased by about 20%; in occipital cortex a 10.5% ChAT decrease was obtained. No significant changes in ChAT activity were obtained in striatum or hippocampus.

The brain regional concentrations of dopamine and its metabolite DOPAC were not altered by the MNBF lesion. The brain regional concentrations of serotonin and its metabolite 5-HIAA were not altered in parietal cortex or hippocampus. In frontal cortex, the serotonin

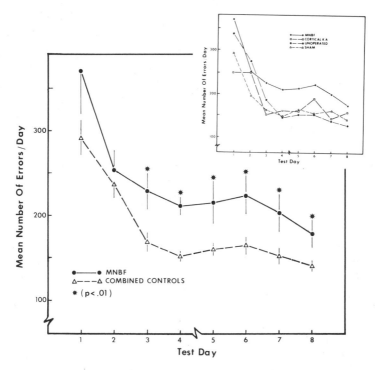

Figure 1. Mean number (± S.E.M.) of daily errors made in a serial reversal discrimination task. The dashed line shows the data for the combined control groups (n = 22) and the solid curve shows the data for MNBF-lesioned rats (n = 13). Asterisks indicate a significant performance difference between controls and MNBF-lesioned rats on days 3 through 8 (p < 0.01, Student's t-test). The inset shows data for the individual control groups and the MNBF-lesioned group.

concentration was decreased by about 14% (p < 0.05, df = 3, 31, 1-Way ANOVA); however, no change in 5-HIAA was obtained in this region.

DISCUSSION

Bilateral KA lesions in rats, placed ventral to the globus pallidus, in an area homologous to the nucleus basalis of Meynert in humans, produced a significant loss of ChAT in the cerebral cortex. The ChAT loss was restricted to the cortex, as no concomitant ChAT loss was observed in either striatum or hippocampus. The biochemical deficits were accompanied by a significant decrease in 24 hr retention of a passive avoidance task and a significantly slower rate of acquisition of a serial discrimination reversal.

Table 1

Choline Acetyltransferase

Brain Areas

Groups	Frontal Cortex	Parietal Cortex	Occipital Cortex	Hippo-campus	Striatum
Combined Controls (n = 22)	54.95 $\pm$ 0.56	51.53 $\pm$ 1.19	46.06 $\pm$ 0.56	69.41 $\pm$ 1.10	194.49 $\pm$ 5.50
MNBF (n = 15)	44.67* $\pm$ 1.54	41.09* $\pm$ 0.97	41.23* $\pm$ 0.96	67.49 $\pm$ 2.12	204.34 $\pm$ 8.01
% ChAT Reduction	-18.7	-20.3	-10.5	- 2.8	+ 5.1

*$p < 0.01$, 1-Way ANOVA; Difference from control. ChAT activity was measured under saturating conditions for both choline and [^{14}C]-Acetylcoenzyme A. ChAT values are expressed as mean $\pm$ S.E.M. nmoles/mg protein/hr.

Although the MNBF-lesioned rats were clearly impaired in retention of the passive avoidance task, there was nevertheless evidence of a learning capacity. This could be seen in the retention latency of the MNBF-lesioned group which was significantly greater than the training latency. Caudate damage is known to impair passive avoidance acquisition and retention (25, 26, 29). It seems unlikely, however, that caudate damage mediated the retention deficit in this experiment as we found no loss of striatal ChAT activity. Direct KA damage of the striatum would have surely reduced striatal ChAT activity (25). Neither are differences in shock sensitivity between MNBF-lesioned and control animals likely to have caused performance differences (8). We conclude therefore that the cortical ChAT deficit was responsible for the impaired passive avoidance retention, although we have not ruled out the possibility that subcortical alterations in other neurotransmitter substances may partly mediate the behavioral deficit.

The performance of the MNBF-lesioned group on the serial discrimination reversal task improved over the course of testing sessions, although these rats made substantially more daily errors than did the controls. This, again, suggests that MNBF-lesioned rats were capable of acquiring new learning. Errors made by the MNBF-lesioned group were perseverative in nature. This is characteristic of a type of behavior we noted previously in a mazerunning spontaneous alternation task (17, 18) and it suggests inflexibility of response patterns.

MNBF-lesioned rats did not differ from controls in the active avoidance task. This could also reflect a residual learning capacity. Alternatively, the lack of differences between the groups might reflect the generally poor performance of rats on lever-press avoidance tasks (24), which would obscure any experimental effects. More importantly, the behavioral processes mediating active and passive avoidance are different, and electrolytic brain lesions have often affected these two types of avoidance behavior in opposite directions (1). Therefore, it may not be unreasonable to expect excitotoxic MNBF lesions to have different effects on both active and passive avoidance tasks.

Changes in motor activity resulting from MNBF lesions could conceivably account for behavioral differences in learning tasks. We found no significant change in the activity levels of MNBFlesioned rats, however. Therefore, it is unlikely that behavioral deficits reflect non-associative, secondary differences in activity.

As for the biochemical indices we measured, the cortical ChAT deficit was most clearly correlated with the behavioral deficits which we found. Of the neurotransmitter indices, only ChAT was consistently lower in the MNBF-lesioned group when compared to the controls. Furthermore, this ChAT decrease was localized in cortical regions and was not found in either hippocampus or striatum. This suggests that our KA lesions were specific to the MNBF and did not produce damage in other non-cortical areas. The absence of a ChAT change in hippocampus and striatum is important because the integrity of these two areas is essential for normal learning and memory processes. Direct damage of striatum or hippocampus would have substantially depleted ChAT (6, 12).

Furthermore, neither behavioral deficits nor cortical ChAT activity changes could have been produced by accidental KA-induced damage to the cortex. It has been demonstrated by others (6), and our data corroborate, that direct cortical injections of KA do not reduce cortical ChAT activity. The CORTICAL KA group in this study did not differ in either behavioral or biochemical data from the UNOPERATED and SHAM-OPERATED controls. The integrity of the CORTICAL KA group confirms previous findings that cholinergic innervation of cortex is derived predominantly from subcortical origins (16).

The behavioral deficits we assume to be indicative of learning and memory problems were accompanied by a 20% decrease in cortical ChAT. By contrast, post-mortem analyses of cortical ChAT in AD patients show up to 80% ChAT loss (4). It is possible that when memory losses are first noted, early in the AD syndrome, the magnitude of the ChAT loss approximates that obtained in our study in rats. Important issues for further validation of this animal model are whether or not larger cortical ChAT decreases can be created with sufficiently restricted lesions and without increased mortality; and whether these large ChAT deficits would produce more pronounced behavioral changes.

Excitotoxic lesioning of the MNBF, with its consequent neurochemical and behavioral deficits, provides a useful means of examining the cholinergic substrates of learning and memory. The effects of the lesion mimic two characteristics of Alzheimer's Disease: cholinergic deficit and cognitive impairment. Hence, this method may prove useful in researching the disease and in developing new pharmacological treatment strategies.

Acknowledgements. We thank Mr. George Vincent and Mr. John Warner for their invaluable assistance, and Dr. Kevin Keim for his comments on the manuscript. This work was partly supported by 5T 32MH15137 (B.L.) and USPHS RSDA MH 00208 (E.F.).

REFERENCES

1. Black, A.H., Nadel, L. and O'Keefe, J. (1977): Psychol. Bull. 84:1107-1129.
2. Boff, E., Gamzu, E., Poonian, D. and Zolcinski, M. (1982): Soc. Neurosci. Abstr. 8:320.
3. Coyle, J.T., Price, D. and DeLong, M. (1983): Science 219: 1184-1190.
4. Davies, P. (1979): Brain Res. 171:319-327.
5. Divac, I. (1975): Brain Res. 93:385-398.
6. Fibiger, H.C. and Lehmann, J. (1981): In: Cholinergic Mechanisms: Advances in Behavioral Biology, Vol. 25 (eds) G. Pepeu and H. Ladinsky, Plenum Press, New York, pp. 663-672.
7. Flicker, C., Dean, R.L., Watkins, D.L., Fisher, S.K. and Bartus, R.T. (1983): Pharmacol. Biochem. Behav. 18:973-981.
8. Friedman, E.F., Lerer, B. and Kuster, J. (1983): Pharmacol. Biochem. Behav. 19:309-312.
9. Gamzu, E., Boff, E., Zolcinski, M., Vincent, G. and Verederese, T. (1983): Soc. Neurosci. Abstr. 9:824.
10. Green, R.H., Beatty, W.W. and Schwartzbaum, J.S. (1967): J. Comp. Physiol. Psychol. 64:444-452.
11. Heise, G.A. and Boff, E. (1971): Neuropharmacology 10:259-266.
12. Hughey, D. and Friedman, E. (1983): Soc. Neurosci. Abstr. 9:648.
13. Johnston, M.V., McKinney, M. and Coyle, J.T. (1981): Exp. Brain Res. 43:159-172.

14. Kelly, P.H. and Moore, K.E. (1978):   Exp. Neurol. _61_:479-484.
15. Kuhar, M.J., Sethy, V.H., Roth, R.H. and Aghajanian, G.K. (1973): J. Neurochem. 20:581-593.
16. Lehmann, J., Nagy, J.I., Atmadja, S. and Fibiger, H.C. (1980): Neuroscience 5:1161-1174.
17. Lerer, B. and Friedman, E. (1982): Soc. Neurosci. Abstr. 8:838.
18. Lerer, B. and Friedman, E. (1983):   In: Alzheimer's Disease (ed) B. Reisberg, Macmillan, New York, pp. 421-427.
19. LoConte, G., Bartolini, L., Casamenti, F., Marconcini-Pepeu, I. and Pepeu, G. (1982): Pharmacol. Biochem. Behav. 17:933-937.
20. Lowry, O.H., Rosebrough, N.J., Farr, A.L. and Randall, R.J. (1951): J. Biol. Chem. 193:265-275.
21. Meller, E. and Friedman, E. (1982):   J. Pharmacol. Exp. Ther. 220:609-615.
22. O'Keefe, J. and Nadel, L. (1978):   The Hippocampus as a Cognitive Map, Oxford University Press, Oxford.
23. Perry, E.K., Tomlinson, B.E., Blessed, G., Bergman, K., Gibson, P.H. and Perry, R.H. (1978): Br. Med. J. 52:1457-1459.
24. Riess, D. (1971): Psychonomic Sci. 25:283-286.
25. Sanberg, P.R., Lehmann, J. and Fibiger, H.C. (1978): Brain Res. 149:546-551.
26. Sanberg, P.R., Pisa, M. and Fibiger, H.C. (1979): Pharmacol. Biochem. Behav. 10:137-144.
27. Shute, C.C.D. and Lewis, P.R. (1967): Brain 90:497- 520.
28. Wenk, H., Bigl, V. and Mayer, V. (1980): Brain Res. 203:295-316.
29. Winocur, G. (1974): J. Comp. Physiol. Psychol. 86: 432-439.

CLINICAL STUDIES OF THE CHOLINERGIC

DEFICIT IN ALZHEIMER'S DISEASE

C.A. Johns, R.C. Mohs, E. Hollander, B.M. Davis,
B.S. Greenwald, M. Davidson, T.B. Horvath and K.L. Davis

Department of Psychiatry
Bronx Veterans Administration Medical Center
130 West Kingsbridge Road
Bronx, New York   10468
and
Departments of Psychiatry and Pharmacology
Mount Sinai School of Medicine of the
City University of New York
One Gustave L. Levy Place
New York, New York   10029
U.S.A.

INTRODUCTION

Alzheimer's disease (AD) is a major mental health problem which will reach epidemic proportions as the percentage of elderly in the population increases. Three to four percent of the population of the United States above age 65 are estimated to be afflicted by this progressive dementia (31, 52), for which at present neither treatment nor definitive antemortem diagnostic tests exist. As current research elucidates the neurochemical and histologic changes that characterize AD, biologic markers useful in diagnosis and rational treatments may be developed.

The most consistent finding in neurochemical studies of brain tissue obtained both from autopsy and biopsy of AD patients is a large reduction of choline acetyltransferase (ChAT) activity. ChAT, an enzyme located almost exclusively within cholinergic neurons, serves as a biochemical marker for these cells. This decrease of ChAT activity has been demonstrated in the hippocampus and cortex (4, 8, 16-18, 45-48, 53, 60, 68) areas of the brain believed to be associated with learning and memory. These brain areas are also the site of the largest concentration of neurofibrillary tangles and neuritic plaques, the classic histologic hallmarks of AD (29).

The reduction of ChAT activity in these brain areas is quantitatively correlated both to plaque number and to severity of dementia (48), as measured by cognitive and behavioral rating scales. The decrease of cortical ChAT activity is likely to be due in part to selective degeneration of cholinergic neurons which originate in the nucleus basalis of Meynert (nbM). Large acetylcholinesterase-rich neuronal bodies in the nbM project widely to the cerebral cortex, and similar neurons in the adjacent diagonal band of Broca and medial septum project to the hippocampus (40), providing the major source of extrinsic cholinergic input to these areas. A decrease of up to 80 percent of these neurons has been demonstrated in the nbM of AD patients (50, 69-71). In contrast, AD brains do not differ from age-matched controls in their ability to bind muscarinic antagonists, indicating a preservation of post-synaptic cholinergic receptor sites (2, 9, 15, 43-48, 53, 68). Taken together, the histologic data strongly suggest that a selective loss of cholinergic neurons from the nbM is a major, specific pathologic mechanism in AD. Changes in concentration of other neurotransmitters have also been demonstrated, and a pronounced degeneration of the locus coeruleus seen in some AD patients (6, 7, 49, 66) suggests that a noradrenergic deficiency may also be an underlying pathologic factor in a subgroup of AD patients. Deficits in somatostatin concentrations have alos been found (17). None of these changes, however, are as well correlated with both the histopathologic lesions and clinical symptomatology of AD as the cholinergic deficits.

Pharmacologic studies of the effects of cholinergic agents on human memory lend further support to the hypothesis that the cholinergic deficit seen in AD mediates at least some of the cognitive impairment seen clinically. Administration of the antimuscarinic agent scopolamine to healthy young subjects transiently impairs their ability to store new information in long-term memory (LTM) while leaving their short term memory and their ability to retrieve perviously-learned information intact (26). This selective deficit is similar to that seen in naturally aged subjects and in patients with early AD (26, 27). Scopolamine's effects on memory can be partially reversed by the cholinesterase inhibitor physostigmine but not by amphetamines (25), specifically implicating the anticholinergic action of scopolamine in impairing cognition rather than its more general sedating effects which impair attention and alertness. Conversely, memory enhancement is seen when either physostigmine (20) or arecoline, a direct short-acting muscarinic receptor agonist (61), is administered to young normal subjects.

These pharmacologic and neurochemical studies, taken together, strongly suggest that a rational treatment approach in AD might be to enhance central cholinergic activity. The development of practical cholinergic therapies has, however, been hampered by the lack of safe, long-acting oral cholinomimetic agents. Response of AD to cholinergic manipulation has been assessed by administering intravenous physostigmine, and results suggesting that memory is enhanced by this agent have encouraged trials of the clinically more practical oral route of physostigmine administration.

Preclinical studies also raise the possibility that a clinical test useful in the diagnosis of AD might be established if an in vivo marker that accurately reflects central cholinergic activity could be identified; such a marker might also be of help in predicing which patients would be most likely to respond to cholinomimetic therapy. Various central nervous sytem (CNS) and neuroendocrine measures were investigated as potential indices of central cholinergic activity. Cerebrospinal fluid (CSF) acetylcholine (ACh) and choline levels, as well as other neurotransmitter metabolites, were determined and compared with severity of dementia in AD patients. Plasma concentrations of cortisol and growth hormone (GH) were quantified. Cortisol secretion is increased by systemic or intrahypothalamic administration of centrally-acting cholinomimetic drugs and is decreased by central but not peripheral anticholinergic drugs (14, 27, 35), suggesting that fluctuations of this hormone may reflect central cholinergic deficits in patients with AD. The GH peak which normally occurs after sleep onset is also highly sensitive to muscarinic blockade (39), and basal GH secretion has been shown to be raised by peripherally-acting cholinomimetics (5, 37) and lowered by peripheral anticholinergic drugs (24). Thus, a cholinergic synapse outside the blood-brain barrier may stimulate GH secretion.

METHODS

Patients in the following studies had at least a 1-year history of progressive memory loss with insidious onset, were in good physical health, and were screened with the aid of CT scans, skull films, EEG, CSF analysis, serum analysis, detailed history and physical examination to rule out other causes of dementia, particularly multi-infarct dementia and depression. Cognitive and behavioral functioning were rated using the Memory and Information Test (MIT) and Dementia Rating Scale (DRS), respectively, scales which are significantly correlated with extent of histopathologic changes seen on autopsy (48). Patients with MIT <10 or DRS> 4 were studied as "definite AD" patients in accordance with long-term follow-up and autopsy studies which confirm that these criteria select patients with a high probability of having AD. Patients with moderate degrees of dementia whose MIT or DRS were within three points of the above criteria were also studied but classified as "equivocal AD" as such patients are statistically more likely to have dementias of various etiologies, especially depression, misclassified as AD (32). Their results were analyzed and reported separately. No patients were psychotic or grossly agitated and all could cooperate with testing procedures. Most patients had never received any psychotropic drugs, and all were completely drug-free except for an occasional chloral hydrate for at least two weeks prior to and throughout the experimental period.

Scales used to identify response to pharmacologic intervention included the Alzheimer's Disease Assessment Scale (ADAS) and selected memory tests such as the Recognition Memory Task (RMT). The ADAS has a broad scope, measuring behavioral disorders, mood states, and cognitive functions such as language, praxis, and memory, and was designed specifically for use in AD. Test-retest reliability was: Wilcoxin r=0.97 (p<.01); inter-rater reliability was: Spearman r=.99 order (p<.001); and the total ADAS score was significantly correct with the Sandoz Clinical AssessmentGeriatric (SCAG) score (r=.89, p<.01) (55). The RMT, a subtest of the ADAS which was also used as an independent measure of LTM in the following studies, was designed for its specificity in testing LTM encoding, its sensitivity to changes in performance, and its comprehensibility to the subject. The test consists of showing patients 12 pictures or words which they are asked to describe or read. These are then shown to the patient again intermixed with 12 unfamiliar words or pictures and the patient is asked to state whether or not he recognizes each item. Three trials are completed. More severely demented patients are shown pictures, less severely demented patients are tested with words. All activity tests were conducted in a dimly lit room between 9 am and 1 pm with a background white noise stimulus.

STUDY I: CSF ACh AND CHOLINE IN AD

Lumbar punctures were performed at 0900h after 10 hours of complete bedrest in 11 male and 4 female AD patients, in whom degree of dementia had been rated using the MIT. Patients ranged in age from 54 to 80 with a mean age of 65 (S.D.=8.7 years). The first through fifth milliliters of CSF obtained were analyzed for ACh and Choline by gas chromatography/mass spectrometry (GC/MS; 28); the sixth through twelfth milliliters were assayed for homovanillic acid (HVA), 5-hydroxyindole acetic acid (5-HIAA) and 3-methoxy-5-hydroxphenylethylene glycol (MHPG), also by GC/MS (1, 30).

Patients with greater memory impairment tended to have lower CSF ACh (Pearson correlation coefficient of MIT with ACh level was r=0.79, p<0.001) (Figure 1). There was a nonsignificant trend for higher choline levels to be associated with greater severity of dementia. The ratio of choline to ACh was inversely correlated to MIT score (r=-0.79,p<0.001). There was no significant correlation between severity of dementia and level of the other neurotransmitter metabolites measured (23).

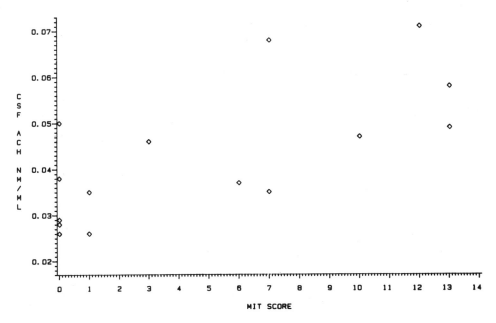

Figure 1. Correlation of cognitive impairment measured by the Memory and Information Test (MIT) with concentration of ACh in cerebrospinal fluid. Subjects were 15 patients with Alzheimer's Disease (r=.70, p=0.004).

STUDY II: SLEEP STUDY

Sixteen male and 9 female patients with definite and probable AD (mean age 66.6 $\pm$ 1.6, relative weight 1.03 $\pm$ 0.02) and ten normal volunteers (7 male, 3 female, mean age 59.1 $\pm$ 2.3, relative weight 1.20 $\pm$ 0.05) participated in sleep studies. Normals spent one, and patients, at least two adaptation nights in the sleep-room with mock intravenous (IV) and blood pressure cuff applied all night. Polyethylene catheters were inserted into a forearm vein at 2100h and blood sampled every half hour from 230h to 1200h the next day. A crude index of sleep was obtained by nurses' observations of the patient, as sleeping or awake, at each half hour. Breakfast was served at 0800h. Plasma was separated and frozen immediately at -4°C, until analysis. Cortisol and growth hormone were measured by radioimmunoassay (24). Not all subjects completed the entire protocol.

Hormonal values were compared using an analysis of variance with time as a within-subject variable, and diagnosis as a between-subjects variable. Cortisol values were log-transformed and age and weight corrected where appropriate, for analysis. Correlations were performed using data from patients having all hormonal values from 2300 to 1030h (N=21 A.D. subjects, 8 normal subjects), except where otherwise indicated. Individual cortisol values, and peak-to-trough differences, were compared using Student's t-test.

    Cortisol concentrations during the secretory period from 0200 to 0830 were higher in AD patients than in controls [F=4.8, df=1,28, (age-corrected); p=.04, Figure 2]. The difference between highest and lowest cortisol values for each individual, a measure of the magnitude of the diurnal peak, tended to be greater in AD than in controls (17 $\pm$ 1.0 vs 14.5 $\pm$ 1.4, p=0.07). Because age and relative weight were correlated (r=.56, n=35, p=.003), and the AD patients were older than the controls, an age-matched subset of AD patients was selected (mean age 60.2 $\pm$ 1.2). In the agematched group, mean nocturnal cortisol did not correlate with age (r=.10, n=18, p=n.s.), or relative weight (r=.22, n=18, p=n.s.).

    As the degree of dementia increased (i.e. the MIT decreased) mean cortisol increased (r=.53, n=29, p=.003, Figure 3). In the group of subjects containing the normals and age-matched controls,

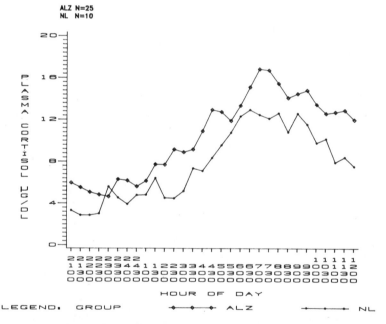

Figure 2. Plasma cortisol concentrations during the night and morning in AD patients and normal controls. Mean cortisol during the secretory period 0200 to 0830 was significantly greater in patients than in controls (F=4.8, df 1, 28 [age-corrected], p=.04).

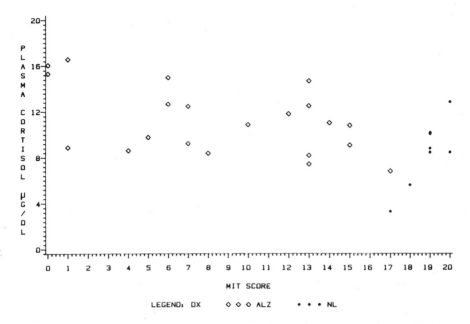

Figure 3. Mean nocturnal cortisol versus degree of dementia in AD patients and normal controls. Plasma cortisol increased with decreasing MIT score (r=.53, n=29, p=.003).

neither age (r=-.11, n=23, p=n.s.) nor relative weight (r=.13, n=23, p=n.s.) correlated with the MIT.

In all normals and patients mean cortisol also increased as the longest continuous period of apparent sleep shortened (r=.46, n=26, p=.02, Figure 4). Among just normal subjects having sleep observations, the correlation between continuous sleep period and cortisol was (r=.72, n=7, p=.07). Neither age nor weight correlated with the sleep measure. Partial correlations between cortisol (1) and severity of dementia (2), adjusting for the effects of sleep (3), age (4) and relative weight (5), were performed. Under these conditions cortisol was still significantly correlated with severity ($r_{12.345}$=-.41,p=.03). When this analysis was repeated for the relationship between cortisol and dichotomous group membership (AD vs. normal), the adjusted correlation became marginal ($r_{12.345}$=-.32, p=.07). Mean cortisol was not significantly related to CSF ACh, HVA or 5-HIAA. A trend suggesting an inverse correlation with CSF MHPG emerged after correction for age and relative weight, (r=.61, df=5, p=.07).

Nocturnal GH concentrations were not different in patients and normals (Figure 5) nor did GH correlated with the severity

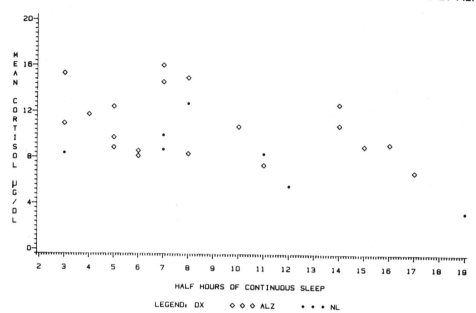

Figure 4.   Mean nocturnal cortisol versus the longest continuous
period of apparent uninterrupted sleep.   As the sleep period lengthened, mean cortisol decreased (r=.46, n=26, p=.02).

of dementia, age, or weight.   Mean nocturnal GH Concentrations increased with increasing CSF MHPG (r=.63, n=12, p=.03).

STUDY III: INTRAVENOUS PHYSOTIGMINE IN AD

The cognitive ability of 13 male and 7 female AD patients, 3
of whom were diagnosed "equivocal AD", was measured during IV infusion
of physostigmine.   The drug was tested in two phases.   In the first,
or dose-response phase, subjects were each given 0.0 mg, 0.125 mg,
0.25 g and 0.5 mg of physotigmine delivered in 100cc of saline over
30 minutes, in random order under double-blind conditions.   Each
infusion was preceded by 2.5 mg Probanthine, IV.   For the last six
subjects, the 0.125 mg dose was replaced by a testing day on 0.0
mg physostigmine and 0.0 mg Probanthine to provide a control condition
for the saline/Probanthine day of testing in order to rule out untoward
effects of the anticholinergic drug.   In the second, or replication
phase, the dose of physostigmine associated with the subject's best
performance was repeated as was a placebo infusion, again randomly
and under double blind conditions, preceded by 2.5 mg of IV Probanthine.   Trials were separated by two to four days and were done at
the same time each day.

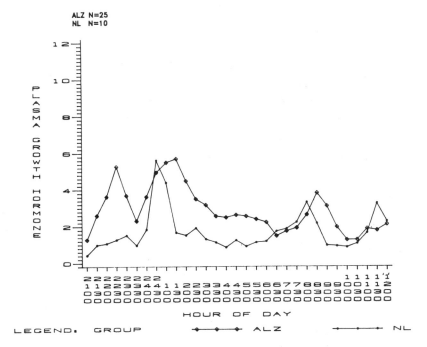

Figure 5.   Plasma growth hormone concentrations during the night and morning in AD/SDAT patients and controls.   Repeated measures ANOVA, covarying for age, showed no significant difference between groups (F=.08, df 1, 24, p=n.s.).

Cognitive testing consisted of: (1) Famous Faces test (LTM retrieval); (2) Digit Span Task (STM); and (3) Recognition Memory Task (RMT) for either 12 words or pictures, or a word recall task wherein the least demented patients were shown ten words and asked to repeat them from memory over three trials.

During the dose-finding phase, all except 2 of the 17 "definite AD" patients had their best response on RMT on some dose of physostigimine rather than placebo, although the optimum dose of physostigimine varied from patient to patient.   No adverse effect of Probanthine was noted in these patients who were tested on two placebo days with and without Probanthine.   One equivocal AD patient performed best on 0.5 mg physostigmine and two showed no improvement over placebo.   During the replication phase given to the 16 patients (15 definite AD and 1 equivocal AD) who demonstrated a best response to physostigmine in the dose-finding phase, twelve patients performed better on LTM storage tests on IV physostigmine than they did on placebo (p=0.02, paired t-test, two-tailed).   When the equivocal AD patient, who responded to physostigmine, was eliminated from

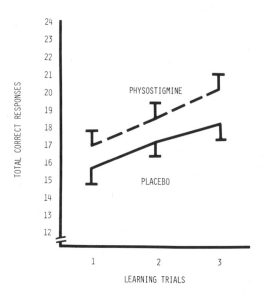

Figure 6. Mean number of correct responses on three trials of the
visual recognition memory task for 13 patients in the replication
phase of the intravenous physostigmine study. The standard error
of the mean is shown for each point.

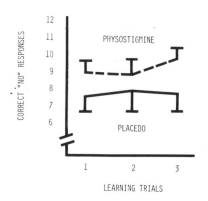

Figure 7. Mean number of correct "no" responses to new items on
the visual recognition memory task for 13 patients in the replication
phase of the intravenous physostigmine study. The standard error
of the mean is shown for each point.

statistical analysis differences remained significant at p=0.02. Degree of improvement of the twelve patients who responded to IV physostigmine ranged from 1.3% to 18.1% (mean 9.4%) (22).

There was a trend for all definite AD patients who demonstrated a response to physostigmine in the dose-finding phase to have more correct responses on physostigmine than on placebo on each of the three RMT trials during the replication phase (Figure 6). These patients also made significantly fewer false positive errors on physostigmine (p=0.02, ANOVA) (Figure 7: analysis done as "correct no", which is equal to "incorrect yes" answers subtracted from 12 and statistically equivalent). In contrast to its effect on RMT, IV physostigmine did not alter performance on either digit span or famous faces tasks.

STUDY IV: ORAL PHYSOSTIGMINE IN AD

Eleven male and five female patients, ten of whom met MIT or DRS criteria for "definite AD" and six of whom wre designated "equivocal AD" wre given oral physostigmine and placebo in a two phase protocol as in Study II. In the dose-finding phase patients were given varying doses of physostigmine and placebo, each dose for four days, every two hours from 7 a.m. to 11 p.m. Cognitive testing was done on the final day of each dose trial. The replication phase consisted of four days of best-dose administration randomized with four days of placebo, with cognitive testing done on the last day of drug or placebo administration. On the last night of each drug condition in the replication phase patients participated in an all night sleep study during which blood was drawn every half hour and assayed for cortisol. To insure that drug effects on early morning changes in these hormones could be detected, administration of oral physostigmine or placebo was continued until 1100h on the morning following the sleep study.

The first 4 patients (3 definite AD and 1 equivocal AD) were given physostigmine in dosages of 0.0 mg, 0.25 mg, 0.5 mg, 0.75 mg and 1.0 mg every two hours. This low dose was chosen in order that any potential adverse reactions might be demonstrated before higher doses were administered in subsequent trials. Testing consisted of a visual memory task. Three of the 4 patients had their best performance on a dose of physostigmine; the one who did not was the one "equivocal AD" patient. These same 3 patients again performed better on physostigmine than on placebo during the replication phase; the improvement ranged from 6% to 12%. The "equivocal AD" patient who showed no improvement on any dose of physostigmine during the dose-finding phase was given physostigmine 1.0 mg and placebo during the replication phase and again performed better on placebo.

As no patients experienced significant side effects to physostigmine at these doses, the maximum dose was raised to 2.0 mg for the next twelve patients, seven of whom had "definite AD." Methodology of drug administration remained the same; doses of 0.0 mg, 0.5 mg, 1.0 mg, 1.5 mg and 2.0 mg were now given. The procedure was further modified by expanding cognitive testing to the ADAS which includes but is not limited to the recognition memory task used initially.

All "definite AD" patients and all but one "equivocal AD" patient had their best performance on a dose of physostigmine; this best dose varied from patient to patient. During the replication phase, given to those eleven patients who demonstrated a response to physostigmine in the dose-finding phase, six of the seven "definite AD" patients again showed improvement on physostigmine; this ranged from 3.5%-20.8% improvement (Figure 8).

One of the four "equivocal AD" patients showed a mild improvement on physostigmine; two others worsened, particularly on depression/irrability and cognitive/language tasks. The fourth patient became delusional and reported visual hallucinations while receiving the 0.5 mg dose of physostigmine. Although it seemed likely that these symptoms were secondary to the patient's advancing disease, intercurrent and unrelated medical problems, or possibly the stress of hospitalization, the study was discontinued and the patient's data were excluded from further analysis. No other patient experienced any serious side effects during the study.

When overnight cortisol measurements were analyzed, a significant positive correlation was found (r=.88, p < 0.01) between percent

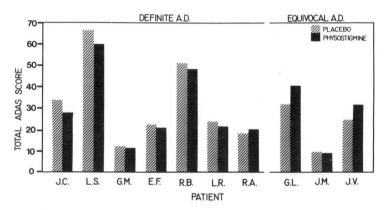

Figure 8. Total ADAS scores for the 10 patients who participated in the replication phase of the oral physostigmine study.

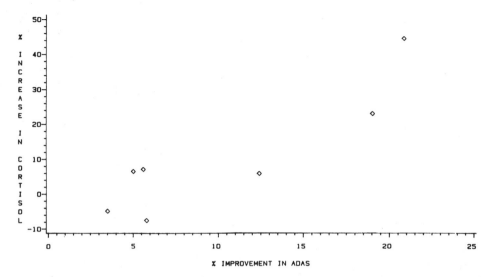

Figure 9. Correlation of symptom change with mean nocturnal cortisol change for seven patients who had complete sleep studies in both drug conditions of the replication phase. Symptom change is measured by % improvement on the ADAS during physostigmine administration. Cortisol change is % increase in mean nocturnal cortisol during physostigmine administration (r=.88, n=7, p < .01).

increase in cortisol and symptom improvement (Figure 9) for the seven patients who had complete sleep studies during the replication phase. This suggests that patients' symptoms were improved only to the extent that the drug enhanced central cholinergic transmission.

Both the clinical staff and families of the 3 "definite AD" patients with the best responses noted improvement in the patients, manifested by less disorientation, better relatedness and generally more appropriate behavior. Although these patients remained quite demented, families actively encouraged further treatment.

DISCUSSION

The studies presented in this paper investigated biological markers that might be used clinically to identify cholinergic or other neurotransmitter abnormalities in patients with AD, and investigated the effects of intravenous and oral physostigmine in patients with AD. In CSF, decreased concentrations of ACh were highly correlated with degree of cognitive impairment but metabolites of dopamine, norepinephrine and serotonin were not. Neuroendocrine studies indicated that patients with AD had higher mean levels of cortisol than nondemented controls but their mean levels of GH and the diurnal variation in GH secretion did not differ from controls.

Intravenous physostigmine enhanced memory in most but not all patients with clinically diagnosed AD, and these improvements were found to be statistically reliable. Oral physostigmine produced overall symptomatic improvement in 6 of 7 "definite AD" patients and was relatively free of side effects. Correlational analysis indicated that patients improved on oral physostigmine only to the extent that the drug had a cholinomimetic effect as measured by an increase in mean cortisol concentration.

An assessment of the potential importance of these studies requires that they be placed in the context of recent findings on the diagnosis, neurochemistry, and neuropathology of AD. Prior to 1976, when the reports of a selective loss of cholinergic neurons in AD first appeared, very few studies of the clinical psychopharmacology or clinical neurochemistry of this disease were being conducted. The finding that cholinergic neurons are affected to a greater extent than are many other neuronal systems in AD suggested that cholinomimetic drugs might be used to treat the disease and that peripheral markers of cholinergic activity might be useful in the diagnostic process. However, recent findings have also made it abundantly clear that the search for pharmacologic treatments and neurochemical markers will not be easy or simple and that several major problems must be addressed if these efforts are to succeed. The present findings represent an initial step in addressing these problems.

The first major problem results from the difficulties involved in making the diagnosis of AD. Clinically, the diagnosis is made by documenting that patients have a history and current mental status consistent with AD and by then performing laboratory and other tests to rule out other possible causes for the dementia. The hope is that such procedures will identify a group of patients with neuropathological and neurochemical characteristics qualitatively similar to those found in patients with AD confirmed at autopsy. Unfortunately, long-term follow-up studies and biopsy studies indicate that as many as 30% of all clinically diagnosed AD patients may not have these characteristics. In follow-up studies covering 1-15 years, between 15% and 30% of all patients originally given a clinical diagnosis of AD were found either to be not demented at follow-up or to have another neurological disorder (38, 54). In a biopsy study of 20 clinically suspected cases of AD only 14 actually had histopathological changes characteristic of AD and of those only 10 had a dramatic loss of ChAT activity (10). Thus, it is to be expected that in studies conducted with clinically diagnosed patients even a valid peripheral marker for central cholinergic activity will be only moderately correlated with mental state and that the effects of cholinergic drugs will be quite variable from patient to patient.

A second major problem to be addressed in clinical trials in-

volving patients with AD is that at least some of these patients
are likely to have deficits other than those in the cholinergic
system which may limit the utility of cholinomimetic therapies.
Noradrenergic cells of the locus coeruleus are lost in some patients
with AD, particularly those with presenile onset. The neuropeptide
somatostatin is also reduced in patients with AD (19, 56-58, 72).
Procedures to identify individuals with pronounced deficits in neurotransmitters other than ACh should ultimately be useful for identifying
patients who are less likely to respond to cholinomimetic therapy.
The present results, admittedly derived from a small sample, suggest
that dysregulation of cortisol in patients with AD may reflect degeneration of another neurotransmitter system, possibly noradrenergic.
Since noradrenergic cells of the locus coeruleus tonically inhibit
the release of cortisol (36), loss of such cells could account for
the hypersecretion of cortisol found in AD patients. Similar studies
in depression have suggested a link between measures of noradrenergic
function and cortisol (21, 59).

A third problem to be addressed in clinical trials of cholinomimetic drugs is how to determine that these drugs actually increase
cholinergic activity. Initial attempts to treat AD with precursors
to ACh were unsuccessful, at least in part, because these substances
do not actually enhance cholinergic transmission except possibly
under conditions when cholinergic neurons are firing at an increased
rate (42). Physostigmine given intravenously has reliable central
cholinomimetic effects (34) but its effects when given orally are
likely to be more variable. In addition, some patients with AD
may have such a severe loss of cholinergic neurons that no amount
of cholinesterase inhibition could enhance central cholinergic transmission. Conclusions about the efficacy of any potential cholinomimetic therapy can only be made with certainty if some biological
measure of the drug's cholinomimetic effects is obtained along with
measures of symptom improvements. The present results suggest that
an increase in mean nocturnal cortisol secretion may be such a measure.

Another potentially useful indicator of oral physostigmine's
cholinomimetic effect has been suggested by Thal et al. (65) who
found that oral physostigmine improved memory only in those patients
who showed substantial cholinesterase inhibition in CSF. Eventually
it may be that a safe, long acting cholinomimetic drug will be developed which has reliable effects on cholinergic transmission in
the great majority of patients. However, the limited number and
uncertain efficacy of available cholinergic drugs makes it almost
imperative that clinical trials of potential cholinomimetics combine
biological markers of neurotransmitter activity with measures of
cognitive and behavioral effects.

Given the problems mentioned above it is not surprising that
results of published studies on the clinical effects of cholino-

mimetics in patients with AD present some inconsistencies. As previously mentioned nearly all of the published studies on the effects of precursors to ACh have found no effect on memory or other symptoms [see Bartus et al. (4) for a review]. Apart from the present study the largest series of patients receiving physostigmine was reported by Christie et al. (12) who found a small but reliable improvement in memory; these investigations also found improvements in memory following arecoline administration. Other smaller series (51, 63, 65) and reports of individual cases (62) including one patient who had a marked improvement in constructional praxis (41), have supported the conclusion that physostigmine produces mild to moderate improvement. Two published studies found no effect of physostigmine but one (11) involved patients with uncertain diagnoses and the other (3) did not seek an appropriate dose for each patient. Two studies with the longer-acting cholinesterase inhibitor, THA, have been conducted in patients with advanced cases of AD (33, 64). Modest global improvement was reported following THA administration in both studies.

Combining careful biological assessment and clinical diagnosis with attention to the unique pharmacological properties of cholinergic drugs may ultimately identify a group of AD patients who can derive meaningful clinical benefit from cholinesterase inhibitors or cholinergic agonists. However, it must be recognized that even when these conditions are met the utility of such drugs could be severely limited. It has been pointed out, for example, that cholinergic cells have relatively few post-synaptic projections which are also non-overlapping so that it is difficult for surviving cholinergic cells to compensate functionally for lost cells by increasing their firing rate (13). In addition, none of these drugs may be able to duplicate the phasic action of cholinergic cells in transmitting information. If such considerations prove to be true, then any pharmacotherapy of AD may have to depend on manipulation of transmitters and co-transmitters operating at sites post-synaptic to the dying neurons. Whatever these pharmacologic methods may be, considerations such as those described in this paper will be important in the design and interpretation of clinical studies.

REFERENCES

1. Anderson, G.M., Young, J.G. and Cohen, D.J. (1979): J. Chromatogr. 164:501-505.
2. Antuono, P., Sorbi, S., Bracco, C., Fusco, T. and Amaducci, L. (1980): In Aging of the Brain and Dementia (Aging, Vol. 13) (eds) L. Amaducci, A.N. Davison and P. Antuono, New York, Raven Press, pp. 151-158.
3. Ashford, J.W., Soldinger, S., Schaeffler, J., Cochran, L. and Jarvik, L.F. (1981): Am. J. Psychiat. 138:829-830.
4. Bartus, R.T., Dean, R.L., Beer, B. and Lippa, A.S. (1982): Science 217:408-417.

5. Bicknell, R.J., Young, P.W. and Schofield, J.G. (1967): Mol. Cell. Endocrinol. 13:167-180.
6. Bondareff, W., Mountjoy, C.Q. and Roth, M. (1981): Lancet 1: 783-784.
7. Bondareff, W., Mountjoy, C.Q. and Roth, M. (1982): Neurology 32:164-168.
8. Bowen, D.M., Spillane, J.A., Curzon, G., Meier-Ruge, W., White, P., Iwangoff, P. and Davison, A.N. (1979): Lancet i:11-14.
9. Bowen, D.M. and Davison, A.N. (1980): Psychol. Med. 10:315-319.
10. Bowen, D.M., Benton, J.S., Spillane, J.A., Smith, C.C.T. and Allen, S.J. (1982): J. Neurol. Sci. 57:191-202.
11. Caine, E.D. (1980): New Engl. J. Med. 303:585-586.
12. Christie, J.E., Shering, A., Ferguson, J. and Glen, A.I.M. (1981): Brit. J. Psychiat. 138:46-50.
13. Coyle, J.T., Price, D.L. and DeLong, M.R. (1983): Science 219: 1184-1190.
14. Cozanitis, D.A. (1974): Anaethesia 29:163-168.
15. Davies, P. (1978): In Alzheimer's Disease, Senile Dementia and Related Disorders (Aging, Vol. 7) (eds) R. Katzman and R. Terry, New York, Raven Press, pp. 453-459.
16. Davies, P. (1979): In Cogenital and Acquired Cognitive Disorders (ed) R. Katzman, New York, Raven Press, pp. 153-160.
17. Davies, P. (1979): Brain Res. 171:319-326.
18. Davies, P. and Maloney, A.J.F. (1976): Lancet ii:1403.
19. Davies, P., Katzman, R. and Terry, R.D. (1980): Nature 288: 279-280.
20. Davis, K.L., Mohs, R.C., Tinklenberg, J.R., Hollister, L.E., Pfefferbaum, A. and Kopell, B.S. (1978): Science 201:274-276.
21. Davis, K.L., Hollister, L.E., Mathe, A.A., Davis, B.M., Rothpearl, A.B., Faull, K.F., Hsieh, J.Y.,.K., Barchas, J.D. and Berger, P.A. (1981): Am. J. Psychiat. 138:1555-1562.
22. Davis, K.L. and Mohs, R.C. (1982): Am. J. Psychiat. 139:1421-1424.
23. Davis, K.L., Hsieh, J.Y.K., Levy, M.I., Horvath, T.B., Davis, B.M. and Mohs, R.C. (1982): Psychopharm. Bull. 18:193-195.
24. Davis, B.M., Mathe, A.A., Mohs, R.C., Levy, M.I. and Davis, K.L. (1983): Psychoneuroendocr. 18:103-107.
25. Drachman, D.A. (1977): Neurology 27:783-790.
26. Drachman, D.A. and Leavitt, J. (1974): Arch. Neurol. 30:113-121.
27. Hillhouse, E.W., Burden, J. and Jones, M.T. (1975): Neuroendocr. 17:1-11.
28. Jenden, D.J., Roch, M. and Booth, R.A. (1973): Anal. Biochem. 55:438-448.
29. Jervis, G.A. (1971): In Pathology of the Nervous System, Vol. 2 (ed) J. Minckler, New York, McGraw Hill, pp. 1385-1395.
30. Karoum, F., Gillin, J.C., Wyatt, R.J. and Costa, E. (1975): Biomed. Mass Spectrom. 2:183-189.
31. Katzman, R. (1976): Arch. Neurol. 33:217-218.
32. Kay, D.W.K. (1977): In Cognitive and Emotional Disturbances in

the Elderly (eds) C. Eisdorfer and R.O. Friedel, Chicago, Year Book Medical Publishers, pp. 11-26.

33. Kaye, W.H., Sitaram, N., Weingartner, H., Ebert, M.H., Smallberg, S. and Gillin, J. (1982): Biol. Psychiat. 17:275-280.

34. Ketchum, J.S., Sidell, F.R., Crowell, E.B., Aghajanian, G.K. and Hayes, A.H. (1973): Psychopharmacologia 28:121-145.

35. Krieger, D.T., Silverberg, A.I., Rizzo, F. and Krieger, H.P. (1968): Am. J. Physiol. 215:959-967.

36. Lancranjan, I., Ohnhaus, E. and Girard, J. (1979): J. Clin. Endo. Metab. 49:227-230.

37. Leveston, S.A. and Cryer, P.E. (1980): Metabolism 29:703-706.

38. Marsden, C.D. and Harrison, M.J.G. (1972): Br. Med. J. 1:249-252.

39. Mendelson, W.B., Sitaram, N., Wyatt, R.J. and Gillin, J.C. (1978): J. Clin. Invest. 61:1683-1690.

40. Mesulam, M., Mufson, E.J., Levey, A.I. and Wainer, B.H. (1982): In: Banbury Report No. 15. Biological Aspects of Alzheimer's Disease, R. Katzman (ed), pp. 79-93.

41. Muramoto, O., Sugishita, M., Sugita, H. and Toyokura, Y. (1979): Arch. Neurol. 36:501-503.

42. Mohs, R.C., Davis, K.L. and Levy, M.I. (1980): Life Sci. 29: 1317-1323.

43. Nordberg, A., Adolfsson, R., Aquilonius, S.M., Marklund, S., Oreland, L. and Winblad, B. (1980): In Aging of the Brain and Dementia (Aging, Vol. 13) (eds) L. Amaducci, A.N. Davison and P. Antuono, New York, Raven Press, pp. 169-171.

44. Perry, E.K. (1980): Aging 9:1-8.

45. Perry, E.K., Gibson, P.H., Blessed, G., Perry, R.H. and Tomlinson, B.E. (1977): J. Neurol. Sci. 34:247-265.

46. Perry, E.K., Perry, R.H., Blessed, G. and Tomlinson, B.E. (1977): Lancet 1:189-191.

47. Perry, E.K., Perry, R.H., Gibson, P.H., Blessed, G. and Tomlinson, B.E. (1977): Neurosci. Lett. 6:85-89.

48. Perry, E.K., Tomlinson, B.E., Blessed, G., Bergman, K., Gibson, P.H. and Perry, R.H. (1978): Brit. Med. J. 2:1457-1459.

49. Perry, E.K., Tomlinson, B.E., Blessed, G., Perry, R.H., Cross, A.J. and Crow, T.J. (1981): J. Neurol. Sci. 51:279-287.

50. Price, D.L., Whitehouse, P.J., Struble, R.G., Clark, A.W., Coyle, J.T., DeLong, M.R. and Hedreen, J.C. (1982): Neurosci. Comment. 1:84-85.

51. Peters, B.H. and Levin, H.S. (1979): Ann. Neurol. 6:219-221.

52. Plum, F. (1979): Nature 279:372-373.

53. Reisine, T.D., Yamamura, H.I., Bird, E.D., Spokes, E. and Enna, S.J. (1978): Brain Res. 149:447-481.

54. Ron, M.A., Toone, B.K., Garralda, M.E. and Lishman, W.A. (1979): Br. J. Psychiat. 134:161-168.

55. Rosen, W.G., Mohs, R.C. and Davis, K.L. (1984): Am. J. Psychiat. 141 (No.11):1356-1364.

56. Rossor, M.N., Emson, P.C., Mountjoy, C.Q., Roth, M. and Iverson, L.L. (1980): Neurosci. Lett. 20:373-377.

57. Rossor, M.N., Fahrenberg, J., Emson, P., Mountjoy, C.Q., Iverson, L.L. and Roth, M. (1980): Brain Res. 201:249-253.
58. Rossor, M.N., Emson, P.C., Mountjoy, C.Q., Roth, M. and Iverson, L.L. (1981): Life Sci. 29:405-410.
59. Sachar, E.J. (1976): Ann. Rev. Med. 27:389-396.
60. Sims, N.R., Smith, C.C.T., Davison, A.M., Bowen, D.M., Flack, R.H.A., Snowden, J.S. and Neary, D. (1980): Lancet 1:333-336.
61. Sitaram, N., Weingartner, H. and Gillin, J.C. (1978): Science 201:274-276.
62. Smith, C.M. and Swash, M. (1979): Lancet 1:42.
63. Sullivan, E.V., Shedlack, K.J., Corkin, S. and Growden, G.H. (1982): In Alzheimer's Disease: A Report of Progress (eds) S. Corkin, K.L. Davis, J.H. Growden, E. Usdin and R.J. Wurtman, New York, Raven Press, pp. 361-368.
64. Summers, W.K., Viesselman, J.O., Marsh, G.M. and Candelora, K. (1981): Biol. Psychiat. 16:145-153.
65. Thal, L.J., Fuld, P.A., Masur, D.M. and Sharpless, N.M. (1983): Ann. Neurol. 13:491-496.
66. Tomlinson, B.E., Blessed, G. and Roth, M. (1970): J. Neurol. Sci. 11:205-242.
67. Torak, R.M. (1978): The Pathological Physiology of Dementia, Berlin, Springer-Verlag.
68. White, P., Goodhardt, M.J., Kent, J.P., Hiley, C.R., Carrasco, L.H., Williams, J.E. and Bowen, D.M. (1977): Lancet i:668-671.
69. Whitehouse, P.J., Price, D.L., Clark, A.W., Coyle, J.T. and DeLong, M.R. (1981): Ann. Neurol. 10:122-126.
70. Whitehouse, P.J., Price, D.L., Struble, R.G., Clark, A.W., Coyle, J.T. and DeLong, M.R. (1982): Science 215:1237-1239.
71. Wilcock, G.K., Esiri, M.M., Bowen, D.M. and Smith, C.C.T. (1983): Neuropath. and Applied Neurobiol. 9:175-179.
72. Yates, C.M., Harmar, A.J., Rosie, R., Sheward, J., Sanchez, J., deLevy, G., Simpson, J., Maloney, A.F.J., Gordon, A. and Fink, G. (1983): Brain Res. 258:45-52.

PROTECTION AND THERAPEUTIC EFFECT OF CDP CHOLINE

IN HYPOCAPNIC NEURONS IN CULTURE

S. Mykita, D. Hoffmann, H. Dreyfus,
L. Freysz and R. Massarelli

Centre de Neurochimie du CNRS,
and U.44 de l'INSERM
5 rue Blaise Pascal, 67084 Strasbourg Cedex France

INTRODUCTION

A low oxygen supply to the nervous tissue induced by hypoxia or anoxia in vitro or by ischemia in vivo is known to produce a severe modification of phospholipid metabolism (2) and to be at the origin of the well known post mortem effect which has hindered for years the correct determination of free choline concentrations in the brain (7). Recently the stimulation of phospholipid metabolism and especially of phosphatidylcholine,through the use of its precursors has opened up new ways of understanding the metabolism of nerve cell membranes and, in some instances, to produce new therapeutic approaches for various diseases such as Alzheimer's disease, senile dementia, or for protection in aging cells (4). The relevance of CDP-choline as a precursor of phosphatidylcholine has been known since the establishment of the essential steps of the Kennedy pathway, and its use in medical and clinical practice has opened up new useful therapeutic treatments. Traumatic cerebral oedemas, cerebral anoxia, local ischemic damages, producing in some cases hemiplegia, can be treated with some success with CDPcholine (14,10).

The reduction of PO_2 in nervous tissue produced by anoxia or ischemia should equally affect in vivo the PCO_2 in the blood, and in the cells. Nerve cell cultures are grown normally in a growth medium containing a phosphate bicarbonate buffer which requires the presence of an atmosphere rich in CO_2 (5% CO_2/95% air). It is therefore quite feasible to reduce the CO_2 content to atmospheric values and observe the effects upon the neuronal morphology and neurochemistry. This report will present some preliminary results on the protection and possible therapeutic use of CDP-choline in hypocapnic neurons in culture.

MATERIALS AND METHODS

Exclusively neuronal cultures from 8-day-old chick embryo cerebral hemispheres were obtained following standard procedures (13). Briefly, mechanically dissociated cerebral tissue was seeded into plastic Petri dishes (100 or 60 mm diameter) precoated with L-Polylysine. The morphological and biochemical development of these cells has been the object of several reports (11, 6). Between the second and third day in culture (synaptogenesis starts at this age in these cultures) the cells, grown in Dulbecco's modified Eagle's medium supplemented with 20% of fetal calf serum, were put in an incubator under a normal air atmosphere (with approximately 0.03% CO_2), 90% humidity and at 37°C. Six to twelve hours later (the time was a function of the rapidity of the gas exchange between the growth medium and the atmosphere, and was determined by the inclusion of trypan blue as a sign of cell death) the cultures were put back into an incubator under 5% CO_2. Pictures were taken one and four days afterwards and: (a) the number of viable cells; (b) the number of cellular clumps (the size of these varied from a few cells indistinguishable from one another, to several tens of neurons); (c) the number of primary neuronal processes; and (d) the number of clearly identifiable glioblasts were measured per photographic field. The cells were grown with CDP-choline 10^{-6}M and 10^{-8}M from zero d.i.c. (days in culture) to seven d.i.c., to study the protective effect of the compound. Alternatively, neurons were grown for 5 days under normal conditions, then kept in hypocapnia for 4 hours in Petri dishes of 60 mm diameter containing 4 ml of growth medium. This was followed by changing the medium, which now contained 10^{-6}M CDP-choline. To study the possible effects of mononucleotides, CMP, AMP and GMP 10^{-4}M and 10^{-6}M were added at zero d.i.c.

Choline metabolism was followed by incubating neurons with radioactive [^{3}H]choline (final concentration: 50 μM, 2 μCi/ml of incubation medium, specific activity: 15 Ci/mmole, Amersham) for various time intervals in Krebs Ringer phosphate solution and homogenizing the cells after addition of 0.4 N HCl. After centrifugation (4,000 g x 20 min) the radioactive content of the acid soluble fraction and the pellet were counted with 10 ml of Rotiszint 22 (Roth).

RESULTS

Control cultures grown under standard conditions (two 8-day-old embryonal cerebral hemispheres per 3 Petri dishes of 100 mm diameter) showed a higher number of neurons per microscopic field (59.52 ± 7.49, mean ± S.D., Table 1), when compared to cultures grown in the presence of 10^{-6}M CDP-choline (47.48 ± 6.89; 2P < 0.001). Conversely less cellular clumps were observed in control (7.02 ± 2.65) compared to treated cultures (12.92 $\pm$ 3.59; 2P < 0.001) while more glial cells were observed in the control cultures (3.68 $\pm$ 1.92 compared to 2.32 $\pm$ 1.52) even if the latter values were not significantly different using Mann and Whitney U test and Student t-test.

After the hypocapnic treatment, however, control cultures appeared to be full of debris and the cells which remained attached to the polylysine film were vacuolized (Fig. 1A). CDP-choline treated cultures, on the other hand, appeared normal in all aspects under similar hypocapnic conditions (compare Fig. 1B with 1C and 1D which had been kept in normal CO_2 atmosphere). An analysis of the pictures taken showed that (Table 1): CDP-choline treated cultures contained more cells after hypocapnic treatment than the corresponding controls (90.35 $\pm$ 9.5 cells per microscopic field compared to 24.9 $\pm$ 4.99, mean $\pm$ S.D.; 2P < 0.001). Moreover, the number of clumped cells was also significantly increased, even if to a smaller extent (7.62 $\pm$ 2.76 compared to 5.53 $\pm$ 2.35, 2P < 0.001). In addition, a small increase in the number of primary processes was also observed in the treated cells (3.84 $\pm$ 1.05 in treated cells and 3.12 $\pm$ 1.79 in the control cultures; 2P < 0.01).

Table 1

Effect of CDP-Choline on Neuronal Morphometry
In The Presence and The Absence of CO_2

	Nø/mf	Nc/mf	Npp/c	Ng/mf
Control (+ CO_2)	59.52$\pm$7.49 (25)	7.02$\pm$2.65 (42)	4.03$\pm$2.00 (32)	3.68$\pm$1.92 (22)
CDP-Ch 10^{-6}M (+ CO_2)	47.48$\pm$6.89 (27)	12.92$\pm$3.59 (52)	*3.9$\pm$1.97 (59)	*2.32$\pm$1.52 (34)
Control (- CO_2)	24.90$\pm$4.99 (69)	5.53$\pm$2.35 (71)	3.19$\pm$1.79 (158)	3.59$\pm$1.89 (56)
CDP-Ch 10^{-6}M (- CO_2)	90.35$\pm$9.50 (74)	7.62$\pm$2.76 (74)	3.84$\pm$1.05 (134)	*3.86$\pm$1.96 (65)

Neuronal cultures were analyzed at 6 d.i.c. mf: microscopic field; Nø: number of cells; Nc: number of clumps; Npp: number of primary processes per cell; Ng: number of glial cells. The number of determinations is shown in parenthesis. The values represent means $\pm$ S.D. *Statistically <u>not</u> significant when compared to the controls. All other values in treated cultures compared to controls were significantly different by 2 p < 0.001.

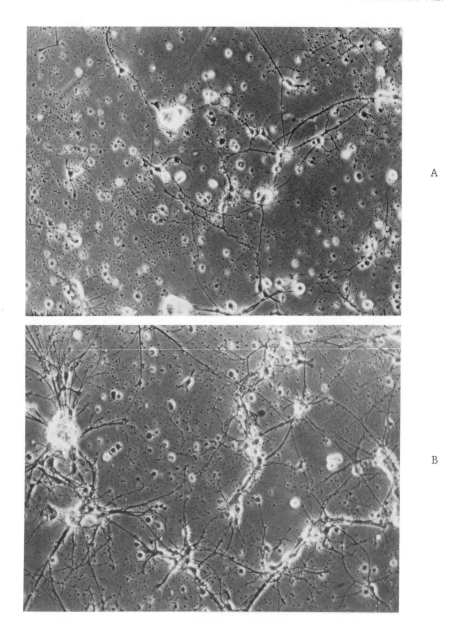

Figure 1.   Effect of low CO_2 on pure neuronal cultures (6 d.i.c.) and the protective effect of CDP-choline.   A: Cells were kept for 6 hours (with 0.03% CO_2) to 3 d.i.c. and the picture was taken at 6 d.i.c.   Neurons appear vacuolized and severely damaged. B: Cells grown with 10^{-6}M CDP-choline from 0 d.i.c. to 3 d.i.c.   After 6 hours of hypocapnia, cultures were grown to 6 d.i.c. in a medium containing CDP-choline.

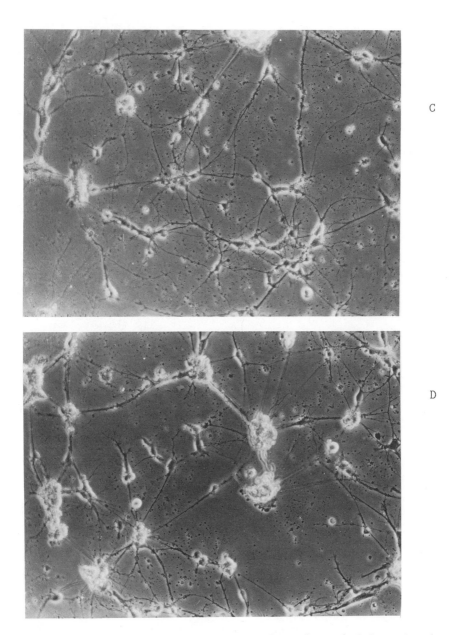

C

D

C and D: Control cells and cells grown from 0 to 6 d.i.c. in the presence of CDP-choline respectively.

Concerning the amount of glial cells per microscopic field in treated and untreated hypocapnic and control cells it is interesting to note that in control cultures (+ CO_2) glioblasts represented 5% of the total cells present. After hypocapnia their content was raised up to 13%. CDP-choline exposure appeared instead to maintain a low amount of support cells similar to controls (4.8% in the presence of CO_2 and 4.2% in the absence of CO_2). However, no difference was seen in the number of glial cells appearing per microscopic field under the two conditions of treatment (controls vs. CDP-choline treated).

Following our observation that CDP-choline could successfully protect cells against hypocapnic treatment, we next wanted to establish whether CDP-choline might have a therapeutic effect on neurons after hypocapnic treatment. Neuronal cell cultures were therefore incubated without CO_2 at 5 d.i.c. for 6 hrs and the medium was thereafter changed with fresh medium containing $10^{-6}M$ CDP-choline. Two days later morphological observations (Figs. 2A and B) indicated that CDP-choline preserved neurons in a satisfactory shape. To determine whether the entire molecule or only parts of it were responsible for the observed effects, cultures were next treated with CMP, GMP and AMP. No effect was observed with any of these nucleotides at $10^{-4}M$ or $10^{-6}M$ (results not shown).

The incorporation of $[^3H]$choline showed between 10 and 20 min an increased metabolism of this compound in the acid soluble extract of cells treated with CDP-choline (+ CO_2), when compared to the control (+ CO_2) (Fig. 3A), while no apparent difference was observed in the total lipid fractions (Fig. 3B). Hypocapnic control cells (Fig. 3A) metabolized more choline than treated cells (Fig. 3A); this was observed in both acid soluble and total lipid fractions (compare Fig. 3A with Fig. 3B).

DISCUSSION

Several reports have indicated that an increased supply of choline to the nervous tissue may influence the metabolism of ACh; notably by increasing its synthesis. Choline itself, deanol and choline glycerophospholipids have been used with uncertain success to test new possible ways to stimulate ACh synthesis via increased supply of the precursor (4). The neuronal cell cultures which have been used in this study are poorly cholinergic, and the above mentioned compounds have been used in various trials to stimulate the synthesis of the neurotransmitter. One of the compounds used is CDP-choline, which is the direct precursor of phosphatidylcholine in most organs, included the brain. CDP-choline has retained the attention of several research groups, since it was shown that it had an important therapeutic effect in hypoxia (5) and anoxia (1), and that it could be used with some success in cases of cerebrovascular injuries (15, 9).

Figure 2. Therapeutic effect of CDP-choline on neurons (7 d.i.c.).
Cells were grown normally from 0 to 5 d.i.c. after which the CO_2
content was brought to atmospheric value for 6 hours, and treated
afterwards with 10^{-6}M CDP-choline. **A**: Control cells full of debris
and showing evident signs of cell death. **B**: Cells treated with
CDP-choline (compare with Fig. 1C).

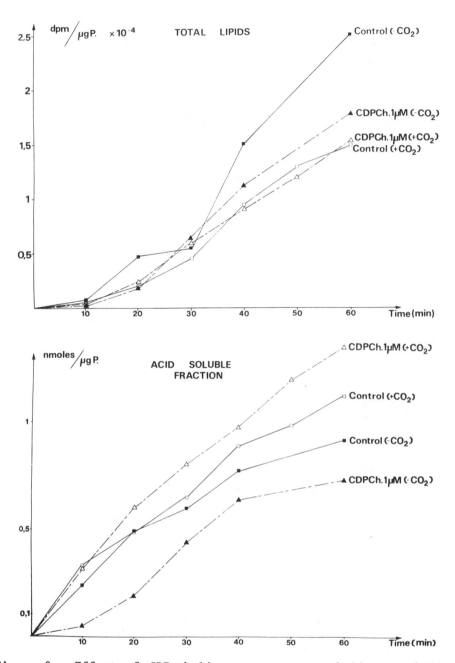

Figure 3. Effect of CDP-choline treatment on choline metabolism in normal and hypocapnic neurons. A: Acid-soluble fraction representing the supernatant of an HCl homogenate of the cells. B: Total lipids fraction representing the radioactive content of the pellet.

The present data indicate that CDP-choline can have a protective and a therapeutic effect on neurons grown in culture and subjected to a hypocapnic treatment.

CDP-choline added to the growth medium under normal conditions (5% CO_2), slightly decreased the number of isolated cells but significantly increased the number of cellular clumps. The actual amount of cells measured after trypsinization and at 6 d.i.c. was in fact higher in treated cells than in controls (9.5 x 10^5 vs 8.25 x 10^5). After hypocapnic treatment, considerable cellular death was observed. Pre-treatment with CDP-choline however greatly enhanced the survival of cells, and kept the amount of cellular clumps and of primary processes closer to the value found for the controls which had been maintained continuously in the presence of CO_2. The higher percentage of glial cells found in control (- CO_2) probably indicates that glioblasts are much more resistant to hypocapnic conditions than neurons.

The question to be raised is why CDP-choline exerts such protective and therapeutic effects. We have observed that isolated nucleotides at 10^{-6}M or at higher concentrations (10^{-4}M) do not have any morphological influence on cell cultures under the present condition (results not shown). It is also known from the work of the group of Porcellati (8) that CDP-choline does not cross the cellular membrane as such and that, with great probability, it is broken down by several enzymes to the nucleoside, phosphate and choline. It is also possible that the half life of the drug in the growth medium is reduced by the presence of hydrolytic enzymes in fetal calf serum. Notwithstanding this, it remains that CDP-choline has an effect on the cell membranes also in normal growth conditions, since it stimulated the aggregation of cells (number of clumps), and it protected isolated cells against the effects of a low PCO_2. The first result of a reduction in CO_2 in these cultures is a rapid rise in the pH which, in the present experimental conditions, goes from 7.0 - 7.5 to a maximum of 8.4. Such a pH (8.4) is known to activate a large variety of enzymes, including hydrolytic enzymes. CDP-choline molecules may conceivably protect the cells against such an activation, and must act in this case at the membrane level.

The results on the aggregating effect and on the antihypocapnic action suggest moreover that, because of its nature, CDP-choline should influence phospholipid synthesis in stimulating, perhaps, the production of choline phosphoglycerides (3). However, the incubation with radioactive choline and its metabolic fate in treated and untreated cells indicate that the action of CDP-choline is certainly of a more complex nature than just that described above.

In the presence of CO_2 more radioactivity was found in the acid soluble fraction of neurons treated with CDP-choline, when compared to their control cells, while no difference was seen in the lipid fraction.   In the absence of CO_2 less radioactivity was contained in the acid soluble fraction of CDP-choline treated cells, and still less in their lipid fraction.   It thus appears that, in the presence of CO_2, CDP-choline increases the amount of the water soluble precursors of choline phospholipids, while the opposite is seen in the absence of CO_2.   Under these conditions, control cells incorporate radioactive choline into their lipid fraction, while treated cells are less prone to do so.   Is this because treated cells are stimulated to make choline by other means (for example by increasing methylation of ethanolamine, which we know can occur in these cells; 12)?

It is of course also possible that the cellular survival effect is but an epiphenomenon of the action of CDP-choline on the structure and constitution of membranes.   In any case, CDP-choline may well reveal itself in the future as being a valuable tool for studying cell membrane dynamics; particularly in cholinergic neurons.

Acknowledgements.   The secretarial assistance of Ms. C. Thomassin-Orphanides is greatly appreciated.   This work was partly supported by a NATO grant (#75.82).

REFERENCES

1. Alberghina, M., Viola, M., Serra, I., Mistretta, A. and Giuffrida, A.M. (1981): J. Neurosci. Res. 6:421-433.
2. Ansell, G.B. and Spanner, S. In Cholinergic Mechanisms (eds) G. Pepeu and H. Ladinsky, Plenum Press, New York (1981): pp. 393-403.
3. Arienti, G., Corazzi, L., Mastrofini, P., Montanini, I., Tirillini, B. and Porcellati, G. (1979): Italian J. Biochem. 28:39-45.
4. Blusztajn, J.K. and Wurtman, R.J. (1983): Science 221:614-621.
5. Dorman, R.V., Dabrowiecki, Z. and Horrocks, L.A. (1979): J. Neurochem. 40:276-279.
6. Dreyfus, H., Louis, J.C., Harth, S., Durand, M. and Massarelli, R. In Neural Transmission, Learning and Memory (eds) R. Caputto and C. Ajmone Marsan, Raven Press, New York (1983): pp. 21-33.
7. Dross, K. and Kewitz, H. (1972): N.S. Arch. Pharmacol. 274:91-105.
8. Floridi, A., Vecchini, A., Palmerini, C.A., Binaglia, L. and Porcellati, G. (1981): Italian J. Biochem. 30:317-319.
9. Horrocks, L.A., Dorman, R.V., Dabrowiecki, Z., Goracci, F. and Porcellati, G. (1981): Progr. Lip. Res. 20:531-534.
10. Kondo, Y. (1963): Arch. Japan Chir. 32:489-511.
11. Louis, J.C., Dreyfus, H., Wong, T.Y., Vincendon, G. and Massarelli, R. In Neural Transmission, Learning and Memory (eds) R. Caputto and C. Ajmone Marsan, Raven Press, New York (1983): pp. 49-61.
12. Massarelli, R., Dainous, F., Freysz, L., Dreyfus, H., Mozzi, R., Floridi, A., Siepi, D. and Porcellati, G. In Basic and Clinical Aspects of Molecular Neurobiology (eds) A.M. Giuffrida Stella,

G. Gombos, C. Benzi and H.S. Bachelard Fond. Intern. Menarini (1982): pp. 147-155.

13. Pettman, B., Louis, J.C. and Sensenbrenner, M. (1979): Nature 281:378-380.

14. Tanakamaru, S. (1964): Kumamoto Igaku Zasshi 38:145-163.

15. Trovarelli, G., De Medio, G.E., Dorman, R.V., Piccinin, G.L., Horrocks, L.A. and Porcellati, G. (1981): Neurochem. Res. 6: 821-833.

THE AGING OF CHOLINERGIC SYNAPSES: ONTOGENESIS

OF CHOLINERGIC RECEPTORS

A. Nordberg

Department of Pharmacology
University of Uppsala
Box 573
S-751 23 Uppsala, Sweden

INTRODUCTION

Cholinergic synapses in brain undergo dynamic changes during development, maturation and aging. During the life span we can assume that there is a continuous interregulation between presynaptic and postsynaptic activity. In recent years several investigators have shown for example that both muscarine- and nicotine-like receptor binding sites in brain can be modulated by a continuous drug treatment (13, 23, 31, 32, 34, 37). Signs for a compensatory "upregulating" mechanism for muscarinic binding sites have been reported in pathological states such as dementia of Alzheimer type (29) and Parkinson's Disease (35).

Studies on the ontogenesis of muscarine- and nicotine-like receptor binding sites in brain and their correlation to the development of presynaptic cholinergic activity might provide a better insight into the neurochemical basis for cholinergic function in both immature and mature brain. They might also yield information concerning possible early susceptibility for neurotoxic exposure and involvement later in life in certain pathological states and neurological diseases.

Most studies on ontogenesis have been performed in experimental animals while very few studies have been reported on human postmortem material (6, 15). The enzyme choline acetyltransferase (ChAT) has often been used as a marker for presynaptic cholinergic activity (7, 12, 14) and more infrequently acetylcholine (ACh) content and choline (Ch) uptake (7). In addition, the activity of acetylcholinesterase (AChE) has been followed, although it must be considered

165

to be a rather nonspecific parameter for cholinergic activity (3, 12). For the muscarinic receptors which are considered to be located mostly postsynaptically, although presynaptic autoreceptors have been reported (39), the muscarinic antagonists quinuclidinyl benzilate (QNB), scopolamine and the agonist oxotremorine have been utilized as ligands in ontogenesis studies (7, 9, 10, 12, 14, 18, 42). Alfa-bungarotoxin (Btx) has, in some studies, been chosen to measure the development of nicotine-like binding sites (3, 14, 16, 40) and recently also another antagonist, tubocurarine (14), and nicotine itself (21), have been in use.

This chapter summarizes our current status of knowledge re-garding the development of cholinergic muscarinic and nicotinic binding sites in animal and human brains.

ANIMAL STUDIES

Development of presynaptic cholinergic activity. According to Coyle and Yamamura (7) measurable amounts of endogenous ACh can be detected in rat brain at 15 days of gestation (22% of adult level) when hardly any activity of ChAT is measurable (1.4% of adult level). The high affinity uptake of Ch by the cholinergic nerve terminal was first detectable one week post-partum (7). From that time point there was a parallel continuous increase in ChAT activity, ACh content and Ch uptake (7). A similar linear increase in ChAT activity up to adult level has been found in brain of chicken (12) and mouse (14), including a pattern of caudalrostral maturation (14).

Development of muscarine-like binding sites. The muscarine-like binding sites start to develop in the rodent about one week before birth (3, 7) and then increase linearly up to 15-20 days of age (3, 10, 14). The cerebellum shows the lowest number of muscarinic binding sites (less than 25% of the number in other mouse brain regions) and also attains maturity at an earlier time (14). Yavin and Harel (42) have studied the development of muscarine-like binding sites in the rabbit brain. They found that in the cerebellum the number of muscarinic binding sites was comparable to the number in other brain regions at birth. In contrast to other brain regions, however, cerebellar muscarinic binding sites then decreased progres-sively in number of binding sites as the animal reached adulthood. A similar developmental pattern was also observed for the lower brain stem (42).

The use of ligands of agonist type has mede it possible to detect at least two major types of muscarinic binding sites with different affinities for the agonist in brain (4). A 6-7 day lag in the appearance of high affinity binding sites following the appear-ance of the low affinity sites was demonstrated by Kuhar et al. (18) in rat cortex. The significance of this finding is unknown. In the adult animal, however, a regional variation of the low affinity sites seem to exist while the proportion of high affinity sites is rather constant (5). A coupling between low affinity binding sites and functional activity has also been discussed (5).

As shown in Fig. 1 the development of muscarinic binding sites in mouse brain parallels but precedes that of ChAT in all regions studied, when the values are expressed as percentage of levels found in the mature animal. Since, as has been pointed out above, the content of endogenous ACh is remarkably high early in the animal's development, when the activity of its synthesizing enzyme ChAT is low, we can also conclude that the development of muscarinic receptors precedes that of the animal's capacity to synthesize ACh (Fig. 1). Unfortunately, no studies have yet been reported on the turnover

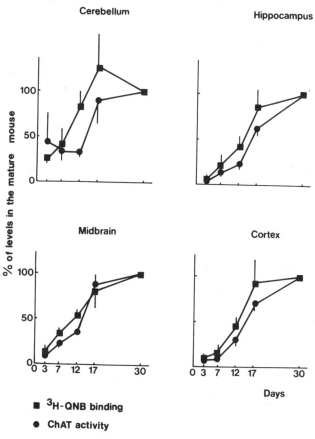

Figure 1. Development of choline acetyltransferase (ChAT) activity and number of labelled quinuclidinyl benzilate ([^{3}H]-QNB) binding sites in different areas of mouse brain expressed as percentage of levels in mature mice. Each point indicates mean value $\pm$ SE (from Falkeborn et al., 14).

rate of ACh during development. Coyle and Yamamura (7) suggested that a low activity of AChE could explain for the early, rather high level of ACh. One might also speculate whether some release (leakage) of ACh from the presynaptic terminal is needed and triggers the formation of receptor binding sites.

Development of nicotine-like binding sites. The use of Btx as a ligand for nicotine-like binding sites in brain is somewhat controversial (25, 26, 28). In contrast to muscarinic binding sites the binding of Btx alternates with maturation (1, 3, 14, 16, 26, 40). As seen in Fig. 2 the Btx binding is substantial at day 3, attains a peak between day 7 and 12 and then decreases subsequently towards values seen in adulthood, in all regions except cerebellum. Two developmental peaks for Btx (day 7-10; day 30) in the superior colliculus have been reported (26). Interestingly, a higher density of Btx binding sites in mouse amygdala of male in comparison to female was observed throughout postnatal life (1).

A variable age-dependent development of nicotine-like binding sites has also been found using tubocurarine (TC) as ligand (Fig. 2). Fig. 2 shows a continuous decrease in the number of TC binding sites per protein, from early age towards maturity, in the cerebellum, hippocampus and cortex.

As seen in Fig. 2 the profiles for the developmental curves of the two ligands Btx and TC are not identical. Previous studies have indicated both high and low affinity binding sites for TC, while only one high affinity site for Btx has been found in the hippocampus of adult mice (19). These findings might explain some of the differences in the developmental pattern between the two ligands.

The picture becomes more complicated, however, if we also add data using highly labelled nicotine (NIC) and acetylcholine (ACh) as ligands. Fig. 3 shows the time course of specific NIC binding from 3 to 30 days of age. As seen in this figure, the pattern is somewhat different between the regions studied. The number of NIC binding sites is rather high at day 3 and increases approximately two fold up to day 30 except in the cortex, where the increase in binding is more than 7-fold that measured on day 3. A drop in the binding is noticed, mostly on day 12 and 7. The reason for this phenomenon is unknown. Studies by Diamond and Miledi (8) have shown a receptor loss around the time of innervation of the muscle. It is therefore tempting to suggest that the decrease in NIC binding could have something to do with time of functional innervation. Kobayashi et al. (17) followed the EEG pattern in mice and noticed that it matures at 16-17 days of age. Another interesting observation is that the acute toxicity of nicotine to mice is lower in 12-day old animals when compared to both 3 day old animals and adult mice (38). Both these observations may conceivably be related to the

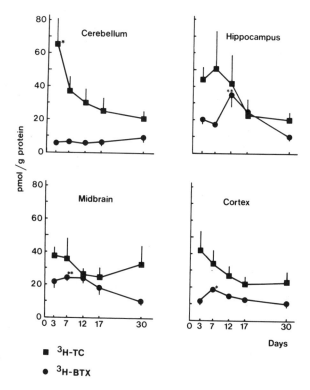

- ■ ³H-TC

- ● ³H-BTX

Figure 2. Development of labelled tubocurarine ([³H]-TC) and alfa-bungarotoxin ([³H]-Btx) binding sites in different areas of mouse brain. Each point indicates mean value ± SE. Levels of significance (t-test) of difference between young and 30 day old mice are given on the curves (**p < 0.01; *p < 0.05). (From Falkeborn et al., 14.)

drop in binding described above. The ACh binding, measured in the presence of atropine to prevent binding to muscarinic receptors, was unchanged with age (Fig. 3).

NIC binding studies in adult hippocampus indicate two binding sites with a twenty-fold difference in affinity constants. Moreover, the constants (Bmax; K_d) for the high affinity site for NIC and TC and the binding site for Btx and ACh are very similar (20, 21, Larsson and Nordberg, this book). To further understand the complexity in the ontogenesis of nicotine-like binding sites a study of time course in changes of subpopulations of binding sites and affinity constants is necessary.

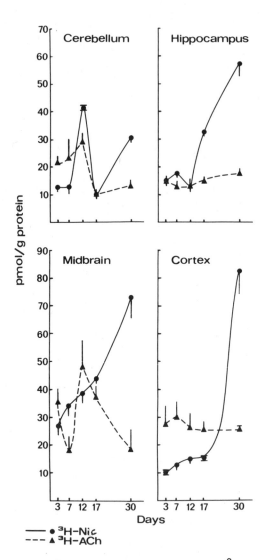

Figure 3. Development of labelled nicotine ($[^3H]$-Nic) and acetyl-choline ($[^3H]$-ACh) binding sites in different areas of mouse brain. Each point indicates mean value $\pm$ SE (from Larsson et al., 21)

Effect of early drug treatment on ontogenesis of cholinergic receptors. In adult animals treatment with anticholinergics such as scopolamine and atropine can increase the number of muscarinic binding sites (3, 23, 41) while cholinesterase inhibitors and oxotremorine decrease muscarinic receptor binding (9, 11, 36). Recently it has been reported that similar treatment can also influence the ontogenesis of muscarine-like binding sites (9, 22) 24). The consequence of such an up/down regulation of binding sites during maturation is still unknown. Commonly used substances such as nicotine and ethanol have recently been shown to affect nicotine-like (13, 37) and muscarine-like (31) binding sites in adult animals, respectively. Their influence on ontogenesis needs to be investigated.

HUMAN BRAIN STUDIES

Studies on the ontogenesis of neurochemical markers in human brain must be encountered by a large variance in results in comparison to animal studies where the experimental conditions can be more controlled. Although human brain is considered to be more mature at birth than rodent brain an extensive synapse formation and normal differentiation is considered to occur also after birth (2).

Fig. 4 shows our data on muscarine-like binding sites obtained in the hippocampus of the fetus, infant (4 hours - 3 years of age), and adult brain (17 - 96 years of age). From birth there seem to be a slight, but continuous decrease in number of muscarinic

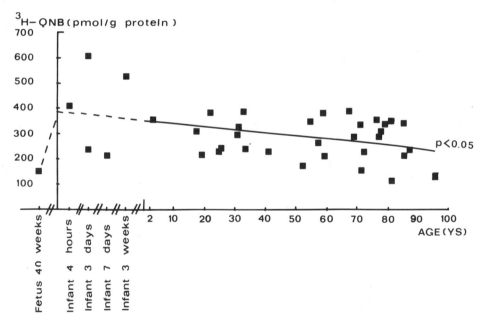

Figure 4. Number of labelled quinuclidinyl benzilate ([^{3}H]-QNB) binding sites in fetus (20 weeks), infant (4 hours - 3 years) and adult (17 - 96 years). Each point represents data from one individual. Regression analysis indicates: r = 0.38; p < 0.05, df 35. Data taken from Nordberg et al., (27); Marcusson, Nordberg and Winblad, unpublished observations.

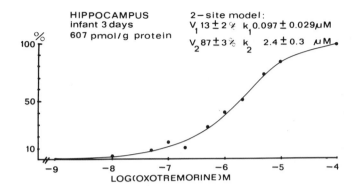

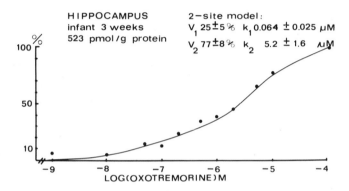

Figure 5. High and low affinity muscarinic binding sites in the hippocampus of infants 3 days and 3 weeks of age, respectively. Competition studies were performed using labelled quinuclidinyl benzilate ([³H]-QNB) and unlabelled oxotremorine. The competition data were fitted to a 2-site receptor model by non-linear least square analysis. The points represent the experimental data obtained and the solid line the best fit of these experimental data.

binding sites with increasing age (p < 0.05). These findings in the hippocampus are in agreement with the findings of Brooksbank et al. (6) in the cerebral cortex, where the highest number of muscarinic binding sites was obtained in the perinatal period, slightly preceding the maximal capacity of ChAT (6). In the cerebellum, on the other hand, Brooksbank et al (6) measured the highest number of muscarinic binding sites in the fetal period.

Figure 5 shows that in hippocampus from infants (3 days, 3 weeks of age) one can discern a subdivision of muscarinic binding sites into two distinct populations with different affinity constants. These studies were performed as competition experiments

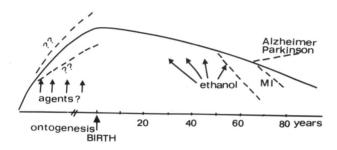

Figure 6. A schematical drawing of the life span of muscarinic receptors in human brain.

using unlabelled oxotremorine. The constants for high and low affinity binding sites given in Fig. 5 are comparable to those obtained in adult hippocampus (33).

Our knowledge concerning the development of cholinergic synapses and function in human brain is still very fragmentary, and further studies are necessary, especially in human fetal brain.

Finally, to speculate, Fig. 6 shows a schematic drawing on the life span of muscarinic receptors in human brain. From our own human postmortem brain studies we know that a normal decline in number of receptors with increasing age can be accelerated in chronic alcoholism (30) and multi-infarct dementia (27). In dementia of Alzheimer type (29) and Parkinson's disease (35) on the other hand, signs for receptor "hypersensitivity" with increased number of binding sites have been observed. As stated in the introduction it is known from animal studies that treatment with cholinergic drugs during ontogenesis may have an effect both in a positive or negative direction on the number of cholinergic receptor sites in brain. Whether in the human an early fetal exposure of drugs might influence the susceptibility of brain for aging processes, toxic substances and different pathological diseases later in life, is still an open question.

It can be concluded from this chapter that the time course of development of muscarinic and nicotinic binding sites in discrete areas of the brain is quite different. Moreover, the opposite finding in binding data using NIC and TC/Btx/ACh as ligands respectively, might reflect different patterns of ontogenesis of subpopulations of nicotinic receptor sites whose functional role in brain is still unknown.

Acknowledgement. This study was supported by the Swedish Medical Research Council (project No. 5817) and the Swedish Tobacco Company.

REFERENCES

1. Arimatsu, Y. and Seto, A. (1982): Brain Research 234:27-39.
2. Balazs, R. and Brooksbank, B.W.L. (1981): In: Fetal Disorders - Recent Approaches to the Problem of Mental Deficiency (eds) B.S. Hetzel and R.M. Smith, Elsevier/North Holland Biomedical Press, pp. 67-96.
3. Ben-Barak, J. and Dudai, Y. (1979): Brain Research 166:245-257.
4. Birdsall, N.J.M., Burgen, A.S.V. and Hulme, E.C. (1978): Molec. Pharmacol. 14:723-736.
5. Birdsall, N.J.M., Hulme, E.C. and Burgen, A. (1980): Proc. R. Soc. Lond. 207:1-12.
6. Brooksbank, B.W.L., Martinez, M., Atkinson, D.J. and Balazs, R. (1978): Dev. Neurosci. 1:267-284.
7. Coyle, J.T. and Yamamura, H.I. (1976): Brain Research 118: 429-440.
8. Diamond, J. and Miledi, R. (1962): J. Physiol. Lond. 162:393-408.
9. Dudai, Y., Ben-Barak, J., Silman, I. and Gazit, H. (1980): In: Neurotransmitters and Their Receptors (eds) U.Z. Littauer, Y. Dudai, I. Silman, V.I. Teichberg and Z. Vogel, John Wiley and Sons Ltd., pp. 217-239.
10. Egozi, Y., Kloog, Y. and Sokolovsky, M. (1980): In: Neurotransmitters and Their Receptors (eds) U.Z. Littauer, Y. Dudai, I. Silman, V.I. Teichberg and Z. Vogel, John Wiley and Sons Ltd., pp. 201-215.
11. Ehlert, F.J., Kokka, N. and Fairhurst, A.S. (1980): Molec. Pharmacol. 17:24-30.
12. Enna, S.J., Yamamura, H.I. and Snyder, S. (1976): Brain Research 101:177-183.
13. Falkeborn, Y., Larsson, C. and Nordberg, A. (1981): Drug and Alcohol Dependence 8:51-60.
14. Falkeborn, Y., Larsson, C., Nordberg, A. and Slanina, P. (1983): Int. J. Devl. Neurosci. 1:187-194.
15. Gale, J.S. (1977): Brain Research 133:172-176.
16. Hunt, S. and Schmidt, J. (1979): Neuroscience 4:585-592.
17. Kobayashi, J., Inman, O., Buno, W. and Himwich, H.E. (1963): Rec. Adv. Biol. Psychia. 293-308.
18. Kuhar, M.J., Birdsall, N.J.M., Burgen, A.S.V. and Hulme, E.C. (1980): Brain Research 184:374-383.
19. Larsson, C. and Nordberg, A. (1983): In: Neurotransmitters and Their Receptors (eds) U.Z. Littauer, Y. Dudai, I. Silman, V.I. Teichberg and Z. Vogel, John Wiley and Sons Ltd, pp. 297-301.
20. Larsson, C. and Nordberg, A. (1984): J. Neurochem. (in press).
21. Larsson, C., Nordberg, A., Falkeborn, Y. and Lundberg, P.-A. Brain Research (submitted).
22. Levy, A. (1981): Life Sci. 29:1065-1070.
23. Majocha, R. and Baldessarini, R.J. (1980): Eur. J. Pharmacol. 67:327-328.

24. Meyer, M.R., Gainer, M.W. and Nathanson, N.M. (1982):   Molec. Pharmacol. 21:280-286.
25. Morley, B. (1981):   Pharmac. Ther. 15:111-122.
26. Morley, B.J. and Kemp, G.E. (1981):   Brain Research Reviews 3:81-104.
27. Nordberg, A., Adolfsson, R., Marcusson, J. and Winblad, B. (1982): In: The Aging Brain: Cellular and Molecular Mechanisms of Aging in the Nervous System (eds) E. Giacobini, G. Filogamo, G. Giacobini and A. Vernadakis, Raven Press, New York, pp.   231-245.
28. Nordberg, A. and Larsson, C. (1981):   In: Cholinergic Mechanisms: Phylogenetic Aspects, Central and Peripheral Synapses and Clinical Significance (eds) G. Pepeu and H. Ladinsky, Plenum Press, New York, pp. 819-824.
29. Nordberg, A., Larsson, C., Adolfsson, R., Alafuzoff, J. and Winblad, B. (1983):   J. Neural. Transmission 56:13-19.
30. Nordberg, A., Larsson, C., Perdahl, E. and Winblad, B. (1982): Drug and Alcohol Dependence 10:333-344.
31. Nordberg, A. and Wahlstrom, G. (1982):   Life Sci. 31:277-287.
32. Nordberg, A., Wahlstrom, G. and Larsson, C. (1980):   Life Sci. 26:231-237.
33. Nordberg, A. and Winblad, B. (1981):   Life Sci. 29: 1937-1944.
34. Overstreet, D.H. and Yamamura, H.I. (1979):   Life Sci. 1865-1878.
35. Ruberg, M., Ploska, A., Javoy-Agid, F. and Agid, Y. (1982): Brain Research 232:129-139.
36. Schiller, G.D. (1979):   Life Sci. 24:1159-1164.
37. Schwartz, R.D. and Kellar, K.J. (1983):   Science 220:214-216.
38. Stalhandske, T., Slanina, P., Tjalve, H. and Hansson, E. (1969): Acta Pharmacol. Toxicol. 27:363-380.
39. Szerb, J.C. (1977):   In: Cholinergic Mechanisms and Psychopharmacology (ed) D.J. Jenden, Plenum Press, New York, pp. 49-60.
40. Wade, P.D. and Timiras, P.S. (1980):   Brain Research 181:381-389.
41. Westlind, A., Grynfarb, M., Hedlund, B., Bartfai, T. and Fuxe, K. (1981):   Brain Research 225:131-141.
42. Yavin, E. and Harel, S. (1979):   Febs Letters 97:151-154.

AGING OF CHOLINERGIC SYNAPSES:

FICTION OR REALITY?

E. Giacobini, I. Mussini[*] and T. Mattio

Department of Pharmacology
Southern Illinois University School of Medicine
Springfield, Illinois 62708   U.S.A.

[*]National Research Council
Unit for Muscle Biology
Institute of General Pathology
University of Padova, Italy

INTRODUCTION

Cholinergic system and aging of the human brain. The normal process of aging of the human brain seems to be associated with a loss of both cortical and subcortical neurons (Table 1). A central problem is whether this decrease in number of neurons is related to neuro-transmitter-specific systems or whether it involves unspecifically all neuronal systems. Recent autopsy and biopsy studies in humans (Table 1) indicate that aging of the brain is indeed accompanied by a loss of neurons in certain areas, mainly in specific cortical regions. However, no significant changes in cell number are found in other areas, such as brainstem nuclei, cranial nerve nuclei, pons-hypothalamic areas or the medulla oblongata (Table 1). A review of the recent morphological and biochemical literature (Table 1) suggests that the number of brain regions known to escape neuronal loss in the course of the aging process has grown. Therefore, it seems that in the aging human brain two populations of neurons coexist side by side. One undergoing regressive changes leading to cell death and a second surviving and undergoing continuous and dynamic growth of processes. In pathological conditions, such as in Parkinson and Alzheimer diseases, some decreases are found in corresponding areas but not in others, indicating a variance from normal aging (Table 1). The most pronounced difference is seen in the substantia innominata (nucleus basalis of Meynert) of Alzheimer patients, with a 75-80% loss in neurons (2, 19, 20). It should be noted that in the same nucleus a 66% decrease in cell density is seen also in

Parkinson patients (Table 1) (1). A decrease in number of cells in Alzheimer brains is seen also in non-cholinergic nuclei such as the locus coeruleus (Table 1) (10, 18). In spite of the fact that neuronal loss in Parkinson disease may be the same or greater than in Alzheimer disease (Table 2), no extensive cortical abnormalities either morphological or neurochemical (ChAT, choline acetyltransferase activity) have been reported in the former disease (1). A similar situation is found in the locus coeruleus where neuronal loss in Alzheimer brains is not correlated with a reduction in cortical activity of noradrenergic enzymes (10).

The mechanism of neuronal loss may be different in the two diseases as suggested by the differences in pathological changes observed in perikarya and neuropil. In fact, there seems to be no apparent relationship between the formation of characteristic neuro-fibrillary changes in one case and the presence of Lewy bodies and neuroaxonal spheroids in the other. It looks, therefore, unlikely that the two abnormalities might represent different manifestations of the same disease. Recent results on changes in cell density, total number of cells, ChAT and AChE (acetylcholinesterase) activity in the nucleus basalis of human controls and SDAT (senile dementia of Alzheimer type) brains are reported in Table 2. It can be seen that all four parameters are strongly affected in SDAT patients, however, contrasting differences between results of various investigators are seen in cell number and cell density (1, 11, 19, 20).

In conclusion, in the case of Alzheimer disease, both morphological and neurochemical results suggest a primary degeneration of cholinergic axons and terminals projecting to cortex and a secondary loss of cholinergic neurons in subcortical nuclei. Whether the irreversible neuronal loss is large or moderate and its nature remains to be demonstrated by further studies, possibly at the ultrastructural level. Another fundamental question is whether Alzheimer disease is primarily a "cell body" or a "terminal disease" and what is the time course and sequence of the neuronal damage (Table 3). Other important questions related to brain aging in general and to SDAT in particular are reported in Table 3.

Choline acetyltransferase is synthesized in the cell body and then transported to the terminal of the cholinergic neurons while acetylcholine (ACh) can be synthesized at both locations. A defect could occur at both places in developing, adult or aging cholinergic neurons (5). It is, therefore, important to develop models of aging which make it possible to study cell bodies as well as terminals in the same population of cholinergic neurons throughout the entire life of the animal. Such a model is described in the next section.

Table 1

Changes In Number of Neurons with Age in
Normal and Pathological Humans*

Structure	Age (years)	NORMAL	PATHOLOGICAL Parkinson	Alzheimer
Cerebral cortex (Inf. frontal and sup. temporal)	90	45% decr.		17-18% large neurons decr.
Hippocampal cortex	45-90	27-80% decr.		47% decr.
Cerebellar cortex (Purkinje cells)	60-100	25% decr.		
Parietal and occipital cortex	80	unchanged		no differ.
Telencephalon (putamen)	80	decr.		
Subst. nigra		(dopam.)	(dopam.)	
	Newborn	400,000		
	60	250,000	60,000	decreased
Nucleus coeruleus	Young Adult	(norep.) 19,000		
	60-80	17,000- 11,000		43% decr.
Nucleus bas. Meynert's (subst. innominata)	Young Adult 50-66	450,000 unchanged(?)	66% decr.	75-80% decr.
Brainstem nuclei (ventral coc., inf. olive, mamm. bodies)		unchanged		
Cranial Nerve Nuclei (trochlear, abducens, facial)		unchanged		unchanged
Pons-hypothalamus		unchanged		
Medulla oblongata		unchanged		

*For ref. see text.

Table 2

Changes in Cell Density and Cholinergic Enzymes Activity in
Nucleus Basalis of Control and SDAT[X] Brains

	CONTROL	SDAT	% DECREASE	AUTHOR/YEAR
CELL NUMBER nr cells/grid	5.9 ± 0.9	1.6 ± 0.3	73	Whitehouse et al.,1981,1982 (19,20)
total nr cells	344.6 ± 64.6	72.4 ± 14.6	79	Whitehouse et al.,1982 (20)
	-----	-----	<50	McGeer et al., 1983 (8)
CELL DENSITY mean area per neurone m x 10^3	10	15	33	Candy et al., 1983 (1)
	10	30[XX]	66[XX]	Candy et al., 1983 (1)
	-----	-----	33	Perry et al., 1982 (11)
ChAT[XXX] POSITIVE CELLS (immunohistochem.)	104	35	66	Nagai et al., 1983 (9)
	-----	-----	<50	McGeer et al., 1983 (8)
ChAT ACTIVITY (nmol/h/mg)	-----	-----	89	Perry et al., 1982 (14)
	-----	-----	70	Rossor et al., 1982a (13)
	196.8 ± 88	13.4 ± 4.4	94	Candy et al., 1983 (1)
	81.8[XXXX]	25.9[XXXX]	64	Rossor et al., 1980; 1982b (12,14)
AChE ACTIVITY[XXXXX] (histochemistry)	INTENSE STAINING	-----	--	Rossor et al., 1982b(14)
(mol/min/mg protein)	676.8 ± 101.2	141.9 ± 33.9	79	Candy et al., 1983 (1)

[X]Senile dementia of Alzheimer type; [XX]Parkinson patients; [XXX]Choline acetyltransferase;
[XXXX]Medium value; [XXXXX]Acetylcholinesterase

Table 3

Some Basic Questions Which Remain Unanswered

---

1. Are biochemical changes selectively localized to certain brain nuclei or are they distributed to all cholinergic synapses in the CNS?

2. Are changes related to the normal cerebral aging process, i.e., are they mechanisms of enzymatic adaptation or are they specific for senile dementia? How important is the age range of the controls? How important is the severity of the disease?

3. Which is the primary target for the chemical damage and the neuronal degeneration? Does the aging process involve both pre- and postsynaptic structures? Does the process involve cholinergic terminals firstly and perikarya secondly?

4. Are cholinergic neurons in the peripheral nervous system (PNS) and central nervous system (CNS) equally affected?

5. Is there a relationship between the reduction in cholinergic cortical innervation and the pathogenesis of plaques?

THE AVIAN IRIS AS A MODEL OF AGING

Aged chickens are defined as birds of age 36 months or older. However, as in mammals, incipient signs of senescence may appear in the avian nervous system at earlier stages of life. In the chicken, these early signs can be identified morphologically and biochemically in the PNS of the chicken after 24 months of age (5). Approximately 90% of adult weight as well as sexual maturity are reached at 200 days (7 months) of age. The rate of growth of the brain is very rapid at first and then gradually decreases as the body weight increases, becoming very small beyond 300 days of age. Among degenerative signs related to old age, the close similarity of spontaneous avian atherosclerosis to the human disease has been recognized since the beginning of the twentieth century (15, 17). A combination of causal factors which remind us of human pathogenesis, such as hemodynamic conditions, blood lipids (including elevated plasma cholesterol) and thrombogenic mechanisms has been demonstrated in avians (17).

In addition, in the chicken as an effect of age, blood pressure increases significantly in males and females from age 10-14 months to age 42-54 months. It is interesting to note that most of the drugs and hormones which are pressor or depressor in mammals have the same effect in birds (17).

Additional advantages in using the avian nervous system as a model of aging are: a) the relatively long life span which allows to observe gradual changes. In rodents, age-related modifications occur almost abruptly at the end of life; b) possibility of studying discrete populations of homogenous (cholinergic) neurons such as in the ciliary ganglion, maintaining a constant number of cells throughout life span (5); c) possibility of investigating separately, cell body and terminals belonging to the same neuron (ciliary ganglia and iris); d) possibility of parallel physiological and pharmacological testing of pupillary functions in vivo; and e) a good correlation between biochemistry and morphology feasible through morphometric electron microscopic studies and quantitative cytochemical analysis (5).

The ciliary ganglion preparation has proved to be a useful and simple neuronal model of cholinergic synapses since its introduction in studies of synaptic development and plasticity (3-5).

In a series of studies which has been recently summarized by Giacobini (5), we have made use of the ciliary ganglion iris preparation of the aging chicken as a model of senescent peripheral cholinergic synapses. In humans, one predictable aging marker has been emphasized: i.e., pupil function as judged by pupil size. Senile miosis, a reduction in pupil size seems to contribute a reliable sign of aging as the diameter of the pupil correlates closely to age for both the dark-adapted and light-adapted eye (7). Seventy-five percent of subjects 55 or older show 3 mm pupil diameter while all those under 55 have pupils at least 4 mm in diameter. In addition, pupillary responses to cholinergic drugs are impaired in the elderly (16). Thus, we believe that the the ciliary ganglion preparation is an appropriate system to study, when investigating age-related phenomena.

Based on the studies performed on the iris, an hypothesis of aging of the cholinergic synapse has been suggested by us (6). This hypothesis contemplates age related changes in carrier mediated mechanisms and in molecular and physical properties of neuronal membranes leading to a "chemical denervation" of the cholinergic synapse. The concept of neuronal aging as an integration of progressive and regressive cell changes, involving specific membrane mechanisms in selective parts of cholinergic neurons has been previously discussed (5, 6). In order to firmly establish the cholinergic hypothesis of neuronal aging in humans, several basic questions still remain to be answered (Table 3). Experimentally, a close integration of biochemical and ultrastructural data and the development of new tests to detect early signs of pathological aging of the cholinergic system are necessary. In the next section we will describe an attempt in this direction.

ABILITY OF THE AGING CHOLINERGIC SYNAPSE TO TAKE UP CHOLINE, RELEASE ACETYLCHOLINE AND TO DEPLETE AND REFORM SYNAPTIC VESICLES

As illustrated in Fig. 1A, neuromuscular junctions (NMJ) in the iris of an aging chicken (2 year) show polymorphic signs of damage such as, reduction and polymorphism of synaptic vesicles, and an increase of neurofilaments and mitochondria. Accumulations of cytoplasmic organelles and lysosomes are seen in the axoplasm

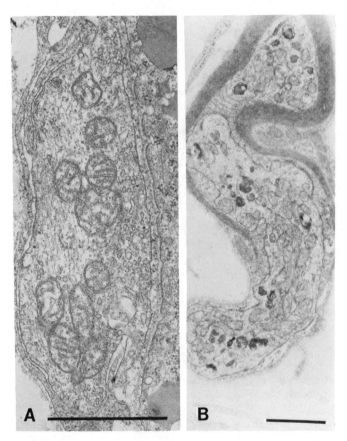

Figure 1. Iris muscle of a 2-year-old chicken. The neuromuscular junction (A) shows reduction and polymorphism of synaptic vesicles, and an increase of microfilaments. (B) Note the accumulation of mitochondria and lysosomes in the axoplasm of a nerve fiber.. Calibration bars = 1 μm.

of the nerve fiber (Fig. 1B). At later stages the nerve ending is enveloped by Schwann cells infiltrating and partially filling the synaptic cleft. Quantitative changes in the ratio describing the relation between volumes of terminals and volumes of synaptic vesicles showing a trend to shift to the left are seen from 4 m to 9 year (Fig. 2). This indicates a progressive decrease in the volume occupied by synaptic vesicles (Fig. 2) and a possible functional deficit.

In order to establish the nature of such a deficit, we have examined the ability of cholinergic synapses in the iris at various ages to take up the precursor [^{3}H]-choline (Ch) and release the formed [^{3}H]-ACh in response to high K$^+$ (115 mM) depolarization.

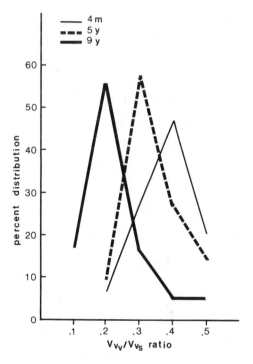

Figure 2. Summary of preliminary results of morphometric analysis of nerve endings and synaptic components in the iris of young adult (4 mo.) and aged (5 and 9 years) chickens. Percent distribution of V_{VV}/V_{VS} ratio, where V_{VS} = synaptic bouton volume fraction; V_{VV} = synaptic vesicle volume fraction. The results obtained by the analysis of the iris NMJ of one animal/age group are the average of 15-20 samples/age. As seen from the figure, there is a trend of the peaks to shift to the left from 4 m to 9 y, demonstrating an age-related decrease in the relative volume of synaptic vesicles.

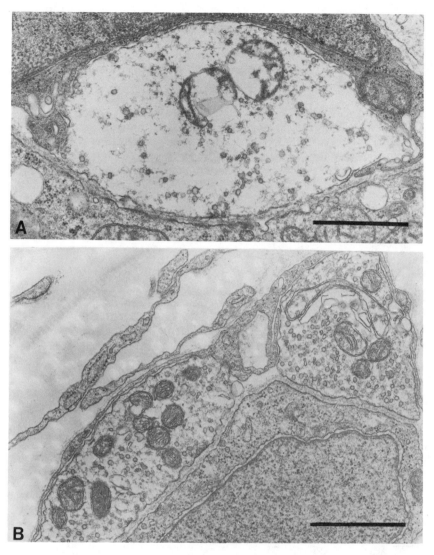

Figure 3. Iris muscle of a 2-month-old chicken. Following prolonged depolarization with high-K^+ concentration (A) the nerve ending appears almost depleted of synaptic vesicles. After reincubation in physiological medium (B) a nearly total retrieval of synaptic vesicles is achieved. Calibration bars = 1 μm.

We have observed that following release of ACh, exocytosis clearly prevails over endocytosis and a nearly total depletion of vesicles is present (Fig. 3A). This depletion is reversible; after 60 min incubation in a physiological medium the morphological recovery is almost complete (Fig. 3B).

The scheme of a typical experiment is reported in Fig. 4.

The first experiment (Type I) was designed to assess the effect of age on the K^+ stimulated release of ACh from cholinergic terminals of the iris. Iris samples were pre-incubated in chick Tyrode for 10 min and then placed in 1 μM choline spiked with 1% [^{3}H]-choline for 10 min to label ACh pools. After a 20-min wash, samples were placed in a stimulation chamber with high K^+ (115 mM) flowing at a rate of 1 ml/min for 60 min. The perfusate was collected in 1 ml fractions and radioactivity was determined by scintillation counting. The second experiment (Type II) was similar to the Type I experiment except that incubation was preceded by 60 min of high K^+ (depletion), followed by 30 min of [^{3}H]-choline uptake, a 20-min wash, and then high K^+ and determination of [^{3}H]-ACh

Experiment I:

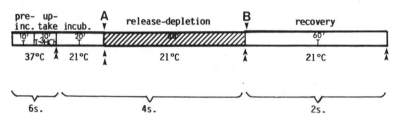

6s.   4s.   2s.

Total duration of Experiment I = 2 hours and 40 min.

Experiment II:

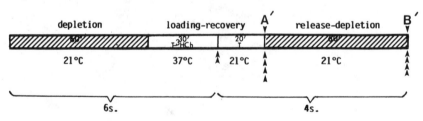

6s.   4s.

Total duration of Experiment II = 2 hours and 50 min.

LEGEND

T = Tyrode solution   Λ = 1 sample fixed for E.M.

s. = samples   ▨ = high [K^+] stimulation

Figure 4.   Scheme of Release-Recovery-Depletion.

release. This experimental protocol was designed to determine the ability of the 3-year iris to undergo depletion-reloading-release when compared to the 4-month tissue.

These data were analyzed by a two-way analysis of variance to determine age versus experiment type. This analysis showed no interaction between age and experimental type for any of the four parameters. That is, the release of [^{3}H]-ACh from either 4-month or 3-year tissues was independent of whether the tissues were treated by a Type I or Type II scheme. The main effects for age showed the following for the four parameters. The 3-year tissue (Table 4) released significantly less [^{3}H]-ACh than the 4-month tissue as determined by the area under the release curve (peak area) (Fig. 5). Also, the 3-year tissue showed a lower peak release of [^{3}H]-ACh than the 4-month tissue. The time needed for the 3-year tissue to reach its peak release was significantly longer than at 4-month and its rate of release (rate of increase) was significantly slower (Table 4).

These neurochemical results demonstrate a decreased ability, dependent only on age and not experimental design, of the 3-year tissue to release [^{3}H]-ACh when compared to a 4-month tissue. This correlates well with the morphological data, which demonstrate that two important features for neurotransmitter release (vesicular volume and length of the appositional membrane) were decreased in the 5-year (or 9-year) tissue (Fig. 2). Our results support the hypothesis that cholinergic transmission in the chicken iris is age-dependent and declines with increasing age. This decline is related to modifications of presynaptic mechanisms of release and uptake of the neurotransmitter and its precursor (5).

TABLE 4

K^+-Stimulated Release of Acetylcholine in Iris Preparations

Age	Peak Area	Peak Height	Time to Peak	Rate of Increase
4 M	281.18 ± 18.68	1.62 ± 0.33	7.50 ± 0.68	0.26 ± 0.06
3 Yr	202.38 ± 20.04**	0.71 ± 0.12*	9.62 ± 0.69*	0.07 ± 0.01**

Mean values (± SEM) for [^{3}H]-ACh K^+-stimulated release from 4-month and 3-year iris samples. The 3-year tissue showed a decreased [^{3}H]-ACh release for all parameters studied.
**p < .02, *p < .05

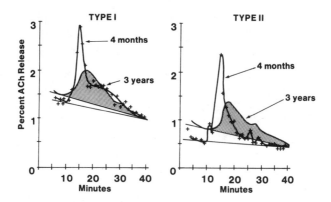

Figure 5.   Typical release curves in Type I and Type II experiments. ACh release is represented as the percent [3H]-ACh released of the total [3H]-ACh present in the tissue at the time of release.

SUMMARY

In conclusion, experiments testing the uptake and release capacity of the peripheral cholinergic synapse under acute conditions of stimulation suggest that severe changes may occur at later stages of life. If a sufficiently long period of recovery is allowed, aging· iris terminals are still capable of responding with an adequate depletion of ACh following two subsequent periods of release and reloading. However, the 3-year iris shows a decreased ability to release [3H]-ACh when compared to a 4-month iris. The results of the release experiments correlate well with the morphometric results showing decreased vesicular volume in older iris. Our results support the hypothesis that cholinergic transmission in the chicken iris is age-dependent and declines with increasing age. This decline is related to modifications of pre-synaptic mechanisms of release and uptake of the neurotransmitter and its precursor (5). Our experiments seem to point out for the first time a specific functional defect in the cholinergic synapse during aging.

Acknowledgements. The investigations carried out in the author's (E.G.) laboratory were supported by AFOSR Grants 81-9229 and 83-0051; by grants from the Nowatski Eye Research Foundation and by the E.F. Pearson foundation. Support from the National Research Council of Italy to the Unit for Muscle Biology (I.M.), Institute of General Pathology, University of Padova, Italy is acknowledged. The authors wish to thank the Department of Pathobiology (Dr. L. Pierro, Chairman), University of Connecticut, Storrs, CT and the University of Connecticut Research Foundation for providing funds and assistance in order to maintain aged birds at their facilities.

REFERENCES

1. Candy, J.M., Perry, R.H., Pery, E.K., Irving, D., Blessed, G., Fairbairn, A.F. and Tomlinson, B.E. (1983): J. Neurological Sci. 54:59:277-289.
2. Coyle, J.T., Price, D.L. and DeLong, M.R. (1983): Amer. Assoc. Adv. Science 219:1184-1190.
3. Giacobini, E. (1959): Acta Physiologica Scandinavica 45:1-45.
4. Giacobini, E. (1970): In: Biochemistry of Simple Neuronal Models (eds) E. Costa and E. Giacobini, Raven Press, 2:9-64.
5. Giacobini, E. (1983a): Adv. in Cell Neurobiology 3:173-214.
6. Giacobini, E. (1983b): In: Aging of the Brain (eds) D. Samuels, S. Algeri, S. Gershon, V. Grimm and G. Toffano, Raven Press, New York, pp. 197-210.
7. Kornzweig, A. (1954): Sight Saving Review 24:138-140.
8. McGeer, P.L., McGeer, E.G., Peng, J.H., Kimura, H., Pearson, T., Suzuki, J. and Dolman, C. (1983): J. Neurochem. Abst. 41(Suppl.):S20D.
9. Nagai, T., McGeer, P.L., Peng, J.H., McGeer, E.G. and Dolman, C.E. (1983): Neurosci. Lett. 36:195-199.
10. Perry, E.K., Tomlinson, B.E., Blessed, G., Perry, R.H., Cross, A.J. and Crow, T.J. (1981): J. Neurol. Sci. 51:279-287.
11. Perry, R.H., Candy, J.M., Perry, E.K., Irving, D., Blessed, G., Fairbairn, A.F. and Tomlinson, B.E. (1982): Neurosci. Lett. 33:311-315.
12. Rossor, M.N., Fahrenkrug, J., Emson, P., Mountjoy, C., Iversen, L. and Roth, M. (1980): Brain Res. 109:249-253.
13. Rossor, M.N., Garrett, N.J., Johnson, A.L., Mountjoy, C.Q., Roth, M. and Iversen, L.L. (1982a): Brain 105:313-330.
14. Rossor, M.N., Svendsen, C., Hunt, S.P., Mountjoy, C.Q., Roth, M. and Iversen, L.L. (1982b): Neurosci. Lett. 28:217-222.
15. Siller, W.G. (1965): In: Comparative Atherosclerosis (eds) J.C. Roberts, Jr., R. Straus and M.S. cooper, Harper and Row, New York, pp. 66.
16. Sitaram, N. and Pomara, N. (1981): Pathology 9:409-410.
17. Sturkie, P.D. (1976): In: Avian Physiology (ed) P.D. Sturkie, Springer-Verlag, pp. 76-101.
18. Vijayashankar, N. and Brody, H. (1979): J. Neuropath. Expl. Neurol. 38:490-497.
19. Whitehouse, P.J., Price, D.L., Clark, A.W., Coyle, J.T. and DeLong, M.R. (1981): Ann. Neurol. 10:122-126.
20. Whitehouse, P.J., Price, D.L., Struble, R.G., Clark, A.W., Coyle, J.T. and DeLong, M.R. (1982): Science 215:1237-1239.

(NOTE ADDED AFTER COMPLETION OF MANUSCRIPT)

Three laboratories reported independently at the meeting on Dynamics of Cholinergic Function (which is the basis for this book) some new results which confirm the loss of cholinergic cells in forebrain structures of human SDAT patients. McGeer et al. reported a decline in cholinergic neurons from 450,000 in the young adult to 50-100,000 in SDAT patients. It should be noted that similar changes in cell number were observed in two noncholinergic areas (locus coeruleus and substantia nigra) by the same authors. According to these authors age-related conditions and senile dementia follow the same pattern, however, in SDAT the insult is more pronounced. According to Bigl et al. (see T. Arendt, V. Bigl, A. Arendt and A. Tennstedt, Acta Neuropathol., 61:101-108, 1983), the neuronal loss is 70% in Alzheimer, 75% in Parkinson (paralysis agitans) and 50% in Korsakoff patients. Etienne et al. found a 60-90% decrease in cell counts. These three recent observations underline the necessity of answering the questions reported in Table 3 before reaching a final conclusion about SDAT and involvement of the cholinergic system.

INTERACTIONS OF CALCIUM HOMEOSTASIS, ACETYLCHOLINE METABOLISM, BEHAVIOR AND 3,4-DIAMINOPYRIDINE DURING AGING

G.E. Gibson and C. Peterson

Department of Neurology
Cornell University Medical College
Burke Rehabilitation Center
785 Mamaroneck Avenue
White Plains, New York U.S.A.

AGING AND MENTAL FUNCTION IN MAN

Normal aging is often associated with a reduction in mental abilities. However, the functions that are altered and the age when they begin to decline is controversial. For example, decreased performance on various aspects of the Wechsler Adult Intelligence Scale (WAIS) occurs at different ages. The full scale WAIS is composed of a verbal scale (6 tests) and a performance scale (5 tests). Although some reductions in the full scale performance are apparent by age 40, the largest decline occurs after the age of 70. The performance tests show greater age-related deficits than the verbal tests (15, 62). The molecular mechanisms that lead to the decline in brain function with aging and their relation to the biochemical basis of the debilitating deficits of Alzheimer's disease are unknown. The similarities between aging and Alzheimer's disease are detailed elsewhere (18).

THE CHOLINERGIC SYSTEM DURING AGING IN MAN AND ANIMALS

Considerable evidence implicates the cholinergic system in the production of geriatric cognitive deficits. Cholinergic dysfunction diminishes learning and memory in animals (14). In young adults, the muscarinic blocker scopolamine leads to psychological changes that resemble those in aging and Alzheimer's disease (17). Furthermore, the acetylcholinesterase inhibitor, physostigmine, or the cholinergic agonist, arecoline, improve the memory of cognitively impaired aged humans (16) or non-human primates (4). Aging causes a 10-20% loss in the activities in brain of the degradative enzyme,

191

acetylcholinesterase (9, 31, 44, 49, 50, 73) and the synthetic enzyme, choline acetyltransferase (43, 44), as well as in high affinity choline uptake (67). However, aging does not alter acetylcholine concentrations (45, 72). The magnitude of these changes is small when compared to the large alterations in behavior that accompany aging.

IN VIVO ACETYLCHOLINE SYNTHESIS AND ITS BEHAVIORAL IMPORTANCE DURING AGING

Acetylcholine synthesis declines with aging in both whole brain and in various brain regions. Since neither enzyme activities, nor acetylcholine concentrations, accurately reflect the dynamics of the cholinergic system, in vivo acetylcholine formation was measured. Incorporation of [U-^{14}C]glucose or [^{2}H$_4$] choline into whole brain acetylcholine decreases from 100% (3 months) to approximately 50% (10 months) or 25% (30 months) in two strains (C57BL and BALB/c) of mice (25). The diminished synthesis is apparently not due to a lack of precursor availability because [U-^{14}C]glucose and [^{2}H$_4$] choline entry into the brain is similar at all ages. The decline in acetylcholine formation is similar in the hippocampus, striatum and cortex (Fig. 1; 22), even though striatal synthesis is much higher than that in the other two regions. This uniform, age-related

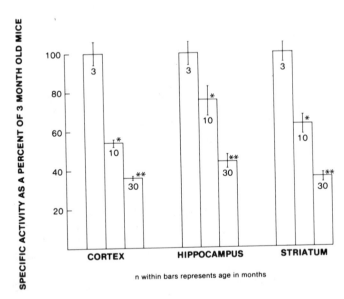

Figure 1. Regional acetylcholine synthesis during aging in vivo. Acetylcholine specific activity was determined one minute after injection of [U-^{14}C]glucose. *Denotes value differs from 3 months and ** denotes value differs from 3 and 10 months (p < 0.05; 22).

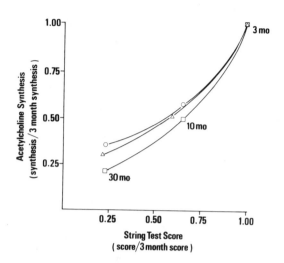

Figure 2. Tight rope test performance and in vivo acetylcholine
synthesis with aging. Acetylcholine formation was measured from
[U-14C]glucose (△ , C57B1; □ Balb/c) or [2H4]choline (○ , Balb/c).
Values at 3, 10 and 30 months differ significantly from each other.
The figure is reprinted with permission of Science (25).

decline of in vivo regional acetylcholine synthesis contrasts with
the alterations in choline acetyltransferase activity which increases
in cortex (+22%), decreases in the striatum (-19%) and is unaffected
in the hippocampus (68).

     Decreased acetylcholine synthesis with aging appears to be
behaviorally important. Tight rope test performance, which quanti-
tates a rodent's ability to maneuver along an elevated taut string,
declines with age and appears to reflect attention. In thiamin
deficiency (3) and hypoxia (21), deficits in tight rope test performance indicate central cholinergic dysfunction. Passive avoidance
(5), 8-arm radial maze (10) and tight rope test performance (21,
51) show similar age-related decreases. The incorporation of [U-14C]-
glucose or [2H4]choline into acetylcholine is highly correlated
(r = 0.98) to reduced tight rope test performance by aged mice (Fig.
2; 25). Although correlation does not prove a cause-effect relationship, these findings are consistent with the hypothesis that
deficits in the cholinergic system underlie age-related behavioral
changes.

ACETYLCHOLINE SYNTHESIS DURING HYPOXIA AND AGING

     Depressed acetylcholine synthesis may make the aged brain more
vulnerable to metabolic insults. During normoxia (Fig. 3, open

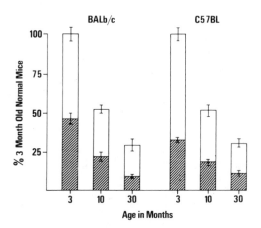

Figure 3. In vivo acetylcholine synthesis in aged mice during hypoxia. [U-^{14}C]Glucose incorporation into acetylcholine was examined in mice of various ages that were made hypoxic by injection of sodium nitrite. Values are the % of acetylcholine synthesis by 3 month old mice. The open and hatched portions represent rates of synthesis during normoxia and hypoxia, respectively (26). The figure is reprinted with permission of Raven Press (27).

bars) and hypoxia (Fig. 3, hatched bars), the percent reduction in acetylcholine formation is nearly the same at 3, 10 and 30 months. However, the residual rate of synthesis during hypoxia in the aged mouse is only one tenth of that by young non-hypoxic animals (26). Thus the age-related reduction in acetylcholine synthesis may interfere with the ability of the senescent brain to cope with additional metabolic insults.

OTHER NEUROTRANSMITTERS DURING AGING

The cholinergic system is not selectively depressed by aging, since other neurotransmitter systems are also altered. Aging also depresses the synthesis of dopamine (60, 61) and the amino acid neurotransmitters (26). Increasing the age from 3 to 30 months, decreases [U-^{14}C]glucose incorporation into the amino acids that are derived from the tricarboxylic acid cycle (gamma-aminobutyrate [GABA; -46%], aspartate [-23%], glutamine [-45%] and glutamate [-31%]) as well as from glycolysis (alanine [-20%] and serine [-12%]). However, the magnitude of the decline in their formation is less than that seen with acetylcholine synthesis and is more difficult to interpret because of metabolic compartmentation.

Although the multiplicity of neurotransmitter changes demonstrates that aging is a dynamic complex process, the experiments that are described below were designed to determine the molecular basis of only the age-related decrease of in vivo acetylcholine synthesis.

IN VITRO ACETYLCHOLINE SYNTHESIS WITH AGING

The age-induced reduction in acetylcholine synthesis persists in vitro, which suggests that the decline is not due to indirect factors that occur in vivo. Alterations in cerebral blood flow, vasculature (e.g. amyloid deposition), substrate availability (e.g.-altered blood brain barrier), peripheral responses, or other neurotransmitter systems may contribute to the in vivo decline. In vitro, with excess substrate and oxygen, these factors are minimized. [U-^{14}C]glucose incorporation into acetylcholine by brain slices from aged mice declines more in low potassium than in high potassium media (Fig. 4; 24), but the magnitude of the change is still less than in vivo. These findings suggest that the age-related deficit in acetylcholine metabolism is not secondary, is partially reversible by potassium depolarization and its molecular basis can be elucidated by in vitro analysis.

In vitro carbohydrate metabolism declines during aging, but the deficits are not enough to account for the large in vivo alterations in acetylcholine metabolism. Oxidative metabolism, which is closely linked to acetylcholine synthesis (20), decreases during senescence in vivo in humans (8) and animals (19) as well

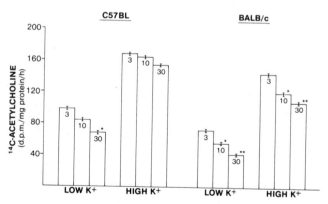

Figure 4. In vitro acetylcholine synthesis during aging in low (5 mM) and high (31 mM) potassium media. Brain slices were prepared from aged C57Bl or Balb/c mice and incubated with excess glucose, oxygen and choline. The numbers within the bars represent the age in months. *Denotes the value differs from 3 months. **Denotes value differs from 3 and 10 months (p < 0.05; 24).

as in vitro with brain slices (51, 53), homogenates (59) or synaptic and non-synaptic mitochondria (11, 12). Brain slices [^{14}C]CO_2 production from [U-^{14}C]glucose (a measure of overall oxidative metabolism), from [3,4-^{14}C]glucose (a measure of pyruvate oxidation in the presence of excess pyruvate) show age-related changes. In low potassium media, aging (3, 10, 30 months) depresses the oxidation of [U-^{14}C]glucose (100, 80, 61%), [3,4-^{14}C]glucose (100, 87, 76%) or [1-^{14}C]pyruvate (100, 88, 87%) (Fig. 5; 24). Depolarizing concentrations of potassium diminish the age-related deficits in oxidative metabolism, just as they did with acetylcholine synthesis, which suggests that the decline in oxidative metabolism is not due to a reduction in the activity of a critical enzyme. Decreased pyruvate oxidation in the presence of saturating pyruvate concentrations does not support the hypothesis of an age-related deficit in the mitochondrial pyruvate transport system (12, 13). Nor can changes in the total activity or the activation state of pyruvate dehydrogenase account for the age-related alterations in oxidative metabolism (34).

NEUROTRANSMITTER RELEASE MECHANISMS DURING AGING

Depressed acetylcholine release may underlie the age-related in vivo decline in synthesis. In brain slices, the potassium-stimulated calcium-dependent release of acetylcholine that was prelabeled with [U-^{14}C]glucose declines as the age increases from 3 months (100%) to 10 months (57%) or 30 months (26%; Fig. 6; 56). Non-calcium

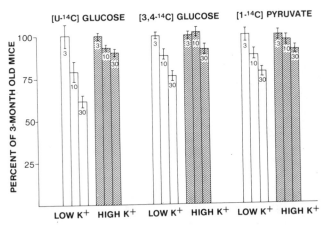

Figure 5. Oxidative metabolism during aging in low (5 mM) and high (31 mM) potassium media. Slices from the whole brain of aged mice were prepared and incubated with excess glucose or pyruvate. The numbers within the bars represent the age in months (24).

dependent release is unchanged. The magnitude of the decline in calcium-dependent release parallels the age-related in vivo decrease in acetylcholine synthesis. A similar age-related decline in acetyl-choline release occurs with high frequency electrical stimulation (52). The possibility of alterations with senescence in the release of norepinephrine, dopamine, GABA and choline/acetylcholine release. Non-calcium dependent release of dopamine, choline/acetylcholine and norepinephrine increased. The inconsistencies between these results and the current studies may have been due to the shorter age-span, the use of a non-purified synaptosomal preparation and measurement of choline/acetylcholine release rather than only acetyl-choline.

CALCIUM HOMEOSTASIS DURING AGING

Altered calcium homeostasis may lead to cholinergic dysfunction since only the calcium-dependent release of acetylcholine declines during aging. If altered calcium metabolism is a primary underlying cause of senescent-related changes, therepautic approaches to the mental deficits that accompany aging will have to change. Several lines of evidence indirectly suggest that aging alters calcium me-tabolism. Calcium activated processes in hippocampal neurons are less responsive in aged animals (35). Tangle formation in Alzheimer's disease (65, 66) and inhibition of neuritic extensions and synapse formation (69) may be due to altered calcium uptake. Selective cell death in ischemia (58) and other disorders may also be associated with altered calcium metabolism.

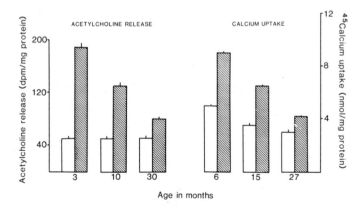

Figure 6. Age-related changes in calcium-dependent acetylcholine release and calcium uptake. Calcium-dependent acetylcholine release was measured after preincubation of slices from whole mouse brain with [U-^{14}C]glucose (56). Calcium uptake was measured in Ficoll purified synaptosomes from aged rats (55). Values in high potassium (hatched portion) differ from each other (p < 0.05). All incubations are for 10 min.

Calcium homeostasis can be examined directly in synaptosomes from aged animals. Synaptosomal calcium-45 uptake is an equilibrium measurement between calcium influx and efflux. In rats, aging (3, 6, 15, 27 months) reduces synaptosomal calcium uptake (106, 100, 68, 46%; 55) in parallel with the decline in acetylcholine release (Fig. 6). Previous studies of calcium uptake during aging found no effect (32). However, these experiments made no distinction between uptake and superficial binding (70). The latter is another indicator of calcium homeostasis, which may also reflect altered membrane fluidity. Superficial binding is the total calcium that is associated with the synaptosome minus the calcium that is transported across the plasma membrane. Aging (3, 6, 15 and 27 months) increases (105, 100, 129 and 167%) the superficial binding of calcium by synaptosomes from rats (55).

THERAPEUTIC APPROACHES TO AGE-RELATED COGNITIVE DISORDERS

There have been numerous attempts to treat geriatric cognitive disorders, although they have been unsuccessful. One strategy has been to try and stimulate the cholinergic system. Chronic choline supplementation improves passive avoidance behavior by aged mice (5), but has been disappointing in the treatment of cognitively impaired humans (6). Since calcium uptake and calcium-dependent release of acetylcholine are diminished by aging, compounds that interact with calcium metabolism may conceivably be more efficacious than choline treatment. The calcium antagonists (nimodipine, nitrendipine, verapamil and nifedipine) have been used for age-related cardiovascular disorders (57), but only nimodopine is effective in the central nervous system (1). The calmodulin antagonists trifluoperazine (74) and prenylanine (33) increase intracellular calcium, but their clinical importance is unknown. The calcium ionophore, A23187, increases the physiologically active pool of calcium (2) and stimulates the calcium-dependent release of acetylcholine (7, 8). The aminopyridines interact with calcium, although their precise mechanism of action is unknown (Table 1; 28, 71). They stimulate the calcium-dependent release of acetylcholine at the neuromuscular junction (48) and from central nervous system preparations such as brain slices or synaptosomes (23, 54). 3,4-Diaminopyridine (54) is more potent than 4-aminopyridine (23) but it is less toxic in man (40) and animals (36). The aminopyridines are ineffective in the absence of extracellular calcium (23, 37).

3,4-Diaminopyridine appears to partially reverse the age-related deficit in calcium homeostasis. No previous studies directly examined the effects of the aminopyridines on calcium homeostasis in the central nervous system. In synaptosomes from 3 and 6 month old rats, 3,4-diaminopyridine or potassium stimulate calcium uptake to a similar extent (55). This may reflect a direct action of 3,4 diaminopyridine on the calcium channel, since outwardly directly potassium channels would not be involved. 3,4-Diaminopyridine

Table 1

Aminopyridine Actions at the Neuromuscular Junction

Action	Reference
Inhibits repolarization by blocking outward potassium currents to increase calcium influx	(46,47,63)
Decreases calcium binding to intracellular structures	(36)
Excites presynaptic fibers	(39,42)
Acts directly on voltage sensitive calcium channels to facilitate calcium entry	(38)
Increases the number of synaptic vesicles at release sites	(30)

reduces the age-related deficits in calcium metabolism. In synaptosomes from 15 and 27 month old rats, 3,4-diaminopyridine stimulates uptake more than potassium depolarization and thus diminishes the age-related deficit (Fig. 7). 3,4-Diaminopyridine also reduces the excess superficial calcium binding that occurs at 27 months (from 167% to 136%) and 15 months (from 129% to 118%), but does not alter superficial calcium binding in young animals.

Age-related alterations in calcium metabolism may underlie the deficits in acetylcholine metabolism. 3,4-Diaminopyridine stimulates acetylcholine release by brain slices in low but not high potassium media and is ineffective in the absence of calcium. Furthermore, the beneficial effects of 3,4-diaminopyridine on acetylcholine release during aging closely parallel its stimulation of calcium uptake (Fig. 7).

RELATION OF ALTERED CALCIUM AND ACETYLCHOLINE METABOLISM DURING AGING TO BEHAVIORAL CHANGES

If alterations in calcium uptake, superficial binding or acetylcholine release are primary changes during aging, then 3,4-diaminopyridine should also alleviate age-related behavioral deficits. 3,4-Diamonpyridine, at a wide range of dosages and times after injection, improves tight rope test performance. The most effective dosage and timing is 10 pmole/kg at 5 minutes after administration (Fig. 8). Although the tight rope test measures some aspect of central cholinergic function, it does not necessarily reflect memory. Performance on the 8-arm radial maze, which directly tests memory, declines with age and can be improved with 3,4-diaminopyridine treatment (Table 2; 10).

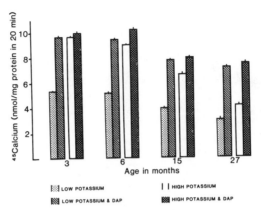

Figure 7. 3,4-Diaminopyridine (DAP) reverses the age-related deficit in calcium uptake. Ficoll purified synaptosomes were prepared from aged rats and incubated in either low (5 mM) or high (31 mM) KCl media with or without DAP. The 10 min calcium-45 uptake incubations were terminated by EGTA/ruthenium red and filtration (55).

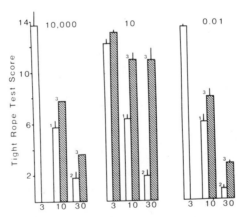

Figure 8. Age-related deficits in behavior are reversed by 3,4-diaminopyridine (hatched portion) 5 min before testing. 1: Denotes value differs from 3 months; 2: Denotes value differs from 3 and 10 months; 3: Denotes significant improvement with 3,4-diaminopyridine (p < 0.05; 55).

Table 2

8-Arm Maze Radial Performance During Aging and
3,4-Diaminopyridine Treatment

Age in Months	Number of Correct Arm Choices
3 months	7.1 ± 0.1
12 months	6.7 ± 0.2
24 months	$6.1 \pm 0.2^*$
24 months + 3,4-diaminopyridine	$6.8 \pm 0.2^{**}$

*Denotes value differs from 3 months; **Denotes significant improvement
with 3,4-diaminopyridine ($p < 0.05$; 10).

CONCLUSIONS

In conclusion, altered brain calcium homeostasis during aging
may underlie the deficits in acetylcholine metabolism, as well as
those in behavior. Diminished calcium uptake during aging parallels
the decline in the calcium dependent release of acetylcholine. This
deficit in release may lead to, at least by correlation, the age-
related reduction in acetylcholine synthesis and behavior. Further
evidence that the calcium deficit is important in this chain of
events is that reversal of the decline in calcium uptake partially
ameliorates the age-induced deficits in release, synthesis and behavior. Thus, therapies that interact with calcium metabolism may
overcome geriatric cognitive dysfunction.

REFERENCES

1. Allen et al. (1983): New Engl. J. Med. 308:619-624.
2. Akerman, K.E.O. and Nicholls, D.G. (1981):   Eur. J. Biochem.
    115:67-73.
3. Barclay, L.L., Gibson, G.E. and Blass, J.P. (1981): Pharm. Bio-
    chem. Behav. 14:153-157.
4. Bartus, R.T. and Dean, R.L. (1980): Neurobiol. Aging 1:145-152.
5. Bartus, R.T., Dean, R.L., Goas, J.A. and Lippa, A.S. (1980):
    Science 209:301-303.
6. Bartus, R.T., Dean, R.L., Beer, B. and Lippa, A.S. (1982): Science
    217:408-417.
7. Casamenti, F., Mantovani, P. and Pepeu, G. (1978): Br. J. Phar-
    mac. 63:259-265.
8. Dastur, D.K., Lane, M.H., Perlin, S. and Sokoloff, L. (1963):
    In Human Aging: A Biological Study (eds) J.E. Birrin, R.N. Butler,
    S.W. Greenhouse, L. Sokoloff and M.R. Yarrow, U.S. Gov't. Printing
    Office, Washington, D.C. pp. 59-75.

9.   Davies, P. and Maloney, A.J.F. (1976):   Lancet $\underline{II}$:1403.
10. Davis, H.P., Idowu, A. and Gibson, G.E. (1983):   Exper. Aging Res. $\underline{9}$:211-214.
11. Eeshmukh, D.R., Owen, D.E. and Patel, M.S. (1980):   J. Neurochem. $\underline{34}$:1219-1224.
12. Deshmukh, D.R. and Patel, M.S. (1980):   Mech. Ageing Devel. $\underline{13}$:75-81.
13. Deshmukh, D.R. and Patel, M.S. (1982):   Mech. Ageing Devel. $\underline{20}$:343-351.
14. Deutsch, J.A. (1970):   Science $\underline{174}$:788-794.
15. Dopplet, J.E. and Wallace, W.L. (1955):   J. Abn. Soc. Psych. $\underline{51}$:312-330.
16. Drachman, D.A., Noffsinger, D., Sahakian, B.J., Krudziel and Fleming, R. (1980): Neurobiol. Aging $\underline{1}$:39-43.
17. Drachman, D.A. (1981):   In Brain Neurotransmitters in Aging and Age-Related Disorders Vol. 17 (eds) S.J. Enna, T. Samorajski and B. Bernard, Raven Press, New York, pp. 255-268.
18. Drachman, D.A. (1983):   In Aging of the Brain Vol. 20 (eds) D. Samuel, E. Giacobini, G. Filogamo, G. Giacobini and A. Vernadakis, Raven Press, New York, pp. 19-31.
19. Ferrendelli, J.A., Sedgwick, W.G. and Suntziff, V. (1971):   J. Neuropath. Exp. Neurol. $\underline{30}$:638-649.
20. Gibson, G.E. and Blass, J.P. (1983):   In Handbook of Neurochemistry Vol. 3, 2nd edition (ed) A. Lajtha, Plenum Press, New York, pp. 633-651.
21. Gibson, G.E., Pelmas, C.J. and Peterson, C. (1983):   Pharm. Biochem. Behav. $\underline{18}$:909-916.
22. Gibson, G.E. and Peterson, C. (1981):   Age (abstr.) 56.
23. Gibson, G.E. and Peterson, C. (1982):   Biochem. Pharm. $\underline{32}$:111-115.
24. Gibson, G.E. and Peterson, C. (1981):   J. Neurochem. $\underline{37}$:978-984.
25. Gibson, G.E., Peterson, C. and Jenden, D.J. (1981):   Science $\underline{213}$:674-676.
26. Gibson, G.E., Peterson, C. and Sansone, J. (1981):   Neurobiol. Aging $\underline{2}$:165-172.
27. Gibson, G.E. and Peterson, C. (1982):   In Aging of the Brain Vol. 20 (eds) D. Samuel, E. Giacobini, G. Filogamo, G. Giacobini and A. Vernadakis, Raven Press, New York, pp. 107-122.
28. Glover, W.E. (1982):   Gen. Pharmacol. $\underline{13}$:259-285.
29. Haycock, J.W., White, W.F., McGaugh, J.L. and Cotman, C.W. (1977):   Exp. Neurol. $\underline{57}$:873-882.
30. Heuser, J.E. (1977):   In Approaches to the Cell Biology of Neurons (eds) W.M. Conan and J.A. Ferrendelli, Society of Neurosci. Symp. $\underline{2}$:215-239.
31. Hollander, J. and Barrow, C.H. (1968):   J. Geront. $\underline{23}$:174-181.
32. Jones, T.W.G. and Beaney, J. (1980):   Mech. Ageing Devel. $\underline{14}$:417-426.
33. Karaki, H., Murakahmi, K., Nakagawa, K., Osaki, H. and Urakawa, N. (1982):   Br. J. Pharmac. $\underline{77}$:661-666.
34. Ksiezak-Reding, H.J., Peterson, C. and Gibson, G.E. (1984):   Mech. Ageing Devel. $\underline{26}$:67-73.

35. Landfield, P.W., Pitler, T.A., Applegate, M.D. and Robinson, J. H. (1983): Soc. Neurosci. Abstr. 923.

36. Lamiable, D. and Millart, H. (1983): J. Chromat. 272:221-225.

37. Lundh, H., Leander, S. and Thesleff, S. (1977): J. Neurobiol. 32:29-43.

38. Lundh, H. and Thesleff, S. (1977): Eur. J. Pharmac. 42:411-412.

39. Lundh, H. (1978): Brain Res. 153:307-318.

40. Lundh, H., Nilsson, O. and Rosen, I. (1983): J. Neurol. Neurosurg. and Psych. 46:684-685.

41. Mantovani, P. and Pepeu, G. (1981): Pharmacol. Res. Comm. 13: 175-184.

42. Marshall, I.G., Lambert, J.J. and Durant, N.N. (1979): Eur. J. Pharmac. 54:9-14.

43. McGeer, P.L. and McGeer, E.G. (1976): In Nutrition and the Brain Vol. 5 (eds) A. Barbeau, J.H. Growdon and R.J. Wurtman, Raven Press, New York, pp. 177-199.

44. McGeer, E.G., Gibinger, H.C., McGeer, P.L. and Wickson, V. (1971): Exp. Gerontol. 6:391-396.

45. Meek, J.L., Bertilsson, L., Cheney, D.L., Zsilla, G. and Costa, E. (1977): J. Gerontol. 32:129-131.

46. Molgo, J., Lemeignan, M. and Lechat, P. (1975): C.R. Acad. Sci. (Paris), Serie D. 281:1637-1639.

47. Molgo, J., Lemeignan, M. and Lechat, P. (1977): J. Pharm. Exp. Ther. 203:653-663.

48. Molgo, J., Lundh, H. and Thesleff, S. (1980): Eur. J. Pharmac. 61:25-34.

49. Moudgil, V.K. and Kanungo, M.S. (1973): Biochem. Biophys. Acta. 329:211-215.

50. Ordy, J.M. and Schjeide, O.A. (1973): In Progress in Brain Research (ed) D.H. Ford, Elsevier, New York, pp. 25-52.

51. Parmacek, M.S., Fox, J.H., Harrison, W.H., Garron, D.C. and Swenie, D. (1979): Gerontol. 25:185-191.

52. Pedata, F., Slavikova, J., Kotas, A. and Pepeu, G. (1983): Neurobiol. Aging 4:31-35.

53. Peng, M.T., Peng, Y.I. and Chen, F.N. (1977): J. Gerontol. 32:517-522.

54. Peterson, C. and Gibson, G.E. (1982): J. Pharm. Exp. Ther. 222:578-582.

55. Peterson, C. and Gibson, G.E. (1983): J. Biol. Chem. 258:11482-11486.

56. Peterson, C. and Gibson, G.E. (1983): Neurobiol. Aging 4:25-30.

57. Rahwan, R.G. and Witiak, D.T. (eds) (1982): Calcium Regulation by Calcium Antagonists, ACS Sympos. Ser. #201, American Chemical Society, Washington, D.C.

58. Raichle, M.E. (1983): Ann. Neurol. 13:2-10.

59. Reiner, J.M. (1947): J. Gerontol. 2:315-320.

60. Reis, D.J., Rosa, R.A. and Joh, T.H. (1977): Brain Res. 136: 465-474.

61. Samorajski, T. (1981): In Brain Neurotransmitter and Receptors

in Aging and Age-Related Disorders Vol. 17 (eds) S.J. Enna,
T. Samorajski and B. Bernard, Raven Press, New York, pp. 1-12.
62. Schaie, K.W. (1959): J. Gerontol. 14:208-215.
63. Schauf, C.L., Colton, C.A., Colton, J.S. and Davis, F.A. (1976):
J. Pharm. Exp. Ther. 197:414-425.
64. Severson, J.A. and Finch, C.E. (1981): Soc. Neurosci. Abstr.
11:186.
65. Schlaepfer, W.W. (1983): In Biochemical Aspects of Alzheimer's
Disease (ed) R. Katzman, Cold Spring Harbor Laboratory, New York,
pp. 107-116.
66. Selkoe, D.J., Ihara, Y. and Salazar, F.J. (1982): Science 215:
1243-1245.
67. Sherman, K.A., Kuster, J.E., Dean, R.L., Bartus, R.T. and Friedman, E. (1981): Neurobiol. Aging 2:99-104.
68. Strong, R., Hicks, P., Hsu, L., Bartus, R.T. and Enna, S.J.
(1980): Neurobiol. Aging 1:59-63.
69. Suarez-Isla, B.A., Pelto, D.J. and Rapoport, S.I. (1983): Age
(abstr.) 13:5.
70. Sun, A.Y. and Seaman, R.N. (1977): Exp. Aging Res. 3:107-116.
71. Thesleff, S. (1981): Neurosci. 5:1413-1419.
72. Vasko, M.R., Domino, L.E. and Domino, E.F. (1974): Eur. J. Pharmacol. 27:145-147.
73. Valcana, T. and Timiras, P.S. (1969): In Proc. 8th Int'l. Congress
of Geront. Vol. II, Int'l. Assoc. Gerontol., Washington, D.C.
pp. 24-29.
74. Wada, A., Yanagihara, N., Izumi, F., Sakurai, S. and Kobayashi,
H. (1983): J. Neurochem. 40:481-486.

CHOLINERGIC DYSFUNCTION AND MEMORY;
IMPLICATIONS FOR THE DEVELOPMENT OF ANIMAL MODELS
OF AGING AND DEMENTIA

M.J. Pontecorvo, C. Flicker and R.T. Bartus

Department of CNS Research
Medical Research Division
American Cyanamid Company
Lederle Laboratories
Pearl River, New York   10965
and
Department of Psychiatry
New York University Medical Center
New York, New York 10016

INTRODUCTION

Aged humans show a pattern of cognitive impairments in which recent memory deficits are superimposed on a more global cognitive deterioration which typically includes deficits in attention, decreased processing speed, cognitive rigidity and interference from previous associations (1, 10). Research extending over the last decade has shown that this pattern of impairments, particularly the decline in recent memory, is a common characteristic of aging in many mammalian species including humans, nonhuman primates, rats and mice (2, 3, 17, 22). In humans these impairments are exacerbated by senile dementia of the Alzheimer type (SDAT), which affects 10% of all persons over 65 years of age and 30% of those over 85. Although the neurochemical changes responsible for the cognitive loss in aged or demented subjects are not completely understood, extensive evidence (5, 9, 11) suggests that a dysfunction of central cholinergic transmission plays an important role.

First, significant changes in a number of brain cholinergic markers, including choline uptake, choline acetyltransferase (ChAT) activity, muscarinic receptor binding, and the response of hippocampal pyramidal cells to iontophoretically applied acetylcholine (ACh), may accompany the cognitive impairments in aged humans, SDAT patients and animals (5, 9). The magnitude of the ChAT decrease

205

has been correlated with the severity of cognitive impairments and with the density of senile plaques in aged humans (25, 28). Recent evidence (26) demonstrating the presence of acetylcholinesterase (AChE) and ChAT in monkey senile plaques further suggests a cholinergic role in this age related neuropathology.

Although SDAT patients may exhibit similar changes in other neurotransmitter systems, correlations between these changes and cognitive deterioration have yet to be established. Furthermore, it is unclear whether these changes exist in all SDAT patients, or whether they may appear partly, as a result of confounding problems involving subject selection and diagnosis (11). In contrast, the decrease in ChAT activity in SDAT, and the correlation with cognitive deficits is remarkably consistent, and has come to be regarded as a neurochemical hallmark of SDAT.

Additional support for this cholinergic hypothesis of geriatric cognitive dysfunction comes from psychopharmacological studies with young and aged subjects. For example, administration of the muscarinic blocker scopolamine to young humans and monkeys produces memory/cognitive impairments similar to those seen in normal aged subjects (3, 14). That comparable deficits are not produced in monkeys by pharmacological blockade of dopaminergic, beta adrenergic or nicotonic receptors (3, 12) suggests a specific involvement of muscarinic cholinergic transmission in these memory/ cognitive functions.

Furthermore, as would be predicted from this hypothesis, the effects of scopolamine can be partially, but reliably reduced by administration of the AChE inhibitor physostigmine, but not by CNS stimulants, such as methylphenidate and amphetamine (3, 13). Thus, it is unlikely that the retention deficit induced by scopolamine in either human or nonhuman primates can be related to more general effects on motivation, arousal and other sedativelike properties.

Finally, AChE inhibitors (physostigmine, tetrahydroaminoacridine) as well as direct cholinergic agonists (arecoline, oxotremorine) can also partially reverse the memory deficits in normally aged monkeys (3, 4). Although this reversal is not complete, cholinergic agents appear more effective than any other class of compounds tested so far including CNS stimulants, neuropeptides, nootropics, antidepressants, and direct-acting dopaminergic GABAergic and alpha-adrenergic agonists. Similar results have also been obtained for aged humans and SDAT patients. Thus, neuropeptides and nootropics are reported to have weak ameliorative effects, at best (8, 15), whereas arecoline and physostigmine, in the appropriate dose, have been reported to produce reliable, and possibly even clinically relevant improvements in cognitive performance in aged and demented subjects (8).

Taken together, these studies indicate that an important relationship may exist between age-related changes in muscarinic cholinergic neurotransmission and geriatric cognitive impairments. These studies further suggest a therapeutic strategy (i.e., treatment via cholinergic agents) that has shown some promise in clinical tests with aged humans and SDAT patients. However, it should be noted that even the most effective of the currently available compounds have limited success in reversing the severe degenerative deficits produced by SDAT (8). Thus, there remains a clear need to continue the search for an effective therapeutic compound.

It is likely that an extensive multidisciplinary effort involving clinical and basic scientists will ultimately be required to produce a satisfactory understanding and treatment for SDAT. However, an important preliminary step, useful in testing hypotheses and potential therapeutic approaches is the development of an animal model that more closely mimics the neurochemical and neuropathological characteristics of SDAT. Although aged animals and humans suffer cognitive deficits and changes in cholinergic transmission, these deficits are typically not as severe as those seen in SDAT, and they are not accompanied by the classic neuropathological signs of the disease. Specifically, brains from normally aged subjects do not show neurofibrillary tangles; nor do they show the dramatic decrease in ChAT activity and correlated increase in senile plaque density that accompanies the cognitive loss in SDAT.

Recent evidence suggests that degeneration in cholinergic basal forebrain nuclei (diagonal band, medial septum, nucleus basalis) may account for the loss of ChAT activity in SDAT. These data are reviewed in detail elsewhere in this book (e.g., 26). Briefly, combined retrograde labeling and immunohistochemical techniques indicate that the basal forebrain is a major source of cholinergic input to the hippocampus, cortex and olfactory bulb. Lesions of the basal forebrain in young monkeys and rodents produce decreases in ChAT activity analogous to those seen in SDAT. Finally, large, presumably cholinergic cell bodies in the basal forebrain degenerate with SDAT.

It has been suggested that this degeneration may also account for the cognitive loss in SDAT. If it can be demonstrated that destruction of these cell bodies can produce severe memory/cognitive deficits analogous to those in SDAT, this hypothesis would be strengthened. It might then be possible to use animals with basal forebrain lesions to study treatment alternatives for the major neurobehavioral symptoms of SDAT.

RESULTS

Little information presently exists that can be used to predict the functional consequences of degeneration or destruction of basal forebrain nuclei. Animals with lesions of the medial septal nucleus have been shown to be impaired on a variety of learning and discrimination tasks (see reference 20 for a recent review). However, only a few recent studies have specifically examined the effects of septal lesions on memory performance. Furthermore, most studies have utilized electrolytic lesions and thus, may be confounded by damage to fibers of passage.

Still less is known about the consequences of lesions of the nucleus basalis and diagonal band. In recent months, a number of groups, including our own, have begun to investigate the effects of lesions of these basal forebrain nuclei (6, 16, 18, 23, 24). In our initial study (16), ibotenic acid (.4 μl @ 6 μg/μl) was infused bilaterally into the ventromedial globus pallidus, the basal forebrain area that is most homologous to the nucleus basalis in humans and monkeys. Ibotenic acid was chosen for lesioning purposes because it causes excitotoxic destruction of dendrites and cell bodies while selectively sparing fibers of passage (27). Also, it is less likely than alternative neurotoxins (e.g., kainic acid) to produce seizure activity and lesions distal to the site of injection (21).

Histological evaluation (Figure 1) confirmed that the lesions destroyed the intended cell bodies. ChAT activity in the frontal cortex of the nucleus basalis lesioned animals was approximately 66% of that in the controls.

Two weeks following surgery all rats were tested on a battery of behavioral tasks. Four different psychomotor tasks, intended to measure muscle strength, stamina, and coordination revealed no effects of the lesion. Furthermore, no differences were observed in shock sensitivity or initial latency to step from a bright to a dark compartment in a one trial passive avoidance task. However, the nucleus basalis lesioned rats were severely impaired at retention of the passive avoidance response (Figure 2). Similar results have also been reported by other labs (18, 23, 24).

Because our nucleus basalis lesioned animals exhibited deficits at both 1 hour and 24 hour retention intervals, it was unclear whether these deficits reflected problems in learning, memory or both. Further, the passive avoidance procedure is recognized as a relatively crude behavioral paradigm, and its results are often open to multiple alternative interpretations. For these reasons we performed a second experiment, in a new group of rats (6). Rats were first trained to obtain food reward by visiting each of 8 arms of a radial arm maze. Repeat visits to an arm were never reinforced and were scored as errors. Following several months of training and establishment of near-perfect performance, the rats received sham lesions or nucleus basalis lesions similar to those described above.

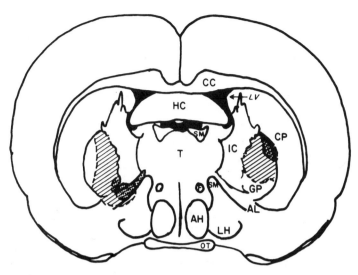

Figure 1. Maximum (hatched) and minimum (stippled) extent of the
lesions produced by infusion of ibotenic acid into the ventromedial
globus pallidus (nucleus basalis - left) or dorsolateral globus
pallidus (control - right), shown at the level of the maximum extent
of the nucleus basalis lesion. Abbreviations: AH, anterior hypothalamus; AL, ansa lenticularis; CC, corpus callosum; CP, caudate-
putamen; F, fornix; GP, globus pallidus; HC, hippocampal commissure;
IC, internal capsule; LH, lateral hypothalamus; LV, lateral ventricle; OT, optic tract; SM, stria medullaris; T, thalamus. From
Flicker, et al., (15).

Two weeks following surgery, the rats were retested on the
radial arm maze task. During this initial retesting period there
were no differences in performance between the sham and nucleus
basalis lesioned rats. Thus, both groups continued to perform the
task as originally trained, going to each of the eight arms only
once, without returning to any of the arms during the session.

By contrast, when a retention interval was interposed between
the selection of the first four arms and the remaining four arms,
a profound, time-related decrement in performance by the nucleus
basalis lesioned animals was revealed (Figure 3). This time-related
deficit suggests that nucleus-basalis-lesioned rats may suffer recent
memory deficits analagous to those seen in human SDAT patients,
and is consistent with the hypothesis that degeneration of basal
forebrain nuclei contributes to the cognitive impairments of SDAT.

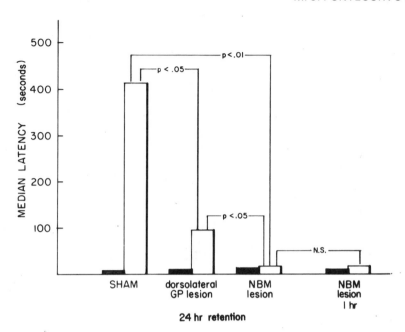

Figure 2.   Passive avoidance retention by nucleus basalis (NBM), sham and dorsolateral globus pallidus (GP) lesioned rats. The solid bars show the latencies to step through to the dark chamber on the first exposure to the apparatus (training). There were no differences among the groups. The open bars show the latencies to step through on the retention test (1 or 24 hours). The NBM lesioned animals showed a retention deficit relative both to sham and globus pallidus lesioned controls. From Flicker et al. (15).

DISCUSSION AND SUMMARY

These findings could have important implications for the development of therapeutic approaches to SDAT. Specifically, they suggest that the most efficient treatment for the cognitive deficits in SDAT might be to compensate for the loss of cholinergic input to the cortex and hippocampus by stimulating post synaptic receptors with direct cholinergic agonists. This suggestion is consistent with previous results in our laboratory using normally aged monkeys (4), and in other laboratories with humans (7), that showed direct cholinergic agonists to be more effective than other cholinergic agents in overcoming memory impairments in aged and demented subjects.

It should be noted, however, that the present data are considered to be preliminary. A number of additional experiments are essential both to provide a more complete understanding of the function

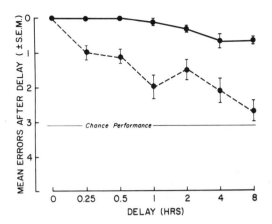

Figure 3.   Retention of spatial location in the radial arm maze by
nucleus basalis (broken line) and sham lesioned (solid line) rats.
The chance line was empirically determined by placing trained animals
in the maze and arbitrarily designating 4 arms as correct and 4
incorrect.   Thus, it approximates the performance that could be
expected from animals with no memory for choices made prior to the
retention interval.

of basal forebrain nuclei in cognitive processes, and to refine the
model for use in evaluating potential therapeutic approaches for SDAT:

First, studies of the effects of basal forebrain lesions in
nonhuman primates are clearly needed.   The basal forebrain nuclei
(particularly the nucleus basalis) are poorly defined in many sub-
primate species; only the monkey has a well-defined cell population
that has been demonstrated to be clearly homologous to the nucleus
basalis of Meynert in humans (19).   The similarity of structure
and organization of the basal forebrain nuclei in human and nonhuman
primates, combined with the relative similarity of behavioral reper-
toire and cognitive test capabilities in these species greatly reduces
the assumptions necessary to extrapolate from the effects of basal
forebrain lesions to the functional consequences of degeneration
of the basal forebrain nuclei in aged demented humans.

Second, it would be useful to compare the effects of nucleus
basalis lesions to the effects of lesions of other basal forebrain
nuclei (e.g., medial septum) since these nuclei have also been implicated in learning and memory processes, and are also reported to
degenerate in SDAT.   A wide range of behaviors should be examined
to allow the comparison of points of similarity and difference between
the effects of each of these lesions and the impairments in human
SDAT patients.

Finally, it is necessary to demonstrate that the response of basal forebrain lesioned animals to pharmacological treatments is similar to that of human SDAT patients. Although there are no recognized standard and prototypical cognition enhancing compounds, certain cholinergic agents (arecoline, physostigmine), and also some nootropic compounds have been reported to have limited ameliorative effects in aged humans and SDAT patients. Such drugs might therefore serve as reference compounds in lieu of more suitable standards.

REFERENCES

1. Albert, M.S. and Kaplan, E. (1980): In: New Directions in Memory and Aging (eds) L.W. Poon, J.L. Fozard, L.S. Cermak, D. Arenberg and L.W. Thompson, Erlbaum, Hillsdale, New Jersey, pp. 403-432.
2. Bartus, R.T. (1979): In: Aging: Sensory Systems and Communication in the Elderly, Vol. 10 (eds) J.M. Ordy and K. Brizzee, Raven Press, New York, pp. 85-114.
3. Bartus, R.T. (1980): In: Aging in the 1980's: Psychological Issues (ed) L.W. Poon, American Psychological Association, Washington, D.C., pp. 163-180.
4. Bartus, R.T., Dean, R.L. and Beer, B. (1983): Psychopharmacology Bulletin 19:168-184.
5. Bartus, R.T., Dean, R.L., Beer, B. and Lippa, A.S. (1982): Science 217:408-417.
6. Bartus, R.T., Dean, R.L., Flicker, C., Pontecorvo, M.J., Figueiredo, J.C. and Fisher, S.K.: In Preparation.
7. Christie, J.E. (1982): In: Aging: Alzheimer's Disease: A Report of Progress in Research, Vol. 19 (eds) S. Corkin, K.L. Davis, J.H. Growdon, E. Usdin and R.J. Wurtman, Raven Press, New York, pp. 413-420.
8. Corkin, S., Davis, K.L., Growdon, J.H., Usdin, E. and Wurtman, R.J. (1982): (eds) Aging: Alzheimer's Disease: A report of Progress in Research, Vol. 19, Raven Press, New York.
9. Coyle, J.T., Price, D.L. and DeLong, M.R. (1983): Science 219: 1184-1190.
10. Craik, F.I.M. (1977): In: Handbook of Psychology of Aging (eds) J.E. Birren and K.W. Schaie, Van Nostrand Reinhold, New York, pp. 384-419.
11. Davies, P. (1981): In: Strategies for the Development of an Effective Treatment for Senile Dementia (eds) T. Crook and S. Gershon, Mark Powley Associates, New Canaan, Connecticut, pp. 19-34.
12. Dean, R.L., Beer, B. and Bartus, R.T. (1982): Neuroscience Abstracts 8:87.16.
13. Drachman, D.A. (1977): Neurology 27:783-790.
14. Drachman, D.A. and Leavitt, J. (1974): Archives of Neurology 30:113-121.

15. Ferris, S.H., Reisberg, B. and Gershon, S. (1980): In: Aging in the 1980's: Psychological Issues (ed) L.W. Poon, American Psychological Association, Washington, D.C., pp. 212220.

16. Flicker, C., Dean, R.L., Watkins, D.L., Fisher, S.K. and Bartus, R.T. (1983): Pharmacology, Biochemistry and Behavior 18:973-981.

17. Flicker, C., Bartus, R.T., Ferris, S.F. and Crook, T.: In preparation.

18. Friedman, E., Lerer, B. and Kuster, J. (1983): Pharmacology, Biochemistry and Behavior 19:309-312.

19. Gorry, J.D. (1963): Acta Anatomica 55:51-104.

20. Gray, J.A. and McNaughton, N. (1983): Neuroscience and Biobehavioral Reviews 7:119-188.

21. Guldin, W.O. and Markovitch, H.J. (1982): Journal of Neuroscience Methods 5:83-93.

22. Kubanis, P. and Zornetzer, S.F. (1981): Behavioral and Neural Biology 31:115-172.

23. Lerer, G., Gamzu, E. and Friedman, E.: This book.

24. LoConte, G., Bartolini, L., Casamenti, F., Marconcini-Pepeu, I. and Pepeu, G. (1982): Pharmacology Biochemistry and Behavior 17:933-937.

25. Perry, E.K., Tomlinson, B.E., Blessed, G., Bergmann, K., Gibson, P.H. and Perry, R.H. (1978): British Medical Journal 2:1457-1459.

26. Price, D.L., Kitt, C.A., Hedreen, J.C., Whitehouse, P.J., Struble, R.G., Cork, L.C., Walker, L.C., Mobley, W.C., Salvaterra, P.M. and Wainer, B.H.: This book.

27. Schwarcz, R., Hokfelt, T., Fuxe, K. Jonsson, G., Goldstein, M. and Terenius, L. (1979): Experimental Brain Research 37:199-216.

28. Wilcock, G.K., Esiri, M.M., Bowen, D.M. and Smith, C.C.T. (1982): Journal of the Neurological Sciences 57:407-417.

RELATIONSHIPS BETWEEN CHOLINE ACETYLTRANSFERASE AND MUSCARINIC BINDING IN AGING RODENT BRAIN AND IN ALZHEIMER'S DISEASE

E.D. London[1,2] and S.B. Waller[2]

[1]Addiction Research Center
National Institute on Drug Abuse
and
[2]Gerontology Research Center
National Institute on Aging
c/o Francis Scott Key Medical Center
Baltimore, Maryland 21224 U.S.A.

INTRODUCTION

The literature concerning age effects on cholinergic neuro-chemical parameters in the brain presents a heterogeneous picture, both within and among species. Whereas some studies of human post-mortem material have demonstrated age-dependent declines in choline acetyltransferase (ChAT) activity in various brain regions (5, 12, 19), others have revealed no ChAT losses in some of the same regions (5, 10, 12, 19, 28). In brain regions from C57BL/6J and A/J mice, we have observed age-dependent increases in ChAT (26); whereas, other studies have revealed no changes or decreases in the brains of C57BL/6J mice (23, 25), other mouse strains (24), and rats (2, 23, 25).

There also is some disagreement regarding age effects on cerebral muscarinic binding. Davies and Verth (6) reported no age-dependence of muscarinic binding in a number of human brain areas; however, others have noted declines in the cerebral cortex and hippocampus (17, 28). In the C57BL/6J mouse, Freund (7) observed age-related decreases in samples of whole brain less the cerebellum; and Strong et al. (23) reported decrements in the cerebral cortex and striatum, but not in the hippocampus. Age-dependent declines in muscarinic binding have been observed in striata of Sprague-Dawley and Long-Evans rats and the cerebral cortices of Sprague-Dawley rats, but not in the hippocampi of either strain (15, 23).

Although variation due to genetic strain and species is a factor which must be considered, some of the aforementioned discrepancies may be inherent in assay techniques as well as the experiences of the animals prior to their being killed for assay (e.g., whether animals were naive or received behavioral training and testing; whether they were virgins or retired breeders). In order to assess the contributions of strain and species differences to age effects on ChAT and muscarinic binding, we used standardized assay procedures to determine ChAT activity and muscarinic binding in brain regions from C57BL/6J mice, Fischer-344 rats and Wistar rats.

Because of evidence from studies in rodents and in tissue culture systems that muscarinic binding is modulated by cholinergic input (1, 13, 21), it seemed possible that age-induced alterations in ChAT might be associated with changes in muscarinic binding. Furthermore, evidence from the noradrenergic system indicates that presynaptic compensation could accompany postsynaptic receptor deficits (22). Thus, it is plausible that ChAT activity could be affected by age-dependent changes in muscarinic binding. We therefore have examined how the relation between ChAT and muscarinic binding might be affected by aging in mouse and rat brains. Preliminary data also are presented regarding this relation in postmortem cerebral cortex samples from human subjects who died with Alzheimer's disease (AD) and from age-matched controls.

MATERIALS AND METHODS

All animals used in these studies were naive, virgin males. C57BL/6J mice, aged 4-, 12-, 18-, or 24 months (mo) were obtained from the rodent colony of the Gerontology Research Center (GRC), National Institute on Aging (NIA). The maximum life span of C57BL/6J mice is from 30 to 35 mo, with a median of 22 to 27 mo (11, 20). Wistar rats, at 6- and 24 mo of age, also were obtained from the GRC colony. The median lifespan for male Wistar rats in the GRC colony is 24 mo. Fischer-344 rats, at 6-, 12- and 24 mo of age, were obtained from Charles River Breeding Laboratories. The Fischer-344 rat has a maximal survival time of approximately 35 mo and a mean survival of 29 mo (4).

Mice were housed six to eight per cage; Wistar rats were housed singly, and Fischer-344 rats were housed two per cage. The animals were maintained in a vivarium with controlled temperature (25 - 27°C) and alternating 12 hr light (0500-1700 hr) and dark cycles. Food (24% protein, NIH formula) and tap water were available at all times. At the designated time, animals were decapitated, and their brains rapidly removed and dissected into regions while on an aluminum plate cooled by ice. Data are presented for samples of cerebral cortex, corpus striatum and hippocampus from mice, and the cerebral cortex and hippocampus from the two strains of rats. Brain samples were stored at -70°C until assayed.

Postmortem cerebral cortex tissue from individuals with AD and from age-matched controls was provided by Dr. M.J. Ball (University of Western Ontario). Tissue samples were either frozen immediately after dissection, and stored at -70°C until the time of assay; or they were sonicated in 20-100 volumes (w/v) of Tris HCl buffer, pH 7.4, frozen as the sonicates, and stored at -70°C. Storage in these different states had no apparent effects on ChAT or muscarinic binding.

For assays of ChAT activity, rodent tissue samples were sonicated in 20 volumes (w/v) of 50 mM Tris-HCl buffer, pH 7.4, containing 0.2% (v/v) Triton X-100. Enzyme activity was determined according to the method of Bull and Oderfeld-Nowak (3). The effect of acetyl coenzyme A ([1-^{14}C]acetyl coenzyme A, 2-5 mCi/mmole; Amersham Corp.) concentration on enzyme activity was determined by varying the concentration of acetyl coenzyme A (1 to 100 μM) in the assay medium, while the concentration of choline chloride was maintained at 60 mM. The effect of choline chloride concentration on enzyme activity was determined by varying the concentration of choline chloride (0.05 to 60 mM), while maintaining the acetyl coenzyme A concentration at 100 μM. Values of K_m and V_{max} were obtained by the Eadie-Hofstee method.

Assays of muscarinic binding were perfomed on tissue sonicates diluted with 20 mM Tris-HCl buffer, pH 7.4, using [^{3}H]quinuclidinyl benzilate ([^{3}H]QNB) as the ligand, and a modification of the method of Yamamura and Snyder (29). The sonicates were incubated with L-[^{3}H]QNB (33.1 Ci/mmol, New England Nuclear Corp.), for 90 min at room temperature. The final incubation volume was 1 ml, and incubations were terminated by filtration through pre-wetted Whatman GF/B glass microfibre filters. Nonspecific binding, obtained in the presence of 100 μM oxotremorine, was subtracted from total binding at the various concentrations of ligand. Variation in total specific binding among triplicate samples (or duplicates in human studies) was always less than 10% of the mean. The percent of [^{3}H]QNB bound was always less than 5% of the free ligand concentration.

Using the rodent brain samples, estimates of total muscarinic receptor density (B_{max}) and the dissociation constant of [^{3}H]QNB from the binding sites (K_{QNB}) were obtained by Eadie-Hofstee analysis of saturation isotherm data, obtained from incubations in which the radioligand concentration was varied from 10^{-12} to 10^{-8} M. In the mouse brain, estimates of the percentage of total specific [^{3}H]QNB binding that was associated with muscarinic agonist high affinity binding sites (B_{Hi}), and the affinities of agonist for high- and lower affinity sites ($K_{Hi-Carb}$ and $K_{Lo-Carb}$) were obtained from assays of [^{3}H]QNB binding in the presence of 10^{-2} to 10^{-10} M carbamylcholine chloride (Sigma), in aliquots of the same tissue sonicates used to obtain binding saturation isotherm data for [^{3}H]QNB. The concentration of [^{3}H]QNB used in these assays was 0.35 nM. Theoretical binding curves of the competitive displacement of $[^3H]QNB$ by the agonist carbamylcholine were analyzed using least squares and non-linear regression analysis. Data were fitted to a simple mass action expression, defined for the case of two unique binding sites for a single ligand (carbamylcholine), as described by Ikeda et al. (8). In the human brain samples, estimates of total muscarinic receptor binding (B_{max}) were calculated using an "abbreviated" assay method described by McKinney and Coyle (13), with a $[^3H]QNB$ concentration of 0.35 nM. The value of K_{QNB} was taken to be 0.098 nM, as determined in studies of human control superior temporal gyrus. Protein concentrations were determined by the method of Lowry et al. (9).

For brain regions obtained from rats, significance of age differences were assessed by one-way analysis of variance and Bonferroni $\underline{t}$ statistics (14). Differences in ChAT activity and binding site density from human postmortem samples between diagnostic groups were assessed separately by region using an analysis of covariance, controlling for any effect associated with either the subject's age or sex or the group in which the sample was included for shipping and neurochemical determination (18). Unless indicated otherwise, the criterion for all significance statements was $p < 0.05$.

RESULTS

Data obtained in Fischer-344 rats of three ages are shown in Table 1. In the cerebral cortex, K_{QNB} and B_{max} showed significant differences at each age point. K_{QNB} was increased by 51% at 12 mo as compared with 6 mo, and was reduced by approximately the same magnitude at 24 mo. B_{max} was reduced progressively with age, showing decrements of 26% and 23% at 12 mo and 24 mo, respectively, as compared with the immediately preceding age groups. There were no age-related differences in cortical ChAT.

In the hippocampus, there were no age differences in B_{max}, but a transient increase in K_{QNB} at 12 mo, a decline in $K_{Hi-carb}$ between 12 and 24 mo, and a decline in the percentage of $[^3H]QNB$ binding to high affinity agonist sites between 6 and 12 mo. ChAT showed a transient decline in the K_m for acetyl coenzyme A at 12 mo, and a decline in V_{max} between 6 and 12 mo.

In Wistar rats (Table 2), the cortex showed an age-dependent decline in ChAT activity and an increase in K_{QNB}, but no change in B_{max} for $[^3H]QNB$. In the hippocampus, K_{QNB} and B_{max} declined between 6 and 24 mo; whereas, ChAT was unaltered with age.

In cortical, hippocampal and striatal samples from C57BL/6J mice of 4 ages, the only age differences noted were in B_{max} for $[^3H]QNB$ and V_{max} for ChAT (Table 3). In the cortex, there was a 32% decline in B_{max} at 18 mo as compared with 12 mo; a similar decline

Table 1

Effect of Age on Muscarinic Binding and ChAT in the Fischer-344 Rat

	CEREBRAL CORTEX			HIPPOCAMPUS		
	6 mo	12 mo	24 mo	6 mo	12 mo	24 mo
Muscarinic Binding (n=6)						
K_{QNB} (pM)	70 ± 8.1	106 ± 8.3*	61 ± 7.9*	32 ± 4.9	60 ± 6.4*	44 ± 7.3
B_{max} (fmol/mg protein)	2301 ± 200	1700 ± 110*	1311 ± 110*	1300 ± 160	1250 ± 120	1486 ± 240
K_{Hi}-Carb (uM)	1.6 ± 0.2	1.2 ± 0.1	1.3 ± 0.2	0.88 ± 0.1	0.76 ± 0.1	0.33 ± 0.1*
K_{Lo}-Carb (uM)	456 ± 53	399 ± 61	299 ± 51	260 ± 15	233 ± 47	215 ± 31
B_{Hi} (%)	44 ± 2	35 ± 3	34 ± 2	47 ± 3	38 ± 4*	32 ± 1
ChAT (n=8)						
K_m (uM acetyl coenzyme A)	13 ± 1.0	14 ± 2.1	12 ± 7.8	15 ± 1.0	12 ± 0.8*	16 ± 1.0*
V_{max} (nmol ACh/hr/mg protein)[a]	80 ± 3	78 ± 3	66 ± 5	110 ± 2	96 ± 3*	99 ± 3

Each value is the mean ± S.E.M. for the number of animals indicated in parenthesis. [a]V_{max} was determined in incubations where the concentration of choline chloride was held constant and the concentration of acetyl coenzyme A was varied. *Significantly different from previous age, $P \leq 0.05$.

Table 2

Effect of Age on Muscarinic Binding and ChAT in the Wistar Rat

	CEREBRAL CORTEX		HIPPOCAMPUS	
	6 mo	24 mo	6 mo	24 mo
Muscarinic Binding				
K_{QNB} (pM)	80 ± 5.0	123 ± 15[*]	450 ± 50	193 ± 17[*]
B_{max} (fmol/mg protein)	1360 ± 121	1410 ± 30	2430 ± 182	1540 ± 126[*]
ChAT				
(nmol ACh/hr/mg protein)	40 ± 2.2	34 ± 1.9[*]	49 ± 3.4	42 ± 3.3

ACh value is the mean ± S.E.M. for four animals. [a]V_{max} was determined in incubations where the concentration of choline chloride was held constant and the concentration of acetyl coenzyme A was varied. [*]Significantly different from 6 mo, P ≤ 0.05.

Table 3

Effect of Aging on Muscarinic Binding and ChAT in the C57B1/6J Mouse

Brain Region	4-6	12	18	24
	B_{max} (fmol/mg protein)			
Cerebral cortex	964 ± 27	877 ± 39	597 ± 31[*]	408 ± 10[*]
Hippocampus	506 ± 8	485 ± 22	485 ± 24	565 ± 14[*]
Striatum	820 ± 45	702 ± 36	785 ± 28	583 ± 11[*]
	V_{max} (nmol ACh/hr/mg protein)[a]			
Cerebral cortex	33.2 ± 3	39.4 ± 4	51.0 ± 4[*]	61.4 ± 4[*]
Hippocampus	71.6 ± 5	70.0 ± 3	70.2 ± 4	97.6 ± 2[*]
Striatum	252 ± 31	264 ± 18	282 ± 47	477 ± 41[*]

Data are from Waller and London (27). Each value is the mean ± S.E.M. for 7-8 mice. [a]V_{max} was determined by Eadie-Hofstee analysis of enzyme activities obtained in incubations with varying concentrations of acetyl coenzyme A. Values did not differ from those obtained when the concentration of acetyl coenzyme A was held constant but the choline chloride concentration was varied. [*]Significantly different from previous ages, P ≤ 0.05.

was noted between 18 and 24 mo. V_{max} for cortical ChAT was increased by 29% at 18 mo, and again by 20% at 24 mo. As in the cerebral cortex, striatal B_{max} for [^{3}H]QNB was reduced concomitant with an increase in the V_{max} for ChAT between 18 and 24 mo. In the hippocampus, the V_{max} for ChAT was increased between 18 and 24 mo of age. Unlike the effects observed in the cortex and striatum, the hippocampus showed an increase in B_{max} between 18 and 24 mo of age.

In postmortem samples of four brain areas from people who had AD, ChAT activity was significantly reduced by 38% to 77% as compared with corresponding control samples (data not shown). Preliminary studies (Fig. 1) indicated that B_{max} for [^{3}H]QNB was not significantly different between AD and control samples in the temporal, frontal, and postcentral-parietal areas, although B_{max} generally was lower in the AD samples. In the hippocampal area, B_{max} was 77% greater in the AD samples as compared with control. This difference did not reach statistical significance ($p \leq 0.10$).

DISCUSSION

The results obtained in these studies demonstrate that age-related effects on cholinergic neurochemical parameters vary among species. Effects on muscarinic binding and ChAT in rat brains generally are consistent with age-dependent regional declines in cholinergic neurotransmission. The major effects in this regard were age-related decreases in B_{max} for [^{3}H]QNB and in V_{max} for ChAT. Discrepancies between effects in Fischer-344 and Wistar rats occurred when there was a decline in one strain and no change in the other. These findings in rats sharply contrast with those obtained in C57BL/6J mice. In this mouse strain, B_{max} for [^{3}H]QNB was reduced in the cortex and striatum and elevated in the hippocampus. V_{max} for ChAT was increased in all three brain regions.

The data obtained in the mouse brain demonstrate a regional variation in the effect of age on relations between muscarinic binding and ChAT. Elevations in ChAT as well as [^{3}H]QNB binding in the aged hippocampus may reflect compensation for decrements in acetylcholine release, other aspects of cholinergic neurotransmission, or noncholinergic aspects of hippocampal function. The increase in cortical and striatal ChAT associated with an agedependent decline in [^{3}H]QNB binding may reflect presynaptic compensatory mechanisms to maintain cholinergic transmission despite decreased densities of muscarinic receptors.

Conversely, the tendency for an increased density of muscarinic binding sites in the hippocampal area and not in the temporal, frontal or postcentral-parietal areas of AD brain suggests an upregulation to compensate for presynaptic losses. Evidence for muscarinic receptor upregulation in the rodent brain has been presented by McKinney and Coyle (13). They reported that samples of neocortex from rats

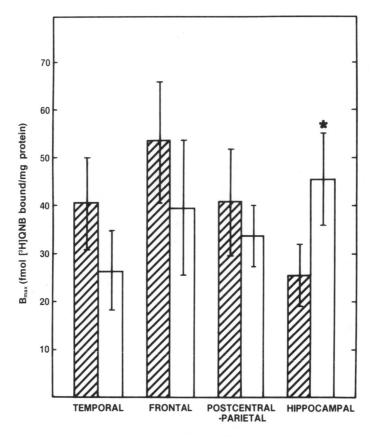

Figure 1. Muscarinic binding (B_{max} for [^{3}H]QNB) in postmortem samples of four brain areas from people who had Alzheimer's disease (open bars, n = 20-25) and from age-matched controls (hatched bars, n = 13-16). *$P \leq 0.10$.

with ibotenic acid lesions of the nucleus basalis magnocellularis showed decrements in neocortical ChAT and an increased proportion of high affinity muscarinic agonist binding sites. In addition, Nordberg et al. (16) noted that, although there was no difference in mean densities of muscarinic binding sites in AD as compared with control human hippocampal samples, ChAT was negatively correlated with muscarinic binding in the AD hippocampus. These observations suggest an upregulation of muscarinic receptors in the AD hippocampus.

CONCLUSIONS

In summary, age effects on central neurochemical parameters and on the relations between these parameters vary when comparing two rat strains with the C57BL/6J mouse. Preliminary studies of postmortem samples from individuals who had AD present still another picture. These inconsistencies demonstrate the speciesspecificity of age effects, and support the view that extrapolations from data obtained in rodents to generalizations about cerebral senescence or senile dementia may yield inaccuracies.

REFERENCES

1. Ben-Barak, Y., Gazit, H. and Dudai, Y. (1980): Brain Res. 194:-249-253.
2. Briggs, R.S., Petersen, M.M. and Cook, P.J. (1982): Neurobiol. Aging 3:259-261.
3. Bull, G. and Oderfeld-Nowak, B. (1971): J. Neurochem. 19:935-947.
4. Coleman, G.L., Barthold, L.S., Osbaldiston, G.W., Foster, S.J. and Jones, A.M. (1977): J. Gerontol. 32:258-278.
5. Davies, P. (1979): Brain Res. 171:319-327.
6. Davies, P. and Verth, A.H. (1977): Brain Res. 138:385-392.
7. Freund, G. (1980): Life Sci. 26:371-375.
8. Ikeda, S.R., Aronstam, R.S. and Eldefrawi, M.E. (1980): Neuropharmacol. 19:575-585.
9. Lowry, O.H., Rosebrough, N.J., Farr, A.L. and Randall, R.J. (1951): J. Biol. Chem. 193:265-275.
10. MacKay, N.V., Davies, P., Dewar, A.J., and Yates, C.M. (1978): J. Neurochem. 30:827-839.
11. Maker, H.S., Lehrer, G.M., Siliders, D.J. and Weiss, C. (1973): Prog. Brain Res. 40:293-307.
12. McGeer, E.G. and McGeer, P.L. (1976): In Neurobiology of Aging (eds) R.D. Terry and S. Gershon, Raven Press, New York pp. 389-403.
13. McKinney, M. and Coyle, J.T. (1982): J. Neurosci. 2:97-105.
14. Miller, R.G., Jr. (1966): Simultaneous Statistical Inference, McGraw-Hill, New York, pp. 67-70.
15. Morin, A.M. and Wasterlain, C.G. (1980): Neurochem. Res. 5:-301-308.
16. Nordberg, A., Larsson, C., Adolfsson, R., Alafuzoff, I. and Winblad, B. (1983): J. Neural Transmission 56:13-19.
17. Nordberg, A. and Winblad, B. (1981): Life Sci. 29:1937-1944.
18. Pedhazur, E.J. (1982): Multiple Regression in Behavioral Research, CBS College Publishing, New York, p. 103.
19. Perry, E.K., Gibson, P.H., Blessed, G., Perry, R.H. and Tomlinson, B.E. (1977): J. Neurol. Sci. 34:257-265.
20. Russell, E.S. (1966): In Biology of the Laboratory Mouse (ed) E.L. Green, McGraw-Hill, New York, pp. 511-519.
21. Simon, R.G. and Klein, W.L. (1979): Proc. Natl. Acad. Sci., U.S.A. 76:4141-4145.
22. Snyder, S.H. (1976): In Basic Neurochemistry (eds) G. Siegel, R. Albers, R. Katzman and B. Agranoff, Little, Brown and Company, Boston, p. 203-217.
23. Strong, R., Hicks, P., Hsu, L., Bartus, R.T. and Enna, S.J. (1980): Neurobiol. Aging 1:59-63.

24. Unsworth, B.R., Fleming, L.H. and Caron, P.C. (1980): Mech. Aging
     Develop. 13:205-217.
25. Vijayan, V.K. (1977): Exp. Neurol. 12:7-11.
26. Waller, S.B., Ingram, D.K., Reynolds, M.A. and London, E.D.
     (1983): J. Neurochem. 41:1421-1428.
27. Waller, S.B. and London, E.D. (1983): Exp. Gerontol. 18:419-425.
28. White, P., Goodhardt, M.J., Keet, J.P., Hiley, C.R., Caraso,
     L.H., Williams, J.E.J. and Bowen, D.M. (1977): Lancet 1:668-671.
29. Yamamura, H.I. and Snyder, S.H. (1974): Proc. Nat. Acad. Sci.
     U.S.A. 71:1725-1729.

CHOLINERGIC HYPOFUNCTION AND DEMENTIA: RELATION OF NEURONAL LOSS IN DIFFERENT PARTS OF THE NUCLEUS BASALIS TO CORTICAL NEUROPATHOLOGICAL CHANGES

V. Bigl[1], T. Arendt[1], A. Tennstedt[2], and A. Arendt[3]

[1]Department of Neurochemistry
Paul Flechsig Institute of Brain Research
Karl Marx University
Leipzig, G.D.R.

[2]Pathological Institute
District Hospital of
Neurology and Psychiatry
Muhlhausen-Pfafferode, G.D.R.

[3]Department of Neuropathology
Pathological Institute
Karl Marx University
Leipzig, G.D.R.

INTRODUCTION

The severe deficiency in cases with senile and presenile dementia of the Alzheimer's type (SDAT) of the cholinergic cortical projection system from the population of large neurons in the basal forebrain referred to collectively by many authors as the Nucleus basalis of Meynert (NbM) (11) is now well established (18, 2). Loss or almost complete disappearance of these neurons in SDAT is reflected in the concomitant reduction of cholinergic cortical parameters (6). The NbM is involved in a number of behavioral reactions and lesions of the NbM show some similarities to the impairment of learning and memory (9) as well as the reduction of total electrical activity described in patients with SDAT (15). Residual cortical cholinergic activity seems to correlate well with the residual intellectual capacity revealed by psychometric tests (13).

From a neuropathological point of view neurofibrillar changes and formation of neuritic plaques together with severe brain atrophy and granulovacuolar degeneration of neurons are the classical diagnostic criteria in SDAT (1). Quantitative studies of neuritic plaques in several brain areas revealed a significant correlation between an index of the number of plaques in the brain and psychometric measures of the degree of dementia (4).

The present study was undertaken in order to investigate the possible relationship between these two parameters crucially involved in SDAT, loss of cholinergic neurons in the basal forebrain, and formation of neuritic plaques. Quantitative evaluation of the number of neuritic plaques in five cortical areas was correlated with the loss of neurons in different parts of the NbM, taking into consideration the topography of these projections as revealed in monkey (10) and rat (3).

MATERIAL AND METHODS

The five cases of SDAT (mean age 61.7 $\pm$ 1.2 years) were selected postmortem on the basis of their clinical history and neuropathological diagnosis.

Tissue blocks containing the NbM were formaldehyde-fixed, paraffin-embedded, cut in the coronal plane (20 μm), and stained with cresyl violet by standard techniques. Frozen sections (20 μm) of the cerebral cortex were stained by the von Braunmühl technique (5). In addition, tissue blocks of cerebral cortex were paraffin-embedded, cut (20 μm) and stained according to a Palmgreen modification (12). The NbM was divided into subpopulations, adapting Mesulam's classification in monkeys (10) to man (Fig. 1). The Ch1 group corresponds to the medial septal nucleus, the Ch2 group to the nucleus basalis neurons of the substantia innominata. The anterior part (Ch4a) appears posterior to the tuberculum olfactorium. Its rostro-caudal extent overlaps with the crossing of the anterior commissure. A rarefication of neural density and in many instances a vascular structure provides a demarcation between the medial (Ch4am) and lateral (Ch4al) part. The intermediate (Ch4i) subpopulation is characterized by the presence of the ansa peduncularis and the posterior sector (Ch4p) is situated posterior to the ansa peduncularis.

The total number of nucleolated NbM neurons was counted in the six subdivisions of the NbM in every 10th section throughout the entire length of the cell population, and expressed as per cent of control (five cases without mental symptoms, mean age 62.1 $\pm$ 1.6 years).

Number of plaques was counted within the hippocampus, the fronto-basal (area 11 according to Brodmann), frontodorsal (area 8), parietal (area 17), temporal (area 20) and occipital (area 17) cortex of each hemisphere. Twenty random fields measuring 1mm x 1mm were examined in each section.

RESULTS

The individual cases showed large variations in the overall neuronal loss in the NbM, in the total number of neuritic plaques, as well as in the distribution of plaques in the different cortical

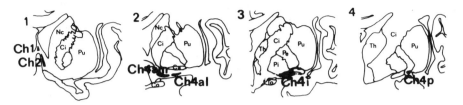

Figure 1.   Schematic frontal sections through the human forebrain.
1.   Section through the junction of the caudate and lenticular nuclei containing the nucleus septi medialis (Ch1) and the vertical limb of the nucleus of the diagonal band of Broca (Ch2).
2. Section through the anterior commissure containing the anterior medial (Ch4am) and anterior-lateral (Ch4al) part of the NbM in the substantia innominata.
3. Section through the cephalic limit of the thalamus containing the intermediate subdivision (Ch4i) of the NbM in the substantia innominata.
4.   Section situated posterior to the ansa peduncularis containing the posterior part of the NbM (Ch4p).
The nomenclature of the NbM subdivisions was adapted to the human brain based on studies in monkey by Mesulam (10).

Abbreviations:	Ca	Commissura anterior	Pe	Pallidum externum
	Ci	Capsula interna	Pi	Pallidum internum
	Co	Chiasma opticum	Pu	Putamen
	Nc	Nucleus caudatus	Th	Thalamus
			To	Tractus opticus

areas (Figs. 2 and 3).   In some cases differences in both, neuronal loss and plaque counts, were seen between the two hemispheres (Fig. 3).

The mean plaque count as calculated from samples of the five cortical areas investigated and the hippocampus showed a significant correlation with the reduction in number of the ipsilateral NbM neurons (Fig. 2).

The extent of the loss of NbM neurons was markedly different in the different subpopulations of the NbM complex (Fig. 3).

The frequency of neuritic plaques in the different cortical areas correlates significantly with the neuronal loss of those subpopulations of the NbM which, according to studies in monkey (10) and rats (3), give rise to the cholinergic innervation of the affected cortical areas (Figs. 4 and 5).   In the individual cases of SDAT the following relationship was apparent between the cortical area with the highest frequency of neuritic plaques and the subpopulation of the NbM with predominant neuronal loss:

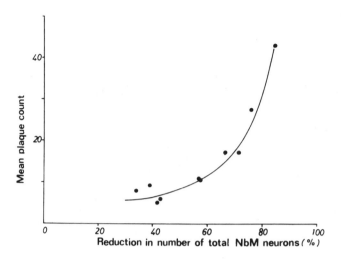

Figure 2. Relation between mean plaque count of the cerebral cortex
and reduction of the ipsilateral NbM neurons. The correlation was
statistically significant (quasilinear regression of ln y = ln b +
ax; r = 0.93; p < 0.001).

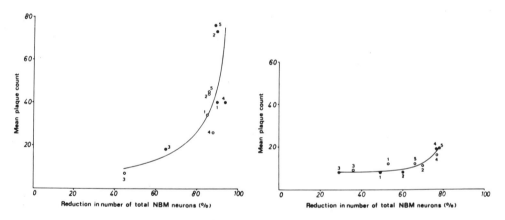

Figure 3.   Relation between plaque counts in different cortical
areas and neuronal loss in corresponding subpopulations of the NbM
within two individual cases of SDAT.
1. Frontobasal cortex -   Ch4al   right hemispheres
2. Frontodorsal cortex-   Ch4i   left hemispheres
3. Parietal cortex   -   Ch4am
4. Temporal cortex   -   Ch4p
5. Hippocampus   -   Ch1 and Ch2
The correlation was statistically significant (case 1: r = 0.8855;
p < 0.01; case 2: r = 0.7528; p < 0.05).

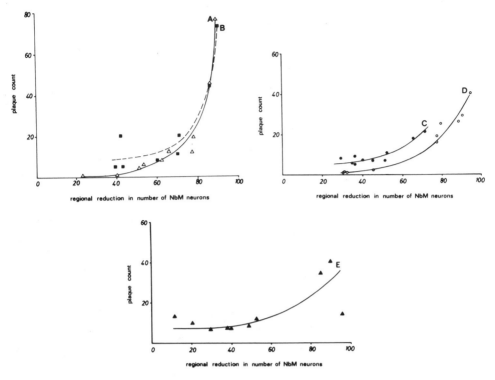

Figure 4.   Relation between plaque counts in different cortical areas (number of plaques x mm^{-2}) and reduction in number of neurons in the corresponding subpopulations of the NbM (in percent of controls).

A. Correlation between plaque count in the hippocampus and neuronal loss in Ch1 and Ch2 (△);
B. Correlation between plaque count in the frontodorsal cortex (area 8) and neuronal loss in Ch4i (■);
C. Correlation between plaque count in the parietal cortex (area 7) and neuronal loss in Ch4am (●);
D. Correlation between plaque count in the temporal cortex (area 20) and neuronal loss in cH4P (○);
E. Correlation between plaque count in the frontodorsal cortex (area 11) and neuronal loss in Ch4al (△).

The correlation is statistically significant (quasilinear regression analysis) with the following correlation coefficients (r): A: r = 0.69, p < 0.05; B: r = 0.87, p < 0.01; C: r = 0.88, p < 0.01; D: r = 0.94, p < 0.01; E: r = 0.65, p < 0.05.

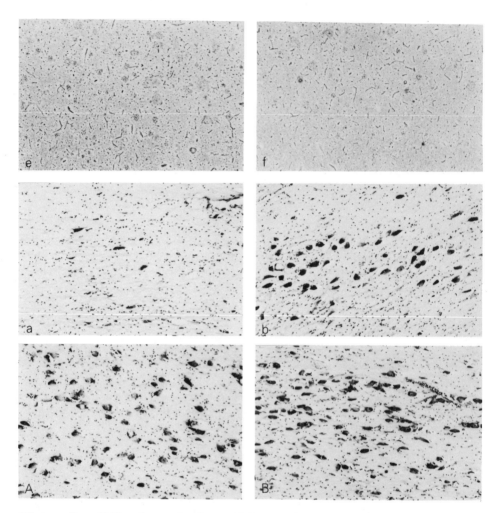

Figure 5. NbM subpopulations Ch4am and Ch4al and distribution of
plaques in their corresponding cortical target areas in a patient
with SDAT and in an age-matched control patient.

  A - D: Ch4am and Ch4al in a nondemented age-matched control
           patient
  a - d: Ch4am and Ch4al in a patient with SDAT

A,a Ch4am right hemisphere; B,b Ch4al right hemisphere; C,c Ch4am
left hemisphere; D,d Ch4al left hemisphere. Nissl stain x 80.

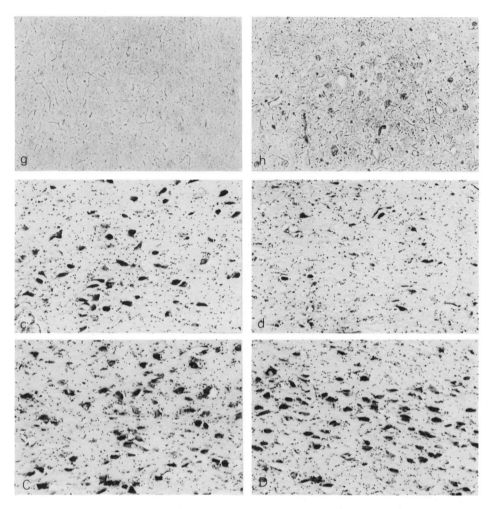

e - h: neuritic plaques in parietal and frontobasal cortex, which receive their cholinergic innervation mainly from Ch4am and Ch4al respectively, in a patient with SDAT.

e - right parietal cortex; f - right frontobasal cortex; g - left parietal cortex; h - left frontobasal cortex. von Braunmühl stain x 40. No plaques were demonstrated in the cerebral cortex of the control patient. Note the different involvement of the two hemispheres.

Hippocampus - nucleus septi medialis (Ch1) and vertical limb of the nucleus of the diagonal band of Broca (Ch2);

Parietal cortex (area 7) - anteriomedial part of NbM neurons in the substantia innominata (Ch4am);

Frontobasal cortex (area 11) - anteriolateral part of the NbM neurons in the substantia innominata (Ch4al);

Frontodorsal cortex (area 8) - intermediate subdivision of the NbM neurons in the substantia innominata (Ch4i);

Temporal cortex (area 20) - posterior part of the NbM (Ch4p).

The neuritic plaque count of the occipital cortex, which was widely differently affected in the cases of this study, could not be matched with neuronal loss of either one of the NbM subpopulations.

DISCUSSION

Our results demonstrate a significant correlation between the number of cortical neuritic plaques and the loss of cholinergic neurons in the ipsilateral NbM in cases of SDAT.

As was shown in the rat (3) and monkey (10), the cholinergic projection from the NbM is topographically well organized, and distinct sets of NbM neurons innervate predominantly certain cortical areas. The loss of neurons in these subpopulations of the NbM also correlates highly significantly with the number of neuritic plaques in the corresponding cortical target areas. Only the plaque counts in the occipital cortex which in the rat receives a dual innervation from different groups of NbM neurons (3) could not be correlated to neuronal counts in any of the different parts of the NbM studied here.

These results, suggesting a link between plaque formation and loss of cholinergic cortical innervation, give further support to the hypothesis of a cholinergic pathogenesis of neuritic plaques.

Electron microscopy has revealed that the plaques are composed of dystrophic neurites (19) intermingled with bundles of amyloid filaments and with non-neuronal reactive cells (17, 20). Most of the distended neurites have the morphological features of presynaptic terminals (8). Within the neuritic plaques AChE activity was demonstrated and a relation between the amount of histochemically demonstrable AChE activity within the plaques and the normal pattern of enzyme activity has been suggested (7). In addition, neuritic plaques in aged monkeys seem to react with monoclonal antibodies against the specific cholinergic marker choline acetyltransferase (14). The dystrophic neurites may, therefore, be regarded as degenerating intracortical terminals of NbM axons. The observation of a decrease in the initially high AChE activity during maturation of senile plaques in aged monkeys gives further support to this suggestion (16).

The nonlinearity of the relation between plaque count and neuronal loss within the NbM, which is typical for a degeneration process, may reflect two different modes of plaque formation. Either plaque formation is a self perpetuating process with an increasing rate, depending on the number of plaques already formed; or additional mechanisms might be induced during plaque formation. The slope of the correlation function is different for the various cortical regions and their corresponding NbM neurons, suggesting a different dependency of neuritic plaque formation from the neuronal loss in the NbM. This might reflect different density of cholinergic fibres within these areas, a different degree of collateralization of the fibres, or other factors not yet known.

Our results give no direct evidence regarding the nature of the plaques or their mode of formation. However, the significant correlation between the number of cortical plaques in SDAT and loss of neurons in the NbM, together with the well known correlation of both parameters with the psychometrically evaluated degree of dementia, suggest that both plaque formation and loss of cholinergic innervation are closely linked.

Thus, loss of cholinergic neurons in the NbM in cases of SDAT might account for both formation of neuritic plaques, the classical neuropathological lesion, and impairment of cognitive and intellectual function, due to loss of cortical cholinergic innervation.

REFERENCES

1. Alzheimer, A. (1907): Allg. Z. Psych. 64:146.
2. Arendt, T., Bigl, V., Arendt, A. and Tennstedt, A. (1983): Acta neuropathol. 61:101-108.
3. Bigl, V., Woolf, N.J. and Butcher, L.L. (1982): Brain Res. Bull. 8:727-750.
4. Blessed, G., Tomlinson, B.E. and Roth, M.B. (1968): Br. J. Psychiatry 114:797-811.
5. von Braunmühl, A.Z. (1929): Z. ges. Neurol. 122:317-322.
6. Davies, P. and Maloney, A.J.F. (1976): Lancet II, 1403.
7. Friede, R.L. (1965): J. Neuropath. exp. Neurol. 24:477-491.
8. Gonatas, N.K., Anderson, W. and Evangelista, I. (1967): J. Neuropath. exp. Neurol. 26:29-39.
9. LoConte, G., Casamenti, F., Bigl, V., Milaneschi, E. and Pepeu, G. (1982): Arch. Ital. Biol. 120:176-188.
10. Mesulam, M.-M., Mufson, E.J., Levey, A.I. and Wainer, B.H. (1983): J. comp. Neurol. 214:170-197.

11. Meynert, T. (1872): In: Strickers Handbuch der Lehre von den Geweben des Menschen Vol. 2, Engelmann, Leipzig.

12. Palmgreen, A. (1960): Acta Zool. 41:239-265.

13. Perry, E.K., Tomlinson, B.E., Blessed, G., Bergmann, K., Gibson, P.H. and Perry, R.H. (1978): Br. J. Med. 2:1457-1459.

14. Price, D.L., Kitt, C.A., Hedreen, J.C., Whitehouse, P.J., Struble, R.G., Cork, L.C., Walker, L.C., Mobley, W.C., Salvaterra, P.M. and Wainer, B.H. (1983): This book.

15. Soininen, H., Partanen, J.V., Puranen, M. and Riekkinen, P. J. (1982): J. Neurol. Neurosurg. Psychiatry 45:711-714.

16. Struble, R.G., Cork, L.C., Whitehouse, P.J. and Price, D.L. (1982): Science 216:413-415.

17. Terry, R.D. and Wisniewski, H.M. (1970): In: Alzheimer's Disease and Related Conditions (eds.) Y.E.W. Wolstenholme and M. O'Connor, Churchill, London, pp. 145-168.

18. Whitehouse, P.J., Price, D.L., Struble, R.G., Clark, A.W., Coyle, J.T. and DeLong, M.R. (1982): Science 215:1237-1239.

19. Wisniewski, H.M., Sinatra, R.S. and Iqbal, I.Y. (1981): In: Aging and Cell Structure (ed) J.E. Johnson, Vol. 1, Plenum Press, New York, pp. 105-142.

20. Wisniewski, H.M. and Terry, R.D. (1973): In: Progress in Neuropathology (ed) H. M. Zimmerman, Vol. 2, Grune and Stratton, New York, pp. 1-26.

BASAL FOREBRAIN CHOLINERGIC SYSTEMS IN PRIMATE BRAIN: ANATOMICAL ORGANIZATION AND ROLE IN THE PATHOLOGY OF AGING AND DEMENTIA

D.L. Price[*], C.A. Kitt[*], J.C. Hedreen[*], P.J. Whitehouse[*], R.G. Struble[*], L.C. Cork[*], L.C. Walker[*], W.C. Mobley[*], G.D. Crawford[**], P.M. Salvaterra[**] and B.H. Wainer[***]

[*]Neuropathology Laboratory
The Johns Hopkins University
School of Medicine
Baltimore, Maryland U.S.A.

[**]Division of Neuroscience
City of Hope Research Institute
Duarte, California U.S.A.

[***]Department of Pathology
University of Chicago
Chicago, Illinois U.S.A.

INTRODUCTION

Over the past several years, there have been substantial advances in our understanding of cholinergic systems in the primate central nervous system. Cholinergic neurons in the medial septum, diagonal band of Broca (dbB), and nucleus basalis of Meynert (nbM) provide the major cholinergic innervation to the amygdala, hippocampus, and neocortex. These cholinergic cell groups, recently termed the Ch1-4 system (26), are affected in Alzheimer's disease (AD) and related disorders. In this review, we first discuss: the anatomical organization of the Ch1-4 system; evidence implicating this system in the pathology of AD and related disorders; and hypothetical models by which dysfunction and, eventually, death of these cells may account for some of the neurochemical/neuropathological changes observed in the brains of individuals with AD and related dementias.

Ch SYSTEM IN NORMAL PRIMATE BRAIN

Cytology. In the primate, the medial septum, nucleus of the dbB, and nbM have high levels of choline acetyltransferase (ChAT) activity (24) and contain moderately large bipolar and multipolar neurons which show acetylcholinesterase (AChE) activity and ChAT immunoreactivity (17, 21, 26, 28, 44, 45, 50). Based on cytoarchitectonic

235

features and patterns of connectivity, these neuronal populations have been divided by Mesulam and his coinvestigators into four cell groups - Ch1-4 (26). Cholinergic neurons contained within the medial septum (Ch1) and vertical limb of the dbB (Ch2) are the most rostral component of this system, cholinergic neurons of Ch3 are present in the horizontal limb of the dbB, and the largest part of the system is located in the nbM (Ch4).

Efferents. In primates, neurons of the Ch1-4 system project directly to cortex (13, 18, 19, 25, 27) and are topographically organized (26, 30, 49). Cholinergic projections have been studied by the use of labeling techniques in which markers are injected into targets of the Ch1-4 system and, subsequently, are retrogradely transported to cell bodies of origin. Neurons of Ch1 and Ch2 project predominantly to hippocampus, Ch3 projects to the olfactory bulb, and Ch4 projects to amygdala and neocortex (10, 13, 18, 19, 22, 26, 49). Rostral portions of Ch4 project to prefrontal neocortex, while more caudal cells project to more posterior areas of neocortex; medial groups project to medial neocortex, and lateral groups project to lateral cortex. Caudolateral portions of Ch4 project to temporal lobe, while neurons in the posterior part of Ch4 project to visual neocortex.

The topography of Ch1-4 projections has also been analyzed by injecting [^{3}H]amino acids in proximity to cell bodies of the Ch1-4 cell group (18). Incorporated into newly synthesized proteins, the radiolabel is delivered by anterograde axonal transport to distal terminals, and autoradiography allows visualization of trajectories and terminal fields of labeled Ch1-4 system axons. Using this approach, Kitt and coworkers (22) have delineated four major Axonal trajectories of the Ch1-4 system: Ch1 and Ch2 project via the fornix to the hippocampus; lateral Ch4 neurons project to the amygdala by ventral pathways; medial Ch4 axons pass through the septum and cingulum to reach the medial neocortex; and the main body of Ch4 gives rise to axons which course beneath the basal ganglia and ascend through the external capsule into the corona radiata before innervating the dorsolateral neocortex. Immunocytochemistry using monoclonal antibodies against ChAT has confirmed these anatomical tracing methods; there are cholinergic axons in the septum, extreme capsule, cingulum, corona radiata, and within the laminae of neocortex (16, 21). Within the cortex, there is some regional variability in ChAT activity: the temporal lobe and motor cortex have the highest levels of enzyme activity; prefrontal cortex has intermediate levels with variation depending on the gyri sampled; and occipital neocortex has the lowest activities (45).

Afferents. Synapses on Ch1 neurons come from a variety of sources (40) including those from nerve cells in the lateral septum (47). Ch2 and Ch4 neurons receive afferents from the peripeduncular nucleus of the midbrain, the hypothalamus, the amygdala, and a variety of brainstem nuclei (18) (Kitt, Mitchell, DeLong and Price, personal

observation). In vitro receptor labeling techniques following autoradiography (55) can be used to map receptors of these various inputs
in the Ch system. Ultrastructural studies (50) of synapses afferent
to Ch4 neurons have shown that distal axondendritic and axospinous
synapses are frequently asymmetrical and have large round vesicles,
while axosomatic synapses on Ch4 neurons tend to be symmetrical
with either flat or pleomorphic vesicles.

Ch SYSTEM IN ALZHEIMER'S DISEASE AND RELATED DISORDERS

Ch1-4 neurons play important roles in regulating the activity
of neurons in the amygdala, hippocampus, and neocortex, structures
which are known to be important in memory, cognition, and behavior
(9, 35). Current evidence indicates that alterations in the Ch1-4
system contribute to the cognitive and amnestic deficits occurring
in individuals with AD (35, 37, 52) and in some patients with Parkinson's disease (3, 8, 34, 39, 51).

Markers of the Cholinergic System. Although a variety of trans-
mitter-associated abnormalities have been described in AD (12),
current evidence suggests that reductions in cholinergic markers
(activities of ChAT and AChE, high-affinity uptake of choline, and
synthesis of acetylcholine from [^{14}C]glucose) in the neocortex appear
to be the most severe, consistent, and perhaps earliest transmitter-
specific abnormalities occurring in these brains (4, 6, 7, 11, 38,
41).

Quantitative Analysis of Cholinergic Neurons. Studies from several
laboratories indicate that dysfunction/death of neurons are responsible, in part, for reductions in cholinergic markers in AD and related
disorders (3, 29, 35, 36, 48, 51, 53, 54).

Senile Plaques and Cholinergic Innervation. Senile plaques are
often present in high density (>15/mm^2) in the amygdala, hippocampus,
and neocortex of individuals with AD but are also seen in low density
(<5 per mm^2) in these regions in aged monkeys and aged humans (42,
43, 56). Plaques measure up to 200 Ìm in diameter and are composed
of a variety of elements: argentophilic neurites (axons and nerve
terminals frequently containing filaments and degenerating membranous
organelles); extracellular amyloid (polypeptide fibrils in a B-pleated
sheet configuration) which stains with Congo red dye and thioflavin-T;
and variable numbers of astroglia and microglia in the surrounding
neuropil. Plaques appear to represent foci of axonal/synaptic re-
organization with ongoing degeneration (and possible sprouting) of
axon terminals.

In individuals with AD (15, 32, 33, 46) and in aged nonhuman
primates (46), some neurites in plaques contain AChE activity.
Because AChE is enriched in axons and neurons of the Ch1-4 complex
and because lesions of the Ch1-4 complex result in a marked decrease

in cortical AChE activity, we hypothesized that some AChE-containing neurites in plaques may be derived from the Ch1-4 system. To test this hypothesis, we have begun to examine the character, topography, sources, and transmitter specificities of neurites in plaques which first appear in rhesus monkeys 17-20 years of age. The earliest plaques are rich in AChE-containing neurites and contain little amyloid. Amyloid plaques show fewer neurites and abundant amyloid (42). Sources of neurites in plaques can be assessed by injection of [^{3}H]amino acids into the basal forebrain (or other regions), followed by autoradiography to identify labeled neurites. If abnormal axons are derived from injected regions, radiolabel should be present in these neurites. The transmitter specificity of neurites can be assessed by immunocytochemical methods, and preliminary results suggest that ChATimmunoreactive neurites are present in plaques of aged nonhuman primates (20, 23).

In AD, there are correlations between cognitive deficits, plaque densities, and reductions in ChAT activity (5, 31). In monkeys, plaque densities appear to be better predictors of impaired performance on specific tasks than chronological age (Struble, Price Jr., Cork, and Price, personal observations). It is not surprising that neuro-pathological abnormalities, such as plaques, may be more important in the expression of behavioral abnormalities than age per se. Since plaques appear at different rates in different regions of the brain in individual animals (42), differences in plaque densities among individual animals may account for some of the variabilities in performance in individual animals of a given age.

MODELS OF PATHOLOGY INVOLVING THE Ch SYSTEM IN DEMENTIA

The Ch1-4 system appears to be important in normal cognitive and memory functions. When the system is altered in diseases, such as AD, these cognitive and memory processes are compromised. Pre-liminary studies in animals with partial Ch1-4 lesions are consistent with this concept - animals show memory abnormalities which are more readily demonstrated when animals are treated with agents which block muscarinic cholinergic receptors (1, 14). In AD and other forms of dementia associated with cortical cholinergic deficiencies, processes leading to dysfunction and death of neurons in the Ch1-4 system are not known. Several models of pathogenesis have been suggested: destruction of Ch1-4 nerve cells with little evidence of chronic structural changes, e.g., plaques and tangles; ongoing dysfunction of nerve cells manifest in distal terminals (neurites of plaques) and in the cell body (neurofibrillary tangles); and primary disease involving forebrain target neurons with associated changes of trophic influences on other neuronal populations which show secondary degeneration.

- In the first model, cholinergic neurons (and other cell groups) dysfunction and die, and there is little

evidence of their participation in the formation of tangles
or plaques. This type of pathological process occurs in
some demented individuals with Parkinson's disease (3,
8, 39, 52).

- In the second model, the disease process selectively
affects certain neurons, including those in the Ch1-4
system. An important expression of dysfunction in these
cells is the formation of abnormal axons and nerve terminals
(distal axonopathy) in the forebrain. As these neurons
begin to function abnormally, they may show impaired abilities to synthesize or deliver (by axonal transport) specific
proteins destined for distal axons/terminals. Alterations
in the biology of the cytoskeleton of affected neurons
are associated with the formation of neurites, and ongoing
degenerative/regenerative changes in axons are the structural
substrate of this distal axonopathy. With time, some of
these neurites disappear; if sprouting of remaining axons
cannot compensate for these changes, specific transmitter
markers are reduced in the affected area. Eventually,
dysfunctional cells, which may show neurofibrillary tangles,
die. This model would explain some of the changes occurring
in AD and in some cases of Parkinson's disease.

- In the third model, changes in axon terminals of this
system are secondary to primary processes occurring in
cortex. A pathogenetic process (e.g., local toxin) or
transmissible agent could cause focal damage to axons/
terminals --the result is the formation of a senile plaques.
A consequence of damage to distal axon terminals is eventually associated with retrograde changes (reduction in
size and number) in neurons projecting to cortex. This
model would also explain some of the abnormalities occurring
in AD.

- In the fourth model, the disease could primarily affect
target neurons, causing these cells to alter production
of trophic signals, which, in turn, cause changes in nerve
cells dependent on these trophic factors (2). Depending
on the nature of the trophic signal, cells dependent on
signals could show degenerative or regenerative responses.
Since cortical neurons are reduced in AD, a process of
this type could contribute to the changes in subcorti-
cal/cortical systems which occur in AD.

Whatever the mechanism, in Alzheimer-type dementia, the eventual
result appears to be an alteration in synaptic interactions in the
amygdala, hippocampus, and neocortex. Clarification of the nature
and evolution of these processes is an essential prerequisite for

understanding the pathogenesis of AD and certain other types of dementia.

Acknowledgements. The authors thank Drs. John W. Griffin, Joseph T. Coyle, John Lehmann, Arthur W. Clark, Susan J. Mitchell, and Russell T. Richardson for helpful discussions of this work. Ms. Nancy Cook provided excellent assistance in the preparation of this manuscript.

These studies were supported by grants from the U.S. Public Health Service (NIH NS 15721, NS 10580, AG 03359, NS 07179, NS 17661 and RR), The Commonwealth Fund, the McKnight Foundation, the Alfred P. Sloan Foundation, and the Alzheimer's Disease and Related Disorders Association. Dr. Cork is the recipient of a Research Career Development Award (NS 00488).

REFERENCES

1. Aigner, T., Aggleton, J., Mitchell, S.J., Price, D., DeLong, M. and Mishkin, M. (1983): Soc. Neurosci. Abstr. $\underline{9}$:826.
2. Appel, S.H. (1981): Ann. Neurol. $\underline{10}$:499-505.
3. Arendt, T., Bigl, V., Arendt, A. and Tennstedt, A. (1983): Acta Neuropathol. (Berl) $\underline{61}$:101-108.
4. Bird, T.D., Stranahan, S., Sumi, S.M. and Raskind, M. (1983): Ann. Neurol. $\underline{14}$:284-293.
5. Blessed, G., Tomlinson, B.E. and Roth, M. (1968): Br. J. Psychiatry $\underline{114}$:797-811.
6. Bowen, D.M., Benton, J.S., Spillane, J.A., Smith, C.C.T. and Allen, S.J. (1982): J. Neurol. Sci. $\underline{57}$:191-202.
7. Bowen, D.M., Smith, C.B., White, P. and Davison, A.N. (1976): Brain $\underline{99}$:459-496.
8. Candy, J.M., Perry, R.H., Perry, E.K., Irving, D., Blessed, G., Fairbairn, A.F. and Tomlinson, B.E. (1983): J. Neurol. Sci. $\underline{59}$:277-289.
9. Coyle, J.T., Price, D.C. and Delong, M.R. (1983): Science $\underline{219}$:1184-1190.
10. Daitz, H.M. and Powell, T.P.S. (1954): J. Neurol. Neurosurg. Psychiatry $\underline{17}$:75-82.
11. Davies, P. and Maloney, A.J.F. (1976): Lancet $\underline{2}$:1403.
12. Davies, P. and Terry, R.D. (1981): Neurobiol. Aging $\underline{2}$:9-14.
13. Divac, I. (1975): Brain Res. $\underline{93}$:385-398.
14. Flicker, C., Dean, R.L., Watkins, D.L., Fisher, S.K. and Bartus, R.T. (1983): Pharmacol. Biochem. Behav. $\underline{18}$:973-981.
15. Friede, R.L. (1965): J. Neuropathol. Exp. Neurol. $\underline{24}$:477-491.
16. Hedreen, J.C., Bacon, S.J., Cork, L.C., Kitt, C.A., Crawford, G.D., Salvaterra, P.M. and Price, D.L. (1983): Neurosci. Lett. $\underline{43}$:173-177.
17. Hedreen, J.C., Struble, R.G., Whitehouse, P.J. and Price, D.L. (1984): J. Neuropathol. Exp. Neurol. $\underline{43}$:1-21.

18. Jones, E.G., Burton, H., Saper, C.B. and Swanson, L.W. (1976): J. Comp. Neurol. 167:385-420.

19. Kievit, J. and Kuypers, H.G.J.M. (1975): Science 187:660-662.

20. Kitt, C.A., Becher, M.W., Wainer, B.H., Mobley, W.C. and Price, D.L. (in preparation): Trajectories of cholinergic axons derived from neurons in the basal forebrain.

21. Kitt, C.A., Hedreen, J.C., Bacon, S.J., Becher, M.W., Salvaterra, P.M., Levey, A.I., Wainer, B.H. and Price, D.L. (1983): Soc. Neurosci. Abstr. 9:966.

22. Kitt, C.A., Price, D.L., DeLong, M.R., Struble, R.G., Mitchell, S.J. and Hedreen, J.C. (1982): Soc. Neurosci. Abstr. 8:212.

23. Kitt, C.A., Price, D.L., Struble, R.G., Cork, L.C., Becher, M.W., Wainer, B.H. and Mobley, W.C. (in press): Evidence for the cholinergic innervation of neuritic plaques.

24. McKinney, M., Struble, R.G., Price, D.L. and Coyle, J.T. (1982): J. Neurosci. 7:2363-2368.

25. Mesulam, M.-M. and Van Hoesen, G.W. (1976): Brain Res. 109: 152-157.

26. Mesulam, M.-M., Mufson, E.J., Levey, A.I. and Wainer, B.H. (1983): J. Comp. Neurol. 214:170-197.

27. Mesulam, M.-M., Mufson, E.J., Levey, A.I. and Wainer, B.H. (1983): In: Banbury Report 15: Biological Aspects of Alzheimer's Disease (ed) R. Katzman, Cold Spring Harbor Laboratory Publications Vol. 15, pp. 79-93.

28. Nagai, T., McGeer, P.L., Peng, J.H., McGeer, E.G. and Dolman, C.E. (1983): Neurosci. Lett. 36:195-199.

29. Nakano, I. and Hirano, A. (1982): J. Neuropathol. Exp. Neurol. 41:341.

30. Pearson, R.C.A., Gatter, K.C., Brodal, P. and Powell, T.P.S. (1983): Brain Res. 259:132-136.

31. Perry, E.K., Perry, R.H., Gibson, P.H., Blessed, G. and Tomlinson, B.E. (1977): Neurosci. Lett. 6:85-89.

32. Perry, E.K., Tomlinson, B.E., Blessed, G., Bergmann, K., Gibson, P.H. and Perry, R.H. (1978): Br. Med. J. 2:1457-1459.

33. Perry, R.H., Candy, J.M. and Perry, E.K. (1983): In: Banbury Report 15: Biological Aspects of Alzheimer's Disease, (ed) R. Katzman, Cold Spring Harbor Laboratory Publications, Vol. 15, pp. 351-361.

34. Price, D.L., Struble, R.G., Whitehouse, P.J., Kitt, C.A. and Cork, L.C. (in press): Drug Dev. Res.

35. Price, D.L., Whitehouse, P.J., Struble, R.G., Clark, A.W., Coyle, J.T., DeLong, M.R. and Hedreen, J.C. (1982): Neurosci. Comment. 1:84-92.

36. Price, D.L., Whitehouse, P.J., Struble, R.G., Coyle, J.T., Clark, A.W., DeLong, M.R., Cork, L.C. and Hedreen, J.C. (1982): Ann. NY Acad. Sci. 396:145-164.

37. Price, D.L., Whitehouse, P.J., Struble, R.G., Price, D.L. Jr., Cork, L.C., Hedreen, J.C. and Kitt, C.A. (1983): In: Banbury Report 15: Biological Aspects of Alzheimer's Disease, (ed)

R. Katzman, Cold Spring Harbor Laboratory Publications Vol. 15, pp. 65-77.

38. Rossor, M.N., Emson, P.C., Iverson, L.L., Mountjoy, C.Q., Roth, M., Fahrenkrug, J. and Rehfeld, J.F. (1982): In: Alzheimer's Disease: A Report of Progress in Research (Aging, Vol. 19) (eds) S. Corkin, K.L. Davis, J.H. Growdon, E. Usdin and R.J. Wurtman, Raven Press, New York, pp. 15-24.

39. Ruberg, M., Ploska, A., Javoy-Agid, F. and Agid, Y. (1982): Brain Res. 232:129-139.

40. Segal, M. and Landis, S.C. (1974): Brain Res. 82: 263-268.

41. Sims, N.R., Bowen, D.M., Allen, S.J., Smith, C.C.T., Neary, D., Thomas, D.J. and Davison, A.N. (1983): J. Neurochem. 40: 503-509.

42. Struble, R.G., Cork, L.C., Price, D.L. Jr., Price, D.L. and Davis, R.T. (1983): Soc. Neurosci. Abstr. 9:927.

43. Struble, R.G., Cork, L.C., Whitehouse, P.J. and Price, D.L. (1982): Science 216:413-415.

44. Struble, R.G., Hedreen, J.C., Whitehouse, P.J. and Price, D.L. (Submitted for publication): The topography of acetylcholines- terase- rich neurons in the human basal forebrain.

45. Struble, R.G., Lehmann, J., Mitchell, S.J., Cork, L.C., Coyle, J.T., Price, D.L. and DeLong, M.R.: This book.

46. Struble, R.G., White, C.L., Clark, A.W., Whitehouse, P.J., Cork, L.C. and Price, D.L. (1982): J. Neuropathol. Exp. Neurol. 41:364.

47. Swanson, L.W. and Cowan, W.M. (1979): J. Comp. Neurol. 186: 621-655.

48. Tagliavini, F. and Pilleri, G. (1983): Lancet 1:469-470.

49. Tigges, J., Tigges, M., Cross, N.A., McBride, R.L., Letbetter, W.D. and Anschel, A. (1982): J. Comp. Neurol. 209:29-40.

50. Walker, L.C., Tigges, M. and Tigges J. (1983): J. Comp. Neurol. 217:158-166.

51. Whitehouse, P.J., Hedreen, J.C., White, C.L. III, and Price, D.L. (1983): Ann. Neurol. 13:243-248.

52. Whitehouse, P.J., Hedreen, J.C., Jones, B.E. and Price, D.L. (1983): Ann. Neurol. 14:149-150.

53. Whitehouse, P.J., Price, D.L., Clark, A.W., Coyle, J.T. and DeLong, M.R. (1981): Ann. Neurol. 10:122-126.

54. Whitehouse, P.J., Price, D.L., Struble, R.G., Clark, A.W., Coyle, J.T. and DeLong, M.R. (1982): Science 215:1237-1239.

55. Whitehouse, P.J., Wamsley, J.K., Zarbin, M.A., Price, D.L., Tourtellotte, W.W. and Kuhar, M.J. (1983): Ann. Neurol. 14: 8-16.

56. Wisniewski, H.M. and Terry, R.D. (1973): Prog. Brain Res. 40:167-186.

CELL COUNTS IN THE NUCLEUS BASALIS OF MEYNERT AND

THE SUPRAOPTIC NUCLEUS IN ALZHEIMER-DISEASED BRAINS

P. Etienne, Y. Robitaille[*] and A. Parent[**]

Douglas Hospital Research Centre
6875 Lasalle Boulevard
Verdun, Quebec, Canada H4H 1R3

[*]Montreal Neurological Institute
3801 University Street
Montreal, Quebec
Canada H3A 2B4

[**]Laboratoire de Neurobiologie
Hospital de l'Enfant-Jesus
1401, 18e rue
Quebec, Quebec
Canada G1J 1Z4

INTRODUCTION

In autopsy studies of Alzheimer's disease (AD) cases, a profound reduction of choline acetyltransferase (ChAT) activity in cerebral cortex and hippocampus brain homogenates has been observed by several groups. The loss of ChAT activity correlates with intellectual impairment and degree of histopathological change. In brain biopsy studies, Bowen and his group have shown that the ChAT deficiency was accompanied by loss of acetylcholine synthesizing ability regardless of substrate concentration (11). Catecholaminergic and GABAergic systems have also been implicated in this condition although their functional alterations are so far generally considered to be smaller and only found in a fraction of AD pa- tients. Recently, Bowen and his group reported that the uptake of serotonin, the content of indoleamines and the number of serotonin receptors are reduced in Alzheimer brains (3). Finally, somatostatin-like immunoreactivity is reduced in brain (4, 9) and CSF (17) of Alzheimer patients. Clearly, a cholinergic functional deficit in AD is now well-established but its importance relative to other neurotransmitter deficits is now being questioned.

Cell loss is quite a widespread phenomenon in AD. It is very prominent in the neocortex (12) and hippocampus (1). The locus ceruleus is affected in a large number of advanced cases of AD (2,

243

13). There may also be loss of 5-HT containing nerve cells, situated in the raphe nuclei of the brain stem (5). However, the precise relationship between functional cortical changes in NE or 5-HT parameters and loss of neuronal cell bodies is not yet clear. Similarly, a precise relationship between biochemical cholinergic changes and morphological abnormalities of the cholinergic system is still not clear. It has been shown, at least in animals, that the principal neuronal sources of cholinergic projections to the neocortex and hippocampus are the substantia innominata (including the nucleus basalis of Meynert [NBM]) and the medial septal nuclei (MSN)/nucleus of the diagonal band of Broca (NDB), respectively. The NBM contains clusters of neurons that can be recognized on the basis of their large size, abundant Nissl material and acetylcholinesterase (AChE) activity (16). Degeneration and loss of large neurons within the region of the SI in AD had been observed as early as 1954 (14) but more systematic investigations were only undertaken recently.

Whitehouse et al. first reported a 90% cell loss of large neurons of the NBM in a patient with a familial form of AD (15) and later a cell loss of over 75% in the NBM of 5 AD patients who were compared to age-matched controls (16). Using Nissl stained 15 μ thick sections they did total counts of large NBM neurons within recognizable anatomical boundaries and also estimated maximum density of those large neurons. The criteria for counting were: diameter > 30 μ, abundant Nissl substance and visible nucleolus.

Perry et al. reported only a 33% cell loss in the NBM of 6 AD patients compared to 5 controls (8). Their results are difficult to compare with those of Price's group because: 1) their patients were different (mean age 63.8 versus 82.5); 2) their counting method was different (total counts versus a T square sampling procedure); and 3) their counting criteria, as described in their report, were not explicit.

Nagai et al., using a new immunocytochemical technique, reported a 66% loss of ChAT positive cells (total counts) in NBM-NDB of 3 AD cases, when these were compared with 3 controls (6). The degree of cell loss was comparable to that reported by Price but the total number of cells cannot be compared because only a fraction of large NBM neurons are positively stained by immunocytochemistry. They also reported that there was no loss of cells positively stained for ChAT in the putamen.

These three groups did not mention the severity of Alzheimer histopathological change, nor did they correlate it with cell loss in NBM or NDB.

In this study, we have attempted to evaluate the relationship between cell loss in the NBM and several variables: 1) severity of Alzheimer classical neuropathological changes; 2) severity of cell loss in a geographically close structure: the supraoptic nucleus (SON); 3) laterality; and 4) age of onset of illness.

MATERIALS AND METHODS

Morphometric studies were performed on 9 control cases (no history of dementia or neurologic illness; mean age $\pm$ SD = 75.8 $\pm$ 9.9 years) and 10 AD cases (mean age $\pm$ SD = 78.3 $\pm$ 10.5). Of these 10 AD cases, 3 had the characteristic early onset and rabid course of presenile dementia. The diagnosis of AD was confirmed by the presence of abundant senile plaques and/or neurofibrillary tangles in the hippocampus and neocortex, and granulovacuolar degeneration in hippocampus. Substantia nigra was examined in all cases but locus ceruleus cell counts are not yet available at the time of this writing. The severity of Alzheimer changes (global evaluation of frequency of plaques and tangles) and the severity of cerebro-vascular changes was rated on a 0 - 3 scale (none, light, moderate, severe). Two late onset AD cases also had histopathological changes compatible with a diagnosis of multiinfarct dementia (MID).

A brain slab of about 3 cm. thickness was dissected out from 1 hemisphere through a transverse section made 1 cm. rostral to the level of the anterior aspect of the optic chiasm, and another along the posterior end of the mammillary bodies. This brain slab contained the septal nuclei, the nucleus of the diagonal band of Broca, the nucleus basalis of Meynert and a large portion of the basal ganglia. It was kept in a fixative containing 10% formaldehyde, 0.9% NaCl, 3% sucrose and 10% dimethylsulfoxide for 24 to 36 hr. as suggested by Rossor et al (10). After rinsing in 10% sucrose and sinking in 30% sucrose, the tissue block was sectioned in a transverse plane with a freezing microtome. Two to three 20 μ thick sections were taken at 400 μ intervals; one served for the demonstration of AChE pattern according to the histochemical procedure of Karnovsky and Roots (results not presented here), whereas the other two were stained with cresyl violet and Bielschowsky respectively.

Cell counts were performed by four observers, not blind to the diagnosis. For NBM cell counts, they examined all sections obtained but only counted large neurons in the sections caudal to the anterior commissure crossing the mid line. The area examined is limited medially by the optic tract and laterally by the posterior limb of the anterior commissure. All neurons with a diameter larger than 30 μ, a visible nucleolus and abundant "Nissl" substance were counted according to Whitehouse and his coworkers. The correlation between observers was > .75 and the correlation between two obser-vations by the same observer was > .90. One observer systematically achieved higher consistency between observations and only her results are presented here. For comparisons between cases, two methods were used: 1) comparing sections with the highest cell numbers; and 2) comparing averages of the four richest consecutive sections (to compensate for angularity of the "transverse" plane section). For SON cell counts, all sections with a visible supraoptic nucleus were examined and counted. This structure was found to be more vulnerable than the NBM to manipulation accidents (shearing, and so forth) so that we could not often examine more than 2 or 3 consecutive sections. All neurons with a diameter larger than 20 μ and a visible nucleolus were counted. For comparison between cases, two methods were used: 1) comparing sections with the highest cell numbers; and 2) comparing averages of the four richest (not necessarily consecutive) sections (to compensate for angularity of the "tranverse" plane).

RESULTS

In the NBM cell loss is striking in Alzheimer cases and always over 50%. There is no overlap between Alzheimer cases and controls. For total counts in the richest sections, the mean $\pm$ SEM was 88 $\pm$ 58 in AD cases and 447 $\pm$ 74 in controls (p < .001) and for averages of counts in the four richest consecutive sections, the mean $\pm$ SEM was 71 $\pm$ 46 in AD cases and 398 $\pm$ 84 in controls (p < .001). There does not seem to be any difference in cell loss severity in NBM between early and late onset cases when Alzheimer type changes are moderate to severe. The cell loss is striking in both left and right hemispheres (3 cases). We also counted the cells which did not meet Whitehouse's criteria but resembled typical large neurons. In AD cases, the percentage of large neurons which did not meet Whitehouse's criteria ranges from 75 to 125% of the "Whitehouse's criteria fitting" neurons but is much smaller in controls (from 10 to 30%).

In the supraoptic nucleus, cell density was comparable in AD and control cases: for total counts in the richest sections, the mean $\pm$ SEM was 202 $\pm$ 102 in Alzheimer cases and 219 $\pm$ 79 in control cases and for averages of counts in the four richest sections the mean $\pm$ SEM was 165 $\pm$ 74 in Alzheimer cases and 195 $\pm$ 64 in control cases. Tables 1 and 2 summarize the results obtained.

We also compared severity of Alzheimer changes with cell counts in the NBM. The results presented in Table 3 suggest that NBM cell loss increases with severity of classical AD neuropathological changes.

DISCUSSION

Our results are completely compatible with those of Whitehouse et al. We found that cell loss in NBM was symmetrical, and that

Table 1

Controls

Age	Alzheimer Changes	Vascular Changes	Nucleus Basalis of Meynert		Supraoptic Nucleus	
			Total Count Richest Section	$\overline{X}$+S.E.M. 4 Richest Consecutive Sections	Total Count Richest Section	$\overline{X}$+S.E.M. 4 Richest Sections
80	0	0	509	519±31 Left	384	316±27
82	1	0	514	467±29 Left	239	211±10
61	0	1	551	501±23 Left	125	120
63	0	0	409	346±27 Left	184	142±18
81	0	0	483	418±23 Left	---	---
88	1	3	311	270±17 Left	161	136±16
84	1	1	398	328±31 Right	225	212±10
60	0	0	437	380±32 Right	252	238± 7
77	0	0	415	352±28 Left	186	186

Table 2

Alzheimer Cases

Age	Illness Duration In Years	Alzheimer Changes	Vascular Changes	Neuro-Pathological Diagnosis	Nucleus Basalis of Meynert		Supraoptic Nucleus	
					Total Count Richest Section	$\overline{X}$±S.E.M. 4 Richest Consecutive Sections	Total Count Richest Section	$\overline{X}$±S.E.M 4 Richest Sections
68	> 10	3	0	Alz.	125	96±10 Left	132	123± 4
81	> 5	3	2	Alz.	36	24± 6 Right	145	124±12
60	> 10	3	0	Alz.	40	39± 2 Left	161	146± 5
75	> 5	3	0	Alz.	50	41± 5 Left	79	77± 4
83	> 10	3	1	Alz.	54	49± 2 Left	158	135±13
90	> 5	2	2	Alz.	136	121± 7 Right	238	196±18
80	> 5	3	0	Alz. Park.	23	12± 4 Left	335	265±24
					67	45±12 Right	264	260± 3
82	---	3	1	Alz.	198	129±34 Left	178	121±19
95	> 5	3	3	Alz. SNa. MID	142	135± 5 Left	---	---
					163	102±23 Right	200	182± 8
69	> 5	3	3	Alz. SNa. MID	80	61± 7 Left	396	303±34
					82	60± 5 Right	230	202±13

NOTE: 1) SNa = atrophy of substantia nigra; 2) MID = multi-infarct dementia; 3) Parkinson's disease (SNa + Lewy Bodies)

Table 3

NBM Cell Density (Mean of 4 Richest Sections)
Versus Severity of Alzheimer Changes

Severity	0	1	2	3
	519	467	121	96
	501	270		24
	346	328		39
	418			41
	380			12
	352			129
				135
				61
				49
$\bar{X}\pm SD$	419±75	355±101	121	65±45

a noncholinergic structure geographically close to the NBM is relatively well-preserved in that condition.

Large intrinsic cholinergic neurons which exist in human neostriatum are not affected in AD (6, 7, 10). This indicates that
AD does not equally affect all cholinergic cell groups. Also, relative
preservation of SON indicates that AD does not equally affect geographically close structures in basal forebrain. These findings
may have implications concerning the etiology of the illness.

Acknowledgements. The authors express their sincere gratitude to
P. Lemoyne and C. Csonka for their excellent technical help. This
research was made possible through grants MT-5781 and PG-22 of the
Medical Research Council of Canada.

REFERENCES

1. Ball, M.J. (1977): Acta Neuropathologica 37:111-118.
2. Bondareff, W., Mountjoy, D.Q. and Roth, M. (1981): Lancet I:
   783-784.
3. Bowen, D.M., Allen, S.J., Benton, M.J., Goodhardt, M.J., Haan,
   E.A., Palmer, A.M., Sims, N.R., Smith, C.C.T., Spillane, J.A.,
   Esiri, M.M., Neary, D., Snowden, J.S., Wilcock, G.K. and Davison,
   A.N. (1983): J. Neurochem. 41:266-272.
4. Davies, P. and Terry, R.D. (1981): Neurobiol. Aging 2:9-14.

5. Mann, D.M., Yates, P.O. (1983): J. Neurol. Neuro surg. and Psych. 46:96-98.
6. Nagai, T., McGeer, P.L., Peng, J.H., McGeer, E.G. and Dolman, C.E. (1983): Neurosci. Lett. 36:195-199.
7. Parent, A., Csonka, C. and Etienne, P. (1984): Brain Research 291:154-158.
8. Perry, R.H., Candy, J.M., Perry, E.K., Irving, D., Blessed, G., Fairbairn, A.F. and Tomlinson, B.E. (1982): Neurosci. Lett. 33:311-315.
9. Rossor, M.N., Emson, P.C., Mountjoy, C.Q., Roth, M. and Iversen, L.L. (1980): Neurosci. Lett. 20:373- 377.
10. Rossor, M.N., Svendsen, C., Hunt, S.P., Mountjoy, C.Q., Roth, M. and Iversen, L.L. (1982): Neurosci. Lett. 28:217-222.
11. Sims, N.R., Bowen, D.M., Davison, A.N. (1981): Biochemical Journal 196:867-876.
12. Terry, R.D., Peck, A., DeTeresa, R., Schechter, R., Horonpian, D.S. (1981): Annals of Neurology 10:184-192.
13. Tomlinson, B.E., Irving, D. and Blessed, G. (1981): J. Neurol. Sci. 49:419-428.
14. Von Buttlar-Brentano, K. (1954): Journal fur Hirnforschung Bd. 1, Heft 4/5, p. 407.
15. Whitehouse, P.J., Price, D.L., Clark, A.W., Coyle, J.T. and De Long, M.R. (1981): Ann. Neurol. 10:122-126.
16. Whitehouse, P.J., Price, D.L., Struble, R.B., Clark, A.W., Coyle, J.T. and De Long, M.R. (1982): Science 15:1237-1239.
17. Wood, P.L., Etienne, P., Lal, S., Gauthier, S., Cajal, S. and Nair, N.P.V. (1982): Life Sci. 31:2073-2079.

SYMPATHETIC INGROWTH: A RESULT OF CHOLINERGIC

NERVE INJURY IN THE ADULT MAMMALIAN BRAIN

J.N. Davis

Neurology Research Laboratory
V.A. Medical Center
and
Departments of Medicine and Pharmacology
Duke University Medical Center
Durham, North Carolina 27705 U.S.A.

INTRODUCTION

The brain is capable of dramatic and vigorous responses to injury. We have been studying one such rearrangement, sympathetic ingrowth, which takes place in the adult rat brain after injury to certain central cholinergic pathways. The purpose of this chapter is to describe sympathetic ingrowth, its regulation and function. Although sympathetic ingrowth appears to be an unusual and perhaps anomalous form of neuronal plasticity, its study has led to a better understanding of the molecular mechanisms that probably underlie the regulation of other neuronal rearrangements. The ability to regulate neuronal rearrangements such as sympathetic ingrowth is likely to lead to the development of pharmacological agents that may be useful to patients with brain injury, stroke or dementia.

DISCUSSION

Peripheral Synmpathetic Nerves Appear in the Brain. Sympathetic nerves were first seen in the rat hippocampal formation independently in two different laboratories (19, 34) after anterior hippocampal or septal injury. These fibers were clearly sympathetic since removing the superior cervical ganglia resulted in their disappearance. Furthermore sympathetic fibers had a different microscopic appearance than the noradrenergic locus coeruleus fibers in the hippocampus. Sympathetic fibers were thick, knobby, and easily identified under low power microscopy while central fibers were

fine, beaded, and usually required the high power objective of a
fluorescent microscope to be detected.

Sympathetic nerves are usually found in the intracranial cavity
on the surface of the brain innervating pial blood vessels lying
outside the blood brain barrier. After septal or hippocampal injury,
however, the nerves appear to grow out from the surface blood vessels
across the hippocampus to reach certain parts of the dentate and
hippocampal gyrus (6, 20). Sympathetic fibers are outside of the
blood-brain barrier in the normal brain, but after growing into
the hippocampus, the nerves come to lie inside this barrier (22).

Cholinergic Injury Initiates Sympathetic Ingrowth. Four pieces of
evidence suggested that sympathetic ingrowth in the rat hippocampus
is initiated by an injury to cholinergic septo-hippocampal neurons
(6). First, sympathetic ingrowth was caused by a lesion in the
medial septal nucleus or anywhere along the fornix and anterior
hippocampus where the septo-hippocampal pathway was found. Second,
lesions of other known afferents to the hippocampal formation did
not elicit sympathetic ingrowth. Third, the topography of sympathetic
ingrowth resembles the topography of the terminal projections of
septal neurons, suggesting a replacement of these fibers by the
ingrowing peripheral nerves. The final and most compelling evidence
came from animals with partial lesions of the medial septum and
only a partial loss of septo-hippocampal fibers. In these cases,
sympathetic fibers were found only in those areas of the hippocampus
where the septal projections had been lost.

Sympathetic ingrowth occurs in other areas besides the hippocampal formation. The same septal lesions that elicited sympathetic
ingrowth in the hippocampus also elicited sympathetic ingrowth into
the medial habenulae (8). The medial septum is known to send a
cholinesterase-containing projection to the medial septum by way
of the stria medularis. There is substantial evidence to suggest
that this pathway is cholinergic. Cholinesterase-containing neurons
in the substantia innominata of the rat have been shown to project
to the neocortex (25). Although widely scattered in the rodent,
they seem analogous to the more compact nucleus basalis in man.
Electrolytic lesion of these neurons in the rat is made difficult
because of their scattered locations. Nonetheless, large lesions
of the substantia innominata led to the appearance of sympathetic
fibers in the neocortex of the rat (4).

Yet not all central cholinergic pathways are involved in sympathetic ingrowth when lesioned. We recently studied the alteral
dorsal tegmental nucleus of the rat, a set of neurons lying just
medial and slightly rostral to the locus coeruleus containing abundant
acetylcholinesterase. These neurons seem to be cholinergic since
their lesion leads to the loss of choline acetyltransferase in the
thalamus and lateral septum to which the neurons are known to project

(31). Lesion of this nucleus does not lead to sympathetic ingrowth in the thalamus or the septum (Haring and Davis, unpublished observations).

Sympathetic ingrowth does not appear to be related to central noradrenergic neurons. Lesion of the locus coeruleus does not elicit sympathetic ingrowth nor does sympathetic ingrowth seem to be affected by the presence of central noradrenergic fibers. Since locus coeruleus neurons are highly collateralized and project to most brain regions in the rat, it is hard to place a lesion that does not interrupt some locus coeruleus fibers. These neurons also undergo a form of neuronal rearrangement after injury, but one that differs significantly from sympathetic ingrowth. Locus coeruleus fibers appear to expand after injury by a process that we and others have called "pruning", in which the injury to a collateral axon elicits expansion of remaining collaterals from the injured neuron. The pruning of locus coeruleus neurons takes about four to six months in the rat in contrast to the one month for sympathetic ingrowth to become apparent (24).

Thus the best evidence at present suggests that an injury to these cholinergic neurons found in the basal forebrain initiates sympathetic ingrowth. It is intriguing to speculate on possible relationships between these basal forebrain cholinergic neurons and sympathetic neurons. Some sympathetic nerves are cholinergic, those that innervate sweat glands, for example. Sympathetic nerves from neonatal rats can become either noradrenergic or cholinergic in vitro depending on the conditions under which they are grown (28). The cholinergic neurons innervating peripheral parasympathetic ganglion have their origin in the central nervous system. Basal forebrain cholinergic neurons could thus be related to parasympathetic structures such as the vagus or sacral parasympathetic neurons either in development or function.

Despite careful observations, however, there is no evidence that the sympathetic nerves growing into the central nervous system in apparent replacement of basal forebrain cholinergic nerves ever contain or release acetylcholine. Choline acetyltransferase levels do not increase in the hippocampus with the ingrowth of sympathetic nerves (Madison, Fox, and Davis, unpublished observations) and acetylcholinesterase staining is not present on the sympathetic fibers (Crutcher and Davis, unpublished observations).

Evidence that a Target Tropic Factor Regulates Sympathetic Ingrowth.
Although cholinergic neuronal injury initiates sympathetic ingrowth, a target factor seems to regulate the ingrowth. The first evidence for such a target factor came from an experiment in which the preganglionic input to the superior cervical ganglion was cut (5). Sympathetic neurons in the ganglia were not damaged but were deprived of any afferent input. A subsequent septal lesion was placed. Since there are projections from the septum to the lateral hypothalamus

and from there to the intermediolateral column of the spinal cord, it is possible that the signal regulating ingrowth may come from the septum, through afferent input to the superior cervical ganglia. However sympathetic ingrowth occurred even after the preganglionic nerves were cut, strongly suggesting that ingrowth was drawn into the brain by some factor from the target hippocampal formation.

There are two possible sources for a target factor regulating sympathetic ingrowth; the degenerating cholinergic fibers, or cellular elements in the target hippocampus. In order to differentiate between these two possibilities, Crutcher and Davis (9) studied rats who had been subjected to a prior entorhinal cortex lesion. Entorhinal cortex lesions were known to produce an expansion of the septo-hippo-campal pathway in the outer third of the molecular layer of the dentate gyrus (21, 35). Sympathetic ingrowth was not found, however, in the outer third of the molecular layer in septal lesioned animals. If degeneration of the cholinergic fibers were responsible for regulating ingrowth, then ingrowth into the outer molecular layer should have been seen in animals with a prior entorhinal lesion after a septal lesion. This experiment provided strong evidence that a cellular element in the hippocampal formation was responsible for producing the tropic factor that regulated sympathetic ingrowth.

Perhaps the best studied of tropic factors is Nerve Growth Factor (NGF). Proteins with NGF activity have been isolated from male mouse submaxillary gland, snake toxins, prostate, and seminal fluid (36). These proteins support and promote neurite outgrowth of sympathetic neurons and certain sensory neurons in vitro. When injected into animals, NGF elicited sympathetic ingrowth into the area of injection (18). One hypothesis to explain sympathetic ingrowth after cholinergic neuronal injury is that a NGF-like protein is produced by one of the cells in the hippocampus and is normally transported away by the cholinergic septo-hippocampal neurons. After injury to these neurons, the concentration of this factor would increase in the hippocampal formation drawing in sympathetic fibers. Evidence for this hypothesis is based on the observation that $[^{125}I]$-NGF injected into the normal rat hippocampus is transported away only by the septo-hippocampal neurons and not by locus coeruleus neurons (33). Coincidentally, submaxillary gland NGF is complexed with high concentrations of zinc (29), and high concentrations of zinc are found in the hippocampus (3).

Source of Cells Producing Sympathetic Ingrowth Regulating Factor.
There are several candidates for the cells producing sympathetic ingrowth regulating factor. Glial cells have been suggested (2) since glial cells have been shown to produce NGF-like activity in vitro. However, the characteristic pattern of ingrowth would not be explained by a glial factor since glial cells are ubiquitous throughout the hippocampus. The interneurons of the hippocampal

formation may be the physiological target of the septo-hippocampal projection could produce the factor (30). Their anatomy would fit reasonably well with the topography of ingrowth and would be consistent with the evidence that cholinergic lesions initiate the ingrowth. A final candidate is the mossy fiber axons of dentate granule cells.

It is not unreasonable to think that the mossy fibers might produce a NGF-like protein. They may have high concentrations of zinc. Furthermore, sympathetic ingrowth, while correlating with the cholinergic neuronal topography, correlates even better with the location of mossy fibers in the hippocampus (10). Granule cells and their mossy fibers can be selectively lesioned with colchicine. There has been conflicting evidence on the results of colchicine injection in the hippocampal formation. Crutcher and Davis (10) found that animals given a septal lesion one month after intra-hippocampal injection of colchicine did not show sympathetic ingrowth. Furthermore, if colchicine was administered after sympathetic ingrowth had occurred, a prompt regression was observed (Davis, unpublished observations). Conflicting results have been found by Loy and her colleagues (30) when colchicine was administered two weeks before the septal lesion. In that case ingrowth was still observed.

One possible explanation for the discrepancy is the difference in time between the colchicine administration and the septal lesion. Histochemical stains for heavy metals consistently show persistence of some mossy fiber staining for several weeks after colchicine. In our experiments ingrowth was seen in areas distant from the colchicine injection, where islands of granule cells and their mossy fibers had not been injured. This seems to suggest that a complete lesion is needed to abolish ingrowth. In these experiments (10, 30) toxic damage to other cellular constituents with kainic acid did not prevent ingrowth after septal lesions, suggesting that tissue damage alone did not prevent ingrowth.

Another experiment supporting the mossy fiber candidacy for the source of tropic factor was carried out in rats treated with thyroxine for two weeks at birth (12). When they reached adulthood, these thyroxine-treated animals had a number of central nervous system anomalies, one of which was an unusual mossy fiber pathway in the hippocampal gyrus. Normal rats have mossy fibers "above" the pyramidal cells in stratum radiatum of the hippocampal gyrus, while thyroxine-treated animals have two mossy fiber pathways, one above and one below the pyramidal cells. Sympathetic ingrowth does not occur normally in the area below the pyramidal cells, but in these thyroxine-treated animals sympathetic ingrowth occurred both above and below the pyramidal cells (12). This anomalous ingrowth closely paralleled the presence of anomalous mossy fibers, lending further indirect support to the candidacy of these fibers as a source of tropic factor.

It should be noted that these experiments are indirect. At present, the source of a target tropic factor remains elusive. Yet the evidence implicating the factor in the regulation on ingrowth seems good. In fact, NGF-like activity was recently detected in the rat hippocampal formation (7). Sympathetic ingrowth is one of the best examples of a tropic factor-regulated neuronal rearrangement in the adult brain.

Regulation of Sympathetic Ingrowth by Afferent Input, Age, and Sex. Although section of the preganglionic afferent input did not prevent sympathetic ingrowth, it had distinct effects on the histofluorescent appearance of the ingrowing fibers and their norepinephrine content. After deafferentation, sympathetic fibers growing into the hippocampal formation had less norepinephrine content and were not as easily visualized by catecholamine histofluorescent stains (23). However, if the preganglionic input was allowed to reinnervate the ganglion, norepinephrine content and histofluorescent appearance rapidly returned to the levels observed in animals without preganglionic interruptions. Furthermore, the ability of synaptosomes prepared from the hippocampus to take up norepinephrine remained the same in animals with deafferentation despite the lowered norepinephrine levels. Uptake is a function of the nerve terminal membrane while content is dependent on the presence of storage vesicles and to a certain extent on the amount of nerve impulse flow in the sympathetic neuron. The preservation of uptake during ganglion deafferentation strongly suggested that ingrowth is not affected by deafferentation, but that the ingrowing sympathetic nerves had less neurotransmitter per nerve terminal because of the reduced nerve impulse flow.

This dissociation of norepinephrine content from actual presence of nerve terminals is important to emphasize. Many experiments from our laboratory and others have relied solely on histofluorescence to study sympathetic ingrowth. While this is usually valid in a qualitative sense, it has serious limitations when quantitative results are seen. For example, after deafferentation the decrease in the number of fibers seen with histofluorescence is clearly due to the decrease in norepinephrine content and does not reflect a decrease in sympathetic ingrowth.

There has been one study of old rats after septal lesions in which, using histofluorescent staining, sympathetic ingrowth was observed in some but not all of the animals (32). While it seems reasonable to interpret the data as indicating that sympathetic ingrowth diminishes as a function of age, caution should be utilized until the studies are confirmed by measurements of norepinephrine uptake.

A study of sex differences in sympathetic ingrowth is also based on histofluorescent observations. Female rats were noted to have greater ingrowth than male rats when studied together (27).

Again, the data must be carefully interpreted because of the qualitative nature of histofluorescent observations. For example, Davis
and Martin did not see a difference in the histofluorescent appearance
of sympathetic ingrowth between male and female animals (12). We
also found no difference between males and females in measurements
of whole hippocampal norepinephrine content (Davis and Fox, unpublished
observations), but measurements of micro-disected dentate gyrus
may be necessary to detect small differences in sympathetic ingrowth
(23). In any event, the reason for the discrepancy in finding sex
differences with sympathetic ingrowth is not clear. It may be due
to different suppliers of the rat used in these studies.

Sympathetic Ingrowth and Recovery of Function. Sympathetic ingrowth
represents a replacement of cholinergic with noradrenergic nerves,
suggesting that this neuronal rearrangement may not replace the
lost function of the septo-hippocampal pathway. There have been a
number of studies looking at behavior in rats after septal lesions
with and without subsequent ganglionectomies (11, 13, 16). To date
only one of these has shown an effect of ganglionectomy (13). The
exception is a study of radial arm maze performance. This behavior
is sensitive to lesions of the hippocampal formation and its afferents. Medial septal lesions produce a transient disruption of this
behavior lasting two to three weeks. By contrast, animals with
superior cervical ganglionectomies at the time of the septal lesions
experience a shorter disruption lasting only one to two weeks (13)
suggesting that sympathetic ingrowth may have slowed "relearning"
of maze performance after septal injury.

It is quite possible that there are other hippocampal functions
normally mediated by the cholinergic septal input that can be replaced
by sympathetic ingrowth. In the periphery muscarinic cholinergic
stimulation produces some of the same cellular events as $alpha_1$
adrenergic stimulation (e.g., calcium entry, smooth muscle contraction). Thus the presence of norepinephrine in the hippocampus could
replace acetylcholine at physiological targets of the septo-hippocampal
fibers that have both muscarinic cholinergic and $alpha_1$ adrenergic
receptors. Evidence for this type of replacement is suggested by
a recent study of the spontaneous firing of CA_3 pyramidal neurons
in acutely anesthetized animals (1). No change in firing was observed
after septal lesions when sympathetic ingrowth had taken place.
However removal of the superior cervical ganglion resulted in a
dramatic increase in firing rates, and a decrease in inter-spike
intervals. The sympathetic fibers may play a role in maintaining
spontaneous firing of CA_3 neurons.

We have studied $[^3H]$-2-deoxyglucose (2-DOG) uptake in the hippocampal formation after septal lesions (14). In the first week after
lesion there is a small increase in metabolism in the areas of the
hippocampus where cholinergic nerve terminals are degenerating,
followed over the next few weeks by a general decrease in 2-DOG

uptake throughout the hippocampal formation. By three months after the lesion, 2-DOG uptake has returned to control levels and remains there for a least four months. These experiments demonstrate an adaptation of 2-DOG uptake, probably reflecting an adaptation of glucose metabolism in the hippocampal formation. Preliminary experiments suggest that the septo-hippocampal fibers do influence 2-DOG uptake throughout the hippocampal formation in normal animals (15). If sympathetic ingrowth also can influence 2-DOG uptake, this could provide further evidence for an adaptive role of this noradrenergic replacement of cholinergic neurons.

CONCLUSIONS

A Sympathetic Ingrowth Hypothesis of Alzheimer's Disease. Sympathetic ingrowth takes place in the rat brain and represents an apparent replacement of central cholinergic nerves by peripheral noradrenergic nerves. There is strong evidence that this neuronal rearrangement may not be as anomalous as once thought, but rather that it results from the presence of a hippocampal tropic factor in adult rat brain. This form of neuronal plasticity is the best studied example of target-factor regulated rearrangement in the intact animal. There is some evidence to suggest that afferent input to the superior cervical ganglion, the age of the animal and its sex may all influence sympathetic ingrowth, but it has been difficult to separate regulation of the rearrangement from regulation of neurotransmitter content in the rearranging nerves. Finally, it appears that sympathetic ingrowth has both deleterious and beneficial effects on the brain. It may disrupt a "relearning" behavior, but may act to regulate firing and metabolism in the hippocampus in a similar manner to the cholinergic input in an uninjured animal.

Sympathetic ingrowth has been described in guinea pigs (17) and in dogs (26). It seems likely that ingrowth also occurs in primates. Although a stroke could injure forebrain cholinergic neurons, the one condition where forebrain cholinergic neurons are most clearly affected is Alzheimer's Disease. Considering the disruption of "relearning" in rats with sympathetic ingrowth, it is possible that ingrowth may partly contribute to the deficit seen in Alzheimer's patients. Demonstrating that sympathetic ingrowth occurs in primates would be a logical first step towards testing this sympathetic ingrowth hypothesis of Alzheimer's disease.

REFERENCES

1. Barker, D.J., Howard, A.J. and Gage, F.H. (1984): Brain Res. 291:357-363.
2. Björklund, A. and Stenevi, U. (1981): Brain Res. 229:403-428.
3. Crawford, J.D. and Connor, J.D. (1972): J. Neurochem. 19:1451-1458.
4. Crutcher, K.A. 91981): Exp. Neurol. 74:324-329.

5. Crutcher, K.A., Brothers, L. and Davis, J.N. (1979):   Exp. Neurol. 66:778-783.
6. Crutcher, K.A., Brothers, L. and Davis, J.N. (1981):   Brain Res. 210:115-128.
7. Crutcher, K.A., Kessner, R.P. and Novak, J.M. (1983):   Brain Res. 262:91-98.
8. Crutcher, K.A. and Davis, J.N. (1980): Exp. Neurol. 70: 187-191.
9. Crutcher, K.A. and Davis, J.N. (1981): Brain Res. 204:410-414.
10. Crutcher, K.A. and Davis, J.N. (1982): Exp. Neurol. 75:347-359.
11. Crutcher, K.A. and Collins, F. (1982): Science 217:67-68.
12. Davis, J.N. and Martin, B. (1982): Brain Res. 247:1245-1248.
13. Harrell, L.E., Barlow, T.S. and Davis, J.N. (1983): Exp. Neurol. 82:379-390.
14. Harrell, L.E. and Davis, J.N. (1984): Exp. Neurol. 85: 128-138.
15. Harrell, L.E. and Davis, J.N. (1985): Neurosci., in press.
16. Kimble, D.P., Anderson, S., BreMiller, R. and Dannen, E. (1979): Physiol. Behav. 22:461-466.
17. LaForet, G. and Davis, J.N. (1983): Neurosci. Abstr. 9: 1223.
18. Levi-Montalcini, R., Chen, M.G.M. and Chen, J.S. (1978): Zoon. 6:201-212.
19. Loy, R. and Moore, R.Y. (1977): Exp. Neurol. 57:645-650.
20. Loy, R., Milner, T.A. and Moore, R.Y. (1980): Exp. Neurol. 67:399-411.
21. Lynch, G.S., Matthews, D.A., Mosko, S., Parks, T. and Cotman, C.W. (1972): Brain Res. 42:311-318.
22. Madison, R.A., Crutcher, K.A. and Davis, J.N. (1981): Brain Res. 213:183-187.
23. Madison, R.A. and Davis, J.N. (1983): Brain Res. 270:1-9.
24. Madison, R.A. and Davis, J.N. (1983): Exp. Neurol. 80:167-177.
25. McKinney, M., Coyle, J.T. and Hedreen, J.C. (1983): J. Comp. Neurol. 217:103-121.
26. McNicholas, L.F., Martin, W.R., Sloan, J.W. and Nozaki, M. (1980): Exp. Neurol. 69:383-394.
27. Milner, T.A. and Loy, R. (1980): Anat. Embryol. 161:159-166.
28. Patterson, P.H., Reichardt, L.F. and Chen, L.L.Y. (1976): Cold Spring Harbor Symp. Quant. Biol. 40:389-397.
29. Pattison, S.E. and Dunn, M.F. (1975): Biochemistry 14:2733-2739.
30. Peterson, G.M. and Loy, R. (1983): Brain Res. 264:21-29.
31. Rotter, A. and Jacobowitz, D.W. (1981): Brain Res. Bull. 6: 525-529.
32. Scheff, S.W., Bernardo, L.S. and Cotman, C.W. (1978): Science 202:775-778.
33. Schwab, M.E., Otten, U., Agid, Y. and Thoenen, H. (1979): Brain Res. 168:473-483.
34. Stenevi, U. and Bjorklund, A. (1978): Neurosci. Lett. 7:219-224.
35. Storm-Mathisen, J. (1974): Brain Res. 80:181-197.
36. Thoenen, H. and Barde, Y-A. (1980): Physiol. Rev. 60:1284-1335.

DETERMINANTS OF ACETYLCHOLINE LEVELS IN AGED RATS

D.O. Smith

Department of Physiology
University of Wisconsin
Madison, Wisconsin 53706 U.S.A.

INTRODUCTION

Aged rats exhibit altered capabilities to sustain synaptic transmission at the diaphragm neuromuscular junction. For example, synaptic depression is enhanced in older animals (15). Moreover, there are associated changes in end-plate structure. In senescent animals, there are more nerve terminal branches and less sprouting and degeneration per end plate (17; cf. 4). In addition, the number of synaptic vesicles per terminal increases with age (17).

The mechanisms underlying such age-related phenomena are poorly defined. One plausible cause which has not yet been thoroughly investigated is reduced availability of the transmitter, acetylcholine (ACh), and corresponding changes in ACh release. Thus, the objective of this study was an assessment of ACh content and release in aged rats. Levels of ACh were found to decrease in older animals. Possible reasons for this decline were also examined. Portions of this study have been reported in Smith (16).

MATERIAL AND METHODS

General Procedures. All experiments were performed on the phrenic nerve-diaphragm muscle of Fischer 344 rats obtained from the aged animal colonies at the Charles River Breeding Laboratories. Males aged 10 and 28 months were chosen. During each series of experiments, data obtained from animals of each age group were compared statistically using routine 2-tailed t-tests.

After dissection, each hemidiaphragm and the associated phrenic nerve was pinned down at resting length in a small chamber containing 5 ml of saline solution. The composition of the solution is described by Liley (8); the pH was 7.4. Fresh saline aerated with 95% O_2-5% CO_2 and maintained at 37°C, was circulated over the tissue at a rate of 10 ml/min; oxygen saturation was maintained at >80% (608 mm Hg).

Biochemical Assays. Tissue to be utilized for biochemical assays of ACh were dissected and then allowed to equilibrate for 20 min in circulating saline. The muscle was cut into strips which ran along the length of the muscle fibers. Phrenic nerve innervation was left intact in only one of the strips. Three pieces of tissue were obtained from each strip: (i) a region containing the phrenic nerve terminals (innervated), (ii) a sample of about the same size but with no nervous tissue (noninnervated), and (iii) a small piece which was subsequently viewed under a microscope and in which the number of muscle fibers was counted. Each of these three pieces was composed of about 1500 to 2500 muscle fibers.

Morphological studies (17) have shown that at least 98% of the muscle fibers in each age group have a single end plate. Moreover, there are an average ($^+_-$ S.E.) of 13.1 ($^+_-$ 0.9) and 17.5 ($^+_-$ 1.0) nerve terminals per end plate in the 10-month and the 28-month animal, respectively. Thus, the number of muscle fibers in a tissue sample was used as an estimate of the number of end plates in the sample. The number of nerve terminals assayed was also calculated from these data. The protein content of the assayed samples was determined using the Folin reagent method of Lowry as modified by Peterson (11).

ACh. To determine the amount of ACh released during nerve stimulation, the anticholinesterase physostigmine (60 μM) was added to the bath saline. After 20 min, bath circulation was stopped, and the nerve was stimulated (1000 impulses at 3 Hz). Following stimulation, the preparation was allowed to sit in the bath for at least 15 min to allow diffusion of released ACh into the saline; the bath was aerated during this time. The entire bath volume was then collected for subsequent analysis.

To assay ACh, the radiochemical method developed by McCaman and coworkers (5, 10) was used. Technical aspects and control experiments specific to this study are documented by Smith (16). The ACh content of noninnervated tissue was seldom more than 5% of the levels in innervated tissue. Thus, the contribution of ACh in the underlying muscle was generally ignored.

Choline Uptake. Choline uptake was also assayed in innervated and noninnervated tissue. To minimize choline leakage from cells into the bath saline, tissue was first allowed to equilibrate in circulating choline-free saline for 2 to 3 hrs prior to the beginning of the experiment. During the final 3 min of this period, 60 μM physostigmine was added to the saline to prevent hydrolysis of released ACh, thus further reducing the amount of free choline entering the bath from tissue sources.

Radiolabeled incubation medium was next substituted for the bath saline. The incubation medium consisted of the bath saline containing a known amount of unlabeled choline, [methyl-^{3}H] choline chloride (80 Ci/mmol), [U-^{14}C]sucrose (435 Ci/mol), and physostigmine (30 μM). During this incubation period, the nerve was stimulated repetitively at 25 Hz. Oxygen saturation and temperature were maintained at >80% and 37°C, respectively, throughout this phase of the assay. A 100 μl sample of the incubation medium was collected for subsequent determination of the specific activities of the bath.

Accumulation of intracellular choline was found to be linear over 5 min of incubation. Thus, to determine the kinetics, choline uptake was measured after a 4-min incubation. Unlabeled choline chloride and [^{3}H]choline were added to the saline to obtain final concentrations ranging from 0.25 to 12.0 μM. In separate tracer wash-out experiments, error in these concentrations due to leakage of endogenous choline was found to be less than 6%. After incubation in the labeled saline, the tissue was quickly removed, placed in a Buchner funnel, and rinsed thoroughly in ice-cold saline. They were then homogenized and centrifuged, and the supernatant was collected for counting on the scintillation counter, using a double-label program for [^{14}C] and [^{3}H].

Activity due to choline in the extracellular space was estimated from the [^{14}C]sucrose space for each sample. Intracellular choline activity was then obtained by subtracting the amount of choline in the sucrose space from the total [^{3}H]choline activity of the sample. Uptake by the nerve-terminal region of innervated tissue samples was then calculated by subtracting activity associated with the non-terminal component (on the basis of relative protein content) from the total intracellular activity.

ChAT. In addition, choline acetyltransferase (ChAT) activity was measured. The liquid extraction assay of Rand and Johnson (13), in which [^{3}H]choline is converted to [^{3}H]ACh, was used without modification.

Electrophysiology

Recording Techniques. Membrane potentials, membrane input resistance, miniature end-plate potentials (m.e.p.p.s), and end-plate potentials (e.p.p.s) were recorded intracellularly in curarized muscles (10^{-8}M d-tubocurarine; dTC) using KCl-filled microelectrodes with 20- to 30-megohm tip resistances. Technical details are described by Smith (16).

ACh Leakage.   Leakage of cytoplasmic ACh from the phrenic nerve terminals was determined indirectly by measuring the hyperpolarizing shift following addition of dTC to the bath saline.   The basic technique has been described by Katz and Miledi (6), although it was modified slightly in these experiments (16).   After locating a recording site quite close to the end plate, the "irreversible" acetylcholinesterase (AChE) inhibitor echothiophate (10^{-7}M) was added to the bath saline.   Following 20 min, fresh saline was circulated over the tissue, and the echothiophate was rinsed away.   This reduced the possibility of nonspecific effects due to the anticholinesterase.

To establish membrane sensitivity to the transmitter, ACh was added to the bath, and the resulting membrane depolarization was recorded.   The ACh was then washed away.   This procedure was repeated for ACh concentrations in the bath ranging from 10^{-10} to 10^{-6} M.   Dose-response curves were then constructed from these data.   After the added ACh had been removed from the bath and the membrane potential had returned to resting levels, dTC (10^{-9} to 10^{-6} M) was introduced, and the resulting hyperpolarization was recorded.

The concentration and the amount of ACh leakage from the nerve terminals was then determined by comparing the hyperpolarization due to addition of dTC with the membrane potential shifts observed in response to known amounts of ACh.   The rate of leakage, or flux F, was also calculated from these data (16; cf. 6).

RESULTS

ACh Levels

Tissue Content.   Biochemical assays, summarized in Table 1, indicated less ACh in motor nerve terminals of the aged than the young adult rats.   In nonstimulated and stimulated tissue, the aged rats contained 59% and 47%, respectively, as much ACh/end plate as the younger animals.   The differences between the younger and the older animals become even larger when the results are expressed per nerve terminal. It can be concluded that within nonstimulated and stimulated single nerve terminals the total amount of ACh is significantly less in the older animals.

ACh Released.   During stimulation, greater amounts of ACh were released into the bath by the 28-month animals.   As shown in Table 1, 22% more ACh per end plate was released by the older rats.   This difference is significantly different at the 0.05 level.   However, when these data were further normalized to the number of nerve terminals per end plate, the age-related difference was no longer evident; the amount released per nerve terminal was not significantly different for the two age groups.   Thus, the increased amount of ACh released into the bath by the older animals may be attributed to the greater number of nerve terminals per end plate.

Table 1

ACh Content ($^{\pm}$S.E.) of Phrenic
Nerve-Diaphragm Muscle Neuromuscular Junction.
Adapted from Smith (16).

Age (mos)	Tissue Type	ACh/end plate (fmol/end plate)	ACh/terminal (fmol/terminal)
10	Nonstimulated	[a]29.3 $^{\pm}$ 4.1	2.2 $^{\pm}$ 0.4
	Stimulated	[c]27.1 $^{\pm}$ 2.3	2.1 $^{\pm}$ 0.2
	Released	[b] 3.2 $^{\pm}$ 0.2	0.24 $^{\pm}$ 0.02
28	Nonstimulated	[a]17.2 $^{\pm}$ 2.4	1.0 $^{\pm}$ 0.2
	Stimulated	[c]12.8 $^{\pm}$ 3.8	0.7 $^{\pm}$ 0.2
	Released	[b] 3.9 $^{\pm}$ 0.2	0.22 $^{\pm}$ 0.02

Stimulation: 1000 impulses at 3 Hz [a,b]: 0.05 level [c]: 0.01 level Number of rats = 8 from each age group

These results indicate that the fraction of the resting ACh content which is released during nerve stimulation increases with age. During stimulation (1000 impulses delivered at 3 Hz), 10.92% and 22.76% of the ACh assayed in the nonstimulated tissue was released in the 10- and the 28-month animals, respectively. These values represent an increase of 108% in the older animals.

ChAT Activity. Reduced amounts of ACh in the older animals could be due to lower activity of ChAT, which catalyzes synthesis of ACh from the substrates choline and acetyl-CoA. To test this possibility, the activity of ChAT was assayed in both innervated and non-innervated tissue. There was no significant nonneuronal ChAT activity in any of the samples tested.

The ChAT activities measured in tissue obtained from the 10- and the 28-month animals did not differ by a statistically significant amount. Specifically, average ($^{\pm}$S.E.) values of V_{max} were 21.7 ($^{\pm}$1.8) and 18.5 ($^{\pm}$1.5) fmol/min/terminal. This confirms similar results reported by Tuček and Gutmann (19).

Choline Uptake. Limited availability of the choline substrate could also reduce ACh levels and affect synaptic transmission (12). The supply of this choline is associated with a high-affinity transport system (21). Thus, uptake by the terminals was assayed.

Accumulation of [^{3}H]choline was observed in tissue from both age groups. As shown in Fig. 1, the amount was linearly dependent on the duration of incubation with labeled choline for at least 5 min. Moreover, uptake was reduced by 70 to 80% when hemicholinium

(HC-3; $2x10^{-5}M$) was added to the incubation saline. It was similarly reduced when Na^+ was deleted from the medium. Such inhibition by HC-3 and dependence upon extracellular Na^+ is characteristic of the high-affinity choline uptake system in brain tissue (14, 21).

The kinetics of [^3H]choline uptake were measured in experiments with extracellular choline concentrations ranging from 0.25 to 12 μM. The results, summarized in Fig. 2, were fit to the basic rectangular hyperbolic Michaelis-Menton relationship using an iterative, least-squares computer routine kindly provided by Dr. W.W. Cleland (2). Estimates of the kinetic parameters are presented in Table 2.

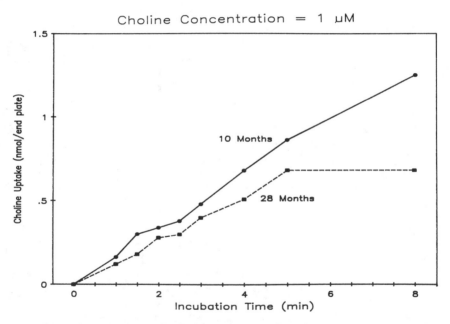

Figure 1. Time-course of choline uptake by phrenic nerve terminals in rats aged 10 and 28 months. Each point represents the average value of 8 determinations. Temperature was maintained at 37°C. The volume of extracellular space was determined by co-incubation in [^{14}C]sucrose; the amount of choline in this space was calculated and subtracted from the total choline activity of the tissue. Uptake by the underlying muscle tissue was subtracted on a per-mgprotein basis.

Table 2

Choline Uptake ($^+$S.E.) by Phrenic Nerve Terminals

Age	K_m	V_{max}(nmol/4 min)	
(mos)	(μM)	per end plate	per terminal
10	1.54 $^+_-$ 0.06	[a]1.63 $^+_-$ 0.02	[b]0.124 $^+_-$ 0.009
28	1.58 $^+_-$ 0.09	[a]1.35 $^+_-$ 0.03	[b]0.077 $^+_-$ 0.005

Number of rats = 12 from each age group   [a],[b]:   0.05 level

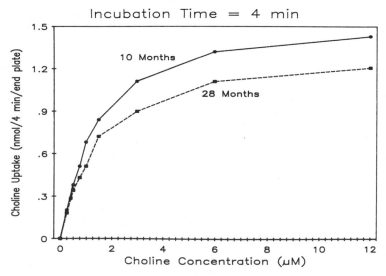

Figure 2.   Choline uptake by phrenic nerve terminals in rats aged 10 and 28 months as a function of choline concentration.   Each point represents the average value of between 4 and 8 determinations; temperature was kept at 37°C.   Uptake was linear for at least 5 min, so the incubation time was 4 min.   As in the time-course experiments (Fig. 1), the total uptake data were corrected for choline entry into the extracellular ([^{14}C]- sucrose) space and the underlying muscle tissue.

There were no age-related changes in the values of K_m, but V_{max} was significantly lower in the aged rats. This difference was even more pronounced when the data are expressed per terminal. These results are consistent with similar reductions in choline uptake observed in chicken iris (9).

Leakage of Cytoplasmic ACh. Continuous leakage of ACh located in cytoplasmic pools has been shown in similar neuromuscular preparations to cause minor depolarization of the postsynaptic membrane (6, 20). Substantial leakage in aged animals could cause ACh levels to decrease and affect synaptic transmission. Therefore, leakage of cytoplasmic ACh was tested by applying dTC to anticholinesterasetreated preparations and recording the resulting hyperpolarization (16).

Initially, the sensitivity of the postsynaptic membrane to ACh was tested by adding ACh in varying concentrations to the bath saline and recording the resulting depolarization. Doseresponse data from each site were plotted, providing calibration curves with which to evaluate the effects of subsequent dTC application (16). The hyperpolarizing responses to dTC were not significantly different for the two age groups, as shown in Table 3. However, due to an age-related decrease in sensitivity to ACh (16), these results indicate greater leakage and higher cleft concentrations of ACh in the aged preparations. Indeed, the recorded hyperpolarizations correspond to 2.6×10^{-11} and 1.0×10^{-10} mol of ACh in the cleft region at the end plates of 10- and 28-month rats, respectively (16). The corresponding values of ACh flux were calculated and are shown in Table 3. In the older rats there is 5 times greater ACh flux per nerve terminal.

Release of Vesicle-Bound ACh. Vesicle-bound ACh is released spontaneously, as well as in response to presynaptic action potentials, resulting in the characteristic postsynaptic responsesthe e.p.p. and the m.e.p.p. The amount of ACh released by this route was further examined by recording spontaneous and evoked end-plate potentials, and then calculating quantal release statistics from these data. Using techniques documented by Smith (16; cf. ref. 7), the average quantal release, m, was determined.

The number of quanta released spontaneously (m.e.p.p.s) and per action potential was significantly greater in the aged rats. During stimulation at 20 Hz, the value of m was 54% greater in the older animals, as shown in Table 4. Qualitatively similar results were obtained during stimulation at rates ranging from 5 to 50 Hz. However, when these data were divided by the average number of nerve terminals per end plate, the calculated average m.e.p.p. rate and quantal content per nerve terminal were not significantly different in the two age groups (Table 4). Thus, the age-related increase in vesicular

Table 3

Leakage of ACh ($\pm$S.E.) from cytoplasmic
sources in rat phrenic nerve terminals
Adapted from Smith (16).

Age	ACh in Cleft	ACh Flux (amol/s)	
(mos)	(pmol)	per end plate	per terminal
10	26 $\pm$ 15	0.21 $\pm$ 0.12	0.016 $\pm$ 0.009
28	101 $\pm$ 51	1.39 $\pm$ 0.71	0.079 $\pm$ 0.041

Number of rats = 7 from each age group

release recorded from the end plate may be attributed to the corresponding increase in the number of nerve terminals per end plate in
the aged rats.

DISCUSSION

The results of this study demonstrate an age-related decline
in the amount of transmitter, ACh, in nerve terminals at the rat
neuromuscular junction. This is associated with increased spontaneous leakage of ACh from cytoplasmic sources and decreased kinetics
of choline uptake and ChAT activity in the older animals.

Under steady-state conditions, ACh synthesis and degradation
are balanced. Thus, assuming first-order kinetics, intracellular
ACh levels may be described by the following relationship:

$$[ACh] = \frac{K_s \, [Choline] \, . \, [AcetylCoA]}{P + K_b}$$

where K_s is the synthesis rate constant, P is the membrane permeability to ACh (which reflects leakage from cytoplasmic sources),
and K_b is the degradation rate constant. Accordingly, decreased
ChAT activity and choline uptake and increased leakage observed in
aged rats may all contribute to lower ACh levels. Membrane permeability to cytoplasmic ACh may play a major role in ACh reductions,
for it is considerably greater in the aged rats. Specifically,
data in Table 2 indicate that:

$$P_{10}[ACh]_{10} = (.079/.016) \cdot P_{28}[ACh]_{28}$$

Table 4

Quantal Content and Rate of Miniature
End-Plate Potentials ($\pm$S.E.) at the Diaphragm.
Adapted from Smith (16).

Age	Average Quantal Content (m)		m.e.p.p.s/second	
(mos)	per end plate	per terminal	per end plate	per terminal
10	[a]74.7 ± 10.0	5.7 ± 0.8	[b]0.79 ± 0.09	0.060 ± 0.007
28	[a]116.1 ± 12.6	6.6 ± 0.7	[b]1.51 ± 0.27	$0.086 +0.015$

Number of rats = 12 (m) and 8 (m.e.p.p.s) from each age group
[a],[b]: 0.05 level

and therefore, that $P_{10}/P_{28}=10.9$. Quantitative evaluation of these relative contributions however require more detailed information about age-related changes in rates of synthesis and degradation.

Measures of transmitter release obtained using electrophysiologic techniques, which are comparable to those reported in previous studies of this preparation (1), were consistent with those based upon biochemical assays. Age-related increases in mean quantal content correspond to the correlated rise in ACh release. Similarly, both characteristics may be attributed to the more extensive terminal arborization in the older animals. Moreover, enhanced synaptic depression in the aged rats reflects the increased fraction of ACh released per action potential. The two phenomena must be causally related if the fraction of transmitter available for release is proportional to the total ACh content of the terminal, since synaptic depression has been shown to result from progressive depletion of the fraction of transmitter available for release (3, 18; cf.1).

Acknowledgements. This work was supported by NIH grants AG01572 and NS00380. Technical assistance of C. Gibson and C.D. Johnson is gratefully acknowledged.

REFERENCES

1. Christensen, B.N. and Martin, A.R. (1970): J. Physiol. (Lond.) 210:933-945.
2. Cleland, W.W. (1967): Adv. Enzymol. 29:1-32.
3. Elmqvist, D. and Quastel, D.M.J. (1965): J. Physiol. (Lond.) 178:505-529.
4. Fahim, M.A. and Robbins, N. (1982): J. Neurocytol. 11:641-656.
5. Goldberg, A.M. and McCaman, R.E. (1973): J. Neurochem. 20:1-8.

6. Katz, B. and Miledi, R. (1977): Proc. R. Soc. Lond. B 196:59-72.
7. Katz, B. and Thesleff, S. (1957): J. Physiol. (Lond.) 137: - 267-278.
8. Liley, A.W. (1956): J. Physiol. (Lond.) 132:650-666.
9. Marchi, M., Hoffman, D.W., Giacobini, E., and Fredrickson, T. (1980): Brain Res. 195:423-431.
10. McCaman, R.E. and Stetzler, J. (1977): J. Neurochem. 28:669-671.
11. Peterson, G.L. (1977): Anal. Biochem. 83:346-356.
12. Potter, L.T. (1970): J. Physiol. (Lond.) 206:145-166.
13. Rand, J.B. and Johnson, C.D. (1981): Anal. Biochem. 116:361-371.
14. Simon, J.R., Atweh, S., and Kuhar, M.J. (1976): J. Neurochem. 26:909-922.
15. Smith, D.O. (1979): Exptl. Neurol. 66:650-666.
16. Smith, D.O. (1984): J. Physiol. (Lond.) 347:161-176.
17. Smith, D.O. and Rosenheimer, J.L. (1982): J. Neurophysiol. 48: 100-109.
18. Thies, R. (1965): J. Neurophysiol. 28:427-442.
19. Tuček, S. and Gutmann, E. (1973): Exptl. Neurol. 38:349-360.
20. Vyskočil, F. and Illes, P. (1977): Pflugers Arch. 370:295-297.
21. Yamamura, H.I. and Snyder, S.H. (1973): J. Neurochem. 21:1355-1374.

ELEVATED RED CELL TO PLASMA CHOLINE RATIOS

IN ALZHEIMER'S DISEASE

J.P. Blass[1], I.Hanin[2], L. Barclay[1],
U. Kopp[2] and M.J. Reding[1]

[1]Altschul Laboratory for Dementia Research
Burke Rehabilitation Center
785 Mamaroneck Avenue
White Plains, New York 10605 U.S.A.

[2]Western Psychiatric Institute and Clinic
Department of Psychiatry
University of Pittsburgh School of Medicine
Pittsburgh, Pennsylvania 15213 U.S.A.

INTRODUCTION

Alois Alzheimer originally described the disease that bears his name as a form of premature aging and degeneration of the nervous system (1). Since then, most studies have focused on it as a disease of neurones - most recently, of specific populations of large cholinergic neurones in the nucleus basalis and of larger neurons in association cortex (7, 39). In the last five years, however, a number of reports have appeared of abnormalities in non-neural cells from patients with Alzheimer disease (DAT). These include lymphocytes (32, 33, 35), leukocytes (26), cultured skin fibroblasts (2), and red blood cells (3, 13, 15, 23). Four groups have reported that some patients with Alzheimer's disease have abnormally high levels of choline in their red cells (3, 13, 15, 23). Values were expressed as a ratio of red blood cell to plasma choline, to allow for dietary variation. We now confirm this finding in a prospective, double-blind study of 118 subjects, and discuss the implications of these and other abnormalities in non-neural tissues for the pathophysiology of neuronal cell damage in Alzheimer's disease.

PATIENTS

Patients were evaluated in a subspeciality outpatient demential clinic (14). They were routinely examined by a neurologist, an internist, a psychiatrist, and a social worker, each with a special interest in geriatrics; had neuropsychological testing; and had a variety of laboratory tests, normally including a CAT scan. Diagnosis was made by staff in conference, using all available data. The diagnosis of all forms of dementia required evidence of global cognitive impairment of greater than three months' duration in a clear sensorium. In this report we refer to several categories of subjects. Patients with Alzheimer's disease had a progressive dementia of insidious onset with no evidence of stroke or other specific cause adequate to explain the dementia. Research criteria for Alzheimer's disease were those of Eisdorfer and Cohen (9). Patients with clinical Alzheimer's disease met the inclusion criteria of Eisdorfer and Cohen (9) but had complicating conditions (most often cardiovascular) which precluded the diagnosis by research criteria. Four of the patients had dementia complicating Parkinson disease, a combination in which the dementia is well documented to be associated with the typical progressive neuropathological changes of Alzheimer's disease (5, 17, 31). Patients with multiinfarct dementia had a relatively sudden onset of dementia, a stuttering downhill course, evidence for generalized vascular disease, often evidence of one or more frank strokes, and modified Hachinski ischemia scores of 4 or more (9, 36). Patients with mixed dementia had characteristics of both Alzheimer's disease and multi-infarct dementia. Patients with other dementias had other specific causes adequate to account for their disabilities, such as chronic schizophrenia or neuro-syphilis. Depression was diagnosed clinically, on the basis of a depressed affect and vegetative signs. Intellectually intact subjects included individuals who were normal on intellectual and behavioral testing, even if they had evidence of other types of nervous system dysfunction. Indeed, the two individuals in this group with abnormally high red cell/plasma choline ratios both had evidence of central nervous system disease. Both were 73-year-old men with episodic confusion of unknown etiology, responding in one case to dilantin. Neither showed enough progression over two to three years to justify the diagnosis of Alzheimer's disease or other progressive dementia.

Diagnoses were established in White Plains before the choline measurements were done.

METHODS

Choline measurements. Blood was obtained by venipuncture into evacuated glass (Vacutainer®) tubes containing sodium heparin as an anticoagulant. Blood samples were then centrifuged at 2,500 rpm for 30 min at 4°C. The plasma was transferred to separate tubes; the buffy coat and a superficial layer of red cells were quantitatively removed by gentle suction, and discarded.   One ml duplicate aliquots of either plasma or of compressed red blood cells were deproteinized by addition to 3 ml ice-cold 0.4 N perchloric acid.   Samples were then mixed thoroughly, and centrifuged for 20 min at 4°C and 32,000 g.   The supernatants were next transferred quantitatively to poly-ethylene tubes, and the pH of these solutions adjusted to 4.2 to 4.4 by addition of 120 μl or 190 μl of ice-cold 7.5 N potassium acetate per ml of red cells or plasma, respectively.   Deproteinization and pH adjustment are essential for stability of the samples, par-ticularly red cells.   Unless the samples are promptly deproteinized and the pH adjusted to 4.2-4.4 prior to freezing of the samples, their lipids will gradually break down, even at subzero temperatures, ultimately resulting in artificially elevated and unreliable levels of free choline (20).

Samples were stored frozen at -20° to -80°C until the time of analysis.   At that stage, samples were thawed slowly at room tempera-ture, mixed, and centrifuged at 32,000 g for 20 min at 4°C.   The supernatants (2.5 ml) were transferred to screw-capped centrifuge tubes (15 ml).   Internal standard, consisting of deuterium-labelled choline (1 nmol) was added to each tube, and choline then precipitated from the supernatants and processed for GC/MS analysis as described by Hanin and Skinner (18).

Choline analysis was done in Pittsburgh without knowledge of the clinical characteristics of the patients.

Analyses of results were by conventional statistical methods (10).

RESULTS

Overall, the Alzheimer's disease and non-DAT groups were reasona-bly matched for age, sex and degree of cognitive and behavioral impairment (Table 1).   As expected, patients with multiinfarct dementia tended to be somewhat older, more likely to be male, and to have higher Hachinski scores (16, 36).   Therefore, the whole group of demented controls tended to be older (75 $\pm$ 1 vs. 70 $\pm$ 1, p < 0.001), more male (49% male vs. 24%) and to have higher Hachinski scores (5.2 $\pm$ 0.5 vs. 1.3 $\pm$ 0.2, p < 0.001) (16).   However, the demented controls were comparable to patients with Alzheimer's disease in degree of mental disability [mental status questionnaire (MSQ) of 3.4 $\pm$ 0.5 vs. 4.0 $\pm$ 0.4, p > 0.05; behavior score 16.8 $\pm$ 1.3 vs. 15.2 $\pm$ 1.2, p > 0.05].   The intellectually intact controls differed (by definition) from both demented controls and Alzheimer's patients on mental testing [MSQ, mini-mental scale (MMS), and behavioral scale].   They also had a significantly higher Hachinski score than the Alzheimer's patients, reflecting the relative incidence of vascular disease and other focal disease in the two groups.

Table 1

Various Characteristics of the Subjects Involved in This Study

		Age	Female/Male	Hachinski Score	MSQ	MMS	Behavioral Score
Controls	(55)	74 ± 2	25/30	4.6 ± 0.4***	5.2 ± 0.5***	13.4 ± 1.9	12.6 ± 1.3***
Intellectually intact	(16)	71 ± 4***	5/11	3.2 ± 0.7***	9.4 ± 0.2***	27.3 ± 1.0	2.6 ± 0.8***
MID	(20)	77 ± 2***	8/12	7.5 ± 0.6	3.4 ± 0.6	9.6 ± 2.4	15.4 ± 1.8
Mixed	(9)	74 ± 4	8/1	4.1 ± 0.7	3.1 ± 0.9	12.7 ± 3.5	15.5 ± 2.7
Other dementias	(10)	71 ± 2***	4/6	2.2 ± 0.6	4.0 ± 1.2	6.2 ± 3.6*	20.5 ± 2.6
[All demented controls	(39)]	[75 ± 1***]	[20/19]	[5.2 ± 0.5***]	[3.4 ± 0.5]	[9.6 ± 1.7]	[16.8 ± 1.3]
Alzheimer	(50)	70 ± 1	33/17	1.3 ± 0.2	4.0 ± 0.4	15.0 ± 1.8	15.2 ± 1.2
Research Criteria	(9)	72 ± 3	7/2	0.3 ± 0.2	4.2 ± 0.9	17.0 ± 3.2	12.0 ± 1.8
Clinical Criteria	(37)	70 ± 1	26/11	1.5 ± 0.2	3.8 ± 0.6	15.4 ± 2.1	16.2 ± 1.4
Parkinson	(4)	72 ± 4	0/4	2.8 ± 0.8	5.8 ± 0.9	16; 5	13.5 ± 5.6
Depressed	(13)	72 ± 2	8/5	2.2 ± 0.5	7.5 ± 1.0***	21.5 ± 2.8	4.6 ± 1.5***

Values represent mean ± SEM, except for female/male ratios. Numbers in parentheses are number of subjects. All subjects were not on all scales.
* Indicates $p < 0.05$; *** indicates $p < 0.001$ vs. all Alzheimer patients.

The depressed patients were comparable to the controls in age and sex, but as expected were less cognitively impaired than the demented controls (Table 1).

The value for red cell/plasma choline ratios was significantly higher in the patients with Alzheimer's disease than the controls (Table 2). The differences were significant for all three subgroups of Alzheimer's patients, but were more marked for the better defined groups of research criteria Alzheimer's disease and Alzheimer's-disease-complicating Parkinson disease. Differences were significant using Student's "t" test (10). The increase appeared to reflect red cell choline concentrations. Mean values for red cell choline concentrations were almost twice as high for Alzheimer's patients as for controls. The variations within each group were, however, large, and the differences did not reach statistical significance. The differences between control and Alzheimer's populations reflect a shift in distributions; values for individual subjects overlapped (Fig. 1).

Red cell/plasma choline ratios were higher for the 13 depressed patients than for the controls (p < 0.05), but lower than for the Alzheimer's patients (Table 2, Fig. 1).

DISCUSSION

Elevations of red cell/plasma choline ratios have been reported in Tourette syndrome (6, 20), in mania (28), in depression (21, 23), and in reports by four groups on Alzheimer's disease (3, 13, 15, 23). The data presented in this paper confirm the existence of these abnormalities in Alzheimer's disease in a larger, prospective study.

The mechanism(s) underlying these increases need to be studied. The changes in red cell/plasma choline are clearly not specific for any single neuropsychiatric disorder. They appear to be due to increases in red cell choline and not decreases in plasma choline. They might prove to be relatively specific for choline, or the abnormalities in choline distribution might reflect more generalized abnormalities in the transport and distribution of a number of materials. They might reflect an abnormality in red cell membranes, in some metabolic process which provides energy for, or is in some other way related to transport. They might be extrinsic to the red cell - for instance, related to some abnormality in lipoproteins which exchange constituents with red cells. Fortunately, obtaining red cells and plasma from living subjects is relatively easy and risk-free so that these and other possibilities can be tested experimentally.

Table 2

Red Cell and Plasma Choline Values

		RBC	Plasma	Ratio Value	Ratio >2.8
Controls					
Intellectually intact	(55)	23.3 ± 3.2	11.5 ± 0.6	1.97 ± 0.24	6/55
	(16)	24.9 ± 9.0	11.5 ± 0.8	1.86 ± 0.50	2/16
MID	(20)	23.9 ± 4.9	11.6 ± 1.0	2.11 ± 0.47	2/20
MIXED	(9)	24.2 ± 5.9	9.5 ± 2.5	2.14 ± 0.50	1/9
Other dementia	(10)	18.8 ± 3.1	11.8 ± 1.2	1.67 ± 0.33	1/10
Alzheimer					
Research Criteria	(50)	40.3 ± 9.1	10.4 ± 0.8	3.83 ± 0.49***	23/50
	(9)	38.7 ± 12.1	9.3 ± 1.1	4.18 ± 1.34***	5/9
Clinical Criteria	(37)	40.0 ± 12.1	10.6 ± 1.0	3.56 ± 0.55**	14/37
Parkinsonian	(4)	47.0 ± 8.7	10.0 ± 1.4	5.51 ± 2.04***	4/4
Depressed	(13)	33.4 ± 8.5	10.0 ± 0.7	3.26 ± 0.80*	7/13

Values are mean ± SEM, for the number of subjects in parentheses. The last column indicates the fraction of patients with ratios above 2.8. *p < 0.05, **p < 0.005, ***p < 0.001, vs. all controls by t-test.

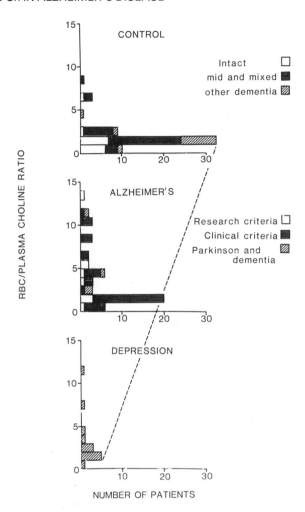

Figure 1. Red cell (RBC) to plasma choline ratios measured in controls, patients with Alzheimer's disease, and depressed patients.

Chemical or functional abnormalities have now been described in a number of non-neural tissues in Alzheimer's disease. These might be secondary to the neurological disease - for instance, to a neuroendocrine abnormality. On the other hand, they might reflect a generalized biochemical abnormality in Alzheimer's disease, present in all cells but only leading to clinical disease in certain populations of neurones. The evidence which indicates a genetic factor in Alzheimer's disease is in accord with the latter possibility (12, 30). The abnormal gene(s) would be present in all cells at some point in their life time, even if expressed and functionally important only in specialized cell types. Other disorders in which generalized but subtle biochemical abnormalities lead to the death of nerve cells are known. Classic examples include Tay-Sachs disease due to a generalized deficiency of hexosaminidase A (37) and spino-cerebellar system degnenerations with deficiencies of specific mito-chondrial (4, 8, 34, 38) or other (27) enzymes.

Non-neural tissues are obviously easier to obtain from living patients than are brain biopsies. Whatever turns out to be the basis of the abnormalities in non-neural tissues in Alzheimer's disease, study of such tissues may be a useful approach to studying the pathophysiology of cell damage in this devastating and common condition.

Acknowledgements. Supported by grants from the General Foods Cor-poration, the Will Rogers Institute, the Winifred Masterson Burke Relief Foundation, the NIA (AG 03853), and the NIMH (MH 26320). We thank Mrs. Louise Orto for her help with obtaining and preparing the samples, and our colleagues at Burke Rehabilitation Center and New York Hospital-Westchester Division whose careful clinical evalua-tions of the patients made these studies possible.

REFERENCES

1. Alzheimer, A. (1907): Cbl. Nervenheilk Psychiat. 18:177-179.
2. Andia-Wallenbaugh, A.M. and Puck, T.T. (1977): J. Cell. Biol. 75:279a.
3. Barclay, L.L., Blass, J.P., Kopp, U. and Hanin, I. (1982): N. Engl. J. Med. 307:501.
4. Blass, J.P. (1981): Current Neurol. 3:66-91.
5. Boller, F., Mizutani, T., Roessman, U. and Gambetti, P. (1980): Ann. Neurol. 7:329-335.
6. Comings, D.E., Gursey, B.T., Avelino, E., Kopp, U. and Hanin, I. (1982): In Gilles de la Tourette Syndrome, eds., A.J. - Friedhoff and T.N. Chase, Raven Press, New York, pp. 255-258.
7. Coyle, J.T., Price, D.L. and DeLong, M.R. (1983): Science 219: 1184-1190.
8. Djikstra, U.J., Willems, J.L., Joosten, E.M.G. and Gabreels, F.J.M. (1983): Ann. Neurol. 13:325-327.
9. Eisdorfer, C. and Cohen, D. (1980): J. Fam. Pract. 11:553-557.
10. Ferguson, C. (1959): Statistical Analysis in Psychology and Education, McGraw-Hill, London.
11. Folstein, M.F., Folstein, S.E. and McHugh, P.R. (1975): J. Psychiatr. Res. 12:189-198.
12. Folstein, M.F. and Breitner, J.C.S. (1981): Johns Hopkins Med. J. 149:145-147.
13. Friedman, E., Sherman, K.A., Ferris, S.H., Reisberg, B., Bartus, R.T. and Schneck, M.K. (1981): N. Engl. J. Med. 304:1490-1491.
14. Garcia, C.A., Reding, M.J. and Blass, J.P. (1981): J. Am. Geriatr. Soc. 29:407-410.

15. Greenwald, B.S., Edasery, J., Mohs, R.C., Horvath, T.B., Shah, N., Trigos, G.G. and Davis, K.L. (1983): Proc. Am. Psych. Assoc. May:NR 127.
16. Hachinski, V.C., Lassen, N.A. and Marshall, J. (1974): Lancet 2:207-210.
17. Hakim, A.M. and Mathieson, G. (1978): Lancet 2:729.
18. Hanin, I. and Skinner, R.F. (1975): Anal. Biochem. 66:568-583.
19. Hanin, I., Kopp, U., Zahniser, N.R., Shih, T.-M., Spiker, D.G., Merikangas, J.R., Kupfer, D.J. and Foster, F.G. (1978): In Cholinergic Mechanisms and Psychopharmacology, ed., D.J. Jenden, Plenum Press, pp. 181-195.
20. Hanin, I., Merikangas, J.R., Merikangas, K.R. and Kopp, U. (1979): N. Engl. J. Med. 301:661.
21. Hanin, I., Kopp, U., Spiker, D.G., Neil, F., Shaw, D.H. and Kupfer, D.J. (1980): Psychiatry Res. 3:345-355.
22. Hanin, I., Cohen, B.M., Kopp, U. and Lipinski, J.F. (1982): Psychopharmacol. Bull. 18:186-190.
23. Hanin, I., Reynolds, C.F. III, Kupfer, D.J., Kopp, U., Taska, L.S., Hoch, C.C., Spiker, D.G., Sewitch, D.E., Martin, D., Marin, R.S., Nelson, J.P., Zimmer, B. and Morycz, R. (1984): Psych. Research 13:167-173.
24. Haycox, J. (1982): Neurosci. Abstr. 8:629.
25. Haycox, J. (1984): J. Clin. Psychiatry 45:23-24.
26. Jarvik, L.F., Matsuyama, S.S., Kessler, J.O., Fu, T.-K., Tsai, S.Y., Clark, E.O. (1982): Neurobiol. Aging 3:93-99.
27. Johnson, W.G. (1981): Neurology 31:1453-1456.
28. Jope, R.S., Jenden, D.J., Ehrlich, B.E., Diamond, J.M. and Gosenfeld, L. (1980): Proc. Natl. Acad. Sci. USA 77:6144-6148.
29. Kahn, R.L., Goldfarb, A.I., Pollack, M. and Peck, A. (1960): Am. J. Psychiatry 117:326-328.
30. Larsson, T., Sjogren, T. and Jacobson, G. (1963): Acta Psychiatr., Scand. Suppl. 167:39.
31. Lieberman, A., Dzietolowski, M., Kupersmith, M., Serby, M., Goodgold, A., Korein, J. and Goldstein, M. (1979): Ann. Neurol. 6:355-359.
32. Miller, A.E., Neighbour, P.A., Katzman, R., Aronson, M. and Lipkowitz, R. (1981): Ann. Neurol. 10:506-510.
33. Nordenson, I., Adolfson, R., Beckman, G., Bucht, G. and Winblad, B. (1980): Lancet 1:481-482.
34. Plaitakis, A., Nicklas, W.J. and Desnick, R.J. (1980): Ann. Neurol. 7:297-303.
35. Robbins, J.H., Otsuka, F., Tarone, R.E., Polinsky, R.J., Brumback, R.A., Moshell, A.N., Nee, L.E., Ganges, M.B. and S.J. Cayeux (1983): Lancet 1:468-469.
36. Rosen, W.G., Terry, R.D., Fuld, P.A., Katzman, R. and Peck, A. (1980): Ann. Neurol. 7:486-488.
37. Sloan, H.R. and Fredrickson, D.S. (1972): In The Metabolic Basis of Inherited Disease, 3rd edition, eds., J.B. Stanbury, J.B. Wyngaarden, D.S. Frederickson, McGraw-Hill, New York, pp. 615-638.

38. Stumpf, D.A., Parks, J.K., Egwen, L.A. and Haas, R. (1982):
     Neurology 32:221-228.
39. Terry, R.D., Peck, A., DeTeresa, R., Schechter, R. and Horoupian,
     D.S. (1981): Ann. Neurol. 10:184-192.

PLASMA AND RED BLOOD CELL CHOLINE IN AGING:

RATS, MONKEYS AND MAN

E.F. Domino, B. Mathews[*] and S. Tait[*]

Department of Pharmacology
University of Michigan
Ann Arbor, Michigan 48109, U.S.A.

[*]Lafayette Clinic
Detroit, Michigan 48207, U.S.A.

INTRODUCTION

Plasma and red blood cell choline concentrations have been suggested as possible peripheral markers of brain cholinergic dysfunction in various human diseases such as major affective disorders, tardive dyskinesia, Tourette's syndrome and senile dementia of the Alzheimer's type (1, 2, 4-8, 10). The present study was initiated to explore these possible relationships in normal human volunteers and patients of similar age. The patients were diagnosed as having tardive dyskinesia or dementia of the Alzheimer's type. These patients were relatively old and their diets were difficult to control. Therefore, additional studies were done in rats and monkeys of various age groups where dietary factors could be held constant. The results obtained indicate that plasma and red blood cell choline levels do not correlate with brain dysfunction in either humans or monkeys. However, in these same species, with increasing age, plasma choline levels increase. Interestingly, increased plasma choline was not observed in aging rats.

MATERIALS AND METHODS

A total of 20 male Holtzman rats were studied. Ten were young adults, approximately 90 days old, and 10 were geriatric, approximately 730 days old. All were on a standard Purina rodent chow diet. Each was decapitated approximately 8:00-10:00 A.M. on the day of the experiment and the exsanguinated blood used for choline assays.

283

A total of 18 adult Macaca mulatta monkeys of mixed sex were used. Nine were approximately 6-7 years, and 9 were 18 plus years old. All were on a standard monkey chow diet. Venous blood samples were collected from each restrained animal between 8:00-10:00 A.M. on the day of the experiment without the use of any anesthesia. These animals were used in a delayed matching to sample box test and had previously been given various drugs and chemicals thought to improve memory. No drugs were given in the week prior to withdrawal of the venous blood sample in which blood choline levels were compared to the animal's behavioral performance (11).

The normal human volunteers (18 to 86 years, both sexes, 58 total) were obtained from various sources including newspaper ads, word of mouth contacts, hospital staff from the University Hospital in Ann Arbor, and patients with various medical diseases from the Turner Geriatric Clinic in Ann Arbor. The patients with tardive dyskinesia (40 to 86 years, mixed sexes, 18 total) were obtained from various sources throughout the State and referred to the Lafayette Clinic in Detroit and from the Ypsilanti Regional Psychiatric Hospital in Ypsilanti. The patients with dementia (50 to 86 years, mixed sexes, 35 total) were obtained from the Lafayette Clinic. All of the volunteers (with the exception of the staff personnel from University Hospital) were given complete histories, physical and mental examinations and were selected as being normal. The patients with tardive dyskinesia and dementia were all diagnosed by at least one neurologist and one psychiatrist. Complete medical and neurological workups were conducted. The diagnosis of tardive dyskinesia was made only in patients with a history of long term (years) neuroleptic medication. Most of the patients were chronic schizophrenics. To be included in the study, each patient had to show involuntary or semivoluntary movements of a choreiform (tic-like) nature of which some had an athetotic or dystonic component. Oral-lingual-masticatory movements were present in each selected patient. In addition to movements of the tongue and facial muscles, some patients had neck and extremity muscle involvement. The motor movements were not relieved by anticholinergic medication. Specifically excluded diseases included Huntington's, Meig's, Wilson's, brain neoplasms, Gilles de la Tourette and Fahr's syndromes.

The diagnosis of dementia was made in a completely different group of patients who had a history of memory impairment of insidious onset, smooth progression, and of at least six months duration. The memory impairment was verified with the Wechsler Memory Scale (WMS) employing normative data of Hulicka. Exclusion criteria were as follows: history of schizophrenia, major affective disorder, alcoholism or other central nervous system (CNS) disorder; abnormalities in laboratory studies, including urinalysis, CBC, SMA-17, B-12, folate, T3, or T4; Hamilton Depression Scale score greater than 12; evidence of a focal or mass lesion on the basis of neurological examination, neuropsychological evaluation, CT scan, or

EEG; severe sensory impairment. Patients with transient ischemic attacks, or any dementing process not consistent with the Alzheimer's type were specifically ruled out as possible participants.

All normal and patient volunteers were maintained on a normal diet in which high sources of choline, including lecithin supplements, were specifically excluded. Patient compliance was more of a problem than with the normal volunteers in which fasting A.M. blood samples were drawn for choline assay. This research was approved by the Institutional Review Boards for human research at the Lafayette Clinic, University Hospital, and Ypsilanti Regional Psychiatric Hospital on various therapeutic protocols of which some of the baseline choline data are reported herein. Patients on any medication known to alter blood choline (such as lithium or dietary supplements) were specifically excluded. Plasma and red blood cell choline levels were measured by a gas chromatographic assay using chemical demethylation (3, 9). A nitrogen phosphorus (NP) detector was used to measure endogenous levels of choline. 3-Hydroxy-N,N,N-trimethylproponominium iodide was used as an internal standard for each sample. This assay involves deproteinization with formic acid-acetone, and extraction from the aqueous sample with dipicrylamine. Both the choline and the internal standard were esterified and then chemically demethylated with sodium benzenethiolate prior to GC NP analysis.

RESULTS AND DISCUSSION

As illustrated in Figure 1, the only consistent change observed was an increase in plasma choline concentration as a function of age in normal monkeys (P <.05) and humans (P <.001). This was not observed in rat plasma. There was no significant change in red blood cell choline concentrations as a function of old age in any of the three species studied.

When mean plasma and red blood cell choline levels were compared between the normal volunteers to a comparable age group of patients with dementia of the Alzheimer's type or patients with tardive dyskinesia, no statistical differences were observed (Figure 2). Among the normal volunteers (19 to 86 years old) plasma choline was highly correlated with age (Figure 3) but red blood cell choline was not (Figure 4).

The results of the present study are consistent with our previous report (11) that plasma and red blood choline levels do not correlate with behavioral performance on a memory test in young adult and geriatric monkeys. Previous reports in the literature have emphasized the value of the ratio of red blood cell to plasma choline in relation to disease states including major affective disorders, tardive dyskinesia, Tourette's syndrome and senile dementia of the Alzheimer's type (1, 2, 4-8, 10). To date, we have not made a detailed analysis of the relationship of the red blood cell to plasma choline ratio to disease states but are in the process of doing so. This will be the subject of future reports. However, the present data indicate quite clearly that plasma and red blood cell choline alone are of no value as peripheral markers of brain pathology for patients with dementia of the Alzheimer's type or for patients with tardive dyskinesia. It is not clear why the present study is so negative, especially since other investigators have reported an elevated red blood cell choline level, for example, in Alzheimer's disease. Perhaps differences in methods of diagnosis and severity of the illness are crucial factors. The Alzheimer patients we studied were, in general, severely demented.

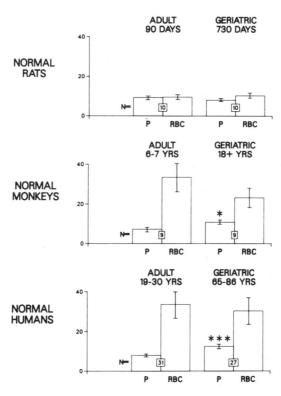

Figure 1. Comparative baseline plasma and red blood cell choline levels in young adult and geriatric rats, monkeys and humans. In this and the subsequent figures, the plasma and red blood cell choline levels are expressed as mean $\pm$ S.E.M. in nmol/ml. Group comparison Student "t" tests were run in which: *p < .05, **p < .01 and ***p < .001.

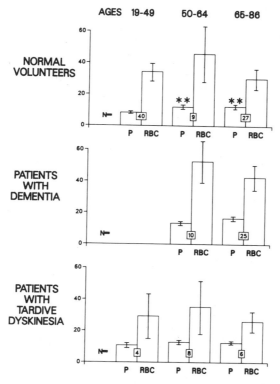

Figure 2. Baseline plasma and red blood cell choline levels in young adult, middle age, and geriatric subjects. The subjects included normal volunteers, patients with dementia of the Alzheimer's type and patients with tardive dyskinesia. The number (N) of subjects in each group is given in the small square inserts in the bar graph. The only significant finding is an increase in plasma choline with age.

The fact that plasma choline increases as a function of age in two species of primates, humans and monkeys, is of considerable interest with regard to the biology of aging. Theoretically, plasma choline may be elevated with old age even if dietary choline is constant because there may be greater absorption of choline from the gastrointestinal tract, greater breakdown of phosphatidylcholine to free choline, greater synthesis of choline, less urinary excretion of choline, or less utilization and turnover of choline by various body tissues. Recently, we have studied the pharmacokinetics of $[^2H]_4$-choline in monkeys following its intravenous administration. The data obtained indicate that geriatric monkeys do not clear $[^2H]_4$-choline from the blood as readily as young adult animals (12). This

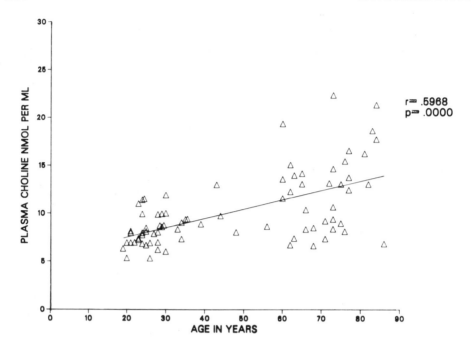

Figure 3.   Linear regression relationship between plasma choline
and age in normal volunteers. The fasting A.M. plasma choline concentration of each subject studied is plotted as a △ .   The correlation
coefficient between age in years and plasma choline was r = +.5968.
This was highly significant (p < .001).

is reduced in geriatric monkeys and possibly in humans. Obviously,
further research is needed to assess the significance of this finding
in the biology of the aging process.

Acknowledgements.   Portions of this research were supported by funds
from the Michigan Department of Mental Health to the Lafayette Clinic,
Wayne State University and the University of Michigan, as well as
the Psychopharmacology Research Fund (Domino).   The authors would
like to acknowledge Drs. Duff, May and Pomara for referring some
of the patients and/or volunteers used in this study.

REFERENCES

1.   Barclay, L., Blass, J., Kopp, U. and Hanin, I. (1982): New Engl.
     J. Med. 307:501.
2.   Blass, J., Hanin, I., Barclay, L., Kopp, U. and Reding, M. (1983):
     This book.
3.   Freeman, J., Choi, R. and Jenden, D. (1975): J. Neurochem. 24:
     729-734.

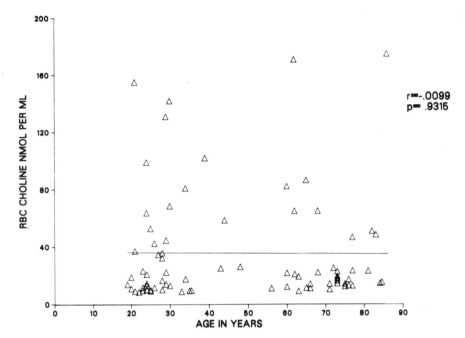

Figure 4. Lack of a linear regression relationship between red blood cell choline and age in normal volunteers. The fasting A.M. red blood cell choline concentration of each subject studied is plotted as a Δ similar to their plasma concentrations shown in Figure 3. Note that there is no relationship between red blood cell choline and age.

4. Friedman, E., Sherman, K., Ferris, S., Reisberg, B., Bartus, R. and Schneck, M. (1981): New Engl. J. Med. 304:1490-1491.
5. Greenwald, B., Edasery, J., Mohs, R., Horvath, T., Shah, N., Trigos, G. and Davis, K. (1983): Proc. Amer. Psychiat. Assoc., New Research Abstracts:NR 127.
6. Hanin, I., Merikangas, J., Merikangas, K. and Kopp, U. (1979): New Engl. J. Med. 301:661-662.
7. Hanin, I., Kopp, U., Spiker, D., Neil, J., Shaw, D. and Kupfer, D. (1980): Psychiat. Res. 3:345-355.
8. Hanin, I., Cohen, B., Kopp, U. and Lipinski, J. (1982): Psychopharm. Bull. 18:186-190.
9. Jenden, D. and Hanin, I. (1974): In: Choline and Acetylcholine Handbook of Chemical Assay Methods (ed.) I. Hanin, Raven Press, New York, pp. 135-150.
10. Jope, R., Jenden, D., Ehrlich, B., Diamond, J. and Gosenfeld, L. (1980): Proc. Nat. Acad. Sci. USA 77:6144-6146.
11. Marriott, J.G., Domino, E.F., Tait, S.K., Mathews, B.N., Fucek, F. and Abenson, J.S. (1981): Soc. Neurosci. Abst. 7:950.

12. Domino, E.F., Marriott, J.G., Mathews, B.N., Domino, S.E. and
     Tait, S.K. (1982): Proceedings Abstracts: 13th CINP Congress,
     Jerusalem, Israel 1:174.

HUMAN RED BLOOD CELL CHOLINE IN AGING AND SENILE DEMENTIA:

EFFECTS OF PRECURSOR THERAPY

K.A. Sherman[1] and E. Friedman[2]

[1]Department of Pharmacology
Southern Illinois University School of Medicine
Springfield, Illinois 62708 U.S.A.
[2]Department of Psychiatry and Pharmacology
New York University School of Medicine
New York, New York 10016 U.S.A.

INTRODUCTION

The decline of cognitive function associated with aging is thought to involve deficits in brain cholinergic mechanisms (4, 8, 16, 21, 28). Cognitive deterioration is greatly exacerbated in senile dementia of the Alzheimer's type (SDAT). In this disease the activity of choline acetyltransferase (ChAT), the synthetic enzyme for acetylcholine (ACh), is markedly reduced in cortex and hippocampus, when compared to age-matched controls (5, 8, 22, 25, 26). The decrease of ChAT activity in cortex is closely correlated with the severity of cognitive impairment and the extent of neuropathology (5, 22). High affinity uptake of choline (Ch) and in vitro synthesis of ACh in cortical tissue biopsied from SDAT patients were reduced (29, 30). These biochemical changes in Alzheimer's patients appear to reflect a loss of cholinergic innervation to hippocampus and cortex (31, 34), which subserve an important role in memory and cognition (4, 11). Therefore, many of the recent strategies for treatment of SDAT reflect attempts to reverse this deficit in cholinergic function.

The hypothesis that peripheral availability of choline influences the synthesis and release of ACh in brain (6) prompted a number of clinical trials of choline or lecithin administration in SDAT patients (7, 13; reviewed in 4). However, these trials have not resulted in any consistent pattern of cognitive improvement.

Recent evidence has suggested that the availability and metabolism of glucose may also be an important determinant of brain ACh synthesis rate (9, 16). Hence, deficits in brain glucose metabolism may exacerbate the cholinergic deficits in aging and Alzheimer's disease (15, 16). In light of these data, rats were treated with a combination of choline and piracetam, a drug reported to enhance brain metabolism (20). The piracetam/ choline combination led to a dramatic reversal of age-related memory deficits, whereas either choline or piracetam alone had little effect (3).

In a parallel, preliminary study of 15 SDAT patients, we found no significant improvement after one week of choline/piracetam treatment. However, four patients improved in subjective and objective measures of memory and cognitive performance (13). This subgroup of "responders" was distinguished in having significantly higher red blood cell (RBC) Ch levels and higher RBC:plasma Ch ratios, both before onset of treatment, and after the week of therapy (14). In a similar vein, Barclay et al. (2) reported an increase in the RBC:plasma Ch ratios of patients with primary degenerative dementia.

In the present investigations, we have characterized the effects of age and SDAT on blood choline measures, and have investigated the influence of various precursor loading strategies on blood Ch.

METHODS

Venous blood (10 ml) was collected in heparinized green top Vacutainer® tubes, placed immediately on ice, and prepared according to the protocol of Hanin et al. (14, 18). Ch content was determined by gas chromatography (19, 28).

Elderly subjects were diagnosed for SDAT and the severity of symptoms was rated according to the Global Deterioration Scale (GDS) described by Reisberg et al. (24). Elderly controls had no objective evidence of cognitive deterioration (GDS 1 or 2) and were often spouses of SDAT patients. Volunteers from N.Y.U. Medical Center served as young controls.

RESULTS

Effect of Age and SDAT on Blood Choline. Both plasma and RBC Ch content were significantly elevated in elderly subjects when compared to young controls (Table 1). Patients diagnosed as having SDAT did not differ significantly from unimpaired elderly controls with regard to plasma or RBC Ch content, or the RBC: plasma Ch ratio (Table 1).

In all groups, the RBC Ch concentrations varied over a wide range (from 4 to 90-113 nmol/ml). Therefore, we examined the frequency distribution of RBC Ch levels in an enlarged sample of controls

Table 1

Plasma and Red Blood Cell Choline Content as a Function of
Age and Cognitive Impairment[a]

	Young Controls	Elderly Controls (nmol/ml)	Senile Dementia
PLASMA	6.2 ± 0.4 (22)	10.2 ± 0.7^{b} (29)	10.2 ± 0.6 (33)
RBC	17.8 ± 4.3 (22)	30.6 ± 4.7^{b} (34)	31.7 ± 5.5 (34)
RBC:PLASMA	2.8 ± 0.6 (22)	3.4 ± 0.4 (29)	3.1 ± 0.6 (33)

[a]Blood Ch content (in nmol/ml) was determined in young controls (16-50 yrs.), unimpaired elderly controls (GDS 1 or 2, 60-80 yrs.), and SDAT patients (GDS $\geq$ 4, 60-80 yrs.). Results are expressed as group mean Ch in nmol/ml (or ratio) $\pm$ S.E.M. for N subjects. [b]$p <$ .05 Student's $\underline{t}$-test compared to young controls.

and SDAT patients. As shown in Figure 1, the percentage of unimpaired elderly subjects with low RBC Ch values (<10 nmol/ml) was significantly lower (19%) compared to young controls (37%), and conversely, the incidence of very high RBC Ch values (>50 nmol/ml) in the elderly subjects was much greater (29%) than in young controls group (2.6%; X^2 = 12.5, p < .01). No significant difference in the distribution of RBC Ch values was observed between SDAT patients and age- and sex-matched controls (X^2 = 4.42, NS).

The stability and reliability of RBC Ch measured in elderly subjects was assessed by taking repeated samples at 3 to 35 month intervals. The correlation between initial and retest RBC Ch values was significant for the overall group N = 18, r = 0.98, p < .01), and for the subgroups with high RBC Ch (> 50 nmol/ml; N = 7, r = 0.86, p < .05), or low RBC Ch (<25 nmol/ml; n = 11, r = .64, p < .05).

Effect of Precursor Treatment on Blood Choline. Treatment with Ch chloride (12 g/d, t.i.d.) for 6 weeks resulted in significant elevations of plasma and RBC Ch content of SDAT patients (Table 2). The RBC:plasma Ch ratios during treatment with this precursor were compared to the baseline ratios for each patient from this study. Figure 2 shows the striking correlation between baseline and treatment RBC:plasma Ch content (N = 15, r = .90, p < .01). Patients characterized by high baseline ratios maintained high ratios during Ch therapy, although the treatment ratio tended to plateau in patients with the highest baseline values. RBC Ch content at baseline and level during treatment were also significantly correlated (N = 15, r = 0.83, p < .01). However, there was no significant relationship between plasma Ch concentration at baseline and during Ch-loading

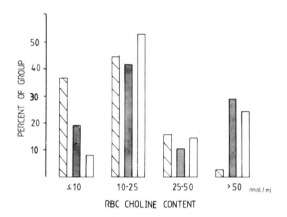

Figure 1. Distribution of RBC choline concentrations in young (striped) and elderly (striped) controls, and in SDAT patients (open bars). RBC Ch was analyzed in groups of: 1) young Ss (N = 38; 16-50 years old, $\bar{X}$ = 29.9; 65% male); 2) unimpaired elderly Ss (N = 69; GDS 1 or 2; 56-94 years of age, $\bar{X}$ = 71 $\pm$ 1; 56% male), and 3) SDAT patients (N = 62; GDS $\geq$ 5; 56-87 years old, $\bar{X}$ = 72 $\pm$ 1; 56% male). The groups were significantly different in distribution of RBC values (X^2 = 20.4, p < .01) and group mean Ch content (ANOVA: F(2,166) = 5.1, p < .01). Mean RBC Ch was elevated in elderly controls (33.4 $\pm$ 3.4) compared to young controls (17.7 $\pm$ 2.7, p < .005, Student's t-test); RBC Ch in SDAT patients (33.4 $\pm$ 43.8) was not different from age- and sex-matched controls..

Ch-loading (r = .23). Similar pre- to post-treatment relationships were observed when piracetam was coadministered with Ch for one week. RBC Ch and RBC:plasma Ch ratio during piracetam/Ch treatment were correlated with the baseline values (Table 3).

In other studies we have examined lecithin (Phospholipon-100) administration as a means of raising Ch availability. The time course of changes in plasma and RBC Ch following acute and repeated (4 weeks) lecithin treatment are shown in Figure 3. Plasma and RBC Ch concentrations were elevated about three-fold from 2 to 6 h after acute lecithin and approached control values by 24 h. The pharmacokinetic pattern was similar after repeated treatment. However, levels of Ch in both plasma and RBC were higher after the chronic treatment regimen than after a single administration of the same dose of lecithin.

Lecithin administration (Phospholipon-100, 34-53 g/d) resulted in significant increases in the plasma and RBC Ch concentrations in SDAT patients (Table 2). The RBC Ch content during lecithin

Table 2

Effect of Precursor Loading on Plasma and
Red Blood Cell Choline[a]

Study/ Treatment	Plasma (nmol/ml)	RBC
A.		
PLACEBO	10.7 ± 1.1 (15)	37.7 ± 7.6 (15)
CHOLINE	31.2 ± 2.4^b (15)	83.3 ± 11.0^b (15)
B.		
PLACEBO	9.7 ± 5.5 (13)	42.3 ± 11.2 (13)
LECITHIN	65.8 ± 5.5^b (13)	156.6 ± 22.1^b (13)
LEC/PIR	64.2 ± 9.0^b (12)	142.2 ± 23.3^b (12)
C.		
PLACEBO	10.2 ± 0.9 (13)	36.0 ± 7.5 (13)
LEC/PIR	31.6 ± 2.0^b (12)	138.2 ± 24.6^b (12)

[a]Plasma and RBC Ch concentrations were determined in: A) outpatients treated with choline chloride (12 g/d, t.i.d., 6 weeks) (ref. 13); B) outpatients treated with lecithin (34-53 g/d, t.i.d., 2 weeks) alone or coadministered with piracetam (4.8 g/d); and C) inpatients treated with lecithin (25 g/d) and piracetam (7.2 g/d) as a maintenance dose for 4 weeks, with bloods drawn 16 h after the last dose of lecithin (ref. 27). Results are expressed as group mean choline concentration in nmol/ml $\pm$ S.E.M. for N subjects.
[b]$P < .05$ compared to placebo, Student's t-test.

treatment was significantly correlated with the patient's baseline value (Table 3). Likewise, the ratio of RBC:plasma Ch attained during lecithin administration was correlated with the pre-treatment ratio, but the levels of Ch in plasma were not significantly related (Table 3). Piracetam coadministration (4.8 g/d) with the lecithin did not alter the elevations of plasma and RBC Ch induced (Table 2). Significant correlations were found between RBC Ch content or RBC:plasma Ch ratio at baseline and during piracetam/lecithin combination treatment in two separate studies (Table 3). In both studies, plasma Ch levels at baseline and during treatment were not significantly correlated.

DISCUSSION

The present results demonstrate a wide range of RBC Ch content in the normal adult human population. Higher mean RBC Ch content

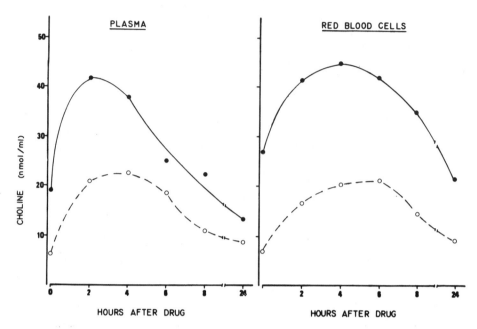

Figure 2. The relationship of RBC:plasma Ch ratio during precursor loading to baseline value. The ratio of RBC to plasma Ch during treatment is plotted against the baseline ratio for fifteen SDAT patients treated with 12 g/day choline chloride (t.i.d.) for six weeks. The correlation between ratio during treatment and at baseline was significant ($r = 0.9$, $P < .01$).

and a greater incidence of high RBC Ch content were observed in the elderly subgroup of the population. We have also found an age-related increase in plasma Ch concentrations consistent with previous reports (12, 18). No alterations in these blood Ch parameters were observed in patients with senile dementia of the Alzheimer's type as compared to age- and sex-matched control subjects.

The mechanism and functional significance of these age-related changes in blood Ch remain to be understood. Elevated free plasma Ch may be related to the increase of CSF Ch content which has been reported in elderly human subjects (17). It is conceivable that these age associated increases in Ch result from altered turnover of choline-containing phospholipids. In the rat brain, a decreased synthesis of choline-phosphoglycerides has been reported with increasing age (23). It would be of interest to examine the effect of age on bound forms of Ch in blood, since these may also contribute to ACh synthesis in brain.

Table 3

Relationship Between Baseline and Treated Blood Choline Values:
Summary of Precursor Loading Studies at N.Y.U.[a]

Study/Treatment	RBC Ch	Plasma Ch	Ratio RBC:Plasma Ch
CHOLINE	.83[b] (15)	.22 (15)	.90[b] (15)
CHOLINE/PIR	.72[b] (10)	.11 (10)	.95[b] (10)
OUTPATIENT:			
LECITHIN	.68[b] (13)	.02 (13)	.61[b] (13)
LEC/PIR	.83[b] (12)	.13 (12)	.68[b] (12)
INPATIENT:			
LEC/PIR	.71[b] (12)	.43 (11)	.58[b] (10)

[a]Pearson correlation coefficient (r) calculated for comparison of baseline (or placebo) and treated value for the SDAT patient reported in Table 2 and ref. 14. Number of patients is in parentheses.
[b]$p < .01$.

     The present results clearly emphasize the need for controlling for age and for utilizing a large number of subjects in studies comparing RBC Ch in different populations. The relative stability of baseline RBC Ch reported previously in young subjects (18) was shown also in our longitudinal analysis of elderly subjects, including those with high RBC Ch. Moreover, when plasma Ch concentrations were elevated three- to six-fold by choline or lecithin treatment, the inter-individual differences in the RBC Ch content achieved and the distribution of Ch across the RBC membrane (indicated by the RBC:plasma Ch ratio) remained highly correlated with the differences observed at baseline. These obervations suggest that inherent properties in the dynamic function of the RBC membrane, such as Ch transport, may be responsible for the differences in RBC Ch among subjects. We observed no differentiation between SDAT patients and unimpaired elderly controls with regard to RBC Ch content or in the incidence of subjects with high Ch levels. Glen et al. (17) found no difference in RBC Ch uptake in SDAT patients. These data argue that RBC Ch does not directly reflect on the pathophysiological mechanisms involved in Alzheimer's disease, although RBC Ch may be altered in more severe cases of the disease than were studied here (2). Nonetheless, the greater prevalence of high RBC Ch in elderly subjects might be related to changes in CNS function which underlie the predisposition to SDAT or to affective disorders. Increased

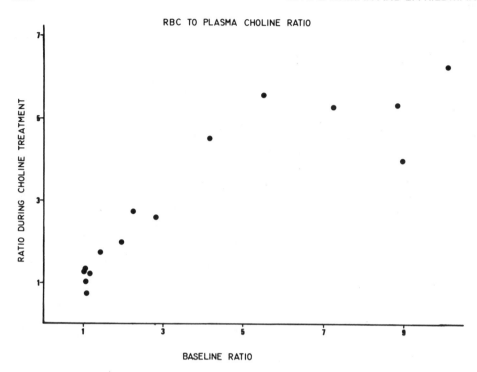

Figure 3.   Time course of changes in plasma and RBC choline after
lecithin: acute (o--o) versus chronic (o—o) administration.   Blood
choline was measured in a patient treated with 15.8 g/70kg lecithin
(Phospholipon-100;95% phosphatidyl Ch) before or after a chronic
treatment regimen (32.5 g/70kg, b.i.d., for 4 weeks).   The lecithin
challenge was given 16 h after the last chronic dose.

RBC Ch level has been reported in certain affective disorders (18)
and it is well known that the incidence of depression increases
with age.

    Choline or lecithin loading has now been widely tested as a
therapeutic approach for SDAT and several other CNS disorders.
Psychotropic effects consistent with central cholinergic stimulation
have been reported, e.g. improved memory, reduction of dyskinetic
movements and mood changes.   However, these effects have not been
consistently achieved (33), and the results of precursor loading
studies have generally been discouraging (4, 13).   A major difficulty
in interpreting these negative findings is the absence of a suitable
measure of the efficacy of such treatments in altering brain Ch
levels and cholinergic function.   Differences in plasma Ch elevation
do not appear to adequately explain variability in therapeutic response
(14).   In our studies, we have measured RBC Ch levels in humans as

a possible indicator of intracellular accumulation of Ch in brain. Preliminary results suggest that while the majority of SDAT patients do not respond to precursor therapy, those patients who accumulated higher levels of Ch in RBCs were more likely to improve (14). Domino et al. (10) also reported a correlation between improvement of cognitive task performance and RBC Ch levels during lecithin treatment. In patients with tardive dyskinesias, we have found no reliable clinical improvement or changes in EEG measures during treatment with lecithin (1, 32). However, when lecithin was administered in conjunction with lithium, much higher RBC Ch levels were achieved and a reduction of dyskinetic movements in 5 of 9 patients ensued (32). Thus, RBC Ch may offer a valuable measure in attempting to understand the variable consequences of precursor treatment.

Acknowledgements.   We thank Drs. Steven Ferris and Barry Reisberg of the N.Y.U. Geriatric Study and Treatment Program, Dr. Michael Serby of the Manhattan Veteran's Administration, and Dr. Jan Volavka of the Manhattan Psychiatric Center for their very generous cooperation in making these studies of blood Ch possible, and providing diagnoses and treatment.

These studies were supported in part by U.S.P.H.S., RSDA, MH00208, MHCRC 35976 and MH29590.

REFERENCES

1.   Anderson, B., Friedman, E., Sherman, K., Banay-Schwartz, M., O'Donnell, J., Carlton, M. and Volavka, J. (Submitted).
2.   Barclay, L.L., Blass, J.P., Kopp, U. and Hanin, I. (1981):   New Engl. J. Med. 307:501.
3.   Bartus, R.T., Dean, R.L., Sherman, K.A., Friedman, E. and Beer, B. (1981):   Neurobiol. Aging 2:105-111.
4.   Bartus, R.T., Dean, R.L., Beer, B. and Lippa, A.S. (1982): Science 217:408-417.
5.   Bowen, D.A., Smith, C.B., White, P. and Davison, A.M. (1976): Brain 99:459-496.
6.   Cohen, E.L. and Wurtman, R.J. (1976):   Science 191:561-562.
7.   Corkin, S., Davis, K.L., Growdon, J.H., Usdin, E. and Wurtman, R.J. (Eds) (1982):   Alzheimer's Disease: A Report of Progress in Research (Aging V. 19), Raven Press, New York.
8.   Davies, P. (1979):   Brain Res. 171:319-327.
9.   Dolezal, V. and Tuček, S. (1982):   Brain Res. 240:285-293.
10. Domino, E.F., Minor, L., Duff, I.F., Tait, S. and Gershon, S. (1982):   In Alzheimer's Disease: A Report of Progress in Research (eds) S. Corkin, K.L. Davis, J.H. Growdon, E. Usdin and R.J. Wurtman, Raven Press, New York, pp. 393-397.
11. Drachman, D.A. (1977):   Neurology 27:783-790.
12. Eckernas, S.A. and Aquilonius, S.M. (1977):   Scand. J. Clin. Lab. Invest. 37:183-187.

13. Ferris, S.H., Reisberg, B., Crook, T., Friedman, E., Schneck, M.K., Sherman, K.A., Corwin, J., Gershon, J. and Bartus, R.T. (1982): In Alzheimer's Disease: A Report of Progress in Research (Aging V. 19), (eds) S. Corkin, K.L. Davis, J.H. Growdon, E. Usdin and R.J. Wurtman, Raven Press, New York, pp. 475-481.

14. Friedman, E., Sherman, K.A., Ferris, S.H., Reisberg, B., Bartus, R.T. and Schneck, M.K. (1981): New Engl. J. Med. 304:1490-1491.

15. Gibson, G.E., Peterson, C. and Sarsone, J. (1981): Neurobiol. Aging 2:165-172.

16. Gibson, G.E. and Peterson, C. (1982): In The Aging Brain: Cellular and Molecular Mechanisms of Aging in the Nervous System (Aging V. 20), (eds) E. Giacobini, G. Filogano, G. Giacobini and A. Vernadakis, Raven Press, New York, pp. 107-122.

17. Glen, A.M., Yates, C.M., Simpson, J., Christie, J.E., Shering, A., Whalley, L.J. and Jellinek, E.H. (1981): Psychol. Med. 11:469-476.

18. Hanin, I., Kopp, U., Spiker, D.S., Neil, J.F., Shaw, D.H. and Kupfer, D.J. (1980): Psychiat. Res. 3:345-355.

19. Jenden, D.J. and Hanin, I. (1974): In Choline and Acetylcholine: Handbook of Chemical Assay Methods (ed) I.Hanin, Raven Press, New York, pp. 135-150.

20. Nicholson, V.J. and Wolthuis, O.L. (1976): Biochem. Pharmacol. 25:2241-2244.

21. Perry, E.K., Perry, R.H., Gibson, P.H., Blessed, G. and Tomlinson, B.E. (1977): Neurosci. Letters 6:85-89.

22. Perry, E.K., Tomlinson, B.E., Blessed, G., Bergmann, K., Gibson, P.H. and Perry, R.H. (1978): Brit. Med. J. 42:1457-1459.

23. Porcellati, G., Gaiti, A. and Brunetti, M. (1982): In The Aging Brain: Cellular and Molecular Mechanisms of Aging in the Nervous System (Aging V. 20), (eds) E. Giacobini, G. Filogamo, G. Giacobini and A. Vernadakis, Raven Press, New York, pp. 77-86.

24. Reisberg, B., Ferris, S.H., deLeon, M.J. and Crook, T. (1982): Am. J. Psychiat. 139:1136-1139.

25. Reisine, T., Yamamura, H.I., Bird, E.D., Spokes, E. and Enna, S.J. (1978): Brain Res. 159:477-481.

26. Rossor, M.N., Rehfeld, J.F., Emson, P.C., Mountjoy, C.O., Roth, M. and Iversen, L.L. (1981): Life Sci. 29:405-410.

27. Serby, M., Corwin, J., Rotrosen, J., Ferris, S.H., Reisberg, B., Friedman, E., Sherman, K.A., Jordan, B. and Bartus, R. (1983): Psychopharm. Bull. 19:126-129.

28. Sherman, K., Kuster, J., Dean, R.L., Bartus, R. and Friedman, E. (1981): Neurobiol. of Aging 2:99-104.

29. Sims, N.R., Bowen, D.M. and Davison, A.N. (1981): Biochem. J. 196:867-876.

30. Spillane, J.A., White, P., Goodhardt, M.J., Flack, R.H.A., Bowen, D.M. and Davison, A.w. (1977): Nature 266:558-559.

31. Tagliavini, F. and Pilleri, P. (1983): Lancet 1:469-470.

32. Volavka, J., Lifshiftz, K., Friedman, E., Sherman, K.A. and Banay-Schwartz, M. (1983): Biol. Psychiat. 18:1175-1179.

33. Vroulis, G.A., Smith, R.C., Brinkman, S., Schodar, J. and Gordon, J. (1981): Psychopharm. Bull. 17;127-128.
34. Whitehouse, P.J., Price, D.L., Struble, R.G., Clark, A.W., Coyle, S.T. and deLong, M.R. (1982): Science 215:1238-1239.

THE USE OF POSITRON EMISSION TOMOGRAPHY FOR THE EVALUATION

OF CHOLINE METABOLISM IN THE BRAIN OF THE RHESUS MONKEY

S.-Å. Eckernäs, S.-M. Aquilonius, K. Bergström,
P. Hartvig, A. Lilja, B. Lindberg, H. Lundqvist,
B. Långström, P. Malmborg, U. Moström and K. Någren

Departments of Neurology, Diagnostic Radiology,
Gynecology and Pharmacy
University Hospital
S-751 85 Uppsala, Sweden
and
Department of Organic Chemistry
and
Gustaf Werner Institute
Box 531, S-751 21 Uppsala, Sweden

INTRODUCTION

Positron emission tomography (PET) offers an unique opportunity to study regional distribution of different compounds noninvasively. After i.v. injection of substances labelled with short-lived isotopes such as $[^{11}C]$, $[^{13}N]$, $[^{15}O]$, $[^{18}F]$ or $[^{68}Ga]$ the distribution of the radioactive label is measured as a function of time by means of a tomographic technique, based on the annihilation radiation produced during the process of positron emission.

Up to now studies on brain choline (Ch) metabolism have been performed in small animals using $[^3H]$ or $[^{14}C]$- isotopes (4, 9, 12) while for human studies no accurate methods have been available. The present paper reports from an ongoing project, and suggests a possible method for the study of brain Ch-metabolism in vivo in primates using $[^{11}C]$-labelled Ch. Initial experiments in dogs have been summarized earlier (3) and one monkey experiment using $[^{11}C]$-Ch was recently reported by Friedland and collaborators (8).

MATERIAL AND METHODS

The [^{11}C] was obtained as [^{11}C]-carbon dioxide. By using [^{14}N](p,α) [^{11}C]-reaction on the van der Graaff accelerator at the Tandem Accelerator Laboratory (University of Uppsala), [^{11}C]-methyl iodide was prepared. The corresponding N-demethylated analogue of choline (dimethylaminoethanol) was alkylated (10). The specific radioactivity varied between 5-30 mCi/Ìmol and about 2-5 mCi (75-180 MBq) was injected i.v. in each experiment. [^{68}Ga]-EDTA was obtained from a commercially available ion exchange column (NEN).

Four female rhesus monkeys (Macaca mulatta) weighing 6-10 kg were used after an overnight fast. Three of the monkeys (A, B, C) were anaesthetized with ketamine, 5 mg/kg, about 30 min before administration of [^{11}C]-Ch. Ketamine 5-10 mg/kg were thereafter given i.m. every 60 to 120 min as required. In one monkey (D) pentobarbital 30 mg/kg was used instead of ketamine for anaesthesia. The labelled compounds were given via a venous catheter in a hindleg with the monkey lying in a fixed position in the positron emission tomograph. In most animals a second injection of [^{11}C]-Ch or [^{68}Ga]-EDTA was given after 2-3 hours (Table 1) when the radioactivity after the first injection had decayed. Blood samples were collected in heparinized tubes from a venous catheter in the other hindleg before, and at regular time intervals after the injection of the compound. The total radioactivity in the blood was measured using a scintillation counter.

In two monkeys (B and C) 1-1.5 mg of atropine were given i.v. one hour prior to the second [^{11}C]-Ch injection.

Immediately after administration of the radiolabelled compounds imaging of the head of the monkey was started. The positron emission tomograph used is equipped with two rings of detectors giving three simultaneous images with a slice thickness of 13.5 mm and an in-plane resolution of about 8 mm (PC 384-3B, AB Scanditronix, Uppsala, Sweden) (6). Images were recorded continuously for 4 x 10 and 200 sec periods at predetermined intervals.

Regions of interest corresponding to total brain, cortical areas and white matter, temporal muscle and the intranasal tissue were analyzed. To ensure correct anatomical localization of the regions of interest horizontal cryosections (2) corresponding to the PET sections were taken from the head of one rhesus monkey using a heavy cryomicrotome (see Fig. 2).

The measured radioactivity was corrected for physical decay to the time of administration of the radioactive dose. The uptake of radioactivity within the tissue is expressed as an "uptake index" which represents the radioactivity in relation to activity given

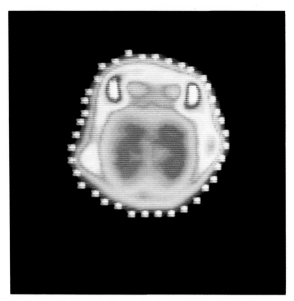

Figure 1. One typical summation image including the images obtained during a [^{11}C]-Ch experiment.

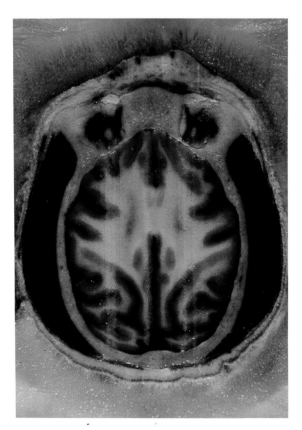

Figure 2. A cryosection corresponding to the upper level of the 13.5 mm thick PET slice shown in Figure 1.

Table 1

Monkey ID	Inj No	Isotope	Age yrs	Drugs*	MRT min (^{11}C–ACh + ^{11}C–Ch) Total Brain	Cortex	White substance
A	1	11C–Ch	11	k	12.3	17.0	15.5
A	2	11C–Ch	11	k + a	13.7	----	----
B	1	11C–Ch	25	k	13.4	14.1	13.2
B	2	68Ga–EDTA	25	k	----	----	----
C	1	11C–Ch	8	k	13.6	14.5	12.8
C	2	11C–Ch	8	k + a	13.5	14.3	12.9
C	3	68Ga–EDTA	8	k + a	----	----	----
D	1	11C–Ch	30	p	----	15.5	14.3

* k = ketamine, a = atropine, p = pentobarbital

per gram body weight. Thus, an uptake index of 1.0 means an even distribution of the tracer in the body.

The radioactivity within brain blood vessels can be corrected for by the use of $[^{68}Ga]$ EDTA. The $[^{68}Ga]$-EDTA complex does not pass the intact blood brain barrier and by dividing the $[^{68}Ga]$ activity found within different brain regions with the blood activity of $[^{68}Ga]$, an approximative value of the blood volume within different regions was obtained.

It is assumed that the radioactivity in the brain 40 min after the $[^{11}C]$-Ch injection mainly represents other Ch-containing compounds (ChM) than acetylcholine (ACh) and free Ch. The radioactivity of ChM is assumed to increase linearly from zero time and a curve supposed to reflect mainly $[^{11}C]$-ACh + $[^{11}C]$-Ch was constructed by subtraction of the extrapolated ChM curve (see Fig 6 and justification for this approach in the Discussion section). The mean residence time (MRT, equal to the turnover time) of what is assumed to be $[^{11}C]$-ACh + $[^{11}C]$-Ch can be calculated by a technique used in pharmacokinetics and described by Yamamoko et al (14).

The area under the curve of data shown in Fig 6 is described by the following equation:

$$AUC = {}_0\!\int^{\infty} (uptake\ index) \times dt$$

The area under the first statistical movement curve is defined as:

$$AUCM = {}_0\!\int^{\infty} t \quad (uptake\ index) \times dt$$

The MRT is obtained by dividing AUCM with AUC:

$$MRT = \frac{AUCM}{AUC}$$

RESULTS

Figure 1 shows one typical "summation image" which includes all images taken after one injection of $[^{11}C]$-Ch. These images are used for the definition of regions of interest. The cryosection in Fig. 2 corresponds to the upper part of the 13.5 mm PET section.

A relatively small fraction was found within the brain and the total brain uptake of Ch represents about 1% of the injected dose, a figure similar to that previously found in the mouse (13). In the nasal tissue a high uptake was found which declined in parallel with the blood activity. In brain and muscle the radioactivity was found to increase slowly with time (Fig. 3).

In the total brain area the calculated blood volume was about 10% and the corresponding figure in the cortex was 6%. The fraction of the uptake index corresponding to radioactivity of [^{11}C]-Ch within brain blood vessels was subtracted and the net results are shown in Fig. 4 and 5. After a rapid initial uptake phase during the first 2 min a slow almost linear increase of radioactivity follows as long as the activity can be traced (about 2 hours). As can be seen in Fig. 5 there is a higher uptake in cortical areas in comparison to white matter.

The MRT for [^{11}C]-ACh + [^{11}C]-Ch in ketamine anaesthetized monkeys was found to be 15.2 $\pm$ 1.6 min in the cortex and 13.8 $\pm$ 1.5 min in white matter (Table 1). There were no obvious differences between monkeys of different ages (8-30 years). The MRT for [^{11}C]ACh + [^{11}C]-Ch was not conclusively affected by the injection of 1 mg atropine or by anaesthesia with pentobarbital (Table 1). The only effect seen after atropine treatment was a decrease in the uptake in cortical areas while the uptake in white matter was unchanged.

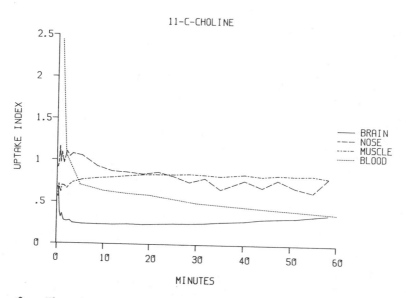

Figure 3. The observed uptake index in the total brain area, nose and temporal muscle in one ketamine anaesthetized monkey.

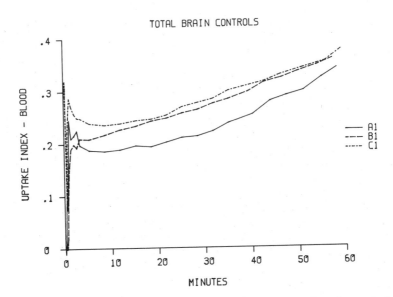

Figure 4. The uptake of radioactivity with time in the total brain area in 3 ketamine anaesthetized monkeys (A, B, C). The calculated activity within brain blood vessels has been subtracted.

DISCUSSION

PET is a non-invasive technique, which permits the measurements of labelled compounds distributed to discrete areas in the brain or other organs. However, it is fundamental that the technique only provides measurement of total radioactivity and its change with time. This means that the images represent the sum of all metabolic products of Ch within the brain including activity local-ized in brain blood vellels. Total brain blood volume as well as blood volumes within areas of interest were measured by means of [68Ga] EDTA. Reasonable estimates of intravascular activity could then be obtained by detecting the changes in blood radioactivity with time following each i.v. injection of [11C]-Ch.

The similarity in total brain [11C]-Ch uptake kinetics in monkeys A, B and C is striking (Fig. 4) indicating a constancy in general Ch-metabolism during ketamine anaesthesia. The further analysis of the uptake kinetics and calculation of the MRT for [11C]-ACh + [11C]-Ch has been based on knowledge from experiments with [3H] and [14C]-labelled Ch performed in small laboratory animals. It is well known from such studies (1, 11, 13) that brain radioactivity during the first 5-10 min following i.v. injection of labelled Ch is mainly in the form of ACh and Ch while phosphorylcholine accounting for about 50% of total radioactivity at 20 min successively increases.

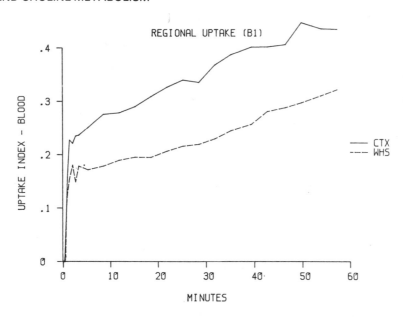

Figure 5.   The uptake of radioactivity with time in cortical (CTX) and white matter (WHS) areas in one ketamine anaesthetized monkey (B) injected with [^{11}C]-Ch.   The calculated activity within brain blood vessels has been subtracted.

Labelled phospholipids are found after longer time periods (5). If these metabolic sequences are similar in the rhesus monkey it seems reasonable to subtract the extrapolated ChM curve to construct an initial phase (Fig. 6) supposed to reflect mainly [^{11}C]-ACh + [^{11}C]-Ch.   The extrapolation is, however, not entirely correct and at this point only hypothetical, since the exact time course of ChM is not known.

The physiological validity of this calcualtion principle is not easily tested in the anaesthetized monkey.   The PET technique employed has a spatial resolution in the plane of about 8 mm.   In the image slice of 13.5 mm thickness the separation of regions and differentiation of gray from white matter must be subject to substantial errors.   Nevertheless, the higher uptake of [^{11}C]-Ch in the cortex as compared to the white matter (Fig. 5) can probably not be explained by differences in blood flow only, and thus indicates a more pronounced Ch-uptake in gray matter.

In rat and mouse it has been shown that pentobarbital decreases (10), and atropine increases (4) ACh turnover in the cortex.   In the present experiments, however, using a relatively low dose of atropine, no clear cut changes in MRT for [^{11}C]-ACh + [^{11}C]-Ch could

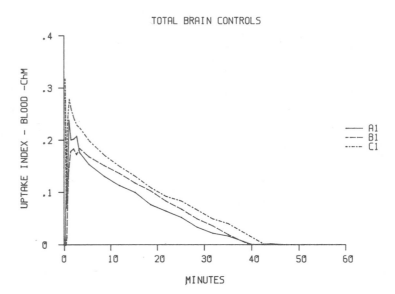

Figure 6. The uptake of radioactivity with time in the total brain area in 3 ketamine anaesthetized monkeys (A, B, C) after subtraction of ChM (see text), as well as the radioactivity within blood vessels.

be seen. Changes in ACh turnover will, however, not necessarily change the total sum of $[^{11}C]$-ACh + $[^{11}C]$-Ch since an increased Ch-uptake could be balanced by an increased release of ACh. The effect of ketamine anaesthesia during the experiments is not known at present.

Thus, this approach does not seem to be amenable to the measurement of changes in brain ACh turnover in the monkey by PET. Regional $[^{11}C]$-Ch kinetics measured by this approach, however, may represent an overall Ch-metabolizing capacity of the brain. As it seems impossible to study the unanaesthetized monkey with the PET technique, it might be more fruitful to test the validity of this suggested method by comparing the uptake of $[^{11}C]$-Ch with that of $[^{11}C]$-labelled Ch analogues known to be taken up by the brain Ch-transporting systems, but not further metabolized (see 7).

It is evident that a method allowing estimation of different processes of brain Ch metabolism in man would be of great potential in clinical research. About an 8-times larger brain volume, and unanaesthetized experimental conditions, are two important factors favouring experimental conditions in man. The $[^{11}C]$-Ch approach used in these studies, in combination with the PET analysis still, however, requires further refinement in animals before it could be applied safely and effectively in man.

Acknowledgement. This work was supported by the Swedish Medical Research Council (grant No 7151) and by grants from the Tore Nilson Foundation for Medical Research and the Swedish Society for Medical Sciences.

REFERENCES

1. Ansell, G.B., Spanner, S. (1975): In: Cholinergic Mechanisms (ed) P.G. Waser, Raven Press, New York, pp. 117-129.
2. Aquilonius, S.-M., Eckernäs, S.-Å., and Gillberg, P.-G. (1983): In: IBRO/Handbook Services: Methods in Neurosciences (ed) A.C. Cuello, John Wiley Sons, London, pp. 155-170.
3. Bergström, K., Aquilonius, S.-M., Bergson, G., Berggren, B.-M., Bergström, M., Brismar, T., Eckernas, S.-Å, Ehrin, E., Eriksson, I., Greitz, T., Gillberg, P.-G., Lagerkranser, M., Litton, J., Lunqvist, H., Långström, B., Malmborg, P., Sjöberg, S., Stalnacke, C.-G. and Widen, L. (1981): J. Comp. Ass. Tomogr. 5:938.
4. Eckernäs, S.-Å. (1977): Acta Physiol. Scand. Suppl. 449:1-62.
5. Eisenstedt, M. and Schwartz, J.H. (1975): J. General. Physiol. 65:293-313.
6. Eriksson, L., Bohm, C., Bergström, M., Ericson, K., Greitz T., Blomqvist, G., Litten, J., Widen, L., Hansen, P., Holte, S. and Stjernberg, H. (1983): In: Positron Emission Tomography of the Brain (eds) W.D. Heiss and M.E. Phelps, Springer Verlag, Berline, Heidelberg, New York, pp. 38-45.
7. Fisher, A. and Hanin, I. (1980): Life Sci. 27:1615- 1634.
8. Friedland, R.P., Mathis, C.A., Budinger, T.F., Moyer, B.R. and Rosen, M. (1983): J. Nucl. Med. 24: 812-815.
9. Kewitz, H. and Pleul, O. (1981): In: Cholinergic Mechanisms (eds) G. Pepeu and H. Ladinsky, Plenum Press, New York, pp. 405-413.
10. Långström, B. (1980): Acta Universitatis Upsaliensis 555.
11. Nordberg, A. (1977): Acta Physiol. Scand. Suppl. 445.
12. Schuberth, J., Sparf, B. and Sundwall, A. (1970): J. Neurochem. 17:461-468.
13. Sparf, B. (1973): Acta Physiol. Scand. Suppl. 397.
14. Yamamoko, K., Nakagawa, T. and Uno, T. (1978): J. Pharmacokinetic Biopharmaceutic. 6:547-558.

CHOLINERGIC MECHANISMS IN SPINAL CORD AND MUSCLE

S.M. Aquilonius, H. Askmark and P.G. Gilberg

Department of Neurology
University Hospital, S-751 85
Uppsala, Sweden

INTRODUCTION

Neuropsychiatric diseases mainly affecting intracerebral mechanisms e.g. dementia, psychosis and movement disorders have in the last years been the objects of intense research in clinical cholinergic neuropharmacology. In contrast, studies devoted to spinal cholinergic mechanisms in man have been few and, if neurophysiological studies are excluded, corresponding investigations in human muscle tissue are sparse.

In the present paper current knowledge regarding the distribution of acetylcholinesterase (AChE) cholineacetyltransferase (ChAT) and cholinergic receptors in the spinal cord will be presented as well as changes in these markers coupled to the degenerations in amyotrophic lateral sclerosis (ALS). Furthermore, the principal changes in ChAT and nicotinic receptors in rat hindleg muscles during denervation and reinnervation will be discussed as a background for quantitative studies in human muscle biopsies.

AChE, ChAT AND CHOLINERGIC RECEPTORS IN THE SPINAL CORD

Animal studies. Several decades ago the presence (27) and synthesis (11) of acetylcholine (ACh) in the spinal cord was demonstrated in animal experiments and the first applications of AChE histochemistry (13, 26) depicted intense staining of the motor neurons. Subsequent analysis of the AChE distribution in relation to the structure of cat spinal cord (36) demonstrated heavy staining also in the dorsal horn, especially corresponding to laminae III and IV of Rexed (34). Accumulation of the AChE following spinal hemisections (16) have indicated the existence of AChE transporting neurons within the white

313

matter, and parallel changes in ChAT analyzed in small dissected samples suggest the presence of both ascending and descending cholinergic fibres in the cord (17). A precise identification of these tracts is still lacking.

In all species studied the highest spinal ChAT activity has been found in the ventral horn (1, 17, 23) while a relatively high activity in the dorsal horn has also been demonstrated in spinal cord obtained from cat (17), cow (1) and man (see below). It is quite possible that a small area of high ChAT activity in an apical part of the dorsal horn may be missed during dissection in the rat. Subcellular studies performed on spinal cord from cow revealed that the the major proportion of dorsal horn ChAT was present in a particulate form, indicating a main localization, in nerve terminals (1). In the ventral horn and especially in the white matter most of the ChAT activity was soluable. Using an immunohistochemical technique to demonstrate ChAT, Kan et al. (22) found no staining in the dorsal horn of the rabbit spinal cord while the perikarya of the ventral horn motor neurons were stained. However, probably due to improved sensitivity of the immunohistochemical method, intense terminal staining of ChAT has now been demonstrated in substantia gelatinosa of the dorsal horn in the cat (25). ChAT activity within spinal gray matter did not change following rostral hemisection in the cat (23) or following lesion of the sensory input in the rat (20), but restricted changes might have been hard to detect with the techniques used.

Radioligand binding techniques have enabled the demonstration and characterization of cholinergic receptors within the spinal cord. Lately, consequently, new data have rapidly been accumulated, especially regarding muscarinic receptors. In this field the introduction of in vitro autoradiographic procedures (35, 43) have been of utmost importance, and such methods seem to be replacing binding studies in homogenates from the spinal cord (15, 24). The first published autoradiograph on spinal cord muscarinic receptors (39) was from the rat, depicting an intense binding of radiolabeled quinuclidiny benzilate ([^{3}H]-QNB) in the ventral horn, and especially in an apical part of the dorsal horn claimed to correspond to substantia gelatinosa. Different proportions of muscarinic receptor subtypes has been shown to exist in the substantia gelatinosa and in the ventral horn of the rat (42), which is of special interest with regard to the changes found in patients suffering from ALS (see below). Following spinal transection in the rat [^{3}H]-QNB binding decreased in cord homogenates from below the lesion (10) but spinal lesions and autoradiographic studies of muscarinic receptors have not been combined as yet.

The distribution of α-[^{125}I] bungarotoxin (α-Btx) has been studied by in vitro autoradiography in the spinal cord of mouse (6) and rat (32). It is interesting to note that in none of these studies was accumulation of grains associated to the motor neurons, while certain small cell bodies and their processes were labelled in the ventral horn. In the mouse a diffuse labelling of laminae I, II and III was reported while, in the rat, low levels of binding sites were found in lamina II of the dorsal horn. However, as discussed by Fibiger (12), at present, there is as yet no conclusive evidence that α-Btx labels reliably nicotinic cholinergic receptors in the CNS.

Human studies. AChE histochemistry on human spinal cord sections (19) has shown low activity in all fiber tracts except the ventral roots. In the gray matter of the cervical cord (Fig 1B) intense staining has been demonstrated in the different motor cell groups of the ventral horn, and in the substantia gelatinosa of the dorsal horn. The correspondence in disbribution of AChE, ChAT and muscarinic receptors (Fig 1) in the cervical human spinal cord is striking, and suggest that AChE may be a marker of cholinergic innervation in this tissue.

When the detailed distribution of ChAT was studied with the "punch technique" of tissue sampling, a high activity in the ventro-lateral part of the ventral horn could be traced into the ventral root region (1). The technique permitted demarcation of a high activity area also in the dorsal horn (Fig. 1C).

The first autoradiographs on muscarinic receptors in human spinal cord were published by Whitehouse and collaborators (41), and the distributions found were in perfect agreement with Fig.1D from our laboratory (15). Further comparison with Fig. 1A demonstrates the close correspondence of high receptor density to motor neuron cell groups, and to the substantia gelatinosa. In the latter region where specific [^{3}H]-etorphine binding sites are similarly distributed (Fig. 1G) cholinergic regulation of pain might take place. Morphine induced analgesia seems to be enhanced by physostigmine (40). Dis-placement studies with carbachol that binds preferentially to high-affinity "agonist" cholinergic sites have demonstrated that these sites dominate in the ventral horn while in the dorsal about equal numbers of high- and low-affinity sites seem to exist (41).

Changes in ALS. As ALS is characterized by progressive degeneration of the cortico-spinal tracts and the lower motor neurons, several groups have found it of value to investigate the involvement of spinal cholinergic mechanisms in this disorder. Nagata and collaborators (31) found pronounced reductions of ChAT in the ventral gray matter of ALS spinal cords, with less pronounced changes in AChE activity. It is interesting, in relation to the known degenerations in the pyramidal tracts, that in lumbar and thoracic parts, ChAT activity was reduced also in the lateral white matter. Further, a tendency to lower ChAT activity in dorsal gray matter was found. In our study of ALS spinal cords, (14) pronounced and patchy reductions in ChAT

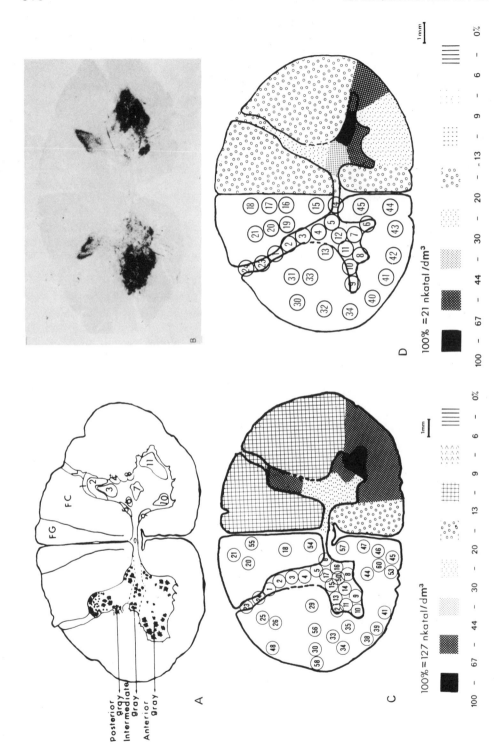

Figure 1A. Cervical segment C8 schematically [from Chambers and Liu (9)]. FG, fasciculus gracilis; FC, fasciculus cuneatus L, fasciculus dorso-lateralis (Lissauer). 1, Nucleus posterior marginalis; 2, substantia gelatinosa of Rolando; 3, nucleus proprius cornu dorsalis; 4, spino-reticular nucleus; 5, nucleus cornu commissuralis posterior; 6, nucleus magnocellularis basalis; 7, intermediomedialis (intermediate nucleus of Cajal); 8, intermediolateral cell column; 9, nucleus cornu commissuralis anterior; 10, anterior medial nuclear column; 11, anterior lateral nuclear column.

Figure 1B. AChE staining of human cervical segment C5 [from Ishii and Friede (19)].

Figure 1C. ChAT distribution in human cervical segment C6.
For details see Aquilonius, Eckernäs and Gillberg (1).

Figure 1D. ChAT distribution in cervical segment C8 in a case of ALS.

activity in the ventral horn was found, probably corresponding to an uneven reduction of motor neurons (Fig 1 D). The reductions were not restricted to the motor neuron area alone, and the normally high ChAT activity area in the dorsal horn was considerably reduced.

When muscarinic receptors were analyzed in homogenates from spinal cord tissue samples, both unchanged (18) and reduced numbers (30) of [^{3}H]-QNB binding sites have been reported in cords from ALS patients.

Using the in vitro autoradiographic technique Whitehouse and collaborators (41) have shown pronounced reductions in muscarinic receptors in the motor neuron areas, and less pronounced reductions in other laminae, especially those in the dorsal horn. The changes were caused solely by reduction in high-affinity agonist sites. Fig. 1F illustrates the extent of the changes in [^{3}H]-QNB binding, as depicted in an autoradiograph from our laboratory.

There are presently no definitive indications of a specific involvement of cholinergic structures in ALS. Increased knowledge regarding transmitter dysfunction might, however, create ideas of symptomatic drug therapy. To get further insight into functional aspects, it seems important to perform studies with the above-mentioned techniques in spinal cord sections from patients dying with other well defined spinal and supraspinal lesions.

ACh-SYNTHESIZING ACTIVITY AND CHOLINERGIC RECEPTORS IN MUSCLE

Animal studies. One aim of this project has been to evaluate the potential application of ACh-synthesizing (ACh-s) activity, as well as receptor quantitation, for the purposes of diagnosis in clinical muscle biopsies. Theoretically in diseases such as ALS, involving pronounced denervation-reinnervation phenomena, changes in muscular ACh-s activity as well as receptor changes would be expected.

It has earlier been shown (33, 37) that most but not all of the ACh-s activity of skeletal muscle disappears following complete denervation. Also it has been demonstrated that the residual ACh-s activity in muscle from frog (28) and rat (38) is different from ChAT, and probably referable to carnitine acetyltransferase.

We have followed the changes in ACh-s activity in rat hind limb muscles for 32 days after complete denervation (3), and for up to 6 months after a cryolesion of the sciatic nerve (7). In parallel the binding of the nicotinic ligands [^{3}H]-α-Btx and [^{3}H]-d-tubocurarine ([^{3}H]-dTC) to a muscle homogenate was studied. The results obtained during denervation-reinnervation are schematically illustrated in Fig. 2. It is interesting to note that normalization of α-Btx binding sites seems to be an earlier sign of reinnervation than normalization of ACh-s activity. The number of dTC binding sites was found to be

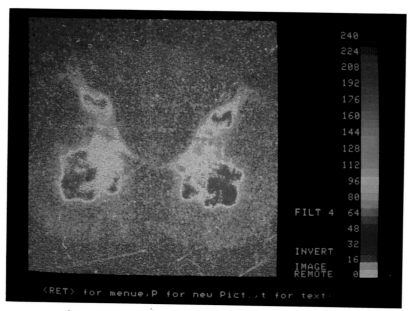

Figure 1E. [³H]-QNB binding sites in human cervical segment C_6. The density of grains is increasing with colour in the sequence: blue-green-yellow-red-white. The _in vitro_ autoradiographs were colour-coded using an OSIRIS-system (Department of Physics IV, Royal Institute of Technology, Stockholm, Sweden).

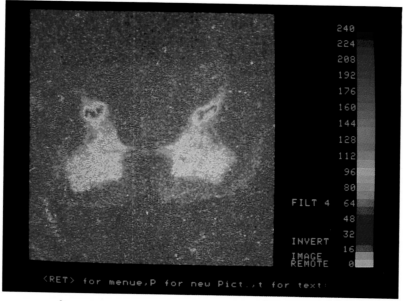

Figure 1F. [³H]-QNB binding sites in cervical segment C_6 in a case of ALS.

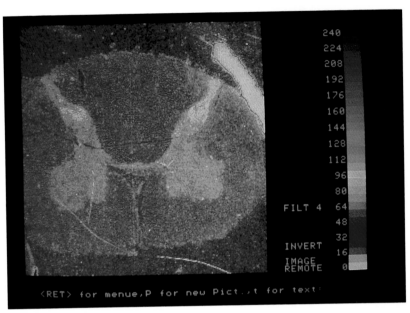

Figure 1G. [^{3}H]-etorphine binding sites in a section adjacent to that used in Fig. 1E.

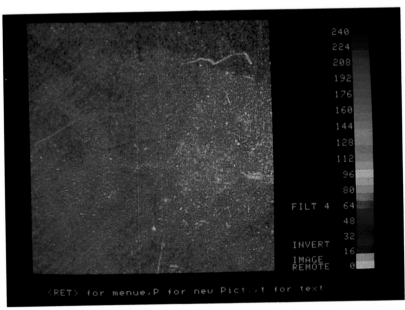

Figure 1H. [^{3}H]-etorphine binding in the presence of levorphanol (10^{-4}M).

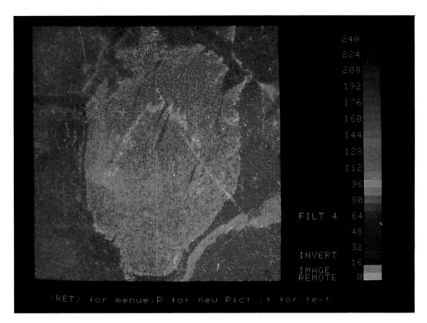

A. Control

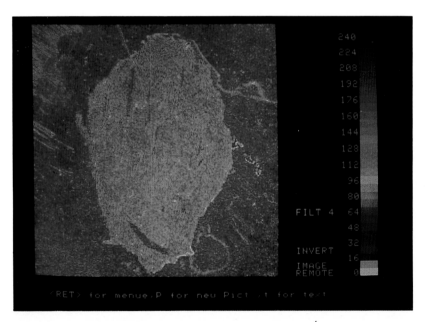

B. As A but displaced (dTC, 10^{-4}M)

Figure 3. Colour-coded (see Fig 1E) autoradiographs of $[^{3}H]$-α-Btx binding to rat tibialis anterior muscle.

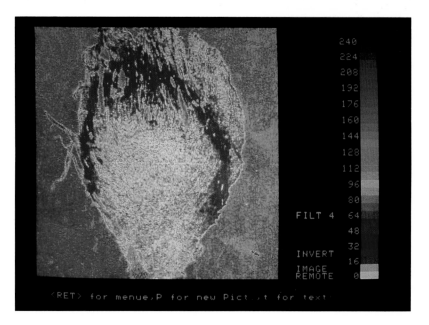

C. 8 days following cryolesion of the sciatic nerve

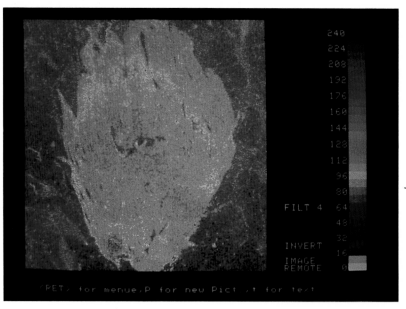

D. As C but displaced (dTC, 10^{-4}M)

considerably higher than α-Btx sites in normal muscle, but the dTC was relatively unchanged following nerve lesion. The preferential increase in α-Btx binding indicates the special characteristics of the newly formed receptors. In order to determine how these binding sites are distributed within the muscle, in vitro autoradiography in cryosections was performed. At a time of increased α-Btx binding in a homogenate (Fig. 2) an intense α-Btx binding was seen all over the muscle (Fig. 3). This phenomenon which has been known for a long time (for reference see 21) seems not to have been visualized before in a whole limb muscle.

Human studies. At operations such as thymectomy and thoracotomy, rib to rib biopsies of intercostal muscles have been obtained for determination of ACh concentration and ChAT activity (8, 29, 30). The ACh-s activity related to ChAT was localized at the end-plate region, which is in accordance with aforementioned animal studies from the same laboratory. Increased ChAT activity was found in intercostal muscles from patients suffering from myasthenia gravis; a finding which was explained as a compensatory mechanism to enable increased ACh release (29).

As the intercostal muscle is unsuitable for biopsy in the ambulatory patient, we have tried to analyze ACh-s activity in homogenates from routine biopsies of biceps, tibialis anterior and quadriceps muscles. When the ChAT inhibitor bromoacetylcholine (38) was used as a tool to identify "true" ChAT no ACh-s activity referable to ChAT was found in the biopsies, while in rat hindleg muscles processed

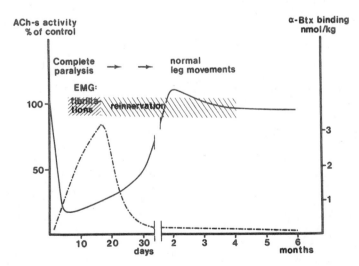

Figure 2. Schematic illustration of ACh synthesizing activity (———), [^{3}H]-α-Btx binding (—·—·) and EMG in rat peroneus longus muscle following cryolesion of the sciatic nerve.

in parallel, more than 50% of the ACh-s activity was regarded as ChAT. These negative results are probably explained by the fact that only a few motor end-plates were included in the biopsies used.

Since available detailed knowledge is sparse concerning the topographical distribution of end-plates in human muscles, horizontal sections of whole muscles were made by means of a large cryostat microtome (4) and stained for AChE. With this technique a topographical mapping of motor end-plate distribution in the biceps brachi (Fig. 4) and in the tibialis anterior muscles have been made (2, 5) such that recommendations concerning place for end-plate rich biopsies in these muscles could be given (5).

SUMMARY

By means of microdissection in combination with ChAT analysis, radioligand binding techniques and in vitro receptor autoradiography

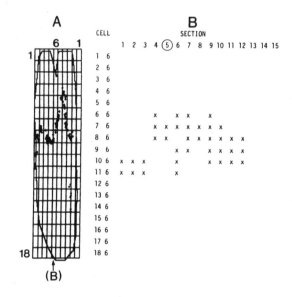

Figure 4. Distribution of end-plates in cryosections of whole human biceps muscle (5). Construction of a longitudinal section in a plane perpendicular to the original section. To the left, is shown a printout of the original muscle after digitization, corresponding to section 5 in a horizontal plane. To the right, perpendicular section (sagittal plane) along the middle of the muscle, corresponding to "cell" 6, which is also indicated to the left. The original and constructed sections are oriented in the same way but have different scales. Proximal end upwards. "X" in the constructed part indicates only the presence of endplates, and is not a quantitative indication of the number of end-plates.

the cholinergic representation within spinal cord and muscle has been studied in detail. In the human spinal cord there is a close correspondence in the distribution of AChE, ChAT and of muscarinic receptors, with predominance in the motor neuron areas and in the substantia gelatinosa of the dorsal horn.

In ALS, muscarinic receptors are markedly reduced in the motor neuron areas, while the reduction in ChAT is more widespread within the spinal section. There are not indications of a specific involvement of cholinergic structures in ALS but increased knowledge regarding transmitter dysfunction might create ideas of symptomatic drug therapy.

The changes in ChAT activity and in density and distribution pattern of α-Btx binding sites have been studied in rat hind limb muscles during the denervation-reinnervation process. "True" ChAT disappears after denervation while α-Btx binding sites increase and spread to the whole muscle. During reinnervation these markers are normalized. In human biceps and tibialis anterior muscle the topographical distribution of motor end-plates has been mapped by staining whole muscle cryosections for AChE. Recommendations concerning the site for end-plate-rich biopsies thus could be given on the basis of such observations. Using guidelines such as those described in this chapter, it might be possible to adopt ChAT analysis and quantitation of α-Btx binding as diagnostic measures of denervation and reinnervation in clinical research.

Acknowledgements. Studies in the laboratory of the authors were supported by the Swedish Medical Research Council (project No. 4373).

REFERENCES

1. Aquilonius, S.M., Eckernäs, S.Å. and Gillberg, P.G. (1981): Brain Res. 211:329-340.
2. Aquilonius, S.M., Arvidsson, B., Askmark, H. and Gillberg, P. G. (1982): Muscle Nerve 5:418.
3. Aquilonius, S.M., Askmark, H., Eckernäs, S.Å. and Nordberg, A. (1982): Acta Physiol. Scand. 114:345-350.
4. Aquilonius, S.M., Eckernäs, S.Å. and Gillberg, P.G. (1983): In IBRO Handbook Series: Methods in Neurosciences (ed) A. C. Cuello, John Wiley & Sons, London, pp. 155-170.
5. Aquilonius, S.M. Askmark, H. Gillberg, P.G., Nandekar, S., Olsson, Y. and Stalberg, E. (1984): Muscle & Nerve 7:287-293.
6. Arimatsu, Y., Seto, A. and Amano, T. (1981): J. Comp. Neurol. 198:603-631.
7. Askmark, H., Aquilonius, S.M., Fawcett, P., Nordberg, A. and Eckernäs, S.Å. (1982): Acta Physiol. Scand. 116: 429-435.
8. Braggaar-Schaap, P. (1979): J. Neurochem. 33:389-392.
9. Chambers, W.W. and Liu, C.N. (1983): In The Spinal Cord (ed) G.M. Austin, Ikagu-Shoin, New York, Tokyo, pp. 3-44.

10.   Charlton, G., Brennan, M.J.W. and Ford, D.M, (1981):   Brain
       Res. 218:372-375.
11.   Feldberg, W. and Vogt, M. (1948):   J. Physiol. 107:372-381.
12.   Fibiger, H.C. (1982):   Brain Res. Rev. 4:327-388.
13.   Giacobini, E. and Holmstedt, B. (1958):   Acta Physiol. Scand.
       42:12-27.
14.   Gillberg, P.G., Aquilonius, S.M., Eckernäs, S.Å., Lundqvist,
       G. and Windblad, B. (1982):   Brain Res. 250:394-397.
15.   Gillberg, P.G., Nordberg, A. and Aquilonius, S.M. (1984):
       Brain Res. 300:327-333.
16.   Gwyn, D.C. and Wolstencroft, J.H. (1966):   Science 153:1543-
       1544.
17.   Gwyn, D.G., Wolstencroft, J.H. and Silver, A. (1972):   Brain
       Res. 47:289-301.
18.   Hayashi, H., Suga, M., Satake, M. and Tsubaki, T. (1981):
       Ann. Neurol. 9:292-294.
19.   Ishii, T. and Friede, R.L. (1967):   In International Review
       of Neurobiology (eds) Pfeiffer and Smythies, Academic Press,
       New York, pp. 231-275.
20.   Jessell, T., Tsunoo, A., Kanazawa, I. and Otsuka, M. (1979):
       Brain Res. 168:247-259.
21.   Jones, R. (1979):   Brain Res. 49:391-402.
22.   Kan, K.S. K., Chao, L.P. and Eng, L.F. (1978):   Brain Res.
       146:221-229.
23.   Kanazawa, O., Sutoo, D., Oshima, I. and Saito, S. (1979):
       Neurosci. Lett. 13:325-330.
24.   Kayaalp, S.O. and Neff, N.H. (1980):   Brain Res. 196:429-436.
25.   Kimura, H., McGeer, P.L., Peng, J.H. and McGeer, E.G. (1981):
       J. Comp. Neurol. 200:151-201.
26.   Koelle, G.B. (1954):   J. Comp. Neurol. 100:211-235.
27.   Macintosh, F.C. (1941):   J. Physiol. 99:436-442.
28.   Molenaar, P. C. and Polak, R.L. (1980):   J. Neurochem. 35:1021-
       1025.
29.   Molenaar, P.C., Newsom-Davis, J., Polak, L. and Vincent,
       A. (1981):   J. Neurochem. 37:1081-1088.
30.   Molenaar, P.C., Newsom-Davis, J., Polak, L. and Vincent,
       A. (1982):   Neurology 32:1062-1065.
31.   Nagata, Y., Okuya, M., Watanabe, R. and Honda, M. (1982):
       Brain Res. 244:223-229.
32.   Ninkovic, M. and Hunt, S.P. (1983):   Brain Res. 272:57-69.
33.   Ranish, N.A., Kiauta, T. and Dettbarn, W.D. (1979):   J.
       Neurochem.32:1157-1164.
34.   Rexed, B. (1954):   J. Comp. Neurol. 100:297-379.
35.   Rotter, A., Birdsall, N.J.M. Burgen, A.S.V., Field, P.M.,
       Hulme, E.C., and Raisman, G. (1979):   Brain Res. Rev.
       1:141-165.
36.   Silver, A. and Wolstencroft, J.H. (1971):   Brain Res. 34:
       205-227.
37.   Tuček, S. (1973):   Exp. Neurol. 40:23-35.
38.   Tuček, S. (1982):   J. Physiol. 322:53-69.

39.   Wamsley, J.K., Lewis, M.S., Young, W.S. and Kuhar, M.J. (1981):
      J. Neurosci. 1:176-191.
40.   Weinstock, M., Davidson, J.T., Rosin, A.J. and Schneiden,
      H. (1982):   54:429-434.
41.   Whitehouse, P.J., Wamsley, J.K., Zarbin, M.A., Price, D.L.,
      Tourtellotte, W.W. and Kuhar, M.J. (1983):   Ann. Neurol.
      14:8-16.
42.   Yamamura, H.I., Wamsley, J.K., Deshmukh, P. and Roeski, W.
      R. (1983):   Eur. J. Pharmacol. 91:147-149.
43.   Young, W.S. and Kuhar, M.J. (1979):   179:255-270.

CENTRAL ACETYLCHOLINE AND STRESS INDUCED CARDIOVASCULAR, NEUROENDOCRINE AND BEHAVIORAL CHANGES: EFFECTS OF PHYSOSTIGMINE AND NEOSTIGMINE

D.S. Janowsky, S.C. Risch, J.C. Gillin,
M. Ziegler, B. Kennedy and L. Huey

Department of Psychiatry
San Diego Veterans Administration Medical Center and
University of California at San Diego
La Jolla, California 92093 U.S.A.

INTRODUCTION

A growing body of data indicates that central neurotransmitters and neuromodulators, including norepinephrine, serotonin, GABA, dopamine, vasopressin, and the opioid polypeptides, are likely to have a major role in the regulation of stress (1). In the following paragraphs, we will review and present evidence suggesting that central acetylcholine may also have an important role in the regulation of stress. Supportive of this, Gilad et al. (7) have observed stress-induced increases in central nervous system acetylcholine turnover and neuronal choline uptake, and down-regulation of muscarinic receptors in rats; and these effects are exaggerated in stress-sensitive rats. Furthermore, there is evidence from a number of studies and from our own work that many of the manifestations of naturally occurring stress are mimicked by administration of centrally acting cholinomimetic agents in both animals and in man.

Cholinergic Regulation of Stress-Sensitive Neuroendocrines. A variety of preclinical studies suggest that acetylcholine is important in stress and non-stress induced release of ACTH and cortisol. Several of these show that ACTH is released from rat hypothalamic preparations in a dose dependent manner by acetylcholine and other cholinomimetic agents, and that this effect, as well as stress induced ACTH release, is reversed by atropine (10, 18). There is also evidence that beta-endorphin is an important stress sensitive neurohormone (9). As with ACTH, which shares a common precursor molecule with beta-endorphin, and as with serum cortisol, serum beta-endorphin levels have been found to increase in animals (23) following physostigmine and arecoline administration, and these effects closely parallel ACTH release.

Prolactin is also highly sensitive to being released by stress. Although animal experiments generally suggest that serum prolactin is decreased following cholinomimetic administration (8), some animal data suggests that cholinomimetics do increase serum prolactin levels. Furthermore, all human studies performed to date using arecoline and physostigmine have demonstrated increased prolactin release (24).

Animal studies also show that growth hormone is released following administration of such cholinomimetics as acetylcholine, physostigmine, and pilocarpine, and that this effect is reversed by anticholinergic agents (4), and appears to be mediated peripherally (22).

Central Cholinergic Mechanisms in the Regulation of Pulse and Blood Pressure. A major effect of acute stress is its ability to increase pulse rate and blood pressure. Exposure of the central nervous system to virtually all muscarinic cholinomimetic agents increases blood pressure in a variety of animal species. This effect is reversible with centrally acting, but not peripherally acting antimuscarinic agents, thus suggesting that cholinomimetics increase blood pressure via a central, muscarinic mechanism. This central cholinergic mediation of blood pressure appears to occur via the muscarinic activation of central noradrenergic neurons, leading to increased peripheral sympathetic neuronal outflow (3). Several animal studies have also demonstrated that acetylcholine may also be of significance in the regulation of hypertension. Mean blood pressure is more dramatically increased by physostigmine in spontaneously hypertensive rats (SHR), a phenomenon which occurs in young SHRs even before hypertension has become manifest (5, 20). Conversely atropine but not methylatropine, can lower blood pressure in the SHR, an effect which also occurs after administration of the acetylcholine synthesis inhibitor, hemicholinium-3 (2).

Central Cholinergic Regulation of Sympatho-Adrenomedullary Function. Peripheral release of catecholamines, such as epinephrine and norepinephrine, regularly occurs following acute physical and/or mental stress. As with other pituitary and adrenalcortical neurohormones, epinephrine and norepinephrine are released in animals following administration of physostigmine and other cholinomimetics, and these effects are blocked by administration of centrally acting, but not peripherally acting anticholinergic agents (19).

Cholinergic Effects on Behavior, Mood and the Perception of Pain. It is well known that stress has profound effects on behavior, cognition, and perception. Specifically, there is much evidence that stress can cause increases in analgesia, anxiety, aggression and depression in appropriate animal models, as well as in man, and that centrally acting cholinomimetic agents produce similar effects.

With respect to the phenomenon of analgesia, intermittent, prolonged, uncontrollable stress causes a naloxone-reversible analgesia, which is likely to be due to the activation of opioid polypeptides (21). Several lines of evidence suggest that this analgesia may be mediated by acetylcholine, since in animals it can be blocked by the centrally acting anticholinergic drug scopolamine, and since cholinomimetic administration causes atropine reversible analgesic effects in animals and in man (21, 26). Similarly, major parallels exist between the emotional effects of stress and those of cholinomimetic agents.

Acetylcholine-GABA-Bendodiazepine Interactions. Finally, there is evidence that acetylcholine and the presumed antistress neurochemical, GABA, are antagonistic. GABA has been shown to block acetylcholine induced release of corticotropin releasing factor (CRF) from isolated rat hypothalami (17). Furthermore, in man, several studies have reported that physostigmine is an effective agent for reversing the soporific effects of benzodiazepine overdosage, an effect postulated to occur via antagonism of the GABA system (6).

METHODS

To test the stress-like effects of cholinomimetic drugs in man, we have compared the effects of physostigmine and neostigmine. Subjects were 12 adult male psychiatric patients of mixed diagnoses who were in good health and between the ages of 18 and 55. Written, informed consent from each subject was obtained in full accordance with institutional requirements. Subjects stayed food- and fluid-free overnight, and off psychoactive drugs for at least one week. To protect from peripheral cholinergic effects, on the two days of the experiment, each subject was given an intramuscular injection of 0.75 mg methscopolamine, a non-centrally acting anticholinergic agent. Thirty minutes later, the first 10 subjects received, over a 10 minute period, on a double-blind basis, in counterbalanced order, using a crossover design, .022 mg/kg (up to 2.0 mg) physostigmine salicylate (Antilerium) on one day and equipotent dose of the non-centrally acting cholinesterase inhibitor, neostigmine (.011 mg/kg) two or more days before or later. The next two subjects received a .022 mg/kg dose of neostigmine instead of the .011 mg/kg dose of neostigmine. Atropine 1.0 mg intramuscular was given 20 minutes after the physostigmine and neostigmine infusions ended to abort central effects.

In addition, to further test the effects of physostigmine on stress relevant parameters, 54 psychiatric patients, after receiving methscopolamine pretreatment as described above, received physostigmine, .022 mg/kg, on one occasion and saline placebo, instead of neostigmine, on another.

All subjects had sitting diastolic and systolic blood pressure measured, using a standard pressure cuff, and pulse rates were evaluated before administration of the anticholinergic medications, at -15 and -1 minutes before physostigmine, neostigmine, or placebo infusion (after the anticholinergic agent had taken effect); and at +10, +20, +30, +45 and +60 minutes after the infusion had ended.

Behavior was measured before and after drug infusion, as described previously (23, 24), using the Profile of Mood States, BPRS, and Janowsky-Davis Activation-Inhibition Scale.

Blood samples were drawn in these first 10 patients at -30, -15, -1 and at +10, +20, +30 minutes after each drug or placebo infusion. Samples were later analyzed for serum epinephrine and norepinephrine levels using the method of Ziegler et al. (27), and for serum ACTH, beta-endorphin, cortisol, growth hormone, and prolactin by previously described methods (23, 25). In the first ten subjects, maximum nausea and vomiting were evaluated by rating the patients' status using a 0-3 point scale, in which 0=no nausea; 1=mild nausea; 2=nausea with gagging; and 3=vomiting.

ANOVA techniques were utilized to determine the main effects of physostigmine versus neostigmine and versus placebo on behavior, blood pressure, pulse rate, serum catecholamines, and serum neuro-endocrines. Data are presented in tabular form, comparing the subjects treated with both physostigmine and neostigmine, and presented descriptively on the patients receiving placebo and physostigmine.

RESULTS

Cardiovascular Changes. Physostigmine, to a significantly greater extent than neostigmine, caused an increase in pulse rate and systolic blood pressure in the 12 subjects in whom it was studied. These effects are shown in Table 1. In the physostigmine and placebo treated groups, pulse rate, systolic and diastolic blood pressures also increased significantly more following physostigmine adminis-tration than after placebo administration.

Neuroendocrine Changes. As shown in Table 2, physostigmine alone caused larger increases in serum prolactin, cortisol, beta-endorphin, and ACTH. Similar changes were noted in the psychiatric patients in whom physostigmine and placebo were compared. Physostigmine, but not neostigmine dramatically increased serum epinephrine levels, but not norepinephrine, as shown in Table 3.

Behavioral Effects. As shown in Table 4, physostigmine, in contrast to neostigmine, caused an increase in behavioral inhibition, a decrease in activation, and an increase in negative affect. These results are consistent with the results of the experiment with 54 psychiatric patients who received physostigmine on one occassion and placebo on another. However, in these 54 patients, additional changes in self-rated behavior occurred, with decreases in self-rated

Table 1

Comparison of the effects on pulse rate and
blood pressure of physostigmine and neostigmine
in twelve methscopolamine pretreated patients

	Maximum Pre-Infusion Values	Maximum Post-Infusion Values
Systolic BP (mmHg)		
Physostigmine	117.8 ± 2.6	126.7 ± 3.1*
Neostigmine	120.6 ± 2.5	121.0 ± 2.9
Diastolic BP (mmHg)		
Physostigmine	78.1 ± 1.9	84.1 ± 3.0
Neostigmine	78.1 ± 2.0	81.7 ± 2.2
Pulse (Beats/Min)		
Physostigmine	106.5 ± 2.7	119.7 ± 4.4*
Neostigmine	103.7 ± 3.0	104.2 ± 2.8

*$p < .05$ comparing neostigmine and physostigmine maximum pre-to post-infusion values

activation on the Activation-Inhibition Scale and friendliness on the POMS Scale, and with increases in fatigue and hostility occurring on the POMS Scale to a significantly greater extent following physostigmine.

Nausea-Emetic Effects. Physostigmine caused a variable response with respect to nausea and emesis. No changes in nausea-emesis ratings occurred after neostigmine administration. Physostigmine caused some degree of nausea in 9 of the 10 subjects. Four subjects had nausea ratings of 2 and five had nausea ratings of 1. Similar changes occurred in the 54 psychiatric patients in whom physostigmine and placebo were compared. Here, physostigmine caused a variable response in nausea-emesis, while placebo had no such effects. Ten subjects received nausea-emesis ratings of 0, 28 subjects had ratings of 1, six had ratings of 2, and ten had ratings of 3 (i.e., emesis).

DISCUSSION

The above information offers evidence that most of the major manifestations of naturally occurring stress are mimicked in both animals and in man by the centrally active muscarinic cholinomimetic, physostigmine, but not by non-centrally acting neostigmine, and

Table 2

Effects of intravenous physostigmine (.022 mg/kg) and neostigmine (.011 mg/kg) on stress sensitive neurohormones in ten psychiatric patients

	Physostigmine***		Neostigmine		F-Value
	Pre	Post	Pre	Post	
Growth Hormone	2.5 ± 0.2	3.2 ± 0.6	2.2 ± 0.2	1.8 ± 0.0	2.08
Prolactin	9.9 ± 3.6	42.7 ± 10.4	9.4 ± 3.5	8.7 ± 3.1	8.65*
Beta-Endorphin	15.2 ± 1.4	29.6 ± 5.0	13.4 ± 2.0	13.1 ± 2.0	6.99*
ACTH	39.4 ± 18.4	148.6 ± 35.4	17.5 ± 1.2	19.6 ± 2.2	9.24*
Cortisol	6.4 ± 0.7	10.8 ± 1.5	5.9 ± 0.3	5.9 ± 0.4	14.71**

*p < .05; **p < .01; ***pre = average of all pre-physostigmine samples; ***post = average of all +10 and +20 minute post-physostigmine samples

Table 3

Effects of intravenous physostigmine (.022 mg/kg) and neostigmine (.011 mg/kg) on stress sensitive adrenal-medullary hormones in ten psychiatric patients

	Physostigmine**		Neostigmine		F-Value
	Pre	Post	Pre	Post	
Epinephrine	56.4 ± 12.3	346.9 ± 83.4	44.3 ± 6.5	41.8 ± 9.8	14.00*
Norepinephrine	370.0 ± 55.4	375.7 ± 41.1	334.6 ± 47.4	218.3 ± 42.2	.18

*p < .01; **pre = average of all pre-physostigmine samples; **post = average of +10 and +20 minutes post-physostigmine samples

Table 4

Effects of intravenous physostigmine (0.22 mg/kg) and neostigmine (.011 mg/kg) on behavioral measures in ten psychiatric patients

Behavioral Scales	Physostigmine**		Neostigmine		F-Value
	Pre	Post	Pre	Post	
BPRS Scale					
Anergia	5.5 ± 0.2	8.3 ± 0.7	5.3 ± 0.3	5.5 ± 0.3	10.5*
Depression	8.1 ± 0.7	10.8 ± 0.7	7.6 ± 0.8	8.2 ± 0.6	10.3*
Total	33.3 ± 1.4	42.8 ± 2.3	33.4 ± 1.5	33.6 ± 1.1	12.3*
A-I Scale-Observer rated					
Inhibition	9.0 ± 2.1	25.7 ± 3.8	8.2 ± 2.2	10.1 ± 1.6	14.5**
Activation	7.5 ± 0.8	4.2 ± 0.9	7.5 ± 0.8	7.1 ± 0.4	6.5*
Dysphoria	2.3 ± 0.6	4.0 ± 0.7	1.5 ± 0.4	1.3 ± 0.3	5.7*
A-I Scale-Self rated					
Inhibition	25.5 ± 4.3	35.9 ± 5.0	21.8 ± 3.7	15.4 ± 3.8	8.3*
Activation	8.8 ± 1.4	6.7 ± 1.5	8.9 ± 1.2	9.2 ± 1.0	1.9
Dysphoria	4.9 ± 1.2	5.6 ± 1.4	3.6 ± 0.9	3.4 ± 1.2	.3
POMS Subscales					
Tension/Anxiety	8.3 ± 1.5	14.5 ± 3.4	7.6 ± 1.9	6.5 ± 1.2	5.7*
Depression	20.9 ± 4.8	21.5 ± 5.2	16.5 ± 3.9	13.7 ± 3.9	.4
Anger/Hostility	5.8 ± 2.8	9.0 ± 4.3	6.9 ± 2.0	4.3 ± 1.4	4.2
Vigor	8.9 ± 2.0	4.6 ± 1.5	9.9 ± 1.7	9.7 ± 1.8	8.9*
Fatigue	6.8 ± 2.3	9.8 ± 2.2	5.1 ± 1.8	3.8 ± 1.4	3.5
Confusion	9.3 ± 1.8	12.1 ± 2.1	8.3 ± 1.7	9.7 ± 1.4	.4
Elation	5.2 ± 1.4	3.6 ± 1.5	5.3 ± 0.9	6.0 ± 1.0	6.2*
Friendliness	15.4 ± 2.6	11.7 ± 2.6	15.0 ± 2.6	16.3 ± 1.4	4.9

*p < .05; **p < .01

that these effects are probably blocked by centrally acting anticholinergic drugs. Our results in man are similar to previous observations of the effects of cholinomimetics in animals and our own
previous studies comparing placebo and physostigmine (11, 13-16,
23, 25). Although acetylcholine may be regulating stress, it is
also possible that cholinomimetics may be causing discomfort, and
thus may be non-specifically activating central stress mechanisms.
However, although it is clear that the cholinomimetic syndrome generally is not pleasant, one study has reported the existence of physostigmine-induced behavioral, blood pressure, pulse rate, cortisol,
and prolactin increases in selected subjects who showed no nausea
and minimal anergia. Also, effects of physostigmine may precede
the nauseating effects by 5-10 minutes (12).

Like the current data, earlier studies with arecoline and physostigmine have been found to show cholinomimetic induced increases
in pulse rate and blood pressure levels in man, an effect which is
exaggerated in prehypertensives (13). Similarly, in man, epinephrine
release, presumably of adrenal medullary origin, has been shown to
increase dramatically following physostigmine infusion, an effect
which is not associated with increases in norepinephrine (16).
Finally, like the effects noted above, a variety of centrally acting
cholinesterase inhibitors and cholinergic agonists, including diiso-
propylflurophosphate (DFP), physostigmine and arecoline have been
shown to cause anxiety, depression, hostility, anergia, and fatigue
in normals, and to exaggerate these effects in patients with affective
disorders (14).

In conclusion it is not inconceivable that acetylcholine itself
is a major or primary moderator of stress. Indeed, in considering
the various central neurotransmitters and neuromodulators implicated
in stress regulation, such as norepinephrine, serotonin, GABA, opioid
polypeptides, vasopressin, epinephrine, and angiotensin, we know
of none except acetylcholine which simultaneously activates the
emotional, analgesic, hypertensive, tachycardiac, pituitary-adreno-
cortical and adrenal medullary manifestations of stress.

Acknowledgement. This work was supported by NIMH Grant #MH 30914.

REFERENCES

1. Anisman, H. and Zacharko R.M. (1983): Behav. and Brain Sci.
   5(1):89-137.
2. Brezenoff, H.E. and Caputi, A.P. (1980): Life Sci. 26:1037-
   1045.
3. Brezenoff, H.E. and Giuliano, R. (1982): Ann. Rev. Pharmacol.
   Toxicol. 22:341-381.
4. Bruni, J.F. and Meites, J. (1978): Life Sci. 23:1351-1358.
5. Buccafusco, J.J. and Spector, S. (1980): J. Cardiovasc. Pharmacol. 2:347-355.

6. Ghoneim, M.M. (1980): Anesthesiology 52:372.
7. Gilad, G.M., Rabey, J.M. and Shenkman, L. (1983): Brain Research 267(1):171-174.
8. Grandison, L. and Meites, J. (1976): Fed. Proc. 35:306.
9. Guillemin, R., Vargo, T., Rossier, J. et al. (1977): Science 197:1367-1369.
10. Hillhouse, E.W., Burden, J. and Jones, M.T. (1975): Neuroendocrinology 17(10):1-11.
11. Janowsky, D.S., El-Yousef, M.K. and Davis, J.M. (1974): Psychosom. Med. 36(3):248-257.
12. Janowsky, D.S., El-Yousef, M.K., Davis, J.M. and Sekerke, H.J. (1973): Arch. Gen. Psychiatry 28:542-547.
13. Janowsky, D.S., Risch, S.C., Huey, L.Y., Judd, L.L. and Rausch, J. (1983): Psychopharm. Bull. 19(4):675-681.
14. Janowsky, D.S., Risch, S.C., Judd, L.L., Parker, D.C., Kalin, N.H. and Huey, L.Y. (1983): In Treatment of Depression: Old Controversies and New Approaches (eds) P. Clayton and J.E. Barrett, Raven Press, New York, pp. 61-74.
15. Janowsky, D.S., Risch, S.C., Ziegler, M., Kennedy, B., Judd, L.L. and Huey, L.Y. (1984): In Catecholamines (eds) E. Usdin, A. Carlsson, A. Dahlstrom and J. Engel, Alan R. Liss, Inc., New York, pp. 327-335.
16. Janowsky, D.S., Risch, S.C., Ziegler, M., Kennedy, B., Judd, L.L. and Huey, L.Y. (in press): In Stress: The Role of Catecholamines and Other Neurotransmitters (ed) E. Usdin, Gordon and Breach, New York.
17. Jones, M.T. and Hillhouse, E.W. (1977): In ACTH and Related Peptides: Structure, Regulation and Action (eds) D.T. Kreiger and W.R. Ganong, New York Academy Sciences, New York, pp. 536-560.
18. Kaplanski, J. and Smelik, P.G. (1973): Acta Endocrinology 73:651-659.
19. Kaul, C.L. and Grewal, R.S. (1968): J. Pharm. Sci. 57:1741.
20. Kubo, T. and Tatsumi, M. (1979): Naunyn Schmiedebergs Arch. Pharmacol. 306:81-83.
21. Lewis, J.W., Terman, G.W. and Liebeskind, J.C. (1982): Abstracts of the 13th Collegium Internationale Neuro-Psychopharmacologicum Congress, Jerusalem, Israel, June 1982.
22. Mendelson, W.B., Lantingua, R.A., Wyatt, R.J., Gillin, J.C. and Jacobs, L.S. (1981): J. Clin. Endo. Metab. 52(3): 309-415.
23. Risch, S.C., Cohen, R.M., Janowsky, D.S., Kalin, N.H. and Murphy, D.L. (1980): Science 209:1545-1546.
24. Risch, S.C., Kalin, N.H. and Janowsky, D.S. (1981): J. Clin. Psychopharmacology 1(4):186-192.
25. Risch, S.C., Kalin, N.H., Cohen, R.M., Weker, J., Insel, T.R., Cohen, M.L. and Murphy, D.L. (1981): Peptides 2:95-97.
26. Sitaram, N. and Gillin, J.C. (1979): In Brain Acetylcholine and Neuropsychiatric Disease (eds) P. Berger and K. Davis, Plenum Press, New York, pp. 311-343.

27. Ziegler, M.G., Durrett, L.R. and Milano, A. (in press): In _Norepinephrine_ (eds) M.G. Ziegler and C.R. Lake, Williams and Wilkins, Baltimore.

MUSCARINIC SUPERSENSITIVITY OF ANTERIOR PITUITARY ACTH

AND β-ENDORPHIN RELEASE IN MAJOR DEPRESSIVE ILLNESS

S.C. Risch, D.S. Janowsky and J.C. Gillin

Department of Psychiatry, M-003
University of California, San Diego
La Jolla, California 92093 U.S.A.

INTRODUCTION

Studies from a number of centers have suggested that there may be an "up-regulation" or supersensitivity of muscarinic cholinergic receptor function in major depressive illness. We have previously demonstrated that major affective disorder patients may be behaviorally supersensitive to the dysphoric-inhibitory effects of centrally active cholinesterase inhibitors (18). In addition, Sitaram, Gillin and co-workers (22) have reported that early rapid eye movement (REM) latency, a major biological marker for depressive illness, may reflect muscarinic supersensitivity of the REM axis in sleep. Janowsky and co-workers (10) have recently reported that affective disorder patients may also have an increased sensitivity to the cardiovascular effects of centrally active cholinergic agents. Finally, Nadi and co-workers (13) have recently reported an increased number of muscarinic receptors on fibroblasts in patients with major depressive illness, which may possibly be a genetically transferrable trait marker for depressive illness. Consequently, a number of investigations have implicated muscarinic supersentitivity in major depressive illness.

Numerous studies have suggested that central acetylcholinergic activity may stimulate hypothalamic-pituitary adrenal (HPA) function. Thus, in vitro (2, 5) in vivo animal (1, 3, 6, 11, 14, 15, 23, 24) and human (3, 4, 7, 9) investigations suggest that cholinergic agonists may activate the HPA axis to increase adrenal cortisol secretion. We have recently demonstrated that centrally active cholinergic agonists also increase plasma concentrations of ACTH (21) and β-Endorphin (16, 17, 19, 20, 21).

Since, as reviewed above, numerous physiological systems display muscarinic receptor supersensitivity in major depressive illness, we have hypothesized that anterior pituitary release of ACTH and β-Endorphin immunoreactivity may also be muscarinically supersensitive in depression.

METHODS

These studies were conducted under an FDA-approved IND and with local human subjects committee approval. All subjects gave written informed consent, were males, and there was no significant difference in age composition among experimental and control groups. All subjects also were off medications for at least 2 weeks prior to participating in the studies, and had received a Schedule of Affective Disorder and Schizophrenia (SADS) and Research Diagnostic Criteria (RDC) diagnosis, based on an interview by an experienced diagnostician and collateral information obtained from previous hospitalizations, medical records, and interviews with family members.

Differences in cholinergically stimulated hypothalamic-pituitary-adrenal plasma β-Endorphin, ACTH and cortisol concentrations among individuals were determined as follows. After an 8 AM insertion of an indwelling heparin lock venous catheter and subsequent plasma samplings, each subject received methoscopolamine, 1.0 mg IM, 20 min prior to physostigmine or placebo infusions to induce partial peripheral cholinergic blockade. Each individual received physostigmine salicylate (22 μg/kg) and placebo (sterile saline), 10 cc infused over 10 min, in a randomized, double-blind, counterbalanced paradigm. Physostigmine and placebo infusions were separated by 3 to 7 days. Plasma samples were obtained from the indwelling venous heparin lock catheters at 10, 20, 30, 45, and 60 min subsequent to the physostigmine and placebo infusions. The average of the three base-line preinfusion plasma β-Endorphin, ACTH and cortisol concentrations were subtracted from peak postinfusion plasma concentrations, to determine each individual's physostigmine- and placebo-associated increase in concentrations of plasma β-Endorphin, ACTH and cortisol immunoreactivity.

Four subject groups were compared with respect to cholinergically stimulated hypothalamic-hypophyseal plasma β-Endorphin, ACTH and cortisol increases. The groups compositions were SADS/RDC: Normals, Major Affective Disorder; Depressed, Nonaffective Disorder psychiatric patients; and Schizoaffective Disorder, Depressed.

All bloods were immediately placed on ice; cortisol bloods in glass heparinized Vacutainer® tubes, and β-Endorphin and ACTH bloods in polypropylene tubes containing EDTA (20 mM); and spun at 3,000 rpm in a refrigerated centrifuge.   The resultant plasma was immediately placed in a -80°C freezer until assayed one week to three months later.

Plasma β-Endorphin and cortisol determinations were performed using radioimmunoassay kits developed by New England Nuclear Company, with antiserum from rabbits prepared against synthetic human βEndorphin and cortisol 21-succinyl bovine albumin, respectively.   The antibody for β-Endorphin demonstrates a 50% cross-reactivity with β-lipotropin, but less than 0.01% with α-Endorphin, α-MSH, and less than 0.004% with leucine enkephalin and methionine enkephalin.   All samples for plasma β-Endorphin and cortisol immunoreactivity were assayed within the same assays, and blindly by the investigators.   Within assay variability for β-Endorphin is approximately 5% and sensitivity is 5 pg/ml.   Within assay variability for cortisol is approximately 4% and sensitivity is 2 μg/dl.

Plasma ACTH concentrations were determined as follows: All samples were assayed in duplicate within the same assay, using an equilibrium radioimmunoassay utilizing a rabbit anti-porcine ACTH antibody from Immunoclear Corporation.   The antibody has less than 0.1% cross reactivity with α-MSH, β-Endorphin, β-lipotropin, leucine and methionine enkephalin, bombesin, parathyroid hormone, prolactin, somatostatin, neurotensin, thyroid stimulating hormone, calcitonin, follicle stimulating hormone, leuteinizing hormone, human growth hormone, and substance P.   Within-assay variability for ACTH is approximately 10% and sensitivity is 2 pg/ml.   Significance of changes in plasma cortisol, ACTH and βEndorphin immunoreactivity after physostigmine and saline were determined by repeated measures analysis of variance.

RESULTS

In all subject groups, physostigmine (relative to placebo) caused significant increases in plasma concentrations of cortisol, ACTH, and β-Endorphin immonoreactivity.

Physostigmine caused significantly greater increases in plasma concentrations of ACTH immunoreactivity [p < 0.001 (Figure 1)] and β-Endorphin immunoreactivity, [p < .0001 (Figure 2)] in subjects with Major Affective Disorder or Schizoaffective Disorder, than in Normal Control subjects or in control psychiatric subjects without Major Affective disorder.

There were no significant differences in the effects of physostigmine on plasma cortisol concentrations in the different nosological groups (Figure 3).

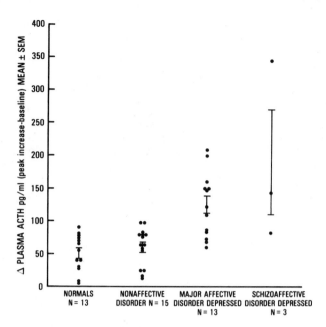

Figure 1. Scatter diagram of maximal physostigmine-associated increases in ACTH immunoreactivity in four nosological groups of subjects.

DISCUSSION

These results parallel our earlier reports that centrally active
acetylcholinesterase inhibitors stimulate the HPA axis, resulting
in increases in plasma ACTH, β-Endorphin and cortisol immunoreac-
tivities.

These results are consistent with the possibility that subjects
with major depressive illness may have muscarinic supersensitivity
of the hypothalamic-pituitary (ACTH - β-Endorphin) axis, and that
such supersensitivity might potentially represent a "biological
marker" for depressive disorders, as well as implicate muscarinic
mechanisms in the pathophysiology of affective disorders.

It is of interest that the cholinomimetic associated supersensitivity of anterior pituitary ACTH release in patients with
major depressive illness (Figure 1) was not paralleled by supersensitive adrenal cortisol release (Figure 3). We have recently
reported a poor and nonsignificant correlation between cholinomimetic
associated increases in peak plasma ACTH and cortisol concentrations
(21).

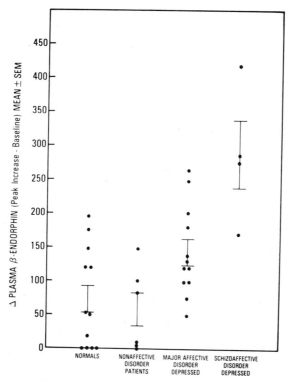

Figure 2. Scatter diagram of maximal physostigmine-associated increases in β-Endorphin immunoreactivity in four nosological groups of subjects.

In fact, over 10 years ago Krieger and associates reported that while there was a close temporal correlation between plasma ACTH and corticosteroid peaks, there was no apparent proportionality between spontaneously released plasma ACTH and corticosteroid levels, either within or among individuals (12). This lack of proportionality may be reflective of differing adrenal receptor sensitivities to ACTH, or differences in metabolic clearance rates of plasma ACTH and cortisol. Krieger and coinvestigators have also suggested that the magnitude of corticosteroid response to ACTH is in part dependent on the recent history of prior adrenal exposure to ACTH. In addition, Holaday and associates (8) have reported a synchronized ultradian cortisol rhythm in monkeys which persists during supramaximal infusions of ACTH, and have suggested therefore that bursts of cortisol secretion are not entirely dependent upon an immediately preceding release of ACTH.

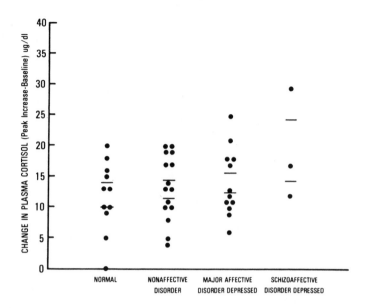

Figure 3. Scatter diagram of maximal physostigmine-associated increases in cortisol immunoreactivity in four nosological groups of
subjects.

       In summary, our data suggest that central muscarinic mechanisms
may be important in the physiology of HPA function, and that patients
with major depressive illness may have a muscarinic receptor supersensitivity of this peptidergic axis.

Acknowledgements.   Supported by Research Scientist Award, 1 K01
MH00393-01 and PHSMH 30914-06 and San Diego Veterans Administration
Medical Center.

REFERENCES

1. Balfour, D.J.K., Khyllar, A.K. and Longden, A. (1975):   Phar-
    mac. Biochem. Behav. 3:179-184.
2. Bradbury, M.W.B., Burden, J., Hillhouse, E.W. and Jones, M.
    T. (1974):   J. Phyisol. Lond. 239:269-283.
3. Cozanitis, D.A. (1974):   Anaesthesia 29:163-168.
4. Davis, B.M. and Davis, K.L. (1980):   Biol. Psychiat. 15:303-
    310.
5. Edwardson, J.A. and Bennett, G.W. (1974):   Nature 251:425-427.
6. Endroczi, E., Schreiberg, G. and Lissak, V. (1963):   Acta Physi-
    ol. Hung. 24:211-221.
7. Hill, P. and Wyndber, E.S. (1974):   Am. Heart J. 87:491-496.

8. Holaday, J.W., Martinez, H.M. and Natelson, B.H. (1977): Science 198:56-58.
9. Janowsky, D.S., Risch, S.C., Parker, D., Huey, L.Y. and Judd, L. (1980): Psychopharmacology Bull. 16:29-31.
10. Janowsky, D.S., Risch, S.C., Huey, L.Y., Judd, L.L. and Rausch, J.L. (1982): Physostigmine-Induced Cardiovascular Changes. Behavioral and Neuroendocrine Correlations. Presented at the Annual Meeting of the American College of Neuropsychopharmacology, San Juan, Puerto Rico.
11. Krieger, H.P. and Krieger, D.T. (1970): Am. J. Physiol. 218: 1632-1641.
12. Krieger, D.T., Allen, W., Rizzo, F. and Krieger, H.P. (1971): J. Clin. Endocr. Metab. 32:266-284.
13. Nadi, N.S., Nurnberger, J.I. and Gershon, E.S. (1982): Presented at the Annual Meeting of the American College of Neuropsycho-pharmacology. San Juan, Puerto Rico.
14. Naumenko, E.V. (1968): Brain Res. 11:1-10.
15. Otsuka, K. (1966): Tohoku J. Exp. Med. 88:165-170.
16. Risch, S.C., Cohen, R.M., Janowsky, D.S., Kalin, N.H. and Murphy, D.L. (1980): Science 209:1545-1546.
17. Risch, S.C., Kalin, N.H., Cohen, R.M., Weber, J., Insel, T.R., Cohen, M.L. and Murphy, D.L. (1981): Peptides 2:95-97.
18. Risch, S.C., Kalin, N.H. and Janowsky, D.L. (1981): J. Clin. Psychopharm. 1(4):180-185.
19. Risch, S.C. (1982): Biological Psychiatry 17(10):1071-1080.
20. Risch, S.C., Janowsky, D.S., Judd, L.L., Rausch, J.L., Huey, L.Y., Beckman, K.A., Cohen, R.M. and Murphy, D.L. (1982): Peptides 3:319-322.
21. Risch, S.C., Kalin, N.H., Janowsky, D.S., Cohen, R.M., Pickar, D. and Murphy, D.L. (1983): Science 222:77.
22. Sitaram, N., Nurnberger, J.I. and Gershon, E.S. (1980): Science 208:200-202.
23. Suzuki, T., Hirai, K., Toshio, H., Kurouji, I. and Hirose, T. (1964): J. Endocr. 31:81-82.
24. Suzuki, T., Abe, K. and Hirose, T. (1975): Neuroendocrinology 17:75-82.

CHOLINERGIC RECEPTOR BINDING IN THE

FRONTAL CORTEX OF SUICIDE VICTIMS

M. Stanley

Departments of Psychiatry and Pharmacology
Wayne State University School of Medicine
and
Division of Pharmacology
LaFayette Clinic
Detroit, Michigan 48207 U.S.A.

INTRODUCTION

A number of studies using a variety of experimental approaches have suggested that changes in the cholinergic system may be related to affective disorders. Patients with a diagnosis of affective disorder are reported to have increased sensitivity to cholinergic agonists as measured by changes in REM sleep latency (18). Physostigmine reduces the manic symptoms of manic-depressives and exacerbates the symptoms of depression in depressives (9, 10). Additionally, there have been reports that physostigmine induces depressive symptoms in normals (16). The latter observation is also consistent with reports that exposure to organophosphate insecticides (irreversible cholinesterase inhibitors) is associated with depressive symptoms (8). More recently increases in the density of muscarinic cholinergic receptors in skin fibroblasts of manic-depressive patients has been reported (14). In sum, these findings are suggestive of an increased sensitivity in the cholinergic systems of individuals with an affective disorder (as well as the introduction de novo of depressive symptoms in normals).

Because there is a high incidence of individuals diagnosed as having an affective disorder who subsequently commit suicide (1, 6, 17) we thought it would be of interest to determine QNB binding in the brains of a large sample of suicide victims, and to compare these findings with a well-matched control group.

Table 1

Patient Characteristics

		SUICIDES						CONTROLS			
Age	Sex	Cause of Death	Postmortem Delay (min)	B_{max}	K_d	Age	Sex	Cause of Death	Delay (min)	B_{max}	K_d
46	M	Hanged	1440	191	17	45	M	Gunshot wound	1650	232	8
13	M	Gunshot wound	1140	197	15	21	M	Gunshot wound	1205	242	17
15	M	Hanged	555	677	13	22	M	Cardiovascular	750	735	14
25	M	Gunshot wound	1560	387	28	20	M	Gunshot wound	1570	411	23
33	M	Gunshot wound	1020	773	15	31	M	Cardiovascular	880	423	11
55	M	Jumped from height	1140	388	16	47	M	Cardiovascular	1200	526	13
25	M	Hanged	1365	450	10	28	M	Auto accident	735	439	8
30	M	Hanged	1320	640	11	18	M	Gunshot wound	805	649	19
34	M	Jumped from height	1320	454	13	39	M	Fell from height	1245	510	11
22	M	Gunshot wound	555	671	10	30	M	Auto accident	460	527	10
25	M	Drowned	1005	498	8	26	M	Fell from height	1350	652	12
80	M	Gunshot wound	1335	555	24	53	M	Cardiovascular	860	393	36
18	M	Gunshot wound	1055	505	16	24	M	Cariovascular	1305	485	12
30	M	Jumped from height	1260	553	9	23	M	Gunshot wound	1035	563	10
37	M	Overdose	795	605	11	39	M	Knife wound	600	597	11
64	M	Overdose	1110	574	16	40	M	Gunshot wound	735	606	16
43	M	Gunshot wound	460	583	16	33	M	Gunshot wound	435	593	13
65	M	Jumped from height	1110	406	10	82	M	Fell from height	770	382	8
30	F	Gunshot wound	1290	621	12	23	F	Gunshot wound	865	640	10
72	F	Overdose-Aspirin/ Valium	1185	339	17	45	F	Knife wound	1020	592	19
79	F	Overdose-Nembutal	1080	543	12	73	F	Cardiovascular	1440	279	10
18	F	Jumped from height	600	229	9	50	F	Auto accident	630	343	10

METHOD

Brain samples were obtained at autopsy from 44 individuals (22 suicide victims and 22 controls). Frontal cortex samples were dissected, frozen immediately on dry ice, and stored at $-80^{\circ}C$ until assayed. The average elapsed time between death and tissue collection for all individuals (postmortem delay) was 17.1 hours. The samples were from 18 men and 4 women in each group. Complete patient characteristics are shown in Table 1. The mean ages were 39 years for the suicides and 36.9 for the controls. The postmortem delay for the suicides and controls was 17.9 and 16.3 hours, respectively. There were no significant differences between the two groups with respect to age and postmortem delay. Data on either the diagnosis or prior medications of the individuals included in this study were not available.

Binding of $[^3H]$-QNB was determined according to the method described by Wastek and Yamamura (19). Samples of tissue homogenate were incubated in duplicate at $37^{\circ}C$ with 10 concentrations of $[^3H]$-QNB (2.5-300 pM final concentration). Specific binding was determined in the presence or absence of 10 μM atropine.

RESULTS

The binding of $[^3H]$-QNB to frontal cortex was saturable at the concentrations used in this study (Figure 1). Scatchard analysis of the binding data indicated that there were no significant differences in the mean number of binding sites B_{max} for the suicide (493 fmoles/mg protein) or control groups (492 fmoles/mg protein) (Figure 2). Nor were there any significant differences in affinity (K_d) between the means of the two groups (suicides 14 pM and control 13.69 pM) (Figure 3).

Correlations between B_{max} or K_d for either suicides or controls were not significantly related to factors such as age and postmortem delay. However, when both groups were combined, B_{max} was significantly correlated with postmortem delay ($r = 0.35$; $p < 0.2$).

Comparisons between the B_{max} values of suicide victims who died by violent means (gunshot wounds, hangings, or jumping from height) and controls who had died either by violent or by non-violent methods, revealed no significant differences. Since variations in muscarinic cholinergic binding as a function of the time of day that individuals died have been reported (15), we also examined the B_{max} values for our combined samples (suicides and controls) at eight separate 3 hour intervals (one way ANOVA). None of the intervals significantly differed from each other (data not shown).

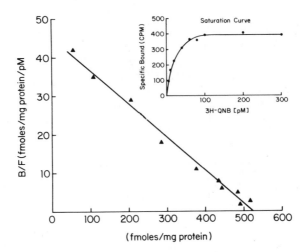

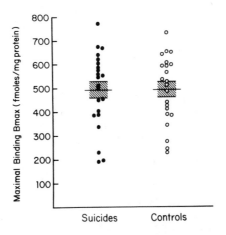

Figure 1. Results of a typical binding experiment showing a saturation curve (inset) and Scatchard curve of the same data (r = 0.97;
B_{max} = 526 fmoles/mg protein; K_d = 13 pM).

Figure 2. Number of binding sites (B_{max}) for suicide victims and
controls.

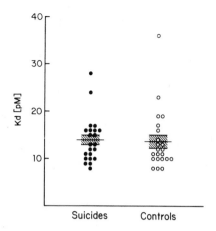

Figure 3.   Receptor affinities (K_d) for suicide victims and controls.

DISCUSSION

The principal finding of this study is that QNB binding in the frontal cortex of suicide victims, who had died in a determined manner, was not significantly different from a well-matched control group. Therefore, our findings do not support the hypothesis that there is increased tone in the cholinergic systems of individuals suffering from affective disorders. Unlike a previous study, which had reported a significant increase in QNB binding in 8 suicide victims (13), our larger matched sample was controlled for factors such as age, sex and postmortem delay. Also, a recent study by Kaufman et al. (11) reported results similar to ours, i.e., that there were no significant differences between QNB binding in both the cortex and hypothalamus of suicide victims and controls.

While these findings do not offer support for the cholinergic hypothesis of depression, they do not by necessity invalidate the impressive amount of _in vivo_ data that have been gathered which indicate that the mood of individuals with affective disorders as well as normals can be altered by drugs affecting the cholinergic system (9, 10, 16).

Does the absence of a finding of increased number of cholinergic receptors in the brains of suicide victims necessarily mean that cholinergic receptors are unchanged in affective disorders? There are several possible explanations suggested by our findings. One possible explanation of the data from this study pertains to the failure of QNB binding to reflect alterations in the functional activity of presynaptic cholinergic neurons. For example, in Alzheimer's disease there is a well characterized loss of presynaptic cholinergic neurons as determined by levels of choline acetyltransferase (4) and histological examinations (20). Based on these findings one would assume that the brains of individuals with Alzheimer's disease would have a significant compensatory increase in muscarinic binding sites; however, postmortem binding studies have revealed no such change (5). Additionally, a recently developed animal model using a compound, AF64A, has been shown to result in significantly decreased levels of choline acetyl- transferase and acetylcholine, but does not induce changes in QNB binding (7). These findings are in contrast to what occurs in response to chronic treatment with a direct acting muscarinic antagonist (e.g., atropine) in which there is a significant increase in QNB binding (12). Thus, the muscarinic cholinergic receptor seems to be more resistant to change following presynaptic manipulation than the receptors associated with serotonergic and dopaminergic systems (2, 3).

A second explanation may be related to suicide itself. Studies which have examined the diagnosis of individuals who have committed suicide have reported the incidence of affective disorder to range from approximately 28 to 75 percent (1, 6, 17). Thus, it is possible that in our sample this diagnosis may have been under-represented, and may therefore have obscured findings directly related to affective disorders.

In conclusion, the results of our study confirm that of Kaufman and colleagues (11) in which no changes were noted in suicide brains. While these findings neither disprove nor support the cholinergic hypothesis of depression, they do suggest that the neurochemical basis for the in vivo observations of increased responsivity of depressed individuals to muscarinic cholinergic agents might not involve changes in receptors estimated by QNB binding. As previously mentioned, alterations in presynaptic cholinergic input or postsynaptic receptor mediated responses need not be accompanied by changes in QNB binding.

Acknowledgements.   I would like to thank Drs. Nunzio Pomara and Bernard Lerer for their helpful comments, Ms. Sandi Tait and Sandy Demetriou for expert technical assistance, and Ms. Carol Dixon and Ms. Mary Ratza for secretarial assistance.   This work was supported in part by NIMH grants MH40210 and MH40048.

REFERENCES

1. Barraclough, B., Bunch, J., Nelson, B. and Sainsbury, P. (1974): Brit. J. Psychiat. 125:355-373.
2. Brunello, N., Chang, D.M. and Costa, E. (1982): Science 215: 1112-1115.
3. Creese, I., Burt, D.R. and Snyder, S.H. (1977): Science 197: 596-598.

4. Davies, P. (1979): Brain Res. 171:319-327.
5. Davies, P. and Verth, A.H. (1978): Brain Res. 138:385-392.
6. Dorpat, T.L. and Ripley, H.S. (1960): Compr. Psychiatry 1:349-359.
7. Fisher, A., Mantione, C.R., Abraham, D.J. and Hanin, I. (1982): J. Pharmacol. Exp. Ther. 222:140-145.
8. Gershon, S. and Shaw, F.H. (1961): Lancet II:1371-1374.
9. Janowsky, D.S., El-Yousef, M., Davis, J.M. and Seherke, H.J. (1973): Arch. Gen. Psychiat. 28:542-547.
10. Janowsky, D.S., Judd, L.L. and Groom, G. (1982): Proceedings of the 13th Collegium Internationale Neuropsychopharmacologicum Congress, Jerusalem, Israel.
11. Kaufman, C.A., Gillin, J.C., Hill, B., O'Laughlin, T., Phillips, I., Kleinman, J.E. and Wyatt, R.J. (1983): New Research Abst. #66, Amer. Psychiat. Assn., New York.
12. Lerer, B., Stanley, M., Demetriou, S. and Gershon, S. (1983): J. Neurochm. 41:1680-1683.
13. Meyerson, L.R., Wennogle, L.P., Abel, M.S., Coupet, J., Lippa, A.S., Rauh, C.E. and Beer, B. (1982): Pharmacol. Biochem. and Behav. 17:159-163.
14. Nadi, N.S., Nurnberger, J.I. and Gershon, E.S. (1982): Abstract, Amer. Coll. of Neuropsychopharmacol., p. 14.
15. Perry, E.K., Perry, R.H. and Tomlinson, B.E. (1977): Neurosc. Lett. 4:185-189.
16. Risch, S.C., Cohen, P.M., Janowsky, D.S., Kalin, N.H. and Murphy, D.L. (1981): J. Psychiat. Res. 4:89-94.
17. Robins, E., Murphy, G.E., Wilkinson, R.H., Gassner, S. and Kayes, J. (1959): Am. J. Public Health 49:888-899.
18. Sitaram, N., Nurnberger, J.I., Gershon, E.S. and Gillin, J.C. (1980): Science 208:200-202.
19. Wastek, G.H. and Yamamura, H.I. (1978): Mol. Pharmacol. 14:768-780.
20. Whitehouse, P.J., Price, D.L., Struble, R.G., Clark, A.W., Coyle, J.T. and Delon, M.R. (1982): Science 215:1237-1239.

MEMBRANE TRANSPORT OF CHOLINE BY THE HUMAN
ERYTHROCYTE: RELATIONSHIP TO INTRACELLULAR
CHOLINE CONTENT, AND EFFECT OF TREATMENT WITH LITHIUM

A.G. Mallinger, U. Kopp, J. Mallinger, R.L. Stumpf,
C. Erstling and I. Hanin

University of Pittsburgh School of Medicine
Western Psychiatric Institute and Clinic
Pittsburgh, Pennsylvania 15213   U.S.A.

INTRODUCTION

Choline (Ch) is the metabolic precursor of acetylcholine, and therefore may have an important role in brain function. Because there are a number of similarities in Ch handling between brain synaptosomes and erythrocytes (RBCs) (23), some investigators have felt that the RBC might potentially serve as a model for neuronal Ch transport. Thus, the Ch content of RBCs has been studied in a variety of psychiatric and neurologic conditions, including major depression (10), acute mania (13, 16, 24), Gilles de la Tourette syndrome (3, 9), and dementia (2, 4).

The mechanisms that regulate intracellular Ch content are not well understood. However, previous investigations have demonstrated that lithium treatment of psychiatric patients can produce significant elevations of RBC Ch content (6, 12, 15, 20, 24), as well as inhibition of Ch transport across the RBC membrane (11, 14, 17, 18, 21). Thus, lithium appears to exert its effect on RBC Ch content at the level of the cell membrane, via inhibition of Ch transport (11, 21). Moreover, it is possible that Ch transport processes could also have a role in the normal physiological regulation of RBC Ch content, in subjects who are not being treated with lithium.

The studies described in this chapter were conducted in order to investigate the relationship between in vivo RBC Ch content and Ch transport across the RBC membrane (measured in vitro). For the in vitro studies, we utilized a method for the simultaneous measurement of Ch influx and efflux based on stable isotope methodology and gas chromatography/mass spectrometry (GC/MS). We measured Ch transport

351

and Ch content in RBCs from a variety of subject groups, including drug-free controls, bipolar patients (both prior to and during lithium treatment), pregnant individuals, and samples of cord blood (which are representative of the fetus at term).

Our findings suggest that: 1) membrane Ch transport may have a role in the regulation of endogenous RBC Ch content; 2) lithium treatment can produce varying degrees of transport inhibition in different subjects; and 3) maternal lithium treatment affects Ch transport by fetal RBCs. Moreover, our data indicate that fetal metabolism of Ch may differ from that of adults.

MATERIALS AND METHODS

Subjects. The patients who participated in these investigations were diagnosed as having bipolar 1 or bipolar 2 affective disorder, according to the Research Diagnostic Criteria (25). The findings from physical examinations and routine clinical laboratory tests were normal, except that one patient had uncomplicated essential hypertension and adult-onset diabetes mellitus. Pre-lithium studies were performed only in patients who had been free of this agent for at least 6 months, because of the known long-term effect of lithium on RBC Ch content (11). Patients did receive other medications during the study (neuroleptics, antidepressants), but the drugs and doses were kept constant during the pre-lithium and lithium-treatment periods.

Control subjects were employees at the University of Pittsburgh, who had no history of psychiatric or medical illness, and who were drug-free for at least 2 weeks prior to being studied. None of the control subjects had ingested lithium during the year prior to their participation in this investigation.

All patients and control subjects gave informed consent for participation in these studies, after the procedures had been fully explained.

Measurements of In Vivo Ch Content. These measurements were performed using a method described previously (8). Briefly, venous blood was collected, and the RBCs were separated from plasma by centrifugation. One ml duplicate aliquots of both plasma and RBCs were deproteinized by addition of perchloric acid and then centrifuged. The supernatants were collected, and the pH adjusted to 4.2-4.4 using potassium acetate. Samples were stored at -20°C until the time of analysis; they were then centrifuged, mixed with an internal standard consisting of deuterium labeled Ch, and subsequently processed for analysis using gas chromatography/mass spectrometry (GC/MS) (7).

In Vitro Measurements of RBC Ch Transport. These experiments were designed to simultaneously measure <u>influx</u> of deuterium-labeled Ch (D_4Ch; four hydrogen atoms in ethylene chain replaced by 4 deuterium atoms) and <u>efflux</u> of nondeuterated endogenous Ch (termed D_0Ch). The method has been described in detail previously (21). Briefly, RBCs were collected and washed twice in a physiologic buffer solution (19), then added to an incubation solution that was identical to the washing solution, except for the addition of 10 μM D_4Ch (a concentration that is similar to plasma Ch levels typically found <u>in vivo</u>). Cells were incubated at 17°C with gentle shaking. Samples were taken at the time that the RBCs were initially mixed with the incubation solution (zero time), and at equal intervals of 6 or 8 minutes over the next 24 minutes. Levels of D_0Ch and D_4Ch in the samples were determined using GC/MS.

The amount of hemolysis occurring during the incubations was always less than 1%. Changes in RBC D_0Ch and D_4Ch concentrations were linear with time during the 24-minute incubations, so that D_0Ch efflux and D_4Ch influx could be calculated directly from the slopes of the concentration versus time curves. Each of these curves contained measurements from at least 4 (usually 5) time points.

RESULTS

Incubation Conditions and Characteristics. Several previous studies of RBC Ch transport have utilized Ch concentration versus time curves that were derived from incubations performed at 37°C (11, 14). At this temperature, however, Ch transport can be very rapid, so that within 30 to 45 minutes, the two Ch isotopes reach an equilibrium distribution across the membrane and net transport is no longer measurable (see upper portion of Figure 1). Moreover, the initial linear segments of the RBC Ch concentration versus time curves (used to determine unidirectional net fluxes of the Ch isotopes) cannot be measured accurately, because for practical reasons samples cannot be collected more often than every four minutes.

In an effort to improve the quantitative assessment of RBC Ch transport, we have conducted studies using an incubation temperature of 17°C. At this lower temperature, the observed rate of Ch transport is slower, and remains linear over a 24-minute interval, so that unidirectional net flux values can be calculated directly from the slopes of the concentration versus time curves. A typical incubation using this technique is illustrated in the lower portion of Figure 1.

The reproducibility of this method for measuring RBC Ch transport was assessed by performing three successive incubations of RBCs from a drug-free control subject at 10 to 14 day intervals (21). The observed coefficients of variation were 8.5% for D_0Ch efflux and 10.7% for D_4Ch influx. In comparison, the coefficient

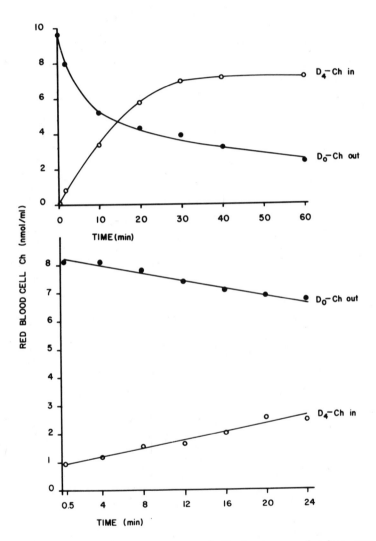

Figure 1.   Changes in RBC content of Ch isotopes during typical in vitro incubations.   Upper portion illustrates findings from cells incubated at 37°C.   Lower portion shows linear transport that occurs when temperature is lowered to 17°C.   Lines were obtained by nonlinear curve-fitting (upper portion) or by the method of least squares (lower portion).   From Mallinger, et al. (21).

of variation for measurements of _in vivo_ RBC Ch content (determined at the time of each incubation) was 15.4%.

Relationship Between RBC Ch Transport and Content in Drug-Free Control Subjects. Ch transport studies were conducted using RBCs from seven drug-free control subjects, who were selected because they represented a wide range of endogenous RBC Ch levels. Two of these subjects were studied twice, and one subject was studied on five occasions. The findings from these incubations are illustrated in Figure 2. Replicate studies in the same individuals are shown connected by lines. The D_4Ch influx data from one subject was lost in a technical mishap.

The figure demonstrates that an inverse, linear relationship exists between each of the _in vitro_ measures of RBC Ch transport (D_4Ch influx, D_0Ch efflux) and the natural logarithm of _in vivo_ RBC Ch content. Logarithmic transformation was performed because

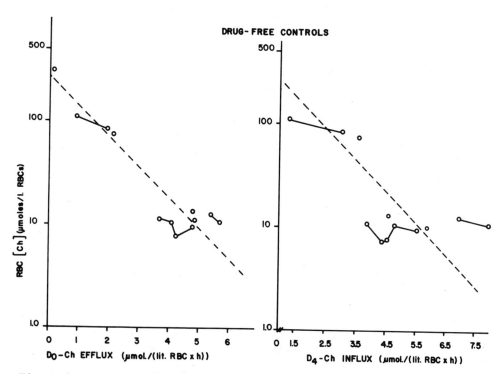

Figure 2. Semi-logarithmic scatter diagrams of _in vivo_ RBC Ch content versus D_0Ch efflux (left portion) and D_4Ch influx (right portion) in 7 drug-free control subjects. Solid lines connect repeated measurements performed with the same subject on different days. From Mallinger et al. (21).

scatterplots of the untransformed data appeared to be approximately exponential. When the mean values for each subject were analyzed, there was a significant inverse linear correlation between ln RBC Ch content and D_OCh efflux (r=-0.97, p<0.001); in addition, the correlation between ln RBC Ch content and D_4Ch influx closely approached statistical significance (r=-0.81, p<0.06).

Effect of Lithium Treatment on RBC Ch Transport and Content in Patients with Bipolar Affective Disorder. Five bipolar patients participated in this phase of the investigation. Three of these subjects were studied both prior to, and during treatment with lithium carbonate. The findings from these studies are illustrated in Figure 3. Lines connecting points represent studies performed in the same subject prior to and during lithium treatment.

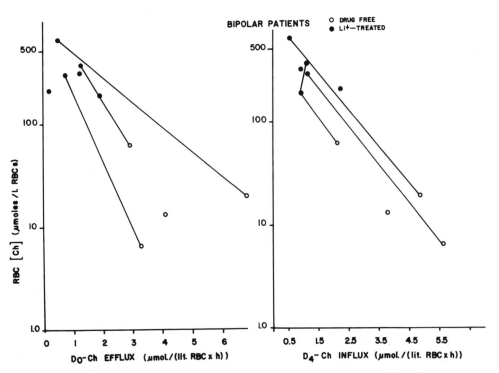

Figure 3. Semi-logarithmic scatter diagrams of *in vivo* RBC Ch content versus D_OCh efflux (left portion) and D_4Ch influx (right portion) in patients with bipolar affective disorder. Lines connect measurements performed in the same subject prior to (open circles) and during (closed circles) treatment with lithium carbonate. From Mallinger et al. (21).

Table 1

Percent Inhibition of Erythrocyte Choline Transport in
Three Bipolar Patients Studied Prior to and During
Treatment with Lithium Carbonate

Patient I.D.	Inhibition of Choline Transport[a,b] (Percent of pretreatment value)	
	D_oCh Efflux	D_4Ch Influx
121	77.8	80.5
137	92.2	99.0
237	46.6[c]	51.0[c]

[a] Calculated from flux data shown in Figure 3.

[b] Percent inhibition =

$$\frac{(\text{Pre-treatment flux value}) - (\text{Post-treatment flux value})}{\text{Pre-treatment flux value}} \times 100$$

[c] Calculated using the mean of 2 post-treatment determinations.

Individuals treated with lithium had lower rates of D_oCh and D_4Ch transport, and higher levels of RBC Ch, than untreated patients. As in the studies of drug-free control subjects described above, there was an inverse relationship between Ch flux and the natural logarithm of Ch content in RBCs; however, lithium-treated patients generally had higher levels of RBC Ch than control subjects (see Figures 2 and 3 for comparison).

Table 1 presents data from the three bipolar patients who were studied both prior to and during lithium treatment. There was con-siderable intersubject variation in the degree of Ch transport in-hibition produced during treatment (46.6 to 99.0%). In each indi-vidual, however, inhibition of D_oCh efflux was similar to that of D_4Ch influx. In fact, there was a statistically significant corre-lation between the percent inhibition values for D_oCh efflux and D_4Ch influx (r=0.997, p<0.05), although this finding should be in-terpreted with reservation because of the small sample size.

The general relationship between RBC Ch content and membrane Ch transport appeared to be similar in both control subjects and bipolar patients. Figure 4 illustrates the RBC Ch content and D_oCh efflux values for all control subjects (drug-free) and patients (drug-free or lithium-treated) who were studied. Mean values are

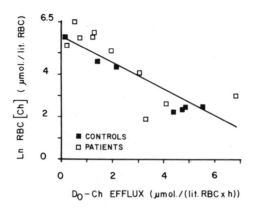

Figure 4. Scatter diagram of in vivo ln RBC Ch content versus D_0Ch efflux for all control subjects (drug-free) and patients (drug-free or lithium-treated) shown in Figures 2 and 3. Mean values are shown for those control subjects who had repeated measurements performed. Line was fitted by the method of least squares.

shown for those control subjects who had repeated measurements per-formed. The apparent similarity of the relationship between RBC Ch content and membrane Ch transport in both subject groups suggests that the same mechanism which accounts for physiological variations of RBC Ch content in drug-free subjects could also be involved in the Ch-elevating effects of lithium treatment.

Relationship Between D_0Ch Efflux and D_4Ch Influx. As the data pre-sented in Table 1 suggests, there appears to be a close correlation between D_0Ch efflux and D_4Ch influx in the RBC. This relationship is independent of the actual rate of Ch transport, or the endogenous RBC Ch content. Figure 5 presents mean values for D_0Ch efflux and D_4Ch influx in the six drug-free control subjects for whom both of these flux measures are available (see Figure 2). There was a sig-nificant correlation between these two transport parameters ($r=0.91$, $p<0.02$), despite large differences in the ratio of intracellular D_0Ch to extracellular D_4Ch.

Maternal and Fetal Blood Ch Measures in a Lithium-Treated Bipolar Patient and in Drug-Free Control Subjects. Because little is known about the effects of treatment with lithium on Ch metabolism during pregnancy, we conducted a series of preliminary studies with maternal and fetal (cord) blood. The findings of these studies are presented in Table 2. In a bipolar patient who was treated with lithium car-bonate during the final month of pregnancy, we measured the Ch levels of plasma and RBCs in both maternal and fetal blood (20). Similar measurements were performed using maternal and fetal blood from

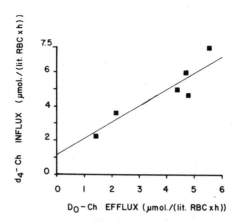

Figure 5.   Scatter diagram of D_4Ch influx versus D_oCh efflux in
RBCs from the six drug-free control subjects for whom both flux
measures are available (see Figure 2).   Mean values are shown for
those subjects who had repeated measurements performed.   Line was
fitted by the method of least squares.

drug-free control subjects.   In addition, Ch transport studies were
performed with cord RBCs from one of the control subjects.

     The observed plasma Ch levels of maternal blood did not differ
from those typically found in blood from non-pregnant subjects (12).
However, plasma Ch levels measured in cord blood from either lithium-
treated or drug-free subjects were somewhat higher than those of
maternal blood, in agreement with previous observations (26).

     In the lithium-treated subject, maternal RBC Ch levels were
elevated, and were similar to those observed in non-pregnant, lithium-
treated patients (see Figure 3).   Moreover, RBC Ch content increased
as the duration of treatment became longer, in accordance with previous
observations (11).   RBCs from the cord blood of this subject, which
are representative of the fetus at term, also had elevated Ch content.
In addition, the serum lithium concentration of cord blood was similar
to that of maternal blood.

     Cord RBCs from two drug-free subjects also had slightly increased
Ch content, as compared to maternal cells.   Nevertheless, Ch levels
measured in cord RBCs from these two subjects were approximately
sixfold lower than the values observed in the lithium-exposed fetus.

     Fetal Ch transport was measured in vitro using cord RBCs from
one drug-free subject (see footnote in Table 2).   The measured value
in this single case were 10.4 μmoles/(liter RBCs x h) for D_oCh efflux,

Table 2

Maternal and Fetal Blood Choline Measures in a Lithium-Treated
Bipolar Patient, and in Drug-Free Control Subjects

	Duration of Lithium Treatment (weeks)	Choline (μM) Plasma	RBC	Serum Lithium Level (mM)
LITHIUM-TREATED:				
MATERNAL BLOOD				
Pregnant	2	13.2	316	0.48
Postpartum	12	6.6	824	0.67
CORD BLOOD	4	50.5	259	0.49
DRUG-FREE[a]:				
MATERNAL BLOOD (Subject 1)				
Pregnant	0	11.6	13.6	----
CORD BLOOD (Subject 1)	0	33.8	34.3[b]	----
CORD BLOOD (Subject 2)	0	52.6	38.6	----

[a]Findings from both maternal and cord blood in Subject 1, and from cord blood only in Subject 2. [b]Transport: D_0Ch efflux = 10.4 μmoles/ (liter RBCs x h); D_4Ch influx = 9.4 μmoles/(liter RBCs x h).

and 9.4 μmoles/(liter RBCs x h) for D_4Ch influx. In contrast, substantially lower flux values have previously been observed in adult control subjects (see Figures 2 and 5: D_0Ch efflux < 6 μmoles/(liter RBCs x h); D_4Ch influx < 8 μmoles/(liter RBCs x h)). Thus, our preliminary findings suggest that in the fetus, RBC Ch content and transport, as well as plasma Ch levels, may be increased as compared to those of adult subjects.

DISCUSSION

It has previously been shown that the Ch which is transported into RBCs is neither metabolized nor incorporated into phospholipids (1, 22). Moreover, there are a number of important similarities between RBCs and synaptosomes (from guinea pig brain) in terms of the characteristics of their Ch transport systems (5, 23): 1) both preparations contain a saturating, low-affinity Ch transport system that does not require added monovalent cations and is not directly dependent on metabolic energy; 2) Ch influx is inhibited in both preparations by external Na, K and Li ions, by substances which attack sulfhydryl groups, and by hemicholinium; and 3) the Ch concentration giving half-maximum flux is similar in both preparations (83 μM in guinea pig synaptosomes and 20 to 30 μM in human RBCs). Thus, the RBC may represent a potentially useful model for studying brain-related Ch transport processes.

The findings that are described in this chapter were obtained using an experimental method that simultaneously measures net D_0Ch efflux and D_4Ch influx, in cells that are incubated in vitro at $17^\circ C$. One advantage of this method is that it utilizes physiologically relevant amounts of intracellular and extracellular Ch for the in vitro incubations. A second advantage of this method is that Ch transport remains linear during the 24-minute incubations, so that net flux values can be accurately determined from the slopes of RBC Ch concentration versus time curves. This contrasts with RBC Ch transport measured at $37^\circ C$ (11, 14), which is so rapid that the concentration versus time curves can become nonlinear in less than eight to ten minutes (see ref. 11), as the two isotopes begin to equilibrate across the membrane. This comparatively rapid loss of linearity can reduce the accuracy of measurements of the initial transport rates, because sampling frequency is limited by the time consuming procedures needed for processing the cell suspensions.

When studies were performed with RBCs from psychiatrically normal, drug-free control subjects, we found a significant inverse correlation between D_0Ch efflux and ln RBC Ch content, as well as a nearly-significant inverse correlation between D_4Ch influx and ln RBC Ch content. These findings suggest that elevated in vivo RBC Ch levels in drug-free subjects are associated with reduced Ch transport across the RBC membrane. Thus, elevations of RBC Ch content that have previously been reported to occur in various psychiatric and neurologic conditions (see INTRODUCTION) could reflect alterations in the physiological handling of this important neurotransmitter precursor at the level of the cell membrane. Investigations of RBC Ch transport might thus provide useful additional information regarding such conditions.

It should be noted that individuals with relatively high RBC Ch levels are unusual among control subjects. Although a previous investigation of 48 blood bank donors found that RBC Ch levels ranged from 6.7 to 144.2 μmoles/liter RBCs (12), the median value was 12.8 μmoles/liter RBCs, and Ch levels exceeded 50 μmoles/ liter RBCs in fewer than 10% of subjects. Three of the control subjects who participated in this investigation (see Figure 2) were selected because they had unusually elevated RBC Ch levels (> 50 μmoles/liter

RBCs), despite having no known medical or psychiatric disorders, and no history of treatment with lithium or other drugs. They were included in this investigation in order to help elucidate the relationship between RBC Ch content and transport over a wide range of values.

Earlier investigations have demonstrated that lithium treatment results in greatly elevated RBC Ch content (6, 12, 15, 24), presumably as a consequence of inhibition of Ch transport across the RBC membrane (11, 14). In the present investigation, we made the additional observation that varying degrees of transport inhibition (ranging from approximately 50% to 90%) can be produced in different subjects (see Table 1). Future studies to determine whether variations in the degree of transport inhibition could be related to clinical factors (such as treatment response) would be of interest.

Our preliminary findings suggest that maternal lithium treatment may produce elevated RBC Ch content in the fetus at term. Lithium is known to cross the placenta, and in this case was present in cord serum in significant amounts. This is a likely means by which lithium could exert its effect on RBC Ch in the fetus. The clinical interpretation of this finding must await further research. However, because the lithium-induced elevation of RBC Ch content is known to be accompanied by profound inhibition of Ch transport across the RBC membrane, a similar effect on Ch transport in other tissues could have potentially important consequences on fetal development, and might be a factor in the suspected teratogenicity of lithium.

Even in drug-free subjects, fetal Ch handling may differ from that of adults, with greater intracellular and extracellular concentrations of Ch, and a correspondingly increased rate of Ch transport. This finding, if confirmed in larger studies, could relate to differences in Ch metabolism in the developing fetus.

SUMMARY

In conclusion, this chapter has described an improved method for quantitatively studying Ch transport by human RBCs in vitro. Using this method, we have been able to investigate the relationship between in vitro Ch transport and in vivo (endogenous) Ch content in adult and fetal cells, and we have been able to quantitate varying degrees of Ch transport inhibition produced by lithium treatment. The method described in this chapter could be readily applied to clinical studies of Ch transport in various psychiatric or neurologic disorders. Moreover, studies using this method could potentially help to clarify the relationship between lithiuminduced Ch transport alterations and the clinical actions of this drug. In this regard, the effects of other psychotherapeutic agents on Ch transport would also be of interest. Hopefully, future studies of RBC Ch transport will help to increase our understanding of cholinergic function in specific psychiatric or neurologic disorders.

Acknowledgement. Supported by NIMH grant #MH26320. Our thanks to Joyce O'Leary for assisting with the preparation of this manuscript.

REFERENCES

1. Askari, A. (1966): J. Gen. Physiol. 49:1147-1160.
2. Barclay, L.L., Blass, J.P., Kopp, U. and Hanin, I. (1982): New Eng. J. Med. 307:500.
3. Comings, D.E., Gursey, B.T., Avelino, E., Kopp, U. and Hanin, I. (1982): In Gilles de la Tourette Syndrome (eds) A.. Friedhoff and T.N. Chase, Raven Press, New York, pp. 255-258.
4. Davis, B., Mohs, R.C., Levy, M.I., Greenwald, B., Rosen, W.G., Mathe, A.A., Deutsch, S.I., Horvath, T.B. and Davis, K.L. - (1982): Proc. Am. Coll. of Neuropsychopharmacology, December 15-18:44.
5. Diamond, I. and Kennedy, E.P. (1969): J. Biol. Chem. 244: - 3258-3263.
6. Domino, E.F., Mathews, B., Tait, S.K., Demetriou, S. K. and Fucek, F. (1981): In Cholinergic Mechanisms (eds) G. Pepeu and H. Ladinsky, Plenum Press, New York and London, pp. 891-900.
7. Hanin, I. and Skinner, R.F. (1975): Anal. Biochem. 66:568-583.
8. Hanin, I., Kopp, U., Zahniser, N.R., Shih, T.M., Spiker, D.G., Merikangas, J.R., Kupfer, D.J. and Foster, F.G. (1978): In Cholinergic Mechanisms and Psychopharmacology (ed) D.J. Jenden, Plenum Press, New York, pp. 181-195.
9. Hanin, I., Merikangas, J.R., Merikangas, K.R. and Kopp, U. (1979): New Eng. J. Med. 301:661-662.
10. Hanin, I., Kopp, U., Spiker, D.G., Neil, J.F., Shaw, D.H. and Kupfer, D.J. (1980): Psychiatry Res. 3:345-355.
11. Hanin, I., Mallinger, A.G., Kopp, U., Himmelhoch, J.M. and Neil, J.F. (1980): Comm. Psychopharmacol. 4:345-355.
12. Hanin, I., Spiker, D.G., Mallinger, A.G., Kopp, U., Himmelhoch, J.M., Neil, J.F. and Kupfer, D.J. (1981): In Cholinergic Mechanisms (eds) G. Pepeu and H. Ladinsky, Plenum Press, New York and London, pp. 901-919.
13. Hanin, I., Cohen, B.M., Kopp, U. and Lipinski, J.F. (1982): Psychopharmacol. Bull. 18:186-190.
14. Jenden, D.J., Jope, R.S. and Fraser, S.L. (1980): Comm. Psychopharmacol. 4:339-344.
15. Jope, R.S., Jenden, D.J., Ehrlich, B.E. and Diamond, J.M. (1978): New Eng. J. Med. 299:833-834.
16. Jope, R.S., Jenden, D.J., Ehrlich, B.E., Diamond, J.M. and Gosenfeld, L.F. (1980): Proc. Natl. Acad. Sci. 77:6144-6146.
17. Lee, G., Lingsch, C., Lyle, P.T. and Martin, K. (1974): Br. J. Clin. Pharmacol. 1:365-370.
18. Lingsch, C. and Martin, K. (1976): Br. J. Pharmacol. 57:323-327.

19. Mallinger, A.G., Kupfer, D.J., Poust, R.I. and Hanin, I. (1975):
     Clin. Pharmacol. Ther. 18:467-474.
20. Mallinger, A.G., Hanin, I., Stumpf, R.L., Mallinger, J., Kopp,
     U. and Erstling, C. (1983): J. Clin. Psychiatry 44:381-384.
21. Mallinger, A.G., Kopp, U. and Hanin, I. (1984): J. Psychiatric
     Res. 18:107-117.
22. Martin, K. (1968): J. Gen. Physiol. 51:497-516.
23. Martin, K. (1974): In Drugs and Transport Processes (ed) B.H.
     Callingham, MacMillan, London, pp. 275-286.
24. Shea, P.A., Small, J.G. and Hendrie, H.C. (1981): Biol. Psy-
     chiatry 16:825-830.
25. Spitzer, R.L., Endicott, J. and Robins, E. (1978): Arch.
     Gen. Psychiatry 35:773-782.
26. Zeisel, S.H., Epstein, M.F. and Wurtman, R.J. (1980): Life
     Sci. 26:1827-1831.

STUDIES ON THE SYNAPTIC FUNCTION IN MUSCLE

AFTER NERVE CRUSH IN NORMAL AND DIABETIC RATS

R. Siliprandi, S. Consolo[*], F. Di Gregorio,
H. Ladinsky[*], C. Scozzesi[*], R. Zanoni and A. Gorio

Department of Cytopharmacology
Fidia Research Laboratories
Via Ponte della Fabbrica 3/A
Abano Terme, Italy

[*]Cholinergic Neuropharmacology
Mario Negri Institute for Pharmacological Research
Milan, Italy

INTRODUCTION

Spontaneous diabetes can develop a neuropathy that affects function and morphology of the peripheral nerves both in men and in animals (7, 14, 16, 22). Diabetes can also be induced by a single injection of Alloxan or Streptozotocin (15), drugs that exert a toxic effect on the beta-cells of the Langerhans islets. In rats treated with either Alloxan or Streptozotocin a decrease in nerve conduction velocity is evident within a few weeks after intoxication (3, 20).

A variety of changes are associated with the neuropathy: reduction in axon diameter (9, 21), segmental demyelination (1, 17) and changes in internodal myelin membrane structure (5). In addition, several authors report alterations in the slow component of axonal transport (8, 13, 19, 23).

In this context it was of interest to study whether the regenerative capacity of the neuron is affected by the disease. Neuronal regeneration, indeed, requires modifications in metabolism to synthesize required materials for axonal elongation and sprouting (4). The elongation velocity seems to be correlated to the b component of the slow axonal transport as shown by Lasek et al. (11).

In this work we report our studies on muscle reinnervation after sciatic nerve crush in alloxan treated rats. Results are compared with those obtained in a similar study performed on normal rats (2, 6).

MATERIALS AND METHODS

The experiments were carried out on adult male Sprague-Dawley rats weightin 250-300 g. Diabetes was induced by a single injection of alloxan monohydrate at the dose of 100 mg/Kg. Animals showing glucose levels above 250 mg/100 cc 30 days after the treatment were included in the experiments.

Denervation of the extensor digitorum longus (EDL) muscle was performed on ether anesthetized rats by crushing the sciatic nerve at the level of the last gluteal nerve.

EDL muscle reinnervation was monitored electrophysiologically by standard intracellular recordings and morphologically by light and electron microscopy (for a detailed description of the methods see ref. 2 and 6).

ACh was determined by radioenzymatic assay as described by Consolo et al. (this book).

The data are presented as the means $\pm$ standard errors.

RESULTS

We studied the course of muscle reinnervation by monitoring both spontaneous and nerve evoked electrical activities recorded at the neuromuscular junction. For several days after the crush no nerve evoked electrical activity could be recorded. Only at day 19 were we able to record the first signs of reinnervation: a weak contraction of muscle fibers after nerve stimulation, and minia-ture end plate potentials (m.e.p.p.s) of very low frequency (less than 1/min).

Table 1 shows that m.e.p.p.s were recorded consistently 15 days after crush in non-diabetic rats, and only about 21 days after crush in alloxan-treated rats. Recovery was faster, too, in normal rats. This was indicated by the finding that m.e.p.p. frequency was less that 50% recorded on day 40 after crush, and complete recovery was not obtained during the 90 day experimental period.

Reinnervation was characterized, in both normal and diabetic rats, by alterations of m.e.p.p. amplitude distribution (Figs. 1 and 2). In healthy rats, the shape of the distribution was normal within day 50 from crush, while in non-diabetic animals the m.e.p.p. amplitude distribution showed a trend towards nondenervated values

Table 1

Miniature End Plate Potential Frequency
(m.e.p.p.s/60 sec)

Time After Crush (days)	Non Diabetic Rats	Alloxan Treated Rats
non denervated	123.9 ± 16.5	132.2 ± 8.8
15	1.1 ± 0.3	/
21	17.6 ± 2.0	3.9 ± 1.0
30	53.4 ± 5.0	16.5 ± 2.3
40	124.0 ± 11.2	49.1 ± 8.7
50	151.1 ± 19.9	56.6 ± 7.7
70	/	86.2 ± 4.9
90	/	112.1 ± 7.8

only on day 90 from crush. In later stages of diabetic rat reinnervation, the main feature of m.e.p.p. distribution was the presence of a large fraction of m.e.p.p.s with unusually high amplitudes. "Giant" m.e.p.p.s, with amplitudes greater that 3.0 mV were often recorded.

In order to better understand these unusual findings, we performed a series of experiments to measure the input resistance of muscle fibers. Table 2 shows, as expected, that the input resistance in reinnervating muscle was higher than in control. Accordingly, the calculated radius was reduced.

The degree of synaptic maturation during the various phases of reinnervation was analyzed by recording the stimulation of m.e.p.p. frequency induced by calcium-dependent and calcium- independent processes. In the first case, potassium concentration in the bathing solution was raised to 30 mM. The high K^+ depolarized the nerve terminal and activated the Ca^{++} channels, provoking a massive release of ACh. In the second case the osmolarity of the perfusing solution was raised by adding 50% sucrose. The results are shown in Tables 3 and 4. In diabetic animals both K^+-stimulated and hypersmolarity-stimulated m.e.p.p. frequency was very low during the early period of reinnervation but then recovered to control values at the same rate as it did in nondiabetic animals. These findings suggest that both the Ca^{++}dependent and the Ca^{++}-independent stimulated release mechanisms are not affected by the disease.

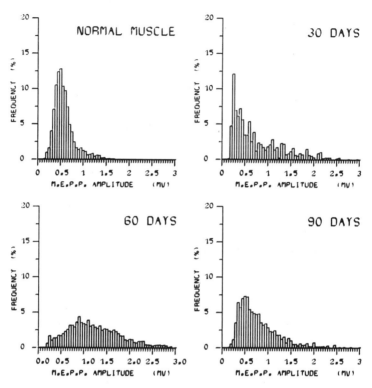

Figure 1. Amplitude distribution of miniature endplate-potentials in EDL muscle fibers in alloxan-treated rats as a function of time after crush.

Table 2

Input Resistance and Calculated Radius

	Input Resistance (MΩ)	Radius (μ)
diabetic non denervated rats	0.269 ± 0.024	26.75 ± 1.04
diabetic rats 30 days from crush	1.368 ± 0.163	11.85 ± 0.56
diabetic rats 60 days from crush	0.948 ± 0.08	13.16 ± 1.26
diabetic rats 90 days from crush	0.563 ± 0.20	18.85 ± 2.24

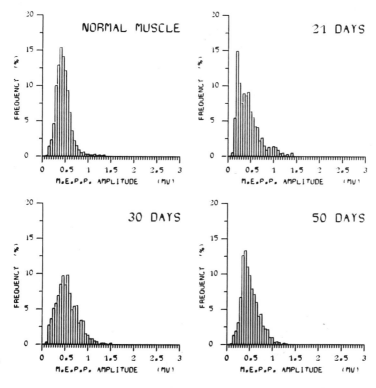

Figure 2. Amplitude distribution of miniature endplate potentials in EDL muscle fibers of non-diabetic rats as a function of time after crush.

During nerve regeneration, neurite sprouting causes enlargement and overlapping of newly formed motor units. As a result many muscle fibers are innervated by two or more nerve terminals until synaptic repression re-establishes the normal one-to-one pattern of innervation between muscle fibers and axon terminals. The degree of polyinnervation during nerve regeneration was estimated under conditions where nerve evoked ACh release was reduced in high Mg^{++} (12.5-18 mM) solutions. This permits the recording of subthreshold end plate potentials (e.p.p.s) without inducing muscle e.p.p.s after electrical stimulation of the nerve. Table 5 compares the data obtained in diabetic and non-diabetic rats. In both cases a maximum value of about 60% was observed, which was reached within about 10 days after the onset of reinnervation. Despite the different delay periods to the beginning of reinnervation in the two groups, the time courses of polyinnervation and synaptic repression were similar in the normal and the diabetic rats; nevertheless, in some cases we were able to detect polyinnervation and axon sprouting even 90 days from crush.

Table 3

Miniature End Plate Potential Frequency Stimulated by 30 mM K+
(m.e.p.p.s/sec)

Time After Crush (days)	Non Diabetic Rats	Alloxan Treated Rats
non denervated	225.1 ± 8.8	354.6 ± 27.1
15	2.9 ± 0.4	/
21	43.1 ± 4.5	3.6 ± 1.1
30	/	79.7 ± 22.5
40	258.7 ± 10.5	173.85± 35.7
50	307.3 ± 22.6	286.2 ± 33.1
70	/	342.5 ± 22.5
90	/	361.8 ± 33.2

Table 4

Miniature End Plate Potential Frequency Stimulated by
Hypertonic Sucrose Solution
(m.e.p.p.s/sec)

Time After Crush (days)	Non Diabetic Rats	Alloxan Treated Rats
non denervated	51.6 ± 4.7	43.9 ± 5.5
15	/	/
21	2.9 ± 0.3	0.2 ± 0.2
30	10.5 ± 1.6	9.1 ± 5.6
40	29.7 ± 3.2	12.1 ± 1.7
50	53.2 ± 1.8	21.5 ± 4.8
70	/	41.3 ± 3.3
90	/	47.8 ± 16.8

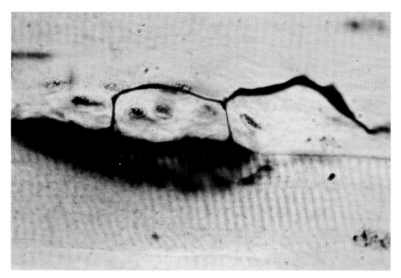

Figure 3. Low magnification light micrograph of an EDL muscle end plate 90 days from crush in a non-diabetic rat.

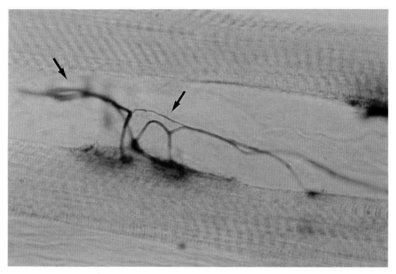

Figure 4. Low magnification light micrograph of an EDL muscle end plate 90 days after crush in a diabetic rat. Two nerve terminals (arrows) are still present at the same end plate.

Figs. 3 and 4 are low magnification light micrographs showing the pattern of reinnervation of EDL muscle 90 days from crush in a normal and in a diabetic rat. It is evident that in the alloxan-treated rat, the end plate is still polyinnervated.

Radioenzymatic assay was carried out to measure the recovery of ACh content in reinnervating EDL muscle. The results are shown in Fig. 5. ACh content fell sharply in both groups to about 20% of control levels within 10 days after crush. The ACh content in non-diabetic animals recovered completely by 30-40 days post crush. Onset of recovery was delayed and the recovery rate was slower in the diabetic rats. In these animals, ACh content only returned to about 75% of control by 40 days post-crush.

DISCUSSION

The rate of both nerve regeneration and synaptic maturation is decreased in alloxan treated animals, while sprouting and synaptogenesis are normal. Our results show that the onset of muscle reinnervation begins with a delay of 5 days in diabetic animals suggesting a reduced elongation velocity of the growing neurites. This alteration may be correlated to the slowing of the axonal transport rate observed in diabetes (8, 13, 19, 23). Conceivably, among the molecules whose anterograde flow is impaired, may be included the cytoskeletal components required for elongation, which are synthetized in the cell body (24).

Table 5

Polyinnervation and Repression
(% of fibers polyinnervated)

Time After Crush (days)	Non Diabetic Rats	Alloxan Treated Rats
15	25.9 $\pm$ 5.8	/
21	55.4 $\pm$ 5.0	23.10 $\pm$ 6.5
25	58.3 $\pm$ 7.1	48.6 $\pm$ 3.0
30	36.6 $\pm$ 2.6	61.9 $\pm$ 5.1
40	23.4 $\pm$ 4.0	33.1 $\pm$ 5.8
50	23.3 $\pm$ 6.6	34.0 $\pm$ 5.1
60	8.3 $\pm$ 3.4	22.3 $\pm$ 3.8
70	/	16.7 $\pm$ 1.6
90	/	13.8 $\pm$ 1.2

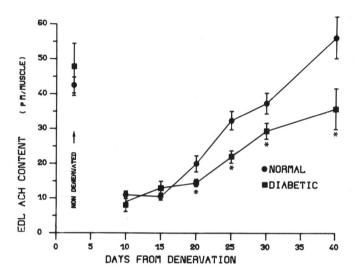

Figure 5. Recovery of acetylcholine content in EDL muscle of normal and diabetic rats as a function of time after sciatic nerve crush. The data represent the means $\pm$ S.E.M. (vertical bars) of 6-9 animals per group. * = $p < 0.001$ vs normal rats of the same time period (Student's t-test). The arrow indicates acetylcholine content in non-denervated muscles.

The slower recovery of m.e.p.p. frequency found in diabetic rats correlates with a slower growing rate of nerve terminals (10). Accordingly, electron microscopy investigations performed on EDL muscle show a high degree of immaturity of the end plates even 60, 70 days from denervation (18). This late recovery is corroborated by ACh assay performed on muscles during reinnervation which demonstrates a delayed onset of recovery as well as a slower rate of recovery of ACh levels in diabetic animals.

Interestingly, the recovery of Ca^{++}-dependent and Ca^{++}-independent release mechanisms are not altered by the disease, as was shown by the experiments in which m.e.p.p. frequency was stimulated by high K^{+} or hyperosmolarity.

Diabetes, furthermore, does not appear to affect neurite sprouting. In alloxan diabetic rats, the time course of polyinnervation and its consequent repression was almost equal to the pattern found in non-diabetic rats, except for the shift in time due to the delay in the onset of reinnervation. The maximum increase in motor size (expressed as maximum of polyinnervation) in the normal and pathological condition were almost identical, suggesting that neurite branching is modulated primarily by the environment of the terminal, rather than by neuronal metabolism.

Despite the almost normal pattern of polyinnervation, the muscle is not entirely unaffected by the disease. Input resistance values were higher than in non-denervated fibers even after long intervals of 60 and 90 days from crush. In a similar study performed on the reinnervation of normal rat muscle (12), the input resistance returned to control value much earlier. These data indicate that muscle fibers recover more slowly from atrophy caused by denervation in diabetic rats, perhaps as a consequence of insulin deficit on muscle metabolism. Long lasting muscle alterations could be correlated with the presence of axon sprouts noticed even 3 months after nerve crush.

We can conclude that, beside a delayed muscle reinnervation and a slower synaptic maturation, diabetes also seems to affect the physiological properties of the reinnervated muscle. Neurite sprouting, on the contrary, is unaltered by the disease.

REFERENCES

1. Brown, M.J., Sumner, A.J., Greene, D.A. et al. (1980): Ann. Neurol. 8:168-178.
2. Carmingnoto, G., Finesso, M., Siliprandi, R. and Gorio, A. (1983): Neuroscience 8:393-401.
3. Elliasson, S.G. (1964): J. Clin. Invest. 43:2353-2359.
4. Frizell, M. and Sjostrand (1981): In PostTraumatic Peripheral Nerve Regeneration (eds) A. Gorio, H. Millesi and S. Mingrino, Raven Press, New York, pp. 59-70.
5. Fukuma, M., Carpentier, J.L., Orci, L., Greene, D.A. and Winegrad, A.I. (1978): Diabetologia 15:65-72.
6. Gorio, A., Carmignoto, G., Finesso, M., Polato, P. and Nunzi, M.G. (1983): Neuroscience 8:403-416.
7. Gregerson, G. (1967): Neurology 17:972-980.
8. Jacobsen, J. and Sidenius, R. (1980): J. Clin. Invest. 66:292-297.
9. Jakobsen, J. (1976): Diabetologia 12:547-553.
10. Kuno, M., Turkanis, S.A. and Weakly, J.R. (1971): J. Physiol. 213:545-556.
11. Lasek, J.R., McQuarrie and Wujek, J.R. (1981): In PostTraumatic Peripheral Nerve Regeneration (eds) A. Gorio, H. Millesi and S. Mingrino, Raven Press, New York, pp. 59-70.
12. McArdle, J.J. and Albuquerque, X.E. (1973): J. Gen. Physiol. 61:1-23.
13. McLean, W.W.J. and Meiri, K. (1980): J. Physiol. 301:57P.
14. Moore, S.A., Peterson, R.G., Felten, D.L., Cartwright, T.R. and O'Connor, N.L. (1980): Exp. Neurol. 70:548-555.
15. Mordes, J.P. and Rossini, A.A. (1981): The Am. Med. 70:353-360.

16. Norido, F., Cannella, R., Zanoni, R. and Gorio, A. (1984): Exp. Neurol. 83:221-233.
17. Powell, H., Knox, D., Lee, S. et al. (1977): Neurology 27:60-66.
18. Schiavinato, A.: Unpublished data.
19. Schmidt, R.E., Matschinsky, R.M., Godfrey, D.A., Williams, A.D. and McDougal Jr., D.B. (1975): Diabetes 24:1081-1085.
20. Sharma, A.K. and Thomas, P.K. (1974): J. Neurol. Sci. 23:1-15.
21. Sugimura, K., Windebank, M.D.A., Natarajan, V., Lambert, E., Schmid, H.H.O. and Dyck, P. (1980): J. Neuropathol. Exp. Neurol. 39:710-721.
22. Sima, A.A.F. and Robertson, D. (1978): Acta Neuropathol. 41:85-89.
23. Vitadello, M., Couraud, J.Y., Hassig, R., Gorio, A. and Di Giamberardino, L. (1983): Exp. Neurol. 82:143-147.
24. Vitadello, M., Filiatreau, G., Dupont, B.L., Hassig, R, Gorio, A. and Di Giamberardino, L.: Submitted to J. Neurochem.

CHOLINERGIC REM SLEEP INDUCTION RESPONSE AS

A MARKER OF ENDOGENOUS DEPRESSION

D. Jones, S. Kelwala, S. Dube,
E. Jackson and N. Sitaram

Lafayette Clinic
Wayne State University
951 E. Lafayette
Detroit, MI 48207

INTRODUCTION

At the present time the diagnosis of mental disorders is defined by operational criteria, which are primarily based on the presenting sysmptoms, i.e., Research Diagnostic Criteria (RDC) (18); Diagnostic and Statistics Manual III (2). This is also true for treatment decisions and prognosis. There is, however, a growing body of evidence that specific biological abnormalities that have both pathophysiological as well as diagnostic significance may be present, in disorders such as manic depressive illness.

Sleep EEG abnormalities have been consistently reported in patients with major depressive disorders (MDD), especially primary and endogenous subtypes (1, 8, 12, 19). The above studies using sleep electroencephalographic (EEG) recordings have found abnormality in Rapid Eye Movement sleep (REM) architecture. Vogel et al. (19), in summarizing the REM abnormalities identified: a) short REM latency; b) high density of eye movements during REM sleep; and c) abnormal temporal distribution of REM sleep with preponderance in the early part of the night. Feinberg et al. (6) and Rush et al. (14) found short REM latency to be more sensitive but less specific than non-suppression of serum cortisol using the Dexamethason Suppression test (4), a commonly reported biological marker for endogenous depression. There is evidence that short REM latency, enhanced REM density and disturbances of sleep continuity may result from enhanced cholinergic and/or decreased catecholaminergic functional activity (7, 9, 11). There is pharmacological evidence that the onset of REM sleep in man may be mediated by cholinergic mechanisms (15).

Using the REM sleep induction response seen after intravenous infusion of a muscarinic agonist (arecoline) during sleep as a potential marker of CNS cholinergic function, we have shown that patients with primary affective disorders in their euthymic state had a supersensitive REM induction response when compared with normal controls (16, 17). Preliminary evidence in seven depressed patients also indicated a supersensitive response.

In this study we evaluated the cholinergic REM induction response in a larger sample of depressed patients with MDD, as well as non-affective psychiatric control patients. Our objectives were: (a) to replicate our original findings (15, 16) in a larger, independent collected sample of depressed patients, tested in a different research setting; (b) to test the specificity of the putative marker to affective disorders as compared to other nonaffective disease states, since such a demonstration would have obvious significance for our underlying affective illness; and (c) to evaluate its relative sensitivity and specificity in differentiating clinical subtypes of depression such as Endogenous versus Nonendogenous, Primary versus Secondary, and Unipolar versus Bipolar MDD.

METHODS

Sixty inpatients and outpatients admitted to the study with RDC diagnosis of MDD (30 Endogenous and 19 Nonendogenous), and 11 nonaffective psychiatric control patients participated in this study. All patients were administered a schedule for Affective Disorders and Schizophrenia (SADS) interview (5). The SADS was administered by experienced research clinicians, with good interrater reliability as previously reported (3). Both current and lifetime diagnostic assignments of patients were based on RDC criteria, using information from SADS part I and II respectively. Raters were blind to all sleep data. The research assistant who scored the polygraphic records was in turn unaware of any clinical data.

All patients were medically cleared before they were allowed to participate in the study. Every patient received the following: a complete physical and neurological exam, EKG, blood chemistry (SMA-17, T_3, T_4), CBC with differential, and urinalysis. They were free of all psychotropic and other drugs which might affect the patients' mood or polygraphic data, for at least two weeks. Informed consent was obtained from the patients. Two patients refused to participate in sleep studies after SADS interview. Three patients with definite MDD (1 endogenous and 2 nonendogenous) who underwent sleep infusion studies had to be dropped from data analysis: two subjects because of excessive and prolonged spontaneous awakenings during the second non REM period, and one subject because of our inability to perform venipuncture due to extreme difficulty in finding a forearm vein (see below).

SLEEP

Subjects underwent 3 to 4 nights of sleep recordings during which their EEG (C3-A2), EMG (submental electromyogram) and EOG (electrooculogram) were monitored as per the standard criteria of Rechtschaffen and Kales (13). Electrocardiogram (lead II) was also monitored throughout the night. The first night served to acclimatize the patient to sleeping with electrodes. No polygraphic recording was obtained for the majority of patients. However, since the publication of data by Akiskal (1) indicating that comparison of variability between nights one and two may be useful in differentiating primary and secondary depressives, we have begun to routinely obtain polygraphic recordings on night one on all subjects.

During the second night a regular polygraphic recording was obtained ("Baseline" night). On nights three and four, in addition to placement of polygraphic electrodes, subjects also slept with a 21-gauge scalp vein needle placed in a forearm vein attached to a 10-foot polyethylene tube extending into an adjacent room. A continuous infusion of normal saline was maintained at the rate of approximately 0.25 ml/minute (using an IVAC infusion pump) throughout the night, to prevent blood coagulation. The intravenous set-up as described above permits discrete pharmalogical challenges to be carried out during specific sleep stages or cycles, safely and without disturbing the sleeping subject. At the end of the first REM period glycopyrrolate, a peripherally acting anticholinergic agent was administered in a dose of 0.1 mg. on both infusion nights. Approximately twenty-five minutes after the end of the first REM period, the patient received an infusion of either 0.5 mg arecoline or placebo (in random order on nights three and four). All infusions were administered over a three minute period.

RESULTS

Overall t-test (2 tailed) comparisons between all MDD patients (N = 49) on one hand, and Nonaffective controls (N = 11) on the other, revealed significant differences in age [MDD = 37.9 $\pm$ 12.8 (mean $\pm$ S.D.); Nonaffectives = 29.5 $\pm$ 4.1; t = 2.14; p = 0.037; Df = 58], REM latency [MDD = 58.6 $\pm$ 30.7; Nonaffectives = 80.8 $\pm$ 21.8; t = -2.26; p = 0.028] and Arecoline response [MDD = 30.9 $\pm$ 24.5; Nonaffectives = 48.1 $\pm$ 20.4; t = -2.15; p = 0.036]. No differences were seen with respect to duration or density of the first REM period.

The subject population was then analyzed after division into 3 groups: Endogenous MDD, Nonendogenous MDD, and Nonaffectives. Table 1 shows the demographic and diagnostic subtype breakdown between the groups. No differences in age or sex distribution between the groups was found, although the mean age of the Nonaffective controls (29.5 years) was almost a decade lower than the Endogenous patients (38.8 years).

Table 1

Demographics and Breakdown of Clinical Diagnosis

	MDD Endogenous (N=30)	MDD Non-Endogenous (N=19)	Nonaffective Controls (N=11)
Inpatient	22	13	7
Outpatient	8	6	4
Age (Mean ± S.D.)	38.8 ± 14.6	36.5 ± 9.7	29.5 ± 4.1
Sex:			
Males	13	8	4
Females	17	11	7
A. MDD Patients Subtypes (SADS-RDC)			
Primary	27	10	
Secondary**	3	9	
Unipolar	24	19	
Bipolar I	2	0	
Bipolar II	4	0	
Psychotic	4	1	
Nonpsychotic	26	18	
B. Principal Diagnosis: Nonaffective Controls			
Panic		3	
Obsessive-Compulsive			2
GAD		1	
Schizophrenia		4	
Alcoholism		1	

**Primary diagnoses in patients with secondary MDD were: Panic disorder
- N=7, Obsessive Compulsive disorder - N=2, Schizophrenia - N=2

Table 2 indicates the results of one-way ANOVA with post-hoc Scheffe's test between the 3 groups on a number of selected sleep parameters. REM latency and arecoline REM-induction latency (Inf-REM2 latency) were significantly reduced in Endogenous as compared to both Nonendogenous and Nonaffective patients. The density of the first REM period was significantly increased in Endogenous as compared to Nonaffective patients. The Nonendogenous patients did not differ from either of the other two groups.

A discriminant function analysis (SPSS; 10) using the four sleep variables was next undertaken to separate the endogenous from the nonendogenous group. Table 3 shows the results using: (a) REM latency alone as a discriminant variable; (b) arecoline response alone; and (c) four sleep measures, consisting of REM latency (RL), effect of arecoline, duration of first REM period (RT) and density of first REM period (RD). As shown in Table 3 the performance of arecoline REM-induction as a diagnostic predictor (Sensitivity = 70%, Specificity = 79%, percentage correctly classified = 74%) was more robust than REM latency (Sensitivity = 70%, Specificity = 53%, percentage correctly classified = 63%). It can be seen that while achieving the same sensitivity the arecoline response had a much greater specificity than REM latency. In the third analysis, among the four sleep measures only two satisfied the criteria (probability of 0.05 or less) to be considered for entry into the discriminant function equation. Arecoline was entered at Step 1 (Wilks Lambda = 0.69, F = 26.9, Df 1,47) and REM latency at Step 2 (Lambda = 0.61, F = 14.9, Df 2,46). Together the two variables achieved a sensitivity of 86.2%, specificity of 75% and correct classification of total cases of 81.6%.

DISCUSSION

In the current study we have attempted to test the validity of supersensitive cholinergic REM sleep induction as a putative biological marker of endogenous depression, using a much larger sample than had been done previously. In addition, in contrast to our prior studies (15-17), where we had compared affective disorder patients with normal controls, here we have extended the investigation to include nonaffective psychiatric controls as well. This control group is critical to assess the specificity of the response to major affective disorder as compared to other psychiatric diseases.

Our results showing shortened REM latency and increased REM density of the first period in endogenous patients corroborate findings of other investigators (1, 6-8, 12, 14, 19). Our study thus extends the important body of data on EEG sleep physiological abnormalities in depression by the above investigators into the pharmacological domain.

Table 2

Breakdown of Selected EEG Sleep Physiological and
Pharmacological (Arecoline Response) Parameters in
Depressed and Nonaffective Psychiatric Controls

	Endogenous MDD (N=30)	Non-Endogenous MDD (N=19)	Nonaffective Controls (N=11)	ANOVA Oneway & Scheffe's Test
REM Latency (Minutes)	53.4 (21.8)	72.7 (27.3)	81.1 (21.8)	p < .05 GRP 1 < 2, 3
Duration of 1st REMP (Minutes)	16.6 (13.4)	19.8 (9.5)	12.4 (4.5)	p = NS
REM Density of 1st REMP Units (0-8)	2.2 (1.0)	1.97 (0.98)	1.37 (0.66)	p < .05 GRP 1 > 3
Arecoline Response (INF-REM2 Latency)	23.1 (23.2)	43.5 (21.8)	48.3 (20.4)	p < .002 GRP 1 < 2, 3

**Time elapsed from sleep onset (Stage 2) to the onset of first REM period. Awake time within this period was not subtracted. Sleep onset defined as the first minute of Stage 2 sleep followed by at least 9 additional minutes of sleep uninterrupted by no more than 1 minute of awake time.

Table 3

Discriminant Analysis: Sleep Variables as
Predictors of MDD Endogenous Subtype Diagnosis

Patient Population (Major Depressive Disorder)
Classification   Number of Patients used in Analysis

Endogenous   30
(SADS-RDC)
Non-Endogenous   19
(SADS-RDC)

A. Variable(s) Used: REM Latency
                        Predicted Group Membership
                  Endogenous   Non-Endogenous
Endogenous   21 (70%)   9 (30%)
Non-Endogenous   9 (47.4%)   10 (52.6%)
                  Cases correctly classified: 63.3%

B. Variable(s) Used: Arecoline REM Induction
                        Predicted Group Membership
                  Endogenous   Non-Endogenous
Endogenous   21 (70%)   9 (30%)
Non-Endogenous   4 (21.1%)   5 (78.9%)
                  Cases correctly classified: 73.5%

C. Variable(s) Used: RL, Arecoline, Duration of First
   REM Period (RT) and Density of First REM Period (RD)
                        Predicted Group Membership
                  Endogenous   Non-Endogenous
Endogenous   25 (86.2%)   4 (13.8%)
Non-Endogenous   4 (25%)   15 (75%)
                  Cases correctly classified: 81.6%

Variables Entered: Wilks Lambda   F   DF
Step 1   Arecoline   0.69   26.9   1, 47
Step 2   RL   0.61   14.9   2, 46

Standardized Canonical Correlation:
Arecoline   0.84
RL   0.35

Group Centroids (Canonical DF at Group Means)
1.   -0.65   Endogenous Diagnosis
2.   0.95   Non-Endogenous Diagnosis

From a practical point of view this leads us to believe that the arecoline REM induction response may, in addition to other sleep and neuroendocrine markers, prove to be of some value as an external objective validator of endogenous MDD diagnosis. Insofar as endogenicity has therapeutic and prognostic implications (1), the development of biological markers may have a limited role in clinical situations, where the manifestation of underlying depression may be masked or confounded by superimposed personality, or panic disorders, or in depressions occurring in the aged (i.e., "pseudodementia").

We caution, however, that the results presented here represent at best _statistically_ _significant_ differences between arecoline REM induction response in Endogenous versus Nonendogenous patients. There is considerable overlap between the two groups, as indicated by the large standard deviations (Table 2). As such, pending extensive and thorough field testing, the cholinergic REM induction response should remain just that: a "response" marker and not be construed as a "test". At this time we are more intrigued by the role of acetylcholine in the pathophysiology and possible genetic vulnerability to affective disorders, rather than in encouraging premature clinical application.

REFERENCES

1. Aikiskal, H.S. (1980): J. Clin. Psychiatry 41:6-14.
2. American Psychiatric Association (1980): Diagnostic and Statistical Manual of Mental Disorders. 3rd Ed.
3. Bell, J., Lycaki, H., Jones, D., Kelwala, S. and Sitaram, N. (1983): Psychiatry Research 9:115-123.
4. Carroll, B.J., Feinberg, M., Greden, J.F., Tarika, J., Albala, A.A., Haskett, R.F., James, N.M., Kronfol, Z., Lohr, N., Steiner, M., de Vigne, J.P. and Young, E. (1981): Arch. Gen. Psychiatry 38:15-22.
5. Endicott, J. and Spitzer, R.L. (1978): Arch. Gen. Psychiatry 35:837-843.
6. Feinberg, M., Gillin, J.C., Carroll, B.J., Greden, J.F. and Zis, A.P. (1982): Biol. Psychiatry 17:327-341.
7. Gillin, J.C., Mendelson, W.B., Sitaram, N. and Wyatt, R.J. (1978): Ann. Rev. Pharmacol. Toxicol. 18:563-569.
8. Gillin, J.C., Duncan, W.C., Pettigrew, K., Frankel, B.L. and Snyder, F. (1979): Arch. Gen. Psychiatry 36:85-90.
9. Hobson, J.A., McCarley, R.W., McKenna, T.M. (1976): Prog. Neurobiol. 6:280-376.
10. Hull, C.H. and Nie, N.G. (1979): SPSS Update. New Procedures and Facilities for Releases 7 and 8. New York:McGraw-Hill.
11. Karczmar, A.G., Longo, V.G., Scott de Carolis (1970): Physiol. Behav. 5:175-182.
12. Kupfer, D.J., Foster, F.G., Coble, P., McPartland, R. and Ulrich, R. (1978): Am. J. Psychiatry 135:69-74.

13. Rechtchaffen, A. and Kales, A. (1968): A Manual of Standard Techniques and Scoring System for Sleep States of Human Subjects. NIH Publication #204, U.S. Govt. Printing Office, Washington, D.C.

14. Rush, A.J., Giles, G.E., Roffworg, H.P. and Parker, C.R. (1982): Biol. Psychiatry 17:327-341.

15. Sitaram, N. and Gillin, J.C. (1980): Biol. Psychiatry 15:925-955.

16. Sitaram, N., Nurnberger, J.I., Gershon, E.S. and Gillin, J.C. (1980): Science 208:200-202.

17. Sitaram, N., Nurnberger, J.I., Gershon, E.S. and Gillin, J.C. (1982): Am. J. Psychiatry 139:571-576.

18. Spitzer, R.L., Endicott, J. and Robins, E. (1978): Research Diagnostic Criteria for a selected group of functional disorders. d.3 New York:Biometrics Research Division, New York State Psychiatric Institute.

19. Vogel, G.W., Vogel, F., McAbee, R.S. and Thurmond, A.J. (1980): Arch. Gen. Psychiatry 17:305-316.

STEREOSELECTIVITY OF SOME MUSCARINIC AND ANTIMUSCARINIC
AGENTS RELATED TO OXOTREMORINE

R. Dahlbom[1], B. Ringdahl[2], B. Resul[1] and D.J. Jenden[2]

[1]Department of Organic Pharmaceutical Chemistry
Biomedical Center
Uppsala University
P.O. Box 574
S-751 23 Uppsala, Sweden

[2]Department of Pharmacology, School of Medicine
University of California
Los Angeles, California 90024, U.S.A.

INTRODUCTION

Oxotremorine, N-(4-pyrrolidino-2-butynyl)-2-pyrrolidone ([1]), is a muscarinic agent equal in potency to acetylcholine (3, 7). Its extraordinary potency is surprising in view of its structural dissimilarity to other muscarinic agents. Whereas most powerful muscarinic agents possess a quaternary trimethyl ammonium group, oxotremorine is a tertiary amine with no N-methyl groups at all. Moreover, it contains an acetylenic bond at the position in the molecule where strong muscarinic agents have an oxygen atom. The structural requirements for muscarinic activity are rather specific and even slight changes in the structure lead to loss of the muscarinic activity or to a change of the type of activity from agonistic to antagonistic (4, 14). The only amino congeners which show agonistic activity of the same order of magnitude as oxotremorine are the dimethylamino ([2]) and the azetidino ([3]) analogs. Compound [3] is in fact the most potent tertiary muscarinic agent known to date (see structures on next page).

A great number of compounds related to oxotremorine and its succinimide analog has been prepared in our laboratories (4) with the object of finding potent central oxotremorine antagonists with maximal specificity relative to peripheral anticholinergic activity.

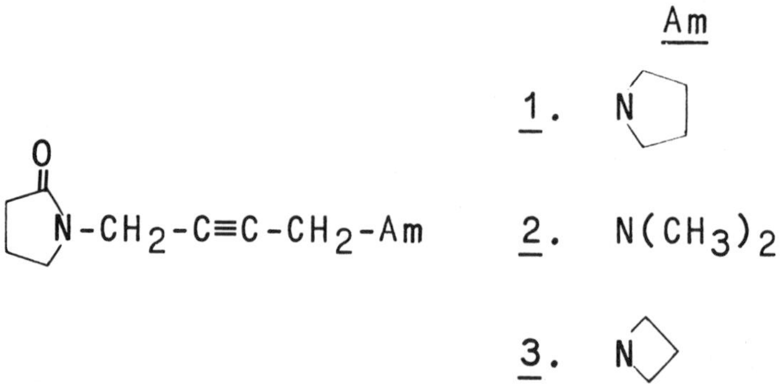

Am

1. N⬡ (pyrrolidine ring)

$N-CH_2-C{\equiv}C-CH_2-Am$

2. $N(CH_3)_2$

3. N⬦ (azetidine ring)

Many of the compounds synthesized have marked selectivity as central anticholinergic agents. The most active compounds have the same intermediate chain connecting the two nitrogens in the molecule.

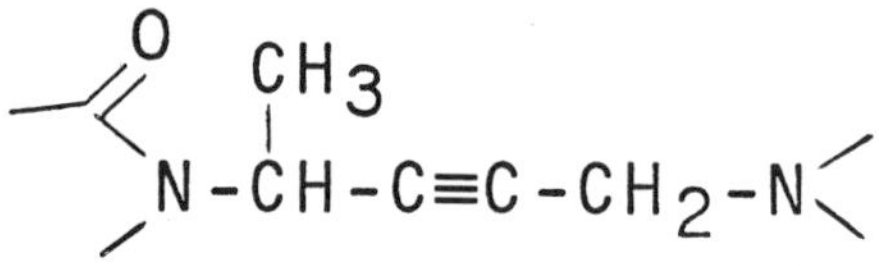

$$N-CH-C{\equiv}C-CH_2-N{<}$$
(with O and CH_3 substituents)

As this chain contains an asymmetrically substituted carbon atom, we decided to prepare the enantiomers of compounds [4-9] in order to investigate possible stereoselectivity of action (6, 12). The enantiomers were tested for antagonism to oxotremorine- induced tremors in mice, and for antagonistic activity towards carbachol on the isolated guinea-pig ileum.

Table 1 shows that the R-isomers are uniformly more active than the S-isomers. The enantiomeric potency ratio increases with increasing activity of the R-isomer. This is in agreement with Pfeiffer's rule (1, 8), which states that the enantiomeric potency ratio of a more active compound is higher than that of a less active one.

It is quite remarkable to note that higher activity is displayed by analogs with the smaller substituents, the most active being compound [4], whose structure differs from that of oxotremorine by only one additional methyl group. It is unique among muscarinic agents in that this small change in structure reverses the type of activity and converts a potent agonist into a potent antagonist.

We found it also of interest to investigate compounds with a chiral center in the amine part of the molecule and introduced a methyl group in the 2- and 3-positions of the pyrrolidine ring of oxotremorine and its succinimide analog (10, 11). In both cases potent oxotremorine antagonists were obtained. Only the enantiomers of the 2-substituted derivatives show stereoselectivity, the S-enantiomers being the most active. However, this stereoselectivity is less pronounced than that of the enantiomers with the chiral center at the 1-position of the butynyl chain (Table 2).

In view of these findings we found it worthwhile to investigate oxotremorine analogs with both these chiral centers. We prepared eight optical isomers of this type (15), and the results of the pharmacological tests are shown in Table 3.

As might be expected, the isomer having methyl groups at both chiral centers and the "right" configuration (R,S) at both centers is the most active, and is the most potent oxotremorine antagonist presently known. The potency ratios of the enantiomeric pairs of [14] and [15] increase with increasing activity of the more potent isomer of each pair in agreement with Pfeiffer's rule. Large differences in activity are also shown by epimers of [14] and [15] with the same configuration in the pyrrolidine ring but opposite configuration at the chiral center in the butynyl chain. The epimeric potency ratios clearly increase with increasing activity of the more active isomer (Table 4).

Table 4 also shows a comparison between the epimeric pairs of [14] and [15] with the opposite configuration at the chiral center in the pyrrolidine ring but with the same configuration in the butynyl chain. In this case a small potency ratio is observed. The conclusion which can be drawn from these data is that only the chiral center in the butynyl chain is critical to stereoselectivity, since inversion leads to large differences in potency. The configuration of the chiral center in the pyrrolidine ring is of much less importance, since inversion causes only small changes in the epimeric potency ratios.

As we now had a good picture of the stereochemical requirements for central and peripheral antimuscarinic activity in this class of compounds, we found it desirable to find a corresponding chiral agonist to compare its active configuration with that found for the antagonists. However, substitution at the C-1 position of the butynyl chain of oxotremorine invariably leads to a substantial loss of muscarinic activity or to antagonistic properties. This difficulty was circumvented by investigating an analogous series of compounds, the first of which were described by Bebbington et al. (2). N-(4-Pyrrolidino-2-butynyl)-N-methylacetamide was reported to be a tremorogenic agent about half as active as oxotremorine,

Table 1

Pharmacological Data for N-(1-alkyl-4-pyrrolidino-2-butynyl)-
Substituted 2-Pyrrolidones and Succinimides (6,12,13)

4 - 6   7 - 9

Compd	R	Tremorolytic potency[a] (μmol/kg)	Enantiomeric potency ratio R/S	Parasympatholytic potency[b] pA_2	Enantiomeric potency ratio R/S
R-4		0.26		7.55	
	CH_3		77		257
S-4		20		5.14	
R-5		0.52		7.09	
	C_2H_5		52		47
S-5		27		5.42	
R-6		3.5		6.79	
	C_3H_7		15		14
S-6		51		5.63	
R-7		0.48		7.07	
	CH_3		58		49
S-7		28		5.38	
R-8		1.2			
	C_2H_5		25		
S-8		30			
R-9		8.2			
	C_3H_7		8		
S-9		68			

[a]Dose required to double the dose of oxotremorine inducing a predetermined tremor intensity in 50% of the mice.   [b]In vitro antimuscarinic activity determined on guinea-pig ileum against carbachol.

Table 2

Pharmacological Data for N-[4-(methylpyrrolidino)-2-butynyl]-
Substituted 2-Pyrrolidones and Succinimides (10,13)

10, 11   12, 13

Compd	R	R'	Tremorolytic potency (μmol/kg)	Enantiomeric potency ratio S/R	Parasympatholytic potency pA_2	Enantiomeric potency ratio S/R
R-10			56		5.17	
	CH_3	H		22		5.2
S-10			2.6		5.89	
R-11			6.5		5.69	
	H	CH_3		1.1		1.0
S-11			5.8		5.68	
R-12			17.4			
	CH_3	H		2.1		
S-12			8.3			
R-13			20			
	H	CH_3		1.2		
S-13			17.2			

and the dimethylamino analog also displayed considerable activity.
Chiral derivatives of these compounds proved to have interesting
properties (9). Introduction of a methyl group at C-1 in the butynyl
chain of the pyrrolidino compound ([16]) yielded an antagonist to
oxotremorine in the CNS with partial agonist properties in the guinea--
pig ileum. The dimethylamino analog ([17]) was a pure agonist both
centrally and peripherally.

Table 3

Pharmacological Data for N-[1-alkyl-4-(2-methylpyrrolidino)-
2-butynyl]-2-pyrrolidones (13,15)

Compd*	R	Tremorolytic potency (μmol/kg)	Enantiomeric potency ratio R/S	Parasympa-tholytic potency pA$_2$	Enantiomeric potency ratio R/S
RS-14		0.10		7.80	
	CH$_3$		200		646
SR-14		20		4.99	
RR-14		0.50		7.35	
	CH$_3$		24		78
SS-14		12		5.46	
RS-15		1.5		6.96	
	C$_3$H$_7$		16		17
SR-15		24		5.73	
RR-15		4.2		6.68	
	C$_3$H$_7$		2.1		5.8
SS-15		8.9		5.92	

*The first configurational symbol refers to the configuration of the
chiral center in the butynyl chain; the second symbol to the configuration of the chiral center in the pyrrolidine ring.

Table 4

The Epimeric Potency Ratios for N-[1-alkyl-4-(2-methylpyrrolidino)-2-butynyl]-2-pyrrolidones (13,15)

Compd	R	Tremorolytic potency (μmol/kg)	Epimeric potency ratio	Parasympatholytic potency pA_2	Epimeric potency ratio
RS-14		0.10		7.80	
	CH$_3$		120		219
SS-14		12		5.46	
RR-14		0.50		7.35	
	CH$_3$		40		229
SR-14		20		4.99	
RS-15		1.5		6.96	
	C$_3$H$_7$		5.9		11
SS-15		8.9		5.92	
RR-15		4.2		6.68	
	C$_3$H$_7$		5.7		8.9
SR-15		24		5.73	
RS-14		0.10		7.80	
	CH$_3$		5.0		2.8
RR-14		0.50		7.35	
SS-14		12		5.46	
	CH$_3$		1.7		2.9
SR-14		20		4.99	
RS-15		1.5		6.96	
	C$_3$H$_7$		2.8		1.9
RR-15		4.2		6.68	
SS-15		8.9		5.92	
	C$_3$H$_7$		2.7		1.5
SR-15		24		5.73	

Table 5

Pharmacological Properties of N-(4-amino-1-methyl-2-butynyl)-
N-methylacetamides

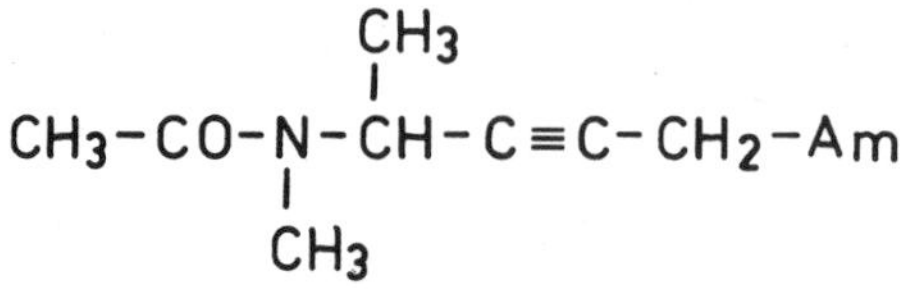

$$\text{CH}_3\text{-CO-N-CH-C}\equiv\text{C-CH}_2\text{-Am}$$

with CH_3 on the N and CH_3 on the CH

Compd	Am	Guinea-pig ileum pD_2	Enantiomeric potency ratio R/S	Tremorogenic potency ED_{50} (μmol/kg)	Tremorolytic potency (μmol/kg)
R-16		6.63*			0.35
			13.2		
S-16		5.51*			10
R-17		6.50		12	
	$N(CH_3)_2$		16.2		
S-17		5.29		inactive	
R-18		6.80			
	$N^+(CH_3)_3$		33.1		
S-18		5.28			
Oxotremorine		7.50		0.5	
Carbachol		7.15			

*Partial agonist.

We next prepared the optical isomers of the two compounds and
the N-methyl quaternary salt of [17] (5). The pharmacological tests
showed that the R-isomers were more potent than the S-isomers both
in vivo and in vitro, regardless of whether the compounds were agonists, partial agonists or antagonists (Table 5). The active configuration is depicted in the structure below:

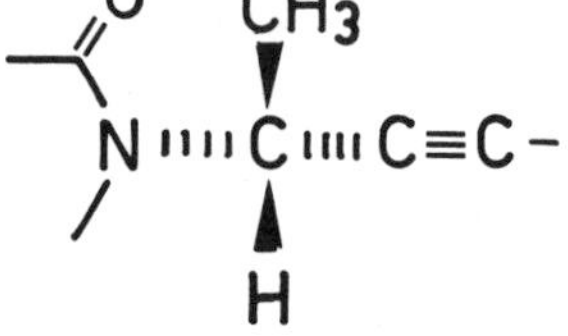

Thus, in the oxotremorine series the stereochemical requirements for muscarinic and antimuscarinic activity appear to be similar. This finding and the close structural similarity between agonists and antagonists suggests that agonists and antagonists interact with a common receptor site. This observation is in contrast to classical muscarinic antagonists, which are believed to bind also to accessory receptor areas differing from, but located close to the agonist binding site.

Acknowledgements. This work was supported in part by USPHS grant MH-17691 and the IF Foundation for Pharmaceutical Research.

REFERENCES

1. Ariëns, E.J. (1966): Advan. Drug Res. 3:235-285.
2. Bebbington, A., Brimblecombe, R.W. and Shakeshaft, D. (1966): Brit. J. Pharmacol. 26:56-67.
3. Cho, A.K., Haslett, W.L. and Jenden, D.J. (1962): J. Pharmacol. Exp. Ther. 138:249-257.
4. Dahlbom, R. (1981): In Cholinergic Mechanisms: Phylogenetic Aspects, Central and Peripheral Synapses, and Clinical Significance (eds) G. Pepeu and H. Ladinsky, Plenum Press, New York, pp. 621-638.
5. Dahlbom, R., Jenden, D.J., Resul, B. and Ringdahl, B.(1982): Brit. J. Pharmacol. 76:299-304.
6. Dahlbom, R., Lindquist, A., Lindgren, S., Svensson, U., Ringdahl, B. and Blair, M.R. Jr. (1974): Experientia 30:1165-1166.
7. George, R., Haslett, W.L. and Jenden, D.J. (1962): Life Sci. 1:361-363.
8. Pfeiffer, C.C. (1956): Science 124:29-31.
9. Resul, B., Dahlbom, R., Ringdahl, B. and Jenden, D.J. (1982): Eur. J. Med. Chem. 17:317-322.
10. Ringdahl, B. and Dahlbom, R. (1978): Experientia 34:1334-1335.
11. Ringdahl, B. and Dahlbom, R. (1978): Acta Pharm. Suec.15:255-263.
12. Ringdahl, B. and Dahlbom, R. (1979): Acta Pharm. Suec. 16:13-20.
13. Ringdahl, B. and Jenden, D.J. (1983): Mol. Pharmacol. 23:17-25.
14. Ringdahl, B. and Jenden, D.J. (1983): Life Sci. 32:2401-2413.
15. Ringdahl, B., Resul, B. and Dahlbom, R. (1979): J. Pharm. Pharmacol. 31:837-839.

AFFINITY AND EFFICACY OF OXOTREMORINE ANALOGS

AT ILEAL MUSCARINIC RECEPTORS

B. Ringdahl

Department of Pharmacology
School of Medicine
University of California
Los Angeles, California 90024 U.S.A.

INTRODUCTION

The pharmacological activity of a muscarinic agonist, as measured for example by contractile responses in smooth muscle preparations, is a function of both its affinity for the receptor and its efficacy in producing a stimulus that leads to a response (26). Because of the existence in many tissues of a receptor reserve for muscarinic agonists, ED_{50} values for pharmacological responses often are poor estimates of affinities. More satisfactory estimates of agonist affinities may be obtained from receptor binding studies. Although the relationship between binding parameters and ED_{50} values of muscarinic agonists is not yet fully understood, radioligand binding techniques have given valuable information on the structural demands for agonist binding at the muscarinic receptor (3, 12). In contrast, the structural features of muscarinic agonists that cause them to activate the receptor are less readily assessed.

I have resolved the muscarinic activities in the isolated guinea pig ileum of several analogs of oxotremorine into affinity and efficacy components. The pharmacological method used allows assessment of the structural requirements not only for receptor binding but also for receptor activation. Pharmacologically determined affinities and efficacies of muscarinic agonists should be helpful in further elucidating the relationship between receptor binding data and pharmacological responses. Efficacy is intimately associated with the ability of an agonist to convey binding into receptor activation, presumably through a specific conformational change in the receptor protein (4). Assessment of the structural requirements underlying efficacy will permit the design of highly efficacious agonists useful as probes for investigation of the mechanism of activation of the receptor-effector system.

PHARMACOLOGICAL ESTIMATION OF AFFINITY AND EFFICACY OF PARTIAL AGONISTS

Furchgott's method (10, 11) for estimation of dissociation constants (K_A) of agonist-receptor complexes involves analysis of dose-response data before, and after fractional inactivation of receptors with dibenamine. It has been widely used for agonists acting at α-adrenergic receptors (2, 23) and also for selected muscarinic agonists (11, 20, 24, 27). However, the use of irreversible antagonists such as dibenamine has been criticized (25) because of their possible interactions with calcium channels (8). I have used a modification of Furchgott's method to determine the K_A values and relative efficacies at muscarinic receptors in the isolated guinea pig ileum of some oxotremorine analogs. Instead of dibenamine with its multiplicity of actions, I used the more specific irreversible muscarinic antagonist propylbenzilylcholine mustard (PrBCM). This modified method, referred to below as Method 1, has been described in detail elsewhere (17).

In view of the doubts raised about the validity of Furchgott's method (8, 25), I first studied some partial agonists for which alternative pharmacological procedures for K_A determinations are available, so that it would be possible to test for internal consistency. Among these methods, comparison of the dose-response curve of a partial agonist with that of an agonist (Method 2) (28) and use of the partial agonist as a competitive antagonist after irreversible occlusion of a major fraction of the receptors (Method 3) (11) are the most commonly used. Table 1 summarizes the dissociation constants and relative efficacies of four partial muscarinic agonists determined by the three independent methods referred to above (17). The results obtained with Method 2 deserve special attention since this method does not make use of an irreversible antagonist. The important assumption underlying this method is that the agonist, with which the partial agonist is compared, has high efficacy (28). Carbachol was found to satisfy this criterion (see below).

The good agreement between the K_A values and relative efficacies determined by Methods 1 and 2 (Table 1) appears to justify the use of PrBCM for determining agonist affinities and efficacies. Method 1, as applied to partial agonists, requires occlusion of only a minor fraction (<50%) of the receptors with PrBCM. This contrasts with the high degree (usually >90%) of receptor inactivation required for K_A determinations of strong agonists using Method 1. Method 3 employed PrBCM to occlude 90-95% of the receptors. At this degree of receptor alkylation, the responses to the partial agonists were completely abolished whereas carbachol still produced a maximal response. The excellent agreement between the K_A values obtained with this method and Methods 1 and 2 (Table 1) suggests that inactivation even of a major fraction of the receptors with PrBCM does not affect the affinity of the compounds studied for the remaining receptors.

Table 1

Summary of Parameters Characterizing the Muscarinic Activity of Some Partial Agonists in the Isolated Guinea Pig Ileum

Parameter	R-I	S-I	R-II	S-II
ED$_{50}$ (μM)	2.85 ± 0.27	0.88 ± 0.17	0.094 ± 0.009	1.70 ± 0.12
E$_{max}$[a]	0.72 ± 0.05	0.55 ± 0.04	0.83 ± 0.03	0.91 ± 0.03
K$_A$ (μM):				
Method 1	5.02 ± 0.80	0.66 ± 0.22	0.13 ± 0.025	5.50 ± 1.27
Method 2	5.50 ± 0.73	1.23 ± 0.13	0.20 ± 0.030	3.23 ± 0.35
Method 3	4.39 ± 0.29	1.11 ± 0.090	0.11 ± 0.009	4.00 ± 0.25
Relative efficacy[b]:				
Method 1	0.099 ± 0.012	0.062 ± 0.011	0.090 ± 0.008	0.15 ± 0.020
Method 2	0.076 ± 0.008	0.033 ± 0.004	0.095 ± 0.010	0.11 ± 0.009

All values are given as means $\pm$ S.E. 4-7 preparations were used for each compound
[a]Maximum response relative to the maximum for oxotremorine which equals 1.00.
[b]Efficacy relative to that of oxotremorine which equals 1.00.

This suggestion was confirmed by estimating the dissociation constants of the enantiomers of the competitive antagonist III (Compound III) in pairs of ileal strips, one of which had been pretreated with PrBCM to alkylate about 90% of the receptors (17). The pA_2 values, determined against carbachol, of R-III and S-III, respectively, were 7.34 and 5.02 in untreated and 7.30 and 5.06 in PrBCM-treated tissue. These results are consistent with the observation of Birdsall et al. (3) that alkylation of up to 95% of the muscarinic binding sites in homogenates of rat cerebral cortex with PrBCM did not alter the IC_{50} or Hill coefficient for competition between carbachol and [^{3}H]-propylbenzilylcholine at the remaining sites.

OXOTREMORINE   COMPOUND III

PHARMACOLOGICAL ESTIMATION OF AFFINITY AND
EFFICACY OF STRONG AGONISTS

The results obtained with the partial agonists suggest that Method 1 is valid also for agonists requiring a high degree of receptor inactivation before a depression of the maximum response is observed, i.e., for strong agonists. However, no alternative pharmacological methods for K_A determinations of such agonists are readily available to confirm the validity of this suggestion.

For oxotremorine and carbachol, a large body of pharmacological and receptor binding data have accumulated in the literature (Table 2) from which information may be obtained about their affinities and relative efficacies at muscarinic receptors. The K_A values of oxotremorine and carbachol estimated by Method 1 (Table 2) agreed with those previously determined pharmacologically using dibenamine (11, 27), and also with the IC_{50} values for inhibition of [^{3}H]-antagonist binding to muscarinic receptors in homogenates of the guinea pig ileum (29) and of the rat brain (3). These IC_{50} values (Table 2) should approximate the dissociation constants of oxotremorine and carbachol, since the concentrations of the [^{3}H]-antagonists used were well below their own apparent dissociation constants (3, 29).

There was also good agreement between the K_A values of oxotremorine and carbachol and the low affinity dissociation constants (K_L) of these agonists determined from inhibition of [^{3}H]-antagonist binding to ileal (7) and brain (3) muscarinic receptors. Agreement between K_A and K_L values has previously been noted for a series of muscarinic agonists (3).

Table 2

Comparison of Dissociation Constants and Efficacies of Oxotremorine and Carbachol Obtained from Pharmacological and Receptor Binding Techniques

Parameter	Tissue	Method (Reference)	Oxotremorine	Carbachol
K_A (μM)	a	Method 1	0.68	16.4
K_A (μM)	a	Furchgott's (27)	0.5	-
K_A (μM)	b	Furchgott's (11)	-	15.9
IC_{50} (μM)	a	[^{3}H]QNB displ. (29)[e]	0.5 - 0.8	20 - 30
IC_{50} (μM)	c	[^{3}H]PrBCh displ. (3)[f]	0.46	15
K_L (μM)	d	[^{3}H]QNB displ. (7)[e]	1.0	12
K_L (μM)	c	[^{3}H]PrBCh displ. (3)[f]	0.83	89.1
K_A/ED_{50}	a	Method 1	28	204
K_L/K_H	d	[^{3}H]QNB displ. (7)[e]	116	313
K_L/K_H	c	[^{3}H]PrBCh displ. (3)[f]	11	195

[a]Guinea pig ileum; [b]rabbit stomach fundus; [c]rat cerebral cortex; [d]rat ileum; [e]displacement of [^{3}H]-3-quinuclidinyl benzilate from homogenates; [f]displacement of [^{3}H]-propylbenzilyl-choline from homogenates.

Finally, the ratio K_A/ED_{50}, which is a measure of agonist efficacy, was in qualitative agreement with the ratio of the low (K_L) and high (K_H) affinity dissociation constants obtained from binding studies (3, 7). The latter ratio has also been suggested to reflect agonist efficacy (3). Collectively, these results show that Furchgott's method in the modification employed here provides a good estimate of dissociation constants and relative efficacies of muscarinic agonists.

AFFINITY AND EFFICACY OF SOME OXOTREMORINE ANALOGS

Previous results have shown that structural modifications of the amino group of oxotremorine and of its highly potent acetamide analog (Compound Va, Table 3) often preserve and sometimes enhance the muscarinic activity (16). Structural alterations in other parts of these molecules almost always are accompanied by a decrease in efficacy since the resulting compounds, with only a few exceptions, are partial agonists or antagonists (5, 6, 15, 17, 20-22).

Therefore, in a search for highly efficacious agonists, it seemed logical to focus on analogs of oxotremorine, and Compound Va, modified only in the amino group (Table 3).

The ED_{50} values for contraction of the isolated guinea pig ileum of some analogs of oxotremorine and Compound Va are summarized in Table 3 (18, 19). All of the compounds produced the same maximum response as carbachol, indicating full agonist activity. The dissociation constants and relative efficacies of the compounds were determined by Method 1 referred to above. Oxotremorine had the highest affinity among the compounds. Substitution of the pyrrolidine ring in oxotremorine with a four-membered azetidine ring (Compound IVb) reduced the affinity 2-fold. The dimethylamino analog IVc had only about 3% of the affinity of oxotremorine. N-Methylation of Compound IVc increased the affinity nearly 8-fold (Compound IVd) whereas N-methylation of oxotremorine decreased the affinity 10-fold (Compound IVe). Similar changes in affinity were observed (18) on structural modification of the amino group of the acetamide Va (Table 3). With the exception of Compound IVe, all the compounds studied had higher relative efficacies than oxotremorine. The dimethylamino analogs (IVc and Vb) and oxotremorine-M (IVd) had about the same relative efficacies as carbachol, which is regarded as a highly efficacious muscarinic agonist (3). The trimethylammonium derivative, Vc, appeared to be somewhat more efficacious than carbachol. Table 3 also summarizes the percentage receptor occupation that each compound requires for half-maximal response.

The relationship between pharmacological response and fractional receptor occupancy in the guinea pig ileum for oxotremorine, carbachol and Compound Vc is illustrated graphically in Figure 1. There is clearly a non-linear relationship between receptor occupancy and response, and a large receptor reserve exists for the compounds with respect to contractile responses in this tissue.

In the present study, the parameter chosen as effect (contraction) is situated late in the sequence of events constituting the receptor-effector coupling. Generally, the closer the measured effect is related to the receptor activation, the smaller the receptor reserve will be, presumably because the number of amplification steps involved is smaller (1). Thus oxotremorine, which has a substantially smaller receptor reserve than carbachol when contraction is considered (Figure 1), behaves like a partial agonist with respect to carbachol when stimulation of phospholipid turnover is used as the end-point (9). Stimulated phospholipid turnover is believed to be closely coupled to receptor activation (13). Whereas the spare receptor capacity and the behavior as partial or strong agonist depend on the end-point chosen as effect, the order of efficacies of the compounds in a series like the one in Table 3 should be independent of the end-point.

Table 3

Parameters Characterizing the Muscarinic Activity of some Oxotremorine Analogs in the Isolated Guinea Pig Ileum

IV V

Compound	Am	ED$_{50}$(M)	K$_A$(M)	Relative affinity[a]	Relative efficacy	% Occupancy for 50% response[b]	n
IVa	N	0.025 ± 0.002	0.68 ± 0.19	1.00	1.00	3.5	6
IVb	N	0.011 ± 0.0004	1.37 ± 0.18	0.50	4.38 ± 0.51	0.79	6
IVc	N(CH$_3$)$_2$	0.13 ± 0.006	22.9 ± 2.7	0.030	6.51 ± 0.71	0.55	5
IVd	Ṅ(CH$_3$)$_3$	0.013 ± 0.0007	2.92 ± 0.14	0.23	7.28 ± 0.30	0.46	5
IVe	Ṅ CH$_3$	0.32 ± 0.014	6.74 ± 1.03	0.10	0.78 ± 0.10	4.5	5
Va	N	0.069 ± 0.002	2.18 ± 0.31	0.31	1.14 ± 0.13	3.1	6
Vb	N(CH$_3$)$_2$	0.39 ± 0.028	66.9 ± 12.4	0.0097	6.46 ± 1.14	0.55	6
Vc	Ṅ(CH$_3$)$_3$	0.037 ± 0.001	10.1 ± 1.5	0.067	9.50 ± 1.26	0.36	5
Carbachol		0.080 ± 0.005	16.4 ± 3.1	0.042	7.20 ± 1.19	0.48	6

Values are given as means ± S.E. [a]Calculated by dividing the dissociation constant of oxotremorine by the dissociation constant of each compound. [b]Calculated from the law of mass action.

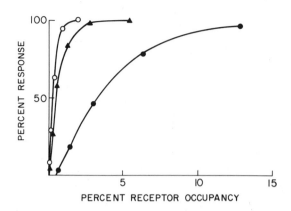

Figure 1. Contractile responses of oxotremorine (●), carbachol
(▲) and Compound Vc (○) as a function of receptor occupancy in the
isolated guinea pig ileum. Percent receptor occupancy at each concentration was calculated from the law of mass action employing
the K_A values listed in Table 3.

From the data in Table 3 it appears that structural modifications of oxotremorine or of its acetamide analog that increase
efficacy also decrease affinity. The converse is not always true,
however, since for example Compounds R-I, S-II (Table 1) and IVe
(Table 3) have lower efficacies and affinities than oxotremorine.
Therefore, no simple relationship exists between the affinities of
these oxotremorine analogs for the receptor and their ability to
activate the receptor. In fact, most evidence suggest different
structural requirements for these two parameters of drug action
(17-20).

CONCLUSIONS

The significance of the pharmacological method used lies primarily
in the fact that it provides an estimate of agonist efficacy, since
this parameter is not readily available by other methods. Efficacy
is not solely a characteristic of the agonist-receptor pair in question. It also depends among other things on receptor density which,
for a given type of receptor, may vary from tissue to tissue. Therefore, the efficacy of a particular agonist may vary among tissues
even if they contain identical receptors. We have previously suggested
that central specificity of some oxotremorine analogs may be due
to differences in efficacy at central and peripheral sites of action
rather than to differences in affinity (21). The observation (14)
that Compound II (Table 1) acts simultaneously as a presynaptic
antagonist and as a postsynaptic agonist suggests that differences
in efficacy also may be important in discriminating between pre-
and postsynaptic muscarinic receptors. Further studies of the struc-

tural requirements underlying efficacy of oxotremorine analogs in different tissues seem likely to uncover more selective muscarinic ligands.

Acknowledgements. This work was supported by USPHS grant MH-17691. I wish to thank Dr. Donald J. Jenden for his helpful suggestions and enthusiastic support of this work and Dr. Nigel J.M. Birdsall for kindly providing PrBCM.

REFERENCES

1. Ariëns, E.J. (1979): Trends Pharmacol. Sci. 1:11-15.
2. Besse, J.C. and Furchgott, R.F. (1976): J. Pharmacol. Exp. Ther. 197:66-78.
3. Birdsall, N.J.M., Burgen, A.S.V. and Hulme, E.C. (1978): Mol. Pharmacol. 14:723-736.
4. Burgen, A.S.V. (1981): Fed. Proc. 40:2723-2728.
5. Dahlbom, R. (1981): In Cholinergic Mechanisms: Phylogenetic Aspects, Central and Peripheral Synapses, and Clinical Significance (eds) G. Pepeu and H. Ladinsky, Plenum Press, New York, pp. 621-638.
6. Dahlbom, R., Jenden, D.J., Resul, B. and Ringdahl, B. (1982): Brit. J. Pharmacol. 76:299-304.
7. Ehlert, F.J., Kokka, N. and Fairhurst, A.S. (1980): Biochem. Pharmacol. 29:1391-1397.
8. El-Fakahany, E. and Richelson, E. (1981): Mol. Pharmacol. 20:519-525.
9. Fisher, S.K., Klinger, P.D. and Agranoff, B.W. (1983): J. Biol. Chem. 258:7358-7363.
10. Furchgott, R.F. (1966): Adv. Drug Res. 3:21-55.
11. Furchgott, R.F. and Bursztyn, P. (1967): Ann. N.Y. Acad. Sci. 144:882-899.
12. Jim, K., Bolger, G.T., Triggle, D.J. and Lambrecht, G. (1982): Can. J. Physiol. Pharmacol. 60:1707-1714.
13. Michell, R.H. and Kirk, C.J. (1981): Trends Pharmacol. Sci. 2:86-89.
14. Nordström, Ø., Alberts, P., Westlind, A., Unden, A. and Bartfai, T. (1983): Mol. Pharmacol. 24:1-5.
15. Resul, B., Dahlbom, R., Ringdahl, B. and Jenden, D.J. (1982): Eur. J. Med. Chem. 17:317-322.
16. Resul, B., Ringdahl, B., Dahlbom, R. and Jenden, D.J. (1983): Eur. J. Pharmacol. 87:387-396.
17. Ringdahl, B. (1984): J. Pharmacol. Exp. Ther. 229:199-206.
18. Ringdahl, B. (1984): Brit. J. Pharmacol. 82:269-274.
19. Ringdahl, B. (1985): J. Pharmacol. Exp. Ther. 232:67-73.
20. Ringdahl, B. and Jenden, D.J. (1983): Mol. Pharmacol. 23:17-25.
21. Ringdahl, B. and Jenden, D.J. (1983): Life Sci. 32:2401-2413.
22. Ringdahl, B., Resul, B., Jenden, D.J. and Dahlbom, R. (1982): Eur. J. Pharmacol. 85:79-83.
23. Ruffolo, R.R., Jr. (1982): J. Auton. Pharmacol. 2:277-295.

24. Sastry, B.V.R. and Cheng, H.C. (1972): J. Pharmacol. Exp. Ther. 180:326-339.

25. Siegel, H. and Triggle, D.J. (1982): Life Sci. 30:1645-1652.

26. Stephenson, R.P. (1956): Brit. J. Pharmacol. 11:379-393.

27. Takeyasu, K., Uchida, S., Wada, A., Maruno, M., Lai, R.T., Hata, F. and Yoshida, H. (1979): Life Sci. 25:1761-1772.

28. Waud, D.R. (1969): J. Pharmacol. Exp. Ther. 170:117-122.

29. Yamamura, H.I. and Snyder, S.H. (1974): Mol. Pharmacol. 10:861-867.

IN VIVO AND IN VITRO STUDIES ON A MUSCARINIC PRESYNAPTIC

ANTAGONIST AND POSTSYNAPTIC AGONIST: BM-5

O. Nordström, A.Undén, V. Grimm*, B. Frieder*,
H. Ladinsky** and T. Bartfai

Department of Biochemistry
Arrhenius Laboratory
S-106 91 Stockholm, Sweden

*Isotope Department
Weizmann Institute of Science
Rehovot, 76100   Israel

**Cholinergic Neuropharmacology
Mario Negri Institute for Pharmacological Research
Milan, 20157, Italy

INTRODUCTION

The oxotremorine analog, compound BM-5: N-methyl-N(1-methyl-4-pyrrolidino-2-butynyl)acetamide (11) (Fig. 1) has been studied with respect to its muscarinic actions (9, 11).   Jenden and colleagues found that the compound inhibited oxotremorine induced tremor in mice (11) while it was a partial agonist producing contraction of the ileum (9, 11).   In parallel studies we found that compound BM-5 behaved as a presynaptic antagonist enhancing the evoked release of acetylcholine from the myenteric plexus and from synaptosomes prepared from rat hippocampus (9).   In other presumable postsynaptic tests, such as the atropine sensitive stimulation of cyclic 3'5'-guanosine monophosphate (cGMP) synthesis in human lymphocytes - a noninnervated "postsynaptic model tissue" - or in hippocampal slices, compound BM-5 behaved as a muscarinic agonist since it enhanced cGMP synthesis (9).

We report here further in vitro and in vivo studies with this interesting compound which, at proper dosage, could have great potential since it could enhance cholinergic transmission by being a presynaptic antagonist and postsynaptic agonist.

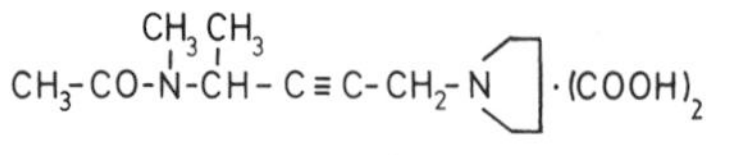

COMPOUND BM-5

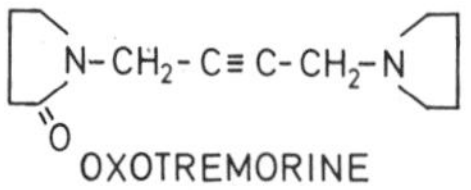

OXOTREMORINE

Figure 1. The structural formula of compound BM-5. Oxotremorine is shown for comparison.

MATERIALS AND METHODS

Binding studies were carried out on homogenates of some brain regions or rat heart (glass teflon homogenizer (695 rpm, 10 up and down strokes)). Homogenates (10% w/v) were prepared in 0.32 M sucrose buffered with 20 mM Hepes: pH 7.0. The homogenates were diluted 4-fold with a Hepes buffer (20 mM, pH 7.0) containing 10 mM NaCl and 10 mM $MgCl_2$ according to Berrie et al. (1). Various concentrations of BM-5 were added ranging from (10^{-9} to 10^{-6} M) at a constant concentration of the labelled muscarinic antagonist $[^3H]$-4-NMPB (0.5 nM). The incubations were carried out at 37°C for 20 min and were terminated by filtration over Whatman GF/C filters which were then rinsed with 10 ml ice cold buffer. The filters were dried and counted with 10 ml scintillation coctail in a Beckman counter at a counting efficiency of 34%.

Chronic Treatment with Compound BM-5 and Atropine. Adult male Sprague-Dawley rats about 60 days of age received i.p. injections of either BM-5 (4 mg/kg) or atropine (20 mg/kg). The injections were given around 9-10 a.m. daily for fourteen days. The treatments were then withdrawn and the animals were killed by decapitation 47 h later. The brains were rapidly removed and dissected on ice and a crude mitochondrial pellet was prepared from the selected brain regions of 5 rats belonging to the vehicle, atropine (20 mg/kg/day), or BM-5 (4 mg/kg/day), treated groups. The membranes were stored at -70°C until assayed.

Determination of Acetylcholine Content. Female CD-COBS rats, 180-200 g body weight, were treated with saline or compound BM-5, and killed by fast focussed microwave irradiation to the head (1.3 KW at 2.45 GHz for about 4 sec.). ACh was determined in striatum by the radio-enzymatic method of Saelens et al. (13) with modifications (7).

Open Field Behavior and Memory Consolidation. Adult, male, naive Wistar rats were used. The animals were 45-60 days old. Drug treated and control pairs were aged matched. All rats were tested 15 min after i.p. injection with BM-5 or saline.

In the open field test, the floor surface (100 x 100 x 41 cm) was divided into 16 squares. The number of squares crossed and the number of rearings (i.e. rat standing upright on its hind legs) were counted during a three min period.

In the experiments on memory consolidation, the rats were tested in a discrimination learning task in the six unit black-white discrimination maze.

The first group of animals were given 4 trials in the maze. After the fourth time of arrival to the goal box, each animal was injected either with 2 mg/kg or an equal volume of saline. Twenty-four hours later the whole group was given another four trials in the same maze. The number of errors was the behavior measured. The experiment was repeated with new animals which were treated with 5 mg/kg of compound BM-5.

The experiments thus involved different strains of rats: Wistar in Israel, CD-COBS rats in Italy, and Sprague-Dawley rats in Sweden, due to local regulations. Attempts are being made now to carry out the follow-up studies in the same strain. Comparison of data within this paper is admitedly made difficult by the strain differences.

Compound BM-5 was synthesized (9) and generously donated by Professor Richard Dahlbom, Department of Organic Pharmaceutical Chemistry, Biomedical Center, Uppsala, Sweden. The compound was stored desiccated at $4^{\circ}C$ when not being used.

Solutions were made up freshly every day.

$[^{3}H]$-4-N-methylpiperidinyl benzilate: $[^{3}H]$-4-NMPB (77 Ci/mmole) was synthesized and generously donated by Professor   Mordechai Sokolovsky, Department of Biochemistry, Tel Aviv University, Tel Aviv, Israel. $[^{3}H]$-3-QNB (33 Ci/mmole) was purchased from NEN. All other chemicals were of reagent grade.

RESULTS AND DISCUSSION

Binding Studies. $[^{3}H]$-4-NMPB, a muscarinic antagonist, was displaced by compound BM-5 in membranes from striatum, cerebral cortex, cerebellum and hippocampus. Table 1 summarizes the binding data, which for each brain area involved 86-92 data points evaluated by means of nonlinear regression methods. Compound BM-5 recognized in each brain region a population of high and a population of low affinity binding sites; both of which were labelled with $[^{3}H]$4-NMPB. The proportion of high and low affinity binding sites varied from brain

Table 1

Binding of Compound BM-5 to [^{3}H]-4-NMPB Labelled Sites in
Membrane from the Rat Cerebral Cortex, Hippocampus,
Striatum and Cerebellum

Brain region	Data points (i)	Hill coeff. (n)	Portion of high affinity sites (%)	K_D high affinity (nM)	K_D low affinity (M)
Cerebral cortex	81	0.85	14 ± 5	5.0 ± 1.4	0.21 ± 0.01
Hippo-campus	90	0.88	12 ± 2	7.5 ± 2.6	0.22 ± 0.01
Striatum	86	0.87	10 ± 1	4.0 ± 1.5	0.24 ± 0.01
Cerebellum	86	0.66	29 ± 4	2.3 ± 6	0.07 ± 0.01

region to brain region as was found with other muscarinic agonists (cf. for reviews 2, 6, 14).

The Hill coefficients were less than unity in each case; another phenomenon previously found with muscarinic agonists but not with antagonists (2, 14).

In rat heart, too, the specific binding of [^{3}H]-4-NMPB to muscarinic receptors was displaced by compound BM-5 and yielded a curve which best fits a two-site model (Fig. 2). The figure also shows that 5'-quanylylimidophosphate [Gpp (NH)p], a nonhydrolyzable analog of GTP, at 50 mM concentration, shifted the displacement curve to the right. This phenomenon has been observed when studying other muscarinic agonists (1), but not antagonists in the rat heart.

The in vitro binding studies therefore suggest that compound BM-5 may behave as an agonist, and that it is a more potent agonist in the cerebellum and in the heart than in the hippocampus, striatum or cerebral cortex. The Hill coefficients observed also indicate that the most agonist-like action can be expected in the cerebellum.

The effect of chronic treatment with compound BM-5 on the number of muscarinic receptors in rat cerebral cortex is shown in Table 2. The compound induced a 60% increase in the number of muscarinic

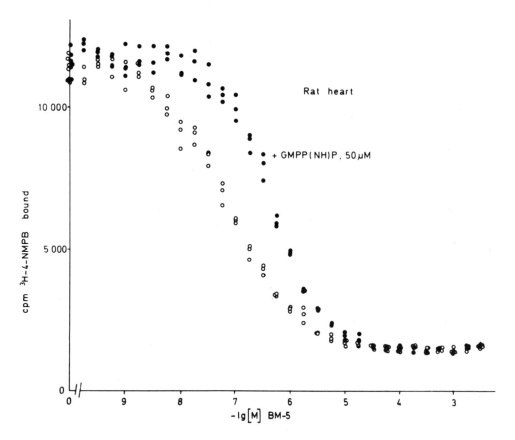

Figure 2. The effect of [Gpp (NH)p] (50 μM) on the displacement of [^{3}H]-4-NMPB binding to muscarinic receptors in membranes of rat heart by compound BM-5. The concentrations of compound BM-5 (31 points run in triplicate) are shown on the abscissa and the amount of [^{3}H]-4-NMPB bound (in cpm) is given on the ordinate. In the absence of the GTP analog the curve best fits a two-site model as analyzed by nonlinear regression methods.

binding sites, an increase which was just as large as that produced by chronic treatment with a higher dose of atropine (Table 2 and ref. 17) the classical antagonist.

The results are consistent with the possibility that the chronic administration of BM-5 (4 mg/kg) brings out an antagonist-like profile of the drug. The continuous blockade of muscarinic receptors led to the development of muscarinic receptor supersensitivity as reflected by the increase in the number of specific binding sites for [^{3}H]-3-QNB (Fig. 2).

Table 2

Chronic Treatment with BM-5 or Atropine Causes Muscarinic
Supersensitivity in Rat Cerebral Cortex

Treatment	Muscarinic receptors* (%)
Vehicle	100 ± 4
Compound BM-5 (4 mg/kg/day)	161 ± 9**
Atropine (20 mg/kg/day)	161 ± 7**

*The number of receptors were measured by determining the specific binding (1 μM atropine inhibitable) of [^{3}H]-3-QNB (8 nM) to a washed crude mitochondrial membrane fraction from the cerebral cortex. Since there still may have been some residual atropine and BM-5 present despite the 48 h washout period, high [^{3}H]-3-QNB concentration was used to fully saturate all muscarinic receptors even in the presence of these competitive inhibitors.   **$p < 0.01$ vs the vehicle treated group; Student's t-test.

Table 3

Dose-Dependent Effect of Compound BM-5 on Rat
Striatal Acetylcholine Content

Drug (mg/kg)	Striatal ACh (nmoles/g wet wt)
Saline	68.0 ± 1.9 (5)
BM-5 1	67.1 ± 3.1 (5)
BM-5 10	60.3 ± 3.2 (5)
BM-5 30	50.7 ± 3.0 (5)*

BM-5 was dissolved in distilled water and given i.p.   The rats were killed after 30 min.   The data are means and S.E.M. (n).   * = $p < 0.01$ vs saline treated group; Dunnett's test.

Table 4

Effect of Compound BM-5 on Open Field Activity of Naive Rats

Dose of BM-5	Saline		BM-5		BM-5 (% of control)	
	squares crossed	rearing number	squares crossed	rearing number	squares crossed	rearing number
1 mg/kg	41.8(2.3)	18.3(1.6)	34.4(3.5)	11.6(1.8)*	82	63
2 mg/kg	48.8(5.1)	9.1(0.9)	14.5(1.05)**	2.5(0.4)**	30	27
5 mg/kg	40.0(4.4)	9.3(2.1)	4.0(1.1)**	0.8(0.4)**	10	9

The data are means(S.E.M.); n = 8; *p < 0.05; **p < 0.01 with respect to the saline-treated group.

Dose Response Effect of Compound BM-5 on Rat Striatal Acetylcholine Content. Table 3 shows that compound BM-5 did not effect striatal ACh content at 1 mg/kg, tended to decrease the level at 10 mg/kg and gave a significant 25% decrease in the ACh content at the dose of 30 mg/kg. Such an effect is reminiscent of decreases in striatal ACh content produced by high doses of the muscarinic receptor antagonist trihexphenidyl (3) and atropine (4).

EFFECT OF COMPOUND BM-5 ON OPEN FIELD ACTIVITY

Naive Rats. The effect of compound BM-5 on open field activity in naive rats is shown in Table 4. The drug (1-5 mg/kg) inhibited motor activity (rearing) and open field behavior in a dose-dependent manner. It is known that muscarinic agonists as well as the acetylcholinesterase inhibitor physostigmine depress locomotor activity (cf. for review 5, 8, 10). Thus BM-5 has certain cholinergic agnosit-like properties with respect to this behavior.

EFFECT OF COMPOUND BM-5 ON MEMORY CONSOLIDATION

Compound BM-5, in the doses tested, 2 mg/kg and 5 mg/kg, did not significantly improve performance in the discrimination learning test in the black and white maze (Fig. 3) It is notable, however, that the lower dose of the drug provoked a tendency toward improved performance, reminiscent of muscarinic receptor agonists or physostigmine (15, 19). Namely rats injected with 2 mg/kg BM-5 after the first training session (trials 1-3) did commit fewer or equal number of errors with the vehicle injected group, when retested 24 hrs later in session 2 (trails 4-7) upper right panel of Figure 3. On the other hand, the higher dose of compound BM-5 (5 mg/kg) showed a

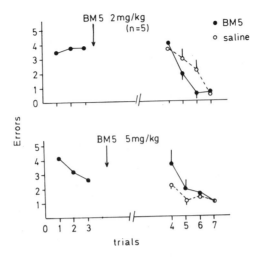

Figure 3. The effects of BM-5 on memory consolidation in rats trained in a discrimination learning task in a six unit black and white maze.

tendency to impair performance, as do the muscarinic receptor antagonists (5, 16, 18): at this dose the number of errors commited in the second session was higher for the BM-5 than for the vehicle injected rats (lower right panel, Fig. 3).

DISCUSSION AND SUMMARY

In conclusion, compound BM-5 causes muscarinic cholinergic agonist-like effects such as redness of the eye, increased motility of the gut, and impairment of locomotor behavior. However, it behaves at the same or higher doses as an antagonist with a tendency to impair memory consolidation. It also produces muscarinic super-sensitivity upon chronic treatment, and decreases rat striatal ACh content by acute treatment.

These data could be explained by assuming that different centers, located in different brain regions, have muscarinic receptors which recognize BM-5 as an agonist or an antagonist. This assertion is in line with the in vitro findings indicating regional differences in the affinity of BM-5 for muscarinic receptors, as well as indicating that the compound behaves more like an agonist in the cerebellum, a center for motor activity, than in other brain areas.

More detailed studies using lower (<1 mg/kg) doses of compound BM-5 in acute and chronic tests are required to evaluate its poten-tial as a centrally acting cholinergic agonist.

Acknowledgements. We are indebted to Professor Richard Dahlbom for the supply of BM-5. This study was supported by grants from the Swedish Medical Research Council and from the National Institute of Mental Health, Bethesda, U.S.A. (#MH31107-05).

REFERENCES

1. Berrie, C.P., Birdsall, N.J.M., Burgen, A.S.V. and Hulme, E.C. (1979): Biochem. Biophys. Res. Commun. 87:1000-1005.
2. Birdsall, N.J.M. and Hulme, E.C. (1976): J. Neurochem. 27:7-16.
3. Consolo, S., Ladinsky, H. and Garattini, S. (1974): J. Pharm. Pharmacol. 26:275-277.
4. Garattini, S., Forloni, G.L., Tirelli, S., Ladinsky, H. and Consolo, S. (1984): Psychopharmacology 82:210-214.
5. Hingtgen, J.N. and Aprison, M.H. (1976): In Biology of Cholinergic Function (eds.) A.M. Goldberg and I. Hanin, Raven Press, New York, pp. 515-566.
6. Kloog, Y., Egozi, Y. and Sokolovsky, M. (1979): FEBS Lett. 97:265-268.
7. Ladinsky, H. Consolo, S., Bianchi, S. and Jori, A. (1976): Brain Res. 108:351-361.
8. Maayani, S., Egozi, Y., Pinchasi, I. and Sokolovsky, M. (1978): Biochem. Pharmacol. 27:203-214.
9. Nordström, Ö., Alberts, P., Westlind, A., Undén, A. and Bartfai, T. (1983): Mol. Pharmacol. 24:1-5.
10. Pradhan, S.N. and Dutta, S.N. (1971): Int. Rev. Neurobiol. 14:173-231.
11. Resul, B., Dahlbom, R., Ringdahl, B. and Jenden, D.J. (1983): Eur. J. Med. Chem. 17:317-322.
12. Russell, R.W. (1977): In Cholinergic Mechanisms and Psychopharmacology (ed.) D.J. Jenden, Plenum Press, New York, pp. 709-731.
13. Saelens, J.K., Allen, M.P. and Simke, J.P. (1970): Arch. Int. Pharmacodyn. Ther. 186:279-286.
14. Sokolovsky, M. and Bartfai, T. (1982): Trends Biochem. Sci. 6:303-305.
15. Stratton L.D. and Petranovich, L. (1963): Psychopharmacologia 5:47-54.
16. Takeyashu, K., Uchida, S., Noguchi, Y., Fujita, N., Saito, K., Hata, F. and Yoshida, H. (1979): Life Sci. 25:585-592.
17. Westlind, A., Grynfarb, M., Hedlund, B., Bartfai, T. and Fuxe, K. (1981): Brain Res. 225:131-141.
18. Whitehouse, J.M. (1964): J. Comp. Physiol. Psychol. 57:13-15.
19. Whitehouse, J.M. (1966): Psychopharmacologia 9:183-188.

INTERACTIONS OF ALAPROCLATE, A SELECTIVE 5HT-UPTAKE BLOCKER, WITH MUSCARINIC RECEPTORS: IN VIVO AND IN VITRO STUDIES

E. Danielsson, A. Undén, Ö. Nordström,
S.O. Ögren[*] and T. Bartfai

Department of Biochemistry
Arrhenius Laboratory
S-106 91 Stockholm, Sweden

[*]Department of Pharmacology
Astra Läkemedel AB
S-151 85 Södertalje, Sweden

INTRODUCTION

Cholinergic mechanisms play an important role in higher brain functions such as learning and memory (6). It is hoped that drugs which improve cholinergic transmission would be therapeutically effective in e.g. senile dementia (1, 6). The search for such drugs has led to introduction of precursor: choline (or lecithin) (cf. 1) therapy; to the testing of directly acting cholinergic agonists such as arecoline; or to the use of the acetylcholinesterase inhibitor physostigmine (cf. 3, 5). All of these attempts have met only with partial success to date. Thus the search goes on.

In 1982, Ogren, while testing a selective 5HT uptake blocker alaproclate (8) on the behavior of mice, discovered that this non-tricyclic, potential antidepressant agent, enhanced the tremor evoked by submaximal doses of oxotremorine, the potent muscarinic agonist. Here we report more extensive studies in vivo in mice and rats, on the alaproclate mediated potentiation of muscarinic cholinergic responses such as tremor, salivation and hypothermia (7). The results of biochemical studies on the interaction of alaproclate with the muscarinic system are also summarized.

MATERIALS AND METHODS

Animals.   Adult mice (NMRI Strain, Anticimex, Stockholm) weighing
20-22 g were used in all but one _in vivo_ experiment.   Adult male
rats (Sprague-Dawley 150-200 g) were used in one _in vivo_ experiment,
and in all _in vitro_ experiments.

Chemicals and isotopes.   [^{3}H]-4-N-methylpiperidinylbenzilate ([^{3}H]-
4-NMPB, 77 Ci/mmole) was generously donated by Professor Mordechai
Sokolovsky, Tel-Aviv University.   Methyl [^{3}H]-choline chloride (84
Ci/mmole) was bought from Amersham, UK.   Alaproclate, the alanine
ester of 2(4-chlorophenyl)1-1-dimethylethanol, batch F20, was syn-
thesized by Lindberg _et al_. (8) at Astra Läkemedel AB, Södertälje.
Oxotremorine semibisfumarate and teraphenylboron were bought from
EGA AG, Steinheim, FRG.

Methods.   Potentiation of oxotremorine induced tremor was studied
in the following manner: Alaproclate dissolved in saline was injected
i.p. 30 min prior to administration of oxotremorine, which was injected
into the neck of the mice or s.c. into rats.   Tremor was scored
for 60 min according to the following scale: score 0, no tremor;
score 1, moderate discontinuous tremor of head and fore limbs; score
2, strong tremor involving the whole body.

Salivation and hypothermia were studied in mice following the
same schedule of injection: alaproclate preceding the injection
of oxotremorine by 30 min.

Measurement of muscarinic receptors was carried out _in vitro_
using a membrane preparation (10,000 g x 1 hr) from selected brain
areas of the rat.   [^{3}H]-4-NMPB concentrations were varied (0.05-7
nM) at various alaproclate concentrations (0, 10, 50, 100 μM).
Incubations were carried out at 37°C for 60 min and were terminated
by filtration on Whatman GF/C filters.

[^{3}H]-ACh release was studied from synaptosomes as described
previously by Nordström and Bartfai (9).

RESULTS

Salivation, tremor and hypothermia caused by oxotremorine are
classical, centrally mediated muscarinic responses (7).   Alaproclate (0-60 mg/kg) when injected 30 min prior to oxotremorine did
not evoke any of these responses in mice or rats.   Figure 1 shows
that alaproclate potentiated the oxotremorine (200 μg/kg) evoked
tremor in mice in a dose dependent manner.

The cholinergic response of salivation to oxotremorine was
not potentiated, but was significantly prolonged by i.p. injections

of alaproclate (Figure 2). An additional, very strong muscarinic

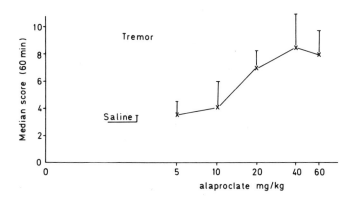

Figure 1. Potentiation of the oxotremorine (200 μg/kg) induced tremor
in mice by alaproclate at different doses injected prior to oxotremorine.

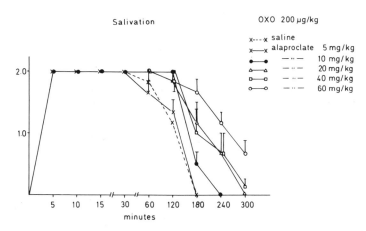

Figure 2. Oxotremorine (200 μg/kg) induced salivation and its prolongation by alaproclate (5-60 mg/kg) injected i.p. 30 min prior to
oxotremorine. The ordinate is in arbitrary units. The experiments
were carried out with 12 animals.

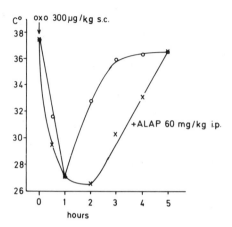

Figure 3. Oxotremorine (300 μg/kg) induced hypothermia in mice and its prolongation by alaproclate (60 mg/kg) injected 30 min prior to oxotremorine.

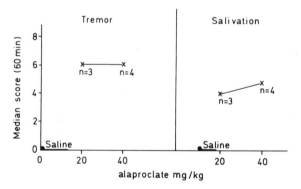

Figure 4. Potentiation of the oxotremorine (200 μg/kg) caused tremor and salivation in rats by alaproclate (20-40 mg/kg) injected i.p. 30 min prior to oxotremorine.

response in mice is that of hypothermia. Figure 3 shows that alaproclate (60 mg/kg) injected 30 min prior to oxotremorine (300 μg/kg)
prolonged the hypothermic response to oxotremorine for 2-3 hours.

Since it is known that there are large differences between
mice and rats in the metabolism of certain drugs and also in sensitivity to certain drugs, we have examined the tremor and salivation
response in rats using the same paradigm as in the above experiments
with mice. Figure 4 shows that alaproclate markedly potentiated
both the oxotremorine induced tremor and salivation in rats, similarly to that seen in mice.

In order to examine whether a possible pharmacokinetic interaction between alaproclate and oxotremorine could account for the
potentiation of muscarinic effects, we have studied the effects of
alaproclate on physostigmine induced tremor in mice. This muscarinic

Table 1

Alaproclate effects on the oxotremorine (100 μg/kg)
physostigmine (200 μg/kg) induced tremor in mice

Alaproclate (40 mg/kg) 30 min prior to cholinergic stimulation	Cholinergic stimulation	Atropine (1 mg/kg)	Metergoline (1 mg/kg)	Tremor score for 60 min
+	-	-	-	0
-	oxo (100 μg/kg)	-	-	0.5
+	oxo	-	-	4.0**
-	physostigmine (200 μg/kg)	-	-	2.0**
+	physo	-	-	5.0**
+	oxo	+	-	0
+	physo	+	-	0
+	-	-	+	0
-	-	-	+	0
-	oxo	-	+	1.5**
+	oxo	-	+	6.0**

**$p < 0.05$ (n=12)

response to inhibition of the acetylcholinesterase was also significantly potentiated (Table 1).

It is also known that activation of serotonin type 2 receptors
may lead to tremor, thus the muscarinic versus serotonergic character
of the tremor when both alaproclate and oxotremorine were injected
was examined by use of serotonin antagonist: metergoline and the
classical muscarinic antagonist, atropine (Table 1).

Hexamethonium, the ganglionic blocker (2 mg/kg), did not block
the oxotremorine induced tremor or its potentiation by alaproclate.

In vitro studies. Binding of alaproclate to muscarinic receptors
in the rat cerebral cortex, hippocampus, and striatum was examined
by measuring competition between the [^{3}H]-labeled antagonist [^{3}H]-
4-NMPB and alaproclate in equilibrium binding experiments. Table
2 indicates that alaproclate binds weakly K_D (28-40 μM) to muscarinic
receptors. The effects of alaproclate on muscarinic agonist: acetylcholine and oxotremorine binding, were also examined. Alaproclate
(10-100 μM) did not "potentiate" the unlabeled muscarinic agonists
to displace [^{3}H]-4-NMPB.

Table 2

Summary of the results of in vitro experiments with
alaproclate on rat brain cholinergic mechanisms

Assay	Effect
Binding to muscarinic receptor	Yes; K_D (28-40 μM)
"Agonist"-like effects - "GTP-effect"[a]	No
[^{3}H]-choline uptake	No
[^{3}H]-ACh release	inhibition at 10^{-4} M
cGMP synthesis	inhibition at 10^{-5} M
Acetylcholinesterase	No

[a]The binding of alaproclate to muscarinic receptors in the rat heart
or cerebellum is unaffected by the presence of GTP (1 mM) which is
known to shift the binding affinity of several muscarinic agonists
(2).

In vitro studies on the uptake of $[^3H]$-choline into nerve terminals, studies on its conversion into $[^3H]$-acetylcholine that was determined before release experiments, or in release experiments, indicate that alaproclate does not affect these presynaptic cholinergic mechanisms (Table 2).

Alaproclate at high (10^{-4} M) concentration inhibited the evoked release of $[^3H]$-acetylcholine from $[^3H]$-choline loaded synaptosomes from the cerebral cortex, hippocampus and striatum.

Alaproclate (10^{-5} M) also inhibited the depolarization evoked increases in cyclic GMP levels in tissue slices from the rat hippocampus, similarly to the effect of atropine on this preparation.

DISCUSSION

The selective 5HT-uptake blocker alaproclate (8, 11) has been shown to potentiate or prolong the effects of muscarinic cholinergic stimulation either by oxotremorine or by physostigmine. These interactions cover the whole range of muscarinic responses such as salivation, tremor and hypothermia, both in rats and mice. Thus there is reason to assume that even cholinergic potentiation of learning and memory consolidation could be assisted by the drug.

The responses to alaproclate could only be observed when triggered by a muscarinic stimulus such as oxotremorine or physostigmine, and could be fully blocked by atropine, indicating that these were ultimately mediated via muscarinic receptors.

The participation of serotonergic mechanisms in eliciting tremor (4) is well documented. Nevertheless, the alaproclate effects observed in conjunction with oxotremorine are not primarily due to the serotonin accumulation, since atropine fully blocks this tremor.

The mode of alaproclate action in potentiating and prolonging muscarinic responses is not known. The previous in vitro studies (10) and those summarized here do not give a clue to the molecular mechanism underlying the strong, muscarinic potentiation by this drug in vivo. The metabolites of alaproclate, alanine and 2(4 chlorophenyl)1-1 dimethylethanol do not produce tremor.

Further investigations are necessary to elucidate the interactions, at the molecular level, which are responsible for the strong and apparent effects of alaproclate, seen in mice and rats, in potentiating both physostigmine and oxotremorine induced muscarinic responses.

Acknowledgements. This study was supported from grants by the Swedish Medical Research Council.

REFERENCES

1. Bartus, R.T., Dean, R.L. III, Beer, B. and Lippa, A.S. (1982): Science 217:408-417.
2. Berrie, C.P., Birdsall, N.J.M., Burgen, A.S.V. and Hulme, E.C. (1979): Biochem. Biophys. Res. Commun. 87:1000-1005.
3. Christie, J.E., Shering, A., Ferguson, J. and Glen, I.M. (1981): Br. J. Psychiatry 138:46-50.
4. Costall, B., Fortune, D.H., Naylor, R.J., Marsden, C.D. and Pycock, C. (1975): Neuropharmacology 14:859-868.
5. Davis, K.L., Mohs, R.C., Tinklenberg, J.R., Pfefferbaum, R., Hollister, L.E. and Kopell, B.S. (1978): Science 201:272-274.
6. Hingtgen, J.N. and Aprison, M.M. (1976): In Biology of Cholinergic Function (eds) Goldberg, A.M. and Hanin, I., Raven Press, New York, pp. 515-566.
7. Koelle, G.B. (1975): In The Pharmacological Basis of Therapeutics (eds) Goodman, L.S. and Gillman, A., MacMillan Publishing Co., New York, pp. 467-476.
8. Lindberg, U.H., Thorberg, S.-O., Bengtsson, S., Renyi, A.L., Ross, S.B. and Ögren, S. O. (1978): J. Med. Chem. 21:448-456.
9. Nordström, Ö. and Bartfai, T. (1980): Acta Physiol. Scand. 103:347-353.
10. Nordström, Ö., Danielsson, E., Peterson, L.-L., Ögren, S.O. and Bartfai, T. (1985): J. Neural. Transmission. 61:1-20.
11. Ögren, S.O., Holm, A.C., Hall, H. and Lindberg, U.H. (1984): J. Neural. Transmission. 59:265-288.

PRESYNAPTIC MUSCARINIC RECEPTORS: CHANGE OF SENSITIVITY

DURING LONG-TERM DRUG TREATMENT

M. Marchi and M. Raiteri

Institute of Pharmacology and Pharmacognosy
University of Genova
Viale Cembrano 4   16148
Genova, Italy

INTRODUCTION

It is widely accepted that the sensitivity of neurotransmitter receptors can be modified during long-term activation or blockade of these receptors by specific drugs. Desensitization of receptors occurs in general after long-lasting exposure to agonists, whereas reduced supply of the agonist transmitter to its receptors, as in the case of a chronic treatment with antagonist drugs, appears to cause receptor supersensitivity. These modifications of receptor sensitivity are likely to have important implications, both in the ethiology of some diseases of the nervous system and in the mechanism of action of drugs used in the treatment of these disorders. In view of these considerations, it is important to ascertain whether and in what conditions a given receptor system can undergo adaptive changes, since neurotransmitter receptors may not be all equally adaptable.

One useful approach to this problem may be that of studying the receptors regulating transmitter release, using this process as a functional parameter. Although limited to release-regulating receptors, this approach allows comparisons between receptors of the same type regulating either the release of different transmitters (heteroreceptors) or the release of the same transmitter, but in different areas of the brain (autoreceptors) (1, 3, 10, 11).

In the case of the cholinergic system there are some presynaptic muscarinic receptors which can be studied by means of this functional parameter, in particular the autoreceptors mediating inhibition of acetylcholine (ACh) release (4, 5, 6, 7, 13) and the muscarinic

423

presynaptic heteroreceptors mediating potentiation of striatal dopamine
(DA) release (2, 9).

In the present research we have investigated some of the characteristics of auto- and heteroreceptors from different brain areas,
and have monitored their alteration in sensitivity following chronic
drug treatment.

MATERIALS AND METHODS

Adult male Sprague-Dawley rats were used. The animals were
sacrificed by decapitation and the brain was dissected out immediately. Crude synaptosomal fractions (P_2) were prepared from cortex,
striatum and hippocampus. In the experiments aimed to study changes
in sensitivity of auto- or heteroreceptors following chronic drug
treatment, the rats were injected subcutaneously once daily for 11
days with scopolamine (10 mg/Kg) or paraoxon (0.15 mg/Kg on day 1
and 2 followed by 0.07 mg/Kg on day 3 through 11). Control animals
were injected with saline. The animals were killed 48 hours after
the last injection. The synaptosomes were prelabeled with [^{3}H]-choline
(0.1 μM; 4 min) or with [^{3}H]-dopamine ([^{3}H]-DA) and the release of
[^{3}H]-Acetylcholine ([^{3}H]-ACh) and of [^{3}H]-DA was studied in superfusion
as previously described (8).

The effects of the muscarinic drugs used was calculated from
the release curves obtained, collecting 1 min-fractions of the superfusion medium: the fractional rate of the basal efflux was subtracted
from the fractional rate of the K^+-evoked release at the peak point
of the release curves obtained in the absence, and in the presence
of the drug tested. Results are therefore expressed as percent of
inhibition ([^{3}H]-ACh) or potentiation   ([^{3}H]-DA) of the K^+-evoked
release. Each point is the average ($^\pm$ S.E.M.) of at least 6 experiments run in triplicate.

RESULTS

Figure 1 shows that exogenous ACh can inhibit, in a concentration-dependent manner, the K^+-evoked release of [3H]-ACh from
rat hippocampal and cortical synaptosomes prelabeled with [^{3}H]choline;
atropine (1 μM) completely counteracted the action of ACh. As was
expected using superfused synaptosomes, atropine per se did not
cause the increase of [^{3}H]-ACh release generally found in slices
(12, 13) or in "incubated" synaptosomes (7). Both oxotremorine
and carbachol inhibited the K^+-evoked release of [^{3}H]-ACh, although
less potently than ACh. Their action was antagonized by atropine
or scopolamine (not shown). Interestingly, striatal cholinergic
nerve endings seem to differ from hippocampal or cortical nerve
endings; in fact, muscarinic autoreceptors inhibiting the K^+-evoked
[^{3}H]-ACh release were much less effective in rat striatal synaptosomes.

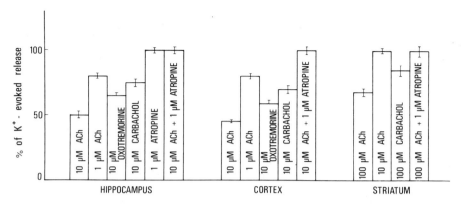

Figure 1. Muscarinic autoreceptors on nerve endings of rat brain. Synaptosomes from hippocampus, cortex and striatum were labeled with [³H]-choline and superfused as previously described (8). Depolarization was obtained with 15 mM KCl. The fractions collected during superfusion were analyzed for their content of [3H]-ACh and fractional release curves were obtained. Each value represents the mean ± S.E.M. of 6 - 8 experiments run in triplicate.

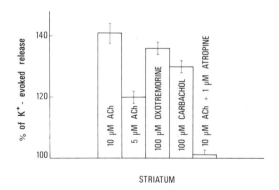

Figure 2. Potentiation of [³H]-DA release mediated by muscarinic heteroreceptors in rat striatal synaptosomes. Rat striatal synaptosomes, prelabeled with [³H]-DA, were superfused as previously described (8). Depolarization was obtained with 15 mM KCl. The fractions collected during superfusion were analyzed for their content of [³H]-DA. The data reported represent the percent of K+-evoked release at the peak point of the release curves. Each value represents the mean ± S.E.M. of results from 12 experiments run in triplicate.

Figure 2 shows the effect of various agonists and antagonists on the K^+-evoked release of $[^3H]$-DA in rat striatum. The release of $[3H]$-DA induced by 15 mM KCl from superfused rat striatal synaptosomes, previously labelled with the radioactive amine, was increased in a concentration dependent manner by exogenous ACh. Atropine (0.1 μM) totally antagonized the stimulatory action of 10 μM ACh. Oxotremorine and carbachol were less potent than ACh as stimulators of $[^3H]$-DA release. Their effect was antagonized by atropine (not shown). The potentiation by ACh if the K^+-evoked $[^3H]$-DA release decreased with increasing K^+ concentrations and totally disappeared when synaptosomes were depolarized with 55 mM KCl (Table 1).

Figure 3 shows the effect of long-term treatment with paraoxon or scopolamine on the sensitivity of presynaptic muscarinic auto- and heteroreceptors regulating, respectively, $[^3H]$-ACh release from hippocampal and cortical synaptosomes, and $[3H]$-DA release from striatal nerve endings. The results of experiments on muscarinic autoreceptors show that the inhibitory potency of exogenous ACh (1 or 5 μM) on $[^3H]$-ACh release differed, depending on the drug treatment received by the animals. In synaptosomes from rats treated chronically (but not acutely) with paraoxon, the release of $[^3H]$-ACh was significantly less inhibited than in synaptosomes from saline treated animals (autoreceptor subsensitivity). The opposite was true in the case of rats chronically treated with scopolamine (autoreceptor supersensitivity). This effect was evident both in hippocampal and cortical synaptosomes.

In contrast to the results obtained with autoreceptors, the experiments on heteroreceptors show that the sensitivity of the muscarinic presynaptic heteroreceptors mediating potentiation of $[^3H]$-DA release in striatal synaptosomes remained unchanged during chronic administration of paraoxon or scopolamine.

Table 1

K^+ concentration (mM)	% increase
15	41.4 ± 3.9 * (12)
35	25.6 ± 4.2 (4)
45	6.8 ± 1.3 (4)
55	0 (4)

Potentiation of $[^3H]$-DA release mediated by muscarinic receptors in rat striatal synaptosomes: effect of different depolarizing concentrations of KCl. The results are expressed as average $\pm$ S.E.M. In parentheses is the number of experiments. Experimental details are as in the legend for Fig. 2.

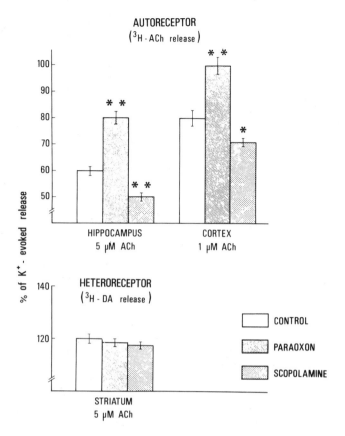

Figure 3. Effect of long-term treatment with paraoxon or scopolamine on the sensitivity of presynaptic muscarinic auto- and heteroreceptors regulating, respectively, [³H]-ACh release from hippocampal and cortical synaptosomes and [³H]-DA release from striatal nerve endings. Rats were treated as described in the text. The action of ACh was evaluated in synaptosomes obtained 2 days after the last injection. The data represent the percent of the K⁺-evoked release at the peak point of the release curves (see legend to Fig. 1 and Fig. 2 for more details). Each value represents the mean ± S.E.M. of 5 - 9 quadruplicate experiments. Asterisks indicate level of significance compared to the acutely treated animals (Student's t test). * p < 0.05; ** p < 0.005

CONCLUSIONS

1. In rat striatal nerve terminals, the inhibitory muscarinic autoreceptors have markedly reduced release-modulating capacity as compared to the autoreceptors in cortical or hippocampal nerve endings.

2.   Muscarinic presynaptic receptors regulating dopamine release
     are present on striatal dopaminergic nerve terminals.
     Their activation mediates increase of the depolarization-
     evoked release of DA when moderate releasing stimuli are
     used.

3.   The results obtained following long-term treatment with
     paraoxon or scopolamine suggest that muscarinic presynaptic
     autoreceptors may be more susceptible than muscarinic
     presynaptic heteroreceptors to changes of sensitivity
     elicited by long-term in vivo administration of agonist
     or antagonist drugs.

The difference in susceptibility between auto- and heteroreceptors
with respect to changes of sensitivity may represent a further criterion to discriminate between muscarinic receptor subtypes.

Acknowledgement. This work was supported by Grants from the Italian
C.N.R. (Progetto Finalizzato Chimica Fine e Secondaria and CT 82.-
02052.04) and from the Italian Ministry of Education.

REFERENCES

1.   Frankhuyzen, A.L. and Mulder, A.H. (1982):   Eur. J. Pharmacol.
     81:97-106.
2.   Lehmann, J. and Langer, S.Z. (1982):   Brain Res. 248:61-69.
3.   Lehmann, J., Smith, R.V. and Langer, S.Z. (1983):   Eur. J. Phar-
     macol. 88:81-88.
4.   Molenaar, P.C. and Polak, R.L. (1970):   Br. J. Pharmacol. 40:
     406-417.
5.   Marchi, M., Paudice, P. and Raiteri, M. (1981):   Eur. J. Phar-
     macol. 73:75-79.
6.   Marchi, M., Paudice, P., Caviglia, A. and Raiteri, M. (1983):
     Eur. J. Pharmacol. 91:63-68.
7.   Nordstrom, O. and Bartfai, T. (1980);   Acta Physiol. Scand.
     108:347-353.
8.   Raiteri, M., Angelini, F. and Levi, G. (1974):   Eur. J. Pharma-
     col. 25:411-414.
9.   Raiteri, M., Marchi, M. and Maura, G. (1982):   Eur. J. Pharma-
     col. 83:127-129.
10. Raiteri, M., Maura, G. and Versace, P. (1983):   J. Pharmacol.
     Exp. Ther. 224:679-684.
11. Raiteri, M., Leardi, R. and Marchi, R. (1984):   J. Pharmacol.
     Exp. Ther. 228:209-214.
12. Rospars, J.P., Lefresne, P., Beaujouan, J.C. and Glowinski,
     J. (1977):   Naunyn-Schmied. Arch. Pharmacol. 300:153-161.
13. Szerb, J.C. (1977):   In Cholinergic Mechanisms and Psychophar-
     macology (ed) D.J. Jenden, Plenum Press, New York, pp. 49-60.

CHARACTERIZATION OF [^{3}H]-NICOTINE BINDING IN RODENT BRAIN AND COMPARISON WITH THE BINDING OF OTHER LABELLED NICOTINIC LIGANDS

C. Larsson and A. Nordberg

Department of Pharmacology
Uppsala University Biomedical Center
Box 591
S-751 24 Uppsala, Sweden

INTRODUCTION

Nicotine can induce diverse pharmacological effects on the central nervous system (CNS), including alterations of spontaneous motoric activity, behavior and memory, antinociceptive effects, hypothermia, antidiuresis and convulsions (11, 15, 23). It is also considered to be a dependence-producing drug (4) and it is widely assumed that the nicotine content of tobacco smoke may play an important role in the development of the tobacco smoking habit (3, 5). Although it is usually assumed that at least some of these effects of nicotine are mediated by nicotinic receptors on neurons in the CNS, there is today no direct evidence for this assumption. It is of interest to note, however, that in some recent experimental studies a change in nicotine-like binding sites in the brain has been observed following chronic administration of nicotine (8, 26).

Alfa-bungarotoxin (α-BTX) is a well-documented ligand for nicotinic acetylcholine receptors in skeletal muscle and electric fish (9), a fact which has motivated the use of α-BTX as a probe for central nicotinic receptors (18-21, 24). Several studies have indicated that α-BTX binds specifically to brain homogenates (7, 11, 13). The inability of α-BTX to block cholinergic function in the CNS and autonomic ganglia has, however, raised questions about the physiological relevance of α-BTX binding sites in the CNS (19).

For investigation of the receptor through which nicotine exerts its central actions, radioactively labelled nicotine itself would, by definition, be the best probe in biochemical _in vitro_ binding studies. Tritium-labelled nicotine ([^{3}H]NIC) with high specific activity has also become available recently, making it possible to use nicotine itself as a ligand. As seen in Table 1, several investigators have now measured the in vitro binding of radiolabelled nicotine to brain homogenate. The discrepancies in the findings are remarkable, however, with dissociation constant (Kd) values varying 1000-fold. These differences may at least partly be a result of technical difficulties associated with in vitro agonist binding studies. They may also be due to differences in procedure and in incubation conditions (e.g. incubation temperature, time, buffer composition and protein concentration). In this in vitro binding study we have therefore tried to define the conditions yielding optimal specific binding of [^{3}H]-NIC to brain homogenate.

Table 1

A Comparison of Binding Data for [^{3}H]-Nicotine Maximum Binding (B_{max}) and Dissociation Constants (K_d), Reported in the Literature

Kd nM	Bmax pmol/g protein	Brain area and species	Incubation conditions time (min)	temp.(^{o}C)	Ref.
0.6 16,000	2.9 112	whole rat brain	30	25	30
6.0	20	whole rat brain	30	25	2
28 460	3.2[a] 10.2[a]	whole rat brain	40	37	25
400 23,000	170 6700	whole mouse brain	20	21	27
59[b]	88[b]	whole mouse brain	5	37	16
60[c]		whole rat brain	5	2	17
0.19[c] 1.7	5.1 29	whole rat brain	15	4	1
23.7 590	76 646	whole rat brain	30	37	6
6 125	60 230	mouse hippocampus	10	4	12

a) pmol/g wet tissue; b) at 4°C also a low affinity site; c) (-)[^{3}H]-Nicotine

METHODS

Aliquots of brain homogenate (P2 fractions, 10) were, as a rule, incubated with 20 nM $\pm$[³H]-nicotine (71.2 Ci/mmol) for 10 min at 4°C in plastic microcentrifuge tubes containing 50 mM Tris-HCl buffer (pH 8.0). The incubation was terminated by centrifugation in a Microfuge® (Beckman) for 5 min at 4°C. After removal of the supernatant by another centrifugation, the surface of the pellet was washed with 50 mM Tris-HCl buffer. The tips of the tubes were cut and the pellet was dissolved in 1 ml of Soluene®-toluene (1:3) overnight. After addition of 4 ml of Instafluor®, the radioactivity was measured by liquid scintillation. For additional information concerning the technical procedure in the binding assays, see (12).

RESULTS

A low incubation temperature was found to favor [³H]NIC binding to mouse hippocampus, as seen in Fig. 1. The effect of temperature on [³H]NIC binding was further analyzed by saturation-binding experiments at different temperatures. Increasing concentrations of [³H]NIC were incubated with P2 fractions of mouse hippocampus at 4 and 25°C for 10 min. The specific [³H]NIC binding was found to be approximately four times higher (when expressed as pmol/g protein) at 4°C than at 25°C (Fig. 2). The rate of association of [³H]NIC to brain homogenate was rapid, with maximal binding after only 5-10 min of incubation, followed by a reduction (Fig. 3). Thus, the present data indicate that the [³H]NIC binding and/or dissociation is so rapid that to be able to measure adequate binding the incubation temperature must be low and the incubation time short. These incubation conditions are different from those used for in vitro binding with other labelled nicotinic ligands [³H]-α-BTX, [³H]-tubocurarine ([³H]TC) (12, 22). The temperature dependence of [³H]NIC binding observed in this study is in agreement with the results of Martin and Aceto (17), and Marks and Collins (16). The latter authors were unable to detect a low affinity site unless the incubation temperature was reduced to 4°C. The above mentioned studies on the temperature dependence of [³H]NIC binding are, as far as we know, the only ones reported. As seen in Table 1, most other studies have been performed at higher temperatures, resulting in a reduced binding capacity of [³H]NIC.

Very few studies have been concerned with the time course of [³H]NIC binding to brain homogenate. Our findings (Fig. 3) are in agreement with Martin and Aceto (17), who reported that the association kinetics for [³H]NIC binding to brain homogenate were very rapid, with maximal binding after only 5-10 min of incubation, followed by a continuous decrease. This result is in contrast with those of Marks and Collins (16) and Costa and Murphy (6). At present it is difficult to explain these discrepancies between findings. The Scatchard plot (Fig. 4) of the saturationbinding data for [³H]NIC (shown in Fig. 3) indicates a nonlinear curve as has also been found

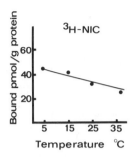

Figure 1. Temperature dependence of [³H]NIC binding to mouse hippocampus. Each point is the mean value of two independent experiments
performed in triplicate. Data taken from Larsson and Nordberg (12).

previously for [³H]TC binding (11). The nonlinearity of the Scatchard
plots suggests that [³H]NIC and [³H]TC bind to two different populations of binding sites (12), in contrast to [³H]-α-BTX binding,
for which only one binding site has been detected (7, 11, 13). As
seen in Table 1, whole brain studies reported in the literature
indicate both one and two binding sites for [³H]NIC. The experimental
conditions may probably play a role in these discrepancies between
results.

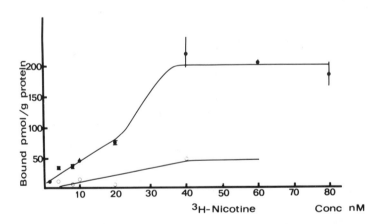

Figure 2. Saturation analysis of binding of [³H]NIC binding to mouse
hippocampus at different temperatures. Increasing concentrations
of [³H]NIC were incubated with P2 fractions of mouse hippocampus at
4°C (●) and 25°C (○) for 10 min. Results are given as the mean
(± SE) of three determinations at 4°C and as the mean of two determinations at 25°C. Data taken from Larsson and Nordberg (12).

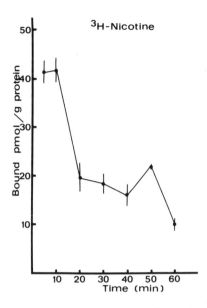

Figure 3. Time course of association of [³H]NIC to brain homogenate.
[³H]NIC (20 nM) was incubated with P2 fractions of rat cerebral cortex
at the times indicated. Each point is the mean (± SE) of four independent experiments. Data taken from Larsson and Nordberg (12).

        The possible existence of two binding sites for [³H]NIC and
[³H]TC and one for [³H]-α-BTX fits well with displacement data obtained
previously in our laboratory (11, 22). We compared the inhibitory
effects on the binding of [³H]-α-BTX, [³H]NIC and [³H]TC to brain
homogenate by the same serial dilutions of unlabelled nicotine and
tubocurarine (11). If the inhibitory activites of an unlabelled
competitor on two radiolabelled markers are compared, the values
representing percentage remaining specific binding of the radio-
indicators will fall on a straight line for any competitor tested
when the two radioligands occupy identical binding sites. If the
two radiolabelled indicators occupy different sites, there will be
a deviation from the straight line (28). When the inhibitory effects
of nicotine on [³H]NIC and [³H]TC binding to mouse hippocampus were
compared, the remaining binding of [³H]TC and [³H]NIC were found
to be the same, indicating that the binding sites for [³H]TC and
[³H]NIC share common properties. However, in comparisons of the
inhibitory activities of unlabelled nicotine against [³H]-α-BTX/[³H]TC
and [³H]-α-BTX/[³H]NIC, the values representing percentage remaining
specific binding fell on a curve which deviated from a straight
line (11), indicating less preference for the [³H]-α-BTX binding
sites.

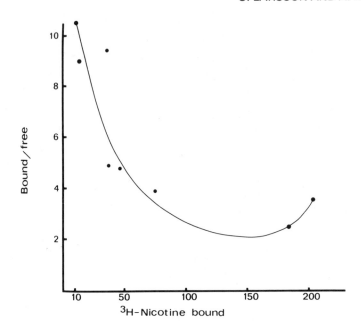

Figure 4.   Scatchard plot of [3H]NIC saturation binding to mouse
hippocampus at 4°C (shown in Fig. 3). Receptor ligand concentration
(bound) is expressed in pmol/g protein and ligand concentration (free)
in nM. Analysis of the data revealed the following parameters: High
affinity site: Kd=6 nM and B_{max}=60 pmol/g protein; low affinity
site: Kd=125 nM and B_{max}=230 pmol/g protein.

The regional distribution of [3H]NIC binding to brain homogenates has also been studied by several investigators (6, 12, 16,
17, 27, 29). As seen in Table 2, [3H]NIC binding to brain homogenate
seems to show a differential regional distribution. Although the
values differ between research groups the highest concentration of
binding of [3H]NIC has been noted in the cortex, hypothalamus and
midbrain. A marked increase in the [3H]NIC binding in the cortex
in comparison with other brain regions has also been observed during
ontogenesis (Larsson et al., 1983 in manuscript; see also Nordberg,
this book).

CONCLUSIONS

In conclusion, this study indicates that [3H]NIC binding and/or
dissociation is so rapid that to be able to measure binding adequately
the incubation temperature must be low (+4°C) and the incubation
time short (10 min). These conditions are different from those
found for other labelled nicotinic ligands (11, 22). The findings
for [3H]NIC in the hippocampus indicate two binding sites. The

Table 2

Regional Distribution of [³H]-Nicotine Binding: A Comparison of Reported Data

References	Cerebellum	Cortex	Hippocampus	Hypothalamus	Midbrain	Medulla Oblongata	Striatum
			pmol/mg protein ± SE				
29 a	25	25	90	95	65	–	–
17 b,c	147	221	323	409	217	209	213
27	36.5 ± 2.2	70.0 ± 6.1	67.2 ± 2.6	51.7 ± 4.6	55.5 ± 4.9	32.7 ± 2.7	43.8 ± 1.5
16	7.9 ± 0.9	23.5 ± 2.8	20.7 ± 3.7	24.9 ± 2.9	48.7 ± 8.0	–	38.3 ± 5.6
6 a	5	10	35	40	–	–	30
12	23 ± 7	110 ± 22	41 ± 7	24 ± 5	54 ± 10	53 ± 7	68 ± 10

a: Values calculated from a histogram; b: (-)-[³H]Nicotine; c: Values calculated from % values of the binding in hypothalamus (409 pmol/g protein). References 6, 17, and 29 have studied the regional distribution of [³H]NIC in rat brain. References 12, 16, and 27 have studied the regional distribution of [³H]NIC in mouse brain.

data for the high affinity sites of this ligand are comparable with those obtained for the high affinity site of [^{3}H]TC and the binding site of [^{3}H]-α-BTX and [^{3}H]-acetylcholine (12, 22, 26). In addition, there appears to be a regional distribution for [^{3}H]NIC binding in the CNS.

Acknowledgement. This study was supported by the Swedish Medical Research Council (Project no. 5817) and the Swedish Tobacco Company.

REFERENCES

1. Abood, L.G., Grassi, S. and Costanza, M. (1983): FEBS Lett. 157:147-149.
2. Abood, L.G., Reynolds, D.T. and Bidlack, J.M. (1980): Life Sci. 27:1307-1314.
3. Armitage, A.K., Hall, G.H. and Morrison, C.F. (1968): Nature 217:331-334.
4. Balfour, D.J.K. (1982): Pharmac. Ther. 15:239-250.
5. Battig, K. (1981): Trends in Pharmacol. Sci. 2:145-147.
6. Costa, L.G. and Murphy, S.D. (1983): J. Pharmacol. Exp. Ther. 226:392-397.
7. Eterovic, V.A. and Bennet, E.L. (1974): Biochim. et Biophys. Acta 362:346-355.
8. Falkeborn, Y., Larsson, C. and Nordberg, A. (1981): Drug and Alcohol Dependence 8:51-60.
9. Fambrough, D.M. (1979): Physiol. Rev. 59:165-226.
10. Gray, E.G. and Whittaker, V.P. (1962): J. Anat. 96:79-89.
11. Larsson, C. and Nordberg, A. (1980): In Neurotransmitters and their Receptors (eds) U.Z. Littauer, Y. Dudai, I. Silman, V.I. Teichberg and Z. Vogel, John Wiley & Sons Ltd., pp. 297-301.
12. Larsson, C. and Nordberg, A. (1985): J. Neurochem. 45:24-31.
13. Lowy, J., McGregor, J., Rosenstone, J. and Schmidt, J. (1976): Biochem. 15:1522-1527.
14. Mangan, G.L. and Golding, J.F. (1984): In: The Psychopharmacology of Smoking, Cambridge University Press, pp. 97-140.
15. Mansner, R. (1972): Ann. Med. Exp. et Biol. Fenniae 50: 205-212.
16. Marks, M.J. and Collins, A.C. (1982): Mol. Pharmacol. 22 554-564.
17. Martin, B.R. and Aceto, M.D. (1981): Neurosci. Biobehav. Reviews 5:473-478.
18. Morley, B.J. (1981): Pharmac. Ther. 15:111-122.
19. Morley, B.J., Farley, G.R. and Javel, E. (1983): Trends in Pharmacol. Sci. 4:225-227.
20. Morley, B.J. and Kemp, G.E. (1981): Brain Research 3:81-104.
21. Morley, B.J., Kemp, G.E. and Salvaterra, P. (1979): Life Sci. 24:859-872.
22. Nordberg, A. and Larsson, C. (1981): In Cholinergic Mechanisms:

Phylogenetic Aspects, Central and Peripheral Synapses and
Clinical Significance (eds) G. Pepeu and A. Ladinsky, Plenum
Press, New York, pp. 819-824.

23. Nordberg, A. and Sundwall, A. (1983): Acta pharmacol et
     toxicol. 52:341-347.
24. Oswald, R.E. and Freeman, J.A. (1981): Neuroscience 6:1-14.
25. Romano, C. and Goldstein, A. (1980): Science 210:647-649.
26. Schwartz, R.D. and Kellar, K.J. (1983): Science 220:214-216.
27. Sershen, H., Reith, M.E.A., Lajtha, A. and Gennaro, J. Jr.
     (1981): J. Recept. Research 2:1-15.
28. Terenius, L. and Wahlström, A. (1976): Eur. J. Pharmacol. 40:
     241-248.
29. Yoshida, K. and Imura, H. (1979): Brain Research 172:453-459.
30. Yoshida, K., Engel, J. and Lilijequist, S. (1982): Arch.
     Pharmacology 321:74-76.

THE CHOLINERGIC LIGAND BINDING MATERIAL OF AXONAL MEMBRANES

H.G. Mautner, J.E. Jumblatt and R. Coronado*

Department of Biochemistry and Pharmacology
Tufts University School of Medicine
136 Harrison Avenue
Boston, Massachusetts 02111 U.S.A.

*Section of Biochemistry, Molecular and Cell Biology
Cornell University
Ithaca, New York 14583 U.S.A.
Now: Department of Pharmacology
University of North Carolina
Chapel Hill, North Carolina 27514 U.S.A.

INTRODUCTION

Evidence has accumulated that acetylcholine (ACh) might play a role in axonal as well as in synaptic membranes. The postulate of Nachmansohn (23) that ACh triggers axonal as well as synaptic permeability changes has been a subject of controversy. However, cholinergic agonists and antagonists affect conduction at the nodes of Ranvier where permeability barriers to quaternary ammonium compounds are minimal (4, 12). Furthermore, choline acetyltransferase and acetylcholinesterase, the enzymes responsible for the synthesis and the hydrolysis of ACh, are present in nerve fibers. Also, in crustacean peripheral nerves release of ACh from cut nerve fibers could be demonstrated (13). Moreover, Denburg and his coworkers (10, 11) have prepared closed membrane vesicles from lobster walking leg nerve plasma membrane, and have demonstrated saturable binding of cholinergic agonists and antagonists to such membranes. Our laboratory has thus been engaged in studying this "axonal cholinergic binding material," and in elucidating its functions.

MATERIALS AND METHODS

Axonal membrane vesicles were prepared from lobster (Homarus americanus) walking leg nerve bundles, following the procedure of Marquis et al. (18). For planar bilayer reconstitution, they were

439

equilibrated overnight in 0.4 M sucrose and 5 mM Tris-HCl (pH 7.4). Bilayers of the Mueller-Rudin type (22) were formed on Teflon cups having an aperture of 0.8 mm. in diameter. The phospholipid mixture (Avanti Chemicals) was composed of 60% phosphatidylethanolamine and 40% acidic phospholipid (phosphatidylserine: phosphatidylinositol:diphosphatidylglycerol) at a molar ratio of 1:1:0.4). The presence of phosphatidylserine and calcium were found to be essential for fusion. Lipids were dissolved in n-decane and applied to the cup at 35 mg/ml. The aqueous phase contained 100 mM KCl, 0.1 mM EGTA, 5 mM Hepes-Tris (pH 7.4). Measurements were carried out at 21-23°C, under voltage-clamp conditions. Cis defines the side of the bilayer to which axonal vesicles are added; trans is the opposite side.

After formation of the bilayer in 100 mM KCl buffer the bilayer was voltage-clamped to +30 mV. $CaCl_2$ was then added to the cis chamber to a final concentration of 1.8 mM, followed by addition of axonal vesicles. Protein concentration in the chamber was 20 mg/ml. Incorporation was stopped by addition of 2 mM EDTA to the cis chamber.

Reversal potentials were measured under bi-ionic conditions consisting of 100 mM KCl on the trans side and 100 mM XCl on the cis side. After fusion of the vesicles under the usual conditions, the cis chamber was perfused with the test cation solution and reversal potentials were measured (6).

Labelling with [³H]-MBTA and [³H]-NEM, following disulfide reduction, was carried out similarly to the procedure of Karlin and Cowburn (1, 20).

RESULTS

In extending Denburg's work, we had found earlier that the binding of [³H]-nicotine to lobster nerve plasma membranes was antagonized by a series of cholinergic ligands as well as by a series of local anesthetics (18), and that this preparation was capable of binding [125I]-α-bungarotoxin (α-Bgt) (19), a ligand widely believed to be a specific label for nicotinic ACh receptor (AChR). A conjugate of α-Bgt and horseradish peroxidase could be used to visualize binding sites not only on membrane vesicles but also on the axolemma of crustacean nerves. Binding was blocked by pretreatment with d-tubocurarine or native α-toxin (5). Freedman and Lentz (14) subsequently showed the above conjugate to be localized at nodes of Ranvier of rat sciatic nerve. After pretreatment with collagenase and trypsin, binding of the conjugate could be seen in all areas of the axolemma where the myelin sheath had separated from the axon. Schwartz et al. (25) visualized binding on goldfish optic nerve both with a conjugate of α-Bgt and rhodamine, and with antibodies against eel AChR. Dahlström et al. (9) showed rat sciatic nerves to contain material capable of cross-reacting with polyclonal and monoclonal antibodies prepared against nicotinic AChR as well as with human myasthenia IgG, making it likely that this biopolymer is indeed related to the synaptic AChR.

A careful study of the binding of cholinergic ligands to axonal membranes, as contrasted to synaptic membranes, showed several differences. Axonal binding sites are capable of the tight binding of both nicotinic and muscarinic antagonists (11, 16). On the other hand, the binding of [^{125}I]-α-Bgt to axonal membranes is much looser, more reversible, and much more sensitive to the presence of salts than binding to synaptic membranes (16).

To obtain more specific labelling we studied the interaction of axonal membrane fragments with [^{3}H]-4-(N-maleimido) benzyltrimethylammonium (MBTA), a reagent claimed to be capable, after disulfide reduction, of specifically alkylating the α-subunit of the nicotinic AChR (17). Following reduction with dithiothreitol (DTT) of axonal membrane fragments, MBTA labelling occurred largely on a peptide with an estimated molecular weight of 50,000 (20). Labelling could be largely prevented by pretreatment with d-turbocurarine, and to a lesser degree with atropine, nicotine and procaine.

While these findings support the presence, in axonal membranes, of an unusual AChR, they do not prove the presence of such a receptor, and provide no information about its functions. Therefore,

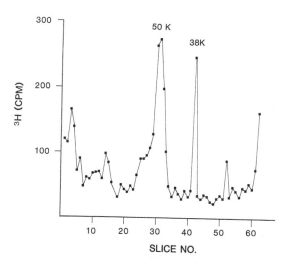

Figure 1. Labelling of several peptide bands with [^{3}H]-NEM following reduction of membrane disulfides with DTT.

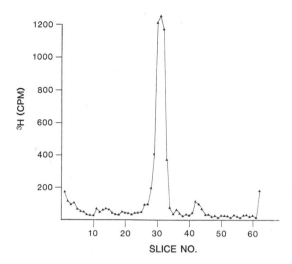

Figure 2. Labelling of 50 K peptide band with [³H]-MBTA following membrane disulfide reduction. Molecular weight was calibrated with phosphorylase B, bovine serum albumin, carbonic anhydrase and soybean trypsin inhibitor.

we investigated whether cholinergic ligands could alter the cation permeability of lobster axonal membranes. Villegas (1) had reported that using lobster axonal membrane vesicles, $[^{22}Na]^+$ efflux was reduced by tetrodotoxin (TTX) and increased by veratridine and α-Bgt (1), and similar results were shown for liposomes into which axonal membrane fragments had been incorporated (27). We encountered considerable difficulties in carrying out cation flux measurements with lobster axon plasma membrane fragments since, because of their leakiness, experimental error was considerable. Fast efflux, showing no ion selectivity, was followed by a "slow" efflux phase (from 40 to 100 minutes), showing first order kinetics. $[^{86}Rb]^+$ efflux was twice as fast as that of $[^{22}Na]^+$. The efflux of $[^{22}Na]^+$ in our hands, was not affected by veratridine, TTX or carbamylcholine. Rb^+ efflux, on the other hand, was reduced not only by the K^+ channel blocking agents 4-aminopyridine and tetraethylammonium (TEA), but also, slightly but reproducibly, by d-tubocurarine, procaine and eserine (15).

It proved relatively easy to incorporate axonal plasma membrane fragments into planar bilayers. Addition of vesicles preloaded with 0.4 M sucrose, in the presence of Ca^{++}, to planar bilayers of the Mueller-Rudin type (22), resulted in an increase of membrane conductance by two to three orders of magnitude (6). The incorporated channels proved to be highly selective for K^+ with permeability for Na^+ and Li^+ being very low. Blockade of K^+ conductance

Table 1

Cation Selectivity of Axonal Membrane Channels
Incorporated in Planar Bilayers

Cation (M^+)	Permeability$_{M^+}$/Permeability$_{K^+}$
Tl^+	2.02
K^+	1.00
Rb^+	0.65
NH^+	0.30
Cs^4	0.23
Na^+	0.12
Li^+	0.07

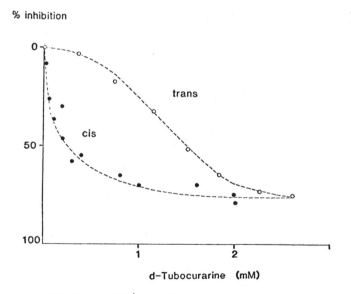

Figure 3.   Inhibition of K^+ conductance increase, induced by the
fusion of axonal membrane fragments to planar bilayers by the addition
of d-tubocurarine, when added on the <u>cis</u> or <u>trans</u> side.

was achieved with d-tubocurarine, atropine, decamethonium and procaine. Since these compounds blocked the conductance increase much
more effectively when added on the _cis_ side (the side on which the
vesicles had been added) rather than on the _trans_ side of the bilayer,
the K^+ channels must have been introduced asymmetrically. Steady-state
K^+ conductance was voltage-independent in the 0-60 mV range (6).
TEA, 4-aminopyridine, carbamylcholine and α-Bgt did not block K^+
conductance.

In bilayers produced by the monolayer assembly method of Montal
and Meuller (21), in addition to the voltage-independent K^+ channels
described above, insertion of axonal membrane fragments, introduced
voltage-sensitive channels with even higher selectivity for K^+ (8).
These latter channels were subject to voltage-sensitive blockade
by TEA and Cs^+, and behaved like the K^+ channels involved in the
delayed rectification of axonal membranes. Coronado and Latorre
have also observed lobster axon K^+ channels, using bilayers formed
with patch clamp pipettes (7).

DISCUSSION

The synaptic AChR, likely to be a water filled pore inserted
in the membrane, shows little ion selectivity or selectivity to
applied electric fields. It is shielded by few permeability

barriers slowing access by cholinergic agonists and antagonists
(generally quaternary ammonium compounds). Thus, it is relatively
easy with synaptic membranes to carry out studies of ligand binding
and to relate these to studies of ligand induced permeability changes.
On the other hand, access to axonal channels is hindered by a variety
of complex permeability barriers. Furthermore, in considering the
permeability changes taking place during the conduction of the action
potential, one has to consider a variety of axonal membrane channels,
several of which show great ion selectivity and great voltage sensitivity. It is not surprising, therefore, that our understanding
of axonal membrane channels has lagged behind our understanding of
the nicotinic AChR, now well defined in terms of subunit structure,
the arrangement of subunits with respect to each other, and subunit
amino acid sequences.

Binding studies using cholinergic ligands and AChR-antibodies
suggest the presence of biopolymers related to the AChR in axonal
membranes. These ligand binding studies also suggest that this
biopolymer can be classified neither as a nicotinic nor as a muscarinic AChR. The possibility has to be considered that one is
dealing with a proreceptor undergoing axonal transport. Indeed,
Dahlstrom et al. (9) showed both orthograde and retrograde transport
of the biopolymer capable of binding AChR antibodies in the peri-
karyon of motor nerves. However, it seems unlikely that biopolymers
localized on the axolemma would be undergoing transport, and there

is evidence both for the presence of α-Bgt-horseradish peroxidase binding material (5, 19) and of acetyl-cholinesterase (3) on axolemmal membranes both at nodes of Ranvier and in internodal sections.

Unfortunately, we have as yet little information about the functions of the axonal cholinergic receptor-like molecule present not only at nodes of Ranvier but also in the paranodal and internodal regions of rat sciatic nerve (14). Since saltatory conduction is presumed to be taking place in myelinated nerves, it seems unlikely that this biopolymer is involved in the conduction of the action potential.

It is suggestive that the localization of the α-Bgt-horseradish peroxidase binding material (14) and of acetylcholinesterase (3) corresponds to the localization of K^+ channels in myelinated nerves (2). These are also found in paranodal and internodal regions and, again, are unlikely to be needed for axonal impulse propagation.

The observation that insertion of axonal membrane fragments into planar bilayers results in the asymmetric introduction of K^+ channels that are not voltage-sensitive and that are capable of blockade by cholinergic antagonists, raises the possibility that these fragments contain a biopolymer involved in control of the axonal resting potential long known to be generated largely by the leakage of K^+ ions. This hypothesis is supported by the claim that the ACh content of Schwann cells greatly exceeds that of axons (26). However, it should be noted that while we have found cholinergic antagonists to be capable of blocking K^+ conductance, we have not been able to increase K^+ conductance with cholinergic agonists (6). Components of the AChR cycle are distributed widely not only in nervous, but also in non-nervous tissue, including unicellular organisms and plants (24). K^+ channels controlling the resting potential of cells seem logical candidates for being targets of primordial cholinergic systems.

Acknowledgement. We are indebted to the National Science Foundation for support of this work (BNS-81-04175).

REFERENCES

1. Barnola, F.V. and Villegas, R. (1976): J. Gen. Physiol. 67:81-90.
2. Brismar, T. (1982): Trends Neurosci. 5:179-181.
3. Brzin, M. (1966): Proc. Nat. Acad. Sci. 56:1560-1563.
4. Chang, C.C. and Lee, C.J. (1966): Brit. J. Pharmacol. 28:172-182.
5. Chester, J., Lentz, T.L., Marquis, J.K. and Mautner, H.G. (1979): Proc. Nat. Acad. Sci. 76:3542-3546.
6. Coronado, R., Huganir, R.L. and Mautner, H.G. (1981): FEBS Letters 131:355-358.
7. Coronado, R. and Latorre, R. (1983): Biophys. J. 41:56.

8. Coronado, R., Latorre, R. and Mautner, H.G. (1984): Biophys. J. 45:288-290.

9. Dahlström, A.B. et al. This book.

10. Denburg, J.L. (1972): Biochim. Biophys. Acta 282:453-458.

11. Denburg, J.L. and O'Brien, R.D. (1972): J. Med. Chem. 16:57-60.

12. Dettbarn, W.D. (1960): Nature 186:891-892.

13. Dettbarn, W.D. and Rosenberg, P. (1966): J. Gen. Physiol. 50:447-460.

14. Freedman and Lentz, T.L. (1980): Comp. Neurol. 193:179-185.

15. Jumblatt, J.E. and Mautner, H.G. (1980): Abstr. Am. Chem. Soc. Meet., Biol. Chem. Div. :96.

16. Jumblatt, J.E., Marquis, J.K. and Mautner, H.G. (1981):J. Neurochem. 37:392-400.

17. Karlin, A. and Cowburn, D.A. (1973): Proc. Nat. Acad. Sci. 70:3636-3640.

18. Marquis, J.K., Hilt, D.C. and Mautner, H.G. (1977): Proc. Nat. Acad. Sci. 74:2278-2282.

19. Marquis, J.K., Hilt, D.C. and Mautner, H.G. (1977): Biophys. Res. Comm. 78:476-482.

20. Mautner, H.G., Cole, J.L. and Belew, M.A. (1983): Biochem. Biophys. Res. Comm. 111:61-66.

21. Montal, M. and Mueller, P. (1972): Proc. Nat. Acad. Sci. 69:3561-3566.

22. Mueller, P. and Rudin, D.O. (1969): In Laboratory Techniques in Membrane Biophysics (eds) H. Passow and R. Stampfli, Springer-Verlag, Berlin, New York, pp. 141-156.

23. Nachmansohn, D. (1953): Harvey Lect., Ser. 49:57-99.

24. Rama Sastry, B.V. and Sadavongvivad, C. (1978): Pharmacol. Revs. 30:65-132.

25. Schwartz, M., Axelrod, D., Feldman, E.L. and Agranoff, B.W. (1980): Brain Res. 194:171-180.

26. Villegas, J. and Jenden, D.J. (1979): J. Neurochem. 32:761-766.

27. Villegas, R., Villegas, G.M., Barnola, F.V. and Racker, E. (1977): Biochem. Biophys. Res. Comm. 79:210-217.

DO MOTOR NEURONS CONTAIN FUNCTIONAL

PREJUNCTIONAL CHOLINOCEPTORS?

G.G. Bierkamper[*], E. Aizenman and W.R. Millington

[*]Department of Pharmacology
University of Nevada School of Medicine
Reno, Nevada 89557 U.S.A.

Department of Environmental Health Sciences
The Johns Hopkins University
Baltimore, Maryland 21205 U.S.A.

INTRODUCTION

Among the many mechanisms which may regulate acetylcholine (ACh) release from mammalian motor nerve terminals is the possibility of negative feedback by the neurotransmitter itself on prejunctional cholinoceptor sites. Autoregulation of neurotransmitter release has been documented in adrenergic systems (21) and in central and autonomic cholinergic systems (12, 15, 22). Negative feedback of released ACh through presynaptic muscarinic receptors has been shown in cerebral cortex (22), hippocampus (15), guinea pig ileum (12), sympathetic ganglion of the bull frog (13) and rat atrium (24).

Considerable controversy surrounds the existence, function, and pharmacological typing of the putative prejunctional cholinoceptors on skeletal motor neurons (6, 7, 17, 18, 20). Histochemical evidence indicates the presence of nicotinic binding sites on the nerve terminal (14, 23) as well as axonal transport of these "receptor" sites in sciatic nerves and ventral roots (see Aizenman et al. and Dahlstrom et al., this book). However, other investigators view histochemical staining of prejunctional sites as artifactual (11).

Pharmacological studies have both supported (1) and dismissed (9, 10) the functional significance of prejunctional muscarinic receptors on motor neurons as a mechanism for regulating ACh release. Prejunctional nicotinic receptors have been placed in the role of modulating ACh release by positive feedback (6, 20) or by negative

447

feedback (7, 17, 20, 25). Recent evidence suggests that motor nerve terminals have a population of nicotinic-like cholinoceptors which function to decrease the quantity of evoked ACh release when activated by elevated levels of ACh in the junctional cleft (7, 20, 25); and allow an increase in evoked ACh release when blocked with a nicotinic antagonist (16, 17).

This hypothesis, i.e., negative feedback by nicotinic cholino-ceptors (nAChR) on motor nerve terminals, has been the focus of our recent investigations. The principal aims of these studies are: 1) proving the existence of the receptor, 2) determining its pharmacologic characteristics, and 3) demonstrating that it can, indeed, alter ACh release. The results of our preliminary experiments are the subject of this report.

METHODS

Acetylcholine release from the vascular perfused rat phrenic nerve-hemidiaphragm preparation has been assessed by three methods: 1) measurement of force of contraction (FC); 2) direct assay of released ACh by radioenzymatic assay; and 3) intracellular recording from the endplate region of the myofiber.

Vascular perfused hemidiaphragms from rats [male, Long-Evans descent, hooded (Blu: (LE)), Blue Spruce Farms, Altamont, New York, 275-375 gm] were prepared as previously described (3, 5). For FC measurements the central tendon area was attached to a plastic frame connected by a chain to a force transducer (Grass Model FT 03); the rib edge was fixed with pins to the silicon rubber base of a vertical chamber (5). The preparation was stimulated indirectly via the phrenic nerve by suction electrode (0.2ms rectangular pulses, 0.8 - 1.2V, supramaximal) or directly through the medium using two opposing electrodes (1ms, 90 - 100V, supramaximal; from a Grass Model SD9 stimulator). Transducer output was recorded on a Beckman Dynograph type R411 recorder. Preparations were perfused at 39 1/min with HEPES-buffered medium (4), pH 7.4, at 30°C, through a PE 10 cannula inserted in the inferior phrenic vein. Choline chloride (10 M) supplementation was present in all experiments (4). The chamber was filled with medium for transmural stimulation.

Direct measurement of ACh in the perfusate was accomplished in other vascular perfused preparations mounted in a closed chamber as previously described (3). Perfusate samples were collected every 15 min during rest or stimulation of the phrenic nerve, then extracted and assayed for ACh by the radioenzymatic method of Goldberg and McCaman (8), as modified for this preparation by Bierkamper and Goldberg (4). Neostigmine (10^{-5}M) was used, except where noted, as the acetylcholinesterase (AChE) inhibitor.

Intracellular recordings were performed by conventional techniques using glass microelectrodes (5-10 Mohm) filled with 3M KCl. Resting membrane potentials (RMP), miniature endplate potentials (mepp) and endplate potentials (epp) were amplified by a Neuroprobe electrometer (A-M Systems, Inc., Everett, Washington), visualized by oscilloscope, and recorded on an FM tape recorder. Experiments were subsequently analyzed on a Smartscope waveform analyzer/averager (T.G. Branden Corp., Portland, Oregon). Drugs were administered by superfusion as well as through the cannula. Muscle twitch was prevented with low calcium (1 mM) and high magnesium (6-7 mM). Carbachol was applied by pressure ejection (Picospritzer II, General Valve Corp., East Hanover, New Jersey) through a separate micropipette positioned extracellularly at the recording site. A single cell was impaled and held throughout the experiment so that control and drug data could be easily compared (2). Only cells with a stable RMP were used.

These three techniques were employed to test the hypothesis that nicotinic agonists or excess ACh in the cleft would activate the prejunctional nAChR and depress ACh release, FC and quantal content. Conversely, nicotinic antagonists would, in theory, block the prejunctional nAChR and result in an enhancement of ACh release, FC and quantal content. The agonists tested were nicotine, carbachol, ACh (by anti-AChE enhancement with neostigmine, physostigmine, and DFP). The antagonists included alpha-bungarotoxin and d-tubocurarine. All drugs were obtained from Sigma Chemical Co., St. Louis, Missouri.

RESULTS

When continuously stimulated (10 Hz) via the phrenic nerve and in the presence of an AChE inhibitor, the release of ACh into the perfusate from the vascular perfused hemidiaphragm reaches a steady plateau level after approximately 1 hr of stimulation (5). Normal release of ACh at 10 Hz averaged 10.5 pmol/minute/hemidiaphragm. The addition of nicotine (10^{-5} M) to the perfusion medium during the plateau of release resulted in a 25.5% decline in ACh after 30 min (Fig. 1). Washout of nicotine with normal medium restored stimulated release of ACh to the pre-drug level. Similar experiments with carbachol (10^{-5} M) revealed a statistically significant decrease (23%) in ACh release (10 Hz), and a partial washout to restore release levels statistically to the pre-drug control. These results suggest a negative influence on stimulated ACh release by nicotinic receptor agonists.

The classical nicotinic receptor antagonist, d-tubocurarine (d-TC), caused a dramatic inhibition of ACh release at the high dosage of 10 mM (36% decrease within 30 min), an effect which was opposite to what was predicted by our hypothesis (Table 1). However, lower doses (10^{-5} M and 10^{-6} M) of d-TC resulted in a 16% and 43%

Table 1

Effect of D-Tubocurarine on Stimulated Acetylcholine Release

Treatment	ACh Release
Control[a]	8.91 ± 0.81 (13)[c]
d-TC[b] (10^{-4} M)	5.75 ± 0.90 (6) *
(10^{-5} M)	10.32 ± 0.07 (4)
(10^{-6} M)	12.74 ± 1.42 (3) *

[a] Value represents steady release level obtained by continuous stimulation (7 Hz) in the absence of drug (75-90 min). [b] Value represents the level of release reached after 30 min of perfusion with d-TC. [c] Values are given as pmol ACh released/min/hemidiaphragm $\pm$ SEM as measured by radioenzymatic assay. The number of preparations is given in parentheses. The experimental design consisted of 60 min of equilibration, 30 min resting release, 90 min of stimulated release; then the introduction of d-TC for the remainder of the experiment with uninterrupted stimulation (7 Hz). Stimulated ACh in the presence of d-TC was measured 30 min after the introduction of the drug through the cannulated vascular supply. Only one concentration of d-TC was used in each preparation. * P (t) < 0.05 by the student's t-test.

increase in stimulated ACh release, respectively (Table 1), 30 min after drug introduction.

The administration of a single dose of alpha-bungarotoxin (50 ug) through the cannulated vascular system of the hemidiaphragm preparation provoked a delayed but remarkable increase in ACh release, and thus, followed the prediction of our hypothesis (Fig. 2). Similar results were obtained by Haggblad and Heilbronn (this book) using a purified alpha-bungarotoxin at a lower dosage.

The effects of nicotinic agonists and antagonists were also studied by measuring FC in contracting hemidiaphragm preparations as a means of circumventing the necessity for total AChE inhibition when directly measuring the release of endogenous ACh by biochemical assay. In these preparations, nicotine (10^{-5} M) suppressed maximum FC at 20 Hz presumably via enhanced activity of the prejunctional AChR (Fig. 3). To test this theory, d-TC (10^{-8} M) was administered in a subsequent trial with nicotine (10^{-5} M). This concentration of d-TC, which had no effect on FC alone, partially restored the FC at 20 Hz in the presence of nicotine. Direct stimulation of the preparations revealed little drug effect on FC and thus supports

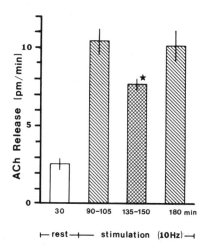

Figure 1. Acetylcholine release from the perfused rat hemidiaphragm as measured by radioenzymatic assay of perfusate samples collected every 15 min. Neostigmine (10^{-5} M) allowed complete recovery of released ACh at rest or during 10 Hz phrenic nerve stimulation. Nicotine (10^{-5} M), administered through the cannulated vascular system, significantly depressed ACh release at 135-150 min [*p (t) < 0.05; n = 6]. Washout of nicotine restored release (180 min) to pre-drug levels (90-105 min).

prejunctional rather than postjunctional receptor interaction or muscle contractile alterations. Carbachol (10^{-5} M) slightly depressed FC at 20 Hz; however, in this case d-TC (10^{-8} M) had no effect in reversing the depression. Atropine (10^{-6} M - 10^{-3}M), a muscarinic receptor antagonist, had no effect on FC alone, nor did it antagonize the effects of nicotinic agonists.

A similar design was applied to test the effects of elevated levels of residual ACh in the synaptic cleft. By administering neostigmine (10^{-6} M) and stimulating at 20 Hz, it was predicted that residual ACh in the cleft would act on the prejunctional nAChR and depress FC by reducing the amount of ACh released. Indeed, neostigmine caused a dramatic depression of FC at 20 Hz. The co-administration of d-TC (5×10^{-8} M) with neostigmine partially blocked this rapid depression of FC (Fig. 3).

Further experiments with physostigmine (PS; 10^{-5} M) explored the frequency dependence of the apparent negative feedback as assessed by FC measurements (Fig. 4). At 5 Hz and above, FC in the presence of PS decreased significantly as compared to control; at 20 Hz, FC rapidly decreased toward baseline.

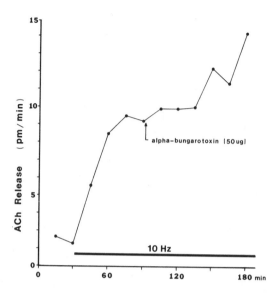

Figure 2. Acetylcholine release from the vascular perfused rat hemi-diaphragm as measured by radioenzymatic assay. Each point represents the average release in pm/min/hemidiaphragm from two experiments. Alpha-bungarotoxin (50 mg) was infused as a single bolus through the cannulated vascular system (arrow).

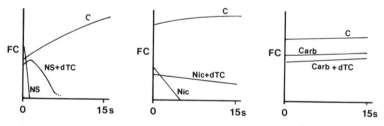

Figure 3. A facsimile of typical force of contraction (FC) experiments in the perfused hemidiaphragm. Curves represent maximum FC over 15 sec of 20 Hz phrenic nerve stimulation for control (C), neo-stigmine (NS, 10^{-6} M), nicotine (Nic, 10^{-5} M), carbachol (Carb, 10^{-5} M) or d-tubocurarine (d-TC, 5×10^{-8} M). Curare antagonized the rapid depression of FC induced by NS and Nic. Carb. had a slight depressive effect which was unaffected by d-TC.

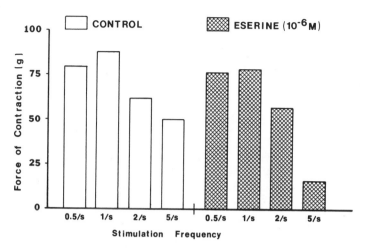

Figure 4. Typical experiment using the vascular perfused rat hemidiaphragm to measure FC at different frequencies of nerve stimulation
with and without the anticholinesterase agent, physostigmine (eserine,
10^{-6} M). FC measurements represent maximum tension developed 5 min
after initiating stimulation at 0.5, 1, 2, or 5 Hz; a 5 min rest
was allowed between trials. Reversing the order of frequencies of
stimulation gave the same results. Neostigmine (10^{-5} M) also yielded
similar results.

        The dose-response relationship of neostigmine and physostigmine
to quantal content of ACh release at 10 Hz was studied by intracellular
recording from the vascular perfused hemidiaphragm preparation. At
10^{-5} M concentration both drugs provoked a decrease in quantal content;
at 10^{-4} the quantity of ACh was significantly decreased even further
(Table 2).

        Additional intracellular recording experiments were performed
in the hemidiaphragm to test the effect of the agonist, carbachol.
Carbachol (10^{-3} M) was pressure ejected by a picospritzer through
a micropipette onto the endplate region from which recordings were
made. It is estimated that the effective concentration in the recording area was briefly 10^{-4} M or less. Epp's were recorded during
trains-of-5 stimulation (50 Hz burst of 5 pulses every 5 sec) of
the phrenic nerve. Application of carbachol caused a decrease in
epp amplitude in the 3rd and 4th events of the train-of-5. Further
ejections of carbachol every two minutes reduced the amplitude of
the 1st, 2nd, and 3rd events. Partial washout of carbachol revealed
a rebound of epp amplitudes (Fig. 5).

Table 2

Effect of Acetylcholinesterase Inhibition on Quantal Content
of Acetylcholine Release

Treatment	Quantal Content	Quantal Content
	At 1 Hz	At 10 Hz
Control (N = 3)	3.01	3.21
Eserine (10^{-6} M)	3.26	3.10
Eserine (10^{-5} M)	2.59	2.50
Eserine (10^{-4} M)	1.88**	1.90**
Control (N = 1)	1.91	2.47
Neostigmine (10^{-6} M)	1.91	2.37
Neostigmine (10^{-5} M)	1.43	1.93
Neostigmine (10^{-4} M)	*	*

Intracellular recordings were made from single cells in a rat hemidiaphragm preparation (2). The same cell was held throughout the
experiment in the presence of choline (10 μM) and at 30°C. Quantal
content was determined by dividing mean epp amplitudes by mean mepp
amplitudes (signal averaging of at least 100 events each). RMP was
normalized to -75 mV; actual RMP averaged -64 mV. Similar decreases
in quantal content were observed at 7 Hz. ** P (t) < 0.05. * mepp
and epp amplitudes decreased to below measurable levels.

DISCUSSION

Our preliminary experiments have tested the premise that cholinergic motor neurons contain nicotinic-like receptors which function
to modulate ACh release. These putative receptors comprise an autoinhibition mechanism whereby increased neurotransmitter levels in
the cleft tend to decrease further release in order to keep the
neurotransmitter output per pulse in a reasonably constant range
commensurate with the physiological demand.

Experimental conditions which increase the cleft levels of
ACh (Fig. 3, Table 2) or activate the prejunctional nAChR, e.g.,
via nicotinic agonists (Fig. 1), demonstrate an enhancement of negative
feedback as measured by the decreased output of endogenous ACh,
depression of FC and reductions in quantal content and epp amplitudes.
Conversely, nicotinic receptor antagonists appear to block the prejunctional nAChR's in a dose-dependent manner and provoke an increase
in ACh release (Table 1, Fig. 2). Moreover, the presence of low

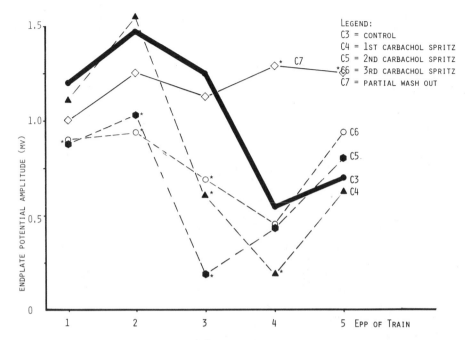

Figure 5. Trains-of-5 (5 pulses spaced 20 ms apart) were applied to the phrenic nerve every 5 sec; endplate potentials were recorded by intracellular electrodes in the isolated rat hemidiaphragm with low Ca^{++} (1 mM) and elevated Mg^{++} (7 mM). The bold curve represents the average epp amplitudes of 21 Control (C3) sweeps. Carbachol (10^{-3} M; C4, C5, C6) was applied extracellularly to the recording site by pressure injection (picospritzer) through a micropipette. Washout (C7) was accomplished for 5 min by normal superfusion of buffered medium over the preparation. * Indicates statistical significance by ANOVA.

concentrations of a nicotinic receptor antagonist, d-TC, produced a consistent antagonism of the enhanced negative feedback provoked by nicotine or excessive cleft levels of ACh. Therefore, the results of our experiments can be interpreted to support the hypothesis that cholinergic motor neurons contain functional nicotinic-like receptors which act to auto-regulate ACh release.

Anticholinesterase drugs cause residual ACh to remain briefly in the synaptic cleft after each stimulus as reflected by the prolongation of the decay phase of endplate potentials (2). Rapid stimulation would be expected to increase cleft levels of residual ACh and thus activate the negative feedback cholinoceptor on the nerve terminals. Wilson (25) used this principle in an electrophysiological study aimed at determining the prejunctional effects

of ACh. Neostigmine (6 mM) and physostigmine (3 mM) caused a substantial decrease in quantal content of epps at 20 and 50 Hz. Curare
(0.14 mM) produced the opposite effect. Wilson ruled out postjunctional receptor desensitization and saturation as a possible mechanism,
and also demonstrated that the effect was not due to elevated K^+ levels
or a direct effect of the anticholinesterase drugs. Our studies
corroborate Wilsons' results with the additional evidence that the
negative feedback by residual ACh attenuates the performance of
the muscle by decreasing FC.

Our preliminary results indicate a frequency-dependence for
negative feedback to exert an influence on muscular function. Low
frequencies of stimulation or single pulses cause no change in FC
in the presence of an AChE inhibitor, whereas, higher frequencies
(>5 Hz) or trains of stimuli (10-50 Hz) reduce FC, presumably via
negative feedback. Since low doses of d-TC antagonized the negative
feedback, it appears likely that at higher rates of stimulation,
residual ACh in the cleft reaches concentrations which provoke substantial negative feedback. Thus, it was surprising that further
negative feedback could be induced during experiments with nicotine
in which endogenous ACh was recovered from the hemidiaphragm during
maximal AChE inhibition; a condition which would be expected to
cause high concentrations of residual ACh.

The mechanism by which prejunctional cholinoceptors act to
limit ACh release is unknown. Our experiments indicate however,
that the events are very rapid. This observation suggests an alteration of the release mechanism, or terminal membrane excitability,
rather than an alteration in choline uptake or synthesis of ACh.

The relative physiological importance of the prejunctional
nAChR is equally uncertain. At near resting or low demand situations
the prejunctional nAChR may have little effect on ACh release and
thus muscular function. However, in disease states, cholinesterase
poisoning or high demand situations, the nAChR may play a significant
role in regulating ACh release to ensure optimal muscular performance
and conservation of neurotransmitter.

One must be cautious in interpreting all our preliminary data
as negative feedback through a prejunctional nicotinic cholinoceptor.
The complexity of the modulatory mechanisms at the nerve terminals
raises the possibility of other mechanisms acting in concert with
the nAChR (6, 7, 17, 19, 20). Nevertheless, when taken together,
our results suggest the presence of a pharmacologically active nicotinic-like site on motor nerve terminals which can alter stimulated
ACh release.

Acknowledgements. The dedicated technical assistance of Frances
Davenport, Jane Bowman, Fred Jacka, Libby Hoops and Dr. Nikou
Amirhessami is gratefully acknowledged. We thank Shou-Lin Lin for

typing the manuscript. These experiments were funded, in part, by NINCDS NS 20456 and DAMD 17-83-C-3190.

REFERENCES

1. Abbs, E.T. and Joseph, D.N. (1981): Br. J. Pharmacol. 73:481-483.
2. Bierkamper, G.G. (1981): Europ. J. Pharmacol. 73:343-348.
3. Bierkamper, G.G. and Goldberg, A.M. (1978): J. Electrophysiol. Tech. 6:40-46.
4. Bierkamper, G.G. and Goldberg, A.M. (1979): In Nutrition and the Brain, Vol. 5 (eds) A. Barbeau, J.H. Growdon and R.J. Wurtman, Raven Press, New York, pp. 243-251.
5. Bierkamper, G.G. and Goldberg, A.M. (1982): In Progress in Cholinergic Biology: Model Cholinergic Synapses, (eds) I. Hanin and A.M. Goldberg, Raven Press, New York, pp. 113-136.
6. Bowman, W.C. (1980): Anesthesia and Analgesia 59:935-943.
7. Foldes, F.F. and Vizi, E.S. (1980): In Modulation of Neurochemical Transmission (ed) E.S. Vizi, Pergamon Press, New York, pp. 335-382.
8. Goldberg, A.M. and McCaman, R.E. (1973): J. Neurochem. 20:1-8.
9. Gundersen, C.B. and Jenden, D.J. (1980): Br. J. Pharmacol.70:8-10.
10. Haggblad, J. and Heilbronn, E. (1983): Br. J. Pharmacol. 80:-471-476.
11. Jones, S.W. and Salpeter, M.M. (1982): Abstr. Soc. Neurosci. 8:134.3.
12. Kilbinger, H. and Wessler, I. (1980): Naunyn-Schmiedberg's Arch. Pharmacol. 314:259-266.
13. Koketsu, K. and Yamada, M. (1982): Br. J. Pharmacol. 77:075-082.
14. Lentz, T.L., Mazurkiewicz, J.E. and Rosenthal, J. (1977): Brain Res. 132:423-442.
15. Marchi, M., Paudice, P. and Raiteri, M. (1981): Europ. J. Pharmacol. 73:75-79.
16. Miledi, R., Molenaar, P.C. and Polak, R.L. (1978): Nature 272:-641-643.
17. Molenaar, P.C. and Polak, R.L. (1980): Prog. Pharmacol. 3:39-44.
18. Ramaswamy, S., Geetha, V.A., Nazimudeen, S.K. and Kameswaran, L. (1978): Europ. J. Pharmacol. 52:197-200.
19. Shain, W. and Carpenter, D.O. (1981): Intl. Rev. Neurobiol. 22:205-250.
20. Standaert, F.G. (1982): Br. J. Anaesth. 54:131-145.
21. Starke, K. (1981): Ann. Rev. Pharmacol. Toxicol. 21:7-30.
22. Szerb, J.C. (1979): In Presynaptic Receptors (eds) S.Z. Langer, K. Starke and M.L. Dubocovich, Pergamon Press, New York, pp. 293-298.
23. Tsujihata, M., Hazama, R., Ishii, N., Ide, Y. and Takamori, M. (1980): Neurology 30:1203-1211.
24. Wetzel, G.T. and Brown, J.H. (1983): Circulation 68:Abtr. 246.
25. Wilson, D.F. (1982): Am. J. Physiol. 242(Cell. Physiol. 11): C366-C372.

PHARMACOLOGICAL CHARACTERIZATION OF AXONALLY TRANSPORTED

[^{125}I]-α-BUNGAROTOXIN BINDING SITES IN RAT SCIATIC NERVE

E. Aizenman[o], W.R. Millington[+], M.A. Zarbin[*],
G.G. Bierkamper[++] and M.J. Kuhar[*]

[*]Departments of Environmental Health
Sciences[o] and Neuroscience[*]
The Johns Hopkins University
615 N. Wolfe Street
Baltimore, Maryland 21205 U.S.A.

[+]NIH-NINCDS, Experimental
Therapeutics Branch
Building 10, Room 5C-205
Bethesda, Maryland 20814 U.S.A.

[++]University of Nevada
School of Medicine
Department of Pharmacology
Howard Building
Reno, Nevada 89557 U.S.A.

INTRODUCTION

Cholinergic receptors are known to occur at the postsynaptic membrane of the mammalian neuromuscular junction. It has been proposed that cholinergic drugs may also act at the motor nerve terminal (rev. 5, 20, 27, 28). However, there is as yet no conclusive evidence that demonstrates that these drug effects occur via an acetylcholine receptor on the presynaptic membrane (4, 6, 14, 17, 30).

One of the principal difficulties in resolving this question is the problem of identifying a presumed low density of presynaptic cholinoceptive sites from a very high density of postjunctional receptors (10). This has confused morphological techniques aimed at visualizing the receptor by specific ligand binding [alpha-bunga-rotoxin (α-BuTX) (7) conjugated to a marker or radio-ioidinated; 16, 18] and physiological techniques directed at differentiating the pre- and postsynaptic receptors on the basis of their pharmacological characteristics (i.e., 13, 26).

In the present investigation we have attempted to circumvent the same problem by labeling the putative receptors as they are axonally transported in peripheral nerves (1). With the use of an innovative autoradiographic technique (32), this approach has enabled us to investigate the pharmacological properties of the toxin-binding site interaction.

MATERIALS AND METHODS

Autoradiographic procedures as described by Young and Kuhar (32) have been modified for the study of receptors undergoing axonal transport (29, 33, 34). Adult male Long-Evans rats were anesthetized and their sciatic nerves exposed. The nerves were ligated with non-absorbable silk suture at a location approximate to half the distance between the sciatic notch and the knee, as measured from the notch itself. The animals were allowed to survive for a specified amount of time, after which a 1-2 cm segment from each nerve was removed under anesthesia, with the ligature at the center of the segment. The nerve segments were frozen in liquid nitrogen onto a cryostat chuck, longitudinally sectioned (8 μm), and thaw mounted onto gelatin coated microscope slides.

Tissue sections were preincubated at room temperature for 20 min in 50 mM phosphate buffer saline (PBS) pH 7.4 alone or in the presence of either curare, atropine, decamethonium, or nicotine ('displacers'). This was followed by a 60 min incubation at room temperature in PBS containing 2 nM [^{125}I]-α-BuTX (200 Ci/mmol; Amersham) with or without displacer. Sections were then washed in cold (4°C) buffer for an additional 15 min.

Glass cover-slips coated with photographic emulsion (Kodak, NTB-3) were apposed to the nerve section-mounted slides and the complex was stored in the dark at 4°C for 4 weeks. After this time, the emulsion was developed and the tissue stained with toluidine blue. Quantification of autoradiograms was performed by grain counting. It has been demonstrated that with this technique grain densities are proportional to time of exposure and amount of radioactivity in the tissue (32).

RESULTS

Figure 1 shows a bright field micrograph as well as a dark field autoradiograph of a ligated (12 hr) rat sciatic nerve section incubated with [^{125}I]-α-BuTX. Note that the silver grains are clustered at both sides of the ligature. This is presumably due to toxin binding sites that have accumulated both proximally (toward spinal cord) and distally (toward periphery) to the ligature as a result of bi-directional axonal transport.

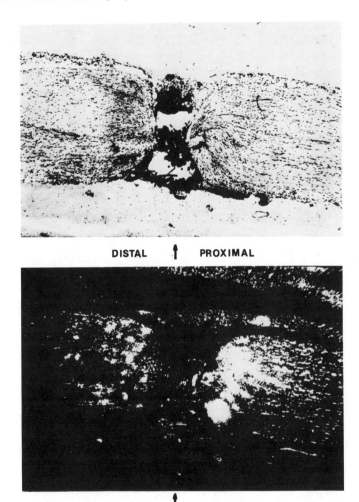

DISTAL ↑ **PROXIMAL**

Figure 1. Bright field micrograph and dark field autoradiograph of a ligated rat sciatic nerve section (12 hour accumulation) that was incubated with [^{125}I]-α-BuTX. Silver grains are clustered at both sides of the ligature (arrow). Proximal = toward spinal cord; Distal = toward periphery.

A toxin binding saturation analysis for both proximally and distally accumulating sites allowed for the estimation of dissociation constants (K_d's) for the ligand-binding site interaction, and of maximum binding site concentration (B_{max}) values. The data are summarized in Table 1-A. Binding saturation was established by incubating nerve sections in increasing amounts of ligand. Specific binding was determined by substracting non-displaceable grain densities (with 10^{-3}M curare) from total (i.e., no displacer present) values. To estimate K_d and B_{max} parameters it was necessary to approximate bound ligand values from silver grain densities. This was done by counting the grains produced by a known amount of radioactivity (in the form of bound $[^{125}I]$-α-BuTX) contained in a standard amount of tissue (homogenized brain frozen sections).

Furthermore, displacement of the toxin-binding site interaction was carried out by increasing the displacer concentration in logarithmic fashion in the preincubation and incubation steps (see Methods). This enabled us to determine inhibitory constants (K_i's) for the various displacers (Table 1-A). Data in Table 1-A represent averages for at least 3 animals (3 nerve sections per animal, each section counted in 3 different randomly selected regions at 0.23 mm distance from the ligature).

Results obtained indicate that: 1) proximally and distally accumulating sites have similar pharmacological properties; 2) the toxin-binding site interaction may be reversible; and 3) curare and nicotine are the most effective displacers, followed by decamethonium, which is only partially effective. Atropine only displaces at very high concentrations. In addition, B_{max} values indicate that after a 12 hour period 50% more sites have accumulated proximally than distally at the ligature.

DISCUSSION

It has been repeatedly demonstrated that treatment of mammalian neuromuscular preparations with cholinergic agents results in, not only postsynaptic, but also presynaptic effects. For example, facilitatory and depolarizing agents can cause spontaneous and/or stimulus-induced repetitive antidromic activity in motor fibers; an effect abolished by curare at sub-paralytic doses (8, 9, 14, 19, 27, 31). This action of cholinergic drugs has been explained on the basis of two hypotheses: the activation of a presynaptic cholinergic receptor, and a potassium effect on the nerve terminal.

The first hypothesis (rev. 5, 20, 27, 28) supports the view that cholinergic facilitatory drugs (i.e., edrophonium, physostigmine), and depolarizing (i.e., decamethonium, succinylcholine) and non-depolarizing (i.e., curare, galamine blocking agents act on a receptor that may be analogous to the postsynaptic acetylcholine receptor (4, 15).

An alternative explanation was proposed by Katz (17). He suggested that cholinomimetics repeatedly activate the receptors in the muscle fiber endplate and may significantly elevate the potassium concentration in the synaptic cleft. The potassium in turn can bring the nerve terminal to threshold, generating backfiring. For a time this theory was abandoned, mainly because it was shown that

Table 1

Summary of Pharmacological Observations

α-BuTX Binding Site Source	K_D(nM)(c)	B_{max}(fmol/mg. prot)(d)	Cur.	$K_I(\mu M)$(e) Nic.	C_{10}	Atr.
-A-						
Rat Sciatic Nerve						
Anterograde(a)	0.970	184	0.36	0.68	21.52	>200
Retrograde(b)	0.530	131	0.16	0.74	9.97	>200
-B-						
(From Barnard and Dolly – see ref. 2)						
Optic Lobe (chick)	0.130	---	0.90	0.30	340	---
Brain (rodent)	0.300	---	1.00	0.10	500	---
Muscle (chick)	0.001	---	0.20	1.40	0.02	---

-A- Pharmacological properties of ^{125}I-α-BuTX binding sites in ligated rat sciatic nerve. Data were obtained by quantification of autoradiograms. (a) Values obtained from toxin binding sites proximal to a ligature (12 hr accumulation); (b) values from distal sites; (c) K_d=dissociation constant of the toxin-binding site interaction; (d) B_{max}=maximum binding site concentration at a distance of 0.23 mm from the ligature after a 12 hr accumulation; and (e) K_i=IC$_{50}$/1+(L/K_d), IC$_{50}$ is the concentration of drug which displaces the toxin binding by 50%, **L** is the toxin concentration used in the experiments (2nM), and K_d is from (c) above. Cur. =curare, Nic.=nicotine, C_{10}=decamethonium, Atr.=atropine.

-B- Properties of putative cholinergic receptors from optic lobe, brain, and muscle, presented for comparison.

transversly cutting the muscle at both sides of the end-plate did
not permanently abolish retrograde activity (3, 25). However, this
has recently been contested by Hohlfeld and co-workers (4) who did
not detect antidromic action potentials after cutting the muscle,
in support of the potassium hypothesis.

The histological data do not clarify the controversy. Lentz,
et. al. (18), and Freedman and Lentz (11) have reported that a horseradish peroxidase-α-BuTX conjugate binds to axonal membranes both
at the motoneuron terminals and at the nodes of Ranvier. In contrast,
Jones and Salpeter (16) failed to show [^{125}I]-α-BuTX binding in
motor nerve terminals using high resolution autoradiography.

Since it is generally believed that in nerve cells only the
soma can synthesize proteins, any receptors destined for the nerve
terminal must be translocated to their functional sites via axonal
transport. We have previously reported that [^{125}I]-α-BuTX binding
sites undergo axonal flow in rat sciatic nerve (1). Ninkovic and
Hunt (22) have reported a similar finding. The binding sites accumulate both proximally and distally at a ligature in a linear, time
dependent fashion for the initial 8 hr after ligation (Millington
and Aizenman, unpubl. obs.). At later times, sites continue to
accumulate linearly at the proximal end only. This is possibly
due to lack of de novo synthesis distal to the ligature. Ninkovic
and Hunt (22) observed that maximum accumulation at the proximal
end of the ligature occurred at 18-24 hr, with subsequent decline
in grain densities at 48 and 72 hr after ligation.

Ninkovic and Hunt have argued against the possibility that the
toxin binding sites undergoing flow in the sciatic nerve are of
motoneuron origin; suggesting that the sites are being transported
in sensory afferents. This argument is based on existing physiological evidence on the action of acetylcholine and related drugs
on sensory nerves (see refs. in 22), and on these authors' observation of lack of any significant binding in the ventral horn of
the spinal cord when compared to the dorsal root ganglion. However,
since binding site densities in motoneuron cell bodies may be too
low to detect above background, putative cholinergic receptors in
motor fibers cannot be discounted; at least in part; until α-BuTX
binding is proven absent in ligated ventral roots or in mixed nerves
in which the sensory component has been allowed to degenerate.

The results of the present investigation provide basic pharmacological information regarding the nature of [^{125}I]-α-BuTX binding
to sites which are transported in peripheral nerves. Using information
compiled by Barnard and Dolly (2; see also 12, 23, 24), one can
compare the data in Table 1-A to pharmacological parameters obtained
with toxin binding studies using cholinergic receptors from various
tissues (Table 1-B). Based on dissociation and inhibitory constants,
the axonally transported toxin binding sites are more similar to

brain, when compared to muscle receptors. Thus, while muscle receptors bind [^{125}I]-α-BuTX almost irreversibly, rodent brain, chick optic lobe, and rat peripheral nerve axonally transported toxin binding sites may interact with the ligand reversibly. Furthermore, decamethonium is the most powerful inhibitor of the toxin-muscle receptor interaction, while it is not very effective or a very poor displacer at the neural sites.

If we presume that axonally transported [^{125}I]-α-BuTX binding sites correspond to receptors whose destination is the presynaptic membrane, then the data presented in this study may provide a pharmacological basis for differentiating pre- and postsynaptic sites of action of cholinergic drugs on the mammalian neuromuscular junction.

Acknowledgements. This word was funded in part by USA DAMD Contract 17-82-C-2128, NIH Awards MH 25951 and MH 00053, and NIEHS Training Grant ES-07094.

REFERENCES

1. Aizenman, E., Millington, W., Zarbin, M., Bierkamper, G. and Kuhar, M. (1983): Fed. Proc. 42:1147.
2. Barnard, E. and Dolly, J. (1982): T.I.N.S. 5:325-327.
3. Barstard, J. (1962): Experientia 18:579-580.
4. Blaber, L. (1970): J. Pharm. Exp. Ther. 175:664-672.
5. Bowman, W. (1980): Anaesth. Analg. 59:935-943.
6. Bowman, W. and Webb, S. (1976): Clin. Exp. Pharm. Phys. 3:545-555.
7. Chang, C. and Lee, C. (1963): Arch. Int. Pharmacodyn. 144:241-257.
8. Eccles, J., Katz, B. and Kuffler, S.W. (1942): J. Neurophysiol. 5:211-230.
9. Feng, T. and Li, T. (1941): Chin. J. Physiol. 16:37-50.
10. Fertuck, H. and Salpeter, M. (1976): J. Cell Biol. 69:144-158.
11. Freedman, S. and Lentz, T. (1980): J. Comp. Neurol. 193:179-185.
12. Fumagali, L., de Renzis, G. and Miani, N. (1976): J. Neurochem. 27:47-52.
13. Glavinovik, M. (1979): J. Physiol. 290:499-506.
14. Hohlfeld, R., Sterz, R. and Peper, K. (1981): Pflugers Arch. 391:213-218.
15. Hubbard, J., Schmidt, R. and Yokota, T. (1965): J. Physiol. 181:810-829.
16. Jones, S. and Salpeter, M. (1983): J. Neurosci. 3:326-331.
17. Katz, B. (1962): Proc. Roy. Soc. Lond. 155:455-477.
18. Lentz, T. Mazurkiewicz, J. and Rosenthal, J. (1977): Brain Res. 132:423-442.
19. Masland, R. and Wigton, R. (1940): J. Neurophysiol. 3:269-275.
20. Miyamoto, M.D. (1978): Pharm. Rev. 29:221-247.
21. Morley, B. and Kemp, G. (1981): Brain Res. Rev. 3:81-104.

22. Ninkovic, M. and Hunt, S. (1983): Brain Res. 272:57-69.
23. Oswald, R. and Freeman, J. (1981): Neurosci. 6:1-4.
24. Polz-Tejera, G., Hunt, S. and Schmidt, J. (1980): Brain Res. 195:223-230.
25. Randič, M. and Straughan, D. (1964): J. Physiol. 173:130-148.
26. Rang, H. (1981): Postgr. Med. J. 57(Suppl. 1):89-97.
27. Riker, W. and Okamoto, M. (1969): Ann. Rev. Pharmacol. 9: 173-208.
28. Standaert, F. (1982): Br. J. Anaesth. 54:131-144.
29. Wamsley, J., Zarbin, M. and Kuhar, M. (1981): Brain Res. 217: 155-161.
30. Webb, S. and Bowman, W. (1974): Clin. Exp. Pharm. Phys. 1: 123-134.
31. Werner, G. (1960): J. Neurophysiol. 23:171-187.
32. Young, W. and Kuhar, M. (1979): Brain Res. 179:255-270.
33. Young, W., Wamsley, J., Zarbin, M. and Kuhar, M. (1980): Science 210:76-78.
34. Zarbin, M., Wamsley, J. and Kuhar, M. (1982): J. Neurosci. 2:934-941.

NICOTINIC CHOLINERGIC RECEPTORS LABELED WITH [3H]ACETYLCHOLINE IN BRAIN: CHARACTERIZATION, LOCALIZATION AND IN VIVO REGULATION

R.D. Schwartz[1] and K.J. Kellar

Department of Pharmacology
Georgetown University Schools of Medicine and Dentistry
Washington, D.C.   20007   U.S.A.

[1]Current Address: Section on Molecular Pharmacology
Clinical Neuroscience Branch
National Institute of Mental Health
Building 10, Room 4N214
Bethesda, Maryland   20205   U.S.A.

INTRODUCTION

There is evidence from behavioral (21), electrophysiological (15) and pharmacological (23) studies that mammalian brain contains a type of nicotinic cholinergic receptor (nAChR). The utility of alpha-bungarotoxin (α-BTX) for labeling and measuring nAChR in electroplax, skeletal muscle, and certain peripheral neuronal tissues has encouraged similar investigations in the mammalian central nervous system. Although α-BTX binding sites in brain have been extensively characterized in vitro (for review, see 20), the relationship between these binding sites and nAChR is unclear since nicotinic cholinergic drugs are relatively weak competitors for the α-BTX binding sites (16, 24, 30) and the toxin does not appear to block cholinergic function in various neuronal tissues (6, 18).

[3H]Nicotine has been used to label a high affinity site in rat brain in vitro (1, 12, 22). This high affinity site appears to have the characteristics of nAChR (12, 22). We have recently used   [3H]acetylcholine ([3H]ACh) of high specific radioactivity to characterize binding sites in several rat brain areas (29). In the presence of atropine to block muscarinic cholinergic receptors, [3H]ACh binds with high affinity and selectivity to a single class of sites with characteristics of nAChR. This [3H]ACh binding site and the high affinity [3H]nicotine binding site appear to be similar

in several respects. Most notable is the finding that in both cases
nicotinic cholinergic agonists such as nicotine, cytisine and carbachol
have high affinity for the sites, while most nicotinic cholinergic
antagonists such as tubocurarine, hexamethonium and mecamylamine
have relatively low affinity for the sites (22, 29). In addition,
α-BTX does not compete very effectively for either [^{3}H]nicotine or
[^{3}H]ACh binding sites (22, 29).

[^{3}H]Acetylcholine appears to label the recognition site for
the endogenous ligand of the nAChR. We have used the [^{3}H]ACh binding
assay to investigate the pharmacology (29), chemical bond requirements
and modification (26) and in vivo regulation (25) of the nAChR recognition sites. In addition, lesion studies have indicated that the
[^{3}H]ACh binding sites are located on catecholamine and serotonin
(5-HT) terminals in certain brain areas of the rat (25), thus supporting the concept that the nAChR plays a functional role in regulating release of neurotransmitters (2) and hormones (8) in brain.

MATERIALS AND METHODS

Binding Assays. The [^{3}H]ACh binding assays in rat brain were carried
out using [^{3}H]ACh (80 Ci/mmol) synthesized by acetylation of [^{3}H]-
choline (80 Ci/mmol, New England Nuclear) as previously described
(29).

In **Vitro** Modifications. Cerebral cortical homogenates were incubated
with dithiothreitol (DTT), p-chloromercurobenzoic acid (PCMB) or
5,5-dithiobis (2-nitrobenzoic adic) (DTNB) for 30 min at 0°C (pH
8.5) as previously described (26). Incubations were terminated by
dilution with cold buffer and centrifugation at 49,000 g for 10
min, and the pellets were resuspended and either incubated with
another agent or washed twice more by centrifugation in preparation
for the [^{3}H]ACh binding assay.

Lesion Experiments. Rats were infused intraventricularly with 200
μg 6-hydroxydopamine (6-OHDA) or 5,7-dihyroxytryptamine (5,7-DHT,
which followed pretreatment with 25 mg/kg desmethylimipramine, i.p.) as
previously reported (28). The nucleus basalis magnocellularis (nBM)
was lesioned by infusion of 15 μg ibotenic acid into the nucleus
as previously described (28). Rats were sacrificed 7 or 10 days
after administration of 6-OHDA or 5,7-DHT, respectively, and 4 days
or 4 weeks after administration of ibotenic acid.

In the striatum, 6-OHDA decreased high affinity [^{3}H]dopamine
uptake by approximately 50% and 5,7-DHT decreased serotonin content
by a similar amount. In the hypothalamus, 6-OHDA and 5,7-DHT decreased
[^{3}H]norepinephrine ([^{3}H]NE) and [^{3}H]5-HT uptake, respectively, by
75-80%. In the cortex, the uptake of [^{3}H]NE and [^{3}H]5-HT was reduced
40-60% by 6-OHDA and 5,7-DHT, respectively. Ibotenate lesions of

the nBM reduced choline acetyltransferase activity in the cortex by 36-46%.

In **Vivo** Treatments. Diisopropylfluorophosphate (DFP) was injected s.c. for 10 days (1 mg/1kg on day 1 and 0.2-0.4 mg/kg on subsequent days). Nicotine-tartrate (equivalent to 1.0 mg/kg = 6 μmol/kg (-)nicotine) dissolved in water and adjusted to pH 7.0, was injected twice daily for 1, 5, 10 or 21 days. Control rats for both treatments received vehicle (water) injections. At sacrifice, 18 hrs after the last injection, the brains were removed, dissected, immediately frozen on dry ice and stored at -80°C until assayed for [³H]ACh binding sites.

In all experiments, treated and control tissues were assayed in parallel. Scatchard plots were based on a one-site model and fit by linear regression analysis. Differences between groups were analyzed statistically by Student's t test or Duncan's new multiple range test when more than 2 groups were compared.

RESULTS

General Characteristics of the [³H]ACh Binding Site in Brain. [³H]Acetylcholine binds rapidly, reversibly and with high affinity (K_D = 12-15 nM) to rat brain membranes (Figure 1 and ref. 29). The binding sites are distributed unevenly throughout the brain. The highest densities of binding sites are found in the thalamus, cerebral cortex, superior colliculus and striatum, while the hippocampus and pyriform cortex appear to contain a very low density of sites (29).

Nicotinic agonists have much higher affinity for the [³H]ACh binding sites than do antagonists (Table 1). In fact, with the exception of dihydro-β-erythroidine, nicotinic antagonists have very low affinities for the [³H]ACh binding site (micromolar range).

In **Vitro** Modification. Nicotinic receptor functions in electroplax and peripheral tissues are altered by treatments which disrupt tissue disulfide bonds (9). The [³H]ACh binding site in brain can also be modified by procedures which affect disulfide bonds (26). After reduction of disulfide bonds with DTT, the density of [³H]ACh binding sites was reduced in a dose dependent manner (Figure 2). This could be reversed by reoxidation of sulfhydryl groups with DTNB (Table 2). The reoxidation by DTNB could be prevented by complexing the reduced sulfhydryl groups with PCMB (Table 2).

While the reduction of disulfide bonds decreased the number of binding sites it did not alter the apparent affinities of nicotinic agonists or antagonists for the remaining sites, indicating that these remaining binding sites retain the native properties which determine the affinity of agonists and antagonists (26).

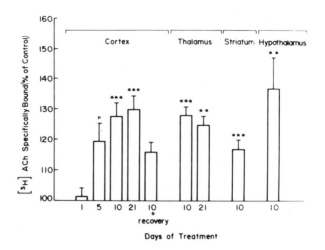

Figure 1. Scatchard plot of [^{3}H]ACh specific binding to cerebral cortex. Cortical homogenates were incubated with varying concentrations of [^{3}H]ACh (2.5 - 37 nM) for 40 min at 0°C as described previously (29). The apparent K_D = 12.6 nM and the B_{max} = 4.6 pmol/g tissue. Inset: saturation isotherm of [^{3}H]ACh specific binding (•--•). Non-specific binding (o--o) was determined in the presence of 100 μM carbachol.

Thus, the reversible decrease in the number of [^{3}H]ACh binding sites following treatment with DTT suggests that membrane disulfide bonds are required for the binding of acetylcholine to nAChR in brain (26).

Presynaptic Location of [^{3}H]ACh Binding Sites. Nicotinic receptors may be involved in the regulation of release of neurotransmitters such as NE, DA and 5-HT in various brain areas (for review see ref. 2). To determine if the [^{3}H]ACh binding sites are located on catecholamine or 5-HT axons, lesion studies were carried out.

[^{3}H]Acetylcholine binding in the striatum and hypothalamus was significantly decreased following either 6-OHDA or 5,7-DHT lesions (Table 3). No change in [^{3}H]ACh binding in the cortex or thalamus was observed following either lesion (Table 3). Scatchard plots of [^{3}H]ACh binding revealed that both lesions decreased the density of [^{3}H]ACh recognition sites without significantly altering the affinity of the remaining sites (28). In rats lesioned with ibotenate in the nBM there was no apparent change in cortical [^{3}H]ACh binding sites when measured either 4 days or 4 weeks after the lesion (Table 3). Thus, while [^{3}H]ACh binding sites appear to be located on catecholamine and 5-HT in the striatum and hypothalamus, we found no evidence for a presynaptic location on cholinergic axons which project from the nBM to the cerebral cortex.

Table 1

Inhibition of [³H]ACh Binding in Rat Cerebral Cortex
By Various Drugs

	K_I (nM)		
Cholinergic agonists			
cytisine	1.3	±	0.5
(-)nicotine	6.4	±	0.5
acetylcholine	7.6	±	1.4
carbachol	13.4	±	2.5
(+)nicotine	146	±	28.5
succinylcholine	6,700	±	1,900
piperidine	8,200	±	2,600
decamethonium	9,700	±	2,300
methacholine	84,000	±	4,400
choline	134,000	±	63,400
cotinine	175,000	±	16,400
Cholinergic antagonists			
dihydro-β-erythroidine	111.3	±	20.9
d-tubocurarine	28,500	±	4,400
gallamine	157,000	±	1,600
hexamethonium	182,000	±	37,000
atropine	476,000	±	189,000
pancuronium	556,000	±	144,000
trimethaphan	610,000	±	166,000
chorisondamine	686,000	±	187,000
mecamylamine	822,000	±	273,000
α-bungarotoxin (Sigma)	displaced <20% at 40 μM		
α-bungarotoxin (2.2 fraction)	displaced 50% at 40 μM		
Other drugs[1]			
hemicholinium-3	80,000	+	9,900
spiperone	260,000	±	42,500
4-aminopyridine	284,000	±	12,500
phencyclidine	>1,000,000		
diisopropyl fluorophosphate	>1,000,000		

[1]The following drugs displaced < 30% at 100 μM: thioridazine, piperazine, amitriptyline and diazepam.

Cortical homogenates were incubated with 10 nm [³H]ACh and 4-7 concentrations of competing drugs for 40 min at 0°C as described previously (29). The inhibitory constant (K_I) was derived from the IC_{50} by using the equation $K_I = IC_{50}/$ (1+([³H]ACh/K_D)). The K_I reported is the mean + S.E.M. obtained from at least 3 experiments. [Adapted from Schwartz et al. (29)]

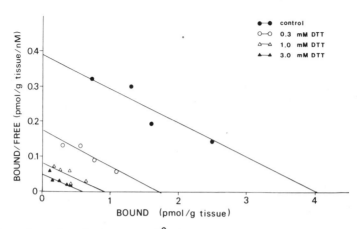

Figure 2. Scatchard plots of [^{3}H]ACh binding in DTT-treated cortex.
Cortical homogenates were incubated with DTT for 30 min at 0°C, washed
and assayed as described in Figure 1. The data are representative
of 2-7 experiments [from Schwartz and Kellar (26)].

Table 2

Effects of Disulfide Bond and Sulfhydryl Reagents
on [^{3}H]ACh Binding in Cerebral Cortex

Pretreatment	Percent of Control
control	100.0 ± 3.8
a. 1 mM DTT	24.4 ± 8.0*
b. 1 mM DTNB	90.0 ± 8.0
c. 1 mM DTT followed by 1 mM DTNB	90.6 ± 4.1
d. 1 mM PCMB	107.6 ± 3.7
e. 1 mM DTT followed by 1 mM PCMB	33.7 ± 1.2*
f. 1 mM DTT followed by 1 mM PCMB followed by 1 mM DTNB	71.4 ± 5.7**

Cortical homogenates were incubated with 1 mM DTT, DTNB or PCMB
and subsequently incubated with [^{3}H]ACh (10 nM) as described previously
(26). Control binding = 2.11 ± 0.08 pmol/g tissue. Data are expressed
as the mean ± S.E.M. for 4 experiments [from Schwartz and Kellar
(26)]. *p<0.001 compared to control; **p<0.01 compared to control
and to treatments c and e.

Table 3

[³H]ACh Binding in Brain Regions From Rats Lesioned
With 6-OHDA or 5,7-DHT Injected Intraventricularly or
With Ibotenic Acid Injected Into the nBM.

Brain area	Specific Binding (fmol/mg protein)		
	Control	6-OHDA	5,7-DHT
Cortex	38.4 ± 2.6 (4)	36.2 ± 1.9 (4)	36.0 ± 2.1 (4)
Thalamus	53.7 ± 1.2 (4)	56.6 ± 2.1 (4)	54.0 ± 2.1 (4)
Striatum	47.3 ± 2.2 (8)	33.0 ± 3.2* (6)	33.2 ± 1.7** (7)
Hypothalamus	41.5 ± 3.5 (5)	16.9 ± 1.1** (6)	23.6 ± 3.6** (6)

	Control	Ibotenic acid
Cortex 4 days	35.0 ± 1.5 (6)	33.1 ± 1.7 (6)
28 days	22.3 ± 1.6 (6)	22.6 ± 2.1 (6)

Binding assays were conducted as described previously (29) using 10 nM [³H]ACh for the 6-OHDA and 5,7-DHT experiments and using 11.6 and 9.2 nM [³H]ACh for the 4 day and 28 day ibotenic experiments, respectively. Values are the mean ± S.E.M. from number of animals indicated in parentheses. *p<0.01; **p<0.001 [from Schwartz et al. (28)].

In Vivo Regulation. The effects of a sustained increase in synaptic concentrations of ACh in vivo on [³H]ACh binding sites in brain were studied by treating rats with the cholinesterase inhibitor, DFP. After 10 days of treatment, [³H]ACh binding in the cerebral cortex, thalamus, striatum and hypothalamus was decreased by 20-38% (Table 4). In the cortex, the reduced binding was due to a decrease in the number of binding sites while the affinity of the sites was unchanged (25).

Table 4

Effect of Repeated DFP Administration (10 days)
on [³H]ACh Binding In Various Areas of Rat Brain

Brain area	Control	DFP
Cortex	2.2 ± 0.1 (13)	1.7 ± 0.1** (11)
Thalamus	3.2 ± 0.2 (9)	2.6 ± 0.2* (10)
Striatum	2.2 ± 0.1 (6)	1.5 ± 0.1** (5)
Hypothalamus	0.7 ± 0.1 (4)	0.4 ± 0.1* (4)

Brain homogenates were prepared from rats injected with DFP, as described in the Methods section. Homogenates were assayed for [³H]ACh binding (10 nM) as previously described (29). Cholinesterase activity was inhibited in the DFP-treated rats by 82.1 ± 1.8%. The data are expressed as the mean ± S.E.M. for the number of animals shown in parentheses. *p<0.05, **p<0.001 [from Schwartz and Kellar (25, 27)].

To determine the effects of repeated exposure to a direct acting nicotinic agonist on nicotinic sites labeled with [³H]ACh, binding was measured in several brain areas following repeated administration of nicotine. [³H]ACh binding was increased in all 4 brain areas tested (Figure 3). The time-course of the nicotine-induced increase in [³H]ACh binding was examined in the cerebral cortex. There was no change in binding either 1 hr after a single injection (data not shown) or after 1 day of treatment (Figure 3). However, following 5 days of nicotine administration [³H]ACh binding was increased by approximately 20% (Figure 3). This binding appeared to be increased maximally by 10 days of nicotine treatment and this increase (28-30%) was maintained following 21 days of treatment (Figure 3). When rats were sacrificed 7 nicotine-free days after a 10-day nicotine treatment period, [³H]ACh binding appeared to be returning to control values and was significantly decreased compared to the binding in rats sacrificed 1 day following 10 days of nicotine treatment (Figure 3). Following 10 days of nicotine administration [³H]ACh binding was also increased (15-30%) in the thalamus, striatum and hypothalamus (Figure 3). Scatchard analyses indicated that repeated nicotine

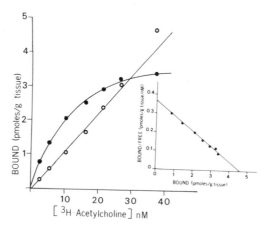

Figure 3. Effect of repeated nicotine administration for various times on [³H]ACh binding in different brain areas. Tissue homogenates were prepared from rats which were injected twice daily with nicotine (1 mg/kg s.c.), for the number of days indicated. [³H]ACh binding (10 nM) was assayed as described previously (29). Data are expressed as the mean $\pm$ S.E.M. (N = 6-16). *p<0.05, **p<0.01, ***p<0.001 [adapted from Schwartz and Kellar (27)].

administration increased the density of [³H]ACh binding sites (30-40%) without altering the affinity of the sites (Table 5).

The selectivity of the effect of repeated nicotine adminis-tration was examined by measuring muscarinic receptor binding sites using [³H]quinuclidinyl benzylate ([³H]QNB). Following nicotine administration for 21 days, no changes in [³H]QNB binding in the cerebral cortex, thalamus or striatum were observed (25, 27).

DISCUSSION

The [³H]ACh binding site described here has properties of a nicotinic cholinergic receptor. However, the brain receptor may be different from peripheral nicotinic receptors in ganglia and striated muscle, and different from the brain site that binds α-BTX. Evidence for this is found in studies of the pharmacologic charac-teristics and the brain regional distribution of the [³H]ACh binding site (3, 29). The lack of correlation between the distribution of [³H]ACh and α-BTX sites indicates that these two binding sites are not located on the same molecule (3, 29). Nicotinic antagonists, as compared to agonists, have relatively low affinities for the [³H]ACh binding site (29). This has also been observed in the case of [³H]nicotine binding (12, 22). The very low affinities of antago-

Table 5

Effect of Repeated Nicotine Administration on
Density (B_{max}) and Affinity (K_D) of [^{3}H]ACh Binding Sites
In Various Areas of Rat Brain

Brain area	B_{max} (pmol/g tissue)	K_D
Cortex		
Control	4.1 ± 0.3	8.3 ± 0.9
Nicotine	5.3 ± 0.1**	8.0 ± 0.4
(21 days)		
Thalamus		
Control	6.9 ± 0.6	9.1 ± 1.0
Nicotine	9.0 ± 0.3*	10.9 ± 1.4
(10 days)		
Striatum		
Control	4.6 ± 0.3	10.6 ± 2.0
Nicotine	6.4 ± 0.4**	11.1 ± 1.8
(10 days)		

Brain homogenates were prepared from rats which were injected twice daily with 1 mg/kg nicotine as described in the Methods section. The homogenates were incubated with [^{3}H]ACh as described in Figure 1. The data are expressed as the mean ± S.E.M. for 4 experiments. *$p<0.05$, **$p<0.01$ [from Schwartz and Kellar (27)].

nists may be due to an agonist-induced shift of the [^{3}H]ACh binding site to a high affinity, agonist-selective or desensitized state (26, 29), but as yet there is no evidence for this.

The decrease in density of [^{3}H]ACh binding sites in striatum and hypothalamus after 6-OHDA and 5,7-DHT lesions indicates the existence of nAChR on catecholamine and 5-HT axon terminals. Since not all the DA or 5-HT terminals in the striatum were destroyed by the chemical lesions (see Methods), it can be concluded that at least 30% of the [^{3}H]ACh binding sites in the striatum are located on DA terminals and 30% on 5-HT terminals. However, in the hypothalamus, where neuronal damage was more complete, it appears that virtually all of the [^{3}H]ACh binding sites may be presynaptic on catecholamine or 5-HT terminals.

It has been well established that presynaptic receptors may play an important role in the modulation of neurotransmitter release (9). Nicotinic cholinergic modulation of neurotransmitter release has been studied in several brain areas (for review see ref. 2). Thus, the presence of nAChR labeled by [³H]ACh on catecholamine and 5-HT nerve terminals in specific areas of rat brain suggests that regulation of neurotransmitter release may be a functional property of these sites. The implications of these findings may contribute to a better understanding of presynaptic nicotinic cholinergic mechanisms involved in striatal neurotransmission as discussed by McGeer et al. (13), and in the secretion of several hypothalamic hormones (8, 17, 31).

Neurotransmitter receptors may undergo apparent changes in number in response to long term changes in stimulation in vivo. This up- or down-regulation of receptor number is usually reciprocal to the change in stimulation and is consistent with compensatory adaptation (5). The nAChR labeled by [³H]ACh in brain can be regulated in vivo in such a manner, as evidenced by the down-regulation which follows chronic cholinesterase inhibition (25, 27). Similar results have been obtained for nicotinic receptors labeled with [³H]nicotine (4) and for muscarinic receptors labeled with [³H]QNB in brain (7, 14). This indicates that [³H]ACh binding sites are responsive to the synaptic concentration of ACh in vivo, and suggests that the sites are innervated by cholinergic neurons. In addition, it indicates that the regulation of these [³H]ACh recognition sites is similar to that of other neurotransmitter and hormone receptor binding sites which down-regulate in response to increased agonist concentration.

In contrast to the effects of DFP, repeated administration of nicotine increases the number of [³H]ACh recognition sites in several areas of rat brain (25, 27). Although nicotine is a cholinergic agonist, it is also considered a functional antagonist because of its capacity to stimulate and then to block (via depolarization blockade and/or desensitization) cholinergic function (32). Nicotine has a short half-life in vivo, and we injected it only twice each day. The increase in [³H]ACh recognition sites in each of the brain regions examined after this schedule of treatments could indicate that a functional antagonism of the nAChR in these brain areas outlasts the presence of the drug. This phenomenon has also been discussed by Marks et al. (12) who observed that chronic infusion of nicotine also increases [³H]nicotine binding sites in several brain areas of mice. This is consistent with the concept that [3H]ACh and [³H]-nicotine bind to the same site in brain. In fact, a recent comparison of the distribution of these binding sites throughout the rat brain using autoradiographic techniques indicates that the distributions of these two sites are virtually identical (3). Thus, up-regulation of nAChR labeled by two different ligands has been observed following repeated exposure to nicotine in two species. The up-regulation, possibly resulting from a functional antagonism of the nAChR, might be involved in mechanisms underlying nicotine addiction.

CONCLUSIONS

The [^{3}H]ACh binding site in brain described here has characteristics of a nAChR. Initial studies of this site have resulted
in several interesting observations:

1. The pharmacology of the site suggests that the brain
   nAChR may be different from either of the peripheral
   nAChR found in skeletal muscle and autonomic ganglia.

2. In at least two areas of the brain, the striatum and
   the hypothalamus, the [^{3}H]ACh binding site is located
   on catecholamine and 5-HT axons, suggesting a role in
   the release of neurotransmitters.

3. The down-regulation of [^{3}H]ACh binding sites following
   repeated exposure to DFP (and thus to increased synaptic
   concentrations of acetylcholine) suggests that these
   sites are innervated by cholinergic neurons.

4. The nAChR recognition site in brain up-regulates in
   response to repeated nicotine exposure. This suggests
   that nicotine acts as a functional antagonist in brain,
   and this could be related to nicotine addiction.

These observations may provide a basis for further investigation
of the functional roles of nAChR in the central nervous system.
[^{3}H]ACh should prove to be a useful ligand for probing these receptors.

Acknowledgements. This research was supported by the Scottish Rite
Schizophrenia Research Program, NMJ, USA and the US Army Medical
Research and Development Command. RDS was supported by NIH Pre-
Doctoral Training Grant GM07443, and by NIDA National Research Service
Award DA05229.

REFERENCES

1. Abood, L.G., Reynolds, D.T. and Bidlack, J.M. (1980):   Life
   Sci. 27:1307-1314.
2. Balfour, D.J.K. (1982): Pharmacol. Ther. 16:269-282.
3. Clarke, P.B.S., Schwartz, R.D., Paul, S.M., Pert, C.B. and Pert,
   A. (1985): J. Neurosci. 5:1307-1315.
4. Costa, L.G. and Murphy, S.D. (1983): J. Pharmacol. Exp. Ther.
   226:392-397.
5. Creese, I. and Sibley, D.R. (1981): Ann. Rev. Pharmacol. Toxicol. 21:357-391.
6. Duggan, A.W., Hall, J.G. and Lee, C.Y. (1976): Brain Res. 107:
   166-170.

7. Ehlert, F.J., Kokka, N. and Fairhurst, Á.S. (1980): Mol.Phar-macol. 17:24-30.
8. Hillhouse, E.W., Burden, J. and Jones, M.T. (1975): Neuroendo-crin. 17:1-11.
9. Karlin, A. (1973): Fed. Proc. 32:1847-1853.
10. Langer, S.Z. (1977): Br. J. Pharmacol. 60:481-497.
11. Marks, M.J., Burch, J.B. and Collins, A.C. (1983): J. Phar-macol. Exp. Ther. 226:291-302.
12. Marks, M.J. and Collins, A.C. (1982): Mol. Pharmacol. 23:554-564.
13. McGeer, P.L., McGeer, E.G. and Innanen, V.T. (1979): Adv. Neur-ol. 24:225-233.
14. McKinney, M. and Coyle, J.T. (1982): J. Neurosci. 2:97-105.
15. McLennan, H. and Hicks, T.P. (1978): Neuropharmacol. 17:329-334.
16. McQuarrie, C., Salvaterra, P.M., DeBlas, A., Routes, J. and Mahler, H.R. (1976): J. Biol. Chem. 251:6335-6339.
17. Meites, J. (1981): In Brain Neurotransmitters and Receptors in Aging and Age-Related Disorders, Aging, Vol. 17 (ed)S.J. Enna, Raven Press, New York, pp. 107-115.
18. Misgeld, U., Weiler, M.H. and Bak, I.J. (1980): Brain Res. 39:401-409.
19. Morley, B.J., Lorden, J.F., Brown, G.B., Kemp, G.E. and Bradley, R.J. (1977): Brain Res. 134:161-166.
20. Oswald, R.E. and Freeman, J.A. (1981): Neurosci. 6:1-14.
21. Pratt, J.A., Stolerman, I.P., Garcha, H.S., Giardini, V. and Feyerabend, C. (1983): Psychopharmacol. 81:54-60.
22. Romano, C. and Goldstein, A. (1980): Science 210:647-649.
23. Sakurai, Y., Takano, Y., Kohjimoto, Y., Honda, K. and Kamiya, H. (1982): Brain Res. 242:99-106.
24. Schmidt, J. (1977): Mol. Pharmacol. 13:283-290.
25. Schwartz, R.D. and Kellar, K.J. (1983): Science 220:213-216.
26. Schwartz, R.D. and Kellar, K.J. (1983): Mol. Pharmacol. 24:387-391.
27. Schwartz, R.D. and Kellar, K.J. (1985): J. Neurochem. In press.
28. Schwartz, R.D., Lehmann, J. and Kellar, K.J. (1984): J. Neuro-chem. 42:1495-1498.
29. Schwartz, R.D., McGee, R., Jr., and Kellar, K.J. (1982): Mol. Pharmacol. 22:56-62.
30. Segal, M., Dudai, Y. and Amsterdam, A. (1978): Brain Res. 148:105-119.
31. Sladek, C.D. and Joynt, R.J. (1979): Endocrinol. 104:659-663.
32. Thesleff, S. (1955): Physiol. Scandinav. 34:218-231.

RESTING RELEASE OF ACETYLCHOLINE AT THE MOTOR ENDPLATE

P.C. Molenaar[1] and R.L. Polak[2]

[1]Department of Pharmacology of the University of Leiden
Sylvius Laboratories
Wassenaarsweg 72
2333 AL Leiden, the Netherlands

[2]Medical Biological Laboratory/TNO
P.O. Box 45
2280 AA Rijswijk, the Netherlands

INTRODUCTION

When skeletal muscle is incubated after inactivation of the cholinesterase, significant amounts of acetylcholine (ACh) diffuse from the preparation into the incubation medium even when the nerve is not stimulated. It is now well established that only a few percent of the ACh released under resting conditions ("resting release") can be accounted for by the spontaneous discharge of ACh quanta responsible for the miniature endplate potentials (min.e.p.ps) (2, 11, 15, 18, 22). Apparently, the ACh from resting muscle is released in a predominantly non-quantal way. The origin, mechanism and physiological significance of non-quantal ACh release in skeletal muscle are still uncertain. The present paper reviews our recent work on resting release of ACh in frog, rat and mouse skeletal muscle (8, 11, 13, 17, 18).

WHAT IS THE ORIGIN OF RESTING RELEASE IN SKELETAL MUSCLE?

As shown in Fig. 1, both the total ACh content and the resting release of ACh in the rat diaphragm are decreased by about 80% in the period of 16-18 h after section of the phrenic nerve near its entrance into the muscle. At this time the motor nerve terminals - but not the axons - are known to have degenerated (14) and it is therefore likely that about 80% of the total ACh content of the diaphragm as well as of the resting release of ACh is located in an originates from the motor nerve terminals. Fig. 1 shows in addition

481

that at a later stage after denervation - the motor axons degenerate within 24-48 h (14) - neither the ACh content nor the resting release decrease further. This suggests that the <u>axons</u> contain and release only little ACh. Apparently the remaining 20% is contained in and released from the <u>muscle</u> <u>fibers</u>.

In similar experiments on frog sartorius muscle, in which the motor nerve terminals degenerate within 6-8 days (1), we found that they contain about 70% of the total amount of ACh in the muscle (10) and that they are the source of about half the resting release of ACh, the remaining part originating from the muscle fibers (11).

IS NON-QUANTAL ACh RELEASE INCREASED BY DEPOLARIZATION OF THE PRESYNAPTIC MEMBRANE?

The fact that the resting release of ACh is mainly non-quantal seems to imply that it originates directly from the cytosol of the motor nerve terminal, where it probably is synthesized (3, 4). If the ACh diffuses outward along its electrochemical gradient through

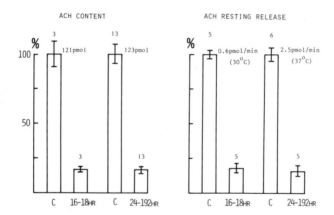

Figure 1. **Effect of denervation on ACh content and resting release of ACh in the rat diaphragm.** Muscles were isolated 16-18 h or 24-192 h after transthoracal section of the left phrenic nerve from male 5-6 week old Wistar rats. Contralateral hemidiaphragms served as controls. The muscles were either extracted immediately for determination of the ACh content or incubated in Ringer solution at 30°C or 37°C for measurement of the ACh release after cholinesterase inhibition by Soman. ACh was measured by mass fragmentography. Means ± S.E.; the figures above the bars give the number of observations. The data have been taken from a study by Miledi et al. (12), except the values at 16-18 h with their controls (Molenaar and Polak, unpublished).

pores in the presynaptic membrane, one would expect that non-quantal release of ACh^+ ions increases when the motor nerve terminal is depolarized. In fact, Vyas and Marchbanks (23) demonstrated that the efflux of positive ions from brain synaptosomes incubated in the absence of Ca^{2+} ions is enhanced by depolarization, and they postulated that this is also true for ACh. However, Vizi and Vyskočil (22) did not find an increase of non-quantal ACh release from the mouse diaphragm incubated in a medium without Ca^{2+} ions when the motor nerve terminals were depolarized by 14 mM KCl. Moreover, Katz and Miledi (7) failed to see even a trace of a non-quantal endplate response in mammalian and amphibian muscle when the motor nerve was stimulated in Ringer lacking Ca^{2+} ions to prevent the release of ACh quanta. This implies that non-quantal ACh release is not due to electrochemical diffusion across a leaky presynaptic membrane.

A small fraction of the min.e.p.ps at the motor junctions consists of potentials with abnormal time course and great amplitude (giant min.e.p.ps). The origin of giant min.e.p.ps is uncertain but the phenomenon seems to be caused by the release of large packets of ACh. In search of a possible different mechanism of resting release we wondered whether it could be caused by the release of giant ACh "quanta." Since 4-aminoquinoline strongly enhances the frequency of giant min.e.p.ps (16), we tested whether this drug enhanced the resting release of ACh from the rat extensor digitorum longus muscle and from the rat diaphragm. However, since we found that 4-aminoquinoline did not cause an increase of the resting release of ACh (17) this possibility was rejected. Apparently, the ACh release associated with the increase of giant min.e.p.ps was below the level of chemical detection.

We (18) and others (5) have observed that C. botulinum A toxin decreased, but did not block, non-quantal ACh release in rat muscle (but see ref. 19). This effect of botulinum toxin and a recent finding that both evoked and resting release of ACh were decreased in diaphragms from mice injected with IgG from patients suffering from the Eaton-Lambert myasthenic syndrome (8), suggested that Ca^{2+} ions may be involved in non-quantal ACh release. This was tested in rat and mouse diaphragms. The results from the experiments with mice, illustrated in Fig. 2, show that the resting release of ACh was depressed when the normal Ringer solution was replaced by Ringer without added $CaCl_2$ and with an increased $MgCl_2$ concentration. A similar effect was seen in the rat diaphragm (not illustrated).

Because of the effect of Ca^{2+} on resting release it was of interest to know whether non-quantal release of ACh is increased under conditions of increased Ca^{2+} influx, viz. during depolarization of the nerve terminals. The following results (for technical reasons obtained on frog muscle at $4^{\circ}C$) suggest that this is not the case. As shown in Fig. 3 depolarization of the motor nerve terminals by

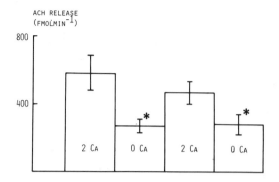

Figure 2. **Effect of Ca^{2+} ions on the resting release of ACh in the
mouse diaphragm.** The hemidiaphragms were incubated at 22°C in successive 30 min periods in medium whose composition alternated between
that of normal Ringer, containing 2 mM $CaCl_2$ and 1 mM $MgCl_2$ (2 Ca),
and of Ringer without added $CaCl_2$ containing 2 mM $MgCl_2$ (0 Ca).
The cholinesterase was inactivated by Soman. Means $\pm$ S.E. from 4
observations. *Significantly different ($P_2 < 0.05$, paired t-test)
from release in preceding 30 min period. Data taken from a study
by Lang et al. (8).

K^+ ions led to an increase of both min.e.p.c. frequency and chemically
detectable ACh release. The relation between log KCl concentration
and log min.e.p.c. frequency (slope 6.2) was similar to that between
log KCl concentration and log ACh release (slope 6.6). This implies
that non-quantal ACh release either did not increase at all or increased as much as quantal release. However, the second possibility
was difficult to reconcile with the results presented in Table 1:
when the KCl concentration was raised from 2 to 12 mM, total ACh
release (quantal + non-quantal) did not increase significantly,
whereas quantal release increased more than 25-fold at 10 mM KCl.
(The ACh release associated with min.e.p.cs was still below detection.) Consequently it appears unlikely that, in the experiments
of Fig. 3, KCl in concentrations between 2 and 20 mM induced a significant non-quantal release of ACh. In this context it is perhaps
relevant that the ACh contained in muscle fibers from frog and rat
is not liberated by depolarization with 50 mM KCl (11-13). Of course
the present observations do not exclude the possibility that the
much larger depolarization caused by the nerve action potential
releases some non-quantal ACh in a normal, Ca^{2+} containing, Ringer.

IS A CARRIER INVOLVED IN NON-QUANTAL ACh RELEASE?

Katz and Miledi (7) suggested that this may be so. It is well
known that Ca^{2+} independent ACh release is strongly enhanced by
inhibition of the membrane Na^+-K^+ ATPase and greatly reduced by
activation of the enzyme (20-22). The paper by Vizi and Vyskočil

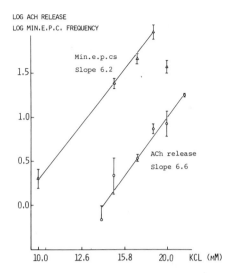

Figure 3. Effect of KCl on ACh release (circle symbols) and min.
e.p.cs (triangle symbols) in frog sartorius muscle incubated at 5°C.
Cholinesterase was inactivated by diethyl-dimethylpyrophosphonate
(DEPP). Double ^{10}log plots of KCl concentration in mM, on the abscissa; min.e.p.c. frequency in sec^{-1} and ACh release in fmol h^{-1}
per endplate, on the ordinate. Means $^{+}_{-}$ S.E. The regression lines
were calculated from observations obtained form 8 to 17 muscles
(ACh) and 10-33 endplates (min.e.p.cs). The data at KCl concentrations above 20 mM were not used for the lines since at these concentrations both min.e.p.c. frequency and ACh release were not well
sustained in the time necessary to sample sufficient endplates and
collect measurable amounts of ACh. With 15-18.5 mM KCl min. e.p.c.
frequency and ACh release gradually increased during the first 30
min, and thereafter became stable for at least 60 min. Data taken
from a study by Miledi et al. (13).

(22) demonstrates that the ouabain effect is likely to be caused
through the inhibition of the ATPase proper and certainly not due
to depolarization of the motor nerve terminals which results from
ATPase inhibition after some time, since in the presence of Ca^{2+}
ions the ouabain-induced increase of total ACh release occurred at
a time before the frequency of min.e.p.ps began to rise. However,
the function of ATPase in non-quantal ACh releaes is not well understood. It is clear that ATPase itself is not responsible for the
transport of ACh out of the motor nerve terminal, since its activation causes a decrease rather than an increase of the release of
ACh.

Table 1

ACh Release and Min.e.p.c. Frequency at 2, 10 and 12 mM KCl

KCl mM	Min.e.p.c. Frequency sec^{-1}	KCl mM	ACh Release $fmol\ min^{-1}$
2	0.13 ± 0.015 (36)	2	15 ± 3.5 (18)
10	3.2 ± 0.83 (16)[a]	12	18 ± 5.2 (10)

ACh release and min.e.p.c. frequency were measured in frog sartorius muscles at 5°C. Means $\pm$ S.E. with number of observations in parentheses (min.e.p.cs: number of endplates; ACh release: number of muscles). [a]3.2 min.e.p.cs sec^{-1} correspond to only 3 fmol min^{-1} in the whole muscle (assuming 10^4 ACh molecules per quantum and 10^3 endplates per muscle). Data taken from a study by Miledi et al. (13).

We considered the possibility that the high affinity uptake carrier for choline might be responsible for non-quantal ACh release. The idea was that under resting conditions the carrier turns at a low, basal rate exchanging choline which is transported into the terminal, against other cations, including ACh, which are transported outward. If this idea is true one would expect that hemicholinium-3, a potent inhibitor of the high affinity uptake of choline, would inhibit the resting release of ACh. However, this compound had no effect on the output of ACh from frog and rat skeletal muscle under normal resting conditions (Molenaar and Polak, unpublished). Consequently, the choline carrier is normally not invoved in non-quantal ACh release.

POSSIBLE PHYSIOLOGICAL SIGNIFICANCE OF NON-QUANTAL RELEASE OF ACh

The ACh which is released non-quantally has no effect on the postsynaptic membrane unless the cholinesterase is inactivated (6, 24). Apparently, ACh released in dilute form cannot penetrate through the barrier of cholinesterase present in the synaptic cleft (see ref. 9). Consequently it seems very unlikely that the non-quantal resting release of ACh serves an important physiological function, such as for instance a trophic effect on the muscle fibers. It appears more likely that such a task is assigned to the quanta of concentrated ACh, which give rise to spontaneous min.e.p.ps.

One wonders, then, whether it is worthwhile to put much effort into studying non-quantal ACh release if it has no significant function. However, the phenomenon cannot be ignored in studies of the synthesis, storage and release of ACh, simply because its share in total turnover of ACh is so large, especially in mammalian muscle.

Acknowledgements. We thank Mrs. J.W.M. Tas and Mr. A. van der Laaken for excellent technical assistance. Financial support by FUNGO/ZWO and the Wellcome Trust is gratefully acknowledged.

REFERENCES

1. Birks, R., Katz, B. and Miledi, R. (1960): J. Physiol. London 150:145-168.
2. Fletcher, P. and Forrester, T. (1975): J. Physiol. London 251: 131-144.
3. Fonnum, F. (1967): Biochem. J. 103:262-270.
4. Fonnum, F. (1968): Biochem. J. 109:389-398.
5. Gundersen, C.B. and Jenden, D.J. (1983): J. Pharmac. Exp. Ther. 224:265-268.
6. Katz, B. and Miledi, R. (1977): Proc. R. Soc. Lond. B196:59-72.
7. Katz, B. and Miledi, R. (1981): Proc. R. Soc. Lond. B212:131-137.
8. Lang, B., Molenaar, P.C., Newsom-Davis, J. and Vincent, A. (1984): J. Neurochem. 42:658-662.
9. Matthews-Bellinger, J. and Salpeter, M.M. (1978): J. Physiol. London 279:179-213.
10. Miledi, R., Molenaar, P.C. and Polak, R.L. (1977): Proc. R. Soc. Lond. B197:285-297.
11. Miledi, R., Molenaar, P.C. and Polak, R.L. (1981): In Cholinergic Mechanisms (eds) G. Pepeu and H. Ladinsky, New York, Plenum Press., pp. 205-214.
12. Miledi, R., Molenaar, P.C. and Polak, R.L. (1982): Proc. R. Soc. Lond. B214:153-168.
13. Miledi, R., Molenaar, P.C. and Polak, R.L. (1983): J. Physiol. London 334:245-254.
14. Miledi, R. and Slater, C.R. (1970): J. Physiol. London 207:507-528.
15. Mitchell, J.F. and Silver, A. (1963): J. Physiol. London 165: 117-129.
16. Molgó, J. and Thesleff, S. (1982): Proc. R. Soc. Lond. B214: 229-247.
17. Molgó, J., Gomez, S., Polak, R.L. and Thesleff, S. (1982): Acta Physiol. Scand. 115:201-207.
18. Polak, R.L., Sellin, L.C. and Thesleff, S. (1981): J. Physiol. London 319:253-259.
19. Stanley, E.F. and Drachman, D.B. (1983): Brain Res. 261:172-175.
20. Vizi, E.S. (1972): J. Physiol. Lond. 226:95-117.
21. Vizi, E.S. (1978): Neuroscience 3:367-384.
22. Vizi, E.S. and Vyskočil, F. (1979): J. Physiol. London 286:1-14.
23. Vyas, S. and Marchbanks, R.M. (1981): J. Neurochem. 37:1467-1474.
24. Vyskočil, F. and Illès, P. (1978): Physiologia bohemoslov. 27:449-455.

EVOKED RELEASE OF ACETYLCHOLINE AT THE MOTOR ENDPLATE

R. Miledi[1], P.C. Molenaar[2] and R.L. Polak[3]

[1]Department of Biophysics
University College London
Gower Street
London WC1E 6BT   U.K.

[2]Department of Pharmacology of the University of Leiden
Sylvius Laboratories
Wassenaarseweg 72
2333 AL Leiden, the Netherlands

[3]Medical Biological Laboratory/TNO
P.O. Box 45
2280 AA Rijswijk, the Netherlands

INTRODUCTION

According to the hypothesis of vesicular release of acetylcholine (ACh) neuromuscular transmission is caused by the simultaneous exocytosis of a large number of synaptic vesicles from the nerve terminals upon the arrival of the action potential (2).   At present there is general agreement that nerve stimulation causes a synchronized discharge of ACh packages (quanta), but whether these quanta derive from vesicles is still uncertain.   In fact, this has been challenged by results from biochemical experiments, notably those on the electric organ of <u>Torpedo</u> (11, 27).

We undertook a combined electrophysiological and biochemical study on ACh in the neuromuscular junction with the aim, among other things, to obtain more insight regarding the question of vesicular ACh release.   The present paper briefly reviews our results bearing on this matter.

CHOICE OF PREPARATION

We used the sartorius muscle of the frog, because this preparation is very suitable for electrophysiological experiments, and is sufficiently rich in ACh to allow chemical determinations of the transmitter (19). In the course of the study it became apparent that our choice had been fortunate because the experiments would have been very difficult to perform on mammalian muscle. Among the important advantages of frog muscle one can point to the low non-quantal resting release of ACh, and the sensitivity of the motor endplate to La^{3+} ions.

1. **Does Stimulation Release Exclusively ACh Quanta?** This is an important question because the ACh output from resting muscle is mainly non-quantal, its rate being about 20 times greater than the ACh release associated with spontaneously occurring miniature endplate potentials (min.e.p.ps) (5, 8, 20-22, 25). On the other hand, the non-quantal resting release of ACh does not lead to electrical effects at the postsynaptic membrane, unless the cholinesterase (AChE) is inhibited, in which case it gives rise to a minute depolarization (12, 30). If depolarization of motor nerve terminals were to cause some increase of non-quantal release of ACh, in addition to that of quantal release, this could have serious consequences for the biochemical analysis of the ACh stores, even if such an effect would be unimportant from an electrophysiological point of view.

In the frog sartorius about 50% of the non-quantal resting release of ACh originates from the nerve terminals and 50% from the muscle fibers, as demonstrated in denervation experiments (22) (muscle fibers have the capacity to synthesize and store some ACh (19, 22, 26, 28)). We found that the release of ACh from denervated muscle fibers did not increase upon direct depolarization by 50 mM KCl (22). Other experiments, on innervated preparations, showed that moderate depolarization of the nerve terminals with 10 mM KCl failed to induce non-quantal release, whereas it caused a more than 10-fold increase of the rate of quantal discharge, an increase which by itself was not great enough for chemical detection (24). All these data taken together strongly suggest that non-quantal ACh release remains constant, and predict that there should be a linear correlation between the frequency of miniature endplate currents (min.e.p.cs) and the amount of ACh collected in the organ bath, under conditions of evoked transmitter release. The results from experiments in which transmitter release was stimulated by K^+ or by hypertonic NaCl medium, shown in Fig. 1, indicate that this was indeed the case.

A difficulty may arise if the nerve terminal were to release substances, other than ACh, causing an electric effect at the end-plate. However, in the frog sartorius, electrical stimulation of the motor nerve does not lead to a postsynaptic response when quantal ACh release is prevented by omission of Ca^{2+} ions (13), or when the effect of ACh is blocked by α-bungarotoxin (14). Only under special conditions, i.e., hyperpolarization of the muscle fiber, a small endplate current, up to 1 nA, may be observed; probably caused

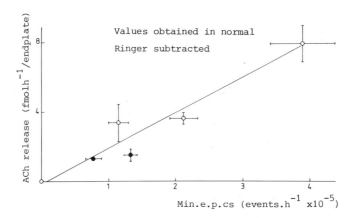

Figure 1. Correlation between chemically detectable ACh release (ordinate) and min.e.p.cs (abscissa) in DEPP-treated frog sartorius muscle. Temperature was 5°C. The frequency of min.e.p.cs was increased by KCl (15, 17, and 18.5 mM: open symbols) or by hypertonic NaCl (216 and 241 mM: filled symbols). Means $\pm$ S.E. (From ref. 24, with permission of the J. of Physiol.)

by the depolarization-induced release of K^+ ions from the nerve terminal (14).

2. Is Evoked ACh Release Abolished After Depletion of Synaptic Vesicles by La^{3+} Ions? Lanthanum was an important tool in the present study because it causes the virtual disappearance of min. e.p.ps and the irreversible loss of synaptic vesicles, without causing other conspicuous alterations of the ultrastructure of the motor nerve terminals (9). We found that upon addition of La^{3+} ions there was a great, but transient, outburst of min.e.p.ps and chemically detectable ACh which lasted for about one hour (20, 21, 24). Interestingly, the ACh content, after a transient decrease during the first hour, "recovered" to control values at a time when the vesicles were practially totally depleted (21, 24). We concluded that the synthesis and storage of ACh after La^{3+} treatment took place in the cytoplasm of the nerve terminals, among other things because the phenomenon did not occur in denervated muscles (21). [The increase of cytoplasmic ACh after La^{3+} differs from the permanent depletion of transmitter caused by black widow spider venom, a toxin also causing vesicle depletion (1, 7).] Consequently, the fact that min.e.p.ps disappear after exposure of the muscle to La^{3+} cannot be explained by shortage of transmitter in the nerve terminals.

Keeping this in mind, we tried to evoke transmitter release after La^{3+} by direct depolarization of the nerve terminals with K^+ ions, or by use of the calcium ionophore A 23187, in order to bypass Ca^{2+} channels which perhaps could be inactivated at a later stage

of the incubation period (in these experiments La^{3+} was washed away
after 15 min, but its effect on ACh in the muscle continued in the
same way as in the continuous presence of La^{3+}). The results, illustrated in Fig. 2, show that 50 mM KCl or A 23187 induced ACh release
in normal untreated muscles, but had no statistically significant
effect in muscles about 4 h after La^{3+} treatment, in which a second
addition of La^{3+} was also ineffective. However, when 50 mM KCl
was given in the period between 45 and 60 min after the first addition
of La^{3+}, i.e. at a time when min.e.p.ps were still observed at a
high frequency (21), there still was an increase of ACh release
(from 11 $\pm$ 2.2, n = 9, to 19 $\pm$ 3.0 pmol, n = 6, P_2 = 0.05; results
not illustrated in Fig. 2).

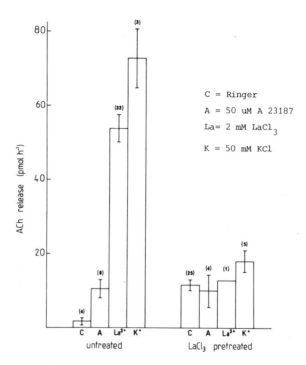

Figure 2. Effect of pretreatment with $LaCl_3$ on evoked ACh release
from frog sartorius muscle. After 15 min of incubation with or without
La^{3+}, the muscles were incubated for a period of 4 h. Thereafter
ACh was collected during 30 min in normal Ringer, with 50 μM A 23187,
with a second dose of La^{3+}, or with 50 mM KCl. In untreated muscles
the KCl concentration was 25 mM (in one experiment the release caused
by 50 mM KCl was about 15 pmol/h during the first 15 min, but ACh
release was then exhausted on prolonged incubation with KCl; not
illustrated). Means $\pm$ S.E. (From ref. 24, with permission of the
J. of Physiol.)

3. **Does Vesicular ACh Decrease Preferentially During a Short Period of Intensive Stimulation of the Nerve Terminals?** To study this question a method was needed for measuring vesicular ACh, rapid enough to prevent possible redistribution of ACh between the vesicular and cytoplasmic pools of ACh. Dunant et al. (3, 4) used such a method for the electric organ of Torpedo, based on a homogenization technique originally devised for brain tissue (see for review ref. 31). The principle of the method is simple: the tissue is homogenized in physiological saline, a procedure causing the instantaneous hydrolysis of extravesicular ("free") ACh by the AChE in the electric tissue; the ACh surviving this treatment, the so-called "bound" ACh, is presumably protected against AChE by being occluded in synaptic vesicles (17); the "free" ACh is then calculated as the difference between "total" acid-extractable) and "bound" (Saline-extractable) ACh.

We found that the same homogenization technique could be applied to frog muscle, provided that excess of purified AChE from electrical eel was added to ensure rapid hydrolysis of free ACh (23). Fortunately the measurement of bound ACh in frog muscle turned out to be a rather specific estimation of vesicular ACh:bound ACh predominantly derived from vesicles, and was practically not contaminated by cytoplasmic ACh, as shown by the observation that bound ACh in La^{3+}-treated muscles was reduced to post-denervation levels, i.e., 30% of controls. Apparently, the large amount of cytoplasmic ACh stored after La^{3+} was completely hydrolyzed by the added AChE (23).

Subsequently, we studied the effects of a brief but intensive stimulation on the levels of free and bound ACh in the sartorius muscle. To this end we used a chloride-free, isotonic potassium propionate medium to obtain total depolarization of the nerve terminals, a procedure known to cause a great reduction of synaptic vesicles in the sartorius within 5 min (6). The synthesis of ACh was blocked by hemicholinium-3 (HC-3), as a precaution against the possibility that the effect of K^+ on ACh was masked by rapid replenishment of the stores. [Actually, this precaution was not really necessary because ACh synthesis in frog muscle is a very slow process (26), and moreover, ACh synthesis was found to be blocked during depolarization with high K^+ (Molenaar and Polak, unpublished).] The results in Table 1 show that K^+-treatment for 5 min reduced total ACh by 16 pmol. Practically all this loss could be attributed to a decrease of bound ACh (which was 70% of the decrease caused by denervation). On the other hand, the free ACh remained unchanged after high K^+. In unpublished experiments, in which no HC-3 was used, we incubated the muscles for 5 min in high K^+ and allowed the muscles to recover for 60 min in normal Ringer containing 50 μM choline-d_9. Neuromuscular transmission returned already after 20 min. Synthesis of ACh did occur during the recovery period since the ACh in the muscle was labeled by choline-d_9. The specific labeling, the ratio ACh-d_9/ACh-d_0, was nearly four times higher in

Table 1

The Effect of Potassium Propionate on Total, Bound and Free ACh

	Total	Bound	Free
Content at t = 0	50 ± 3.4 (14)	20 ± 2.8 (7)	30 ± 2.8 (7)
Content after 5 min incubation in high-K^+	34 ± 3.5 (14)*	8 ± 1.6 (7)*	31 ± 5.2 (7)
Content after 15 days' denervation	8 ± 2.0 (8)	3 ± 1.5 (7)	6 ± 1.3 (6)

Muscles were incubated for 30 min in 10 μM HC-3, thereafter for 5 min in K^+ propionate medium (6), and then fractioned. No AChE inhibitor was used. Means ± S.E. with number of muscles in parentheses. *Significantly different from values at t = 0, $P_2 < 0.01$, Student's t-test.

free ACh than in bound ACh, indicating that ACh synthesis took place in the cytoplasm and that there was not a rapid exchange of ACh between cytoplasmic and vesicular ACh. However, when recovery was allowed to proceed for 3 h, the levels of ACh in the fractions were back to normal, and there was no significant difference in the degree of labeling of ACh in the fractions.

4. **Does the ACh Quantum Contain the Same Amount of ACh as the Vesicle?** In the experiments concerned with the effect of La^{3+} on total ACh, the compartment of vesicular ACh was estimated to be 25 pmol (21). In these experiments it was assumed that the depletion of ACh caused by La^{3+}, after inhibition of ACh synthesis by HC-3, was due to a loss of the vesicular ACh store. Recycling of vesicles seems to be little or nonexistent after La^{3+} (29), but it is possible that during the initial stage of the La^{3+}-treatment there was a little recycling of vesicles, leading to a reduction of cytoplasmic ACh, and consequently, to an overestimation of ACh in the vesicular store.

In the experiments in which free and bound ACh were measured, La^{3+} caused an 18 pmol reduction of bound ACh (23). This value clearly provides a lower limit, because some vesicular ACh might have been lost during isolation of the bound ACh fraction, and because cytoplasmic ACh cannot have biased this estimate. Taken together, the effects of La^{3+} on total and bound ACh indicate that the vesicular compartment contains more than 18, but less than 25 pmol ACh, or between 11,000 and 15,000 molecules per vesicle [assuming 10^3 endplates in the muscle and 10^6 vesicles in the endplate (9)].

In the experiment summarized in Fig. 1, the min.e.p.c. corresponded to the release of 12,000 molecules of ACh, indicating that the ACh quantum, on the average, consists of 12,000 molecules. This value is somewhat higher than previous estimates, on other muscles, for the ACh content of a quantum (5, 15).

DISCUSSION

Since we obtained a positive answer to the four questions posed here, the vesicle hypothesis still stands as a likely mechanism for ACh release in skeletal muscle. But does the present and other evidence, such as the incorporation of the vesicular membrane into the membrane of the nerve terminal after La^{3+}, while being a necessary corollary of the hypothesis, exclude other more or less plausible mechanisms?

The Gate Hypothesis. The possibility that evoked ACh release occurs by diffusion from the cytoplasm through gates in the presynaptic membrane, a mechanism proposed by Marchbanks (16), implies among other things that the size of the quantum is dependent on the concentration of cytoplasmic ACh and the electrochemical gradient across the membrane. This is not the case and other objections have been raised against this hypothesis (see for instance ref. 27). The present observation that after La^{3+} min.e.p.ps of small amplitude occur, albeit at a very low frequency, is also difficult to reconcile with this hypothesis, since cytoplasmic ACh is not decreased, but increased, after La^{3+}.

The Operator Hypothesis. Another hypothesis by Israël and Dunant (10) for the ACh release in the electric organ states that ACh release is brought about by an "operator" in the presynaptic membrane, which would recruit fixed amounts of ACh directly from the cytoplasmic store. According to this hypothesis the present results could be interpreted as follows. La^{3+} ions would first activate the operator molecules, and at a later time synchronously with the disappearance of the vesicles, gradually poison them. Treatment with high-K^+would lead to loss of cytoplasmic ACh via the operator and this loss would remain unnoticed because the cytoplasmic store would be replenished instantaneously by vesicular ACh, functioning as a large buffer store of ACh. In conclusion then, the present results do not contradict the operator hypothesis. Yet, as far as skeletal muscle is concerned, there is not clearly a raison d'être for this hypothesis.

Acknowledgements. We thank Mrs. J.W.M. Tas for excellent technical assistance. The financial support by FUNGO/ZWO and the Wellcome Trust is gratefully acknowledged.

REFERENCES

1. Clark, A.W., Hurlbut, W.P. and Mauro, A. (1972): J. Cell Biol. 52:1-14.
2. del Castillo, J. and Katz, B. (1957): In Microphysiologie comparée de éléments excitables, Colloques du Centre National de la Recherche Scientifique, Paris, Gif/Yvette, no. 67, pp. 245-258.
3. Dunant, Y., Gautron, J., Israël, M., Lesbats, B. and Manaranche, R. (1972): J. Neurochem. 19:1987-2002.
4. Dunant, Y., Israël, M., Lesbats, B. and Manaranche, R. (1977): Brain Res. 125:123-140.
5. Fletcher, P. and Forrester, T. (1975): J. Physiol. (London) 251:131-144.
6. Gennaro, J.F., Nastuk, W.L. and Rutherford, D.T. (1978): J. Phys iol. (London) 280:237-247.
7. Gorio, A., Hurlbut, W.P. and Ceccarelli, B. (1978): J. Cell Biol. 78:716-733.
8. Hebb, C.O., Krnjevic, K. and Silver, A. (1964): J. Physiol. (London) 171:504-513.
9. Heuser, J.E. and Miledi, R. (1971): Proc. R. Soc. London B179:247-260.
10. Israël, M. and Dunant, Y. (1979): In Progress in Brain Res., Vol 49, (ed) S. Tuček, Amsterdam, Elsevier, pp. 125-139.
11. Israël, M., Dunant, Y. and Manaranche, R. (1979): Progr. Neurobiol. 13:237-275.
12. Katz, B. and Miledi, R. (1977): Proc. R. Soc. London B196:59-72.
13. Katz, B. and Miledi, R. (1981): Proc. R. Soc. London B212:131-137.
14. Katz, B. and Miledi, R. (1982): Proc. R. Soc. London B216:497-507.
15. Kuffler, S.W. and Yoshikami, D. (1975): J. Physiol. (London) 251:465-482.
16. Marchbanks, R.M. (1975): In Metabolic Compartmentation and Neurotransmission (ed) S. Berl, New York, Plenum Press, pp.609-620.
17. Marchbanks, R.M. and Israël, M. (1971): J. Neurochem. 18:439-448.
18. Matthews-Bellinger, J. and Salpeter, M.M. (1978): J. Physiol. (London) 279:197-213.
19. Miledi, R., Molenaar, P.C. and Polak, R.L. (1977): Proc. R. Soc. London B197:285-297.
20. Miledi, R., Molenaar, P.C. and Polak, R.L. (1978): In Cholinergic Mechanisms and Psychopharmacology (ed) D.J. Jenden, New York, Plenum Press, pp. 377-386.
21. Miledi, R., Molenaar, P.C. and Polak, R.L. (1980): J. Physiol. London 309:199-214.
22. Miledi, R., Molenaar, P.C. and Polak, R.L. (1981): In Cholinergic Mechanisms (eds) G. Pepeu and H. Ladinsky, New York, Plenum Press, pp. 205-214.
23. Miledi, R., Molenaar, P.C. and Polak, R.L. (1982): J. Physiol. London 333:189-199.
24. Miledi, R., Molenaar, P.C. and Polak, R.L. (1983): J. Physiol. London 334:245-254.

25. Mitchell, J.F. and Silver, A. (1963): J. Physiol. London 165: 117-129.
26. Molenaar, P.C. and Polak, R.L. (1980): J. Neurochem. 35:1021-1025.
27. Tauc, L. (1979): Biochem. Pharmac. 27:3493-3498.
28. Tuček, S. (1982): J. Physiol. London 322:53-69.
29. von Wedel, R.J., Carlson, S.S. and Kelly, R.B. (1981): Proc. Natl. Acad. Sci. 78:1014-1018.
30. Vyskočil, F. and Illès, P. (1978): Physiologia Bohemoslov. 27:-449-455.
31. Whittaker, V.P. (1969): Progr. Brain Res. 31:211-222.

DEPOLARIZATION INDUCED HYDROLYSIS OF

CYTOPLASMIC ACh IN MOUSE BRAIN

P.T. Carroll

Department of Pharmacology
Texas Tech University Health Sciences Center
Lubbock, Texas
U.S.A

INTRODUCTION

Acetylcholine (ACh) release occurs from cholinergic nerve endings in brain by different processes. Spontaneous ACh release, for example, appears to occur primarily from the cytosol by a Ca^{2+} independent process (5-7) and may be synthesized by a soluble, cytoplasmic form of choline-O-acetyltransferase (EC 2.3.1.6, ChAT) (3). In contrast, the evoked form of ACh release appears to occur from a vesicular fraction by a Ca^{2+} dependent process (6-9, 18) and it may be synthesized by a membrane-bound form of ChAT closely associated with the vesicular fraction (1-3, 17). Recent reports suggest that the Ca^{2+} dependent and Ca^{2+} independent forms of ACh release differ in yet another respect. Depolarization of brain tissue only stimulates the Ca^{2+} dependent, not the Ca^{2+} independent form of ACh release (9, 13-14). That depolarization of brain tissue is unable to stimulate the Ca^{2+} independent release of ACh is difficult to reconcile with the finding that high K^+ induced depolarization of brain tissue not only lowers the ACh content of the vesicular fraction but also lowers the ACh content of the cytoplasmic fraction (16). One might expect that elevated K^+ would exclusively lower the level of vesicular ACh since it only stimulates the Ca^{2+} dependent form of ACh release. One possible explanation for the K^+ induced reduction of cytoplasmic ACh in the absence of a corresponding stimulation of Ca^{2+} independent ACh release is that high K^+ may cause the hydrolysis of cytoplasmic ACh and thereby lower its level.

An approach to testing this possibility would be to depolarize brain tissue in the absence of extracellular Ca^{2+} so that ACh release would not be augmented, and then to determine whether depolarization

499

had caused a selective disappearance of cytoplasmic ACh. If so, and extracellular Ca^{2+} were required for depolarization to stimulate ACh release and lower the level of vesicular ACh, then it would appear likely that depolarization did effect the hydrolysis of cytoplasmic ACh rather than a stimulation of its release. Experiments to test this possibility were done by incubating minces of mouse forebrain in a Ca^{2+} free Krebs solution containing either elevated K^+ (35 mM) or veratridine (50 μM), and determining the effects of these agents on the subcellular levels of ACh. Similar experiments were also done in the presence of extracellular Ca^{2+} and compared with the Ca^{2+} free experiments.

The results of these experiments indicated that both depolarizing agents caused a selective breakdown of cytoplasmic ACh. To ascertain whether this effect might be mediated by an intraterminal form of AChE, minces were initially pretreated in a normal Krebs solution with various AChE inhibitors (neostigmine, physostigmine, paraoxon) judged to differ in their rates of penetration into central cholinergic nerve endings by their relative abilities to expand the cytoplasmic pool of ACh. Then these pretreated minces were subsequently exposed to high K^+ or veratridine in the absence of Ca^{2+}, to determine if these agents would then stimulate the Ca^{2+} independent release of ACh; also whether the magnitude of the stimulation would be correlated with the degree of intraterminal AChE inhibition prior to depolarization. Results obtained in the study suggested that the choline formed from the veratridine-induced hydrolysis of cytoplasmic ACh might be recycled to form a portion of the ACh released from the vesicular fraction. To test this possibility, minces were pretreated to hydrolyze cytoplasmic ACh and the effect of this reduced level of hydrolyzable ACh then determined on the veratridine induced release of ACh over time.

This chapter describes some of the experiments which we conducted in the context of the above mentioned studies.

MATERIALS AND METHODS

To test the effect of veratridine on the levels of choline and ACh in subcellular fractions of mouse forebrain under conditions where the Ca^{2+} independent release of ACh would not be stimulated, minces were incubated for 5 min in a Ca^{2+} free Krebs solution (no added Ca^{2+} + 0.1 mM EGTA) in the absence or presence of veratridine (50 μM) and paraoxon (3 μM), and subcellular fractions prepared according to previously described procedures (11-12, 15-16). The amounts of choline and ACh in the cytoplasmic (S_3) and vesicle-bound (P_3) fractions were then assayed by a previously described method (15).

To determine the effect of tetrodotoxin (TTX) on the veratridine induced decrease in S_3 ACh content, minces were initially washed

twice with 5 ml of ice-cold Ca^{2+} free Krebs containing TTX (1 μg/ml).
Then the minces were incubated for 5 min in a Ca^{2+} free Krebs solution
containing TTX (1 μg/ml) in the absence or presence of veratridine
(50 μM).

The influence of elevated K^+ (35 mM) on the levels of choline
and ACh in S_3 and P_3 fractions of mouse forebrain minces incubated
without Ca^{2+} was determined in the identical way as the veratridine
experiments. To maintain osmolality, the sodium chloride concentration was reduced.

To test the effect of high K^+ and veratridine on the subcellular
levels of ACh under conditions where these agents would stimulate
the Ca^{2+} dependent release of ACh, the same experiments as those
described above were performed except that extracellular Ca^{2+} was
present (2.5 mM) and EGTA was not.

In other experiments, minces were incubated for 30 min in normal
Krebs or not incubated and the effect of incubation on the subcellular
levels of choline and ACh determined. Also, the effects of neostigmine
(1 x 10^{-4}M), physostigmine (1 x 10^{-4}M) and paraoxon (3 μM) were
determined on the subcellular levels of ACh in minces incubated
for 30 min in normal Krebs. To ascertain whether these various
pretreatments would enable high K^+ or veratridine to stimulate the
Ca^{2+} independent release of ACh, minces were initially pretreated
for 30 min in normal Krebs with or without a particular AChE inhibitor,
washed twice with ice cold Ca^{2+} free Krebs containing 0.1 mM EGTA,
and then incubated for 5 min in a Ca^{2+} free Krebs solution containing
0.1 mM EGTA in the absence or presence of either high K^+ or veratridine. To determine how quickly paraoxon (3 μM) might penetrate
central cholinergic nerve endings and inhibit an intraterminal form
of AChE, minces were incubated for either 5 or 10 min periods in
normal Krebs with or without paraoxon (3 μM) and the effect of paraoxon
on the subcellular levels of ACh determined.

To ascertain whether a reduction in hydrolyzable ACh would
lead to a reduction in evoked ACh release, minces were pretreated
for 5 min in a Ca^{2+} free Krebs solution to maintain the S_3 level
of ACh or a Ca^{2+} free high K^+ Krebs to hydrolyze S_3 ACh. Then both
sets of minces were washed twice with ice cold Krebs containing
paraoxon (3 μM) and ultimately incubated in normal Krebs with or
without veratridine (50 μM) for 5 or 10 min periods and the veratridine-induced release for the 2 sets of pretreated minces determined and compared.

Release of choline and ACh from mouse brain minces into the
various incubation media was determined according to previously
described procedures (5-10).

Table 1

Effect of elevated K[+] and veratridine on the subcellular levels of ACh in mouse forebrain minces incubated without added Ca[2+] [a]

	TREATMENT Krebs	35mMK[+]	Δ
S_3	14.0 ± 1.0	8.4 ± 1.0[b]	-5.6 ± 1.0
P_3	7.2 ± 1.1	6.6 ± 0.9	-0.6 ± 0.8
Media	9.8 ± 0.5	9.6 ± 0.9	-0.2 ± 0.4
	Krebs	Veratridine	Δ
S_3	10.9 ± 0.8	6.3 ± 0.8[c]	-4.6 ± 0.7
P_3	5.8 ± 0.4	5.2 ± 0.4	-0.6 ± 0.3
Media	10.8 ± 0.9	11.7 ± 0.6	+0.9 ± 0.3

[a]Minces of mouse forebrain were incubated for 5 min in either Krebs, 35mMK[+] Krebs, or Krebs with veratridine (50 μM) in the absence of added Ca[+] and presence of 0.1 mM EGTA. All incubation media contained paraoxon (3 μM). The amount of ACh released into the incubation media as well as the amount of ACh in cytoplasmic (S_3) and vesicular (P_3) fractions were determined for an n of 6 in the high K[+] experiments and an n of 5 in the veratridine experiments in nmoles/gram wet wt/5 min. Results are expressed as the mean ± S.E.M. in nmoles/-gram wet wt/5 min. [b]Results differ significantly from Krebs treatment at $p < 0.05$ (paired Student t-test). [c]Results differ significantly from Krebs treatment at $p < 0.05$ (ANOVA).

RESULTS

Previously reported results indicated that both high K[+] and veratridine stimulate the release of ACh from mouse forebrain minces in a dose-dependent fashion when extracellular Ca[2+] is present, but not when it is absent (9). In the present study, both agents caused a substantial loss of ACh from the S_3 fraction when minces were incubated without Ca[2+], which could not be attributed to a gain in P_3 or medium ACh (Table 1). However, when Ca[2+] was present in the incubation media, then both agents not only stimulated ACh release but also lowered the ACh content of the P_3 fraction (9). They also significantly reduced the ACh content of the S_3 fraction when Ca[2+] was present in the incubation medium.

Table 2

Effects of tetrodotoxin on the veratridine-induced loss of S_3 ACh and the veratridine-induced gain in P_3 choline during incubation of mouse forebrain minces in the absence of Ca^{2+}

| TREATMENT | SUBCELLULAR FRACTIONS | | | |
| | S_3 | | P_3 | |
	ACh	Choline	ACh	Choline
Krebs	10.9 ± 0.8	42.0 ± 2.7	5.8 ± 0.4	12.5 ± 1.5
Veratridine (50 μM)	6.3 ± 0.8^b	39.0 ± 2.6	5.2 ± 0.4	18.5 ± 1.1^b
Krebs + TTX (1 μg/ml)	10.7 ± 0.9	44.6 ± 2.1	6.3 ± 0.5	23.1 ± 2.1
Krebs + TTX (1 μg/ml) + veratridine (50 μM)	9.3 ± 0.8	41.2 ± 1.2	5.1 ± 0.7	21.9 ± 2.6

[a]Minces of mouse forebrain were incubated for 5 min in the Ca^{2+} free Krebs in the absence or presence of veratridine (50 μM), or in Ca^{2+} free Krebs with TTX (1 μg/ml), or in Ca^{2+} free Krebs with TTX (1 μg/ml) and veratridine (50 μM). In the TTX experiments, minces were initially washed with ice-cold Krebs containing TTX and then incubated. All incubation media contained paraoxon (3 μM). Results are expressed as Mean $\pm$ S.E.M. in nmoles/gram wet wt for an n of 4 or 5. [b]Results differ significantly from Krebs at $p < 0.05$ (ANOVA).

Veratridine, but not high K^+, increased the choline content of the P_3 fraction by an amount which closely matched its decrease of the ACh content in the S_3 fraction (Table 2). Addition of TTX to the Ca^{2+} free incubation media to obviate veratridine's stimulation of sodium influx prevented both the veratridine induced decrease in the ACh content of the S_3 fraction, and its increase in the choline content of the P_3 fraction. Incubation of minces in a normal Krebs solution for 30 min elevated the ACh content of the S_3 fraction by over 300% and that of the P_3 fraction by approximately 200% relative to non-incubated control minces (Table 3). Addition of AChE agents such as neostigmine, physostigmine, and paraoxon to a 30 min incubation of minces in normal Krebs all raised the ACh content of the S_3 fraction still further, but by much different magnitudes, and more so than incubation alone had raised it. None of these inhibitors significantly changed the ACh content of the P_3 fraction (data not shown). A 5 min incubation of minces with paraoxon (3 μM) did not change the ACh content of the S_3 fraction, but a 10 min incubation of minces with paraoxon raised it approximately 23%.

When minces were pre-incubated in normal Krebs for 30 min to expand the S_3 pool of ACh by over 300%, only veratridine was then able to stimulate the Ca^{2+} independent release of ACh from minces during subsequent incubation (Table 3). When AChE inhibitors were included during the 30 min pre-incubation, then both high K^+ and veratridine were able to stimulate the Ca^{2+} independent release of ACh; the magnitude of these stimulations was correlated with the magnitude of S_3 ACh expansion caused by the respective inhibitors prior to depolarization (Table 3).

When minces were pretreated in a Ca^{2+} free high K^+ Krebs solution to reduce the level of S_3 ACh, and thereby make less available for hydrolysis during subsequent depolarization, veratridine released less ACh than it did from minces pretreated in a Ca^{2+} free Krebs to maintain the level of S_3 ACh (Table 4). This reduction in veratridine-induced ACh release closely matched the loss of hydrolyzable S_3 ACh caused by the Ca^{2+} free high K^+ pretreatment of minces. However, the loss of hydrolyzable ACh caused by the pretreatment only affected the amount of ACh released by veratridine during a 10 min post-incubation period, not during a 5 min post-incubation period.

DISCUSSION

The results obtained in this study corroborate those of others suggesting that depolarization of brain tissue normally does not stimulate the Ca^{2+} independent, spontaneous release of ACh (9, 13-14). The inability of depolarization to do so appears to be due to its hydrolytic influence on the subcellular source of the Ca^{2+} independent, ACh release, that is on the cytoplasmic pool of ACh. Both depolarizing agents used in the study caused a breakdown of cytoplasmic ACh rather than a stimulation of its release. This breakdown appeared to be mediated by an intraterminal form of AChE; inhibition of this enzyme form by various inhibitors prior to depolarization enabled both depolarizing agents to stimulate the Ca^{2+} independent release of ACh. In contrast, expansion of the cytoplasmic pool did not; only veratridine stimulated the Ca^{2+} independent release of ACh from minces pretreated in normal Krebs without an AChE inhibitor. The results also indicated that the inclusion of paraoxon in a 5 min incubation of minces did not prevent the depolarization induced hydrolysis of cytoplasmic ACh, since it apparently is unable to penetrate cholinergic nerve endings and inhibit an intraterminal form of AChE that rapidly.

Table 3

Relationship between the expansion of S_3 ACh caused by incubation of mouse forebrain minces with various AChE inhibitors and the respective abilities of high K^+ and veratridine to augment the Ca^{2+} independent, spontaneous release of ACh.[a]

PRETREATMENTS	S_3 ACh % of Control	POST-INCUBATIONS			POST-INCUBATIONS		
		Krebs	35mMK+	Δ	Krebs	Veratridine	Δ
1) Krebs	313 ± 31	6.1 ± 0.4	5.3 ± 0.6	-0.8	4.9 ± 0.4	7.3 ± 0.5[b]	+2.3
2) Krebs + neostigmine	113 ± 6	N.D.	N.D.	N.D.	7.9 ± 0.3	13.6 ± 1.0[b]	+5.7
3) Krebs + physostigmine	128 ± 4	9.6 ± 3.2	14.3 ± 2.1[b]	+4.7	N.D.	N.D.	N.D.
4) Krebs + paraoxon	178 ± 11	8.0 ± 0.6	14.3 ± 1.7[b]	+6.3	7.5 ± 0.5	20.6 ± 2.1[b]	+13.1

[a]Minces of mouse forebrain were pre-incubated for 30 min in normal Krebs in the absence or presence of neostigmine (1 x 10^{-4}M), physostigmine (1 x 10^{-4}M), or paraoxon (3 μM), and the effects of these treatments determined on the levels of ACh in S_3 and P_3 fractions. Only incubation in Krebs (treatment #1) elevated the ACh content of the P_3 fraction. All AChE inhibitors caused a selective expansion of S_3 ACh when compared with their individual control groups. In the release experiments, following the pretreatments, minces were washed twice in ice cold Ca^{2+} free Krebs solution containing 0.1 mM EGTA and paraoxon (3 μM), then incubated for 5 min in either 35mMK$^+$ Krebs or veratridine (50 μM) or 4.7 mM K$^+$ Krebs, all of which lacked Ca^{2+} and contained 0.1 mM EGTA. N.D.=not determined. Results are expressed as the Mean ± S.E.M. in nmoles/gram wet wt/5 min. [b]Results differ significantly from minces post-treated in Krebs without Ca^{2+} at $p < 0.05$ (ANOVA, or paired Student t-test).

Table 4

Relationship between the amount of cytoplasmic ACh available
for hydrolysis and the amount of ACh released by veratridine[a]

| | ACh RELEASE | |
	POST-TREATMENT Krebs without Ca^{2+}	PRETREATMENTS 35mM K^+ Krebs without Ca^{2+}
5 min		
Krebs	8.0 ± 1.1	7.7 ± 0.9
Veratridine (50 μM)	17.6 ± 2.0^{b}	15.1 ± 1.0^{b}
Δ	$+ 9.6 \pm 1.3$	$+ 7.4 \pm 1.0$
10 min		
Krebs	13.0 ± 1.5	9.4 ± 1.2
Veratridine (50 μM)	29.9 ± 1.1^{b}	19.9 ± 1.1^{b}
Δ	$+17.0 \pm 1.6$	$+10.5 \pm 0.8^{c}$

[a]Minces of mouse forebrain were pretreated for 5 min in either Krebs without Ca^{2+} to maintain the cytoplasmic pool of ACh or 35mM K^+ Krebs without Ca^{2+} to hydrolyze it and make it unavailable as a source of choline during subsequent depolarization, washed twice with 5 ml of ice cold Krebs containing paraoxon (3 μM), and then incubated for 5 or 10 min periods in Krebs or Krebs with veratridine (50 μM) and paraoxon (3 μM). Results are expressed as the Mean $\pm$ S.E.M. for an n of 8 animals. Data are presented as nmoles ACh released/gm wet wt. [b]Results differ significantly from control Krebs at p < 0.05 (Student t-test paired). [c]Results differ significantly from Krebs without Ca^{2+} pretreatment at p < 0.05 (two sample rank test).

The results obtained in this study suggest that depolarization-induced hydrolysis of cytoplasmic ACh may play an important role in providing substrate for the formation and release of ACh from vesicles. For example, the choline formed from the veratridine induced hydrolysis of cytoplasmic ACh appeared to be transferred to the vesicular fraction. This was not the case in the high K^+ experiments and this difference between veratridine and high K^+ might reflect differences in how these agents influence sodium influx across vesicular membranes. Additionally, veratridine was unable to release as much ACh from minces that had been pretreated in a high K^+ Krebs without Ca^{2+}, to hydrolyze cytoplasmic ACh, and thereby to reduce the source of transferable choline, as it was able to release from minces which had been pretreated in a Ca^{2+} free Krebs to maintain the cytoplasmic pool of ACh. That this diminution of veratridine induced ACh release only occurred during a 10 min and not a 5 min post-incubation period may be due to an initial stimulation in the release of preformed ACh from the vesicular fraction by veratridine, followed by a stimulation in the release of newly synthesized ACh, formed from hydrolyzed cytoplasmic ACh, from the same fraction.

Acknowledgements.   This work was supported in part by NSF grant BNS-811 7975 and NINCOS grant NS212189-01.   I wish to thank Eri Thomas for manuscript preparation and Tricia Craig for technical assistance.

REFERENCES

1. Badamchian, M. and Carroll, P.T.:   J. Neurosci. (in press).
2. Benishin, C.G. and Carroll, P.T. (1981):   J. Neurochem. 36: 732-740.
3. Benishin, C.G. and Carroll, P.T. (1983):   J. Neurochem. 41: 1030-1039.
4. Benishin, C.G. and Carroll, P.T. (1984):   J. Neurochem. 43: 885-887.
5. Carroll, P.T. (1983):   Neurochem. Res. 8:1271-1283.
6. Carroll, P.T. (1984):   Brain Res. 321:55-62.
7. Carroll, P.T. and Aspry, J.M. (1980):   Science 210:641-642.
8. Carroll, P.T. and Aspry, J.M. (1981):   Neuroscience 6:2555-2559.
9. Carroll, P.T. and Benishin, C.G. (1984):   Brain Res. 291: 261-272.
10. Carroll, P.T. and Goldberg, A.M. (1975):   J. Neurochem. 25: 523-527.
11. Carroll, P.T. and Nelson, S.H. (1978):   Science 199:85-86.
12. Collier, B., Poon, P. and Salehmoghaddam, S. (1972):   J. Neurochem. 51-60.
13. Gibson, G.E. and Peterson, C. (1981):   J. Neurochem. 978-984.
14. Meyer, E.M. and Cooper, J.R. (1981):   J. Neurochem. 37:1186-1192.
15. Nelson, S.H., Benishin, C.G. and Carroll, P.T. (1980):   Biochem. Pharmacol. 29:1949-1957.
16. Salehmoghaddam, S.H. and Collier, B. (1976):   J. Neurochem. 27:71-76.
17. Smith, C.P. and Carroll, P.T. (1980):   Brain Res. 185:363-371.
18. Suszkiw, J.B. and O'Leary, M.E. (1982):   J. Neurochem. 38:1668-1675.

SEPARATION OF RECYCLING AND RESERVE SYNAPTIC VESICLES FROM CHOLINERGIC NERVE TERMINALS OF THE MYENTERIC PLEXUS OF GUINEA-PIG ILEUM

D.V. Ágoston, J.W. Kosh[1], J.Lisziewicz,
P.E. Giompres[2] and V.P. Whittaker

Abteilung Neurochemie
Max-Planck-Institut für Biophysikalische Chemie
Postfach 2841
D-3400 Gottingen, FR Germany

[1]Present Address:
  College of Pharmacy
  University of South Carolina
  Columbia, SC 29208 U.S.A.

[2]Present Address:
  Department of Human and
   Animal Physiology
  University of Patras
  Patras, Greece

INTRODUCTION

Our understanding of the dynamics of vesicle recycling and transmitter synthesis, storage and release at a cholinergic synapse has been greatly increased by the discovery that synaptic vesicles isolated from stimulated cholinergic electromotor nerve terminals of Torpedo marmorata can be separated into two subpopulations (designated VP_1 and VP_2) by methods utilizing differences in size (6) or density (18, 19). Studies utilizing dextran particles which are recognizable in the electron microscope but are unable to diffuse through lipoprotein membranes and whose presence within the lumen of synaptic vesicles could only have occurred during a cycle of exo- and endocytosis, have enabled the VP_2 vesicles to be identified as the recycling subpopulation and as the smaller vesicles seen in whole tissue sections near the presynaptic plasma membrane which accumulate in the terminal cytoplasm as stimulation continues (17, 18). Work with radioactive precursors of acetylcholine have shown that newly synthesized acetylcholine is preferentially taken up into these smaller, denser vesicles (6, 17) and that on the restimulation of already labelled tissue blocks, the acetylcholine released has a specific radioactivity not significantly different from that in the VP_2 fraction (13). Finally, false transmitters that distribute

509

themselves differently from newly synthesized acetylcholine and from each other between cytoplasm and VP$_2$ vesicles are released on restimulation in a ratio equal to that in which they are stored in the VP$_2$ subpopulation and not in that in which they occur in the cytoplasm (9). Thus VP$_2$ vesicles have been identified, in this terminal, as the immediate source of released transmitter.

The question now arises as to the generality of these findings. An analogous phenomenon is seen in mammalian cortical cholinergic nerve terminals from which two synaptic vesicle fractions can also be isolated: fraction D, comprising monodisperse vesicles; and fraction H, comprising vesicles adhering to the presynaptic plasma membrane (1). The latter became preferentially labelled with newly formed acetylcholine (1, 14) and acetylcholine analogues (14) and can similarly be implicated as the source of released transmitter (14). However, so far, it has not proved possible to isolate H-fraction vesicles free from presynaptic plasma membrane.

The successful preparation from the myenteric plexus of the guinea pig - a relatively rich source of cholinergic nerve terminals - of highly purified cholinergic synaptic vesicles (2) without the need (unlike brain cortex) to resort to hypo-osmotic lysis of nerve terminals suggested that this tissue would be a suitable one with which to test whether recycling and reserve vesicles from a mammalian cholinergic synapse could be separated. This paper describes the work we have done so far on this problem.

METHODS

The Myenteric Plexus-Longitudinal Muscle Preparation

Stimulation and isotopic labelling of tissue. Myenteric plexus-longitudinal muscle (MPLM) strips were prepared from the guinea-pig ileum as previously described and incubated for 45-70 min in Krebs-Ringer solution containing either deuterated (D$_4$) choline (50 μM) or [^{3}H]acetate (2 μCi.ml^{-1} of specific radioactivity 200 μCi. μmol^{-1}). Stimulation, if applied, was after an initial 30 min incubation period and was for 10 min with potassium (by replacing the KCl and NaCl of the normal Krebs-Ringer solution with 50 mM KCl and 87 mM NaCl for the duration of the stimulus) or electrically at 0.1, 1 or 10 Hz delivered through a coaxial Pt electrode by a Grass Medical Instruments (Quincy, MA, USA) S48 stimulator (11). The stimuli were of 1 msec duration and 10 V/cm intensity. At 5, 10 or 30 min after stimulation (or an equivalent period of stimulus-free incubation) the tissue was briefly washed in 3 changes of ice-cold label-free Krebs-Ringer solution to remove extracellular label, dried between hardened filter paper, weighed and homogenized (for details see below). Samples (2 ml) containing D$_4$ acetylcholine were next extracted with ice-cold formic acid-acentonitrile for 30 min (4), for gas chromatography-mass spectrometry (GCMS)

(15 and references cited therein) using a Hewlett-Packard model 5992 B instrument. Samples containing [^{3}H]acetylcholine were submitted (3 x 0.7 ml) to liquid ion-exchange extraction (3), and the radio-activity determined in a scintillation counter (Bertold model LB 5004).

Homogenization of tissue. This was carried out as previously described (2) except that the pellet and pellicle obtained after the first centrifugation were combined, rehomogenized with 15 up-and-down passes with a 5 min cooling break after the first eight passes, and the homogenate centrifuged at 1000 g for 20 min. The supernatant so obtained was added to the first supernatant. This procedure eliminated zonal peak III and usually zonal peak II (for nomenclature see (2)) as well. It also increased the yield of acetylcholine in peak I to 60-70% of homogenate.

Isolation of vesicles. Using centrifugal density gradient centri-fuging in a zonal rotor synaptic vesicles were isolated from the combined supernatants derived from homogenates of strips stimulated at 1 Hz for 10 min with 5 min recovery, or from unstimulated strips incubated for the same length of time. The procedure was essentially as previously described (2) except that a 300 ml Beckman Ti-60 rotor was used, the first dense sucrose was 0.6 M, the second 1.45 M and the change-over was after 240 ml of 0.6 M sucrose had been delivered. This gave a shallower gradient and greater resolving power in the region of the vesicle peak. The sample volume was 30 ml (equivalent to 9-11 g of tissue from 5-6 animals). Fractions were 5 ml in volume. Protein was measured as described (12). Acetylcholine was determined by bioassay on thin slips of leech muscle (16) using 1 ml samples acidified to pH 4.0 with 0.1 M HCl or by GCMS or scintillation counting as described above. The leech assays were useful in identifying the peak of vesicular acetylcholine and agreed well with the values for $D_0 + D_4$ acetylcholine as determined by GMCS.

Calculations

Vesicle Density. Theoretical calculations of vesicle density were made by assuming a diameter of 50 nm for myenteric cholinergic vesi-cles, a membrane thickness of 4.3 nm (5) and a membrane density of 1.13 g.ml^{-1} as measured for _Torpedo_ electromotor synaptic vesicles (5). The increase in density attendant on a 15% reduction in core volume was then calculated. This is less than that observed between VP_1 and VP_2 vesicles in _Torpedo_ (7).

Isotopic ratio. The isotopic ratio for deuterated acetylcholine was calculated as D_4 acetylcholine /($D_0 + D_4$ acetylcholine) and is a measure of the incorporation of label into the endogenous acetyl-choline pool analogous to the specific radioactivity for a radioactive label.

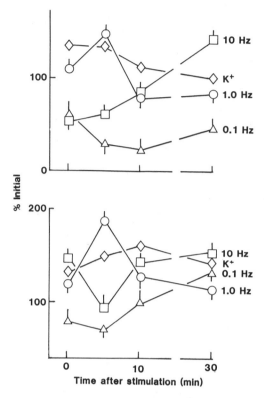

Figure 1. Time-course of incorporation of [^{3}H]acetate (upper diagram) or D_4 choline (lower diagram) into the acetylcholine of MPLM strips in response to electrical or chemical stimulation. Values are means $\pm$ range of two experiments and are expressed as percentages of the values at the beginning of stimulation. These were (per g of tissue, means of 5 expts): for [^{3}H], 38,600 $\pm$ 7,000 dpm; for [^{2}H], 8.16 $\pm$ 1.82 nmol; for endogenous acetylcholine, 40 $\pm$ 5 nmol.

Relative specific concentration (RSC). This was calculated in the usual way as the D_0 or D_4 acetylcholine recovered in a fraction expressed as a percentage of the total recovered D_0 or D_4 acetylcholine, divided by the protein content of the fraction expressed as a percentage of the total recovered protein.

RESULTS

Effect of stimulation parameters on the labelling of homogenate acetylcholine. Some preliminary experiments were carried out to determine the optimum conditions for the incorporation of labelled acetylcholine precursors into vesicles. Fig 1 shows the effect of various modes of stimulation on the incorporation of [^{3}H]acetate (upper diagram) or D_4 choline (lower diagram) into the acetylcholine of homogenates of MPLM strips.   Since no anticholinesterase was added to the homogenization medium and most of the homogenate acetylcholine was subsequently recovered, on zonal centrifugation, in a single vesicle peak, the homogenate acetylcholine presumably mainly represents vesicular acetylcholine.

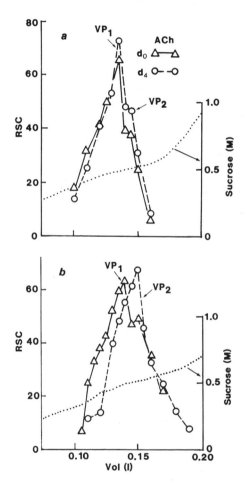

Figure 2. Zonal separation of cholinergic synaptic vesicles isolated from MPLM strips after exposure to D_4 choline. (a) Unstimulated control; (b) preparation stimulated at 1 Hz for 10 min with 5 min recovery. Note the presence in the preparation from stimulated tissue of two vesicle peaks with different isotopic ratios, the denser of which (VP_2) is more labelled than the lighter (VP_1). ACh, acetylcholine.

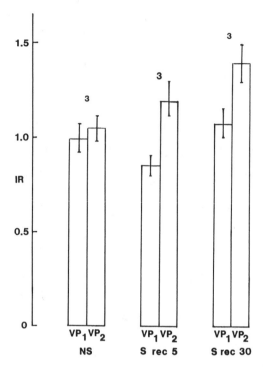

Figure 3. Mean isotopic ratios (IR) of the two vesicle populations equilibrating to regions of the gradient with refractive indices of 1.065 and 1.070, respectively. In a paired t test the VP_2 fraction is significantly more labelled than the VP_1 under stimulation (s) conditions [1 Hz for 10 min, with recovery (rec) for 5 or 30 min] ($p < 0.01$). The bars are SEMs or range (2 experiments). The number of experiments is shown above the blocks. NS indicates nonstimulated control ($p < 0.5$).

The two precursors behaved similarly: insufficient experiments were done to establish whether the differences between their rates of incorporation are significant, though in view of differences between them in the mechanism of uptake and dilution by the endogenous pool it would not be surprising if they were. The results show that incorporation of label (i.e. de novo synthesis) of acetylcholine is stimulus-dependent. Much more incorporation of label took place 5 min after stimulation at 1 Hz for 10 min than under any of the other conditions tested (except K^+ stimulation). Vesicle purification was therefore undertaken from tissue stimulated under these conditions.

Zonal separation of two vesicle populations from stimulated tissue. Fig 2 shows the distribution, in a representative experiment, of endogenous (D_0) and newly synthesized (D_4) acetylcholine in a zonal density gradient separation of vesicles from (above) unstimulated, and (below) stimulated tissue. The RSCs are plotted as a function of gradient volume. It will be seen that at rest one main peak of vesicular acetylcholine is obtained; there is, however, a shoulder in the denser region of the gradient which, indeed was also visible in earlier results (2, Fig 2) and this is somewhat more highly labelled than the main peak. On stimulation (lower figure) this shoulder becomes a recognizable peak and is even more highly labelled. Experiments with [^{3}H]acetate gave comparable results (not shown). Fig 3 shows the isotopic ratio for the two peaks in two control and six stimulation experiments. Under stimulation conditions, the labelling of the denser peak is significantly higher than that of the lighter peak. Under resting conditions there were insufficient results to establish a significant difference.

The metabolic heterogeneity of the vesicle peak is quite like that seen with vesicles from Torpedo electromotor nerve terminals, except that the difference between the MPLM VP_1 and VP_2 populations is less than that between the corresponding Torpedo populations both in labelling and density. The following section will show that the greater density of the VP_1 population and the smaller difference in density between VP_1 and VP_2 as compared to the Torpedo vesicles is to be expected in view of the smaller diameter and probable lower acetylcholine and ATP content of the MPLM vesicles (mammalian brain vesicles are estimated to have a much lower acetylcholine content than Torpedo vesicles) (10).

Density differences between vesicle populations. Table 1 compares the density of the peak fraction of vesicles from unstimulated tissue with those of the corresponding fractions of the two vesicle populations from stimulated tissue. The less highly labelled (VP_1) vesicle population from stimulated tissue has essentially the same density as the vesicles from unstimulated tissue and may thus be regarded as the reserve population that has either not been involved in vesicle recycling or has been fully recharged. The more highly labelled (VP_2) vesicle population, like its counterpart in Torpedo electromotor terminals, is denser than VP_1. However the density of the VP_1 population is greater than that of VP_1 vesicles from Torpedo and the difference in density between the VP_1 and VP_2 populations is considerably less (about one-third). Calculations show that this could be accounted for by the smaller size of the resting vesicle and a smaller loss of core water during recycling.

DISCUSSION

The results show that a mammalian cholinergic synapse, when stimulated, generates a vesicle population which, like that in the electromotor nerve terminals of Torpedo, is metabolically and biophysically heterogeneous. As in Torpedo the more highly labelled

Table 1

The Densities of the Two Vesicle Populations
Isolated from Stimulated Strips of Myenteric
Plexus-Longitudinal Muscle of Guinea-Pig Ileum

Conditions (No. of Expts)	Vesicle Fraction	Density (G/L) Found	Calcd*
Unstimulated (2)	VP	$1.0655 + 0.0010$	--
Stimulated (3)	VP_1	$1.0653 + 0.0016$**	1.0654
	VP_2	$1.0700 + 0.0016$**	1.0701
(Density Difference)		$0.0046 + 0.0011$	0.0047

*Vesicle initially 50 nm diam with 4.3 nm thick membrane of density
1.13 losing 15% of its core water and solutes (Density 1.016) as a
result of recycling. **Differences significant ($p < 0.05$). (Values
are Means of 2 Expts + Range or 3 Expts $\pm$ S.E.M.)

vesicles generated by stimulation are denser than their less highly
labelled precursors and the observed increase in density can be
accounted for by a lower transmitter and water content. Differences
in density between either population and the corresponding populations in stimulated electromotor terminals can be accounted for
by the smaller size of the mammalian vesicles and the higher proportion
of the total vesicle volume and mass contributed by the membrane.

It is thus clear that vesicle metabolic heterogeneity is not
a feature peculiar to Torpedo. However, the dynamics of the MPLM
system appear to be faster than those of Torpedo. The turnover of
acetylcholine appears to be faster and there may be more impulse
traffic and spontaneous activity in the MPLM system than in the
electromotor system. Recycling vesicles may recover the biophysical
characteristics of resting vesicles more rapidly and the two refilling mechanisms postulated to occur in Torpedo, a fast partial
refilling during recycling and a slower "topping up" process during
rest (8) may differ less in their kinetics in the MPLM system. A
careful comparison of the kinetics of the two systems will be necessary
to settle these points. Further work will also examine whether
false transmitters preferentially enter the VP_2 compartment and
whether this compartment is, as in electromotor terminals, the preferred source of released transmitter on restimulation.

REFERENCES

1. Barker, L.A., Dowdall, M.J. and Whittaker, V.P. (1972): Biochem. J. 130:1063-1080.
2. Dowe, G.H.C., Kilbinger, H. and Whittaker, V.P. (1980): J. Neurochem 35:993-1003.
3. Fonnum, F. (1975): J. Neurochem. 24:407-409.
4. Freeman, J.J., Choi, R.I. and Jenden, D.J. (1975): J. Neurochem 24:729-734.
5. Giompres, P.E., Morris, S.J. and Whittaker, V.P. (1981): Neuroscience 6:757-763.
6. Giompres, P.E., Zimmermann, H. and Whittaker, V.P. (1981): Neuroscience 6:765-774.
7. Giompres, P.E., Zimmerman, H. and Whittaker, V.P. (1981): Neuroscience 6:775-785.
8. Giompres, P.E. and Whittaker, V.P. (1984): Biochim. Biophys. Acta 770:166-170.
9. Luqmani, Y.A., Sudlow, G. and Whittaker, V.P. (1972): Neuroscience 5:153-160.
10. Nagy, A., Baker, R.R., Morris, S.J. and Whittaker, V.P. (1975): Brain Res. 109:285-309.
11. Paton, W.D.M. and Zar, M.A. (1968): J. Physiol. 194:11-33.
12. Peterson, G.L. (1977): Biochemistry 83:346-356.
13. Suszkiw, J.B., Zimmermann, H. and Whittaker, V.P. (1978): J. Neurochem. 30:1269-1280.
14. von Schwarzenfeld, I. (1979): Neuroscience 4:477-493.
15. Weiler, M., Roed, I.S. and Whittaker, V.P. (1982): J. Neurochem 38:1187-1191.
16. Whittaker, V.P. and Barker, L.A. (1972): Meth. Neurochem. 2:1-52.
17. Zimmermann, H. and Denston, C.R. (1977): Neuroscience 2:695-714.
18. Zimmermann, H. and Denston, C.R. (1977): Neuroscience 2:715-730.
19. Zimmermann, H. and Whittaker, V.P. (1977): Nature, London, 267:633-635.

BIOCHEMICAL IMPLICATIONS OF THE SYNAPTIC VESICLE

ORGANIZATION IN QUICK FROZEN ELECTRIC ORGAN NERVE TERMINALS

A.F. Boyne and T.E. Phillips

Department of Pharmacology and Experimental Therapeutics
University of Maryland School of Medicine
Baltimore, MD 21201 U.S.A.
and
Department of Pharmacology
Northwestern University School of Medicine
Chicago, IL 60611 U.S.A.

INTRODUCTION

The elegant application of the quick freezing technique to neuromuscular junctions by John Heuser and Thomas Reese and their collaborators (8) has put an end to the era of reliance on aldehyde fixations in the correlation of ultrastructure and quantitative details of electrophysiology. We have attempted to apply the same technique to _Torpedine_ electric organ nerve terminals in the hope of resolving some of the biochemical controversies associated with that system (18). This chapter describes recent observations which lead to a re-evaluation of discrepancies in the literature vis-a-vis the vesicle hypothesis. The discussion emphasizes a critical need for information on how the acetylcholine (ACh) balance is affected when synaptic vesicles pump calcium.

MATERIALS AND METHODS

We have used a liquid N_2 based quick freezing technique (1) with the "Gentleman Jim" instrument (Pelco International, Tustin, CA 92681). Cylinders of electric organ were punched from the organ of anesthetized or pithed rays and the cylinders were sliced to produce a 0.6 mm thick sheet of stacked electrocytes with their intervening layers of nerve terminals. Individual stacks were dissected from the sheet by cutting along their collagenous margins and were mounted on a piece of filter paper glued to a rectangle of foam in a freezing probe. The tissue was delivered _without bounce_

519

to the dry surface of a highly polished copper bar which extended into liquid N_2. Under these conditions, the surface 12-15 microns is frozen without visible ice crystals.

There are various options in the next step (18). For the results discussed here, the ca. 15 mg tissues were dropped into and stored in tetrahydrofuran (THF; 16) containing 1% OsO_4 at -78°C for 3 days. The tissue was then slowly warmed to room temperature, which caused it to become osmicated. It was then washed in pure THF and embedded in plastic.

RESULTS

At low magnification, quick frozen material can look less appealing than aldehyde-treated tissue. The crystallinity of the deeper cells attracts attention as an evident artefact. However, at higher magnifications, in the well frozen surface, an improvement in fixation is dramatically evident. Although slice damage often causes nerve terminals in this region to lose their synaptic vesicles and fill with vacuoles (18), the electrocyte stacking permits examination of deeper terminals which were protected from mechanical damage by an overlying electrocyte. Unexpectedly, we found that nerve terminal cytoplasm crystallizes less readily than the extracellular fluid; this serves to increase the depth to which one can find well preserved nerve terminals. (This is also true for mammalian neuromuscular junctions; Meshul, Boyne and Albuquerque; unpublished).

In these deeper terminals, the occasional mitochondria are free of the inclusions seen in damaged surface terminals. Moreover, they are filled with synaptic vesicles which display a greater orderliness and resistance to clumping than is evident in aldehyde fixed tissues.

The most striking feature of the quick frozen nerve terminals is the presence of rows of synaptic vesicles aligned along the presynaptic membrane (Figure 1). A comparison with a cross-sectioned nerve terminal from tissue fixed in an aldehyde mixture containing 10 mM Ca^{++} is provided in Figure 2. As reported previously (2), the Ca^{++} content determines the frequency of synaptic vesicle fusion and bulging against the presynaptic membrane (e.g., at arrow in Figure 2). However, rows of such fusions were not found after Ca^{++}-containing fixatives were used. More recently, highly variable rows of vesicle-membrane alignments without bulging fusions have been found in tissues chemically fixed in the absence of Ca^{++} (by using a vehicle with 90 mM Na oxalate (17)). These may be analagous to the quick frozen arrangements.

When the pattern of distribution was quantitated in quick frozen blocks from four fish, the mean proportion of vesicles attached to the membrane was 18.3% $\pm$ 3.2 (S.D.) (18).

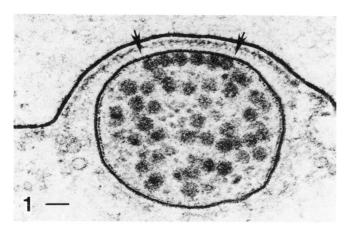

Figure 1.   Cross section of <u>Narcine brasiliensis</u> electric organ cholinergic nerve terminal prepared by quick freezing and tetra-hydrofuran freeze-substitution.   Note that a substantial proportion of the synaptic vesicle population (average=18%) is aligned with the presynaptic membrane and faces the postsynaptic receptors. The nature of the material in the vesicle core which becomes osmi-cated during this form of freeze-substitution is not yet established. Scale bar=100 nm.   [Reproduced from reference (18) with permission from Alan R. Liss, Inc.].

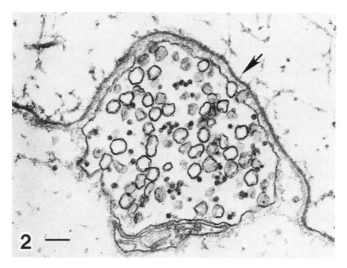

Figure 2.   Similar cross section to Fig. 1, except that this terminal was in a tissue prepared by chemical fixation in a solution containing 4% formaldehyde, 5% glutaraldehyde, and a hypotonic buffer and salt mixture which included 10 mM $CaCl_2$ [see ref. (2) for further details]. Occasional vesicles can be seen to be firmly attached to the nerve terminal membrane and bulging against it (e.g., at arrow) but orderly rows of such attachments are absent.   Scale bar=100 nm.

We also noted that the axons which pass between the electrocytes contain occasional vesicles associated with their neurotu-
bules. However, just prior to the terminal proper, the neurotu-
bules depolymerize and synaptic vesicles become more closely packed
(18). This close packing was not in itself a sufficient condition
for vesicle attachment to the axonal plasma membrane. Only when
the nerve terminal attained a position against the basal lamina
covered electrocyte did one see vesicle attachements to the presynaptic face. The apparent requirement for a basal lamina coating
was further tested by counting the frequency of vesicle attachements
over the post-junctional folds which, in this preparation, are so
wide that they leave a 'bare' patch on the nerve terminal membrane.
We found that 30.1 $\pm$ 4.2% (S.D.; n=4) of the basal lamina coated
membrane bore attached synaptic vesicles but that the frequency
dropped to 2.1 $\pm$ 1.4% (S.D.; n=4) over the middle of post-junctional
folds.

DISCUSSION

The basal lamina is known to be able to aggregate nicotinic
receptors (23). It seems, therefore, reasonable to assume that
the strands of basal lamina which can be seen to touch the postsynaptic membrane, are associated with the receptors in that surface. The observed arrangement of 18% of the vesicles relative to
receptors suggests that this extracellular structure provides a
signal which positions vesicles in such a way that, if they discharge,
cholinergic receptors are appropriately positioned to register the
event. We emphasize that such an interpretation is directly suggested
by the images of quick frozen material and is not dependent on prior
acceptance of the vesicle hypothesis.

Rows of membrane attached vesicles are not generally seen in
electric organ fixed in standard, Ca^{++}-containing fixatives. What
happens to them? Two possibilities are apparent: (1) they exocytose
during aldehyde fixation, i.e., they are structurally labile, or
(2) they detach from the membrane and pull back into the cytoplasm,
i.e., they are positionally labile. Although membrane-attached
vesicles may be labile in both senses, the following observations
suggest that the former possibility is a more likely explanation
for their disappearance.

(1) Chemical fixation has been shown to trigger bursts
of quantal transmitter release from neuromuscular junctions
(25). Whether vesicle exocytosis is the source of quanta
or is an epiphenomenon associated with quantal release,
this observation predicts vesicle loss during aldehyde
fixation.

(2) We have found that use of 90 mM Na oxalate (a
Ca^{++} precipitator which can cross cell membranes) in the

fixative does lead to some preservation of membrane associated vesicles (17). This suggests a Ca^{++} dependence of the loss during aldehyde fixation.

(3)   Plattner has reported that exocytotic loss of paramecium trichocysts during aldehyde fixation is also Ca^{++} dependent (20).

The simplest interpretation is that membrane attached vesicles are structurally labile and exocytotically evert into the plasma membrane during aldehyde fixation. If a tendency to exocytotic destruction is also expressed during homogenization, the ACh in this population would be hydrolyzed and therefore attributed to the cytoplasmic pool. One might argue that the inclusion of EDTA during homogenization should block a Ca^{++} dependent event. However, we have observed that it takes 20 min for a Ringers solution with Mg^{++} substituted for Ca^{++} to block transmission in a 15 mg, 0.6mm thick stack of tissue and a further 20 min to reverse the block (17). We doubt that the EDTA included in the incubation medium at the time of homogenization can penetrate the synaptic clefts sufficiently rapidly to block the loss of these vesicles during depolarizations triggered by the mechanical disruption.

The next question to consider is whether these labile, membrane associated vesicles, represent the hypothesized "membrane operators" (9) or "vesigates" (26, 27) which accumulate cytoplasmic ACh and discharge it in a quantal manner.

Vesicles have both an ACh uptake system (3, 11, 14) and a high affinity Ca^{++} pump (10, 15, 22). They may be expected to accumulate some of the Ca^{++} influx associated with nerve terminal activity. Is there a conflict between these two activities? There may be for the following reason. It has been shown that when Ca^{++} enters vesicles in vitro, ACh is released (5, 19). Since active zone vesicles would be exposed to the highest concentrations of incoming Ca^{++} (28), they would be most likely to be involved in Ca^{++} buffering (it has been suggested that the Ca^{++} gates are actually in the active zone (21)). The loss of ACh from these vesicles to the cytoplasm would result in a progressive reduction of the quantal amplitude as they accumulate more and more Ca^{++} and lose more and more ACh with repetitive stimulation.

Since reduction of quantal size with repetitive stimulation does not appear to occur in most preparations unless ACh synthesis is blocked, adherence to the vesicle hypothesis requires new but fairly specific postulates about vesicle behavior. For example, either (i) vesicles are able to release their Ca^{++} load and regain ACh in the intervals between action potentials, or (ii) the number of copies of the Ca^{++} pump protein is less than the number of vesicles so that only a small proportion of the vesicle population can carry out Ca^{++} buffering. If neither of these conditions apply, then even the newly defined population of mambrane associated vesicles does not provide a complete explanation of the quantal release process and it would be necessary to postulate the existence of another quantized releasing system in the nerve terminal, as has been done by Israël et al., (9) and Tauc (26).

The above considerations emphasize the importance of answering the following questions.

(1) How does the vesicle handle the osmotic stresses involved in electroneutral inward pumping of Ca^{++}? If Ca^{++} is taken up along with an anion, the vesicle core would become hypertonic and the internal pressure would rise as water moved in. Although an increase in osmotic pressure is an attractive way of priming the vesicles for discharge during the next action potential, it implies that the probability of release during trains of stimuli would go through massive oscillations as the starting population of active zone vesicles reached a similar level of osmotic turgor at the same time. Although cytoplasmic ACh has been reported to oscillate in electric organ (7), only small oscillations occur in cholinergic endplate potentials (13).

On the other hand, evidence that Ca entering vesicles drives out ACh^+ (5, 19), suggests that an electroneutral exchange of 2 x ACh^+ for 1 x Ca^{++} may be possible. Since an exchange mechanism could automatically reverse during the time between action potentials (when the Ca^{++} level outside the membrane-attached vesicles falls to resting levels), then there is at least one way in which vesicles could buffer Ca^{++} while avoiding osmotic stress and maintaining the constancy of quantal amplitude.

(2) It has been reported that vesicles isolated from frozen electric organ are loaded with ca. 150 mM Ca^{++} (24). These are the same vesicles that have now been shown to be able to accumulate ACh^+ (3, 11, 14). It would be of great interest to answer a corollary of question (1): When vesicles accumulate ACh^+ in vitro, in a low Ca^{++} medium, do they lose their internal Ca^{++} to the medium?

Israël et al., (9) and Tauc (27) have discussed how the concepts of "membrane operators" and "vesigates", could explain apparent conflicts in the interpretation of current data. We will consider the explanatory value of a population of membrane associated vesicles, the ACh of which is continually cycling through the cytoplasm as Ca^{++} is absorbed during an action potential and released in between action potentials.

The model readily explains the preferential release of newly synthesized ACh (6) since the vesicles at the membrane would be preferentially involved in Ca^{++} buffering and enhanced ACh turnover simply because they would be exposed to a higher concentration of Ca^{++} than other vesicles (28). Vesicles discharging during any given action potential would also be expected to be drawn from this pool because of their location at the membrane. Accelerated local ACh turnover would also explain the gradual but selective incorporation of false transmitters into the secreted material (12) prior to their incorporation into the general vesicle transmitter pool (4).

When AChE is injected into neuronal cell bodies, transmission is blocked before there is a measurable loss of vesicular ACh (27). This has appeared to be a powerful argument against the role of vesicles in neurotransmission; it suggests that the released ACh is cytoplasmic in origin. However, the result would be expected if the active zone vesicles must cycle ACh back and forth from the cytoplasm. Elimination of cytoplasmic ACh would eliminate their ability to refill.

In several preparations used in developmental studies, it has been reported that quantal neurotransmission can be detected prior to the appearance of synaptic vesicles in the associated nerve terminals. This conclusion assumes that aldehyde fixation would preserve the first vesicles to arrive in the maturing nerve terminals. A recent observation with quick frozen, 7 day old <u>Torpedo</u> <u>omata</u> is relevant to this assumption. Although the nerves are well developed and full of vesicular and cisternal elements, the terminals are small and do not cover as much of the innervated membrane as they will in adulthood. Furthermore, in the one ray analyzed to date, 48% of the vesicles were found attached to the terminal membrane. Although clearly preliminary, this suggests that the first vesicles to arrive have an increased likelihood of being found in the presynaptic membrane attachment sites. It would not be surprising if they were selectively destroyed by aldehyde fixation, giving the false impression that quantal transmission preceded the appearance of synaptic vesicles.

We thus conclude that the outstanding difficulties with the vesicle hypothesis become predictable consequences if (1) active zone vesicles are more unstable than cytoplasmic vesicles and (2) vesicular Ca^{++} loading can be reversed <u>in</u> <u>situ</u>. We have provided evidence for the first condition and look forward with great interest to data from other laboratories on the second.

Acknowledgement. This work was supported by NINCDS grants NS 13043 and 16167 and by the U.S. Army Res. and Development Command (Contract 17-81-C-1279). We thank Alan R. Liss Inc., for permission to reproduce Figure 1.

REFERENCES

1. Boyne, A.F. (1979):   J. Neurosci. Methods 1:353-364.
2. Boyne, A.F., Bohan, T.P. and Williams, T.H. (1974):   J. Cell Biol. 63:780-785.
3. Carpenter, R.S. and Parsons, S.M. (1978):   J. Biol. Chem. 253: 326-328.
4. Carroll, P.T. and Aspry, J.M. (1980):   Science 210:641-642.
5. Diebler, M.F. (1982):   J. Neurochem. 38:1405-1411.
6. Dunant, Y., Gautron, J., Israël, M., Lesbats, B. and Manaranche, R. (1972):   J. Neurochem. 19:1987-2002.
7. Dunant, Y., Israël, M., Lesbats, B. and Manaranche, R. (1977): Brain Res. 125:123-140.
8. Heuser, J.E., Reese, T.S., Jan, Y., Jan, L. and Evans, L. (1979): J. Cell Biol. 81:275-300.
9. Israël, M., Dunant, Y. and Manaranche, R. (1978):   Progr. Neurobiol. (Oxford) 13:237-275.
10. Israël, M., Manaranche, R., Marsal, J., Meunier, F.M., Frachon, P. and Lesbats, B. (1980):   J. Physiol. (Paris) 76:478-485.
11. Koenigsberger, R. and Parsons, S.M. (1980):   Biochem. Biophys. Res. Comm. 94:305-312.
12. Large, W.A. and Rang, H.P. (1978):   J. Physiol. (London) 285: 24-34.
13. Meiri, H. and Rahamimoff, R. (1978):   J. Physiol. (London) 278: 513-523.
14. Michaelson, D.M. and Angel, I. (1981):   Proc. Nat. Acad. Sci. (U.S.A.) 78:2048-2052.
15. Michaelson, D.M., Ophir, I. and Angel, I. (1980):   J. Neurochem. 35:116-124.
16. Ornberg, R.L. and Reese, T.s. (1981):   J. Cell Biol. 91:387a.
17. Phillips, T.E. (1982):   "Utilization of quick freezing and freeze-substitution, in conjunction with chemical fixation, in the ultrastructual analysis of Narcine brasiliensis electric organ nerve terminals." Ph.D. Thesis. Northwestern University, Chicago, IL 60611, U.S.A.
18. Phillips, T.E. and Boyne, A.F. (1984):   J. Electron Microsc. Tech. 1:9-23.
19. Pinchasi, I., Michaelson, D.M. and Sokolovsky, M. (1978):   Febs. Letts. 109:188-192.
20. Plattner, H. (1981): Cell Biol. Int. Rep. 5:435-458.
21. Pumplin, D.W., Reese, T.S. and Llinas, R. (1981):   Proc. Natl. Acad. Sci. (U.S.A.) 78:7210-7213.
22. Rephaeli, A. and Parsons, S.M. (1982):   Proc. Natl. Acad. Sci. 78:5783-5787.
23. Rubin, L.L., Gordon, A.S. and McMahan, U.J. (1980):   Neurosci. Abstr. 120:330.
24. Schmidt, R., Zimmermann, H. and Whittaker, V.P. (1980):   Neurosci. 5:625-638.
25. Smith, J.E. and Reese, T.S. (1980):   J. Exp. Biol. 89:19-29.
26. Tauc, L. (1978):   Biochem. Pharmacol. 27:3483-3488.

27. Tauc, L. (1982): Physiol. Rev. 62:857-893.
28. Zucker, R.S. and Stockbridge, N. (1983): J. Neurosci. 3:1263-
     1268.

PRESYNAPTIC CHANGES ACCOMPANYING TRANSMISSION OF A SINGLE NERVE IMPULSE: AN INTERDISCIPLINARY APPROACH USING RAPID FREEZING

Y. Dunant, L.M. Garcia-Segura[*], G.J. Jones,
F. Loctin, D. Muller and A. Parducz

Départements de Pharmacologie et de Morphologie[*]
C.M.U.
1211 Genève 4, Switzerland

INTRODUCTION

The electric organ of the fish <u>Torpedo</u> is an extremely favour-able tool to study presynaptic aspects of cholinergic transmission, because its profuse synaptic innervation allows biochemical analysis of transmission to be combined with both electrophysiological and morphological studies. The synaptic transmission, however, as at other cholinergic synapses, is very rapid. As judged from the electrophysiological postsynaptic response, release of the transmitter acetylcholine (ACh) occurs after a delay of a few (about two) ms and lasts for only 2-4 ms. To study the presynaptic changes accompanying transmission, it is therefore necessary to act very quickly, or alternatively, to prolongate (as it were, to slow down) the processes involved. Two separate strategies can be employed to achieve these goals:

1) By rapid freezing of single isolated prisms (columns of electroplaques), we have been able to study changes in presynaptic ACh stores with a time resolution about the same as the duration of a burst of 20 single impulses at 100Hz. (Such bursts are very similar to the activity of the electric organ of <u>Torpedo</u> in its natural habitat.) Freeze-fracture morphology of the presynaptic membrane during and after a single impulse is also possible for the rapidly frozen tissue, and preliminary results indicate changes in the membrane during this time.

2) It is well known that 4-aminopyridine (4-AP) has a strong potentiating action at cholinergic synapses.

For the <u>Torpedo</u> electric organ this action is predominantly one of increasing the duration of the electrical response, rather than its amplitude. The duration may be increased more than 100 times. Because this duration is much longer than the time resolution of our freezing experiments, we have been able to use rapid freezing to study both bio-chemical and morphological changes during the whole time course of the greatly prolonged single impulse which occurs in the presence of 4-AP.

Results of some of our recent work using these strategies are presented in this chapter.

MATERIALS AND METHODS

Stimulation and Rapid Freezing. The experiments described were done using single isolated prisms, dissected from slices of electric organ excised from <u>Torpedo</u> anesthetized by tricaine methane sul-phonate (MS 222, Sandoz; 1g/3l sea water). The isolated prisms were maintained in an oxygenated elasmobranch saline medium and allowed to recover during a time period of at least one hour. Further details, including also details of radioactive labelling of ACh in these tissue fragments, and of the techniques used for assaying ACh and ATP, may be found in Dunant et al. (6, 7).

For stimulation and rapid freezing of the specimens, a machine has been constructed on which up to 24 single prisms are held in place by their stimulating and recording electrodes. Stimulation of the prisms is controlled electronically and synchronized with the opening of a magnetic lock. A precise stimulation pattern can be applied to the prisms, which are then all frozen at the same time as the cartwheel head holding the prisms falls into a large volume of cryogenic liquid. In early experiments, the cryogenic liquid was isopentane cooled to -130°C with liquid nitrogen. More recently, we have used a mixture of 40% isopentane in propane as coolant, which has the advantage that it can be cooled to liquid nitrogen temperature without freezing (18).

Freeze-fracture replicas of the presynaptic membrane were obtained using conventional techniques (21) by carefully fracturing and shadow-ing the very surface regions ($< 10 \mu m$ depth; 10). Only these surface regions are frozen sufficiently fast that there is no ice damage when, as in these experiments, there is no chemical fixation or cryoprotection.

Rate of Freezing. At present, we do not have a definite measure of the freezing times for the specimens in our experimental con-ditions. Such measurements are difficult and, especially, difficult to interpret (cf. ref. 2). However, theoretical calculations and extrapolation from measurements on larger samples of <u>Torpedo</u> electric organ tissue indicate that the time for freezing half the volume of single prisms is about 50 ms (7, 8). For the surface regions studied morphologically, Heuser et al. (12) and van Harreveld and Trubatch (22) have measured freezing times of 2-3 ms. The more recent work of Escaig (9) and theoretical calculations (19) suggest that these regions are frozen in times of 1 ms or less.

RESULTS

Changes Accompanying Transmission of a Burst of Impulses. In these experiments our aim was to study changes in ACh levels and compartments during transmission of a short burst of impulses designed so as to mimic the natural activity of the electric organ (1). The extractable ACh is thought to be predominantly presynaptic in the electric organ (14) and exists in two forms: a "bound" form, which is contained in synaptic vesicles and a "free" form which is probably cytoplasmic in the nerve terminals. These forms may be separated, because only the former is obtained when the plasma membrane is disrupted and the free ACh hydrolyzed by the endogenous acetylcholinesterase activity.

Fig. 1a shows the changes observed in total and bound ACh during the short burst of impulses (see also ref. 7). From these results it is clear that it is the free ACh (defined as the difference between total and bound) which is utilized and reformed during the transmission - there is no significant change in bound ACh. Furthermore, by using radioactive labelling of the ACh pools, it could be shown that there is no significant exchange between the two pools of ACh during release (3).

Changes in Tissue ACh During Transmission of a Single Giant Impulse. The enormous potentiation of evoked ACh release on stimulation in the presence of 4-AP may be seen by comparing the electrical recordings of Figs. 1a and 1b. With 4-AP, the electroplaque potential (e.p.p.) is slightly increased in amplitude, and greatly prolonged. This giant e.p.p. has an initial peak which may last up to 100 ms, and is followed by a second peak, the late rebound. The overall duration is between 500 and 1000 ms, depending on the 4-AP concentration and, particularly, the temperature, mainly because the latter affects the amplitude and time course of the late rebound (3).

Fig. 1b shows the changes in total ACh during the giant e.p.p. The greater scatter in these measurements reflects the difficulties of keeping the conditions of 4-AP treatment and temperature constant. However, the results show clearly changes in ACh similar to those for a burst of impulses (cf. Fig. 1a). Interestingly, it appears from these results that the late rebound of the giant e.p.p. is associated with the rise in ACh after the initial fall. Although not shown in Fig. 1b, we have also measured the ACh content at later times, after the end of the giant e.p.p. Total ACh remains at a

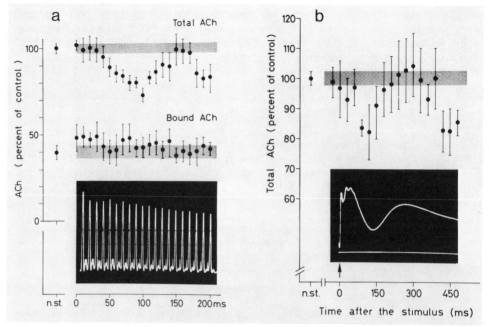

Figure 1.   Changes in ACh during a short burst of impulses (a), and
during a single giant e.p.p. due to 4-AP treatment (b). The biochemical data and electrical records have the same time scale.

reduced level (about 80% of control) for at least 1.5 s. The level
of bound ACh again does not change during the course of the giant
e.p.p. and in the following second (3).

Freeze-Fracture Morphology During the Giant e.p.p. We have obtained
replicas of freeze-fracture synaptic terminals frozen at 30 ms intervals during the whole time course of the giant e.p.p., with the
aim of looking for morphological changes which might be associated
with ACh release. The typical appearance of terminals cryofixed
and examined in this way is shown in Fig. 2.

Our first measurements were of the number of pits in the membrane,
as an index of vesicle-membrane interaction (12, 13). However,
only a relatively small number of pits were found and, although
there was a detectable increase during the giant e.p.p., this occurred
at a late time (between the initial peak and the late rebound);
there was no increase during the initial rise when ACh release must
be very high (Fig. 3).

In a second series of measurements, the density and size distribution of intramembrane substance (IMP) were evaluated. This

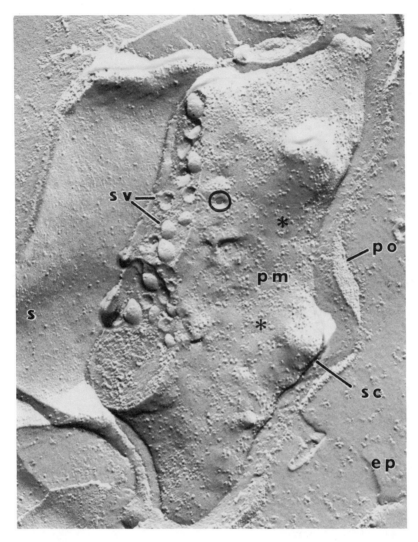

Figure 2. Freeze-fracture replica of a <u>Torpedo</u> electric organ synapse. PM, presynaptic membrane (P-face); SV, synaptic vesicles; PO, postsynaptic membrane; SC, synaptic cleft; EP, electroplaque cytoplasm; S, Schwann cell membrane (E-face). The asterisks show IMP on the presynaptic membrane. Circled is an example of a pit in the presynaptic membrane (x 60,000).

was suggested by the finding (15, 16, 17) that ACh release from synaptosomes of <u>Torpedo</u> electric organ appears to be accompanied by IMP changes. We found a very significant increase in IMP density during the giant e.p.p., which recovered with a similar time course

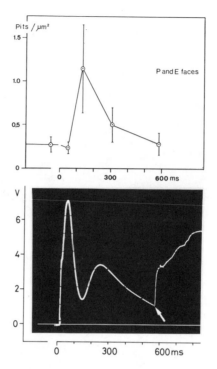

Figure 3. Changes in the density of pits in the presynaptic membrane P-face during the giant e.p.p. The increase in density at 150 ms is just significant ($0.2 < p < 0.5$).

to the electrical response (see Fig. 5 and ref. 10). The increase in IMP density occurred for both the P- and E-faces of the presynaptic membrane and was predominantly due to an increase in the density of large (> 10 nm diameter) particles (Fig. 4).

We have tested one experimental condition for which ACh release is prevented and found that this also does not produce a rise in IMP density. The condition tested was a low-Ca^{++}, high-Mg^{++} solution, which is well known to block ACh release (11, 20). Fig. 5 shows that, in this solution, even in the presence of 4-AP, ACh release was blocked, and there was no increase in IMP density.

DISCUSSION

The results of our experiments using rapid freezing confirm those of earlier work with <u>Torpedo</u> electric organ, using much longer periods of stimulation (5, 7). It appears that the pool of ACh which is used and reformed during stimulation is the "free" ACh, which is probably cytoplasmic, implying that transmitter release

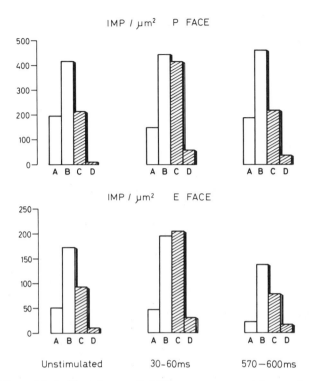

Figure 4. Size distributions of IMP on presynaptic membranes during the giant e.p.p., on the rising phase (30-60 ms) and on recovery (570-600 ms). Groupings A-D are diameters < 7, 7-10, 10-15 and > 15 nm.

is from this compartment and that the bound, i.e., vesicular, ACh is not immediately involved in transmitter release (14). It also appears that, during potentiation of the e.p.p. by 4-AP treatment, which most likely is due to a greatly increased amount of rate and quantity of transmitter released, it is the cytoplasmic pool of ACh which is the source of the transmitter (3).

Obviously, these biochemical results are difficult to reconcile with the vesicle hypothesis of ACh release (4), which implies that the release is due to exocytosis of ACh from the vesicular store. Indeed, we do not find, in our morphological studies on the Torpedo presynaptic membrane, evidence in the form of pits indicating vesicle exocytosis, such as those seen at the frog neuromuscular junction (12). It might be, of course, that our rapid freezing technique is not fast enough to catch such events, and that the increase in IMP that we see represents late events in the process of vesicle fusion (cf. ref. 13). We think this is unlikely, however, since

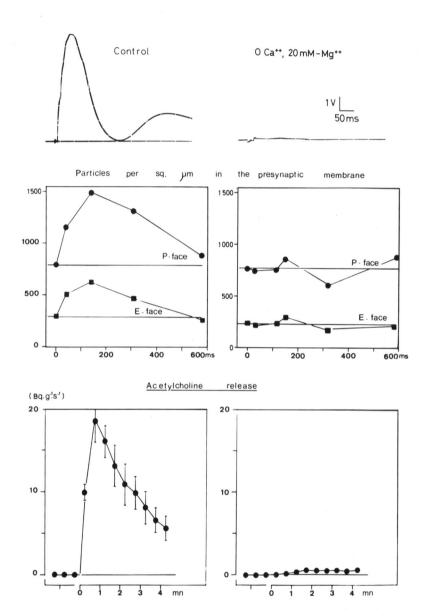

Figure 5. Low-Ca⁺⁺, high Mg⁺⁺ solution and giant e.p.p. Reduced Ca⁺⁺ abolishes the electrical response, the increase in IMP on both faces of the presynaptic membrane, and the overflow of radioactivity due to ACh release from prisms labelled with [^{14}C]-choline.

we can find clear images of pits indicating vesicle-membrane interaction, and a change in their density at a later time during the
giant e.p.p. (see Figs. 2 and 3). Also for the, admittedly small,
number of cross-fractures of nerve terminals that we have obtained
from stimulated prisms, there is no clear evidence for a significant
reduction in the number of synaptic vesicles, which would be expected
during the massive ACh release produced by 4-AP treatment, as indicated
by our biochemical measurements.

At present, almost all studies of the mechanism of presynaptic
transmission have used either repetitive stimulation for a long
time or 4-AP treatment to enhance release. It is evident that the
next step is to study transmission of a single impulse in normal
conditions. The biochemistry will be very difficult, since the
time resolution in freezing even of very 'small samples is expected
to be significantly longer than the duration of a single normal
e.p.p. Morphological studies are more promising, and we now have
results from six experiments in which single prisms were frozen
with successive delays of 1 ms during stimulation. These preliminary
results clearly indicate an increase in IMP density on the presynaptic
membrane P-face occurring during the e.p.p. and, again, do not give
evidence for vesicle exocytosis in the form of pits.

More work will obviously be needed with this type of experiment
because of uncertainties in the time at which the terminals were
frozen. We can conclude, however, that the hypothesis of a membrane
"operator" responsible for the release of cytoplasmic ACh seems to
be the most likely mechanism operating at the nerve-electroplaque
junction of the Torpedo (14, 17).

Acknowledgements. This work is supported by grants from the Swiss
FNRS. We are indebted to A. Masiero, F. Pillonel, J. Richez, P. Sors,
P. Vadi and N. Collet for their excellent technical and secretarial
assistance. Dr. A. Parducz was on leave from the Institute of Biophysics, Hungarian Academy of Sciences, Szeged, Hungary, with a
fellowship from the "Roche Foundation," Basle, Switzerland.

REFERENCES

1. Balbenoit, P. (1970):   Z. Vergl. Physiol. 67:205-216.
2. Bald, W.B. (1983): J. Microsc. 131:11-13.
3. Corthay, J., Dunant, Y. and Loctin, F. (1982):   J. Physiol.
    (London) 325:461-479.
4. Del Castillo, J. and Katz, B. (1957): In Microphysiologie Comparée
    des Elements Excitables. Colloques du CNRS, No. 67, pp. 245-258,
    Ed. CNRS: Paris (1957).
5. Dunant, Y., Gautron, J., Israël, M., Lesbats, B. and Manaranche,
    R. (1972): J. Neurochem. 19:1987-2002.
6. Dunant, Y., Eder, L. and Servetiadis-Hirt, L. (1980): J. Physi-
    ol. (London) 298:185-203.

7. Dunant, Y., Jones, G.J. and Loctin, F. (1982):   J. Physiol. (London) 325:441-460.
8. Dunant, Y., Jones, G.J. and Schaller-Clostre, F. (1982): J. Physiol. (Paris) 78:357-365.
9. Escaig, J. (1982):   J. Microsc. 126:221-229.
10. Garcia-Segura, L.M., Muller, D., Jones, G.J. and Dunant, Y. (1985):   Neuroscience (Submitted).
11. Harvey, A.M. and MacIntosh, F.C. (1940):   J. Physiol. (London) 97:408-416.
12. Heuser, J.E., Reese, T.S., Dennis, M.J., Jan, Y., Jan, L. and Evans, L. (1979):   J. Cell. Biol. 81:275-300.
13. Heuser, J.E. and Reese, T.S. (1981):   J. Cell. Biol. 88:564-580.
14. Israël, M., Dunant, Y. and Manaranche, R. (1979):   Prog.Neurobiol. 13:237-275.
15. Israël, M., Manaranche, R., Morel, M., Dedieu, J.C., Gulik-Krzywicki, T. and Lesbats, B. (1981):   J. Ultrastruct. Res. 75:162-178.
16. Israël, M., Lesbats, B., Manaranche, R., Morel, N., Gulik-Krzywicki, T. and Dedieu, J.C. (1982):   J. Physiol. (Paris) 78:348-356.
17. Israël, M. et al. (1984):   This book.
18. Jehl, B., Bauer, R., Dorge, A. and Rick, R. (1981):   J. Microsc. 123:307-309.
19. Jones, G.J. (1984):   J. Microsc. 136:349-351.
20. Katz, B. (1969):   The Release of Neural Transmitter Substances, Liverpool University Press, Liverpool, pp. 56-60.
21. Moor, H. and Muhlethaler, K. (1963):   J. Cell. Biol. 17:609-628.
22. Van Harreveld, A. and Trubatch, J. (1979):   J. Microsc. 115:243-256.

TENTATIVE IDENTIFICATION OF THE CHOLINE TRANSPORTER IN CHOLINERGIC PRESYNAPTIC PLASMA MEMBRANE PREPARATIONS FROM TORPEDO ELECTRIC ORGAN

I. Ducis

Abteilung Neurochemie
Max-Planck-Institut für Biophysikalische Chemie
Postfach 2841
D-3400 Gottingen, FR Germany[1]

INTRODUCTION

Sealed vesiculated membrane fragments, isolated from various tissues (1, 11, 14, 22), have proved extremely useful for studying the transport of many substances across membranes. Recently, neurotransmitter uptake has been studied in resealed plasma membrane fragments after osmotic lysis of synaptosomes (3, 10, 14, 17, 21). Isolated vesiculated synaptic membranes are a useful means of studying neuronal uptake mechanisms since energy sources can be readily controlled and there is no subsequent intracellular storage and metabolism of the substance under investigation. Although resealed membrane fragments have been used to study uptake processes in mammalian presynaptic terminals, only the transport of putative amino acid neurotransmitters has been reported (14, 17, 21).

The reason for the absence of studies on choline uptake in mammals, with two exceptions (15, 19), could well be the complex and heterogenous nature of mammalian brain. Ideally, in order to obtain membranes with high choline transport or binding activity, a preparation enriched in cholinergic nerve endings should be utilized. One such tissue is the purely cholinergic (7, 8) electromotor system of Torpedo marmorata with its abundance of synaptic contacts. Plasma membrane vesicles derived from Torpedo synaptosomes have been used to study release of acetylcholine (12) and the mem-

---

[1]Current Address: Section of Biochemistry, Molecular and Cell Biology, Cornell University, Division of Biological Sciences, Wing Hall, Ithaca, New York 14853, U.S.A.

branes, fractionated on sucrose gradients after disruption of synaptosomes, have been biochemically characterized (20, 23), but so
far no binding or transport studies have been carried out.

In this study, specific high-affinity choline transport into
resealed membrane fragments from <u>Torpedo</u> was demonstrated, the amount
of bound choline to various subfractions of synaptosome lysate was
estimated, and tentative identification of the choline transporter
was made.

METHODS AND MATERIALS

Preparation of membrane fractions. Synaptosomes from <u>Torpedo</u> were
prepared by a method similar to (13). The homogenate from 120-150
g of electric tissue was layered over twelve discontinuous Ficoll
density gradients comprising 12, 14, and 8 ml layers of 12, 8 and
2% (w/v) Ficoll in <u>Torpedo</u> Ringer solution (13). The gradients
were centrifuged in a Beckman SW 27 rotor at 23,000 rev/min for 45
min. Material equilibrating at the 2 to 8% interface was removed,
diluted (1:4) with <u>Torpedo</u> Ringer solution, and centrifuged for 20
min at 17,000 x g. The pellet (synaptosomes) was resuspended in a
small volume of 0.7 M sucrose in a buffer comprising 20 mM imidazole-
HCl (pH 8.5, 2°C) and 0.1 mM $MgCl_2$ (buffer 1 plus sucrose). After
diluting the synaptosomal suspension with buffer 1 to 0.1 mg of
protein per ml, the suspension was gently stirred for 90 min at
2°C before the addition of a 3 M KCl solution, resulting in a final
KCl concentration of 0.3 M. After 20 min, the lysate was centrifuged
at 27,000 x g for 20 min and the pellet was resuspended in 0.5-1.0
ml of a buffer comprising 20 mM imidazole-HCl (pH 8.0, 2°C), 300
mM KCl and 0.1 mM $MgCl_2$ (buffer 2) and dialyzed for 16-20 h against
1.5 l of buffer 2 at 2°C. When ion effects were being investigated,
the pellet was resuspended in a buffer with the appropriate loading
ions and dialyzed for 16-20 h against 1.5 l of the resuspension
buffer at 2°C. Before use, membranes were centrifuged at 10,000
x g for 10 min before resuspension in an appropriate amount of loading
buffer.

To subfractionate the synaptosome lysate (parent fraction),
the pellet was resuspended in 6 ml of buffer 2 and layered over
two discontinuous density gradients comprising 7 ml layers of 12,
10, 8, 6 and 4% (w/v) Ficoll in buffer 2. The gradients were centrifuged in a SW 27 rotor at 20,000 rpm for 90 min. Material equilibrated
at the interfaces was removed, diluted with buffer 2 (1:4), and
centrifuged at 27,000 x g for 20 min. After resuspension of the
pellets in 0.5-1.5 ml of buffer 2, they were either used immediately
or refrigerated and used for uptake, binding or biochemical characterization studies the following day.

Uptake assay. Membranes were preincubated for 20 min at 25°C before
uptake measurements were initiated. Valinomycin (2.5 μM) was present

during the preincubation where noted. Uptake was initiated by adding 50 μl of membranes to 400 μl of incubation medium. The incubation media were: buffer 2 (pH 7.4, 25°C) containing radiolabelled choline (2 μM, 10 μCi. mol^{-1}) or a buffer comprising 20 mM imidazole-HCl (pH 7.4, 25°C), 300 mM NaCl and 0.1 mM MgCl$_2$ (buffer 3), containing radiolabelled choline in the presence or absence of 10 μM hemicholinium-3. At specified time intervals, a 50 μl sample was removed from the incubation medium and diluted in either 1 ml of ice-cold buffer 2 or buffer 3, as appropriate. The diluted sample was immediately mixed and applied to a 0.45 μm cellulose filter (Millipore HAWP) and the membranes were washed with 4 ml of ice-cold buffer 2 or buffer 3. The filters were removed, solubilized in Bray's solution (2) and assayed for radioactivity in a Berthold LB 5004 liquid scintillation spectrometer. The uptake measurements were performed in triplicate or quadruplicate. The standard errors of the mean are indicated in the figures if they exceed the symbol size.

To estimate the specific binding of choline, membranes were preincubated for at least 45 min at 25°C in the appropriate hyperosmotic medium before uptake measurements were initiated. The incubation media were buffer 2 or 3, containing appropriate amounts of sucrose to give the final desired osmolarity. Reactions were terminated by diluting a 50 μl sample of incubation medium into 1 ml ice-cold buffer 2 or 3 of the same osmolarity as the incubation medium, applying it to a 0.45 μm cellulose filter, and washing it with 4 ml of appropriate ice-cold hyperosmotic buffer.

Enzyme assays. Acetylcholinesterase was measured as described in (6). Choline acetyltransferase was measured by the method of (9). Ouabain-sensitive (Na$^+$-K$^+$)ATPase activity was assayed by a modification of the method described in (18). 10-50 μl of sample were pipetted into tubes on ice. An appropriate amount of a stock solution of 30 mM MgCl$_2$, 200 mM KCl and 900 mM NaCl was premixed with distilled water and added to the tissue samples, resulting in a total volume of 250 μl. 100 μl of a stock solution (250 mM) of imidazole-HCl, ph 7.0, and 100 μl of a 0.05% saponin solution were next added. To appropriate tubes, 25 μl of water or a ouabain solution (2 mM) were added. The sample tubes, after mixing, were preincubated at 25°C in a shaking water bath. After 15 min, the tubes were placed on ice, allowed to cool and 25 μl of an ATP stock solution (60 mM) were added. After mixing, the samples were incubated for 40 min at 25°C in a shaking water bath. Final volume in each reaction tube was 500 μl. Final concentrations of all components were (mM): MgCl$_2$, 3; KCl, 20; NaCl, 90; ATP, 3; imidazole-HCl, 50; ouabain, 0.1; saponin, 0.01%. To stop the reaction, samples were placed on ice and ice-cold 1.2 M HClO$_4$ was added (335 μl). After 10 min, the samples were centrifuged at 10,000 x g for 10 min. 700 μl of the protein-free supernatant were removed and 700 μl of a freshly prepared solution of 144 mM FeSO$_4$ in 8.15 mM NH$_4$ - molybdate/0.58 M H$_2$SO$_4$ were added, mixed and left at room temperature for 20 min.

Absorbance at 700 nm was determined against distilled water blanks.
Ouabain-sensitive ATPase activity was calculated by subtracting
ouabain-insensitive activity from total ATPase activity after subtraction of appropriate blanks.

5'-nucleotidase (α,β-methylene adenosine diphosphate-sensitive
phosphatase) activity was assayed by a method exactly like that
described for ouabain-sensitive (Na^+-K^+)ATPase with the following
exceptions: saponin was not present, and the final concentrations
of the components in the incubation medium were (mM): $MgCl_2$, 5;
KCl, 5; NaCl, 250; Tris-HCl, 100 (pH 7.4); AMP, 1.5; α,β-methylene
ADP, 0.375.

Polyacrylamide gel electrophoresis and immune blotting. Sodium
dodecyl sulphate polyacrylamide gel electrophoresis (SDS-PAGE) and
immune blotting were performed exactly as described in (24).

Protein. This was determined as previously described (5).

Animals. Live Torpedo marmorata were obtained from the Institut
de Biologie Marine, Arcachon, France. They were housed in tanks
of running sea water (16°C) until required.

Chemicals. [^{3}H]choline chloride (80 Ci/mmol) and [^{3}H]acetylCoA
(705 mCi/mmol) were purchased from New England Nuclear (Boston,
MA, USA). Valinomycin and all specific reagents needed for enzyme
assays were obtained from Sigma (München, FRG). Ficoll 400 was
purchased from Pharmacia (Uppsala, Sweden). All other chemicals
used were of highest grade and obtained from Merck (Darmstadt, FRG).

RESULTS

The uptake of choline by plasma membrane vesicles is Na^+ and
Cl^- specific. Although optimal uptake of choline is absolutely
dependent on the presence of a Na^+-gradient (out > in), it is only
partially Cl^--gradient (out > in) dependent (Fig. 1). When Li^+,
Rb^+, K^+ or Cs^+ were substituted for Na^+, uptake of choline was significantly reduced (data not shown); when Br^-, SCN^- or NO_3^- were
substituted for Cl^-, uptake of choline was also significantly reduced
although Br^- could partially substitute for Cl^- (Fig. 1). The uptake
of choline as a function of time under various experimental conditions
is shown in Fig. 2. In the presence of an inwardly directed Na^+-
gradient choline is rapidly taken up during an initial 3-5 min period
over its equilibrium concentration but is thereafter lost more slowly
from the membrane vesicles (overshoot). By contrast, membranes
incubated at K^+ equilibrium show little uptake. When membranes
are incubated in the presence of an inwardly directed Na^+-gradient
and also in the presence of 10 μM hemicholinium-3, the Na^+ dependent
uptake of choline is completely eliminated and resembles the uptake
curve observed at K^+ equilibrium. Valinomycin potentiates the Na^+

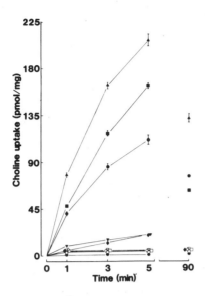

Figure 1. Uptake of 2 μM [^{3}H]choline by vesiculated fragments of
<u>Torpedo</u> electromotor presynaptic plasma membranes. Membranes, pre-
loaded with 150 mM K_2SO_4, 20 mM imidazole-H_2SO_4, 0.1 mM $MgSO_4$ and
150 mM mannitol were incubated in the following media buffered in
20 mM imidazole-H_2SO_4 and 0.1 mM $MgSO_4$: (▲) 300 mM NaCl, Na^+ and
Cl^- gradient; (●) 300 mM NaBr, Na^+ and Br^- gradient; (▼) 300 mM
$NaNO_3$, Na^+ and NO_3^- gradient; (□) 300 mM NaSCN, Na^+ and SCN^- gradient;
(○) 300 mM KCl, Cl^- gradient; (◇) 150 mM Na_2SO_4, 150 mM mannitol,
Na^+ gradient; (◉) 150 mM K_2SO_4, 150 mM mannitol, no gradient. Membranes, preloaded with 300 mM KCl, 20 mM imidazoleH_2SO_4 and 0.1 mM
$MgSO_4$ were incubated in the following media buffered in 20 mM imida-
zoleH_2SO_4 and 0.1 mM $MgSO_4$: (◆) 150 mM Na_2SO_4, 150 mM mannitol,
Na^+ gradient and Cl^- gradient (out < in); (■) 300 mM NaCl, Na^+ gradient, no Cl^- gradient. Membranes, preloaded with 150 mM Na_2SO_4,
20 mM imidazole-H_2SO_4, 0.1 mM $MgSO_4$ and 150 mM mannitol, were incubated
in 300 mM NaCl, 20 mM imidazole-H_2SO_4 and 0.1 mM $MgSO_4$ (▽), Cl^-
gradient, no Na$^+$ gradient. Uptake was carried out in all cases at
pH 7.4, 25°C. Points are means of triplicates or quadruplicates;
bars, the standard errors of the mean.

dependent uptake of choline but is without effect at K^+ equilibrium.
The insert of Fig. 2 shows that 10 μM hemicholinium-3, contrary to
observations made with membrane vesicles, does not completely block
choline uptake into synaptosomes.

The time course of choline uptake was found to be linear for
60 sec when the uptake of 1 and 10 μM choline was observed (data

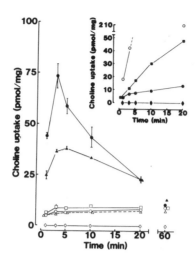

Figure 2. Uptake of 2 μM [^{3}H]choline by vesiculated fragments of
Torpedo electromotor presynaptic plasma membranes loaded with 300
mM KCl, 20 mM imidazole-HCl and 0.1 mM MgCl$_2$. Membranes were incubated in 300 mM NaCl, 20 mM imidazole-HCl, 0.1 mM MgCl$_2$ and (a) in
the presence (●) and absence (▲) of 2.5 μM valinomycin, or (b) in
the presence of 10 μM hemicholinium-3 (□). Membranes were incubated
in loading buffer in the presence (o) and absence (△) of 2.5 μM
valinomycin. The insert shows the uptake of 2 μM choline into
Torpedo synaptosomes at 2°C (●) or at 25°C in the presence (■) or
absence (o) of 10 μM hemicholinium-3. ◇ and ◆ show the choline
associated with membranes and synaptosomes, respectively, after
samples were diluted into appropriate buffer containing 0.1% Triton
X-100 after maximum uptake had occurred. Points are means of triplicates or quadruplicates; bars, the standard errors of the mean.

not shown). In the experiments determining K_m, V_{max} and K_i of hemi-
cholinium-3 inhibition, incubations of 40 sec were used. For determination of K_m and V_{max} (16) which were, respectively, 1.7 μM
and 96 pmol. min^{-1}.mg of protein^{-1} (Fig. 3), membranes were incubated
at 25°C in the presence of various concentrations of choline (1-10
μM) either in the presence of an inwardly directed Na$^+$-gradient,
or at K$^+$ equilibrium. The specific uptake of choline was taken as
the difference between uptake in the presence of an inwardly directed
Na$^+$-gradient and that observed at K$^+$ equilibrium. The radioactivity
associated with membranes at K$^+$ equilibrium was assumed to be of a
non-specific nature. This assumption seems valid from the experimental
observations made in Fig. 2, since hemicholinium-3 is a potent inhibitor of high-affinity choline uptake.

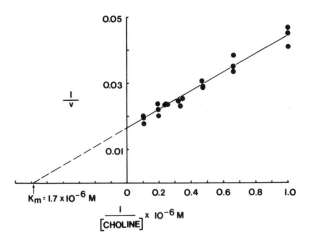

Figure 3. Double-reciprocal plot of the initial uptake of [3H]choline into resealed membranes in the presence of an inward directed Na$^+$-gradient. Velocity (v) was expressed as picomoles of choline accumulated per mg protein per 40 sec. An equation for the line was determined by the method of least squares.

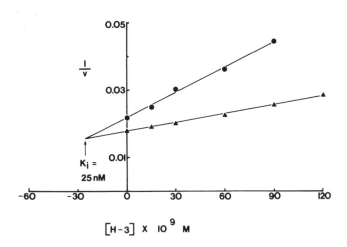

Figure 4. Estimation of K_i for hemicholinium-3 by the method of Dixon (4). Plots of the reciprocal of the rate of [3H]choline uptake (1/v) by membrane vesicles vs hemicholinium-3 concentration revealed hemicholinium-3 to be a purely competitive inhibitor with a K_i of 2.5 x 10^{-8} M. Velocity (v) is expressed in pmol of [3H]choline accumulated per mg protein per 40 sec. ● represent values determined at a [3H]choline concentration of 2.1 μM; ▲ represent values determined at a [3H]choline concentration of 5.5 μM. Equations for the lines were determined by the method of least squares.

When the K_i of hemicholinium-3 was determined according to the method of Dixon (4), membranes were incubated in the presence of various concentrations of hemicholinium-3 and either 2.1 or 5.5 μM choline. The membranes were incubated for 40 sec in the presence of an inwardly directed Na^+-gradient or at K^+ equilibrium. Hemicholinium-3 inhibited high-affinity uptake of choline in a purely competitive manner with a K_i of 25 nM (Fig. 4).

When the synaptosome lysate was subfractionated on a Ficoll gradient six fractions (A-F) were obtained. In order to estimate the Na^+ specific binding of choline, membrane fractions were exposed to increased osmolarity of the incubation medium during transport. Radioactivity associated with membranes at K^+ equilibrium was subtracted. Extrapolation of the curves to infinite osmolarity was used as an estimate of choline binding sites (Fig. 5). Correlations of the number of binding sites with enzyme marker activities showed the best positive correlation ($r^2 > 0.993$) with acetylcholinesterase activity (Fig. 6), although a good positive correlation ($r^2 > 0.987$) was also found with 5'-nucleotidase and a good negative correlation ($r^2 > -0.915$) with ouabain-sensitive ATPase. SDS-PAGE analysis of the proteins of the membrane fractions showed seven components that increased with estimated enrichment in carrier concentration and had approximate M_rs of 200, 140, 68 (doublet), 57, 54 and 28 KDa. Three of these (200, 140, 68 doublet) were tentatively identified as myosin, 5'-nucleotidase and choline acetyltransferase. Acetylcholinesterase was positively identified as a 68 KDa component by immune blotting (data not shown). The 57, 54 and 28 KDa components, therefore, remain as candidates for the choline carrier. However, since the 54 KDa shows a better correlation ($r^2 > 0.958$) with acetylcholinesterase (Fig. 7), a marker for presynaptic plasma membranes, this component has been tentatively identified as the choline transporter.

DISCUSSION AND SUMMARY

In the resealed membranes derived from <u>Torpedo</u>, and described in this chapter, Na^+/choline cotransport was routinely demonstrated. Since the membranes are relatively impermeable to both Na^+ and choline (the impermeability of membranes to choline can be estimated by measuring the uptake of choline at KCl equilibrium, where the only driving force for the choline transporter is the 2 μM choline gradient), a transient Na^+-gradient-dependent increase in the intramembranal choline concentration over the equilibrium value can be demonstrated. However, optimal Na^+/choline cotransport is not observed in the absence of Cl^-. Since significantly lower Na^+/choline cotransport is observed in the presence of SCN^-, NO_3^- and Br^-, anions which are more permeant to membranes than Cl^-, this suggests a specific role for Cl^- in the transport process beside its role as a counterion. The evidence, then, suggests that the primary role of Cl^- is to somehow "activate" the transporter.

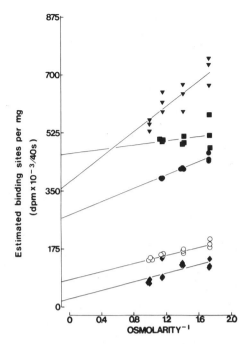

Figure 5. Radioactivity associated with different membrane fractions after exposure to 2 μM [^{3}H]choline as a function of the osmolarity of the incubation medium. The radioactivity associated with membrane vesicles decreased after 40 sec of incubation with 2 μM [^{3}H]choline at 25°C when the osmolarity of the incubation medium was increased. The equations for the relationships between radioactivity associated with membrane vesicles and osmolarity of the incubation medium was computed by the method of least squares. The lines associated with different membrane fractions intersect the ordinate at different values. They are as follows (x 10^{-3}): A ($\blacklozenge$), 16.5; B (not shown), 52.3; C (o), 76.0; D (not shown), 120.1; E ($\blacktriangledown$), 357.1; F ($\blacksquare$), 458.3 and the parent fraction ($\bullet$), 268.0.

Although a systematic study has not yet been made, there was some indication that the fairly large differences between preparations in the absolute uptake of choline were due to age and seasonal variations between fish. However, the following properties for the high affinity uptake of choline could be routinely demonstrated in my resealed synaptic plasma membrane fragments: 1) the absolute dependency of choline uptake on the Na$^+$-gradient (out > in); 2) the stimulation of choline uptake by an inside negative diffusion potential generated by the addition of valinomycin, a specific K$^+$ ionophore; 3) the dependency of choline uptake on the presence (external) of Cl$^-$, and only partially on its (out > in) gradient; and,

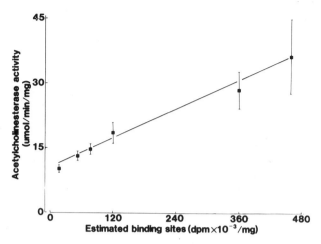

Figure 6. Correlation of acetylcholinesterase activity of the different membrane fractions with estimated binding sites. The equation for the line, y = 0.0549x + 10.4, was computed by the method of least squares, with a correlation coefficient greater than 0.993. Points represent the average of six separate measurements of acetylcholinesterase activity; bars, the standard errors of the mean.

4) observation of high affinity hemicholinium-3 inhibition of high affinity choline transport.

When the synaptosome lysate was subfractionated on a Ficoll gradient, it was found that all fractions exhibited the ability to transport choline to varying degrees. However to correlate transport ability with the number of carriers present did not seem justified. Since morphological analysis of lysed synaptosomes revealed many structures besides intact resealed membranes, it was possible that the choline transporter could be even more abundant in unsealed membrane fragments which would contribute little or nothing to an assay measuring Na$^+$/choline cotransport. Therefore, the specific binding of choline to the various subfractions was estimated by exposing membranes to hyperosmotic conditions during short (40 sec) periods of choline uptake. This method may have several advantages over methods using equilibrium and low temperature conditions to measure binding: choline binding, like choline transport, could be temperature and Na$^+$-gradient dependent and there is no assumption that diffusion or transport did not occur. However, this method does not provide a way to calculate binding constants, only relative specific binding capacity.

Relative specific binding measurements of choline showed that the estimated concentration of choline carriers increased down the

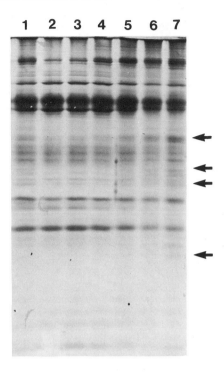

Figure 7. SDS-PAGE analysis of the proteins present in the different membrane fractions after fractionation of the synaptosome lysate (1.) on a 4-12% Ficoll - 300 mM KCl - 20 mM imidazole - 0.1 mM MgCl$_2$ (pH 8.0 at 2°C) gradient. The various gel patterns represent the following membrane fractions recovered, respectively, from top to bottom of the gradient: (2.), A; (3.), B; (4.), C; (5.), D; (6.), E; (7.), F. The position of acetylcholinesterase (68 KDa) and the three tentative candidates for the choline transporter (57, 54 and 28 KDa) are represented by arrows. 100 μg of protein were applied to each gel.

gradient, which showed high positive correlation with both 5'-nucleotidase and acetylcholinesterase activities. Ouabain-sensitive ATPase activity, on the other hand, was highest at the top of the gradient and was lowest at the bottom, resulting in a high negative correlation with increased estimated enrichment in choline carriers. Since acetylcholinesterase was directly identified on gels immunochemically, and the best positive correlation was found between its activity and estimated number of binding sites, it was assumed that the choline transporter gel pattern would be similar to that of acetylcholinesterase. This rationale resulted in the tentative identification of a 54 KDa protein as the choline carrier. However, direct identification of the choline carrier will come only through fractionation and reconstitution of membrane components or, perhaps, affinity labelling experiments.

REFERENCES

1. Aronson, P.A. and Sactor, B. (1974): Biochim. Biophys. Acta 356:231-243.
2. Bray, G.A. (1960): Anal. Biochem. 1:279-285.
3. Breer, H. (1983): FEBS Letters 153:345-348.
4. Dixon, D.M. (1953): Biochem. J. 55:170-171.
5. Ducis, I. and Koepsell, H. (1983): Biochim. Biophys. Acta 730: 119-129.
6. Ellman, G.L., Courtney, K.D., Andres, Jr., V., Featherstone, R.M. (1961): Biochem. Pharmacol. 7:88-95.
7. Feldberg, W., Fessard, A., Nachmansohn, D. (1940): J. Physiol. (Lond.) 97:3P.
8. Feldberg, W. and Fessard, A. (1942): J. Physiol. (Lond.) 101: 200-216.
9. Fonnum, F. (1975): J. Neurochem. 24:407-409.
10. Gordon, D., Zlotkin, E. and Kanner, B. (1982): Biochim. Biophys. Acta 688:229-236.
11. Hopfer, V., Nelson, K., Purotto, J. and Isselbacher, K.S. (1973): Biol. Chem. 248:25-32.
12. Israël, M., Lesbats, B. and Manaranche, R. (1981): Nature, Lond., 294:474-475.
13. Israël, M., Manaranche, R., Mastour-Frachon, P., Morel, N. (1976): Biochem. J. 160:113-115.
14. Kanner, B.T. (1978): Biochemistry 17:1207-1211.
15. King, R.G. and Marchbanks, R.M. (1980): Nature 287:64-65.
16. Lineweaver, H. and Burk, D. (1934): J. Amer. Chem. Soc. 56:658-666.
17. Marvizón, J.G., Mayor, Jr., F., Aragon, M.C., Gimenez, C. and Valdivieso, F. (1981): J. Neurochem. 37:1401-1406.
18. Medzihradsky, F., Nandhasri, P.S., Idoyaga-Vargas, V., Sellinger, O.Z. (1971): J. Neurochem. 18:1599-1603.
19. Meyer, E.M. and Cooper, J.R. (1982): Science 217:843-845.
20. Morel, N., Manaranche, M., Israël, M., Gulik-Krzywicki (1982): J. Cell Biology 93:349-356.
21. Roskoski, R., Jr. (1981): J. Neurochem. 36:544-550.
22. Rudnick, G. (1977): J. Biol. Chem. 252:2170-2174.
23. Stadler, H. and Tashiro, T. (1979): Eur. J. Biochem. 101:171-178.
24. Walker, J.H., Obrocki, J., Zimmermann, C.W. (1983): J. Neurochem. 41:209-216.

A RECONSTITUTED PRESYNAPTIC MEMBRANE EQUIPPED WITH PROTEIN STRUCTURES WHICH PERMIT THE CALCIUM DEPENDENT RELEASE OF ACETYLCHOLINE

M. Israël, N. Morel, R. Manaranche, B. Lesbats,
T. Gulik-Krzywicki and J.C. Dedieu

Département de Neurochimie
Laboratoire de Neurobiologie Cellulaire and
Centre de Génétique Moléculaire
C.N.R.S.
91190 Gif-sur-Yvette, France

INTRODUCTION

Presently most authors accept the view that there is, in the nerve terminal, a genuine cytosolic free acetylcholine (ACh) compartment (20 to 50% of the total ACh; 8, 31) and that the enzyme choline acetylase, which synthesizes it, is also located in the cytosol (5). Free ACh decreases and is renewed in the course of stimulation of electric organ slices (1, 2, 3, 7, 8). Starting with the whole tissue it was shown that the characteristic decay curve of the electrical discharge of a stimulated electric organ slice was associated with a characteristic variation of the free ACh compartment (1, 7, 8). The correlation was followed down to a few stimuli (3).

The release of free ACh from isolated cholinergic synaptosomes has more recently been obtained (11). This approach enables the cytosolic and vesicular pools to be directly evaluated and physically separated from each other. This study was performed with the aid of the chemiluminescent procedure developed by Israël and Lesbats (11, 12). In a reaction mixture containing cholinesterase, choline oxidase, peroxidase and luminol, the presence of ACh gives rise to a light emission. If cholinergic synaptosomes isolated from Torpedo electric organ (19, 25) are frozen-thawed in such a reaction mixture, there is an initial light emission giving the size of the cytosolic compartment (40%). Then the light measured by opening the synaptic vesicles with a detergent gives the size of the bound ACh pool (11). It has also been possible to confirm with the chemiluminescent method all previous observations showing that the cytosolic pool (i.e., free ACh) could be depleted by stimulation (11).

The release of cytoplasmic ACh could conceivably be performed by a special population of vesicles, "operative vesicles," which would be submitted to a rapid recycling process at the active zone (8). The phenomenon would be related to a process called "futile recycling" in the model proposed by Suszkiw (28) and Weiler et al. (31). Such a model explains at least the great stability and slow turnover of a majority of the vesicular pool; a stability which was quite unexpected when it was first described (1). Another possibility could be that the release of cytoplasmic ACh is not performed through a special population of vesicles, but rather results from a typical property of the presynaptic membrane. Such a mechanism should be seriously considered, since it has indeed been found that synaptosomal sacs refilled with ACh released it in proportion to their internal ACh concentration, and in proportion to the calcium influx triggered by a calcium ionophore (13).

The stability of the vesicular population, and the fact that the few endo-exocytosis pits found in some conditions (26) were not systematically associated with the release of ACh (15, 16, 18) suggested that the packet of transmitter might well cross the membrane through some other structure. Several investigations have already studied the effect of stimulation on presynaptic intramembrane particles (4, 27, 29, 30, 32). In most cases variations of densities have been reported, but they have not been studied in parallel with the release of ACh. The only common ultrastructural change which was observed was the appearance of a category of large 8 to 18 nm particles, while smaller particles (5 to 11 nm) disappeared (15, 16, 18). The large particles could be pinched-off with the E or P faces of the membrane, according to the conditions used. The hypothesis that calcium causes the appearance of large particles in the membrane and that these particles ensure the calcium dependent translocation of ACh has been proposed in this context by Israël et al. (16, 18).

Since intact tissues, synaptosomes, and synaptosomal sacs were shown to release their cytoplasmic or free ACh content upon calcium entry, it appeared possible to go one step further and try to reconstitute a proteoliposome which would incorporate the presynaptic membrane elements that might ensure the calcium dependent ACh permeability.

In our recent works (14, 17), we have shown that a functional presynaptic membrane can be reconstituted from a lyophilized presynaptic membrane powder, mixed with lipid in an organic solvent. The reconstituted proteoliposome is equipped with a calcium dependent ACh release mechanism, which seems to depend on specific presynaptic membrane proteins. Intramembrane particles which are formed in the proteoliposomal membrane, while it releases ACh, might well be directly involved in the calcium dependent translocation of the transmitter.

The reconstitution of a functional presynaptic membrane appears to be a valuable tool to solve several problems dealing with the uptake and release of transmitter. Several reports by us and others have already described positive results in this area of investigation (14, 17, 20, 23, 24).

METHODS

Biochemical Methods. Torpedo synaptosomes were isolated as previously described (19, 25). The procedure involves a gradual comminution of the tissue by filtrations through calibrated metallic grids and a non equilibrium centrifugation of the filtrate on a density gradient. The physiological medium used was: 280 mM NaCl, 3 mM KCl, 1.8 mM $MgCl_2$, 3.4 mM $CaCl_2$, 1.2 mM Na phosphate buffer (pH 6.8), 5.5 mM glucose, 300 mM urea, and 100 mM sucrose. When equilibrated with O_2, $NaHCO_3$ (4-5 mM) was added to adjust the solution to pH 7.0 - 7.1.

Preparation of Presynaptic Membrane Proteoliposomes. After their isolation, synaptosomes were osmotically shocked, their membrane was centrifuged at a low speed (12,000 g x 30 min) resuspended in 300 μl H_2O, and lyophilized as previously described (25). The lyophilized membrane powder derived from 15 to 20 g tissue (1 to 2 mg presynaptic proteins) was mixed with 3.5 to 4 mg lecithin (1-phosphatidyl choline, dipalmitoyl) and dissolved in 1 ml 1butanol. The organic solvent was evaporated under a stream of N_2. Then the material was resuspended in 0.5 ml of a solution consisting of: 100 mM potassium succinate, 10 mM Tris buffer pH 7.2 and 10^{-5}M phospholine (ecothiopate iodide). After 15 minutes, necessary to block esterase, 50 mM ACh chloride was added to the suspension (the potassium succinate was prepared by neutralization of the acid with KOH). The suspension was then sonicated at room temperature (15 s with the 1 cm probe of a Bioblock France sonicator, set at maximum power scale 4). The proteoliposomes were cleaned from external ACh by two gel filtrations done in succesion on 5 ml of Sephadex G-50 (coarse) columns. The Sephadex beads were in a solution consisting of 150 mM Tris buffer, pH 8.4 and 50 mM NaCl. In this solution, the proteoliposomes are stable for many hours and retain amounts of ACh in the order of 50 pmol per μl suspension.

Acetylcholine Release from Proteoliposomes. The chemiluminescent procedure for measuring ACh was used (9-13, 17). In this method acetylcholinesterase hydrolyzes ACh, the choline generated is oxidized by choline oxidase, and the H_2O_2 produced gives a light emission in the presence of luminol plus peroxidase. We first prepare a mixture consisting of: 100 μl choline oxidase (Wako) (250 units ml^{-1}), 50 μl horseradish peroxidase (Sigma type II) (2 mg ml^{-1}), and 100 μl luminol (Merck) (1 mM in 0.2 M Tris buffer, pH 8.6). Each assay tube receives 15 μl of the mixture, 3 μl of acetylcholinesterase (Boehringer; 1000 units ml^{-1}; cleaned on Sephadex G-50), and 250 μl of a solution consisting of 100 mM potassium succinate in 10 mM Tris buffer pH 8.4. This assay mixture is constantly stirred with a small magnet. The release of ACh is measured as follows: first the calcium ionophore A 23187 is added (1 to 16 μM final concentration) (the stock solution, 20 or 50 mM is dissolved in dimethylsulfoxide). Five to 20 μl of proteoliposomes are then added (this corresponds to 1 nmol of entrapped ACh). When all traces of choline in the external solution are oxidized, the light emission returns to baseline. The release of ACh is then triggered.

Morphological Studies. Freeze-fracture studies were performed as previously described (15, 26). A thin layer of suspension is set between two copper plates and frozen in liquid freon. It is then fractured, shadowed and replicated according to the "sandwich technique" of Gulik-Krzywicki and Costello (6) and examined in a Siemens Elmiskop 102 electron microscope.

Protein Analysis. SDS-polyacrylamide gel electrophoresis was carried out as described by Laemmli (21). The stacking gel and the separating gel (3 and 7.5% acrylamide, respectively) each contained 0.1% SDS. Samples (50-60 μg protein) were heated at 100°C for 3 min in the presence of 2.5% SDS and 5% 2-mercaptoethanol. Electrophoresis was carried out at 40 mA (constant current) for about 4 h. Gels were stained and fixed in 0.0125% Coomassie Brilliant Blue, 20% methanol, 7% acetic acid.

RESULTS

Properties of Proteoliposomes Obtained from Lyophilized Presynaptic Membranes. The first property we expected from these proteoliposomes was that they would entrap a significant amount of ACh. This was indeed the case, since a volume of 1 μl of the fraction collected after the final gel filtration step liberated upon addition of Triton X-100 an amount of ACh of 56 $\pm$ 5 pmol of ACh (n = 10). Since the proteoliposomes were formed in a solution of 50 mM ACh (i.e. 50 pmol/nl) it is concluded that the proteoliposomal volume in 1 μl fraction is 56/50 = 1.12 nl (0.1% of the fraction volume). For comparison, 1 μl of the initial synaptosomal fraction represents a synaptosomal volume of 1.5 nl. One ml of the proteoliposomes is associated to 300 μg sedimentable protein. Thus, the reconstituted proteoliposomes prepared with lyophilized presynaptic membranes have an internal volume and a protein content close to the native synaptosomes.

The reconstituted proteoliposomes (Fig. 1) show in their convex and concave faces intramembrane particles resembling the ones found

Figure 1. Reconstituted presynaptic membrane. The proteoliposomes obtained do not contain any internal organelles. Their convex and concave faces show numerous intramembrane particles.

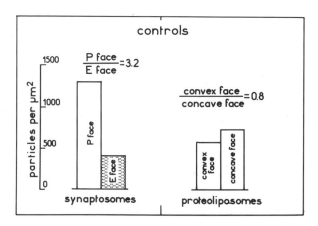

Figure 2. Comparison of intramembrane particles in the native and reconstituted membrane. The non-symetrical distribution of particles between the P and E faces of the synaptosomal membrane (P/E = 3) is lost in the reconstituted membrane (convex/concave ratio: 0.8).

in the P and E faces of the synaptosomes. However, the ratio of
the number of particles in the two faces is quite different in the
reconstituted and synaptosomal membranes; it is close to 0.8 for
proteoliposomes and 3.2 for synaptosomes (Fig. 2). Another difference
is that proteoliposomes do not contain any visible structures (Fig. 1)
while synaptosomes contain numerous synaptic vesicles.

Acetylcholine Release from Proteoliposomes. A neurotoxin extracted
from the venom glands of an annelid Glycera convoluta (22) triggers
a large increase of the miniature end-plate frequency at neuromuscular
junctions or electric organ synapses (22), and a substantial ACh
release from synaptosomes. When the toxin was applied to proteoliposomes filled with ACh it induced the release of ACh (Fig. 3a).
When the buffer without toxin was added, the record remained flat
(Fig. 3b).

It is also possible to trigger the release of ACh from the
proteoliposomes by adding them to a solution already containing
the venom (Figure 4a). In this case all traces of external choline
in the proteoliposomes were first oxidized by incubating them with
a few μl of the choline-oxidase reaction mixture. Fig. 4b shows
the blank value obtained by adding these proteoliposomes to mixture
without venom; Fig. 4a shows the release of ACh obtained after adding
these proteoliposomes to a reaction mixture containing Glycera venom.

We have attempted to establish whether or not the Glycera neurotoxin induced release is calcium dependent. Fig. 5 shows that the
slope of ACh release measured when proteoliposomes were incubated
in the presence of venom was considerably increased by the addition
of calcium, confirming the calcium dependency of venom action.

In other experiments, we have shown that the proteoliposomes
are able to release ACh when a calcium influx is generated. The
ionophore A 23187 was first incorporated into the system. The subsequent addition of calcium triggered the efflux of ACh (Fig. 6a).
In the absence of ionophore, no ACh release occurred upon calcium
addition (Fig. 6b).

We have also checked whether the ACh content of the proteoliposomes decreases after ACh release. This was performed by bursting
proteoliposomes in the reaction mixture with a detergent (Triton
X-100) added before, or after triggering the release of ACh. In
general, 20 to 50% of the proteoliposomal ACh content was released
within a few minutes. The release was slower after neurotoxin action
than with the calcium ionophore.

In order to evaluate the efficiency of the reconstitution procedure, the release of ACh from proteoliposomes (Fig. 3, 4, 5 and
6) was compared to the release of ACh from stimulated synaptosomes

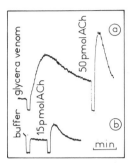

Figure 3. Acetylcholine release from proteoliposomes. The release
of ACh was continuously recorded with the chemiluminescent ACh assay.
In a, the Glycera venom triggered the release of ACh which was evaluated by the injection of an ACh standard. In b, control values
obtained by injecting buffer instead of venom.

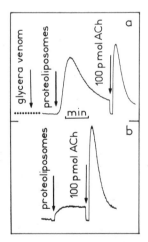

Figure 4. Acetylcholine release from proteoliposomes. a: Release
of ACh triggered by injecting the proteoliposomes in a reaction mixture
containing the Glycera venom (in this case the proteoliposomes were
previously treated with a small amount of choline oxidase reaction
mixture to eliminate all traces of external choline).   b: Control
values obtained by injecting the proteoliposomes in a reaction mixture
which did not contain the venom.

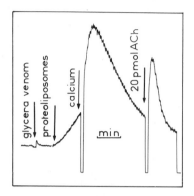

Figure 5. Acetylcholine release from proteoliposomes: Calcium dependency of venom action. After having added the Glycera venom to the proteoliposomes in the absence of calcium, the release of ACh was very small. The addition of calcium changed immediately the slope of the recorded light emission.

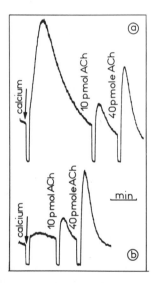

Figure 6. Calcium influx and acetylcholine release from proteoliposomes. In a, the release of ACh was triggered by the addition of calcium to proteoliposomes which had been treated with the calcium ionophore A23187. In b, the ionophore was omitted, and the calcium addition failed to induce a release of ACh comparable to the record shown.

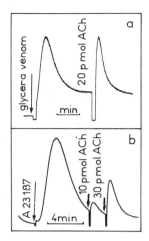

Figure 7. Acetylcholine release from synaptosomes. This figure is shown to compare the release of ACh from synaptosomes to the release capabilities of the reconstituted proteoliposomes shown. In all cases the ACh contents were comparable.

(Fig. 7a,b). The amount of occluded ACh in the stimulated synaptosomes, as with proteoliposomes, was 1 nmol. Moreover, the concentrations of Glycera venom or ionophore used were comparable to those previously shown. It can be seen from such a comparison that the reconstituted membrane is efficiently equipped with the native release mechanism.

Role of Presynaptic Membrane Proteins in the Release of ACh. In two experiments we treated the lyophilized presynaptic membrane powder with pronase before the reconstitution of the membrane. It was found that these pronase treated liposomes lost most of their ability to release ACh. Furthermore, the protein content of the reconstituted structure was reduced to 30% of the control.

In order to extract the protein structures involved in the calcium dependent ACh translocation through the presynaptic membrane, the lyophilized powder was treated with a mild detergent solution (1% sodium cholate in 10 mM Tris buffer pH 8). The suspension was submitted to gel filtration (ultragel ACA 22) on a column equilibrated with the same solution. Four experiments gave similar results. The elution profile (Fig. 8a) of the column was monitored by the absorption at 280 nm. The turbidity recovered in fraction 1 corresponds to material above 600,000 daltons. Fraction 2 corresponds to elements of apparent molecular weight situated between 669,000 and 230,000 daltons, and fraction 3 is situated between 230,000 and 68,000 daltons. The detergent was removed by dialysis, then the material was lyophilized, and proteoliposomes tested for their

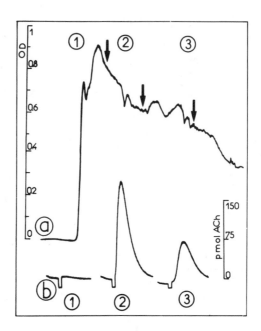

Figure 8. Separation of a membrane component which carries the ACh release capabilities. A cholate extract of lyophilized synaptosomal membranes was separated on ultragel ACA 22. The record gives the optical density of the eluate at 280 nm. Three fractions were collected (1, 2, and 3), and incorporated to the liposomes. Fraction 2 proteoliposomes had the greatest ACh release capability, as shown by the ACh release curves.

ability to release ACh. In spite of a similar ACh content, it was found that most of the calcium dependent ACh release activity was recovered in fraction 2, although some activity was also present in fraction 3. The ACh release curves for the three fractions are shown in Fig. 8b.

In the experiment shown, the slope of ACh release expressed in pmol ACh released per second and per mg protein was 442, 12,217 and 4,598, respectively, for the three fractions. Since fraction 2 was the most active, we compared its protein composition with that of the other two. The proteoliposomes were boiled in 2.5% NaDodSO$_4$ plus 5% 2-mercaptoethanol, and protein subunits analyzed by polyacrylamide gel electrophoresis. The protein pattern published elsewhere was simpler than that of the whole presynaptic membrane. Four major bands were always found in fraction 2 (92,000, 60,000, 42,000 and 36,000 daltons). These bands were only found after boiling in the presence of 2-mercaptoethanol, suggesting that they derive from heavier protein complexes, which is in accordance with the gel filtration results. The 68,000 daltons band specific of pre-synaptic membrane was also detected with immunological methods, and often visible after silver staining of the gels.

Intramembrane Particles Associated with Release. Since it was previously found that a category of large particles appear in the synaptosomal membrane whenever ACh is released, we tried to find out if these particles appear in the reconstituted proteoliposomal membrane when it releases ACh subsequent to the influx of calcium. Fig. 9 shows that this is indeed the case, since a significant increase in particle density occurred in the calciumplus-ionophore condition. Fig. 9 also shows for comparison the same phenomenon previously described on stimulated synaptosomes. These large particles are essentially pinched off with the E face of the synaptosomal membrane, and in the convex face of the proteoliposomal membrane. The histogram of particle sizes in the convex face of controls and stimulated proteoliposomal membranes is given in Fig. 10. The moderate ACh release induced by Glycera venom is to be compared to a more intense release triggered by the calcium ionophore. The release curves are shown below each histogram. It can be seen that the 10-to-13 nm particles become more numerous in the two release conditions. The mean particle density in the convex face is indicated above each histogram. The opposite concave face (not shown) remained equal to the control.

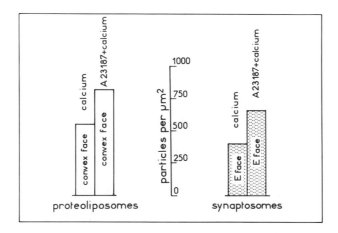

Figure 9. Effect of stimulation on intramembrane particles. The number of intramembrane particles increases in one of the membrane faces of stimulated proteoliposomes or synaptosomes. These particles are pinched off with the convex face of the proteoliposome or with the E face of the synaptosome. ACh release was triggered with A 23187 plus calcium.

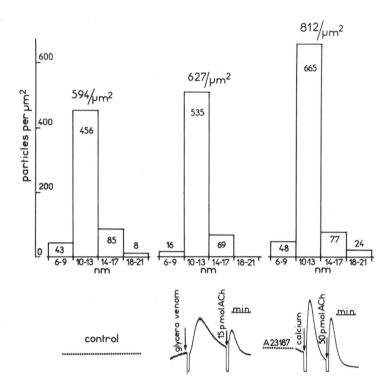

Figure 10. Histogram of intramembrane particles diameters in stimulated proteoliposomes. Particles of 10 to 13 nm become more numerous
in the convex face of proteoliposomes as a result of stimulation
with Glycera venom or ionophore. The increase in density seems to
follow the intensity of ACh release given below each histogram.
The opposite concave face (not shown) remains constant.

DISCUSSION

Since the only common and constant ultrastructural change found
for a variety of stimulation conditions leading to the release of
ACh from nerve terminals is a decrease in the number of small particles
(5 to 11 nm) and an increase of larger (8 to 18 nm) particles, it
was natural to consider the possibility that these particles are
the structure through which the cytoplasmic ACh transmitter is released. This view is strengthened by noting that stimulation occurs
either through the natural calcium channel, or through short-circuiting
of the channel, as is the case with the calcium ionophore.

The fact that large particles appeared in the reconstituted
proteoliposomal membrane at the peak of ACh release supports the

view that these particles appear in the membrane independently of synaptic vesicles, which were not present within the proteoliposomes.

Moreover, a protein component of high molecular weight was extracted from the membrane of synaptosomes, and it was shown that this element gives the proteoliposome its ability to release ACh whenever a calcium influx is generated.

In several previous reports, the term "operator" was used to identify the element of the presynaptic membrane which permits the translocation of ACh. It is probable that this "operator" was extracted and inserted in the reconstituted presynaptic membrane, since it was able to release the transmitter. It is also probable that this presynaptic constituent has a rather high molecular weight and that it is formed by several protein subunits.

The results obtained with the gels point toward several possible candidates for this constituent. It is however still premature to propose the composition of this substance. This structure is, however, no longer a hypothetical element, since it has become an experimental object. The term "operator" was initially chosen to designate this substance when it was only a working hypothesis. This was not a very good choice of terminology and we shall probably have to change it, if we are able to purify it. What is certain at this point is that an extractable element can be inserted in a membrane, and it ensures transmitter release upon calcium action.

REFERENCES

1. Dunant, Y., Gautron, J., Israël, M., Lesbats, B. and Manaranche, R. (1974): J. Neurochem. 23:635-643.
2. Dunant, Y., Gautron, J., Israël, M., Lesbats, B. and Manaranche, R. (1972): J. Neurochem. 19:1987-2202.
3. Dunant, Y., Jones, G.J. and Loctin, F. (1982): J. Physiol. (London) 325:441-460.
4. Fesce, R., Grohovaz, F., Hurlbut, W.P. and Ceccarelli, B. (1980): J. Cell. Biol. 85:337-345.
5. Fonnum, F. (1968): Biochem. J. 101:389-398.
6. Gulik-Krzywicki, T. and Costello, M.J. (1978): J. Microsc. 112:103-113.
7. Israël, M. and Dunant, Y. (1979): In Progress in Brain Research (ed) S. Tuček, Elsevier, vol. 49, pp. 125-139.
8. Israël, M., Dunant, Y. and Manaranche, R. (1979): Prog. Neurobiol. 13:237-275.
9. Israël, M. and Lesbats, B. (1980): C.R. Acad. Sci. (Paris) 291:-713-716.
10. Israël, M. and Lesbats, B. (1981): Neurochem. Int. 3:81-90.
11. Israël, M. and Lesbats, B. (1981): J. Neurochem. 37:1475-1483.
12. Israël, M. and Lesbats, B. (1982): J. Neurochem. 39:248-250.

13. Israël, M., Lesbats, B. and Manaranche, R. (1981): Nature (London) 194:474-475.
14. Israël, M., Lesbats, B., Morel, N., Manaranche, R., Gulik-Krzywicki, T. and Dedieu, J.C. (1984): Proc. Nat. Acad. Sci. (USA) 81:277-281.
15. Israël, M., Manaranche, R., Lesbats, B. and Gulik-Krzywicki, T. (1981): J. Ultrastruct. Res. 75:162-178.
16. Israël, M., Lesbats, B., Manaranche, R., Morel, N., Gulik-Krzywicki, T. and Dedieu, J.C. (1982): J. Physiol. (Paris) 78:348-356. 19-20.
17. Israël, M., Lesbats, B., Manaranche, R. and Morel, N. (1983): Biochim. Biophys. Acta 728:438-448.
18. Israël, M., Manaranche, R., Lesbats, B. and Gulik-Krzywicki, T. (1982): In Advances in Biosciences (ed) P. Lechat, Pergamon Press, vol. 35, pp. 173-182.
19. Israël, M., Manaranche, R., Mastour-Frachon, P. and Morel, N. (1976): Biochem. J. 160:113-115.
20. King, R.G. and Marchbanks, R.M. (1982): Biochem. J. 204:565-576.
21. Laemmli, U.K. (1970): Nature (London) 227:680-685.
22. Manaranche, R., Thieffry, M. and Israël, M. (1980): J. Cell. Biol. 85:446-458.
23. Meyer, E.M. and Cooper, J.R. (1982): Science 217:843-845.
24. Meyer, E.M. and Cooper, J.R. (1983): Neuroscience 3:987-994.
25. Morel, N., Israël, M., Manaranche, R. and Mastour-Frachon, P. (1977): J. Cell. Biol. 75:43-45.
26. Morel, N., Manaranche, R., Gulik-Krzywicki, T. and Israël, M. (1980): J. Ultrastruct. Res. 70:347-362.
27. Pumplin, D.W. and Reese, T.S. (1977): J. Physiol. (London) 273:443-457.
28. Suszkiw, J.B. (1980): Neuroscience 5:1341-1349.
29. Takano, Y. and Hamiya, H. (1979); Experientia 35:1076-1078.
30. Tokunaga, A., Sandri, C. and Akert, K. (1979): Brain Res. 174:207-219.
31. Weiler, M., Roed, I.S. and Whittaker, V.P. (1982): J. Neurochem. 38:1187-1191.
32. Venzin, M., Sandri, C., Akert, K. and Wyss, U.R. (1977): Brain Res. 130:393-404.

STORAGE AND SECRETION OF ACETYLCHOLINE

AND DOPAMINE BY PC12 CELLS

B.D. Howard and W.P. Melega

Department of Biological Chemistry
School of Medicine
University of California, Los Angeles
Los Angeles, California 90024 U.S.A.

INTRODUCTION

PC12 is a clonal line of rat pheochromocytoma cells that upon exposure to nerve growth factor extend neurites and acquire the morphology of neurons (7, 25). Both undifferentiated and nerve growth factor treated cells store and secrete dopamine and acetylcholine (9, 14, 16, 20, 23). This property makes PC12 useful for pharmacological studies in that one can compare the effects of a particular drug on two different transmitter systems in the same cell.

Of particular interest to us is 2-(4-phenylpiperidino)cyclohexanol (AH5183) which has been reported to cause neuromuscular blockade by inhibiting the evoked release of acetylcholine from motor nerve terminals (15). The electrophysiological characteristics of this blockade have been interpreted to indicate an AH5183-induced decrease in synaptic vesicle stores of acetylcholine from motor nerve terminals (15). The electrophysiological characteristics of this blockade have been interpreted to indicate an AH5183 only slightly affected the synthesis of ACh in intact PC12 cells while markedly inhibiting the loading of acetylcholine into storage vesicles. Anderson et al. (2) found that AH5183 inhibited the transport of acetylcholine into isolated synaptic vesicles from Torpedo with an IC_{50} of 40 nM. In this chapter we describe additional studies on the effects of AH5183 on the metabolism of acetylcholine and dopamine in PC12. We also report on the ability of various pharmacological agents to inhibit secretion of these compounds by PC12.

The evoked secretion of many compounds from a variety of secretory cells and nerve terminals requires ATP (3, 10, 11, 17, 22). In

565

contrast, the secretion of dopamine and acetylcholine from undifferentiated PC12 cells was found not to be dependent on ATP; depletion
of PC12 ATP pools to as little as 10% of control values did not
result in any inhibition of dopamine or acetylcholine secretion
evoked by 56 mM K^+ or by carbachol (21). Reynolds et al., (21)
proposed that evoked secretion could occur from ATP-depleted PC12
cells because the cells had circumvented or already accomplished
the ATP-dependent reaction(s) involved in secretion from other secretory systems. If this were indeed the case, PC12 would be useful
for identifying those reactions.

One postulated role for ATP in secretion is as a substrate
for protein kinase; certain proteins are known to be phosphorylated
in secretory systems that have been stimulated to secrete (1, 6,
12, 24). Perhaps the best characterized of these proteins are M_r
86,000 and M_r 80,000 proteins called proteins Ia and Ib, respectively
(12). They are found in nerve terminals of mammalian brain and
peripheral nervous systems (4, 5).

Because PC12 does not require intact stores of ATP for secretion, we have examined protein phosphorylation in PC12 cells that
have been induced to secrete. We report that both proteins Ia and
Ib are present in PC12 cells but protein is not phosphorylated.

MATERIALS AND METHODS

[^{125}I]-Labelled protein A was a gift from Dr. Jim McGinnis.
Proteins Ia and Ib purified from bovine brain, rabbit antiserum
against these proteins, and preimmune serum were gifts from Dr. Paul
Greengard and Dr. Wilson Wu. AH5183 was a gift from Allen and Hanbury.

Culture and incubation of PC12 cells in defined buffers. PC12 cells
were cultured on Petri dishes as described. The cells were incubated
at 37°C while still attached to dishes in a cell incubation buffer
consisting of 5.6 mM glucose, 125 mM NaCl, 4.8 mMKCl, 1.3 mM $CaCl_2$,
1.2 mM $MgSO_4$, 1.2 mM KH_2PO_4, and 25 mM Hepes, pH 7.4. Where indicated,
the KCl concentration was raised to 56 mM to provide a "high-K^+"
buffer for depolarization of the cells. In these cases, the NaCl
concentration was correspondingly lowered to maintain isoosmolarity.
When Ba^{2+} was used as a releasing agent, Ca^{2+} was omitted from the
buffer. A storage vesicle fraction was obtained as described (20).

Acetylcholine and dopamine were measured as described (21).

Polyacrylamide gel electrophoresis. Protein samples were dissolved
in sodium dodecyl sulfate (NaDodSO$_4$) sample buffer (2% NaDodSO$_4$,
62 mM Tris-HCl, pH 6.8, 10% glycerol, 0.77% dithiothreitol, 0.01%
bromphenol blue) and boiled for 3 min. The samples were subjected
to NaDodSO$_4$-slab gel electrophoresis (13) using a stacking gel of
2.5% acrylamide and a resolving gel of 8.5% acrylamide. The proteins

were stained with Coomassie brilliant blue G-250, destained, dried
and exposed to X-ray film. The following proteins were used as
molecular weight standards: phosphorylase a, bovine serum albumin,
catalase, and proteins Ia and Ib from bovine brain.

Immunoblots. A PC12 protein extract was subjected to $NaDodSO_4$ polyacrylamide gel electrophoresis and subsequently transferred to nitrocellulose paper by electroblotting as described (27). The protein-
containing nitrocellulose paper was incubated at $40^{\circ}C$ for 30 min
in Tris-saline buffer (0.9% NaCl, 10 mM Tris-HCl, pH 7.4) containing
5% bovine serum albumin. The nitrocellulose paper was transferred
to fresh buffer containing a 1:35 dilution of antiserum to bovine
protein I and incubated overnight at room temperature. The nitrocellulose paper was washed stepwise at room temperature in Tris-saline
buffer (10 min), 2 changes of Tris-saline buffer containing 0.05%
Nonedit P-40 (20 min each), and Tris-saline buffer (10 min). After
the washing, it was incubated for 3.5 hr at room temperature in
Tris-saline containing 5% bovine serum albumin and $[^{125}I]$-labelled
protein A (3 x 10^6 cpm/ml) and then washed again in stepwise fashion
as described above. Finally, the paper was washed 10 min in H_2O,
dried, and exposed to X-ray film for 4 days at $-70^{\circ}C$ using an intensifying screen.

**Immunoprecipitation of PC12 protein with antibody to bovine protein
I.** PC12 cells were incubated for 21 hr in phosphate-free Dulbecco's
modified Eagle's medium (GIBCO) supplemented with 5% dialyzed fetal
calf serum and 10% dialyzed horse serum, and containing 0.35 mCi
of $[^{32}P_i]$ per ml. The cells were lysed and treated with nucleases
as described (19) and the lysate was incubated for 1 hr at $4^{\circ}C$ with
a 1.5% suspension of Pansorbin. This preadsorption with Pansorbin
is to remove proteins that nonspecifically bind to Pansorbin. The
suspension was centrifuged and the supernate was incubated overnight
at $4^{\circ}C$ with a 1:10 dilution of preimmune serum or antiserum raised
against bovine brain protein I, and then again for 1 hr with a 1.5%
suspension of Pansorbin. The suspension was centrifuged, the pellet
was washed as described (18) and then dissolved and boiled in $NaDodSO_4$
sample buffer. The Pansorbin was removed by centrifugation and
the proteins in the supernate were separated by $NaDodSO_4$ polyacrylamide
gel electrophoresis. The gel was dried and exposed to X-ray film
for 24 hr.

RESULTS

Table 1 shows that in PC12 AH5183 at 4 x 10^{-7} M only partially
inhibits the synthesis of acetylcholine (63% of control) but almost
completely inhibits the loading of newly synthesized $[^2H_4]$ acetylcholine into storage vesicles. The levels of unlabelled acetylcholine in the cells and vesicles were not significantly affected
by the drug. AH5183 did not inhibit the uptake of choline by the
cells, but it did markedly inhibit the transport of dopamine into

the cells. Dopamine loading into the vesicles was only partially inhibited.

Table 2 shows that AH5183 does not inhibit K^+-evoked or Ba^{2+}-evoked release of newly accumulated [^{3}H] dopamine. These results further indicate that AH5183 does not block dopamine loading into vesicles.

Several agents were examined for their effect on transmitter release from PC12. Pretreatment of PC12 cells for 5 min with 10 μM trifluoperazine, 10 μM chlorpromazine, or 25 μM diphenylhydantoin resulted in the inhibition of K^+-evoked release of both acetylcholine and dopamine. No effect on K^+-evoked release was observed after a similar preincubation with 40 μM AH5183, 10 μM oxotremorine, or 100 μM hexamethonium, or after a 1 hr preincubation with 1 μg/ml of β-bungarotoxin.

Table 3 shows the details of the effect of trifluoperazine on the evoked secretion of acetylcholine and dopamine from PC12. The ability of trifluoperazine and chlorpromazine to block release from PC12 indicates that calmodulin is involved in the release process.

Protein Phosphorylation Experiments. When PC12 cells were preincubated with [^{32}P$_i$] and then stimulated by exposure to 56 mm K^+ or 1 mM carbachol, we did not detect any increase in the phosphorylation of proteins corresponding to proteins Ia and Ib. However, PC12 does have proteins similar to proteins Ia and Ib. As shown in the immunoblot of Figure 1A, antibody raised against proteins Ia and Ib from bovine brain cross reacts strongly only with M_r 80,000 and M_r 86,000 proteins from PC12. PC12 cells contain more of the M_r 86,000 protein than the M_r 80,000 protein. This situation resembles that in rat and bovine brain (12, 28).

Although present in PC12 cells the M_r 80,000 protein does not appear to be phosphorylated. We have examined this matter in the following way. Undifferentiated cells were grown for 21 hr in medium containing [^{32}P$_i$]. A cell lysate was incubated with antiserum against protein I and Pansorbin (protein A-producing Staph. aureus cells). The immunoprecipitated protein was subjected to NaDodSO$_4$ polyacrylamide gel electrophoresis and the gel was analyzed by autoradiography. As shown in Figure 1B, a M_r 86,000 phosphoprotein was present in the immunoprecipitate but not when preimmune serum was used. In neither case was a [^{32}P]-labelled M_r 80,000 protein detected.

The lack of [^{32}P]-labelled M_r 80,000 protein was not due to a failure of antibody and Pansorbin to immunoprecipitate a PC12 M_r 80,000 protein. Using a lysate of unlabelled PC12 cells, this immunoprecipitation procedure was found to selectively precipitate both the M_r 80,000 protein and the M_r 86,000 protein; in fact, the immunoprecipitate, which was analyzed for proteins Ia and Ib by

Table 1

Effect of AH5183 on the metabolism of
[$^2H_4$] acetylcholine (ACh) and [^3H] dopamine (DA)

	Parameters Measured (% of Control Values)		
[AH5183], M	[$^2H_4$] Choline Uptake, Cells	[$^2H_4$] ACh Synthesis	[$^2H_4$] ACh Loading into Vesicles
4×10^{-8}	120 ± 4	92 ± 7	22 ± 2
4×10^{-7}	107 ± 1	63 ± 5	4 ± 1
4×10^{-6}	117 ± 4	73 ± 8	< 2
4×10^{-5}	124 ± 5	61 ± 8	< 2

[AH5183], M	[^3H] DA Uptake, Cells	[^3H] DA Loading into Vesicles
4×10^{-5}	28 ± 2	78 ± 4

Cells were incubated for 30 min with 10 μM [$^2H_4$] Choline or for 10 min with 1 μM [^3H] dopamine (0.2 μC$_i$/ml) in the presence of varying concentrations of AH5183. Then the levels of [$^2H_4$] choline, acetylcholine, and vesicle function were determined. The results are means ± mean deviation for duplicate incubations. Control cells contained, per mg of protein, 1.5 ± 0.1 nmol of [$^2H_4$] choline, 1.5 ± 0.05 nmol of acetylcholine and 0.25 ± 0.02 nmol of [$^2H_4$] acetylcholine; control vesicles contained, per mg of protein, 1.1 ± 0.1 nmol of acetylcholine and 0.1 ± 0.01 nmol of [$^2H_4$] acetylcholine. The radioactive dopamine accumulated by control cells and granules corresponded to 14,100 ± 300 dpm and 7,600 ± 100 dpm, respectively, per mg of protein.

Table 2

Effect of AH5183 on the evoked release of
newly accumulated [^{3}H]dopamine (DA)

Releasing Agent	[AH5183], M	DA Accumulated dpm/plate	(% Control)	DA Released % DA Released	(% Control)
56 mM K$^+$	0	146,300 ± 8,200		15.0 ± 0.5	
"	4 x 10^{-8}	150,100 ± 7,500	(103)	16.4 ± 1	(109)
"	4 x 10^{-7}	135,400 ± 4,100	(93)	17.2 ± 2	(115)
"	4 x 10^{-6}	101,100 ± 2,600	(69)	16.0 ± 0.5	(107)
2 mM Ba^{2+}	0	111,700 ± 9,300		62.5 ± 4	
"	4 x 10^{-7}	88,100 ± 4,300	(79)	63.0 ± 2	(101)
"	4 z 10^{-6}	56,600 ± 3,400	(51)	57.1 ± 3	(91)
"	4 x 10^{-5}	25,100 ± 1,600	(22)	55.4 ± 2	(89)

Cells were incubated for 30 min with [^{3}H]dopamine in the presence of AH5183, washed and exposed
for 5 min to 56 mM K$^+$ or 2 mM Ba^{2+}.

Table 3

Effect of trifluoperazine (TFP) on secretion from PC12

Conditions		Release		
		Undifferentiated PC12		NGF-Treated PC12
K$^+$ (mM)	TFP (μM)	Dopamine	Acetylcholine	Acetylcholine
6	0	1.0 ± 0.1	1.5 ± 0.1	1.6 ± 0.2
6	10	0.8 ± 0.1	1.2 ± 0.1	1.2 ± 0.2
6	100	1.4 ± 0.1	2.1 ± 0.1	1.2 ± 0.2
55	0	21.0 ± 2.0	17.0 ± 3.0	19.0 ± 2.0
55	10	11.0 ± 1.0	8.0 ± 1.0	10.0 ± 1.0
55	100	5.0 ± 1.0	7.0 ± 1.0	6.0 ± 1.0

Cells were preincubated for 5 min in low K$^+$ buffer containing the indicated
concentration of TFP (trifluoperazine) indicated, washed and incubated for
an additional 5 min in buffer containing the indicated concentrations of K$^+$
and TFP. The release is expressed as the percentage of total dopamine or
acetylcholine present in the medium at the end of the incubation period.
The values are means ± S.D. for triplicate incubations. Cells were induced
to extend neurites by a 14 day treatment with (NGF) nerve growth factor prior
to measurement of acetylcholine release. For undifferentiated cells, total
dopamine and acetylcholine were 1.9 ± 0.3 and 2.2 ± 0.2 nmol, respectively,
per mg of protein. For nerve growth factor-treated cells total acetylcholine
was 3.3 nmol per mg of protein.

essentially the same technique as described in Figure 1A, contained
more of the M_r 80,000 protein than the M_r 86,000 protein (Mayumi
Koide, unpublished results).

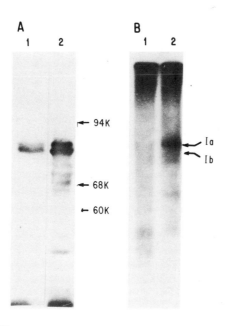

Figure 1.   Autoradiogram of immunoblot and immunoprecipitate of PC12
proteins. (A) PC12 proteins were labelled with antibody to bovine
brain protein I and then with $[^{125}I]$-labelled protein A.   Lane 1
contains proteins from PC12 cells (450 μg of protein applied to
gel).   Lane 2 contains purified protein I (Ia plus Ib) standard
from bovine brain. The bovine brain protein I standard shows heavily
labelled bands at 86,000 and 80,000 daltons with multiple bands at
lower molecular weights.   These other bands in lane 2 are degradation products of protein I.   The positions of molecular weight
standards are indicated.   (B) PC12 cells were incubated with $[^{32}P_i]$
for 21 hr, immunoprecipitated with antibody to bovine protein I
and subjected to NaDodSO$_4$ polyacrylamide gel electrophoresis.   Lane
1 has proteins incubated with preimmune serum and Pansorbin.   Lane
2 has proteins immunoprecipitated with anti-protein I antibody and
Pansorbin.   In each case the protein added to the gel corresponded
to 40% of the sample obtained from one flask of cells.   The positions
of bovine brain proteins Ia and Ib standards are indicated by arrows.
Note that a M_r 86,000 phosphoprotein is pronounced in lane 2 but
that there was no M_r 80,000 phosphoprotein present.

DISCUSSION

AH5183 causes a marked inhibition of the loading of acetyl-choline into storage vesicles and only a slight inhibition of acetyl-choline synthesis. Choline transport across the PC12 plasma membrane, a carrier mediated process (16), is not blocked by AH5183. In contrast, AH5183 blocks dopamine transport across the plasma membrane but not across the vesicle membrane. The fact that AH5183 did not decrease the level of endogenous unlabelled acetylcholine in PC12 storage vesicles indicates that the drug does not act by lysing the vesicles or making them leaky.

PC12 cells contain two proteins that appear to be equivalent, respectively, to proteins Ia and Ib from bovine brain. Results presented here demonstrate that these PC12 proteins comigrate with the bovine proteins Ia and Ib on NaDodSo$_4$ polyacrylamide gel electrophoresis, and the two PC12 proteins specifically cross react with antibody against the bovine proteins. Another similarity is their solubilities under certain conditions of pH shifts. Bovine brain proteins Ia and Ib remain soluble when the pH of a crude protein extract is adjusted to pH 3.0 and then to 6.0 (28). This property is also exhibited by the PC12 M_r 80,000 and M_r 86,000 proteins (M. Koide, unpublished results). In light of these findings it seems justifiable to refer to the two PC12 proteins in question as proteins Ia and Ib.

The phosphorylation of PC12 proteins Ia and Ib differs from that of the corresponding proteins in mammalian brain. Unlike the case with mammalian brain proteins (12), we did not detect an increased phosphorylation of any M_r 80,000 or M_r 86,000 protein when undifferentiated or nerve growth factor PC12 cells were stimulated with 56 mM K^+ or 1 mM carbachol. In itself this finding is not particularly remarkable because the content of proteins Ia and Ib in PC12 may be too low to detect an increased phosphorylation under the conditions used. However, what is unusual is that when the cells were PC12 cells were incubated in [$^{32}P_i$] for 21 hr and proteins Ia and Ib purified by immunoprecipitation, only protein Ia exhibited phosphorylation. This finding occurred in spite of the fact that there is more protein Ib than Ia in PC12 (Fig. 1A) and in the immunoprecipitate (M. Koide, unpublished results).

In mammalian brain nerve terminals, both proteins Ia and Ib undergo some phosphorylation even in the absence of stimulation of the terminals (12). The 21 hr period used for the labeling of PC12 in this experiment was greater than half of the cell doubling time, so there should have been substantial synthesis and phosphorylation of protein Ib during this period.

The unusual phosphorylation pattern of protein I in PC12 makes PC12 an attractive system for determining whether protein I functions in transmitter release. It is tempting to speculate that the lack of detectable phosphorylation of protein Ib is related to the fact that the evoked release of dopamine and acetylcholine from undifferentiated PC12 cells is independent of ATP stores (21). An interesting possibility is that some modification of protein Ib in PC12 not only prevents phosphorylation of the protein, but allows protein Ib to perform whatever function it has in transmitter release even in the absence of phosphorylation. A speculation by Reynolds et al. (21) may be relevant in this regard. They found that, unlike the case with undifferentiated PC12 cells, acetylcholine release from nerve growth factor-treated PC12 was partially inhibited when cellular ATP pools were depleted. They suggested, therefore, that the protein modification, which they postulated to account for the ATP independence of release from undifferentiated PC12 cells, may be present to a lesser extent at release sites in differentiated cells.

Acknowledgements. We are very indebted to Dr. Paul Greengard and Dr. Wilson Wu for the gift of bovine brain protein I and antiserum to protein I and to Dr. Jim McGinnis for the gift of [^{125}I]-labelled protein A. We also thank Dr. Donald Jenden for assistance in the measurement of acetylcholine.

REFERENCES

1. Amy, C.M. and Kirshner, N. (1981): J. Neurochem. 36:847-854.
2. Anderson, D.C., King, S.C. and Parsons, S.M. (1983): Molec. Pharmacol. 24:48-54.
3. Baker, P.F. and Knight, D.E. (1979): Trends Neurosci. 2:288-290.
4. Bloom, F.E., Ueda, T., Battenberg, E. and Greengard, P. (1979): Proc. Natl. Acad. Sci. U.S.A. 76:5982-5986.
5. DeCamilli, P., Ueda, T., Bloom, F.E., Battenberg, E. and Greengard, P. (1979): Proc. Natl. Acad. Sci. U.S.A. 76:59775981.
6. DeLorenzo, R.J. and Freedman, S.D. (1977): Biochem. Biophys. Res. Commun. 77:1036-1043.
7. Greene, L.A. and Tischler, A.S. (1976): Proc. Natl. Acad. Sci. U.S.A. 73:2424-2428.
8. Greene, L.A. and Rein, G. (1977): Brain Res. 129:247-263.
9. Greene, L.A. and Rein, G. (1977): Nature 268:349-351.
10. Johansen, T. (1979): Eur. J. Pharmacol. 58:107-115.
11. Kirshner, N. and Smith, W.J. (1966): Science 154:422-423.
12. Krueger, B.K., Forn, T. and Greengard, P. (1977): J. Biol. Chem. 252:2764-2773.
13. Laemmli, U.K. (1970): Nature 227:680-685.
14. Lee, K., Sabban, E., Goldstein, M., Seeleg, P.-J. and Greene, L.A. (1982): Neurosci. Soc. Abs. 8:886.
15. Marshall, I.G. (1970): Br. J. Pharmacol. 38:503-516.
16. Melega, W.P. and Howard, B.D. (1981): Biochemistry 20:4477-4483.
17. Nelson-Krause, D.C. and Howard, B.D. (1978): Brain Res. 147:91-105.

18. Opperman, H., Levinson, A.D., Varmus, H.E., Levintow, L. and Bishop, J.M. (1979): Proc. Natl. Acad. Sci. U.S.A. 76:1804-1808.
19. Pawson, T., Guyden, J., Kunz, T.S., Radke, K., Gilmore, T. and Martin, G.S. (1980): Cell 122:767-775.
20. Rebois, R.V., Reynolds, E.E., Toll, L. and Howard, B.D. (1980): Biochemistry 19:1240-1248.
21. Reynolds, E.E., Melega, W.P. and Howard, B.D. (1982): Biochemistry 21:4795-4799.
22. Rubin, R.P. (1970): J. Physiol. 206:181-192.
23. Schubert, D. and Klier, F.G. (1977): Proc. Natl. Acad. Sci. U.S.A. 74:5184-5188.
24. Sieghart, W., Theoharides, T.C., Alper, S.L., Douglas, W.W. and Greengard, P. (1978): Nature 275:329-331.
25. Tischler, A.S. and Greene, L.A. (1978): Lab. Invest. 39: 77-89.
26. Toll, L. and Howard, B.D. (1980): J. Biol. Chem. 255:1787-1789.
27. Towbin, H., Staehlin, T. and Gordon, J. (1979): Proc. Natl. Acad. Sci. U.S.A. 76:4350-4364.
28. Ueda, T. and Greengard, P. (1977): J. Biol. Chem. 252:5155-5163.

EFFECTS OF EXOGENOUS CHOLINE ON ACETYLCHOLINE

AND CHOLINE CONTENTS AND RELEASE IN STRIATAL SLICES

J-C. Maire[*], J.K. Blusztajn and R.J. Wurtman

Department of Nutrition and Food Science
Massachusetts Institute of Technology
Cambridge, Massachusetts 02139, U.S.A.
and
[*]Departement of Pharmacology
Centre Médical Universitaire
Geneva, Switzerland

INTRODUCTION

The synthesis of acetylcholine (ACh) requires choline as the immediate precursor. The free extracellular choline can enter cholinergic neurons through a high affinity and a low affinity transport system (10, 26). ACh content of various tissues has been reported to increase after administration of choline (4, 7, 11, 13, 19), although this effect was not always observed in animal brain in other studies (8, 14, 18, 22). In spite of these discrepancies, results often support the hypothesis that the level of exogenous choline regulate ACh synthesis during periods of increased neuronal demand for the precursor or when choline supply to the neurons is reduced (16, 21, 23).

Choline is present in all cells as an abundant constituent of membrane choline-phospholipids. It has been found that brain neurons can synthesize choline by methylating phosphatidylethanolamine to phosphatidylcholine (PC) (1, 3, 5). The subsequent hydrolysis of PC would generate free choline, but it is not yet clear whether choline originating from the brain's choline- phospholipids can be utilized directly as a source for brain ACh synthesis (2, 12, 25).

This study was undertaken to determine whether exogenous choline supplied to an in vitro preparation of striatal neurons could affect intracellular pools of choline and ACh, as well as the release of ACh during periods of increased neuronal activity. Some experiments were designed to test the possibility that choline, liberated from the cell membranes, might be used for the synthesis of ACh.

METHODS

Slices (0.3 mm) were prepared from rat striatum with a McIlwain tissue chopper. The tissue was then continuously superfused (0.5 ml/min; 37°C) with physiological solution (PhS) bubbled with a mixture of 95% O_2/5% CO_2 (pH=7.4). The composition of the PhS was (mM): NaCl, 120; KCl, 3.5; $CaCl_2$, 1.3; $MgSO_4$, 1.2; NaH_2PO_4, 1.2; $NaHCO_3$, 25; glucose, 10; and eserine salicylate, 0.02 (a cholinesterase inhibitor). After a period of equilibration (30 min), the effluent was collected at 4-min intervals. During the collection period, the slices were subjected to an electrical field stimulation at 15 Hz for 30 min, using $Ag/AgCl_2$ electrodes.

Efflux experiments were conducted as follows: slices were maintained at rest for 12 min after the beginning of the collection period, then they were stimulated for 30 min and finally allowed to rest for a 18 min period of time.

Choline and ACh were extracted from the effluent by a liquid cation exchange procedure using 1 ml of tetraphenylboron in heptanone-3 (10 mg/ml). An aliquot (0.75 ml) of the organic phase was then back-extracted into 0.5 ml of 0.4 N HCl. Aliquots of the HCl extract were dried under vacuum, and choline and ACh were quantitated in the residue by the radio-enzymatic method of Goldberg and McCaman (9).

The tissue contents of choline and ACh were also determined in slices removed from the perfusion apparatus at the beginning ("initial content") or at the end ("final content") of the collection period. The tissue (20-50 mg) was homogenized in 0.5 ml 1N formic acid/acetone (15% v/v). After centrifugation, the formic acid/acetone extract was dried under vacuum and the residue was reconstituted in 2 ml PhS. Choline and ACh were then assayed as described above. Protein content of the pellet obtained after centrifugation was assayed according to Lowry et al. (15).

When the effect of hemicholinium-3 (HC-3) or choline was studied, these substances were added to the PhS from the beginning of the superfusion period. In experiments where choline concentrations were much greater than those of ACh, it was possible that failure to phosphorylate all of the choline might lead to artefactually high ACh concentration; hence, in this circumstance, we assayed the ACh by initially converting it to choline (by incubating samples with acetylcholinesterase, 1.5 U) and then measuring choline levels with and without this prior hydrolysis.

Table 1

Spontaneous and Electrically Evoked Release of ACh

Conditions	Spontaneous (pmol/mg prot/min)	Evoked (pmol/mg prot/stim. period)
PhS, control	7.5 ± 1.3	767 + 177
+ choline 5 μM	13.2 ± 3.0*	773 ± 128
+ choline 20 μM	22.7 ± 5.7*	1121 ± 201*
+ HC-3 5 μM	3.8 ± 2.2*	335 ± 55*

Mean of 3 separate experiments (± S.D.), ACh assay run in duplicate, evoked release is the difference between total ACh released during a stimulation period (30 min. at 15 Hz) minus the anticipated spontaneous efflux.
*Significant difference ($p<0.05$) from control value in PhS (t-test)

RESULTS

Figure 1 shows the rate of ACh efflux slices superfused with PhS or PhS containing choline (5 and 20 μM) or HC-3 (5 μM). Exogenous choline accelerated the spontaneous efflux of ACh, while inhibition of high affinity choline uptake by HC-3 slowed this efflux (Table 1).

Electrical stimulation of the slices also accelerated ACh release. This higher rate was maintained constant throughout the stimulus period, except in the presence of HC-3, when the rate quickly declined to that characteristic of spontaneous efflux (Figure 1). The release of ACh evoked during the stimulus period was increased when the PhS contained 20 μM of choline and was decreased by addition of HC-3 (5 μM) to choline-free PhS (Table 1).

When slices were superfused with choline-free PhS, choline appeared in the effluent at a rate of 55 pmol/mg prot/min. In the presence of HC-3, there was a 1.5 fold increase in the rate of choline efflux. Electrical stimulation had no significant effect on this efflux.

Preincubation for 30 min with HC-3 lowered tissue ACh levels (Table 2); these were further reduced (to about 20% of control levels) during the stimulation period. Tissue choline levels were not affected by preincubation with or without choline, however, after stimulation,

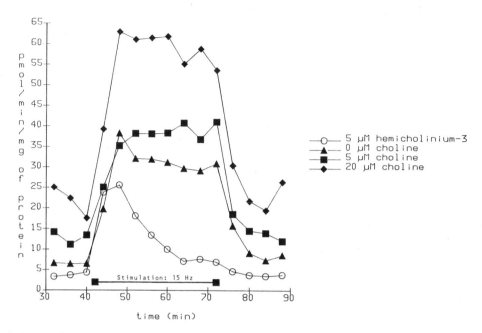

Figure 1. Rate of ACh efflux from striatal slices. The bar immediately above the x-axis represents the stimulation period. Each point is the mean of 3 separate experiments (ACh assay run in duplicate).

choline levels tended to be higher in samples incubated with choline (Table 2).

In tissues superfused without exogenous choline [e.g., experiments involving PhS and (PhS + HC-3) groups], it was possible to compare the decrease in intracellular (choline + ACh) during the collection period, with the measured total efflux of (ACh + choline). From Table 2 we calculate a combined loss of (choline + ACh) of 15 and 9.6 pmol/mg prot/min PhS and (PhS + HC-3) groups, respectively. During the same period, the mean combined effluxes of ACh and choline were, respectively, 75 and 94 pmol/mg prot/min. Hence, the decreases in intracellular ACh and choline could be responsible for, at best, only 20% and 10% of the amounts of these compounds recovered in the effluent.

DISCUSSION

The concentrations of choline in the media superfusing the slices failed to affect basic tissue choline and ACh levels (i.e., prior to stimulation) (Table 2), an observation similar to that of Millington and Goldberg who studied cortical slices (16). In the sympathetic ganglion, choline availability failed to affect ACh content studied _in vitro_, however, administration of choline _in_

Table 2

Initial and Final Contents of ACh and Choline (nmol/mg prot)

| Conditions | ACh | | Choline | |
	Initial	Final	Initial	Final
PhS, control	1.96 ± 0.29	1.33 ± 0.40	1.05 ± 0.13	0.78 ± 0.16
+ choline 5 μM	1.37 ± 0.52	1.51 ± 0.62	1.00 ± 0.35	0.95 ± 0.10
+ choline 20 μM	1.86 ± 0.48	1.36 ± 0.45	1.16 ± 0.35	1.32 ± 0.45
+ HC-3 5 μM	0.62 ± 0.14*	0.28 ± 0.07*	1.35 ± 0.33	1.12 ± 0.21

Mean of 3-4 separate experiments (± S.D.), assays run in triplicate, elapsed time between initial and final measurements: 1 hr. *Significant difference (p<0.05) from control value obtained in PhS (t-test)

vivo did increase the ganglionic ACh levels (17). However, HC-3, which impairs choline uptake, did markedly decrease tissue ACh (Table 2), while choline levels remained as high as, or higher than, in controls. This latter observation may be explained by the fact that HC-3 was applied to the slices before the tissue had fully recovered from the stress of the preparation the [choline levels have been found to be increased post-mortem (6, 20)]: inhibition of the bidirectional transport system will tend to conserve a high level of intracellular choline.

The rate of spontaneous release of ACh was increased by the presence of exogenous choline in the medium. This rate was least when the high affinity choline uptake was inhibited by HC-3 (Figure 1). Since tissue choline and ACh levels failed to be similarly affected, (Table 2), our observations suggest that spontaneous ACh turnover depends more upon the choline concentration of the medium, than on that of the tissue.

The amounts of ACh released during a long period of electrical stimulation were similar in choline-free PhS and in PhS + choline 5 μM. The evoked release was increased by about 50% when choline levels in the media approached their physiologic range (20 M), and decreased by 44% when choline uptake was blocked (by HC-3) (Table 1). In fact, evoked ACh release was practically abolished at the end of the stimulation period in the presence of HC-3. These results support the hypothesis that the total amount of transmitter liberated during neuronal activity can depend upon the availability of extra-cellular choline. However, it should be noted that, even in the absence of added choline in the PhS, (i.e., in the control group), a portion of the choline liberated from the slices apparently is reuptaken from the extracellular space in sufficient quantities to maintain a constant release of ACh throughout the stimulation period.

The fact that ACh and choline are recovered in the effluent in much higher amounts than expected from the decrease in their intracellular pools suggests a large efflux of choline from another tissue source, most likely the membrane choline-phospholipids. Both the ACh content of the slices and the release of ACh into the superfusate were diminished when choline uptake was inhibited by HC-3. In contrast, the choline content of the slices and superfusate was not decreased. This observation indicates that, in the event that some of the choline originating from the breakdown of phospho- lipids remains intracellularly, the synthesis of ACh from this par- ticular choline pool is not efficient enough to prevent the depletion of tissue ACh (see Table 2). Thus, maintaining a steady level of choline in the slices is not sufficient to assure a steady level of ACh. This affirms the importance of a functional choline uptake mechanism, and of the presence of extracellular choline in sufficient levels, to maintain normal ACh levels and steady ACh release.

This study shows that the striatal cholinergic neuron preferen- tially used, for ACh synthesis, choline entering the cell through its uptake system, rather than the existing intracellular pool of free choline. Whether one concludes that the most important choline for this purpose is that entering via high-affinity or low-affinity uptake depends, in part, on the specificity that one attributes to HC-3 in blocking the high-affinity process.

In summary, this study shows that, in our preparation, ACh synthesis depends upon availability of extracellular choline. The amount of ACh released during increased neuronal activity is also dependent upon the exogenous choline level. Furthermore, choline originating from an intracellular source other than free choline or ACh, probably the membrane choline- phospholipids, can be a source of extracellular choline, and may be important in maintaining suf- ficient ACh synthesis when neuronal activity is increased and cir- culating plasma choline levels are inadequate. This suports the hypothesis (12, 14, 17, 24) that the activity of the cell may determine its ability to respond to exogenous choline by making more ACh.

Acknowledgements. These studies were supported in part by NIMH grant MH-28783. J-C. Maire holds a fellowship from the Swiss Foun- dation for Fellowships in Experimental Medicine and Biology.

REFERENCES

1. Blusztajn, J.K. and Wurtman, R.J. (1981): Nature, 290: 417-418.
2. Blusztajn, J.K. and Wurtman, R.J. (1983): Science, 221:614-620.

3. Blusztajn, J.K., Zeisel, S.H. and Wurtman, R.J. (1979): Brain Res., 179:,319-327.

4. Cohen, E.L. and Wurtman, R.J. (1976): Science, 191:561-562.

5. Crews, F.T., Hirata, F. and Axelrod, J. (1980): J. Neurochem. 34:1491-1498.

6. Dross, K. and Kewitz, H. (1972): Naunyn- Schmiedeberg's Arch. Pharmakol., 274:91-106.

7. Eckernäs, S.Å., Sahlström, L. and Aquilonius, S-M. (1977): Acta Physiol. Scand., 101:404-410.

8. Flentge, F. and Van den Berg, C.J. (1979): J. Neurochem., 32: 1331-1333.

9. Goldberg, A.M. and McCaman, R.E. (1973): J. Neurochem., 20:1-8.

10. Haga, T. and Noda, H. (1973): Biochem. Biophys. Acta, 291: 564-575.

11. Haubrich, D.R., Wedeking, P.W. and Wang, P.F.L. (1974): Life Sci., 14:921-927.

12. Jenden, D.J., Weiler, M.H. and Gunderson, C.B. (1982): In: Alzheimer's Disease: A Report of Progress, (eds) Corkin, S., Davis, K.L., Growdon, J.H. and Wurtman, R.J., Raven Press, New York, pp. 315-326.

13. Kuntscherová, J. (1972): Physiol. Bohemoslov., 21:655-660.

14. London, E.D. and Coyle, J.T. (1978): Biochem. Pharmacol., 27: 2962-2965.

15. Lowry, O.H., Rosebrough, N.J., Farr, A.L. and Randall, R.J. (1951): J. Biol. Chem., 193:265-275.

16. Millington, W.R. and Goldberg, A.M. (1982): Brain Res., 243: 263-270.

17. O'Regan, S. and Collier, B. (1981): J. Neurochem., 36:420-430.

18. Pedata, F., Wieraszko, A. and Pepeu, G. (1977): Pharmacol. Res. Commun., 9:755-761.

19. Racagni, G., Trabucchi, M. and Cheney, D.L. (1975): Naunyn-Schmiedeberg's Arch. Pharmacol., 290:99-105.

20. Stavinoha, W.B. and Weintraub, S.T. (1974): Science, 183: 964-965.

21. Trommer, B.A., Schmidt, D.E. and Wecker, L. (1982): J. Neurochem., 39:1704-1709.

22. Wecker, L., Dettbarn, W.D. and Schmidt, D.E. (1978): Science, 199:86-87.

23. Wecker, L. and Goldberg, A.M. (1981): In: Cholinergic Mechanism: Phylogenetic Aspect, Central and Peripheral Synapses, and Clinical Significance, (eds) Pepeu, G. and Ladinsky, H., Plenum Press, New York, pp. 451-461.

24. Weiler, M.H., Bak, I.J. and Jenden, D.J. (1983): J. Neurochem., 41:473-480.

25. Wurtman, R.J. and Zeisel, S.H. (1982): In: Alzheimer's Disease: A Report of Progress, (eds) Corkin, S., Davis, K.L., Growdon, J.H. and Wurtman, R.J., Raven Press, New York, pp. 303-313.

26. Yamamura, H.I. and Snyder, S. (1973): J. Neurochem., 21:1355-1374.

ACETYLCHOLINE AND CHOLINE CONTENTS OF RAT SKELETAL MUSCLE

DETERMINED BY A RADIOENZYMATIC MICROASSAY: EFFECTS OF DRUGS

S. Consolo, M. Romano, C. Scozzesi,
A.C. Bonetti[*] and H. Ladinsky

Mario Negri Institute for Pharmacological Research
Via Eritrea 62
20157 Milan, Italy

INTRODUCTION

Skeletal muscle contains a small amount of acetylcholine (ACh) relegated primarily to the nerve terminals in the endplate region (10), and a large quantity of choline (3). The ratio of the concentrations of choline:ACh in muscle is at least two orders of magnitude greater than it is in brain. Due to these particular characteristics some of the recently devised radio-chemical and chemical methods are unsuitable for the determination of ACh in this tissue since they lack either sufficient sensitivity or the necessary capacity to separate quantitatively the ACh from the high choline. Gas chromatography-mass fragmentography methodology (5, 11) has the required sensitivity and separative capacity, and is currently used for muscle ACh assay. Mass fragmentography is, however, expensive and not available to everybody. Hence, the relative dearth of information on drug effects on muscle ACh content, although such studies are warranted to complement extensive knowledge on drugs affecting the neuromuscular junction, obtained by physiological and electrophysiological means.

This paper describes a specific, radioenzymatic microassay for measuring picomole quantities of ACh in skeletal muscle. The levels of ACh as well as choline in the extensor digitorum longus (EDL), soleus and gastrocnemius muscles of the rat are reported after focussed microwave irradiation to the hind limb to prevent

---

[*]Permanent address: Fidia Research Laboratories, Abano Terme, Italy

post-mortem changes in ACh and choline contents. Particular attention was directed to the effect of some drugs potently affecting neuro-muscular transmission at the end plate.

MATERIALS AND METHODS

Tissue samples. Female CD-COBS rats (Charles River, Italy) weighing 180-200 g were used in these studies. The lightly ether anesthetized animals were placed in a modified perspex rectangular holder from which the constrained head and left hind limb protruded. First the leg was subjected to focussed microwave irradiation (1.3 KW at 2.45 GHz for 4 sec); then the holder was quickly rotated and the head was microwaved. The EDL, soleus and gastrocnemius muscles were rapidly excised from the irradiated paw, frozen on dry ice, weighed and pulverized in liquid nitrogen in a specially devised mortar and pestle (8).

Extraction. The frozen muscle powder was added to 1.5 ml Beckman polyethylene conical tubes containing 700 l of an ice-cold mixture of 1N formic acid:acetone 15:85, V/V). The mixture was shaken in ice for 30 min, centrifuged at low speed and the precipitate dis-carded. The supernatant was washed twice with heptane:chloroform (4:1, V/V) and the organic phase was discarded after each wash. For the ACh assay, a known amount (usually 120 μl) of the aqueous phase was placed into a 450 μl Beckman polyethylene microtube and freeze-dried overnight in an Edwards Mini-Fast Lyophilizer.

Electrophoresis. The residue obtained after lyophilization was suspended in 25 μl of water and briefly centrifuged at high speed in a Beckman microfuge. An 18 μl aliquot of the clear aqueous extract was electrophoresed on paper as described previously (1, 13). ACh was separated from choline, other choline analogs, as well as potentially interfering substances, by this technique. Using appropriate labels of ACh and choline, it was found that more than 90% of the ACh was recovered in the ACh band and less than 2% of the choline overlapped onto the ACh band.

After electrophoresis, the papers were dried in an air-circulated oven at 100°C for 5 min. The tetraethylammonium (TEA) band containing the ACh was lightly stained with iodine vapor and the outer edge carefully delineated by a thin pencil line. 2.3 cm measured from the outer edge of the TEA band towards the origin was carefully cut out, folded and placed into microtubes as described by Saelens et al. (14). The ACh was quantitatively eluted with 2 x 60 μl volumes of a methanol: 100 μM HCl (7:1, V/V) mixture. The eluant was evapor-ated in a thermostatically controlled water bath at 70°C for 7 min and then at 95°C for 20 min. The acidity of the mixture (pH about 5) protected the ACh against hydrolysis during evaporation.

Phosphorylation. To the dried tubes, placed in an ice bath, was added a 20 μl portion of an incubation mixture (buffer substrate A) consisting of (all at final concentrations): 0.15 M sodium phosphate buffer, pH 8, 1 mM ATP (neutralized to pH 7), 6 mM $MgCl_2$ and 180 μg choline kinase lyophilizate (Boehringer Mannheim). The microtubes were gently "buzzed" on a Beckman micromixer, centrifuged for 5 sec and incubated at 37°C for 15 min; the choline present in the medium was completely phosphorylated. The samples were then placed at 95°C in a water bath for 5 min and centrifuged. The heat denaturation of the choline kinase produced a white precipitate which did not interfere with the subsequent steps. The ACh was stable during these steps.

Acetylcholine assay. After cooling the microtubes, 1 μl of acetylcholinesterase (eel Type V, Sigma Chemical Co., St. Louis, MO), approximately 0.03 U, was added to each sample, which were then incubated at 37°C for 15 min.

A 26 μl portion of a second incubation mixture (buffer substrate B) containing 635 pmoles of [^{3}H]acetylcoenzyme A (0.7-1.3 Ci/mmole, New England Nuclear), 26 nmoles of physostigmine sulfate and 14 μl of a partially purified choline acetyltransferase solution (see preparation of choline acetyltransferase) were added. The microtubes were "buzzed" and incubated for 60 min at 37°C. Forty μl of the total 46 μl sample were transferred to a scintillation vial containing 5 ml of 0.1 M sodium phosphate buffer, pH 7.4; 2 ml of acetonitrile containing 10 mg sodium tetraphenylborate (Merck) were then added. After gentle shaking, 10 ml of toluene scintillator were added to the vials. The radioactivity was measured in a Packard Tri-Carb 300 liquid scintillation spectrometer.

Choline assay. Choline, which migrates immediately ahead of the ACh in the electrophoretic separation, is contained in the 2.3 cm band delineated from the outer edge of the TEA band away from the origin. However, due to the highly concentrated lyophilized extract used for ACh and the large amount of choline present in muscle, it is instead preferable to electrophorese 10 μl of the original unlyophilized extract and to proceed by the method of Saelens et al. (14) as modified by Ladinsky et al. (7).

ENZYMATIC CONVERSION EFFICIENCIES

These were checked in each experiment by the addition of appropriate substrates as described below.

Phosphorylation of choline. An excess of choline (200 pmoles) is added to blanks carried through electrophoresis and elution. The subsequent phosphorylation and acetylation reactions should yield blank counts.

Hydrolysis of acetylcholine. 41 pmoles of ACh dissolved in 1 μl of 0.1 mM HCl are added to microtubes containing 20 μl of buffer substrate A just after the phosphorylation step. The subsequent acetylcholinesterase and acetylation reactions should yield an equivalent amount of ACh as [^{3}H]ACh.

Acetylation of choline. 41 pmoles of choline dissolved in 1 μl of 0.1 mM HCl are added to microtubes containing 20 μl of buffer substrate A just before the addition of buffer substrate B. The subsequent acetylation reaction should yield equimolar ACh as [^{3}H]ACh.

PREPARATION OF CHOLINE ACETYLTRANSFERASE

The enzyme was prepared from mouse brains (minus cerebella) by the method of Ladinsky and Consolo (6). The crude enzyme was partially purified and concentrated by isoelectric precipitation with 1 M acetic acid. After centrifugation at 20,000 g for 20 min, the pellet was suspended in 0.1 M sodium phosphate buffer, pH 7 (4 ml per g brain) and the cloudy solution centrifuged at 20,000 g for 30 min. The clear supernatant was adjusted to pH 6 with glacial acetic acid. The solution was subjected to fractional ammonium sulfate precipitation and the 35%-70% fraction containing the enzyme was collected by centrifugation at 20,000 g for 15 min. The pellet was dissolved in 25 mM sodium phosphate buffer, pH 7.4 (3 ml per g brain) and the solution was dialyzed overnight against 2 liters of 50 mM sodium phosphate buffer, pH 7.4. Physostigmine sulfate, 5 mg/ml, was added in a volume of 133 μl/ml of dialysate to give a final concentration of 1.9 mM. The solution was spun at low speed to remove any insoluble material and the dialysate was placed in 5 ml Liovac flasks in 500 μl aliquots and lyophilized. The protein concentration before lyophilization was usually 5-6 mg/ml which yielded a 3-4 fold higher purification as compared with the previous method (6). The lyophilized vials were sealed and stored at -20°C. In this condition, the enzyme was found to be stable for up to 3 months. Before use the enzyme was dissolved in 700 μl of 0.1M sodium phosphate buffer.

RESULTS AND DISCUSSION

A flow sheet giving an overview of the analytical steps involved in the method is shown in Fig. 1. The electrophoretic step largely separates ACh from choline and choline analogs, and further presents the advantage of removing from the tissue extract many kinds of administered drugs which could potentially interfere with the assay. The phosphorylation reaction is able to convert the choline contaminating the ACh band, as well as related compounds and possible traces of interfering substances in the reagents, paper strips etc., to unlabeled phosphate esters. The adaptation of combining electrophoresis and phosphorylation thus affords an effective means of

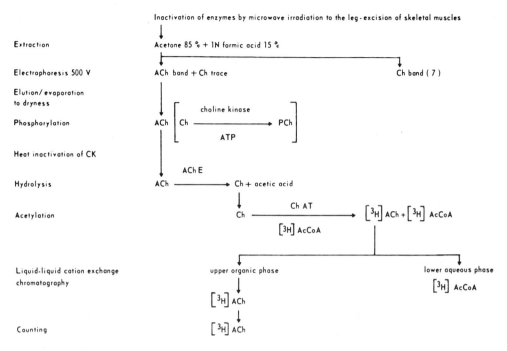

Figure 1. Flow sheet for the enzymatic microradioassay of ACh in skeletal muscle.

quantitatively removing the choline (>99.9%) while maintaining high recovery of ACh (86%). We noted that without prior electrophoresis the phosphorylation reaction does not go to completion, probably because the concentrated muscle extract is too acidic to be brought to the alkaline pH at which the phosphorylation reaction must be carried out.

Complete recovery of radiolabeled ACh was obtained in the extraction and lyophilization. An approximately 10% loss of ACh occurred in the electrophoretic step while the enzymatic reactions went virtually to completion as determined by the appropriate checks of enzymatic conversion efficiencies. The overall running recovery of ACh was 86%.

A number of structural analogs of ACh, some of which are present in muscle, did not interfere with the assay in amounts up to 200 pmoles when carried through the entire procedure either with or without the electrophoretic step. These substances included choline (200 pmoles), phosphorylcholine (200 pmoles), dimethylaminoethanol (82 pmoles), monomethylaminoethanol (82 pmoles), tetraethylammonium (140 pmoles), D,L-carnitine (165 pmoles), D,L-acetylcarnitine (165 pmoles).

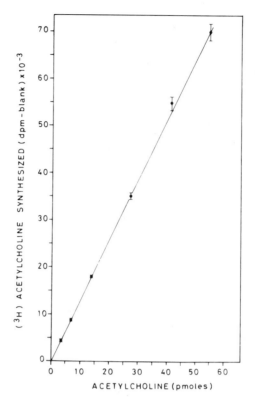

Figure 2. **Internal standard curve for ACh.** Standards of ACh iodide
(3.4, 6.8, 13.7, 27.3, 41.1 and 54.8 pmoles) in 10 μl solutions were
electrophoresed and assayed. The ordinate shows the amount of radiolabelled ACh formed in dpm after subtraction of blanks. The quantities
of standard are shown on the abscissa. The points represent the
means and S.E.M. (vertical bars) of 7 replicates. The slope is
1300 dpm/pmole ACh. The specific activity of [^{3}H]AcCoA in this
experiment was 0.7 Ci/mmole. A parallel curve was obtained in the
presence of tissue extract with the intercept on the ordinate yielding
the ACh amount in the extract. The coefficient of variation of
the replicate samples of tissue extract or of ACh standards was
less than 5%. The sensitivity of the method is 4 pmoles.

Fig. 2 shows the proportionality of dpm of product to ACh added
before electrophoresis over the range of 3.4-54.8 pmoles. The lower
sensitivity obtained is approximately 4 pmoles which is roughly
equivalent to that achieved by the radiochemical assay of Goldberg
and McCaman (2) who used [^{32}P]ATP of high specific activity to label
the product. The use of [^{3}H]acetylcoenzyme A has averted the in-
convenience inherent in the [^{32}P]-labelled substrate, a high energy
β-emitting isotope of short half-life and rapid decay character-

istics. The major drawback of using tritiated acetylcoenzyme A is the relatively high blank value. This, however, represents less than 1% of the total radioactivity added per tube and is likely caused by tritium exchange with water, an unavoidable phenomenon.

The ACh and choline content of rat EDL (white), gastrocnemius (mixed) and soleus (red) muscles is shown in Table 1. In freshly excised EDL, the ACh content was 33.5 ± 1.7 pmoles/muscle. Polak et al. (12) have reported similar values for EDL muscles excised from rats under continuous flow of oxygenated Krebs-Ringer solution. In microwaved EDL muscle the ACh content was 25% greater (41.8 ± 2.4 pmoles) while the amount of choline was reduced by almost 50%, from 4.1 ± 0.2 to 2.3 ± 0.1 nmoles. The ratio of choline content to ACh content was 122 in the fresh EDL and 55 in the microwaved EDL. Microwaving produced a 10% reduction in muscle weight which is due to water loss (from 78.9% to 70.9%).

The soleus muscle contained similar quantities of ACh (35.4 ± 2.7 pmoles) and choline (2.09 ± 0.2 nmoles) as did the EDL while the gastrocnemius possessed 5-fold higher ACh (168.0 ± 6.5 pmoles) and 10-fold higher choline (24.2 ± 1.1 nmoles). Therefore, the ratio of choline/ACh in this muscle is 143, likely indicating that much of the choline is extra-neuronal.

Surgical denervation reduced the ACh content of EDL within 10 days to less than 10% of innervated muscles. The ACh remained depressed at that level for at least 30 days. After denervation the muscle weight fell by 40% and 58% at the two times studied (Fig. 3). These results are in keeping with the findings of Polak et al. (12) who had used pyrolysis-mass fragmentography for their ACh determinations.

Table 1

Acetylcholine and Choline Content in
Freshly Excised Extensor Digitorum Longus and in
Some Microwaved Muscles of the Rat Hind Limb

	Acetylcholine (pmoles/muscle)	Choline (nmoles/muscle)	Choline/Acetylcholine (ratio)	Muscle mass (mg)
Fresh EDL	33.5 ± 1.7 (7)	4.09 ± 0.2 (5)	122	110.1 ± 2.4 (17)
Microwaved EDL	$*41.8 \pm 2.4$ (8)	$*2.29 \pm 0.1$ (5)	55	98.7 ± 1.6 (22)
Microwaved gastrocn.	168.0 ± 6.5 (6)	24.17 ± 0.1 (6)	143	899.8 ± 62.6 (6)
Microwaved soleus	35.4 ± 2.7 (7)	2.09 ± 0.2 (7)	59	85.7 ± 6.6 (7)

The data are the means $\pm$ S.E.M. (n); $*$ = p<0.01 vs fresh muscle, Student's t-test

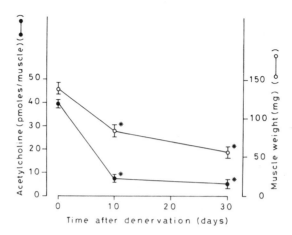

Figure 3. The effect of sciatic nerve degeneration on the ACh content
and fresh muscle weight of the rat extensor digitorum longus muscle
as a function of time. The EDL muscle was totally denervated by
resection of the sciatic nerve just after the last gluteal nerve
buds. The time after section, in days, is given on the abscissa
and the ACh content (pmoles/muscle) (●—●) and muscle wet weight
(mg) (o—o) is shown on the ordinate. The data are the means and
S.E.M. of 6 replicates. * p < 0.01 vs sham-operated controls, 0
time.

The dose-dependence of procaine, a local anesthetic having
neuromuscular blocking properties, on EDL ACh content, 15 min after
i.m. administration is shown in Table 2. The drug gave no significant
effect at 15 mg/kg and then produced an increase of 78% at the dose
of 30 mg/kg and doubled the ACh content at 60 mg/kg. All the drug
doses induced muscular paralysis of the limb. The choline content
remained unchanged at all three doses. The ACh increasing effect
of procaine is likely due to a blockade of ACh release. Indeed,
there is substantial evidence suggesting that local anesthetics
bind specifically to Na^+ channels at the nerve membrane, preventing
the normal Na^+ flux and as a result causing a nerve conduction block
(4, 15). In accordance, procaine and other local anesthetics reduce
release of [^{14}C]ACh evoked by high K^+ from cerebral cortex synaptosomes of guinea pig (16).

Previous in vitro studies have shown that ACh content increased
with time in either innervated or denervated EDL muscles incubated
with Soman, an organophosphate cholinesterase inhibitor, indicating
that the muscle tissue is capable of synthesizing ACh (12). On the
other hand, the carbamate cholinesterase inhibitor, physostigmine,
0.5 mg/kg and 1 mg/kg, i.m. was not able to produce an effect on
EDL ACh content up to 60 min of its administration, while it decreased
choline content by about 45% (Table 2).

Table 2

Effect of Drugs Administered Intramuscularly on the Acetylcholine
and Choline Content of Rat Extensor Digitorium Longus Muscle

Treatment	Dose (mg/kg)	Time (min)	Acetylcholine (pmoles/muscle)	Choline (nmoles/muscle)
Saline	--	--	39.7 ± 2.0	1.8 ± 0.2
Procaine	15	10	55.1 ± 4.6	1.9 ± 0.1
Procaine	30	15	70.8 ± 8.0*	1.7 ± 0.1
Procaine	60	15	76.9 ± 8.0**	1.8 ± 0.1
Saline	--	--	44.3 ± 3.1	2.0 ± 0.1
Physostigmine	0.5	30	47.1 ± 2.7	1.3 ± 0.2**
Physostigmine	0.5	60	41.6 ± 1.1	1.1 ± 0.05**
Physostigmine	1	10	42.3 ± 2.4	1.1 ± 0.2**
Saline[o]	--	--	56.2 ± 2.1	2.4 ± 0.3
α-BTX	100 μg/rat	45	43.0 ± 2.3**	3.8 ± 0.1**

The data are means ± S.E.M. (n = 5-7);   [o]Larger rats were used in
this experiment; *p < 0.05; **p < 0.01 VS saline treated group

α-Bungarotoxin (BTX), a nicotinic receptor blocker, decreased
the ACh content by about 30% at 100 μg/rat, 45 min after its i.m.
administration and increased choline content by 58%. Accordingly,
BTX doubled the amount of ACh collected from rat diaphragm after
phrenic nerve stimulation (9). The basal content of ACh, 56.2 ± 2
pmoles per muscle, was higher in this group of rats, which is a
reflection of the larger size of the animals used in this experiment
(Table 2).

The intraperitoneal administration of several drugs of various
classes did not alter the ACh content of EDL muscle. These drugs,
doses and times were: pentylenetetrazol, 80 mg/kg, 30 min; methadone
10 mg/kg s.c., 60 min; choline chloride, 300 mg/kg, 20 min; halop-
eridol, 2 mg/kg, 20 min; physostigmine, 1 mg/kg, 20 min; dichlorvos,
20 mg/kg, p.o., 20 min and isoniazide, 1 mg/kg, 30 min. Methadone
markedly increased the EDL choline content by 53%, from 3.0 ± 0.2
to 4.8 ± 0.1 nmoles/muscle; choline chloride administered at the
high dose increased EDL choline content 7-fold. The large changes,
both increases and decreases, in the choline content of EDL muscle
induced by the various drugs is of interest, since such effects
are rarely seen in brain. Whether this metabolic change is receptor
mediated and related to cholinergic activity remains to be investigated.

The radioenzymatic method described here has proved to be specific and sensitive enough to detect changes in ACh content of EDL muscle after different treatments. The method is also suitable for the measurement of ACh in other peripheral tissues, as well as in punches of small brain regions.

Acknowledgements. This work was partially supported by a grant from the Gustavus and Louise Pfeiffer Research Foundation, Los Angeles, CA., U.S.A.

REFERENCES

1. Feigenson, M.E. and Saelens, J.K. (1969): Biochem. Pharma col. 18:1479-1486.
2. Goldberg, A.M. and McCaman, R.E. (1973): J. Neurochem. 20:1-8.
3. Gundersen, C.B., Jenden, D.J. and Newton, M.W. (1981): J. Physiol. (Lond.) 310:13-35.
4. Hille, B. (1977): J. Gen. Physiol. 69:497-515.
5. Jenden, D.J. and Hanin, I. (1974): In Choline and Acetylcholine: Handbook of Chemical Assay Methods (ed.) I. Hanin, Raven Press, New York, pp. 135-150.
6. Ladinsky, H. and Consolo, S. (1974): In Choline and Acetyl choline: Handbook of Chemical Assay Methods (ed.) I. Hanin, Raven Press, New York, pp. 1-17.
7. Ladinsky, H., Consolo, S., Bianchi, S. and Jori, A. (1976): Brain Res. 108:351-361.
8. Ladinsky, H., Consolo, S., and Sanvito, A. (1972): Anal. Biochem. 49:294-297.
9. Miledi, R., Molenaar, P.C. and Polak, R.L. (1978): Nature 272:641-643.
10. Miledi, R., Molenaar, P.C., Polak, R.L., Tas, J.W.M. and Van der Laaken, T. (1982): Proc. R. Soc. Lond. B 214:153-168.
11. Polak, R.L. and Molenaar, P.C. (1979): J. Neurochem. 23:1295-1297.
12. Polak, R.L., Sellin, L.C. and Thesleff, S. (1981): J. Physiol. 319:253-259.
13. Potter, L.T., and Murphy, W. (1967): Biochem. Pharmacol. 16:1386-1388.
14. Saelens, J.K., Allen, M.P. and Simke, J.P. (1970): Arch. Int. Pharmacodyn. 186:279-286.
15. Strickartz, G. (1976): Anaesthesiology 45:421-441.
16. Vyas, S. and Marchbanks, R.M. (1983): Biochem. Pharmacol. 32:2827-2829.

PRE- AND POSTSYNAPTIC EFFECTS OF MUSCARINIC ANTAGONISTS

IN THE ISOLATED GUINEA PIG ILEUM

H. Kilbinger, W. Weiler and I. Wessler

Pharmakologisches Institut der Universität Mainz
Obere Zahlbacher Str. 67
D-6500 Mainz (FRG)

INTRODUCTION

Several compounds have been claimed to differentiate between subtypes of muscarinic receptors (2). In the experiments described in this chapter we have studied in the guinea-pig ileum whether the presynaptic muscarinic receptors of the cholinergic nerves differ from the postsynaptic muscarinic receptors of the longitudinal muscle in their affinities for several muscarinic antagonists (methylatropine; trihexyphenidyl; clozapine; DAMP). Inhibition by oxotremorine of the evoked release of acetylcholine (ACh) was used as a parameter of presynaptic activity, and the increase by oxotremorine of smooth muscle tension as a postsynaptic parameter. The affinity constants (pA_2 values) of the antagonists were determined by constructing complete concentration response curves for pre- and postsynaptic effects of oxotremorine, in the absence and presence of different concentrations of the respective antagonist. Differences in the affinities of a given antagonist to either pre- or postsynaptic receptors should result in different pre- and postsynaptic pA_2 values.

METHODS

The method of measuring the release of [^{3}H]-ACh from the myenteric plexus-longitudinal muscle preparation of the guinea-pig ileum in the absence of a cholinesterase inhibitor has been described in detail (9). In short, two longitudinal muscle strips were incubated in a 2 ml organ bath with [^{3}H]-choline (1 μM; 5 μCi/ml) and subsequently superfused (1 ml/min) at 37°C with Tyrode solution (composition in mM: NaCl 137; KCl 2.7; $CaCl_2$ 1.8; $MgCl_2$ 1.0; NaH_2PO_4 0.4; $NaHCO_3$ 11.9; glucose 5.6; choline 0.001). After a 70 min washout period the superfusate was collected in 3 min samples. Tritium was determined by liquid scintillation spectrometry. Results are expressed as pmol of labelled ACh, based on the specific radioactivity of $[^3H]$-choline used for labelling.

Strips were stimulated by field stimulation four times (S_1-S_4) at a frequency of 1 Hz. 120 square wave pulses (1 ms) were applied by two platinum electrodes, that were positioned parallel to the strips. Stimulation periods started 9 min (S_1), 39 min (S_2), 66 min (S_3), and 93 min (S_4) after the washout period. The stimulation-evoked outflow of $[^3H]$-ACh was obtained from the difference of the total tritium outflow during stimulation and the following 13 min, and the calculated spontaneous outflow. The spontaneous outflow of tritium was assumed to decline exponentially from the 9 min period before stimulation to the period 15-24 min after the onset of stimulation.

Cumulative concentration-response curves for the inhibitory effect of oxotremorine were obtained by increasing, in a stepwise manner, the concentration of oxotremorine in the superfusate by a factor of 10. On any single strip a complete concentration-response curve for oxotremorine was established. Each concentration of oxotremorine was superfused from 24 or 21 min before, until 6 min after the respective stimulation period (S_2, S_3, S_4). The effects of oxotremorine on the outflow of $[^3H]$-ACh were expressed as the ratio between the outflow of $[^3H]$-ACh evoked by S_2, S_3, or S_4, and that evoked by S_1. The ratios of $[^3H]$-ACh outflow obtained in the presence of oxotremorine were expressed as the percentage of the equivalent ratio obtained in control experiments:

$$\% \ [^3H]\text{-ACh outflow} = \frac{(S_X/S_1) \ \text{oxotremorine}}{(S_X/S_1) \ \text{control}} \times 100$$

where S_X indicates either S_2, S_3, or S_4.

For interaction experiments muscarinic antagonists were added to the superfusate 50 min before S_1 and remained in the medium throughout the experiment. EC_{50} values for oxotremorine (either in the absence or presence of an antagonist) were calculated from concentration-response curves obtained on a single muscle strip. Dose ratios were obtained from the ratios of individual EC_{50} values for oxotremorine in the presence of an antagonist, and the mean EC_{50} value for oxotremorine from all control experiments. pA_2 values for the presynaptic effects of muscarinic antagonists were calculated from Schild plots, by linear regression analysis, according to Furchgott (6).

Contractions of the longitudinal muscle strips were recorded isometrically. EC_{50} values were obtained from individual concentration-response curves to oxotremorine, and dose ratios were calculated for each experiment at the level of the EC_{50}. Muscarinic antagonists were added 45 min before the effects of oxotremorine were tested. pA_2 values for the postsynaptic effects of muscarinic antagonists were determined, from Schild plots, by regression analysis.

Statistics.   Given are means $\pm$ S.E.M.   Significance of difference was calculated by t-test.   N, number of experiments.   Regression lines of Schild plots were calculated by the least square method. The slope of a regression line was tested for difference from unity according to Sachs (11).   The 95% confidence limits of the pA_2 values were calculated according to Documenta Geigy (3).

Drugs.   Oxotremorine sesquifumarate (Aldrich); N-methylatropinium bromide (Merck, Darmstadt); trihexyphenidyl hydrochloride (Lederle, Munchen); clozapine base (Sandoz, Basel); 4-diphenyl-acetoxy-N-methylpiperidine methiodide (DAMP, synthesized according to Abramson et al. (1) by Dr. G. Lambrecht, University of Frankfurt).

RESULTS AND DISCUSSION

In control experiments the stimulation-evoked outflow of $[^3H]$-ACh declined from S_1 through S_4.   Oxotremorine decreased the evoked outflow in a concentration-dependent manner (Table 1).   From the concentration-response curve a presynaptic EC_{50} of 33 ± 5 nM was calculated.

Muscarinic antagonists were superfused from 50 min before S_1 onwards.   Neither of the antagonists significantly enhanced the outflow during S_1 (cf. Table 1 with Table 2 and Table 3).   In the presence of the antagonists alone the decline in evoked outflow

Table 1

Effects of Oxotremorine on Stimulation-Evoked Outflow of
$[^3H]$-ACh from Longitudinal Muscle Strips
Preincubated with $[^3H]$-choline.

	S_1 (pmol/g)	S_2 / S_1	S_3 / S_1	S_4 / S_1	N
Control	34 ± 11	0.88 ± 0.09	0.72 ± 0.03	0.62 ± 0.08	4
Oxo-tremorine	34 ± 3	0.72 ± 0.02	$0.13 \pm 0.04**$	$0.05 \pm 0.02**$	5

Each strip was stimulated four times at 1 Hz, 120 pulses (S_1 - S_4). The concentrations of oxotremorine were 0.01 μM (S_2), 0.1 μM (S_3) and 1 μM (S_4). $**p < 0.01$ vs control experiments.

Table 2

Effects of Oxotremorine on Stimulation-Evoked Outflow of
[^{3}H]-ACh in the Presence of Methylatropine

Methylatropine (nM)	Oxotremorine (μM) during S2; S3; S4	S1 (pmol/g)	S2/S1	S3/S1	S4/S1	N
3	-	35±7	0.79±0.06	0.61±0.05	0.49±0.05	4
3	0.01; 0.1; 1.0	50±5	0.80±0.08	0.31±0.05**	0.09±0.02**	4
10	-	39±10	0.83±0.11	0.73±0.05	0.55±0.05	4
10	0.01; 0.1; 1,0	31±5	0.77±0.05	0.57±0.06**	0.21±0.01**	4
30	-	43±15	1.09±0.09	0.83±0.07	0.62±0.05	3
30	0.1; 1.0; 10.0	36±9	0.76±0.01**	0.40±0.01**	0.11±0.01**	4
100	-	69±18	0.83±0.09	0.63±0.03	0.45±0.04	4
100	0.1; 1.0; 10.0	80±13	0.80±0.03	0.54±0.01**	0.19±0.02**	4

Significance of difference from control experiments: *$p < 0.05$;
**$p < 0.01$.

Table 3

Effects of Oxotremorine on Stimulation-Evoked Outflow of [^{3}H]-ACh
in the Presence of DAMP

DAMP (nM)	Oxotremorine (μM) during S2; S3; S4	S1 (pmol/g)	S2/S1	S3/S1	S4/S1	N
10	-	28±8	0.86±0.05	0.69±0.05	0.59±0.05	4
10	0.01; 0.1; 1.0	38±3	0.81±0.05	0.42±0.04**	0.07±0.01**	4
30	-	37±8	0.90±0.08	0.79±0.09	0.58±0.08	4
30	0.01; 0.1; 1.0	31±4	0.91±0.04	0.50±0.03**	0.10±0.02**	4
100	-	28±6	0.93±0.07	0.78±0.06	0.68±0.05	4
100	0.1; 1.0; 10.0	20±2	0.89±0.10	0.29±0.06**	0.08±0.03**	4
300	-	27±4	0.85±0.08	0.65±0.05	0.49±0.07	4
300	0.1; 1.0; 10.0	47±2	0.80±0.05	0.45±0.04**	0.11±0.01**	4

Significance of difference from control experiments: *$p < 0.05$;
**$p < 0.01$.

was similar to that in the control experiments. The inhibitory effects of oxotremorine were diminished in the presence of the antagonists. Typical examples for the interaction between oxotremorine and the antagonists are given in Tables 2 and 3. Methylatropine, trihexyphenidyl, DAMP and clozapine behaved as competitive antagonists of oxotremorine, causing parallel shifts of the oxotremorine concentration-response curve to the right without affecting the maximal inhibitory effect. From the dose ratios pA_2 values for the presynaptic effects of the antagonists were calculated. The slopes of the regression lines did not differ significantly from unity, the theoretical value for an interaction between an agonist and a competitive antagonist.

The four antagonists studied caused a concentration-dependent inhibition of the contractile effects of oxotremorine on the smooth muscle. The concentration-response curves for the postsynaptic effects of oxotremorine were shifted to the right in a parallel manner, and the maximum response was unaffected.

The pre- and postsynaptic pA_2 values of the muscarinic antagonists studied in the present experiments are compared in Figure 1. Included are the published values for pre- and postsynaptic effects of pirenzepine and scopolamine (7). It is evident that neither of the six antagonists shows a preferential affinity to presynaptic neuronal or postsynaptic ileal muscarinic receptors. The results suggest

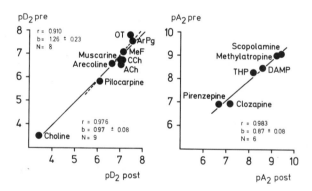

Figure 1. Correlation between pre- and postsynaptic potencies of muscarinic agonists and antagonists in the guinea-pig ileum (from Ref. 8). Left panel: pD_2 values of muscarinic agonists. Presynaptic response, inhibition by drugs of [^{3}H]-ACh outflow evoked by electrical stimulation (1 Hz, 180 pulses). Postsynaptic response, increase by drugs in longitudinal muscle contraction. ArPg, arecaidinepropargylester. OT, oxotremorine. MeF, methylfurmethide. CCh, carbachol. Interrupted line, data without choline, N = 8. Solid line, data including choline, N = 9. Data taken from Ref. 9 and 10. Right panel: Pre- and postsynaptic pA_2 values of muscarinic antagonists. THP, trihexyphenidyl.

that the pharmacological properties of pre- and postsynaptic muscarinic receptors in the guinea-pig ileum are not different. This suggestion is corroborated by previous experiments in which we have found that the relative potencies for pre- and postsynaptic effects of muscarinic agonists in the ileum were similar (9, 10).

Figure 1 summarizes the results of experiments in which pre- and postsynaptic effects of muscarinic drugs were studied in the isolated guinea-pig ileum. It is obvious that there are highly significant correlations between pre- and postsynaptic pD_2 values of agonists, and pre- and postsynaptic pA_2 values of antagonists.

From all these results it is thus concluded that the muscarinic receptors located on both sites of the cholinergic synapse in the ileum are a homogeneous group. The situation is comparable to that in the adrenergic system where the muscarinic receptors mediating both inhibition of cardiac performance and inhibiton of noradrenaline release possess similar properties (4, 5).

Acknowledgement. This work was supported by the Deutsche Forschungsgemeinschaft.

REFERENCES

1. Abramson, F.B., Barlow, R.B., Franks, F.M. and Pearson, J.D.M. (1974): Brit. J. Pharmacol. 51:81-93.
2. Birdsall, N.J.M. and Hulme, E.C. (1983): Trends Pharmacol. Sci. 4:459-463.
3. Documenta Geigy, Wissenschaftliche Tabellen (1969): 7th edition, Ciba-Geigy Ltd., Basel, p. 178, equation (672).
4. Fozard, J.R. and Muscholl, E. (1972): Brit. J. Pharmacol. 45:616-629.
5. Fuder, H., Meiser, C., Wormstall, H. and Muscholl, E. (1981): Naunyn-Schmiedeberg's Arch. Pharmacol. 316:31-37.
6. Furchgott, R.F. (1972): In Handbook of Exp. Pharmacol., Vol. 33 (eds) H. Blaschko and E. Muscholl, Springer, Berlin, Heidelberg, New York, pp. 283-335.
7. Halim, S., Kilbinger, H. and Wessler, I. (1982): Scand. J. Gastroenterol. 17(Suppl. 72):87-93.
8. Kilbinger, H. (1984): Trends Pharmacol. Sci. 5:103-105.
9. Kilbinger, H. and Wessler, I. (1980): Naunyn-Schmiedeberg's Arch. Pharmacol. 314:259-266.
10. Kilbinger, H. and Kruel, R. (1981): Naunyn-Schmiedeberg's Arch. Pharmacol. 316:131-134.
11. Sachs, L. (1969): Statistische Auswertungsmethoden (2. Aufl.), Springer, Berlin, Heidelberg, New York.

PRESYNAPTIC MODULATION OF ACETYLCHOLINE RELEASE

J.R. Cooper, J.H. Reese and H. Park

Department of Pharmacology
Yale University School of Medicine
New Haven, Connecticut 06510   U.S.A.

INTRODUCTION

Modulation of synaptic transmission, both pre- and postsynaptically is gaining considerable recognition as a means of fine tuning
the nervous system, a requisite for understanding behavioral changes.
In this laboratory we are focusing on presynaptic modulation, with
an ultimate goal of investigating the molecular mechanisms that
may be involved.

To this end we are utilizing the myenteric plexus-longitudinal
muscle of the guinea-pig ileum, a preparation with a long history
in identifying neuroactive agents that regulate acetylcholine (ACh)
release.   Among these agents are serotonin, γ-Aminobutyric acid,
norepinephrine, oxotremorine, opioids, somatostatin, substance P,
neurotensin, and CCK-8. Another reason for utilizing this preparation
is its similarity to the central nervous system with its unmyelinated
axons, a blood-myenteric plexus barrier, non-neuronal cells which
resemble glia more than Schwann cells, its integrative nervous activity
and the fact that, like the brain, the myenteric plexus exhibits
tolerance and physical dependence following the in vivo administration
of morphine to the guinea-pig.

Since as noted above, we are primarily interested in elucidating
the mechanisms involved in modulation, we developed a synaptosomal
preparation of the myenteric plexus-longitudinal muscle (2).   Subsequently we demonstrated that when this preparation was preloaded
with [^{3}H]choline in order to generate [^{3}H]ACh, the transmitter was
released in a calcium-dependent manner on depolarization with K^+,
veratridine, or the nicotinic agonist dimethyl phenyl piperazinium
(DMPP).   We further showed that modulation was considerable more

marked when release was evoked with DMPP as compared to K^+ or veratridine (3, 6). To date, we have identified a muscarinic receptor
(3), a purinergic receptor (A_1 or R_i type) (6) and an adrenergic
receptor (α_2 type) (7), all of which result in an inhibition of evoked
ACh release; as well as a nicotinic excitatory receptor.

The present communication concerns the effects of cyclic nucleotides and forskolin on ACh release and modulation, using the synaptosomal preparation of the myenteric plexus-longitudinal muscle.

METHODS

Preparation of P_2 fraction. Preparation of a crude synaptosomal
fraction was as described elsewhere (2). Briefly, the ileum from
one Hartley guinea-pig was removed and placed in Krebs Ringer Bicarbonate (KRB) buffer consisting of NaCl (118 mM), KCl (4.7 mM),
$CaCl_2$ (2.5 mM) $MgSO_4$ (1.2 mM), NaH_2PO_4 (1.2 mM), $NaHCO_3$ (25 mM)
and dextrose (10 mM), and gassed with 95% O_2-5% CO_2. Longitudinal
muscle strips were minced and homogenized in 5 ml of 0.32 M sucrose
containing 3 mM sodium phosphate buffer (pH 7.2). Following centrifugation at 1000 x g for 10 min the supernatant was collected and
centrifuged for 20 min at 17,000 x g. The pellet (P_2) produced
was used for the release studies.

Synthesis and release of [^{3}H] ACh. The tissue preparation containing
the synaptosomes was incubated in 3 ml of KRB containing 3.0 μM [^{3}H]-
choline (5.0 x 10^3 Ci/mol) for 30 min at 37°C. After centrifugation
for 10 min at 5000 x g the pellet was washed in fresh KRB containing
10 μM physostigmine. The tissue was then suspended in 10 ml of the
KRB plus eserine, and aliquots (200 μl) were added to 10 x 75 mm
tubes which contained the appropriate drugs for a final volume of
220 μl. The tubes were incubated for 10 min at 37°C under 95% P_2-5%
CO_2. After returning the tubes to the ice bath, the tubes were
centrifuged at 3600 x g for 10 min and the supernatants were collected.

[^{3}H]ACh determination. The choline kinase-ion pair extraction method
(5) was used to separate [^{3}H]ACh from [^{3}H]choline.

MATERIALS

The materials used included: 80 x 10^3 Ci/mol (methyl-[^{3}H])
choline chloride (New England Nuclear; physostigmine sulfate (ICN
Pharmaceuticals); butyronitrile (Aldrich Chem. Col) glycylglycine
(Calbiochem-Behring); adenosine (Boehringer Mannheim); ATP, choline
chloride, ATP: choline phosphotransferase (EC 2.7.1.32, choline
kinase), DMPP iodide, 8-bromo-cAMP, 8-bromo-cGMP, IBMX, EGTA and
tetraphenylboron (Sigma Chem. Co.).

RESULTS

Effects of cyclic nucleotides on release of [³H]ACh. Increasing concentrations of 8-bromo-cAMP caused a marked increase in the release of [³H]ACh from the guinea-pig ilial myenteric plexus synaptosomes (Fig. 1a). The addition of DMPP (10 μM) produced an increased release of [³H]ACh to levels above those induced by 8-bromo-cAMP alone. When 8-bromo-cGMP was evaluated, this cyclic nucleotide stimulated the release of [³H]ACh only at high concentrations (Fig. 1b). DMPP similarly stimulated the release above the levels produced by 8-bromo-cGMP.

The reduction of the DMPP-induced release in the presence of 8-bromo-cAMP (i.e., the difference between the release in the presence of DMPP plus the 8-bromo-cAMP and the release caused by the cyclic nucleotide itself), suggested that there was an interaction between the release caused by the two agents (Fig. 1c). In contrast, with 8-bromo-cGMP, the increase in release caused by DMPP was constant at all concentrations of the cyclic nucleotide tested, suggesting no obvious interaction.

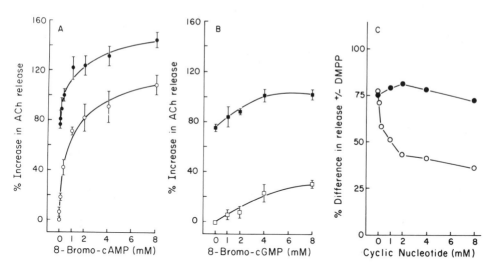

Figure 1. Stimulation of release of [³H]ACh from synaptosomes by cyclic nucleotides and DMPP. (a) Effect of increasing concentrations of 8-bromo-cAMP in the absence ○ or presence ● of DMPP. Release and determination of [³H]ACh were as in the Methods section. The data are the means ± S.E.M. of 3 experiments. (b) Effect of increasing concentrations of 8-bromo-cGMP in the absence □ and presence ■ of DMPP. (c) The difference in release of [³H]ACh in the presence and absence of DMPP with 8-bromo-cAMP (from a) ○ and with 8-bromo-cGMP (from b) ● .

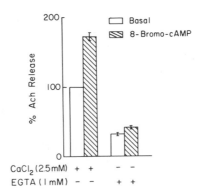

Figure 2. Effect of Ca^{2+} on the basal and 8-bromo-cAMP-induced release
of [³H]ACh from the synaptosomes. All steps prior to the evaluation
of release were carried out in KRB containing Ca^{2+}. The data are
the means ± S.E.M. of 3 experiments.

Effect of Ca^{2+} on the 8-bromo-cAMP induced release. In the presence
of 2.5 mM Ca^{2+}, 8-bromo-cAMP (4mM) increased the release of [³H]ACh
from the synaptosomes (Fig. 2). In the absence of exogenous Ca^{2+}
and the presence of 1 mM EGTA both the basal and 8-bromo-cAMP-stimu-
lated release were markedly reduced. Thus, in contrast to the effect
of 8-bromo-cAMP on protein phosphorylation which is independent of
Ca^{2+} (4), the release of ACh by this agent is dependent on the presence
of the cation.

Effect of IBMX on the release of [³H]ACh. The ability of IBMX, a
phosphodiesterase inhibitor, to promote release was evaluated in
the presence and absence of DMPP. IBMX stimulated the release of
[³H]ACh from the synaptosomes (Fig. 3) but not to the level produced
by 8-bromo-cAMP. At lower concentrations of IBMX, DMPP stimulated
the release above the level caused by IBMX alone. As IBMX was increased, however, the ability of DMPP to cause additional release
was reduced.

Effects of neuromodulators on the 8-bromo-cAMP-stimulated release.
Release by 8-bromo-cAMP was evaluated in the presence of adenosine
(10 μM), clonidine (10 μM), and oxotremorine (100 μM). As can be
seen in Fig. 4a, these substances all caused reductions of the basal
release and the total release in the presence of the nucleotide.
However, when the 8-bromo-cAMP-stimulated release (the difference
between the total release and the basal release) was determined,
no significant modulation was observed (Fig. 4b).

Effect of Forskolin on the [³H]ACh release. The addition of forskolin
(50 μM), an activator of adenylate cyclase (8), increased the release
of [³H]ACh by 60% when compared to the basal release (Fig. 5).

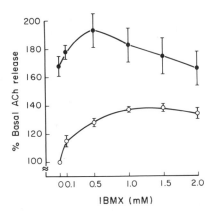

Figure 3. Effect of IBMX on the release of [³H]ACh in the absence ○ and presence ● of DMPP. Release and determination of [³H]ACh were as in the Methods section. The data are the means ± S.E.M. of 3 experiments.

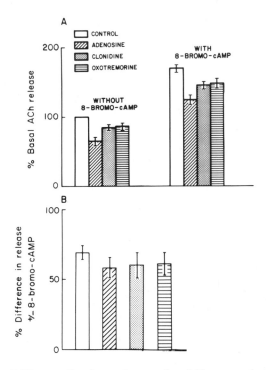

Figure 4. (a) Effect of adenosine, clonidine, and oxotremorine on the release of [³H]ACh in the presence or absence of 8-bromo-cAMP. The data are the means ± S.E.M. of 3 experiments. (b) The difference in the release of [³H]ACh in the presence or absence of 8-bromo-cAMP with adenosine, clonidine or oxotremorine (from a).

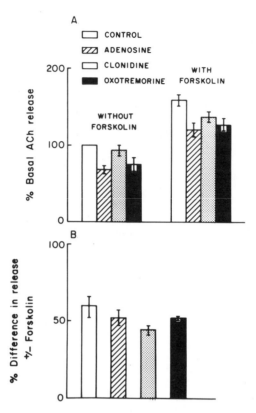

Figure 5. (a) Effect of adenosine, clonidine, and oxotremorine on the release of [³H]ACh in the presence or absence of 50 μM forskolin. The data are the means ⁺ S.E.M. of 3 experiments. (b) The difference in the release of [³H]ACh in the presence or absence of forskolin with adenosine, clonidine or oxotremorine (from a).

While each of the modulators adenosine (10 μM), clonidine (10 μM) and oxotremorine (100 μM) reduced the basal release somewhat, the forskolin-induced release remained unchanged (Fig. 5b).

DISCUSSION

Since Alberts and Stjarne (1) first demonstrated that the electrically stimulated release of ACh from the intact guinea-pig ileum myenteric plexus was increased by 55% in the presence of 0.5 mM 8-bromo-cAMP, we have entertained the possibility that cAMP might act as a second messenger in this system. This membrane permeable cAMP analogue in fact induced release of [³H]ACh at concentrations up to 8 mM. Concentrations of this magnitude were evaluated in our studies since Forn and Greengard (4) showed that these concentrations of 8-bromo-cAMP stimulated maximally the phosphorylation of membrane proteins.

DMPP, which maximally stimulates the nicotinically-induced release of [³H]ACh (3), caused an additional increase in [³H]ACh release at each concentration of 8-bromo-cAMP used. At the higher concentrations of 8-bromo-cAMP however, the nicotinically-induced increase in release was much reduced. This finding suggests a relationship between the release caused by 8-bromo-cAMP and that caused by DMPP. In contrast, 8-bromo-cGMP promoted release of [³H]ACh, only at high concentrations; with this release however, the effect of DMPP was constant at all concentrations of 8-bromo-cGMP tested. This suggests that there was no direct relationship between the two types of release.

Alberts and Stjarne (1) reported no effect of 8-bromo-cGMP on the release of [³H]ACh. While the specific reason for the different findings in these two studies is not clear, Alberts and Stjarne were using only 0.5 mM to 1.0 mM 8-bromo-cGMP and, in addition they used the intact longitudinal muscle-myenteric plexus preparation.

Since we have previously shown that the DMPP-induced release from ileal synaptosomes is calcium dependent (3), we were interested to determine whether the release caused by 8-bromo-cAMP required the presence of this cation. Calcium was indeed required for the 8-bromo-cAMP-induced release, in contrast to the calcium-independent effect of the cyclic nucleotide in promoting protein phosphorylation. This observation implies that if protein kinase activity is involved in the release process, a different step is involved, where the presence of extracellular Ca^{2+} is mandatory. Further support of a second messenger role for cyclic nucleotides is provided by experiments with IBMX a phosphodiesterase inhibitor. In our study IBMX stimulated the release of [³H]ACh from the synaptosomes, suggesting that 8-bromo-cAMP may have been acting through a cAMP-dependent mechanism.

We have previously reported that adenosine, clonidine, and oxotremorine reduce the DMPP-induced release of [³H]ACh from guinea-pig ileal synaptosomes (3, 6, 7). The three modulators reduced the basal release but left unaffected the release caused by the cAMP analog. The results using 8-bromo-cAMP reported above indicate that the inhibition of the nicotinically-induced release caused by adenosine, clonidine and oxotremorine must act at some point prior to that of cAMP, and subsequent to protein phosphorylation activity.

Similarly, forskolin, which directly stimulates the catalytic subunit of adenylate cyclase causing an increase of intracellular cAMP (8) stimulated the release of [³H]ACh. However, neither adenosine, clonidine nor oxotremorine reduced this forskolin-stimulated [³H]ACh release. If DMPP in fact requires cAMP as a second messenger for the release of [³H]ACh, then the inability of the modulators of the DMPP-induced release to affect the forskolin-induced release again suggests that the modulation must occur at some point prior to the actual production of cAMP by adenylate cyclase. Whether the modulation is a direct interaction on calcium channels or at the receptor level is as yet unknown.

In this chapter we have demonstrated that [^{3}H]ACh can be released from guinea-pig ileal myenteric plexus synaptosomes by agents (IBMX, forskolin) which raise intracellular cAMP levels, and by the direct action of the membrane permeable cAMP analog, 8-bromo-cAMP. Since protein phosphorylation involves protein kinase activity, our results suggest that a cAMP-dependent protein kinase is involved. On the other hand, we have observed (unpublished) that trifluoperazine inhibits the DMPP-induced release of ACh in our preparation. This observation adds an additional level of complexity to the release mechanism, with the implication of calmodulin stimulating a calmodulin-dependent protein kinase. At any rate, regardless of the protein phosphorylation mechanism, it is clear from our results that this mechanism is not involved in modulatory activity; our previously described modulators exert their inhibitory activity only when ACh release is evoked by DMPP, and not by the agents which lead to protein phosphorylation.

REFERENCES

1. Alberts, P. and Stjarne, L. (1982): Acta Physiol. Scand. _115_: 269-272.
2. Briggs, C.A. and Cooper, J.R. (1981): J. Neurochem. _36_:1097-1108.
3. Briggs, C.A. and Cooper, J.R. (1982): J. Neurochem. _38_:501-508.
4. Forn, J. and Greengard, P. (1978): Proc. Nat. Acad. Sci. U.S.A. _75_:5195-5199.
5. Nemeth, E.F. and Cooper, J.R. (1979): Brain Res. _165_:166-170.
6. Reese, J.R. and Cooper, J.R. (1982): J. Pharm. Exp. Therap. _223_:612-616.
7. Reese, J.R. and Cooper, J.R. (1984): Biochem. Pharmac. _33_:3007-3011.
8. Seamon, K. and Daly, J.W. (1981): J. Biol. Chem. _256_:9799-9801.

EFFECT OF DIISOPROPYLFLUOROPHOSPHATE ON SYNAPTIC TRANSMISSION AND ACETYLCHOLINE SENSITIVITY IN NEUROBLASTOMA-MYOTUBE CO-CULTURE

M. Adler, F.-C. T. Chang, D. Maxwell,
G. Mark, J.F. Glenn and R.E. Foster

Neurotoxicology Branch
United States Army Medical Research Institute
of Chemical Defense[*]
Aberdeen Proving Ground, Maryland 21010 U.S.A.

INTRODUCTION

The toxicity of organophosphorous cholinesterase inhibitors is believed to result primarily from inhibition of acetylcholinesterase (AChE) and consequent increases in acetylcholine (ACh) lifetime (10). At the vertebrate neuromuscular junction, blockade of AChE leads to increases in the amplitude and time course of spontaneous miniature endplate potentials (MEPPs), evoked endplate potentials (EPPs) and their underlying currents (7, 9, 11). In mammalian but not amphibian preparations, anticholinesterase agents also cause repetitive antidromic firing in the presynaptic axon in response to a single conditioning stimulus (12). This phenomenon has been attributed to nerve terminal depolarization due to accumulation of ACh (17) or K^+ (12) in the synaptic cleft following blockade of ACh hydrolysis.

Although the above effects can be explained by AChE blockade, several actions of a less well-understood nature have also been reported for AChE inhibitors. Thus, paraoxon has been found to increase MEPP frequencies and to depolarize muscle fibers, even at relatively low concentrations (13). Neither of these actions are explained adequately by AChE blockade. In high concentrations, several cholinesterase inhibitors have been shown to reduce the

---

[*]"The opinions or assertations contained herein are the private views of the authors and are not to be construed as official or as reflecting the views of the U.S. Army or the Department of Defense."

amplitude of EPPs and their associated currents, convert their decays from single to multi-exponential functions, and enhance the rate of desensitization to ACh (11, 16). These appear to be caused largely by direct actions of the inhibitors on the ACh receptor complex.

We have investigated the effects of the irreversible organo-phosphorous cholinesterase inhibitor, DFP, on clonal G8-1 myotubes co-cultured with ACh- secreting NG108-15 neuroblastoma x glioma hybrid cells. The myotube AChE activity increases gradually after fusion of myoblasts and attains maximal levels after 2-3 weeks in culture (19). Since myotube AChE is developmentally regulated, the co-cultures provide a favorable preparation for examining the direct effects of cholinesterase inhibitors, as well as for inves-tigating the role of AChE in synaptic transmission.

Our results indicate that DFP depresses the amplitude of ion-tophoretically-induced ACh potentials and spontaneous miniature synaptic potentials (MSPs) that result from quantal secretion of NG108-15 ACh. DFP also enhances the rate of desensitization of the myotube to ACh but has little or no effect on the resting potential or cable properties of either clonal cell type. These findings suggest that DFP interferes with nicotinic receptor activation. Co-cultures incubated with DFP and subsequently washed with physio-logical solution showed no increases in the synaptic potential ampli-tude or time course, both of which are indicative of AChE blockade in adult neuromuscular preparations. The absence of such potentiation implies that G8-1 AChE activity is inadequate to influence the ACh lifetime or that transmitter diffusion is sufficiently rapid to terminate the action of ACh at these clonal synapses.

MATERIALS AND METHODS

Culture techniques. Experiments were performed on clonal myogenic G8-1 cells and hybrid NG108-15 cells generated by the fusion of neuroblastoma N18TG2 with glioma line C6BU-1. The cells were gener-ously supplied by Dr. Marshall Nirenberg. G8-1 myoblasts (subcultures 4 to 19) were grown in Dulbecco's modified Eagle's medium (DMEM) supplemented with 5% fetal bovine serum in 250 ml Falcon flasks. The cultures were harvested at 40-50% confluence with phosphate buffered saline containing 0.1% trypsin and 0.05 mg/ml DNAse. NG108-15 cells (subcultures 12 to 26) were grown in DMEM supplemented with 5% fetal bovine serum, 10^{-4}M hypoxanthine, 10^{-6}M aminopterin and 1.6×10^{-5}M thymidine in 250 ml flasks. Myotube cultures were prepared for electrophysiological recording by plating 10^{5} myoblasts in collagen-coated 35 mm tissue culture dishes. The cultures were maintained in DMEM and 5% fetal bovine serum until 70% confluent and subsequently switched to DMEM with 10% horse serum to promote fusion of myoblasts. Nonfused dividing cells were eliminated by addition of 10^{-5}M 5'-fluorodeoxyuridine and 10^{-4}M uridine for 2-3 days beginning one day after addition of horse serum. Myotube cultures remained viable for over 30 days. Co-cultures were established by seeding 8-10 day old myotube dishes with 5×10^4 NG108-15 cells. Co-cultures were maintained in DMEM containing 10% horse serum, 10^{-4}M hypoxanthine, 1.6×10^{-5}M thymidine and 10^{-3}M dibutyryl cyclic-AMP. Recordings were carried out after at least three days in co-culture.

Electrophysiology. Intracellular recordings were performed at 35°C by standard electrophysiological procedures using 3M KCl filled microelectrodes (30-40 megohm). Stimulation and recording techniques were similar to those described by Atlas and Adler (3). ACh was applied to high density receptor patches of G8-1 myotubes with iontophoretic pipettes (120-160 megohm) containing 2M ACh. Braking currents of 2-4 nA were usually sufficient to prevent background leakage. For brief focal applications, the ACh microelectrode was positioned to produce optimal ACh sensitivity with a pulse duration of 0.2-1.0 msec. For steady state responses, the ACh pipettes were positioned 10-15 um above the myotube and the pulse duration was increased to 5-30 sec in order to produce ACh potentials with a stable plateau.

Determination of AChE activity. For measurements of AChE activity, myotubes treated similarly to those used for electrophysiological studies were washed twice with phosphate buffered saline to remove traces of horse serum proteins and detached from the substrate with the aid of 0.1% trypsin and 0.05 mg/ml DNAse. The cells were centrifuged at 300 x g for 5 min. The supernatant was removed and the pellet was frozen at -20°C for 2 weeks. The cells were subsequently thawed and homogenized in 1% Triton X-100 using a Potter Elvehjem homogenizer with a Teflon pestle. The homogenate was centrifuged at 15,000 x g for 10 min. The resultant supernatant was assayed for AChE activity at 37°C in 0.1M potassium phosphate buffer (pH 7.4) using the radiometric method of Siakotos et al. (18). Protein was analyzed by the method of Bradford (4) using bovine serum albumin as a standard. Experiments in which cells were harvested by scraping instead of trypsin revealed no significant differences in AChE activity.

Solutions and Drugs. The control solution for electrophysiological studies was either DMEM (GIBCO) or a physiological solution of similar composition. The bath was perfused at a rate of 0.5 to 2 ml/min. Tetrodotoxin was added as necessary to reduce spontaneous myotube action potentials and contractions. Diisopropylfluorophosphate (DFP) was purchased from Sigma and working dilutions were made fresh daily.

RESULTS

Determination of myotube cholinesterase activity. The AChE activity of G8-1 cultures is developmentally regulated. This is demonstrated

in Fig. 1 where the enzyme activity is plotted as a function of
time in culture. The zero time point was obtained from myoblasts
maintained in logarithmic growth for three successive passages.
Such cells were found to have low AChE activity (0.03 nmole [^{14}C]ACh/
min/mg protein). The enzyme activity remained low over the next
four days. At the end of this time, the cultures were nearly confluent
with myoblasts but contained less that 2% multinucleated myotubes.
The AChE activity increased gradually after horse serum was added
to the growth medium to promote myotube formation, reaching a maximum
of 1.1 nmole [^{14}C]ACh/min/mg protein on day 15. This represents
an approximate 35-fold enhancement in activity. By use of the specific
inhibitors, BW-284C51 and ISO-OMPA, 83% of the total activity was
determined to be true AChE, and only 16% to be butyrylcholinesterase.
Thus the ratio of AChE to butyrylcholinesterase in G8-1 cells is
similar to that in adult muscle (20).

The increase in cholinesterase activity appears to be due, at
least in part, to an actual induction of enzyme activity as well
as to increases in the myotube/myoblast ratio. Evidence for the
former is based on results with standard histochemical staining
procedures for cholinesterase (8). The myotubes in a 6-day culture

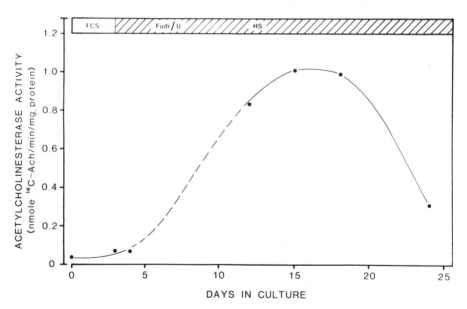

Figure 1. Development of acetylcholinesterase activity in G8-1
cultures. The cells were maintained in DMEM with fetal calf serum
(FCS) for 4 days, then switched to horse serum (HS) for the remaining
period. 5'-Fluorodeoxyuridine (10^{-5}M) and uridine (10^{-4}M) (Fudr/U)
were added for the indicated time to inhibit further myoblast proliferation.

were only faintly stained whereas 15 day cultures exhibited intense staining (Adler, Chang and Foster, unpublished observations). The developmental sequence is similar to that reported by Sugiyama (19) although the absolute maximum activities reported here are lower.

Since G8-1 cells have inducible AChE and can form synaptic contacts with cholinergic neuronal cell lines (5, 19), it is possible to use this system to study the actions of cholinesterase inhibitors. To achieve this aim we investigated the effects of DFP on synaptic transmission in co-cultures of G8-1 myotubes and NG108-15 neuroblastoma x glioma hybrid cells. NG108-15 cells synthesize, store and release ACh, and form functional synapses with myotubes in culture (15). NG108-15 ACh is stored in synaptic vesicles from where it is released in quantal fashion (14).

Effect of DFP on spontaneous transmitter release. Three to five days after seeding G8-1 myotube cultures with NG108-15 cells, spontaneous MSPs were observed in approximately thirty percent of the myotubes impaled. The MSPs resembled spontaneous MEPPs of the adult vertebrate neuromuscular junction. However, unlike the MEPP, the MSP generally exhibited a hyperpolarizing component following the decay of the depolarizing phase (Fig. 3). Similar afterhyperpolarizing potentials (AHPs) have been reported by previous investigators in NG108-15/myotube co-cultures (5).

The origin of the hyperpolarization is presently unknown. Since the hyperpolarization may complicate assessment of the action of DFP on the MSP time course, it was of interest to characterize this component in more detail. Fig. 2 shows a series of iontophoretically-elicited ACh potentials at membrane potentials from -40 to -80 mV. As is clear from Fig. 2, ACh potentials were also accompanied by AHPs. This excludes the possibility that the hyperpolarization is due to the release of putative neurotransmitters other than ACh from NG108-15 cells. The AHP magnitude was observed to decrease as the membrane potential was shifted from -40 to -80 mV and reversed at approximately -75 mV. The reversal potential of -75 mV is near the expected Nernst potential for K^+ and Cl^-, indicating that the AHP is carried by one or both of these ions. The AHP persisted when extracellular Cl^- was replaced by the impermeant isethionate anion, suggesting that Cl^- ions are apparently not involved in its generation. Additional experiments in which Ca^{2+} ions were varied suggested that the AHP is carried by a Ca^{2+}dependent K^+ current (data not shown). Since this current is difficult to block without altering the synaptic potentials, the AHPs were eliminated by holding the membrane near the K^+ equilibrium potential in experiments where alterations of the MSP decay were of interest (See Fig. 4).

Fig. 3 shows the effect of DFP on MSPs from a 14 day old myotube. Under control conditions, the MSP frequency varied from 2 to 51 min^{-1} (mean $\pm$ SEM = 17.2 $\pm$ 2.1 min^{-1}) and depended on temperature,

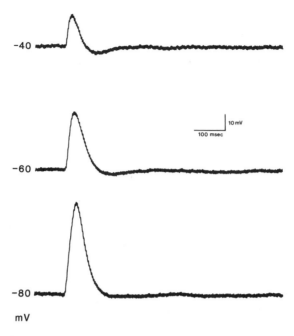

Figure 2. ACh potentials elicited by 1 msec iontophoretic pulses at the indicated membrane potentials to illustrate the reversal of the after hyperpolarization.

Ca^{2+} concentration, age of the co-cultures, the extent of differentiation of the NG108-15 cells and hybrid cell density. The addition of DFP led to a reduction in the MSP frequency and a depression in its amplitude. Within 5 min of perfusion with a 10^{-3}M DFP solution, the MSP amplitude and frequency was markedly reduced (Fig. 3B). The cell was bathed in DFP for a total of 30 min to ensure complete inactivation of AChE and then washed with drug-free solution. The MSP amplitude and frequency recovered to control values within 10 min of wash and remained stable thereafter (Fig. 3C). No potentiation in the MSP amplitude was evident even after washing for 50 min. The dose-response curves in Fig. 3D reveal that DFP inhibits synaptic transmission in this preparation at concentrations above 10^{-5}M.

The decrease in the MSP frequency may stem from an impairment in the release process, or may be secondary to the blockade of the MSP amplitude. Although presynaptic effects cannot be ruled out, the DFP-induced depression of MSP amplitudes may be expected to cause the smaller classes of MSPs to disappear within the baseline noise and give rise to a lower apparent frequency. Evidence consistent with this mechanism is demonstrated by the histograms of MSP amplitude distributions in Fig. 4. The MSP amplitudes show a skewed distribution under both control conditions and after washout

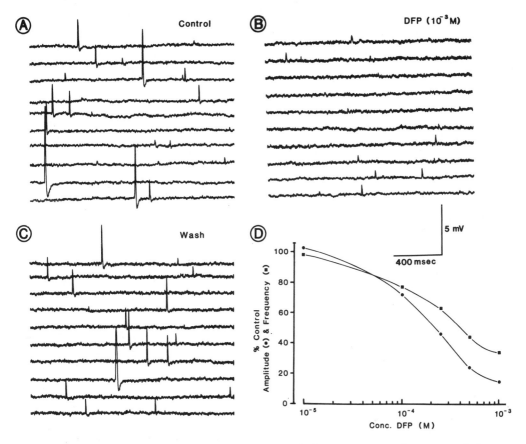

Figure 3.Effect of DFP on spontaneous miniature synaptic potentials in NG108-15/G8-1 co-cultures. The data were obtained from a 14 day old myotube under the indicated conditions. A. & B. The traces in A and B are continuous. C. The first 5 traces were taken 10 min after the onset of wash; the remaining 5 traces were taken 40 min later. MSP after-hyperpolarizations were observed in most cells and reversed at approximately -75 mV. D. Dose response curve showing the DFP-induced inhibition in the MSP amplitude and frequency.

of excess DFP. In the presence of DFP (10^{-4}M), MSP amplitudes were depressed on average by approximately 30% and up to 50% in some experiments. This is seen in the histogram as a shift to the left in the amplitude distribution. Since the baseline noise is relatively high (0.3 mV), due to the necessity of using high resistance micro-electrodes, the reduction in the MSP amplitudes causes the smaller MSPs to merge within the instrument noise level and escape detection.

Fig. 4 also illustrates the distribution of MSP rise and half-decay times from the same cell before, during and after perfusion with 10^{-4}M DFP. The MSP time course appears to be more normally distributed than the MSP amplitudes. DFP (10^{-4}M) produced no significant alterations in the MSP rise and half-decay times, even though this concentration was found to inhibit G8-1 AChE completely. Similarly, no significant alterations were detected in the MSP kinetics after extensive washout of excess inhibitor. The absence of potentiation in the MSP amplitude or prolongation in its time course, after treatment by DFP, suggests that hydrolysis is not rate limiting in terminating transmitter action at this synapse.

Effect of DFP on acetylcholine potentials. To determine the mechanism of depression of MSP amplitudes by DFP, experiments were performed on ACh potentials elicited by microiontophoretic application of ACh. Fig. 5 shows a dose-response curve for the effect of DFP on ACh potential amplitudes. As indicated, DFP depresses the ACh potential amplitude in a concentrationdependent fashion. The IC_{50} (2.2×10^{-4}M) estimated from this curve is close to that obtained for the depression of MSP amplitudes. These data suggest that DFP interferes with activation of the ACh receptor-channel complex in G8-1 myotubes. DFP was also found to reduce the slope and maxima of ACh potentials elicited by varying the intensity of the current delivered to the iontophoretic pipette (Fig. 6). This implies that the DFP blockade is noncompetitive.

In the presence of relatively high concentrations, DFP was found to enhance the rate of desensitization to ACh. This is illustrated in Fig. 7 where ACh potentials were elicited repetitively at 0.3 Hz. The first impulse in each train shows the steady-state level of inhibition produced by DFP. During repetitive application of ACh there was a rapid depression in the amplitudes of successive ACh potentials to a plateau level reached by the 10th pulse. Comparison of the control and DFP records reveals that the desensitization rate increases with DFP concentration. The recovery from desensitization was rapid and nearly complete after a 10 sec quiescent period as shown by the single traces following the trains (Fig. 7). Washout of DFP led to a complete restoration of ACh potential amplitudes within 10 min but the altered desensitization rates did not recover fully even after a 1 hr wash. The mechanisms responsible for depression of the ACh potential amplitude and the enhancement in the desensitization rate are thus different.

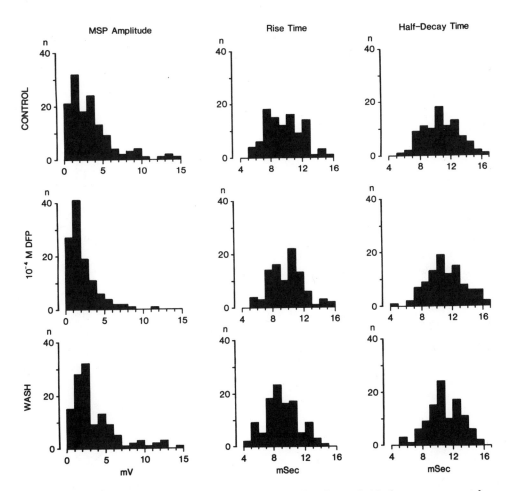

Figure 4. Histograms illustrating distribution of miniature synaptic potential (MSP) amplitudes, rise and half-decay times in NG10815/G8-1 co-cultures. The data were obtained from the same myotube under control conditions, 8 min after perfusion with 10^{-4}M DFP and 20 min after washout of excess DFP. Recordings were made on 19 day old myotubes co-cultured for 7 days with NG108-15 cells. All histograms contain the same number of events to facilitate comparisons.

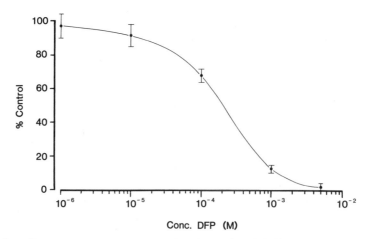

Figure 5. Dose-response curve showing the inhibition of ACh potential amplitudes by DFP. The symbols represent the mean $\pm$ SEM of responses obtained from G8-1 myotubes (n=6-8). ACh potentials were elicited by brief iontophoretic pulses (0.3-2 msec). The control ACh sensitivity was 1040 $\pm$ 82.1 mV/nC.

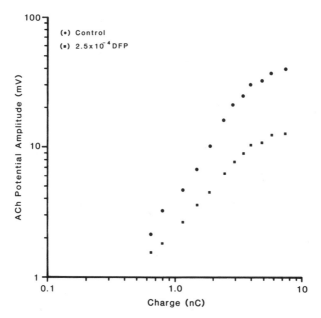

Figure 6. Dose-response curve showing the effect of DFP on the steady-state ACh potential of G8-1 cells. Acetylcholine potentials were elicited with 20 sec iontophoretic pulses from pipettes containing 2M ACh positioned approximately 10 μm above the cell.

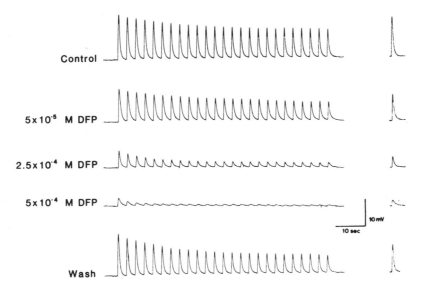

Figure 7. ACh potentials from a G8-1 myotube elicited by trains of 200 msec iontophoretic pulses of ACh delivered at 0.3 Hz. DFP produced a marked enhancement in the desensitization rate (middle three traces) that was only partially reversed by 30 min of wash (bottom trace). The single pulses following the trains were obtained after a 10 to 15 sec. quiescent period.

DISCUSSION

The results of the present investigation demonstrate that DFP reduces the amplitude and frequency of spontaneous synaptic potentials in NG108-15/G8-1 co-cultures. DFP also depresses the amplitude of the iontophoretically elicited ACh potential and enhances the desensitization rate of G8-1 myotubes to ACh. Although the DFP concentrations used in the study were relatively high (10^{-6}-10^{-3}M), the actions of the organophosphate were found to be confined to the synaptic region: even in the presence of 10^{-3}M DFP, little or no change was detected in the resting potential, input resistance or action potential generation of either G8-1 or NG108-15 cells (2).

Depression in the amplitudes of the MSP (Fig. 3) and ACh potential (Fig. 5) appear to result from a noncompetitive blockade of the ACh receptor-ion channel complex. This is based on the finding that increases in the ACh concentration enhanced rather than relieved the blockade (Fig. 6). It is not clear whether the DFP effect is exerted on the ACh-gated channel since we didnot observe voltage-dependent alterations in the synaptic current- voltage relationship, or in the kinetics of the synaptic responses. The former is indicative of a closed channel block while the latter is generally associated with open channel blockade (1). The failure of excess DFP to modify the MSP decay differs from its action on frog sartorius muscle where the organophosphate produced double exponential endplate current decays (11). It is of interest, however, that in both systems high DFP concentrations cause a marked depression in the macroscopic synaptic conductance.

The enhancement of desensitization observed in the presence of DFP (Fig. 7) is similar to that reported recently for pyridostigmine at the vertebrate neuromuscular junction (16). The ability of DFP to alter the rate of desensitization in G8-1 cells is most likely a direct effect not related to AChE blockade. Evidence that desensitization is not correlated with AChE activity comes from findings that (1) desensitization rates increased when the DFP concentration was raised from 2.5×10^{-4}M to 5×10^{-4}M, although AChE was completely inhibited at the lower concentration (Fig. 7); and (2) that the ability of DFP to enhance desensitization was similar in 6 and 15 day old myotubes, although the AChE activities differed widely at these two ages (Fig. 1).

Blockade of AChE by DFP did not result in increases in the MSP amplitude or lengthening of its decay, both of which are typical manifestations of AChE inhibition at the vertebrate neuromuscular junction. Potentiation of the MSP was absent even when recordings were made during the peak of G8-1 AChE activity (Fig. 1). The absence of synaptic potentiation implies that the ACh lifetime is limited by diffusion of transmitter out of the synaptic cleft rather than by hydrolytic cleavage. There are at least two possible reasons for this. The first is that the AChE activity is too low, even during maximal induction, to become rate limiting. This is suggested by comparison of G8-1 AChE activities with those of mammalian muscle (20). In addition to being low, our preliminary histochemical data indicate that G8-1 AChE is not localized to regions in contact with NG108-15 neurites. The second possibility is that the NG108-15/G8-1 synapse has fewer diffusion barriers due to its relatively simple morphology and absence of junctional folds (14). Katz and Miledi (9) have proposed that AChE blockade allows transmitter molecules to interact repetitively with receptors, thus delaying their diffusion out of the cleft region. The extent of this delay depends on the ACh receptor density and the length of the diffusion pathway. For the reasons outlined above, the diffusion pathway is expected to be shorter for NG108-15/G8-1 synapses, and this might allow the transmitter diffusion rates to approach the the theoretical maximum (6).

REFERENCES

1. Adler, M., Albuquerque, E.X. and Lebeda, F.J. (1978): Mol. Pharmacol. 14:514-529.

2. Adler, M., Pascuzzo, G.J., Maxwell, D., Glenn, J.F. and Foster, R.E. (1983): Soc. for Neurosci. (Abstr) 9:1138.
3. Atlas, D. and Adler, M. (1981): Proc. Natl. Acad. Sci. 78: 1237-1241.
4. Bradford, N.K. (1976): Anal. Biochem. 72:248-254.
5. Christian, C.N., Nelson, P.G., Peacock, J. and Nirenberg, M. (1977): Science 196:995-998.
6. Eccles, J.C. and Jaeger, J.C. (1958): Proc. Roy. Soc. Lond. B. 148:38-56.
7. Fatt, P. and Katz, B. (1952): J. Physiol. (Lond) 117:109-128.
8. Karnovsky, M. and Roots, L. (1964): J. Histochem. Cytochem. 12:219-221.
9. Katz, B. and Miledi, R. (1973): J. Physiol. (Lond) 231:549-574.
10. Koelle, G.B. (1975): In: The Pharmacological Basis of Therapeutics (eds) L.S. Goodman and A. Gilman, Macmillan Publishing Co., Inc., New York, pp. 445-466.
11. Kuba, K., Albuquerque, E.X., Daly, J. and Barnard, E.A. (1974): J. Pharmacol. Exp. Ther. 189:499-512.
12. Hohlfeld, R., Sterz, R. and Peper, K. (1981): Pflugers Arch. 391:213-218.
13. Laskowski, M.B. and Dettbarn, W.D. (1979): J. Pharmacol. Exp. Ther. 210:269-274.
14. Nelson, P.G., Christian, C.N., Daniels, M.P. Henkart, M., Bullock, P. Mullinax, D. and Nirenberg, M. (1978): Brain Res. 147:245-259.
15. Nelson, P., Christian, C. and Nirenberg, M. (1976): Proc. Natl. Acad. Sci. 73:123-127.
16. Pascuzzo, G.J., Akaike, A., Maleque, M.A., Shaw, K-P., Aronstam, R.S., Rickett, D.L. and Albuquerque, E.X. (1983): Mol. Pharmacol. 25:92-101.
17. Riker, W.F. Jr. and Standaert, F.G. (1966): Ann. N.Y. Acad. Sci. 135:163-176.
18. Siakotos, A.N., Filbert, M. and Hester, R. (1969): Biochem. Med. 3:1-12.
19. Sugiyama, H. (1977): FEBS Letters 84:257-260.
20. Vigny, M., Koenig, J. and Rieger, F. (1976): J. Neurochem. 27:1347-1353.

ACETYLCHOLINE RELEASE FROM RAT DIAPHRAGM:

A SEARCH FOR RELEASE REGULATORY MECHANISMS

J. Häggblad and E. Heilbronn

Unit of Neurochemistry and Neurotoxicology
University of Stockholm
Enköpingsvägen 126, S-172 Sundbyberg, SWEDEN

INTRODUCTION

During the last five years there have been a number of reports concerned with possible muscarinic mechanisms at the motor nerve terminals of the phrenic nerve. Both positive and negative muscarinic feedback autoregulation of acetylcholine (ACh) release has been suggested, as well as an absence of any regulatory muscarinic receptor. The designs of the previously performed experiments differ from each other and from ours mainly in the technique used for ACh analysis. There are also differences in the use of anticholinesterases. It is possible that these differences in approach may have led to the divergent results. We decided to reinvestigate the possible involvement of a presynaptic muscarinic autoreceptor in ACh release from motoneurons. For this purpose a chemiluminescence technique for the quantitation of ACh released from mammalian neuromuscular preparations was developed. Nerve stimulation was accomplished either by 50 mM KCl or by electrical stimulation of the phrenic nerve.

In this report the ACh quantitation method and the results of its application to the analysis of drug mediated responses on ACh release from rat hemidiaphragm preparation are described. The results obtained suggest the absence of a muscarinic feedback of ACh release at the neuromuscular junction of the rat.

METHODS

Diaphragm Preparation. Rat left hemidiaphragms were prepared as previously described (5). The tissue was kept in 5 ml of oxygenated medium (114 mM NaCl, 3.5 mM KCl, 2,5 mM CaCl$_2$, 1 mM MgCl$_2$, 11 mM glucose, 6 mM Hepes, pH 7.4), pinned to a silicone rubber layer at the bottom of a thermostated well in a perspex apparatus. Experiments were performed at 25°C. Acetylcholinesterases were inhibited by preincubation, before the release experiment, with 10 μM sarin during 45 minutes. Sarin, 1 μM, was also present during the release experiments. Acetylcholine release was induced either by electrical stimulation (5 Hz, 0.2 ms) of the phrenic nerve with suction electrodes connected to a Grass stimulator or by an increase in potassium concentration to 50 mM. (Osmolarity was kept constant by an equimolar decrease in sodium.) When high potassium was used as a mean to evoke release tetrodotoxin (TTX), 1 μM, was used to prevent backfiring caused by Sarin. Fractions for analysis of ACh content were collected during 30 minute periods.

Purification and Quantitation of ACh. As some of the intermediates in the luminescence reactions are free radicals, radical scavengers in the sample have to be removed. Furthermore, the sample has to be concentrated in order to fit the analysis protocol. For this purpose the potassium periodide precipitation method was used (13). Samples were placed in ice-water. Internal standard [(methyl-[^{3}H])-ACh], coprecipitant (tetramethylammonium bromide) and the KI_3-reagent was added and the samples were kept at 0°C for 10 minutes. Subsequently, they were centrifuged at 1000g x 10 minutes at 4°C. The resultant pelleted precipitate was dissolved in diethyl ether. To the ether phase 0.1 mM HCl was added. The phases were mixed and the diethyl ether phase was discarded. The aqueous phase was further washed three times with diethyl ether and then boiled to dispose of the diethyl ether and remaining traces of iodine. After purification the samples were either measured directly or stored for some days at -20°C before quantitation.

Acetylcholine was measured with a chemiluminescence technique, modified (5) from that described by Israël and Lesbats (8). The samples were added to a reaction mixture containing choline oxidase, luminol and horseradish peroxidase. Initially, a large light emission was recorded, mainly due to choline released from the preparation. The signal subsequently declined to a steady baseline. Electrophorus acetylcholinesterase was added and a peak due to hydrolysis of ACh and subsequent oxidation of formed choline was recorded. The peak was quantitated by addition of appropriate amounts of ACh (Fig. 1). Peak heights were routinely used as a measurement of ACh since it was found that peak height and peak area were closely correlated (corr = 0.988, 5).

RESULTS

The hemidiaphragm preparation released spontaneously 0.5 pmol ACh/min x hemidiaphragm. Addition of 50 mM K$^+$ led to a 3.5 fold increase in release. Electrical stimulation of the phrenic nerve at 5 Hz, 0.2 ms, augmented ACh release by a factor of approximately

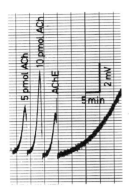

Figure 1. Luminometer recording from a quantitation of ACh released from a hemidiaphragm preparation. Addition of acetylcholinesterase (AChE) leads to the hydrolysis of ACh to choline, which subsequently is oxidized. The light emission produced is quantitated by addition of known amounts of ACh.

5 (Table 1). Evoked release was taken as the difference between basal and stimulated release. A test of agents known to affect ACh release was made. Veratrine augmented both basal and stimulated release while 4-aminopyridine increased only stimulated release (Table 1). These results are consistent with the general view on synaptic transmission (6, 12).

In subsequent experiments effects of muscarinic drugs were investigated. Addition of the muscarinic agonist oxotremorine to the release media altered neither basal nor stimulated release. The muscarinic antagonists atropine and 3-quinuclidinylbenzilate (3QNB) were also ineffective in altering basal and stimulated release (Table 2). Some experiments were performed (data not shown) with permeable cyclic nucleotide derivatives in order to mimic autoreceptor activation (11). 8-Br-cGMP (2×10^{-4} M) did not alter ACh release while 8-Br-cAMP (2×10^{-4} M) somewhat enhanced basal release ($p < 0.05$), although evoked release remained unaltered. In order to increase the amounts of possible muscarinic autoreceptors rats were treated long-term with atropine (20 mg/kg, i.p., 20 days). This treatment is known to cause muscarinic supersensitivity in brain (15) and would thus probably lead to an amplification of a possible muscarinic autoreceptor action also at the neuromuscular junction. No changes in ACh release were, however, recorded there due to the atropine treatment (data not shown). In the same experiments a binding study of [3H]-3QNB to cerebral cortex homogenates from the atropine treated rats revealed the expected 30-40% increase in muscarinic receptor amounts while attempts to demonstrate any [3H]-3QNB bound to diaphragm homogenates of the atropine treated, supersensitized rats, revealed no specific binding of the muscarinic ligand.

Release experiments were also performed with the nicotinic antagonist α-bungarotoxin (α-BGT). Preincubation of hemidiaphragms with the toxin resulted in an approximately 40% increase in the potassium evoked release (Table 2).

Table 1

Effect of High Potassium, Electrical Stimulation, 4-AP and
Veratrin on Acetylcholine Release from Rat Left Hemidiaphragm

Conditions	Basal Release	Evoked Release[1]
no drug (50 mM K^+ evoked release)	$0.5^{\pm}0.13$ (10)	$1.4^{\pm}0.31$ (8)
no drug (electrical stimulation, 5 Hz, 0.2 ms)	$0.5^{\pm}0.13$ (6)	$2.2^{\pm}0.16$ (6)
10^{-4} M 4-AP (electrical stimulation, 5 Hz, 0.2 ms)	$0.6^{\pm}0.08$ (3)	$3.2^{\pm}0.25$ (3)
280 μg veratrine (electrical stimulation, 5 Hz, 0.2 ms)	$2.1^{\pm}0.22$ (3)	$1.0^{\pm}0.24$ (3)

Values are expressed as pmoles ACh released/min x hemidiaphragm,
mean $\pm$ S.D. Figures within parenthesis show number of observations.
(1) Evoked release was taken as the difference between basal and
stimulated release. 4-AP, 4-aminopyridine.

Table 2

Effects of Cholinergic Drugs on ACh Release From
Rat Left Hemidiaphragm

Conditions	Basal Release	Evoked Release [3]
no drug[1]	$0.5^{\pm}0.09$ (9)	$1.3^{\pm}0.28$ (7)
10^{-6} M oxotremorine[1]	$0.6^{\pm}0.09$ (4)	$1.5^{\pm}0.24$ (4)
10^{-5} M oxotremorine[1]	$0.4^{\pm}0.05$ (5)	$1.3^{\pm}0.24$ (5)
10^{-8} M 3QNB[1]	$0.5^{\pm}0.06$ (7)	$1.3^{\pm}0.14$ (5)
10^{-7} M 3QNB[1]	$0.5^{\pm}0.06$ (4)	$1.4^{\pm}0.07$ (4)
10^{-6} M 3QNB[1]	$0.5^{\pm}0.03$ (4)	$1.4^{\pm}0.09$ (4)
5 μg α-BGT[1] (preincubated)	$0.5^{\pm}0.04$ (4)	$1.8^{\pm}0.12$ (4)**
no drug[2]	$0.5^{\pm}0.13$ (6)	$2.2^{\pm}0.16$ (6)
10^{-5} M atropine[2]	$0.4^{\pm}0.11$ (4)	$2.2^{\pm}0.22$ (4)

Values are expressed as pmoles ACh released/min x hemidiaphragm,
mean $\pm$ S.D. Figures within parenthesis show number of observations.
(1) 50 mM K^+ evoked release, 1 μM TTX added. (2) Electrical stimulation. (3) Evoked release was taken as the difference between basal
and stimulated release. **; $p < 0.01$ as compared to no drug. 3QNB;
3-quinuclidinylbenzilate, α-bgt; α-bungarotoxin.

DISCUSSION

The data presented suggest the absence of a muscarinic auto-receptor activity at the neuromuscular junction of the rat.   No effects on ACh release were recorded by the use of muscarinic drugs. Previous reports that concern muscarinic autoregulation of ACh release (chemically measured) from mammalian skeletal motoneurons present results both in favor of, and against involvement of an autoreceptor mechanism.   These results fall mainly into three categories: first, those who claim a positive feedback mechanism e.g. Das et al. (2); secondly, those who claim a negative feedback (1), like the mechanism described for brain and smooth muscle; and lastly, those who suggest the absence of a muscarinic autoreceptor mechanism (this report, 3, 4).   Seen in the light of the fast all-or-none character of neuro-muscular transmission a relatively slow process such as muscarinic feedback control of transmitter release may not be of much use at this site of ACh release.

There have accumulated results that suggest other types of cholinergic release regulatory mechanisms at the neuromuscular junc-tion.   Miledi et al. (10) reported that preincubation of hemidiaphragms with α-BGT increased the release of ACh from the preparation.   This observation is verified by our results.   It was also mentioned in the same paper that high concentrations of d-tubocurarine had a similar effect.   The observations could be interpreted in a number of ways, one of them being the presence, at the neuromuscular endplate, of a nicotinic analog to the concept of negative muscarinic feedback control.   A positive feedback of ACh release from motoneurons can also be observed under certain conditions.   ACh can enhance its own release in anticholinesterase-treated preparations by inducing repetitive firing.   The mechanism is probably of nicotinic nature since d-tubocurarine inhibits the effect (9).   Thus, if there exists a presynaptic autoreceptor mechanism, it seems most likely to be of the nicotinic type.   Its precise nature remains to be clarified.

It may also be appropriate to discuss whether a negative trans-synaptic feedback may exist, and if this mechanism in some way is mediated through activation of the postsynaptic nicotinic ACh recep-tor.   Such a mechanism would explain why a block of the synaptic transmission with nicotinic antagonists (e.g., α-BGT) enhanced evoked ACh release, as the assumed transsynaptic mechanism also would be blocked in that situation.   Possible candidates for this transsynaptic mediation are ATP or other adenosine nucleotides.   ATP has been reported to decrease both the quantal content of endplate potentials, and the frequency of miniature endplate potentials in mammalian neuromuscular preparations (14).   Furthermore, in the Torpedo electro-plax system ATP, which has been suggested to be released from the postsynaptic side, was found to decrease evoked release of ACh, probably by interfering with presynaptic, release coupled calcium channels (7).   In these experiments the agonist induced release of

ATP was blocked by nicotinic antagonists. It is our intention to
further investigate the matter.

Acknowledgement. This work was supported by grants from the Swedish
medical research council to E.H.

REFERENCES

1. Abbs, E.T. and Joseph, P.N. (1981): Br. J. Pharmac. 73:481-483.
2. Das, M., Ganguly, D.K. and Vedasiromoni, J.R. (1978): Br. J.
    Pharmac. 62:195-198.
3. Gundersen, C.B. and Jenden, D.J. (1980): Br. J. Pharmac. 70:
    8-10.
4. Häggblad, J. and Heilbronn, E. (1983): Br. J. Pharmac. 80:
    471-476.
5. Häggblad, J., Eriksson, H. and Heilbronn, E. (1983): J.
    Neurochem. 40:1581-1584.
6. Illes, P. and Thesleff, S. (1978): Br. J. Pharmac. 64:623-627.
7. Israël, M., Lesbats, B., Manaranche, R., Meunier, F.M. and
    Frachon, P. (1980): J. Neurochem. 34:923-932.
8. Israël, M. and Lesbats, B. (1981): Neurochem. Int. 3:81-90.
9. Masland, R.L. and Wigton, R.S. (1940): J. Neurophysiol. 3:
    269-375.
10. Miledi, R., Molenaar, P.C. and Polak, R.L. (1978): Nature
    272:641-643.
11. Nordström, O. and Bartfai, T. (1981): Brain Res. 213:467-471.
12. Narahashi, T. (1974): Physiol. Rev. 54:813-889.
13. Polak, R.L. and Molenaar, P.C. (1974): J. Neurochem. 23:
    1295-1297.
14. Ribeiro, J.A. and Walker, J. (1975): Br. J. Pharmac. 54:
    213-218.
15. Westlind, A., Grynfarb, M., Hedlund, B., Bartfai, T. and
    Fuxe, K. (1981):225:131-141.

DISTRIBUTION OF PRESYNAPTIC RECEPTORS

MODULATING ACETYLCHOLINE RELEASE

E.S. Vizi and G.T. Somogyi

Institute of Experimental Medicine
Hungarian Academy of Sciences
P.O.B. 67
H-1450 Budapest, Hungary

INTRODUCTION

In addition to neurotransmitter substances acting at close range in chemical synaptic neurotransmission, there exist chemical interactions between neurones without any close synaptic contact: interneuronal modulation of transmission which operates over some distance (17-19). This would be an intermediary form between classical neurotransmission and the broadcasting of neuroendocrine secretion. Transmission of information can, however, also be modulated at synaptic junctions pre- and postsynaptically, as well as at axon terminals and target cells (soma, dendrites, axon hillock) which are not in synaptic contact with the release site where the modulator/transmitter comes from.

Since the release from varicosities devoid of typical synaptic specialization has long been known (e.g., peripheral nervous system) the question arises as to whether noradrenaline and/or ATP/adenosine released from non-synaptic varicosities can actually reach their target cells and suppress neuronal firing inhibiting the release of genuine transmitters. All available neurochemical evidence seems to favor this hypothesis (19).

Firstly, the surgical or neurochemical removal of the inhibitory pathway results in an enhanced release of the transmitter. Secondly, administration of the modulator of the stimulation of the inhibitory axon leads to a reduction of the genuine transmitter release and the neuronal firing. In the Auerbach plexus the stimulation of the sympathetic fibres and the subsequent release of noradrenaline from varicosities for example leads to an inhibition of acetylcholine release (20, 17). Since the synaptic contact between noradrenergic and cholinergic neurons is very sparse it is likely that noradrenaline does not exert its effect solely on a one-to-one basis, on restricted areas of postsynaptic specializations, but can diffuse to reach distant targets.

The long-distance message can be monitored only by those cells that are equipped with receptors sensitive to modulators. The present study was therefore carried out to examine the distribution of pre-synaptic cholinergic receptors: is each varicosity branch equipped with these receptors, or does each modulator have a population of neurons equipped with only one receptor?

METHOD AND RESULTS

The release of $[^3H]$acetylcholine (24) was measured from the longitudinal muscle strip with attached Auerbach plexus, as described by Vizi et al. (22). Table 1 shows typical data obtained following loading with methyl-$[^3H]$choline (1.44×10^5 Bq/ml).

Evidence has been obtained by us, as well as others, that the cholinergic varicosities are equipped with presynaptic inhibitory receptors (17, Table 2). The release of acetylcholine (ACh) thus

Table 1

Longitudinal Muscle Strip (n=6) Attached with Auerbach Plexus Prepared According to Paton and Vizi (1969) (9)

	no eserine	with eserine	p^x
Tritium content (Gq/g)			
before:	142981 ± 11996	123910 ± 14996	
after:	105305 ± 9010	96400 ± 8410	
Fractional release at rest ($\times 10^{-3}$)	14.95 ± 1.76	8.35 ± 1.53	<0.02
at simulation (1 Hz, 300 shocks)	48.69 ± 5.50	22.15 ± 3.80	<0.01
release (S_1) at 1 Hz	9517 ± 815	5253 ± 610	<0.01

[x]Significance between the values obtained in the presence and absence of eserine. Note that in the presence of eserine the release is under muscarinic feedback inhibition.

Table 2

Substances with Inhibitory Effect on Acetylcholine Release

Substances	Receptor	References
1-noradrenaline	α_2	Vizi (1968)(13) Paton & Vizi (1969)(9) Beani et al (1969)(1) Vizi & Knoll (1971)(20) Vizi (1972)(14) Vizi (1974)(15) Drew (1977)(3) Beani et al (1978)(2)
acetylcholine	m (muscarinic)	Szerb & Somogyi (1973)(12) Vizi (1974) (15) Kilbinger & Wagner (1975)(7) Sawynok & Jhamandas (1977)(11) Forsbraey & Johnson (1980)(4)
adenosine	P_1	Vizi (1975)(16) Vizi & Knoll (1976)(21) Hayashi et al (1978)(6) Gustafsson et al (1978)(5) Reese and Cooper (1982)(10)

can be inhibited by stimulation of presynaptic α-(1,8,9,13), which later turned out to be α_2-(3,16,17,21), opiate (μ)- and muscarinic receptors (12).

In our experiments when the cholinesterase was not inhibited, both theophylline (170 μM) and atropine (1 μM) enhanced the release of ACh, indicating an endogenous tonic control of ACh release by adenosine and/or by ACh (Tables 3 and 4). The finding that atropine enhanced the release of ACh raises the question of whether endogenous ACh might control, via muscarinic receptors, the release of ACh, and if the released ACh measured is in fact already modulated. Therefore, a possible explanation for the lack of effect of adenosine in the presence of eserine on the release of ACh might be that the output is controlled by ACh. In addition, the observation, in the

Table 3

Effect of Adenosine and α-methyl-noradrenaline on
[^{3}H]-acetylcholine Release. Size of Population of
P_1 and α_2-adrenoceptors

	S_2/S_1 (1 Hz, 300 shocks)	% inhibition
Control	0.780 ± 0.032 (12)	
Atropine, 3×10^{-7}M	3.086 ± 0.245 (6)	
+adenosine, 10^{-7}M	1.806 ± 0.186 (4)	41.5
+adenosine, 2×10^{-5}M	0.905 ± 0.035 (4)	70.7
+adenosine, 2×10^{-5}M; α-m-NA, 10^{-6}M	0.780 ± 0.065 (4)	74.7
+α-m-NA, 10^{-6}M	0.960 ± 0.050 (4)	68.9

α-m-NA, α-methyl-noradrenaline. For method, see Table 4.

presence of eserine, that theophylline failed to enhance the release
of ACh (Table 4), raises the further question of the size of the
neuron population equipped with adenosine and/or muscarinic receptors.

The question, then, is whether each cholinergic branch is equipped
with these receptors, or whether each neuron possesses only one
type of inhibitory receptor. If the former is the case, the stimulation of one type of receptor will produce an inhibition of ACh
release without leaving space for another presynaptic modulator to
further reduce the release. On the other hand, if each neuron has
only one recognition site, then the total release should be reduced
to the extent of the size of population.

The inhibitory effect of adenosine at a concentration of 20
μM was considered as (near)-maximal, therefore this concentration
was chosen in our experiments. In the absence of muscarinic modulation (i.e., in the presence of atropine) (Table 4) adenosine (20
μM) produced an inhibition of 70.7 percent. The residual release
was not affected by α-methyl-noradrenaline (10^{-6}M) (Table 3). These
findings indicate that there is no population of cholinergic neurons
which has only one type of receptor. Similar conclusions have been
drawn from our data obtained with striatal slices (22).

DISCUSSION

From these findings it seems likely that interneuronal modulation of both peripheral and central cholinergic neurons is likely

Table 4

Effect of Different Drugs on [^{3}H]-acetylcholine Release
from Longitudinal Muscle Strip Attached with
Auerbach Plexus (Eserinized preparation 2×10^{-6}M)

		Eserine sulfate, 2×10^{-6}M
Control	0.685 ± 0.043 (6)	0.761 ± 0.048 (6)
Adenosine 2×10^{-5}M	0.343 ± 0.078^x (5)	0.680 ± 0.063 (4)
Theophylline 1.7×10^{-4}M	1.060 ± 0.050^x (6)	0.740 ± 0.075 (4)
Atropine 3×10^{-7}M	0.951 ± 0.062^x (4)	3.086 ± 0.245^x (6)

Mean $\pm$ S.E.M. S_2/S_1. Number of experiments in brackets.
^{x} Significant at the level of 0.05. S_2/S_1 = ratio between the overflow
of tritium evoked by S_2 and S_1. Krebs solution. The preparation
was incubated (45 min) in Krebs solution containing 1.44×10^5 Bg/ml
methyl-[^{3}H]-choline chloride. The bath was bubbled with O_2/CO_2 (95/5).
The tissue was perfused at a rate of 1 ml/min with a Krebs solution
containing hemicholinium (10^{-5}M).

to be mediated via different receptors, and that these different
recognition sites are located in each neuron (Fig. 1). Their proportional involvement in the modulation of neuronal function is time
and space dependent; the type of modulatory influence depends on
the regional localization of the non-synaptic and/or synaptic inhibitory input.

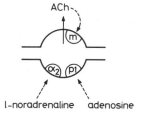

Figure 1. Schematic representation of a
cholinergic varicosity equipped with muscarinic
(m), α_2- and P_1-receptors.

The inhibition of ACh release by adenosine and acetylcholine appears, according to our studies, to be via independent mechanisms, and mediated via different receptors. Similarly, evidence has been obtained (22, 23) that opioid peptides and adenosine do not share a common pathway in inhibiting ACh release.

REFERENCES

1. Beani, L., Bianchi, C. and Crema, A. (1969): Br. J. Pharmacol. 36:1-17.
2. Beani, L., Bianchi, C., Giacomelli, A. and Tamberi, F. (1978): Eur. J. Pharmac. 48:179-193.
3. Drew, G.M. (1977): Eur. J. Pharmac. 72:123-130.
4. Fosbraey, P. and Johnson, E.S. (1980): Br. J. Pharmac. 69: 145-149.
5. Gustafsson, K., Hedqvist, P., Fredholm, B.B. and Lundgren, G. (1978): Acta Phys. Scand. 104:469-478.
6. Hayashi E., Kunitomo, M., Mori, M., Shinozuka, K. and Yamada, S. (1978): Eur. J. Pharmacol. 48:297-307.
7. Kilbinger, H. and Wagner, P. (1975): Naunyn-Schmiedeberg's Arch. Pharmac. 287:47-60.
8. Kosterlitz, H.W., Lyden, R.J. and Watt, A.J. (1970): Br. J. Pharmac. 287:47-60.
9. Paton, W.D.M. and Vizi, E.S. (1969): Br. J. Pharmacol. 35: 10-29.
10. Reese, J.H. and Cooper, J.R. (1982): J. Pharmacol. 223:612-616.
11. Sawynok, J. and Jhamandas, K.H. (1977): Can. J. Physiol. Pharmac. 55:909-916.
12. Szerb, J.C. and Somogyi, G.T. (1973): Nature, 241:121-122.
13. Vizi, E.S. (1968): Naunyn-Achmiedeberg's Arch. Exp. Path. Pharmac. 259:199-200.
14. Vizi, E.S. (1972): J. Physiol. 226:99-117.
15. Vizi, E.S. (1974): J. Neural. Transm. 11:61-78.
16. Vizi, E.S. (1975): In: Subcortical Mechanisms and Sensorimotor Activities, (ed) T.L. Frigyesi, Hans Huber, Bern., pp. 63-87.
17. Vizi, E.S. (1979): Progr. in Neurobiol. 12:181-290.
18. Vizi, E.S. (1980): Trends Pharm. Sci. 2:172-175.
19. Vizi, E.S. (1983): In: Dale's Principle and Communication Between Neurones, (ed) N.N. Osborne, Pergamon Press, Oxford and New York, pp. 83-110.
20. Vizi, E.S. and Knoll, J. (1971): J. Pharm. Pharmacol. 23:918-925.
21. Vizi, E.S. and Knoll, J. (1976): Neuroscience 1:391-398.
22. Vizi, E.S., Somogyi, G.T. and Magyar, K. (1981): J. Auton. Pharmac. 1:413-419.
23. Vizi, E.S., Somogyi, G.T. and Magyar, K. (1983): In: Physiology and Pharmacology of Adenosine Derivatives, (eds) J. W. Daly, Y. Kuroda, J.W. Phillis, H. Shimizu and M. Ui, Raven Press, New York, pp. 209-217.
24. Wikberg, J.E.S. (1978): Nature 173:164-166.

MUSCARINIC RECEPTOR ACTIVATION INCREASES
EFFLUX OF CHOLINE FROM ISOLATED HEART AND RAT CORTEX
IN VIVO. INTERACTIONS WITH FORSKOLIN AND IBMX

K. Löffelholz, R. Brehm and R. Lindmar

Department of Pharmacology
University of Mainz
Obere Zahlbacher Strasse 67
D-6500 Mainz, Federal Republic of Germany

INTRODUCTION

Muscarinic receptor activation modulates functions of the heart
and neurotransmission in the peripheral and central nervous system.
Moreover, muscarinic agonists produce changes in the metabolism
of, for example, heart tissue, such as inhibition of beta-adreno-
ceptor-mediated cAMP accumulation, glycogenolysis and lipase activation.

The recent observation that cholinesterase inhibitors and muscarinic agonists increased cardiac and cortical efflux of choline
(Ch) (2, 3) indicated that muscarinic receptor activation increases
hydrolysis of Ch-containing phospholipids, possibly of phosphatidylcholine or sphingomyelin. At first, this phenomenon has been envisaged
as a possible autoregulatory mechanism controlling the availability
of the precursor Ch for acetylcholine biosynthesis (3). It may
also be that effects on the metabolism of Ch-containing phospholipids
as the most abundant constituents of the phospholipid bilayer are
linked to intracellular mechanisms by which certain membrane properties
associated with ion channels, membrane-bound enzymes or receptor-
proteins are regulated.

In the present study, therefore, the role of cAMP in the effects
of muscarinic receptor activation on Ch efflux was investigated.

MATERIALS AND METHODS

Chickens were stunned by a blow to the head and bled from the
neck vessels. The hearts or heart atria were quickly removed and
perfused or incubated at 37°C. For perfusion and incubation Tyrode's

solution (mmol/l) was used; containing: Na^+ 149.3; K^+ 2.7; Ca^{2+} 1.8; Mg^{2+} 1.05; Cl^- 145.5; HCO_3^- 11.9; $H_2PO_4^-$ 0.4; and (+)-glucose 5.6. The perfusion rate was adjusted to 20 ml/min by a peristaltic pump and the solutions were gassed with a mixture of 95% O_2 + 5% CO_2. Atrial force of contraction was recorded isometrically.

Adult male Wistar rats weighing 200 to 260 g were anesthetized with urethane. The efflux of cortical Ch was investigated by the "cortical cup technique" (for details see Corradetti et al., this book).

Ch was determined by two different methods: 1) the chemolumi-nescent method described by Israël and Lesbats (8); and 2) the radio-enzymatic method described by Goldberg and McCaman (6).

RESULTS

A. **Resting Ch efflux.** Table 1 shows that the Ch efflux from the heart had a rate of around 1000 pmol g^{-1} min^{-1}, which was constant for the duration of the experiments (80 min).

The phospholipase A2 inhibitor mepacrine reduced the resting Ch efflux from the perfused chicken heart in a concentration-dependent way (Fig. 1). Half-maximal inhibition (EC_{50}) was observed by approxi-mately 8 x 10^{-6} mol/1 mepacrine. Maximum inhibition (3 x 10^{-5} mol/1) was restricted to about 50% of the basal Ch efflux, i.e., the remaining efflux was mepacrine-insensitive under the present conditions.

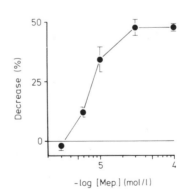

Figure 1. **Decrease of cardiac Ch efflux by mepacrine (Mep.), a phos-pholipase A_2 inhibitor.** Ordinate, decrease in rate of Ch efflux in % of efflux before Mep. addition (0% = 1075 $\pm$ 62 pmol g^{-1} min^{-1}). Abscissa, concentration of Mep. (mol/1). Note, that the maximum effect was limited to a 50% reduction. Presented are means $\pm$ SE of 3-4 experiments; each concentration was tested on a different preparation.

Table 1

Effects of low temperature, mepacrine and atropine on resting
Ch efflux and its increase caused by acetylcholine (ACh)[a]

| | N | Efflux of Choline (pmol/g min) | | |
		Before ACh	During ACh	ACh-evoked increase
Control (37°C)	4	1060 ± 90	2088 ± 276	948 ± 234
17°C	6	202 ± 24[b]	372 ± 48	171 ± 37
Control	3	1094 ± 52	1699 ± 153	605 ± 104
Mepacrine 10⁻⁵mol/1	4	694 ± 128[b]	1245 ± 174	552 ± 70
Control	4	1060 ± 90	2088 ± 276	948 ± 234
Atropine 3 x 10⁻⁶mol/1	5	1099 ± 90[b]	1255 ± 147	241 ± 93

[a]ACh, 3 x 10^{-7}mol/1 (temperature and atropine) or 10^{-6}mol/1 (mepacrine), was infused for 40 min. [b]Ch efflux rates (pmol/g min) before reducing the temperature and infusion of mepacrine or atropine were 1030 ± 137, 898 ± 155 or 1014 ± 140, respectively.

B. Muscarinic stimulation of Ch efflux. Infusion of acetylcholine (3 x 10^{-7} or 10^{-6} mol/1) evoked a gradual increase in Ch efflux during the 40 min period of infusion, and finally about doubled the Ch efflux rate (Table 1; 3). In these experiments, Ch originating from acetylcholine cleavage was subtracted from the total Ch efflux. Throughout the 40 min infusion of acetylcholine at 3 x 10^{-7} or 10^{-6} mol/1, hydrolysis of acetylcholine calculated from the arteriovenous difference, 37 ± 6% (N = 4) or 21 ± 3% (N = 6), respectively, was constant. Similar results on hydrolysis were obtained previously (5). In the presence of 3 x 10^{-6} mol/1 atropine, acetylcholine (3 x 10^{-7} mol/1) failed to enhance cardiac Ch efflux rate (Table 1), whereas hydrolysis of acetylcholine was unchanged (43 ± 7%, N = 5).

Mobilization of cellular Ch was studied in the rat cortex in vivo using the "cup-technique" (2). At 37°C, Ch efflux rate was constant for 3 h. In contrast, when using Ringer solution kept at room temperature (about 23°C) and changing the cup solution every 10 min, the efflux declined from 60 to 15 pmol cm^{-2} min^{-1} within 3 h (see Corradetti et al., this book). Under these conditions, mobilization of cellular Ch by muscarinic receptor activation was found (3).

In the chicken heart, Ch efflux was temperature- sensitive like that of the rat cortex. Table 1 shows that a reduction of the temperature from 37°C to 17°C caused a decrease in the rates of resting and acetylcholine-evoked Ch efflux to about 20%. In contrast, the two kinds of efflux had different sensitivities to 10^{-5} mol/l mepacrine which reduced the resting efflux to about 70% of the control (Tables 1 and 2), but left the acetylcholine-evoked increase in Ch efflux unchanged (Table 1).

C. **Ch efflux and cAMP.** In order to study a possible involvement of cAMP in the muscarinic stimulation of Ch efflux in the heart, we used forskolin, an adenylate cyclase activator; and 3-isobutyl-1-methylxanthine (IBMX); a cAMP phosphodiesterase inhibitor. As expected, both substances increased the force of contraction of incubated chicken atria driven electrically at 4 Hz (Fig. 2). Both substances also increased the Ch efflux rate in perfused chicken hearts; a 50% increase was observed with about 3×10^{-7} mol/l forskolin, and with 3×10^{-4} mol/l IBMX.

We found that DFP (10^{-4} mol/l for 30 min; 10 min washout; not illustrated), an irreversible cholinesterase inhibitor, reduced the increase by forskolin from 554 ± 59 to 229 ± 49 pmol g^{-1} min^{-1} (N = 4 each). It is unclear whether, in the forskolin-DFP experiments, the increase of the Ch efflux during forskolin infusion was due to the muscarinic stimulation of Ch efflux (due to DFP treatment) or to residual forskolin-induced Ch efflux. The responses to forskolin occurred after a latency of several min and were still increasing at the end of the 40 min period of infusion.

Table 2

Effect of mepacrine on resting and forskolin-evoked Ch efflux[a]

| | | Choline Efflux (pmol/g min) | | |
	Before drugs	Mepacrine	Forskolin + Mepacrine Total efflux	Increase
Control	13	1401 ± 81	---	2420 ± 136	1032 ± 100
Mepacrine 10^{-5}mol/l	4	1022 ± 118	685 ± 98	1195 ± 115	510 ± 34
Mepacrine 3×10^{-5}mol/l	4	2522 ± 166	1768 ± 74	1661 ± 61	-107 ± 91

[a]Forskolin, 10^{-6}mol/l, was infused for 40 min.

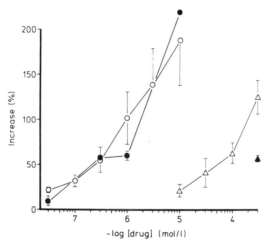

Figure 2. **Forskolin and IBMX: positive inotropic effects in in-cubated chicken atria and increase in Ch efflux in perfused chicken heart.** The concentration-response curves for the inotropic effects were obtained using electrically driven (4 Hz) atria exposed to cumulative concentrations of forskolin (open circles; N = 4) and of IBMX (open triangles; N = 3), whereas the effects of forskolin (closed circles) and of IBMX (closed triangles) on Ch efflux were obtained using different hearts for each concentration (N varies between 3 and 10 experiments). Presented are means $\pm$ SE.

Forskolin (10^{-6} mol/l), unlike acetylcholine, caused an increase of the Ch efflux that was reduced by 10^{-5} mol/l mepacrine to 49% and blocked by 3×10^{-5} mol/l (Table 2). This effect occurred at a concentration that eliminated the mepacrine-sensitive fraction of the resting Ch efflux (Fig. 1).

D. Ch efflux, coronary resistance and contractions. Coronary resistance, force of contraction and heart rate were measured in perfused chicken hearts under the conditions of the Ch efflux experiments (Fig. 3). It was found that forskolin and IBMX did not affect coronary resistance, while they increased force of contraction, heart rate and, as shown in Fig. 2 and 3, Ch efflux. Atropine (3×10^{-7} mol/l) slightly increased the forskolin-induced stimulation of Ch efflux but reduced the mean effects of forskolin on force of contraction and on heart rate. Acetylcholine alone increased cardiac Ch efflux, but unlike forskolin and IBMX it increased coronary resistance and decreased force of contraction and heart rate. These data suggest that the facilitation of the cardiac Ch efflux by muscarinic receptor activation, on the one hand, and by forskolin and IBMX, on the other hand, were not indirectly caused by changes in coronary resistance, force of contraction and heart rate.

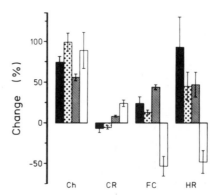

Figure 3. Comparative presentation of changes in Ch efflux (Ch), coronary resistance (CR), force of contraction (FC) and heart rate (HR) of isolated perfused chicken hearts by forskolin (10^{-6} mol/l; black columns), atropine plus forskolin (3 x 10^{-7} and 10^{-6} mol/l, respectively; stippled columns), IBMX (5 x 10^{-4} mol/l; hatched columns) and acetylcholine (3 x 10^{-7} mol/l; open columns). For Ch experiments N varies between 3 and 10 experiments. Changes of CR, FC and HR were measured in 4 hearts for each drug. Given are decreases or increases in % of the values measured before drug addition (means $\pm$ SE). These values (0%) were 33 $\pm$ 1 mm Hg (CR), 3.8 $\pm$ 0.3 g/g wet weight (FC) and 145 $\pm$ 7 beats/min (HR) (N = 16).

DISCUSSION

Ch-containing phospholipids play an important structural role in the biomembrane phospholipid bilayer. Although integrity of biomembranes is critically dependent on the maintenance of an intact bilayer, the phospholipids undergo a continuous metabolic turnover. On the other hand, circumscribed hydrolysis of phospholipids may be important for cell membrane functions such as receptor-mediated regulation of ion permeability (18, 15), activity of membrane-bound enzymes (13) and agonist-receptor interactions (7).

So far, receptor-mediated effects on phospholipid metabolism have been concentrated around two phenomena: 1) hydrolysis of phosphatidylinositol enhanced by muscarinic receptor activation (11) or adrenoceptor activation (16) (so-called "PI - effect"); and 2) phospholipid methylation (methylation of phosphatidylethanolamine to phosphatidylcholine) enhanced by beta-adrenoceptor activation (7). Clearcut effects of receptor activation on hydrolysis of phosphatidylcholine have not been observed yet, possibly because studying overall tissue content or overall turnover of phosphatidylcholine is inadequate to detect metabolic effects on a small fraction of phosphatidylcholine involved in cellular responses to receptor activation. Moreover, phosphatidylcholine in a certain cellular fraction may be degraded by various pathways with different susceptibility to receptor activation.

For the above reasons we decided to approach the problem by measuring the formation of Ch, a metabolic product of phosphatidyl-choline, rather than looking at overall phosphatidylcholine in the tissue. It was found that physostigmine caused an atropinesensitive facilitation of the Ch efflux in the perfused chicken heart and the rat cortex in vivo (2). Acetylcholine infused into the heart caused a gradual increase in cardiac Ch efflux (Ch derived from acetylcholine was subtracted) which was still increasing at the end of the 40 min-period of observation and was blocked by atropine (3; present investigation).

In order to study an involvement of cAMP in the muscarinic mobilization of cellular choline, the effects of two substances were studies which are known to increase cAMP through different mechanisms: forskolin increases cAMP by a direct activation of adenylate cyclase (17) and IBMX by inhibition of the cAMP phospho-diesterase (10, 9). As expected, both substances increased the force of contraction in chicken atrial preparations driven electrically (Fig. 2). In the spontaneously beating perfused chicken heart both substances had positive inotropic and chronotropic effects that faded within 10 min. There were no changes of the coronary resistance during the period of observation (40 min) (Fig. 3). It was found that forskolin and IBMX enhanced the Ch efflux rate causing a 50% increase at about 3×10^{-7} mol/l forskolin and 3×10^{-4} mol/l IBMX. Positive inotropic effects were also observed on the atrial preparation and on the perfused heart at these concentrations (Fig. 2 and 3); a result that would be expected if both increase of Ch efflux and of force of contraction were mediated by cAMP.

Acetylcholine, 3×10^{-7} mol/l, had an effect on Ch efflux similar to that of forskolin, 10^{-6} mol/l, whereas its effects on coronary resistance, force of contraction and heart rate were qualitatively different. Acetylcholine increased coronary resistance and exhibited negative chronotropic and inotropic effects. Thus, it can be concluded that the effects of muscarinic receptor activation and of forskolin were not subsequent to changes of coronary resistance or of cardiac mechanics.

The present results indicate that cAMP stimulates cellular mobilization of Ch. This was initially difficult to reconcile with the muscarinic mobilization of Ch, because muscarinic receptor acti-vation is known to decrease, rather than to increase, forskolin-induced adenylate cyclase activity (12). The apparent discrepancy was solved when we found that the cholinesterase inhibitor DFP antagonized, and atropine facilitated, the forskolin-evoked stimulation of the Ch efflux. Thus, it appeared that muscarinic receptor activation had two opposite effects on cardiac efflux: 1) Facilitation of Ch efflux that is independent of cAMP and seems to be present in heart and brain; and 2) inhibition of Ch efflux that is controlled by the cAMP system.

The effects of low temperature and of mepacrine, a phospholipase A_2 inhibitor, were also studied, on resting and evoked Ch efflux. Reduction of the perfusion temperature from $37^\circ C$ to $17^\circ C$ decreased resting and ACh-evoked efflux of Ch to the same extent, indicating a similar energy-dependency for both processes. Moreover, mepacrine caused inhibition of only the resting and forskolin-evoked Ch efflux, whereas the efflux caused by acetylcholine apparently was unchanged. Since mepacrine maximally reduced the Ch resting efflux to 50%, a mepacrine-sensitive Ch formation can be distinguished from an equally fast Ch formation that is mepacrineinsensitive. In other words, the phospholipase A_2 activity is responsible possibly for part of the resting Ch efflux and the forskolin-evoked activation, and seems to be irrelevant for the effect of acetylcholine on Ch efflux.

Although it has been shown that the Ch efflux from the heart is not simply efflux of free cellular Ch, but is associated with the liberation of Ch from bound Ch (3), presumably from phosphatidyl-choline, the precise mechanisms involved in the resting and evoked Ch efflux are virtually unknown.

In general, Ch efflux could be increased by enhancing phospholipid degradation or by inhibiting Ch incorporation. Since the significance of the CDP-pathway for the incorporation of Ch into phosphatidylcholine has been shown in a perfused heart preparation (19) and since this incorporation may be inhibited by cAMP (14), it seems possible that forskolin and IBMX facilitate Ch efflux by restricting re-incorporation of Ch. Alternatively, phospholipases A_2 or D, which hydrolyze phospha-tidylcholine to free Ch and phosphatidic acid, may be involved. These enzymes have been detected in rat central nervous system axolemma (4).

If phospholipase D is involved in the muscarinic enhancement of Ch efflux, we have to consider the possibility that the same mechanism also caused the muscarinic reduction in cAMP-mediated enhancement of the Ch efflux, since phosphatidic acid was found to reduce adenylate cyclase activity (1). The muscarinic inhibition of the activation of the cAMP system may be related to phospholipase D-generated phosphatidic acid. Whether this is true or not, the presence of phospholipase D in cellular membranes (4) may have func-tional implications in the regulation of essential proteins of the plasma membrane. Receptor-mediated changes of Ch efflux, as shown in the present study, may be related to the presence of phospholipases in cell membranes.

Acknowledgements. The experiments described herein are part of the Ph.D. thesis of R. Brehm. We are grateful for the technical assistance of Ulrike Stieh-Koch. The study was supported by a grant from the Deutsche Forschungsgemeinschaft.

REFERENCES

1. Clark, R.B., Salmon, D.M. and Honeyman, T.W. (1980): J. Cyclic Nucl. Res. 6:37-41.
2. Corradetti, R., Lindmar, R. and Löffelholz, K. (1982): Europ. J. Pharmacol. 85:123-124.
3. Corradetti, R., Lindmar, R. and Löffelholz, K. (1983): J. Pharmacol. Exp. Ther. 226:826-832.
4. DeVries, G.H., Chalifour, R.J. and Kanfer, J.N. (1983): J. Neurochem. 40:1189-1191.
5. Dieterich, H.A. and Löffelholz, K. (1977): Naunyn-Schmiedeberg's Arch. Pharmacol. 296:143-148.
6. Goldberg, A.M. and McCaman, R.E. (1974): In: Choline and Acetylcholine, Handbook of Chemical Assay Methods (ed) I. Hanin, Raven Press, New York, pp. 47-61.
7. Hirata, F. and Axelrod, J. (1980): Science 209:1082-1090.
8. Israël, M. and Lesbats, B. (1982): J. Neurochem. 39: 248-250.
9. Korth, M. (1978): Naunyn-Schmiedeberg's Arch. Pharmacol. 302:77-86.
10. McNeill, J.H., Brenner, M.J. and Muschek, L.D. (1973): Recent Adv. Stud. Card. Struct. Metab. 3:261-273.
11. Michell, R.H. (1975): Biochim. Biophys. Acta 415:81-147.
12. Nemeček, G.M. and Honeyman, T.W. (1983): Fed. Proc. 42(4):5001.
13. Panagia, V., Michiel, D.F., Dhalla, K.S., Nijjar, M.S. and Dhalla, N.S. (1981): Biochim. Biophys. Acta 676:395-400.
14. Pelech, S.L., Pritchard, P.H. and Vance, D.E. (1982): Biochim. Biophys. Acta 713:260-269.
15. Philipson, K.D., Frank, J.S. and Nishimoto, A.Y. (1983): J. Biol. Chem. 258:5905-5910.
16. Putney, J.W., Jr. (1983): In: Adrenoceptors and Catecholamine Action - Part B (ed) G. Kunos, John Wiley & Sons, Chichester, New York, pp. 51-64.
17. Seamon, K.B. and Daly, J.W. (1983): TIPS 4:120-123.
18. Stahl, W.L. (1973): Arch. Biochem. Biophys. 154:56-67.
19. Zelinski, T.A., Savard, J.D., Man, R.Y.K. and Choy, P.C. (1980): J. Biol. Chem. 255:11423-11428.

LONG TERM POTENTIATION OF CHOLINERGIC TRANSMISSION

IN THE SYMPATHETIC GANGLION

C.A. Briggs and D.A. McAfee

Division of Neurosciences
Beckman Research Institute
City of Hope
1450 East Duarte Road
Duarte, California 91010 U.S.A.

INTRODUCTION

Long term potentiation (LTP), like the better known posttetanic potentiation (PTP), is a use-dependent increase in synaptic efficacy. Both potentiations are induced after repetitive synaptic stimulation for a few seconds. But while posttetanic potentiation decays away in a few minutes after the tetanic stimulation, long term potentiation continues for hours or even longer (7, 14, 27). The striking disparity between the endurance of long term potentiation and the brevity of its induction suggests a metabolic modulation of synaptic transmission.

Initial observations of LTP were made more than 15 years ago. In 1966, Lomo (26) first indicated that a long-lasting facilitation of synaptic transmission followed repetitive stimulation in the hippocampus. Shortly thereafter, Dunant and Dolivo (17, 18) reported a few experiments in which an apparently similar phenomenon was seen in the rat superior cervical ganglion. However, the preliminary observations made in the ganglion were not pursued and several years later detailed accounts of LTP in the hippocampus in vivo (8, 9) and in vitro (33) appeared. Further studies of LTP have concentrated on the hippocampus. Recently, however, this laboratory in collaboration with Dr. Thomas H. Brown affirmed that LTP does indeed occur in the rat superior cervical sympathetic ganglion (11, 13, 28). Descriptions of LTP in areas of the central nervous system outside the hippocampus (20, 23, 24, 32) and even in the invertebrate neuromuscular junction (5) have also appeared recently.

This paper discusses ganglionic LTP with reference to new observations which bear on the mechanism of this durable increase in synaptic efficacy.

RESULTS AND DISCUSSION

An example of the potentiation of ganglionic transmission at $22^{\circ}C$ which follows tetanic stimulation of the preganglionic nerve is shown in Figures 1 and 2. To quantitate this effect, the degree of potentiation I(t) at time t after the tetanus is expressed as a fractional increase such that

$$I(t) = (V_t - V_c)/V_c, \qquad (1)$$

where V_t is the response amplitude at time t after the tetanus and V_c is the pretetanic control response amplitude. At $22^{\circ}C$, the potentiation decays with time in a manner that can be described by the sum of two exponential terms

$$I(t) = P[\exp(-t/\tau p)] + L[\exp(-t/\tau_L)]. \qquad (2)$$

The time constants τ_p and τ_L quantify the duration of the rapidly-decaying and the slowly-decaying components, respectively. The coefficients P and L represent the initial magnitude of each component. Values for these parameters can be determined by standard linear regression analysis and peeling techniques (Figure 1B). Figure 2 demonstrates that the total potentiation is closely described by Equation 2.

The time constant for the rapidly decaying component (τp) ranged from 1 minute to 6 minutes in different experiments. As one might expect, this phase accords with PTP (22, 40). The other component always had a much slower time constant (τ_L) - 20 minutes to 250 minutes. We identify this slow phase as long term potentiation (LTP) because its time constant is at least 10-fold greater than that of PTP while it is induced by a tetanus lasting only 5-20 seconds. Ganglionic LTP is apparently even longer lasting at $32^{\circ}C$ than at $22^{\circ}C$ (11).

In order to measure LTP with extracellular recording of compound action potentials, ganglionic transmission must first be made submaximal. In lieu of submaximal stimulation of the preganglionic nerve, submaximal transmission can be achieved by partial nicotinic antagonism, by reducing the concentration of Ca^{2+} or increasing the concentration of Mg^{2+}, by adding procaine, or by partially cutting the preganglionic nerve between the stimulating electrode and the ganglion. LTP is induced by preganglionic tetanic stimulation with all of these paradigms (11).

Muscarinic facilitory processes have been known for a number of years in the sympathetic ganglion (2, 12, 25, 31, 36, 37). However, LTP apparently is not <u>induced</u> by either muscarinic or nicotinic

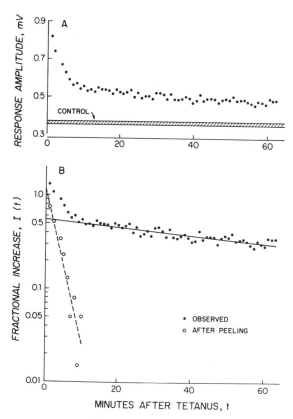

Figure 1. **Short-term and long-term potentiation of ganglionic transmission.** Atropine (2 μM) and d-tubocurarine (100 μM) were present throughout. The preganglionic nerve was stimulated once every 60 seconds with a supramaximal pulse (0.5 msec duration square wave) and the resultant postganglionic compound action potential was measured. **(A):** The amplitude of the postganglionic response is shown as a function of time after the 20 Hz, 20 second preganglionic tetanus (solid circles). Also shown is the average control response preceding the tetanic stimulation (striped bar; mean $^{\pm}$ one standard deviation, n = 5). **(B):** The total potentiation of the response (solid circles) was computed from the data in **(A)** using Equation 1. The potentiation was quantified as a function of time according to Equation 2, using standard linear regression analysis. This provided estimates for P (1.2, or 220% of control), τp (2.5 minutes), L (0.55, or 155% of control) and τL (105 minutes). Reproduced from reference 11 with permission from The <u>Journal of Physiology</u>.

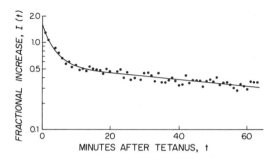

Figure 2. Summation of post-tetanic potentiation and long-term potentiation. The data are the same as shown in Figure 1B (solid circles). The solid line was drawn using Equation 2 and the values for P, τp, L, and τ_L determined in Figure 1(B). The actual data are closely described by this analysis. Reproduced from reference 11 with permission from The Journal of Physiology.

receptor activation (11). Transient exposure of the ganglion to carbachol (10 μM to 1 mM for 3 minutes) induced no potentiation of subsequent ganglionic transmission while, in the same ganglia, LTP was evident following preganglionic tetanic stimulation. Furthermore, LTP induced by tetanic stimulation was not reduced by atropine nor by blocking postganglionic discharge during the tetanic stimulation with a nicotinic antagonist (Figure 3).

The induction of LTP is, however, dependent upon extracellular Ca^{2+} (11). A paradigm similar to that of Figure 3 was used. No long term potentiation occurred following tetanic stimulation in the absence of Ca^{2+} and presence of elevated (8 mM) Mg^{2+} while the same ganglia showed prominent LTP after similar tetanic stimulation in normal medium (2.2 mM Ca^{2+}, 1.2 mM Mg^{2+}). A similar calcium dependency has been found in the hippocampus (19, 35). Perhaps LTP is induced after the release of some modulator. If so, it would appear that a noncholinergic agent is released from preganglionic afferents. The induction of LTP by preganglionic stimulation was not blocked by blocking doses of atropine plus hexamethonium, and LTP was not induced by nonsynaptic stimulation of postganglionic neurons with antidromic stimulation (13, but see 21), with the intra-cellular microelectrode (10), or with bath-applied carbachol (11). Alternatively, Ca^{2+} influx could initiate a metabolic modulatory process in the cholinergic afferents.

LTP is not due to altered excitability of preganglionic axons (13, 17, 18), leaving increased synaptic efficacy or increased ex-citability of postganglionic neurons as possible mechanisms of the potentiation. Intracellular studies have provided direct evidence that ganglionic LTP is due to an increase in nicotinic synaptic transmission (10). In 50% of 42 postganglionic neurons, nicotinic

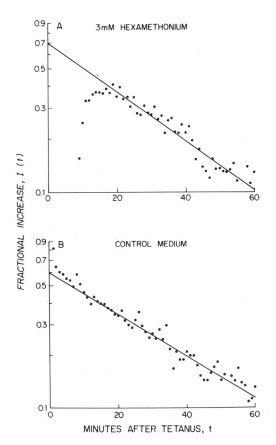

Figure 3. **Nicotinic independence of the induction of long term potentiation.** All data were obtained from the same ganglion. Hexamethonium (200 μM) was used to partially depress synaptic transmission while 2 μM atropine blocked muscarinic transmission. The fractional increase in compound potential amplitude [I(t)] is plotted as a function of time after preganglionic tetanic stimulation (5 Hz for 5 seconds). **(A):** Transmission was blocked with 3 mM hexamethonium during the preganglionic tetanic stimulation and then hexamethonium was returned to the control level (200 μM) at the end of the tetanic stimulation. LTP with the following paramaters was induced: τ_L = 31 min and L = 0.71. In the absence of tetanic stimulation, the response returned to control within 18 min after the start of 3 mM hexamethonium washout and no hysteresis was evident. **(B):** The concentration of hexamethonium was maintained at the control level throughout. This LTP had the following parameters: τ_L = 36 min and L = 0.59. PTP is evident during the first few minutes after the tetanic stimulation. Similar results were obtained in three other experiments. Reproduced from reference 11 with permission from <u>The Journal of Physiology</u>.

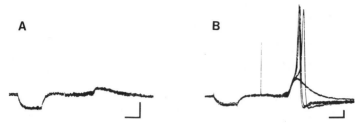

Figure 4. Intracellular recording of long-term potentiation. Standard intracellular techniques were used (29). Once every 30 seconds, the input resistance of the postsynaptic neuron was measured by injection of hyperpolarizing current (to left of each trace; 200 pA, 30 msec square pulse) and a nicotinic fast EPSP was elicited by submaximal stimulation of the preganglionic nerve (to right of each trace). Calibration bars represent 5 mV vertically and 20 msec horizontally. (A): Shown are two superimposed pretetanic control responses. (B): Shown are four superimposed responses to the same stimulus 40 minutes after preganglionic tetanic stimulation at 20 Hz for 20 seconds. Most responses after the tetanic stimulation were synaptically driven action potentials. When fast EPSPs were discernible, they were obviously potentiated. This behavior continued for 80 minutes after the tetanic stimulation, until the impalement was lost. Notice that the response is amplified two-fold in (A) compared to (B).

Before this synaptic tetanic stimulation, the same postganglionic neuron was tetanically stimulated nonsynaptically (20 Hz for 20 seconds) by injection of depolarizing current pulses through the microelectrode. Each pulse during the nonsynaptic tetanic stimulation was 1 nA for 30 msec and elicited a nonsynaptic action potential. This treatment did not potentiate synaptic transmission.

fast excitatory postsynaptic potentials (fast EPSPs) were potentiated after tetanic synaptic stimulation (20 Hz for 20 seconds). In 13 of the cells displaying LTP, fast EPSPs were potentiated by an average of 150% at 15 mintues after tetanic stimulation; pretetanic control responses averaged 5 mV. In the other 8 observations of LTP, the postganglionic neuron fired an action potential in response to all or almost all of the preganglionic test stimuli after the tetanic stimulation, even when the pretetanic response was a fast EPSP of only 2 mV. Nevertheless, it was still possible to determine that the fast EPSP was potentiated. Figure 4 shows one example.

These potentiations were not correlated with changes in membrane potential or input impedance as measured in the soma of the post-ganglionic neuron; similar findings have been reported in the hippo-campus (1, 3, 38, 39). Thus, we do not find evidence that LTP results from a change in passive membrane properties. It appears more likely that LTP, like PTP, is due to an increase in the strength of synaptic transmission. Possible mechanisms include increased postsynaptic receptor efficacy and increased presynaptic transmitter release. These possibilities are not mutually exclusive.

It has been proposed that postsynaptic Ca^{2+} influx during tetanic stimulation leads to an increase in postsynaptic receptors for glutamate in the hippocampus (4) and for acetylcholine in the frog sympathetic ganglion (21). In contrast, our findings indicate that neither electrical nor cholinergic stimulation of postganglionic neurons induces LTP in the rat superior cervical ganglion. The postsynaptic nicotinic receptors might, however, be modified by the action of an extracellular agent released during tetanic stimulation.

Other studies suggest increased neurotransmitter release as a mechanism for LTP. At the crayfish neuromuscular junction, increased quantal content has been correlated with LTP (5) and in the hippocampus a long-lasting increase in the release of aspartate (34) and of glutamate (16) has been noted after tetanic stimulation. In the ganglion, acetylcholine is the primary transmitter and nicotinic transmission undergoes LTP. Prolonged tetanic stimulation of the cat superior cervical ganglion induces a long lasting increase in the content and release of acetylcholine (6, 15). In addition, we in collaboration with Dr. Richard E. McCaman have recently found that brief tetanic stimulation (20 Hz for 20 seconds) of the rat superior cervical ganglion induces an hours-long potentiation of evoked acetylcholine release without increasing the ganglionic content of acetylcholine (30). The potentiation of transmitter release correlated with the potentiation of ganglionic transmission measured in the same experiments. This suggests a presynaptic locus for LTP in the symapthetic ganglion.

SUMMARY

A burst of activity lasting only seconds induces a remarkably durable, hours-long potentiation of nicotinic synaptic transmission in the superior cervical sympathetic ganglion. The induction of this potentiation apparently requires synaptic activation and extracellular calcium, but is independent of cholinergic receptor activation. Increased acetylcholine release occurs in conjunction with the potentiation of synaptic transmission. Thus, it is likely that a presynaptic mechanism underlies ganglionic long term potentiation. Nevertheless, more than one mechanism may underly the phenomenon and possible postsynaptic contributions, such as modification of nicotinic receptors or active membrane properties, have not been ruled out. Future intracellular studies will help resolve these questions.

Acknowledgements. Supported by an Advanced Research Fellowship (766 F1-1) from the American Heart Association - Greater Los Angeles Affiliate, and by a grant from the National Science Foundation (BNS 81 12414).

REFERENCES

1. Andersen, P., Sundberg, S.H., Sveen, O., Swann, J.W. and Wigstrom, H. (1980): J. Physiol (London) 302:463-482.
2. Ashe, J..W. and Libet, B. (1981): Brain Res. 217:93-106.
3. Barrionuevo, G. and Brown, T.H. (1983): Proc. Natl. Acad. Sci.-USA 80:7347-7351.
4. Baudry, M., Siman, R., Smith, E.K. and Lynch, G. (1983): Eur. J. Pharmacol. 90:161-168.
5. Baxter, D.A., Bittner, G.D. and Brown, T.H. (1985): Proc. Natl. Acad. Sci. USA in press.
6. Birks, R.I. (1978): J. Physiol. (London) 280:559-572.
7. Bliss, T.V.P. (1979): Trends in Neurosciences 2:42-45.
8. Bliss, T.V.P. and Gardner-Medwin, A.R. (1973): J. Physiol. (London) 232:357-374.
9. Bliss, T.V.P. and Lomo, T. (1973): J. Physiol. (London) 232:331-356.
10. Briggs, C.A., Brown, T.H. and McAfee, D.A. (1983): Trans. Am. Soc. Neurochem. 14:141.
11. Briggs, C.A., Brown, T.H. and McAfee, D.A. (1985): J. Physiol (London) 359:503-521.
12. Brown, D.A., Constanti, A. and Adams, P.R. (1981): Fed. Proc. 40:2625-2636.
13. Brown, T.H. and McAfee, D.A. (1982): Science 215:1411-1413.
14. Chung, S.H. (1977): Nature 266:677-678.
15. Collier, B., Kwok, Y.N. and Welner, S.A. (1983): J. Neurochem. 40:91-98.
16. Dolphin, A.C., Errington, M.L. and Bliss, T.V.P. (1982): Nature 297:496-498.
17. Dunant, Y. (1969): Prog. Brain Res. 31:131-139.
18. Dunant, Y. and Dolivo, M. (1968): Brain Res. 10:271-273.
19. Dunwiddie, T., Madison, D. and Lynch, G. (1978): Brain Res. 150:-413-417.
20. Gerren, R.A. and Weinberger, N.M. (1983): Brain Res. 265:138-142.
21. Kumamoto, E. and Kuba, K. (1983): Nature 305:145-146.
22. Larrabee, M.G. and Bronk, D.W. (1947): J. Neurophysiol. 10:139-154.
23. Lee, K.S. (1982): Brain Res. 239:617-623.
24. Lewis, D., Teyler, T. and Shashoua, V. (1981): Soc. Neurosci. Abs. 7:66.
25. Libet, B. (1975): Nature 258:155.
26. Lomo, T. (1966): Acta Physiol. Scand. Suppl. 277:128.
27. Lynch, G., Browning, M. and Bennett, W.F. (1979): Fed. Proc. 38:2117-2122.

28. McAfee, D.A. and Brown, T.H. (1981): Soc. Neurosci. Abs. 7:710.
29. McAfee, D.A. and Yarowsky, P. (1979): J. Physiol. (London) 245:-
     447-466.
30. McCaman, R.E., Briggs, C.A. and McAfee, D.A. (1984): Soc. Neuro-
     sci. Abs. 10:195.
31. McIsaac, R.J. (1977): J. Pharmac. Exp. Ther. 200:107-116.
32. Racine, R.J., Milgram, N.W. and Hafner, S. (1983): Brain Res.
     260:217-231.
33. Schwartzkroin, P.A. and Wester, K. (1975): Brain Res. 89:107-119.
34. Skrede, K.K. and Malthe-Sorenssen, D. (1981): Brain Res. 208:-
     436-441.
35. Turner, R.W., Baimbridge, K.G. and Miller, J.J. (1982): Neuro-
     sci. 7:1411-1416.
36. Volle, R.L. (1966): Pharmac. Rev. 18:839-869.
37. Weight, F.F., Schulman, J.A., Smith, P.A. and Busis, N.A. (1979):
     Fed. Proc. 38:2084-2094.
38. Wigstrom, H., McNaughton, B.L. and Barnes, C.A. (1982): Brain
     Res. 233:195-199.
39. Yamamoto, C. and Chujo, T. (1978): Exp. Neurol. 58:242-250.
40. Zengel, J.E., Magleby, K.L., Horn, J.P., McAfee, D.A. and Yarow-
     sky, P.J. (1980): J. Gen. Physiol. 76:213-231.

CONTROL OF THE RELEASE OF [^{3}H]-ACETYLCHOLINE FROM RAT

HIPPOCAMPAL SLICES BY AMINOPYRIDINES AND PHENCYCLIDINE

R.D. Schwarz, C.J. Spencer, A.A. Bernabei
and T.A. Pugsley

Warner-Lambert/Parke-Davis Pharmaceutical Research
Ann Arbor, Michigan 48105 U.S.A.

INTRODUCTION

Release of acetylcholine (ACh) resulting from the depolarization of cholinergic nerve terminals appears to involve the movement of specific ions across the nerve membrane. Experimentally, electrical stimulation and veratridine release ACh by increasing the flux of Na^+ as shown by their sensitivity to the Na^+ channel blocker tetrodotoxin (TTX), while elevated K^+ releases ACh by a TTX insensitive mechanism. In addition, all three methods appear to require the presence of extracellular Ca^{++} in order to release ACh from vesicular stores (3, 9, 11). Electrophysiological data have suggested that blockade of K^+ channels will also enhance neurotransmitter release by increasing Ca^{++} entry or utilization, due to the prolongation of the repolarization period (14).

Aminopyridines (APs) have been suggested to increase the release of various neurotransmitters, including ACh, by blocking voltage-dependent K^+ channels that are linked to Ca^{++} and/or having a direct effect on Ca^{++} itself (16). The ability to increase transmitter release has led to the suggestion that APs may be clinically useful in a variety of disorders, such as: myasthenia gravis, botulism, curare poisoning, Parkinsonism, and Alzheimer's disease.

Phencyclidine (PCP) has also been suggested to increase transmitter release by blocking certain K^+ channels (1, 4). However, PCP may act at a non Ca^{++}-linked channel which is distinct from the AP-sensitive K^+ channel. Those sites at which PCP acts may also be identical to the proposed PCP binding sites labeled by [^{3}H]-PCP.

The purpose of the present experiments was to examine the ability of various APs and PCP to release ACh from hippocampal slices. The hippocampus was chosen since cholinergic neurons terminating within the hippocampus have been extensively studied biochemically, electrophysiologically, and anatomically.

METHODS

The dorsal and ventral portions of the hippocampus were dissected out from the brains of male rats (Long Evans, 200-300 g) following decapitation. Cuboidal slices (0.3 mm X 0.3 mm) were then cut on a McIlwain tissue chopper and dispersed in ice-chilled Krebs-Ringer Hepes buffered media. Normal medium consisted of: NaCl (119 mM), KCl (4.75 mM), $CaCl_2$ (1.25 mM), KH_2PO_4 (1.20 mM), $MgSO_4$ (1.18 mM), Hepes (22 mM), EDTA (0.03 mM), ascorbate (0.6 mM), and glucose (10 mM). The pH was adjusted to 7.2 with 3.0 M Tris HCl. Following a quick centrifugation at 500 X g, the supernatant was discarded and the slices incubated with [^{3}H]-choline (80 Ci/mmole; 0.01 μM final concentration) in 10 ml normal medium for 15 minutes at 37°C. The slices were washed 3X with 5 ml medium and then resuspended in a volume of medium such that 0.2 ml aliquots of this suspension would be equivalent to at least 10-15 mg of tissue. The slices were further incubated for 15 minutes in a final volume of 3 ml in normal medium, or medium with elevated K^+ or veratridine in the presence or absence of test compound. The reaction was terminated by placing the samples on ice followed by the separation of medium and tissue by rapid centrifugation. One ml of 0.4N perchloric acid plus 2.0 ml H_2O were added to the tissue, while 0.5 ml acid was added to the medium. Following tissue homogenization, the amount of radioactivity was measured in both fractions by liquid scintillation counting. Previous results had shown that under stimulated conditions 90% of the [^{3}H] present was due to the presence of [^{3}H]-ACh.

RESULTS

Of the seven APs examined at concentrations of 10^{-7} - 10^{-4}M, 4-AP and 3,4-AP significantly increased the spontaneous (basal) release of [^{3}H]-ACh from rat hippocampal slices at a concentration of 100 μM, while 2-, 3-, 2,3-, 2,5-, and 2,6-AP were inactive. PCP at similar concentrations was also inactive in altering the spontaneous release of [^{3}H]-ACh (Table 1).

Table 2 shows the effects of APs and PCP on K^+-stimulated [^{3}H]-ACh release. The same APs at concentrations which significantly increased spontaneous [^{3}H]-ACh release, markedly decreased the ability of K^+ to stimulate release. While 4-AP and 3,4-AP appeared to be similar in their ability to stimulate the spontaneous release of [^{3}H]-ACh, 3,4-AP was more potent in its ability to decrease K^+-stimulated release than was 4-AP. In addition, PCP failed to alter K^+-stimulated release of [^{3}H]-ACh.

Table 1

The Effect of Aminopyridines (APs) and
Phencyclidine (PCP) on the Spontaneous Release
of [^{3}H]-ACh From Rat Hippocampal Slices

Compound	% Control Release			
	10^{-7}	10^{-6}	10^{-5}	10^{-4} M
2-AP	100.4 ±4.5	99.5 ±3.6	94.1 ±4.1	99.1 ±3.2
3-AP	102.5 ±2.4	105.9 ±3.9	107.4 ±3.9	100.0 ±2.9
4-AP	108.5 ±4.5	104.5 ±4.5	109.5 ±4.0	124.1*** ±4.0
2,3-AP	94.2 ±3.2	100.5 ±3.4	99.5 ±4.2	94.7 ±3.2
2,5-AP	100.0 +2.9	103.8 ±3.4	110.1 ±3.8	105.1 ±4.1
2,6-AP	104.9 ±4.9	107.8 ±3.9	103.4 ±3.9	101.5 ±2.0
3,4-AP	104.1 ±4.1	106.9 ±2.8	109.7 ±5.5	116.6*** ±3.2
PCP	101.5 ±4.0	97.0 ±7.4	95.1 ±6.9	87.6 ±8.9

*$p<0.05$ compared to control release. **$p<0.01$. ***$p<0.001$. Each
value represents the mean ± S.E.M. of 4-8 determinations. Basal
release in 15 min was 20.7 ± 0.4% of total [^{3}H] present.

It was shown in the above results that both 4-AP and 3,4-AP
markedly decreased the ability of K$^+$ to release [^{3}H]-ACh from hippocampal slices (Table 2). Although both veratridine and K$^+$ appear
to release [^{3}H]-ACh by Ca^{++}-dependent processes (Fig. 2), there is
a significant difference in the mechanism by which the two agents
promote release. Both 4-AP and 3,4-AP were tested for their ability
to alter veratridine-induced release and were then compared to their
effect on K$^+$-stimulated release. Figure 1 shows that 4-AP and 3,4-AP

Table 2

The Effect of Aminopyridines (APs) and
Phencyclidine (PCP) on the K^+-Stimulated Release
of $[^3H]$-ACh from Rat Hippocampal Slices

Compound	% K^+-Stimulated Release			
	10^{-7}	10^{-6}	10^{-5}	10^{-4} M
2-AP	90.0 ±11.4	114.3 ±12.8	90.0 ±7.1	100.0 ±17.1
3-AP	87.8 ±5.1	81.6 ±5.1	78.6 ±7.1	85.7 ±4.1
4-AP	95.5 ±2.7	94.6 ±2.7	80.4* ±0.1	67.0** ±6.2
2,3-AP	106.3 ±5.1	92.4 ±9.5	93.0 ±9.5	103.8 ±9.5
2,5-AP	94.7 ±8.5	90.4 ±9.6	94.6 ±9.6	91.5 ±7.4
2,6-AP	87.8 ±12.2	85.7 ±9.2	90.8 ±9.2	94.9 ±12.2
3,4-AP	80.8 ±8.1	82.8 ±6.1	64.6** ±8.0	46.5*** ±6.1
PCP	95.0 ±3.7	95.6 ±5.4	101.3 ±5.0	96.6 ±5.7

*$p<0.05$ compared to K^+ release. **$p<0.01$. ***$p<0.001$. Each value
represents the mean ± S.E.M. of 4-8 determinations. K^+-stimulated
release in 15 min was 30.4 ± 0.2% of total $[^3H]$ present.

did not alter veratridine-induced release of $[^3H]$-ACh, while markedly
decreasing K^+-stimulated release.

Figure 2 shows that the increase in spontaneous release produced
by the APs was Ca^{++} dependent. Omision of Ca^{++} from the incubation
medium significantly reduced the ability of the APs to stimulate
$[^3H]$-ACh release. The depolarizing agents veratridine (5 μM) and
K^+ (20 mM) required the presence of extracellular Ca^{++} for $[^3H]$-ACh

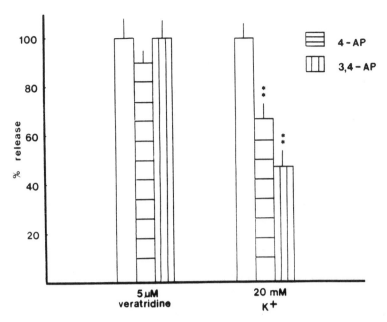

Figure 1. Effect of 100 μM 4-AP and 3,4-AP on veratridine and K$^+$-stimulated release of [^{3}H]-ACh from rat hippocampal slices. **p < 0.01 when compared to treatment in the absence of AP.

release with veratridine also utilizing intracellular Ca^{++}, since it significantly released [^{3}H]-ACh even in the absence of Ca^{++} (Figure 2).

The release of ACh by veratridine has been shown to involve Na$^+$ movement through specific channals which can be blocked by TTX, while release induced by K$^+$ is TTX insensitive. To further test whether APs increase release by acting on Na$^+$ channels in a manner similar to veratridine, 4-AP and 3,4-AP, were incubated with the slices in the presence or absence of TTX. It can be seen that TTX had no statistically significant effect on the release of [^{3}H]-ACh produced by either 4-AP or 3,4-AP (Fig. 3). In comparison, TTX significantly decreased veratridine-induced release, while having no effect on K$^+$-induced release.

                              DISCUSSION

These results confirm previous evidence that APs can actively release [^{3}H]-ACh from brain tissue by a Ca^{++} dependent process (6, 15). The examination of a series of related APs and the finding that only 4-AP and 3,4-AP were active, indicated that there are structural requirements for the activation of hippocampal cholinergic nerve terminals and the subsequent release of ACh. The inactivity

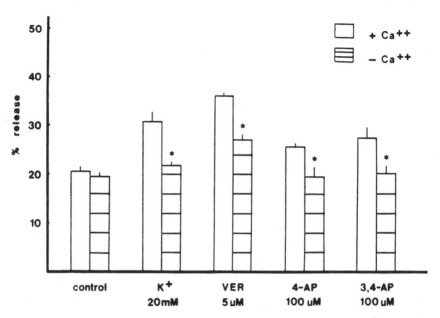

Figure 2. Effect of Ca^{++} (1.25 mM) on the release of $[^3H]$-ACh from rat hippocampal slices. *$p < 0.05$ when compared to respective treatment with Ca^{++} present.

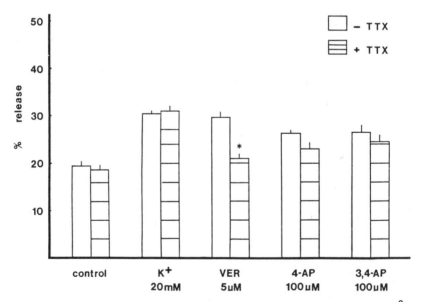

Figure 3. Effect of tetrodotoxin (1 μM) on the release of $[^3H]$-ACh from rat hippocampal slices. $p < 0.05$ when compared to respective treatment with no TTX present.

of PCP on spontaneous or K^+-stimulated release further suggests that AP sensitive K^+ channels are more important in the control of [^{3}H]-ACh released from rat hippocampal slices than PCP sensitive K^+ channels.

The mechanism by which APs release ACh, as well as other neuro-transmitters, is currently unclear. It has been suggested that APs have a direct action on voltage dependent Ca^{++} channels (7). Our results (Figure 2) show that the presence of extracellular Ca^{++} is a vital requirement for the AP-stimulated release of ACh from rat hippocampal slices. However, we did not measure whether APs directly control Ca^{++} channels. APs have also been suggested to block specific K^+ channels, which increases the entry of Ca^{++} by prolonging repolarization (8, 18). The ability of 4-AP and 3,4-AP to decrease K^+-stimulated release (Table 2), but not veratridine-induced release may support the above mentioned mechanism. Further support is given in part by the ability of TTX to block veratridine-induced release, but not that of K^+ or the APs.

These results suggest that APs are more similar to K^+ in their mechanism and also that they do not affect Na^+ channels. There is, however, evidence in the literature that 4-AP and 3,4-AP increase K^+-induced release (12) or the spontaneous release of ACh (6). Our results show a clear increase in spontaneous release and a decrease in K^+-stimulated release. These differences may be accounted for by the fact that the experiments differed as to the species used (mouse versus rat), age of animals (3-30 months versus 6 weeks), origin of the pool of ACh measured ([^{14}C]-glucose versus [^{3}H]-choline), and brain area utilized (forebrain versus hippocampus).

PCP has a variety of pharmacological ations which recently have been correlated to its ability to block K^+ channels, resulting in increased transmitter release and also binding to specific sites within neural tissue (2). However, only a marked increase in dopamine release from rat striatal slices has been reported (17). The present results fail to show an effect of PCP on rat hippocampal cholinergic neurons, as did Murray and Cheney (10) who measured the effect, in vivo, of PCP on ACh turnover in the hippocampus (certain cortical regions did show an increase).

APs may potentially be of great therapeutic benefit. While their actions have been much more widely studied on peripheral neuronal systems (5, 16), the present evidence suggests that they can also influence hippocampal cholinergic function. Additional work in our laboratory indicates that APs may also affect cholinergic systems in the cortex and striatum as well (13).

REFERENCES

1. Albuquerque, E.X., Aguayo, L.G., Warnick, J.E., Weinstein, H., Glick, S.D., Maayani, S., Ickowics, R.K., and Blaustein, M.P. (1981): Proc. Natl. Acad. Sci. $\underline{12}$:7792-7796.
2. Albuquerque, E.X., Aguaya, L.G., Warnick, J.E., Ickowicz, R.K. and Blaustein, M.P. (1983): Fed. Proc. $\underline{42}$:2584-2589.
3. Blaustein, M.P. (1975): J. Physiol. $\underline{247}$:617-655.
4. Blaustein, M.P. and Ickowicz, R.K. (1983): Proc. Natl. Acad. Sci.: $\underline{80}$:3855-3859.
5. Bowman, W.C. (1982): Trends Pharmacol. Sci. $\underline{3}$:183-185.
6. Dolezal. V. and Tucek, S. (1983): Arch. Pharm. $\underline{323}$:90-95.
7. Lundh, H. and Thesleff, S. (1977): Eur. J. Pharm. $\underline{42}$:411-412.
8. Meves, H. and Pichon, Y. (1977): J. Physiol. (Lond) $\underline{268}$: 511-532.
9. Minchin, M.C. (1980): J. Neurosci. Methods $\underline{2}$:111-121.
10. Murray, T.F. and Cheney, D.L. (1981): J. Pharm. Exp. Ther. 733-737.
11. Paton, D.M. (1979) In: The Release of Catecholamines From Adrenergic Neurons, Pergamon Press, NY, 323-332.
12. Peterson, C. and Gibson, G.E. (1983): Neurobiol. Aging $\underline{4}$:25-30.
13. Schwarz, R.D., Spencer, C.J., Bernabei, A.A. and Pugsley, T.A. (1983): Soc. Neurosci. Abstracts $\underline{9}$:433.
14. Siegelbaum, S.A. and Tsien, R.W. (1983): Trends in Neurosci. $\underline{6}$:307-320.
15. Tapia, R. and Sitges, M. (1982): Brain Res. $\underline{250}$:291-299.
16. Thesleff, S. (1980): Neurosci. $\underline{5}$:1413-1419.
17. Vickroy, T.W. and Johnson, K.M. (1983): Neuropharm. $\underline{22}$: 839-842.
18. Yeh, J.Z., Oxford, G.S., Wu, C.H. and Narahashi, T. (1976): J. Gen. Physiol. $\underline{68}$:519-535.

A COMPARISON BETWEEN ENDOGENOUS ACETYLCHOLINERELEASE AND

[^{3}H] CHOLINE OUTFLOW FROM GUINEA-PIG BRAIN SLICES

L. Beani, C. Bianchi, A. Siniscalchi,
L. Sivilotti, S. Tanganelli and E. Veratti

Department of Pharmacology
University of Ferrara
Via Fossato di Mortara, 64/B - 23
44100 Ferrara, Italy

INTRODUCTION

The measure of [^{3}H]Choline ([^{3}H]Ch) efflux from preloaded brain slices is considered a valid method for studying cholinergic function. At variance with the assay of endogenous acetylcholine (ACh) this procedure gives an index of the release process in the absence of esterase inhibition, thus excluding the dampening of the autoreceptor-mediated negative feedback (4, 6, 7).

In order to establish the equivalence of these approaches the two methods have been compared on electrically-stimulated guinea-pig brain slices, kept under the same experimental conditions.

MATERIALS AND METHODS

Slices of guinea-pig cerebral cortex (CC), thalamus (Th) and caudate nucleus (CN), prepared and superfused as recently described (1), were used to investigate the ACh release and [^{3}H]Ch efflux. In general, the tissue was electrically stimulated (alternate pulses of 5 msec duration, strength 30 mA/cm^2) at the 45th (St$_1$) and 75th (St$_2$) min, for 2 min at 2 Hz (5). Some slices were superfused with Krebs solution containing physostigmine, and ACh release was measured by bioassay (1).

Others were preloaded for 30 min with [^{3}H]Ch 0.1 μmol/l (80 Ci/mmol), superfused with normal Krebs solution, and the radioactivity of the samples (taken every 5 min) was determined with a scintillation beta spectrometer (Beckman LS 1800).

The electrically-evoked extra-outflow was calculated by subtracting the presumed spontaneous efflux (6). The fractional rate (5) of tritium release was measured as the ratio between pmol of radioactivity lost in 5 min and pmol of tritium actually present in the slices. The [^{3}H]Ch remaining in the tissue (solubilized with Protosol), was counted at the end of experiment.

Drug treatment started at the 60th min of perfusion. Fresh solutions of commercially available drug were employed. All the values are given as pmol per g of fresh tissue, per min.

Student's $\underline{t}$ and the Wilcoxon, Mann and Witney tests were used for statistical analysis of the results.

RESULTS

The pattern of spontaneous and stimulus-evoked ACh and [^{3}H]Ch outflow is depicted in Fig. 1. It is evident from the different ordinate values that, as expected, CN displays the highest releasing property and Th the lowest. In addition, the resting [^{3}H]Ch outflow, matched with the spontaneous ACh release, is proportionally higher in Th than in CN and CC.

Consequently, the ratio between evoked and resting tritium efflux is unsatisfactory in Th, good in CC and excellent in CN. The tritium loss consistently declines in Th during the entire course of the experiment and also in Cn during St$_2$. On the contrary, the ACh release profile is steady in every area, in agreement with previous reports (2, 3).

The ratio between [^{3}H]Ch and ACh outflow is approximately 4:1000 in CN, Th and CC and the ratio between the two subsequent periods of stimulation (St$_2$/St$_1$) ranges between 0.75 and 1 both with the ACh and the [^{3}H]Ch methods (Table 1).

Thus, apart from some minor differences in the efflux profiles and in the absolute values of endogenous transmitter and tritium release, the equivalence of ACh and [^{3}H]Ch approach seems to be satisfactory. Since in most instances the drug treatment is performed between St$_1$ and St$_2$, the responsiveness of the cholinergic structures to a second stimulus of increased intensity was tested. To this aim, CC slices were stimulated during ST$_2$ at 10 Hz, instead of 2 Hz.

By increasing the stimulation rate from 2 Hz to 10 Hz, the [^{3}H]Ch efflux only changed from 0.128 $\pm$ 0.010 ng^{-1}g^{-1}min^{-1} to 0.253 $\pm$ 0.015 ng/g/min (6 expts). Therefore, the responsiveness of ACh release in comparison to [^{3}H]Ch efflux was proportionately higher when the stimulus intensity increased by a factor of 5.

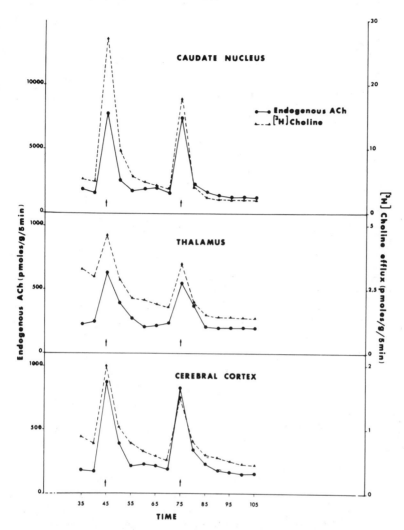

Figure 1.   Outflow of [³H]Choline ([³H]Ch) and endogenous ACh from guinea-pig cerebral cortex, caudate nucleus and thalamus slices. ↑ = Stimulation for 2 min at 2 Hz, except in the thalamus (for 5 min at 2 Hz).

In view of this finding, drugs known both to enhance and reduce the neurotransmitter release were tested on [³H]Ch preloaded slices. At the highest concentration tested (6 mM) GABA reduced the evoked [³H]Ch efflux and increased the spontaneous tritium loss both in CC and CN to nearly the same extent as it did on ACh release (Table 2). On the other hand the amino acid showed a significant effect on such efflux at 0.3 mM, a concentration insufficient to affect the endogenous transmitter release.

Table 1

Comparison between Acetylcholine (ACh) and [^{3}H]Choline ([^{3}H]Ch)
electrically-evoked efflux in guinea-pig
Cerebral Cortex, Thalamus and Caudate Nucleus slices.

Brain Area	Methods	Evoked extra release at the 45th min: pmoles g^{-1} min^{-1}	St$_2$/St$_1$
CEREEBRAL CORTEX	Endogenous ACh	276 ± 12.6	0.95 ± 0.06
	[^{3}H]Ch efflux	1.04 ± 0.07	0.75 ± 0.02
THALAMUS	Endogenous ACh	172 ± 16	0.82 ± 0.07
	[^{3}H]Ch efflux	0.68 ± 0.09	0.96 ± 0.05
CAUDATE NUCLEUS	Endogenous ACh	2762 ± 90	1.00 ± 0.03
	[^{3}H]Ch efflux	10.5 ± 0.56	0.76 ± 0.01

St$_2$/St$_1$ represents the ratio between ACh or [^{3}H]Ch efflux at the
75th(St$_2$) and 45th(St$_1$) min, respectively. The values are the mean
± S.E.M. of 10 experiments.

   Morphine 30 μM significantly inhibited the evoked ACh release
from the Th slices (to 62 ± 4.5 percent, 5 expts.), while the drug
reduced [^{3}H]Ch efflux to an insignificant degree (to 84 ± 19 percent,
5 expts.). In addition the enhancement of ACh release previously
reported in the slices treated with 30 μM morphine plus 10 μM naloxone
(increase to 157 ± 8.4; 2) was no longer seen with the [^{3}H]Ch technique; the efflux remaining close to the control values (96 ± 7.9
percent, 5 expts.).

CONCLUSIONS

   From the above results only a partial equivalence of the two
methods can be recognized. As a rule, the [^{3}H]Ch technique permits
better observation of drug induced inhibition than facilitation.
A drawback of the [^{3}H]Ch approach is the exhaustion or dilution of
tritium stores, so that drug-induced increases of evoked efflux
are minimized. In addition, the involvement of insufficiently labelled

Table 2

Effect of GABA on ACh Release and on [³H]Ch Efflux from
Guinea-Pig Cerebral Cortex and Caudate Nucleus Slices
Kept at Rest and Electrically-Stimulated at 2 Hz.

Brain Area	Experimental Conditions	GABA (Molar concentrations)			
		3×10^{-4}	1×10^{-3}	3×10^{-3}	6×10^{-3}
CEREBRAL CORTEX	Endogenous ACh				
	resting release	----	118 ± 9.4*	141 ± 11.3*	162 ± 13.1**
	evoked-release	----	89 ± 9.1*	78 ± 6.9	61 ± 6.0**
	[³H]Ch				
	resting release	120 ± 4.9*	134 ± 9.7**	143 ± 6.4**	166 ± 12.1**
	evoked-release	89 ± 4.1*	84 ± 9.5*	83 ± 6.8*	78 ± 6.6**
CAUDATE NUCLEUS	Endogenous ACh				
	resting release	----	115 ± 10.2**	133 ± 12.1**	150 ± 14.0**
	evoked-release	----	83 ± 6.8*	71 ± 7.3**	57 ± 6.1**
	[³H]Ch				
	resting release	104 ± 4.5	126 ± 9.1**	143 ± 5.7**	131 ± 5.7**
	evoked-release	87 ± 2.5*	73 ± 2.6**	67 ± 3.8**	66 ± 3.8**

The effect of the amino acid is reported as % of control (mean ± S.E.M., 10 expts.).
Statistically different from control: *P < 0.05; **P < 0.01.

pools cannot be ruled out as site of action for certain drugs. This could explain why morphine facilitation (exerted through naloxone-insensitive receptors; 2) is not seen in Th slices with the [^{3}H]Ch technique.

On the other hand, the [^{3}H]Ch approach permits one to study the influence of drugs or other putative transmitters on cholinergic nerve-endings, without the dampening effect of auto-receptor mediated negative feedback. Thus, the actual sensitivity of the release process to other signalling mechanisms can be precisely assessed. Accordingly, the threshold for GABAergic modulation is three times lower with the [^{3}H]Ch than the ACh approach.

We must conclude, therefore, that until such time as a better understanding of ACh disposition in the nerve endings is reached, parallel investigation with both of the methods tested in this study is necessary.

Acknowledgment. This research was supported by Grant n. 82.02103.04 from Consiglio Nazionale delle Ricerche, Rome and from M.P.I.

REFERENCES

1. Beani, L., Bianchi, C., Giacomelli, A. and Tamberi, F. (1978): Eur. J. Pharmac. 48:89-93.
2. Beani, L., Bianchi, C. and Siniscalchi, A. (1982): Br. J. Pharmac. 76:393-401.
3. Bianchi, C., Tanganelli, S., Marzola, G. and Beani, L. (1982): Arch. Pharmacol. 318:253-258.
4. Dunant, Y. and Walker, A.I. (1982): Eur. J. Pharmacol. 78:201-212.
5. Hertting, G., Zumstein, A., Jackisch, R., Hoffman, I. and Starke, K. (1980): Arch. Pharmacol. 315:111-117.
6. Szerb, J.C. and Somogyi, G.I. (1973): Nature 241:121-122.
7. Szerb, J.C., Hadhazy, P. and Dudar, J.D. (1977): Brain Res. 128:285-292.

RELATIONSHIP BETWEEN CALCIUM ENTRY AND ACh RELEASE IN

K^+-STIMULATED RAT BRAIN SYNAPTOSOMES

J.B. Suszkiw, M.E. O'Leary and G.P. Toth

Department of Physiology
University of Cincinnati Medical Center
Cincinnati, Ohio 45267

INTRODUCTION

The release of acetylcholine (ACh) and other neurotransmitters by potassium depolarization of isolated nerve endings in vitro-(synaptosomes) exhibits a rapid temporal decay (1, 3, 6). This could be due either to the depletion of the releasable transmitter pool (4), inactivation of Ca^{++} entry (7), or a combination of both factors.

In this report we summarize our recent work (10, 11) which examined the pattern of Ca^{++}-dependent ACh release in relation to the kinetics of Ca^{++} entry, and its inactivation in rat brain synaptosomes exposed to 50 mM K_o^+ for short and prolonged durations.

METHODS

Rat cerebrocortical synaptosomes were prepared according to Gray and Whittaker (2). Intrasynaptosomal ACh was radiolabeled from [^{3}H]Choline (1 μM, 30 min incubation) in the presence of 20 μM Paraoxon (diethyl-p-nitrophenyl phosphate) to inhibit the acetylcholinesterase activity. The release of [^{3}H]ACh was studied in superfused synaptosomal beds formed on glass microfiber filters (Whatman, GFF), and by rapid filtration through 0.65 μ Millipore filters under vacuum. Released [^{3}H]ACh in the superfusates or filtrates was assayed as described previously (10, 11). The synaptosomal uptake of [^{45}Ca]$^{++}$ was measured by rapidly quenching the reaction mixture with 10 mM EGTA, passing the synaptosome suspension through 0.65 μ filters, and counting the radioactivity retained with synaptosomes on the filters (11). Inactivation kinetics of K^+-stimulated Ca^{++}-entry (channels) was followed by predepolarizing synaptosomes

in Ca^{++} deficient high K^+ solutions for different durations and
subsequently measuring the influx of $[^{45}Ca]^{++}$ during 1 sec or 10 sec
pulses of 1.2 mM $[^{45}Ca]Cl_2$ injected into the predepolarization medium.
All experiments were performed at 30°C. Tris- (TKR) or Hepes- (HKR)
buffered (pH 7.4), oxygenated Krebs-Ringer solutions were used.
High K^+ (50 mM) solutions were prepared by equimolar replacement
of NaCl with KCl. Due to variations in experimental protocols final
concentrations of high K^+ varied from 47.5 mM in some experiments
to 52.5 mM in others. All data shown represent the "specific" K^+-
elicited effects corrected for background in 5 mM K^+ solutions.

RESULTS

Figure 1 illustrates that intermittent stimulation of superfused
synaptosomal beds by 3-min pulses of 50 mM K^+ (K_1-K_5) evoked decre-
mental output of $[^3H]ACh$ which reached nearly undetectable levels
after the fifth (K_5) stimulus. The omission of Ca^{++} prevented,
and reintroduction of 1.2 mM Ca^{++} restored the K^+-evoked release,
indicating that the measured K^+-stimulated release was a Ca^{++}-dependent
process. The cumulative release, expressed as a percentage of total
$[^3H]ACh$ initially present in synaptosomes, was 27.1 ± 3.6% (mean ±
SD, n = 4) (inset). Since the synthesis of new ACh was prevented
in these experiments by omitting choline from the superfusion medium,
the cumulative percentile release provided an estimate of the relative
size (fraction) of the releasable $[^3H]ACh$ pool in synaptosomes.

Figure 2 shows that $[^3H]ACh$ output from synaptosomal beds stimulated continuously by 52.5 mM K^+ consisted of an early "phasic"
release which was followed by a slower "tonic" release. The absence
of the tonic component of release from synaptosomes stimulated for
3 minutes by high K^+ and then superfused by nondepolarizing (5 mM
K^+) HKR, indicates that the sustained efflux was not an artifact
of the wash-out kinetics. The cumulative Ca^{++}-dependent release
from synaptosomes in suspension, expressed as a percentage of the
initial $[^3H]ACh$ content, was 12.1 ± 1.8% (44% of the releasable
pool) for the phasic component (3-min output), and 1.69% for the
combined phasic and tonic components (30 min output) (Inset, means
± SD, n = 3). Consequently, the decay of release cannot be ascribed
to the depletion of the releasable pool.

Figure 3-A shows the time course of the phasic $[^3H]ACh$ release
from synaptosomes in suspension, stimulated by 47.5 mM K_o^+. At the
indicated times synaptosomes were passed through 0.65 μ filters,
and $[^3H]ACh$ in the filtrates was expressed as a percentage of total
$[^3H]ACh$ content in synaptosomes prior to K^+-stimulation. The cumulative $[^3H]ACh$ output reached a plateau (12.1 ± 1.8%, mean ± SD, n
= 3) within 60 sec of applied stimulus. The approach to plateau
could be fitted by a single exponential (inset) from which the rate
coefficient (0.039 sec^{-1}) and $t_{1/2}$ (≈ 17.8 sec) for the cessation
of transmitter release were estimated. In similar experiments the

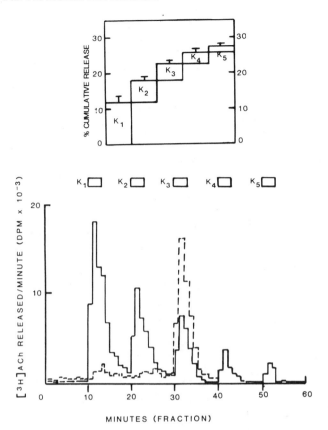

Figure 1. K⁺-evoked release of [³H]ACh from superfused synaptosomal
beds. **Solid trace:** stimulation by a 3-min pulse of 52.5 mM K⁺-HKR.
Dashed trace: continuous 30-min exposure to 52.5 mM. Synaptosomes
on the filters were prewashed for 30 min with Hepes-buffered Krebs-
Ringer solution (HKR). Rate of superfusion was approx. 1 ml/min;
1 ml fractions were collected. **Inset:** Cumulative release of [³H]ACh
from synaptosomes in suspension during 3-min and 30 min K⁺ depolarization. Results are expressed as a percentage of initial [³H]ACh in
synaptosomes prior to stimulation (means ± SD, n = 3).

specific K⁺-activated influx of [⁴⁵Ca]⁺⁺ into synaptosomes rapidly
approached a plateau (Figure 3-B). The estimated rate constant
for this process was 0.12 sec⁻¹ ($t_{1/2} \simeq 5.7$ sec). In summary, the
results illustrated in Figure 3 suggest that the phasic [³H]ACh
release is a consequence of rapid cessation of Ca⁺⁺ influx into
synaptosomes.

In contrast to calcium influx evoked physiologically by presynaptic action potentials, the apparent phasic Ca⁺⁺ entry in K⁺

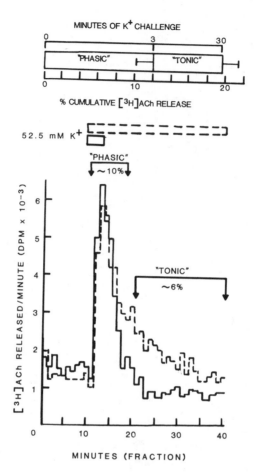

Figure 2. K⁺-evoked release of [³H]ACh from synaptosomal beds stimulated intermittently by 3-min pulses of 50 mM K⁺-TKR (Tris-buffered
Krebs-Ringer) (K_1-K_5). Solid trace: normal (1.2 mM Ca⁺⁺). Dashed
trace: Ca⁺⁺ was omitted during the first 30 min and then added during
the third (K_3) period of stimulation. Inset: Percentile cumulative
Ca⁺⁺-dependent release of labeled ACh (means ± SD, n = 4).

stimulated synaptosomes does not appear to result from repolarization
of synaptosomes back to the resting membrane potential. This is illustrated in Figure 4 which shows that synaptosomes remain depolarized
for at least 5 minutes in high K⁺ (52.5 mM), as determined with
the potential-sensitive dye, 3,3'-dihexyloxacarbocyanine (9). Rather,
the time course of [⁴⁵Ca]⁺⁺ influx in synaptosomes is a function
of the inactivation kinetics of the voltage-dependent Ca⁺⁺ channels.

Figure 5 shows that inactivation kinetics can be resolved into
fast, slow and very slow (essentially "non-inactivating") components.

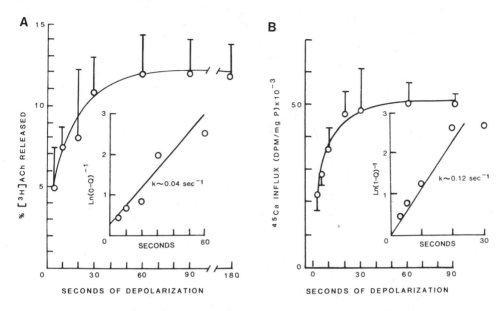

Figure 3. Initial time course of Ca++-dependent [3H]ACh release (A) and [45Ca]++ entry (B) evoked in synaptosomes by 47.5 mM K+-TKR. A: Results (mean ± SD of three experiments) are expressed as a percentage of the initial [3H]ACh n synaptosomes. Inset: semilogarithmic plot of fractional (Q) [3H]ACh release relative to the plateau value at 90 sec. B: Results (mean ± SD, n = 4) represent specific, K+-stimulated [45Ca]++ entry obtained by subtracting [45Ca]++ binding to synaptosomes in 5 mM K+-TKR from total entry at 47.5 mM K+-TKR. Inset: semilogarithmic plot of fractional (Q) [45Ca]++ influx relative to the plateau value at 60 sec.

The experimental points (means ± SD, n = 4) were obtained by first predepolarizing synaptosomes for the indicated durations in Ca++-free 50 mM K+, and then measuring [45Ca]++ influx during a 1 sec pulse of 1.2 mM [45Ca]Cl$_2$ injected into the reaction medium. The influx of [45Ca]++ after different times of predepolarization was expressed as a percentage of [45Ca]++ entry into synaptosomes at zero time of predepolarization. The solid curve fitted to the experimental points in Fig. 5-A is a sum of two exponential components (dashed traces). The fast component (t$_{1/2}$ ≈ 2.4 sec) of inactivation is nearly over by 10 sec of predepolarization. To resolve the slow component, the duration of [45Ca]++ pulse was lengthened to 10 sec and the decline in [45Ca]++ influx with time of predepolarization was expressed relative to the level of [45Ca]++ influx into synaptosomes after 10 sec predepolarization when the fast inactivation process was nearly completed (Fig. 5-B). The solid curve fitted to the experimental points obtained in this way is a sum of the slow (t$_{1/2}$ ≈ 27 sec) and noninactivating (t >> 5 min) components. Noting that "100%"

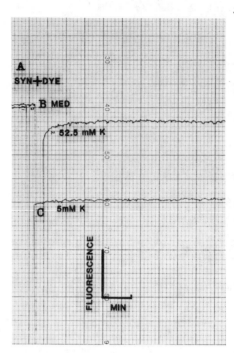

Figure 4. Fluorescence response of 3,3'-dihexyloxacarbocyanine in presence of synaptosomes exposed to 5 mM K^+ and 52.5 mM K^+, Ca^{++}-deficient HKR. Reaction volume was 2 ml and contained 0.5 mg protein and 2 μM dye. Temperature: $30°C$; Excitation: 470 nm; Emission: 500 nm.

in Figure 5-B corresponds to about 20% on the ordinate in Figure 2-A, the dashed traces provide an estimate of the fraction of the fast (0.8), slow (0.1), and noninactivating (0.1) calcium entry pathways at "zero time," and their relative contributions to Ca^{++} entry during longer times of K^+-depolarization.

DISCUSSION

These experiments show that during sustained challenge of synaptosomes by 50 mM K^+ the Ca^{++}-dependent release of preformed [3H]ACh consisted of initial fast "phasic" release which was then followed by a much reduced and slower "tonic" release. By omitting choline from the medium and thus preventing the synthesis of new ACh we estimated that no more than 44% of the releasable [3H]ACh pool was released during the phasic output. Consequently, the rapid decay of the initial [3H]ACh output cannot be ascribed to the depletion of the releasable pool, as has been suggested by some investigators (4).

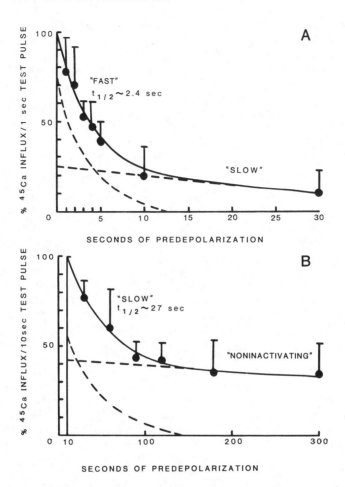

Figure 5. Time course of inactivation of K^+-stimulated $[^{45}Ca]^{++}$ entry in synaptosomes predepolarized in Ca^{++}-deficient 50 mM K^+ solutions. A: Initial time course of inactivation was determined by measuring $[^{45}Ca]^{++}$ influx for 1 sec after predepolarization. Data (means $\pm$ SD, n = 4) are expressed as a percentage of $[^{45}Ca]^{++}$ influx/sec in synaptosomes after zero time of predepolarization. B: Late time course of inactivation was determined by measuring $[^{45}Ca]^{++}$ influx for 10 sec after predepolarization. Data (means $\pm$ SD, n = 4) are expressed as a percentage of $[^{45}Ca]^{++}$ influx/10 sec in synaptosomes after 10 sec predepolarization. Note that 100% on the ordinate in this figure corresponds to 20% on the ordinate in the top figure (A). The best-fit curves (solid) describing the experimental points were obtained by resolving the inactivation process into two first-order "fast" and "slow" components and a linear "noninactivating" component (dashed traces).

In agreement with similar results reported for the release of [^{3}H]dopamine from striatal synaptosomes (1), the initial time course of [^{3}H]ACh release from cerebrocortical synaptosomes employed in this work, appears to be determined by the inactivation kinetics of the K$^+$-activated "fast Ca^{++} channels." We have estimated that these channels may mediate as much as 80% of K$^+$ stimulated Ca^{++} entry in synaptosomes. Their rapid inactivation ($t_{1/2} \simeq 2.4$ sec) effectively terminates the major phase of Ca^{++} influx in synaptosomes within 10 sec of K$^+$-stimulus, and thus accounts for the rapid decay of the Ca^{++}-dependent transmitter output.

Although the latter process occurs at a slower rate ($t_{1/2} \simeq 17.8$ sec) than the inactivation of Ca^{++} channels, this apparent discrepancy does not invalidate the above conclusions. Firstly, the kinetics of [^{45}Ca]$^{++}$ entry and its inactivation in synaptosomes reflect but an average of rates of these processes in many different types of nerve terminals present in synaptosome preparations. Secondly, it is likely that unphysiologically massive influx of Ca^{++} during K$^+$ stimulation overloads the intrasynaptosomal Ca^{++} sequestration and extrusion mechanisms. This would cause a delay in lowering of free Ca^{++} in synaptosomes and, consequently, a delay in transmitter release shut-off. Finally it is also possible that Ca^{++} entry through the small fraction of the "slow Ca^{++} channels" [$t_{1/2}$ (inactivation) $\simeq 27$ sec] contributes to the prolongation of [^{3}H]ACh output under the experimental conditions employed. It can be calculated that the "tonic" output of [^{3}H]ACh during sustained K$^+$ challenge occurred at about 1/40th of the rate of the initial phasic release. This Ca^{++}-dependent release may be a consequence of a sustained elevation of free intrasynaptosomal Ca^{++}, due to its continued entry via the noninactivating pathway(s).

It is interesting to speculate if, and how, the pattern of [^{3}H]ACh release described here may relate to transmitter release during physiological synaptic activation. In situ, the per impulse phasic release of transmitters follows the opening and closure of the voltage-sensitive Ca^{++} channels during a cycle of membrane depolarization and repolarization when action potential invades the terminal. In contrast, the phasic release from synaptosomes is a consequence of inactivation of Ca^{++} channels during unphysiologically long depolarization; this provides for a massive entry of Ca^{++} and, consequently, greatly magnified output of the transmitter. If the phasic release from synaptosomes resembles the stimulus-dependent synchronous release of transmitters in situ, then the tonic output from synaptosomes may be thought of as reflecting an increased rate of the asynchronous i.e., spontaneous quantal release.

A collorary of these, admittedly speculative postulates, would be that the dynamics of ACh in synaptosomes, measured during very short and prolonged periods of K$^+$-stimulation, may reflect different physiological states of nerve terminals. Particularly during prolonged

K$^+$ incubations, the synthesis and release of ACh can be predicted to reach a steady-state not too different from that in preparations "at rest." In this situation the measured transmitter "turnover" may reflect for the most part the synthesis and leakage of the cyto-solic ACh. In agreement with this supposition is the observation that during prolonged depolarization of brain slices in high potassium, the specific activity of ACh synthesized from [^{3}H]Choline was higher in cytoplasmic than in synaptic vesicle fractions (8).

Acknowledgement. This work was supported by NIH grants NS 17442 and NS 17968.

REFERENCES

1. Drapeau, P. and Blaustein, M.P. (1983): J. Neurosci. 3:703-713.
2. Gray, E.G. and Whittaker, V.P (1962): J. Anat. 96:79-88.
3. Haycock, J.W., Levy, W.B., Denner, L.A. and Cotman, C.W. (1978): J. Neurochem. 30:1113-1125.
4. Morel, N., Israël, M. Manaranche, R. and Lesbats, B. (1979): Prog. Brain Res. 49:191-202.
5. Mulder, A.H., Vandenberg, W.B. and Stoof, J.C. (1975): Brain Res. 99:419-424.
6. Murrin, L.C., DeHaven, R.N. and Kuhar, M.J. (1977): J. Neuro-chem. 29:681-687.
7. Nachshen, D.A. and Blaustein, M.P. (1980): J. Gen. Physiol. 76:709-728.
8. Richter, J.A. and Marchbanks, R.M. (1971): J. Neurochem. 18:705-712.
9. Sims, P.J., Waggoner, A.S., Wang, C.H. and Hoffman, J. (1974): Biochem. 13:3315-3330.
10. Suszkiw, J.B. and O'Leary, M.E. (1982): J. Neurochem. 38:1668-1675.
11. Suszkiw, J.B. and O'Leary, M.E. (1983): J. Neurochem. 41:868-873.

ACTIVATION AND INHIBITION OF THE NICOTINIC RECEPTOR:

ACTIONS OF PHYSOSTIGMINE, PYRIDOSTIGMINE AND MEPROADIFEN

E.X. Albuquerque[*], C.N. Allen, Y. Aracava,
A. Akaike, K.P.Shaw and D.L. Rickett[2]

Department of Pharmacology and Experimental Therapeutics
University of Maryland School of Medicine
Baltimore, Maryland 21201 U.S.A.

[2] Medical Chemical Defense Research Program
U.S. Army Medical Research and Development Command
Fort Detrick, Maryland 21701 U.S.A.

INTRODUCTION

The acetylcholine receptor-ionic channnel complex (AChR) is a membrane bound glycoprotein with a molecular weight of about 250,000 daltons (23, 26) that links chemical recognition to the opening of cation channels. The AChR extends from about 50 Å on the extracellular side of the membrane and has a cytoplasmic tail of 15 Å (19, 27). The clusters of these receptors at the muscle endplate assure transmission of the signal from the nerve terminal to the postsynaptic region on the muscle fiber such that the fiber ultimately contracts in response to nerve stimulation. A large number of studies on biochemical aspects of the nicotinic AChR have been undertaken as well as on primary structural, electronmicroscopic, neutron diffraction, and other analyses (see reviews 3, 4, 18 and 31).

The nerve terminal releases 6,000 - 10,000 molecules of the neurotransmitter acetylcholine (ACh) per quantum which diffuse through a synaptic gap of 400 - 600 Å to interact with high-density patches of AChR. Upon colliding with these AChR, brief electrical transients appear across the membrane. Binding of ACh to a recognition site of the AChR causes the ionic channel to open, thereby allowing cationic currents (carried chiefly by sodium under normal conditions) to

---

[*]To whom reprint requests should be sent.

flow down their respective electrochemical potential gradients. The channel spontaneously closes after a few milliseconds and is ready to be reactivated.

Interaction of ACh with its receptor site as well as the kinetics of ionic channel activation may be altered by pharmacological agents (3, 6, 18, 31). A compound may specifically interact with the ACh receptor site by either activating and opening the ion channel (agonist) or by preventing the binding of an agonist molecule (competitive antagonist). The flow of current through the channel can be blocked by an agent which interacts with sites at the ionic channel, either prior to its opening, or once it is in the conducting conformation (non-competitive antagonist). In addition, certain classes of drugs can modify the affinity of ACh for its binding site, affecting the time course of receptor desensitization and the functional properties of the AChR.

Recording techniques such as voltage clamping and noise analysis have traditionally been used to study the kinetic properties of the AChR. These techniques, however, either record macroscopic currents (miniature endplate and endplate currents) which represent the summation of a number of single channel currents, or provide indirect information about the properties of microscopic events. Recording of single channel currents became possible with the development of the patch clamp technique (17, 24). Patch clamp recordings are successful on securely immobilized cells free of connective tissue (Fig. 1). For these reasons, cells grown in culture have been used in the majority of patch clamp studies.

The objective of this study is to identify interactions of physostigmine (PHY), pyridostigmine (PYR) and meproadifen (MEP) with sites on the AChR complex in both adult frog muscle fibers and cultured neonatal rat muscles (myoballs). As discussed, these agents may alter ACh channel activation by interacting either with the agonist recognition site or with ion channel sites of the nicotinic macromolecule.

METHODS AND TECHNIQUES

Single muscle fibers were obtained by enzymatic dispersion of frog toe interosseal or lumbricales muscles. Once isolated, fibers were secured in the recording chamber with a parafilm-paraffin oil adhesive; details of this method are published elsewhere (8). The procedure for myoball culturing was adapted from that reported by Giller et al. (15), and described in detail by Akaike et al. (2). Cells were cultured from hind limb muscles of 1- to 2-day old, neonatal rats and studies were performed with 1- to 2-week old culture. Single channel currents were recorded either from cell-attached or cell-free (inside-out) patches. In this procedure fire-polished micropipettes were filled with ACh solutions (0.05 to 0.4 μM) either

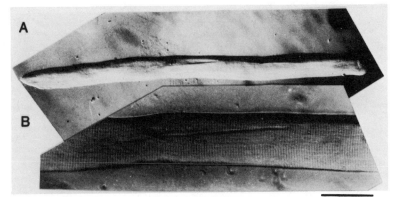

A:240μm
B: 95μm

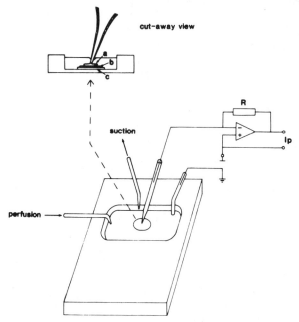

Figure 1. Microphotographs of a single fiber isolated from the interosseal and lumbricales muscles of adult frog. A: Low-power view of a single muscle fiber following enzyme treatment. Note that the endplate region appears as an elongated concavity in the center of the fiber. B: High-power view of the same fiber where the bands are discernible. Diagram of the recording chamber in which the single muscle fibers are immobilized is shown at the bottom of the figure. a indicates the single muscle fibers; b the para-film-paraffin oil adhesive; and c a glass coverslip.

alone or together with different concentrations of drugs under study. The microelectrode was gently pressed against the cell surface and, as a slight suction was applied through the microelectrode, a mechanically stable high resistance (giga-ohm) seal between the cell membrane and the recording pipette was achieved. An LM-EPC-5-patch clamp system (List-Electronics, W. Germany) was used to measure single channel currents which were low-pass filtered to 3 KHz (second order Bessel) and the data were stored on FM magnetic tape for computer-assisted analyses. Automated analysis of patch-clamp data, developed in our laboratory and published in detail elsewhere (2), were performed on a PDP 11/40 (Digital Equipment Corp.) minicomputer. These analyses provided histograms of the distributions of single channel current amplitudes and single channel open times. All experiments were carried out at a temperature of 10°C.

Drugs: ACh chloride (Sigma Chemical Company) solutions were prepared from the crystalline salt for every experiment. Pyridostigmine (PYR) bromide (Hoffman-La Roche), physostigmine (PHY) sulfate (Sigma Chemical Company), PHY methiodide (provided by Dr. John Daly, N.I.H.) and tetrodotoxin (Sigma Chemical Company) were diluted daily from fresh stock solutions. Meproadifen [MEP; 2-(diethylmethyl-aminoethyl)-2,2-diphenylvalerate iodide)] was kindly provided by Dr. J. Cohen (Department of Anatomy, Washington University, St. Louis, MO). A fresh solution was prepared for every experiment performed.

RESULTS

Effects of Acetylcholine on Perijunctional Receptors of Single Adult Muscle Fibers. Single channel currents activated by ACh (0.3-0.4 μM) were recorded from the perijunctional region of frog interosseal or lumbricales muscle fibers. The majority of patches (14 to 16) had single channel openings with a conductance of 32 pS (Fig. 2A), which was similar to previously reported values (14). The current flowing through the open channels was linearly dependent on the membrane potential (Fig. 2B), and the reversal potential was estimated from the current-voltage (I-V) plot to be +12 mV. In a small number (2 of 16) of the patches, channel openings with two different conductances, 32 pS and 20 pS, were observed; the events with the high conductance were the most prevalent. In contrast, in rat myoballs a 20 pS conductance channel opening was predominant while channel openings with 33 pS conductance usually made up only 10% of the total recorded events (2, 12). In addition, low frequency, low-conductance channels (10 pS) were seen, which may represent a sub-conductance state (16), although they have been observed as an independent event (12). These differences in the predominancy of a particular conductance state in frog muscle fibers and rat myoballs may be related to species differences or to differences between adult and embryonic tissues.

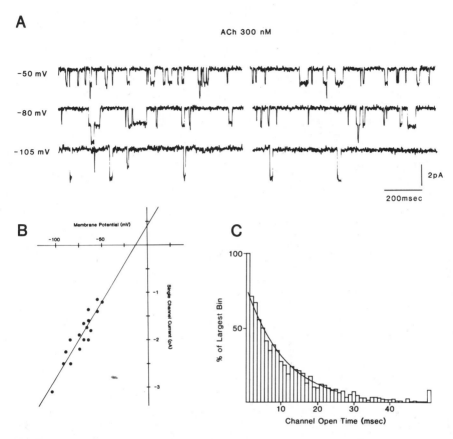

Figure 2. Characteristics of the ACh-activated ion channel currents recorded at the perijunctional region of the interosseal and lumbricales muscle fiber of adult frog. A: Typical records of single channel currents evoked by ACh (0.3 μM) in the recording microelectrode at temperature of 10°C. The downward deflection indicates an inwardly flowing current. A voltage-dependent increase in the amplitude of the single channel currents was observed. Bandwidth: 3 kHz. B: Relationship between the single channel amplitude and the transmembrane potential. The reversal potential is about +12mV. C: Histogram of the distribution of channel lifetimes gathered from a single patch recording at -50mV holding potential. The distribution of channel open times is fit by a single exponential function with a mean of 9.6 msec.

A histogram of the distribution channel open times of the 32 pS events was adequately fit by a single exponential function in 12

of 16 recordings (Fig. 2C). This indicates the existence of a single
open state for these channels. The open time histograms of the
remaining four patches were fit by a double exponential function
due to the presence of a fast component with a mean lifetime of
4.8 msec. The mean channel lifetime was exponentially related to
the membrane potential. For example, at membrane potentials of
-40 mV, -70 mV and -100 mV the respective mean channel lifetimes
were 9.4 $\pm$ 1.0 msec (mean $\pm$ S.E., n = 7), 15.0 $\pm$ 1.7 msec (n = 8),
and 19.8 $\pm$ 3.0 msec (n = 3), respectively. A similar voltage dependence of channel lifetime was previously reported from ACh-induced
endplate current fluctuation analysis (1, 7, 9, 22).

**Effects of Anticholinesterase Agents on the Acetylcholine Receptor
Ion Channel Complex.** The physiological actions of PHY, a tertiary
amine, and PYR, a quaternary compound, are generally attributed to
their inhibition of acetylcholinesterase (AChE). However, voltage-
clamp, noise analysis and patch-clamp experiments demonstrate that
these agents interact directly with the AChR complex (2, 5, 25,
29). PHY decreases the conductance of ACh-activated channels of
the isolated muscle fibers in a dose dependent manner (Fig. 3A).
For example, PHY at concentrations of 100 μM and 200 μM reduced the
conductance of ACh-activated channels from 32 pS to 22.7 pS and
18.6 pS, respectively.

PHY (0.1-100 μM) shortened ACh channel life time and induced
rapid transitions (flickers) between open and closed states. The
flickerings, characterized by many short-lasting movements of the
current level within the channel opening, may be due to an increase
in the rate constant for drug dissociation and/or to a conformational
change of the AChR which results in fast transitions between the
non-conducting and the open species of the nicotinic macromolecule.
At 100 μM PHY shortened the channel open time from 11.1 msec to
4.1 msec at -100 mV. The shortened lifetimes were more pronounced
at hyperpolarized membrane potentials. Similar to control conditions,
the distribution of channel open times in the presence of PHY was
fitted by a single exponential function (Fig. 3B). At PHY concentrations greater than 200 μM, the single channel currents had a more
prolonged burst time and an increased baseline noise during the open
state (Fig. 4).

PHY (0.5-50 μM), alone in the patch electrode, activated channels
with a conductance similar to ACh (Fig. 5). Pretreatment of the
muscle fibers with Naja toxin blocked PHY-induced activation of
the nicotinic AChR, suggesting an agonistic interaction of PHY at
the ACh recognition site. The amplitude of the current flowing
through PHY-activated channels was linearly dependent on transmembrane
potential. The single exponential distribution of the channel open
times suggested a single open state for the channels activated by
PHY.

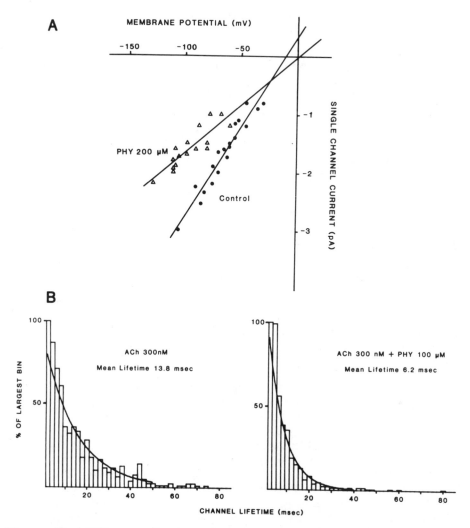

Figure 3. Effects of physostigmine (PHY) on the single channel amplitude and channel open time of the ACh-activated channels at the perijunctional region of the interosseal and lumbricales muscle fiber. A: Relationship between the amplitude of the single channel currents and the transmembrane potential. The patch microelectrode contained either ACh (0.3 μM) alone (control) or combined with PHY (200 μM). The slope conductance was decreased from 32 pS to 18.7 pS in the presence of PHY. B: Histograms of the distribution of channel open times recorded from two different patches with ACh (0.3 μM) or together with PHY (100 μM). Both patches were recorded at holding potential of -60 mV and at temperature of 10°C.

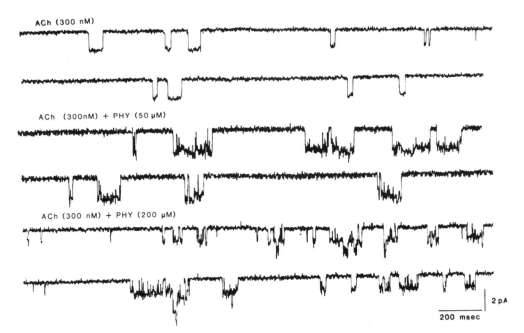

ACh (300 nM)

ACh (300nM) + PHY (50 µM)

ACh (300 nM) + PHY (200 µM)

2 pA

200 msec

Figure 4. The interaction of PHY with the ACh receptor-ion channel complex of the frog muscle fibers. Single channel currents were activated by ACh (0.3 µM) in the absence (control) and presence of PHY (50 and 200 µM). The transmembrane potential was -90 mV and the bath temperature 10°C. Note the increased amount of flickering and open channel noise in the presence of PHY.

PYR, although similar to PHY as an AChE inhibitor, has distinct effects on the AChR complex. In myoballs, PYR (50 µM) in combination with ACh (0.1 µM) inside the patch micropipette induced the appearance of channels with marked flickering (Fig. 6), without changing the mean channel open time (Fig. 7). The frequency of these channel openings increased as a function of time of exposure to both drugs. After 10 min, the frequency of these 21 pS flickering events was gradually decreased and a 10 pS event, which was rarely observed under control conditions (2, 16), became predominant. These low-conductance channel openings showed very little flickering (see Fig. 2 from ref. 2). Similar effects were observed in both cell-attached and inside-out patches when PYR was superfused into the bathing medium. The amplitude of ACh-induced single channel currents was decreased 30 minutes after exposure to 50 and 100 µM PYR to 67% and 48% of the control values, respectively (Fig. 8). A higher concentration of PYR (200 µM) produced a biphasic effect on channel activation. During the initial phase following drug application the number of channel openings increased and irregular waves of

PHY (50 µM)

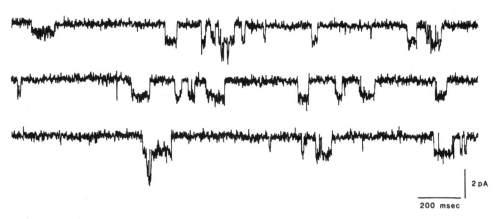

2 pA

200 msec

Figure 5. PHY-activated single channel currents recorded from frog skeletal muscle. Pipette contained PHY 50 µM. The membrane potential was -60 mV and the temperature 10°C.

bursting activity were seen. After cessation of bursting activity, the frequency of channel openings was markedly decreased. These phenomena may indicate the appearance of desensitized receptor-ion channel complexes (2, 28). PYR most likely induced an increase in affinity of the agonist for its receptor site, and altered the rate constant for the transition between the activated but non-conducting conformation and the open state (2, 25).

Similarly to PHY, PYR (100 µM) showed agonist property, activating the opening of low- conductance (10 pS) ionic channels (Fig. 9 and Table 1). The amplitude of PYR-induced channels was voltage dependent with a reversal potential of 0 mV. Pretreatment of the myoballs with α-bungarotoxin blocked activation of these channels. This suggests that PYR interacts with the agonist recognition site at the AChR complex. The activation of ionic channels with low con-ductance and with a lower frequency than those activated by ACh may be partly explained by ascribing weak agonist properties to PYR. However, the frequency of these low-conductance channel openings was higher in the presence of PYR combined with ACh, suggesting that PYR alters the kinetics of the ACh-activated channels by intera-cting with sites distinct from the agonist receptor. Binding studies performed in <u>Torpedo nobiliana</u> electric organs have also demonstrated that PYR interacts with both the ACh receptor sites and sites at the ionic channel of the nicotinic AChR (25).

Modifications of ACh Ion Channel Activation: Enhancement of Receptor Desensitization. Meproadifen (MEP), a quaternary local anesthetic,

ACh 0.1µM   ACh 0.1µM + PYR 50µM

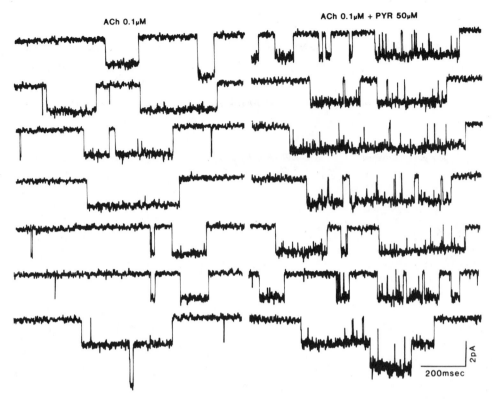

Figure 6. PYR-induced alterations of the single channel currents
activated by ACh on rat myoballs. Recordings of channel currents
activated by ACh (0.1 µM) alone (left traces) or in the presence of
PYR (50 µM) (right traces) at a membrane potential of -60 mV. Note
the increased noise during the open phase in the presence of PYR.

potentiates the rate of agonist induced receptor desensitization,
as demonstrated from voltage clamp and biochemical studies (20,
22). Patch clamp experiments using myoballs showed that MEP did
not change either the mean channel lifetime or the conductance of
ACh-activated channels. Although lacking any effect on properties
of single channel currents, MEP altered the frequency of ACh-activated
channel openings. MEP 0.02 to 0.1 µM, produced a biphasic effect
on opening frequency of ACh channels (10, 11). An initial high
frequency of ACh channel openings was followed by a decreased frequency. At concentrations greater than 0.2 µM the initial phase
was not seen and MEP only produced a concentration-dependent depression
of the frequency of channel openings (Fig. 10). At 5 µM no channel
activity was detected. An interesting feature of MEP action was

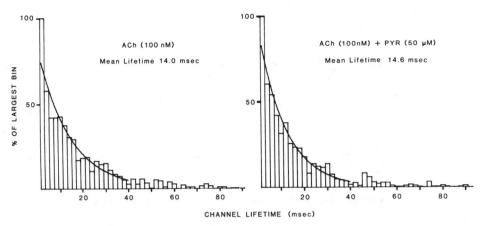

Figure 7.   Open time histograms of ACh-activated channels in the
absence (left) and presence of (50 μM) PYR (right).   Note that PYR
did not change the mean channel lifetime.

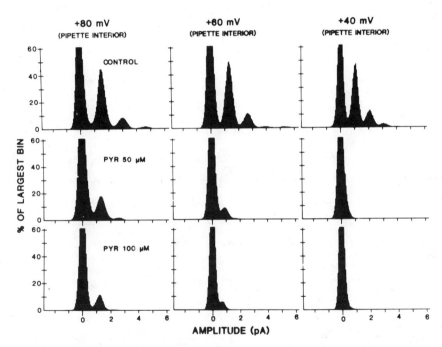

Figure 8.   Total amplitude histograms of channels activated by ACh
(0.1 μM) in the absence (upper figure) and presence of 50 μM (middle
figure) and 100 μM PYR (lower figure).

PYR(100 µM)

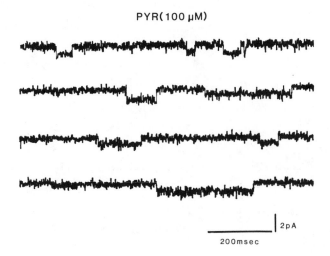

2pA

200msec

Figure 9.   Single channel currents activated by PYR (100 µM) in rat
myoballs. The transmembrane potential was -100 mV. The conductance
was 11.2 pS at temperature of 10°C.

Table 1

Amplitude and Lifetime of Single-Channel Currents Induced by PYR.*

	Cell-attached				Inside-out	
Holding Potential -100Mv	-60mV	-80mV	-100mV	-120mV	-60mV	-80mV
Amplitude (pA)	1.02	1.24	1.48	1.69	0.69	0.94 1.14
Mean lifetime	16.8	26.5	27.5	31.0	9.9	15.4 18.9

*The micropipette solution contained PYR (50-100 µM) without ACh.
Data were obtained from five cell-attached patches and two inside-out
patches.   Each value (channel amplitude and lifetime) is the mean
of 20-30 events per patch. These data were analyzed by hand because
of the low frequency of single-channel events (see ref. 2).

the appearance of events with higher conductance and increased current fluctuations during the open phase at low concentrations of this agent (Fig. 11). This broadening of the baseline during the open state which is similar to that induced by PYR (Fig. 6) may be due to inadequately resolved short closures which induced skewing to the left of the amplitude histogram (see Fig. 5 from ref. 2; 11).

This phenomenon, the cause of which is not clear at the present moment, may be related to perturbations of the nicotinic macromolecule while the channel is open. The lack of effect on channel conductance or mean channel lifetime indicated that MEP interacts with a site at the AChR prior to its opening, suggesting that the drug may be blocking the channel in resting state and also inducing desensitization. The selective application of MEP, either to the exterior of the cells or from the cytoplasmic side of the membrane, allowed an assessment of the location of MEP binding site. The effect on the frequency of ACh channel activation, when MEP together with

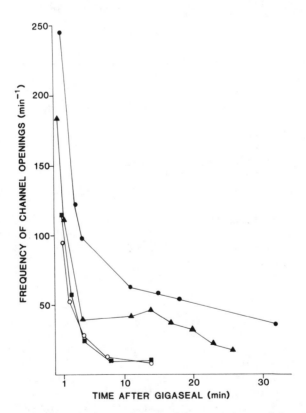

Figure 10. Frequency of channel openings activated by ACh (0.3 μM) in the presence of meproadifen 0.2 (●), 0.5 (▲), 1.0 (■) and 2.5 μM (o) in rat myoballs.

ACh 0.1µM + MEPROADIFEN 0.05µM

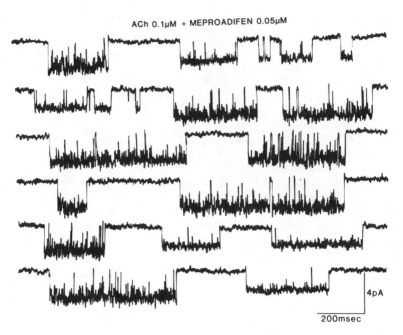

4pA

200msec

Figure 11. Samples of single channel currents activated by ACh (0.1 µM) in the presence of meproadifen (0.05 µM). Note, in addition to the unchanged channel currents, a large number of events with enlarged baseline coupled with large number of flickers during the opening phase.

ACh was in the patch electrode, demonstrated that a MEP binding site is located at the external portion of the AChR. The absence of any change on the activation of ACh channels, when MEP was applied into the bathing medium under inside-out patch clamp conditions, indicated that there is no binding site for this drug on the internal portions of the AChR macromolecule. Superfusion of the recording chamber with this drug after formation of a giga-ohm seal produced no change on ACh channel activation. This suggests that MEP, in contrast to other quaternary agents such as PYR, has no access to its binding site through the cell membrane.

The data presented here have shown that the binding of ACh to the nicotinic receptor and the subsequent activation of its associated ion channel is altered by a variety of pharmacological agents. Certain classes of drugs, such as MEP, increase the affinity of ACh for the receptor. This action may play a role in receptor desensitization. Anticholinesterasic agents, in addition to their inhibition of AChE, interact directly and differentially with sites at the AChR and exhibit agonistic effects, increased agonist affinity, enhancement of the receptor desensitization, and blockade of the open conformation of ACh-activated channels.

DISCUSSION AND SUMMARY

Several important features of drug interactions with the nicotinic AChR are disclosed in this review. First, several factors could contribute to the appearance of the diversified conductance states described in this chapter. These include: rearrangement of the receptor subunits involving changes in composition and organization of the lipid membrane; different conductance states of the same channel; and also different kinetics of channel activation, which may produce a complex distribution of channel open times. In myoballs ACh (0.02 to 0.3 μM) the activated channels with three different conductances. The events with the intermediate conductance (20 pS, at 10°C) were the most common (12). The interosseal and lumbricales muscles of the adult frog has ACh receptor ion channels with two distinct conductances. The low-conductance events of 20 pS conductance were rare while the predominant channel-openings disclosed a conductance value of 32 pS (8). This value is similar to that obtained for ACh channels from noise analysis by Anderson and Stevens (9), by Neher and Sakmann (24) for suberyldicholine (28 pS) and more recently by Gardner et al. (14) for ACh and other agonists (30 pS).

Table 2

Nicotinic Receptor-Ion Channel Complex

RECEPTOR		
Agonists	Partial Agonists	Antagonists
Acetylcholine	Decamethonium	d-Tubocurarine
Carbamylcholine	Nereistoxin	α-Bungarotoxin
Anatoxin-a	Physostigmine	Quinuclidinyl benzilate
Di-isoanatoxin		Tetraethylammonium
Arecoline		Piperocaine
Arecoline Methiodide		Quinidine
Tetramethylammonium		Pumiliotoxin
Cytisine		
Muscarone		
Suberyldicholine		
Pyridostigmine		

Table 3

Ionic Channel Inhibitors

Nicotinic Receptor-Ion Channel Complex Conformations:

Open or Conducting	Resting or Closed	Intermediate-Nonconducting
HTX	HTX	HTX
H_{12}-HTX	H_{12}-HTX	H_{12}-HTX
N-Benzylazaspiro-HTX	Azaspiro-HTX	Azaspiro-HTX
Depentyl-H_{12}-HTX	Depentyl-H_{12}-HTX	
PCP	PCP	PCP
PCP-Methiodide	m-Amino-PCP	m-Amino-PCP
m-Nitro-PCP	m-Nitro-PCP	m-Nitro-PCP
PCC	PCC	PCC
Naltrexone	Naltrexone	
Naloxone	Naloxone	
Levallorphan		
Gephyrotoxin	Gephyrotoxin	Gephyrotoxin
Amantadine & Analogs	Amantadine & Analogs	
Quinidine	Quinidine	
d-Tubocurarine	d-Tubocurarine	
Tetraethylammonium	Tetraethylammonium	
Physostigmine	Physostigmine	Physostigmine
	Pyridostigmine	Pyridostigmine
Decamethonium	Decamethonium	
Piperocaine	Piperocaine	
Bupivacaine	Imipramine	Imipramine
Atropine	Desimipramine	Desimipramine
Scopolamine	Nortriptyline	Nortriptyline
Quinuclidinyl benzilate	Amitriptyline	Amitriptyline
	Meproadifen	Meproadifen

HTX, histrionicotoxin; H_{12}-HTX, perhydrohistrionicotoxin; PCP, phencyclidine; PCC, 1-piperidinocyclohexanecarbonitrile; PCE, N-ethyl-1-phenylcyclohexylamine.

Whatever states of channel conductance are present during ACh action, PYR only activated a low-conductance population of channels (see Table 1). Furthermore, since PYR and ACh together activated these low-conductance openings at a frequency higher than that induced by PYR or ACh alone, it is likely that, in addition to its weak agonistic effect, PYR allosterically modifies the nicotinic receptor-ion channel complex, inducing alteration of the conductance properties of the channels activated by ACh (see Figs. 2 and 9 in ref. 2).

Second, the anticholinesterase agents, PYR and PHY, have differential effects on the AChR. In addition to its marked blockade of the open conformation of the AChR channels, PHY similarly to PYR produces high affinity binding and enhancement of desensitization (2, 5, 25, 30). PHY, similarly to PYR (2), interacts with the AChR at its extracellular surface; neither drug interacts with the AChR at intracellular sites (Aracava et al., unpublished results).

In contrast to the generation of low-conductance channels and weak agonist properties of PYR, PHY at low concentrations induces activation of channels with a conductance apparently similar to that seen with ACh. It is important to recognize that both drugs produce similar depression of endplate decay currents, yet through distinctly different actions; for PYR this is mediated through high affinity binding of ACh and enhancement of receptor desensitization while for PHY, in addition to an effect like that of PYR, a marked blockade of the channel in open conformation accounts for shortening of channel open time and acceleration of the decay phase of endplate currents.

The third aspect disclosed in the present chapter is that the quaternary agent MEP produces a more specific effect on the AChR. MEP does not alter channel lifetime or conductance but it does induce a high affinity binding and desensitization. In combination with ACh, MEP's actions at concentrations as low as $0.02\ \mu M$ increase the frequency of single channel opening followed by depression (11).

Blockade of the neuromuscular junction can thus occur in at least four ways: 1) blockade of the channel in the open state; 2) receptor blockade; 3) drug-induced high affinity binding and desensitization; and 4) the combinations of these above phenomena. There are a number of agents which can be used as a tool to study these various aspects of neuromuscular transmission. The diverse properties of these and other agents are illustrated in Tables 2 and 3 of this chapter.

These tables are not intended to be a complete survey of all the drugs thus far sampled in the literature. The use of these drugs and toxins is a fundamental tool in establishing the basis of modern molecular pharmacology, and has enabled investigators to identify many different states of the activation, desensitization and blockade of the ionic channel of the AChR. Most of the drugs in Table 3, particularly those listed in the second and third columns, increase the affinity of the agonist to its binding site and induce desensitization.

Acknowledgement. This research was supported by U.S. Army Medical Research and Development Command Contract DAMD 17-81-1279.

REFERENCES

1. Adler, M., Oliveira, A.C., Albuquerque, E.X., Mansour, N.A. and Eldefrawi, A.T. (1979): J. Gen. Physiol. 74:129-152.
2. Akaike, A., Ikeda, S.R., Brookes, N., Pascuzzo, G.J., Rickett, D.L. and Albuquerque, E.X. (1984): Mol. Pharmacol.25:102-112.
3. Albuquerque, E.X. and Spivak, C.E. (1984): In Natural Products and Drug Development, Alfred Benzon Symposium (eds) P. Krogsgaard-Larsen, S.B. Christensen and H. Kofod, Munksgaard, Copenhagen, pp. 301-321.
4. Albuquerque, E.X., Adler, M., Spivak, C.E. and Aguayo, L. (1980): Ann. N.Y. Acad. Sci. 358:204-238.
5. Albuquerque, E.X., Akaike, A., Shaw, K.P. and Rickett, D.L. (1984): Fund. Appl. Toxicol. 4:S27-S33.
6. Albuquerque, E.X., Kuba, K. and Daly, J. (1974): J. Pharmacol. Exp. Ther. 189:513-524.
7. Albuquerque, E.X., Warnick, J.E., Aguayo, L.G., Ickowicz, R.K., Blaustein, M.P., Maayani, S. and Weinstein, H. (1983): In Phencyclidine and Related Arylcyclohexylamines: Present and Future Applications (eds) J.M. Kamenka, E.F. Domino and P. Geneste, pp. 579-597.
8. Allen, C.N., Akaike, A. and Albuquerque, E.X. (1984): J. Physiol (Paris) 79:338-343.
9. Anderson, C.R. and Stevens, C.F. (1973): J. Physiol. (London) 235:655-691.
10. Aracava, Y., Ikeda, S.R. and Albuquerque, E.X. (1983): Neurosci. Abs. 9:733.
11. Arcava, Y. and Albuquerque, E.X. (1984): FEBS Letters 174:267-274.
12. Arcava, Y., Ikeda, S.R., Daly, J.W., Brookes, N. and Albuquerque, E.X. (1984): Mol. Pharmacol. 26:304-313.
13. Fertuck, H.C. and Salpeter, M.M. (1974): Proc. Natl. Acad. Sci. USA 71:1376-1378.
14. Gardner, P., Ogden, D.C. and Colquhoun, D. (1984): Nature 309:160-162.
15. Giller, E.L., Jr., Neale, J.H., Bullock, P.N., Schrier, B.K. and Nelson, P.G. (1977): J. Cell Biol. 74:16-29.
16. Hamill, O.P. and Sakmann, B. (1981): Nature 294:462-464.
17. Hamill, O.P., Marty, A., Neher, E., Sakmann, B. and Sigworth, F.J. (1981): Pflugers Arch. 391:85-100.
18. Karlin, A. (1980): In The Cell Surface and Neuronal Function (eds) C.W. Cotman, G. Poste and G.L. Nicolson, Elsevier/North-Holland Biomedical Press, pp. 191-260.
19. Klymkowsky, M.W. and Stroud, R.M. (1979): J. Mol. Biol. 128:319-334.
20. Krodel, E.K., Beckman, R.A. and Cohen, J.B. (1979): Mol. Pharmacol. 15:294-312.
21. Kuffler, S.W. and Yoshikami, D. (1975): J. Physiol. (London) 251:465-482.
22. Maleque, M.A., Souccar, C., Cohen, J.B. and Albuquerque, E.X. (1982): Mol. Pharmacol. 22:636-647.

23. Martinez-Carrion, M., Sator, V. and Raftery, M.A. (1975): Bio-
      chem. Biophys. Res. Commun. 65:129-137.
24. Neher, E. and Sakmann, B. (1976): Nature 260:799-801.
25. Pascuzzo, G.J., Akaike, A., Maleque, M.A., Shaw, K.P., Aronstam,
      R.S., Rickett, D.L. and Albuquerque, E.X. (1984): Mol. Pharma-
      col. 25:92-101.
26. Reynolds, J.A. and Karlin, A. (1978): Biochemistry 17:2035-2038.
27. Ross, M.J., Klymkowsky, M.W., Agard, D.A. and Stroud, R.M. (1977):
      J. Mol. Biol. 116:635-659.
28. Sakmann, B., Patlak, J. and Neher, E. (1980): Nature 286:71-73.
29. Shaw, K.P., Akaike, A. and Albuquerque, E.X. (1983): Neuro-
      sci. Abs. 9:1138.
30. Sherby, S.M., Shaw, K.P., Albuquerque, E.X. and Eldefrawi, M.E.
      (1984): Fed. Proc. Abs. 43:342.
31. Spivak, C.E. and Albuquerque, E.X. (1982): In Progress in Choli-
      nergic Biology: Model Cholinergic Synapses (eds) I. Hanin and
      A.M. Goldberg, Raven Press, New York, pp. 323-357.

MOLECULAR RELATIONSHIP BETWEEN ACETYLCHOLINESTERASE

AND ACETYLCHOLINE RECEPTORS

P. Fossier, G. Baux and L. Tauc[*]

[*]Laboratoire de Neurobiologie Cellulaire
C.N.R.S.
91190 Gif sur Yvette, France

INTRODUCTION

Application of acetylcholinesterase inhibitors (AChEIs; organo-phosphates, carbamates or quaternary ammonium compounds) at the neuromuscular junction leads to an increase in the amplitude and duration of the end-plate potential (1, 16). This effect has been considered to be a consequence of acetylcholinesterase (AChE) inhi-bition and of the resulting increase in concentration of acetylcholine (ACh) in the synaptic cleft (11, 12, 16). At vertebrate ganglionic cholinergic synapses (3), however, the postsynaptic response is not modified when AChE is blocked, indicating that this enzyme might not play a major role in the inactivation of released ACh. It appears that the main reason for this difference might lie in the different geometry of the endplate compared to that of a neuro-neuronal synapse: ACh can accumulate within the cleft of the neuromuscular junction, whereas at a neuro-neuronal synapse it diffuses rapidly into the intercellular space.

We have found yet another mechanism of action of AChE in Aplysia. AChEIs clearly increase the postsynaptic response. Experimental analysis which we present in this chapter indicates that this facili-tation of the postsynaptic response may result from a change in a molecular inhibitory action of AChE on the acetylcholine receptor (AChR).

MATERIALS AND METHODS

Experiments were performed on isolated and desheathed ganglia of Aplysia californica (Pacific Bio Marine Company, California) at room temperature. We used two identified cholinergic interneurones

697

which induce inhibitory Cl⁻-dependent postsynaptic responses (H-type response) (18) in several identified postsynaptic cells in the buccal ganglion (6, 7), and the neurone R15 which is situated in the abdominal ganglion and which has cholinergic receptors opening cationic channels (D-type response) (18).

Current recordings of postsynaptic responses evoked by a pre-synaptic spike or ionophoretic application of the agonist were per-formed using a classical voltage clamp. Because in these central cells it is not possible to record individual postsynaptic currents resulting from the release of quanta from a given interneurone, quantal aspects of the transmission in the buccal ganglion were obtained by an indirect method as previously described by us (14, 15). Similarly, the Cl⁻ channel parameters were calculated using a fast Fourier transform from the responses to prolonged ionophoretic applications of ACh or carbachol onto the somatic receptors (15, 8).

As in most _Aplysia_ central neurones (18), AChRs of the buccal postsynaptic cells are not confined to the synapse but are also distributed on the soma, which is devoid of synaptic contacts. ACh and carbachol were alternately applied at the same membrane spot using a constant current source and a double- barrelled or two single electrodes filled with the agonists at 1 M concentration. Braking current was applied to the electrode not in use.

In all experiments, the Cl⁻ reversal potential of the post-synaptic cell was monitored frequently, as was the temperature. Current amplitudes were routinely transformed into conductances. The preparation was continuously perfused with normal saline or saline to which drugs were added.

AChE inhibitors were: phospholine (Ayerst Laboratories, U.S.A.) at a concentration of 5×10^{-4} M and prostigmine at 5×10^{-6} M. Contrathion (pralidoxime sulfomethylate, Rhone Poulenc Laboratories, France) at 10^{-3} M was used as a reactivator of AChE. Tetrodotoxin (10^{-4} M) and prostigmine were obtained from Sigma (U.S.A.). Triton X-100 (Koch-Light Laboratories Ltd., England) was used in the range 0,0005%-0,00025% and sodium deoxycholate (Merck, U.S.A.), at 10^{-7} to 10^{-9} M. Normal buffer composition was: NaCl 460 mM, KCl 10 mM, $CaCl_2$ 11 mM, $MgCl_2$ 25 mM, $MgSO_4$ 28 mM, and Tris-HCl 10 mM, pH 7.8. The action of each AChE inhibitor was studied on at least 10 preparations (28 for phospholine) for both ACh and carbachol.

RESULTS

When AChEIs were applied on the buccal ganglion preparation, the postsynaptic current (PSC) response to an evoked spike was poten-tiated. With organophosphate inhibitors [which are known to have a reversible curare-like action when present in the bath (4)], it was first necessary to wash the preparation for about 10 min to

remove the depressing effect of the drug on the AChR and to observe
the potentiation of the response (Fig. 1). Also, following bath
application of an oxime, contrathion (10^{-4} M) and washout (with
artificial sea water), it was possible to restore the test amplitude
of the response (Fig. 1). The carbamate prostigmine at the concentration used (5×10^{-6} M) only showed a facilitory effect (Fig. 1).
The postsynaptic response could also be facilitated by the oxime
contrathion (5×10^{-5} M), which upon application first increased the
PSC and then exerted a curare-like depressive action (4).

Using the above described method, quantal analysis of the postsynaptic responses following application of AChEIs showed that the
increase of the postsynaptic response was not due to an increase
in the number of quanta liberated but to an increase in the size
of individual miniature PSCs (4).

Current responses to ionophoretically applied ACh on the somatic
H-receptors were depressed by organophosphate inhibitors. Upon washing

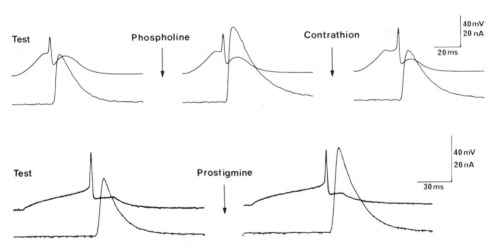

Figure 1. Effects of AChEIs on the evoked postsynaptic current (lower
traces) of the H-type synapse. Upper traces: evoked spikes in the
presynaptic neurone. The postsynaptic cell was voltage-clamped at
-80 mV. The postsynaptic current response was potentiated in the
presence of prostigmine (5×10^{-6} M) or phospholine (10^{-3} M) followed
by washing. The oxime contrathion (10^{-3} M) restored the test amplitude
of the postsynaptic response after phospholine treatment.

the preparation, the amplitude of the response increased to as much
as 300% of the test response.   This response returned to control
level when the preparation was perfused with the reactivator contrathion for 10 min followed by prolonged washing.   The current responses
to ACh ionophoretic application were also potentiated in the presence
of prostigmine and contrathion.

The observed potentiations could be due to the inhibition of
the ACh-hydrolysing function of AChE, thus leading to an increase
in available ACh.   However, when carbachol was applied and the preparation submitted to the same treatments, the current responses
were also potentiated (Fig. 2) (5).   The potentiation of the carbachol
response cannot be due to an increase in available agonist because
carbachol is not hydrolyzed by AChE.   Also, because there is no
synapse on the soma, an indirect effect of AChE resulting in ACh
leakage or activation of some presynaptic endings is unlikely.
Carbachol was applied to the soma, which was freely exposed to the
perfusing fluid, so that the distribution of the drug over the injected
region at different moments was probably identical whether AChE
was active or inhibited.   Furthermore, none of the observed changes
could be attributed to a modification of the Cl⁻ reversal potential,
as it was monitored throughout all experiments.

The potentiation of the responses could not be attributed to
a change in the properties of the ACh receptor (AChR) resulting in

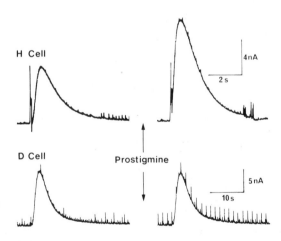

Figure 2.   The AChEI prostigmine (5×10^{-6} M), potentiated the response
to ionophoretic application of carbachol on the H-type cell of the
buccal ganglion (holding potential -80mV) (upper line), whereas it
had no effect on the carbachol response in a D-type cell of the
abdominal ganglion (holding potential -70mV) (lower line).

a change in Cl⁻ channel characteristics. Using noise analysis, we found that the channel parameters were not changed significantly by organophosphorus compounds or prostigmine. Thus, it seems likely that the potentiated response to carbachol following AChE inhibition was due to a change in receptor affinity so that more channels opened for a given dose of agonist.

The potentiating effects of AChE inhibitors were not observed in the R15 neurone, which contains D-type ACh receptors, opening cationic, mostly sodium-selective channels (9, 17, 18). In this cell, the evoked postsynaptic response as well as the response to ionophoretic application of carbachol were not modified by any of the AChE inhibitors used (Fig. 2). This absence of action not only indicates differences in functional receptor properties between R15 and the postsynaptic buccal ganglion neurones, but also could be considered as a good control: it shows that AChE inhibitors did not act non-specifically on the neuronal membrane and that the potentiation was not due to an experimental artefact such as local movements of the ionophoretic micropipettes.

Because all the AChEIs used produce a similar facilitation of the agonist responses in spite of quite different molecular constitution, it seems unlikely that AChEIs act directly on the AChR. Our results are rather in agreement with a hypothesis which would postulate the existence of a molecular relationship between AChE and the neighboring AChR, in which AChE exerts an action on some property of the AChR which modulates its readiness to be activated by ACh. Such an action would differ depending on whether AChE was in an active or inactive state, with active AChE exerting an inhibitory effect.

To test this hypothesis we attempted to change this assumed molecular interaction between AChE and AChR by perturbating the lipidic surroundings in which both macromolecules are embedded.

Detergents at concentrations greater than 0,0005% for Triton X-100 and 10^{-7} M sodium deoxycholate depressed current responses to ionophoretically applied ACh or carbachol for both D or H-type receptors. This depression most probably was due to a direct action of the detergents on AChRs. On the contrary, lower concentrations of the detergents produced a clear increase of the responses, but this increase was seen only for H-type receptors (Fig. 3). Triton X-100 at low concentration also facilitated the evoked postsynaptic current at the H-type synapse but was without effect at D-type synapses.

Moreover, no cumulative effect of AChEIs and detergents was observed: when the potentiating effect of one compound has been obtained, the other compound did not additionally increase the response, whatever the order of application. This result indicates

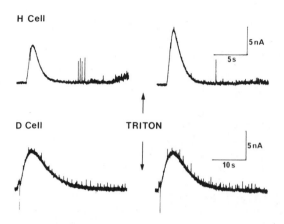

Figure 3. The bath-applied detergent Triton X-100 (0,00025%) increased the amplitude of the response to ionophoretic application of carbachol (upper) on H cells (holding potential -80mV) whereas it did not change the carbachol response in D-type cells (holding potential -70mV).

that both AChE and detergents probably act on the same mechanism. This latter could be the membrane fluidity if AChEIs acted also as detergents. However, with 30 min washing, the facilitatory effect of detergents was reversible, whereas that of organophosphate inhibition was not, unless AChE reactivators were used (5).

DISCUSSION

We consider it unlikely that the facilitatory effect of AChEIs and detergents results from a direct action on the AChR. An alternative explanation is that the probability of the AChR entering into its active state can be modulated by its molecular environment. This may be represented by some protein-to-protein interaction in which the conformational aspect of the AChE molecule in its active state can exert an inhibitory action on the AChR.

The hypothesis assumes that AChE is present in the postsynaptic membrane and that AChE and AChR are arranged in a specific spatial relationship. It is known that AChE is present in Aplysia ganglia, that it is the most potent esterase (2) and that it is present in the somatic membrane of both H and D cells, together with AChR (unpublished observations).

The role of the lipidic environment surrounding the AChE and AChR molecules is certainly of importance, and interactions between the AChR and "boundary lipids" have been described (10, 13). It is not impossible, therefore, that such protein-lipid interactions may also exist between AChE and surrounding lipids.

Two models can be proposed; both conform to our experimental results. In the first model, a conformational change in the AChE molecule induced by an AChEI could induce a local increase in the membrane fluidity, which affects the properties of the neighboring AChR. Detergents would similarly affect the membrane fluidity and thus mimic the effect of AChEIs.

The schematic action would be as follows:

$$AChE \longrightarrow membrane\ fluidity \longrightarrow AChR$$

The absence of action of AChEIs and of detergents on D-type receptors might then be explained by a small dependence of the activability of AChR on membrane fluidity.

In the second model, the lipidic environment would have a relatively passive role: a change in fluidity could increase the distances or change some molecular forces between neighboring AChE and AChR molecules and thus remove the inhibitory action which AChE might normally exert on the AChR. The conformational change of AChE induced by the action of an AChEI would similarly remove the inhibitory effect. The interaction is then more direct:

$$AChE \longrightarrow AChR$$

The absence of effect of AChEI or detergents on D-type receptors may be then due, in addition to the specific properties of D-type receptors, to a difference in AChE distribution or to a difference in positioning of AChE with respect to the AChR.

Whichever of these two models is applicable, our results suggest an experimental model of molecular interaction between neighboring membrane proteins that depends on their states of activity and on their immediate lipidic environment. This kind of behavior is probably a very common property of macromolecular components of the cytoplasmic membrane. However, it is rarely observed, as the experimental demonstration of such interaction is conditional on the presence of a measurable parameter. This is the case here for the proposed relationship between the AChE and AChR molecules of H-type cells.

Acknowledgement. We are grateful for Dr. K. Takeda for helpful comments. This work was supported by grants ATP #950501 and DRET #82/1203 to L.T.

REFERENCES

1. Bois, R.T., Hummel, R.G., Dettbarn, W-D. and Laskowski, M.B. (1980): J. Pharmacol. Exp. Ther. 215:53-59.
2. Dettbarn, W-D. and Rosenberg, P. (1962): Biochim. Biophys. Acta. 65:362-363.

3. Emmelin, B. and MacIntosh, F.C. (1956):   J. Physiol. (Lond.)
    131:477-496.
4. Fossier, P., Baux, G. and Tauc, L. (1983):   Pflugers Arch.
    396:15-22.
5. Fossier, P., Baux, G. and Tauc, L. (1983): Nature 301:710-712.
6. Gardner, D. (1971):   Science 173:550-553.
7. Gardner, D. and Kandel, E.r. (1977):   J. Neurophysiol. 40:
    333-348.
8. Gardner, D. and Stevens, C.F. (1980): J. Physiol. (Lond.) 304:
    145-164.
9. Gerschenfeld, H.M., Ascher, P. and Tauc, L. (1967):   Nature
    213:358-359.
10. Klymkowski, M.W., Heuser, J.E. and Stroud, R.M. (1980): J. Cell
    Biol.: 85-823-838.
11. Laskowski, M.B. and Dettbarn, W-D. (1979): J. Pharmacol. Exp.
    Ther. 210:269-274.
12. Morrison, J.D. (1977): Br. J. Pharmacol. 60:45-53.
13. Neubig, R.R., Krodel, E.K., Boyd, N.D. and Cohen, J.B. (1979):
    Proc. Natl. Acad. Sci. USA 76:690-694.
14. Simonneau, M., Tauc, L. and Baux, G. (1980): Proc. Natl. Acad.
    Sci. USA 77:1661-1665.
15. Simonneau, M., Baux, G. and Tauc, L. (1980): In: Ontogenesis
    and Functional Mechanisms of Peripheral Synapse.   INSERM Sym-
    posium #13, Editor J. Taxi, Elsevier/North Holland Biomedical
    Press, pp. 179-189.
16. Skliarov, A.I. (1980): Gen. Pharmac. 11:89-95.
17. Stinnakre, J. (1970): J. Physiol. (Paris) 62 suppl. 3:452-453.
18. Tauc, L. and Gerschenfeld, H.M. (1962): J. Neurophysiol. 25:
    236-262.

ACETYLCHOLINESTERASE AND THE MAINTENANCE OF NEUROMUSCULAR

STRUCTURE AND FUNCTION

W-D. Dettbarn

Department of Pharmacology
Vanderbilt University School of Medicine
Nashville, Tennessee 37232 U.S.A.

INTRODUCTION AND OVERVIEW

The structural and functional integrity of skeletal muscle depends on the presence and normal function of neuromuscular transmission. Problems may arise from pharmacologic manipulation of neuromuscular transmission and may lead to changes in the development, maintenance, and integrity of the end-plate and muscle fiber. The physiology of neuromuscular transmission can be affected in several ways. The amount of the transmitter acetylcholine (ACh) which is released may be increased or reduced by a wide variety of drugs (2, 6, 7, 11, 28, 29). Post-synaptic sensitivity may be changed by altering either the number of or the affinity of the ACh receptors. Drug induced changes of the input resistance of the muscle fiber membrane can alter the threshold for the generation of muscle action potentials. Variation in one of these factors or in combination with one another can be contributing causes of neuromuscular degeneration (4, 10, 16, 36, 39).

Acetylcholinesterase (AChE), a constituent of the membranes at the synapse, plays an essential role in the inactivation of ACh at the synaptic cleft by hydrolyzing ACh to choline and acetic acid. Inhibition of AChE profoundly modifies neuromuscular transmission. This can be shown by electrophysiological analyses, by studies of the response of the muscle to nerve stimulation, and by observations on muscular activity in the intact animal in the absence of applied nerve stimulation (3, 19, 21, 35, 38, 44).

Anticholinesterases of the carbamate and organophosphate type have a number of characteristic actions at the neuromuscular junction. They produce spontaneous fasciculations of the muscle fibers and,

705

with large doses, muscle weakness or complete neuromuscular block.
In the muscle stimulated through its nerve, the effects depend on
the frequency of stimulation; at low frequencies, the twitch tension
is potentiated due to repetitive firing of the muscle fibers; at
high rates of stimulation, the muscle is unable to maintain a tetanic
contraction.

This report will focus on a review of previous data and more
recent experiments studying the roles of the different molecular
weight forms of AChE in the organophosphate induced myopathy.

Acute AChE Inhibition and Muscle Fiber Degeneration. Reports from
several laboratories have shown that reversible as well as irreversible
AChE inhibitors induce myopathies (1, 14, 15, 34, 40, 42).

Injection (i.p., s.c., or i.v.) of organophosphorus inhibitors
in concentrations that cause fasciculations and other symptoms of
cholinergic hyperactivity produced a progressive myopathy in the
rat diaphragm, EDL, soleus, gastrocnemius, and quadriceps muscles.
The earliest lesions noted were focal areas of abnormality close
to the surface of the muscle fiber (1, 12). With H-E stain, this
area appeared to be more basophilic. The trichrome stain demonstrated
an area of red-staining, and the normal basic pattern of mitochondria,
usually identified with LDH and NADH reactions, was disrupted by
clumping of highly reactive material. These focal changes progressed
to a generalized breakdown of fiber architecture, characterized by
a loss of staining quality followed by phagocytosis. Longitudinal
sections indicated that the early changes in a focal necrosis affected
only a small segment of fiber lengths. The later stages involved
progressively greater lengths of muscle fibers (23).

Ultrastructural Changes Produced by AChE Inhibitors. Motor nerve
terminals showed varying degrees of changes within 30 minutes to 2
hours after injection of soman and paraoxon (24, 34). Some nerve
terminals appeared relatively normal with the exception of swolled
mitochondria. More severely affected nerve terminals displayed
myelin figures, membrane enclosures, and an increase in the number
of large coated vesicles. More obvious changes were seen in the
subsynaptic area and the surrounding muscle fiber. Soman, paraoxon,
and neostigmine initiated the formation of vesicular structures in
the primary and secondary subsynaptic cleft (8, 16, 24, 34). Occa-
sionally, some of these were seen in the sarcoplasm. Many of the
cleft vesicles are similar in density and size to synaptic vesicles
but with considerable variations in diameter. The severity of lesions
in the subsynaptic folds varies even within the same muscle. Normal
subsynaptic clefts with few cleft vesicles were seen side by side
with subsynaptic clefts with many cleft vesicles and a widening of
the cleft itself (23, 24, 34). In general, ultrastructural changes
were seen in almost all nerve terminals and muscle fibers, while

light microscopic changes indicating fully developed necrosis were observed only in a minority of fibers.

The AChE inhibitors, in addition to the changes seen in the region of the end-plate, caused changes in the muscle fiber itself. Muscle surrounding the motor end-plate showed a disruption of cyto-architectural organization.   Initially, the first changes were in the mitochondria, which showed swelling leading to lysis of the central cristae. Myelin figures beneath the end-plate were frequently observed while the region more distal to the end-plate was less affected.   The nucleoli of the muscle cell nucleus were enlarged and moved to the periphery of the nucleus.   There was an increase in the number of sarcoplasmic ribosomes with subsequent dilation of the sarcoplasmic reticulum and loss of striation of the myofibrils, followed by total destruction of the myofilaments and fragmentation of Z bands (23, 24, 34).

Modification of the Myopathy.   In animals undergoing left sciatic or left phrenic nerve transection 4 to 7 days prior to the application of the irreversible inhibitor, the innervated muscles showed a slight increase in the number of lesions, while the denervated muscles were protected against the myopathic process and instead underwent typical denervation atrophy (1, 12, 13, 41).   When animals were pretreated with hemicholinium, an inhibitor of choline acetyltrans-ferase (ChAT) (26), the necrotic inducing action of paraoxon in innervated muscles was significantly reduced (12).   In rats, internal skeletal fixation by pinning the right ankle and knee 3 days before treatment with paraoxon protected the soleus from the paraoxon-induced myopathy and potentiated the myopathy in the soleus of the unfixed limb, which had assumed a greater weight-bearing function (13).

Reactivation of Phosphorylated AChE and Prevention of Phosphorylation by Carbamylation.   When given to rats at a concentration of 0.23 mg/kg, s.c., paraoxon induced an 85% inhibition of AChE of the dia-phragm, and the enzyme remained at this level of inhibition for the next 2 hours.   Administration of Protopam (2-PAM) (60 mg/kg, i.p.), a reactivator of phosphorylated AChE, at various time intervals after the paraoxon injection (10-120 minutes) restored AChE activity to normal. When administered between 10 and 30 minutes after paraoxon, 2-PAM totally prevented the development of the paraoxon myopathy. At longer intervals between paraoxon and 2-PAM administration there was a time related increase in muscle necrosis.   If AChE inhibition proceeded uninterrupted for 2 hours prior to PAM administration, the muscle necrosis, as determined by light microscopy, occurred in 4.2% of the fibers.   The myopathic process depended upon the degree and duration of AChE inhibition (15, 40, 42).

Pretreatment with a single non-necrotizing dose of reversible inhibitors such as neostigmine and physostigmine (0.05-0.3 mg/kg), 5-15 minutes prior to the administration of paraoxon, prevented both enzyme phosphorylation and fiber necrosis otherwise induced by paraoxon. Longer time intervals between pretreatment with the carbamylating inhibitors and the organophosphates (180 minutes) were less effective in protecting AChE and in preventing muscle fiber necrosis.

Reversible AChE Inhibition and Myopathies. Long-term treatment of rats with prostigmine sulfate for 42 to 150 days showed degeneration of post-synaptic folds, mainly in red muscle fiber and less so in white muscle (8, 40). The post-synaptic membrane profile concentration was decreased by 29% in red muscle fibers and by 10% in white muscle fibers. The mean MEPP amplitude was decreased by 29%. Frequency, quantum content, and muscle resting membrane potential was not affected by prostigmine (8).

In acute experiments, prostigmine as well as physostigmine, in concentrations between 0.1-0.6 mg/kg, caused muscle fiber necrosis, similar to that seen with the irreversible inhibitors of AChE. The number of necrotic fibers rose with increasing inhibitor concentrations. The total number of necrotic fibers, however, was less than that caused by paraoxon. Signs of cholinergic intoxication, such as salivation, diarrhea, as well as body tremors and pronounced muscle fasciculations, were seen only for about 30 minutes after reversible inhibition of AChE. The same symptoms were observed for over 2 hours after irreversible inhibition of the enzyme. Repeated application of the reversible inhibitor, i.e., 3 times during a 1.5 hour period, led to an increased number of lesions (22, 42).

Acute Insecticide Poisoning: Myopathic Changes in Human Muscles. Although the clinical symptoms of ACh intoxication due to AChE inhibition are well described, there have been only a few reported cases of myopathic alterations from organophosphorus insecticide exposure. Necrotic lesions similar to those seen in rat muscle have been found in human diaphragm muscle after parathion ingestion, and in intercostal muscle of agricultural workers after being exposed to spray containing malathion and Diazinon (5, 43). In all these cases, the muscle necrosis was established by autopsy. In man, the incidence of acute muscle necrosis is probably larger than reported, if more attention was focused on muscle during autopsy. Whether this necrosis contributes to the overall lethal effect in acute poisoning remains to be seen.

The histological picture from these case-studies was comparable to pathological alterations in rat skeletal muscle after acute exposure to organophosphorus inhibitors of AChE such as parathion, paraoxon, DFP and soman.

Molecular Forms of AChE and the Myopathy. Although the general importance of AChE in cholinergic neurotransmission is well established, the possible physiological role of the multiple molecular forms remains unclear. Recently, differences in molecular species distribution of AChE were observed in rat SOL and EDL muscle following denervation (17). A study of the multiple forms of AChE (16S, 10S and 4S) and their distribution in end-plate and non end-plate regions of rat diaphragm muscle indicated that the 16S form was specifically associated with the end-plate region of muscle and might correspond to the functionally important end-plate enzyme (18). The 4S and 10S enzyme forms were found throughout the muscles and may serve other roles in the cholinergic system. The above findings suggest that different molecular forms to AChE may play specific roles in both normal and abnormal cholinergic neurotransmission processes.

METHODS

Two irreversible inhibitors, the tertiary phospholine and the quaternary phospholine, were used to investigate the localization of the AChE critical to the development. of the necrosis. The quaternary phospholine form is poorly lipid soluble and should inhibit only the extracellularly located AChE. The tertiary phospholine is highly lipid soluble and should inhibit AChE whether intracellularly or extracellularly located. In addition, we have studied the rate of recovery of the individual forms after irreversible inhibition. Changes in the relative proportions of the various forms during prolonged recovery may provide insight into the biosynthetic process of AChE.

For biochemical analyses the left soleus (SOL) and extensor digitorum longus (EDL) muscles were removed. The tissue was weighed, minced and homogenized in phosphate buffer. AChE activity was determined according to previously published procedures (17).

For the analysis of AChE molecular forms, SOL and EDL were homogenized in velocity sedimentation buffer and prepared for linear gradient ultracentrifugation (17). The histochemical analysis of the muscle lesions was performed as previously described (12).

RESULTS

Animals were injected with either the tertiary or quaternary inhibitor and then killed two hours later, the exposure time necessary to cause necrotic changes in muscle. As can be seen from Table 1, the number of lesions caused by the tertiary or quaternary phospholine for a given muscle (EDL or SOL) was similar, an indication that the molecular form of AChE involved in the necrosis is most likely located extracellularly. This is shown in Fig. 1, demonstrating the effect of the inhibitors on the activity of the different molecular forms of AChE.

Table 1

Number of Lesions Following Exposure to Tertiary or
Quaternary Phospholine*
Control AChE Activity is Expressed as μM ACh/G/HR

| TIME PAST INJECTION | PHOSPHOLINE | | | |
| | TERTIARY | | QUATERNARY | |
	EDL	SOL	EDL	SOL
Control	0	0	0	0
AChE	120.90$^\pm$0.18 (100%)	64.23$^\pm$2.97 (100%)	124.68$^\pm$2.51 (100%)	57.59$^\pm$0.99 (100%)
Lesions/ Cross Section of Muscle	84.42$^\pm$11.86 (4%)	55.75$^\pm$2.37 (6%)	84.66$^\pm$13.74 (71%)	54.80$^\pm$3.70 (38%)

*AChE activity in % of control in brackets; 10 animals were used
in each experiment $^\pm$ S.E.

The activity of the individual forms was studied after density
gradient separation. As can be seen from Fig. 1A, in SOL the quaternary inhibitor reduced the activity of all molecular forms to
40% of control, including the 4S form, which has been postulated
to be entirely intracellular. The tertiary phospholine reduced
the activity of all the molecular forms to less than 10% (Fig. 1C).
In the EDL, Fig. 1B, the quaternary inhibitor reduced the 16S and
10S to 40% of their control activity while the activity of the 4S
form was not inhibited significantly. The tertiary inhibitor, on
the other hand, reduced activity of all forms to less than 10%
(Fig. 1D). Regardless of the difference in the degree of inhibition,
the muscle lesions caused by either form was identical, indicating
that reduction to 40% of the extracellularly located enzyme activity
was enough to cause the lesions.

Figure 1. Velocity sedimentation gradient separation of AChE molecular form in SOL and EDL muscle, 120 min. after in vivo injection
of quaternary phospholine (0.1 mg/kg s.c.) A and B after tertiary
phospholine (0.3 mg/kg s.c.) C and D. Δ------Δ = control, o------o
= experimental muscle. AChE activity is expressed in cpm/mg protein
x 10^{-3}. The position of proteins with known sedimentation such as
alkaline phosphatase 6.15 (P), catalase 11.5 (C) and B galactosidase
16.05 (G) are indicated with arrows.

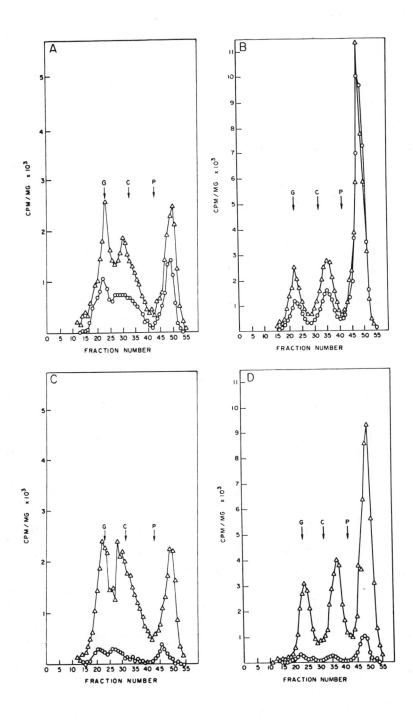

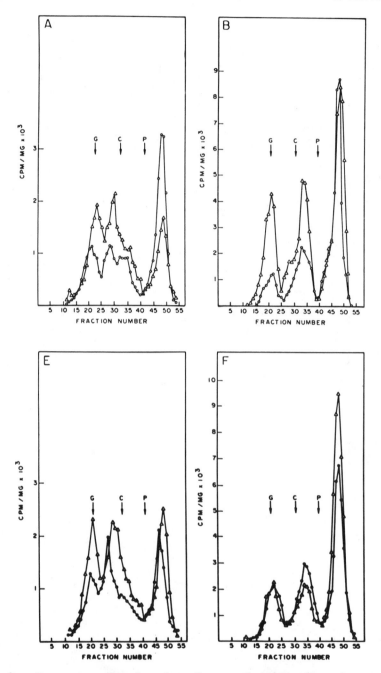

Figure 2. Recovery of molecular forms of AChE after irreversible inhibition with quaternary or tertiary phospholine. AChE recovery in SOL after one day (A) and seven days (E) after quaternary phospholine, one day (C) and seven days (G) after tertiary phospholine.

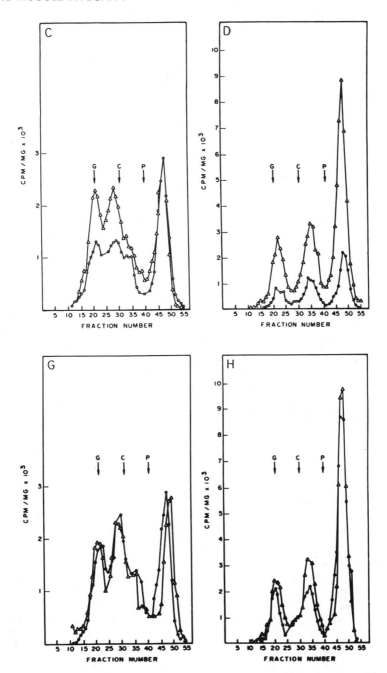

EDL: one day (B) and seven days (F) after quaternary phospholine, one day (D) and seven days (H) after tertiary phospholine. Δ------Δ = control, o------o = experimental muscle. For further explanation see legend of Fig. 1.

The effects of a single dose of either tertiary or quaternary phospholine on the activity of the different molecular forms of AChE and their recovery from inhibition are presented in Fig. 2. Two hours after injection the inhibition was greatest in both muscles (see Fig. 1). It is apparent that the recovery of the heavier molecular forms such as 16S and 10S is slower than that of the 4S form. By day seven most of the molecular forms had recovered to control activity.

DISCUSSION

The role of AChE inhibition as the triggering factor leading to excessive ACh mediated hyperactivity in the muscle and nerve causing the myopathy is strongly supported by the observation that 2-PAM prevents the necrosis; however, other mechanisms could be involved. AChE inhibitors may have a direct toxic effect unrelated to AChE inhibition on the post-synaptic membrane, such as on the ACh receptor or part thereof such as the ion channels and elsewhere along the muscle fiber. The greater number of necrotic fibers in the stimulated, active muscle as compared to the inactive (denervated or immobilized) muscle in the paraoxon treated animals may suggest that the AChE inhibitor causes a "subthreshold myopathy" not seen histologically unless the muscle is stressed by normal or excessive activity.

Recent studies have shown increased Ca^{++} uptake into the end-plate region of muscle stimulated by carbamyl choline (9, 25, 27). The presence of Ca^{++}-activated proteases, which specifically disrupt Z-disks in isolated myofibrils in vitro, could explain the disappearance of Z-disks from myofibrils near the end-plate if prolonged ACh-receptor interaction elevates intracellular Ca^{++} levels sufficiently. The extreme vesiculation and disruption of SR below the end-plate, as well as the swollen mitochondria, may reflect an overloading of the muscle Ca^{++}-binding capacity resulting in high sarcoplasmic Ca^{++} levels. Another observation that argues for elevation of Ca^{++} due primarily to an increased influx is the frequent occurrence of contractures which are restricted to the end-plate region. If these contractures were caused by depolarization-induced release of Ca^{++} from the SR the contracture should extend over a much larger region of the muscle. On the other hand, localized contractures can be produced by focal injections of Ca^{++} and are often used as an indication of such local Ca^{++} influx (20, 31, 37). A similar contracture under the end-plate has been reported after AChE inactivation.

Whether the characteristic contracture seen under the end-plate can itself cause local damage and thus produce the myopathy remains to be seen. From preliminary studies we know that within 2 hours following exposure to paraoxon or DFP, glycogen levels in liver and muscle are reduced by 50%. Whether this is due to direct actions on its synthesis or catabolism or due to increased muscle activity remains to be determined. With increasing duration of muscle activity, induced by the AChE inhibitor, the Ca^{++} uptake may become so excessive that a dangerous fall in glycogen, ATP and creatine phosphate occurs. The energetic state of the hyperstimulated muscle fiber would be comparable to the metabolic situation of a muscle in anoxia and ischemia. In all these cases, muscle function and structural integrity cannot be maintained, because the high-energy phosphate pool is exhausted. Thus, the defect may be caused by the inability of the sarcoplasmic reticulum to sequester the sudden increase in Ca^{++}.

Our results show that exposure of muscles to induced excess of ACh alters the ultrastructure of pre- and post-synaptic elements of the neuromuscular junction and can lead to partial denervation of muscle. Thus, AChE at the neuromuscular junction may, in part, have a protective and regulatory function through its control of free ACh. By limiting the accumulation of ACh and the extent of its interaction with pre- and post-synaptic membranes, it may preserve morphological integrity of nerve terminals and muscle. Whether this is a direct effect of ACh or an indirectly induced process through increased muscle activity is still open to question. An indirect action is supported by findings that proteolytic enzymes are released from muscles being electrically stimulated or treated with ACh or carbamate inhibitors. This release can induce changes at the end-plate and in the muscle (30, 32, 33).

In our studies to determine the involvement of intra- and extracellular pools of AChE in the muscle necrosis, we found differences in the distribution or accessibility of the individual forms between the slow twitch SOL and the fast twitch EDL muscle. In the SOL all forms were inhibited to a larger extent than those of the EDL when the quaternary inhibitor was used (Figs. 1A and B). This indicates that in SOL a greater proportion of all molecular forms is located on or near the extracellular surface. The tertiary inhibitor caused in both muscles significantly greater inhibition of overall activity as well as of that of the individual molecular forms (Figs. 1C and D); however, no difference in the number of necrotic muscle fibers were found when compared to the necrotic action of the quaternary inhibitor (Table 1). This suggests that the inhibition of the 16S and 10S forms seen with the quaternary inhibitor is the trigger mechanism that leads to the necrosis. We can also conclude from these studies that some portion of the 16S and 10S forms must be intracellularly located, since a greater inhibition of these two forms was seen when the tertiary inhibitor was used. The recovery studies (Fig. 2) are in general agreement with other reports of half time rates of recovery after organophosphates. The faster recovery of the 4S form may indicate that this molecular form represents the initial step in the synthesis of the other forms.

In conclusion, AChE inhibitors cause a progressive necrosis in skeletal muscle. The severity of this myopathy depends upon a rapid, critical reduction of AChE activity lasting over a critical time period, involving mainly the extracellularly located 16S and possibly the 12S and 10S molecular forms. This inhibition triggers neurally mediated events, resulting in repeated receptor transmitter interactions causing an increase in the activity of a number of motor units, as indicated by the presence of long lasting muscle fasciculations. It is suggested that the cause for the necrosis is the loss of ATP and creatine phosphate, causing changes in the membrane properties leading to an increase in the intracellular Ca^{++} concentration. This rise in Ca^{++} may activate intracellular proteases, contributing to the observed necrosis. The lesions produced in rat muscle resemble closely those found in human autopsy material of accidental or suicidal organophosphate poisonings.

Acknowledgement. This work was supported by NIH Grants NINCDS 12438-05, EHS 02028-02, Air Force Grant AFOSR 82-0310, and Army Grant DAMD-17-83-C-3244. The author thanks Mrs. Barbara Page for the typing of this manuscript.

REFERENCES

1. Ariens, A.T., Meeter, E., Wolthuis, O.L. and van Benthem, R.M.J. (1969): Experientia 25:57-59.
2. Birks, R. and MacIntosh, F.C. (1961): Can. J. Biochem. Physiol. 39:787-827.
3. Bowman, W.C. and Webb, S.N. (1972): In International Encyclopedia of Pharmacology and Therapeutics (ed) J. Cheymol, Pergamon Press, Oxford, Sec. 14, Vol. 2, pp. 427-502.
4. Chang, C.C., Chen, T.F. and Chuang, S.T. (1973): J. Physiol. (London) 230:613-618.
5. DeReuck, J. and Willams, J. (1975): J. Neurol. 208:309-314.
6. Drachman, D.B. (1972): J. Physiol. (London) 226:619-627.
7. Edstrom, A. and Mattsoon, H. (1972): J. Neurochem. 19:1717-1729.
8. Engel, A.G., Lambert, E.H. and Santa, T. (1973): Neurology 23:1273-1281.
9. Evans, R.H. (1974): J. Physiol. (London) 240:517-533.
10. Fambrough, D.M., Drachman, D.B. and Satyamurti, S. (1973): Science 182:293-295.
11. Feng, T.P. (1937): Clin. J. Physiol. Rep. Ser. 12:177-196.
12. Fenichel, G.M., Kibler, W.B., Olson, W.H. and Dettbarn, W-D. (1972): Neurology 22:1026-1033.
13. Fenichel, G.M., Kibler, W.B. and Dettbarn, W-D. (1974): Neurology 24:2086-1090.
14. Fischer, G. (1968): Histochemie 16:144-149.
15. Fischer, G. (1970): Experientia 26:402-403.
16. Fleming, W.W., McPhillips, J.J. and Westfall, D.P. (1973): Ergeb. Physiol. Biol. Chem. Exp. 68:55-119.

17.   Groswald, D.E. and Dettbarn, W.D. (1983):   Exp. Neurol. _79_:
      519-531.
18.   Hall, Z. (1973):   J. Neurobiol. _4_:343-361.
19.   Hobbiger, F. (1976):   In Handbuch der experimentellen
      Pharmakologie, Erganzungswerk (ed) E. Zaimis, Springer,
      Berlin-Heidelberg-New York, Vol. XLII, pp. 487-582.
20.   Jenkinson, D.H. and Nicholls, J.G. (1961):   J. Physiol.
      (London) _159_:111-127.
21.   Karczmar, A.G. (1967):   Am. Rev. Pharmacol. _7_:241-276.
22.   Kawabuchi, M., Osame, M., Igata, A. and Kanaseki, T. (1976):
      Experientia _32_:623-625.
23.   Laskowski, M.B., Olson, W.H. and Dettbarn, W-D. (1975):
      Exp. Neurol. _47_:290-306.
24.   Laskowski, M.B., Olson, W.H. and Dettbarn, W-D. (1977):
      Exp. Neurol. _57_:13-33.
25.   Leonard, J. and Salpeter, M.M. (1979):   J. Cell Biol. _82_:
      811-819.
26.   MacIntosh, F.C., Birks, R.I. and Sastry, B.P. (1956):   Nature
      _178_:1181-1183.
27.   Miledi, R., Parker, I. and Schalow, G. (1977):   J. Physiol.
      (London) _268_:32-33P.
28.   Otsuka, M. and Endo, M. (1960):   J. Pharmacol. exp. Ther.
      _128_:273-282.
29.   Otsuka, M. and Nonomura, Y. (1963):   J. Pharmacol. exp. Ther.
      _140_:41-45.
30.   O'Brien, R.A.D., Osterberg, A.J.C. and Vrbova, G. (1980):
      Neuroscience _5_:1367-1379.
31.   Parsons, R.L. and Nastuk, W.L. (1969):   Am. J. Physiol. _217_:
      364-369.
32.   Poberai, M. and Savay, G. (1976):   Acta Histochem. _57_:44-48.
33.   Poberai, M., Savay, G. and Osillik, B. (1972):   Neurobiology
      _2_:1-7.
34.   Preusser, H.J. (1967):   Z. Zellforsch. Mikrosk. Anat. _80_:
      436-457.
35.   Riker, W.F., Jr. and Okamoto, M. (1969):   Ann. Rev. Pharmacol.
      _9_:173-208.
36.   Roberts, D.V. and Thesleff, S. (1969):   Eur. J. Pharmacol.
      _6_:281-285.
37.   Salpeter, M.M., Kasprzak, H., Feng, H. and Fertuck, H. (1979):
      J. Neurocytol. _8_:95-115.
38.   Thesleff, S. and Quastel, D.M.J. (1965):   Ann. Rev. Pharmacol.
      _5_:263-284.
39.   Wecker, L. and Dettbarn, W-D. (1975):   Arch. Int. Pharmacodyn.
      Ther. _217_:236-245.
40.   Wecker, L. and Dettbarn, W-D. (1976):   Exp. Neurol. _51_:281-291.
41.   Wecker, L. and Dettbarn, W-D. (1977):   Exp. Neurol. _57_:94-101.
42.   Wecker, L., Kiauta, T. and Dettbarn, W-D. (1978):   J.
      Pharmacol. Exp. Ther. _206_:97-104.
43.   Wecker, L., Mrak, R.E. and Dettbarn, W-D. (1982):   J. Environ.
      Pathol. Toxicol. Oncol. _5_:575-579.

44. Werner, G. and Kuperman, A.S. (1963): In <u>Handbuch der</u>
     <u>experimentellen Pharmakologie, Erganzungswerk</u> (ed) G.B. Koelle,
     Springer, Vol. XV, pp. 570-678.

STRIATAL DOPAMINE-γ-AMINOBUTYRIC ACID-ACETYLCHOLINE INTERACTION

IN ORGANOPHOSPHATE-INDUCED NEUROTOXICITY

I.K. Ho, S.P. Sivam, John C.R. Fernando,
D.K. Lim and B. Hoskins

Department of Pharmacology and Toxicology
University of Mississippi Medical Center
2500 North State Street
Jackson, Mississippi 39216   U.S.A.

INTRODUCTION

The extreme toxicity of organophosphate cholinesterase inhibitors (OP-ChE-Is) is due to their irreversible inactivation of acetyl-cholinesterase (AChE), resulting in cholinergic hyperactivity (19). It has been suggested that various cholinergically-mediated behaviors would be disrupted only if the AChE activity in brain fell below some critical level, i.e., 40% of normal (16, 35). However, some of the behavioral changes induced by OP-ChE-Is do not entirely correlate with the degree of AChE inhibition (4, 11, 18, 21, 42). Therefore, the question of whether all of the toxic symptoms are due to the derangement of cholinergic function or if other neurochemical changes might be, in part, responsible is unclear.

Glisson et al. (14, 15) have shown that acute administration of diisopropylfluorophosphate (DFP) to rabbits significantly increased brain levels of dopamine (DA).   They also showed that DFP-induced elevation of ACh levels in the thalamus and hypothalamus was lowered by pretreatment with the monoamine oxidase inhibitor, JB-835 and dihydroxphenylalanine (DOPA).   When rabbits were pretreated with atropine methylnitrate prior to DFP, the increase in DA induced by DFP was prevented.   In chronic experiments, Freed et al., (13) showed that Mipafox administered to rats daily for 35 days produced ataxia and a reduction of DA levels in the corpus striatum.   Treatment with Leptophos for the same period of time produced slight motor dysfunction and a small but significant decrease in striatal DA levels.   Fenitrothion neither induced motor dysfunction nor changed striatal DA levels.   All three compounds inhibited AChE activity in the corpus striatum. These results led to the suggestion of an involvement of striatal DA in the delayed neurotoxic effects of certain OP-ChE-Is. In a histological study, Dasheiff et al. (7) demonstrated that increased catecholamine stores were found in the locus ceruleus and the zona compacta of the substantia nigra in rats following a 0.33 LD_{50} dose of Sarin. These increases returned to normal after 10 days. In a case report of possible organophosphorus insecticide-induced Parkinsonism, there was the suggestion that chronic exposure of organophosphorus agents altered the cholinergic and dopaminergic activity in the striatum (8).

It has also been suggested that gamma-aminobutyric acid (GABA) may play an important role in organophosphorus insecticide-induced convulsions (28). Compounds (e.g., benzodiazepines) acting via GABAergic mechanisms (5) are potent inhibitors of organophosphorus-induced seizure activity (23, 32). Cyclic-GMP levels appear to reflect the GABAergic inhibitory activity in the cerebellum (27) and it has been shown that Soman elevated cyclic-GMP levels in rat cerebellum (24). Benzodiazepines which block Soman-induced convulsions also block the increase in cyclic-GMP at the onset of convulsions (24). Lundy et al. (25) further showed that diazepam and two GABA-transaminase inhibitors, aminooxyacetic acid and n-dipropyl acetic acid, greatly reduced Soman-induced convulsions. Other organophosphorus agents which cause convulsions and death but do not inhibit AChE (2) are also believed to exert their effects by altering GABAergic function (3). Thus, evidence suggests that organophosphorus-induced convulsions appear to be abolished by altering GABA/ACh balance in an unknown manner, possibly via neuronal depression.

More recently, Sevaljevic et al. (37) reported that inhibition of AChE in Soman- and VX-poisoning coincides with a significant increase in plasma cyclic AMP levels. They further showed that administration of 1-(2-hydroxyiminomethyl-1-pyridino-3-(4-carbamoyl-1-pyridinio)-2-oxapropane dichloride (HI-6), an oxime, provided full protection from poisoning by lethal doses of Soman and VX, however, its effects on restoration of plasma cyclic AMP levels and red blood cell (RBC) AChE activity were dependent upon the agent used. In VX intoxication, HI-6 reactivated RBC AChE and remarkably reduced the level of cyclic AMP in plasma, whereas in Soman-poisoning, HI-6 administration caused neither RBC AChE reactivation nor decreased cyclic AMP. They suggested that, even though HI-6 failed to reactivate RBC AChE, it may reactivate AChE at some other vital site(s) in the body. They therefore speculated that cyclic AMP generation via catecholamine activation of adenylate cyclase may participate in the VX-induced increase of plasma cyclic AMP. This is based on the proposal that induction of catecholamine release may be involved in Soman intoxication (41). Pharmacologically, Stitcher et al. (41) showed that the combination of theophylline and N^2,O^2-dibutyryl cyclic-AMP significantly enhanced the toxicity of Soman in mice. Therefore, it has been suggested that cyclic-AMP participates in

OP-ChE-I poisoning, possibly by increasing the levels of ACh in the brain (1) or by enhancing the release of ACh at neuromuscular junctions (17).

Recently Lundy and Shaw (26) demonstrated elevated cyclic GMP levels in the cerebella of rats which received one or two times the LD_{50} of Soman. These increases in cyclic GMP occurred before the onset of cholinergic symptoms, and prior to convulsions. Pretreatment of animals with an antimuscarinic or antinicotinic agent plus an oxime (HI-6), which attenuated Soman-induced lethality, had little effect on Soman-induced convulsions or cyclic GMP levels. However, pretreatment with clonazepam, which reduced the convulsive activity of Soman, also attenuated the increased cyclic GMP in the cerebellum. Their results also showed that Soman-induced convulsions and the resulting increases in cerebellar cyclic GMP concentrations could not be related mainly to increased cholinergic activity or to muscarinic effects, since pretreatment of poisoned animals with antimuscarinic or antinicotinic compounds or with atropine plus an oxime had little effect on severity of convulsions or on cyclic GMP levels. Therefore they suggested that cyclic GMP may play a role in the initiation and continuation of Soman-induced convulsions which were not related to specific cholinergic receptors but to general neuronal excitation.

The development of tolerance to OP-ChE-Is has since been well established as being a consequence of repeated administration of these agents. During the past two decades, an increase in the number of studies pertaining to OP-ChE-Is has resulted in some understanding of how alterations of cholinergic functional states might be involved in the tolerance phenomenon induced by these agents. These details have been documented by Costa et al., (6).

Although OP-ChE-Is exert their toxicity mainly through the inhibition of AChE, it has been shown that only 2% of AChE activity in brain is sufficient to preserve spontaneous respiration in rats (29). In the literature, reports have consistently shown that when animals had developed tolerance to OP-ChE-Is the AChE activity still remained at 10 to 30% of the normal activities (19, 22, 33, 34, 40).

While susceptibility of cholinergic receptors may play a major role in OP-ChE-Is-induced tolerance, other neuronal activities which interact with cholinergic pathways cannot be overlooked. Russell et al. (33) demonstrated that when control and DFP-tolerant animals were treated with α-methyl-p-tyrosine (which depletes both dopamine and norepinephrine content in the brain), the behaviors of both groups of animals were comparably affected.

The discovery of a variety of neurotransmitters, especially DA and GABA, and their participation in a number of neurological and behavioral disorders (10, 20, 36) suggest that these systems, in addition to the cholinergic system, may be involved in OP-ChE-Is-induced tolerance.   It has been demonstrated that an imbalance of dopaminergic (inhibitory) and cholinergic (excitatory) systems in the basal ganglia is associated with motor dysfunction, particularly Parkinsonism (20).   The basal ganglia are rich in ACh (9, 30).   On the other hand, GABA is one of the major inhibitory neurotransmitters in the mammalian central nervous system (31).   Based on the evidence, then, we believe that other neuronal systems which may be involved in OP-ChE-Is-induced tolerance warrant extensive study.

In this chapter we wish to summarize our recent studies of the effects of acute and subacute administrations of DFP on DA and GABA systems in the rat (12, 22, 38-40).

ACUTE TOXICITY AND THE DEVELOPMENT OF TOLERANCE TO DFP IN THE RAT

DFP toxicity and tolerance in rats were studied (12, 22). Acute injection of DFP showed dose-dependent depressions in body weights and in food and water consumption.   The animals recovered within 72 hours.   Daily injections of DFP caused significant depressions in all three parameters.   However, tolerance to DFP in terms of growth rates, food, and water consumption occurred.   DFP-induced behaviors (e.g. tremors, chewing-movements, hind-limb abduction and hypothermia) increased in a steeply dose-dependent manner; all, except chewing-movements, subsided after 7 hours.   Subacute treatment with DFP for up to one month produced biphasic patterns of change for all the behavioral parameters.   Tremor appeared in a complex spectrum of slow to intense fast types.   Except for chewing-movements, tolerance developed for all these behaviors, but at different rates. The data show dose-dependency of general toxicity during acute and subacute exposure to DFP and of tolerance during subacute exposure.

EFFECTS OF ACUTE AND SUBACUTE ADMINISTRATION
OF DFP ON CHOLINERGIC SYSTEM

The effects of acute and subacute administration of DFP to rats on AChE activity (in striatum, medulla, diencephalon, cortex and cerebellum) and muscarinic receptor characteristics (in striatum) were investigated (40).   After a single injection of (acute exposure to) DFP, the striatal region was found to have the highest degree of AChE inhibition.   After daily DFP injections (subacute treatment), all brain regions had the same degree of AChE inhibition, which remained at a steady level despite the regression of the DFP-induced cholinergic overactivity.   Acute administration of DFP did not affect the muscarinic receptor characteristics; whereas subacute administration of DFP for either 4 or 14 days reduced the number of muscarinic sites without affecting their affinity.   The in vitro addition of DFP to striatal membranes also did not affect muscarinic receptors. Our results also indicate that none of the lethal or sublethal doses had any apparent effect on choline acetyltransferase (38).

EFFECTS OF ACUTE AND SUBACUTE ADMINISTRATION OF DFP ON GABA SYSTEM

The effects of acute and subacute injections of DFP to rats on GABA synaptic function were also investigated in the striatal region of the brain (39, 40). Acute as well as subacute treatments increased levels of GABA and its precursor (glutamate) and decreased GABA uptake and release. However, none of the treatments affected activities of the GABA related enzymes: glutamic acid decarboxylase and GABA-transaminase. Our results also showed that acute administration of DFP increased the number of GABA receptors without affecting their affinity. In subacutely-treated animals, DFP caused an increase in the number of GABA receptors after 14 days of treatment. This increase, however, was considerably lower than that observed after the acute treatment.

EFFECTS OF ACUTE AND SUBACUTE DFP ADMINISTRATION ON DA SYSTEM

The effects of acute and subacute DFP treatment on striatal DA and its metabolite, dihydroxyphenylacetic acid (DOPAC) levels and DA receptor binding characteristics were also studied (12, 40). After acute treatment, striatal DA and DOPAC levels were altered and the DOPAC/DA ratios were consistently increased within the first two hours. After subacute treatment for 4 and 14 but not for 28 days, both DA and DOPAC levels were decreased without a change in their ratios. Our studies further indicated that acute administration of DFP increased the number of DA receptors. Subacute treatment of DFP also caused an increase in the number of DA receptors after 14 days of treatment. However, this increase was considerably lower than that observed after the acute treatment. These data suggest that the changes in DA metabolism arose secondarily to an elevation of brain ACh following AChE inhibition. A prolonged change in the levels or turnover of DA could be responsible for increase of postsynaptic DA receptor density.

CONCLUSIONS

The results indicate an involvement of the GABAergic and dopaminergic systems in the action of DFP. It is suggested that the GABAergic and dopaminergic involvement may be a part of a compensatory inhibitory process to counteract the excessive cholinergic activity produced by DFP.

Based upon the evidence presented, we propose that GABAergic and dopaminergic neurons interact with cholinergic neurons after acute and subacute exposure to OP-ChE-Is, as presented in Fig. 1.

Under normal circumstances (i.e., no exposure to OP-ChE-Is, etc.), a balance between excitatory cholinergic and the inhibitory GABAergic and dopaminergic activities is maintained in the striatum (basal ganglia).

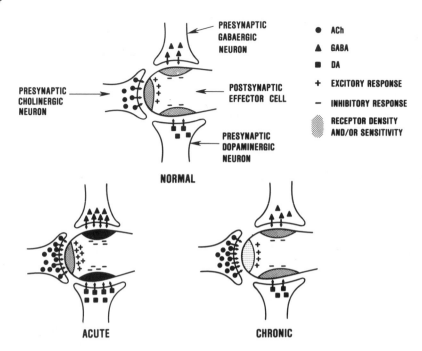

Figure 1.   A model for the interaction of cholinergic, dopaminergic
and GABAergic neurons after acute and chronic exposure to OP-ChE-Is.

After acute exposure to one of the anticholinesterase inhibitors
(DFP) inhibition of AChE results in increased cholinergic activity
which is balanced by increases in dopaminergic and GABAergic activities
due to increases in DA and GABA receptor densities.

After subacute exposure to one of these inhibitors of AChE,
in which synaptic concentrations of ACh are still high, the muscarinic
receptor characteristics are changed such that there is a decrease
in number of receptors as well as in receptor sensitivity to ACh.
Thus, the effect of increased concentrations of ACh in this situation
results in a "normal" cholinergic (excitatory) response that, in
turn, is balanced by "normal" inhibitory responses of GABA and DA
due to a return to normal of the inhibitory receptor populations.

REFERENCES

1.   Askew, W.E. and Ho, B.T. (1975): Canad. J. Biochem. 53:634-635.
2.   Bellet, E.M., and Casida, J.E. (1973): Science 182:1135-1136.
3.   Bowery, N.G., Collins, F.F. and Hill, R.G. (1976):   Nature
      261:601-603.
4.   Burchfield, J.L., Duffy, F.H. and Sim, V.M. (1976):   Toxicol.
      Appl. Pharmacol. 35:365-379.

5. Costa, E., Guidotti, A. and Mao, C.C. (1976): In: <u>GABA in Nervous System Function</u> (eds) E. Roberts, T.N. Chase and D.B. Tower, Raven Press, New York, pp. 413-426.
6. Costa, L.G., Schwab, B.W. and Murphy, S.D. (1982): Toxicol. <u>25</u>:79-97.
7. Dasheiff, R.M., Einberg, E. and Grenell, R.G. (1977): Exptl. Neurol. <u>57</u>:549-560.
8. Davis, K.L., Yesavage, J.A. and Berger, P.A. (1978): J. Nervous Mental Disease <u>166</u>:222-225.
9. Dawson, R.M. and Jarrott, B. (1981): Biochem. Pharmacol. <u>30</u>:2365-2368.
10. Enna, S.J. (1981): Biochem. Pharmacol. <u>30</u>:907-913.
11. Faff, J., Borkoruska, E., Bak, W. (1976): Arch. Toxicol. <u>36</u>:139-146.
12. Fernando, J.C.R., Hoskins, B. and Ho, I.K. (1984): Pharmacology, Biochemistry, Behavior <u>20</u>:951-957.
13. Freed, V.H., Matin, M.A., Fang, S.C. and Kar, P.P. (1976): Eur. J. Pharmacol. <u>35</u>:229-232.
14. Glisson, S.N., Karczmar, A.G. and Barnes, L. (1972): Neuropharmacology <u>11</u>:465-477.
15. Glisson, S.N., Karczmar, A.G., Barnes, L. (1974): Neuropharmacology <u>13</u>:623-631.
16. Glow, P.H. and Rose, S. (1965): Nature <u>206</u>:475-477.
17. Goldberg, A.L. and Singer, J.J. (1969): Proc. Nat. Acad. Sci, USA <u>64</u>:134-141.
18. Green, D.M., Muir, A.W., Stratton, J.A. and Inch, T.D. (1977): J. Pharm. Pharmacol. <u>29</u>:62-64.
19. Holmstedt, B. (1959): Pharmacol. Rev. <u>11</u>:567-688.
20. Hornykiewicz, O. (1975): Biochem. Pharmacol. <u>24</u>:1061-1065.
21. Kozar, M.D., Overstreet, D.H., Chippendale, T.C. and Russell, R.W. (1976): Neuropharmacol. <u>15</u>:291-298.
22. Lim, D.K. Hoskins, B. and Ho, I.K, (1983): Res. Comm. Chem. Pathol. Pharmacol. <u>39</u>:399-418.
23. Lipp, J.A. (1973): Arch. Int. Pharmacodyn. <u>202</u>:244-251.
24. Lundy, P.M. and Magor, G. (1978): J. Pharm. Pharmacol. <u>30</u>:251-252.
25. Lundy, P.M. Magor, G. and Shaw, R.K. (1978): Arch. Int. Pharmacodyn. <u>234</u>:64-73.
26. Lundy, P.M. and Shaw, R.K. (1983): Neuropharmacology <u>22</u>:55-63.
27. Mao, C.C., Guidotti, A. and Costa, E. (1974): Brain Res. <u>75</u>:510-514.
28. Matin, M.A. and Kar, P.P. (1973): Eur. J. Pharmacol. <u>21</u>:217-221.
29. Meeter, E. and Wolthuis, O.L. (1968): Eur. J. Pharmacol. <u>2</u>:377-386.
30. Quastel, J.H. (1962): In: Neurochemistry (K.A.C. Elliot, I.H. Page and J.H. Quastel, eds.), Charles C. Thomas, Springfield, pp. 431-451.
31. Roberts, E., Chase, T.N. and Tower, D.B. (1976): <u>Nervous System Function</u>, Raven Press, New York.

32. Rump, S., Grudzinska, E. and Edelwejn, Z. (1973): Neuropharma-
     cology 12:813-817.
33. Russell, R.W., Overstreet, D.H. Cotman, C.W., Carson, V.G.,
     Churchill, L., Dalglish, F.W. and Vasquez, B.J. (1975): J.
     Pharmacol. Exp. There. 192:73-85.
34. Russell, R.W., Vasquez, B.J., Overstreet, D.H. and Dalglish,
     F.W. (1971): Psychopharmacologia 20:32-41.
35. Russell, R.W., Watson, R.H.J. and Trankenhaeuser, M. (1961):
     Scad. J. Physiol. 2:21-29.
36. Seeman, P. (1981): Pharmacol. Rev. 32:229-313.
37. Sevaljevic, L., Krtolica, K. Poznanovic, G., Boskovic, B. and
     Maksimovic, M. (1981): Biochem. Pharmacol. 30:2725-2727.
38. Sivam, S.P., Hoskins, B. and Ho, I.K. (1984): Fundamen.
     Appl. Toxicol. 4:531-538.
39. Sivam, S.P., Nabeshima, T., Lim, D.K., Hoskins, B. and Ho,
     I.K., (1983): Res. Comm. Chem. Pathol. Pharmacol. 42:51-60.
40. Sivam, S.P., Norris, J.C., Lim, D.K. Hoskins, B. and Ho, I.K.
     (1983): J. Neurochem. 40:1414-1422.
41. Stitcher, D.L., Harris, L.W., Moore, R.D. and Heyl, W.C. (1977):
     Toxicol. Appl. Pharmacol. 41:79-90.
42. Wecker, L., Mobley, P.L. and Dettbarn, W-D. (1977): Biochem.
     Pharmacol. 26:633-637.

POLYMORPHISM OF CHOLINESTERASE: POSSIBLE INSERTION OF THE VARIOUS MOLECULAR FORMS IN CELLULAR STRUCTURES

J. Massoulié

Laboratoire de Neurobiologie
Ecole Normale Supérieure
46, rue d'Ulm
75005 Paris, France

INTRODUCTION

Acetylcholinesterase (EC 3.1.1.7.) and butyrylcholinesterase (EC 3.1.1.8.) are very polymorphic (see review, ref. 20). They occur as a number of molecular forms which are equivalent in their catalytic activity, but differ in their molecular parameters and interactions.

Nomenclature of Molecular Forms. Homologous sets of forms exist in all vertebrates. These forms may be classified as <u>globular</u> (G) and <u>asymmetric</u> (A), depending on the presence of a collagen-like tail. The globular forms can be subdivided into <u>detergent-insensitive</u> and <u>amphipathic</u> molecules, according to their interactions with micelles of non-denaturing detergents, e.g. Triton X 100 (23).

In Electrophorus, it has been shown that the globular forms are monomers (G1), dimers (G2) and tetramers (G4) of the catalytic subunit, and that the asymmetric forms are formed by one (A4), two (A8) and three (A12) tetramers linked to the strands of the triple-helical collagen-like tail (see Figure 1).

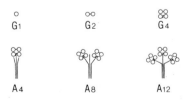

Figure 1

In other species, this general structure seems to be conserved, with however a number of possible modifications. For example, non catalytic globular subunits partially replace the active subunits in Torpedo A forms (14, 15). The catalytic subunits themselves vary widely in size among different species. They are, for example, heavier than 100 kilodaltons in chicken or quail, and about 70K - 80K in mammals. Several types of catalytic subunits in fact occur within each species: in Torpedo, the subunits of asymmetric forms differ from those of the amphipathic globular forms, which themselves present different variants in buffer-soluble and detergent-soluble fractions (16, 19, 31).

Sedimentation coefficients are often used to characterize the different forms. However, they differ markedly between corresponding forms in different species (e.g., A12 is 20S in chicken and 16-17S in mammals), and in the case of amphipathic forms, they change with the presence and nature of detergents (e.g., G1 from bovine caudate nucleus sediments at 4.5S in the absence of detergent and 3.1S in the presence of Triton X 100) (9). The nomenclature proposed for Electrophorus is therefore useful to express homologies between molecular forms which present similar quaternary structures (3, 19, 20).

Molecular Interactions and Solubility. The amphipathic globular forms may be solubilized in low salt buffers. It is generally possible to obtain a fraction which is soluble in the absence of detergent (buffer-soluble fraction) and a fraction which requires detergent (detergent-soluble fraction). The buffer-soluble forms may be associated non-covalently with hydrophobic proteins, as in Torpedo electric organs (4). Their hydrodynamic parameters are clearly modified in the presence of detergents; in Triton X100, the sedimentation coefficient decreases and the Stokes radius increases. The detergent-soluble molecules generally form polydisperse aggregates in the absence of detergent. The activity of extensively purified amphipathic acetylcholinesterase has been shown to depend on hydrophobic interactions with detergents, or with other proteins (7, 20).

The collagen-tailed asymmetric forms interact with polyanionic glycosaminoglycans (2). Therefore, they are fully soluble only in high salt, and in some cases, in the presence of chelating agents (22). Alternatively, they also are soluble in the presence of an excess of heparin sulfate, which probably induces the formation of small aggregates (27).

A fraction of collagen-tailed molecules is solubilized in low salt extracts of Torpedo tissues, in the form of small aggregates (5). This fraction is particularly important in the electric lobes, where the enzyme is mostly found in motoneuron cell bodies (28). This indicates the possibility of different types of insertion of these molecules in cellular structures in vivo.

Different solubility characteristics do not necessarily correspond to distinct cellular compartments. This has been shown clearly in the case of acetylcholinesterase from bovine caudate nucleus; multiple extractions in the absence of detergent do not seem to exhaust a cellular pool of enzyme and increase the total yield of G1, at the expense of G4 (9). This indicates possible artefacts in the proportions of molecular forms. The values obtained after direct solubilization in the presence of an efficient detergent such as Triton X100 or sodium cholate, however, probably reflect a real situation of the enzyme in the cell. This is clearly shown by studies which demonstrate the differential subcellular insertion of molecular forms in living cells.

Subcellular Insertion of Molecular Forms. The classification of cholinesterase forms made according to biochemical criteria has relevance for their cellular location. Whereas the detergent- insensitive globular forms are exclusively soluble, the amphipathic forms may be intracellular, associated with the plasma membrane, or soluble in the extracellular fluid. In particular such forms are secreted in culture media by T28 murine neural cells (13) and occur naturally in Torpedo plasma (19). It is therefore not possible to assume an equivalence of soluble/detergent-insensitive molecules, and of amphipathic/membrane-bound molecules.

The enzyme that is associated with the plasma membrane is accessible to the extracellular medium and is thus physiologically available to hydrolyze released acetylcholine. For example, in the murine T28 neural cell line, the membrane-bound form is G4, while G1 is intracellular (12). In Torpedo organs, the G2 form is found on the membrane of presynaptic nerve endings (17, 21).

The asymmetric forms occur either intracellularly or extracellularly. They are found in both locations in approximately equal amounts in primary cultures of chick muscle cells (Vallette et al., in preparation), as well as in rat muscles _in vivo_ (32). In cultures of the mouse C2 muscle line, they occur almost exclusively as extracellular patches of activity probably associated with basal lamina components (11). In muscles _in vivo_, asymmetric forms are thought to account for the acetylcholinesterase activity that is associated with the synaptic basal lamina (18).

Multiplicity of Metabolic Pools. In T28 cell cultures, we have analyzed the metabolism of acetylcholinesterase G1 and G4 forms by the method of heavy isotope labeling (13). Our main findings are the following:

a) The enzyme is synthesized as an inactive precursor,

b) G1 is renewed much faster than G4; half-replacement of pre-existing light molecules by newly synthesized

heavy molecules being observed after 3.5 hours and
40 hours, respectively,

c)   most of the synthesized enzyme is rapidly degraded
     (approximately 70% of G1 appears to be degraded), and

d)   both G1 and G4 consist of several metabolic pools. For
     example, we have to consider at least three pools of
     G1: a "fast" pool which is mostly degraded, but also
     partially polymerized into G2 and G4, a "medium" pool
     which is at least partially secreted, and a "slow"
     pool which is degraded, with half-lives ranging from 1
     hour to 10 hours.

The metabolic heterogeneity of acetylcholinesterase probably
corresponds to biochemical differences (e.g., in the state of glycosylation) and/or to distinct subcellular compartments.

In studies of chick muscle cultures, using radioactive labeling,
Rotundo has recently observed a large extent of rapid degredation
(24). He has identified sequential changes in the glycosylation
of the enzyme and has shown that the assembly of collagen-tailed
forms takes place in the Golgi apparatus (25).

Distribution and Physiological Regulation of Acetylcholinesterase
Forms: Collagen-Tailed Forms in Skeletal Muscles. The distribution
of collagen-tailed forms in vertebrate tissues is more restricted
than that of globular forms. They appear to be invariably present
at the endplates of fast skeletal muscles, and are subject to a
distinct neural control. Early suggestions that they could be considered as biochemical correlates of nerve-muscle interactions
("endplate-forms") have not been borne out, because of many exceptions. For example, the asymmetric forms disappear from rat (10)
or chicken (26, 28) muscles after denervation, but not from rabbit
slow muscle (1).

CONCLUSION

It seems likely that the complex polymorphism of cholinesterases,
which has been conserved throughout the evolution of vertebrates,
is necessary to accommodate various subcellular locations of these
enzymes, which may be functionally essential in different physiological
situations. The manner in which it is generated and controlled
raises many interesting questions.

REFERENCES

1. Bacou, F., Vigneron, P. and Massoulié, J. (1982):   Nature 296:
    661-664.
2. Bon, S., Cartaud, J. and Massoulié, J. (1978):   Eur. J. Biochem.
    85:1-14.
3. Bon, S., Vigny, M. and Massoulié, J. (1979):   Proc. Natl. Acad.
    Sci. USA 76:2546-2550.
4. Bon, S. and Massoulié, J. (1980):   Proc. Natl. Acad. Sci. USA
    77:4464-4468.
5. Bon, S. (1982):   Neurochem. Int. 4:577-585.
6. Dutta-Choudhury, T.A. and Rosenberry, T.L. (1984):   J. Biol.
    Chem. 259:5653-5660.
7. Frenkel, E.J., Roelofsen, B., Brodbeck, U., Van Deenen, U.M. and
    Ott, P. (1980):   Eur. J. Biochem. 109:377-382.
8. Gómez-Barriocanal, J., Barat, A., Escudero, E., Rodríguez-Borrajo,
    C. and Ramírez, G. (1981):   J. Neurochem. 37:1239-1249.
9. Grassi, J., Vigny, M. and Massoulié, J. (1982):   J. Neurochem.
    38:457-469.
10. Hall, Z.W. (1973):   J. Neurol. 4:343-361.
11. Inestrosa, N.C., Miller, J.B., Silberstein, L., Zisking-Conhaim,
    L. and Hall, Z.W. (1983):   Exp. Cell Res. 147:393-406.
12. Lazar, M. and Vigny, M. (1980):   J. Neurochem. 35:1067-1069.
13. Lazar, M., Salmeron, E., Vigny, M. and Massoulié, J. (1984):
    J. Biol. Chem. 259:3703-3713.
14. Lee, S.L., Heinemann, S. and Taylor, P. (1982):   J. Biol.
    Chem. 257:12282-12292.
15. Lee, S.L. and Taylor, P. (1982):   J. Biol. Chem. 257:12292-12301.
16. Lee, S.L., Camp, S.J. and Taylor, P. (1982):   J. Biol. Chem.
    257:12302-12309.
17. Li, Z.Y. and Bon, C. (1983):   J. Neurochem. 40:338-349.
18. McMahan, U.J., Sanes, J.R. and Marshall, L.M. (1978):   Nature
    271:172-174.
19. Massoulié, J., Bon, S., Lazar, M., Grassi, J., Marsh, D., Méflah,
    K., Toutant, J.P., Vallette, F. and Vigny, M. (1984):   In Cholines-
    terases: Fundamental and Applied Aspects (eds) M. Brzin, E.-
    A. Barnard and D. Sket, W. de Gruyter and Co., Berlin, pp. 73-97.
20. Massoulié, J. and Bon, S. (1982):   Ann. Rev. Neurosci. 5:57-106.
21. Morel, N. and Dreyfus, P. (1982):   Neurochem. Int. 4:283-288.
22. Ramírez, G., Gomez-Barriocanal, J., Barat, A. and Rodríguez-
    Borrajo, C. (1984):   In Cholinesterases: Fundamental and Applied
    Aspects (eds) M. Brzin, E.A. Barnard and D. Sket, W. de Gruyter
    and Co., Berlin, pp. 115-128.
23. Rosenberry, T.L., Scoggin, D.M., Dutta-Choudhury, T.A. and Haas,
    R. (1984):   In Cholinesterases: Fundamental and Applied

Aspects (eds) M. Brzin, E.A. Barnard and D. Sket, W. de Gruy-
terand Co., Berlin, pp. 155-172.

24. Rotundo, R.L. (1984): In Cholinesterases: Fundamental and
    Applied Aspects (eds) M. Brzin, E.A. Barnard and D. Sket, W. de
    Gruyter and Co., Berlin, pp. 203-217.

25. Rotundo, R.L. (1984): Proc. Natl. Acad. Sci. USA 81:479-483.

26. Silman, I., Di Giamberardino, L., Lyles, J., Couraud, J.Y. and
    Barnard, E.A. (1979): Nature 280:160-162.

27. Torres, J.C. and Inestrosa, N.C. (1983): FEBS Lett. 154:265-268.

28. Tsuji, S. (1977): Brain Res. 124:352-356.

29. Vigny, M., Di Giamberardino, L., Couraud, J.Y., Rieger, F.
    and Koenig, J. (1976): FEBS Lett. 69:277-280.

30. Wiedmer, T., Di Francesco, C. and Brodbeck, U. (1979): Eur.
    J.Biochem. 102:59-64.

31. Witzemann, V. and Boustead, C. (1983): EMBO J. 2:873-878.

32. Younkin, S.G., Rosenstein, C., Collins, P.L. and Rosenberry,
    T.L. (1982): J. Biol. Chem. 257:13630-13637.

AN ENDOGENOUS NEUROTROPHIC FACTOR FOR THE MAINTENANCE OF ACETYLCHOLINESTERASE AND BUTYRYLCHOLINESTERASE IN THE PREGANGLIONICALLY DENERVATED SUPERIOR CERVICAL GANGLION OF THE CAT

G.B. Koelle, G.A. Ruch, E. Uchida, R. Davis,
W.A. Koelle, K.K. Rickard and U.J. Sanville

Department of Pharmacology
Medical School/G3
University of Pennsylvania
Philadelphia, PA 19104   U.S.A.

INTRODUCTION

Most studies of neurotrophic activity have been conducted in vitro with tissue culture preparations (e.g., 22). The preganglionically denervated superior cervical ganglion (SCG) of the cat has been found to provide an excellent test-object for demonstrating this type of activity in vivo. It was shown many years ago by Sawyer and Hollinshead (23) that within three days of preganglionic denervation the AChE content of the cat SCG falls to < 20% of its control value, and the BuChE to < 70%; these effects persist for several weeks. When these changes were studied by light microscopic histochemistry it was found that preganglionic denervation results in the total disappearance of AChE from the neuropil; it remains in high concentration in the perikarya of occasional (< 1%, 9) neurons, and in low concentration in the remainder. Butyrylcholinesterase, which is found normally only in the neuropil, shows a moderate fall (10, 15; Fig. 1 in ref. 11). From these and related findings it was concluded that most of the AChE of the normal cat SCG is confined to the preganglionic axons and their terminals, and that BuChE is present only in the capsular glial or Schwann sheath cells.

The development of a reliable electron microscopic histochemical method (12), which permitted direct visualization of the subcellular locations of the enzymes, necessitated considerable revision of these views. Acetylcholinesterase was now found to be localized at the plasma membranes of the dendrites and perikarya of the ganglion cells as well as of the presynaptic axons; BuChE was found chiefly at the identical postsynaptic sites, but not in the axonal terminals,

733

and in minimal amounts in the glial cells (4). Following preganglionic denervation, AChE disappeared from both pre- and postsynaptic sites, and the concentration of BuChE fell (5; Fig. 2 in ref. 11), just as had been noted by light microscopy. This raised the question of why AChE disappeared totally, and BuChE partially, from postsynaptic sites following preganglionic denervation, or conversely, why the integrity of the preganglionic fibers is essential for the normal maintenance of postsynaptic AChE and BuChE. The most likely explanation appeared to be that it is due to the release by the preganglionic fibers of a neurotrophic factor. This possibility was investigated in the studies reported here.

METHODS

A crude extract of the cat brain, spinal cord, and sciatic nerves was prepared by homogenizing these tissues from a single cat in approximately 7 volumes of water with a Waring blender. Sodium chloride solution was added to bring the final concentration to 0.9%, and the homogenate was adjusted to pH 4.5, with rehomogenization at each stage. Following centrifugation (30 minutes at 1,000g) the supernatent was readjusted to pH 7.0, recentrifuged, and the supernatent recovered and stored in the refrigerator overnight. Initially, aprotenin (Trasylol) was added to the homogenates, but this proved to be unnecessary.

Cats were anesthetized with sodium pentobarbital, a one-cm segment was resected from both preganglionic trunks, and the operative wound was sutured. The following day they were reanesthetized and atropinized; the trachea was intubated for artificial respiration, and the saphenous vein catheterized for the infusion of glucose-NaCl solution. The external carotid (EC) and lingual (L) arteries were ligated bilaterally [the SCG of the cat is supplied by the occipital and internal carotid branches of the common carotid artery (3)], and the right common carotid artery was exposed for infusion. The extract was delivered from an ice-water jacketed reservoir by a Harvard Peristaltic Pump at approximately 0.1 ml/min, from approximately 24 to 48 hours following denervation. At the end of that time the SCG were excised, frozen, and homogenized. Acetylcholinesterase and BuChE were assayed by a modification of the method of Ellman et al. (6), and protein was determined by the method of Lowry et al. (16).

RESULTS

The major results of the initial study are presented in Table 1 (13, 14).

It is apparent that infusion of the extract produced a marked decrease in the fall of ganglionic AChE and BuChE that otherwise occurs at 48 hours following preganglionic denervation. It is unlikely

Table 1

Effects of intraarterial infusion of an extract of cat brain,
spinal cord, and sciatic nerves on the AChE and BuChE contents
of preganglionically denervated cat SCG

Procedure	nmol substrate hydrolyzed/mg protein/min	
	AChE	BuChE
1. Normal cat SCG (8)	449 ± 42	590 ± 61
2. Preganglionically denervated SCG; EC and L arteries ligated at 24 hr; ganglia excised at 48 hr post-denervation (8)	167 ± 10	412 ± 29
3. Preganglionically denervated SCG; EC and L arteries ligated and extract infused 24-48 hr post-denervation (6)	354 ± 35*	602 ± 65**

Numbers of SCG in parentheses. *More than preganglionically denervated, arterially ligated controls, $p < 0.001$. **More than preganglionically denervated, arterially ligated controls, $p < 0.05$.

that the enzymes were taken up by the SCG from the extract. In a
previous study in this laboratory, the common carotid artery of
iso-OMPA treated cats, similarly arterially ligated, was perfused
for periods up to 10 hours with normal cat plasma (which contains
over ten times the concentrations of AChE and BuChE of the extract
used here); there was no evidence of uptake by the SCG of either
enzyme (unpublished). it is notable that there was no significant
difference between the AChE or BuChE contents of the right and left
SCG of the infused cats, indicating that there are extensive anasto-
moses between the two internal carotid arteries.

In exploring ways to simplify the assay procedure prior to
attempts to isolate the neurotrophic factor, two modifications were
tested: (1) omission of ligation of the EC and L arteries, and
(2) shortening the period of infusion. The first study led to unexpected results, which are summarized in Figure 1 (14). When SCG
were denervated then excised and assayed 48 hours later, with omission

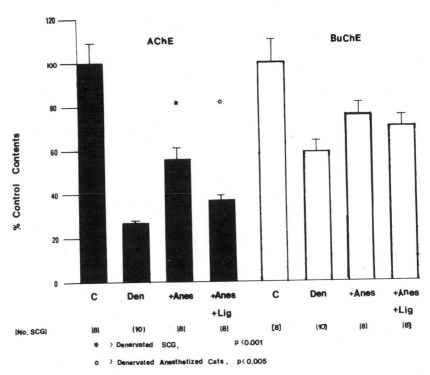

Figure 1. Acetylcholinesterase (solid bars) and BuChE (open bars)
contents of SCG of cats, expressed as percentages of normal controls
(C). All experimental ganglia were excised 48 hours following preganglionic denervation. Vertical lines indicate SEM.

 Den: Denervated.
 +Anes: Denervated; cats kept continually anesthetized
                 with sodium pentobarbitol.
 +Anes+Lig: External carotid and lingual arteries ligated 24
                 hours following denervation.
 ✷ >   Denervated SCG, p < 0.001.
 ☆ >   Denervated, Anesthetized cats, p < 0.005.

of anesthesia and arterial ligation on the intervening day, the
mean AChE contents fell to 29% and the BuChE to 58% of the control
values. When pentobarbital anesthesia was maintained throughout
this period (during which cats received subcutaneous infusions of
glucose-NaCl solution), the mean value for AChE was approximately
twice that of the previous group; BuChE changed in the same direction,
but the difference was not significant. Additionally, cats anesthetized and arterially ligated on the intervening day (the control
group for the earlier series) had mean values for AChE and BuChE
midway between the two preceding groups; for AChE, the difference

between the anesthetized, non-ligated and ligated cats was highly significant (p < 0.005).

In view of the increase in AChE and BuChE contents of denervated SCG produced by continual anesthesia, which was to a great extent negated by the arterial ligations, arterially ligated cats were employed as originally to study the effect of shortening the infusion period. Results are shown in Figure 2 (14). Infusion was begun approximately 24 hours following denervation; as the infusion period

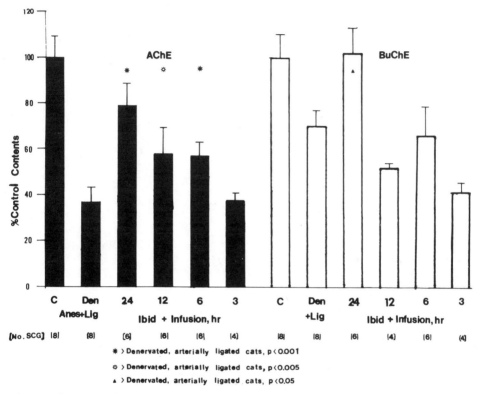

Figure 2. Acetylcholinesterase (solid bars) and BuChE (open bars) contents of SCG of cats expressed as percentages of normal controls (C). All experimental ganglia were exsised 48 hours following pre-ganglionic denervation. Vertical lines indicate SEM.

Den, Anes + Lig: External carotid and lingual arteries ligated
                      24 hours following denervation.

Ibid and Infusion, hr.: Same treatment as preceding group, then
                      infused with nerve tissue extract for number
                      of hours indicated. All cats received approxi-
                      mately the same volume of extract.

✱ >   Denervated, arterially ligated cats, p < 0.001.
☼ >   Denervated, arterially ligated cats, p < 0.005.
▲ >   Denervated, arterially ligated cats, p < 0.05.

Table 2

Effects of 24 Hour-Intraarterial Infusion of Acetylcholine (ACh)
and Nerve Growth Factor (NGF) on the AChE and BuChE Contents
of Preganglionically Denervated Cat SCG. (EC and L Arteries
Ligated at 24 Hours, Ganglia Excised 48 Hours Post-Denervation)

Cat	Infusion	Total Dose	nmol substrate hydrolyzed/mg protein/min	
			AChE	BuChE
T	ACh	2.8 mMol	167	257
V	ACh	15.8 mMol	203	278
	Means, T & V:		185 ± 11	268 ± 9
D	7 S-NGF	0.1 mg	263	263
G	2.5 S-NGF	0.5 mg	305	303
F	2.5 S-NGF	0.5 mg	203	308
	Means G & F:		254 ± 30	306 ± 11

was decreased, the rate was increased so that approximately the
same volume of extract (ca. 180 ml) was delivered in all experiments;
all SCG were excised 48 hours post-denervation. With infusion periods
of 12 and 6 hours, the decreased reduction of AChE was still significant; no effect was obtained with infusion for 3 hours. The
BuChE contents showed such great variation that only at 24 hour's
infusion was the effect significant.

Two authentic compounds have been tested by the foregoing procedure (Table 2). Two cats were infused for 24 hours with acetylcholine bromide (2.8 and 15.8 mMol, respectively). Values for AChE
at 48 hours post-denervation were not appreciably greater than the
controls (Table 1); values for BuChE were lower, which may reflect
a seasonal variation in this enzyme as noted previously (W.A. Koelle,
unpublished).

A cat was infused for the same period with 0.1 mg of 7 S-NGF,
and two additional cats were infused with extremely high doses (0.5
mg) of 2.5 S-NGF. Values for AChE were higher than those of the
controls, but considerably below those of cats infused for 24 hours
with extract (Table 1).

DISCUSSION

These studies have demonstrated the presence of a neurotrophic
factor in crude extracts of cat nerve tissue that opposes the fall
in the AChE and BuChE contents of the cat SCG that otherwise occurs

48 hours post-denervation. It might act by preventing or delaying degeneration of the preganglionic fibers (17); alternatively, it might directly affect the postsynaptic membranes in the same manner as a putative factor released by the presynaptic fibers to maintain the presence there of AChE and BuChE. The site is now under investigation by EM histochemistry (12).

If the neurotrophic factor acts postsynaptically, it is tempting to speculate on its possible mechanism of action. It has been shown that in the SCG of the rat (7, 27) and steer (2) the chief molecular forms of AChE present are G_1, G_4, and a relatively small proportion of A_{12}. The G_1 form is assumed to occur chiefly in the cytoplasm, the G_4 at the plasma membranes, and the A_{12}, by analogy with the motor endplate of skeletal muscle (8), at synaptic sites (19). The light and electron microscope histochemical studies described above have shown that in the cat the cytoplasm (presumably, the rough endoplasmic reticulum) of the vast majority of neurons in the cat SCG contains only traces of AChE; the bulk is present in the neuropil (the preganglionic axonal terminals and dendrites) and at the plasma membranes of the ganglion cells (4, 10). Following preganglionic denervation, the AChE of the cytoplasm appears to remain intact; it disappears from all the remaining sites (5, 15). This suggests that the neurotrophic factor may regulate the conversion of G_1 to G_4, and subsequently to A_{12}. The proposal is consistent with the changes that occur following preganglionic denervation in the SCG of the rat (7), where the relative concentrations of AChE at subcellular sites are quite different from those in the cat (26).

We are also attempting to isolate and characterize the neurotrophic factor. As shown here, it is not acetylcholine. Preliminary findings indicate that it is probably a protein. In view of the limited effects obtained with infusions of high doses of NGF, which is present in relatively low concentration in the mammalian CNS (1, 24, 25), NGF could have had only a minor contributory role in the results obtained with the infusion of nervous system extract.

The effect of continual anesthesia with sodium pentobarbital in opposing the fall in ganglionic AChE and BuChE was unexpected. Barbiturates are known to suppress energy metabolism, and this is probably the basis for their beneficial effects in the treatment of cerebral hypoxia (20), hypotension (21), and brain injury (18). Such an effect may have been operative here, perhaps by delaying degeneration of the preganglionic fibers. It is more difficult to explain how this effect was partially negated by ligation of the EC and L arteries. it is possible that this procedure resulted in an increase in blood flow to the SCG, which may have opposed the action of the barbiturate.

Acknowledgements: We are especially grateful to Dr. Ruth Hoag Angeletti for a generous gift of 2.5 S-NGF. This investigation was supported by Research Grants NS-00282-30,31 from the National Institute of Neurological and Communicative Disorders and Stroke, National Institutes of Health, U.S.A.

Addendum. Since this paper was presented, additional experiments have been completed involving infusions of dialyzed and enzyme treated extracts of the cat central nervous system. They have led to the conclusion that the neurotrophic factor under study is a heat-stable peptide of low molecular weight ($< 1,000$); it is probably present also in the gut, but not in significant concentrations in the liver or skeletal muscle (Koelle, G.B., Sanville, U.J., Rickard, K.K. and Williams, J.E.: Proc. Natl. Acad. Sci. USA 81:6539-6542, 1984).

REFERENCES

1. Barde, Y.A., Edgar, D. and Thoenen, H. (1980): Proc. Natl. Acad. Sci. USA 77:1199-1203.
2. Bon, S., Vigny, M. and Massoulie, J. (1979): Proc. Natl. Acad.-Sci. USA 76:2546-2550.
3. Chungcharoen, D., Daly, M. de B., and Schweitzer, A. (1952): J. Physiol. (London) 118:528-536.
4. Davis, R. and Koelle, G.G. (1978): J. Cell Biol. 78:785-809.
5. Davis, R. and Koelle, G.B. (1981): J. Cell Biol. 88:581-590.
6. Ellman, G.L., Courtney, K.D., Andres, V., Jr. and Featherstone, R.M. (1961): Biochem. Pharmacol. 7:88-95.
7. Gisiger, V., Vigny, M., Gautron, J. and Rieger, F. (1978): J. Neurochem. 30:501-516.
8. Hall, Z.W. (1973): J. Neurobiol. 4:343-361.
9. Holmstedt, B. and Sjöqvist, F. (1959): Acta Physiol. Scand. 47:284-296.
10. Koelle, G.B. (1951): J. Pharmacol. Exp. Ther. 103:153-171.
11. Koelle, G.B., Davis, R., Koelle, W.A., Ruch, G.A., Rickard, K.K. and Sanville, U.J. (1981): In: Cholinergic Mechanisms (eds) G. Pepeu and H. Ladinsky, Plenum Press, New York, pp. 127-132.
12. Koelle, G.B., Davis, R., Smyrl, E.G. and Fine, A.V. (1974): J. Histochem. Cytochem. 22:252-259.
13. Koelle, G.B. and Ruch, G.A. (1983): Proc. Natl. Acad. Sci. USA 80:3106-3110.
14. Koelle, G.B., Ruch, G.A. and Uchida, E. (1983): Proc. Natl. Acad. Sci. USA 80:6122-6125.
15. Koelle, W.A. and Koelle, G.B. (1959): J. Pharmacol. Exp. Ther. 126:1-8.
16. Lowry, O.H., Rosebrough, N.J., Farr, A.L. and Randall, R.J. (1951): J. Biol. Chem. 193:265-275.
17. Lubinska, L. (1982): Brain Res. 223:227-240.
18. Marshall, L.F., Smith, R.W., and Shapiro, H.M. (1979): J. Neurosurg. 50:26-30.
19. Massoulie, J. and Bon, S. (1982): Ann. Rev. Neurosci. 5:57-106.

20. Michenfelder, J.D. and Theye, R.A. (1973): Anesthesiology 39: 510-517.
21. Nilsson, L. (1971): Acta Neurol. Scand. 47:233-253.
22. Perez-Polo, J.R. and de Vellis, J. (eds) (1982): Special Issue on Growth and Trophic Factors. J. Neuroscience Research 8:127-567.
23. Sawyer, C.H. and Hollinshead, W.H. (1945): J. Neurophysiol. 8:137-153.
24. Schwartz, M.J., Ghetti, B., Truex, L. and Schmidt, M.J. (1982): J. Neurosci. Res. 8:205-211.
25. Shine, H.D. and Perez-Polo, J.R. (1976): J. Neurochem. 27: 1315-1318.
26. Uchida, E. and Koelle, G.B. (1983): Proc. Natl. Acad. Sci. 80:6723-6727.
27. Verdiere, M., Derer, M. and Rieger, F. (1982): Developmental Biol. 89:509-515.

PHARMACOKINETICS OF [^{14}C]-SARIN AND ITS CHANGES

BY OBIDOXIME AND PRALIDOXIME

P.G. Waser, R. Sammett, E. Schönenberger,
and A. Chang Sin-Ren

Institute of Pharmacology
University of Zurich
Zurich 8006, Switzerland

INTRODUCTION

Sarin was invented by Schrader in 1937 (6). It was one of the first examples of an organophosphate; it was later stockpiled with other similar compounds (soman, tabun and VX) as nerve gases for warfare. Of all the other 2000 organophosphorus compounds synthesized as insecticides, only several have been used to a major extent in agriculture. For all these compounds their lethal action is based on their blocking of acetylcholinesterase (AChE) (E.C.3.-1.1.7), the specific enzyme in cholinergic transmission which destroys acetylcholine. Death occurring by a highly increased concentration of this neurotransmitter, which depolarizes all synaptic connections in brain and the peripheral nervous system, can be prevented by reactivation of the hydrolyzing enzyme, AChE, by limiting or binding the neurotransmitter output, or by protection of the postsynaptic receptors with atropine or similar antagonists.

In this report we show some of our results in the investigation of the pharmacokinetics of [^{14}C]-labeled sarin in rats and mice (5). We also describe the distribution and interaction of both non-labeled and [^{14}C]-labeled sarin, following the administration of the reactivators of AChE, pralidoxime (2-PAM) (4), and obidoxime (Toxogonin®) (5).

MATERIAL AND METHODS

Synthesis of [^{14}C]-Sarin. Sarin with a [^{14}C]-labeled methyl group (Fig. 1) was synthesized in our laboratory by Dr. Chang Sin-Ren. It had a specific radioactivity of 4.48 - 4.65 mCi/mMol (= 32.0-

32.7 μCi/mg). In order to obtain a positive whole body autoradiography within weeks we had to inject a total activity of 1.0 μCi/20g
mouse, or a total dose of 1.5 mg/kg.

Gas-chromatographic-analysis performed immediately after the
synthesis showed 80% of the substance to be [^{14}C]-sarin. Besides
hydrolytic products, which are formed rapidly (t/2=90 min.) from
0.1 M sarin solution in water at a low pH, 4-10% of diisopropyl-
methyl-phosphonate was present as a side product. This interfering
compound was therefore synthesized by us in pure form, and labeled
with [^{14}C] (4.3 mCi/mMol=23.4 μCi/mg) (Fig. 1) to investigate its
distribution and action.

Synthesis of [^{14}C] Obidoxime and of [^{14}C]-2/PAM. The radiolabeled
carbons of [^{14}C]-obidoxime were synthesized in the methyl groups
of the linking chain (specific activity 4.4 mCi/mMol=12.28 μCi/mg).
Pralidoxime was prepared with a [^{14}C]-N-methyl group at a specific
activity of 3.55 μCi/mg. Both were synthesized by Dr. Chang Sin-
Ren (Fig. 1).

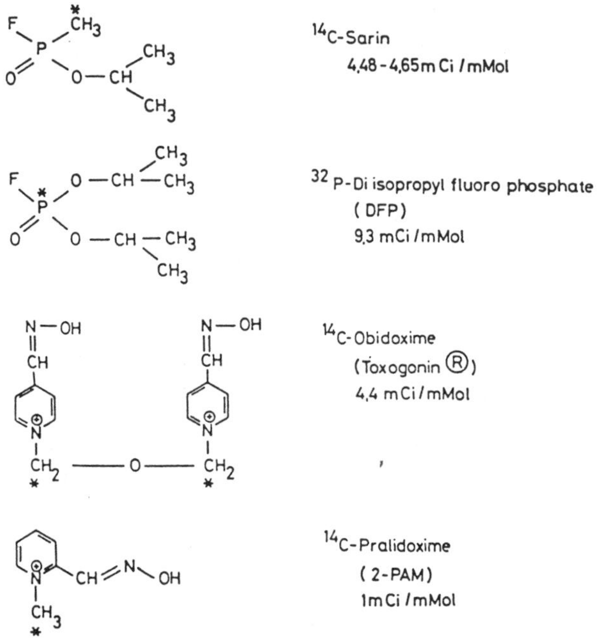

Figure 1. Chemical structure and specific radioactivity of compounds
synthesized by Dr. A. Chang Sin-Ren.

Autoradiographic technique. Narcotized animals were injected intra-venously, and their ECGs recorded continuously. They were artificially ventilated through a tracheal cannula. The experiments were continued after the sarin injection for up to 60 min. The animals were killed by immersion in a hexane/CO_2-mixture (-75°C) and then embedded in carboxy-methylcellulose (5%) and sectioned in a whole body LKB-2250 PMV-Cryo-Microtome 450. The 20 μ thick sections were exposed to Kodak NS or Kodak DEF-5 film for 8 weeks and then developed in a Kodak-Microdol-X-developer; they were later fixed in Kodak-Unifix solution. Some of the sections were stained with haematoxylin-eosin to permit identification of tissues (2).

Radioactivity in the tissues represented by silver grains in the films, was measured using a transmission densitometer TD-504 (Macbeth division of Kollmorgen Corporation). The relative amount of total radioactivity in small areas (0.2 mm) was quantitatively determined by comparison with a series of film emulsion pads containing a scale of 16 dilution concentrations with [¹⁴C]-glucose, on every autoradiograph together with the section through the animal.

The <u>experimental</u> <u>protocol</u> used in these studies on the interaction of sarin and obidoxime is shown in Fig. 2.

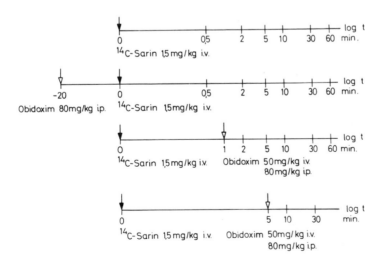

Figure 2. Experimental protocol used in mice, in the study of the interaction of sarin and obidoxime.

RESULTS

[^{14}C]-Sarin.   The most astonishing result obtained was the very
fast uptake of radiolabeled sarin into the central nervous system;
both brain and spinal cord (Fig. 3).   Thirty seconds after the i.v.
injection of 1.5 mg/kg the radioactivity in the CNS, cardiac blood,
and liver was similar; but the suprarenal cortex contained 3 times
the blood concentration.   All other vital organs except the kidneys,
i.e., adrenal medulla, liver, lungs and spleen showed peak concentrations within a minute.   Radioactivity in all organs was subsequently
diminished within the next 60 minutes.   Only the urine in the pelvis
of the kidney showed an increased radioactivity over a long period.
After 60 min the accumulated radioactivity in the nervous tissue
had gradually diminished to one third of the maximal value (Fig. 4).

[^{14}C]-Sarin plus Obidoxime.   As mentioned in the methods section,
sarin and obidoxime were administered in several different ways
and sequences (Fig. 2).   Results obtained are discussed in the sub-
sections which follow.   Some of these are illustrated in Fig. 5
and 6.

   a) Obidoxime, 80 mg/kg, i.p., 20 min before Sarin (prophylactic
application).   Obidoxime produced a remarkable decrease in the concentration of [^{14}C]-sarin in the CNS 30 sec after injection, which

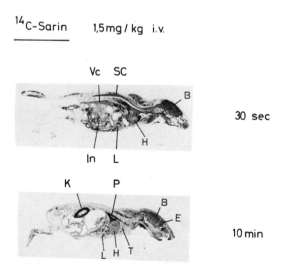

Figure 3. Distribution of [^{14}C]-sarin in mice 30 sec and 10 min after
i.v. injection of 1.5 mg/kg.

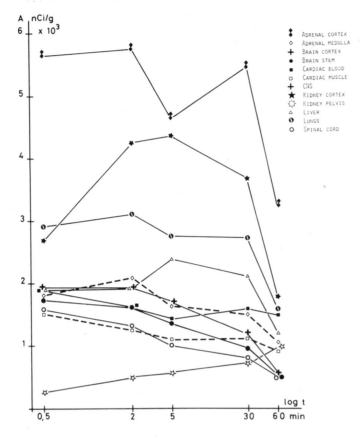

Figure 4. Total radioactivity (μCi/g x 10^3) in different mouse organs at different times after i.v. injection of 1.5 mg/kg of $[^{14}C]$-sarin.

was even more prominent 2 min later (Fig. 5). The concentration in the cardiac blood was 3 times higher than in the cortex, and 1.5 times over that in the cardiac muscle. A decrease followed within 2 min in a few organs, such as lungs and spleen, but not in the liver, kidney, cerebral cortex or adrenal cortex, where we first found an increase within two min, followed by a decrease. From 30 to 60 min after administration of obidoxime much radioactivity was shown to be excreted through the renal cortex and pelvis.

b) Obidoxime, 50 mg/kg, i.v., 1 min after $[^{14}C]$-Sarin (immediate therapeutic application). Two min after sarin administration the radioactivity in the CNS and heart was relatively high, as in a) above (Fig. 5, 6). But after 5-10 min there was a pronounced decrease in radioactivity in the CNS, and contrary to this an increase in liver

14C-Sarin   1,5 mg/kg i.v.

2 min

+ Obidoxim 80 mg/kg i.p. 20 min before

2 min

Obidoxim 50 mg/kg i.v. 1 min after

2 min

Figure 5.   Interaction of obidoxime injected i.p. before or i.v.
afterwards, with [14C]-sarin (1.5 mg/kg i.v.).

and renal cortex and pelvis. Even after 60 min this increase did
not change in comparison with most other organs.

c) Obidoxime, 50 mg/kg, i.p., 1 min after [14C]-Sarin. [14C]-
Sarin distribution was measured 10 min after its i.v. injection,
in order to permit absorption of the i.p. administered obidoxime.
The radioactivity in the CNS was already low, increased in the cardiac
blood and elevated in liver, renal cortex, and adrenal cortex, in
contrast to the previous experiments.

d) Obidoxime, 50 mg/kg, i.v., 5 min after [14C]-Sarin (delayed
therapeutic application). Again the radioactivity in the CNS 10 min
after i.v. injection of sarin was 2 times lower than in the control
experiment, and distinctly higher in the cardiac blood. In the
adrenal cortex and in renal cortex it is very high. Within 30 min
these two organs lost some activity, but it was increased in the liver.

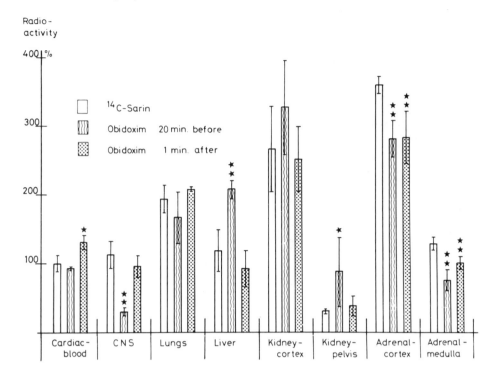

Figure 6.   Influence of obidoxime treatment on [^{14}C]-sarin-radio-activity in different organs 2 min after intravenous injection of sarin.

e)   Obidoxime 80 mg/kg, i.p., 5 min after [^{14}C]-Sarin.   Even 10 min after administration of radiolabeled sarin, [^{14}C]-sarin activity in the CNS was clearly lower when compared to cardiac blood. The highest activity was found in the renal cortex.   During the next 30 min the CNS lost some radioactivity, but in the liver the increase was pronounced.

[^{14}C]-Diisopropyl-methyl-phosphonate i.v.   When compared with the amount of radioactivity incorporated (10%) following 1.5 mg/kg [^{14}C]-sarin administration, the radioactivity accumulated following [^{14}C]-diisopropyl-methyl-phosphonate administration of 0.15 mg/kg was notably low in all organs.   CNS (brain and spinal cord), cardiac muscle, thymus, spleen, and adrenals were equal to the background in the films.   In the cardiac blood radioactivity was low after 1 min, but in the liver and the kidneys and especially in the urinary bladder the concentrations were 10-80 times higher.   Even after a 14 times higher i.v. dose (2.1 mg/kg) the concentration in the CNS was always low and equal to the background.   The highest activity

was found to be in the liver, intestine, and later in the urinary
excretory organs.

[^{14}C]-Obidoxime, i.v., alone (rats).   Obidoxime at a dose of 0.1
g/kg, i.v., after 2-30 min was concentrated mostly in the cardiac
blood, the kidneys, the urinary bladder, and the intestine (Fig. 7).
A lower concentration was found in the liver and spleen. A remarkably
high concentration throughout the 120 min after injection was found
in the cartilage of the skull and other bones, the intervertebral
discs, the ribs and sternum, the subcutaneous tissues, the larynx,
etc. Most impressive, however, was the fact that no traces of [^{14}C]-
obidoxime were found in the CNS, even after 120 min! (8).

[^{14}C]-Pralidoxime, i.v. (mice). [^{14}C]-2-PAM (75 mg/kg) i.v. produced
similar autoradiographs as those seen with obidoxime treatment.
The CNS was blank, and most of the radioactivity was concentrated
in the liver, kidneys, urinary bladder, bile, intestine, and salivary
glands. After 180 min the activity remained mostly in the bile
and intestine (4).

^{14}C-Obidoxim
——————

Distribution of rats after i.v. injection (0,1 g/kg)

2 min

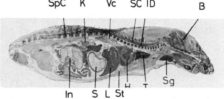

Histological section (20 μ, HE – staining)

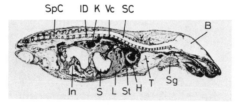

Autoradiography

Figure 7. Distribution of [^{14}C]-obidoxime in rats 2 min after i.v.
injection (0.1 g/kg).

DISCUSSION

[^{14}C]-Sarin was found to be distributed immediately after the injection in the whole body of the mouse and passed readily through the blood-brain barrier, in contrast to radiolabeled diisopropyl-fluoro-phosphonate ([^{32}P]-DFP), which enters only in small amounts into the brain (4), and [^{14}C]-diisopropyl-methyl-phosphonate (5), which does not pass the blood brain barrier. During the first 60 min following injection the radioactivity of the nervous tissue was diminished steadily. As we had to inject 0.75 to 1.5 mg/kg of [^{14}C]-sarin i.v., or 8-21 times its LD$_{50}$ [(70-95 μg/kg, (4, 5)], not much discrimination between different parts of the brain was to be expected. Some autoradiographs did nevertheless show highest concentrations in the cerebellum and the brain stem centers. Much of the injected sarin apparently was not bound to the cholinesterases or proteins, but remained free to be shifted into different parts or compartments of the body. Most of the injected sarin was ultimately excreted through the kidney, although the pelvis contained lower concentrations; the urinary bladder was not recognizable, probably because of maximal cholinergic contraction. The adrenal cortex was always very highly radioactive, probably because of its high vascularization. The accumulation of [^{14}C]-sarin in the liver was not pronounced, and detoxification by its unspecific esterases there-fore is probably not of much importance. Through the biliary system much activity was passed to the intestine. The lungs accumulated a considerable amount of radioactivity, indicating that they provide an important elimination pathway.

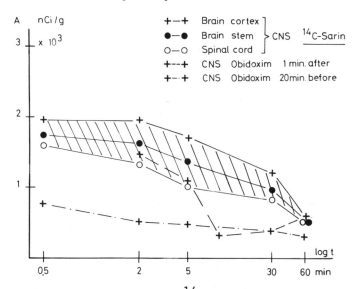

Figure 8. CNS-radioactivity of [^{14}C]-sarin and metabolites alone (hatched area) and of cortex (+), with different interactions of obidoxime.

The prophylactic application of obidoxime i.p., 20 min. before [14C]-sarin, distinctively lowered the radioactivity in the nervous tissues by a factor of 2.3-3.8 (Fig. 8). This protective effect may be explained in different ways:

a) Direct reaction between [14C]-sarin and obidoxime by formation of a complex between the two charged molecules similar to the "Tammelin ester" (7). As obidoxime was 20 times more concentrated in the body than sarin, this radioactive complex may explain the increased radioactivity in the blood, and later in the metabolizing liver, the kidneys, and perhaps the suprarenals.

b) The AChE might be protected in all synaptic regions, reversibly by obidoxime, against the irreversible binding of sarin. As obidoxime is strongly polar (cationic) it might be held in the vascular system, and leave most of the enzymes in the other tissues unprotected, becoming strongly marked by [14C]-sarin.

c) Reactivation of phosphorylated AChE by obidoxime, injected 20 min earlier, may have been possible for well vascularized organs, because the obidoxime was highly concentrated within the blood. The metabolite of hydrolyzed sarin still carried the [14C] labelled methyl group ([14C]-methylphosphoric-monoiso-propyl-ester), and could not be distinguished from sarin by autoradiography.

d) The permeability of the blood brain barrier may have been influenced by obidoxime. It is known that seizures and convulsions, which occur by the action of sarin, enhance the permeability of the blood-brain barrier (1,3). This action might conceivably be diminished by the interaction of obidoxime with sarin in the periphery.

Therapeutic application of obidoxime 1 min after [14C]-sarin i.v. resulted in an immediate decrease of radioactivity in cardiac blood and CNS (Fig. 8). Two to 10 min after the injection the difference is most pronounced (1/4 radioactivity), but after 60 min similar values as in the controls were found in the nervous tissues. At the same time a remarkable increase in the organs of elimination, kidneys and liver, took place as a result of shifting of [14C]-sarin into the CNS from the bloodstream. Within the first 10 min much of the enzymes may have been reactivated and the sarin excreted after formation of an obidoxime/sarin complex. As the blood concentration was diminished, more of sarin might conceivably have returned to the blood stream from the nervous tissues.

The i.p. injection of obidoxime reduced the sarin concentration in the CNS more slowly, probably by limited elimination through the kidneys and biliary system. During the first 10 min, i.v. injection had a more effective and faster action on CNS and cardiac blood than did i.p. injection.

Therapeutic i.v. injection of obidoxime 5 min after i.v. sarin, resulted in decreased radioactivity in the CNS, 10 or 30 min later. At this time the radioactivity of the liver increased continuously. This action was considerably less than after an immediate i.v. injection of obidoxime. Again, i.p. injection 5 min after sarin produced a delayed but marked decrease in radioactivity in the brain, and at the same time high activity in the liver.

Compared to intravenous therapeutic injection, intraperitoneal injection was always more successful in reducing the effects of sarin in mice, and in increasing survival.

$[^{14}C]$-Diisopropyl-methyl-phosphonate as a radioactive byproduct in the injected sarin-solution had evidently no effect on the autoradiographs in most organs, since the slight radioactivity was within the statistical deviations of the $[^{14}C]$-sarin measurements. As a polar and water soluble compound, most of it was immediately eliminated through the kidneys. In the CNS however it was absent, and in other main organs the radioactivity was very low.

Finally the most important finding in these studies was the absence of $[^{14}C]$-obidoxime and $[^{14}C]$-pralidoxime accumulation in the CNS. An explanation for their measurable but rather small and only slightly effective antidotal action can only be found by further investigation of the different states of permeability of the blood brain barrier, a direct chemical complexing effect toward sarin with a resulting increased elimination, or both.

CONCLUSIONS

In conclusion, $[^{14}C]$-sarin has been synthesized by us and used for intravenous injections in mice and rats. It produced known severe symptoms of intoxication at an LD_{50} dose of 70 $\mu g/kg$. Narcotized and artificially ventilated animals were then treated with a prophylactic injection (i.p.) of obidoxime 20 min before sarin, or a therapeutic injection (i.v. or i.p.) of obidoxime 1 or 5 min after sarin. $[^{14}C]$-Diisopropyl-methyl-phosphonate, a contamination product of sarin-synthesis, was synthesized and its distribution in comparison to sarin investigated. Finally, $[^{14}C]$-obidoxime and $[^{14}C]$-pralidoxime were synthesized, and their distribution studied in animals in order to understand their mechanisms of action.

Sarin was found to be distributed rapidly throughout the animal, as seen in the whole-body autoradiographs. It entered the CNS within

15-30 seconds. By 2-10 min the concentration in the CNS diminished continuously, and the excretion organs gained more radioactivity. Within 60 min it was already partly eliminated through the urinary system, the liver-bile pathway, and perhaps through the lungs. The contaminating diisopropyl-methyl-phosphonate has no influence on the distribution of radioactivity due to [^{14}C]-sarin.

The prophylactic administration of obidoxime (80 mg/kg, i.p., 20 min before [^{14}C]-sarin) induced a marked decrease in radioactivity in the CNS. Also the lungs and the adrenal glands showed significantly lower [^{14}C] concentration, whereas the liver, and to a large extent the kidneys gradually accumulated a higher radioactivity count within 60 min.

Therapeutic treatment (i.v. or i.p.) with obidoxime 1 or 5 min after [^{14}C]-sarin did not clearly influence the distribution patterns of [^{14}C]-sarin. Liver and kidney concentrations were higher, and CNS-activity lower than in the controls some minutes after obidoxime administration; showing that hepatic metabolism and renal elimination are increased. [^{14}C]-Obidoxime and [^{14}C]-pralidoxime did not enter the CNS, and were mainly eliminated through the urinary tract.

Finally [^{14}C]-sarin kinetics could only be influenced by prophylaxis with obidoxime, resulting in a decrease of radioactivity in the CNS.

Acknowledgement. We wish to thank the Swiss National Foundation (project 3.544.079) and the Swiss Commission for Military Research in Medicine for financial help. We thank Dr. G. Riggio for analytical help, Dr. W. Hopff for fruitful discussions, Prof. G. B. Koelle, and Prof. I. Hanin for editing the manuscript.

REFERENCES

1. Ashani, Y. and Catravas, G.N. (1981): Biochem. Pharmacol. 30:2593-2601.
2. Cross, S.A.M., Groves, A.D. and Hesselbo, T. (1974): Int. J. Appl. Radiat. Isotopes 25:381-386.
3. Falb, A. and Erdmann, W.D. (1969): Arch. Toxikol. 24:123-132.
4. Gotheil, A.M. (1978): Ganztierautoradiographische Untersuchungen uber die Verteilung von Pyridin-2-Aldoxim-[^{14}C]-Methoiodid in Mausen allein und unter Einwirkung von Methylisopropylfluorophosphonat. Dissertation ETH Zurich, Nr. 6220.
5. Sammet, R. (1983): Kinetik von [^{14}C]-sarin und deren Beeinflussung durch Obidoxime - eine Ganztierautoradiographische Untersuchung an der Maus. Dissertation ETH Zurich, Nr. 7288.
6. Schrader, G. (1952): Monographie Nr. 62, Verlag Chemie, Weinheim.
7. Tammelin, L.E. (1957): Acta Chem. Scand. 11:1340-1349.

8. Waser, P.G. et al. (in preparation): Investigation on [^{14}C]-obidoxime and sarin.

SUCCINYLCHOLINE - A METHOD OF DETERMINATION.

DISTRIBUTION AND ELIMINATION

I. Nordgren

Department of Toxicology
Karolinska Institutet
Box 60400
S-104 01 Stockholm, Sweden

INTRODUCTION

For the analysis of endogenous and exogenous compounds in tissues and body fluids, gas chromatographic methods are very attractive, especially if combined with a mass spectrometer. To be able to be gas chromatographed, the compound must be volatile itself, or be derivatized into a volatile product.

Quaternary ammonium compounds can be gas chromatographed after different kinds of pyrolysis reactions. The Hofmann degradation to form olefins is one example. Under certain conditions pyrolysis leads to the formation of the corresponding tertiary amine by an S_N2 reaction. This was described by Hofmann already in 1851 for the halide salts (4, 5). Szilagyi, Schmidt and Green 1968 (11) used this reaction for the analysis of acetylcholine (ACh) and other esters of choline. They reported that the reaction is also applicable to bisquaternary ammonium compounds, e.g. succinylcholine (SCh) (9).

The demethylated product of ACh can also be obtained by heating with benzenethiolate. The reaction was first described by Shamma, Deno and Remar 1966 (10) as a general reaction for demethylation of quaternary compounds on a preparative scale. This method was adapted for the analysis of ACh in biological material by Jenden, Hanin and Lamb 1968 (3). It was also reported to be applicable to other esters of choline including SCh.

Karlén et al. 1974 (6) combined this demethylation technique with a method for the extraction of ACh as an ion-pair with hexanitrodiphenylamine (DPA) (1). With this technique and the use of

757

gas chromatography - mass spectrometry the limit of detection of
ACh is 20 pg/g tissue. The application of this method to the identification and quantitation of SCh in biological material was described
by Forney et al. 1982 (2) and Nordgren et al. 1983 (8).

The short duration of action of SCh has been ascribed to a
rapid enzymatic hydrolysis by pseudocholinesterase (12). Other
means of disposition seem to have been largely neglected in the
discussion of its evanescent action. To shed further light on this
question, we undertook experiments to study stability, tissue distribution and elimination of SCh from plasma in man and experimental
animals (2, 7, 8). This research was originally prompted by a forensic
case; the experimental results therefore also include studies of
embalmed material.

MATERIALS AND METHODS

Chemicals. All chemicals were of analytical grade. The reagents
were prepared according to Nordgren et al. 1983 (8).

Rat Experiments. Five Sprague Dawley rats received doses of SCh
ranging between 10 and 200 mg/kg by i.m. injection in the gluteal
muscles. Two controls received the same volume of distilled water
(5 ml/kg) also by i.m. injection in the gluteal muscles and were
killed with carbon dioxide. The rats were refrigerated (8°C) for
6 hrs after death and were then embalmed by i.v. and cavity infusion
of FAX (Champion Chemical Co., Ontario, Canada), a commercially
available glutaraldehyde embalming fluid diluted 1:8 with water.
The tissues were homogenized in distilled water with an Ultra Turrax
homogenizer. Perchloric acid (final concentration 0.4 N) and the
internal standard, deuterated SCh, was added. The homogenates were
left in a refrigerator (3°C) for 30 min and were then centrifuged
for 20 min at 100,000 x g and 4°C.

Dog Experiments. Mongrel dogs received doses of SCh ranging between
2 - 106 mg/kg by i.v. or i.m. injection. The dogs were anesthetized
with pentobarbital. In order to study tissue distribution and elimination from plasma, tissues were collected 1-45 min after administration of SCh, and blood was collected by means of an indwelling
catheter in heparinized tubes containing physostigmine (10^{-4}M final
concentration) 0.5 - 36 min after the SCh injection. In reference
to forensic cases some tissues were embalmed by injection and soaking
in FAX, diluted as described above. Perchloric acid was added to
the plasma (final concentration 0.4 N), while the tissues were homogenized in 0.4 N $HClO_4$. Internal standard was added and the samples
were treated as described for rat tissue. Two dogs were kept under
artificial respiration while others were left to die from the dose
of SCh.

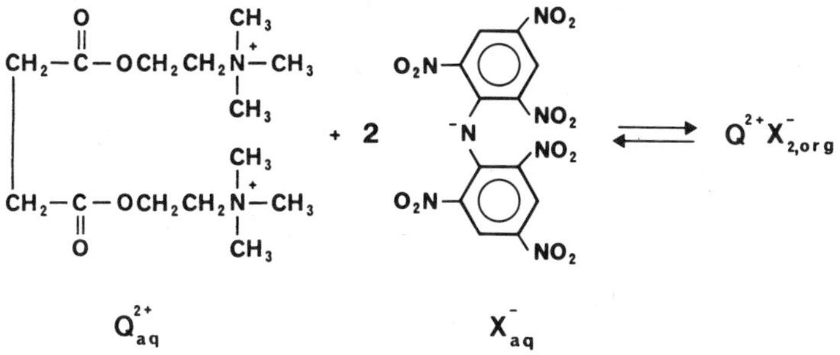

Figure 1. Extraction of SCh (Q^{2+}) into an organic phase as an ion pair with DPA (X^-).

Human Plasma. A male surgical patient received 0.6 mg SCh per kg i.v. Blood samples were collected in heparinized tubes containing physostigmine (final concentration $10^{-4}M$).

Determination of SCh. The analyses were carried out according to Forney et al. 1982 (2) and Nordgren et al. 1983 (8). SCh was extracted from an aqueous phase, a homogenate or plasma, into dichloromethane as an ion pair with DPA'(Fig. 1) and was then demethylated with sodium benzenethiolate to form the corresponding tertiary amine (Fig. 2) which was analyzed by gas chromatography-mass spectrometry. Standard curves were prepared from drug free plasma or tissues. As internal standard we used SCh labelled with 12 or 18 deuteriums.

The gas chromatographic column used was an SE-52 glass capillary column, 25 m in length and with an i.d. of 0.32 mm. The injector and column temperatures were 250°C and 200°C, respectively. Helium was used as carrier gas with a flow rate of 1 ml/min. The mass spectrometer used was an LKB 2091 with computer controlled mass spectra and MID systems. The separator and ion source temperatures were kept at 250°C. The electron energy was 70 eV and the trap current 50 μA.

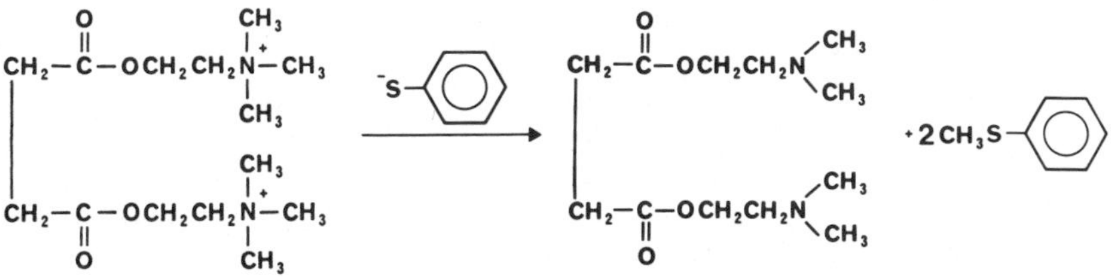

Figure 2. N-Demethylation of SCh with sodium benzenethiolate.

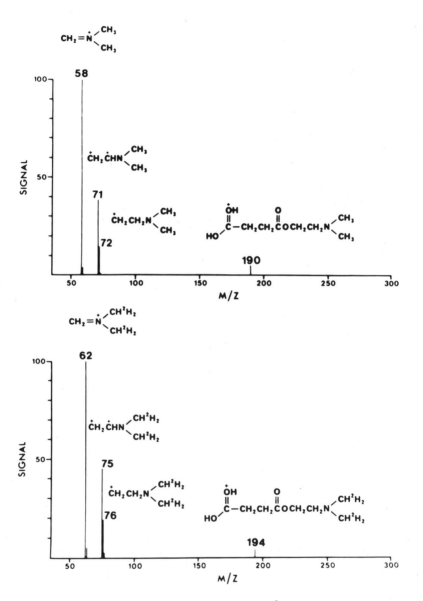

Figure 3.   Mass spectra of demethylated [^{1}H]-SCh (upper spectrum) and [^{2}H]$_{12}$-SCh (lower spectrum).

Fig. 3 shows the mass spectra of demethylated [^{1}H]-SCh and [^{2}H]$_{12}$-SCh and the possible structures of the major fragments. The spectra were obtained at the maximum of the total ion current of the peaks after background subtraction by the computer.

In the quantitative analyses the mass spectrometer was focussed on the most abundant ions formed. After removal of one methyl group the base peaks were m/z 58 for unlabelled SCh, and m/z 62 ([^{2}H]$_{12}$-SCh) and 64 ([^{2}H]$_{18}$-SCh) for the deuterated variants. Fig. 4 shows a mass fragmentogram from plasma.

To confirm the correctness of the analyses, quantitative analyses were also performed at m/z 190 and 194.

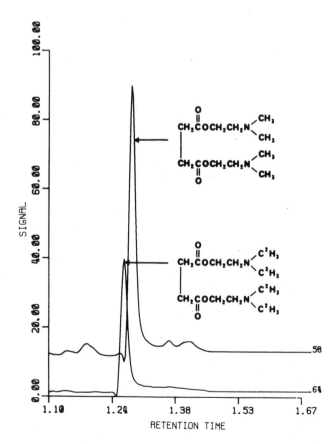

Figure 4. Mass fragmentogram of demethylated SCh (m/z 58) and the internal standard [^{2}H]$_{18}$-SCh (50 ng, m/z 64) in human plasma.

Table 1

SCh in Embalmed Rat Tissues

	SCh dose (mg/kg i.m.)	Time of storage (months)		SCh concentrations (μg/g)		
		8°C	-20°C	Kidney	Liver	Injection site muscle
Control	0	3	3	ND[a]	ND[a]	ND[a]
(CO$_2$)	0	6	0	ND[a]	ND[a]	ND[a]
SCh	10	3	3	-[b]	0.1	0.7
	50	3	3	-[b]	1.9	5.3
	50	6	0	ND[a]	ND[a]	0.3
	50	6	0	0.9	-[b]	0.5
	200[c]	3	3	-[b]	2.4	14.2[c]

[a] ND = Not detected
[b] - = Specimens not analyzed
[c] The rat received 100 mg/kg i.m. in each hind leg muscle. The total dose was 200 mg/kg but only 100 mg/kg was injected into the muscle analyzed.

RESULTS AND DISCUSSION

Table 1 presents the results of analyses of embalmed rat tissues. In order to study the influence of storage temperature on the stability of succinylcholine some of the rats were refrigerated at 8°C for six months before analysis while some of them were refrigerated for three months and then frozen at -20°C for another three months. All animals were analyzed at the same time (2, 8). SCh was still present six months after death. The concentration of the drug in tissues appeared to be dose dependent. Freezing some of the specimens for three months appeared to have preservative value.

Fig. 5 shows the results of analyses of SCh in plasma in a surgical case. SCh was rapidly eliminated.

To study the role of enzymatic hydrolysis as compared to tissue distribution and excretion, we measured the elimination of SCh in plasma from dogs both with and without artificial respiration (7).

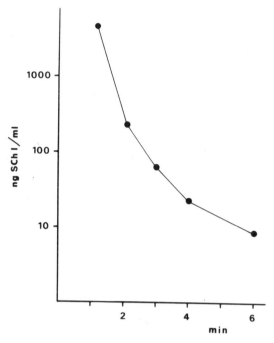

Figure 5. Elimination of SCh from human plasma after i.v. injection of 0.6 mg/kg.

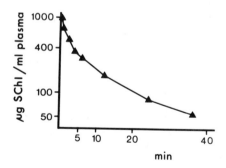

Figure 6. Elimination of SCh from dog plasma. The dog was injected i.v. with SCh 2 mg/kg, 1.5 hrs later followed by a second dose, this time 106 mg/kg. The curve shows the elimination after the second dose. The dog was kept under artificial respiration.

Table 2

Levels of SCh after Administration of 2 mg/kg i.v.,
1.5 hrs Later Followed by a Second Dose of 106 mg/kg i.v.

	Time after the second dose of SCh	μg SChI/g
Kidney	13	1175
"	38	803
Liver	43	5
Diaphragm	45	65
Spleen	42	107

The dog was kept under artificial respiration and the tissues were
embalmed with FAX

Table 3

Levels of SCh after Injection of 10 mg/kg i.v. or i.m.

	Route of Administration	Time after injection (min)	μg SChI/g
Kidney	i.v.	2	219
"	i.v.	10	306
"	i.m.	4	2.6
"	i.m.	8	1.6
Liver	i.v.	1	0.91
"	i.v.	10	0.52
"	i.m.	4	0.022
"	i.m.	8	0.014

The dogs were not artificially ventilated.   The tissues were
embalmed with FAX

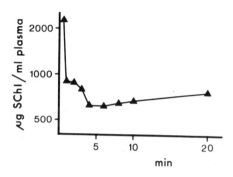

Figure 7. Elimination of SCh from dog plasma after administration
of 67 mg/kg i.v. The dog was not artificially ventilated.

When the dogs were artificially ventilated the SCh was rapidly
eliminated as illustrated by Fig. 6.

In cases where the dogs were not kept under artificial respiration
the decrease in the level of SCh stopped when the animals died and
the circulation failed. This is illustrated in Fig. 7. Cessation
of respiration occurred after 30 sec. Drop in blood pressure and
cessation of regular heart beats occurred after roughly 4 min.

Of the tissues analyzed the highest amounts of SCh were found
in kidney as demonstrated in Table 2.

We found the tissue concentrations to increase in a dose dependent
manner, but the relative distribution between the organs seemed to
be the same irrespective of the dose administered. Tissue concentrations were lower for animals kept under artificial respiration,
when comparable doses were given. I.m. injection led to lower tissue
levels compared to i.v. administration of the same dose (Table 3).

These results show that SCh can be analyzed in tissues and
body fluids from man and experimental animals, and that it is fairly
stable provided that the material is stored under satisfactory conditions.

The results also show that the disappearance of SCh from plasma
is not only due to enzymatic hydrolysis. Tissue distribution and
excretion play an important role. The finding that the elimination
of SCh from plasma in dogs stops upon circulatory failure might
indicate that, at the SCh concentrations used, the enzymatic hydrolysis
is a capacity-limited process. This would lead to an increasingly
important tissue distribution with higher SCh-doses, and also explains
why SCh can be found in tissues from forensic cases.

Acknowledgements. This work was supported by grants from the Swedish Medical Research Council B84-14X-00199 and the Wallenberg Foundation.

REFERENCES

1. Eksborg, S. and Persson, B.A. (1971): Acta Pharmaceutica Suecica 8:205-216

2. Forney, R.B. Jr., Carroll, F.T., Nordgren, I.K., Pettersson, B.-M. and Holmstedt, B.R. (1982): J. Anal. Toxicol. 6:115-119.

3. Jenden, D.J., Hanin, I. and Lamb, S.I. (1968): Analytical Chemistry 40:125-128.

4. Hofmann, A.W. (1851): Annalen der Chemie und Pharmacie 78:253-287.

5. Hofmann, A.W. (1851): Annalen der Chemie und Pharmacie 79:11-39.

6. Karlén, B., Lundgren, G., Nordgren, I. and Holmstedt, B. (1974): In Choline and Acetylcholine: Handbook of Chemical Assay Methods. I. Hanin (ed), Raven Press, New York, pp. 163-179.

7. Nordgren, I., Baldwin, K. and Forney, R. Jr. (1984): Biochemical Pharmacol. 33(15):2519-2521.

8. Nordgren, I.K., Forney, R.B. Jr., Carroll, F.T., Holmstedt, B.R., Jaderholm-Ek, I. and Pettersson, B.-M. (1983): Arch. Toxicol. Suppl. 6:339-350.

9. Schmidt, D.E., Szilagyi, D.I.A. and Green, J.P. (1969): Journal of Chromatographic Science 7:248-249.

10. Shamma, M., Deno, N.C. and Remar, J.F. (1966): Tetrahedron Letters 13:1375-1379.

11. Szilagyi, D.I.A., Schmidt, D.E. and Green, J.P. (1968): Analytical Chemistry 40:2009-2013.

12. Taylor, P. (1980): In The Pharmacological Basis of Therapeutics, Sixth Edition, A.G. Gilman, L.S. Goodman and A. Gilman (eds), Macmillan Publishing Co., New York, Toronto and London, pp. 227, 230.

CHOLINERGIC EFFECTS OF HI-6 IN SOMAN POISONING

T.-M. Shih[1], C.E. Whalley[1], J.J. Valdes[2],
P.M. Lundy[3] and P.A. Lockwood[4]

[1]Basic Pharmacology Branch
U.S. Army Medical Research Institute of Chemical Defense

[2]Toxicology Branch
Chemical Research and Development Center
Aberdeen Proving Ground, Maryland   USA

[3]Biomedical and [4]Chemistry Sections
Defense Research Establishment Suffield
Ralston, Alberta   CANADA

INTRODUCTION

During the past decade considerable advances have been made in the development of therapeutic antidotes against organophosphorus cholinesterase (ChE) inhibitors. Poisoning with most organophosphorus anticholinesterases (AntiChE) is treatable with a combination of an antimuscarinic compound, such as atropine sulfate (ATS), and an oxime, such as pralidoxime chloride (2-PAM) or toxogonin (18, 30, 34). The antidotal effects of oximes are generally thought to be due to their ability to reactivate the inhibited cholinesterase enzyme (10, 25, 26); however, a part of their beneficial effect has been ascribed to actions other than enzyme reactivation (4, 35, 41). Since most of the useful oximes are quaternary in structure and do not readily cross the blood brain barrier, the role played by the oximes in the central nervous system (CNS) is in some doubt (3, 25, 52). Although small quantities of these oximes do enter the central nervous system, the degree of reactivation of brain ChE is not great (11, 15).

Soman (pinacolyl methylphosphonofluoridate), an extremely potent AntiChE, is resistant to standard therapeutic regimens employing anticholinergics and oximes, such as 2-PAM (5, 29). Following its reaction with ChE soman undergoes a chemical change (aging) that makes reactivation by oximes in vivo difficult or impossible (17, 19). Recently a new series of bis-quaternary oximes, designated

767

"H-oximes," has been synthesized (22). These new oximes have been used successfully with an antimuscarinic compound to antagonize soman intoxication (27, 32, 38, 42, 47, 49, 55). Although these new oximes are effective in lowering the toxicity of soman, the mechanism of their antidotal action is not completely understood. The demonstration that HI-6 ([[[(4-aminocarbonyl)pyridino]methoxy]-methyl]-2-[(hydroxyimino) methyl]-pyridinium dichloride), one in the series, could reactivate soman-inhibited ChE in vitro (10, 50, 57) suggested that enzyme reactivation might explain its primary antidotal action in vivo.

We have conducted a series of studies to determine the role of cholinergic mechanisms in the antidotal effects of HI-6 in soman poisoning. We studied the effects of HI-6 in discrete brain areas as well as in blood on soman-induced depression of ChE activity and elevation of acetylcholine (ACh) or choline (Ch) levels in vivo, as well as on muscarinic receptor binding and high affinity Ch uptake (HAChT) in vitro. In similar but separate experiments the effect of 2-PAM was studied on the same cholinergic mechanisms to compare its effects with HI-6.

MATERIALS AND METHODS

Animal Preparation. Male albino Sprague Dawley-Wistar [AMRI, (SDxWI)-BR] rats, weighing between 200 and 250 grams, were used. They were housed as groups of three in plastic cages (45 x 24 x 21 cm) in temperature-controlled animal quarters that were maintained on a 12-h light-dark cycle, with artificial light provided between 0600 and 1800. Laboratory rat chow and water were available ad libitum. At least 6 days were allowed for the animals to become acclimated to the environment of the animal room prior to experimental use. Experiments on animals were carried out between 0900 and 1200 each morning to reduce the possible variation in ChE activity and ACh levels due to circadian rhythms.

For ACh and Ch measurements, rats were killed by microwave irradiation (4.0 KW at 2.45 GHz) focused on the head for 2.5 sec using a Gerling-Moore metabostat microwave system (Gerling-Moore, Inc., Santa Clara, CA). For ChE, muscarinic receptor binding, and synaptosomal HAChT assays, rats were decapitated. In some experiments, blood was separated into plasma and red blood cells (RBC) and the brain was dissected into brainstem, cerebral cortex, hippocampus, midbrain, cerebellum and striatum for subsequent biochemical determinations.

Drugs and Treatments. ATS and atropine methylnitrate (AMN) were purchased from Sigma Chemical Company (St. Louis, MO); pralidoxime chloride (2-PAM) from Ayerst Laboratories, Inc. (New York, NY); (Methyl-[^{3}H]) choline chloride ([^{3}H]-Ch, 80 Ci/mMol) from New England Nuclear (Boston, MA); and [^{3}H]-quinuclidinyl benzilate ([^{3}H]-QNB,

31 Ci/mMol) from Amersham (Arlington Hts, IL). HI-6 was synthesized by the Chemistry Section, Defense Research Establishment Suffield, Canada. Soman with a purity of at least 96% was obtained from the Chemical Research and Development Center or the Defense Research Establishment Suffield. In the in vivo studies, all drugs were dissolved in 0.9% saline and, except for soman, were given as their salts. HI-6 (125 mg/kg, i.p.), 2-PAM (43.2 mg/kg, i.m.), ATS (16 mg/kg, i.m.) or AMN (17 mg/kg, i.m.) were injected immediately following soman or saline (1 ml/kg, s.c.). The doses of oximes were equivalent to one quarter of their respective LD_{50} values (6, 37), while ATS and AMN were equimolar.

Blood and Brain ChE Assay. The blood obtained from each rat was immediately centrifuged and plasma and RBC separated and stored at $4^{\circ}C$ for assay. Each of the six dissected brain areas was placed in a test tube into which 14 vol of 1% Triton X-100 in saline was added for all areas except striatum, which received 30 vol of solubilizer. The brain tissue was then homogenized, using a Caframo tissue homogenizer (setting 6; 10 up and down strokes). The homogenates were centrifuged at $4^{\circ}C$ and the supernatant saved for assay. ChE analysis was performed by a modified automated (21) colorimetric assay (13).

Brain ACh and Ch Assay. ACh and Ch were extracted (31) from these six brain areas and levels quantitatively determined by gas chromatography/mass spectrometry method (51) using deuterated analogues of ACh and Ch as the internal standards.

In Vitro Muscarinic Receptor Binding Assay. Following dissection, the hippocampus, striatum, and cortex were frozen ($-70^{\circ}C$) and stored overnight. Next day the tissues were thawed, and approximately 60 mg were homogenized using a Brinkman polytron (setting 6, for 10 sec) in 5 ml sodium-potassium phosphate-EDTA buffer (pH 7.4). The homogenates were incubated at $30^{\circ}C$ for 15 min and placed on ice. Then, 0.2 ml (approximately 2.5 mg tissue) homogenate was added to a test tube which contained 1.76 ml sodium-potassium-EDTA buffer, 0.02 ml $[^3H]$-QNB (final concentration $2.1 \times 10^{-10}M$) and 0.02 ml of either HI-6 or 2-PAM in one of eleven concentrations (final concentration ranging from 4×10^{-6} to $1 \times 10^{-3}M$), or an ethanol blank. The samples were mixed and incubated for 30 min at $30^{\circ}C$. The reaction was terminated in an ice bath and the contents of the tubes were aspirated onto Whatman GF/B filter strips using a Brandel M24R tissue harvester. The filters were washed three times with 5 ml 0.9% saline. The filters were placed in 5 ml Formula-947 scintillation cocktail (New England Nuclear) and counted in a Packard 300-C scintillation spectrometer.

In Vitro High Affinity Ch Uptake Assay. Following dissection, the striatum and hippocampus were homogenized in 5 ml and the cortex in 20 vol of ice-cold 0.32 M sucrose with a Potter-Elvehjem homogenizer (setting 2; 6 up and down strokes). The homogenates were centrifuged at 1000 x g for 10 min at 4°C to obtain a supernatant fraction (S1) containing the synaptosomes. The S1 supernatant was then centrifuged at 17,000 x g for 15 min to obtain a crude synaptosomal pellet (P2), which was resuspended in 20 vol of 0.32 M sucrose for the uptake study. HAChT was analyzed by adding 0.1 ml of the P2 suspension to 0.86 ml Krebs-Henseleit media (pH 7.4, composition in mM: NaCl, 145; KCl, 5; MgSO4, 1.2; CaCl2, 1.5; D-glucose, 10; and HEPES, 2.5), which also contained one of the following compounds: HI-6 (1×10^{-7} to 1×10^{-2}M) or 2-PAM (1×10^{-6} to 1×10^{-2} M). After adding 0.02 ml of a solution containing [^{3}H]-Ch (0.4 μCi) and unlabeled Ch (final Ch concentration 0.5 μM) to the media, the incubation was carried out for 4 min at 37°C in a Dubnoff shaker or at 4°C in an ice bath. Uptake at 37°C was terminated by transferring the tubes to an ice bath. The synaptosomes containing [^{3}H]-label were collected on Whatman GF/B filter strips and washed 3 times with 5 ml 0.9% saline, utilizing a Brandel M24R tissue harvester. The filters were placed in 10 ml of Ultrafluor (National Diagnostics, Somerville, NJ) and counted in a Beckman LS3800 Liquid scintillation counter. High affinity Ch uptake values were obtained by subtracting those obtained at 4°C from 37°C.

Data Analysis. A 24-hr LD_{50} value was calculated by the probit procedure of Finney (14). The IC_{50}, which is the concentration of unlabeled compound producing 50% inhibition of [^{3}H]-QNB binding or [^{3}H]-Ch uptake, was calculated from a semilog plot of concentration (abscissa) versus % bound or % inhibition (ordinate). Statistical significance was determined by the use of Student's t-test, and differences were considered significant if $p < 0.05$.

RESULTS

Table 1 shows the protective effects of atropine and HI-6 against soman intoxication in rats. The s.c. LD_{50} for soman was determined to be 117 μg/kg. Treatment with HI-6 alone offered some protection against the lethal effect of soman (about 2.5 LD_{50}). Treatment with ATS provided very little protection (about 1.2 LD_{50}), while treatment with AMN, a peripherally acting antimuscarinic compound, alone afforded no protection. Addition of ATS to HI-6 treated rats significantly enhanced the protective ratio from 2.5 to 5.5; however, the addition of AMN failed to increase the protection afforded by HI-6 treatment.

In these same animals, the average time to death (in min) during the first 4 hr immediately following soman in the various antidotal treatment groups is shown in Table 2.

Although ATS increased to some extent the mean time to death following soman administration, HI-6 offered the most significant increase. As the dose of soman was increased from 170 to 216 μg/kg,

Table 1

Protective Effects of HI-6, ATS or AMN
Against Soman Toxicity in Rats

Treatment[a]	LD_{50} of Soman[b] (μg/kg, s.c.)	95% Limits	Protective ratio[c]
Soman	117	105-131	1.0
Soman + HI-6	293	240-330	2.5
Soman + ATS	133	117-148	1.2
Soman + AMN	116	106-127	1.0
Soman + HI-6 + ATS	644	476-795	5.5
Soman + HI-6 + AMN	281	247-327	2.4

[a]HI-6 (125 mg/kg, i.p.), ATS (16 mg/kg, i.m.) and AMN (17 mg/kg, i.m.) were injected immediately following various doses of soman (s.c.).
[b]The LD_{50} values were calculated by the probit procedure based on the fraction of animals that died within 24 hrs of soman administration.
[c]Protective ratio = LD_{50} of soman in protected rats/LD_{50} of soman in unprotected rats.

Table 2

Influence of Antidotal Treatments on
Time to Death in Soman Intoxicated Rats

Treatment[a]	#Dead/#Total[b] (min)	Mean Time to Death[b]	
Soman (170.1)	19/24	9.9	(4-17)
Soman (170.1) + ATS	6/8	30.0	(4-60)
Soman (170.1) + HI-6	1/8	65.0	
Soman (216.0)	16/16	8.0	(2-11)
Soman (216.0) + ATS	8/8	16.1	(7-43)
Soman (216.0) + HI-6	4/16	50.0	(10-62)
Soman (300.0) + HI-6 + ATS	1/16	220.0	
Soman (300.0) + HI-6 + AMN	3/10	78.3	(60-90)

[a]Dose of soman in parentheses (μg/kg, s.c.). Doses for antidotes shown in Table 1.
[b]Observed up to 4 hr following soman administration. Actual range of time in parentheses.

the time to death in each group of animals was decreased. A combination of HI-6 plus ATS therapy provided the most significant increase in time to death, but the combination of HI-6 and AMN was also effective in delaying death.

These results demonstrate that HI-6 by itself is effective in protecting rats exposed to soman. This antidotal effect is even more dramatic when HI-6 is combined with ATS in the therapy regimen. When 2-PAM was administered intramuscularly (43.2 mg/kg equivalent to 1/4 LD_{50}) in a similar study, no protection (protective ratio = 1.0) was observed and the time to death was not increased in soman treated rats.

In a separate series of experiments, the effects of HI-6 and 2-PAM were examined on ChE activity in blood as well as in brain tissues, following soman intoxication. Tables 3a and 3b show that HI-6, 2-PAM or ATS treatment individually did not alter the ChE activity in any tissue, while soman markedly inhibited ChE in all tissues studied. In soman intoxicated rats, the HI-6 treatment group had significantly higher levels of ChE in plasma and RBC than did the group given soman treatment alone; however, 2-PAM was not able to protect against soman-induced ChE inhibition in the CNS. Addition of ATS to the treatment regimen also had no effect upon ChE reactivation. Similar results were obtained from three other brain regions, midbrain, brainstem and cerebellum (data not shown).

Since it had been previously reported that some oximes inhibited the organophosphate-induced rise in brain ACh levels (43, 45), an investigation was undertaken of the effects of oximes on ACh and Ch levels in different brain areas. Representative results from three cholinergic brain areas are presented in Tables 4 and 5. The ACh and Ch values found in the six areas of the brain taken from rats 30 minutes after HI-6 or 2-PAM did not differ significantly from values found in saline-treated control animals. ATS tended to lower the brain level of ACh, as has been reported by others (9, 20, 24), but maintained or elevated the levels of Ch. Thirty minutes following exposure to soman, the levels of ACh and Ch from animals that received HI-6 or 2-PAM immediately after soman were not different from those in brain areas obtained from animals treated with soman alone. Addition of ATS to HI-6 attenuated the striatal levels of ACh or Ch as compared to soman plus HI-6 treatment; but the addition of ATS to 2-PAM was more effective in reducing ACh or Ch levels.

Besides ChE activity or ACh and Ch levels, other cholinergic parameters are possibly affected by these oximes. Since the data suggested the lack of central action of these oximes on ChE activity and on ACh and Ch levels, the effects of HI-6 and 2-PAM on muscarinic receptor binding and HAChT in three cholinergic enriched brain areas were studied in the same in vitro system.

Table 3a

Effects of Soman and its Antidotal Treatments on Blood ChE Activity

Treatment[a]	ChE μmol acetylthiocholine hydrolyzed/ml/min[b]	
	Plasma	RBC
Saline	$0.659 \pm 0.021(13)$[c]	$1.367 \pm 0.057(13)$[c]
Soman	$0.044 \pm 0.006(12)$	$0.163 \pm 0.014(12)$
ATS	$0.623 \pm 0.031(6)$[c]	$1.283 \pm 0.036(6)$[c]
HI-6	$0.686 \pm 0.049(10)$[c]	$1.296 \pm 0.059(10)$[c]
2-PAM	$0.670 \pm 0.014(8)$[c]	$1.353 \pm 0.040(8)$[c]
Soman + ATS	$0.023 \pm 0.006(6)$	$0.182 \pm 0.009(6)$
Soman + HI-6	$0.321 \pm 0.032(10)$[c]	$0.957 \pm 0.093(10)$[c]
Soman + 2-PAM	$0.056 \pm 0.007(8)$	$0.188 \pm 0.023(8)$
Soman + HI-6 + ATS	$0.400 \pm 0.020(10)$[c]	$0.978 \pm 0.041(10)$[c]
Soman + 2-PAM + ATS	$0.041 \pm 0.002(8)$	$0.180 \pm 0.008(8)$

[a]Dose of soman was 100 μg/kg, s.c.. HI-6 (125 mg/kg, i.p.); 2-PAM (43.2 mg/kg, i.m.); ATS (16 mg/kg, i.m.) and AMN (17 mg/kg, i.m.) were injected immediately following soman. [b]Activity expressed as mean $\pm$ SEM (number of animals). [c]$p<0.05$, when compared with soman treatment group.

Table 3b

Effects of Soman and its Antidotal Treatments on Regional Brain ChE Activity

Treatment[a]	ChE μmol acetylthiocholine hydrolyzed/g protein/min[b]		
	Cortex	Hippocampus	Striatum
Saline	$107.81 \pm 4.45(13)$[c]	$112.36 \pm 2.63(13)$[c]	$633.79 \pm 14.24(13)$[c]
Soman	$24.68 \pm 5.87(12)$	$19.39 \pm 7.16(12)$	$269.18 \pm 61.41(12)$
ATS	$127.18 \pm 2.04(6)$[c]	$108.25 \pm 2.77(6)$[c]	$639.77 \pm 15.36(6)$[c]
HI-6	$102.00 \pm 2.12(10)$[c]	$116.27 \pm 2.35(10)$[c]	$582.39 \pm 21.54(10)$[c]
2-PAM	$107.15 \pm 1.96(8)$[c]	$120.00 \pm 1.51(8)$[c]	$628.03 \pm 12.30(8)$[c]
Soman + ATS	$11.05 \pm 4.79(6)$	$9.62 \pm 6.72(6)$	$124.62 \pm 67.81(6)$
Soman + HI-6	$35.76 \pm 6.21(10)$	$22.01 \pm 6.46(10)$	$366.24 \pm 46.16(10)$
Soman + 2-PAM	$18.64 \pm 3.61(8)$	$18.40 \pm 5.50(8)$	$295.00 \pm 57.60(8)$
Soman + HI-6 + ATS	$21.34 \pm 4.85(10)$	$14.67 \pm 5.19(10)$	$254.32 \pm 51.20(10)$
Soman + 2-PAM + ATS	$23.15 \pm 6.07(8)$	$28.83 \pm 9.98(8)$	$367.71 \pm 63.26(8)$

[a]Dose of soman was 100 μg/kg, s.c.. HI-6 (125 mg/kg, i.p.); 2-PAM (43.2 mg/kg, i.m.); AT (16 mg/kg, i.m.) and AMN (17 mg/kg, i.m.) were injected immediately following soman. [b]Activity expressed as mean $\pm$ SEM (number of animals). [c]$p<0.05$, when compared with treatment group.

Table 4

Effects of Soman and its Antidotal Treatments
on Regional Brain Acetylcholine Levels

ACh mean nmoles/g wet wt. $\pm$ SEM (N)

Treatment[a]	Cortex	Hippocampus	Striatum
Saline	19.21 $\pm$ 0.43(29)	20.90 $\pm$ 0.77(29)	62.22 $\pm$ 1.26(27)
Soman	33.52 $\pm$ 2.94(20)[b]	36.30 $\pm$ 2.66(20)[b]	77.68 $\pm$ 3.64(21)[b]
ATS	14.26 $\pm$ 0.38(31)[b]	17.53 $\pm$ 0.63(26)	38.37 $\pm$ 1.32(29)[b]
HI-6	22.45 $\pm$ 1.58(12)	23.37 $\pm$ 1.62(12)	61.79 $\pm$ 2.91(12)
2-PAM	20.74 $\pm$ 0.98(11)	21.59 $\pm$ 1.30(11)	49.90 $\pm$ 3.61(12)[b]
Soman + ATS	47.86 $\pm$ 5.88(14)[b]	44.81 $\pm$ 5.39(14)[b]	63.75 $\pm$ 6.32(13)
Soman + HI-6	33.87 $\pm$ 2.28(12)[b]	29.06 $\pm$ 1.49(10)[b]	83.50 $\pm$ 3.13(11)[b]
Soman + 2-PAM	37.00 $\pm$ 4.39(11)[b]	32.94 $\pm$ 2.56(11)[b]	64.47 $\pm$ 6.56(11)
Soman + HI-6 + ATS	41.27 $\pm$ 4.66(12)[b]	35.31 $\pm$ 2.63(12)[b]	61.89 $\pm$ 7.00(12)
Soman + 2-PAM + ATS	30.98 $\pm$ 2.06(12)[b]	31.12 $\pm$ 2.63(12)[b]	37.49 $\pm$ 2.83(12)[b]

[a]Dose of soman was 100 μg/kg, s.c.. HI-6 (125 mg/kg, i.p.); 2-PAM (43.2 mg/kg, i.m.); AT (16 mg/kg, i.m.) and AMN (17 mg/kg, i.m.) were injected immediately following soman.
[b]$p<0.05$, when compared with saline treatment.

Table 5

Effect of Soman and its Antidotal Treatments
on Regional Brain Choline Levels

Ch mean nmoles/g wet wt. $\pm$ SEM (N)

Treatment[a]	Cortex	Hippocampus	Striatum
Saline	21.78 $\pm$ 0.76(30)	19.45 $\pm$ 0.57(29)	26.03 $\pm$ 0.54(23)
Soman	33.63 $\pm$ 3.94(19)[b]	29.46 $\pm$ 3.91(15)[b]	38.07 $\pm$ 4.95(20)[b]
ATS	22.80 $\pm$ 0.72(22)	23.46 $\pm$ 1.49(32)	34.53 $\pm$ 1.68(31)[b]
HI-6	19.49 $\pm$ 2.70(12)	20.44 $\pm$ 2.33(12)	26.78 $\pm$ 1.63(12)
2-PAM	14.61 $\pm$ 1.42(10)[b]	14.61 $\pm$ 1.42(10)	21.64 $\pm$ 1.36(11)
Soman + ATS	26.80 $\pm$ 2.18(13)	18.39 $\pm$ 1.22(13)	34.42 $\pm$ 3.41(13)[b]
Soman + HI-6	27.62 $\pm$ 2.43(11)[b]	30.41 $\pm$ 4.92(11)[b]	33.63 $\pm$ 1.87(12)[b]
Soman + 2-PAM	33.84 $\pm$ 9.19(10)[b]	29.39 $\pm$ 4.33(9)[b]	38.63 $\pm$ 5.56(9)[b]
Soman + HI-6 + ATS	19.11 $\pm$ 1.06(12)	25.17 $\pm$ 2.57(12)[b]	30.33 $\pm$ 2.32(11)
Soman + 2-PAM + ATS	16.95 $\pm$ 1.29(11)	16.95 $\pm$ 1.29(11)	24.70 $\pm$ 1.52(12)

[a]Dose of soman was 100 μg/kg, s.c.. HI-6 (125 mg/kg, i.p.); 2-PAM (43.2 mg/kg, i.m.); AT (16 mg/kg, i.m.) and AMN (17 mg/kg, i.m.) were injected immediately following soman.
[b]$p<0.05$, when compared with saline treatment group.

Table 6

Effect of HI-6 and 2-PAM on the Binding of
[^{3}H]-QNB to Brain Regional Muscarinic Receptors

| | IC_{50} (M) | | |
Compound	Hippocampus	Striatum	Cortex
HI-6	2.8×10^{-4}	3.5×10^{-4}	3.7×10^{-4}
2-PAM	7.0×10^{-5}	8.0×10^{-5}	9.6×10^{-5}

Table 6 shows the IC_{50} values for the competition of [^{3}H]-QNB binding by these oximes. Overall, 2-PAM exhibited a more potent (approximately 4 times) inhibition of [^{3}H]-QNB binding than did HI-6; however, both oximes showed similar regional order of potency: hippocampus > striatum > cortex.

Similarly, in the three brain regions examined, 2-PAM was more effective in inhibiting HAChT than HI-6. The IC_{50} values for HI-6 were 5.6×10^{-4}, 6.8×10^{-4} and 7.0×10^{-4}M in the hippocampus, striatum and cortex, respectively; while the IC_{50} values for 2-PAM were, respectively, 1.1×10^{-4}, 4.5×10^{-5} and 2.4×10^{-4}M for these three brain areas.

DISCUSSION

Recent studies have demonstrated that some new H-oximes, when combined with ATS, are extremely effective against soman poisoning in mice, rats, dogs and monkeys (6, 28, 32, 38, 42, 49, 53, 56). Among these H-oximes, HI-6, when combined with ATS, was reported to be the least toxic and most efficacious oxime against soman poisoning in mice (6). In the present study, we have also demonstrated that HI-6 treatment alone could protect against soman in rats to a significant degree. Additionally, in those animals that died from this single treatment, the time to death was significantly prolonged.

The most significant protection was afforded by simultaneous treatment with both HI-6 and ATS. This combination appeared to be effective whether ATS was administered after or before soman challenge (40). AMN, on the other hand, provided no further benefit in soman-intoxicated rats that had been treated with HI-6. These results suggest that although HI-6 alone could protect against soman toxicity to a certain degree, the additional central protection provided by ATS administration was beneficial. It should be noted, however,

in this species. In separate experiments soman was very resistant
to conventional therapy with either 2-PAM alone or 2-PAM plus ATS,
which is in agreement with results reported earlier by others (5,
23, 37, 39).

Different routes of administration of HI-6 (i.p.) and 2-PAM
(i.m.) were employed in these studies, based on the published data
from two laboratories (6, 37). The doses of these two oximes were
essentially equimolar and were equivalent to one quarter of their
respective LD_{50} values. Since both oximes were prepared in aqueous
solution, we can assume that both were rapidly absorbed. Because
lethality was examined 24 hrs following soman and oxime administration,
any differences in the effects of oximes on soman toxicity as a
result of absorption rates and tissue distribution would be minimized.

The results of this study have demonstrated that HI-6, when
administered immediately following soman challenge in rats, produces
significant in vivo reactivation of soman-inhibited ChE in the plasma
and RBC, while 2-PAM fails to do so. Clement (6) also reported
that HI-6 produced significant reactivation of soman-inhibited ChE
in the diaphragm and intercostal muscles. However, species differences
occur in the therapeutic efficacy of HI-6 against soman poisoning.
In terms of recovery of neuromuscular transmission due to enzyme
reactivation, HI-6 is effective in in vitro muscle preparations
from mice, rats, guinea pigs and dogs, but not in preparations from
rhesus monkey or humans (55).

Neither oxime was able to reactivate soman-inhibited ChE in
the discrete brain areas. The minimal reactivation of antiChE-
inhibited brain ChE by systemic injection of oximes has been consistently documented in the literature (5, 16, 25, 30, 33, 46) and
was also demonstrated when HI-6 was administered intracerebroventricularly (40). Minimal reaction in the CNS may be due to the
quaternary structure of most oximes, which limits their penetration
of the blood-brain barrier. Nevertheless, some indirect evidence
indicates the presence of oximes in the CNS after systemic injection
(1, 6, 40, 48).

Consistent with our previous findings (40, 51), brain area
ACh and Ch levels rise following soman administration, and only in
the striatum was intramuscularly administered 2-PAM able to block
ACh increase. HI-6 was not able to block the ACh increase and if
other oximes cause the reversal of organophosphate induced elevation
of ACh as has been previously reported (43, 45), the results presented
here argue against the same therapeutic mechanism for HI-6 against
soman poisoning. ATS, when administered alone in unexposed rats,
reduced levels of ACh in the CNS (9, 20, 24). Addition of either
oxime to ATS therapy following soman poisoning did not reverse the
rise of ACh in any brain area except in the striatum.

The hemicholinium-like (4) and the antimuscarinic activity (2, 4, 36) of oximes have also been reported. Our results support these findings and indicate that under identical conditions 2-PAM reduces HAChT more significantly than HI-6 in an in vitro synaptosomal preparation, and also reduces [^{3}H]-QNB muscarinic receptor binding more than HI-6. The extent to which these oximes penetrate the blood-brain barrier and exert these actions in the CNS of the intact organism requires further investigation.

In summary, our present studies, as well as those reported by others, clearly demonstrate the antidotal action of HI-6 against soman poisoning in rats. Since both oximes possess relatively similar effects upon competitive muscarinic receptor binding and upon HAChT, and since 2-PAM neither reactivated ChE in the blood and tissues nor protected rats against soman challenge, it appears that the beneficial effects of HI-6 are most likely due to its ChE reactivating capability, demonstrated by us in plasma and RBC, and by others in peripheral tissues (6, 55). Our data also show that systemically administered HI-6 did not appreciably reactivate central ChE inhibition nor did it attenuate the soman induced increase in the central levels of ACh or Ch; therefore, either HI-6 does not penetrate the blood-brain barrier in significant amounts to be effective, or the central action of HI-6 is questionable (40).

Maintenance by HI-6 of a certain amount of active ChE in the periphery appears to be important for the survival of somanexposed animals (6, 7, 8), although other pharmacologic actions of this oxime can not be ruled out (6, 35, 41). Our results and those reported by others (6, 55) suggest that the reactivation or, possibly, prevention of inhibition by HI-6 of peripheral ChE in soman poisoned animals is important for survival, since neither ATS nor AMN alone afforded any protection. Only in combination with HI-6 treatment did ATS, but not AMN, administration improve therapy.

Since AMN was not effective, the additional beneficial effect of ATS may involve vital central respiratory mechanisms. The lack of antidotal action of 2-PAM against soman in similar experiments is probably due to its very low ChE reactivating potency, combined with its relatively high toxicity (12, 18, 30, 44, 54) in the periphery.

Acknowledgments. The authors express their appreciation to Mr. Bruce Barney, Mr. Otis Smith, Mr. Thomas Koviak, and Mr. Andres Kaminskis for their technical assistance, to LTC Marvin A. Lawson for helpful criticism, and to Ms. Ann E. Sheppard for typing this manuscript.

REFERENCES

1. Aarseth, P. and Barstad, J.A.B. (1968): Arch. Int. Pharmacodyn. Ther. 176:434-442.

2. Amitai, G., Kloog, Y., Balderman, D. and Sokolovsky, M. (1980): Biochem. Pharmacol. 29:483-488.
3. Brown, R.V. (1960): Br. J. Pharmacol. 15:170-174.
4. Clement, J.G. (1979): Eur. J. Pharmacol. 53:135-141.
5. Clement, J.G. (1979): Toxicol. Appl. Pharmacol. 47:305-311.
6. Clement, J.G. (1981): Fundam. Appl. Toxicol. 1:193-202.
7. Clement, J.G. (1982): Biochem. Pharmacol. 31:1283-1287.
8. Clement, J.G. and Lockwood, P.A. (1982): Toxicol. Appl. Pharmacol. 64:140-146.
9. Crossland, J. and Slater, P. (1968): Br. J. Pharmacol. 33:42-47.
10. De Jong, L.P.A. and Wolring, G.Z. (1980): Biochem. Pharmacol. 29:2379-2387.
11. De la Manche, I.S., Verge, D.E., Bouchaud, C., Coq. H. and Sentenac-Roumanou, H. (1979): Experientia 35:531-532.
12. Edery, H. and Shatzberg-Porath, G. (1958): Science 128:1137-1138.
13. Ellman, G.L., Courtney, K.D., Andres, V. and Featherstone, R.M. (1961): Biochem. Pharmacol. 7:88-95.
14. Finney, D.J. (1971): Probit Analysis, 3rd ed., Cambridge University Press, London.
15. Firemark, H., Barlow, C.F. and Roth, L.J. (1964): J. Pharmacol. Exp. Ther. 145:252-265.
16. Fleisher, J.H., Hansa, J., Killos, R.J. and Harrison, C.S. (1960): J. Pharmacol. Exp. Ther. 130:461-468.
17. Fleisher, J.H. and Harris, L.W. (1965): Biochem. Pharmacol. 14:641-650.
18. Fleisher, J.H., Harris, L.W., Miller, G.R., Thomas, N.C. and Cliff, W.J. (1970): Toxicol. Appl. Pharmacol. 16:40-47.
19. Fleisher, J.H., Harris, L.W. and Murtha, E.F. (1967): J. Pharmacol. Exp. Ther. 156:345-351.
20. Giarman, N.J. and Pepeu, G. (1962): Br. J. Pharmacol. 19:226-234.
21. Groff, W.A., Kaminskis, A. and Ellin, R.I. (1976): Clin. Toxicol. 9:353-358.
22. Hagedorn, I. (1977): Offenlegungsschrift 2616481.
23. Harris, L.W., Fleisher, J.H., Vick, J.A. and Cliff, W.J. (1969): Biochem. Pharmacol. 18:419-427.
24. Harris, L.W., Heyl, W.C., Stitcher, D.L. and Moore, R.D. (1978): Life Sci. 22:907-910.
25. Harris, L.W. and Stitcher, D.L. (1983): Drug Chem. Toxicol. 6:235-240.
26. Harris, L.W., Stitcher, D.L. and Heyl, W.C. (1981): Life Sci. 29:1747-1753.
27. Hauser, W., Kirsch, D., and Weger, N. (1981): Fundam. Appl. Toxicol. 1:164-168.
28. Hauser, W. and Weger, N. (1979): Arch. Toxicol. Suppl. 2:393-396.
29. Heilbronn, E. and Tolagen, B. (1965): Biochem. Pharmacol. 14:73-77.
30. Hobbiger, F. (1957): Br. J. Pharmacol. 12:438-446.

31. Jenden, D.J. and Hanin, I. (1974): In Choline and Acetylcholine: Handbook of Chemical Assay Methods (ed) I. Hanin, Raven Press, New York, pp. 135-150.

32. Kepner, L.A. and Wolthuis, O.L. (1978): Eur. J. Pharmacol. 48:377-382.

33. Kewitz, H. and Nachmansohn, D. (1957): Arch. Biochem. Biophys. 66:271-283.

34. Kewitz, H. and Wilson, I.B. (1956): Arch. Biochem. 60:261-263.

35. Kirsch, D., Hauser, W. and Weger, N. (1981): Fundam. Appl. Toxicol. 1:169-176.

36. Kuhnen-Clausen, D. (1972): Toxicol. Appl. Pharmacol. 23:443-454.

37. Lennox, W.J. and Sultan, W.E. (1982): Unpublished findings.

38. Lipp, J. and Dola, T. (1980): Arch. Int. Pharmacodyn. Ther. 246:138-148.

39. Loomis, T.A. and Johnson, D.D. (1966): Toxicol. Appl. Pharmacol. 8:533-539.

40. Lundy, P.M. and Shih, T.-M. (1983): J. Neurochem. 40:1321-1328.

41. Lundy, P.M. and Tremblay, K. (1979): Eur. J. Pharmacol. 60:47-53.

42. Maksimovic, M. Boskovic, B., Radovic, L., Tadic, V., Deljac, V. and Binenfeld, Z. (1980): Acta Pharm. Jugosl. 30:151-160.

43. Mayer, O. and Michalek, H. (1971): Biochem. Pharmacol. 20:3029-3037.

44. Meeter, E. Wolthuis, O.L. and Van Benthem, M.J. (1971): Bull. W.H.O. 44:251-257.

45. Michalek, H. and Bonavoglia, F. (1973): Biochem. Pharmacol. 22:3124-3127.

46. O'Leary, J.F., Harrison, B., Groblewski, G. and Wills, J.H. (1959): Fed. Proc. 18:430.

47. Oldiges, H. and Schoene, K. (1970): Arch. Toxicol. 26:293-305.

48. Rosenberg, P. (1960): Biochem. Pharmacol. 3:212-219.

49. Schenk, J., Loffler, W. and Weger, N. (1976): Arch. Toxicol. 36:71-81.

50. Schoene, K. (1973): Biochem. Pharmacol. 22:2997-3003.

51. Shih, T.-M. (1982): Psychopharmacology 78:170-175.

52. Tong, H.S. and Way, J.L. (1962): J. Pharmacol. Exp. Ther. 138:218-223.

53. Wilhelm, K., Fajdetic, A., Deljac, V. and Binenfeld, Z. (1979): Arch. Hig. Rada Toksikol. 49:415-425.

54. Wills, J.H. (1961): CRDL Special Publication 2-38, Summary of Information on 2-PAM Cl.

55. Wolthuis, O.L., Berends, F. and Meeter, E. (1981): Fundam. Appl. Toxicol. 1:183-192.

56. Wolthuis, O.L., Clason van der Wiel, H.J. and Visser, R.P.I.S. (1976): Eur. J. Pharmacol. 39:417-421.

57. Wolthuis, O.L. and Kepner, L.A. (1978): Eur. J. Pharmacol. 49:415-425.

ACETVLCHOLINE TURNOVER IN MOUSE BRAIN:

INFLUENCE OF CHOLINESTERASE INHIBITORS

B. Karlén[2], G. Lundgren[1], J. Lundin[1] and B. Holmstedt[1]

[1]Department of Toxicology
The Swedish Medical Research Council
Karolinska Institutet
S-104 01 Stockholm, Sweden

[2]KabiVitrum AB, Research Department
S-112 87 Stockholm, Sweden

INTRODUCTION

Evidence has been presented to indicate that in neurons the concentrations of neurotransmitters are maintained at constant levels although the rates of transmitter metabolism are changing (10, 13). However, in cholinergic neurons there is evidence to show that concentrations of acetylcholine (ACh) are accompanied by changes in the rate of metabolism of this transmitter (11).

In catecholaminergic neurons, studies of the regulation of catecholamine synthesis have determined that the concentration of these transmitters in neural tissues is modulated within narrow limits by product inhibition. This process balances the rate of synthesis with the rate of release and utilization of catecholamines (1). It is reasonable to assume that similar regulatory mechanisms may exist also in cholinergic neurons, and that an understanding of the factors regulating ACh biosynthesis may emerge by investigating how cholinoceptive drugs change brain ACh turnover rates. We have therefore studied the influence of drugs injected in mice that perturb the regulation of brain cholinergic neurons and increase (5, 9) or decrease (6) their normal ACh concentration.

We performed this current study to determine whether the irreversible cholinesterase inhibitors soman, sarin or FX, which are thought to increase brain ACh concentration by a mechanism different to that of the muscarinic receptor agonist oxotremorine, also would decrease the turnover rate of brain ACh. We were also interested to see if these organophosphate inhibitors would act in

781

the same way as the previously studied reversible inhibitor physostigmine (7), i.e., to increase the ACh concentration and to decrease
the turnover rate.

The three inhibitors exhibit different degrees of reversibility,
as demonstrated by the speed in which the inhibited phosphorylated
enzymes are transformed to non-reactivateable enzymes. The half
lives of aging for soman, sarin and FX are 2.2 min (2), 5.8 hours
(3) and infinite (12), respectively.

The turnover of ACh in whole brain and striatum of mice was
studied by the pulse injection technique using deuterated choline
(Ch) as precursor (8). The animals were killed by focussed microwave
irradiation and ACh, Ch and their deuterated variants were assayed
by mass fragmentography (4).

MATERIALS AND METHODS

Animals and Drugs. Male NMR, albino mice weighing 20-27 g were
used. N-(2-hydroxyethyl)-N,N,N-tri-[^{2}H$_2$],[^{1}H]methylammonium iodide
(d$_6$-Ch) in saline was given intravenously during 1 sec in a dose
of 20 μmol/kg per volume of 5 ml/kg.

N-(2-hydroxyethyl)-N,N,N-tri-[^{2}H$_3$]methylammonium iodide
(d$_9$-Ch) and N-(2-acetoxyethyl)-N,N,N-tri-[^{2}H$_3$]methylammonium iodide
(d$_9$-ACh) were used as internal standards. N-(2-acetoxyethyl)-
N,N,N,-tri-[^{2}H$_2$],[^{1}H]methylammonium iodide (d$_6$-ACh) was used for
calibration purposes. The synthesis of the deuterium labelled compounds has been described previously (4).

Sarin. Isopropoxymethylphosphoryl fluoride, <u>Soman</u>: (3,3-dimethyl-
2-butoxy)-methylphosphoryl fluoride and <u>FX</u>: 2-(N,N-diisopropylamino)-
eththioethoxymethylphosphineoxide were synthesized in the Research
Institute of National Defence, FOA 4, S-172 04 Sundbyberg, Sweden.

Procedure. After i.v. injection of d$_6$-Ch, its incorporation into
ACh was studied up to 20 min. The mice were killed 0.25, 0.75,
2.5, 5, 10 and 20 min after d$_6$-Ch administration by focussed microwave
irradiation on the head with 2.5 kW for 0.68 sec. (Metabostat, Gerling-Moore, Palo Alto, California, U.S.A.).

Estimation of ACh and Ch in Brain. The whole brains were removed
and homogenized directly in 4 ml 0.4 $\underline{M}$ HClO$_4$ or striata from 1-2
mice were excised, weighed, pooled and homogenized in 2 ml 0.4 $\underline{M}$
HClO$_4$ with an Ultra-Turrax homogenizer in a plastic scintillation
flask (25 ml).

After addition of internal standards (<u>whole brain</u> 0.25 nmol
d$_9$-ACh and 0.60 nmol d$_9$-Ch, <u>striata</u>: 0.10 nmol d$_9$-ACh and 0.30 nmol
d$_9$-Ch) the homogenates were left for 20 min at +4°C and then centri-

fuged for 20 min at 100,000 x g and +4°C. Endogenous ACh and Ch
together with their deuterated variants were extracted with dipicrylamine in dichloromethane as ion pairs. The Ch moieties were
derivatized with propionyl chloride and the resulting mixture of
ACh and propionyl choline derivatives was demethylated with sodium
thiophenoxide and analyzed by mass fragmentography (4). The specific
activities were expressed as the mole ratio between the amount of
labeled compound and the sum of labeled and endogenous compounds.

Calculation of Turnover Rate of ACh. The turnover rate of ACh (TR_{ACh})
was calculated from the specific activity curves of ACh (S_{ACh}) and
Ch (S_{Ch}) between 15 and 45 sec after the injection of d_6-Ch according
to Zilversmit (14). The following equation, where the subscripts
1 and 2 refer to 15 and 45 sec, respectively, was used to calculate
the fractional rate constant, K_a, of ACh synthesis.

$$K_a = \frac{2(S_{ACh2} - S_{ACh1})}{t(S_{Ch1} - S_{ACh1} + S_{Ch2} - S_{ACh2})}$$

TR_{ACh} could then be calculated by multiplying K_a with the endogenous
concentration of ACh.

Statistics. Statistical comparisons of independent sample means
were made using unpaired two-tailed Student's t-test.

RESULTS

The concentrations of Ch, ACh and their deuterated variants
found in whole brain and striatum after pretreatment with saline,
soman, sarin and FX are shown in Tables 1 and 2. Unpaired Student's
t-test of differences between means of ACh and Ch at different times
after d_6-Ch showed them to be the same. The effect of pretreatment
on ACh and Ch was therefore made on the total mean of the individual
values at all time points for animals given the same treatment.

In whole brain the endogenous concentration of ACh (24.3 nmol/g)
was not affected by sarin and only to a slight but significant extent
by FX, while soman increased the level to about 30 nmol/g. All
three substances increased the Ch level in comparison to controls
(32.6 nmol/g).

The specific activity curves of Ch and ACh obtained in whole brain
in animals treated with saline and the three inhibitors are depicted
in Figure 1. The time course for increase and decline of S_{ACh} was
similar in brains of animals treated with saline, sarin and FX,
reaching a peak after 2.5 min. However, in animals treated with
soman the peak was not yet reached after 20 min. The S_{Ch} curves
were affected differently by the inhibitors. Soman and FX caused
an increase of the specific activity of Ch at early times, this

Table 1

Concentration (nmol/g) of Endogenous and d_6-Substituted ACh and
Ch in Whole Brain of Mice Injected i.v. with 20 μmol/kg d_6-Ch
After Pretreatment for 15 Min with Cholinesterase
Inhibitors (Mean $\pm$ S.D.)

Pre-treatment	No. Ani-mals	Time after inj. of d_6-Ch (min)	ACh	d_6-ACh	Ch	d_6-Ch
Saline(1)	7	0.25	24.2$\pm$2.3	0.20$\pm$0.04	34.0$\pm$ 3.4	3.59$\pm$1.09
	8	0.75	24.8$\pm$4.4	0.75$\pm$0.21	32.6$\pm$ 8.1	2.34$\pm$0.54
	8	2.5	23.6$\pm$2.9	0.80$\pm$0.20	33.6$\pm$ 7.1	0.90$\pm$0.23
	4	5	24.5$\pm$3.3	0.73$\pm$0.10	28.5$\pm$ 1.5	0.47$\pm$0.10
	4	10	23.8$\pm$3.9	0.43$\pm$0.90	32.6$\pm$ 8.4	0.43$\pm$0.21
	4	20	25.4$\pm$3.2	0.17$\pm$0.04	32.4$\pm$ 1.5	0.14$\pm$0.02
	35	0.25-20	24.3$\pm$2.7		32.6$\pm$ 7.1	
Sarin(2)	4	0.25	26.0$\pm$2.1	0.19$\pm$0.06	35.6$\pm$ 4.3	3.35$\pm$1.08
330 μg/kg	3	0.75	27.8$\pm$2.4	0.58$\pm$0.14	38.4$\pm$ 4.3	2.11$\pm$0.43
i.p.	7	2.5	24.3$\pm$3.4	0.64$\pm$0.22	54.0$\pm$ 6.2	1.50$\pm$0.52
	4	5	23.6$\pm$3.3	0.56$\pm$0.03	54.4$\pm$ 6.6	1.07$\pm$0.26
	4	10	25.9$\pm$5.6	0.56$\pm$0.13	49.5$\pm$14.7	0.63$\pm$0.17
	4	20	24.6$\pm$4.3	0.33$\pm$0.10	44.9$\pm$10.3	0.29$\pm$0.10
	26	0.25-20	25.1$\pm$3.5		47.2$\pm$10.4	
Soman(3)	4	0.25	30.6$\pm$2.8	0.08$\pm$0.02	48.6$\pm$15.4	7.24$\pm$1.20
150 μg/kg	4	0.75	30.2$\pm$3.8	0.21$\pm$0.09	54.3$\pm$16.0	5.50$\pm$2.53
s.c.	4	2.5	28.5$\pm$5.4	0.32$\pm$0.06	47.0$\pm$14.3	1.64$\pm$0.22
	4	5	31.8$\pm$2.0	0.38$\pm$0.04	49.0$\pm$ 6.3	0.99$\pm$0.10
	4	10	30.4$\pm$1.6	0.46$\pm$0.04	67.7$\pm$ 5.7	0.97$\pm$0.29
	3	20	33.1$\pm$2.5	0.50$\pm$0.14	75.0$\pm$10.5	0.69$\pm$0.37
	23	0.25-20	30.7$\pm$3.2		56.2$\pm$15.0	
Fx(4)	4	0.25	26.4$\pm$3.2	0.10$\pm$0.01	36.7$\pm$ 8.2	4.38$\pm$1.76
17.5	4	0.75	25.9$\pm$3.8	0.26$\pm$0.04	43.5$\pm$ 9.6	3.76$\pm$0.77
μg/kg	4	2.5	27.6$\pm$3.4	0.56$\pm$0.14	47.2$\pm$ 7.7	2.18$\pm$0.99
s.c.	3	5	27.5$\pm$0.2	0.47$\pm$0.08	56.1$\pm$ 9.8	1.27$\pm$0.19
	4	10	29.9$\pm$2.1	0.59$\pm$0.11	57.3$\pm$ 8.3	0.91$\pm$0.17
	4	20	27.0$\pm$0.5	0.21$\pm$0.05	37.3$\pm$ 8.6	0.16$\pm$0.04
	23	0.25-20	27.2$\pm$2.6		45.9$\pm$11.2	

Differences between treatments: ACh mean values (0.25-20 min);
1-2, N.S.; 1-3, P < 0.001; 1-4, P < 0.001. Ch mean values (0.25-20
min); 1-2, P < 0.001; 1-3, P<0.001; 1-4, P < 0.001.

Table 2

Concentration ·(nmol/g) of Endogenous and d_6-Substituted ACh and
Ch in Striatum of Mice Injected i.v. with 20 μmol/kg d_6-Ch
After Pretreatment For 15 Min With Cholinesterase
Inhibitors (Mean ± S.D.)

Pre-treatment	No. Ani-mals	Time after inj. of d_6-Ch (min)	ACh	d_6-ACh	Ch	d_6-Ch
Saline(1)	6	0.25	92.9±9.1	0.36±0.07	67.7±28.8	2.11±0.31
	10	0.75	69.6±9.3	1.43±0.38	45.2±10.2	2.19±0.47
	2	2.5	60.3±12.4	1.73±0.40	43.4±4.4	1.74±0.49
	3	5	65.0±10.7	1.39±0.15	56.4±10.7	1.05±0.15
	6	10	74.9±8.3	1.22±0.16	44.7±5.6	0.79±0.14
	5	20	75.6±7.2	0.85±0.07	55.0±15.9	0.76±0.24
	32	0.25-20	74.5±12.9		51.4±16.8	
Sarin(2)	3	0.25	95.6±12.1	0.44±0.09	69.9±5.3	2.87±0.22
330 μg/kg	4	0.75	93.1±19.9	1.09±0.16	8.4±6.3	2.25±0.27
i.p.	4	2.5	75.6±2.6	1.53±0.14	71.6±8.9	1.80±0.34
	4	5	76.3±14.8	1.70±0.44	62.8±11.4	1.54±0.30
	4	10	92.3±3.8	1.52±0.35	63.5±5.4	0.84±0.16
	4	20	87.1±5.3	0.77±0.08	63.5±5.5	0.81±0.16
	23	0.25-20	86.3±13.0		66.5±7.6	
Soman(3)	3	0.25	73.8±19.4	0.22±0.04	47.1±8.4	5.68±0.87
150 μg/kg	4	0.75	75.9±14.9	0.78±0.18	43.2±2.5	4.42±0.75
s.c.	3	2.5	78.9±2.3	1.67±0.15	53.4±4.5	1.43±0.06
	3	5	75.7±5.7	1.55±0.58	57.6±14.3	1.16±0.36
	4	10	69.8±11.7	1.29±0.07	73.0±17.3	1.28±0.13
	4	20	78.6±6.2	1.15±0.16	66.2±15.8	0.94±0.12
	21	0.25-20	75.3±10.5		57.3±15.4	
Fx(4)	4	0.25	98.4±17.4	0.36±0.10	50.7±14.9	3.76±0.73
17.5	4	0.75	94.0±24.4	0.98±0.25	41.7±3.9	2.61±1.01
μg/kg	4	2.5	94.9±13.9	1.23±0.25	61.2±17.4	1.41±0.52
s.c.	4	5	96.3±17.6	1.34±0.25	60.5±9.1	0.97±0.17
	2	10	89.6±0.7	0.58±0.16	78.1±14.4	0.59±0.03
	4	20	81.1±14.3	0.38±0.08	54.9±3.0	0.39±0.13
	24	0.25-20	93.1±16.5		55.6±14.3	

Differences between treatments: ACh mean values (0.25-20 min);
1-2, $P < 0.005$; 1-3, N.S.; 1-4, $P < 0.0001$. Ch mean values (0.25-20
min); 1-2, $P < 0.001$; 1-3, N.S.; 1 - 4, N.S.

effect being most pronounced after soman. Treatment with sarin did not increase but rather caused a decrease of the S_{Ch} values at early time points. This is probably due to an increase of the endogenous Ch level while the concentration of labelled Ch was unaffected. In all cases the S_{Ch} curves were normalized already after 2.5 min.

The turnover rate of ACh was 15.9 nmol/g/min in control mice (Table 3). The corresponding figures for animals treated with sarin, soman and FX were 11.9, 2.3 and 3.8 nmol/g/min, respectively. The lowest values were obtained after treatment with soman and FX, which drugs also increased the endogenous ACh levels. Sarin, which did not increase the ACh level, still decreased the turnover rate somewhat.

In striatum the effect of the inhibitors on ACh concentrations as well as on turnover rates differed from that in whole brain. The ACh concentration (74.5 nmol/g) in striatum was increased after sarin and FX while it was unaffected after soman. This is roughly the opposite to what occurred in whole brain, where soman and to a

Table 3

Endogenous Levels (nmol/g), Fractional Rate Constants (K_a, min^{-1}) and Turnover Rates (TR_{ACh}, nmol/g/min) of ACh in Mice Pretreated for 15 Min With Saline or Organophosphates. d_6-Ch Was Given i.v. and the Animals Were Killed 15 and 45 Sec Thereafter.

	Whole Brain			Striatum		
Pretreatment	ACh	K_a	TR_{ACh}	ACh	K_a	TR_{ACh}
Saline	24.3	0.656	15.9	74.5	0.820	61.1
Sarin 330 μg/kg i.p.	25.1	0.473	11.9	86.3	0.500	43.2
Soman 150 μg/kg s.c.	30.7	0.075	2.3	75.3	0.150	11.3
Fx 17.5 μg/kg s.c.	27.2	0.140	3.8	93.1	0.210	19.6

The ACh concentration is the mean of all time points collected in Tables 1 and 2, respectively. K_a and TR_{ACh} were calculated as described in Methods.

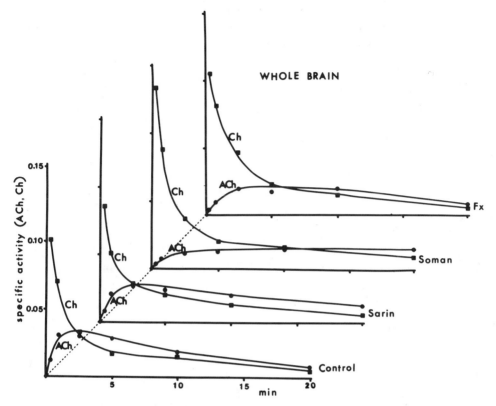

Figure 1. Effect of sarin, soman and FX on the specific activity
of ACh and Ch in whole brain. Mice were injected with saline, sarin
(330 μg/kg, i.p.), soman (150 μg/kg, s.c.) or FX (17.5 μg/kg, s.c.) 15
min before d_6-Ch (20 μmol/kg, i.v.) and killed 0.25 - 20 min later.
The data are means $\pm$ S.D. of at least 3 animals.

lesser extent FX, but not sarin, caused an increase of the ACh level.
The concentration of endogenous Ch (51.4 nmol/g) was increased only
after treatment with sarin. The specific activity curves of Ch
and ACh are depicted in Figure 2. The pattern is the same as that
in whole brain but the increase of the specific activity of Ch at
early times is more pronounced after pretreatment with soman and
FX. Also, there is a decrease after sarin.

The turnover rate of ACh was decreased to about the same extent
in striatum as in the whole brain after treatment with the different
inhibitors (Table 3). However, in contrast to whole brain, the
decrease was not correlated with the ACh concentration, since it
was largest after soman which did not change the ACh concentration
at all.

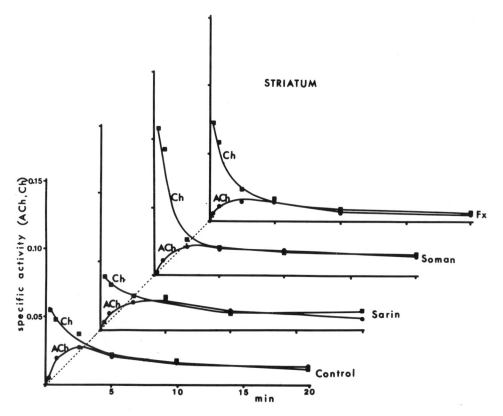

Figure 2.   Effect of sarin, soman and FX on the specific activity
of ACh and Ch in striatum.   Mice were injected with saline, sarin
(330 μg/kg, s.c.), soman (150 μg/kg, s.c.) or FX (17.5 μg/kg, s.c.) 15
min. before d_6-Ch (20 μmol/kg, i.v.) and killed 0.25 - 20 min later.
The data are means ± S.D. of at least 3 animals.

DISCUSSION

        The organophosphates were given in doses corresponding to about
50 percent of the LD 50 doses.   These doses induced in all animals
severe effects, typical of cholinesterase inhibition, e.g. tremor,
salivation, hypothermia and convulsions.   These effects are thought
to be due to an overstimulation of cholinergic receptors by excess
ACh induced by inhibition of the cholinesterases.

        The present results indicate that these inhibitors affect whole
brain and striatum differently, and that they also change ACh levels
and turnover rates to different extents.   Soman, which has the greatest
effect on the level and turnover of ACh in whole brain, has no effect

on the ACh level in striatum, but affects the turnover of ACh more than sarin and FX in this region. These results are evidence against the existence of a clear relationship between change in levels of ACh and of turnover rates of ACh in vivo.

Thus, soman alters the turnover of ACh in striatum even when the ACh level is in the normal range. Theoretically, the level of ACh would depend on whether its rate of synthesis is balanced by its rate of hydrolysis. Normally, cholinesterase inhibitors induce an increase of the ACh levels, and it is thought that the excess ACh reduces its synthesis rate by some feedback inhibition mechanism, e.g. by stimulation of presynaptic receptors. In the case of soman the level of ACh was unaffected in striatum although its rate of synthesis was much reduced. One explanation could be that soman, besides it cholinesterase inhibiting effect, also has a direct action on the ACh synthesizing system; as a result reducing the synthesis rate of the transmitter. The level of ACh would then depend on whether or not the rate of synthesis is balanced by the rate of hydrolysis. In the case of soman it appears that the two processes are affected simultaneously and to the same degree, while in the case of FX and sarin the onset of inhibition of the synthesis rate is comparably slower and possibly develops as a result of the increased ACh level. In the former case the level of ACh would be unchanged, and in the latter it would be increased.

The kinetics of the precursor d_6-Ch in striatum are affected differently following administration of the three inhibitors. Thus, following soman administration the specific activity of the precursor is doubled. In animals treated with soman the rate of ACh synthesis is only 20 percent of its normal rate. The initially high specific activity in these animals might therefore be the result of a decreased synthesis rate in the neuron. If this is true, the specific activity of Ch initially would reflect the degree of inhibition of the ACh synthesis. The initial S_{Ch} levels seem to be correlated with the inhibitory effects of soman, FX and sarin, as reflected by the lowered K_a-values.

The differences seen between the inhibitors is probably not related to the differences in the speed of their aging process; the spontaneous hydrolysis time is so long that a reoccurrence of esterase activity would not take place during the time studied (35 min).

In conclusion, the present study has shown that cholinesterase inhibition caused by "irreversible" phosphates affects the ACh levels to different degrees in the same tissue, as well as in different tissues (whole brain and striatum). The level of ACh is not correlated with the rate of ACh synthesis, thus direct product inhibition seems to be ruled out as the only cause for a reduced synthesis rate. Presynaptic inhibition by production of excess ACh in the synaptic spatium may be in operation, but it is plausible that a direct effect of the inhibitor on the synthetic mechanism is also of importance, especially in the case of soman.

Acknowledgement. Supported by grants from the Swedish Medical Research Council B83-04X-00199-19C and the Research Institute of National Defence FOA 4, Umeå.

REFERENCES

1. Costa, E. (1970): In Advances in Biochemical Psychopharmacology, Vol. 2 (eds) E. Costa and E. Giacobini, Raven Press, New York, pp. 162-204.
2. Fleischer, J.H. and Harris, L.W. (1965): Biochem. Pharmacol. 14:641-650.
3. Harris, L.W., Fleischer, J.H., Clark, J. and Cliff, W.J. (1966): Science 154:404-407.
4. Karlén, B., Lundgren, G., Nordgren, I. and Holmstedt, B. (1974): In Choline and Acetylcholine. Handbook of Chemical Assay Methods (ed) I. Hanin, Raven Press, New York, pp. 163-179.
5. Karlén, B., Lundgren, G., Lundin, J. and Holmstedt, B. (1977): Life Sci. 20:1651-1655.
6. Karlén, B., Lundgren, G., Miyata, T., Lundin, J. and Holmstedt, B. (1978): In Cholinergic Mechanisms and Psychopharmacology (ed) D.J. Jenden, Plenum Publishing Co., New York, pp.643-655.
7. Karlén, B., Lundgren, G., Lundin, J. and Holmstedt, B. (1979): Naunyn-Schmiedeberg's Arch. Pharmacol. 308:61-65.
8. Karlén, B., Lundgren, G., Lundin, J. and Holmstedt, B. (1982): Biochem. Pharmacol. 31:2867-2872.
9. Lundgren, G., Karlén, B. and Holmstedt, B. (1977): Biochem. Pharmacol. 26:1607-1612.
10. Neff, N.H. and Costa, E. (1966): Excerpta. Med. Int. Congr. Ser. 122:23-34.
11. Schuberth, J., Sparf, B. and Sundwall, A. (1969): J. Neurochem. 16:695-700.
12. Sidell, F.R. and Groff, W.A. (1974): Tox. Appl. Pharmacol. 27:241-252.
13. Sedvall, G.C. and Kopin, I.J. (1967): Life Sci. 6:45-51.
14. Zilversmit, D.B. (1960): Am. J. Med. 29:832-848.

CENTRAL VERSUS PERIPHERAL ANTAGONISM OF CHOLINESTERASE

INHIBITOR INDUCED LETHALITY

D.S. Janowsky, S.C. Risch, A. Berkowitz,
A. Turken and M. Drennan

Department of Psychiatry
San Diego Veterans Administration Medical Center and
University of California at San Diego
La Jolla, California 92093 U.S.A.

INTRODUCTION

Concern with the possibility that chemicals may be used in international conflicts has grown in the past several years. Possible use of highly toxic and lethal chemical agents by Soviet forces in Afghanistan, by Soviet supported forces in Southeast Asia, and by Iraq in its war with Iran have been detailed (3, 4, 7, 10, 12, 13).

Although most publications mention that many of the cholinesterase inhibitors stockpiled may cause death by peripheral as well as central mechanisms, little has been written about the specific role of the central nervous system in causing lethality. If, in high doses, central mechanisms were of great importance in causing death, it is possible that a strongly centrally acting anticholinergic drug would be a relatively more effective treatment when compared with a less centrally acting anticholinergic agent.

The present studies compare the effectiveness of the centrally and peripherally acting antimuscarinic drug, scopolamine, with the less centrally acting anticholinergic drug, atropine, and with the non-centrally acting anticholinergic drug, methscopolamine, in preventing lethality by the centrally effective, reversible cholinesterase inhibitor, physostigmine. Also compared is the relative effectiveness of a combination of methscopolamine and scopolamine in preventing physostigmine induced lethality (8), compared to either drug alone.

791

METHODS

Experiment #1. The basic design of Experiment #1 was as follows: 1) saline, or 2) methscopolamine, or 3) scopolamine, or 4) a combination of methscopolamine and scopolamine in varying doses was injected subcutaneously into adult Swiss Webster (Simenson) mice weighing between 22-44 grams. All drugs were diluted in distilled water so that 0.01 ml of solution was injected per 1 gram animal weight.

Twenty minutes later, a 1.8 mg/kg dose of physostigmine (the LD_{100}) or a 6.0 mg/kg dose of physostigmine (3.3 x the LD_{100}) was injected subcutaneously into each mouse. All mice were observed for one hour after the physostigmine injection, at which time a mortality count was performed. Drug doses are outlined in Tables 1 and 2.

Experiment #2. In a replication experiment of Experiment #1, performed three months later, the LD_{100} for physostigmine was found to have decreased to 0.75 mg/kg. Although the reason for this decrease is uncertain, it is possible that litter differences or seasonal variations in susceptibility to cholinesterase inhibitors may have been causative. A dose of 4 times the LD_{100} for physostigmine (3.0 mg/kg) was chosen as the experimental physostigmine dose for this experiment, in which the antidotal effects of methscopolamine, scopolamine, and methscopolamine (3.0 mg/kg) plus scopolamine were compared. Drug doses are outlined in Table 3.

Experiment #3. Varying doses of atropine sulfate, scopolamine, or saline placebo were injected subcutaneously into Swiss Webster (Simenson) mice weighing between 22 and 44 grams. All drugs were diluted in distilled water so that 0.01 ml of solution was injected per 1 gram animal weight, with doses outlined in Table 4. Twenty minutes later, a 0.75 mg/kg dose of physostigmine (the LD_{100}) or a 2.25 mg/kg dose of physostigmine (3 times the LD_{100}) were injected subcutaneously into each mouse. All experimental animals were observed for one hour after the physostigmine injection at which time a mortality count was performed.

RESULTS

Experiment #1. Table 1 compares the effects of pretreatment with scopolamine with that of methscopolamine in mice receiving physostigmine 6.0 mg/kg. A statistically significantly larger number of mice receiving physostigmine (6 mg/kg) survived when low doses of scopolamine were given as a pretreatment than when methscopolamine was used to pretreat the animals. However, at the pretreatment doses were increased to 30 mg/kg and more, the survival rates between the scopolamine and methscopolamine pretreated mice became indistinguishable.

Table 1

Comparisons of the Antagonistic Effects of Methscopolamine, Scopolamine, and Methscopolamine Plus Scopolamine on Lethality Induced by Physostigmine 6.0 mg/kg (3.3 times the LD$_{100}$)

Dose of Methscopolamine or Scopolamine Used (mg/kg)	Various Doses of Methscopolamine Followed by 6 mg/kg Physostigmine		Various Doses of Scopolamine Followed by 6 mg/kg Physostigmine		Methscopolamine 3.0 mg/kg + Various Doses of Scopolamine followed by 6 mg/kg Physostigmine	
	Die	Live	Die	Live	Die	Live
0.025	—	—	14	0	8	6++
0.05	—	—	9	5	5	9
0.10	5	0	7	7	1	13+
0.20	7	0	4	10**	2	11
0.40	9	0	3	11***	2	10
0.60	5	0	—	—	2	0
1.0	9	0	3	14***	1	13
3.0	6	0	3	14***	—	—
5.0	4	0	0	3*	1	13
6.0	—	—	1	13***	—	—
10.0	9	0	0	8***	—	—
30.0	1	4	1	4	—	—
50.0	0	7	1	6	—	—

*p < 0.05; **p < 0.01; ***p < 0.001 – Comparing various doses of scopolamine followed by 6 mg/kg physostigmine with equal doses of methscopolamine followed by 6 mg/kg physostigmine using Fisher's Exact Test
+p < 0.05; ++p < 0.01 – Comparing various doses of scopolamine alone followed by 6 mg/kg physostigmine with equal doses of scopolamine + 3 mg/kg methscopolamine followed by 6 mg/kg physostigmine

With respect to the effects of a combination of methscopolamine and scopolamine, as can also be seen in Table 1, significantly more mice receiving physostigmine (6 mg/kg) survived when a combination of low dose scopolamine (0.025 mg/kg or 0.1 mg/kg) plus methscopolamine (3 mg/kg) was used to pretreat the animals than when scopolamine (0.025 mg/kg or 0.1 mg/kg) or methscopolamine (3 mg/kg) was given alone. Methscopolamine, itself, in the 3.0 mg/kg dose was essentially ineffective in preventing physostigmine induced lethality.

Furthermore, when the physostigmine induced lethality for a composite of all the animals pretreated with 0.025, 0.05, 0.1 and 0.2 mg/kg doses of scopolamine were compared with those pretreated with the above scopolamine doses plus methscopolamine (3.0 mg/kg), a statistically significantly greater number of animals survived when the combination treatment was used (scopolamine + physostigmine = 34 died, 22 lived; scopolamine + methscopolamine + physostigmine = 16 died, 39 lived, $p < 0.01$, X^2 test).

As shown in Table 2, in contrast to the above data using 6 mg/kg doses of physostigmine, at a dose of 1.8 mg/kg physostigmine (1 x the LD_{100}), both methscopolamine and scopolamine appeared relatively effective in blocking the lethal effects of physostigmine in relatively low doses. However, scopolamine did appear several times more effective than did methscopolamine in preventing physostigmine induced lethality.

Table 2

Comparisons of the Antagonistic Effects of Methscopolamine and Scopolamine On Lethality Induced by Physostigmine 1.8 mg/kg (1 times the LD_{100})

Dose of Methscopolamine or Scopolamine Used (mg/kg)	Various Doses of Methscopolamine Followed by 1.8 mg/kg Physostigmine		Various Doses of Scopolamine Followed by 1.8 mg/kg Physostigmine	
	Die	Live	Die	Live
0.005	6	0	6	0
0.01	5	1	4	2
0.025	4	2	1	5
0.05	4	2	0	5
0.10	4	2	1	5
0.20	1	5	-	-
0.40	1	5	-	-
0.60	0	8	-	-
1.0	0	4	-	-
2.0	0	4	-	-

Table 3

Comparison of the Antagonistic Effects of Scopolamine and Methscopolamine + Scopolamine on Lethality Induced by Physostigmine 3.0 mg/kg (4 times the LD_{100})

Dose of Methscopolamine or Scopolamine Used (mg/kg)	Various Doses of Methscopolamine Followed by 3 mg/kg Physostigmine		Various Doses of Scopolamine Followed by 3 mg/kg Physostigmine		Methscopolamine 3.0 mg/kg + Various Doses of Scopolamine followed by 3 mg/kg Physostigmine	
	Die	Live	Die	Live	Die	Live
0.025	–	–	15	0	6	9[++]
0.05	–	–	6	9	4	11
0.10	–	–	10	13	4	21[+]
0.20	–	1	5	10	1	14
3.0	14		–	–	–	–

– indicates no animals studied at these doses. [+]p<0.05; [++]p<0.001 – Comparing various doses of scopolamine alone followed by 3 mg/kg physostigmine with equal doses of scopolamine + 3 mg/kg methscopolamine followed by 3 mg/kg physostigmine

Table 4

Scopolamine and Atropine Effectiveness in Preventing
Lethality Due to Physostigmine Given Twenty Minutes Later[#]

Anticholinergic Drug Dose	Physostigmine ($1 \times LD_{100}$)				Physostigmine ($3 \times LD_{100}$)			
	Scopolamine		Atropine		Scopolamine		Atropine	
	Die	Live	Die	Live	Die	Live	Die	Live
0 (Saline)	20	4	20	4	25	0	25	0
0.025	2	12	5	9	10	1	10	1
0.05	1	13	0	15	10	0	9[*]	1
0.1	0	12	0	12	5	8	9[*]	1
0.2	0	16	0	16	1	9	9[**]	2
4.0	–	–	–	–	0	10	8[***]	1

[*]$p < 0.05$) Fisher's Exact Test, comparing effects of scopolamine and atropine. [**]$p < 0.01$). [***]$p < 0.001$). [#]Atropine or methscopolamine given subcutaneously followed twenty minutes later by subcutaneous physostigmine

Experiment #2. In the replication experiment (Experiment #2), as shown in Table 3, scopolamine, but not methscopolamine, was effective in blocking the lethal effects of physostigmine 3.0 mg/kg (approximately 4 times the LD_{100}) in this group of mice. Furthermore, as in Experiment #1, a combination of methscopolamine (3.0 mg/kg) plus scopolamine (0.025 and 0.1 mg/kg) was more effective than scopolamine or methscopolamine used alone in preventing physostigmine induced lethality.

In addition, as in Experiment #1 when all the 0.025, 0.05, 0.1 and 0.2 mg/kg scopolamine treated animals were compared to those pretreated with the above same scopolamine doses plus methscopolamine (3.0 mg/kg), a significantly greater number of animals survived when the combination treatment was administered (scopolamine + physostigmine = 36 died, 32 lived; scopolamine + methscopolamine + physostigmine = 15 died, 55 lived, $p < .01$, X^2 test). Here again, methscopolamine (3.0 mg/kg) alone was ineffective in preventing physostigmine induced lethality.

Experiment #3. Table 4 compares the effectiveness of varying doses of scopolamine with those of atropine in preventing physostigmine 0.75 mg/kg (the LD_{100}) induced lethality. Scopolamine and atropine were essentially equally potent in preventing lethality. Table 4 also compares the effectiveness of atropine and scopolamine in preventing ultra-high dose physostigmine (2.25 mg/kg = 3 x the LD_{100}) induced lethality. Scopolamine appeared to be at least 100-200 times as potent as atropine in preventing lethality.

DISCUSSION

Our results suggest that the relative effectiveness of atropine, scopolamine, and methscopolamine in preventing physostigmine induced lethality varies with the dose of physostigmine given. With relatively low doses of physostigmine, atropine, scopolamine, and methscopolamine are similar in preventing lethality. At higher physostigmine doses, scopolamine is clearly more effective.

Since atropine and methscopolamine are relatively less effective (8) centrally when compared to scopolamine, our results suggest that at the LD_{100} dose, physostigmine induced lethality probably occurs by predominately peripheral muscarinic mechanisms. However at higher physostigmine doses (i.e., at 3 x the LD_{100} dose), central mechanisms appear to be of great importance, since scopolamine was much more effective than either atropine or methscopolamine in preventing lethality.

The clinical treatment of organophosphorous poisoning has consisted predominantly of the administration of anticholinergic agents, especially atropine, given alone or in combination with a cholinesterase reactivator (8). In man, the suggested dose of atropine used to treat cholinesterase toxicity begins at 1-2 mg intramuscular, with escalation to total doses of 40 mg (approximately 0.75 mg/kg in an averaged sized male) (11).

Although relatively effective when combined with other antidotes (i.e., mecamylamine, cholinesterase reactivators, etc.), atropine alone is ineffective, even in high doses, in counteracting soman and other cholinesterase inhibitor induced lethality when these agents are given in high doses to rodents (1, 2, 5, 6, 9, 14). This ineffectiveness is consistent with our observed data.

Since cholinesterase inhibitors may have central, as well as peripheral lethal effects, our data suggest that at least for some high dose nerve agent poisonings, scopolamine may be a more effective antidote than atropine. This observation is consistent with the work of Bertram et al. (2) who showed that scopolamine plus obidoxine (a cholinesterase reactivator) was more effective than atropine plus obidoxine in preventing DFP induced lethality in mice. Possibly, such differential effects may occur in humans. In addition, even lower doses of scopolamine can be used in combination with moderate doses of methscopolamine to prevent physostigmine-induced lethality, suggesting that treatment using a two compartment model may be advantageous.

This possibility may have clinical relevance for the treatment of cholinesterase inhibitor toxicity. High doses of a peripherally acting agent like methscopolamine, by allowing use of a lower dose of a centrally acting agent, may potentially reduce the occurrence of debilitating psychotoxic episodes following anticholinergic treatment of cholinesterase intoxication, and yet retain and/or enhance antidotal efficacy.

REFERENCES

1. Ashani, Y., Leader, H., Raveh, L., Bruckstein, R. and Spiegelstein, M. (1983): J. Med. Chem. 26:145-152.
2. Bertram, U., Kasten, A., Lullman, H. and Ziegler, A. (1977): Experientia 33(9):1196-1197.
3. Cookson, J. and Nottingham, J. (1969): A Survey of Chemical and Biological Warfare, London, Sheep and Ward.
4. Haggerty, J.J. (1980): Milit. Rev. 60(8):37-44.
5. Harris, L.W., Stitcher, D.L., Heyl, W.C., Lieske, C.N., Lowe, J.R., Clark, J.H. and Broomfield, C.A. (1979): Toxicol. & Appl. Pharmacol. 49:23-29.
6. Harris, L.W., Stitcher, D.L. and Heyl, W.C. (1980): Life Sci. 26:1885-1891.
7. Hoeber, A. and Douglass, J. (1979): Int. Security 3(1):55.
8. Ketchum, J.S., Sidell, R.F., Crowell, E.B., Jr., Aghajanian, G.K. and Hayes, A.H., Jr. (1973): Psychopharmacologia (Berl.) 28:121-145.
9. Lipp, J. and Dola, T. (1980): Arch. Int. Pharmacodyn 246:138-148.
10. Mendelson, M. and Robinson, J.P. (1980): Sci. Amer. 242(2):38-47.
11. Namba, T., Nolte, C.T., Jackrel, J. and Grob, D. (1971): Am. J. Med. 50:475-492.
12. Newhouse, P., Belenky, G. and Del Jones, F. (1980): American Psychiatric Association Syllabus and Scientific Proceedings, 134th Annual Meeting, page 170.
13. Ruehle, H. (1978): In U.S. Army Medical Intelligence and Information Agency, K-9002-A. Washington, D.C.
14. Stitcher, D.L., Harris, L.W., Heyl, W.C. and Alter, S.C. (1978): Drug & Chem. Toxicol. 1(4):355-362.

STRUCTURAL GENE MUTATIONS AFFECTING ACETYLCHOLINESTERASE AND CHOLINE ACETYLTRANSFERASE PERTURB BEHAVIOR AND DEVELOPMENT IN A SIMPLE INVERTEBRATE

R.L. Russell[*],[+] and J.B. Rand[*]

[*]Department of Biological Sciences
and
[+]Department of Psychiatry
University of Pittsburgh
Pittsburgh, Pennsylvania 15260   U.S.A.

INTRODUCTION

For cholinergic transmission, as for many other processes of biological interest, a genetic approach can sometimes offer advantages complementary to those of other approaches.   Most notable amongst these advantages is the ability, by mutation, to alter specifically the products of individual genes which may be involved.   Sometimes such mutations can serve to pinpoint the roles of already identified gene products, and in other cases mutations can serve to point out the unsuspected involvement of other gene products, whose nature remains to be worked out.

Full exploitation of such genetic advantages depends to a considerable degree on the genetic flexibility of the organism studied, a feature which does not always coincide with intrinsic interest. Thus mammals, although obviously of considerable intrinsic interest as regards cholinergic transmission, do not lend themselves well to genetic exploitation.   On the other hand, certain relatively simple invertebrates, most notably the fruit fly Drosophila melanogaster and the soil nematode Caenorhabditis elegans, have considerable genetic flexibility, and thus are well suited to a genetic approach. C. elegans, in addition, exhibits a marked cellular simplicity that is of considerable advantage when studying specific interactions among nerve cells.   For example, the entire C. elegans nervous system consists of only 302 neurons, whose synaptic interactions have been entirely catalogued by serial section electron microscopy (1, 3, 16-18, J.G. White, personal communication).   Moreover, the highly reproducible cell lineages producing these neurons have been followed

799

all the way through development from the fertilized egg to the adult
(13-15).

The work presented below describes one approach toward using
the genetic and cellular advantages of C. elegans for investigating
cholinergic transmission. It includes a summary of previous work,
some of which has been published (2, 4, 5, 10, 7, 8, 10, 11), as well
as a number of new results. In brief, we have been able to identify
two genes affecting acetylcholinesterase (AChE) and a third gene
affecting choline acetyltransferase (ChAT), and we have used mutations
in these genes to investigate the roles played by these important
cholinergic enzymes in nervous function and development.

RESULTS

Acetylcholinesterase. We have previously reported (5) that extracts
of the nematode C. elegans contain multiple molecular forms of AChE.
In fact, five major forms have been described, and each of these
has been separated from the others and characterized with respect
to a variety of properties. The kinetic properties of these five
forms, as summarized in Table 1, make it clear that they belong to
two distinct classes, which we have labeled class A and class B.
In some respects this situation resembles that of the vertebrate
true (or acetyl-) and pseudo- (or butyryl-) cholinesterases, in
that two kinetically distinct cholinesterase classes exist. However,
there is not a detailed homology; the two C. elegans AChE classes
do not exhibit the same constellations of inhibitor and substrate
specificities as the vertebrate enzymes, and thus cannot be called
one true and the other pseudo. Indeed, the relatively high affinities
of both classes for acetylcholine (ACh) suggest that both should be
called true, or acetylcholinesterases (EC 3.1.1.7).

The C. elegans AChE classes also differ from the vertebrate
true and pseudocholinesterases in that there is not a parallel array
of forms in the two classes. As described elsewhere (see Massoulie,
J., this book), both true and pseudocholinesterases from vertebrates
exist as homologous collections of molecular forms, some of which
are globular and others of which are highly asymmetric because they
possess a collagen-like tail. In C. elegans, there are three forms
in Class A, but in class B there are only two. Furthermore, the
collagen-tailed asymmetric AChE forms of vertebrates do not have a
parallel in C. elegans; none is affected by collagenase treatment,
and only one is nearly as asymmetric as the vertebrate tailed forms.
Nonetheless, within each C. elegans class the largest form is quite
insoluble, requiring detergent for solubilization, whereas the smallest
form is quite soluble, and this difference suggests that the smaller
form might be a precursor which is assembled together with other,
less hydrophilic components, to produce a larger form localized on
(synaptic?) membranes.

Table 1

Properties of C. elegans Acetylcholinesterase Forms

| | "Major" | | | | | "Minor" |
| | Class A | | | Class B | | Class C |
	IA	III	IV	IB	IIB	IIC
Sedimentation constant, $S_{20,w}$	5.2	11.3	13.0	4.9	6.7	7
Molecular Weight, Kilodaltons	101	298	357	83	176	------
K_m for Acetylcholine, M	12	15	12	80	67	0.017
Relative activity, BuThCh/AcThCh x 100	64	64	71	18	25	------
"pI50", Eserine (units of 10^{-8}M)	0.29	0.26	0.16	20	20	100,000
"pI50", Aldicarb (units of 10^{-6}M)	1.5	1.7	1.9	27	36	400
"pI50", iso-OMPA (units of 10^{-4}M)	0.47	0.39	0.85	>30	>30	>30
"PI50", Triton X-100 (units of 10^{-4}M)	0.3	0.3	0.3	8	>30	>30

Reproduced from (5) and (8), with permission

The possible functional roles of the two C. elegans AChE classes, A and B, is no less intriguing than the roles of the multiple cholinesterases in vertebrates, and in C. elegans the genetic approach we have taken has helped to shed some light on these roles (2, 4, 7). We have isolated 5 mutations in a gene called ace-1 on the C. elegans X chromosome (Figure 1), and all 5 produce a marked and specific loss of class A AChE, affecting all 3 class A forms. We have also isolated 17 mutations in another gene called ace-2 on chromosome I (Figure 1), and all 17 of these produce a marked and specific loss of class B AChE, affecting both class B forms. Thus the kinetic differences between classes A and B are reinforced by this observation that the two classes are under separate genetic control. Although the evidence in humans is less definitive, it does suggest separate genetic control also for true and pseudo-cholinesterases (9).

Two lines of evidence support the contention that ace-1 and ace-2 are structural genes, respectively, for class A and class B AChE. First, in "gene dosage" experiments, mutant heterozygotes at either locus show activity levels of the corresponding AChE class which are almost exactly half those of wild type homozygotes (Table 2). By analogy with numerous examples in Drosophila melanogaster, this is the result expected for a structural, as opposed to a regulatory, gene. Second, some of the mutant alleles of each gene turn

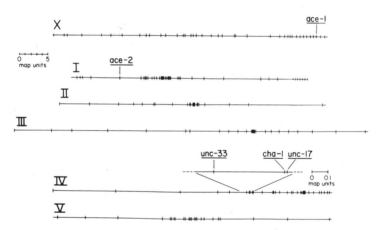

Figure 1. Genetic Map of C.elegans. Each chromosome is indicated by a single horizontal line, and each mapped gene is indicated by a short vertical line at the appropriate location. With the exception of the expanded insert shown for chromosome IV, the scale is as given to the left, with 1 map unit = 1 centimorgan. On limited evidence, one gene is approximately 0.02 centimorgans long. Labelled genes are described in the text. This map is based on the 1983 map distributed by the Caenorhabditis Genetics Center, Columbia, MO, and on our own data.

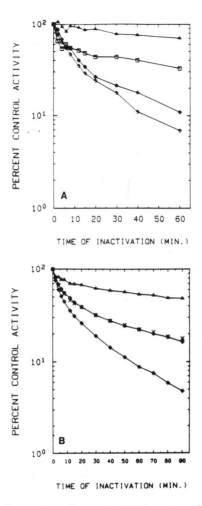

Figure 2. Thermal Inactivation of AChE. In all cases, percentage of surviving AChE activity is plotted against time of thermal inactivation. **A.** Inactivation at 50.3°C of form IV (class A) AChE from wild type N2 (▲) and of that controlled by three mutations in ace-1, the apparent structural gene for class A, p1010 (□), p1008 (♦), and p1004 (◊). **B.** Inactivation at 50.0°C of form IV (class A) AChE from wild type N2 (▲), that controlled by p1008 (♦), and a mixture of the two which was 30% N2 and 70% p1008 (◊). For the mixture, theoretical expectation for completely independent inactivation of wild type and mutant components is also indicated (x). The form IV AChE from the ace-1 mutants was isolated from extracts in which it was present at very low residual levels. Reproduced from (7), with permission.

out to produce only partial deficiencies of the affected AChE class, and in these cases the residual activity of the affected type has been shown to be qualitatively altered; an example of an ace-1 mutant thermolability change in class A AChE is shown in Figure 2. Such qualitative changes are most probably due to primary amino acid sequence changes in some portion of the AChE molecule, such as could only be produced by a structural gene mutation.

From this evidence we have concluded, with reasonable certainty, that ace-1 and ace-2 are indeed structural genes, and this conclusion leads to some interesting sequelae. Since each of the forms within a class disappears when the corresponding gene is mutant, it seems clear that the gene must encode a component shared by all the forms within a class. Kinetically, the forms within a class are indistinguishable, suggesting that they carry a common active site, and thus it seems likely that the active site is carried on the common component encoded by the corresponding structural gene. Together with the solubility differences between forms within a class, this supports the aforementioned notion of assembling within each class an insoluble, membrane-associated large form of AChE, by combining the corresponding smaller form with other, as yet unidentified components.

The behavioral phenotypes of ace-1 and ace-2 mutants are unexpected and informative. At either locus we can identify severe mutants by the criterion that they spare no more than 1% of the corresponding AChE class, and often much less. Surprisingly, even

Table 2

Gene Dosage Effects of ace-1 and ace-2 on
Acetylcholinesterase Activity

| | | Number of | Number of | Acetylcholinesterase Activity | |
| | | ace-1(+) | ace-2(+) | | Percent of Case |
	Genotype	Gene Copies	Gene Copies	Total	with 2 Copies
	ace-2/ace-2 ; ace-1/ace-1	0	0	0.003	1.7
A	ace-2/ + ; ace-1/ace-1	0	1	0.080	47
	+ / + ; ace-1/ace-1	0	2	0.170	(100)
	ace-2/ace-2 ; ace-1/ace-1	0	0	0.004	1.4
B	ace-2/ace-2 ; ace-1/ +	1	0	0.160	57
	ace-2/ace-2 ; + / +	2	0	0.280	(100)

Drawn from (2), with permission.

the most severe ace-1 mutant has no observable behavioral defect, and the same is true of the most severe ace-2 mutant as well. Thus from these results it appears that either class A or class B AChE can be totally dispensed with, or in other words that neither class is uniquely essential for normal behavior. However, lest it be mistakenly concluded that AChE as a whole is unimportant, we should quickly add the further observation that severe double mutants (ace-1; ace-2), lacking both class A and class B AChE are quite severely uncoordinated. Thus AChE is indeed important, but the two AChE classes must be somehow arrayed such that each can functionally suffice for normal behavior in the absence of the other, making it necessary to eliminate both if a behavioral effect is to be produced.

This result suggests quite strongly that the anatomical distributions of class A and class B AChE ought to be very similar, and indeed this is borne out by histochemical staining. Within the nervous system, and with the caveat that observations have so far been made only at the light microscope and not at the electron microscope level, the distribution of class A and class B AChE, as judged by staining of ace-2 and ace-1 mutants, is identical. Both classes are found at all sites of neuromuscular contact, i.e., along the dorsal and ventral nerve cords and around the nerve ring (Figure 3). Thus it seems likely that each cholinergic synapse in C. elegans contains both class A and class B AChE, although no clear and obvious functional reason for this dual presence is forthcoming. In light of this observation, and in view of the presence of homologous collagen-tailed forms among both the true and pseudo cholinesterase of vertebrates, it is tempting to speculate that both types of cholinesterase might be present at some, perhaps even most, vertebrate cholinergic synapses.

Another intriguing, but as yet unexplained parallel between the vertebrate and C. elegans cases concerns AChE outside the nervous system. In C. elegans, only one non-nervous structure stains for AChE; this is a six-celled structure interposed between the pharynx and the intestine, and thought to be a valve preventing regurgitation. We have reconstructed this structure by serial section electron microscopy, and find, to our surprise, that it is not innervated at all. Intriguingly, the AChE which it contains is entirely of class B, providing at least a loose formal parallel with vertebrate serum, where only pseudocholinesterase is found. Perhaps in both cases a better understanding of the presence of two AChE classes in the nervous system would emerge if it were understood why particular extraneuronal sites contain only one class.

The quantitative relationship between AChE level and behavior has been investigated using ace-1 and ace-2 mutants. Animals which are homozygous for a severe mutation at one locus and heterozygous at the other (i.e., ace-2/ace-2; ace-1/+ or ace-2/+; ace-1/ace-1) have approximately 25% of the wild type level of AChE and yet are

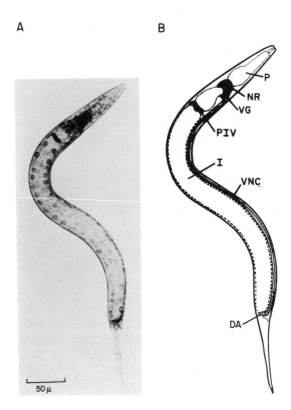

Figure 3. AChE Staining of C. elegans Whole Mount. Late juvenile
stage (L3) animal, permeabilized by a brief acetone treatment and
then stained by the acetylthiocholine direct visualization method.
A. Photomicrograph. B. Schematic drawing. P, pharynx; NR, nerve
ring (major integrative center of nervous system); VG, ventral ganglion; PIV, pharyngeo-intestinal "valve" (may not, in fact, be a
valve); I, intestine; VNC, ventral nerve cord (site of many neuromuscular junctions); DA, depressor ani muscle. Reproduced from
(2), with permission.

behaviorally quite normal. On the other hand, animals homozygous
for a severe mutation at the ace-2 and also homozygous, but for a
less severe mutation, at the ace-1 locus have 5-10% of the wild
type AChE level and exhibit only mild uncoordination. Thus, there
appears to be a behavioral threshold for AChE somewhere in the range
of 10-25% of the normal wild type levels (suggesting a "safety factor"
for AChE of 4-10). Interestingly, even animals homozygous for severe
mutations at both loci are viable, and not noticeably affected in
development, despite having only 1-2% of wild type AChE levels.
Thus, the AChE "safety factor" for viability, as opposed to coordination, is much greater, at least 50-100. Such large safety factors,

if they apply in the vertebrate as well, would certainly cast doubt
on any explanation of the type in which, say, a 2-fold reduction
in AChE levels is posited to be causal for a particular clinical
or animal syndrome.

When first examined, the small amount of AChE activity present
in animals homozygous for severe mutations at both loci was thought
to be residual class B AChE, spared by slightly leaky ace-2 mutant
alleles (7). This thinking was based on the sedimentation constant
(ca. 7S) and marked Triton X-100 resistance of this activity, both
known properties of the class larger B AChE form, form II (Table
1). Upon closer scrutiny, however, this activity has proven to be
quite novel (8). It can be separated from form II (now called form
IIB to distinguish it) either by ion exchange chromatography on
DEAE cellulose, where it elutes later than form IIB, or by affinity
chromatography on N-carboxyphenyldimethyl ethylammonium ion coupled
to Sepharose, to which it binds but form IIB does not. In addition,
its kinetic properties are markedly different from those of form IIB
(and for that matter from those of the class A forms and all other
known cholinesterases as well). For example, its Km for acetylcholine
is about 17nM, about 5000-fold lower than that of form IIB (Table
1), it is about 1000-fold more resistant than form IIB to eserine,
and it has a turnover number of only about 900 per minute, in contrast
to a figure of several hundred thousand per minute for form IIB
and other well characterized acetylcholinesterases (data not shown).

The novel properties and relatively low abundance of this new
AChE, which we have called class C AChE (form IIC), led us to surmise
that a homologous AChE, if present in other systems, could well
have been overlooked. Accordingly, we sought evidence of such a
homologous AChE in rat brain, Torpedo electric organ, and Drosophila
heads. No evidence of such an AChE was obtained in these cases,
but in another nematode species, the AChE-secreting parasite of
swine kidneys, Stephanurus dentatus, a homologous AChE was detected
quite easily (8). Thus, at the moment there is evidence for the
occurrence of this type of AChE only in the class Nematoda. The
functional significance of class C AChE in C. elegans is unclear,
although it does appear essential for viability in ace-2; ace-1
double mutants (these have only class C AChE and can be killed by
specific AChE inhibitors). Class C AChE is present in wild type
C. elegans at levels at least approximately the same as in ace-2;
ace-1 double mutants (8) indicating, along with its novel properties,
that it is probably the product of an as yet unidentified gene (or
genes) different from ace-1 and ace-2. Attempts are now being made
to obtain mutations in this new gene.

Choline acetyltransferase. Fortunately, the molecular aspects of
choline acetyltransferase (ChAT) in C. elegans are, at least initially,
less complex than those of AChE. Only two molecular species of
ChAT have been detected, with molecular weights of 154,000 Kd (6.9S)

and 71,000 Kd (5.2S) (10). These are kinetically indistinguishable, and the larger one can be quantitatively converted to the smaller without activity loss (Figure 4), suggesting a common genetic control. We have isolated 5 mutations in a gene called <u>cha-1</u> on <u>C</u>. <u>elegans</u>

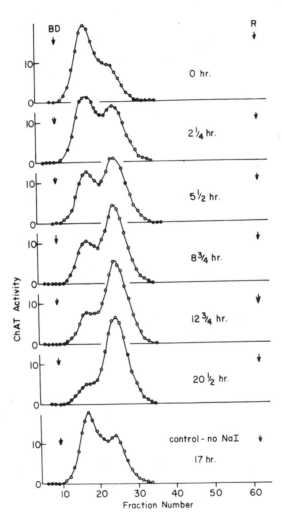

Figure 4.   Interconversion of <u>C</u>. <u>elegans</u> ChAT Forms.   A soluble preparation consisting mostly of the larger ChAT form was incubated in 0.5 M NaI for the indicated times at 4°C, followed by size fractionation using gel permeation chromatography on Sephacryl S200 Superfine.   Progressive conversion to the smaller ChAT form occurred without activity loss.   Left and right arrows mark the positions, respectively, of blue dextran and riboflavin, used as markers. Reproduced from (11), with permission.

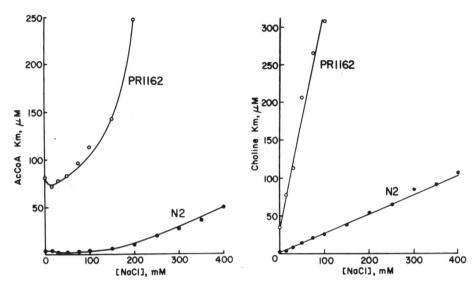

Figure 5. Kinetic Alterations of ChAT in a cha-1 Mutant. At each
of the indicated NaCl concentrations, Km's were determined for each
of the two ChAT substrates, acetyl-CoA and choline, in an excess
of the other substrate. Each Km was determined from the slope of
an Eadie-Hofstee plot (V vs. V/S) involving velocity measurements
at 8-10 concentrations, and the correlation coefficients for the
resulting plots wre from -0.956 to -0.995. Km's were determined
for the small forms of ChAT from wild type N2 (●) and the strain
PR1162, containing the leaky cha-1 mutation p503 (o). The dependence
of the C. elegans ChAT Km's on salt concentration resembles that
seen with other ChAT's, but clearly does not obscure the large difference between wild type and mutant. Reproduced from (10), with
permission.

chromosome IV (Figure 1), and all 5 lead to a marked reduction in
the level of both ChAT forms, substantiating the expected common
genetic control (11). By the same two kinds of criteria applied
to the AChE genes ace-1 and ace-2, cha-1 is almost certainly a structural gene for ChAT. An example of kinetic changes in ChAT in a
partially-defective cha-1 mutant is shown in Figure 5.

The phenotypes of cha-1 mutants, like those of ace-1 and ace-2
mutants, are interesting and informative (Table 3). Four of the 5
cha-1 mutants (b401, p1152, p1154, and p1156) are severe, sparing
only 1-3% of the wild type levels of ChAT, and all four share a
common constellation of properties. First, although viable, they
have a marked, coiling type of uncoordination, quite distinct, for
instance, from that of ace-2; ace-1 double mutants. Second, they

Table 3

Properties of C. elegans cha–1 and unc–17 Mutants

Allele	Complementation Pattern	ChAT Activity, Percent of Control	Uncoordinated?	Slow Growing?	Small as an Adult?	AChE Inhibitor Resistant?
+	(wild-type)	(100)	No	No	No	No
b401	pure cha–1	0.0	Yes, Type A	Yes	Yes	Yes
p1152	pure cha–1	0.8	Yes, Type A	Yes	Yes	Yes
p1154	pure cha–1	1.8	(Yes), Type A	(Yes)	(Yes)	Yes
p503	?	9.3	No	No	No	No
p1156	overlap	2.5	Yes, Type A	Yes	Yes	Yes
e113	overlap	46.0	Yes, Type B	No	No	Yes
e876	overlap	58.0	Yes, Type B	No	No	Yes
p1160	pure unc–17	151.0	Yes, Type A	Yes	Yes	Yes
e245	pure unc–17	90.0	Yes, Type A	Yes	Yes	Yes
7 others	pure unc–17	119.0 (ave)	Yes, Type A	Yes	Yes	Yes

The data presented apply to animals homozygous for the mutant allele given in the left-hand column. The complementation pattern in the second column was derived from the data in Figure 7, which, for technical reasons, does not include p503. ChAT activity represents the mean of 4–6 determinations, and is presented as enzyme activity per animal, as a percent of the wild-type activity. The "A" type of uncoordinated behavior, observed with most alleles, is the coiling, jerky behavior described in the text. The "B" type of uncoordination is seen only with animals homozygous for the e113 or e876 alleles, and is characterized by an omega-shaped posture and the inabilit to propagate waves of body muscle contraction. Drug resistance was determined as the ability to grow and develop in the presence of 0.5 mM Aldi-carb. The parentheses around the p1154 data indicate that, although this allele leads to the same qualitative alter-ations, they are notably less severe than those caused by the cha–1 alleles just above it. Drawn in part from (10), with permission.

are slow growing, especially during early post-embryonic stages
(Figure 6). Third, they are small in the adult stage (Figure 6).
Fourth, they are markedly more resistant than the wild type to several
different inhibitors of AChE. The fact that all four mutants show
the same constellation of properties, coupled with the strong evidence
that cha-1 is a structural gene, leads to the conclusion that all
these properties are secondary consequences of a primary lesion in
ChAT. While the uncoordination and perhaps the inhibitor resistance
could have been expected, the growth defects are somewhat surprising
and suggest either that cholinergic transmission is somehow important
for growth rates and eventual size, or else that acetylcholine might
play a separate and perhaps broader role in growth and development
than its status as simply a neurotransmitter might suggest. In this
regard it is of interest that the most pronounced slowing of growth
in these mutants occurs at a time when many apparently cholinergic
motor neurons are being born by cell division and are forming their
neuromuscular contacts (13-15). The fifth cha-1 mutant (p503) has
only an extremely mild behavioral defect and also contains approximately 10% of the wild type ChAT levels. Thus, as with AChE, there
are apparently different and sizable safety factors for ChAT as
regards viability (50-100x) and coordination (6-10x).

Cha-1 maps extremely close to another "gene", unc-17, previously
identified by mutants with uncoordinated behavior (Figure 1). Unc-17
mutants exhibit the same constellation of properties as those in
cha-1, except for the ChAT deficiency (Table 3). Most unc-17 mutants
complement most cha-1 mutants, but three exceptional "overlap" alleles,

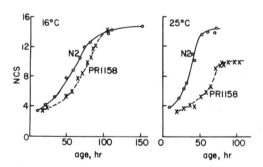

Figure 6. Growth Curves of Wild Type and cha-1 Mutant. Synchronized
populations of N2 and the cha-1 (b401) mutant strain PR1158 were
characterized at the indicated times of development in a sizing
nematode counter, the signal of which (NCS) is proportional to animal
length. Reproductive maturity (first egg laid) occurs for N2 at
47 hr (25°C) or at 90 hr (16°C). Reproduced from (10), with permission.

originally isolated in each of the two genes fail to complement
all alleles of both genes (Figure 7). Surprisingly, two of these
alleles (e113 and e876) exhibit nearly normal ChAT levels (Table
3), and yet fail to complement the cha-1 mutants occurring in the
apparent structural gene for ChAT.   These unusual complementation
properties suggest that cha-1 and unc-17, rather than being separate
genes, may represent different parts of a single complex locus encoding
ChAT.   A working model based on this suggestion in presented in Figure
8.   ChAT in this model is envisaged as a homodimer of a subunit with
two distinct domains; one of these domains (encoded by cha-1) is envisaged as carrying out the catalytic activity and the other (encoded
by unc-17) is envisaged as involved in some as yet unknown additional
function required in vivo (perhaps correct localization of the catalytic activity).   Several biochemical and genetic predictions of
this model are now being tested.

Synapses.   In C. elegans, comparison of detailed neuroanatomy with
that of a larger nematode, Ascaris lumbricoides, allows many identified
synapses to be ascribed a cholinergic status (6, 12).   The numbers
of such apparently cholinergic synapses have been determined by
serial section electron microscopy in wild type and in several severely

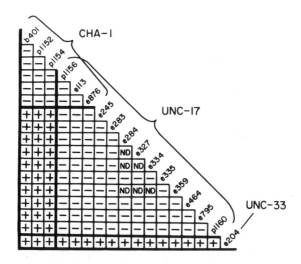

Figure 7.   Complementation Pattern of cha-1 and unc-17 Mutants.
Complementation (+) is indicated by normal coordination in the double
heterozygote for the two alleles involved (e.g., b401/e245).   Lack
of complementation (-) is indicated by uncoordination in the double
heterozygote (e.g., b401/p1152).   ND indicates test not done.   The
gene unc-33 is relatively closely linked (Figure 1) but unrelated,
and serves as a positive control.   Reproduced from (10), with permission.

affected <u>cha-1</u> mutants. As shown in Table 4, the mutants have approximately 30% fewer such synapses. This suggests that the biosynthesis of ACh may be limiting for cholinergic synapse formation and/or maintenance.

DISCUSSION

Results described above provide several examples of the ways in which a genetic approach can be revealing in investigating cholinergic transmission. For example, the distinction of two major AChE classes, while possible on kinetic grounds alone, is rendered much more convincing by the demonstration that these two classes are separately controlled by unlinked genes. Also, the observation that the several forms within each class are all controlled by the same structural gene makes it extremely likely that each form shares at least one common structural subunit with the others. The genetic

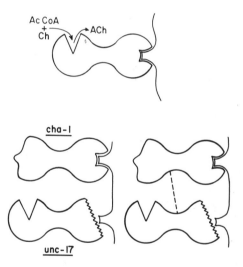

Figure 8. Model for Structure of ChAT. Top drawing indicates two-domain structure postulated for ChAT monomer; left domain catalyzes ACh synthesis, right domain ensures correct localization. Both functions are assumed essential for normal cholinergic transmission. Lower left indicates postulated monomer defects in pure <u>cha-1</u> mutant (upper part) and in pure <u>unc-17</u> mutant (lower part). Neither monomer alone is adequate for normal transmission. Lower right indicates how dimerization between differently altered monomers of <u>cha-1</u> and <u>unc-17</u> could lead to complementation by allowing correct localization of catalytic activity and therefore normal transmission.

Table 4

Frequency of Dorsal Neuromuscular Junctions

Genotype	Number of Series Examined	Total Sections Examined	Total Distance Examined	Junctions Observed			Junctions Per Unit Distance
				DA	DB	Total	
+	4	1786	132.2 m	62	54	116	0.88 m^{-1}
cha-1 (b401)	3	1026	75.9 m	25	22	47	0.62 m^{-1}

Very young animals (2 hr from hatching) were fixed, embedded, and serially sectioned. Series length varied from 200 to 700 sections, with a nominal section thickness of 73 nm, and a loss rate no greater than 3%. Each series was from the anterior part of the animal, just posterior to the retrovesicular ganglion. Each section was photographed and printed, the motor neurons in the ventral and dorsal nerve cords were all identified by standard criteria (18), and the neuromuscular output of each was catalogued. The DA and DB cells are the two classes of presumed cholinergic motor neurons in first-stage C. elegans larvae; their output is exclusively along the dorsal nerve cord. They are believed to be responsible for backward and forward movement, respectively, and apparently represent the entire excitatory neuromuscular output to the body muscles in such animals.

approach was also essential for identifying the third class of C. elegans AChE, since its contribution to total activity is so low as to be obscured in all but the severely affected double mutants of the ace-2; ace-1 type.

Establishing a relationship between AChE level and behavior, while potentially accomplishable by a pharmacological approach, was accomplished more easily and probably with fewer complications by the genetic approach using different mutational alleles. For ChAT, the genetic approach has established that the two molecular species are commonly controlled by a given structural gene, implying that they share a common subunit. And again, establishing a relationship between ChAT level and behavior (or viability) has been made considerably easier by the use of mutants.

Furthermore, two special results of the genetic approach to ChAT would have been quite difficult to obtain by other means. The first is the observation that primary deficienc8ies in ChAT lead to a number of secondary behavioral, growth, and drug resistance phenotypes. The second is the observation, based upon detailed fine-structure genetics, that the structural gene for ChAT is quite probably complex, involving more than a single domain in the resulting gene product.

Finally, we hope that some of the ideas generated by a genetic approach to cholinergic transmission in C. elegans will prove to be testable by other than genetic means in more complicated organisms such as mammals, where the level of intrinsic interest is clearly higher.

Acknowledgements. The work described herein was supported by grants from the U.S. Public Health Service (NS 13749) and from the Muscular Dystrophy Association. We thank our colleagues Jay Bashor, Leslie Cavalier, Leanne Cribbs, Joseph Culotti, Marilyn Culotti, John Duckett, Gunter von Ehrenstein, Paul Gardner, Robert Herman, Carl Johnson, Dennis Kolson, John Lace, Valerie Mazelsky, Philip Meneely, Nancy Peffer, Kristin Peterson, and Brian Stern for assistance, collaborations, discussions, and in some cases the sharing of data before publication. We also thank Donald Riddle and the Caenorhabditis Genetics Center, Columbia, MO, for map information, and we thank Elsevier Biomedical Press, the Genetics Society of America, and Raven Press for permission to reproduce some previously published material.

REFERENCES

1. Albertson, D.G. and Thompson, J.N. (1976): Phil. Trans. Roy. Soc. London B275:299-325.
2. Culotti, J.G., von Ehrenstein, G., Culotti, M.R. and Russell, R.L. (1981): Genetics 97:281-305.

3. Hall, D.H. and Russell, R.L. (1985): Submitted for publication.
4. Johnson, C.D., Duckett, J.G., Herman, R.K., Meneely, P.M. and Russell, R.L. (1981): Genetics 97:261-279.
5. Johnson, C.D. and Russell, R.L. (1983): J. Neurochem. 41:30-46.
6. Johnson, C.D. and Stretton, A.O.W. (1985): J. Neurosci., in press.
7. Kolson, D.L. and Russell, R.L. (1985a): J. Neurogenet., 2:69-91.
8. Kolson, D.L. and Russell, R.L. (1985b): J. Neurogenet., 2:93-110.
9. La Du, B.N. (1972): Fed. Proc. 31:1276-1285.
10. Rand, J.B. and Russell, R.L. (1984): Genetics 106:227-248.
11. Rand, J.B. and Russell, R.L. (1985): J. Neurochem. 44:189-200.
12. Stretton, A.O.W., Fishpool, R.M., Southgate, E., Donmoyer, J.E., Walrond, J.P., Moses, J.E.R. and Kass, S. (1978): Proc. Natl. Acad. Sci. USA 75:3493-3497.
13. Sulston, J.E. (1976): Phil. Trans. Roy. Soc. London B275:287-298.
14. Sulston, J.E. and Horvitz, H.R. (1977): Dev. Biol. 56:110-156.
15. Sulston, J.E., Schierenberg, E., White, J.G. and Thomson, J.N. (1983): Dev. Biol. 100:64-119.
16. Ward, S., Thomson, N., White, J.G. and Brenner, S. (1975): J. Comp. Neurol. 160:313-338.
17. Ware, R.W., Clark, D., Crossland, K. and Russell, R.L. (1975): J. Comp. Neurol. 162:71-110.
18. White, J.G., Southgate, E., Thomson, N. and Brenner, S. (1976): Phil. Trans. Roy. Soc. London B275:327-348.

CHOLINE FLUXES TO AND FROM THE RAT CEREBRAL CORTEX

STUDIED WITH THE "CUP TECHNIQUE" IN VIVO

R. Corradetti, R. Brehm[+], K. Löffelholz[+] and G. Pepeu

Department of Pharmacology
University of Florence
Viale G.B. Morgagni 65
50134 Florence, ITALY

[+]Department of Pharmacology
University of Mainz
Obere Zahlbacher Strabe 67
D-6500 Mainz, Federal Republic Of Germany

INTRODUCTION

Since MacIntosh and Oborin (11) and later Mitchell (12) introduced the "cup technique" as a mean to study acetylcholine release from the cerebral cortex in vivo, this technique has been widely used for investigating the release of various neurotransmitters in anaesthetized as well as unanaesthetized mammals (2, 13, 14). Recently we proposed the "cup technique" as a way for studying the efflux of endogenous choline (Ch) from the rat cerebral cortex (4, 5) and to estimate changes in the extracellular concentration of Ch, if we consider the fluid filling the cup as an extension of the extracellular space.

Ch diffuses from the rat cortex into the cup at a rate of about 60 pmol min^{-1} cm^{-2} (5; Löffelholz et al., this book). This effect is temperature-dependent, and is enhanced by muscarinic agonists. From indirect evidence and from the comparison with results obtained in the isolated perfused chicken heart, we suggested a cellular source for the Ch effluent from the rat cerebral cortex into the cup. The most likely among the possible sources of choline is membrane phospholipid breakdown (Löffelholz et al., this book). However, Ch found in the cup could also originate from the blood plasma. Thus, the observed changes of Ch efflux induced by muscarinic agonists could reflect perturbations of the blood-brain exchanges of Ch, or

817

even of blood flow through the microvessels which is likely to be regulated by muscarinic receptors (7, 9).

Thus, experiments were designed to study the Ch fluxes between brain and cup solution, and vice versa. Moreover, we tackled the problem of whether the muscarinic increase in cortical Ch efflux (5) is caused by effects on Ch flux mechanisms, or on Ch liberation from cortical cells.

MATERIALS AND METHODS

Experiments were carried out on adult male Wistar rats (180-200 g). The head of the anaesthetized animal (urethane 1.25 g/kg i.p.) was positioned in a head holder and two Perspex cups (0.21 cm^2 each) were tightly placed on the exposed cerebral hemispheres after removing the dura mater. Body temperature was kept constant (37°C) by means of a homeothermic blanket with a rectal probe. The cups were filled with Ringer solution (100 or 300 μl; at 37°C unless otherwise stated) of the following composition (mM): Na^+ 155.9, K^+ 5.6, Ca^{++} 1.6, Cl^- 158.8, HCO_3^- 5.9 and (+) glucose 5.5. The Ringer solution was replaced at intervals which varied (2-10 min) according to the experimental design. When the temperature of the cup solution was reduced or drugs were added to the cup solution, one side was kept as control so as to obtain treated and control values from the same rat.

In order to investigate the vessel integrity and the permeability for plasma Ch in control conditions and during local treatments, 10 μCi of (carboxyl-[^{14}C]) - inulin ([^{14}C]-inulin, NEN, 2.6 mCi/g) or 100 μCi of (methyl-[^{3}H]-Ch chloride, [^{3}H]-Ch, NEN, 80 Ci/mmol) was infused in the tail vein (0.09 ml/min for 10 min). [^{14}C]-inulin was chosen since it penetrates the blood-brain barrier (BBB) very slowly and eventually distributes in the extracellullar space (16). At the end of the infusion, blood samples were taken from the heart, and radioactivity was measured after extraction. In the plasma, radioactivity was totally water-soluble, indicating that it represented the free Ch pool.

When the removal mechansisms were investigated trace amounts of [^{3}H]-Ch were added to the Ringer solution in the cup (100 μl).

Ch was determined by two different methods, the chemiluminescence method described by Israēl and Lesbats (10), and the radioenzymatic method described by Goldberg and McCaman (8).

RESULTS

Morphology of the preparation. Since the experiments with the "cup technique" were designed to study Ch fluxes between cortex and cup solution, the tissue under the cup (meninges and neural tissue) was studied by light-microscopy. It appeared from experiments on 8 hemispheres that the removal of the dura mater also took away, at least in part, the outer arachnoid and pial layers and thus opened the arachnoid space, whereas the cortical tissue (under the inner pial layer) did not show any light-microscopical signs of structural damage.

Effects of penicillin on BBB. Evidence that the BBB was perserved after surgery was provided by measuring the appearance of [^{14}C]-inulin in the cup solution during a 10 min infusion of 10 μCi of [^{14}C]-inulin (2.6 mCi/g) in the tail vein (Table 1). Under control conditions, the efflux of labelled inulin into the cup was very small and achieved a plateau after 6-8 min of infusion. The ratio of the plasma concentration at the end of the infusion (4.1 $\pm$ 0.5 x 10^5 dpm/ml; N=4) and the concentration reached in the cup during the plateau phase (8-10 min of infusion) (22 dpm/cup) was found to be about 5000. These experiments indicate that the BBB for inulin was not damaged.

Table 1

Effects of Penicillin Added to the Cup Solution (4%) on the [^{14}C]- or [^{3}H]-Efflux (dpm/cup) from the Cortex During a 10 min Period of [^{14}C]-Inulin or [^{3}H]-Ch Infusion.

	Infusion of [^{14}C]-inulin (10μ Ci/10 min)				
Time(min)	0 - 2	2 - 4	4 - 6	6 - 8	8 - 10
Control(dpm/cup)	1 $\pm$ 1	7 $\pm$ 3	23 $\pm$ 6	21 $\pm$ 9	22 $\pm$ 6
Penicillin(dpm/cup)	10	43	85	148	220

	Infusion of [^{3}H]-Ch (100μ Ci/10 min)				
Time(min)	0 - 2	2 - 4	4 - 6	6 - 8	8 - 10
Control(dpm/cup)	59 $\pm$ 9	152 $\pm$ 20	218 $\pm$ 24	262 $\pm$ 30	308 $\pm$ 37
Penicillin(dpm/cup)	282 $\pm$ 154	743 $\pm$ 231	979 $\pm$ 211	1063 $\pm$ 315	1208 $\pm$ 282

The data represent appearance of radioactivity (dpm/cup) in the cup solution (300 μl).

Preliminary investigations strongly support this conclusion. Topical application of penicillin (4%) to the cup solution 10 min before infusing [^{14}C]-inulin markedly increased the appearance of [^{14}C]-inulin in the cup (220 dpm/cup), thus reducing the "plasma/cup ratio" from about 5000 to 558.

In similar experiments, 100 μCi of [^{3}H]-Ch instead of [^{14}C]-inulin was infused during a 10 min period in the tail vein. Immediately after the infusion, radioactivity in the plasma was 5.6 $\pm$ 0.3 x 10^5 dpm/ml whereas only 308 $\pm$ 37 dpm/cup (N=11) appeared in the cup solution (0.3 ml) during the last sampling period (8-10 min), thus representing a plasma/cup ratio of 544. When unlabelled Ch (10^{-5} mol/l) was added to the cup solution at the beginning of the infusion of [^{3}H]-Ch, the [^{3}H]-flux into the cup (352 $\pm$ 84 dpm/cup, 8-10 min, N=4) was virtually unchanged. In contrast, topical application of penicillin (4%) to the cup solution increased [^{3}H]-flux to 1208 $\pm$ 282 dpm/cup (N=4), thus reducing the plasma/cup ratio from 544 to 139 (Table 1).

Under the conditions of the above flux experiments, the total Ch concentration of the plasma was 11 $\pm$ 1 μmol/l (N=6). From the specific activity of the ^{3}H-Ch in the plasma, it is calculated that the Ch transfer from the plasma into the cup solution was increased by penicillin from 15.0 to 58.7 pmol min^{-1} cm^{-2}, when the radioactivity of the cup was regarded as being Ch. Of course, these figures represent only a fraction of the total cortical efflux of Ch (see below).

Temperature-dependency of Ch fluxes. The rate of endogenous Ch efflux from the cortex into the cup solution was about 60 pmol min^{-1} cm^{-2} at a temperature of 37°C and sampling periods of 10 min (Fig. 1). However, when the cup solution was replaced by Ringer solution kept at 23°C, the Ch efflux rate declined gradually to 15 pmol min^{-1} cm^{-2} within 3 h (Fig. 1) indicating a marked temperature-dependency.

In contrast, the effect of cooling on the transfer of [^{3}H]-Ch from the blood plasma into the cup solution was small. Reducing the temperature of the Ringer solution from 37° to 20°C lowered the appearance of radioactivity in the last sampling period (8-10 min) from 308 $\pm$ 37 (N=11) to 260 $\pm$ 48 dpm/cup (N=4).

Temperature-dependency was also very weak with respect to the flux of [^{3}H]-Ch from the cup to the cortex. A tracer dose of [^{3}H]-Ch (10 nmol/l) was added to the cup solution (100 μl). When the Ringer solution was kept at 0°C, the initial rate of [^{3}H]-Ch flux (removal in % of initial [^{3}H]-Ch concentration) was reduced from 1.1 $\pm$ 0.2 to 0.63 $\pm$ 0.27% per min (N=4). Similar figures (0.74 $\pm$ 0.03 vs. 0.63 $\pm$ 0.08% per min, N=3) were obtained at an

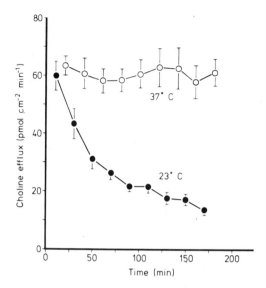

Figure 1.   Temperature dependency of Ch efflux from rat cortex.
Ordinate, rate of Ch efflux into cup solution (300 μl) from a
cortical area of 0.21 cm^2 during periods of 20 min (2 sampling
periods of 10 min).   Abscissa, duration of experiment.   Closed
circles, efflux into cup solution that was replaced every 10 min
by solution kept at 23°C (N=12).   Open circles, efflux into
solution at 37°C (N=7). Means ± SE.

initial concentration of 10 μmol/l of Ch (including the tracer dose
of [^{3}H]-Ch).

Muscarinic receptors and Ch fluxes.   A major goal of the present
study was to find out whether muscarinic receptor activation
alters the [^{3}H]-Ch fluxes as determined in the above experiments
(see Discussion).

      First, the effect of muscarinic stimulants on the transfer of
[^{3}H]-Ch from the blood plasma to the cup solution was investigated
(Fig. 2).   After addition of 3 x 10^{-4} mol/l physostigmine or 5 x
10^{-4} mol/l bethanechol to the cup solution, the appearance of
radioactivity (8-10 min of infusion) (297 ± 42 or 266 ± 22
dpm/cup, respectively, N=3 to 4) was as low as in the controls
(see above).   Likewise physostigmine failed to facilitate the
[^{3}H]-transfer from the plasma to the cup solution (319 ± 138
dmp/cup, N=4), when the reservoir of Ringer solution was kept at
20°C.   (At 23°C, physostigmine and bethanechol increased endogenous
Ch efflux; see Ref. 5).

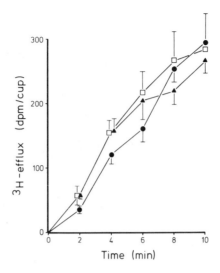

Figure 2.   Tritium efflux from the cortex into the cup solution
during a 10 min period of [^{3}H]-Ch infusion in the tail vein.
Ordinate, efflux of [^{3}H] (dpm/cup) into cup solution (300 μl) at
37°C.   Samples were replaced every 2 min.   Abscissa, time of
infusion.   Open squares, control (N=7).   Triangles, bethanechol, 3
x 10^{-4} mol/l (N=3).   Circles, physostigmine, 5 x 10^{-4} mol/l
(N=4).   Means ± SE.

Second, the effect of physostigmine on the initial flux of a
tracer dose of [^{3}H]-Ch (10 nmol/l) from the cup solution (100 μl)
to the cortex was investigated, which was 1.21 ± 0.07% per min
(N=4) under control conditions.   Again 3 x 10^{-4} mol/l physostigmine
failed to alter the flux rate which was now 1.12 ± 0.09% per min
(N=8).

DISCUSSION

The "cup technique" has been shown to be useful for studying
the efflux of neurotransmitters (13) and also of Ch (4, 5) from
the rat cortex in vivo.   Light-microscopic observations demonstrated
that, in the rat, the arachnoid space is opened by removal of the
dura mater, since the outer layers of the arachnoid and pial
membranes are, at least in part, removed together with the dura.
Hence the cup solution is not separated from the neural cortical
tissue by an intact arachnoid space, but only by inner leptomeningeal
layers which are not expected to represent a serious diffusion
barrier for Ch (6).

The necessity for a closer look at the "cup technique" (11,
12, 13) resulted from the observation that muscarinic receptor
activation increased Ch efflux from the rat cortex into the cup
solution (4, 5).   Although a similar muscarinic effect had been

observed in the chicken heart perfused with Tyrode's solution (4, 5; Löffelholz et al., this book), the possibility is still not excluded, that, in the cortex, the increase in Ch efflux was caused by a muscarinic increase in the BBB permeability for Ch. Evidence for a cholinergic innervation of microvessels isolated from bovine and rat cerebral cortex has been obtained recently (7, 9) suggesting a muscarinic regulation of the cerebral blood flow and capillary permeability.

Exclusion of a direct muscarinic effect on Ch fluxes should strengthen the hypothesis that the metabolism of Ch-containing phospholipids is regulated by muscarinic receptor activation in the heart (Loffelholz et al., this book) and also in the brain. Muscarinic receptors have been detected in cortical neurons (1, 15) besides those found in vessels (7, 9). The present study showed, firstly, the integrity of the BBB for Ch under the cup, i.e., after removal of the dura, and, secondly, a lack of a muscarinic effect on the Ch fluxes between cup solution and brain and vice versa.

The BBB for Ch was studied by intravenous infusion for 10 min of radioactively labelled inulin and Ch, which are known to penetrate the BBB at a slow rate; Ch probably via a low affinity carrier system (3). Indeed, after about 10 min of infusion, the ratios between the plasma concentration and the concentration in the cup solution were about 5000 for $[^{14}C]$-inulin and 500 for $[^{3}H]$-Ch. The amount of radioactivity that appeared in the cup after infusion of $[^{3}H]$-Ch was so small (near the detection limit) that no attempt was undertaken to identify the radioactivity. In these experiments penicillin (4%) was added to the cup solution in order to destruct the BBB (17). It was found that penicillin markedly increased the appearance of radioactivity in the cup, thus reducing the "plasma/cup ratios" from about 5000 to 500 (inulin) and from 544 to 139 (Ch). The result supports the conclusion that the BBB is not pertubated under conditions of the cup technique.

The above results further indicate that the high rate of efflux of endogenous Ch from the cortex into the cup solution (60 pmol min^{-1} cm^{-2}; Fig. 1) was not due to a damaged BBB. The flux of $[^{3}H]$-Ch from the blood into the cup solution not only was small (15 pmol min^{-1} cm^{-2}) but differed from the cortical efflux of Ch in two further points, namely in the temperature-dependency and in the sensitivity to muscarinic stimulants. The cortical Ch efflux was facilitated by physostigmine and bethanechol (5) and was highly temperature-sensitive (Fig. 1), which is indicative of a temperature-dependent process (phospholipid hydrolysis?) rather than of simple diffusion. In contrast, the $[^{3}H]$-flux from the blood to the cup appeared to have a low temperature-sensitivity and was not affected by physostigmine and bethanechol. Likewise the initial rate of $[^{3}H]$-Ch elimination from the cup solution (due to cortical "uptake") was unaffected by physostigmine and showed a poor temperature-dependency as compared to the cortical Ch efflux.

In conclusion, the bulk of the Ch efflux from the rat cortex in vivo originates from the extracellular Ch of the cortical tissue which may be controlled by cellular mobilization of Ch. The muscarinic stimulation of the Ch efflux is not due to an effect on the BBB or on the cerebrospinal fluid-brain "barrier" but rather to an effect on the mobilization of Ch from cortical cells. Finally, phospholipid hydrolysis may be enhanced by muscarinic stimulation, as suggested for similar receptor-mediated changes in Ch efflux from the heart (5; Löffelholz et al., this book).

Acknowledgements. The authors thank Dr. Anna Laura Abbamondi for the histological work. The research project was supported by grants from the Consiglio Nazionale delle Ricerca and from the Deutsche Forschungsgemeinschaft.

REFERENCES

1. Bartus, R.T., Dean, R.L., Beer, B. and Lippa, S.S. (1982): Science 217:408-417.
2. Beani, L. and Bianchi, C. (1970): In Drugs and Cholinergic Mechanisms in the CNS (eds) E. Heilbronn and A. Winter, Res. Inst. of National Defense, Stockholm, pp. 369-386.
3. Cornford, E.M., Braun, L.D. and Oldendorf, W.H. (1978): J. Neurochem. 30:299-308.
4. Corradetti, R., Lindmar, R. and Löffelholz, K. (1982): Europ. J. Pharmacol. 85:123-124.
5. Corradetti, R., Lindmar, R. and Löffelholz, K. (1983): J. Pharmacol. Exp. Ther. 226:826-832.
6. Davson, H. (1976): J. Physiol. (Lond.) 255:1-28.
7. Estrada, C., Hamel, E. and Krause, D.N. (1983): Brain Res. 266:261-270.
8. Goldberg, A.M. and McCaman, R.E. (1974): In Choline and Acetylcholine, Handbook of Chemical Assay Methods (ed) I. Hanin, Raven Press, New York, pp. 47-61.
9. Grammas, P., Diglio, C.A., Marks, B.H., Giacomelli, F. and Wiener, J. (1983): J. Neurochem. 40:645-651.
10. Israël, M. and Lesbats, B. (1982): J. Neurochem. 39:248-250.
11. MacIntosh, F.C. and Oborin, P.E. (1953): Abstr. XIX Int. Physiol. Congr. pp. 580-581.
12. Mitchell, J.F. (1963): J. Physiol. (Lond.) 165:98-116.
13. Moroni, F. and Pepeu, G. (1984): In IBRO Handbook Series: Methods in Neurosciences (ed) C.A. Marsden, John Wiley & Sons Ltd., Chichester, New York, pp. 63-79.
14. Pepeu, G. (1973): Prog. Neurobiol. 2:257-288.
15. Polak, R.L. (1971): Brit. J. Pharmacol. 41:600-606.
16. Reed, D.J. and Woodbury, D.M. (1966): J. Physiol. (Lond.) 169:816-850.

17. Steinwall, O. and Klatzo, I. (1966):   J. Neuropathol. Exp.
     Neurol. 25:542-559.

BIOCHEMICAL EFFECTS OF LECITHIN ADMINISTRATION

R.S. Jope

Department of Pharmacology and
Neuroscience Program
University of Alabama in Birmingham
Birmingham, Alabama 35294 U.S.A.

INTRODUCTION

Lecithin is being administered in large doses to patients with Alzheimer's Disease and other disorders involving impaired CNS cholinergic activity with the goal of stimulating the synthesis and release of acetylcholine. However, there are many questions concerning the biochemical effects of lecithin that still have to be examined. Three questions involving lecithin treatment are addressed in this report: 1) Are there any similarities between the effects of lecithin on blood choline levels in rats and human subjects? 2) What are the effects of chronic versus acute treatment with lecithin on the concentrations of choline and acetylcholine? 3) Does lecithin have significant biochemical effects on metabolic processes besides those involving choline utilization? None of these questions can be answered adequately at this time, but recent results, described in this chapter, should help to formulate further research projects aimed at obtaining a more comprehensive understanding of the metabolic effects of the administration of large doses of lecithin.

Comparison of rat and human blood choline levels following lecithin administration. The increases in rat and human plasma choline following acute, oral administration of lecithin are shown in Figure 1. The time courses of the rise in choline levels are quite similar, especially considering that the rats received much more lecithin on the basis of body weight. The most notable difference is that the choline level remains high longer in rat plasma than in human plasma. Taking into consideration previous dose response results (6) it appears that lecithin can be absorbed and hydrolyzed at a limited rate so that larger doses result in a longer duration of high plasma choline levels while the peak level is limited by the

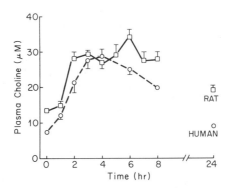

Figure 1. Effects of lecithin on rat and human plasma choline levels.
Rats were administered lecithin (10 mmole/kg; approximately 1.9
gm/rat) by intragastric intubation. Human subjects were fed lecithin
(15 gm/person; approximately 0.29 mmole/kg). Lecithin was 80-90%
phosphatidylcholine. Choline was quantitated by gas chromatography
mass spectrometry. Rat data are from Jope (6) and human data are
from Jope et al. (10). Values are means $\pm$ SEM of 5-10 subjects.

rate of catabolism of lecithin. However, further experiments measuring
the time courses of the effects of several doses of lecithin are required to test this proposal.

During the last few years we have utilized erythrocytes as an
easily available cell type for measuring intracellular choline levels
(6, 8-10). It remains to be determined whether or not this is a
useful indication of choline uptake into other cell types or of
therapeutic response to choline-loading treatments. Figure 2 shows
the effects of oral lecithin administration on choline levels in
rat and human erythrocyes. The time courses of the changes in choline
concentration are similar in the two samples and are similar to
the changes seen in plasma. However, the choline level in rat erythrocytes remained high for a longer period and reached a slightly higher
maximum than did the choline in human erythrocytes. Figure 2 also
compares the change in rat erythrocyte choline level to that in
brain. The time course of the changes in choline concentrations
are similar in brain and erythrocytes. However, the brain choline
levels only rise by approximately 50% whereas the erythrocyte choline
levels rise by 3 to 4-fold. It should be noted that these effects
were obtained in rats given a much higher dose of lecithin (on the
basis of body weight) than human subjects receive. These results
suggest that regulatory mechanisms within the brain or at the blood
brain barrier may act to limit the influence of plasma choline levels
on brain choline levels in the rat. Perhaps of greater importance
than regulation of choline flux, choline levels within the brain
may be regulated by metabolism of free choline to choline derivatives.

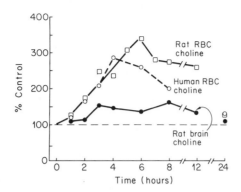

Figure 2.   Effects of lecithin treatment on choline levels in rat and human erythrocytes (RBC) and rat brain. Experiments were carried out as described in the legend to Figure 1. Rat brain choline levels were determined following sacrifice by microwave irradiation focused on the head. Values are presented as percent of control levels which were 15.3 $\pm$ 2.2 μM in human erythrocytes and 11.6 $\pm$ 0.65 μM in rat erythrocytes. Rat brain choline levels are averages of changes measured in the striatum, hippocampus and cortex; control values are given in Table 1.

We previously reported that almost all of the choline entering the brain eventually enters the phosphorylcholine pool (7). Millington and Wurtman (15) observed that choline administration to rats increased the concentration of phosphorylcholine in the brain and the amount of choline entering this pool far outweighed the changes in free choline or acetylcholine. Which, if any, of these regulatory mechanisms operate in human brains to control choline levels and choline utilization remains unknown.

One further comparison warrants discussion. The ratio of erythrocyte choline to plasma choline is lower in rats than in humans and the effect of lecithin administration on this ratio is opposite in the two samples. Following lecithin treatment the ratio increases slightly in rat blood, indicating that intracellular erythrocyte choline responds to changes in plasma choline and actually accumulates choline to some extent (Figure 3). The ratio decreases in human blood, on the other hand, indicating that lecithin increases plasma choline more than it does intracellular erythrocyte choline.

If these results can be applied to other tissue they would suggest that human cells respond less than do rat cells to increased plasma choline levels. This question should be examined further by measuring this ratio using other doses of lecithin and by measuring the effects of lecithin administration on the concentration of choline in other cell types. It is of great interest, in this context,

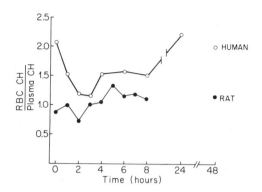

Figure 3. Ratio of erythrocyte to plasma choline concentrations in rats and humans before and during treatment with lecithin. Experiments were carried out as described in the legends to Figures 1 and 2.

that Friedman et al. (4) found that in patients with Alzheimer's Disease there was a better response to treatment with choline plus piracetam in those individuals with higher erythrocyte to plasma choline ratios both before and during treatment. Therefore, this ratio may be of use in predicting and/or confirming response to choline- loading treatments.

Comparison of acute and chronic lecithin treatment. We have compared the effects of acute versus chronic (10 days) oral administration of lecithin (10 mmole/kg) on cholinergic parameters in rats. All measurements were made 6 hours after the last dose of lecithin because that is the time at which plasma choline was at a maximum level. Both plasma and erythrocyte choline levels were the same after chronic treatment as they were after an acute treatment (Table 1). Therefore, there does not appear to be a cumulative effect of 10 successive daily treatments of lecithin. In the brain, chronic treatment resulted in the same levels of choline in the cortex and hippocampus as did acute treatment (Table 1). However, in the striatum the choline level was significantly higher after chronic treatment than following acute treatment. This indicates that chronic lecithin treatment can result in a small, but significant, region-specific accumulation of choline. Other investigators have also reported differences in choline utilization in the striatum compared to other brain regions (16-18). Further research should be aimed at determining whether this difference is due to different rates of acetylation, phosphorylation or transport of choline. There was no effect of either acute or chronic lecithin administration on the concentration of acetylcholine in the cortex, hippocampus or striatum.

Table 1

Effects of Acute and Chronic (10 days) Oral
Lecithin (10 mmole/kg) Administration to Rats on
Choline and Acetylcholine Concentrations

	Control	Acute	Chronic
Choline			
Plasma	11.4 ± 1.1	$34.3 \pm 2.0^*$	$36.7 \pm 2.7^*$
Erythrocytes	11.5 ± 1.2	$39.4 \pm 1.8^*$	$40.0 \pm 2.3^*$
Cortex	20.1 ± 1.6	$31.1 \pm 2.7^*$	$28.4 \pm 1.0^*$
Hippocampus	30.0 ± 2.9	$38.2 \pm 4.6^*$	$41.5 \pm 3.3^*$
Striatum	29.9 ± 3.5	$37.0 \pm 1.9^*$	$50.3 \pm 6.7^*$
Acetylcholine			
Cortex	19.7 ± 0.6	20.8 ± 1.2	18.5 ± 0.4
Hippocampus	30.4 ± 0.8	32.1 ± 1.6	31.5 ± 0.7
Striatum	99.5 ± 3.1	96.9 ± 4.1	89.5 ± 9.2

Choline and acetylcholine levels were measured by combined gas chromatography/mass spectrometry. Data are from Jope et al (11) and represent the mean $\pm$ SEM (n = 8-11 for control and acute; n = 4-5 for chronic). Units are μM for plasma and erythrocytes and nmole/gm for brain regions. *p < 0.05 compared to controls.

We found no effect of chronic lecithin administration on the activities of choline acetyltransferase or acetylcholinesterase or on the apparent affinity or B_{max} of [^{3}H]-1-QNB binding sites in rat cortex, hippocampus or striatum (data not shown).

Effects of lecithin on lipids, fatty acids and liver enzymes. Since high doses of lecithin are being administered chronically to patients, it is obviously important to be aware of the effects that may be produced by this treatment on other processes as well as on the cholinergic system. For example, high doses of lecithin may alter the absorption, distribution or metabolism of other drugs and nutrients. It is possible that such an action could account for the previously reported (6) reduced effect of several drugs that were administered to rats pretreated with lecithin. Zeisel et al. (20) have raised the point that it is also necessary to study the metabolic products of lecithin and also of contaminants often found in lecithin preparations. The interaction between lecithin and other substances is an important area to clarify in future research if lecithin continues to be used therapeutically.

We have begun to study this problem by examining the activity of the hepatic cytochrome P_{450} system and the rate of N-demethylation of a secondary (benzphetamine) and a tertiary (methamphetamine) analog of amphetamine. We have found no difference in these activities in liver microsomes prepared from control rats and rats treated for 10 days with lecithin (10 mmol/kg) (data not shown). Although these results were negative there are many more processes in this area that should be tested.

Lecithin is not just a time-release capsule of choline, although it does seem to fulfill this function. It also contains several other moieties that may have significant metabolic effects. With this in mind, the effect of lecithin on plasma lipids and the fatty acid composition of lipids in the plasma and brain were measured. Ten days of oral lecithin (10 mmol/kg) administration increased the plasma total lipid, free cholesterol and total cholesterol concentrations in rats (Table 2). In this context, Hirsch et al. (5) previously reported that ingestion of a lecithin-containing high choline diet for two days increased human serum triglyceride levels and depressed serum cholesterol.

Lecithin treatment had a large effect on the content of plasma lipid fatty acids. There were large increases in the dienes in plasma phospholipids and triglycerides, and a decrease in the triglyceride monoenes (Table 3). In the brain there was no effect of lecithin treatment on the fatty acid composition of phosphatidyl-

Table 2

Plasma Lipid Levels in Control Rats and
Rats Treated for 10 Days with Lecithin (10 mmole/kg)

	Total lipid (mg/ml)	Free cholesterol (mg/100 ml)	Total cholesterol (mg/100 ml)	Triglyceride (mg/100 ml)
Control	3.2 ± 0.4	3.3 ± 1.1	35.4 ± 2.9	70.7 ± 28.8
Lecithin	4.2 ± 0.2*	7.6 ± 1.8*	49.6 ± 4.6*	104.6 ± 17.6

Lipids were extracted by the method of Folch et al. (3) and quantitated as described by Menon and Dhopeshwarker (13). Data are from Jope et al. (11) and are presented as the mean ± SEM of 4 control and 5 treated rats (*$p < 0.05$).

Table 3

Fatty Acid Composition of Plasma Lipids

	Saturates	Monoenes	Dienes	Tetraenes
Phospholipids				
Control	55.9 ± 1.6	14.5 ± 1.5	20.8 ± 2.3	2.8 ± 1.3
Lecithin	49.6 ± 3.6	12.5 ± 2.3	27.7 ± 3.0*	4.4 ± 2.8
Triglycerides				
Control	30.8 ± 1.4	38.9 ± 3.4	25.5 ± 4.0	---
Lecithin	27.7 ± 3.0	27.7 ± 3.5*	39.9 ± 4.8*	---

Values were obtained as described in the legend to Table 2 ($^*p < 0.05$).

choline, phosphatidylserine or phosphatidylethanolamine (data not shown). These initial results indicate that future investigations should take into consideration the fact that lecithin treatment changes plasma lipid levels and fatty acid compositions as well as the plasma choline level. More detailed studies should be conducted to determine the consequences of these changes on lipid metabolism and lipid-modulated reactions.

CONCLUSIONS

Lecithin is being administered in very high doses to patients with several different disorders, most notably patients with Alzheimer's Disease, with the goal of enhancing cholinergic activity in the brain. Two main problems with this therapeutic regimen are the paucity of data concerning the effects of lecithin, and the ever-present difficulty in translating effects observed in laboratory animals to human patients. The latter problem has been approached by comparing the effeects of lecithin in humans and rats on the plasma and erythrocyte choline concentrations. The observed effects were similar in rats and humans. The most notable difference was the lower ratio of erythrocyte choline relative to plasma choline after lecithin treatment in humans compared to rats. Since we do not yet know the applicability of findings in erythrocytes to other cell types, no conclusions can be drawn from this result. However, if future investigations substantiate the erythrocyte as a good indicator of tissue choline, this result could indicate that it is more difficult to raise cellular levels of choline in humans than in rats by the oral administration of lecithin or choline. This would make the outlook for possible beneficial effects of lecithin rather poor, especially considering the low level to which rat brain choline levels are increased by lecithin.

We have never observed an elevation in rat brain acetylcholine levels following either acute or chronic treatment with lecithin. Other investigators have reported no effect of choline-loading on acetylcholine turnover (1, 2). Therefore, these results do not indicate that increased choline levels will enhance cholinergic activity. However, interpretation of these results faces the problem of possible species differences, as discussed above, as well as possible differences caused by the pathological conditions that are being treated. The latter problem has been approached by examining the effects of choline-loading in rats that have been treated with drugs that alter cholinergic activity. Under these conditions, choline-loading appears to support the synthesis of acetylcholine (6, 12, 19). Therefore the possibility remains open that lecithin may enhance acetylcholine synthesis in patients with impaired cholinergic function.

The few biochemical studies that have been reported concerning the effects of lecithin have been directed towards studying the effects of the cholinergic system. However, as discussed in this chapter, lecithin may have a wide range of other effects. These effects may include alterations of the absorption, distribution and metabolism of nutrients and drugs, alterations of lipid- or fatty acid- regulated enzymes, and alterations of the lipid and fatty acid compositions of plasma and tissues. These effects may contribute to unwanted side effects resulting from chronic lecithin treatment or they may add to the therapeutic efficacy of lecithin. It is also most interesting that Mervis and Bartus (14) found increased dendritic spines in aged mice treated with choline. All these findings point out the need to study the effects of lecithin and of choline on lipid metabolism, in addition to their effect on acetylcholine synthesis.

In conclusion, there is still much to be learned about the metabolic effects of lecithin treatment. Further basic and clinical research should be coordinated to establish the effects of lecithin and its place in today's pharmacopeia.

Acknowledgements. The results presented in this paper were obtained with the generous support of Dr. Donald Jenden and the help of many other collaborators listed in the references. This work was supported by grant #MH17691.

REFERENCES

1. Brunello, N., Cheney, D.L. and Costa, E. (1982): Neurochem. 38:1160-1163.

2. Eckernäs, S.Å., Sahlström, L. and Aquilonius, S.M. (1977): Acta Physiol. Scand. 101:404-410.

3. Folch, J., Lees, M. and Sloan-Stanley, G.H. (1957): J. Biol. Chem. 226:497-508.

4. Friedman, E., Sherman, K.A., Ferris, S.H., Reisberg, B., Bartus, R.T. and Schneck, M.K. (1981): New Engl. J. Med. 304:1490-1491, 1981.

5. Hirsch, M.J., Growdon, J.H. and Wurtman, R.J. (1978): Metabolism 27:953-960.

6. Jope, R.S. (1982): J. Pharmacol. Exp. Therap. 220:322-328.

7. Jope, R.S. and Jenden, D.J. (1979): J. Neurosci. Res. 4:69-82.- Jope, R.S. and Jenden, D.J. (1979): J. Neurosci. Res. 4:69-82.

8. Jope, R.S., Jenden, D.J., Ehrlich, B.E. and Diamond, J.M. (1978): New Engl. J. Med. 299:833-834.

9. Jope, R.S., Jenden, D.J., Ehrlich, B.E., Diamond, J.M. and Gosenfeld, L.F. (1980): Proc. Natl. Acad. Sci. (USA) 77:6144-6146.

10. Jope, R.S., Domino, E.F., Mathews, B.N., Sitaram, N., Jenden, D.J. and Ortiz, A. (1982): Clin. Pharmacol. Therap. 31:483-487.

11. Jope, R.S., Jenden, D.J., Subramanian, C.S., Dhopeshwarker, G.A. and Duncan, J. (1984): Biochem. Pharmacol. 33(5):793798.

12. London, E.D. and Coyle, J.T. (1978): Biochem. Pharmacol. 27: 2962-2965.

13. Menon, N.K. and Dhopeshwarker, G.A. (1980): In: Progress in Lipid Research (Ed) R.T. Holman, pp. 129-138, Pergamon Press, Oxford, England.

14. Mervis, R.F. and Bartus, R.T. (1981): J. Neuropathol. Exp. Neurol. 40:313-321.

15. Millington, W.R. and Wurtman, R.J. (1982): J. Neurochem. 38: 1748-1752.

16. Schmidt, D.E. and Wecker, L. (1981): Neuropharmacol. 20:535-539.

17. Sherman, K.A., Zigmond, M.J. and Hanin, I. (1978): Life Sci. 23:1863-1870.

18. Wecker, L. and Dettbarn, W.D. (1979): J. Neurochem. 32:961967

19. Wecker, L., Dettbarn, W.D. and Schmidt, D.E. (1978): Science 199:86-87.

20. Zeisel, S.H., Wishnok, J.S. and Blusztajn, J.K. (1983): J. Pharmacol. Exp. Ther. 225:320-324.

FACTORS WHICH INFLUENCE THE AVAILABILITY OF CHOLINE TO BRAIN

S.H. Zeisel

Departments of Pathology and Pediatrics
Boston University School of Medicine
Boston, Massachusetts 02118 U.S.A.

INTRODUCTION

Choline is necessary for normal function of the mammalian organism. It is a precursor for the biosynthesis of the choline-phospholipids phosphatidylcholine (PC), sphingomyelin (SM), and choline plasmalogen, which are important constituents of brain membranes [for example, PC contributes eight to fifteen percent of the dry weight of brain (1)]. Free choline is readily accumulated by brain neurons which utilize it as a precursor for the biosynthesis of acetylcholine (ACh; 45). Choline acetyltransferase is not saturated with choline at normal brain concentrations of this amine (41). Therefore, the availability of free choline to neurons directly influences the synthesis of ACh (14, 15, 27). Thus, it is obvious that large numbers of choline molecules must be used by brain for both the maintenance of structural integrity and for cholinergic neurotransmission. Despite this requirement, more unesterified (free) choline leaves the brain, in vivo, than enters it (1, 2, 13, 21). This results in jugular venous levels of free choline that are higher than those of arterial blood entering brain. These concentration differences range from 1 nmol/ml in humans (3) to 1-12 nmol/ml in rats (13, 21, 37). Where does all this free choline come from?

Endogenous Sources of Choline for Brain. Brain can synthesize choline molecules de novo by methylating phosphatidylethanolamine, forming PC (10, 18, 33). Free choline can then be generated when this PC is catabolized (9). We do not know the mechanisms whereby this special pool of PC is degraded. The amount of choline produced by this pathway (approximately 10 pmoles/g/min) is small when compared to net efflux from brain (7 nmoles/g/min; 9, 21). The brain also contains storage pools of choline-esters from which free choline might be derived. The total pool of free choline within normal

837

rat brain (25 nmol/g wet wt) would be quickly exhausted if it were
the sole source of choline efflux (1). The same is true for acetylcholine (20 nmol/g wet wt), CDP-choline (50 nmoles/g wet wt), phosphorylcholine (PCh; 380 nmol/gm wet wt), glycerophosphorylcholine
(GPC; 400 nmol/g wet wt), and choline plasmalogen (600 nmol/g wet
wt; 1, 2). However, a great deal of choline is stored within the
phospholipids SM (3700 nmol/g wet wt), and PC (15,000 nmol/g wet
wt) which might be pools from which free choline could be derived
(1). Over a life-time, even pools as large as these would be insufficient to maintain continuous choline efflux from brain, as
the methylation of PE is the only way that brain can form new choline
molecules to replace those lost (10). These observations lead to
the natural conclusion that significant amounts of choline-containing
compounds must enter brain from the blood (1, 2, 21, 29).

Exogenous Sources of Choline for Brain. Ingestion of choline-containing compounds in the diet results in elevation of blood choline
concentrations (48). We have characterized the kinetics of choline
absorption by rat ileum and jejunum, and found that both a mediated
and a nonsaturable transport mechanism exist (51). The mediated
component has a low Km for choline (0.1-0.3 mM) and it predominates
at low lumenal choline concentrations. It can be inhibited by hemi-
cholinium-3. The nonsaturable component of transport transfers
choline at a rate proportional to this amine's concentration in
the lumen of gut, and may be due to diffusion of choline through
the relatively leaky "tight" junctions of gut endothelium. Choline
that is not absorbed from gut is subject to degradation by gut bacteria
to form methylamines, which are precursors for the formation of
nitrosamines, compounds believed to potentiate carcinogenesis (50).

Most dietary choline is eaten in the form of PC (47, 48).
Both pancreatic secretions, and intestinal mucosal cells contain
enzymes capable of hydrolyzing PC (19, 20, 28, 31, 36). Phospholipase A2 (which cleaves the β-acyl moiety of PC, forming lysoPC is
secreted as a zymogen by the pancreas, and is activated in the gut
by trypsin, calcium, and bile salts (19, 20). LysoPC is absorbed
by enterocytes (28, 31, 36), and is then either acylated, reforming
PC (31, 36), or is converted to glycerophosphorylcholine (GPC) and
free choline (28, 31). Approximately twice as many PC molecules
are absorbed from gut lumen as are reconstituted and secreted into
the lymphatic circulation (31). The choline and GPC formed enter
the portal circulation (28).

The liver influences plasma choline concentration both by synthesizing choline radicals (in PC) via sequential methylation of
phosphatidylethanolamine (11), and by removing choline from the
portal circulation (49). As in the gut, the uptake of choline by
liver is the sum of two processes - one mediated and one nonmediated
(49). Within liver, much choline is irreversibly degraded to form
betaine, the remainder is converted to PC and PCh (49). We believe

that the liver's ample ability to act as a sink for choline accounts
for choline's rapid clearance from plasma. Liver also contributes
PC to plasma, in the form of a phospholipid envelope for lipoproteins:
very low density lipoprotein (VLDL), low density lipoprotein (LDL),
and high density lipoprotein (HDL). Little is known about the fate
of the choline moiety within lipoproteins.

Choline crosses the blood-brain barrier via a specific carrier,
no other endogenous substrate for this transport system has been
identified (35). This carrier is normally not saturated with substrate
at physiologic blood choline concentrations ($Km = 0.44$ mM, plasma
choline concentration $= 0.01$ mM; 47). Neonatal brain may be more
efficient than adult brain at extracting choline from blood (17,
34). As discussed earlier, a net efflux of free choline has usually
been observed for brain. This suggests that choline esters may
cross the blood-brain barrier. Illingworth and Portman (29) also
demonstrated a small influx of lysophosphatidylcholine (lysoPC)
from blood into brain. However, Pardridge (35) determined that
the permeability of the blood-brain barrier to lysoPC or PC was
very low, and, though Spanner et al. (37) observed an arteriovenous
difference (across rat brain) for free choline, they found none
for PC of lysoPC.

Free Choline Release by Rat Brain. Even though estimates of the
relative contribution to brain of exogenous and endogenous. sources
of the choline molecule may be still a matter of controversy, it
is clear that, whatever the derivation of choline may be, mechanisms
must exist within brain for the hydrolysis of choline esters. We
now present data about the factors which may influence these mechanisms.

METHODS

Adult (150 g) male Sprague-Dawley rats (Charles River Laboratories, Wilmington, MA) were used in all studies. They were fed
rat chow and water ad libitum.

For experiments using brain homogenates, rats were killed by
decapitation, their brains were removed and chilled on ice. They
were then homogenized in 4 volumes 0.32 M sucrose. Exactly 30 minutes
after decapitation, aliquots of homogenate were added to Hepes buffer
(pH 7.4) containing additives as noted, and incubations were begun.
Subcellular fractions of brain were prepared using the method of
Whittaker (42). Choline-containing compounds (PC, SM, GPC and PCh)
were isolated using thin layer and ion exchange resin chromatography,
and were assayed by measuring phosphorus content (39).

Choline was measured using a radioenzymatic method (24, 32)
in which it was first isolated from the tissue by liquid cation
exchange chromatography, then combined with AT[^{32}P] in a reaction

catalyzed by choline kinase. The resulting [^{32}P]-choline was isolated using ion exchange chromatography, and quantitated using liquid scintillation spectrophotometry. Recovery for choline exceeded 70%, and was corrected for using an internal standard of [^{3}H]-choline.

RESULTS

When whole brains from decapitated rats were incubated in situ at 37°C, free choline concentration rose 8-fold during a 30 minute period (at a rate of 500 nmoles/g wet weight/hr). A similar increase in choline content was observed in rat brain which was homogenized in pH 7.4 Hepes buffer, and incubated in vitro (10 nmoles/mg protein/hr; Figure 1). Choline was not formed when the homogenates were prepared from brains which had been initially fixed with a beam of microwave irradiation focussed to the head. Free choline production was pH dependent, with maximal activity at pH 8.5. More choline was formed at 37°C than at lower temperatures, and little was produced at 50°C. During incubations of brain homogenates, the concentrations of choline-containing compounds remained relatively constant (Figure 2). However, only a small fraction of the large PC pool needed to be degraded in order to account for the free choline formation that we observed. We were able to demonstrate significant hydrolysis of radiolabeled PC, forming GPC, PCh and free choline (Figure 3). No radiolabel was ever found to accumulate in lysoPC. In similar experiments, we observed that SM ([^{14}C]-methyl choline) was hydrolyzed, forming significant amounts of free choline, and PCh. No radiolabel derived from SM was ever isolated within GPC, PC or lysoPC (Figure 4).

Free choline production occurred at approximately the same rate in all subcellular fractions of brain which contained membranes (total homogenate, P2, myelin, mitochondria, synaptosomes, and microsomes). The soluble fraction formed little free choline, and did not contain factors which enhanced choline formation by the membrane-containing fractions.

Magnesium stimulated the release of choline from brain homogenates (Figure 5). Maximal effect was noted at concentrations greater than 5 mM in the incubation medium. Zinc and nickel were potent inhibitors of choline release. Calcium had little effect upon choline production.

The addition of ATP to homogenates of rat brain inhibited choline accumulation (Figure 6). Pretreatment with magnesium could prevent this effect of ATP. ADP and CTP addition also inhibited choline formation (Figure 6). CMP (3 mM) plus magnesium (15 mM) plus dithiothreitol (80 μM), added so as to optimize conditions for the reversal of the last step in the Kennedy pathway for PC synthesis, had no effect from addition of dinitrophenol, or from incubating under a 100% nitrogen atmosphere.

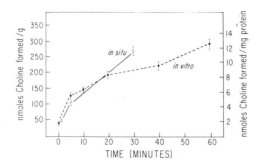

Figure 1. Free choline formation by preparations of rat brain. We studied the formation of free choline by two preparations of rat brain: IN SITU - rats were decapitated and their heads incubated in a sand bath at 37°C for the times indicated. They were then subjected to a beam of focussed microwave irradiation (2450 MHz, 6KW) to stop all enzymatic reactions. IN VITRO - rats were decapitated, and their brains were immediately cooled to 0°C. Brains were then homogenized in 4 volumes 0.32M sucrose. Aliquots of the homogenate were then diluted with an equal volume of 0.2M Hepes buffer (pH 7.4) containing 26mM magnesium chloride. Exactly 30 minutes after the brains were initially collected, brain samples were warmed to 37°C and incubation was begun. At the noted times, reactions were terminated by the addition of 8 volumes chloroform-methanol (1:1 v/v). Choline was assayed using a radioenzymatic method. Data for the in situ experiment are expressed as mean nmoles choline per gram wet weight $^\pm$ SEM. Data for the in vitro experiment are expressed as mean nmoles choline per mg protein $^\pm$ SEM. Six rats were used for each data point.

The supplementation of brain homogenates with choline-containing compounds increased free choline formation. Added PC increased the rate of choline production from 15 to 17 nmoles/mg prot/hr, lysoPC addition had a slightly greater effect (increased rate to 21 nmole/mg prot/hr), while GPC and PCh addition resulted in many-fold increases in choline formation (increased rate to 41 and 80 nmol/mg prot/hr respectively). Added ATP significantly inhibited the increase in choline formation seen after the addition to brain homogenates of exogenous PC, lysoPC, GPC and PCh.

In all experiments we also assessed the rate of PCh and PC synthesis from choline, as such pathways might have been a significant sink for choline. The amount of choline ester that accumulated during incubations was much less than the amount of free choline generated, and treatments which inhibited choline formation did not significantly accelerate PCh and PC synthesis.

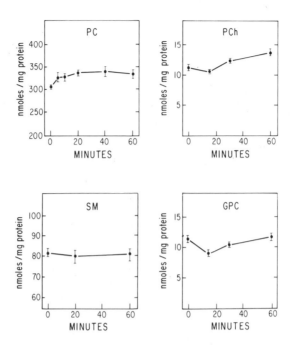

Figure 2. Choline-containing compounds in homogenates of brain. Brains were collected and treated as described for _in vitro_ incubation in Figure 1. At the termination of incubations, chloroform and methanol were added to tubes, and aqueous and organic phases were separated. Phosphatidylcholine (PC) and sphingomyelin (SM) were isolated using ion exchange resin chromatography. Individual compounds were quantitated by measuring phosphorus content. Data are expressed as mean $\pm$ SEM (n=6 rats per point).

DISCUSSION

We have demonstrated that brain can hydrolyze choline esters to form free choline. Similar findings have previously been reported for mouse (30), and rat brain (12, 21, 26). Preparations of heart, ileum, kidney, lung, and liver also produced free choline during _in vitro_ incubations (25, 30). Our data suggest that choline hydrolysis is mediated by an enzyme activity, as we demonstrated temperature dependence and heat inactivation.

We were interested in determining which ester was the initial reservoir from which free choline was produced. We believe that our data (Figure 2) show that only PC or SM could have been the source of choline, as only these two pools are large enough to produce free choline without significantly diminishing overall pool size. In addition, we showed that the soluble fraction of brain contains

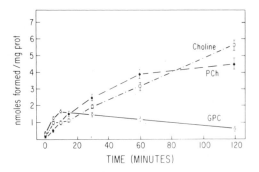

Figure 3. Formation of water-soluble metabolites of radiolabeled phosphatidylcholine. Brains were collected and treated as described for in vitro incubation in Figure 1. ([^{14}C]-methyl)-phosphatidyl-choline was added to each sample. Choline phosphorylcholine (PCh) and glycerophosphorylcholine (GPC) were isolated using ion exchange chromatography. DPM in each metabolite were converted to nmoles/mg protein using the original specific activity of phosphatidylcholine in the incubation mixture. Data are expressed as means $\pm$ SD (n=3 rats per point).

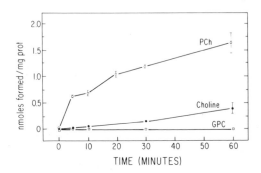

Figure 4. Formation of water-soluble metabolites of radiolabeled sphingomyelin. Brains were collected and treated as described for in vitro incubation in Figure 1. ([^{14}C]-methyl)-sphingomyelin was added to each tube. Choline, phosphorylcholine (PCh) and glycero-phosphorylcholine (GPC) were isolated using thin layer chromatography. DPM within each metabolite were converted to nmoles/mg protein using the original specific activity of sphingomyelin within the incubation mixture. Data are expressed as mean $\pm$ SD (n=3 rats per point).

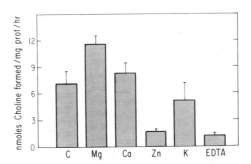

Figure 5. Effects of cations upon choline production by rat brain.
Brains were collected and treated as described for in vitro incubation
in Figure 1.   The incubation buffer was supplemented so that the
final concentration of indicated compounds was: control (C), 13mM
magnesium chloride (Mg), 16mM calcium chloride (Ca), 1mM zinc chloride
(Zn), 100mM potassium chloride with no sodium (K), or 16mM EDTA.
Data are expressed as mean nmoles choline formed per mg protein per
hour ± SD (n=3 rats per point).

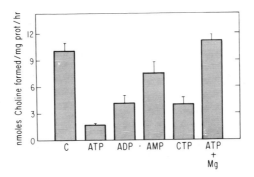

Figure 6.   Effects of ATP and related compounds upon choline production.   Brains were collected and treated as described for in
vitro incubation in Figure 1.   The incubation buffer was supplemented
so that the final concentration of indicated compounds was: control
(C), 10mM adenosine triphosphate (ATP), 10mM adenosine diphosphate
(ADP), 10mM adenosine monophosphate (AMP), 10mM cytidine triphosphate
(CTP), or 10mM ATP plus 10mM magnesium chloride (ATP + Mg). Data
are expressed as mean nmoles choline formed per mg protein per hour
± SD (n=3 rats per point).

no factors which augment choline formation; GPC and PCh are water soluble.

SM is present in brain, and it can be degraded by sphingomyelinase (EC 3.1.4.1.2). Lysosomal sphingomyelinase requires detergent and is activated by free fatty acids (23). The non- lysosomal form of this enzyme is magnesium dependent and is activated by Triton X-100 (23). Optimal SM hydrolysis occurs at pH 7-8, and activity can be found in many subcellular fractions of brain (23). No activity has been reported in brain homogenates in the absence of detergent. We observed significant formation of free choline from radiolabeled SM (Figure 4), therefore we conclude that sphingomyelinase in our brain homogenates was activated (perhaps by fatty acids released).

We were able to demonstrate that radiolabel originally contained within PC made its way into free choline during incubations of brain. There are many possible routes for PC metabolism (Figure 7). It can be deacylated in the 2-position by phospholipase A2 (EC 3.1.1.4), an enzyme activity associated with many membranes which is stimulated by, but does not require, calcium, and which is inhibited by zinc (16, 19, 40, 46). Brain contains a great deal of lysophospholipase A activity (EC 3.1.1.1.5; (29)), which deacylates lysoPC to form GPC. LysoPC can also be metabolized to form free choline directly, without forming GPC as an intermediate. This reaction is catalyzed by brain lysophospholipase D activity which has a pH optimum of 7.2, is stimulated by magnesium and is inhibited by zinc (44). LysoPC is a membrane-toxic compound, and is never present in large amounts in brain (29). We never observed accumulation of significant quantities of lysoPC, and could not even detect its formation when radiolabeled PC was added to incubation solutions. For this reason we conclude that lysoPC was not the primary reservoir from which free choline was produced, though it may have been a transient inter- mediate in the pathway.

Three routes exist for the metabolism of lysoPC formed from PC. It can be reacylated, hydrolyzed to release free choline, or hydrolyzed to form GPC. Reacylation requires ATP to form the acyl-CoA moiety, and is rapid (PC is formed at a rate of 2 nmol/mg protein/min; 22). It is possible that, in the presence of adequate ATP concen- trations, most lysoPC is reacylated, making little available for further degradation to form free choline. This would explain the effects of added ATP upon the degradation of added lysoPC. This phospholipid could then be directly degraded to form free choline by the action of lysophospholipase D, an enzyme activity which is inhibited by zinc, and which is stimulated by magnesium (44).

There are other mechanisms whereby ATP might inhibit choline formation. Addition of magnesium was able to reverse some of ATP's inhibitory influence. It is possible that ATP acts to chelate mag- nesium, and that this ion is absolutely required for choline release

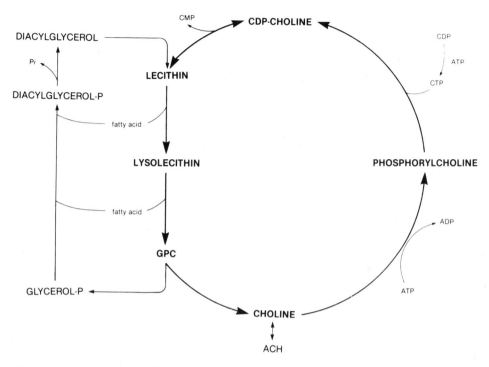

Figure 7. Pathways which might result in the liberation of choline
from phosphatidylcholine. The enzyme activities needed to catalyze
the reactions indicated have all been reported to exist in brain.

from brain tissue. It is also possible that magnesium's stimulation
of choline production could have been mediated by stimulation of
ATPase activity. The inhibitory action of ATP was not mediated by
accelerated utilization of choline for choline ester synthesis;
the addition of ATP to our brain homogenates did increase PCh and
PC synthesis, but not enough to account for the degree of inhibition
of free choline accumulation that we observed. For this reason we
conclude that the major action of ATP is the inhibition of some step
in the pathway of PC catabolism rather than the acceleration of
choline utilization for PC synthesis.

    We have suggested that ATP concentration is an important regulator
of free choline production, and have speculated that it acts, in
part, by influencing the deacylation-acylation cycle of PC metabolism.
If this is the case, then fatty acids (formed during PC deacylation),
as well as free choline, should accumulate after treatments which
lower ATP availability. Several other investigators have reported
that there is a large increase in brain FFA content within 30 seconds
after decapitation or the ligation of the arterial supply of brain

(5, 6, 38, 43). Other treatments associated with seizures and hypoxia, such as electroconvulsive shock, also result in an increase in brain FFA content (4, 7).

SUMMARY

We have discussed various mechanisms which influence the availability of choline to brain. Dietary intake of choline, its absorption from gut, and hepatic metabolism and synthesis of choline each contribute to plasma choline concentration. Transport of choline across the blood-brain barrier does not appear to be limiting, thus brain choline content is, in part, determined by blood choline concentration. Other choline-containing compounds also may enter brain from blood.

Blood tissue is capable of synthesizing choline de novo, but we suggest that this capacity is relatively limited. Much of the free choline used and released by brain must be derived from the hydrolysis of PC and SM. At this time we do not know whether hydrolysis of membrane or blood phospholipid is an important source of choline in vivo. The consistent observation of net efflux of free choline from brain suggests that it is. If this is so, choline ester hydrolysis may be an import factor determining the availability of choline for the synthesis of acetylcholine by neurons. Wurtman, and colleagues (8), have already suggested that cholinergic neurons cannibalize their own membranes to ensure adequate supplies of choline for acetylcholine synthesis.

Acknowledgements: This work was supported, in part, by grants from the National Institutes of Health (#1-R23-HD1627) and from the Whitaker Health Science Fund.

REFERENCES

1. Ansell, G.B. (1973): In: Form and Function of Phospholipids (Ansell, G.B., Hawthorne, J.N. and Dawson, R.M.C., eds.), Elsevier, Amsterdam, pp. 377-422.
2. Ansell, G.B. and Spanner, S. (1982): In: Phospholipids in the Nervous System (Horrocks, L., Ansell, G. and Porcellati, G., eds.), Vol. 1, Raven, New York, pp. 137-144.
3. Aquilonius, S.M., Ceder, G., Lying-Tunell, U., Malmud, H.O., Schubert, J. (1975): Brain Res. 99:430-433.
4. Bazan, N.G. (1970): Biochim. Biophys. Acta 218:1-10.
5. Bazan, N.G. (1971): Lipids 6:211-212.
6. Bazan, N.G., Bazan, H.E.P., Kennedy, W.G. and Joel, C.D. (1971): J. Neurochem. 18:1387-1393.
7. Bazan, N.G. and Turco, E.B.R. (1980): Adv. Neurol. 28:197-205.
8. Blusztajn, J.K., Tacconi, M.T., Zeisel, S.H. and Wurtman, R.J. (1985): In: Physiological Role of Phospholipids in the Nervous System (Horrocks, L., ed.), Raven Press, New York (in press).

9. Blusztajn, J.K. and Wurtman, R.J. (1981): Nature 290:417-418.
10. Blusztajn, J.K., Zeisel, S.H. and Wurtman, R.J. (1979): Brain Res. 179:319-327.
11. Bremer, J. and Greenberg, D. (1961): Biochim. Biophys. Acta 46:205-21.
12. Browning, E.T. (1971): Biochim. Biophys. Res. Commun. 45:1586-1590.
13. Choi, R.L., Freeman, J.J. and Jenden, D.J. (1975): J. Neurochem. 24:735-741.
14. Cohen, E.L. and Wurtman, R.J. (1975): Life Sci. 17:1095-1102.
15. Cohen, E.L. and Wurtman, R.J. (1976): Science 191:561-562.
16. Cooper, M.F. and Webster, G.R. (1970): J. Neurochem. 17:1543-1554.
17. Cornford, E.M., Braun, L.D. and Oldendorf, W.M. (1982): Pediatric Res. 16:324-328.
18. Crews, F.T., Hirata, F. and Axelrod, J. (1980): J. Neurochem. 34:1491-1498.
19. DeHass, G.H., Postema, N.M., Nieuwenhuizen, W. and VanDeenen, L.L.M. (1968): Biochim. Biophys. Acta 159:103-117.
20. DeHass, G.H., Postema, N.M., Nieuwenhuizen, W. and VanDeenen, L.L.M. (1968): Biochim. Biophys. Acta 159:118-129.
21. Dross, K. and Kewitz, H. (1972): N.S. Arch. Pharmacol. 274:91-106.
22. Fisher, S.K., Doherty, F.J. and Rowe, C.E. (1982): In: Phospholipids in the Nervous System (eds) L. Horrocks, G. B. Ansell and G. Porcellati), Vol. 1, Raven Press, New York, pp. 137-144.
23. Gatt, S. (1982): In: Phospholipids in the Nervous System (eds) L. Horrocks, G. B. Ansell and G. Porcellati, Vol. 1, Raven Press, New York, pp. 181-198.
24. Goldberg, A.M. and McCaman, R. (1973): J. Neurochem. 20:1-8.
25. Haubrich, D.R., Gerber, N., Pflueger, A.B. and Zweig, M. (1981): J. Neurochem. 36:1409-1417.
26. Haubrich, D.R., Wang, P.F.L., Chippendale, T. and Proctor, E. (1976): J. Neurochem. 27:1305-1313.
27. Haubrich, D.R., Wang, P.F.L., Clody, D.E. and Wedeking, P.W. (1975): Life Sci. 17:975-980.
28. Houtsmuller, U. (1979): In: Nutrition and the Brain, Vol. 5 (eds) R. Wurtman and J. Wurtman), Raven Press, New York, pp. 83-93.
29. Illingworth, D.R. and Portman, O.W. (1973): Physiol. Chem. Physics 5:365-373.
30. Kosh, J.W., Dick, R.M. and Freeman, J.J. (1980): Life Sciences 27:1953-1959.
31. Lekim, D. and Betzing, H. (1976): Hoppe Seylers Z. Physiol. Chem. 357:1321-1331.
32. McCaman, R. and Stetzler, J. (1977): J. Neurochem. 28:669-671.
33. Mozzi, R. and Porcellati, G. (1979): FEBS Letters 100:363-366.
34. Oldendorf, W.H. and Braun, L.D. (1976): Brain Res. 113:219-223.
35. Pardridge, W.M., Cornford, E.M., Braun, L.D. and Oldendorf, W. (1979): In: Nutrition and the Brain, Vol. 5, (eds) A. Barbeau, J. Growdon and R. Wurtman), Raven Press, New York, pp. 25-34.

36. Parthasarathy, S., Subbaiah, P.V. and Ganguly, J. (1974): Biochem. J. 140:503-508.
37. Spanner, S., Hall, R. and Ansell, G.B. (1976): Biochem. J. 154:133-140.
38. Sun, G.Y., Manning, R. and Strosznajder, J. (1980): Neurochem. Res. 5:1211-1219.
39. Svanborg, A. and Svennerholm, L. (1961): Acta Med. Scand. 169:43-49.
40. Van den Bosch, H. (1980): Biochim. Biophys. Acta 604:191-246.
41. White, H.L. and Wu, J.C. (1973): J. Neurochem. 20:297.
42. Whittaker, V.P. and Barker, L.A. (1972): Meth. Neurochem. 2:2-52.
43. Wieloch, T. and Siesjo, B.K. (1982): Path. Biol. 30:269-277.
44. Wykle, R.L. and Schremmer, J.M. (1974): J. Biol. Chem. 249:1742-1746.
45. Yamamura, H.I. and Snyder, S.H. (1973): J. Neurochem. 21:1355.
46. Zahler, P. and Kramer, R. (1981): Meth. Enzym. 71:690-698.
47. Zeisel, S.H. (1981): Ann. Rev. Nutr. 1:95-121.
48. Zeisel, S.H., Growdon, J.H., Wurtman, R.J., Magil, S.M. and Logue, M. (1980): Neurology 30:1226-1229.
49. Zeisel, S.H., Story, D.L., Wurtman, R.J. and Brunengraber, H. (1980): Proc. Natl. Acad. Sci., U.S.A., 77:4417-4419.
50. Zeisel, S.H., Wishnok, J.H. and Blusztajn, J.K. (1983): J. Pharmacol. Exptl. Therap. 225:320-324.
51. Sheard, N.F. and Zeisel, S.H.: Am. J. Physiol. (submitted).

THE UTILIZATION OF SUPPLEMENTAL CHOLINE BY BRAIN

L. Wecker

Departments of Pharmacology and Psychiatry
Louisiana State University Medical Center
1901 Perdido Street
New Orleans, Louisiana 70112, U.S.A.

INTRODUCTION

The disposition and metabolism of choline and its utilization for the synthesis of acetylcholine (ACh) in brain have been the subject of numerous investigations (for reviews see 2, 3, 4, 18). Although it is well established that following administration, choline is rapidly incorporated into ACh in brain (11, 20, 28, 29), the ability of supplemental choline to increase the steady-state concentration of ACh is questionable. Studies in my laboratory have indicated that the acute administration of choline does not increase the concentration of ACh in brain under "normal" biochemical and physiological conditions, but it does provide choline that can be used to support the synthesis of ACh when there is an increased demand for the precursor, i.e., when the activity of central cholinergic neurons is increased (32, 36). Choline, administered either as a salt or in the form of phosphatidylcholine (PTC), prevents the depletion of ACh in brain induced by atropine, fluphenazine or pentylenetetrazol, when choline is administered acutely prior to these agents that increase cholinergic neuronal activity (14, 21, 27, 31, 33, 38). In addition, prior administration of choline has been shown to prevent opiate withdrawal-induced decreases in the levels of ACh in hippocampus (8). Hence, data suggest that choline enhances the synthesis of ACh during conditions of drug- induced increases in neuronal demand. This idea is further substantiated by studies in the perfused hemidiaphragm preparation, indicating that the release of ACh following phrenic nerve stimulation is enhanced by the presence of choline in the perfusate only when the rate of nerve stimulation is increased (7). Hence, most data, with the exception of one report in which large doses of choline were used that may have decreased the turnover of ACh (30), support the hypothesis that supplemental choline, via acute administration, is used

851

to enhance the synthesis of ACh under conditions of increased neuronal demand.

Although there is an abundance of data on the effects of acute choline administration, clinical interest in the use of choline for the treatment of neuropsychiatric disorders involves administration on a chronic basis. We have previously reported that chronic administration of choline in the diet does not increase the steady-state concentration of ACh in brain, but does alter the behavior of rats (37). In addition, dietary choline supplementation has been shown to protect both rats and mice from the convulsant effects of many pharmacological agents, particularly nicotine (35). Recent findings have indicated that chronic supplementation with choline does not affect the activity of either choline acetyltransferase or acetylcholinesterase, or the density of muscarinic receptors, as measured by both agonist and antagonist binding (34). Hence, the present studies were initiated to determine whether chronic supplementation with choline in the diet could provide choline to support the synthesis of ACh when the demand for precursor was increased.

Effects of Chronic Choline Supplementation on the Induced Depletion of ACh. To determine whether the prophylactic effects of choline measured after acute administration were also present after chronic administration, rats were maintained on either basal choline chow (0.2% choline chloride) or choline-supplemented chow (2.0% choline chloride) for up to 45 days. Plasma and red blood cell levels of choline were determined after 28-32 days on the dietary regimens. The concentration of choline in plasma increased significantly by 117% with supplementation from basal levels of 8.05 $\pm$ 0.25 to 17.5 $\pm$ 1.54 nmoles/ml. Similarly, the levels of choline in red blood cells also increased significantly from 13.9 $\pm$ 1.00 to 22.6 $\pm$ 0.71 nmoles/ml, a 63% increase.

The administration of atropine (7.0 mg/kg, i.p.) to rats maintained on basal choline chow caused a significant 40% decrease in the steady-state concentration of ACh in striatum when measured 30 minutes after injection (Table 1). Atropine did not alter the levels of ACh in striatum from rats maintained on the choline supplemented diet. Similar results, although not statistically significant, were noted in the hippocampus, viz., atropine decreased the concentration of ACh in hippocampus from basal choline rats by 20% while levels in choline-supplemented rats were unaffected.

Administration of the convulsant pentylenetetrazol (30 mg/kg, i.p.) also led to a significant 23% decrease in the concentration of ACh in hippocampus from rats maintained on the basal diet and this effect was prevented by chronic choline supplementation (Table 2). Hence, results indicate that chronic supplementation with choline can support the synthesis of ACh when there is an increased neuronal demand.

Table 1

Effects of Dietary Choline Supplementation on
Atropine-Induced Depletion of Acetylcholine

Brain Region	Dietary Regimen	Acetylcholine (nmoles/g tissue)	
		Saline	Atropine
Striatum			
	Basal	81.1 ± 4.47 (10)	49.8 ± 4.65* (7)
	Supplemented	79.5 ± 3.10 (10)	72.9 ± 4.06 (6)
Hippocampus			
	Basal	28.3 ± 2.20 (9)	22.6 ± 2.86 (8)
	Supplemented	29.9 ± 3.12 (10)	27.6 ± 4.80 (5)

Rats (140-160g initial weight) were maintained on either basal choline chow (0.2% choline chloride) or choline-supplemented chow (2.0% choline chloride) for 28-35 days. Animals were injected (i.p.) with saline (0.10 ml/100g) or atropine sulfate (7.0 mg/kg) and sacrificed by head-focused microwave irradiation 30 minutes following the injection. ACh was analyzed by pyrolysis gas chromatography. Each value is the mean ± SEM. The number of rats is in parentheses. *Significantly less than corresponding saline-injected group, $p < .01$.

Mechanisms Responsible for the Effects of Choline. Since circulating levels of choline were enhanced by dietary supplementation, the effects of the dietary regimens on the concentration of choline in brain were determined. When rats were maintained on the diets for 28-35 days, the concentration of choline in the striatum and hippocampus from basal choline rats was 24.8 ± 3.00 and 20.5 ± 3.28 nmoles/g, respectively. The levels of choline in brains from choline-supplemented rats were not different from control values, viz., 24.7 ± 3.14 and 24.5 ± 4.03 nmoles/g for striatum and hippocampus, respectively. Hence, the prophylactic effects of choline did not appear to be mediated by elevated levels of free choline in brain, results similar to those obtained from acute studies (27, 36).

Table 2

Effects of Dietary Choline Supplementation on
Pentylenetetrazol-Induced Depletion
Of Acetylcholine in Hippocampus

Dietary Regimen	Acetylcholine (nmoles/g tissue)	
	Saline	Pentylenetetrazol
Basal	25.3 ± 0.69	19.4 ± 1.18[*]
Supplemented	23.4 ± 0.78	23.0 ± 0.71

Rats (140-160g initial weight) were maintained on either basal choline chow (0.2% choline chloride) or choline-supplemented chow (2.0% choline chloride) for 35-42 days. Animals were injected (i.p.) with saline (0.10 ml/100g) or pentylenetetrazol (30 mg/kg) and sacrificed by head-focused microwave irradiation 30 minutes following the injection. ACh was analyzed by pyrolysis gas chromatography. Each value is the mean ± SEM of determinations from 7 rats per group. [*]Significantly less than corresponding saline-injected group, p < .01.

Since choline is released from many choline-containing compounds in brain, it was of interest to investigate the possible involvement of esterified sources of choline in mediating the observed effects. To provide a measure of the functional ability of the brain to convert choline esters to free choline, the rate of choline production postmortem was determined (Table 3). Supplementation with choline increased the rate of choline production in striatum, hippocampus, and cortex by 24%, 13%, and 53%, respectively. Hence, results suggested an effect of choline supplementation on the metabolism of bound sources of choline.

Since the production of choline postmortem is thought to be due in part to the release of choline from PTC, the concentration of PTC, as well as total lipid phosphorus (P), was measured in particulate fractions of striata from rats maintained on the two dietary groups. The concentrations of PTC and total lipid P in the synaptosomal membrane fraction isolated from the striata of rats maintained on the basal choline and choline-supplemented groups did not differ (Table 4). However, when lipids were quantitated in the microsomal fraction, a significant increase was apparent in the choline-supple-

Table 3

Effects of Dietary Choline Supplementation on
The Postmortem Production of Choline

Brain Region	Rate of Choline Production (nmoles/g/min)	
	Basal	Supplemented
Striatum	30.2 ± 2.81	37.5 ± 2.18[*]
Hippocampus	24.0 ± 1.85	27.0 ± 2.25
Cortex	18.9 ± 2.39	28.9 ± 2.46[*]

Rats (140-160 g initial weight) were maintained on either basal choline chow (0.2% choline chloride) or choline-supplemented chow (2.0% choline chloride) for 28-32 days. Rats were sacrificed by either head-focused microwave irradiation or decapitation. For the latter, the time between sacrifice and homogenization of tissue was monitored. The rate of choline production is the difference between values obtained from animals sacrificed by microwave irradiation and those obtained from rats sacrificed by decapitation. Choline was analyzed by pyrolysis gas chromatography. Each value is the mean ± SEM of determinations from 7-8 rats per group. [*]Significantly greater than corresponding basal group values, p < .05.

mented group. The concentrations of PTC and total lipid P in the striatum increased significantly with choline supplementation by 12% and 22%, respectively (Table 4).

CONCLUSIONS

Results obtained to date indicate that chronic dietary supplementation with choline does not enhance the synthesis or metabolism of ACh under "normal" physiological and biochemical conditions, but does support neurotransmitter synthesis when the neuronal demand for choline is increased through pharmacological manipulation. Furthermore, these observed effects do not appear to be mediated through an enhanced steady-state concentration of free choline in the brain, but rather may involve alterations in the metabolism of phospholipids.

Whether administered intravenously, intraperitoneally or directly into the brain, choline is metabolized first to the water-soluble compounds ACh and phosphorylcholine (1, 5, 10, 22, 25) followed by conversion to lipid-soluble esters, primarily PTC (1, 13, 15). In

Table 4

Effects of Dietary Choline Supplementation
On Membrane Phospholipids in Rat Striatum

Subcellular Fraction	Dietary Regimen	Phosphatidylcholine	Total Lipid P
		(micromoles/mg protein)	
Synaptosomal Membranes			
	Basal	.3281 ± .0564	.8254 ± .1637
	Supplemented	.3313 ± .0362	.7398 ± .1059
Microsomes			
	Basal	.4895 ± .0151	.9874 ± .0552
	Supplemented	.5498 ± .0142*	1.208 ± .0694*

Rats (140-160 g initial weight) were maintained on either basal choline chow (0.2% choline chloride) or choline-supplemented chow (2.0% choline chloride) for 28-35 days. Animals were sacrificed by decapitation. Subcellular fractions were prepared according to Gray and Whittaker (19). Phospholipids were isolated, separated by two dimensional TLC, and analyzed for phosphorus by spectrophotometry (17, 26). Protein was determined using BSA as a standard (24). Each value is the mean ± SEM of determinations from 5 groups, each group containing striata pooled from 4 rats. *Significantly greater than corresponding basal group values, $p < .05$.

addition to the incorporation of choline into these bound forms, choline can also be released from these in vivo sources (11, 20) to continuously supply free choline for the synthesis of ACh in brain (6, 9, 12, 16, 22, 23).

The involvement of choline esters in mediating the effects of choline supplementation is suggested from evidence of increased concentrations of both PTC and total lipid P in the microsomal fraction of striata, in conjunction with observations of an increased ability of brain from choline-supplemented rats to release choline postmortem. In conclusion, it appears that chronic supplementation with choline enhances the concentration of esterified sources of choline in brain. Furthermore, it appears that choline can be mobilized from these compounds to support the synthesis of ACh under conditions of increased neuronal demand.

Acknowledgements. These studies were supported by USDHHS research grant No. NIMH-33443 and a contract from the USAMRDC No. DAMD 17-83-

C-3012. The author appreciates the assistance of C.J. Flynn, S. Taylor, G. Cawley, S. Rothermel, and B.A. Trommer.

REFERENCES

1. Ansell, G.B. and Spanner, S. (1968): Biochem. J. 110:201-206.
2. Ansell, G.B. and Spanner, S. (1975): In: Cholinergic Mechanisms (ed) P.G. Waser, Raven Press, New York, pp. 117-129.
3. Ansell, G.B. and Spanner S. (1977): In: Cholinergic Mechanisms and Psychopharmacology (ed) D.J. Jenden, Plenum Press, New York, pp. 431-445.
4. Ansell, G.B. and Spanner, S. (1979): In: Nutrition and the Brain, Vol. 5 (eds) A. Barbeau, J.H. Growdon and R.J. Wurtman, Raven Press, New York, pp. 35-46.
5. Aquilonius, S.M., Flentge, F., Schuberth, J., Sparf, B. and Sundwall, A. (1973): J. Neurochem. 20:1509-1521.
6. Bhatnagar, S.P. and MacIntosh, F.C. (1967): Canad. J. Physiol. Pharmacol. 45:249-268.
7. Bierkamper, G.G. and Goldberg, A.M. (1979): In: Nutrition and the Brain, Vol. 5 (eds) A. Barbeau, J.H. Growdon and R.J. Wurtman, Raven Press, New York, pp. 243-251.
8. Botticelli, L.J. and Wurtman, R.J. (1981): Brain Research 210: 479-484.
9. Browning, E.T. (1971): Biochem. Biophys. Res. Comm. 45:1586-1590.
10. Chakrin, L.W. and Whittaker, V.P. (1969): Biochem. J. 113:97-107.
11. Choi, R.L., Freeman, J.J. and Jenden, D.J. (1975): J. Neurochem. 24:735-741.
12. Collier, B., Poon, P. and Salehmoghaddam, S. (1972): J. Neurochem. 19:51-60.
13. Diamond, I. (1971): Arch. Neurol. 24:333-339.
14. Dolezal, V. and Tucek, S. (1982): Brain Research 240:285-293.
15. Dowdall, M.J., Barker, L.A. and Whittaker, V.P. (1972): 130: 1081-1094.
16. Dross, K. and Kewitz, H. (1972): Naunyn-Schmiedeberg's Arch. Pharmacol. 274:91-106.
17. Folch, J., Lees, M. and Stanley, G.H.S. (1957): J. Biol. Chem. 226:497-509.
18. Freeman, J.J. and Jenden, D.J. (1976): Life Sciences 19:949-962.
19. Gray, E.G. and Whittaker, V.P. (1962): J. Anat. (London) 96:79-88.
20. Hanin, I. and Schuberth, J. (1974): J. Neurochem. 23:819-824.
21. Jope, R.S. (1982): J. Pharmacol. Exptl. Therap. 220:322-328.
22. Jope, R.S. and Jenden, D.J. (1979): J. Neurosci. Res. 4:69-82.
23. Kosh, J.W., Dick, R.M. and Freeman, J.J. (1980): Life Sciences 27:1953-1959.
24. Lowry, O.H., Rosebrough, N.J., Farr, A.L. and Randall, R.J. (1951): J. Biol. Chem. 193:265-275.
25. Nordberg, A. (1977): Acta Physiol. Scand. Suppl. 445.

26. Rouser, G., Fleischer, S. and Yamamoto, A. (1970):   Lipids 5: 494-496.
27. Schmidt, D.E. and Wecker, L. (1981): Neuropharmacology 20:535-539.
28. Schuberth, J., Sparf, B. and Sundwall, A. (1969):   J. Neurochem. 16:695-700.
29. Schuberth, J., Sparf, B. and Sundwall, A. (1970):   J. Neurochem. 17:461-468.
30. Sherman, K.A., Zigmond, M.J. and Hanin, I. (1981):   Neuropharmacology 20:921-924.
31. Trommer, B.A., Schmidt, D.E. and Wecker, L. (1982):   J. Neurochem. 39:1704-1709.
32. Wecker, L. and Dettbarn, W-D. (1979): J. Neurochem. 32:961-967.
33. Wecker, L., Dettbarn, W-D. and Schmidt, D.E. (1978):   Science 199:86-87.
34. Wecker, L., Ehlert, F.J., Speth, R.C., Trommer, B.A. and Yamamura, H.I. (1984):   in preparation.
35. Wecker, L., Flynn, C.J., Stouse, M.R. and Trommer, B.A. (1982): Drug-Nutrient Interactions 1:125-130.
36. Wecker, L. and Goldberg, A.M. (1981):   In: Cholinergic Mechanisms: Phylogenetic Aspects Central and Peripheral Synapses, and Clinical Significance (eds) G. Pepeu and H. Ladinsky, Plenum Press, New York, pp. 451-461.
37. Wecker, L. and Schmidt, D.E. (1979): Life Sciences 25:375-384.
38. Wecker, L. and Schmidt, D.E. (1980): Brain Research 184:234-238.

EFFECTS OF THE COMBINED TREATMENT OF RATS WITH FLUPHENAZINE AND CHOLINE OR LECITHIN ON THE STRIATAL CHOLINERGIC AND DOPAMINERGIC SYSTEM

F. Flentge[1], D. Arst[2], B.H. Westerink[3],
M.J. Zigmond[4] and I. Hanin[5]

Departments of Biological Psychiatry[1]
and
Pharmaceutical and Analytical Chemistry[3]
University of Groningen
Groningen, The Netherlands

Departments of Psychiatry[5]
and
Biological Sciences[2,4]
University of Pittsburgh
Pittsburgh, PA 15213 U.S.A.

INTRODUCTION

It has been reported that the acute administration of choline (Ch) or its dietary precursor, lecithin (phosphatidylcholine; PCh) results in an increase in the steady-state concentration of acetylcholine (4, 11-13, 17). This increase was interpreted to be due to an enhanced ACh synthesis. Many clinical experiments have since been performed using Ch or PCh loading, based on the hypothesis that these agents would increase ACh levels, leading to an enhanced ACh release (for review see Bartus et al.; 2). In recent years, however, a number of groups have not been able to confirm the original finding of an increase in ACh levels in the brain or brain regions in rats or mice, after Ch or PCh administration (3, 7, 14, 18-23). Furthermore, no change in the synthesis rate of ACh has been observed after Ch loading (3, 6). Nevertheless, several reports have demonstrated that exogenous Ch or PCh could increase ACh content under conditions of increased cholinergic activity. In these investigations an attempt was made to increase activity by treatment of the animals with drugs like atropine, fluphenazine and pentylenetetrazol. It was observed that the decrease in ACh levels produced by these drugs in one or more regions of the brain could be attenuated by pretreatment of the animals with Ch or PCh (14, 19, 21-23, but see also 20 which failed to confirm these results).

It has been suggested that Ch- or PCh-induced attenuation of the decrease in ACh levels in these cases is a result of enhanced Ch uptake and subsequent ACh synthesis. It is also possible, however, that Ch acts by antagonizing the primary stimulating effect of the mentioned drugs, thus restoring ACh levels but reducing the cholinergic hyperactivity.

For these reasons, we decided to study the effects of fluphenazine on animals pretreated with either Ch or PCh. The primary effect of fluphenazine is probably the blocking of dopaminergic receptors on striatal cholinergic neurons, removing the normal inhibitory influence of dopamine (DA) on these neurons. Dopamine turnover is enhanced as well, resulting in highly elevated levels of the metabolites dihydroxyphenylacetic acid (DOPAC) and homovanillic acid (HVA). We therefore decided to use DOPAC levels as a measure of the primary effect of fluphenazine in experiments similar to those in which it had been shown by Jope (14) that PCh can attenuate the fluphenazine-induced decrease in ACh levels. In subsequent experiments the effect of Ch also was investigated.

METHODS

Some of these experiments were performed in Pittsburgh (Western Psychiatric Institute and Clinic), while the remainder were performed in Groningen (Department of Biological Psychiatry).

Pittsburgh Experiments. Male, Sprague-Dawley rats (Zivic-Miller Laboratories, Allison Park, PA) weighing 200-250g were used. After drug treatment rats were killed by microwave irradiation (2.5 kW, 2.4 MHz) of the head for 4 sec in a modified Litton microwave oven (Medical Engineering Consultants, Lexington, MA). Brain tissue was homogenized in 0.4 N perchloric acid and prepared for the analysis of ACh and Ch content by gas chromatography/mass spectrometry as described by Hanin and Skinner (9). DA, DOPAC and norepinephrine (NE) were assayed in an aliquot of the supernatant by the high pressure liquid chromatographic (HPLC) method with electrochemical detection as described by Keller et al. (15). Fluphenazine hydrochloride (Prolixin®) was obtained from E.R. Squibb and Sons, Inc., Princeton, N.J. and PCh (soybean phosphatidylcholine, 95% pure) was a gift from the American Lecithin Company, Atlanta, GA.

Groningen Experiments. Male Wistar rats (local breed), weighing 200-250 g were killed by microwave irradiation (2.5 kW, 2.4 MHz) for 6 sec in a modified Litton microwave oven (Medical Engineering Consultants, Lexington, MA). The striatum from one hemisphere was homogenized in 1 N formic acid/acetone (15:85 vol/vol), and the supernatant extracted three times with 3 volumes of water saturated ether for the Ch and ACh assays; the striatum from the other hemisphere was used for the determination of DA, DOPAC and HVA, according to the procedure described by Westerink and Mulder (24).

Choline was assayed by a radioenzymatic method described earlier by Flentge and Van den Berg (7). ACh was assayed by a two-step radio-enzymatic assay; firstly, endogenous Ch was converted to phosphorylcholine using choline-kinase as described by Goldberg and McCaman (8). After inactivation of choline-kinase by boiling, ACh was hydrolyzed by acetylcholinesterase. The Ch derived from ACh was subsequently determined by the Ch assay mentioned before, after · inactivation of acetylcholinesterase by boiling. The separation of $[^{14}C]$-ACh from $[^{14}C]$-acetylCoA in the ACh assay was performed by high voltage electrophoresis, using a modification of the approach of Aquilonius et al. (1) (electrophoresis at 4kV, for 30 min, instead of using the separation by Biorad 1-X4 columns in the Ch assay). This modification in the procedure allows one to obtain lower blanks. Fluphenazine hydrochloride was a gift from the Squibb Institute, Princeton, N.J., U.S.A., and PCh (phosphatidylcholine, 89% pure) was a gift from Unilever Research Co., Vlaardingen, The Netherlands.

General. Phosphatidylcholine was suspended in 0.45% NaCl as described by Jope (14), and administered by a single intragastric intubation. Fluphenazine and Ch chloride were injected intraperitoneally. Drug doses and times of administration are indicated in the Results section. Control groups were included in which one or both of the administered drugs were replaced by the vehicle solution.

RESULTS

Effects of PCh Treatment. PCh was administered to animals at a dose and time shown earlier by others (14) to attenuate the fluphenazine induced decrease in striatal ACh levels. Two different batches of PCh were used. Fluphenazine alone decreased ACh levels significantly while PCh alone had no significant effect on the striatal ACh concentration. Pretreatment of the fluphenazine injected animals with PCh (from either source) had no effect at all on the fluphenazine-induced decrease in ACh levels (Table 1).

Some of the animals treated with PCh (American Lecithin Co.) were also used to measure striatal DOPAC, DA and NE levels. As can be seen from Table 2, PCh treatment alone had no effect on either DA or DOPAC levels. Fluphenazine injection, however, increased DOPAC levels 3.5 fold, and slightly reduced DA levels. These effects were unchanged by pretreatment with PCh. None of these treatments had a significant effect on the concentration of NE in the striatum, though a tendency to higher levels could be observed in the PCh pretreated animals.

Table 1

Effect of Phosphatidylcholine (PCh) Treatment on
Fluphenazine-Induced Depletion of Rat Striatal ACh

Treatment	AM. LEC. CO.-PCh		UNILEVER-PCh	
	ACh	Ch	ACh	Ch
Control	69.6 ± 6.3 (9)	42.6 ± 2.9 (7)	80.7 ± 5.9 (6)	37.0 ± 7.0 (4)
Fluphenazine	40.1 ± 2.8* (10)	38.3 ± 2.3 (7)	65.1 ± 3.0*** (6)	39.8 ± 0.9 (5)
PCh	68.7 ± 6.2 (8)	47.6 ± 2.5 (7)	93.4 ± 5.4 (6)	47.7 ± 1.6 (6)
PCh + Fluphenazine	41.9 ± 2.3* (10)	47.5 ± 2.4 (7)	61.0 ± 3.3** (6)	40.8 ± 2.0 (6)

Results are expressed as mean ± S.E.M. in nmoles/g. Number of animals
is in parentheses. *p < 0.001, **p < 0.01, ***p < 0.025, student's
t-test. PCh (10 mmol/kg; intragastric) and fluphenazine hydrochloride
(1 mg/kg; i.p.) were administered 6 h and 1 h before sacrifice,
respectively.

Table 2

Effect of Phosphatidylcholine (PCh) and/or Fluphenazine
Treatment on DOPAC, DA and NE Levels in Rat Striatum

Treatment	DOPAC (μg/g)	DA (μg/g)	NE (ng/g)
Control	1.11 ± 0.01	10.5 ± 0.6	160 ± 14
Fluphenazine	3.73 ± 0.08*	8.7 ± 0.4***	164 ± 15
PCh	1.09 ± 0.05	9.0 ± 0.8	199 ± 16
PCh + Fluphenazine	3.74 ± 0.21*	8.2 ± 0.5**	204 ± 31

Results are expressed as mean ± S.E.M. Number of animals: 5 per
group. *p<0.001, **p<0.02, ***p<0.05, student's t-test. PCh (10
mmol/kg; intragastric) and fluphenazine hydrochloride (1 mg/kg;
i.p.) were administered 6 h and 1 h before sacrifice, respectively.

Effects of Choline Treatment. Next we examined the effect of pretreatment of the animals with Ch on the fluphenazine-induced decrease
in ACh levels and on DOPAC levels. In general pretreatment with
Ch for 15 minutes had no effect on either the fluphenazine-induced
decrease in ACh levels (Fig. 1) or increase in DOPAC-levels (Fig. 2).
In the case of some individual animals of the group sacrificed 45
min after fluphenazine administration the fluphenazine-induced decline
in ACh was not observed. At the same time the increase in DOPAC
levels was also absent in these animals. These observations were
more frequent in the choline-treated group than in the group treated
only with fluphenazine (Fig. 3). In both groups, however, a significant correlation was found between ACh and DOPAC levels (resp. r =
-0.80 and r = -0.78, p < 0.01).

DISCUSSION

The results of the ability of PCh to affect brain ACh levels
are mixed. Hirsch and Wurtman (12) reported a small increase in
whole brain ACh levels of rats that had consumed a meal supplemented
with an impure PCh preparation. Magil et al. (17) also have reported
increased ACh concentrations in the whole brain after dietary administration of an 80% pure PCh preparation. Jope (14), however,

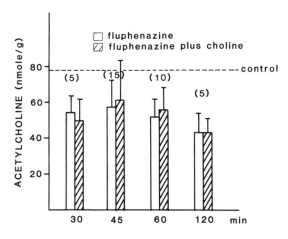

TIME AFTER FLUPHENAZINE

Figure 1. Striatal ACh levels of rats after treatment with fluphenazine or fluphenazine plus choline. Rats were sacrificed by
microwave irradiation at different times after i.p. injection with
fluphenazine hydrochloride (0.5 mg/kg). All groups received an
i.p. injection of either saline (blank columns) or Ch chloride (1
mmole/kg; shaded columns) 15 min before the fluphenazine injection.
Results are expressed as mean ± SD. Figures in brackets denote
the number of animals per column.

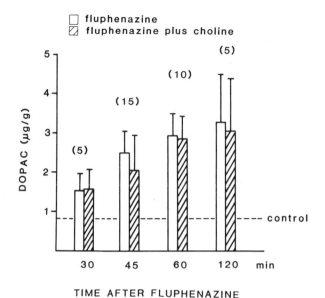

Figure 2. Striatal DOPAC levels of rats after treatment with fluphenazine or fluphenazine plus choline. For details see legend
Fig. 1.

has reported no measured effect of intragastrically administered
80% pure PCh on ACh levels in striatum, hippocampus and cortex.
Our results, as described in this chapter, which were obtained with
two batches of purified PCh from different suppliers, confirm these
latter findings. In these same studies, we were, however, not able
to confirm Jope's finding that the fluphenazine induced depletion
of striatal ACh could be partially antagonized by pretreatment with
PCh. Also we have shown that PCh pretreatment did not influence
the effect of fluphenazine on the dopaminergic system in the striatum,
as measured by an increase in DOPAC levels. The concomitant small
decrease in DA-levels as a result of the fluphenazine-induced enhanced
DA turnover, was also independent of pretreatment with PCh.

The published literature with respect to the question of whether
Ch could antagonize the fluphenazine-induced decrease in striatal
ACh levels also is inconsistent. Sherman et al. (20) were not able
to show such an antagonistic effect, but Trommer et al. (21) did
report that after the combined treatment with Ch and fluphenazine,
ACh levels in the striatum were not significantly different from
those in control animals. These authors, however, did not report
whether the ACh levels after combined treatment (Ch plus fluphenazine)
were significantly higher than in animals treated with fluphenazine
alone.

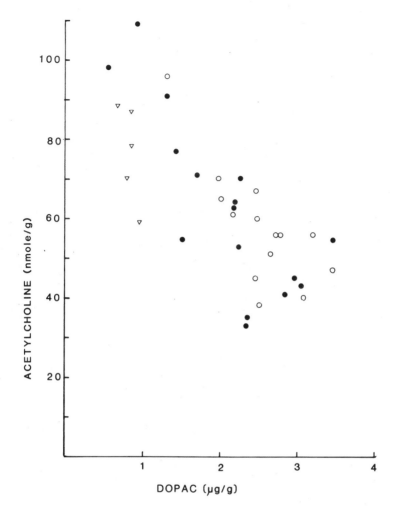

Figure 3.   Relationship between striatal ACh and DOPAC levels 45 min after fluphenazine injection.   Rats received an i.p. injection with saline (o) or choline chloride (1 mmole/kg, ●) 15 min before fluphenazine injection.   Control animals (▽) received only 2 saline injections.

    In general our current experiments on the effect of pretreatment with Ch on the fluphenazine induced decline in ACh levels support the results obtained by Sherman et al. (20). At several time intervals after the administration of fluphenazine (from 30 minutes to 2 hours) we found no difference in ACh levels between groups of rats which had received fluphenazine alone and those groups which were pretreated for 15 minutes with Ch. The same result was obtained with respect

to DOPAC levels, showing that the effect on the the dopaminergic system
was similar under both conditions.

However, the results of the group sacrificed 45 min after fluphenazine administration are of special interest. Only in this
group did we find a highly significant negative correlation between
DOPAC and ACh levels, whether or not the animals were pretreated
with Ch. Those few exceptions in which we observed an apparent
reversal of the fluphenazine-induced decline in ACh levels belonged
to this 45-min group. Also Trommer et al. (21) found an antagonistic
effect of Ch in animals treated for 45 min with fluphenazine. Since
at this time point the effect of fluphenazine is submaximal (Fig. 2),
DA receptor blockade is probably incomplete. In accordance with
these results it has been shown recently that under conditions of
incomplete dopamine receptor inactivation ACh levels show a linear
negative correlation to the degree of receptor occupation (25).

It has been proposed by others that the antagonism by Ch pretreatment is the result of increased uptake of exogenous Ch and
subsequent enhanced ACh synthesis under conditions of increased
cholinergic activity. A condition of this sort could be induced
by fluphenazine, but also by other drugs such as atropine and pentylenetetrazol (14, 19, 21-23). Our finding that the infrequent
reversal of the fluphenazine-induced decline in ACh levels coincided
with a reduction of the effect of fluphenazine on the dopaminergic
system suggests an alternative explanation. Ch might antagonize
or prevent the primary effect of the stimulating drug, i.e., the
stimulation of cholinergic nerve cells itself. In the case of fluphenazine this primary effect is probably the blockade of dopaminergic
receptors on cholinergic cells, removing the normal inhibitory influence of DA on these cells (5, 10). Therefore the increase in
ACh by Ch in these fluphenazine-treated animals might have resulted
from restoration of the normal inhibitory influence of dopaminergic
cells on the cholinergic cells, thus reinstating normal ACh turnover
and normal ACh levels.

Acknowledgements. The authors are indebted for excellent technical
assistance to Ms. A. Russell, Mrs. R. Medema, Mrs. T. Koch and Mr.
F. Postema. These studies were supported in part by NIMH grant
#MH26320.

REFERENCES

1. Aquilonius, S.M., Flentge, F., Schuberth, J., Sparf, B. and
   Sundwall, A. (1973): J. Neurochem. 20:1509-1521.
2. Bartus, R.T., Dean, R.L., Beer, B. and Lippa, A.S. (1982):
   Science 217:408-417.
3. Brunello, N., Cheney, D.L. and Costa, E. (1982): J. Neurochem.
   38:1160-1163.
4. Cohen, E.L. and Wurtman, R.J. (1976): Science 191:561-562.

5. Conner, J.D. (1970): J. Physiol. Lond. 208:691-703.
6. Eckernas, S.A., Sahlstrom, L. and Aquilonius, S.M. (1977): Acta Physiol. Scand. 101:404-410.
7. Flentge, F. and Van den Berg, C.J. (1979): J. Neurochem. 32: 1331-1333.
8. Goldberg, A.M. and McCaman, R.E. (1974): In Choline and Acetyl-choline: Handbook of Chemical Assay Methods. (ed) I. Hanin, Raven Press, New York, pp. 47-63.
9. Hanin, I. and Skinner, R.F. (1975): Anal. Biochem. 66:568-583.
10. Hattori, T., Singh, V.K., McGeer, E.G. and McGeer, P.L. (1976): Brain Res. 102:164-173.
11. Haubrich, D.R., Wang, P.F.L., Clody, D.E. and Wedeking, P.W. (1975): Life Sci. 17:975-980.
12. Hirsch, M.J. and Wurtman, R.J. (1978): Science 202:223-225.
13. Hirsch, M.J., Growdon, J.H. and Wurtman, R.J. (1977): Brain Res. 125:383-385.
14. Jope, R.S. (1982): J. Pharmacol. Exptl. Therap. 220:322-328.
15. Keller, R., Oke, A., Mefford, I. and Adams, R.N. (1976): Life Sciences 19:995-1004.
16. Korf, J., Sebens, J.B., Flentge, F. and Van Der Werf, J.F. (1985): J. Neurochem. 44:314-318.
17. Magil, S.G., Zeisel, S.H. and Wurtman, R.J. (1981): J. Nutri-tion 111:166-170.
18. Pedata, F., Wieraszko, A. and Pepeu, G. (1979): Pharmac. Res. Commun. 9:755-761.
19. Schmidt, D.E. and Wecker, L. (1981): Neuropharmacology 20:535-539.
20. Sherman, K.A., Zigmond, M.J. and Hanin, I. (1981): Neurophar-macology 20:921-924.
21. Trommer, B.A., Schmidt, D.E. and Wecker, L. (1982): J. Neuro-chem. 39:1704-1709.
22. Wecker, L. and Schmidt, D.E. (1980): Brain Res. 184:234-238.
23. Wecker, L., Dettbarn, W.D. and Schmidt, D.E. (1978): Science 199:86-87.
24. Westerink, B.H.C. and Mulder, T.B.A. (1981): J. Neurochem. 36:1449-1462.

MECHANISMS OF ACETYLCHOLINE SYNTHESIS:

COUPLING WITH CHOLINE TRANSPORT

R.J. Rylett

Department of Physiology
Medical Sciences Building
University of Western Ontario
London, Ontario, Canada N6A 5C1

INTRODUCTION

Mechanisms regulating acetylcholine (ACh) synthesis at the cholinergic nerve terminal have not been resolved. Clearly, choline utilized in the synthesis of ACh is derived from an extraneuronal source rather than being synthesized de novo within the cholinergic neuron (12). A number of investigators have demonstrated that the supply of exogenous choline to the nerve ending by sodium-dependent high-affinity choline carriers may be both rate-limiting and regulatory in the synthesis of ACh (4, 5, 9, 14, 29, 34) with a high percentage of the transported choline being metabolized to ACh (5, 35). Contradictory evidence has, however, been presented recently indicating that in guinea-pig brain synaptosomes choline taken into the nerve ending by high- affinity choline carriers makes only a minor contribution as precursor for ACh synthesis (16, 20). Whereas it has been reported that the major source of choline for ACh synthesis is derived from preexisting cytoplasmic choline pools in guinea-pig brain synaptosomes, evidence has been presented that in rat brain synaptosomes synthesis of ACh utilizing a cytoplasmic pool of choline as precursor could not be measured (32).

Controversy exists over the mode of coupling between the immediate source of choline for ACh synthesis and the acetylating enzyme, choline acetyltransferase (ChAT). Two distinct mechanisms have been proposed: a) "direct coupling" (5-7), whereby extraneuronal choline is transported into the nerve ending by specific carriers and metabolized directly to ACh, perhaps by a membrane-associated form of ChAT or some form of carrier-enzyme complex (7), without existing in a free state within the terminal; and b) "kinetic coupling" (15, 31), where choline derived from a small intraterminal pool supplied by the high-affinity choline carriers is utilized for the synthesis of ACh by soluble ChAT which maintains the levels of choline, ACh, acetylCoenzyme A and coenzyme A constant under control by the laws of mass action. In opposition to these hypotheses, Marchbanks and Kessler (20) state that, in guinea-pig brain synaptosomes, choline transport and ACh synthesis are independent mechanisms, with choline transported into the nerve terminal not having privileged access to ChAT.

These mechanisms remain unresolved partly because the subcellular localization of ChAT, especially as it relates to function, has not been established. There is increasing evidence leading to the conclusion that at least some fraction of the enzyme is membrane-bound (8, 30), while other studies have shown that soluble ChAT may be present in multiple forms in a number of animal species and at different stages of ontogeny (1, 2, 10, 18, 19, 22). There appears to be a significant correlation between the presence of a net positively charged form of the enzyme (pI > 8.0) which has appreciable affinity for synaptosomal membrane fragments (10, 11) and the efficient acetylation of choline transported by the high-affinity carriers (1-3). Despite the lack of evidence for differing subcellular localizations of these various molecular forms of ChAT (10, 17), it must be kept in mind that one of the most significant factors affecting metabolic regulation in the cellular microenvironment is the interaction between enzymes and cellular structures (21). A coupling between the high-affinity choline carrier and ChAT may function as an integral component responsible for at least part of ACh synthesized, and may be of especial importance during repetitive neuronal activity. The precise nature of this coupling has not been deduced, nor has its quantitative role in ACh synthesis during different stages of neuronal activity been established (13).

STUDIES WITH CHOLINE MUSTARD AZIRIDINIUM ION (ChMAz)

Inhibition of Synaptosomal ChAT by ChMAz. The nitrogen mustard analogue of choline, ChMAz, is a potent, selective and irreversible inhibitor of synaptosomal high-affinity choline carriers (23, 25-27) and as such has use as a chemical probe at the cholinergic nerve terminal. The interaction of ChMAz with ChAT in rat brain homogenates showed this nitrogen mustard analogue to be a rather weak inhibitor of the enzyme (24). However, subsequent work showed ChMAz to be much more potent as an inhibitor of intrasynaptosomal ChAT observed when rat forebrain synaptosomes were incubated with micromolar concentrations of the compound (26). This time-dependent, irreversible inhibition of synaptosomal ChAT appeared to be associated with alkylation of high-affinity choline carriers; when synaptosomes were incubated in the presence of ChMAz in a sodium-free medium there was no loss of synaptosomal enzyme activity. Similarly, incubation of synaptosomes with monoethylcholine or hemicholinium-3 (HC-3) did not alter ChAT activity. HC-3 protected against the inhibitory effect of ChMAz. Calculations based on the osmotically sensitive volume of the synaptosomes determined the maximum intrasynaptosome concentration of ChMAz to be 20 μM after 7 min, if it were transported at the same velocity as choline (23, 26). This concentration of ChMAz did not alter ChAT activity in rat brain homogenates (24) and it should be stressed that it is unlikely that ChMAz would be transported into synaptosomes at the same velocity as choline, if it is transported at all (23). These results suggest that some fraction of synaptosomal ChAT was depressed in association with alkylation of the high-affinity choline carrier, and subsequent investigations have been directed towards characterizing this inhibition.

Interaction of ChMAz with High-Affinity Choline Carriers. As the first step, we have investigated more fully the irreversible interaction of ChMAz with the high-affinity choline carrier in rat forebrain synaptosomes (27). Based on the β-chloroethylamine structure of ChMAz, it was predicted that inhibitory reactions involving ChMAz would be inherently time-dependent. This prediction was verified experimentally with the observation that both the kinetic characteristics and degree of inhibition of [3H]-choline uptake by ChMAz were a function of time that synaptosomes were exposed to the mustard analogue. In terms of potency, a graded blockade with an overall increase in inhibitory strength of about 12-fold was obtained by increasing the exposure of synaptosomes to ChMAz from 1 to 30 minutes before monitoring high-affinity transport activity at 1 μM [3H]choline. The kinetic characteristics of the inhibition of choline transport also changed relative to the preincubation time of synaptosomes with ChMAz. Increasing the time of exposure of synaptosomes to ChMAz from 1 to 30 minutes before measuring choline transport activity caused the apparent competitive inhibition kinetics observed initially to shift towards noncompetitive inhibition, as indicated by a decreased maximum velocity (V_{max}) of transport.

ChMAz would seem to bind to and inactivate high-affinity choline carriers rapidly. Since it was important to determine how much ChMAz was transported into the nerve terminals to interact with cytoplasmic ChAT, and since we did not have isotopically-labelled ChMAz to facilitate these measurements, we have tried to measure this indirectly by calculating the rate constant for the irreversible bond formation between ChMAz and high-affinity choline carriers. If this bond forms rapidly then most bindings would result in irreversible inactivation of the carriers and very little analogue, and consequently very little choline, would be transported into the nerve terminals.

In determining the rate of alkylation of high-affinity choline carriers, it must be assumed that inhibition of the carrier sites by ChMAz occurs exponentially, at least initially, as a function of time and concentration of ChMAz. The percentage of [3H]-choline uptake in ChMAz-treated synaptosomes compared to controls provides a measure of the uninhibited carrier fraction; the rate of decline of the number of uninhibited choline carriers measured at various ChMAz concentrations allows the calculation of K_3, the rate of formation of the alkylated carriers. As illustrated in Figure 1(a) (inset), a linear relationship was not obtained by plotting the uninhibited carrier fraction semilogarithmically as a function of inhibitor preincubation time over the range 1 to 30 minutes. Although this could imply other inhibitory mechanisms, it could also indicate that the exponential phase of the inhibitory process was extremely fast. As demonstrated in Figure 1(a), exponential inhibition of high-affinity choline carriers was obtained following short (15-60 sec) preincubations of synaptosomes with ChMAz. The rate constants (k_{app}) at each concentration, obtained from the slopes of the lines in Figure 1(a), allow the determination of K_3 by replotting as a function of inhibitor concentration in Figure 1(b). The intercept of this plot shows K_3 to be 5×10^{-2} sec^{-1}, however, since the line intercepts on the y-axis rather than passing through the origin it could indicate that reversible complexes were formed as well as the covalently-bound carriers (33).

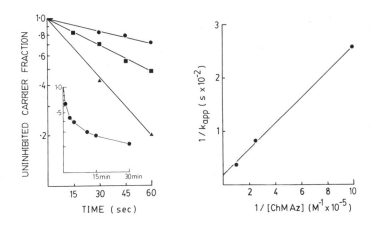

Figure 1. The rate of decline of uninhibited synaptosomal high-affinity choline carriers as a function of ChMAz concentration. A: (Left panel) Synaptosomes were incubated for up to 60 sec with 1 (●), 4 (■) or 10 (▲) μM ChMAz, then washed with inhibitor-free buffer solution before choline transport (1 μM) was measured. The inset shows the nonexponential rate of inhibition of high-affinity choline carriers produced by incubation of synaptosomes with 2 μM ChMAz for up to 30 min. B: (Right panel) Plot of $1/k_{app}$ against $1/[ChMAz]$. The values of k_{app} were obtained from the slopes of the lines in part A. The line represents a regression line. Data published previously, Rylett and Colhoun (27).

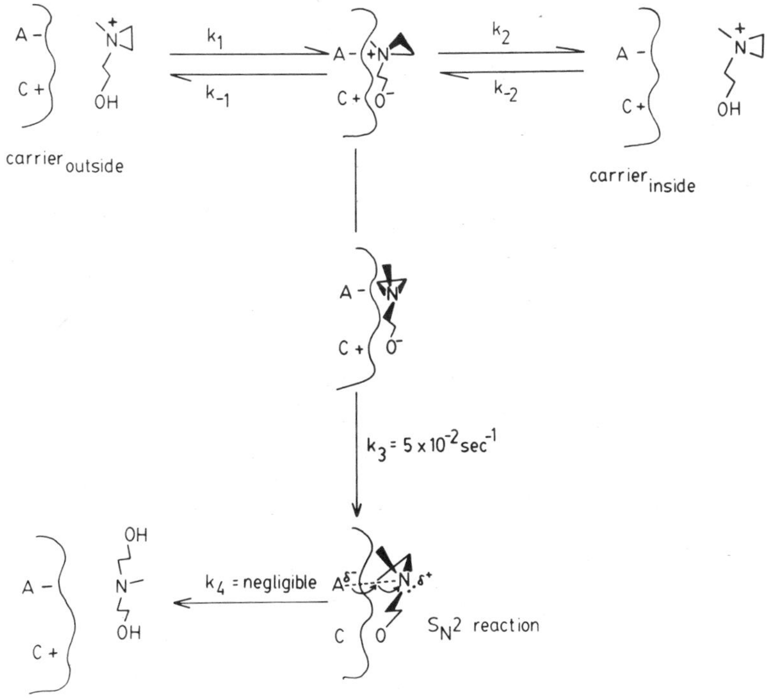

Figure 2. Schematic of proposed interaction of ChMAz with synaptosomal high-affinity choline carriers. ChMAz binds initially by ionic forces. Rearrangement of the molecule could result in irreversible bond formation with a rate constant of 5×10^{-2} sec^{-1}, and thus inactivation of the carrier. Some bindings will result in translocation and dissociation of ChMAz from the carrier at the inner membrane surface. Hydrolysis of the alkylated carrier will occur at some very slow rate (k_3) to yield the free carrier and methyl-diethanolamine. Published previously, Rylett and Colhoun (27).

Reactions between ChMAz and the synaptosomal high-affinity choline carrier, the simplest case, are summarized in Figure 2. It would appear that ChMAz binds to and inactivates high-affinity choline transport rapidly, thereby precluding the transport of the mustard analogue into the nerve terminal. Initially, as a result of a small number of reversible competitive bindings, some translocation of the analogue into the nerve ending might occur, however, it is predicted that the concentration would not be sufficient to mediate significant inhibition of ChAT.

Interaction of ChMAz with Partially Purified ChAT. The interaction of ChMAz with ChAT in rat brain homogenates showed this nitrogen mustard analogue of choline to be a rather weak inhibitor of the enzyme; the IC_{50} was determined to be 1.8 mM after 10 min incubation of the enzyme with the inhibitor (24). Under the conditions of these experiments very little irreversible inhibition of ChAT by ChMAz was detected. The ability of ChMAz to irreversibly inhibit ChAT was tested on a partially purified preparation of the enzyme from rat brain. Comparable inhibition of partially purified ChAT by ChMAz was observed; IC_{50} was found to be 0.7 mM following 30 min incubation with the enzyme. It was determined, as illustrated in Figure 3, that ChMAz could mediate a time-dependent, irreversible alkylation of partially purified ChAT. Reversal of the inhibition of ChAT activity by ChMAz was measured following dialysis of the enzyme-inhibitor complex for 18 hours, and it was found that the degree of irreversible inhibition was a function of the incubation time of enzyme with inhibitor. It was concluded that ChMAz could irreversibly bind to ChAT but that this was significant only with high concentrations of the inhibitor and after prolonged exposure times; these findings do not seem to explain the irreversible inhibition of synaptosomal ChAT produced when rat brain synaptosomes were incubated with micromolar concentrations of ChMAz.

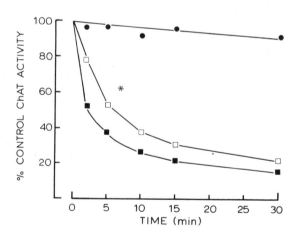

Figure 3. Time-study on inhibition of partially purified ChAT by ChMAz. ChAT activity was tested immediately after a 2 to 30 min incubation with ChMAz at $37^{\circ}C$: 2 mM (■) or 0.02 mM (●) ChMAz, or after 18 hr dialysis of the enzyme-inhibitor complex (2 mM ChMAz, □) to test for reversibility of the inhibition. Inhibition of ChAT by ChMAz was largely reversible after 2 min incubation, however, was irreversible by 30 min exposure. Points represent means of at least four determinations and S.E.M. was about 5% on all points.

Comparative Studies.   Differences between rat and guinea-pig with
respect to molecular forms of ChAT in brain have been reported by
Fonnum and colleagues (10, 11, 18), indicating that there could be
a potential for differences in mechanisms for coupling between the
transport and acetylation of choline in these two species.   Comparative
studies were performed to assess the utilization of choline transported
by synaptosomal sodium- dependent, high-affinity choline carriers
for the synthesis of ACh; it was determined that a significantly
higher percentage of [3H]-choline transported into rat forebrain
synaptosomes was acetylated immediately compared to that of guinea-pig
(72% in rat, 58% in guinea-pig) (28).

Studies were performed to determine whether inhibition of synaptosomal ChAT was produced by incubating guinea-pig brain synaptosomes
with ChMAz, comparable to that observed with rat brain synaptosomes.
Time-dependent inhibition of high-affinity choline transport was
produced in guinea-pig forebrain synaptosomes which was not different
from that produced in rat forebrain synaptosomes (Figure 4).   A
striking difference was observed, however, in the activity of synaptosomal ChAT following exposure of the synaptosomes to the mustard
analogue.   As illustrated in Figure 5, very little inhibition of
ChAT activity was measured in guinea-pig brain synaptosomes compared
to those from rat brain.   That this difference could reflect differing
subcellular localizations of ChAT, and indeed, different relativities
with respect to coupling with choline carriers is speculative and
currently being investigated.

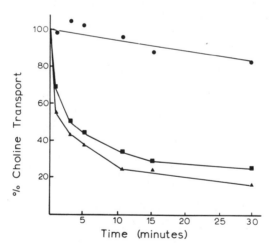

Figure 4.   Time-dependent inhibition of synaptosomal high-affinity
choline transport by 2 μM ChMAz.   No difference was found for inhibition of uptake of 1 μM [3H]-choline into forebrain synaptosomes
from rat (▲) or guinea-pig (■); control transport (●).   Points
represent mean of at least four determinations in duplicate; S.E.M.
was less than 7% on all points.

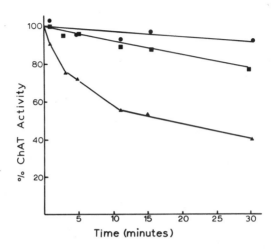

Figure 5. Time-dependent inhibition of synaptosomal ChAT activity by ChMAz. An irreversible decrease in ChAT was found when rat fore-brain (▲) synaptosomes were incubated with 2 μM ChMAz; far less inhibition of synaptosomal enzyme activity was observed when guinea-pig forebrain (■) synaptosomes were tested; control (●). ChAT was assayed radioenzymatically following lysis of the synaptosomes. Points represent the mean of at least four separate determinations, in duplicate, with S.E.M. less than 7% on all points.

In preliminary studies, we have observed that about 22% of the synaptosomal content of ChAT is membrane-bound in rat brain, whereas only minimal enzyme activity appears to be in a bound form in guinea-pig brain synaptosomal membranes. Benishin and Carroll (8) have reported recently that as much as 30% of rat brain synap-tosomal ChAT may be nonionically bound to membrane. In ChMAz-treated rat brain synaptosomes, the irreversible inhibition of ChAT appears to be associated with the membrane-bound form, as well as soluble forms of the enzyme with a pI (isoelectric point) greater than 8.0.

SUMMARY

From the studies reported using ChMAz as a chemical probe, it would appear that some relationship between high-affinity choline carriers and some fraction of ChAT exists in rat brain synaptosomes. Coupling between the high-affinity choline carriers and ChAT could be of prime functional significance in the regulation of ACh syn-thesis. However, the molecular mechanisms underlying this proposed link have not been defined. Future investigations will focus upon further characterization of the ChMAz-mediated inhibiton of synap-tosomal ChAT, and its relevance with regard to ACh synthesis.

Acknowledgements. The author would like to thank Dr. E.H. Colhoun for his assistance and helpful discussion with this work. These studies were supported by a grant from the Medical Research Council of Canada.

REFERENCES

1. Atterwill, C.K. and Prince, A.K. (1977): Brit. J. Pharmacol. 61:111P-112P.
2. Atterwill, C.K. and Prince, A.K. (1978): J. Neurochem. 31:732-740.
3. Atterwill, C.K. and Prince, A.K. (1978): Brit. J. Pharmacol. 62:398P.
4. Atweh, S., Simon, J.R. and Kuhar, M.J. (1975): Life Sci. 17: 1535-1544.
5. Barker, L.A. and Mittag, T.W. (1975): J. Pharmacol. Exp. Ther. 192:86-94.
6. Barker, L.A. and Mittag, T.W. (1976): Molec. Pharmacol. 25: 1931-1933.
7. Barker, L.A., Mittag, T.W. and Krespan, B. (1978): In Cholinergic Mechanisms and Psychopharmacology (ed) D.J. Jenden, New York, Plenum Press, pp. 465-481.
8. Benishin, C.G. and Carroll, P.T. (1983): J. Neurochem. 41:1030-1039.
9. Eisentadt, M.L., Treistman, S.N. and Schwartz, J.H. (1975): J. Gen. Physiol. 65:275-291.
10. Fonnum, F. and Malthe-Sorenssen, D. (1972): In Biochemical and Pharmacological Mechanisms Underlying Behaviour (eds) P.B. Bradley and R.W. Brimblecombe, Prog. Brain Res. 36:13-27.
11. Fonnum, F. and Malthe-Sorenssen, D. (1973): J. Neurochem. 201351-1359.
12. Freeman, J.J. and Jenden, D.J. (1976): Life Sci. 19:949-962.
13. Jope, R.S. (1979): Brain Res. Rev. 1:313-344.
14. Jenden, D.J., Jope, R.S. and Weiler, M.H. (1976): Science 194: 635-637.
15. Jope, R.S. and Jenden, D.J. (1977): Life Sci. 20:1384-1392.
16. Kessler, P.D. and Marchbanks, R.M. (1979): Nature 279:542-544.
17. Kuczenski, R., Segal, D.S. and Mandell, A.J. (1975): J. Neurochem. 24:39-45.
18. Malthe-Sorenssen, D. (1976): J. Neurochem. 26:861-865.
19. Malthe-Sorenssen, D. and Fonnum, F. (1972): Biochem. J. 127: 229-236.
20. Marchbanks, R.M. and Kessler, P.D. (1982): J. Neurochem. 279: 542-544.
21. Masters, C.J. (1978): Trends Biochem. Sci. 3:206-209.
22. Polsky, R. and Shuster, L. (1976): Biochim. Biophys. Acta. 445:25-42.
23. Rylett, R.J. (1980): Ph.D. Thesis, University of Western Ontario, London, Canada. Actions of Choline Mustard Aziridinium Ion at the Cholinergic Nerve Terminal.
24. Rylett, R.J. and Colhoun, E.H. (1979): J. Neurochem. 32:553-558.

25. Rylett, R.J. and Colhoun, E.H. (1980): J. Neurochem. $\underline{34}$:713-719.

26. Rylett, R.J. and Colhoun, E.H. (1980): Life Sci. $\underline{26}$:909-914.

27. Rylett, R.J. and Colhoun, E.H. (1984): J. Neurochem. $\underline{43}$:787-795.

28. Rylett, R.J., Carlton, T.J. and Colhoun, E.H. (1983): J. Soc. Neurochem. $\underline{9}$:970.

29. Simon, J.R. and Kuhar, M.J. (1975): Nature $\underline{255}$:162-163.

30. Smith, C.P. and Carroll, P.T. (1980): Brain Res. $\underline{185}$:363-371.

31. Suszkiw, J.B. and Pilar, G.B. (1976): J. Neurochem. $\underline{26}$:1133-1138.

32. Weiler, M.H., Gundersen, C.B. and Jenden, D.J. (1981): J.Neurochem. $\underline{36}$:1802-1812.

33. Wilson, I.A. and Kitz, R. (1962): J. Biol. Chem. $\underline{237}$:3245-3249.

34. Yamamura, H.I. and Snyder, S.H. (1972): Science $\underline{178}$:626-628.

35. Yamamura, H.I. and Snyder, S.H. (1973): J. Neurochem. $\underline{21}$:1355-1374.

TARGET TISSUE INFLUENCES ON CHOLINERGIC DEVELOPMENT

OF PARASYMPATHETIC MOTOR NEURONS

J.B. Tuttle[*] and G. Pilar

Physiology Section
Biological Sciences Group
The University of Connecticut
Storrs, Connecticut 06268   U.S.A.

[*]Current Address:
Department of Physiology
University of Virginia School of Medicine
Charlottesville, Virginia 22908   U.S.A.

INTRODUCTION

The normal function of neurons in the nervous system depends upon the orderly formation and maintenance of appropriate connections with other neurons and with non-neural target tissues. Knowledge of these processes is thus basic to our understanding of the biology and physiology of the nervous system, and the sequelae to insult or dysfunction among the interacting elements.

In this chapter we address one aspect of this general problem -specifically: having formed an appropriate synapse, how does this interaction influence the subsequent program of neuronal differentiation and survival? At some point in normal development, each neuron makes contact with the appropriate target element. Although most neurons seem to be capable of relatively autonomous development up until synaptogenesis begins, an obvious dependence upon the target element becomes manifest at this time (6, 20).

METHODS

We have employed in our studies neurons from the avian ciliary ganglion and their terminals in the iris. All of the techniques and methods used in this report have been published previously (1, 8, 9, 17, 18) and thus will not be detailed here.

RESULTS

The parasympathetic ciliary ganglion innervates the ciliary and iris muscles, and the choroidal coat muscles of the eye. In vivo, both the synapses in the ganglion and those in the periphery are cholinergic (9). Use of this preparation has the very real advantage that the important developmental landmarks of synaptogenesis, cell death, and the initiation of ganglionic and neuromuscular transmission are well known (5, 8). Some of the relevant aspects of this information are summarized in Figure 1. The ability of neuronal terminals to rapidly synthesize acetylcholine (ACh) under conditions of demand during tonic contracture is basic to their function in this peripheral effector system. Thus, study of the iris terminals during development revealed that neuromuscular transmission in the embryonic iris rapidly fatigued during a tetanus at 20 Hz, whereas soon after functional demands are made upon the system at hatching, this transmission becomes resistant to fatigue (Fig. 1, hatched bars). The failure of transmission at the embryonic iris was shown to be due to a failure to sustain transmitter release at adequate levels, implying an inability of the ACh synthetic system to cope with the demands of sustained transmission (8).

Concomitantly in time with the shift from an embryonic, fatiguable junction to the mature, more secure transmission, there is a large change in the capacity for ACh synthesis measured using radiolabeled substrate (Figure 1) (8). Only at this point in development does one detect an increase in the amount of [^{3}H]ACh synthesized from [^{3}H]Choline ([^{3}H]Ch) in response to a pre-conditioning depolarization (19). This ability to increase transmitter synthesis in response to release demand, appearing cotemporally with mature transmission, suggests causality, and appears to be a useful biochemical index of the mature physiological state.

A study of the ultrastructure of the ciliary ganglion nerve terminals in the iris over this embryonic period might have been expected to reveal a discrete morphological correlate of the shift to adult transmission. However, none was found in the initial survey (1). Instead, a progressive growth and filling of the terminals with clear synaptic vesicles occurred. Either a morphological correlate does not exist, or a finer analysis of terminal structure is needed for its demonstration.

These studies of development in vivo have provided a description of the steps taking place during maturation of a neuromuscular junction. There is a seven-day period between the initiation of stimulus-coupled transmitter release and the formation of a reliable synapse. This implies that the establishment of transmission per se is not sufficient for function at the adult level and that significant developmental events may occur during this lag period. Attainment of secure transmission coincided in time with the appearance of a

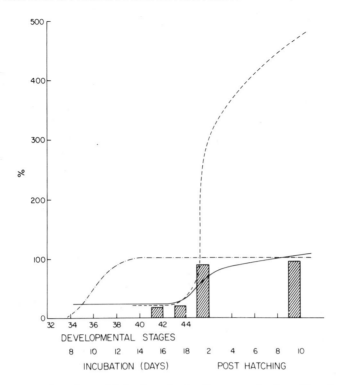

Figure 1. **Functional and biochemical changes during the development of iris neuromuscular junctions.** Data are expressed as percentage of adult values. Neuromuscular transmission (-•-•). Basal ACh synthesis (———). Depolarization stimulated ACh synthesis (----). During electrical stimulation of the ciliary nerve at ST41 and 44, the electrical response of the iris declines by more than 80% (hatched bars) accompanied by mechanical fatigue. The bars represent the amplitude (total area above baseline) of the electrical response 30 sec after the initiation of the stimulation over the control response, expressed as percent. The arrow indicates hatching and the onset of pupillary reflex activity. Near hatching, basal ACh synthesis increases and a conditioning depolarization elicits a large increase in synthesis. Simultaneously, neuromuscular transmission becomes more resistant to fatigue during repetitive activity. Reprinted with permission from (9).

biochemical marker (the increase in ACh synthesis in response to prolonged depolarization), and with the onset of the functional demands associated with a behaving animal. Thus, a timetable of events was clearly defined, but an investigation of the neuron-target interdependency necessitated the use of neurons in cell culture.

In the isolated environment of cell culture, the questions of mechanism are more approachable. Paramount among these is the determination of how the target could convey developmentally relevant influences back to the parent neuron. Studies in other systems and in vivo in the ciliary ganglion have highlighted the potential importance of three types of interaction modes (6). Nerve Growth Factor (NGF) is well known as the first of what may be a class of relatively low molecular weight molecules that support the survival of peripheral neurons in cell culture, and probably influence the development of neurons in vivo (20). Thus one way for the target to influence neuronal maturation is to provide a supply of soluble proteins that are necessary for continued growth. These are then transported from the periphery to the neuronal cell body, where they are available to influence the synthesis of specific proteins or regulate metabolic activity.

Activity through a neuronal pathway has also been shown to support aspects of maturation, and to reinforce initial patterns of connectivity (3, 12, 15). The target may then provide support through this mechanism, by fostering or allowing effective transmission to occur, and thus the developmental consequences of activity to be manifest.

A third possibility is that the contact of neuronal terminals with the membrane of an appropriate target serves as an important event. This would imply that some form of molecular recognition or matching was involved, and that the adequate triggering of this process was necessary and sufficient for the continuation or acceleration of maturation. The reacting neuronal membrane might undergo alterations that not only served to maintain adhesion between the two elements, but also to trigger enzymatic induction or receptor insertion. A specific intracellular messenger may be released or a specific site in the membrane activated. It is conceivable that both specific cell-cell adhesion and synapse formation between neuron and target are mediated by the same process. The initial step involved in these events may be a contact-recognition phenomenon with subsequent consequences for both the neuron and target. This hypothesis embodies the original suggestion Weiss (21) put forth to explain morphogenetic changes during development (see also 11).

Ciliary Ganglion Neuron Cell Culture. Neurons from the chick ciliary ganglion can be removed from the developing embryo prior to neuro-muscular synaptogenesis, dissociated to a suspension of single cells, and successfully maintained in long-term culture (7, 17). All of the ciliary ganglion neurons will survive either in the presence or absence of target tissues or other cell types if a sufficient supply of trophic survival support is provided in the culture medium. It should be recalled that in vivo, up to one-half of the neuronal population ordinarily does not survive through the cell death period (5, 6). Thus, one can release the neurons in culture from their

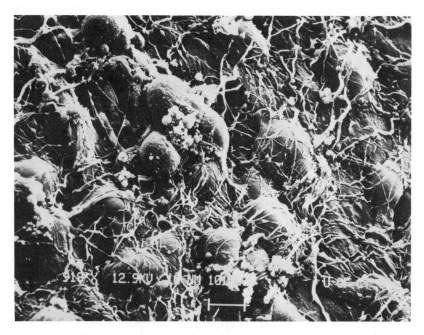

Figure 2. Neurons from the ciliary ganglion in cell culture. Scanning electron microscopy. Calibration bar = 10 μm.

in vivo requirement for target interaction for survival, and subsequently examine developmental status both in the presence and absence of target tissue.

Figure 2 shows such a culture of ganglion cells, prepared for scanning electron microscopy. This image was taken from an area of a culture with a relatively high density of neurons. The dissociated neurons have regenerated new processes, and give all the appearances of having adapted well to the culture environment. Transmission electron microscopy of similar cultures revealed that many of the swellings or boutons formed on the cell bodies of the cultured neurons have all the signs expected for synaptic contacts - including close membrane apposition, membrane densities and aggregations of vesicles (9). However, for reasons detailed below, intracellular recordings revealed virtually no interneuronal synaptic activity. The electrophysiological examination of these cultured neurons did show that they retained many of the characteristics of the ganglionic neurons in vivo, including the details of action potential waveform and time course, afterpotentials, and ability to discharge trains of action potentials in response to sustained depolarization (1).

The cultured neurons also are competent with respect to transmitter synthesis. When incubated in 1 μM [^{3}H]Ch, the neurons take up the tritiated precursor and synthesize labeled ACh (Figure 3). When extracellular Na$^+$ is replaced by Li$^+$ during this incubation, a 60% reduction in synthesis is observed. The partial Na$^+$ dependence of ACh synthesis suggests that much of the labeled substrate available for [^{3}H]ACh synthesis is taken up by the neuron via the Na$^+$-dependent, high-affinity choline transport system that has been described for these cholinergic neurons in vivo (14), and in other cholinergic systems (4). The ability to synthesize ACh is not surprising, as the cultured neurons have been shown to develop substantial levels of choline acetyltransferase (ChAT) activity (18).

The mere competence of the cultured neurons to synthesize ACh, however, does not address our original question, because the ciliary ganglion neurons can achieve this level of function without target interaction. We previously identified a useful index of transmitter synthetic maturation - the ability to respond to depolarization with an increase in synthesis (8). In culture, neurons without target muscle do not increase the amount of ACh synthesized following a depolarization, if they are tested at less than one week in vitro (Figure 3). However, sister cultures plated onto a layer of pre-fused myotubes are capable of a substantial increase in synthesis (Figure 3, C and D). In addition, this responsiveness of synthetic capacity

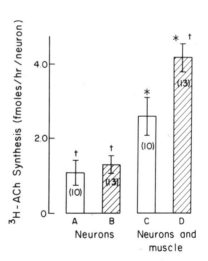

Figure 3. Cultures of ciliary ganglion neurons plated alone (A and B) and in co-culture with myotubes (C, D). [^{3}H]ACh synthesis was assayed in both situations. B and D followed a preincubation in 55 mM[K$^+$]. (*) Differ from neurons alone (A) at P = 0.01; (†) Differ from (C) at P = 0.05. Reprinted with permission from (18).

develops in the neuron-muscle co-cultures at no more than four days
in vitro (2). Thus, the presence of the target tissue has an ac-
celerating influence on the maturation of transmitter synthetic
machinery. When tested at the same time (one week in vitro), the
neurons co-cultured with target also had double the ChAT activity
and approximately double the basal, unstimulated level of ACh synthesis
shown by their counterparts cultured without myotubes (18).

The specificity of this influence was initially examined by
testing fibroblasts and medium conditioned by muscle for a similar
effect on neuronal maturation. No effect of either was seen, suggesting that interaction with muscle was needed. The formation of
effective ciliary neuron-myotube junctions was examined by taking
intracellular recordings from myotubes 4 to 12 days after the neurons
were plated in the culture. At all times it was easy to observe
neuromuscular activity (1). Figure 4A shows one of these neuron-muscle
co-cultures taken with Hoffman-modulation contrast optics. After
plating, some neurons settle between muscle cells on the collagen
substrate, while many of the neurons attach directly to the myotube
surface. Spontaneous junctional potentials were observed in virtually
all myotubes sampled, and were not blocked by tetrodotoxin (TTX),
but were eliminated by curare (1). Thus, these results are a clear
indication that the ciliary ganglion neurons form active junctions
with this muscle, and that ACh is mediating the neuromuscular transmission.

The question of mechanism has not yet been fully addressed,
even though several points are made by the data presented. First,
synapse formation per se is not sufficient for the acceleration of
synthetic capacity. Neurons alone in culture form many interneuronal
synapses but these are both silent and trophically ineffective.
Second, although the conditioned medium studies are not conclusive,
a soluble trophic protein excreted by muscle in high concentration
apparently is not a likely candidate. A final series of experiments
on ACh synthesis seems to rule out muscle activity as the mediative
mechanism. Neurons were plated onto the membrane fragments of myotubes
- fragments that remained after lysis of myotube cultures in distilled
water, and washing of the fragments for at least two hours in distilled
water (16). This treatment leaves no live myotubes (see Figure 4B),
and thus precludes the active transmission that should be required
for activity to play a role in this aspect of maturation. When
the capacity to increase ACh synthesis in response to a pre-conditioning depolarization is measured for neurons on the lysed myotube
substrate, a clear acceleration is seen. At 4-5 days in vitro,
little incremental increase in synthesis occurs. However, by 7
days, a time when neurons plated alone show no stimulation, the
neurons growing on the myotube membranes double ACh synthesis after
the depolarization (Table 1).

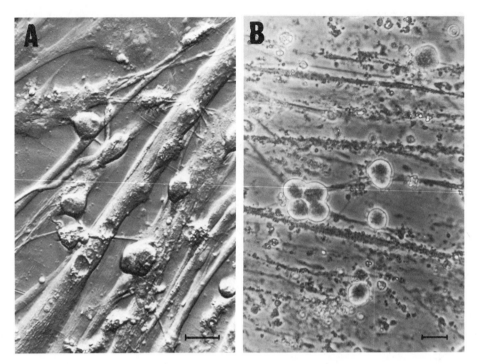

Figure 4. (A) Photomicrograph of a live co-colture of ciliary ganglion neurons and pectoral myotubes taken with Hoffman-modulation contrast optics. (B) Phase contrast micrograph of neuronal culture on lysed myotube membrane remnants. Calibration bar = 10 μm.

The ability of the target membrane to at least partially duplicate the interaction with a live target cell is clear for the case of the maturation of transmitter synthetic machinery toward adult status. It should be recalled that these experiments were conducted on neurons grown in medium that contained levels of trophic survival protein saturating for survival, and that under these conditions, all of the neurons plated have been shown to survive for the duration of these experiments (16). We have not as yet determined if the cultured neurons form terminals on the myotube membrane remnants, and thus if the maturation of transmitter synthetic capacity depends absolutely upon terminal junction structure.

Chemosensitivity of Ciliary Ganglion Neurons in Cell Culture. The synthesis of adequate amounts of neurotransmitter is but one of the prerequisites for normal adult function of the neuron. The neuron must also maintain adequate receptor sensitivity to its input transmitter in order to allow the competent transfer of tonic and reflex activity to the effector organs. As was mentioned above,

Table 1

Effect of Interaction with Myotube Membrane Remnants
on Neuronal Sodium-Dependent [^{3}H]ACh Synthesis

	Days in Culture	
	4-5 days	7 days
Neurons Alone:		
A. Basal Na^+-dependent ACh Synthesis	0.42 ± 0.14	0.60 ± 0.08
B. K^+ Stimulated Na^+-dependent ACh Synthesis	0.32 ± 0.10	0.71 ± 0.19
Neurons on Membrane Remnants:		
A. Basal Na^+-dependent ACh Synthesis	0.46 ± 0.06	0.86 ± 0.31
B. K^+ Stimulated Na^+-dependent ACh Synthesis	0.59 ± 0.16	1.76 ± 0.19

ACh synthesis was measured as tritium-labeled ACh formed during incubation in [^{3}H]choline. Potassium stimulation refers to a pre-incubation in osmolarity-controlled saline with 55mM K^+. Values are mean ± SE. For details see (18).

interneuronal synapses formed in the cultures of neurons plated without myotubes were not transmitting, that is, electrophysiological evidence of interneuronal transmission was not found whereas untrastructural evidence of synaptic contacts was seen in all cases examined (1). Because the ability of the neurons to synthesize ACh has already been demonstrated (vide infra), and other experiments showed an adequate release of transmitter in response to depolarization via the Ca^{++}-dependent mechanism (18), the site of transmission failure at these interneuronal synapses should be postsynaptic. To test this, intracellular recordings were taken from cultured neurons during both ionotophoretic and pressure ejection of ACh, and the cellular responses measured.

Using inotophoretic application of 3M ACh from high resistance (>100 M Ω) pipets, neurons were tested for responsiveness to ACh at three different somal sites per cell, and with at least three different current strengths per site. This allowed quantitation of the response in terms of mV peak depolarization per nanocoulomb charge delivered to be calculated as slopes from plots of response

Table 2

Target Tissue Maintenance of Neuronal Sensitivity to ACh

Culture Condition	Percentage of Neurons Responding at:		
	1 day	6-8 days	>12 days
Collagen - ST 32 ganglia	100(26)	21(40)	17(37)
- ST 37 ganglia	100(17)	0(32)	
Myotube Co-culture		100(26)	100(30)
Myotube Membranes		88(51)	0(7)
Fibroblast Membranes		0(44)	
Muscle Conditioned Medium		0(15)	
Fibroblast Conditioned Medium		0(7)	

Number of neurons tested in parenthesis.   For experimental and culture
details, see (1).

amplitudes.   The minimum response detectable was approximately 1
mV/nC; this sensitivity is 2 to 3 orders of magnitude less than
that found on parasympathetic neurons in vivo (10).   Thus, our method
of detecting and quantitating chemosensitivity allows us to conclude
with certainty whether the neurons possess, or have lost sensitivity,
at physiologically meaningful levels.

Table 2 shows that the neurons cultured without myotube targets
lost ACh sensitivity over the first week in culture.   If neurons
from ST. 32 embryoes were cultured, only about 20% of the neurons
retained any response to ACh at 7 days in vitro, and no neurons of
a large sample were found that were highly sensitive (>100mV/nC).
By contrast, all of the neurons tested that had been grown on live
myotubes retained sensitivity, and many of them approached the 800-
1000mV/nC reported for parasympathetic intraseptal neurons in the
frog (10).   Thus, interaction with a suitable target tissue had a
dramatic influence on the retention of another important physiological
property for normal neuronal function.

Table 2 also details some initial investigation into the inter-
active mechanism in this response.   In these cases, neurons from
older ganglia were cultured, as it was found that neurons taken
from embryoes at a slightly later state of development lost ACh
sensitivity completely in 4 or 5 days in culture.   Medium conditioned
over muscle was without effect, as was medium condition over embryonic
chick skin fibroblasts.   By contrast, culture on the lysed myotube
membrane remnants had a pronounced effect.   Almost all of the neurons
retained sensitivity at high levels for one week.   The subsequent
loss of sensitivity paralleled the loss of membrane remnants from

the culture substrate over time. That the interaction is specific for muscle is suggested by the finding that lysed fibroblasts do not maintain sensitivity.

SUMMARY AND CONCLUSIONS

It is hard to envision a major role for muscle and neuronal activity [all cultured ciliary ganglion neurons are quiescent (1)] or for a soluble trophic protein in the maintenance of ACh sensitivity by ciliary ganglion neurons under the culture conditions we have used. A live target cell is not necessary for the retention of transmitter sensitivity. Instead, the retention of this important functional capability by the neurons requires a specific interaction with either the muscle cell or membrane remnants of the myotubes. Because the membrane remnants are also effective, the operative molecule may be an element of the extracellular matrix, basal lamina, a membrane molecule or an intracellular molecule exposed by the lysis. Cell-to-cell contact interaction is a mechanism proven to be involved in cell and tissue differentiation during embryological development (13). It is an intriguing possibility that a similar mechanism may also be involved in the selective contact between developing neurons and their targets. Contactmediated interactions thus may have significance to both the very early and the later phases of neurodevelopment as they would help to establish not only paths of tissue differentiation, but also to foster specific connections and to support neuronal maturation.

Acknowledgements. The authors would like to acknowledge the participation of the following colleagues in aspects of this research: Dr. K. Vaca and D.B. Gray (ACh synthesis) and Dr. G. Crean (Electron microscopy). Portions of the research discussed in the text were supported by grants from the NIH (NS 10338, Fellowship NS 05382), the Spinal Cord Injury Foundation, The University of Connecticut Research Foundation, the U.S. Army Research Office, and an NIH Career Development Award to JBT.

REFERENCES

1. Crean, G., Pilar, G., Tuttle, J. and Vaca, K. (1982): J. Physiol. 331:87-104.
2. Gray, D.B. and Tuttle, J.B. (1983): Soc. Neurosci. Abstr. 9(2): 846.
3. Harris, W.A. (1981): Ann. Rev. Physiol. 43:689-710.
4. Kuhar, M.J. and Murrin, L.C. (1978): J. Neurochem. 30:15-21.
5. Landmesser, L. and Pilar, G. (1974): J. Gen. Physiol. 73: 605-628.
6. Landmesser, L. and Pilar, G. (1978): Fed. Proc. 37:2016-2022.
7. Nishi, R. and Berg, D.K. (1979): Nature 277:232-234.
8. Pilar, G., Tuttle, J.B. and Vaca, K. (1981): J. Physiol. 321: 175-193.

9. Pilar, G. and Tuttle, J.B. (1982): In Progress in Cholinergic
    Biology: Model Cholinergic Synapses (eds) I. Hanin and A.M.
    Goldberg, Raven Press, New York, pp. 213-247.
10. Roper, S. (1976): J. Physiol. (London) 254:455-473.
11. Roth, S. (1973): Quarterly Rev. of Biol. 48:541-563.
12. Sanes, D.H. and Constantine-Paton, M. (1983): Science 221:
    1183-1185.
13. Saxen, L., Karkinen-Jaaskelainen, M., Lehtonen, E., Nordling,
    S. and Wartiovaara, J. (1976): In The Cell Surgace in Animal
    Embryogenesis and Development (eds) G. Poste and G.L. Nicolson,
    North-Holland Publishing Co., New York, pp. 331-407.
14. Suszkiw, J.B. and Pilar, G. (1976): J. Neurochem. 26:1133-1138.
15. Thompson, W. (1983): Nature 302:614-616.
16. Tuttle, J.B. (1983): Science 220:977-979.
17. Tuttle, J.B., Suszkiw, J. and Ard, M. (1980): Brain Res. 183:161-
    180.
18. Tuttle, J.B., Vaca, K. and Pilar, G. (1983): Dev. Biol. 97:
    255-263.
19. Vaca, K. and Pilar, G. (1979): J. Gen. Physiol. 73:605-628.
20. Varon, S. and Adler, R. (1980): In Current Topics in Develop-
    mental Biology, Vol. 16, Part II (eds) A.A. Moscona, R. Monroy
    and K. Hunt, Academic Press, New York, pp. 206-252.
21. Weiss, P. (1947): J. Biol. Med. 19:235-278.

INHIBITION OF ACETYLCHOLINE SYNTHESIS IN VITRO

J.J. O'Neill, P.H. Doukas, F.. Ricciardi,
B. Capacio, R. Leech and G.H. Sterling*

Department of Pharmacology
Temple University School of Medicine
3420 North Broad Street
Philadelphia, PA 19140

*Department of Pharmacology
Hahnemann University School of Medicine
Broad and Vine Streets
Philadelphia, PA 19102

INTRODUCTION

The concentration of acetylcholine (ACh) in brain is maintained within narrow limits even though the turnover rate is several fold that estimated for other neurotransmitters (16). The precise mechanisms regulating ACh metabolism are complex and there is considerable disagreement as to which factor(s) is rate-limiting to synthesis. Regulation of ACh levels can be separated into several categories, no one of which by itself is controlling but collectively can regulate ACh synthesis: 1) Choline acetyltransferase (ChAT) activity; 2) Availability of its precursors, choline or AcCoA and; 3) Indirectly, by changes in ACh release.

In order to better understand diseases that stem from deficiencies in cholinergic activity, reproducible in vitro and in vivo models displaying cholinergic hypofunction are desirable. This necessitates the availability of specific inhibitors. There are relatively few compounds which are specific inhibitors of high affinity choline transport (HAChT) or of ChAT, which enter the central nervous system from parenteral routes. Of the compounds available, few have been reported to significantly reduce ACh levels in the CNS after systemic administration (14).

Previous investigators have developed ChAT inhibitors sharing similar structural characteristics (See Fig. 1). The present work

891

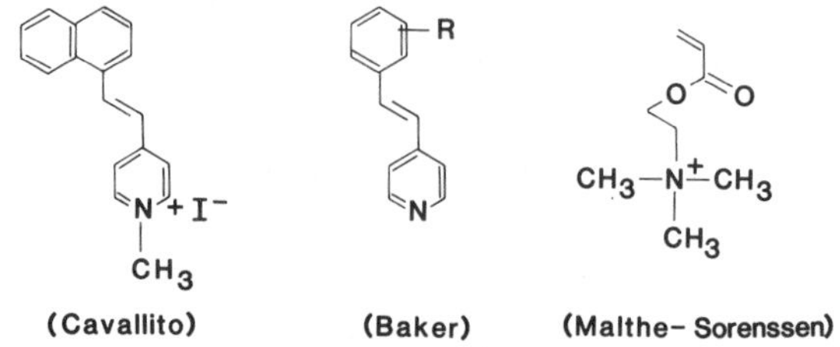

(Cavallito)   (Baker)   (Malthe– Sorenssen)

Figure 1. Inhibitors of choline acetyltransferase reported by Cavallito (Naphthylvinyl pyridinium methyl iodide), Baker (4-Stilbazoles),
and Malthe-Sorenssen (Acryloylcholine). 3',4'-Dichloro-4-stilbazole
was found to be the most potent within that series (1).

involves the design, synthesis and evaluation of quinuclidinyl compounds with structural features similar to those reported earlier
but with certain key differences:

* Quinuclidine is a highly lipophilic tertiary amine,
  able to penetrate the blood brain barrier.

* It is a strong base and is protonated at physiologic
  pH; hence it is capable of forming a charged species
  similar to the reference quaternary ammonium compounds.

* Quinuclidinyl compounds are known to have excellent
  structural characteristics for interaction with cholinergic receptors.

* 3-Quinuclidinol methyl iodide is reported to be an
  inhibitor of HAChT, and is a rigid analog of choline.

The initial phase of this work required structure activity
studies with _in vitro_ assay systems. Results obtained so far are
reported in this chapter.

METHODS

Enzyme Studies. A crude form of ChAT was used in these studies to
allow for non-specific binding, in order to provide better estimates
of potential _in vivo_ activity. Comparable results were also obtained
with Sigma ChAT which had twenty fold higher specific activity,
but this preparation was less stable in solution with storage.
The enzyme was prepared from 3 grams of dog brain caudate nucleus
and extracted with 6 ml of distilled water. The tissue was homogenized
and centrifuged at 1,500 rpm at 4°C for 10 minutes. The supernatant
suspension was diluted to 30 ml (10%) with distilled water, divided
in two equal volumes and stored at -90°C. All enzyme assays employed

a modification of the method of McCaman and Hunt (20). The complete assay system contained: sodium phosphate buffer (pH7.4), 50mM; NaCl, 300mM; EDTA, 0.1mM; choline chloride 1mM; eserine sulfate 2×10^{-4}M; bovine serum albumin, 0.05 percent (w/v); $MgCl_2$, 20mM; acetyl CoA 9×10^{-5}M (specific activity, 19mCi/mmole); and triton X-100, 0.25 percent. This was used when inhibitor A_{50} values were estimated. To determine K_I values, the choline content was varied from 0.25-2mM at constant inhibitor concentration. In a few studies, choline was held constant and acetyl-CoA concentration was varied, but with a constant amount of $[^{14}C]$-Acetyl CoA. At the level of enzyme activity employed, 20 minutes of incubation was chosen, as the activity becomes non-linear at longer times.

To assay routinely for enzyme inhibition, 45 μl of crude caudate extract plus 5 μl of inhibitor were added to 4x450mm glass centrifuge tubes, mixed by vortexing, and placed on ice until all pipetting was completed. Each tube received 20 μl of $[^{14}C]$-Acetylcoenzyme A which was diluted with the components described above as the assay system, and 5 μl of the enzyme-inhibitor mixture. The tubes were vortexed, centrifuged at 1,000 rpm at 4°C for 10 minutes; then incubated for 20 minutes at 37°C. Each tube next received 4 μl of 10mM unlabelled ACh as a carrier. Labelled ACh was separated from other labelled material on preconditioned Bio-Rad Ag 1-X8 anion exchange columns, by the method of Schrier and Shuster (26) as modified by White and Wu (31). The effluents and five 100 μl washes were collected directly into scintillation vials previously filled with 15 ml of Chaikoff's solution, and counted in a Packard Scintillation Counter, Model B2450.

Enzyme inhibition: All activities were converted to international units and expressed as micromoles/gram protein/hour. A software package was developed employing Eadie-Hofstee data plotting to calculate the apparent Km and Vmax (μmoles/g/hr) for ChAT in the absence and presence of inhibitor, and to estimate the inhibitor constant, K_I, in micromoles/liter. This also provided information on the nature of inhibition, i.e., competitive, non-competitive, uncompetitive or mixed-type. A separate program was developed for the TRS-80 to compute inhibitor potency, expressed as A_{50}-values in μmoles/liter.

Cerebral Cortex Slices. Acetylcholine synthesis and CO_2 production from labelled glucose were measured in cerebral cortex slices from male Sprague-Dawley rats, weighing 180-200 g. The animals were sacrificed by decapitation and cerebral cortex slices prepared with an O'Neill-Cummins tissue slicer (21). The incubation procedures were a modification of the method of Sterling and O'Neill, 1978 (29). Two slices were placed in a vial with 1 ml Krebs-Ringer bicarbonate buffer containing D-glucose (10 mM), 5 μCi(U-$[^{14}C]$) glucose, choline chloride (0.8 mM), eserine sulfate (0.4 mM), and potassium chloride (6 mM) w/wo various concentrations of inhibitor. Following aeration

for 10 minutes with $O_2:CO_2$ (95%:5%), samples were incubated at 37°C
for 60 minutes in a Dubnoff metabolic shaker bath at 150/minute.
Potassium chloride (60 mM, final conc.) was added at the 15 minute
mark.   A Kontes cup containing filter paper saturated with Hyamine
hydroxide (200 μl, 1M) was suspended in each vial to capture CO_2.
The incubation was terminated with perchloric acid (0.3 M final conc.),
and samples left on ice for 20 minutes to insure $[^{14}C]O_2$ uptake.
Hyamine was then taken for liquid scintillation counting. $[^3H]$-
Acetylcholine (@ 4,000 dpm) was added to each flask to measure recovery
from the ACh extraction procedure.

Tissue and media were homogenized together and the contents
centrifuged at 9,500 g for 10 minutes.   The supernatant was stored
at -70°C until the following day for extraction.   Acetylcholine
was extracted by ion-pair with dipicrylamine (10, 11, 22).

Calculations.   The nanomoles of ACh and CO_2 produced from $[^{14}C]$-glucose
were calculated from glucose specific activity as previously described
(2, 22).

$$\text{NMOL ACh from } ^{14}C\text{-glucose} = \frac{(\text{DPM ACh})\ (\text{NMOL Glucose})\ (3)}{(\text{DPM Glucose})}$$

$$\text{NMOL } CO_2 \text{ from } ^{14}C\text{-glucose} = \frac{(\text{DPM } CO_2)\ (\text{NMOL Glucose})}{(\text{DPM Glucose})}$$

The numerator was multiplied by 3 in calculating since there
are three times as many radioactive carbons per molecule of glucose
as in ACh.

The data were compared using analysis of variance, followed
by Dunnett's test with $P \leq 0.05$ as the level of significance.

Synthesis of 2-Benzylidene 3-Quinuclidinones.   These compounds were
prepared according to the procedure of Warawa et al. (30).   Quinu-
clidinone free base (0.01 M) and the appropriately substituted benzal-
dehyde (0.01 M) were dissolved in 20 mls. of absolute ethanol.   A
pellet of either NaOH or KOH was added and the mixture refluxed
with stirring for three hours after which it was allowed to stand
at room temperature overnight.   The precipitated solid was removed
by filtration and the filtrate reduced in volume under vacuum and
cooled to yield additional solid.   The combined solids were washed
with water and then allowed to air dry.   Hydrochlorides were prepared
in the usual manner and recrystallized from hot methanol and ether.
Physical constants and spectral data for these compounds are listed
in Table 1.

RESULTS

The studies described were designed to provide information concerning the influence variations in structure have on the inhibition of brain ChAT. As some measure of _in vitro_ potency, inhibitor studies were carried out in which one substrate concentration was varied over an appropriate range while holding the co-substrate constant and initial rates were measured. When choline was held constant and acetyl-CoA varied, vinyl pyridinium compounds produced non-competitive inhibition. In contrast, when choline concentration was varied in the presence of inhibitor, double reciprocal plots of velocity versus substrate concentration became non-linear at high (2mM) choline. Hersh and Peet (15) in a re-evaluation of the kinetic mechanisms of human placental ChAT reaction observed a similar effect employing "dead end" inhibitors. This led to the suggestion that the pattern seemed to change from non-competitive to competitive at high substrate concentration. In order to avoid this complication choline concentrations never exceeded 2 millimolar.

Table 1

2-Benzylidine-3-Quinuclidinones, Melting Points and Spectral Data

R	m.p.[1]	UV/Vis[2] max.	NMR[3]
H	170-173	298 400	1.80-2.20 (m,4H), 2.62 (t,1H), 2.87-3.23 (m,4H), 7.05 (s,1H), 7.27-8.13 (m, 5H).
p-Cl	195-197	307	1.85-2.25 (m,4H), 2.68 (t,1H), 2.91-3.30 (m,4H), 7.11 (s,1H), 7.48 (d,2H), 8.18 (d,(2H).
m-Cl	192-193	292	1.80-2.27 (m,4H), 2.63 (t,1H), 2.80-3.40 (m,4H), 7.00 (s, 1H), 7.17-8.42 (m,4H).
$3,4-Cl_2$	197	298	1.85-2.25 (m,4H), 2.68 (t,1H), 2.86-3.33 (m,4H), 6.92 (s,1H), 7.30-8.40 (m, 3H).
$3,5-Cl_2$	134-136	287	1.83-2.23 (m,4H), 2.63(t, 1H), 2.86-3.28 (m,4H), 6.86 (s,1H), 7.30 (t,1H), 8.05 (d,2H).
$p-OCH_3$	185-188	330	1.85-2.25 (m,4H), 2.68 (t,1H), 2.80-3.40 (m,4H), 3.93 (s,3H), 6.95 (d,2H), 7.05 (s,1H), 8.13 (d,2H).
$m-OCH_3$	163-166	298	1.81-2.23 (m,4H), 2.61 (s,1H), 2.90-3.25 (m,4H), 3.83 (s, 3H), 6.86-7.93 (m,4H), 7.06 (s,1H).
$p-N(CH_3)_2$	226-229	225,391 805	1.83-2.37 (m,4H), 2.68 (t,1H), 2.87-3.33 (m,4H), 3.03 (s,6H), 6.75 (d,2H), 7.07 (s,1H), 8.07 (d,2H).
$m-NO_2$	224-227	278	1.87-2.33 (m,4H), 2.67 (s.1H), 2.80-3147 (m,4H), 7.10 (s,1H), 7.33-8.47 (m,4H).
$m-CH_3$	117-120	298	1.81-2.25 (m,4H), 2.43 (s,3H), 2.63 (t,1H), 2.85- 3.28 (m,4H), 7.05 (s,1H), 7.30 (d,2H), 7.71-8.08 (m,2H).

1. Melting points of HCl salts; 2. Of free base in ethanol; 3.Chemical shift of free base in $CDCl_3$.

In planning for further studies in vivo, a measure of potency (A_{50}) would prove useful. An A_{50} value is the log of the concentration necessary to produce fifty percent inhibition of the enzyme at optimum choline and acetyl-CoA concentrations. For this purpose, a choline concentration $5xK_m$(2 mM) and an acetyl-CoA concentration approximately $5xK_m$(90 μM) were employed. All A_{50} and K_I values were computed employing software developed in our laboratories for this purpose. The values thus obtained for azulylvinyl-2- and azulylvinyl-4-pyridinium and their 3'-bromo-azulyl homologs are presented in Table 2, together with data obtained with the most potent of Cavallito's compounds, the N-hydroxyethyl derivative of Napthylvinyl-pyridine. Substitution of the naphthyl group by azulene did not enhance potency despite its tendency to form charge transfer complexes.

Table 2

A Comparison of Azulene Derivatives with N-Hydroxyethyl
Naphthyl Vinyl Pyridine and Acetylseco-HC3 as Inhibitors of ChAT

COMPOUND	Km+	K_I (μmol/l)	COMMENTS	INHIBITION
Control	0.5 - 1.7	---	Var Choline	
	17 - 62	---	Var AcCoA	
Compound I (N-Methyl Azulylvinyl-2-Pyridinium Iodide)	0.2 - 0.4 6.2	5.7 - 26 27.7 A_{50} = 4.4	Var Choline Var AcCoA Dose-Response*	Noncompetitive
Compound II (N-Methyl Azulylvinyl-4-Pyridinium Iodide)	0.13 2.15 - 2.34	1.1 0.8 - 2.8 A_{50} = 2.4-5.6	Var Choline Var AcCoA Dose-Response*	Uncompetitive
Compound III (N-Methyl-3'-Bromoazulylvinyl-2-Pyridinium Iodide)	0.18	13.7	Var Choline	Noncompetitive
Compound IV (N-Methyl-3'-Bromoazulylvinyl-2-Pyridinium Iodide)	0.52	176.1	Var Choline	Noncompetitive
N-HE-NVP	0.48 3.85	1.7 1.3 A_{50} = 0.9	Var Choline Var AcCoA Dose-Response*	Noncompetitive
AcetylsecoHC3	0.65	15.3 A_{50} = 2.8	Var Choline Dose-Response*	Uncompetitive

Concentrations: Choline = 1 mM; AcCoA = 113.3 μM (140 μM); Inhibitor = 25 μM - 0.025 μM.
+KMCh = mM; kmAcCoA = μM.

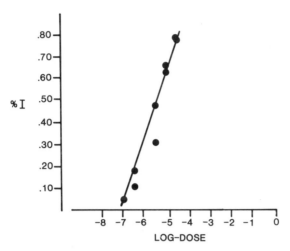

Figure 2.   Inhibition of ChAT as a function of the concentration of acetylseco-HC3 in the presence of sodium phosphate buffer (pH 7.4), 50 mM; NaCl, 300mM; EDTA, 0.1 mM; Choline Chloride, 1 mM; Eserine Sulfate, $2X10^{-4}$M; Bovine Serum Albumin, 0.05 percent (w/v); $MgCl_2$, 20 mM; ^{14}C-Acetyl CoA, $9X10^{-5}$M (Specific activity, 19 mCi/-mmole); and Triton X 100, 0.25 percent.

Also included in Table 2 is the potent HAChT inhibitor, acetyl-secohemicholinium-3 (Acetylseco-HC3). In confirmation of the earlier report of Domino et al. (6), acetylseco-HC3 is a potent inhibitor of ChAT as well, having a K_I value of 15.3 μM.   In Fig. 2 is shown the effect of acetylseco-HC3 concentration on enzyme activity, from which an A_{50} = 2.8 μM was estimated.

With an aim to synthesize ChAT inhibitors with better distribution into the CNS, a series of 2-benzylidene-3-quinuclidinones were synthesized (Table 1) and studied.   The results are presented in Table 3 together with estimates of K_m.   Inhibition in all cases is interpreted as being non-competitive (Fig. 3).   The constancy of Km values in the absence (Km = 0.41 for choline) or presence of inhibitor supports this conclusion.

To study the structure-activity relationship (SAR) of derivatives of 3-hydroxy quinuclidinine directly on ACh synthesis, in vitro, ACh and CO_2 production (as an estimate of respiration) from labelled glucose was measured in brain slices incubated for 1 h with 6 mM KCl.   Potassium stimulation increased production of ACh and CO_2 from glucose; 60 mM (final conc.) yielding maximal stimulation.

Acetylseco-HC3 was used as a reference compound (Fig. 4).   As expected, it produced a dose-related decline in [^{14}C]-ACh synthesis

Table 3

ChAT Inhibitor Action

R-Group	Km+	K_I (μMol/L)	Comments	Inhibition
-H	0.81	17.0	Var. Chol.	Non Competitive (or irreversible)
m-CH₃	0.64	3.1	Var. Chol.	Non Competitive (or irreversible)
m-C	0.35	3.8	Var. Chol.	Non Competitive (or irreversible)
m-OCH₃	0.54	23.0	Var. Chol.	Non Competitive (or irreversible)
p-OCH₃	0.60	13.2	Var. Chol.	Non Competitive (or irreversible)
3,5 di-Cl	0.35	48.0	Var. Chol.	Non Competitive (or irreversible)

Concentrations: Choline = 0.38-2 mM; AcCoA = 113.3 μM.   +KmCh $\doteq$ mM; KmAcCoa = μM.

with increasing concentrations of drug, to 23% of control at 5×10^{-3} M. In contrast, there was no decrease on [^{14}C]O$_2$ production. The quaternary quinuclidinol analog, 3-quinuclidinol methyl iodide, produced a dose-related decrease in [^{14}C]-ACh synthesis to approximately 40% of control values at doses that did not reduce respiration (Fig. 5). In order to determine the necessity of the methyl functional group on the quinuclidinol ring, tissue was incubated with 3-quinuclidinol hydrochloride. In contrast to the quaternary compound, this analog failed to produce a decrease in [^{14}C]-ACh synthesis. With the exception of 3-hydroxy-N-allyl quinuclidinium bromide which produced a modest decrease in [^{14}C]-ACh synthesis to 80% of control, other quinuclidine derivatives tested failed to inhibit synthesis at concentrations up to 10^{-3}M. It should be noted that these quinuclidines inhibit HAChT and should not affect ACh synthesis from

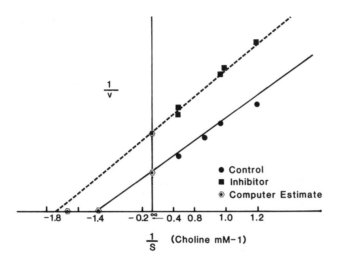

Figure 3.   The non-competitive inhibition of ChAT by m-CH$_3$-2-Benzyl-idene-3-quinuclidinone (3.3 μM) over a range of choline concentrations holding [^{14}C]-AcetylCoA constant (9 X 10^{-5}M).   The values Vmax and Km were computed based on Eadie-Hofstee date treatment.

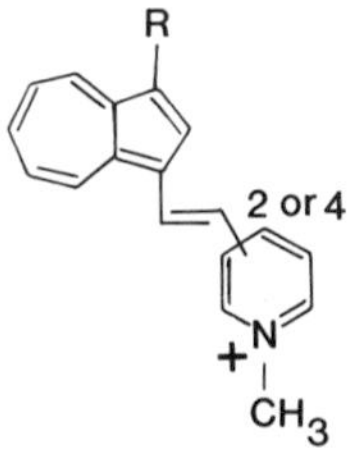

Cpd	R	Pyr. Subs.
I	H	2
II	H	4
III	Br	4
IV	Br	2

Figure 4.   ACh and CO$_2$ production from (U$^-$[^{14}C]) Glucose (5 μCi) was measured in cerebral cortex slices incubated for 1 h, 37°C in modified Krebs, Ringer bicarbonate buffer (KCI, 60mM added at 15 min mark) with various concentrations of acetylseco-HC3.   The values are represented as mean $\pm$ SEM of (N) number of samples.

endogenous choline already present in the nerve terminal; this may explain the observation that no compound completely inhibited ACh synthesis from labelled glucose.

DISCUSSION

The well known studies of Cavallito and Co-workers (3, 4, 28) established a prototypical structure for ChAT inhibitors with high

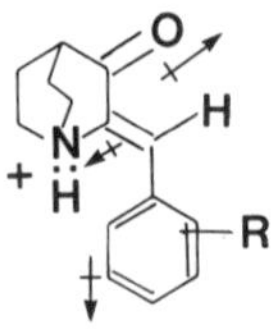

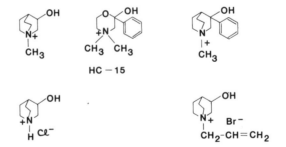

Figure 5. ACh and CO_2 production from $(U^-[{}^{14}C])$ glucose was measured with concentrations of 3-quinuclidinol methyl iodide from 10^{-7} to $5 \times 10^{-3} M$.

in vitro inhibitory potency. They proposed that the stilbazole methyl iodides and quaternized naphthylvinylpyridines provided the proper trans arrangement of π-donor and π-acceptor moieties necessary for the inhibition of ChAT via an interaction that was, in part, determined by hydrophobic and charge transfer forces. Our own interest in charge-transfer interactions (7) and non-benzenoid aromatic compounds, specifically ferrocene and azulene, led us to synthesize some azulylvinylpyridine methyl iodides (AVP) as isomers of the naphthyl systems (Fig. 6). Cavallito had included the ferrocenyl-vinyl-pyridinium system in his series and found it to be poorly active against ChAT. Ferrocene, by virtue of its cylindrical structure, represents a considerable steric departure from benzene and naphthalene and hence the lack of potency may be reasonably explained on this basis. Azulene, on the other hand, may be considered to represent an electronic departure from its isomeric benzenoid counterpart naphthalene, due to the presence of a weak dipole, enhanced reactivity and tendency to form charge transfer complexes. It has been shown to possess steric and lipophilic properties similar to those of naphthalene (18).

Figure 6. Azulylvinyl pyridinium methyl iodides (AVP) substituted either at the 2- or 4- position of the pyridine ring.

Given the foregoing, and previous work in our laboratories with azulylcholine inhibitors of pseudocholinesterase (17), we synthesized several AVP's. The synthesis of these compounds was effected using base catalyzed condensation of the appropriate azulene aldehyde with 2- or 4-picoline methyliodide in methanol. These compounds were shown to possess inhibitory action against ChAT equipotent to the naphthylvinyl-pyridinium agents, and are presented in Table 2. The additional π-donor ability of the azulene does not appear to markedly enhance the inhibitory potency of compounds of this type suggesting a secondary role, if any, for charge transfer on the part of the hydrocarbon ring.

The contribution of lipophilic and stereoelectronic properties to inhibitors was further elucidated by Baker and Gibson (1) with a series of non-quaternary derivatives of 4-stilbazole. In an effort to develop tertiary amines possessing ChAT inhibitory action greater than the few reported by Cavallito et al., they systematically investigated the effects of structural changes in the benzene ring within a large series of compounds. The effort was directed at compensating for the reduction in affinity to ChAT resulting from either the loss of the permanent delocalized positive charge in the quaternized pyridine ring, or the absence of the quaternizing group itself. The 3',4'-dichloro-4-stilbazole derivative was found to be a very potent inhibitor of ChAT in vitro and this was only partly attributed to the lipophilic and stereo electronic contributions of the substituents (1). The authors stressed the importance of the double bond for binding to the enzyme surface and further postulated that its polarization should be in the direction of the pyridine ring, with the benzylic carbon being electron deficient. Further support for this hypothesis was provided in the work of Malthe-Sorenssen et al. (19) and Rowell and Chiou (23), who reported the ChAT inhibitory action of certain quaternized and tertiary aminoalcohol esters of acrylic acid. Both of these studies attributed a crucial role to a polarized double bond in conjugation with a carbonyl function.

Thus, the compounds described above can be seen to share a similar structural theme, namely a conjugated polarized olefinic system separated from an appropriately substituted nitrogen by 3 or 4 atomic units. Several are quaternary ammonium derivatives unable to penetrate cellular membranes with any acceptable facility; others are tertiary amines that are easily protonated in the biological milieau, and would be expected to readily traverse lipoidal membranes. Those compounds that are esters of acrylic acid would be expected to be toxic and vulnerable to hydrolysis in a physiological environment. Most of the reported inhibitors are non-competitive in their action, and it cannot be assumed that they all share a similarity in mechanism at the molecular level. Lastly, none of the compounds reported to date include an asymmetric center.

The high affinity of quinuclidinyl compounds for cholinergic ultra-structures (25, 32), which has led to the use of quinuclidinyl benzilate (QNB) as the specific ligand for muscarinic receptors, the ability of 3-hydroxyquinuclidine methyl iodide to inhibit HAChT (8), and the considerable lipophilicity of this strong base (13), have prompted us to investigate certain derivatives of this rigid bicyclic system as potential inhibitors of either ChAT or HAChT.

2-Benzylidene-3-quinuclidinones (Fig. 7) have been reported in the literature (30) and their ease of synthesis, via aldol condensation of 3-quinuclidinone and appropriately substituted benzaldehydes, led us to prepare a series of these as the initial group of potential inhibitors of ACh synthesis. These compounds have a loose structural similarity to the aforementioned inhibitors in that they possess a double bond in conjugation with a substituted phenyl ring and in which the polarization of the bond is promoted by its position relative to the carbonyl (i.e., an alpha, beta unsaturated ketone). In addition, these compounds are highly lipophilic tertiary amines that are easily protonated at physiologic pH. Lastly, they may be considered oxidized analogs of 3-hydroxyquinuclidine.

The inhibitory constants for the limited series of 2-benzylidene-3-quinuclidinones are presented in Table 3. All of the compounds exhibited non-competitive inhibition against varying concentrations of choline and a representative plot of 1/V vs 1/S is shown in Figure 3. These compounds are 10-100 times more active than the acrylate esters reported by Malthe-Sorenssen et al. (19) and Rowell and Chiou (23). They are within the range of activity demonstrated by the stilbazole analogs reported by Baker and Gibson (1). Although loosely patterned upon the Cavallito and Baker compounds, these agents offer certain departures from those first inhibitor prototypes in that:

* they possess an alicyclic nitrogen heterocycle rather than the aromatic pyridine.

* there appears to be a lack of co-planarity between the double bond and the substituted phenyl ring, due to considerable steric hindrance as evidenced in molecular models. Further evidence as to crowding in the vicinity of the nitrogen derives from the inability to form quaternary methyl iodides.

* the double bond is conjugated to a carbonyl (analogous to the acrylates) rather than to the pyridine ring.

The similarity of the quinuclidines to the Cavallito and Baker compounds, and the acrylates, resides in the polarization of the

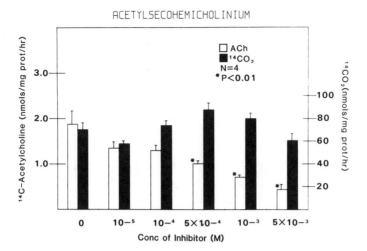

Figure 7. Structure of 2-benzylidene-3-quinuclidinones indicating
the electron withdrawing capacity relative to the double bond of
the keto group, protonated nitrogen, and phenyl ring

double bond (See Fig. 7) and the potential reactivity imparted thereby. In both the vinyl-pyridines and the acrylates the double bond
is polarized by inductive and resonance interactions with electron-
attracting centers, specifically the charged pyridinium in the former
and the carbonyl oxygen in the latter. Several authors (1, 3, 5)
have suggested the presence of an electron-rich center in ChAT,
possibly a sulfhydryl group, that may serve as an anchoring point
for these vinylygous systems. The stabilization of partial charge
separation in the conjugated frameword of the inhibitors is afforded
through resonance distribution. The lack of this resonance possibility in the piperidine analog of the vinyl-pyridines may explain
its reported inactivity (3). Similarly, in our own series of inhibitors, the loss of conjugation via reduction of the keto carbonyl
to a hydroxyl function abolishes activity. The 2-benzylidine-3-
quinuclidinones readily undergo 1,4-Michael additions with appropriate
nucleophiles (30), and one may postulate a similar tendency in the
presence of electron-rich sulfhydryl centers. Although we do not
wish to postulate bond-forming reactions between enzyme and inhibitor
at this early stage in our work, we call attention to this tendency
in the inhibitors (as have previous authors). By focusing on this
reactivity, we may design and test compounds deviating in structure
from the Cavallito and Baker prototypes. This may afford new oppor-
tunities for inhibition of ChAT by a variety of structurally diverse
agents.

The recent excellent review by Fisher and Hanin (9) catalogues
those compounds found to possess inhibitory action on HAChT. The

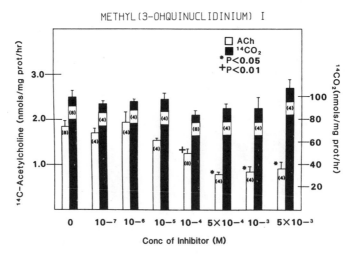

Figure 8. Derivatives of 3-hydroxy quinuclidine compared with HC-15.

reader is referred to this paper, and the references cited therein, for a complete discussion of this particular area. For our own purposes we wish to draw attention to the well known hemicholinium series of compounds, specifically HC-15, HC-3, and acetylseco-HC3, all highly potent inhibitors of HAChT, and to the inhibitory action of 3-hydroxyquinuclidine methyliodide (8). 3-hydroxyquinuclidine methyliodide may be considered a rigid analog of choline and may also be considered the prototype for carbon isosteres of the hemi-choliniums, particularly HC-15.

We have prepared several simple analogs of 3-hydroxyquinuclidine methyl iodide in order to determine the structure activity requirement within this series (Fig. 8). 3-Phenyl-3-hydroxy quinuclidinium methyl iodide can be seen to be the carbon isostere of HC-15 and was prepared by the method of Grob (12). N-Allyl-3-hydroxyquinuclidinium bromide was synthesized as a potential irreversible inhibitor of HAChT for eventual comparison with recently reported mustard analogs of choline (24) and HC-15 (27). 3-hydroxy quinuclidinium methyl iodide, as reported earlier by Dowdall (8) demonstrated inhibition of ACh synthesis (See Fig. 5). The other compounds in this particular series, other than the allyl derivative, did not show inhibition of ACh synthesis.

Acknowledgement. This work was supported by a U.S. ARMY Medical Research and Development Command Contract #DAMD17-82-C-2183. The authors also wish to thank Remus M. Berretta for the artwork and Barbara N. Bush for help in the preparation and typing of the manuscript.

REFERENCES

1. Baker, B.R. and Gibson, R.E. (1971; 1972): J. Med. Chem. 14: 315-322.
2. Browning, E.T. and Shulman, M.P. (1968): J. Neurochem. 15: 1391-1405.
3. Cavallito, C.J., White, H.L., Yun, H.S. and Foldes, F.F. (1970): In Drugs and Cholinergic Mechanisms in the CNS (eds) F. Heilbronn and A. Winter, Stockholm, Sweden, pp. 97-116.
4. Cavallito, C.J., Rittman, A.W. and White, H.L. (1971): J. Med. Chem. 14:230-233.
5. Currier, S.F. and Mautner, H.G. (1974): Proc. Nat. Acad. Sci. 71:3355-3358.
6. Domino, E.F., Mohrman, M.E., Wilson, A.E. and Haarstad, V.B. (1973): Neuropharmacol. 12:549-561.
7. Doukas, P.H. (1975): In Drug Design 5 (ed) E.J. Ariens, Academic Press, pp. 133-167.
8. Dowdall, M.J. (1978): In Cholinergic Mechanisms and Psycho-pharmacology (eds) D.J. Jenden, Plenum, New York, pp. 359-375.
9. Fisher, A. and Hanin, I. (1980): Life Sci. 27:1615-1634.
10. Freeman, J.J., Choi, R. and Jenden, D.J. (1975): J. Neurochem. 24:729-734.
11. Gibson, G.E. and Blass, J.P. (1976): J. Neurochem. 26:1073-1078.
12. Grob, C.A. (1960): Chem. Abstr. 54:8862d.
13. Hansch, C., Leo, A.J. (1979): Substituent Constants for Cor-relation Analysis in Chemistry and Biology, (eds) Wiley, Appendix II, p. 249.
14. Harris, L.W., Stitcher, D.L. and Heyl, W.C. (1982): Life Sci. 30:1867-1873.
15. Hersh, L.B. and Peet, M. (1977): J. Biol. Chem. 252:4796-4802.
16. Jope, R.S. (1979): Brain Res. Rev. 1:313-344.
17. Kim, W.H. (1978): Ph.D. Thesis, Temple University. Non Benzenoid Aromatic Enzyme Inhibitors: Inhibition of Horse Serum Butyrylcholinesterase by Azulene and Ferrocene Derivatives.
18. Leo, A.J., Hausch, C. and Elkins, D. (1971): Chem. Rev. 71:590.
19. Malthe-Sorenssen, D., Andersen, R.A. and Fonnum, F. (1974): Biochem. Pharmacol. 23:577-586.
20. McCaman, R.E. and Hunt, J.M. (1965): J. Neurochem. 12:253-259.
21. O'Neill, J.J., Simon, S.H. and Cummins, J.T. (1963): Biochem. Pharmac. 12:809-820.
22. Peterson, C. and Gibson, G.E. (1982): J. Pharmacol. and Exp. Ther. 222:576-582.
23. Rowell, P.P. and Chiou, C.Y. (1976): Biochem. Pharmacol. 25: 1093-1099.
24. Rylett, B.J. and Colhoun, E.H. (1979): J. Neurochem. 32:553-559.
25. Schulman, J.M., Sabio, M.L. and Disch, R.L. (1983): J. Med. Chem. 26:817-823.
26. Schrier, B.K. and Shuster, L. (1967): J. Neurochem. 14:977-985.
27. Smart, L.A. (1983): J. Med. Chem. 26:104-107.

28. Smith, J.C., Cavallito, C.J. and Foldes, F.F. (1967): Biochem. Pharmacol. 16:2438-2441.
29. Sterling, G.H. and O'Neill, J.J. (1978): J. Neurochem. 31: 525-530.
30. Warawa, E.J., Mueller, N.J. and Jules, R.J. (1974): J. Med. Chem. 17:497-501.
31. White, H.L. and Wu, J.C. (1973): J. Neurochem. 20:297-307.
32. Yamamura, H.I. and Snyder, S.H. (1974): Proc. Natl. Acad. Sci. U.S.A. 71:1725-1729.

THE ROLE OF GLIAL CELLS IN NEURONAL

ACETYLCHOLINE SYNTHESIS

P. Kasa

Central Research Laboratory
Medical University
Somogyi B ut 4, 6720 Szeged, Hungary

INTRODUCTION

Earlier reports have suggested that the choline (Ch) supply for acetylcholine (ACh) synthesis may originate from the blood (1), or be released (5), or synthesized de novo by different enzymes in the brain (9). There are no data, however, on the role of glial cells in neuronal ACh synthesis. Some years ago, Tuček (12) put forward the idea that Ch may be produced in the glial cells, from where it passes into the extracellular fluid, and is then taken up by the high-affinity (16) carrier system into the cholinergic axon terminals. On the basis of biochemical investigations, Ansell and Spanner (2) have suggested that glycerophosphocholine diesterase (GPCD: EC 3.1.4.2) may be important in the release of Ch from the glial cells. It has also been noted that central neurons "fare better" in cultures when in contact with non-neuronal cells (13), and especially glial cells (11). Since neither the fate of the Ch released from the glial cells nor the role of the contact between glial cells and neurons has yet been elucidated, our aim was to investigate these phenomena.

MATERIALS AND METHODS

Cultures. Pure neuronal (E7DIV6), pure glial (E14DIV14), combined glial (E14DIV14) + neuronal (E7DIV7) and glial (E14DIV14) + pure neuronal (E7DIV2) cultures were co-cultured for 6 days by standard methods (3, 10) from embryonic chick brain.

Biochemical experiments. The procedure used to study the transport of Ch from the glial cells to the neurons and its utilization for ACh synthesis was as follows: Glial cells from 14-day-old chick

907

brain were cultured in vitro for 14 days, and were then treated
with 10 μM [^{14}C]-choline solution (spec. act.: 58 mCi/mmol) for 30
min. After this period the cultures were washed 5x with Ch-free
and KCl-free solution. Neurons were then plated from 7-day-old
embryonic chick brain onto the glial cell layer to give combined
cultures. After 2, 5, and 7 days the medium in which the cells
were growing (MEM, Eagle's) was changed and the [^{14}C]-choline influx
into it was determined. The [^{14}C]-choline incorporated into lipids,
phosphocholine, betaine and ACh, as well as the free [^{14}C]-choline,
were determined (8) in the pure glial cell cultures after 24 h,
and in the combined cultures after 7 days.

A second set of experiments was performed to study the role
of the contact between glial cells and neurons in the choline transport. Pure glial cell cultures (E14DIV14) were produced in some
Petri dishes, and pure neuron cultures (E7DIV2) in others. When
the neurons began to develop their processes, the E14DIV14 culture
was treated with [^{14}C]-choline, the surface was washed, and the
E7DIV2 neuron culture was transferred to the Petri dish containing
glial cells in such a way that the neurons could not come into contact
with the glial cells. The [^{14}C]-choline influx into the incubation
medium and the uptake by the neurons were measured. In both cultures
the distribution of [^{14}C]-Choline in lipids (8), the ChAT activity
(4) and the ACh content (6, 7) was determined.

RESULTS

When the tissue of embryonic chick brain (E14) was plated on
collagen-treated Petri dishes, the neurons degenerated and pure
glial cell cultures (Fig. 1) were obtained after 12-14 days in vitro
(DIV12, DIV14).

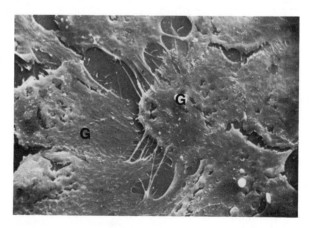

Figure 1. Scanning electron micrograph of a pure glial cell culture.
Note the flat polygonal glial cells (G) and the absence of neuro-
cytes. x 9,000

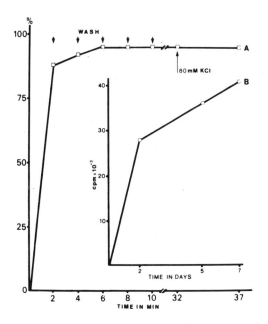

Figure 2. Demonstration of [^{14}C]-choline uptake (A) and release (B) from the tissue cultures. Each point represents the mean of five separate experiments, with a SE of 5-10%.

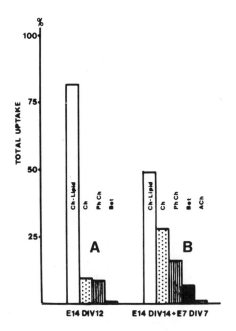

Figure 3. Distribution of radio-activity from [^{14}C]-choline in the different Ch-containing compounds present in pure glial cell cultures incubated for 24 h (A), and in combined cultures incubated for 7 days (B).

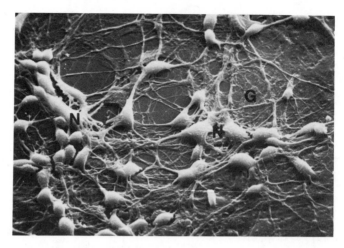

Figure 4.   Scanning electron micrograph of the combined culture.
Note the well-developed neurons (N) on the glial cell (G) layer.
x 6,000

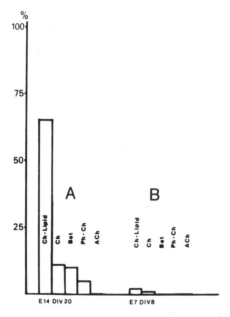

Figure 5.   Distribution of radioactivity from [^{14}C]-choline in the
different Ch-containing compounds present in pure glial (A), and
in pure neuronal (B) cultures.   The neurones were incubated for 6
days in the same Petri dish as that in which the pure glial cells
were treated with the radioactive precursor, but the neurons and
glial cells were separated by a plastic layer.

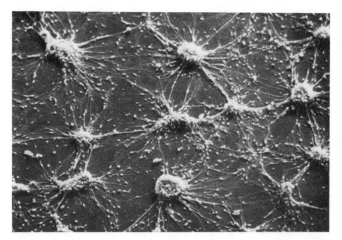

Figure 6. Scanning electron micrograph of a pure neuronal culture
(E7DIV7). Note the absence of glial cells and the presence of well-
developed neurons and their axonal arborizations. x 600

The pure glial cell cultures were treated with 10 μM [^{14}C]-choline
for 30 min, and the excess [^{14}C]-choline was then washed out. Eighty-
eight percent of the radioactivity was eliminated after the 1st,
92% after the 2nd, and 95% after the 3rd washing; subsequently there
was no change (Fig. 2A). When the labelled glial cell cultures
were treated with 80 mM KCl for 5 min, there was no more release
of radioactive Ch into the medium than in the controls. However,
in the incubation medium a continuous release of choline could be
observed during a 7-day period (Fig. 2B). After a 24 h treatment
of the glial cells with [^{14}C]-choline, 80% of the radioactivity
appeared in the lipids and 20% in the acid-soluble fraction. In
the acid-soluble fraction, 9% of the radioactivity was in phosphocholine, 10% in choline and 1% in betaine (Fig. 3A). When the combined
cultures (Fig. 4) were analyzed by TLC, it was found that the radioactivity was shifted from the lipids towards the water-soluble choline
fraction, and radioactivity appeared in the spots where ACh should
be present (Fig 3B). No [^{14}C]-ACh formation could be detected
(Fig. 5A, 5B) in the pure neuron cultures (Fig. 6) when the two
cell populations (glial cells and neurons) were separated by a plastic
layer, although the ChAT activity (1.10 $\pm$ 0.17 nmol/min/mg prot.)
and the non-radioactive ACh content (0.81 $\pm$ 0.07 nmol/mg prot.)
could be measured in these cultures.

DISCUSSION

Our results indicate that the [^{14}C]-choline incorporated into
the glial cells was not only released, but was taken up by neurons,

where [^{14}C]-ACh was synthesized from it if the neurons were in close contact with the glial cells.

Several questions may arise in connection with these experimental results. Firstly, what is the mechanism by which choline is released from the glial cells, and secondly, what is the fate of the released choline?

It has been shown that choline efflux from synaptosomes (14) or neuron cultures (15) may be stimulated by different agents (e.g. choline, ouabain, K$^+$, etc.). On the other hand, Ansell and Spanner (2) suggested that glycerophosphocholine-diesterase, which is localized mainly in the glial cells, can release choline from glycerophosphocholine.

In one series of experiments (not yet published) it has been demonstrated that 5.0 μM ACh in the incubation medium will increase the [^{14}C]-choline efflux from pure glial cell cultures which contain mAChR (31.64 $\pm$ 9.7 fmol/mg prot.) as measured by the method of Yamamura and Snyder (17). It is therefore presumed that the mAChR present in glial cell cultures may be of functional significance during the ACh release from cholinergic axon terminals.

If our results are correct, we may hypothesize that ACh released from the cholinergic neuron may have a dual function: (a) it depolarizes the postsynaptic cholinoceptive cell membrane; and (b) it evokes choline release from the surrounding glial cells, the choline being taken up by the axon terminals and utilized for ACh synthesis. The importance of the close contact between glial cells and neurons must also be emphasized, since no [^{14}C]-ACh was formed when the two cell populations were separated by a plastic layer. The spontaneous outward flux of choline seems therefore not to be an important factor in the synthesis. The mechanism by which the neurons acquire their choline from the glial cells is not yet settled, although the group of Massarelli (15) have shown that the K$^+$ efflux from the neurons and its uptake by the glial cells may trigger the release of the free choline content of these cells.

SUMMARY

The uptake and metabolism of [^{14}C]-choline in different glial and neuronal cultures, and also the role of glial-neuronal contacts in ACh synthesis, were studied. It was shown that the glial cells have the ability first to take up choline from the incubation medium, and in special circumstances to release it, whereby ACh can be synthesized in the cholinergic neurons. Our results demonstrate that no [^{14}C]-ACh synthesis occurred in the neurons when the neuronal and glial cell cultures were separated by a plastic layer. It is suggested that, besides other choline sources, the precursor present in glial cells may play a role in neuronal ACh synthesis if the two cell populations are in intimate contact.

Acknowledgements. This work was supported by the Scientific Research Council, Ministry of Health, Hungary (06/4-20/457).

REFERENCES

1. Ansell, G.B. and Spanner, S. (1971): Biochem. J. 122:741-750.
2. Ansell, G.B. and Spanner, S. (1981): In Cholinergic Mechanism (eds) G. Pepeu and H. Ladinsky, Plenum Press, New York, pp. 393-403.
3. Booher, J. and Sensenbrenner, M. (1972): Neurobiology 2:97-105.
4. Fonnum, F. (1975): J. Neurochem. 24:407-409.
5. Freeman, J.J. and Jenden, D.J. (1976): Life Sci. 19:949-962.
6. Hanin, I. and Jenden, D.J. (1969): Biochem. Pharmacol. 18:837-845.
7. Kilbinger, H. (1973): J. Neurochem. 21:421-429.
8. Marchbanks, R.M. and Israël, M.J. (1971): J. Neurochem. 18:439-448.
9. Mozzi, R. and Porcellati, G. (1979): FEBS Letters 100:363-366.
10. Pettman, B., Louis, J.C. and Sensenbrenner, M. (1979): Nature 281:378-380.
11. Touzet, N. and Sensenbrenner, M. (1978): Dev. Neurosci. 1:159-163.
12. Tuček, S. (1978): In Acetylcholine Synthesis in Neurons (eds) J. Wiley and Sons, New York.
13. Varon, S. and Saier, M. (1975): Exp. Neurol. 48:135-162.
14. Vyas, S. and Marchbanks, M.J. (1981): J. Neurochem. 37:1467-1474.
15. Wong, T.Y., Hoffman, D., Dreyfus, H., Louis, J.C. and Massarelli, R. (1982): Neurosci. Lett. 29:293-296.
16. Yamamura, H.I. and Snyder, S.H. (1973): J. Neurochem. 21:1355-1374.
17. Yamamura, H.I. and Snyder, S.H. (1974): Proc. Natl. Acad. Sci. (Wash.) 71:1725-1729.

THE EFFLUX OF CHOLINE FROM NERVE CELLS: MEDIATION BY IONIC GRADIENTS AND FUNCTIONAL EXCHANGE OF CHOLINE FROM GLIA TO NEURONS

D. Hoffmann, S. Mykita, B. Ferret and R. Massarelli

Centre de neurochimie du CNRS and U44 de l'INSERM
5, rue Blaise Pascal
67084 Strasbourg Cedex, France

INTRODUCTION

The availability of the precursors choline and acetyl Coenzyme A has been considered as one of the essential components for regulation of acetylcholine metabolism (12). One of the main pathways for the supply of choline to neurons for cholinergic metabolism is its transport across the plasma membrane. Recently, however, it has been shown that: a) choline can be synthesized in nerve cells (neurons and glia) by stepwise methylation of ethanolamine, and acetylcholine may be synthesized via this methylation pathway (8); and b) choline may be released from nerve cells following a mechanism which can be stimulated by, and is dependent upon, ions in a manner which is the mirror image of its flow into the cell (13).

The importance of the endogenous methylation process for the regulation of the synthesis of acetylcholine must still be ascertained. The efflux of choline, instead, is somewhat a puzzle. Why should a compound so important for neurons and more particularly for cholinergic neurons be extruded from nerve cells? It should be kept in mind that, in the rat brain, there are at least 20 μmoles/g fresh weight of choline containing compounds (1) and only about 20 nmoles/g of acetylcholine (11) and 5-10 nmoles/g of free choline (3). It is therefore surprising, to say the least, to observe an efflux of choline stimulated by an excess of K^+ ions in the incubation medium (13), suggesting instinctively that choline may normally be "released," and that such a "release" may be provoked by depolarizing conditions.

The present report analyzes the relationship between ions and the efflux of choline, and suggests the possibility of a "balance

915

effect" for choline fluxes (in and out) which is produced and maintained by ionic gradients. It is also suggested that glial cells
may actively exchange choline with neurons during nerve activity,
and that they may function as a choline reservoir for neuronal needs.

MATERIAL AND METHODS

Isolated neurons, grown in the absence of support cells, were
obtained by seeding dissociated cerebral hemispheres, from 8-day-old
chick embryo. They were grown in plastic Petri dishes (60 mm Ø)
precoated with a film of L-Polylysine, according to a previously
published report (10). Glial cell cultures were obtained similarly,
starting from older embryo (14 day-old), and seeded directly on
the bottom of plastic Petri dishes (60 mm Ø) (2).

Cultures were preincubated for about 20 hrs with [^{14}C]methyl-
choline (specific activity: 58 mCi/mmole, Amersham; final concentration: 42 μM; 1 μCi/ml) in their growth medium (Dulbecco's modified
Eagle's medium supplemented with 10% and 20% foetal calf serum in
glial and neuronal cultures, respectively). Previous experiments
had shown that this time was necessary for the saturation of the
various choline containing cellular compartments (8). The cells
were further incubated in Krebs Ringer phosphate solution (NaCl
137 mM, KCl 2.6 mM, CaCl$_2$. 2 H$_2$0 0.7 mM, MgCl$_2$ 0.5 mM, Na$_2$HPO$_4$ 2
H$_2$0 3.2 mM, KH$_2$PO$_4$ 1.4 mM, glucose 10 mM, adjusted to pH 7.4) at
37°C, to which various agents have been added or substituted iso-
osmotically. Aliquots of this incubation medium (normally 100 μl)
were taken every two min starting from 1 min up to 29 min, and an
equal amount of fresh medium was immediately added to maintain the
volume of the incubation medium at 4 ml. To show that the choline
exiting the cells was essentially of free origin both neurons and
glial cells were preincubated overnight with [^{14}C]choline (1 μCi/ml)
to label the pool of choline (7) and, after thorough washing (3
times with 4 ml of 0.147 M NaCl solution), incubated with 4 ml of
Krebs Ringer phosphate solution containing 2.5 μCi/ml of [^{3}H]methyl-
choline (specific activity: 77 Ci/mmol, Amersham; final concentration: 32.4 μM) for 15 min. The freshly uptaken [^{3}H]choline will
essentially label the acid soluble compartments of the choline pool
(7).

The radioactive content present in the various aliquots was
an expression of the outward flux of choline from the cells, and
was measured in scintillation vials by adding 0.5 ml H$_2$0 and 10 ml
of Rotiszint 22 (Roth), using an Intertechnique scintillation spectrometer SL20. Final data are expressed as cpm/μg protein (determined
according to Lowry et al. [5]) and corrected by the following equations:

when x = 1 min, then: $U_{(x)} = 40 \ Q_{(x)}$ (μg protein)$^{-1}$;

if   x > 1 min, then:
$$U_{(x)} = [40 \ Q_{(x)} - 39 \ Q_{(x-2)}] \ (\mu g \ protein)^{-1} + U_{(x-2)}$$

where: x represents the sample analyzed every two minutes starting from 1 min up to 29 min; $Q_{(x)}$ = cpm in a 100 μl aliquot; $U_{(x)}$ = total cpm found in the medium at each time point per μg protein; and $U_{(x-2)}$ = total cpm per μg protein, measured in the preceding aliquots. Statistical significance was measured according to Student's t test.

RESULTS

Neurons and glial cells spontaneously discharge choline into the incubation medium. The exiting choline is essentially of free origin, as can be seen in Figures 1a and 1b, where neurons and glial cells had been prelabelled with [14C]choline overnight, and labelled for 15 min with [3H]choline. The higher amount of [3H]choline exiting the cells, in fact, indicates that it is the freshly labelled choline which is preferentially released. The remaining [14C]Choline exiting the cells corresponds to the free choline of phospholipid origin which amounts to about one third of the total free choline content (7). The efflux of choline was stimulated when Na^+ ions were substituted with iso-osmolar concentrations of Li^+ or Cs^+. This effect was more prominent in glial cells, than in neurons. Cyanide (10^{-3}M) and low temperature inhibited the efflux from both nerve cell types (Table 1).

Table 1

Effect of Na^+ Substitution on the Efflux of Choline

	Neuron	Glia
Control	100.0	100.0
Li^+ Cl^- 137 mM		192.1
Li^+ Cl^- 50 mM		108.3
Cs^+ Cl^- 50 mM		139.2
Cyanide 10^{-3}M	67.9	93.0
Temperature 10°C		53.0

Data, expressed in percent of control values, were measured after 10 min of efflux.

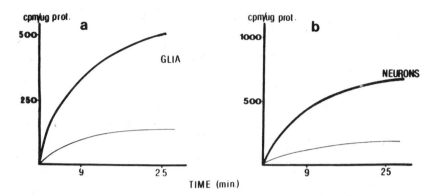

Figure 1.   Efflux of choline from glial cells (a) and neurons (b).
Nerve cells were preincubated with 42 μM [^{14}C]choline (1 μCi/ml) overnight and pulse-labelled with 32.4 μM [^{3}H]choline (2.5 μCi/ml) for
15 min.   The spontaneous efflux was measured as described in Materials
and Methods.

Excess of K^+ ions in the medium was also capable of increasing
the efflux of choline from nerve cells (Table 2); however, 30 mM
KCl were sufficient to produce in glial cells a stimulation similar
to that obtained in neurons only with 100 mM KCl.   The increase
however was not Ca^{++} dependent and the absence of $CaCl_2$ from the
incubation medium actually stimulated the efflux of choline from
neurons and glial cells.   The incubation of glial cells with a medium
containing 50 mM KCl and no $CaCl_2$ was capable of increasing the
efflux of choline to levels reached by an incubation performed without
$CaCl_2$.

The data suggest that choline might be driven out of nerve
cells based upon the K^+ ionic electrochemical gradient.   Alternatively,
it might be cotransported with K^+ ions out of nerve cells.

It is rather difficult to block K^+ channels, although one of
the channels can be blocked by $BaCl_2$.   This ion (0.1 - 1.0 μM) does
not by itself produce any change in the efflux of choline and when
added to the incubation medium it does not affect the efflux stimulated
by the lack of Ca^{++} (results not shown).   However, the K^+ stimulated
efflux can be successfully reduced by the addition of low micromolar
concentrations of Ba^{++} (Table 2).

DISCUSSION

Why is an efflux of choline present in nerve cells?   Choline
synthesis is not sufficient in the liver to supply the whole organism

Table 2

Effect of K$^+$ Stimulation on the Efflux of Choline

	Neuron	Glia
Control	100.0	100.0
K$^+$ 30 mM		133.3
K$^+$ 50 mM		211.1
K$^+$ 100 mM	161.6	
K$^+$ 100 mM; Ba^{++} 0.5 μM	114.7	
K$^+$ 30 mM; Ba^{++} 0.5 μM		119.6
Ca^{++} 0 mM	181.8	170.8
K$^+$ 50 mM; Ca^{++} 0 mM		169.4

Data, expressed as percent of control values, were measured after 10 min of efflux.

with the amount required for normal function. Moreover, even if a de novo synthesis of choline by the stepwise methylation of etha-nolamine has been found in nerve cells (8), the amount thus produced should not be sufficient, at first glance, to supply the needs of nervous tissue for the synthesis and renewal of membrane choline phospholipids.

There is debate at present regarding the number and the function of various mechanisms which apparently mediate the influx of choline. Some authors suggest a specific link between one such mechanism and the synthesis of acetylcholine (4, 14). Others find only one mechanism for choline influx, and little or no specificity as its action for acetylcholine synthesis (6). Based on our current knowledge of factors involved at the cholinergic nerve terminal it appears, in any case, that the neurotransmitter acetylcholine may be synthesized from: a) choline transported from the extracellular environment; and b) choline synthesized de novo within nerve cells. The presence of free choline originating from phospholipids (7) adds a further variable to the possible pathways responsible for the regulation of acetylcholine synthesis.

Our interest in the efflux of choline from nerve cells stems from the results presented ten years ago by the group of Kewitz, when an arteriovenous difference of choline specific activities was found in the rat brain (3), indicating that choline reenters the blood circulation from the brain. A fortiori, then, there was a serious probability that choline might come out of nerve cells, and the present report shows the presence of such a phenomenon both in neurons and glial cells.

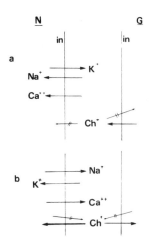

Scheme 1. Movements of ions and of choline in glial cells and
neurons. During an action potential (a) K^+ is extruded into the extracellular space (e.s.) and Na^+ enters into the neuron (N), together
with Ca^{++}. These conditions stimulate Ch efflux from glial cells
(G). After the passage of the depolarizing wave (b), the reestab-
lishment of the normal ionic gradients favors the reuptake of choline.

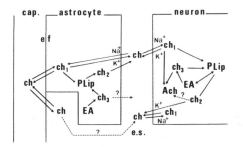

Scheme 2. Exchange and metabolism of choline in nerve cells. Choline
may cross the capillary membrane by the end feet (ef) or into the
extracellular space (e.s.). In astrocytes, choline coming from the
circulation, and hence from extracellular transport (Ch1), is metabolized into phospholipids (PLip) which may themselves produce free
choline (Ch2) (7). Choline (Ch3) may be further synthesized from
ethanolamine (EA) (8). Choline efflux from glia and neurons originates
from Ch1 and Ch2. Ch3 may synthesize in neurons PLip and acetylcholine
but its efflux is only hypothetical. Recent experiments (unpublished)
have shown that, in neurons, Ch2 may synthesize acetylcholine.

The results indicate that choline comes out nerve cells preferentially from the free choline compartment. However, as Figure 1 shows, some [^{14}C]choline efflux was also found in both neurons and glial cells. Such [^{14}C]choline should have been derived from phospholipid metabolism (Ch2 in scheme 2), and it has already been reported that the free choline originating from phospholipids represents 20-30% of total free choline (7). This proportion is roughly similar to that observed in Figures 1a and 1b.

Results from the analysis of the dependence of choline efflux upon ions (summarized in scheme 1) may perhaps advance a possible interpretation of the physiological role of the efflux of choline. During depolarization there is, for a short period of time, an increased concentration of K^+ in the extracellular space and a decrease in Na^+ and Ca^{++} concentrations. These conditions may be closely determined in space and time and, as we have seen, a small deviation in ionic concentrations may affect the efflux of choline much more from glial cells than from neurons. We suggest then that the depolarization of the neuronal membrane produces an increased efflux of choline from glial cells, resulting in an increase in the amount of choline available for neuronal needs, as proposed in scheme 1.

Choline also crosses the blood brain barrier as such (9). What constitutes the blood brain barrier is of considerable importance. Since neurons are not found in close contact with capillaries but astrocytes are (by the end feet), there is a strong possibility that astrocytes may act as the barrier between blood and neurons, and that they may absorb choline, subsequently store it and release it upon neuronal depolarization for neuronal use.

A tentative summary of a possible interrelationship between the fluxes and metabolism of choline in nervous cells, based upon an extrapolation of the data obtained _in vitro_, is presented in scheme 2.

Acknowledgements. The secretarial work of Ms. C. Thomassin-Orphanides is much appreciated. This work was supported by a grant of NATO #75.82/DI.

REFERENCES

1. Ansell, G.B. and Spanner, S. (1978): In Cholinergic Mechanisms and Psychopharmacology (ed) D.J. Jenden, Plenum Press, New York, pp. 431-445.
2. Booher, J. and Sensenbrenner, M. (1972): Neurobiology 2:97-105.
3. Dross, K. and Kewitz, H. (1972): N.S. Arch. Pharmakol. 274:91-106.
4. Kuhar, M.J., Sethy, V.H., Roth, R.H. and Aghajanian, G.K. (1973): J. Neurochem. 20:581-593.

5. Lowry, O.H., Rosebrough, N.J., Farr, A.L. and Randall, R.J. (1951): J. Biol. Chem. 193:265-275.

6. Massarelli, R. and Wong, T.Y. (1981): In Cholinergic Mechanisms (eds) G. Pepeu and H. Ladinski, Plenum Press, New York, pp. 511--520.

7. Massarelli, R., Gorio, A. and Dreyfus, H. (1982): J. Physiol. (Paris) 78:392-398.

8. Massarelli, R., Dainous, F., Freysz, L., Dreyfus, H., Mozzi, R., Floridi, A., Siepi, D. and Porcellati, G. (1982): In Basic and Clinical Aspects of Molecular Neurobiology (eds) A.M.Giuffrida-Stella, G. Gombos, G. Benzi and M.S. Bachelard, pp. 147-155.

9. Pardridge, W.M., Cornford, E.M., Braun, L.D. and Oldendorf, W.-H. (1979): In Nutrition and the Brain (eds) A. Barbeau, J.H. Growdon and R.J. Wurtman, Raven Press, New York, Vol. 5, pp. 25-34.

10. Pettman, B., Louis, J.C. and Sensenbrenner, M. (1979): Nature 281:378-380.

11. Saelens, J.K. and Simke, J.P. (1976): In Biology of Cholinergic Function (eds) A.M. Goldberg and I. Hanin, Raven Press, New York, pp. 661-681.

12. Tuček, S. (1978): Acetylcholine Synthesis in Neurons, Chapman and Hall, London, pp. 62-123.

13. Wong, T.Y., Hoffmann, D., Dreyfus, H., Louis, J.C. and Massarelli, R. (1982): Neurosci. Lett. 29:293-296.

14. Yamamura, H.I. and Snyder, S.H. (1972): Science 178:626.

CHOLINE UPTAKE IN THE HIPPOCAMPUS: INHIBITION OF SEPTAL-HIPPOCAMPAL CHOLINERGIC NEURONS BY INTRAVENTRICULAR BARBITURATES

J.A. Richter

Departments of Pharmacology and Psychiatry
Indiana University School of Medicine
Indianapolis, IN 46223 U.S.A.

INTRODUCTION

Simon et al. (15) first reported that in vivo administration of pentobarbital caused an inhibition of high-affinity, sodium-dependent choline uptake which was measured in vitro in hippocampal synaptosomes. This effect is believed to be related to a decrease in activity in the septal-hippocampal neurons. We have been interested in determining just where in the brain the drug acts to cause this effect and what behavioral consequences result from this particular effect of barbiturates.

We previously demonstrated (12) that injections of pentobarbital directly into the dorsal hippocampus were without effect on choline uptake, even though levels of the drug in the hippocampus were at least as high as those found there when the drug was given i.p. and inhibition of choline uptake was observed. Injections of pentobarbital in the medial septum were also ineffective. Acute lesions of the medial septum (1 hr) inhibited choline uptake and pentobarbital i.p. then had no further effect. We observed a transient recovery of choline uptake at 3 hr after medial septal lesions, but even under these conditions the drug had no effect. These results had suggested to us that the inhibition of choline uptake in the hippocampus could be caused by an action of the drug at some other site, which sends projections to or through the septum. In the experiments reported here we applied phenobarbital to restricted portions of the ventricular system to obtain a general idea of where the barbiturate was acting. Phenobarbital was chosen because its pharmacokinetic properties are more favorable than pentobarbital. We found that the inhibition of choline uptake in the hippocampus occurs when phenobarbital is present near the junction of the lateral and third

923

ventricles, and that this effect is not related to the sedativehyp-
notic effects of this class of drugs.

METHODS AND MATERIALS

The experiments were done in male Wistar rats with chronic
indwelling cannulae so that injections of the drug could be given
in the awake animal. The rats were anesthetized with pentobarbital
and placed in a Kopf stereotaxic frame. The cannula guide (Plastic
Products Co.) was lowered through a burr hole in the skull to the
appropriate coordinates and fixed in place with screws and dental
cement. An obdurator was placed in the guide during the animal's
recovery.

On the day of the experiment an inner cannula was put in the
guide cannula. It was filled with drug or salt solution and attached
by PE tubing to a Hamilton syringe mounted in a Sage syringe pump
which was set to deliver the solution at the desired rate. The
animals were placed in small plastic cages and observed during the
drug administration. In order to place a block in the dorsal portion
of the third ventricle, 4.0 μl of Nivea cream was injected from a
syringe connected by PE tubing directly to the cannula guide positioned
at A5700 μ according to König and Klippel (3). This injection was
made immediately before beginning injection of the drug or control
solution through the cannula located in the lateral ventricle.
This technique was previously described by Herz et al. (2).

Solutions of sodium phenobarbital (J.T. Baker Chemical Co.)
for injection were made in distilled water and control solutions
were made at an equal osmolarity with NaCl (assuming all the phenobarbital was ionized) and brought to pH 9.6 with NaOH to mimic the
pH of the drug solution. Both drug and salt injection solutions
also contained 0.1% Fast Green Dye so that the presence of the injection solution in the ventricles could be verified. In the experiments in which the Nivea cream block was used the drug solution
injected into the lateral ventricle was also spiked with [^{14}C]pheno-
barbital (New England Nuclear) at a final specific activity of 5
dpm/nmole so that a more precise estimate of the spread of drug
solution could be made. In these experiments the remainder of the
brain after cortex and hippocampus were removed was frozen and sliced
coronally in 300 μm sections. The sections were then digested and
the radioactivity determined by liquid scintillation spectroscopy.

At the given time after the onset of injection or at the end
of the infusion the injector cannula was removed and the rat immediately decapitated. The hippocampus was dissected bilaterally and
synaptosomes prepared from both together or each side separately.
Choline uptake was measured in the synaptosomes according to previously
described procedures using 0.5 μM [^{3}H]-choline (New England Nuclear)

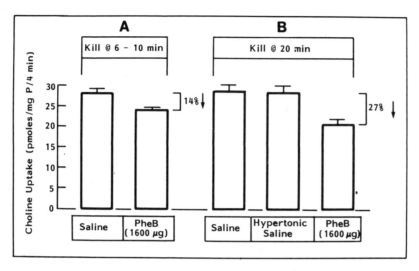

Figure 1. **Bilateral Intraventricular Administration of Phenobarbital (PheB) Inhibits Choline Uptake in the Hippocampus.** In (A) 10 μl per ventricle of 80 μg PheB/μl was injected over 1 min and the rats were decapitated at 6-10 min after beginning the injection. In (B) 5 μl per ventricle of 160 μg PheB/μl was injected over 1 min and the rats were decapitated at 20 min. The hypertonic saline controls were injected with a total of 10 μl of 0.785 M NaCl brought to pH 9.6 with NaOH. After decapitation choline uptake was measured in hippocampal synaptosomes as described in Methods and Materials. Values given are means $\pm$ SEM with the number of rats used indicated in each bar.

(12, 15). Blanks, consisting of choline uptake in the absence of sodium, were subtracted and the resultant high-affinity, sodium-dependent choline uptake is reported as choline uptake in pmoles/mg protein/4 min. Protein was measured by the method of Lowry et al. (5) using bovine serum albumin as the standard.

RESULTS

When a phenobarbital-Fast Green Dye mixture was injected bi-laterally into the lateral ventricles, the dye was found to have spread through the entire ventricular system when the rat was killed 10-20 min later. Choline uptake in the hippocampus was inhibited and the inhibition was apparently greater if 20 min rather than 10 were allowed to elapse after the injection (given over 1 min) (Fig. 1). Since the drug solution was hypertonic, the effects of equally hypertonic saline were also tested in two rats, but no apparent effect was observed. Dye was also included in the normal and hyper-tonic saline solutions so it was not responsible for the effect attributed to the drug.

Injections of phenobarbital were also made unilaterally into one lateral ventricle, and choline uptake was measured separately in the ipsilateral and contralateral hippocampus. Choline uptake was inhibited similarly in both sides, which suggests that the effect of the drug occurs at a periventricular site at, or near to the junction of the lateral and third ventricles (Fig. 2). While the lateralization of the drug distribution was not always complete as estimated from the spread of the dye, there was no indication that the effect was more lateralized when the spread was less. It can also be noted that the degree of inhibition of choline uptake in these experiments was greater than that observed in the bilateral injection experiments (Fig. 1). This may be attributed to the larger amount of drug injected and the fact that it was infused over the 20 min period and not injected in the first 1 min.

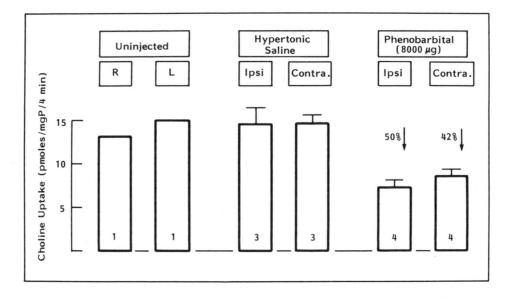

Figure 2. Phenobarbital (PheB) Injected Into One Lateral Ventricle Inhibits Hippocampal Choline Uptake Bilaterally. For PheB injections a solution of 320 μg/μl (1.28 M) was injected into one lateral ventricle (5 μl over 10 sec and then 1 μl/min for 20 min). The hypertonic saline injected rats received a 1.28 M NaCl solution (at pH 9.6) on the same schedule. Uninjected controls were implanted with cannulae but not injected, and were decapitated while the others were being injected. Injected rats were decapitated at the end of the injection. Choline uptake was measured in hippocampal synaptosomes as described in Methods and Materials. Values given are means ± SEM with the number of rats used indicated in each bar.

To determine where along the rostral-caudal extent of the third and fourth ventricles the drug was acting, the caudal spread of the drug was blocked by first injecting Nivea cream at the anterior end of the rostral portion of the third ventricle (see Fig. 3). In this situation, injection of phenobarbital via one lateral ventricle was still effective in inhibiting choline uptake in both hippocampi (Fig. 4). Fig. 3 shows the rostro-caudal distribution of the pheno-barbital. The peak drug levels were usually between A7100 and A6400 and the drug levels usually dropped to very low levels caudal to A5000, i.e., just behind the block of Nivea cream. Injections of

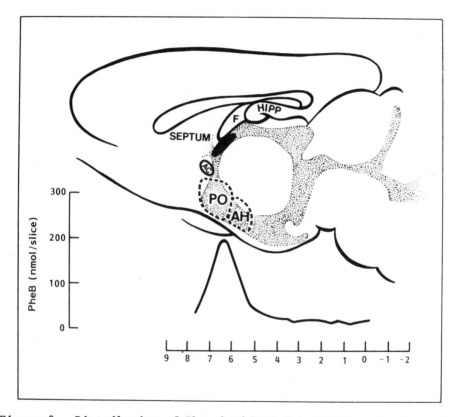

Figure 3. Distribution of Phenobarbital (PheB) After Its Injection Into the Lateral Ventricle with A Block in the Third Ventricle. Just before injection of PheB a block of Nivea cream (4 μl) was injected via a cannula guide in the dorsal portion of the third ventricle. PheB was injected in the left lateral ventricle: 5 μl over 10 sec and then 1 μl/min for 20 min of a 320 μg/μl solution of [14C]-PheB (5 dpm/nmole). The rats were decapitated and the hippo-campus and cortex dissected away. The remainder of the brain was frozen and sliced at 300 μm intervals. The slices were digested and the radioactivity was determined by liquid scintillation spec-troscopy.

phenobarbital were also made into the ventral branch of the third ventricle. These injections caused a significant inhibition of choline uptake in the hippocampus (Fig. 5). After this localized injection of the drug 2/3 of the rats did not lose the righting reflex but they all exhibited vigorous chewing behavior.

DISCUSSION

These results together with our previous findings (12) originally led us to conclude that the site at which the barbiturates act to inhibit choline uptake in the hippocampus was distant from the septal-hippocampal neurons and near the preoptic/anterior hypothalamus. Since it is known that barbiturates can enhance GABAergic inhibition (8, 9) we thought it possible that they had such an effect in this region which was then expressed trans-synaptically on this cholinergic pathway. Indirect effects of other agents on the septal-hippocampal neurons have been found in other studies. Thus, using turnover of ACh in the hippocampus as a measure, Wood et al. (16) have provided evidence that GABAergic interneurons from the lateral septum have an inhibitory effect on the cholinergic cell bodies of the medial septum, and the inhibitory effects of opioid and dopaminergic projections on the septal-hippocampal neurons were found to be mediated by these GABAergic interneurons.

However, more recent work in our laboratory now demonstrates that localized administration of phenobarbital in the medial septum (but not the preoptic/anterior hypothalamus) does inhibit choline uptake in the hippocampus if dose and time are appropriately controlled (Richter and Gormley, in preparation). These more recent results are not inconsistent with the present findings. Even though the drug levels were highest around A6000-A7000 when phenobarbital was injected in the lateral ventricle with a block in the third ventricle (Fig. 3), our calculations indicate that the lower levels of drug nearer the septal region would still be sufficient to be effective in the septal area. Similarly, there would be sufficient spread rostrally from injections of phenobarbital in the ventral third ventricle (Fig. 5) to expose the septum to the drug. Therefore it may be suggested that phenobarbital either enhances GABAergic action on medial septal cholinergic neurons or has a direct GABA-like effect there. Since Brunello and Cheney (1) found that lateral septum to medial septum GABAergic interneurons were apparently not necessary for the effects of systemic pentobarbital on ACh turnover in the hippocampus, it may be more likely that the barbiturates can have a direct GABA-like effect in the medial septum.

Previous results from our laboratory provided some evidence that the inhibition of choline uptake in the hippocampus was not related to the sedative-hypnotic effects of these drugs (14). After i.p. administration of pentobarbital hippocampal choline uptake was not inhibited if the rats were decaptiated just when they lost

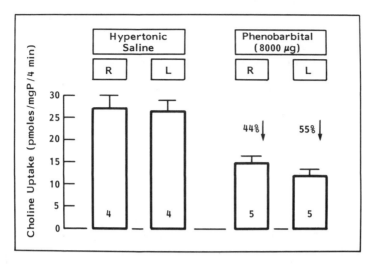

Figure 4. Intraventricular Phenobarbital (PheB) Inhibits Hippocampal Choline Uptake When the Spread of the Drug to the Caudal Ventricular System is Prevented. Just before injection of PheB, a block of Nivea cream was introduced (see Fig. 3). PheB was injected in the left lateral ventricle: 5 μl over 10 sec and then 1 μl/min for 20 min of a 320 μg/μl solution of [^{14}C]-PheB (5 dpm/nmole). Control rats were injected with the Nivea cream and then a solution of hypertonic saline (1.28 M, pH 9.6) on the same schedule. The rats were de-capitated immediately and choline uptake was measured in hippocampal synaptosomes as described in Methods and Materials. Values given are means $\pm$ SEM with the number of rats used indicated in each bar. The rostro-caudal distribution of the [^{14}C]-PheB is shown in Fig. 3.

the righting reflex. However shortly thereafter, when the animals were still unable to right themselves, the inhibition of choline uptake was observed. Using measurements of drug levels in gross brain regions, the lag in the effect on choline uptake could not be attributed to a lower level of drug in the brain. When various convulsants were given and the animals killed during convulsion, hippocampal choline uptake was not always increased. Therefore the changes in choline uptake in the hippocampus are not the result of the alteration of the state of the animal - from hypnosis in the one direction, to convulsions in the other (14).

The present results provide more evidence that the inhibition of choline uptake in the hippocampus is not related to the sedative-hypnotic effects of the barbiturates. While lateral ventricular injections of phenobarbital (with or without the block) caused the majority of the rats to lose their righting reflex, the more localized application of the drug to the ventral portion of the third ventricle

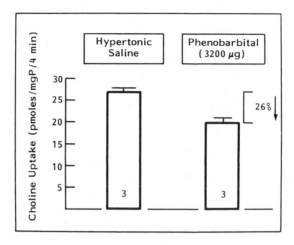

Figure 5. Effect of phenobarbital (PheB) injected in the ventral-rostral third ventricle. [^{14}C]-PheB (320 μg/μl, 5 dpm/nmole) was injected in the ventral portion of the third ventricle (A7000, V2000, L 0) at the rate of 0.5 μl/min for 20 min. Control rats were injected with hypertonic saline (1.28 M, pH 9.6) in the same manner. The rats were decapitated immediately and choline uptake in hippocampal synaptosomes was measured as described in Methods and Materials. Values are means ⁺ SEM (N=3).

did not cause the loss of the righting reflex in the majority of the rats, even though choline uptake in the hippocampus was inhibited. Thus the sedative-hypnotic actions of the barbiturates are apparently mediated by their action at another site and inhibition of cholinergic function in the hippocampus is not necessary for this effect. Systemically administered barbiturates inhibit choline uptake in the cortex as well as the hippocampus (but not in the striatum) (15), and it is likely that inhibition of cholinergic projections to the cortex is also unrelated to the sedative hypnotic effects although this has not yet been tested. The fact that lesions of the medial septum or the nucleus Basalis do not cause loss of the righting reflex, hypnosis or anesthesia is also consistent with this suggestion.

These findings also point to the need to determine what functions of the animal are in fact altered when cholinergic projections to the hippocampus (and probably the cortex) are inhibited by barbiturates. The barbiturates have many other actions including anxiolytic effects, anticonvulsant effects and effects on learning and memory (13). We have, for example, some evidence that the inhibition of choline uptake may be related to an anticonvulsant action of these and some other drugs (Miller and Richter, 6, 7). Moreover, since the loss of cholinergic neurons is a prominent finding in Alzheimer's

Disease (rev. in 10) it is considered quite possible that this loss may be related to the learning and memory deficits so prominent in this disease. In humans with Alzheimer's Disease the loss of ChAT in the cortex has been correlated with the degree of mental impairment (11). In rats, LoConte et al. (4) have recently shown that even unilateral lesions in the cholinergic forebrain nuclei result in decreased cognitive function. Thus it may be suggested that the effects of barbiturates on learning and memory are also mediated through their inhibition of cholinergic function in the hippocampus and cortex.

Acknowledgements. The excellent technical work of Ms. Joanne M. Gormley and assistance of Ms. Sandra Barton are gratefully acknowledged. This research was supported in part by Grant No. DA 00796.

REFERENCES

1. Brunello, N. and Cheney, D.L. (1981): J. Pharmacol. Exp. Therap. 219:489-495.
2. Herz, A., Albus, K., Metys, J., Schubert, P. and Teschemacher, H. (1970): Neuropharmacology 9:539-551.
3. König, J.F.R. and Klippel, R.A. (1974): The Rat Brain. A Stereotaxic Atlas of the Forebrain and Lower Parts of the Brain Stem. Robert E. Krieger Pub. Co., Inc., Huntington, NY.
4. LoConte, G., Bartolini, L., Casamenti, F., Marconcini Pepeu, I. and Pepeu, G. (1982): Pharmacol. Biochem. and Behav. 17: 933-937.
5. Lowry, O.H., Rosebrough, N.J., Farr, A.L. and Randall, R.J. (1951): J. Biol. Chem. 193:265-295.
6. Miller, J.A. and Richter, J.A. (1984): Br. J. Pharmacol., in press.
7. Miller, J.A. and Richter, J.A. (1983): Soc. Neurosci. Abs. 9:971.
8. Nicoll, R. (1978): In: Psychopharmacology: A Generation of Progress, (eds) M.A. Lipton, A. DiMascio and K.F. Killam, Raven Press, NY, pp. 1337- 1348.
9. Olsen, R.W. (1981): J. Neurochem. 37:1-13.
10. Perry, E.K. and Perry, R.H. (1980): In: Biochemistry of Dementia (ed) P.J. Roberts, John Wiley and Sons, Ltd., London, pp. 135-183.
11. Perry, E.K., Tomlinson, B.E., Blessed, G., Bergmann, K., Gibson, P.H. and Perry, R.H. (1978): Brit. Med. J. 2:1457-1459.
12. Richter, J.A. and Gormley, J.M. (1982): J. Pharmacol. Exp. Therap. 222:778-785.
13. Richter, J.A. and Holtman, J.R., Jr. (1982): Prog. Neurobiol. 18:275-319.
14. Richter, J.A., Gormley, J.M., Holtman, J.R., Jr., and Simon, J.R. (1982): J. Neurochem. 39:1440-1445.
15. Simon, J.R., Atweh, S. and Kuhar, M.J. (1976): J. Neurochem. 26:909-922.

16. Wood, P.L., Cheney, D.L. and Costa, E. (1979):   Neuroscience
     $\underline{4}$:1479-1484.

STUDIES ON A cAMP-DEPENDENT PROTEIN KINASE

OBTAINED FROM NICOTINIC RECEPTOR-BEARING MICROSACS

H. Eriksson, R. Salmonsson, G. Liljeqvist and E. Heilbronn

Unit of Neurochemistry and Neurotoxicology
University of Stockholm
Enköpingsvägen 126, S-172 46
Sundbyberg Sweden

INTRODUCTION

The nicotinic acetylcholine receptor (nAChR) of Torpedo californica is composed of four different subunits with apparent relative molecular M_r of 40,000 (α), 50,000 (β), 60,000 (γ) and 66,000 (δ) with 2:1:1:1 stoichiometry in the monomer (4, 14). In the receptor-containing membranes from electric organs of the same fish, a protein kinase is present which is able to phosphorylate the γ- and δ-subunits (3, 6, 7, 15, 17). This phosphorylation has been regarded as cyclic nucleotide independent (6, 15, 17) until recently, when Huganir and Greengard (10) found that cAMP stimulates the endogenous phosphorylation of nAChR from the electric organ of Torpedo californica.

The present report describes work aiming at the isolation and characterization of a cAMP-dependent kinase that phosphorylates the nAChR in the electric organ of Torpedo marmorata. From data on kinase specificity and on the primary structure of the nAChR we also suggest a plausible phosphorylation site on each of the γ- and δ-subunits.

MATERIALS AND METHODS

The nAChR-enriched membranes were prepared from electric organs of Torpedo marmorata essentially as previously described (4). A protease inhibitor mixture was included in all solutions (5). For the preparation of the protein kinase, receptor enriched membranes solubilized in the protease inhibitor mixture containing 2.5% octaethyleneglycol mono n-dodecylether ($C_{12}E_8$; Nikko Chemicals), was layered on top of a sucrose density gradient, 5-20%. The material was centrifuged for 24 h at 36,000 rpm in a Beckman SW 40Ti rotor.

933

Fractions were collected and assayed for protein kinase activity as well as for αbungarotoxin binding (16).

The protein kinase activity in the fractions from the sedimentation velocity centrifugation was assayed with the heavy form of the nAChR as substrate, obtained after the same centrifugation. Each assay tube contained 20 mM Tris-HCl (pH 7.4), 20 mM $MgCl_2$, 1 mM ouabain, 10 mM mercaptoethanol and 0.1% $C_{12}E_8$. The phosphorylation was started by the addition of ATP [2 μM ATP, 1 μCi (γ-^{32}P)ATP; NEN]. The reaction was stopped after 30 s by adding 2.5% SDS, 0.1M mercaptoethanol and 0.1 M Tris-HCl, pH 6.8. When the endogenous phosphorylation of the postsynaptic membranes was studied no exogenous substrate was added, but otherwise the assays were identical. When the effects of cAMP and protein kinase inhibitor were investigated, these were included in the preincubation mixture 5 minutes prior to addition of ATP. The SDS-polyacrylamide gel electrophoresis on 9% gels that followed was according to Laemmli (11). Autoradiograms from dried gels were obtained after exposure to Kodak X-OMAT films.

RESULTS

The endogenous phosphorylation of the nAChR from electric organ has previously been shown to appear on the γ- and δ-subunits of the nAChR (3, 5, 10). The cAMP-dependence of this endogenous nAChR phosphorylation, was investigated. The results (Fig.1) reveal that the phosphorylation of the M_r 66,000 polypeptide was activated approximately twofold by cAMP, while an inhibitor of cAMP-dependent protein kinases reduced the phosphorylation by 80%.

The protein kinase was separated from its substrate (nAChR) by a sedimentation velocity technique (Fig. 2). The protein kinase activity, which still was cAMP-dependent (data not shown) was found to be separated from both receptor forms with only a minor overlap. The small second apparent activity peak observed around fraction 21, in Figure 2, is due to a significant increase in the amount of the substrate (9S form of nAChR) in the assay of these fractions. A sedimentation coefficient of about 6S was estimated for the protein kinase (12), which roughly corresponds to an M_r of 140,000.

DISCUSSION

The phosphorylation of the nAChR, both in a system where the previously separated protein kinase and nAChR were brought together and in solubilized receptor-bearing membranes, was shown to be largely, if not completely due to a cAMP-dependent protein kinase. The observation that the kinase inhibitor reduced the phosphorylation by 80% but was unable to abolish it, at the conditions we used, might be interpreted as if there exists another receptor kinase activity as well.

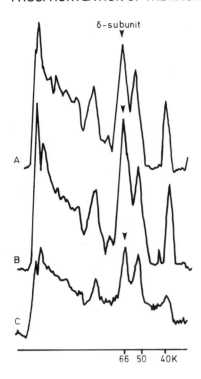

δ-subunit

A

B

C

66 50 40K

Figure 1. Endogenous cAMP-dependent phosphorylation of nAChR containing membranes. The membranes were phosphorylated in the standard assay (A), in the presence of 1 μM cAMP (B)/or 200 μg/50 μl of protein kinase inhibitor (B), as indicated. The results are shown as densitometric scans of autoradiograms produced from SDS gels.

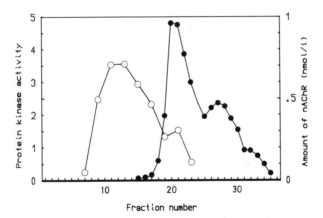

Figure 2. Sedimentation velocity centrifugation of solubilized post-synaptic membranes rich in nAChR. Protein kinase activity (o—o) is expressed in arbitrary units. The amount of nAChR (•—•) is estimated as the amount of α-bungarotoxin binding sites. Fractions are numbered from the top of the gradient. For experimental details see Materials and Methods.

The observation (Fig. 1) that the phosphorylation of most other non-receptor polypeptides (like the M_r 39,000, 52,000 and 94,000 components) were similarily affected by both cAMP and the kinase inhibitor, and that cAMP did activate the protein kinase in both systems used, supports the idea that only a cAMP-dependent protein kinase was present.

The results shown in Figure 2 suggest that the solubilized protein kinase behaves as a tetrameric holoenzyme, still activatable by cAMP. A considerable proportion of free catalytic subunits in the solubilized microsac preparation would have been revealed by a protein kinase activity appearing in fractions of lower number than what actually was observed.

The amino acid sequence for the nAChR subunits (from Torpedo californica; 13), as deduced from the cDNA sequence analysis, may be used for a consideration of probable phosphorylation sites. Models have also been constructed depicting possible transmembrane segments of the receptor subunits (2, 13).

The data in the literature which show that the kinase acts on the cytoplasmic side of the membrane (3) restrict the possible targets for phosphorylation to the sequence intervals 246-251 and 308-449 for the γ-subunit; and 252-256 and 313-455 for the δ-subunit. cAMP-dependent protein kinases phosphorylate preferentially serine residues in peptides or protein substrates. The phosphorylation sites in physiologically significant substrates for cAMP-dependent protein kinases, sequenced so far, often follow the pattern ARG-ARG-X-SER or LYS-ARG-X-X-SER (8, 9) with nonpolar residues on the C-terminal side (1). In the sequences mentioned above only the SER in position 354 (γ-subunit) and in position 362 (δ-subunit) fulfill these requirements. The flanking amino acid sequences are ARG-ARG-ARG-SER-SER(P)-PHE-GLY and ARG-ARG-SER-SER-SER(P)-VAL-GLY. SER 361 (δ-subunit) cannot be excluded as the phosphorylation target, though the residue on the C-terminal side is polar.

The two serine residues that we suggest to be the target for the cAMP-dependent phosphorylation are thus evidently in homologous positions as shown by Noda et al. (13). When comparing the sequences for all the receptor subunits, it is also apparent that these serine residues are present only in the γ- and δ-subunits; and that the α-and β-subunits are lacking homologous sequences to the other two subunits in this part of the structure.

Acknowledgements. This work was supported by the Swedish Natural Science Research Council, The Swedish Medical Science Research Council, and Magn. Bergvalls Stiftelse.

REFERENCES

1. Carlson, G.M., Bechtel, P.J. and Graves, D.J. (1979): Adv. in Enzymol. 50:41-115.
2. Claudio, T., Ballivet, M., Patrick, J. and Heinemann, S. (1983): Proc. Natl. Acad. Sci. (USA) 80:1111-1115.
3. Davis, C.G., Gordon, A.S. and Diamond, I. (1982): Proc. Natl. Acad. Sci. (USA) 79:3666-3670.
4. Eriksson, H., Liljeqvist, G. and Heilbronn, E. (1983): Biochim. Biophys. Acta 728:449-454.
5. Eriksson, H., Salmonsson, R. Liljeqvist, G. and Heilbronn, E. (1983): manuscript submitted.
6. Gordon, A.S., Davis, C.G. and Diamond, I. (1977): Proc. Natl. Acad. Sci. (USA) 74:263-267.
7. Heilbronn, E., Björk, C., Elfman, L., Hartman, A. and Mattsson, C. (1979): In Neurotoxins; Tools in Neurobiology (eds) B. Ceccarelli and F. Clementi, Raven Press, New York, pp. 151-158.
8. Huang, T.S. and Krebs, E.G. (1979): FEBS Lett. 98:66-70.
9. Huang, T.S., Feramisco, J.R., Glass, D.B. and Krebs, E.G. (1979): In From Gene to Protein: Information Transfer in Normal and Abnormal Cells (eds) T.R. Russell, K. Brew, H. Faber and J. Schultz, Academic Press, New York, pp. 449-461.
10. Huganir, R.L. and Greengard, P. (1983): Proc. Natl. Acad. Sci. (USA) 80:1130-1134.
11. Laemmli, U.K. (1970): Nature 227:680-685.
12. Martin, R.G. and Ames, B.N. (1961): J. Biol. Chem 236:1372-1379.
13. Noda, M., Takahashi, H., Tanabe, T., Toyosato, M. Kikyotani, S., Furutani, Y., Hirose, T., Takashima, H., Inayama, S., Miyata, T. and Numa, S. (1983): Nature 302:528-532.
14. Reynolds, J.A. and Karlin, A. (1978): Biochemistry 17:2035-2038.
15. Saitoh, T. and Changeux, J.P. (1980): Eur. J. Biochem. 105:51-62.
16. Schmidt, J. and Raftery, M.A. (1973): Anal. Biochem. 52:349-354.
17. Teichberg, V.I. and Changeux, J.P. (1977): FEBS Lett. 74:71-76.

DIFFERENTIAL EFFECTS OF CARBACHOL AND OXOTREMORINE ON MUSCARINIC RECEPTORS, CYCLIC AMP FORMATION, AND PHOSPHOINOSITIDE TURNOVER IN CHICK HEART CELLS

J.H. Brown and S.B. Masters

Division of Pharmacology
University of California, San Diego
La Jolla, California 92093 U.S.A.

INTRODUCTION

A central theme in current studies of muscarinic receptors is the definition of muscarinic receptor subtypes. Early evidence for the existence of such receptor subtypes came from studies of physiological responses mediated through muscarinic receptors in various tissues (1, 7). More recently, investigators have used radioligand binding studies to look for muscarinic receptor subtypes. Attempts to distinguish subtypes with classical antagonists have not been fruitful since these agents generally recognize a uniform population of binding sites (9). Several non-classical antagonists such as gallamine (8) and pirenzepine (10) do appear to show selectivity in radioligand binding assays, but gallamine probably acts at a site that is different from the primary antagonist binding site (8) and the selectivity of pirenzepine has been disputed (11).

Although there remains some question as to whether muscarinic receptor subtypes can as yet be defined on the basis of antagonist receptor binding properties, there is evidence from receptor binding experiments that subpopulations of muscarinic receptor are differentiated by agonists. Muscarinic agonists generally bind to receptors in membrane preparations with more than a single affinity (9). It has been suggested that agonist binding sites of different affinity result from coupling of a portion of the receptors to their effectors (3).

We have previously shown that cholinergic agonists elicit two distinct biochemical responses through activation of cardiac muscarinic receptors (5, 6). One response is inhibition of catecholamine-stimulated cyclic AMP formation. The other is stimulation of phosphoinositide (PhI) hydrolysis by a phospholipase C. To determine whether these biochemical responses can be used to define distinct populations of muscarinic receptors, we have directly compared the effects of agonists and antagonists on the cyclic AMP and PhI responses in dissociated chick heart cells. Since we are interested in the relationships between receptor binding properties and responses, we have also examined the binding properties of these agents in intact heart cells.

MATERIALS AND METHODS

Hearts from 13-day embryonic chicks were dispersed with 0.25% trypsin, and cells were freed by triturating tissue fragments in Krebs Henseleit medium buffered with 20 mM HEPES (KHB) and containing 1% BSA. Dissociated heart cells were separated from erythrocytes and cellular debris by centrifugation in 30% Percoll. In the final cell preparation 80-90% of the cells excluded trypan blue and this level of viability was maintained for at least 3 hours.

For all assays, cells were incubated at $35^{\circ}C$ in KHB. In cyclic AMP experiments, cells were equilibrated with 100 μM isobutylmethyl-xanthine, a phosphodiesterase inhibitor. Drugs were then added for 2 min and incubations were terminated by centrifugation and immediate addition of 10% TCA. Cyclic AMP was purified and assayed by the competitive binding assay of Gilman. The inhibitory effects of cholinergic agonists and antagonists were assessed in the presence of 3 μM isoproterenol.

Phosphoinositide (PhI) hydrolysis was monitored by measuring accumulation of the hydrolysis product inositol 1-phosphate (Ins1P). Cellular phosphoinositides were labelled by incubating heart cells with [^{3}H]inositol for one hour. Cholinergic agonists and LiCl (10 mM), an inhibitor of Ins1P dephosphorylation (2), were then added for an additional 30 min. At the end of the drug incubation, cells were centrifuged and extracted with chloroform:methanol:water (10: 10:9). The upper (aqueous) phase of this 2-phase extract was frac-tionated by anion exchange chromatography to separate [^{3}H]Ins1P from [^{3}H]inositol (2, 6).

In radioligand binding experiments, intact chick heart cells were incubated for 75 min with various concentrations of [^{3}H]quinu-clidinylbenzilate (QNB) (saturation experiments) or with 0.1 nM QNB and various concentrations of unlabelled agonists or antagonists (competition experiments). Specific binding was defined as that inhibited by 10 μM atropine and was greater than 80% of the total binding. Data from agonist competition experiments and saturation experiments were analyzed with nonlinear least-squares computer curve fitting (12).

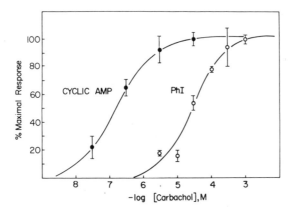

Figure 1. Dose-response relationships for cyclic AMP and PhI responses to carbachol. Cyclic AMP values represent the percent of the maximal inhibition of Iso-stimulated cyclic AMP accumulation by carbachol. PhI values represent the percent of the stimulation above basal of [^{3}H]Ins1P accumulation caused by a maximal dose of carbachol.

RESULTS

The cholinergic agonist carbamylcholine (carbachol) inhibits cyclic AMP formation and stimulates PhI hydrolysis in chick heart cells (Fig. 1). Both responses appear to be mediated through activation of muscarinic receptors since they are blocked by atropine, but not by d-tubocurarine (data not shown). The dose-response relationships for the two responses to carbachol are quite different; the K_{act} value for the cyclic AMP response is 2 x 10^{-7}M, whereas the K_{act} for the PhI response is 2 x 10^{-5}M. Thus, one differential property of the two responses is that carbachol is 100-fold less potent at stimulating PhI hydrolysis than at inhibiting cyclic AMP formation.

The muscarinic agonist oxotremorine also inhibits cyclic AMP accumulation in chick heart cells (Fig. 2). Oxotremorine is about 10 times more potent than carbachol in eliciting a cyclic AMP response, and causes a maximal degree of inhibition of cyclic AMP accumulation which is equivalent to that caused by carbachol. In contrast, oxotremorine is virtually ineffective in stimulating PhI hydrolysis (Fig. 3). Oxotremorine does interact with receptors coupled to PhI hydrolysis since it blocks the stimulatory effect of carbachol on PhI hydrolysis (4). Thus, another differential property of these responses is that both carbachol and oxotremorine inhibit Iso-stimulated cyclic AMP accumulation, but only carbachol stimulates PhI hydrolysis.

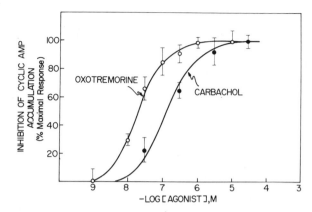

Figure 2. A comparison of the oxotremorine and the carbachol dose-response relationships for inhibition of Iso-stimulated cyclic AMP accumulation.

If an antagonist has selectivity for the receptors coupled to one of the two biochemical responses, we would expect to see differences in the dose-response relationships for response inhibition by the antagonist. The muscarinic antagonist pirenzepine is nearly equally effective at blocking the PhI response and the cyclic AMP response (Fig. 4). There is no difference in the potency of the classical muscarinic blocking agent atropine for inhibiting activation of the receptors that mediate these two biochemical responses (data not shown). Thus, the muscarinic antagonists we studied do not appear to differentiate receptors coupled to the cyclic AMP response from those coupled to the PhI response.

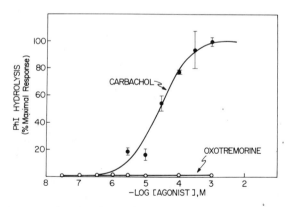

Figure 3. A comparison of carbachol and oxotremorine effects on PhI hydrolysis. Values represent the percent of the maximal carbachol-induced increase in [^{3}H]Ins1P accumulation.

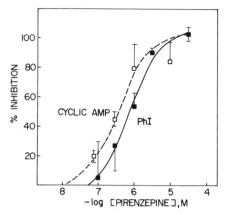

Figure 4. Pirenzepine antagonism of the cyclic AMP and PhI responses to carbachol in chick heart slices. Cyclic AMP values represent the percent decrease in the carbachol-induced inhibition of Iso-stimulated cyclic AMP accumulation. PhI values represent the percent inhibition of the carbachol-induced increase in [^{3}H]Ins1P accumulation. These concentrations of carbachol were chosen for the cyclic AMP and PhI assays were 1 μM and 100 μM, respectively, each approximately 5 times greater than the K_{act} of carbachol for the respective response.

We studied the binding properties of muscarinic receptors on intact chick heart cells under conditions that were identical to those used to assay the two biochemical responses. The radioligand [^{3}H]QNB appears to bind to a homogeneous population of sites with a K_D of 12 pM (Fig. 5). Similarily, both atropine and pirenzepine appear to recognize a single population of receptor binding sites (data not shown). Representative data from a competition experiment with the agonists carbachol and oxotremorine is shown in Figure 6. The data from carbachol competition experiments are best fit by a two-site model while the data from oxotremorine competition curves indicate that oxotremorine has a single high affinity for all of the [^{3}H]QNB binding sites on the intact cell (Fig. 6).

DISCUSSION

The experiments presented here demonstrate that muscarinic agonists have different potencies and efficacies for eliciting cyclic AMP and PhI responses. The agonists carbachol and oxotremorine also differ in their interaction with the muscarinic receptor since carbachol recognizes or induces a low affinity receptor state while oxotremorine binds with a single, relatively high affinity. There appears to be only a single population of binding sites for all three antagonists, QNB, atropine and pirenzepine, and there is no evidence that the PhI and cyclic AMP responses are selectively affected

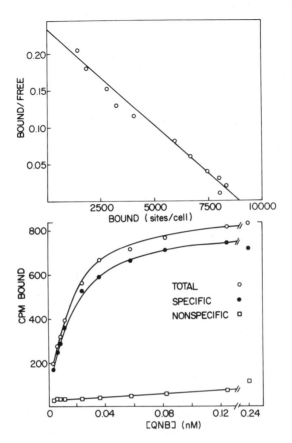

Figure 5. Scatchard plot and saturation curve of [³H]QNB binding to intact chick heart cells. The bottom panel depicts total, specific and nonspecific [³H]QNB binding to 1.8×10^6 cells at various QNB concentrations. Scatchard analysis (top panel) of these data indicated that QNB bound to a single population of sites ($r^2 = 0.98$, Hill slope=1.00) with a K_D of about 12 pM, and that there were approximately 9,000 receptor sites per cell.

by antagonists, even pirenzepine, an antagonist proported to have discriminatory abilities (10).

    The muscarinic receptor is multifunctional because its activation by agonists leads to several distinct biochemical responses, including those we have studied here. One way to think about this multifunctionality is according to a model in which a single type of receptors couples to and thus activates different effectors. Our data suggest that the ability of an agonist to cause receptor-effector coupling may depend on the nature of the effector. Oxotremorine causes the muscarinic receptor to couple to the effector that mediates inhibition

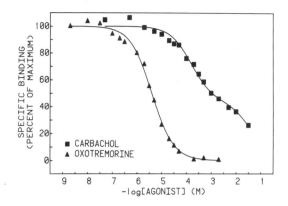

Figure 6. **Oxotremorine and carbachol competition for [³H]QNB binding sites.** The data points represent the percent of the maximal specific [³H]QNB binding. The solid lines are theoretical curves derived from a computer analysis that fits data to a one- or two-site model. The oxotremorine data were best fit by a one-site model while the carbachol data were best fit by a two-site model (p < 0.001).

of Iso-stimulated cyclic AMP formation. Oxotremorine does not, however, cause these receptors to couple to the effector that mediates the PhI response. Carbachol, an agonist that elicits both biochemical responses, can cause the muscarinic receptor to couple to both effectors, but still differentiates these effectors on the basis of potency. Low doses of carbachol initiate coupling to the effector of the cyclic AMP response yet higher doses are required for receptor coupling to the PhI response.

It has been suggested that agonist-induced coupling to cellular effectors alters the agonist binding properties of the receptor (3). The nature of the effector to which receptors couple may also be reflected in the binding properties of agonists. Carbachol recognizes or induces two receptor states in the intact heart cells and also causes both biochemical responses while oxotremorine only recognizes or induces one state and causes one response. The molecular correlate of the inability of oxotremorine to cause receptors to couple to the effector of the PhI response may be the failure of oxotremorine to induce a low agonist affinity state of the receptor (see Fig. 7).

One explanation for these differential effects of agonists may be that different degrees of receptor occupancy are necessary for receptor-effector coupling to the various effectors. In the case of the effector of the PhI response, full receptor occupancy is needed. Thus, high doses of carbachol are required and oxotremorine, a relatively weak agonist, can not occupy enough receptors to cause receptor coupling to this effector.

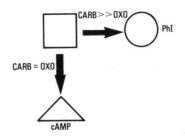

Figure 7. Hypothetical model for differential agonist effects on muscarinic receptor coupling to cyclic AMP and phosphoinositide responses.

In summary, the data we have presented here demonstrate that the agonists oxotremorine and carbachol differentiate muscarinic receptors coupled to adenyate cyclase and to the PhI response, and suggest that the nature of receptor-effector coupling confers specificity upon the activity of the agonist.

Acknowledgements. We thank David Goldstein for his excellent technical assistance. This work was supported by NIH grant HL 28143, and AHA grant 82-S112. JHB is an Established Investigator of the AHA and SLB is an NIH predoctoral trainee.

REFERENCES

1. Barlow, R.B., Berry, K.J., Glanton, P.A.M., Nikolaou, N.M. and Soh, K.S. (1976): Br. J. Pharmacol. 58:613-620.
2. Berridge, M.J., Downes, C.P. and Hanley, M.R. (1982): Biochem. J. 206:587-595.
3. Birdsall, N.J.M., Burgen, A.S.V. and Hulme, E.C. (1978): Mol. Pharmacol. 14:723-736.
4. Brown, J.H. and Brown, S.L. (1984): J. Biol. Chem. 259:3777-3781.
5. Brown, J.H. and Brown, S.L. (1984): Fed. Proc. 43:2613-2617.
6. Brown, S.L. and Brown, J.H. (1983): Mol. Pharmacol. 24:351-356.
7. Burgen, A.S.V. and Spero, L. (1968): Br. J. Pharmacol. 34:99-115.
8. Dunlap, J. and Brown, J.H. (1983): Mol. Pharmacol. 24:15-22.
9. Ehlert, F.J., Roeske, W.R. and Yamamura, H.I. (1981): Fed. Proc. 40:153-159.
10. Hammer, R., Berrie, C.P., Birdsall, N.J.M., Burgen, A.S.V. and Hulme, E.C. (1980): Nature 283:90-92.
11. Laduron, P.M., Leysen, J.E. and Gorissen, H. (1981): Arch. Int. Pharmacodyn. 249:319-321.
12. Munson, P.J. and Rodbard, D. (1980): Anal. Biochem. 107:220-239.

POLYPHOSPHOINOSITIDE AND PHOSPHOPROTEIN RESPONSES TO

MUSCARINIC RECEPTOR ACTIVATION IN BOVINE ADRENAL MEDULLA

J.N. Hawthorne and A-M. F. Swilem

Department of Biochemistry
Medical School
Nottingham NG7 2UH, U.K.

INTRODUCTION

Bovine adrenal medulla responds to nicotinic stimulation by secreting catecholamines, but simultaneous activation of muscarinic receptors in the perfused gland depresses secretion (10). The calcium influx required for secretion is caused by nicotinic activation (3) but the breakdown of inositol phospholipids considered by Michell (8) to be responsible for mobilization of calcium is produced by muscarinic drugs only. This inositol phospholipid (phosphoinositide) effect was previously thought to be initiated by hydrolysis of phosphatidylinositol but recent work suggests that triphosphoinositide (PtdIns $4,5P_2$, Fig. 1), a more labile compound associated with plasma membranes and some secretory granules, is the lipid first involved (reviewed in reference 4). A possible sequence of reactions is given in Figure 2.

Since the bovine adrenal medulla appeared unusual in exhibiting muscarinic receptor-linked phosphoinositide metabolism and no apparent calcium mobilization, we have studied it in more detail using chromaffin cells in culture. Earlier work on adrenal medulla showed changes in labelling of phosphatidylinositol and phosphatidate in response to muscarinic activation (6,11) but more recent work in other tissues (1, 7) indicates that the first lipid change is breakdown of PtdIns $4,5P_2$ as outlined in Figure 2. We have therefore investigated this possibility in the chromaffin cells. It also seemed important to study the requirement for calcium ions in these reactions.

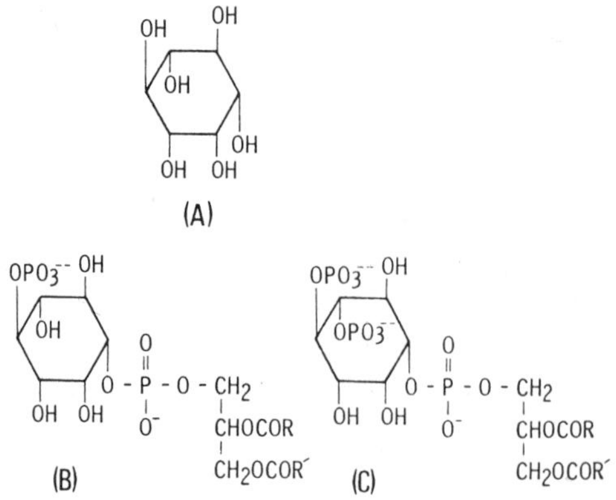

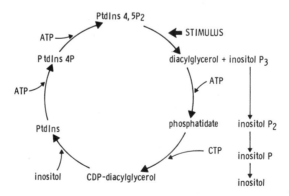

Figure 1. **The major phosphoinositides.** **(A)** Phosphatidylinositol
(PtdIns); **(B)** Phosphatidylinositol 4-phosphate (diphosphoinositide,
PtdIns4P); **(C)** Phosphatidylinositol 4,5-bisphosphate (triphospho-
inositide, PtdIns 4,5P$_2$)

METHODS

Isolation and culture of chromaffin cells.   The method of Fisher
et al. (3) was followed. A Percoll gradient was used for further
purification of the cells and their viability was measured with
Trypan blue (at least 95% of cells were viable). Cells were added
to 16 mm uncoated culture wells (approx. 600,000 per well) containing
1 ml Eagle's minimum essential medium supplemented with 10% heat--
inactivated foetal calf serum, 20 μM 5-fluorodeoxyuridine (to inhibit
cell division and fibroblase proliferation), 100 units/ml penicillin
and streptomycin, and 2.5 μg/ml Fungizone. The cells were used after
2 days culture.

Figure 2.   **Receptor-linked phosphoinositide metabolism.**

Phospholipid labelling studies. The cultured cells were pre-labelled for 75 min in 1 ml Locke's solution containing 50 μCi/ml [^{32}P]orthophosphate and then washed twice with cold Locke's solution to remove excess [^{32}P]. The medium was then replaced with Locke's solution containing cholinergic drug (or control Locke's solution only) and the brief incubation at 25°C terminated by the addition of 1 ml 20% trichloroacetic acid (TCA) and approximately 20 mg wet weight of adrenal medulla homogenate, this latter as "carrier" tissue. The precipitate was centrifuged down, washed once with 1 ml 5% TCA containing 1 mM EDTA and once with 2 ml distilled water.

Lipid was extracted from the pellets with 1.5 ml chloroform-methanol-conc. HCl (100:100:1, by volume). The extraction was repeated with this solvent and a further treatment given with chloroform-methanol-conc. HCl (200:100:1), 1.5 ml of each being used. The extracts were combined and mixed with 1.5 ml chloroform and 1.5 ml 0.1 N HCl. After shaking well the phases were separated by centrifugation and the lower chloroform layer was washed twice with 1.5 ml synthetic upper phase. This lower layer was then dried in a stream of nitrogen and the lipid residue was dissolved in 25 μl chloroform for application to thin-layer plates. The system of Van Rooijen et al. (12) was used for one-dimensional separation of phosphatidate and the three phosphoinositides. The phospholipids were located by iodine vapour staining and also scanning for radioactivity (thin-layer scanner RTLS-1A, Panax Equipment Ltd., Mitcham, Surrey, U.K.). The relevant areas of the plates were scraped off, and radioactivity of the silica gel determined by liquid scintillation counting.

RESULTS

As shown in Table 1, carbachol produced roughly a 20% loss of PtdIns 4,5P$_2$ and of PtdIns 4P in 30 sec when pre-labelled chromaffin cells were incubated with the drug. At the same time there was increased labelling of phosphatidate, probably because the diacylglycerol released from the phosphoinositides was phosphorylated by diacylglycerol kinase (Figure 2). No change was seen in the radioactivity of PtdIns and the same effects were produced by muscarine, confirming that this was a muscarinic and not a nicotinic response.

The loss of inositol lipids did not occur in Ca^{++}-free media. The losses in general are similar to those produced by vasopressin in the hepatocyte (reviewed in reference 4) but the Ca^{++}-dependence of the effect is controversial for that cell. The present results agree with those of Exton's group (9) who also found that the loss of PtdIns 4,5P$_2$ was Ca^{++}-dependent.

DISCUSSION

These results suggest that Michell's theory that receptor-linked phosphoinositide hydrolysis leads to entry of Ca^{++} (8) does not apply

Table 1

Changes in [^{32}P]-Labelling of Phospholipids When Adrenal
Chromaffin Cells are Incubated Briefly with Carbachol

	Ca^{++} (mM)	PtdIns 4,5P$_2$	PtdIns 4P	Phosphatidate
Control	2.2	602 $\pm$ 135	305 $\pm$ 142	202 $\pm$ 47
Carbachol	2.2	481 $\pm$ 112*	231 $\pm$ 83	322 $\pm$ 20**
Carbachol	0	580 $\pm$ 144	325 $\pm$ 175	288 $\pm$ 23
Carbachol	0 (0.5 mM EGTA)	584 $\pm$ 146	320 $\pm$ 162	261 $\pm$ 45

Cells were pre-labelled for 60 min with [^{32}P] and stimulated with
carbachol (3 x 10^{-4} M) for 30 sec. Results are means $\pm$ S.D. from
six experiments and represent total c.p.m. in the lipids. Significantly different from control using Student's t-test:
*p < 0.05; **p < 0.001.

to the adrenal medulla. There are two reasons for this suggestion.
First, muscarinic receptors on the bovine chromaffin cell do not
increase the concentration of intracellular Ca^{++} (3) and second,
the loss of PtdIns 4,5P$_2$ requires extracellular Ca^{++}. The latter
observation might indicate that the phosphoinositide loss is a consequence rather than a cause of Ca^{++} entry, but it might also mean
that external Ca^{++} is required for the interaction between acetylcholine and the receptor.

If the phosphoinositide changes are not responsible for Ca^{++}
mobilization, what physiological function do they have? As yet
there is no direct evidence for a function in the chromaffin cell,
but work with other cells suggests that PtdIns 4,5P$_2$ might activate
the plasma membrane Ca^{++} pump. This is usually measured biochemically
as a Ca^{++}-dependent ATPase which maintains intracellular Ca^{++} concentrations at levels below 10^{-6}M. This Ca^{++} pump ATPase is activated
by PtdIns 4,5P$_2$ in the red cell and in synaptosomes (reviewed in
reference 4). In the chromaffin cell Ca^{++} can be removed from the
cytoplasm by the plasma membrane Ca^{++} pump and also by uptake into
chromaffin granules. In both cases, polyphosphoinositides can be
formed on the cytoplasmic face of the relevant membrane (5). However,
if PtdIns 4,5P$_2$ activates the Ca^{++} pump, the observed loss of this
lipid in response to muscarinic agonists should make the pump less
active and increase the concentration of cytosolic Ca^{++}. Fisher et

al. (3) did not observe such an increase so the role of PtdIns $4,5P_2$ remains enigmatic.

It is also difficult to understand how acetylcholine can both stimulate catecholamine secretion through a nicotinic receptor and reduce it through a muscarinic receptor in the same chromaffin cell. A possible explanation is that the two receptors have different sensitivities to acetylcholine and to Ca^{++}.

Phosphoprotein metabolism has not been studied in the present series of experiments, but work from Nishizuka's group in Japan and Gispen's in Holland (reviewed in reference 4) indicates that phosphoinositide and phosphoprotein metabolism may be inter-related. The diacylglycerol released when PtdIns or PtdIns $4,5P_2$ are hydrolyzed can activate protein C kinase of Nishizuka. In the presence of the diacylglycerol, the kinase appears to be activated by much lower concentrations of Ca^{++} than are required without this lipid. On the other hand, PtdIns 4P kinase, the enzyme responsible for triphosphoinositide synthesis, is regulated by a B50 kinase which may be identical with protein C kinase (2). It is not yet clear how these observations link up, but the final explanation of the receptor-linked phosphoinositide effect may come from a better understanding of the protein kinases.

REFERENCES

1. Abdel-Latif, A.A., Akhtar, R.A. and Hawthorne, J.N. (1977): Biochem. J. 162:61-73.
2. Aloyo, V.J., Zwiers, H. and Gispen, W.H. (1982): Prog. Brain Res. 56:303-315.
3. Fisher, S.K., Holz, R.W. and Agranoff, B.W. (1981): J. Neurochem. 37:491-497.
4. Hawthorne, J.N. (1983): Bioscience Reports 3:887-904.
5. Hawthorne, J.N., Mohd. Adnan, N. and Lymberopoulos, G. (1980): Biochem. Soc. Trans. 8:30-32.
6. Hokin, M.R., Benfrey, B.G. and Hokin, L.E. (1958): J. Biol. Chem. 233:814-817.
7. Kirk, C.J., Creba, J.A., Downes, C.P. and Michell, R.H. (1981): Biochem. Soc. Trans. 9:377-379.
8. Michell, R.H. (1975): Biochim. Biophys. Acta 415:81-147.
9. Rhodes, D., Prpic, V., Exton, J.H. and Blackmore, P.F. (1983): J. Biol. Chem. 258:2770-2773.
10. Swilem, A-M.F., Hawthorne, J.N. and Azila, N. (1983): Biochem. Pharmacology 32:3873-3874.
11. Trifaro, J.M. (1969): Mol. Pharmacology 5:382-393.
12. Van Rooijen, L.A.A., Sequin, E.B. and Agranoff, B.W. (1983): Biochem. Biophys. Res. Commun. 112:919-926.

GABA$_A$ VS. GABA$_B$ MODULATION OF

SEPTAL-HIPPOCAMPAL INTERCONNECTIONS

W.D. Blaker[#], D.L. Cheney[+] and E. Costa

Laboratory of Preclinical Pharmacology
National Institute of Mental Health
Saint Elizabeths Hospital
Washington, D.C. 20032 U.S.A.

[#]Present Address:
Virginia-Maryland Regional
College of Veterinary Medicine
Virginia Polytechnic
Institute and State University
Blacksburg, Virginia 24061 U.S.A.

[+]Director of Neuroscience Research
Neuroscience Subdivision
Ciba Geigy Corporation
Summit, New Jersey 07901 U.S.A.

INTRODUCTION

In rats, intraseptal injection of the GABA receptor agonist muscimol decreases the turnover rate of acetylcholine (TR$_{ACh}$) in the hippocampus in a dose-dependent and bicuculline-reversible manner (15). This is in keeping with the histochemical evidence of a substantial GABAergic innervation of the septal nuclei (11). Furthermore, stimulation of other septal neurotransmitter receptors (e.g., dopamine and beta-endorphin) can decrease the hippocampal TR$_{ACh}$ via an activation of GABAergic interneurons (for review see 4). Conversely, some neurotransmitters, such as substance P, modulate the cholinergic activity of this pathway independently from GABAergic mechanisms.

Recent evidence has pointed to the existence of two classes of high affinity GABA binding sites in the CNS, designated GABA$_A$ and GABA$_B$ (for review see 3). The GABA$_A$ site is part of the "classical" GABA receptor which is linked to Cl$^-$ channels and whose activation increases Cl$^-$ fluxes across cell membranes according to concentration gradients. The GABA$_B$ site is much less characterized with regard to operative molecular mechanisms, but has been implicated in the presynaptic inhibition of excitatory amino acid release.

On the basis of binding, electrophysiological and other studies, muscimol and (-)baclofen have been shown to be rather specific agonists at the $GABA_A$ and $GABA_B$ sites, respectively, while bicuculline is a specific $GABA_A$ antagonist. No specific antagonists of $GABA_B$ sites have been described. The effect of muscimol on the hippocampal TR_{ACh} implicates a septal $GABA_A$ receptor population in the modulation of this cholinergic pathway, but the possibility of modulatory actions mediated by $GABA_B$ receptors has not been investigated.

Numerous studies have been done on the involvement of these limbic structures on various behaviors in the rat (for review see 6). Although lesions of the septum, hippocampus and fimbria-fornix cause a wide range of behavioral alterations, response inhibition is reduced by all three types of lesions. This alteration is expressed by sustained, inappropriate responding during reinforcement schedules involving, for example, extinction. Because intrahippocampal, but not intraseptal, injection of the muscarinic antagonist scopolamine reduces response inhibition and because most, if not all, of the cholinergic input to the hippocampus originates in the septum, it is believed that the cholinergic septal-hippocampal pathway participates in the mediation of such response inhibition.

The studies reported here were undertaken to correlate pharmacologically induced decreases in the hippocampal TR_{ACh} with changes in extinction of a food-reinforced lever press response. We have previously shown that intraseptal muscimol simultaneously and dose-dependently decreases the TR_{ACh} in the hippocampus and increases responding during extinction (2). We shall herein differentiate the behavioral effects elicited by GABAergic vs. non-GABAergic inhibition of hippocampal cholinergic activity as well as show that $GABA_A$ receptor activation in the septum produces a behavioral-biochemical profile different from that elicited by $GABA_B$ receptor activation.

METHODS

The behavioral and surgical methods employed were essentially those described previously by us (2). Briefly, male Zivic-Miller rats were trained on a continuous reinforcement (CRF) schedule for 2-3 sessions (20 min session/day) after which chronic septal guide cannulae were implanted. CRF training continued for another four sessions. On the day of the experiment, appropriate amounts of various drugs dissolved in saline (muscimol, Pierce Chemicals, Rockford, IL; (-) and (+)baclofen, gift from Ciba-Geigy Pharmaceuticals, Summit, NJ; beta-endorphin, Peninsula Laboratories, Belmont, CA; substance P and bicuculline methobromide, Sigma Chemical Co., St. Louis, MO) were injected in a volume of 1 μl over a period of one minute into the area of the medial septum. Fifteen minutes later, the rat was placed in the operant chamber and allowed to lever press under CRF for five minutes, after which the pellet feeder was disconnected and extinction allowed to proceed for ten minutes. The rat was then infused via the lateral tail vein with phosphoryl $(C[^2H]_3)_3$ choline chloride (15 μmole/kg/min) for nine minutes, and killed by head focused microwave irradiation. Brain areas were dissected and analyzed for endogenous and deuterated choline and acetylcholine by gas chromatography-mass spectrometry. The regional content and turnover rate of acetylcholine were determined as described previously (14).

To characterize GABA receptors [^{3}H]GABA binding was performed in rats injected bilaterally with 1 μg kainic acid into the ventral and dorsal hippocampi (4 μg total). After 3 1/2 to 5 weeks, the septal area was dissected, homogenized in 100 volumes distilled water at 0-4°C, centrifuged at 30,000 xg for 15 minutes and the pellet resuspended in 50 mM TRIS-HCl (pH 7.4), 2.5 mM $CaCl_2$. The suspension was frozen, thawed, incubated at 37°C for 20 minutes, and washed five times with the TRIS-Ca^{++} buffer at 0-4°C. Binding conditions included 20 nM [^{3}H]GABA (35 Ci/mmole, New England Nuclear, Boston, MA), 100 μM bicuculline methobromide or (-)baclofen, and approximately 60 μg membrane protein in 500 μl TRIS-Ca^{++} buffer. Incubation at 20°C for five minutes was terminated by centrifugation. Specific binding of [^{3}H]GABA was defined as that displaced by the presence of 1 mM GABA.

RESULTS

Representative cumulative recorder tracings showing the effect of various intraseptal doses of the $GABA_A$ agonist muscimol on extinction after CRF training are shown for one experiment in Fig. 1. The most marked differences between muscimol and saline treated rats were seen in the extinction response patterns. Rats given a non-sedative dose of intraseptal muscimol (i.e., less than 10 nmoles) showed more prolonged and sometimes unabated continuous responding. The combined behavioral results of six independent experiments are shown in Fig. 2A as the average response rates during the five minutes preceding extinction and during the ten minutes with extinction. Doses of 0.3 to 3 nmoles resulted in significant increases in extinction responding, whereas higher doses decreased responding due to sedation. CRF responding also tended to increase but did not consistently reach statistical significance.

The effects of intraseptal muscimol on acetylcholine metabolism in the hippocampus are shown in Fig. 2B. All doses of 0.3 nmoles or higher elicited a decrease in hippocampal TR_{ACh} while doses of 1 nmole or higher, in addition, increased the acetylcholine content. The TR_{ACh} in the parietal cortex, which receives no cholinergic innervation from the medial septum, was decreased only when a sedative dose of 30 nmoles was administered (data not shown). Thus, up to a dose of 3 nmoles, the dose response to reduce the hippocampal TR_{ACh} is similar to that observed to elicit a behavioral response.

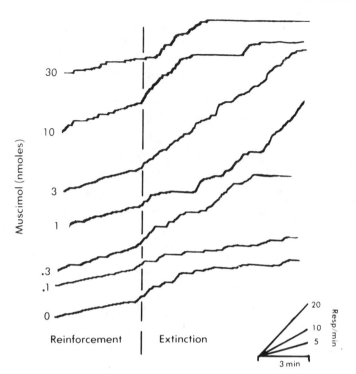

Figure 1.   Cumulative recorder tracings of extinction responding
after acute intraseptal injection of muscimol into CRF-trained rats.
The vertical line indicates the onset of extinction.

        Table 1 shows the behavioral effect of intraseptal bicuculline,
a GABA$_A$ antagonist.   Whereas muscimol increased responding during
extinction, bicuculline led to a decrease in both CRF and extinction
responding.   The dose of bicuculline used produced no gross behavioral
effects such as hyperactivity.   Administration of bicuculline five
minutes before muscimol led to intermediate response rates similar
to those of controls.

        Preliminary testing of neuropeptides in this behavioral paradigm
have shown the following: Intraseptal beta-endorphin, which is
effective in lowering the hippocampal TR$_{ACh}$, produces an increase
in extinction responding (data not shown).   On the other hand, intraseptal substance P produces no behavioral effects when given in
doses which decrease the hippocampal TR$_{ACh}$ to a similar extent as
behaviorally effective doses of muscimol (data not shown).

        To study whether GABA$_B$ receptors mediate behavioral and bio-
chemical effects similar to those of muscimol, the GABA$_B$ agonist
(-)baclofen was tested.   A dose-related modification of extinction

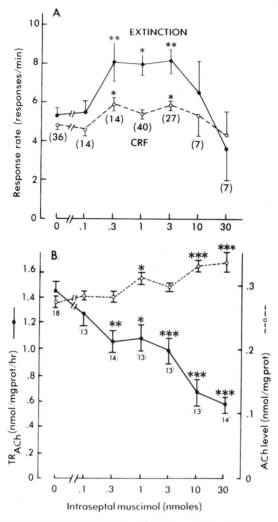

Figure 2. Average response rates during CRF and extinction (A) and hippocampal acetylcholine content and TR_{ACh} (B) after acute intra-septal injection of muscimol. Numbers in parentheses indicate the number of animals in each group and error bars indicate SEM. Significant differences from saline controls by Dunnett's multiple comparison test: $^{*}p < 0.5$; $^{**}p < .02$; $^{***}p < .01$.

was elicited by (-)baclofen (Fig. 3A). This drug increased extinction responding with a biphasic profile similar to that elicited by muscimol. However (-)baclofen is 5-10 times more potent than muscimol on a molar basis. A dose of 0.5 nmoles (+)baclofen was without effect whereas (-)baclofen was active in 0.1 nmole doses. In contrast to muscimol, (-)baclofen failed to decrease hippocampal TR_{ACh} or

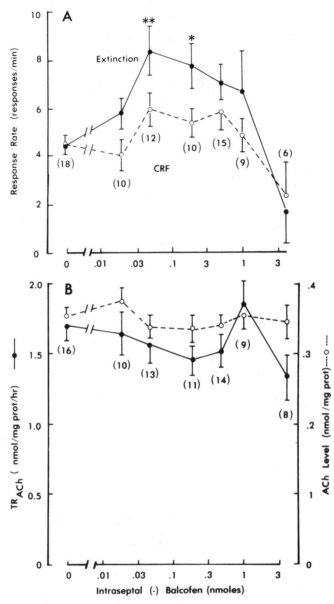

Figure 3. Average response rates during CRF and extinction (A) and hippocampal acetylcholine content and TR$_{ACh}$ (B) after acute intraseptal injection of (-)baclofen. Numbers in parentheses indicate number of animals in each group and error bars indicate SEM. Significant differences from saline controls by Dunnett's multiple comparison test: *p < .05, **p < .02.

Table 1

Effect of Intraseptal Muscimol (1 nmol)
and Bicuculline (1 nmol) on Responding

Treatment	N	Response Rate (Responses/Min) During CRF	During Extinction
Saline	13	5.1 ± .5	6.1 ± .7
Muscimol	16	5.7 ± .2	9.0 ± .8**
Bicuculline	9	2.3 ± .5**	2.9 ± .8*
Muscimol plus Bicuculline	13	4.1 ± .4	7.0 ± .7

Significant difference from saline group: $^*p < .05$; $^{**}p < .02$; $^{***}p < .01$

to increase acetylcholine content, even at near-sedative doses
(Fig. 3B).

To investigate the location of septal $GABA_A$ and $GABA_B$ receptors, the binding of [^{3}H]GABA to septal membranes was analyzed in rats lesioned with kainic acid in the hippocampus. A sufficient length of time was allowed after the lesion to allow for complete degeneration of hippocampal efferents. Table 2 shows that such a lesion decreased $GABA_B$ type (bicuculline-insensitive) binding by approximately 40% while $GABA_A$ binding (baclofen-insensitive) was unaffected.

Table 2

Effect of Intrahippocampal Kainic Acid Lesion
on [^{3}H]GABA Binding to Septal Membranes

Group	N	Specific [^{3}H]GABA binding (fmoles/mg prot) Total	+Baclofen ($GABA_A$)	+Bicuculline ($GABA_B$)	+Baclofen +Bicuculline (Residual)
Sham	22	277 ± 22	189 ± 17	101 ± 13	37 ± 12
KA	23	261 ± 21	163 ± 17	61 ± 9**	28 ± 12

Significant difference from sham group: $^{**}p < .02$

DISCUSSION

Studies involving lesions of hippocampal afferents and localized administrations of cholinergic drugs have contributed a large body of evidence suggesting that the cholinergic septal-hippocampal pathway is capable of regulating extinction as defined by response inhibition consequent to changes in reward patterns (for discussion see 2, 6). We find that intraseptal muscimol in doses that decrease the TR_{ACh} in the hippocampus increase response rates during extinction, indicating that responding during extinction is increased when the cholinergic neurons innervating the hippocampus are firing at a lower average rate. In addition, the trend toward an increase in CRF responding elicited by such doses of intraseptal muscimol is reminiscent of that seen after septal lesions (8).

The behavioral results of preliminary experiments using doses of beta-endorphin and substance P which reduce TR_{ACh} are compatible with the view that increased extinction responding is produced selectively by those decreases in hippocampal cholinergic activity which are mediated by a GABAergic mechanism. Intraseptal beta-endorphin decreases hippocampal TR_{ACh} in a bicuculline-reversible manner (15). This has been interpreted to reflect an activation of septal GABAergic interneurons which in turn inhibit the firing of septal cholinergic neurons. However substance P-induced decrease in the hippocampal TR_{ACh} is unaffected by bicuculline, indicating that this neuropeptide probably elicits an inhibition independent from an activation of GABAergic interneurons. Since we find that beta-endorphin, but not substance P, is effective in increasing extinction responding, this behavioral effect may be selectively mediated by GABAergic inhibition of septal cholinergics. All the above results indicate that a $GABA_A$-type modulation of the cholinergic projections from the septum to the hippocampus may mediate increased responding during extinction.

Experiments were performed to investigate why the $GABA_B$ agonist (-)baclofen produces a behavioral response similar to that of muscimol while failing to reduce hippocampal TR_{ACh}. Considering that the main, if not sole, projection area of the cholinergic neurons located in the medial septum is the hippocampus (10, 13), and that electrical stimulation of this projection disinhibits hippocampal pyramidal cells via a cholinergic mechanism (1, 7), it is probable that reduction of TR_{ACh} reduces pyramidal cell firing. Since the hippocampal pyramidal cells project almost exclusively to the lateral septum (12), and neurochemical studies have shown that these neurons are most likely glutamatergic (5, 9) we conclude that a decrease in hippocampal TR_{ACh} causes a reduction in the activity of these glutamatergic afferents to the lateral septum. Since $GABA_B$ sites have been implicated in the presynaptic inhibition of glutamate release, the neuronal localization of septal $GABA_A$ and $GABA_B$ binding sites was examined. Our results suggest that a sizeable proportion of $GABA_B$ sites in the septum are located presynaptically on afferents from the hippo-campus, whereas few if any of the GABA$_A$ sites can be considered to have such a location.

Thus both intraseptal muscimol and baclofen may disrupt extinction by inhibiting the output of the glutamatergic hippocampal afferents terminating in lateral septum. Muscimol acts by inhibiting the stimulatory cholinergic input to the hippocampal pyramidal cells via the activation of GABA$_A$ receptors presumably located on the cholinergic cell bodies of the medial septum. Baclofen acts by inhibiting presynaptically the glutamatergic hippocampal efferents to the lateral septum via the activation of presynaptic GABA$_B$ re-ceptors. Our results also suggest that a high degree of restraint should be shown when interpreting the behavioral effects of pharma-cological agents in the absence of supporting biochemical data. In fact, these two GABA agonists produce very similar effects on a specific behavior when microinjected into the septum, but do so by different neuronal mechanisms, as documented by the present study of their biochemical effects.

REFERENCES

1. Ben-Ari, Y., Krnjevic, K., Reinhardt, W. and Ropert, N. (1981): Neurosci. 6:2475-2484.
2. Blaker, W.D., Cheney, D.L., Gandolfi, O. and Costa, E. (1983): J. Pharm. Exp. Ther. 225:361-365.
3. Bowery, N.G. (1982): Trends in Pharmacological Sciences 3:400-403.
4. Costa, E., Panula, P., Thompson, H.K. and Cheney, D.L. (1983): Life Sci. 32:165-179.
5. Fonnum, F. and Walaas, I. (1978): J. Neurochem. 31:1173-1181.
6. Grossman, S.P. (1978): In Functions of the Septo-Hippocampal System, Ciba Foundation Symposium, K. Elliot and J. Whelan (eds.) Elsevier/North Holland, New York, pp. 227-273.
7. Krnjevic, K. and Ropert, N. (1981): Can. J. Physiol. Pharma-col. 59:911-914.
8. Lorens, S.A. and Kondo, C.Y. (1969): Physiol. Behav. 4:729-732.
9. Malthe-Sorenssen, D., Skride, K.K. and Fonnum, F. (1980): Neuro-sci. 5:127-133.
10. McKinney, M., Coyle, J.T. and Hedreen, J.C. (1983): J. Comp. Neurol. 217:103-121.
11. Panula, P., Reveulta, A.V., Cheney, D.L., Wu, J.Y. and Costa, E. (1983): J. Comp. Neurol. 222:69-80.
12. Swanson, L.W. and Cowan, W.M. (1976): J. Comp. Neurol. 172:49-84.
13. Swanson, L.W. and Cowan, W.M. (1979): J. Comp. Neurol. 186:621-656.
14. Wood, P.L. and Cheney, D.L. (1979): Can. J. Physiol. Pharma-col. 57:404-411.
15. Wood, P.L., Cheney, D.L. and Costa, E. (1979): Neurosci. 4:1479-1484.

SEPTAL GABAERGIC NEURONS: LOCALIZATION AND POSSIBLE

INVOLVEMENT IN THE SEPTAL-HIPPOCAMPAL FEEDBACK LOOP

D.L. Cheney and P. Panula[*]

Pharmaceuticals Division, CIBA-GEIGY Corporation
Neuroscience Research, Research Department
Summit, New Jersey 07901 U.S.A.

*Department of Anatomy
University of Helsinki
Helsinki, Finland

INTRODUCTION

Much has been learned both histologically and biochemically about the major connections of the distinct areas of the septal complex, the functional interactions with the hippocampus and the chemical character of some of these pathways (7). The cholinergic septal-hippocampal pathway serves as a well-defined link between these two important structures of the limbic system. The study of this pathway has increased our understanding of the synaptic organization within the septal nuclei (7).

It has been shown that cholinergic neurons, regulated by gamma-aminobutyric acid containing (GABAergic) interneurons, originate in the medial septum and diagonal band of Broca and project to the hippocampus where they impinge on what appear to be glutamatergic pyramidal cells. Glutamatergic neurons send collaterals back to the lateral septum where they functionally regulate another set of GABAergic neurons (7). The present study reviews the morphological evidence for GABAergic terminals and cell bodies in the septal nuclei and suggests that there are at least two different GABAergic neuronal systems operating in the septum: a group of large cells in the medial septum and diagonal band, and a population of small cells in the lateral septal nucleus.

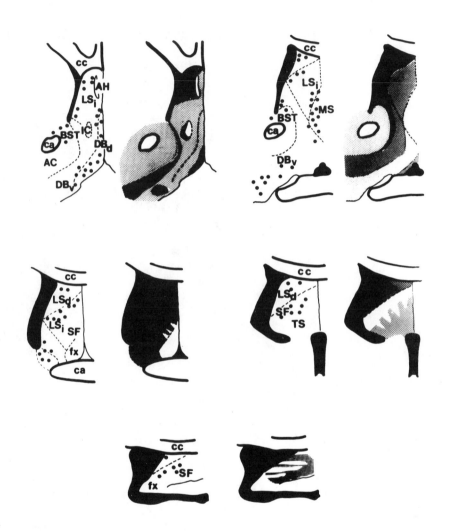

Figure 1. Distribution of GAD-like immunoreactivity in the septal
nuclei after intraseptal colchicine (top left, bottom left, bottom
right) and without colchicine pretreatment (top right). Top left
- #A medium-sized neuron and numerous nerve processes show GAD-like
immunoreactivity in the medial septal nucleus (MS) after colchicine
pretreatment. Bottom left - neuronal cell bodies in the ventral
part of the nucleus of the diagonal band (DB_V) are heavily stained
for GAD. Top right - small dendritic profiles contain GAD-like
immunoreactivity in the dorsal part of the lateral septal nucleus
(LS_D) in normal rat brain. Bottom right - small cells exhibit GAD-like
immunoreactivity after colchicine pretreatment in the intermediate
part of the lateral septal nucleus (LS_I).

Adult male Sprague Dawley rats received a single injection of
saline or colchicine (70 μg/20 μl of saline) 24 to 48 hr before
killing. The animals were perfused with physiological saline followed
by ice-cold 4% formalin. The peroxidase-antiperoxidase method of
Sternberger et al. (38) was used. Free floating coronal sections
of the rat forebrain were incubated with 20% normal swine serum at
room temperature for 20 min; incubated with specific GAD antiserum
diluted 1:5000, 1:1000 or 1:2000 with PBS-T at 4°C for 48 hr; incubated with swine anti-rabbit immunoglobulins diluted 1:50 or 1:100
with PBS-T at room temperature for 30min; incubated with a soluble
complex of horseradish peroxidase-rabbit antihorseradish peroxidase
diluted 1:50 or 1:100 with PBS-T at room temperature for 30 min;
and reacted with 3,3'-diaminobenzidine tetrahydrochloride (50 mg/100
ml) and 0.003% hydrogen peroxide in 0.05 M Tris-HCl buffer, pH 7.6
for 8 min. The sections were then collected on slides, air dried,
embedded in Permount® and studied with brightfield or interference
contrast microscopy.

No staining in septal structures was found in control specimens
(see 23).

THE MEDIAL SEPTAL NUCLEUS AND THE DIAGONAL BAND

Neurons of the medial septal nucleus and the diagonal band project to the hippocampus (8, 9, 34) via the fimbria and dorsal fornix and enter at the level of the striatum oriens (basal dendrites of pyramidal cells) of field CA_3. Immediately after entering the hippocampus, they divide and innervate the pyramidal cells of the stratum oriens of fields CA_1, CA_2 and CA_4 plus the supragranular region of the dentate gyrus (29). Histochemical studies utilizing acetylcholinesterase histochemical techniques (19, 36) and measurement of choline acetyltransferase activity (20) suggest that this pathway is cholinergic. This has been confirmed using measurements of acetylcholine release (12) and acetylcholine turnover (28). Cholinergic cells are distributed throughout the medial septal nucleus and the diagonal band as revealed by immunohistochemical demonstration of choline acetyltransferase (16).

Recently studies using antisera against L-glutamic acid decarboxylase (GAD) have been used to locate GABAergic neurons and nerve terminals in the medial septal nucleus and the diagonal band of rat by using the peroxidase-antiperoxidase method (32). Moderately immunoreactive terminals can be observed in the medial septum. The intensity decreases in the dorsal part of the diagonal band and is weak in the ventral part. Nerve cell bodies can only be demonstrated after intraventricular or intraseptal injections of colchicine. Large GAD-immunoreactive neurons are located in the medial septal nucleus (Fig. 1, top left) and the diagonal band (Fig. 1, bottom left) where they are distributed similarly to the cholinergic neurons (16). Distribution of terminals and cell bodies is shown schematically in Fig. 2. The observation that the distribution of cholinergic cells in the medial septal nucleus and the diagonal band (16) closely resembles the distribution of the cells containing GAD-like immunoreactivity raises the possibility that a close relationship may exist between these two groups of cells.

There is pharmacological evidence that the inhibitory action of dopamine and beta-endorphin on hippocampal cholinergic dynamics, but not that of substance P, is mediated by GABAergic interneurons (5, 44). However, in light of recent demonstrations of choline acetyltransferase and GAD in the same motoneurons (3, 4) the coexistence of these two neurotransmitter synthesizing enzymes in neurons of the medial septal nucleus and the diagonal band cannot be excluded.

THE LATERAL SEPTAL NUCLEUS

The lateral septal nucleus receives its major input from the collaterals of the pyramidal cells in CA_3 of the hippocampus that pass through the fimbria (11, 25, 34, 40). The transmitter of this excitatory projection appears to be either glutamate or aspartate

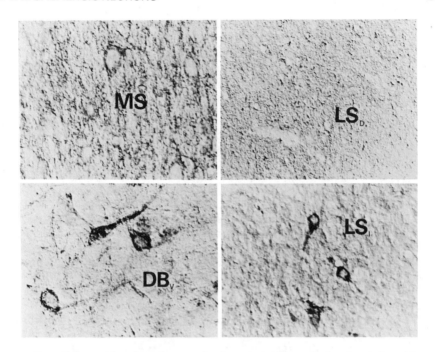

Figure 2. Schematic drawings of the septal nuclei show the distribution of GAD-like immunoreactivity in nerve cell bodies (full circles, left side) and nerve fibers and terminals (right side). The divisions of the septal nuclei (adapted from Swanson and Cowan, 41) are also shown on the left side. Representative sections from the rostral (top left) to the caudal (bottom) part of the septum are shown. The distance from bregma (+ anterior, posterior) are: top left = +0.7 mm, top right = +0.1 mm, middle left = -0.3 mm, middle right = -0.8 mm, bottom = -1.0 mm. Low density terminal staining is shown on the right side by sparsely spaced dots; medium density terminal staining is shown by closely spaced dots; and the high density terminal staining is shown by the dark diagonal lines. Abbreviations are as follows: AC = nucleus accumbens, AH = anterior hippocampal rudiment, BST = bed nucleus of the stria terminalis, ca = commissura anterior, cc = corpus callosum, DB_d = nucleus of the diagonal band of Broca, pars dorsalis, DB_v = nucleus of the diagonal band of Broca, pars ventralis, fx = fornix, IC = island of Calleja magna, SL_d = lateral septal nucleus, pars dorsalis, LD_i = lateral septal nucleus, pars intermedia, MS = medial septal nucleus, SF = septofibrial nucleus, TS = triangular septal nucleus (adapted from 32).

(13, 39, 45). Kainic acid, the potent rigid analog of glutamate, has been used to identify the nature of the cells involved in the feedback regulation of the septal neurons (13). The GAD activity in the septum itself and the choline acetyltransferase activity in the hippocampus remain unchanged after the injection of kainic acid into the medial septum (2). Furthermore, the acetylcholine turnover rate in the hippocampus is not altered (2, 45). Since the cells in the medial septum appear to be resistant to kainic acid, they are probably not innervated by glutamatergic fibers (18, 45). Conversely, the same amout of kainic acid injected into the lateral septum reduces specifically the GAD activity in the ipsilateral septum whereas no significant change in choline acetyltransferase or acetylcholine turnover rate can be detected in the hippocampus after the lesion of the lateral septum (2). This would be consistent with the possibility that GABAergic neurons in the lateral septum are innervated by glutamatergic nerve terminals.

Other afferents to the lateral septal nucleus include noradrenergic fibers from the nucleus locus coeruleus (15, 27, 42), a dopaminergic projection from the ventral tegmental area (21, 26) and serotonergic fibers from the raphe nuclei (1, 6, 17, 35, 37). Studies using antisera to GAD demonstrate that a dense network of terminals containing GAD-like immunoreactivity is present in the lateral septal nucleus, especially in the pars dorsalis, indicating that this area is innervated by GABAergic fibers (Fig. 1, top right). Biochemical lesion studies suggest that these fibers do not originate from neurons located outside the septal area (13, 14, 39). In the lateral septal nucleus of normal rat brain, intense GAD-like immunoreactive staining occurs in terminals whereas no staining of cell bodies is apparent (Fig. 1, top right). After intraseptal colchicine injection numerous uniform neurons are visible in the pars dorsalis, pars intermedia (Fig. 1, bottom right) and pars ventralis. All GAD-immunoreactive cells in the lateral septal nucleus appear to be of medium-size and appear to have one main process which exhibits GAD-like immunoreactivity. Cell bodies and terminals are shown schematically in Fig. 2

The lateral septal nucleus projects to the medial septal nucleus, the diagonal band (34, 41), the medial and lateral preoptic nucleus, the anterior hypothalamic areas (30, 31, 34, 41), the parataenial and paraventricular nuclei of the thalamus and the medial habenular nucleus (41). Little is known about the transmitters that are operative in these projections, since none of the classical neurotransmitters has been demonstrated in the great majority of the lateral septal neurons. GABA could be involved in the functioning of these pathways, and the GABAergic inhibition of the lateral septal cells could be mediated by axon collaterals of the GABAergic projection cells, since several of the receptive areas contain nerve terminals with GAD-like immunoreactivity (33). However, electrophysiological studies indicate that this would not be generally true since the cells of the pars dorsalis of the lateral septal nucleus do not project outside the septum (10).

DISCUSSION

The density of nerve terminals exhibiting GAD-like immunoreactivity in various nuclei (32) is in general agreement with the
distribution of the GAD activity in vitro (13, 43). Thus, the lateral
septal nucleus is densely innervated by GABAergic fibers compared
to the medial septal nucleus (13, 43). The diagonal band contains
considerable enzyme activity (43), but histochemical studies (32)
reveal that most of the GAD-like immunoreactivity is associated
with fibers extending from the area of the olfactory tubercle to
the island of Calleja magna and a thin rim of immunoreactive terminals
between the diagonal band and the nucleus accumbens, which corresponds
to the rostromedial extension of the substantia innominata.

In colchicine-injected rats GAD-like immunoreactivity is widely
distributed in neurons of the lateral and medial septal nuclei,
and the diagonal band (32). Iontophoretic application of GABA to
medial and lateral septal neurons inhibits their firing. This inhibition is antagonized by bicuculline but not strychnine (23). Stimulation of the fimbria elicits prolonged inhibition of both medial
and lateral septal neurons (22). This response is mediated by local
interneurons (22). However, only the lateral septal nucleus neurons
respond to various frequencies of fibrial stimulation (24).

The major output from the septal complex to the hypothalamic
area and the lateral preoptic nucleus is through the medial septal
nucleus and the nucleus of the diagonal band (30, 31, 34, 41).
Moreover, the single most important input to the medial division
of the septal complex is that from the lateral septal nucleus.
Yet, the chemical character of this input remains unknown. Electrophysiological (23) and biochemical (2) results suggest that GABAergic
interneurons may be involved. However, immunochemical studies
suggest that GABA may not be the main neurotransmitter for this
pathway since only scattered GAD-containing cell bodies are found
in the lateral septal nucleus. Indeed, there is a dense network
of GAD-positive fibers and terminals in the lateral septal nucleus
suggesting that, perhaps, many of the GABAergic cell bodies in the
medial septum send axons to the lateral septum. It is evident,
however, that the interactions that occur in the lateral septal
complex between the various hippocampal and medial septal inputs
must be of considerable importance for this feedback loop. Further
studies using the concept of the multiple transmitter coexistence
in the regulation of the turnover rate of acetylcholine in the hippocampus may continue to be useful to increase our understanding of
the neuronal interactions between the septum and hippocampus.

In summary, the immunohistochemical localization of two sets
of GABAergic neurons in the septal nuclei has been highlighted in
this review: (a) a population of large GABAergic cells in the medial
septal nucleus and the diagonal band, and (b) a population of small

GABAergic cells in the lateral septal nucleus. The evidence is consistent with an hypothesis suggesting that (a) GABAergic neurons in medial septum and diagonal band modulate the activity of the septal hippocampal cholinergic pathway, and (b) GABAergic interneurons in the lateral septum modulate the output of the glutamate/aspartate containing collaterals originating in the hippocampus.

REFERENCES

1. Azmitia, E.C., Jr. and Segal, M. J. (1978): J. Comp. Neurol. 179:641-668.
2. Brunello, N. and Cheney, D.L. (1981): J. Pharmacol. Exp. Ther. 219:489-495.
3. Chan-Palay, V., Engle, A.G., Palay, S.L. and Wu, J.Y. (1982): Proc. Natl. Acad. Sci. USA 79:6717-6721.
4. Chan-Palay, V., Engle, A.G., Wu, J.Y. and Palay, S.L. (1982): Proc. Natl. Acad. Sci. USA 79:7027-7030.
5. Cheney, D.L., Robinson, S.E., Malthe-Sorenssen, D., Wood, P.L. Comissiong, J.W. and Costa, E. (1978): In Neuropsychopharmacology, Vol. 5, Advances in Pharmacology and Therapeutics (ed) C. Dumont, Pergamon Press, New York, pp. 241-250.
6. Conrad, L.C.A., Leonard, C.M. and Pfaff, D.W. (1974): J. Comp. Neurol. 156:179-206.
7. Costa, E., Panula, P., Thompson, H.K. and Cheney, D.L. (1983): Life Sci. 32:165-179.
8. Cragg, B.G. and Hamlyn, L.H. (1956): J. Anat. (London) 90: 591-596.
9. Daitz, H.M. and Powell, T.P.S. (1954): J. Neur. Neurosurg. Psychiat. 17:75-82.
10. DeFrance, J.F. (1976): In The Septal Nuclei, Advances in Behavioral Biology, Vol. 20 (ed) J.F. DeFrance, Plenum Press, New York, pp. 185-227.
11. Deolmos, J. and Heinen, L. (1977): Neurosci. Lett. 6:107-114.
12. Dudar, J.D. (1975): Brain Res. 83:123-133.
13. Fonnum, F. and Walaas, I. (1978): J. Neurochem. 31:1173-1181.
14. Fonnum, F., Walaas, I. and Iversen, E. (1977): J. Neurochem. 29:221-230.
15. Fuxe, K. (1965): Acta Physiol. Scand. 64:Suppl.247,39-85.
16. Kimura, H., McGeer, P.L., Peng, J.H. and McGeer, E.G. (1981): J. Comp. Neurol. 200:151-201.
17. Kohler, C., Chan-Palay, V. and Steinbusch, H. (1982): J. Comp. Neurol. 209:91-111.
18. Kohler, C., Schwarcz, R. and Fuxe, K. (1978): Neurosci. Lett. 10:241-246.
19. Lewis, P.R. and Shute, C.C.D. (1967): Brain 90:521-540.
20. Lewis, P.R., Shute, C.C.D. and Silver, A. (1967): J. Physiol. (London) 191:215-224.
21. Lindvall, O. (1975): Brain Res. 87:89-95.
22. McLennan, H. and Miller, J.J. (1974): J. Physiol. (London) 237:607-624.

23. McLennan, H. and Miller, J.J. (1974): J. Physiol. (London)
    237:625-633.
24. McLennan, H. and Miller, J.J. (1976): J. Physiol. (London)
    254:827-841.
25. Meibach, R.C. and Siegel, A. (1977): Brain Res. 124:197-224.
26. Moore, R.Y. (1978): J. Comp. Neurol. 177:665-684.
27. Moore, R.Y., Bjorklund, A. and Stenevi, U. (1971): Brain Res.
    33:13-35.
28. Moroni, F., Malthe-Sorenssen, D., Cheney, D.L. and Costa, E.
    (1978): Brain Res. 150:333-341.
29. Mosko, S., Lynch, G. and Cotman, C.W. (1973): J. Comp. Neurol.
    152:163-174.
30. Nauta, W.J.H. (1958): J. Comp. Neurol. 104:247-271.
31. Nauta, W.J.H. (1956): Brain 81:319-341.
32. Panula, P., Revuelta, A.V., Cheney, D.L., Wu, J.Y. and
    Costa, E. (1984): J. Comp. Neurol. 222:69-80.
33. Perez, de la Mora, M. Rossani, L.D., Tapia, R. Teran, L. Pala-
    cios, R., Fuxe, K., Hokfelt, T. and Ljungdahl, A.(1981): Neuro-
    science 6:875-895.
34. Raisman, G. (1966): Brain 89:317-438.
35. Segal, M. and Landis, S.C. (1974): Brain Res. 82:263-268.
36. Shute, C.C.D. and Lewis, P.R. (1961): Nature (London) 189:
    332-333.
37. Steinbusch, H.W.M. (1981): Neuroscience 4:557-618.
38. Sternberger, L.A., Hardy, P.H., Cuculis, J.J. and Meyer, H.G.
    (1970): J. Histochem. Cytochem. 18:315-333.
39. Storm-Mathisen, J. and Opsahl, M.W. (1978): Neurosci. Lett.
    9:65-70.
40. Swanson, L.W. and Cowan, W.M. (1977): J. Comp. Neurol. 172:
    49-84.
41. Swanson, L.W. and Cowan, W.M. (1979): J. Comp. Neurol. 186:
    621-659.
42. Swanson, L.W. and Hartman, B.K. (1975): J. Comp. Neurol. 163:
    467-506.
43. Tappaz, M.L., Brownstein, M.J. and Palkovits, M. (1976): Brain
    Res. 108:371-379.
44. Wood, P.L., Cheney, D.L. and Costa, E. (1979): Neurosci. 4:
    1479-1484.
45. Wood, P.L., Peralta, E., Cheney, D.L. and Costa, E. (1979):
    Neuropharmacology 18:519-524.

REGULATION OF DRUGS AFFECTING STRIATAL CHOLINERGIC

ACTIVITY BY CORTICOSTRIATAL PROJECTIONS

H. Ladinsky, M. Sieklucka[*], F. Fiorentini,
G. Forloni, P. Cicioni and S. Consolo

Laboratory of Cholinergic Neuropharmacology,
Mario Negri Institute for Pharmacological Research
Milan, Italy

INTRODUCTION

Electrical (30) and biochemical (7, 22) studies have provided evidence for the existence of a massive pathway projecting from the cortex to the striatum. Histologically, the corticostriatal fibers originate from pyramidal cells of layer V (13, 32), and appear to make synaptic contacts with about 60% of all neurons intrinsic to the neostriatum, among which is included the entire population of Golgi type II interneurons associated with choline acetyltransferase-containing cells (12). In addition, the corticostriatal pathway appears to utilize glutamic acid (7, 15, 22, 28), an excitatory neurotransmitter that has been shown to increase acetylcholine (ACh) release from striatal slices in vitro, through N-methyl-D-aspartate and glutamic acid-preferring type receptors, possibly localized on the cholinergic cells (21). This pathway appears to regulate striatal cholinergic activity as the following results indicate: a) cholinergic interneurons are destroyed by the intrastriatal application of kainic acid, a conformationally restricted analog of glutamic acid; and b) striatal cholinergic neurotransmission is compromised after long-term decortication, i.e., the ACh turnover rate is decreased (31) and the sodium-dependent high affinity uptake of choline is reduced (27).

Research outlined in the present communication demonstrates further that the chronic degeneration of the corticostriatal excitatory

---

[*]Permanent address: Department of Pharmacology, Institute of Clinical Pathology, Medical School, Lublin, Poland.

pathway makes the cholinergic neurons of the striatum insensitive to the neuropharmacological action of a number of different drugs.

MATERIALS AND METHODS

Female CD-COBS rats (Charles River, Italy) , body weight 175-200 g were used. Acetylcholine and choline content of striatum was measured by the radioenzymatic method of Saelens et al. (26) with modifications by Ladinsky et al. (17), in rats killed by focussed microwave irradiation to the head (1.3 KW and 2.45 GHz for 4 sec). The ACh synthesis rate was determined by the method of Racagni et al. (24) after the i.v. infusion of [^{3}H]choline precursor as described earlier (29). Choline acetyltransferase activity was measured by the method of Fonnum (9). Striatal noradrenaline, dopamine and serotonin content was measured by electrochemical detection coupled with high pressure liquid chromatography (14, 23). Uptake of [^{3}H]-glutamic acid was estimated by the method of Divac et al. (7).

Frontal decortication by undercutting the cortex was performed in etherized rats as follows: the animals were positioned in a stereotaxic apparatus and the skull was opened laterally to the bregma, 6 mm along the fronto-temporal suture, 2 mm in depth. A horizontal cut of the right hemisphere was made with a glass knife fashioned from a cover slip. In sham-operated animals, the skull was opened but no lesion was made. The experiments were performed 14 days after the lesion.

The data were analyzed statistically by ANOVA (2x2) factorial analysis, Tukey's test, and Tukey's test for unconfounded means.

RESULTS

Transection of the corticostriatal pathway by undercutting the cortex produced neurochemical changes in the striatum. [^{3}H]-glutamic acid uptake was decreased by 55% in the striata ipsilateral to the lesion as compared with striata removed from sham-operated rats [from 5.73 $\pm$ 1.7 to 2.63 $\pm$ 0.13 nmoles glutamic acid/min/mg protein (means and S.E.M.) n=6, p < 0.01], indicating the extensive impairment of the corticostriatal glutamatergic input. At the same time, the content of noradrenaline (102.2 $\pm$ 6.4 ng/g), serotonin (166.5 $\pm$ 9.8 ng/g), dopamine (10.7 $\pm$ 0.4 g/g) and choline (23.0 $\pm$ 1.0 nmoles/g) as well as the activity of choline acetyltransferase (1.9 nmoles/min/mg protein) in the striatum was not altered by the lesion.

Undercutting of the cortex, however, significantly altered the dynamics of the cholinergic neurons, as reflected by reduced ACh turnover. The synthesis rate of ACh was found to be decreased by 30%, from the level of 10.7 $\pm$ 0.8 pmoles ACh synthesized/min/mg wet wt in the sham-operated rats, to 7.5 $\pm$ 0.3 pmoles/min/mg wet wt (n=6; p < 0.01) in the striata of decorticated rats. This result is in accordance with the previous work of Wood et al. (31) in which decortication was produced by ablation of the frontal cortex.

Decortication might have been expected to produce an increase in ACh content as a result of the removal of the tonic positive input. It is unclear at present why this does not occur (see Tables), but it is reasonable to assume from the data that compensatory mechanisms are activated which maintain the cholinergic neurons intrinsic to the striatum in a new, lowered functonal state. Moreover, it was found that the capacity of choline uptake inhibitors to deplete striatal ACh, which has been used as an index of cholinergic neurotransmitter utilization (8), was reduced after decortication (Table 1). The i.c.v. application of hemicholinium-3 (20 μg, 30 min) and the i.p. injection of CM 54903 (40 mg/kg, 30 min), a non-polar analog of dimethylaminoethanol capable of crossing the blood brain barrier (29), induced depletions in ACh content of 57% and 36% in sham-operated rats respectively, whereas in the decorticated rats these reductions amounted to only 44% and 27%. The approximately 25% decrease in ACh utilization in decorticated rats verified by either of the two drugs is in remarkably good agreement with the observed slowed turnover rate.

Chronic, unilateral frontal decortication prevented completely the large increases in striatal ACh content induced by the i.p. administration of several classes of drugs; i.e., the typical dopaminergic agonist apomorphine (1.5 mg/kg, 30 min); the long-acting ergot alkaloids bromocriptine (4 mg/kg, 90 min) and lisuride (0.2 mg/kg, 30 min); the muscarinic agonist oxotremorine (0.5 mg/kg, 20 min); the purinergic drug 2-chloroadenosine (20 μg, i.c.v., 30 min); as well as the opiate, methadone (10 mg/kg, 60 min) (Table 2). On the other hand, the effect of drugs that markedly reduce striatal ACh content; i.e., the typical neuroleptic dopamine receptor antagonist pimozide (1 mg/kg, 240 min); the atypical neuroleptics clozapine (20 mg/kg, 60 min) and l-sulpiride (100 mg/kg, 180 min); and the antimuscarinic agent scopolamine (0.5 mg/kg, 30 min); was not influenced by the lesion (Table 3).

DISCUSSION

The present results provide the first evidence that the effects of a number of drugs capable of depressing cholinergic activity (2, 3, 5, 6, 11, 16, 18, 19) through receptor-mediated responses are operative only if the corticostriatal pathway is integral. It is conceivable that the cholinergic neurons, being already inhibited by the loss of the excitatory input, cannot be further inhibited by drugs which slow down ACh turnover and therefore their activity, whereas these neurons are still responsive to agents that activate them (4). This type of mechanism best explains how such different

Table 1

Efffect of Choline Uptake Inhibitors on Striatal
Acetylcholine Content in Decorticated Rats

Striatal ACh (nmoles/g wet wt)

Drug

Treatment	Hemicholinium-3	CM 54903
Sham	65.4 ± 2.4	66.5 ± 1.9
Decortication	61.4 ± 3.2	66.7 ± 1.7
Drug	28.4 ± 1.2*	42.3 ± 1.1*
Drug + Decortication	34.1 ± 1.0*	48.8 ± 1.3*
Interaction	<0.01	<0.05

Dose and times: hemicholinium-3, 20 μg i.c.v., 30 min; CM 54903,40 mg/kg, i.p. 30 min. The data are the means ± S.E.M. (n > 6). *p < 0.01 vs respective sham-operated groups.

neuropharmacological classes of drugs may share the property of being blocked by the decortication.

Furthermore, we have observed that when the insensitive neurons of the deafferentiated striata were exposed for a longer period of time to the action of a depressant drug, they were actually capable of responding to it. Degeneration of the corticostriatal neurons prevented the ACh-increasing effect of bromocriptine up to 180 min, but not at the longer time of 240 min after administration of the drug (Table 4). At this later time, bromocriptine produced a significant increase in ACh content of 25% which was equivalent to the increase induced in the nonlesioned animals. Based on these data, it is proposed that the decorticating lesion impaired the state of sensitivity of the cholinergic neurons upon which the effects of the drugs appear to be dependent. The overall results shown above also exclude the possibility that the lesion produced some specific damage to striatal cells, at which the agonists, but not the antagonists, might exert their action.

An important feature of this study, which should be considered, is the fact that neuropharmacological responses in the brain appear to be the result of an interaction between several major neurotransmitter systems. Although one transmitter system may be the specific target of a drug, others, too, are essential for the successful completion of a drug's action. Thus, we have attempted to depict

Table 2

Effect of Agonist Drugs of Various Classes on Striatal
Acetylcholine Content in Decorticated Rats

Drug	Striatal ACh (nmoles/g wet wt)				
	Sham	Decortication	Drug	Drug + Decortication	Interaction
Apomorphine	64.4 ± 1.0	64.0 ± 1.8	80.1 ± 2.2*	67.7 ± 2.1	<0.01
Bromocriptine	68.9 ± 1.7	69.4 ± 2.1	80.2 ± 2.3*	72.0 ± 2.1	<0.01
Lisuride	65.1 ± 1.8	66.4 ± 1.4	86.6 ± 2.1*	69.9 ± 1.5	<0.01
Oxotremorine	67.7 ± 1.0	66.4 ± 1.7	90.2 ± 2.1*	69.3 ± 1.9	<0.01
2-Chloro-adenosine	65.1 ± 2.1	65.3 ± 2.6	78.2 ± 1.3*	66.8 ± 3.2	<0.01
Methadone	66.3 ± 1.1	66.9 ± 1.8	82.3 ± 1.5*	69.2 ± 1.4	<0.01

Doses and times: apomorphine, 1.5 mg/kg, i.p., 30 min; bromocriptine, 4 mg/kg, i.p., 90 min; lisuride, 200 μg/kg, i.p., 30 min; oxotremorine, 0.5 mg/kg, i.p., 20 min; 2-chloroadenosine, 20 μg, i.c.v., 30 min; methadone, 10 mg/kg, i.p., 60 min. The data are the means ± S.E.M. (n=6-12)
* = p < 0.01 vs sham-operated group

Table 3

Effect of Typical and Atypical Neuroleptics and Scopolamine on
Striatal Acetylcholine Content in Decorticated Rats

Drug	Striatal ACh (nmoles/g wet wt)				
	Sham	Decortication	Drug	Drug + Decortication	Interaction
Pimozide	66.3 ± 1.4	62.2 ± 3.2	44.7 ± 1.9*	37.8 ± 1.6*	ns
Clozapine	66.5 ± 1.4	62.9 ± 1.8	37.5 ± 1.9*	37.8 ± 1.6*	ns
l-Sulpiride	66.8 ± 1.4	64.2 ± 3.2	49.4 ± 2.9*	47.9 ± 2.0*	ns
Scopolamine	67.4 ± 3.0	65.4 ± 3.1	42.5 ± 4.3*	46.9 ± 2.8*	ns

Doses and times: pimozide, 1 mg/kg, i.p., 240 min; clozapine, 20 mg/kg, i.p., 60 min; l-sulpiride, 100 mg/kg, i.p., 180 min; scopolamine, 0.5 mg/kg, i.p., 30 min. The data are means ± S.E.M. (n=7-8)
* = p < 0.01 vs sham-operated group

Table 4

Time-Course of the Effect of Bromocriptine on Striatal
Acetylcholine Content in Sham-Operated and Decorticated Rats

Time after Bromocriptine (min)	Striatal ACh (nmoles/g wet wt)	
	Sham	Lesioned
Vehicle	66.0 $\pm$ 1.4	66.4 $\pm$ 2.4
60	82.6 $\pm$ 1.4*	66.4 $\pm$ 1.2
120	86.7 $\pm$ 1.6*	69.7 $\pm$ 1.4
180	89.5 $\pm$ 1.9*	73.4 $\pm$ 3.0
240	82.2 $\pm$ 1.3*	82.4 $\pm$ 2.4*

* = $p < 0.01$ vs the vehicle group as analyzed by Dunnett's test.
The data are means $\pm$ S.E.M. (n=6-12)

by the scheme in Fig. 1 the possible interrelationships that exist
among some of the neurotransmitter systems known to be involved in
the regulation of cholinergic neurons. The activity of the cholinergic
neurons is maintained in a state of equilibrium by a balance of
two separate influences: the corticostriatal excitatory input,
which exerts its action at the dendritic level of the cholinergic
interneurons (21), and the nigroneostriatal inhibitory dopaminergic
afferents. Moreover, cholinergic agonists are able to directly
regulate the release of dopamine (10, 20, 25) by a mechanism that
apparently does not involve axoaxonic relationships but rather a
dendroaxonic communicatory process (1). In this manner the cholinergic
interneuron itself is able to interact functionally within the system
to maintain the equilibrium state.

It is generally accepted that any event which alters the activity
of one neurotransmitter system is likely to change the activity of
any other system with which it is linked. Thus, the loss of the
excitatory input in decorticated rats results in a refractoriness
of the cholinergic neurons. As a consequence, according to the
scheme presented in Figure 1, the signal from the cholinergic dendrites
to the dopaminergic terminals may be reduced, leading to lowered
dopaminergic tone. Indeed, we have found in preliminary experiments
(performed with Drs. S. Algeri and F. Ponzio), that the decrease
in the dopamine metabolites homovanillic acid, dihydroxyphenylacetic
acid and 3-methoxytyramine, induced by administering a threshold
dose of apomorphine, is strongly prevented in decorticated rats.

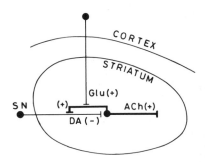

Figure 1. Possible arrangement of the dopaminergic, cholinergic and putative glutamatergic neurons in the striatum. Cholinergic activity may be regulated by a balance between the excitatory and inhibitory neurotransmitter influences. Chronic interruption of the excitatory input alters the steady state conditions and brings the cholinergic interneurons to a new, lowered functional state. In this depressed condition the cholinergic neurons are refractory to dopamine receptor agonists.

The relationship of the cortical pathway with cholinergic and dopaminergic neuropharmacological actions of drugs found here is relevant to clinical evidence. The findings suggest that changes in the activity of these systems are closely associated with the etiology of neurological and psychiatric disorders and with the mode of action of drugs that alleviate the symptoms of such diseases, and thus may suggest new inroads to therapy.

Acknowledgements: Compound CM 54903 was kindly furnished by Clin-Midy, Montpellier, France. This work was supported by a National Research Council Contract (National Pharmacology Group), No. 82.01300.04.

REFERENCES

1. Butcher, L.L. and Wolf, N.J. (1982): Brain Res. Bull. 9:475-492.
2. Carenzi, A., Cheney, D.L., Costa, E., Guidotti, A. and Racagni, G. (1975): Neuropharmacology 14:927-939.
3. Consolo, S., Ladinsky, H. and Garattini, S. (1974): J. Pharm. Pharmacol. 26:275-277.
4. Consolo, S., Ladinsky, H. and Bianchi, S. (1975): Eur. J. Pharmacol. 33:345-351.
5. Consolo, S., Ladinsky, H., Pugnetti, P., Fusi, R. and Crunelli, V. (1981): Life Sci. 29:457-465.
6. Consolo, S., Forloni, G.L., Fisone, G., Sieklucka, M. and Ladinsky, H. (1983): Biochem. Pharmacol. 32:2993-2996.
7. Divac, I., Fonnum, F. and Storm-Mathisen, J. (1977): Nature, Lond. 266:377-378.
8. Domino, E.F. and Wilson, A.E. (1972): Psychopharmacologia 25: 291-298.
9. Fonnum, F. (1975): J. Neurochem. 24:407-409.

10. Giorguieff, M.F., Kemel, M.L. and Glowinski, J. (1977): Neurosci. Letters 6:73-77.
11. Guyenet, P., Euvrard, C., Javoy, F., Herbet, A. and Glowinski, J. (1977): Brain Res. 136:487-500.
12. Hattori, T., McGeer, E.G. and McGeer, P.L. (1979): J. Comp. Neurol. 185:347-354.
13. Hedreen, J.C. (1977): Neurosci. Letters 4:1-7.
14. Keller, R., Oke, A., Mefford, I. and Adams, R.N. (1976): Life Sci. 19:995-1004.
15. Kim, J-S., Hassler, R., Haug, P. and Paik, K-S. (1977): Brain Res. 132:370-374.
16. Ladinsky, H., Consolo, S., Bianchi, S., Samanin, R. and Ghezzi, D. (1975): Brain Res. 84:221-226.
17. Ladinsky, H., Consolo, S., Bianchi, S. and Jori, A. (1976): Brain Res. 108:351-361.
18. Ladinsky, H., Consolo, S., Samanin, R., Algeri, S. and Ponzio, F. (1980): Adv. Biochem. Psychopharmacol. 24:259-265.
19. Ladinsky, H., Consolo, S., Forloni, G. and Tirelli, A.S. (1981): Brain Res. 225:217-223.
20. Lehmann, J. and Langer, S.Z. (1982): Brain Res. 248:61-69.
21. Lehmann, J. and Scatton, B. (1982): Brain Res. 252:77-89.
22. McGeer, P.L., McGeer, E.G., Sherer, U. and Singh, K. (1977): Brain Res. 128:369-373.
23. Ponzio, F. and Jonsson, G. (1979): J. Neurochem. 32:129-132.
24. Racagni, G., Trabucchi, M. and Cheney, D.L. (1975): Naunyn-Schmiedeberg's Arch. Pharmacol. 290:99-105.
25. Raiteri, M., Marchi, M. and Maura, G. (1982): Eur. J. Pharmacol. 83:127-129.
26. Saelens, J.K., Allen, M.P. and Simke, J.P. (1970): Arch. Int. Pharmacodyn. 186:279-286.
27. Simon, J.R. (1982): Life Sci. 31:1501-1508.
28. Spencer, H.J., Gribkoff, V.K., Cotman, C.W. and Lynch, G.S. (1976): Brain Res. 105:471-481.
29. Vezzani, A., Zatta, A., Ladinsky, H., Caccia, S., Garattini, S. and Consolo, S. (1982): Biochem. Pharmacol. 31:1693-1698.
30. Webster, K.F. (1961): J. Anat. 95:532-544.
31. Wood, P.L., Moroni, F., Cheney, D.L. and Costa, E. (1979): Neurosci. Letters 12:349-354.
32. Yeterian, E.H. and Van Hoesen, G.W. (1978): Brain Res. 139:43-63.

EXCITATORY AMINO ACID INFLUENCE ON STRIATAL

CHOLINERGIC TRANSMISSION

B. Scatton and D. Fage

Synthelabo-L.E.R.S.
Biochemical Pharmacology Group
31, avenue Paul Vaillant Couturier
92220 Bagneux, FRANCE

INTRODUCTION

The corpus striatum contains a discrete population of cholinergic interneurons which represents about 1-2% of the total neostriatal neuronal population. The striatal cholinergic interneuron corresponds morphologically to the large aspiny neuron described by Kemp and Powell (7). These nerve cells appear to possess an important functional role in extrapyramidal motor function, and have been particularly implicated in the translation of alterations of nigro-striatal dopaminergic transmission into behavioral patterns (for review see Lloyd, 10).

The transsynaptic regulation of the activity of striatal cholinergic neurons by a number of afferent inputs, e.g. the nigro-striatal dopaminergic (10), the raphe-striatal serotonergic (3) and the intrastriatal GABAergic (17) neurons is now well documented.

Evidence also suggests that the cortico-striatal pathway (which utilizes the excitatory amino acids L-glutamate or L-aspartate as neurotransmitter) may also be involved in the regulation of striatal cholinergic neuron activity. Thus, cortical ablation reduces high-affinity glutamate uptake (13) and concomitantly decreases acetylcholine (ACh) turnover (23) in the striatum. Moreover, an input from the cerebral cortex onto the aspiny dendrites of neurons morphologically similar to the presumed striatal cholinergic interneurons, is suggested by electron microscopic studies (5). Finally, kainate-induced destruction of cholinergic neurons (among others) in the striatum seems to require the integrity of the cortico-striatal pathway (14).

Amino acid receptors mediating neuronal excitation have been broadly classified into 3 subtypes according to their sensitivity to excitatory amino acid agonists and antagonists: the N-methyl-D-aspartate (NMDA), the quisqualate and the kainate-preferring receptors (for recent reviews see 21, 22). The NMDA receptor is preferentially activated by NMDA and antagonized selectively by a number of organic (phosphonate analogs of carboxylic acids) or inorganic (magnesium ions) agents. The second type of receptor is activated by quisqualate, with L-glutamic acid diethyl ester as an antagonist. The third receptor type is activated by kainate, and is insensitive to magnesium ions, with no known specific organic antagonist. The naturally occurring amino acid neurotransmitters L-aspartate and L-glutamate seem to have mixed agonist actions at these proposed receptor sites, those for NMDA and quisqualate probably predominating in the excitation of central neurons by these amino acids when applied iontophoretically (21, 22).

The nature of the excitatory amino acid receptor mediating the cortical excitatory influence on striatal cholinergic neurons is as yet unknown. We have recently attempted to characterize this receptor by investigating the effects of specific excitatory amino acid receptor agonists and antagonists on the release of [^{3}H]-ACh from slices of the rat corpus striatum (9, 18). In this review, we will present and discuss data that suggest the specific involvement of an NMDA-type receptor in the excitatory amino acid influence on striatal cholinergic transmission. Evidence will also be provided to show that these NMDA-type receptors are localized on the dendrites of striatal cholinergic interneurons. Finally the influence of excitatory amino acid agonists on the release of [^{3}H]-ACh in non-striatal ACh-rich brain areas will also be discussed.

INVOLVEMENT OF AN NMDA-TYPE RECEPTOR IN THE EXCITATORY AMINO ACID INFLUENCE ON STRIATAL CHOLINERGIC TRANSMISSION

In order to investigate the pharmacological nature of the receptor(s) mediating the excitatory amino acid influence on striatal cholinergic neurons, the ability of a number of excitatory amino acid agonists to stimulate [^{3}H]-ACh release has been examined in superfusion experiments with rat caudate-putamen slices. Because magnesium ions have been reported to antagonize the L-glutamate evoked neuronal depolarization (1), the release experiments were performed using magnesium-free medium.

As shown in Fig. 1, agonists of excitatory amino acid receptors caused an increase in [^{3}H]-ACh release. Concentration response curves were sigmoidal in shape, suggesting the involvement of a single saturable receptor. The relative order of potency was the following: NMDA > (+)ibotenate $\geq$ N-methyl-DL-aspartate > L-glutamate > L-aspartate $\geq$ cysteate > kainate = quisqualate with EC$_{50}$'s ranging from 15 to 800 μM. More recent studies with quinolinic acid (a rigid analog

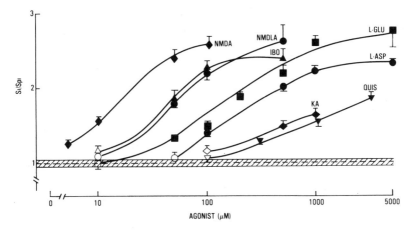

Figure 1. Effects of excitatory amino acid receptor agonists on the outflow of [³H]-ACh from rat striatal slices in magnesium-free medium. For methods see Scatton and Lehmann (18). Each value represents the mean ± S.E.M. of 8 independent determinations. Results are expressed as the ratio of [³H]-ACh outflow following a 2 min pulse of agonist divided by the pre-stimulus spontaneous outflow (S_I/S_{pI}). Filled symbols: statistically significant (p < 0.001) compared with controls. Open symbols: not statistically significant. Abbreviations: IBO, (±)ibotenate; NMDA, N-methyl-D-aspartate; NMDLA, N-methyl-DL-aspartate; L-GLU, L-glutamate; L-ASP, L-aspartate; KA, kainate; QUIS, quisqualate.

of NMDA existing normally in mammals as a metabolite of tryptophan and described as a potent NMDA-type agonist; 20) showed that this compound also evoked a release of [³H]-ACh from striatal slices, although with a relatively low affinity (in preparation).

The rank order of agonists in evoking [³H]-ACh release correlates quite well with the rank order determined by Watkins (21) electrophysiologically for NMDA-type receptors, suggesting that the receptor which mediates the release of [³H]-ACh evoked by excitatory amino acid agonists is akin to the NMDA-type. This view is also supported by experiments performed with inorganic and organic antagonists of excitatory amino acids. Electrophysiological studies on spinal neurons have indicated that NMDA or L-aspartate but not kainate-induced responses are highly sensitive to magnesium ions (1, 21). Similarly, we found that low concentrations of magnesium antagonize the release of striatal [³H]-ACh evoked by N-methyl-DL-aspartate, (+) ibotenate, L-glutamate, cysteate and L-aspartate but not by kainate (9, 18). The EC_{50} of magnesium for this effect was 0.1 mM for both L-glutamate and N-methyl-DL-aspartate.

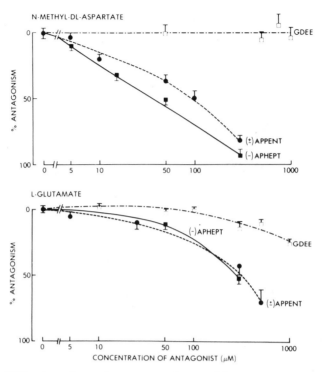

Figure 2. Effects of excitatory amino acid receptor antagonists on N-methyl-DL-aspartate and L-glutamate evoked release of [^{3}H]-ACh from rat striatal slices. Amino acid antagonists were added to the medium 12 min before the 2 min pulse of N-methyl-DL-aspartate (50 μM) or L-glutamate (300 μM). The ratio S_I/S_{pI} was calculated in both the absence and presence of the antagonist and results were expressed as % of this value in respective controls (agonist alone). Results are means $\pm$ S.E.M. of 8 determinations. Filled symbols: statistically significant (p < 0.01) compared with respective controls. Open symbols: not statistically significant.

As suggested by electrophysiological studies, excitatory amino acid receptor subtypes exhibit a differential sensitivity to blockade by specific organic antagonists. The two most potent and selective antagonists of the NMDA-type receptor are (-)2-amino-7-phosphono-heptanoate [(-)APHept] and ($\pm$)2-amino-5- phosphonopentanoate [($\pm$) APPent] whereas, L-glutamic acid diethylester (GDEE) is an antagonist at quisqualate-type receptors (2, 15). As shown in Fig. 2, (-)APHept and ($\pm$)APPent antagonized in a concentration-dependent manner the release of [^{3}H]-ACh evoked by either N-methyl-DL-aspartate (IC$_{50}$'s 40 and 90 μM, respectively) or L-glutamate (IC$_{50}$'s 280 and 320 μM, respectively). The NMDA-type antagonist D-α-aminoadipate as well as phencyclidine, a well known hallucinogenic agent which has been shown electrophysiologically to possess NMDA-type receptor antagonist properties (11), were also found to block the release of $[^3H]$-ACh evoked by N-methyl-DL-aspartate with IC_{50}'s of 1 mM and 0.2 μM, respectively (9; Scatton and Fage, in preparation). In contrast, the quisqualate receptor antagonist GDEE (up to 1 mM) failed to significantly affect the responses to N-methyl-DL-aspartate or L-glutamate. The inhibition by ($\pm$)APPent of NMDA or L-glutamate evoked $[^3H]$-ACh release was found to be purely competitive in nature (18). In contrast, the inhibition caused by magnesium ions appears to be non competitive. This suggests that the inorganic and the organic antagonists of excitatory amino acids act at different membrane sites, a conclusion supported by electrophysiological studies (21, 22). The divalent cation probably acts by interfering with the receptor-ionophore coupling process, while the organic antagonists might block the access of the agonist to the receptor recognition site.

Taken together these experiments clearly indicate that the receptor mediating the excitatory amino acid influence on striatal cholinergic transmission resembles the NMDA-preferring receptor, as previously characterized electrophysiologically according to 3 criteria: 1) rank order of potency of agonists; 2) high sensitivity to inhibition by magnesium ions; and 3) specific antagonism by NMDA-type receptor antagonists. The very weak potency of quisqualate together with the failure of the quisqualate-type antagonist GDEE to block N-methyl-DL-aspartate or L-glutamate evoked $[^3H]$-ACh release indicate that quisqualate-type receptors do not contribute significantly to the excitatory amino acid evoked responses of striatal cholinergic cells. This contrasts with nigro-striatal dopaminergic neurons where quisqualate-type receptors appear to mediate the excitatory amino acid influence on dopamine release (16). A quisqualate-type receptor regulating the release of L-aspartate has also been recently identified in the rat hippocampus (12). It appears, therefore, that the relative involvement of amino acid receptor subtypes in the excitatory influence of amino acids on neuronal cells depends on the nature and/or regional localization of the target cells.

Since NMDA-type receptors mediate the excitatory amino acid influence on striatal cholinergic neurons, these receptors may play a part in transducing information received from the cerebral cortical afferents (which are thought to use L-glutamate or L-aspartate as neurotransmitter) to cholinergic cells. The blockade by magnesium ions, ($\pm$)APPent and (-)APHept but not GDEE, of the effect of L-glutamate on $[^3H]$-ACh release indeed suggests that in the striatum this naturally occurring transmitter interacts predominantly with NMDA-type receptors. Further support for this hypothesis is provided by the lack of additivity of the effects of supramaximal concentrations of NMDA and L-glutamate on the release of $[^3H]$-ACh from striatal slices and occurrences of full cross desensitization between the two amino acid receptor agonists (unpublished data). Electrophysiological studies showing that cortically evoked action potentials in the striatum can be inhibited by doses of the NMDA receptor antagonist D-α-aminoadipate that also affect excitations induced by L-glutamate (19), are also in keeping with this view.

LOCALIZATION OF THE NMDA-TYPE RECEPTORS MEDIATING THE EXCITATORY AMINO ACID INFLUENCE ON STRIATAL CHOLINERGIC TRANSMISSION

The question naturally arises as to the localization of those NMDA-type receptors modulating striatal cholinergic transmission. One possibility is that the excitatory amino acid receptor agonists directly depolarize and initiate action potentials in cholinergic nerve terminals. In fact, some excitatory amino acid receptors have been suggested to exist on nerve terminals (12, 16). However, although a direct depolarization of cholinergic nerve terminals by excitatory amino acid receptor agonists cannot be rigorously ruled out, the failure of N-methyl-DL-aspartate to evoke [^{3}H]-ACh release from slices of hippocampus, interpeduncular nucleus and olfactory tubercle (see below), where cholinergic afferents rather than inter-neurons, are present, strongly argues against this interpretation.

A second possibility is that NMDA-type receptors are located on the dendrites of cholinergic neurons. Indeed, as noted in the Introduction, there is good anatomical evidence for an excitatory input onto the dendrites of striatal cholinergic interneurons; these neurons therefore ought to possess excitatory amino acid receptors on their dendrites. Moreover, it is well known that virtually all neuronal perikarya in the CNS can be excited by application of ex-citatory amino acid receptor agonists, thus suggesting that striatal cholinergic cells may also possess excitatory amino acid receptors. A more direct support for this possibility has been provided by the observation that tetrodotoxin (0.5 μM) abolishes the release of [^{3}H]-ACh evoked by N-methyl-DL-aspartate and L-glutamate in magnesium-free medium (Fig. 3B; 9, 18). These data and the above considerations are consistent with the hypothesis that excitatory amino acid receptor agonists activate receptors located on cholinergic dendrites, causing depolarization and subsequent action potential propagation along the axons of the cholinergic interneurons to the terminals, where [^{3}H]-ACh is released.

The involvement of NMDA-type receptors on cells other than cholinergic neurons which affect the cholinergic neurons via an unknown transmitter cannot, however, be completely excluded. Previous studies have shown that L-glutamate increases the release of dopamine from striatal slices, presumably via a direct presynaptic action at the dopaminergic nerve terminal (16). Since dopaminergic neurons are involved in the transsynaptic regulation of striatal cholinergic neurons, the possibility has to be considered that dopaminergic neurons mediate the excitatory influence on striatal cholinergic neurons. This appears, however, unlikely, since chemical lesion of the nigro-striatal dopaminergic pathway failed to alter the ability

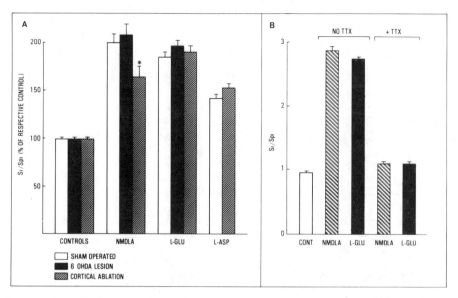

Figure 3.   Effects of excitatory amino-acid receptor agonists on the release of [³H]-ACh after lesion of the nigro-striatal dopaminergic or cortico-striatal glutamatergic pathways (A) or after application of tetrodotoxin into the medium (B).   (A) Animals were killed 2 weeks after 6-hydroxy-dopamine-induced lesion of the nigro-striatal dopaminergic pathway or 10 days following cortical ablation.   (B) Tetrodotoxin (0.5 μM) was added to the superfusion medium 12 min before exposure to amino acid agonists.   Two min long pulses of agonists (100 μM) were applied. Experiments were performed in magnesium-free medium. The ratio S_I/S_{pI} was calculated as described in the legend to Fig. 1.   Results are means $\pm$ S.E.M. of 8 independent determinations.   * p < 0.01 vs sham.   Abbreviations: TTX: tetrodotoxin; NMDLA: N-methyl-DL-aspartate; L-GLU: L-glutamate; -ASP: L-aspartate; 6-OHDA: 6-hydroxydopamine.

of N-methyl-DL-aspartate or L-glutamate to increase [³H]-ACh release from striatal slices (Fig. 3A).   Moreover, since the nigro-striatal dopaminergic system exerts an <u>inhibitory</u> influence on striatal cholinergic neurons (10), it is unlikely to mediate the release of [³H]-ACh evoked by excitatory amino acids.

NMDA-type receptors involved in the modulation of ACh neuron activity may also be (at least partially) located on cortical afferents to the striatum.   In order to test this possibility we have investigated the effect of excitatory amino acids on striatal [³H]-ACh release after cortical ablation.   As shown in Fig. 3A, cortical ablation partially reduced the ability of N-methyl-DL-aspartate to evoke [³H]-ACh release.   These findings may indicate that some NMDA-type receptors are located on the cortical terminals removed by the lesion. Accordingly, the stimulatory influence of N-methyl-DL-aspartate on striatal [^{3}H]-ACh release might be partially effected via: 1) a stimulation of presynaptic NMDA-type receptors located on cortical afferents to the striatum, leading to the release of an endogenous amino acid transmitter which then acts on postsynaptic NMDA-type receptors located on the dendrites of cholinergic cells; and 2) via a direct action on the latter postsynaptic NMDA-type receptors. The absence of effect of cortectomy on the responses to L-glutamate and L-aspartate (Fig. 3A) suggests that these naturally occurring amino acids preferentially stimulate the postsynaptic NMDA-type receptors located on cholinergic cells. Alternatively, an increase in the responsiveness of striatal cholinergic neurons to L-glutamate and L-aspartate due to a decrease in the uptake of these agonists (which normally reduces their synaptic concentrations), caused by cortical ablation, may have masked a diminution of the response due to the destruction of presynaptic NMDA-type receptors located on cortical afferents.

REGIONAL DIFFERENCES IN THE SENSITIVITY OF CHOLINERGIC NEURONS TO N-METHYL-DL-ASPARTATE

Cholinergic neurons are widely distributed within the CNS. In order to test whether NMDA-type receptors also regulate cholinergic neuron activity in non-striatal brain regions, the ability of N-methyl-DL-aspartate to evoke [^{3}H]-ACh release has been examined in a variety of ACh-rich brain areas of the rat. As shown in Fig. 4, N-methyl-DL-aspartate stimulated [^{3}H]-ACh release from slices of olfactory tubercule, septum, nucleus accumbens and frontal cortex, suggesting that NMDA-type receptors may modulate cholinergic transmission also in these areas. In this respect, however, the effect of the excitatory amino acid receptor agonist was less pronounced in these non-striatal regions than in the striatum. This differential regional response may be ascribed to a difference in the relative densities of NMDA-type receptors in these areas.

In other brain areas, e.g. the interpeduncular nucleus, olfactory bulb and hippocampus, N-methyl-DL-aspartate failed to evoke [^{3}H]-ACh release (Fig. 4). Yet, in these regions potassium (25 mM) was able to evoke a calcium-dependent release of [^{3}H]-ACh. The failure of N-methyl-DL-aspartate to evoke [^{3}H]-ACh release from the latter brain areas where cholinergic afferents, but not interneurons, are present (4), supports the view that excitatory amino acid receptor agonists influence cerebral cholinergic neurons by acting on NMDA-type receptors mainly located on their perikarya.

CONCLUSIONS

The present and previous (9, 18) studies provide evidence for an excitatory influence of excitatory amino acid pathways on cholinergic transmission in a variety of brain areas containing cholinergic

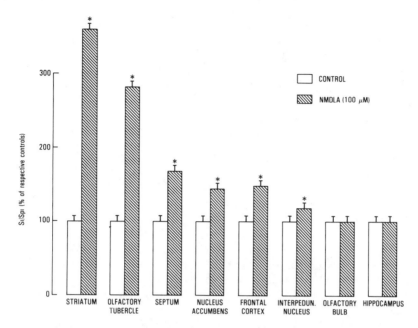

Figure 4. Effect of N-methyl-DL-aspartate (NMDLA) on the release of [³H]-ACh from slices of various ACh-rich brain areas in the rat. Two min long pulses of NMDLA (100 μM) were applied. Slices were perfused with magnesium-free Krebs medium. The ratio S_I/S_{pI} was calculated as described in the legend to Fig. 1. Results are means ± S.E.M. of 5 independent determinations and are expressed as % of respective controls. * p < 0.01 vs related controls.

interneurons (e.g. the striatum, olfactory tubercle, septum and cerebral cortex). This excitatory influence appears to be mediated by NMDA-type receptors. In the striatum, these receptors are located predominantly on the dendrites of cholinergic interneurons, and may play a part in transducing excitatory amino acid neurotransmission carried by cortico-striatal afferents.

Excitatory amino acid neurotransmitters may have an important role in the regulation of motor activities which are strongly associated with striatal cholinergic function (see Introduction). This notion is supported by recent reports showing that intra-striatal application of NMDA causes ipsiversive rotation (6) while intra-globus pallidus application of the compound induces locomotor hyperactivity and dyskinetic reactions (8). Therefore, excitatory amino acid receptor agonists and antagonists easily penetrating the blood brain barrier may prove useful in the treatment of neurological idiopathic or iatrogenic disorders (e.g. Parkinson's disease, tardive dyskinesias) associated with a relative dysfuntion of the striatal cholinergic system.

REFERENCES

1. Davies, J. and Watkins, J.C. (1977): Brain Res. 130:364-368.
2. Davies, J. and Watkins, J.C. (1981): In Glutamate as a Neuro-
    transmitter (eds) G. DiChiara and G.L. Gessa, Raven Press, New
    York, pp. 275-284.
3. Euvrard, C., Javoy, F., Herbet, A. and Glowinski, J. (1977):
    Europ. J. Pharmacol. 41:281-289.
4. Fibiger, H.C. (1982): Brain Res. Reviews 4:327-388.
5. Hassler, R., Chung, J.W., Rinne, U. and Wagner, A. (1978):
    Exp. Brain Res. 31:67-80.
6. Jenner, P., Marsden, C.D. and Taylor, R.J. (1981): Brit. J.
    Pharmacol. 72:570P.
7. Kemp, J. and Powell, T.P.S. (1971): Phil. Trans. Roy. Soc.
    Lond. B. 262:383-401.
8. Kerwin, R.W., Luscombe, G.P., Pycock, C.J. and Sverakova, K.
    (1980): Brit. J. Pharmacol. 68:174P.
9. Lehmann, J. and Scatton, B. (1982): Brain Res. 252:77-89.
10. Lloyd, K.G. (1978): In Essays in Neurochemistry and Neuropharma-
    cology (eds) M.B.H. Youdim, W. Lovenberg, D.F. Sharman and J.R.
    Lagnado, Wiley, New York, pp. 129-207.
11. Lodge, D. and Anis, N.A. (1982): Europ. J. Pharmacol. 77:
    203-204.
12. McBean, G.J. and Roberts, J.P. (1981): Nature 291:593-594.
13. McGeer, P.L., McGeer, E.G., Scherer, U. and Singh, K. (1977):
    Brain Res. 128:369-373.
14. McGeer, E.G., McGeer, P.L. and Singh, K. (1978): Brain Res.
    139:381-383.
15. Perkins, M.N., Stone, T.W., Collins, J.F. and Curry, K. (1981):
    Neurosci. Lett. 23:333-336.
16. Roberts, P.J. and Anderson, S.D. (1979): J. Neurochem. 32:1539-
    1545.
17. Scatton, B. and Bartholini, G. (1981): In Cholinergic Mechanisms
    (eds) G. Pepeu and H. Ladinsky, Plenum Publishing Corporation,
    pp. 771-780.
18. Scatton, B. and Lehmann, J. (1982): Nature 297:422-424.
19. Stone, T.W. (1979): Br. J. Pharmac. 67:545-551.
20. Stone, T.W. and Perkins, M.N. (1981): Europ. J. Pharmacol. 72:
    411-412.
21. Watkins, J.C. (1981): In Glutamate: Transmitter in the Central
    Nervous System (eds) P.J. Roberts, J. Storm-Mathisen and G.A.
    Johnston, John Wiley, Chichester, U.K., pp. 1-24.
22. Watkins, J.C., Davies, J., Evans, R.H., Francis, A.A. and Jones,
    A.W. (1981): In Glutamate as a Neurotransmitter (eds) G.D.
    DiChiara and G.L. Gessa, Raven Press, New York, pp. 263-273.
23. Wood, P.L., Moroni, F., Cheney, D.L. and Costa, E. (1979): Neuro-
    sci. Lett. 12:349-354.

CONTRIBUTION OF THE DORSAL NORADRENERGIC BUNDLE TO THE EFFECT OF AMPHETAMINE ON ACETYLCHOLINE TURNOVER

S.E. Robinson

Department of Pharmacology and Toxicology
Medical College of Virginia
Box 613, MCV Station
Richmond, Virginia   23298   U.S.A.

INTRODUCTION

Central noradrenergic neurons project to many sites in the brain where they may interact with cholinergic neurons. Areas with noradrenergic innervation where this interaction may occur include the cortex, septum, hippocampus, and hypothalamus (3, 7). These areas receive projections from one or more of the following noradrenergic cell groups: A1, A2, A5, A6 (locus coeruleus) and A7.

There is already evidence that cholinergic neurons in the cortex and hippocampus are activated by noradrenergic neurons. Amphetamine and nomifensine, drugs which inhibit the uptake of dopamine and norepinephrine (NE) into nerve endings, increase the turnover of acetylcholine (ACh) in the cortex and hippocampus. These drugs appear to act through a noradrenergic mechanism, because the dopamine agonist apomorphine does not affect the turnover rate of ACh (TR_{ACh}) in the cortex and decreases TR_{ACh} in the hippocampus, while intraseptal administration of the antagonist phenoxybenzamine blocks the action of these drugs (14).

In order to determine the contribution of the noradrenergic projections of the locus coeruleus to the action of amphetamine on cholinergic neurons in several areas of the brain, the dorsal noradrenergic bundle was selectively lesioned by injection of the neurotoxin 6-hydroxydopamine (6-OHDA) (Fig. 1). It was decided to lesion the dorsal noradrenergic bundle, rather than the locus coeruleus, because of the proximity of the locus coeruleus to the fourth ventricle and the possibility of 6-OHDA entering the ventricular system.

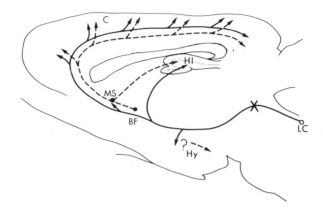

Figure 1. Relationship between dorsal noradrenergic bundle and cholinergic projections in rat brain. X represents lesion site as described
in text; BF, basal forebrain; C, cerebral cortex; HI, hippocampus;
Hy, hypothalamus; LC, locus coeruleus; and MS, medial septum.

METHODS

The dorsal noradrenergic bundles of Equithesin-anesthetized
male Sprague-Dawley rats were lesioned bilaterally by stereotaxic-
ally-placed injections of 6-OHDA (23.6 nmol in 2 μl of 0.2% ascorbic
acid, 0.9% saline solution; 0.5 μl/min) at the coordinates of AP
+1.0, L $\pm$ 0.8, V -0.6, according to the atlas of König and Klippel.
Control animals were injected with the vehicle. A smaller volume
of 6-OHDA was injected than was used by Lidbrink and Jonsson (6);
this was to diminish the possibility of diffusion of neurotoxin to
the ventral noradrenergic bundle.

Ten days after being lesioned, the animals were injected with
either amphetamine (27 μmol/kg; i.p.) or saline, and killed by focussed
microwave radiation 1 hour later. The brains were immediately dissected into appropriate areas. Nine minutes before microwave irradiation
of the rats, constant rate infusion with phosphoryl [^{2}H$_9$] choline
(15 μmole/kg/min) was begun. Animals were killed immediately at
the end of infusion. Levels of ACh and choline and TR$_{ACh}$ were then
determined by a mass fragmentographic technique, as described by
Robinson et al. (15). A 200 μl aliquot of the potassium acetate-
neutralized perchloric acid supernatant was reserved for determination
of levels of NE to confirm lesions of noradrenergic neurons. NE
was determined by an HPLC method with electrochemical detection,
essentially according to the method of Morier and Rips (11). Rats
not exhibiting the proper decrease in NE were excluded from all
data calculations.

RESULTS

Effect of Amphetamine and 6-OHDA on Cortical TR$_{ACh}$ and NE. Amphetamine increases TR$_{ACh}$ in the parietal cortex without changing ACh or choline content. 6-OHDA lesions of the dorsal noradrenergic bundle signifi- cantly decrease the NE content of the parietal cortex, but do not affect TR$_{ACh}$ (Fig. 2). However, the amphetamine-induced increase in TR$_{ACh}$ is reduced to approximately 60% of the control response to amphetamine. This suggests that although noradrenergic neurons from the locus coeruleus do not have a tonic action on cholinergic neurons in or projecting to the cerebral cortex, they do contribute to the excitatory action of amphetamine on cholinergic neurons in this area.

Effect of Amphetamine and 6-OHDA on hippocampal TR$_{ACh}$ and NE. Treat- ment with amphetamine also increases TR$_{ACh}$ in the hippocampus without affecting ACh or choline content. As in the parietal cortex, 6-OHDA

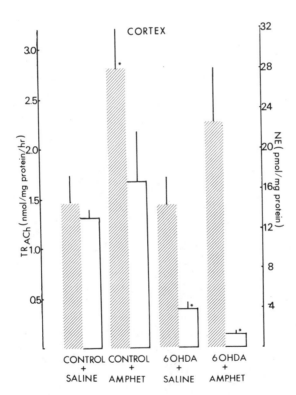

Figure 2. Effect of dorsal noradrenergic bundle lesions and ampheta- mine on TR$_{ACh}$ and concentration of NE in the cortex. Hatched bars represent TR$_{ACh}$, open bars NE content. Vertical lines represent $\pm$ S.E.M. *p<0.05.

lesion of the dorsal noradrenergic bundle has no effect on TR_{ACh}, but causes a significant reduction in hippocampal NE content. As observed in the cortex, 6-OHDA lesions of the dorsal noradrenergic bundle reduce the amphetamine-induced increase in TR_{ACh} in the hippocampus by approximately 40% (Fig. 3). These results suggest that the dorsal noradrenergic bundle does not exert a tonic action on the hippocampal cholinergic innervation, but does contribute to the action of amphetamine on hippocampal cholinergic neurons.

Effect of Amphetamine and 6-OHDA on Hypothalamic TR_{ACh} and NE. Amphetamine increases TR_{ACh} in the hypothalamus (Fig. 4) without changing ACh or choline content (data not shown). 6-OHDA-produced lesions of the dorsal noradrenergic bundle reduce the NE content of the hypothalamus, although the reduction is not as large as that seen in the parietal cortex or hippocampus. The dorsal noradrenergic bundle appears to exert no tonic influence on the cholinergic innervation of the hypothalamus, as 6-OHDA lesions have no effect on TR_{ACh}. However, 6-OHDA lesions of the dorsal noradrenergic bundle

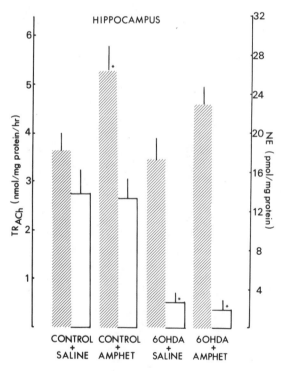

Figure 3. Effect of dorsal noradrenergic bundle lesions and amphetamine on TR_{ACh} and concentration of NE in the hippocampus. Hatched bars represent TR_{ACh}, open bars NE content. Vertical lines represent $\pm$ S.E.M. *$p<0.05$.

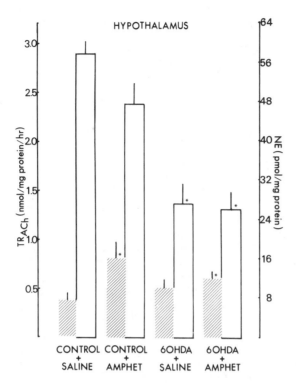

Figure 4.   Effect of dorsal noradrenergic bundle lesions and amphetamine on TR_{ACh} and concentration of NE in the hypothalamus.   Hatched
bars represent TR_{ACh}, open bars NE content.   Vertical lines represent
$\pm$ S.E.M.   *$p<0.05$.

reduce the amphetamine-induced increase in TR_{ACh} in the hypothalamus
(by 50%), although TR_{ACh} after amphetamine and 6-OHDA remains significantly higher than control.

DISCUSSION

        Amphetamine is known to activate cholinergic neurons in multiple
areas of the brain.   This drug has been demonstrated to increase
the rate of release of ACh from the cat cerebral cortex, as well
as to increase TR_{ACh} in the cortex and hippocampus of the rat (4,
12, 14).   Furthermore, this action has been proposed to occur through
noradrenergic mechanisms, since dopaminergic agonists have no effect
on TR_{ACh} in the cortex and decrease TR_{ACh} in the hippocampus (14).
In fact, the actions of amphetamine are blocked by the injection
of the α-antagonist phenoxybenzamine into the septal area, suggesting
that the excitatory action of amphetamine on cortical and hippocampal
neurons occurs through noradrenergic terminals in this area.   In
the present studies, the stimulatory action of amphetamine on TR_{ACh}

in the parietal cortex and hippocampus has been confirmed, as well as extended to the hypothalamus. Knowing the source of the noradrenergic neurons involved in these actions may indicate the functional significance of cholinergic activation in these areas. The locus coeruleus supplies noradrenergic innervation to a wide range of forebrain structures. Therefore, the dorsal noradrenergic bundle was lesioned by 6-OHDA to determine the contribution of the locus coeruleus to this activation.

The locus coeruleus supplies noradrenergic projections to the cerebral cortex where these projections could interact with intrinsic cholinergic neurons, and noradrenergic projections which pass through the medial forebrain bundle and give off collaterals to the lateral hypothalamus, an area containing cholinergic cell bodies projecting to the parietal cortex (2, 5, 9). In addition, a projection from medullary noradrenergic neurons innervates the substantia innominata, the location of additional cholinergic neurons that project to the parietal cortex (8, 9). The fact that phenoxybenzamine, when injected into the septal area, reduces the amphetamine-induced increase in cortical TR_{ACh} suggests that the interaction between cholinergic and noradrenergic neurons occurs in the septal area. Thus, noradrenergic projections from either the locus coeruleus or medullary cell groups could be involved. It appears that the locus coeruleus does contribute to the action of amphetamine on TR_{ACh} in the cortex, as TR_{ACh} after 6-OHDA lesion of the dorsal noradrenergic bundle and treatment with amphetamine is no longer significantly different from control, although it is not totally blocked. This interaction may be of importance to arousal mechanisms, because an increased activity of cholinergic neurons has been postulated to occur during arousal (10).

The locus coeruleus is thought to be the sole source of noradrenergic innervation of the hippocampus. However, the septum and nucleus of the diagonal band, which contain cholinergic cell bodies projecting to the hippocampus, receive noradrenergic innervation from cell groups in the medulla (A1 and A2) as well as from the locus coeruleus (8). As earlier experiments with phenoxybenzamine have suggested that the septal area is the site of amphetamine-enhanced interactions of NE with receptors on interneurons or on cholinergic neurons projecting to the hippocampus, it appears that the source of the NE could be noradrenergic cell groups in either the medulla or the locus coeruleus. Lesions of the dorsal noradrenergic bundle partially reduce the amphetamine-induced increase in hippocampal TR_{ACh}, hence the locus coeruleus probably does play a part in this action of amphetamine. Interestingly, the dorsal noradrenergic bundle has been reported to not be important for atropine-sensitive hippocampal theta activity (16).

The locus coeruleus, although it is not considered the primary source of noradrenergic innervation of the hypothalamus (7), may also be involved in the action of amphetamine on TR_{ACh} in the hypothalamus. Lesion of the dorsal noradrenergic bundle with 6-OHDA reduces, but does not significantly block, the excitatory action of amphetamine on TR_{ACh} in the hypothalamus. Because the locus coeruleus has been implicated in behavioral and physiological responses to changes in blood volume (13), and the hypothalamus is an area important in the control of the autonomic nervous system, this interaction between noradrenergic and cholinergic neurons may be physiologically important. Furthermore, a noradrenergic pressor mechanism has been described in the posterior hypothalamus, an area also known to contain a cholinergic pressor mechanism (1, 17). Dopaminergic involvement in the residual action of amphetamine on TR_{ACh} in the hypothalamus after 6-OHDA lesions of the dorsal noradrenergic bundle has not been ruled out, as the action of dopamine on hypothalamic cholinergic neurons has not been studied. Neurons traveling in the ventral noradrenergic bundle may also be important in this area.

The lack of complete blockade of increases in TR_{ACh} in the brain areas studied may be due to the development of supersensitivity and incomplete lesions. Thus, NE released from any remaining terminals of the dorsal noradrenergic bundle could result in an increased response in the cholinergic neurons. Also, the dorsal noradrenergic bundle, although the most important projection route of the locus coeruleus, does not contain all of the ascending outflow of the locus coeruleus, and specific lesions of this tract could spare locus coeruleus neurons traveling in the dorsal periventricular tract or the central tegmental tract and possibly synapsing with the areas studied (7). Furthermore, the ventral noradrenergic bundle may also supply noradrenergic neurons important in the action of amphetamine on TR_{ACh} in the areas studied.

CONCLUSIONS

Noradrenergic neurons traveling in the dorsal noradrenergic bundle do not exert a tonic action on cholinergic neurons in the cortex, hippocampus or hypothalamus. However, these noradrenergic neurons appear to contribute to the action of amphetamine on cholinergic processes in the parietal cortex, hippocampus, and possibly the hypothalamus.

Acknowledgements. The author wishes to acknowledge the expert technical assistance of Ms. Finley Austin and Mrs. Diane Gibbens, and to thank Mrs. Debbie Slovenec for her help in preparing this manuscript. This work was supported by funds from NIMH grant #R01MH37450 and a Grant-in-Aid from the American Heart Association, with funds contributed in part by the American Heart Association Texas Affiliate, Inc.

REFERENCES

1. Buccafusco, J.J. and Brezenoff, H.E. (1979):   Brain Research 165:295-310.
2. Eckenstein, F. and Thoenen, H. (1983):   Neuroscience Lett. 36: 211-215.
3. Fibiger, H.C. (1982):   Brain Res. Rev. 4:327-388.
4. Hemsworth, B.A. and Neal, M.J. (1968):   Br. J. Pharmacol. 34:543-550.
5. Jones, B.E. and Moore, R.Y. (1977):   Brain Research 127:23-53.
6. Lidbrink, P. and Jonsson, G. (1974):J. Neurochem. 22:617-626.
7. Lindvall, O. and Björklund, A. (1978):   In Handbook of Psycho-pharmacology (eds) L.L. Iversen, S.D. Iversen and S.H. Snyder, Plenum Press, New York, pp. 139-231.
8. Lindvall, O. and Stenevi, U. (1978):   Cell. Tiss. Res. 190:383-407.
9. McKinney, M., Coyle, J.T. and Hedreen, J.C. (1983): J. Comp. Neurol. 217:103-121.
10. Montplaisir, J.Y. (1975):   Electroenceph. Clin. Neurophysiol. 38:263-272.
11. Morier, E. and Rips, R. (1982):   J. Liq. Chromatog. 5:151-164.
12. Pepeu, G. and Bartolini, A. (1968): Eur. J. Pharmacol. 4:254-263.
13. Persson, B. and Svensson, T.H. (1981): J. Neural Trans. 52:73-82.
14. Robinson, S.E., Cheney, D.L. and Costa, E. (1978):   Naunyn - Schmiedeberg's Arch. Pharmacol. 304:263-269.
15. Robinson, S.E., Malthe-Sorenssen, D., Wood, P.L. and Commissiong, J. (1979): J. Pharmacol. Exp. Ther. 208:476-479.
16. Robinson, T.E., Vanderwolf, C.H. and Pappas, B.A (1977):   Brain Research 138:75-98.
17. Zawoiski, E. (1980): Arch. Int. Pharmacodyn., Ther. 247:103-118.

SUBSTANTIA INNOMINATA - CORTICAL CHOLINERGIC PATHWAY:

REGULATORY AFFERENTS

P.L. Wood[*] and P. McQuade[+]

[*]Neuroscience Research
CIBA-GEIGY Corporation, Pharmaceuticals Division
Summit, New Jersey 07901 U.S.A.

[+]Douglas Hospital
Research Centre
Verdun, Quebec H9R 1W3, Canada

INTRODUCTION

The large forebrain neurons of the substantia innominata (SI) extensively innervate the ipsilateral cerebral cortex (2, 6).   In addition, these neurons stain for acetylcholinesterase (7, 9, 13, 18) and lesions of the SI result in 50-80% decreases in cortical choline acetyltransferase (5, 7, 8).   These data strongly support the cholinergic makeup of the substantia innominata-cholinergic pathway (SICP).   In this report, we summarize our studies of the transynaptic regulation of the SICP, with particular focus on potential direct innervations of the cholinergic cell bodies in the SI.   To this end, local injections of defined receptor agonists or antagonists into the SI have been compared with parenteral injections for their actions on cortical cholinergic function.

To measure cholinergic activity, we have utilized acetylcholine turnover (TR_{ACh}) as a biochemical index of cholinergic dynamics (1, 10, 23).   Assay conditions and calculations employed have been described extensively in previous studies (19, 21).

GABAERGIC INPUT

Parenteral GABAergic agents have been found to potently decrease cortical TR_{ACh} (26, 28).   This activity is observed with receptor agonists (THIP, muscimol, kojic amine), benzodiazepines and GABA transaminase inhibitors (Table 1). These actions are also reversed by picrotoxin which alone does not alter cortical TR_{ACh}. It therefore appears that the GABAergic input to the SI, like that in the septum (1), is not tonically active.

Confirmation of a direct GABAergic innervation of the SI (12, 14, 17) was obtained by utilizing direct injections of muscimol into the SI (26). These treatments (Table 2) clearly reduced cortical TR_{ACh} to the same degree as parenteral drug treatment. Anatomical (12) and biochemical (17) studies indicate that the GABAergic input to the SI originates in the N. accumbens. Electrical stimulation of the N. accumbens was therefore examined and found to decrease cortical TR_{ACh} (Table 2).

OPIOIDS

The SICP is potently depressed by μ (mu) and δ (delta) but not k (kappa) opioid receptor agonists (21, 22, 27). These actions are naloxone reversible (Table 3) and in the case of agonists, reversal also occurs with MR-2034 pretreatment (Table 3). However, μ-dependent decreases in cortical TR_{ACh} are not reversed by MR-2034. These data appear to indicate that the SICP possesses a specific μ opiod isoreceptor regulation, namely μ_2 receptors (25), in addition to δ regulation. These actions do not appear to be at the level of the

Table 1

Actions of Parenteral GABAergic Agents on
TR_{ACh} in the Rat Parietal Cortex

Treatment (mg/kg)	Min	ACh (nmol/mg prot)	TR_{ACh} (nmol/mg prot/hr)
Control	30	0.18 ± 0.370	1.10 ± 0.100
Kojic amine (20)	30	0.18 ± 0.160	0.66 ± 0.071[*]
THIP (10)	30	0.20 ± 0.013	0.78 ± 0.072[*]
THIP (20)	30	0.23 ± 0.030	0.54 ± 0.055[*]
AOAA (25)	30	0.18 ± 0.015	0.61 ± 0.016[*]
Muscimol (4)	30	0.20 ± 0.015	0.52 ± 0.040[*]
Muscimol (4)	60	0.21 ± 0.011	0.62 ± 0.098[*]
Muscimol (4)	120	0.19 ± 0.013	0.62 ± 0.023[*]
Muscimol + Picrotoxin (3)	30	0.20 ± 0.018	1.00 ± 0.077
Picrotoxin (3)	30	0.18 ± 0.011	1.00 ± 0.033
Diazepam	30	0.17 ± 0.019	0.67 ± 0.073[*]

[*]$p < 0.05$; mean ± S.E.M. (n = 5-9)

Table 2

Actions of Local Injections of Muscimol in the
Substantia Innominata and of Electrical Stimulation of the
Nucleus Accumbens on Cortical TR_{ACh}

Treatment	Min	ACh (nmol/mg prot)	TR_{ACh} (nmol/mg prot/hr)
A) Control		0.18 ± 0.330	$1.20 \pm .098$
B) Intra S.I. muscimol[1]			
0.1 μg	30	0.18 ± 0.021	0.87 ± 0.061*
0.8 μg	30	0.19 ± 0.027	0.53 ± 0.049*
C) N. Acc. Stimulation[2]			
10 Hz	20	0.17 ± 0.017	0.86 ± 0.020*
30 Hz	20	0.18 ± 0.008	0.50 ± 0.033*

[1]0.5 μl (A 6.5, L 2.5, V-2.0); [2]200 μA, 1.5 msec, 20 sec "on"/20 sec "off", (A 9.0, L 1.1, V-0.6); *$p < 0.05$ (n = 7-10)

cholinergic cell bodies in the SI (Table 4; 27) nor the cholinergic nerve endings in the cerebral cortex (4, 16, 27). It therefore appears that a subcortical opioid site is involved in the actions of opiates on cortical TR_{ACh}. The present hypothesis (Fig. 1) is that the SI receives a tonic excitatory input which in turn is innervated by an enkephalinergic synapse (27). This multi-synaptic regulatory input to the SI for opioids clearly requires neurochemical and anatomical verification.

Neither naloxone (27) nor the enkephalinase inhibitor, thiorphan (24) affects this opioid loop indicating a lack of tonic opioid transmission (Table 4).

SEROTONERGIC AGENTS

The serotonin antagonists mianserin and metergoline as well as the agonists quipazine and m-trifluoropiperazine do not alter cortical TR_{ACh} (Table 5). These data do not support a serotonergic regulation of the SI.

Table 3

Actions of Parenteral and Intraventricular Opiates on
TR_{ACh} in the Rat Parietal Cortex (30 min)

Treatment	mg/kg	TR_{ACh} % Control
Morphine (μ)	16	55[*]
	16 + N (2)	95
	16 + MR (4)	53[*]
MR-2034 (k)	4	100
	32	106
DADLE[**] (δ)	20 μg (IVt)	55[*]
	20 μg + N (5)	98
	20 μg + MR (4)	84

[*]$p < 0.05$; N, naloxone; MR, MR-2034; [**]D-Ala-D-Leu enkephalin

Table 4

Lack of Effect of Local Substantia Innominata Opiate
Injections on Cortical TR_{ACh} (30 min)

Treatment	μg	TR_{ACh} % Control
Morphine	10	105
DADLE[*]	21	95
Thiorphan	100 (IVt)	94

[*]D-Ala-D-Leu enkephalin

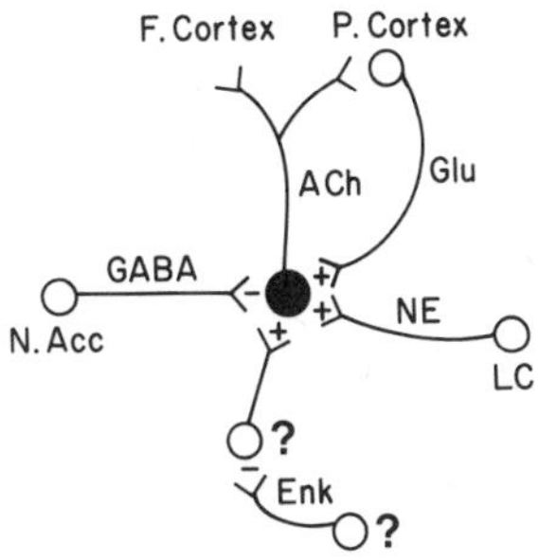

Figure 1. Tentative transsynaptic inputs to the cholinergic neurons of the rat substantia innominata.

ADRENERGIC AGENTS

Previous studies of the noradrenergic regulation of the septal hippocampal cholinergic pathway have demonstrated a positive α-adrenergic input to the septum (1). In these studies amphetamine was also found to increase TR_{ACh} in the parietal cortex. We have extended these initial observations by using defined adrenergic receptor agonists and antagonists (Table 6). In these studies, the β agonists isoproterenol and salbutamol as well as the β antagonist propranolol did not alter cortical TR_{ACh}. Similarly, α_2 agonists and antagonists were without effect on TR_{ACh} in the cortex. However, the α_1 antagonist prazosin decreased cortical TR_{ACh}, indicating that a possible tonic α_1 input regulates these cholinergic neurons (Table 6). Whether this is at the level of the SI or the cortex remains to be defined.

Table 5

Lack of Effect of Parenteral Serotonergic Agents on
TR_{ACh} in the Parietal Cortex of the Rat

Treatment (mg/kg)	TR_{ACh} (nmol/mg prot/hr)
Control	1.2 ± 0.060
Mianserin (20)	1.2 ± 0.058
Metergolin (10)	1.2 ± 0.069
Quipazine (10)	1.3 ± 0.062
m-Trifluoropiperazine (2)	1.2 ± 0.050
m-Trifluoropiperazine (10)	1.3 ± 0.046

Table 6

Actions of Parenteral Adrenergic Agents on Cortical TR_{ACh}

Treatment	mg/kg	TR_{ACh} (nmol/mg prot/hr)
Control		1.10 ± 0.058
Isoproterenol (β)	5.0	1.10 ± 0.100
Salbutamol (β_2)	5.0	1.30 ± 0.100
Propranolol (β)	5.0	1.20 ± 0.098
Clonidine (α_2)	0.1	1.30 ± 0.130
	0.2	1.10 ± 0.091
Piperoxan (α_2)	1.0	1.00 ± 0.047
CP-14304-18 (α_2)	1.0	0.98 ± 0.059
Prazosin (α_1)	0.5	$0.78 \pm 0.065^*$
	1.0	$0.65 \pm 0.041^*$
	5.0	$0.62 \pm 0.031^*$

$^*p < 0.05$

ADENOSINE

The adenosine agonist 2-chloroadenosine decreases cortical TR_{ACh} in a dose-dependent manner (3, 11). This action probably is at the level of the cortex (15). Analysis of possible tonic regulation will await the development of more potent and specific adenosine antagonists.

GLUTAMATE

The cholinergic neurons of the SI are sensitive to the neurotoxic actions of kainic acid (5, 7, 8) indicating a possible glutamatergic input to this nucleus. This input probably originates from the neocortical pyramidal cells and is similar to the pyramidal cell feedback loop to the septum from the hippocampus (1, 20).

SUMMARY

In summary, as shown in figure 1, the SICP receives a potent inhibitory GABAergic input from the N. accumbens. This input appears to be direct innervation of the SI while the inhibitory opioid regulation appears to originate from a more caudal site. A positive glutamatergic feedback from the neocortex and a possible α_1 input from the brainstem also appear to regulate the activity of the cholinergic neurons in the SI.

REFERENCES

1. Costa, E., Panula, P., Thompson, H.K. and Cheney, D.L. (1983): Life Sci. 32:165-179.
2. Divac, I. (1975): Brain Res. 93:385-398.
3. Haubrich, D.R., Williams, M., Yarbrough, G.G. and Wood, P.L. (1981): Can. J. Physiol. Pharmacol. 59:1196-1198.
4. Jhamandas, K., Hron, V. and Sutuk, M. (1975): Can. J. Physiol. Pharmacol. 50:57-62.
5. Johnston, M.V., McKinney, M. and Coyle, J.T. (1979): Proc. Natl. Acad. Sci. (USA) 76:5392-5396.
6. Jones, E.G., Burton, H., Saper, C.B. and Swanson, L.W. (1976): J. Comp. Neurol. 167:385-420.
7. Lehmann, J. and Fibiger, H.C. (1979): Life Sci. 25:1939-1947.
8. Lehmann, J., Nagy, J.I., Atmadja, S. and Fibiger, H.C. (1980): Neuroscience 5:1161-1174.
9. Mesulam, M.M. and Van Hoesen, G.W. (1976): Brain Res. 109:152-157.
10. Moroni, F., Malthe-Sorenssen, D., Cheney, D.L. and Costa, E. (1978): Brain Res. 150:333-341.
11. Murray, T.F., Blaker, W.D., Cheney, D.L. and Costa, E. (1982): J. Pharmacol. Exp. Ther. 222:550-554.
12. Nauta, W.J.H., Smith, G.P., Faull, R.L.M. and Domesick, V.B. (1978): Neuroscience 3:385-401.
13. Parent, A., Gravel, S. and Oliver, A. (1979): Adv. Neurol. 24:1-11.
14. Perez de la Mora, M., Possan, L.D., Tapin, R., Terun, L., Palacios, R., Fuxe, K., Hökfelt, E. and Ljungdahl, A. (1981): Neuroscience 6:875-895.
15. Stone, T.W. and Taylor, D.A. (1980): Exp. Neurol. 70:556-566.
16. Szerb, J. (1974): Eur. J. Pharmacol. 29:192-194.
17. Walaas, I. and Fonnum, F. (1979): Brain Res. 177:325-336.
18. Wenk, H., Bigl, V. and Meyer, U. (1980): Brain Res. Rev. 2:295-316.
19. Wood, P.L. and Cheney, D.L. (1978): Can. J. Physiol. Pharmacol. 57:404-411.
20. Wood, P.L., Peralta, E., Cheney, D.L. and Costa, E. (1979): Neuropharmacology 18:519-524.
21. Wood, P.L. and Stotland, L.M. (1980): Neuropharmacology 19:975-982.

22. Wood, P.L. and Rackham, A. (1981): Neurosci. Lett. 23:75-80.
23. Wood, P.L., Cheney, D.L. and Costa, E. (1981):   Adv. Behav. Biol. 25:715-722.
24. Wood, P.L. (1982): Eur. J. Pharmacol. 82:119-120.
25. Wood, P.L., Richard, J.W. and Thakur, M. (1982):   Life Sci. 31:2313-2317.
26. Wood, P.L. and Richard, J.W. (1982): Neuropharmacology 21:969-972.
27. Wood, P.L., Stotland, L.M. and Rackham, A. (1984): In Dynamics of Neurotransmitter Function (ed) I. Hanin, Raven Press, New York, pp. 99-107.
28. Zsilla, G., Cheney, D.L. and Costa, E. (1976): Naunyn-Schmiedebergs Arch. Pharmacol. 294:251-255.

ENKEPHALINERGIC MODULATION OF CHOLINERGIC TRANSMISSION IN

PARASYMPATHETIC GANGLIA OF THE CAT URINARY BLADDER

W. C. deGroat, M. Kawatani, and A.M. Booth

Department of Pharmacology
School of Medicine
University of Pittsburgh
Pittsburgh, Pennsylvania 15260 U.S.A.

INTRODUCTION

Dense networks of enkephalin containing nerve terminals have been identified in mammalian sympathetic and parasympathetic ganglia (1, 12, 20-22, 25, 35). Evidence from various sources indicates that enkephalins are contained in, and released by axons that arise from neurons extrinsic to the ganglia, and that enkephalins have an inhibitory effect on ganglionic transmission (2, 12, 23, 25, 26, 30, 36, 39).

This paper, which will focus on the role of leucine-enkephalin in parasympathetic ganglia of the urinary bladder of the cat, will summarize the immunohistochemical and pharmacological data suggesting that leucine-enkephalin is an inhibitory transmitter released by axons in the sacral parasympathetic preganglionic pathways.

MATERIALS AND METHODS

The distribution of leucine-enkephalin (L-ENK) in sacral pre-ganglionic neurons and in nerve terminals in vesical ganglia was examined with indirect immunohistochemical techniques using the fluorescein isothiocyanate (FITC) or the peroxidase antiperoxidase method (37). Rabbit anti-L-ENK serum (Lot No. 35381, INC Corp) was used. Spinal cord and ganglia were removed from animals perfused with 4% paraformaldehyde fixative (in 0.1 M phosphate buffer, pH 7.2). Tissue was sectioned in a cryostat. Appropriate controls using antisera preabsorbed with antigen were included in the protocol. Preganglionic neurons were identified by retrograde labeling with a fluorescent dye (true blue or propidium iodide) (5, 22, 27, 34)

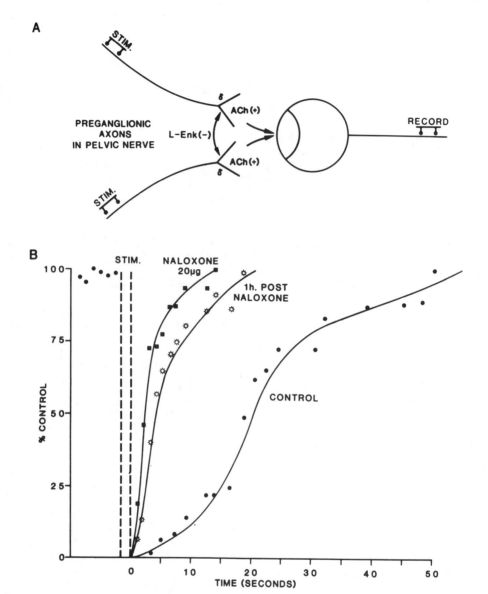

Figure 1.   Fluorescence photomicrographs showing the distribution
of leucine-enkephalin immunoreactivity (L-ENK-IR) in the sacral spinal
cord (A, C, & D) and in a bladder ganglion (B).   A: L-ENK-IR in
neurons in the lateral band of the sacral parasympathetic nucleus
from an animal injected intrathecally with colchicine 48 hours prior
to sacrifice.   B: L-ENK-IR in varicosities surrounding neurons in
a vesical ganglion.   C: High power photomicrograph of four sacral
preganglionic neurons labelled with propidium iodide following retrograde axonal transport from the pelvic nerve.   Cells were identified

by red fluorescence when visualized by light at 450-490 nm excitation
wave length. D: Photomicrograph of the same section as in C showing
two of the labelled cells (filled arrow) exhibiting L-ENK-IR and
two cells (open arrow) which did not exhibit L-ENK-IR. L-ENK varicosities are also shown distributed throughout the nucleus. L-ENK-IR
was identified by green fluorescence when illuminated with light
at 530-560 nm excitation wave length. Animal was pretreated with
colchicine intrathecally. E: Approximate locations of spinal sections
shown in photomicrographs A (no. 1) and C-D (no. 2). Calibration
represent 125 μm in A, C & D 100 μm in B.

---

which was applied to the preganglionic axons in the pelvic nerve.
After an appropriate transport time (7-14 days), colchicine (150-200
μg, injected intrathecally) was applied to the spinal cord to increase
the concentration of peptides in the preganglionic neurons. The
animals were then sacrificed 24-48 hours later. Vesical ganglia
were obtained from normal animals and from animals in which the
sacral ventral roots had been transected unilaterally 1-38 weeks
earlier to eliminate the preganglionic outflow to one side of the
bladder.

Pharmacological experiments were conducted in chloralose or
dial urethane anesthetized cats using multiunit recording on postganglionic nerves in situ, or intracellular recording in ganglia
maintained in a tissue bath in vitro. In the former experiments,
the urinary bladder was exposed by a midline abdominal incision;
ganglia were identified on the surface of the bladder and postganglionic nerves were prepared for monophasic recording (6, 7). Stimulating
electrodes were placed on preganglionic nerves (pelvic nerves).
Drugs and neuropeptides were administered by close intra-arterial
injection into the inferior mesenteric artery. Intracellular recording
was obtained with glass micropipettes from isolated ganglia superfused with oxygenated Krebs-Ringer solution. More detailed descriptions of the electrophysiological technique have been included in
recent papers (8, 36).

RESULTS

Immunohistochemical studies revealed that parasympathetic ganglia
on the surface of the urinary bladder exhibited a dense innervation
by L-ENK-containing axons and varicosities which formed pericellular
baskets surrounding the ganglion cells (Fig. 1B) (24). The L-ENK
varicosities were eliminated by transection of the sacral ventral
roots, indicating that they were a component of an efferent pathway
projecting from the sacral segments of the spinal cord. Sacral
preganglionic neurons were identified as the likely source of this
efferent pathway. As noted by other investigators (17, 18) and
confirmed in the present experiments, neurons in the region of the

sacral parasympathetic nucleus exhibited L-ENK immunoreactivity (Fig. 1A) when the animals were pretreated with colchicine (150-200 μg, intrathecally) 24-48 hours before sacrifice. However, L-ENK immunoreactivity was not detected in this population of cells in animals without colchicine treatment.

Retrograde dye tracing combined with immunohistochemistry (5, 22, 34) established that the neurons containing L-ENK were parasympathetic preganglionic neurons (10). In these experiments, propidium iodide or true blue (fluorescent dyes) were applied to the pelvic nerve to label the entire sacral preganglionic outflow by retrograde transport. As shown in Figure 1C & D, many labelled preganglionic neurons in the lateral band of the sacral parasympathetic nucleus exhibited L-ENK-immunoreactivity. Neurons in this region of the nucleus innervate the urinary bladder (9, 28, 29). It should be noted that some labelled cells did not exhibit L-ENK-immunoreactivity, whereas some unlabelled neurons within the nucleus contained L-ENK. These findings could be attributed to methodological problems such as failure to detect low concentrations of peptide or incomplete retrograde labelling, but also would be compatible with the notion that different populations of neurons exist in the lateral band of the nucleus (9, 29).

In view of prominent enkephalinergic projection from the spinal cord to vesical ganglia, pharmacological studies were undertaken to analyze the effects and possible physiological role of L-ENK in these ganglia. The intraarterial administration of L-ENK to bladder ganglia in situ depressed the postganglionic action potentials elicited by electrical stimulation of the preganglionic pathway in the pelvic nerve (2, 11, 12, 36). The depression was dose dependent in the range between 0.1 and 20 μg, i.a., had a duration of 5-10 minutes and was antagonized by the administration of naloxone (10-40 μg, i.a.), an opiate antagonist (Fig. 2). The magnitude of the enkephalinergic inhibition was also dependent upon the frequency of stimulation. For a particular dose of L-ENK the inhibition was maximal at low frequencies of preganglionic stimulation (0.25 - 0.5 Hz), was reduced by 50% at frequencies of 2-3 Hz and was negligible at frequencies of 5-7 Hz even with large doses of L-ENK.

An analysis of other opiate receptor agonists revealed that methionine-enkephalin was approximately equipotent with L-ENK, whereas selective delta opiate receptor agonists, DSLET ([DSer^2, Leu^5] enkephalyl-thr) or DADLE ([D-Ala^2, D-Leu^5] enkephalin) (0.05 - 2 μg, i.a.) (31) were more potent than L-ENK. Other agents such as ethylketocyclazocine, a kappa-mu receptor agonist and morphine, an agent with 20-50 times greater affinity for mu than delta receptors (31), were relatively weak ganglionic depressants. These findings indicate that delta opiate receptors mediate enkephalinergic inhibition of transmission in vesical ganglia.

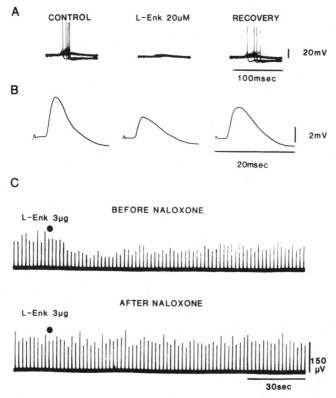

Figure 2. Depressant effects of leucine-enkephalin (L-ENK) on transmission in vesical ganglia. A: Intracellular recording of seven
superimposed sweeps showing EPSPs and firing elicited by electrical
stimulation of a preganglionic nerve (0.5 Hz). L-ENK (20 μM) depressed EPSP amplitude and completely blocked firing. Data obtained
from an in vitro preparation. B: Computer-average of 10 individual
EPSPs from another cell under the same conditions. L-ENK reversibly
depressed the amplitude of the EPSPs. C: Depression of compound
action potentials on a vesical postganglionic nerve by the intra-
arterial administration of L-ENK (3 μg). Action potentials were
elicited by stimulation of a preganglionic nerve (0.5 Hz). The
depressant effects of L-ENK were blocked (lower traces) by the intra-
arterial administration of naloxone (30 μg). Data obtained from a
preparation in situ.

Experiments on isolated ganglia in vitro using intracellular
recording examined the mechanisms involved in enkephalinergic inhibition (2, 36). As illustrated in Fig. 2, L-ENK administered in
the superfusion solution in a concentration of 15-20 μM depressed
synaptically evoked firing, decreased the amplitude of excitatory
postsynaptic potentials (EPSPs) and increased the number of EPSP

failures without producing a consistent change in resting membrane potential or resistance. These effects were observed in 10 of 20 cells tested and were reversible in 10-15 minutes after elimination of L-ENK from the superfusion fluid.

Since vesical ganglia receive an enkephalinergic innervation and are inhibited by the administration of exogenous enkephalins experiments were undertaken to determine whether endogenously released enkephalins would also depress ganglionic transmission. For these experiments we have used a heterosynaptic inhibition paradigm that was developed in previous studies in our laboratory (7). Vesical ganglia receiving inputs from two or more preganglionic nerves were identified in situ (Fig. 3A). Electrical stimulation of one nerve at a low frequency (0.5 Hz) elicited consistent postganglionic action potentials, which were markedly depressed by repetitive stimulation of another preganglionic nerve using trains of stimuli (1-10 sec train duration) at high intratrain frequencies (10-30 Hz). The magnitude (range, 30-80%) and duration (range, 20-50 sec.) of the heterosynaptic inhibition varied with the frequency and duration of the stimulation. Naloxone administered in doses (50-100 μg, i.a.) which antagonized the ganglionic depressant effects of exogenous enkephalins reduced the magnitude and duration of heterosynaptic inhibition in a dose dependent manner (Fig. 3B) (11). The effect of naloxone persisted for 1-3 hr. Naloxone did not alter adrenergic inhibition in vesical ganglia (6) or the postganglionic firing elicited by single stimuli. Therefore it would appear that under the conditions of these experiments, enkephalinergic inhibition does not occur with low levels of preganglionic activity but only at high rates of firing which mimic the normal preganglionic discharge pattern occuring during micturition (9).

DISCUSSION

In the present investigation, we have used various anatomical and electrophysiological techniques to examine the origin and function of leucine-enkephalin-containing pathways to parasympathetic ganglia of the cat urinary bladder. It is clear from these studies that bladder ganglion cells receive an extensive enkephalinergic innervation which is part of an efferent system originating in the sacral segments of the spinal cord. Unilateral transection of the sacral ventral roots or the pelvic nerve led within 2-3 weeks to the disappearance, ipsilaterally, of the dense network of enkephalin-varicosities surrounding the bladder ganglion cells. These varicosities did not reappear in the bladder ganglia over a considerable range of survival times (1-9 months) (24) even though it has been noted that long term survival (1-8 months) after unilateral decentralization is accompanied by axonal sprouting in the periphery and reinnervation of the denervated bladder ganglion cells by cholinergic axons in the hypogastric nerve or by cholinergic axons from the contralateral side of the bladder (13). This finding suggests that only certain efferent pathways which have the potential for innervating autonomic ganglion cells are enkephalinergic.

The most likely source of L-ENK inputs to bladder ganglia are the cholinergic preganglionic neurons in the lateral band of the sacral parasympathetic nucleus (29). HRP tracing techniques showed that neurons in this region innervate the urinary bladder (28). Neurons in the same region also exhibit L-ENK immunoreactivity, although this is usually only detectable if the peptide concentration in the cells is increased by ligating the ventral roots (17, 18) or by administering colchicine to the spinal cord (10, 12, 17, 18). More direct evidence for the presence of L-ENK in sacral preganglionic neurons was obtained in the present study when immunohistochemistry was combined with dye tracing (10). When preganglionic neurons were labelled by retrograde axonal transport with dye applied to the pelvic nerve, it was noted that many dye labeled cells also exhibited L-ENK immunoreactivity in colchicine treated animals. Similar results have been obtained in the lumbar sympathetic pathways of the guinea pig (5) and in the preganglionic pathways to the avian ciliary ganglion (15, 16). It has not been directly established that L-ENK and acetylcholine coexist in the same neurons, although based on the large numbers of sacral preganglionic neurons exhibiting L-ENK-immunoreactivity it would seem likely that the coexistence does occur.

The presence of L-ENK in the sacral preganglionic pathways raises the possibility that enkephalinergic mechanisms both in the spinal cord and in peripheral ganglia may be important in the regulation of bladder function. This view is supported by the demonstration of dense networks of L-ENK-containing varicosities in the sacral autonomic nucleus (10, 12, 17, 18) as well as in bladder ganglia. The origin of the enkephalin-terminals in the spinal autonomic nuclei is uncertain. However, it should be noted that recurrent inhibition is very prominent in the sacral autonomic outflow (9) and therefore the presence of enkephalinergic terminals could represent axon collaterals of the enkephalinergic preganglionic neurons. The role of enkephalins in recurrent inhibition has not been examined, although it is known that the intrathecal administration of exogenous opioid peptides depresses bladder reflexes via activation of delta opiate receptors and that this depression is blocked by naloxone (19). In addition, the administration of naloxone enhances bladder activity suggesting that the bladder reflex pathways are subject to a tonic enkephalinergic inhibition (12, 19, 33, 38).

Enkephalinergic inhibition in peripheral ganglia is also mediated by delta opiate receptors and is readily reversed by naloxone. Mu and kappa receptor agonists (morphine and ethylketocyclazocine, 31) which have relatively weak depressant effects on ganglia have similar weak effects on spinal micturition pathways. Thus there

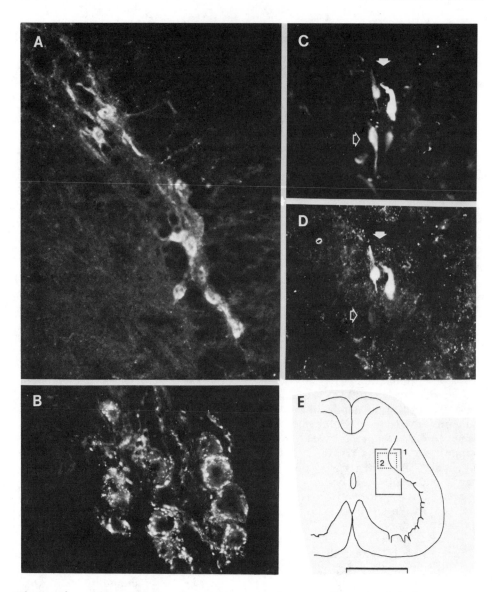

Figure 3. The effect of naloxone on heterosynaptic inhibition in a vesical ganglion. **A**: Diagram showing the experimental preparation. Two bundles of preganglionic fibers innervating the same ganglion were isolated for stimulation (STIM). A postganglionic nerve was isolated for recording evoked postganglionic action potentials. **B**: Graph showing that electrical stimulation of one preganglionic nerve elicits postganglionic potentials which were reduced in magnitude by repetitive stimulation of another preganglionic nerve (STIM between vertical dashed lines, 30 Hz for 2 sec). Abscissa, time in seconds; ordinate, amplitude of evoked postganglionic

potentials. Naloxone (20 μg, intra-arterially) reduced the duration of the inhibition. It is concluded as shown in upper diagram (A) that L-ENK released from one preganglionic nerve can act presynaptically to depress acetylcholine release from adjacent nerve terminals.

———————

are a number of similarities in the effects of opioid drugs on the peripheral and spinal components of the sacral outflow to the bladder.

The mechanism of enkephalinergic inhibition in the spinal autonomic pathways in unknown. However, in the bladder ganglia, it is clear that L-ENK has a predominately presynaptic inhibitory action. L-ENK decreased EPSP amplitude and the number of EPSP failures, without altering resting membrane potential or resistance. L-ENK depression was also frequency dependent, which is consistent with a presynaptic inhibitory mechanism. Other investigators have also noted presynaptic depression by opioid peptides at cholinergic synapses in sympathetic ganglia (23, 25, 26) and the neuromuscular junction (3) as well as at adrenergic synapses (14). Postsynaptic inhibitory effects of enkephalins have also been reported in the peripheral and central nervous system (4, 30, 32, 39, 40).

An inhibitory action of endogenously released enkephalins is suggested by studies of heterosynaptic inhibition in bladder ganglia and the inferior mesenteric ganglia (26). Earlier experiments in bladder ganglia had indicated that heterosynaptic inhibition induced by repetitive stimulation of preganglionic axons was related to postexcitatory depression or refractoriness. However, the present experiments revealed that heterosynaptic inhibition was reduced in magnitude and duration and in a reversible, dose-dependent manner by naloxone. Since naloxone did not have a direct effect on excitatory transmission elicited by low frequency preganglionic nerve stimulation, it is reasonable to conclude that the drug acted selectively to block the inhibitory effects of endogenous opioid peptides which were released by high frequency stimulation.

Based on the above observations, it is tempting to speculate, as indicated in Figure 3A, that L-ENK and acetylcholine are released at the same terminals, and that under conditions of high frequency preganglionic stimulation, L-ENK reaches sufficient concentrations at the synapse to activate delta receptors on the cholinergic terminals to suppress the release of acetylcholine. Enkephalinergic inhibition therefore represents another type of synaptic modulatory mechanism which may regulate transmission in vesical parasympathetic ganglia (6-8).

Acknowledgements. We thank Susan Erdman and Mary Backes for excellent technical assistance. Supported by NSF grant BNS 82-08348, NIH grants AM 31788, and NS 18075, and clinical research grant MH30915.

REFERENCES

1. Burnstock, G., Hökfelt, T., Gershon, M.D., Iversen, L.L., Kosterlitz, H.W. and Szurszewski, J.H. (1979): Neurosci. Res. Prog. Bull., 17:424-459.
2. Booth, A.M., Ostrowski, N., McLinden, S., Lowe, I. and deGroat, W.C. (1981): Soc. Neurosci, Abst. 7:214.
3. Bixby, J.L. and Spitzer, N.C. (1983): Nature, 301:431-432.
4. Carette, B. and Poulain, P. (1982): Regulatory Peptides 3:125-133.
5. Dalsgaard, C.J., Hökfelt, T., Elfvin, L.G. and Terenius, L. (1982): Neuroscience 7:647-654.
6. deGroat, W.C. and Saum, W.R. (1972): J. Physiol. 220:297-314.
7. deGroat, W.C. and Saum, W.R. (1976): J. Physiol. 256:137-158.
8. deGroat, W.C. and Booth, A.M. (1980): Fed. Proc. 39:2990-2996.
9. deGroat, W.C., Booth, A.M., Milne, R.J., and Roppolo, J.R. (1982): J. Auton. Nerv. Syst. 5:23-43.
10. deGroat, W.C., Kawatani, M., Booth, A.M., Lowe, I.P. and Zug, D. (1982): Neuroscience 7:S-50.
11. deGroat, W.C. and Kawatani, M. (1982): Soc. Neurosci. Abst. 8:552.
12. deGroat, W.C., Kawatani, M., Hisamitsu, T., Lowe, I.P., Morgan, C., Roppolo, J.R., Booth, A.M., Nadelhaft, I., Kuo, D. and Thor, K. (1983): J. Auton. Nerv. Syst. 7:339-350.
13. deGroat, W.C. and Kawatani, M. (1983): Soc. Neurosci, Abstr. 9:938.
14. Dubocovich, M.L. and Langer, S.Z. (1980): Brit. J. Phamacol. 70:383-393.
15. Ericksen, J.T., Reiner, A. and Karten, H.J. (1982): Nature 295:407-410.
16. Ericksen, J.T., Karten, H.J., Eldred, W.D. and Brecha, N.C. (1983): J. Neurosci. 2:994-1003.
17. Glazer, E.J. and Basbaum, A.I. (1980): Science 208:147-148.
18. Glazer, E.J. and Basbaum, A.I. (1981): J. Comp. Neurol. 196:377-389.
19. Hisamitsu, T. and deGroat, W.C. (1984): Brain Res. 298:51-65.
20. Hökfelt, T., Johansson, O., Ljungdahl, A., Lundberg, J.M. and Schultzberg, M. (1980): Nature 284:515-521.
21. Hökfelt, T., Schultzberg, M., Elde, R., Nilsson, G., Terenius, L., Said, S. and Goldstein, M. (1978): Acta Pharmacol. et. Toxicol. 43 II:79-89.
22. Hökfelt, T., Skirboll, L., Rehfeld, J.F., Goldstein, M., Markey, K. and Dann, O. (1980): Neuroscience 5:2093-2124.
23. Jiang, Z.G., Simmons, M.A. and Dun, N.J. (1982): Brain Res. 235:185-191.
24. Kawatani, M., Lowe, I.P., Booth, A.M., Backes, M.G., Erdman, S.L. and deGroat, W.C. (1983): Neurosci. Lett. 39:143-148.
25. Konishi, S., Tsunoo, A. and Otsuka, M. (1979): Nature 282:515-516.
26. Konishi, S., Tsunoo, A. and Otsuka, M. (1981): Nature 294:80-82.

27. Kuypers, H., Bentivoglio, M., Van der Kooy, D. and Catsman-Berrevoets, C.E. (1977): Neurosci. Lett. 6:127-135.
28. Morgan, C., Nadelhaft, I. and deGroat, W.C. (1979): Neurosci. Lett. 14:189-194.
29. Nadelhaft, I., deGroat, W.C. and Morgan, C. (1980): J. Comp. Neurol. 193:265-281.
30. North, R.A., Katayama, Y. and Williams, J.T. (1980): In Neural Peptides and Neuronal Communications (eds) E. Costa and M. Trabucchi, Raven Press, New York, pp. 83-91.
31. Peterson, S.J., Robson, L.E. and Kosterlitz, H.W. (1983): Br. Med Bull, 39:31-36.
32. Pepper, C.M. and Henderson, G. (1980): Science, 209:394-396.
33. Roppolo, J.R., Booth, A.M. and deGroat, W.C. (1983): Brain Res. 264:355-358.
34. Sawchenko, P.E. and Swanson, W. (1981): Brain Res. 210:31-51.
35. Schultzberg, M., Hökfelt, T., Terenius, L, Elfvin, L.G., Lundberg, J.M. Brandt, J., Elde, R.P. and Goldstein, M. (1979): Neuroscience 4:249-270.
36. Simonds, W.F., Booth, A.M., Thor, K.B., Ostrowski, N.L., Nagel, J.R. and deGroat, W.C. (1983): Brain Res. 217:365-370.
37. Sternberger, L.A. (1979): Immunocytochemistry, John Wiley and Sons, New York.
38. Thor, K.B., Roppolo, J.R., deGroat, W.C. (1983): J. Urol. 129:202-205.
39. Williams, J.T., Egan, T.M. and North, R.A. (1982): Nature 299:74-77.
40. Zieglgansberger, W. and Bayerl, H. (1976): Brain Res. 115:111-128.

VASOACTIVE INTESTINAL POLYPEPTIDE-MUSCARINIC

CHOLINERGIC INTERACTIONS

B. Hedlund, J. Abens, A. Westlind and T. Bartfai

University of Stockholm
Department of Biochemistry
Arrhenius Laboratory
S-106 91 Stockholm, Sweden

INTRODUCTION

Coexistence of a classical neurotransmitter and a peptide neurotransmitter in the same neuron in the peripheral or central nervous system has been described in a large number of cases (cf. for review (10)). The peptide which has been described to occur together with acetylcholine in the parasympathetic neurons innervating the cat submandibular salivary gland is VIP, the vasoactive intestinal polypeptide (cf. 12). VIP, a 28 aminoacid long peptide with a C-terminal amide, belongs to the secretin family. It was isolated by Said and Mutt (17) from porcine gut and it appears that the aminoacid sequence of the porcine and human VIP is identical.

Based on results of immunohistochemical methods it is likely that VIP and acetylcholine may also coexist in the bipolar cells of the cerebral cortex. Many of these cells contain VIP (8) and some show choline acetyltransferase immunoreactivity (4).

As a natural extension of studies on muscarinic cholinergic transmission in the peripheral and central nervous system we have studied acute and chronic drug effects on the acetylcholine-VIP interactions (1, 9, 13). Some of these data together with new information on ACh-VIP interactions are summarized below.

MATERIALS AND METHODS

Male Sprague-Dawley rats (150-200 g) from Anticimex, Stockholm, were used.

Purified VIP was a generous gift from Professor Viktor Mutt, Department of Biochemistry, Karolinska Institutet, Stockholm.

[^{3}H]-4-N-methylpiperidinylbenzilate ([^{3}H]-4-NMPB) was a generous gift from Professor Mordechai Sokolovsky, Department of Biochemistry, Tel Aviv University, Tel Aviv.

Na[^{125}I] (carrier free) was purchased from New England Nuclear Co., Boston, Mass., U.S.A.

BSA was of RIA grade, and was purchased from Sigma, St. Louis; Charcoal (Norit A) was from Sigma, St. Louis, Missouri. Dextran T70 was from Pharmacia, Uppsala, Sweden. All other chemicals were purchased of reagent grade.

Antiserum to VIP was raised in Stockholm, or was a gift from either Dr. Graham Dockray, University of Liverpool or Dr. Jan Fahrenkrug, Copenhagen.

Iodination of VIP was carried out with the chloramine T procedure with a slight modification of the procedure of Staun-Olsen et al. (18) as described previously (1). The products were separated by HPLC using a Bondapack C18 column and by developing the chromatogram with a gradient 20%-40% (v/v) of acetonitrile in 0.1% trifluoroacetic acid in double distilled water, using a flow rate of 1 ml/min.

VIP levels were determined in tissue which had been dissected, and frozen at -70°C. Frozen tissue was weighed before boiling in 10 volumes of water for 10 min. After cooling, glacial acetic acid was added to make the samples 1.6 M with respect to acetic acid. The samples were subsequently homogenized with a Polytron homogenizer (setting 5) for 45 seconds on ice. The homogenates were kept at room temperature for 30 min and then centrifuged (1000 g x 5 min). The supernatants were lyophilized and resuspended in the RIA buffer [NaH_2PO_4, Na_2HPO_4 20 mM pH 7.0, 0.002% sodium azide, 0.2% bovine serum albumin (BSA)] prior to assay.

The VIP radioimmunoassay procedure was as follows (6, 7): Samples were incubated with rabbit VIP antibodies (66/82, 1:2000) for 24 hours in silanized glass tubes in 1 ml of the buffer described for the extraction procedure (see above), in the presence of Aprotinin (500 KIU) per tube. 3000 cpm of the [^{125}I]-labeled antigen were then added to the tubes. 24 hours later the free [^{125}I]-antigen (see above) was adsorbed by the addition of 200 μl of a charcoal suspension (50 g/l) containing dextran (5 g/l) and 10% (v/v) human blood plasma. The tubes were briefly agitated on a vortex and centrifuged at 1500 x g for 10 min at 4°C. The pellets and supernatants were counted in a Packard gamma counter, and VIP levels were determined by means of a standard curve.

The number of muscarinic ACh receptors was measured as described previously (9). Adenylate cyclase activity was determined according to Robberecht et al. (15) using a one-minute incubation time, and a Krebs-Ringer's buffer without $CaCl_2$, but with 1 mM EGTA present. The reaction mixture contained: ATP 0.5 mM, $MgCl_2$ 6.5 mM (including 1.05 mM in the buffer), phosphoenol pyruvate 2 mM, pyruvate kinase 30 μg/ml, eserine 10 μM, and d-tubocurarine 10 μM. Isobutyl methyl xanthine (0.5 mM) was also included in the buffer. The cAMP measurements were carried out using incubation mixtures without added protein as blanks in the standard curves. cAMP was determined according to Brown et al. (3). Protein was determined according to Lowry et al. (11).

RESULTS

Short term interactions between VIP and acetylcholine. In membranes from the cat submandibular salivary gland the nonequilibrium binding (2 min, 37°C) of both acetylcholine and carbamylcholine was profoundly affected by the presence of VIP (10 nM) (Table 1). Similar experiments using membranes from the rat salivary gland have not yet demonstrated such an effect of VIP on muscarinic agonist binding.

The binding of $[^{125}I]$-VIP to membranes of rat cerebral cortex, a tissue which is rich in those receptors to permit reliable and easy measurements, was found to be inhibited when the muscarinic receptors were occupied with atropine (1 μM) (Table 1).

The VIP-activated adenylate cyclase in membranes from the cat submandibular salivary gland was further activated in the presence of muscarinic agonists (Table 1). Similar experiments using membranes from rat salivary gland have not yet demonstrated any such effect in a convincing manner.

Chronic atropine treatment and its effects on VIP levels and muscarinic and VIP receptors. Chronic treatment of rats with atropine (20 mg/kg/day) for 14 days led to an increase in the number of muscarinic receptor sites in both the rat cerebral cortex and rat salivary gland (Table 2). The number of $[^{125}I]$-VIP binding sites was also increased significantly as a result of this treatment (Table 2). Moreover, tissue levels of VIP were decreased both in the rat salivary gland, and in rat cerebral cortex at the end of the atropine treatment (Table 2).

DISCUSSION

Studies on the cat submandibular salivary gland have indicated that VIP enhances the acetylcholine induced salivary secretion in a synergistic manner (12, 14). One of the biochemical mechanisms behind those effects may be the shift in the affinity of acetylcholine for the muscarinic receptors in the presence of VIP (10 nM)

Table 1

Some short term interactions between muscarinic ligands
and the vasoactive intestinal polypeptide

Tissue	Measurement	Effector	IC_{50} (M)	Activity
Cat submandi- bular salivary gland	ACh binding	--	5×10^{-3} [a] (24 points)	
	"	VIP (10 nM)	8×10^{-8} (24 points)	
Cat submandi- bular salivary gland	Adenylate cyclase	--		100 ± 7 [b]
	"	VIP (1 nM)		275 ± 17** (6)
		ACh (0.5 μM)		242 ± 21** (6)
	"	VIP (1 nM)+ ACh (0.5 μM)		404 ± 15** (6)
Rat cerebral cortex	$[^{125}I]$-VIP binding	--		100 ± 5 (6) [c]
"	"	atropine (1 μM)		73 ± 9 (6)*

*p < 0.1, **p < 0.05. [a]Concentration of acetylcholine required to displace 50% of the bound [^{3}H]-4-NMPB (3 nM) at 37°C, 2 min incubation time. [b]Adenylate cyclase activity was determined as described in Methods and expressed as pmoles/min x mg protein. [c][^{125}I]-VIP binding was measured as described in Methods.

[Table 1 and (13)]. This effect was proven to be specific in that other peptides of the secretin family were inactive. The effect is transient and abolished once VIP is degraded by peptidases, a process which takes 4-7 min at 2 mg/ml protein concentration of membranes from the cat submandibular salivary gland.

Acetylcholine acting at muscarinic receptors is known to decrease the affinity of β-receptors and thus to inhibit adenylate cyclase activated by, e.g., β-receptors in the heart (19). In membranes from the cat submandibular salivary gland however an activation (242% of control) was observed. Furthermore this stimulatory effect was almost additive with the activation caused by VIP (1 nM) (Table 1).

In whole dispersed cells of the cat submandibular, salivary gland Enyedi et al. (5) have made similar observations. Specifically, carbamylcholine potentiated the VIP stimulation of an elevation of cyclic AMP levels. Truly, experiments using whole cells one may argue that carbamylcholine acted indirectly, by causing the release of another substance which further activated the VIP-stimulated

Table 2

Effect of chronic atropine treatment (20 mg/kg/day) for 14 days
on muscarinic receptor binding, [^{125}I]-VIP binding sites, and
VIP-levels in the rat cerebral cortex and rat salivary gland

Tissue	Muscarinic receptors ([^{3}H]-4-NMPB binding) % of control	[^{125}I]-VIP binding sites % of control	VIP like immuno-reactivity % of control
Cerebral cortex	126 ± 7 (5)**	175 ± 11 (5)**	80 ± 15 (4)*
Salivary gland	204 ± 11 (18)**	207 ± 7 (19)**	29 ± 24 (18)**

*p < 0.1, ** p < 0.05. The number of rats is shown in parentheses. The receptors
and VIP levels were measured as described in Methods. 100% of VIP receptors represent
32.000 cpm [^{125}I]-VIP specifically bound/mg protein in the cerebral cortex, and
8.400 cpm in the salivary gland. 100% VIP immunoreactive material in the cerebral
cortex represents 58 picomoles/g wet weight, and 22 picomoles/g wet weight in the
salivary gland. 100% muscarinic receptors in the cerebral cortex corresponds to
2.8 picomoles [^{3}H]-4-NMPB/mg protein, and to 0.12 picomoles/mg protein in the salivary
gland.

adenylate cyclase. This argument is not easily applied to a hypotonic
membrane preparation used in the above experiments, however. The
nature of interaction whereby occupancy of the muscarinic receptor
leads to activation of adenylate cyclase (or VIP-stimulated adenylate
cyclase) is not yet known, but it is likely that the GTP-binding
protein (16) is involved. GTP regulates interconversion of high and
low affinity agonist binding forms of muscarinic receptors in several
tissues (2) and it is also an integral part of the hormone or neurotransmitter sensitive adenylate cyclase complex.

The inhibition of [^{125}I]-VIP binding by high concentrations
of atropine is a finding that is presently being followed up. If
this effect is also observed at lower atropine concentrations, it
may partly explain the VIP-receptor increase as discussed below.
The atropine concentrations in the brain during the applied regimen
were 2-3 nM as determined by [^{3}H]-atropine tracer, or by indirect
estimation of a residual muscarinic ligand [cf. Westlind et al.
(20)]. The results shown in Table 2 illustrate that "a pure muscarinic" drug such as atropine, upon chronic treatment, alters not
only the number of muscarinic receptors as shown previously in several
tissues (cf. 20 and references therein) but also the number of [^{125}I]-
VIP receptors in the tissues where acetylcholine and VIP coexist,
as in the salivary gland.

There is as of yet no direct study to show that acetylcholine and VIP coexist in the rat cerebral cortex, but it is known that a large portion (≈ 20-30%) of bipolar neurons contain VIP-like immunoreactive material(s), and new data using antibodies to choline acetyltransferase (4) indicate that some bipolar cells contain this synthetic enzyme. Whether there is an overlap between these subpopulations of bipolar cells is not yet known. That this indeed may be the case has however recently been shown by Dr. L. Butcher (personal communication). The effects of chronic atropine treatment however are similar in the cerebral cortex and salivary gland.

The large increase in the $[^{125}I]$-VIP binding sites may be a reflection of the fact that VIP levels are decreased upon such atropine treatment. It is hypothesized that muscarinic autoreceptors inhibit release of VIP under normal conditions, and that chronic atropine treatment renders this feedback inhibition of VIP release by acetylcholine ineffective. Consequently VIP levels drop in the nerve terminal, since axonal flow of VIP cannot keep pace with the unhibited release. This phenomenon is more apparent in the salivary gland, where our preparation contains nerve terminals but not cell bodies, while when measuring VIP levels in the cerebral cortex, we measure the peptide content of both the soma, and the terminals, of the bipolar cells.

The large changes described above underlie the importance of taking into account the possibility of coexisting neurotransmitter or peptide when examining the effects of a drug. Thus, changes in VIP metabolism and VIP receptor numbers and sensitivity may well become part of the "cholinergic study" of tissues where acetylcholine and VIP coexist.

Acknowledgements. This study was supported by grants from the Swedish Medical Research Council.

REFERENCES

1. Abens, J., Westlind, A. and Bartfai, T. (1984): Peptides 5:375-377.
2. Berrie, C.P., Birdsall, N.J.M., Burgen, A.S.V. and Hulme, E.C. (1979): Biochem. Biophys. Res. Comm. 87(4):1000-1005.
3. Brown, B.L., Ekins, R.P. and Albano, J.D.M. (1972): Adv. Cycl. Nucl. Res. 2:25.
4. Eckenstein, F. and Thoenen, H. (1983): Neurosci. Lett. 36:211-215.
5. Enyedi, P., Fredholm, B.B. and Lundberg, J.M. (1982): Eur. J. Pharmacol. 79:139-143.
6. Fahrenkrug, J. (1979): Digestion 19:149-169.
7. Fahrenkrug, J. and Schaffalitzky de Muckadell, O.B. (1976): J. Lab. Clin. Med. 89:1379-1388.

8. Fuxe, K., Hökfelt, T., Said, S.I. and Mutt, V. (1977): Neurosci. Lett. <u>5</u>:241-246.
9. Hedlund, B., Abens, J. and Bartfai, T. (1983): Science <u>220</u>: 519-521.

10. Hökfelt, T., Johansson, O., Ljungdahl, A., Lundberg, J.M. and Schultzberg, M. (1980): Nature <u>284</u>:515-521.
11. Lowry, O.H., Rosebrough, N.J., Farr, A.L. and Randall, R.J. (1951): J. Biol. Chem. <u>193</u>:265-275.
12. Lundberg, J.M. (1981): Acta Phys. Scand. Suppl. <u>496</u>.
13. Lundberg, J.M., Hedlund, B. and Bartfai, T. (1982): Nature <u>295</u>:147-149.
14. Lundberg, J.M., Ånggård, A., Fahrenkrug, J., Hökfelt, T. and Mutt, V. (1980): Proc. Natl. Acad. Sci. USA <u>77</u>:1651-1655.
15. Robbererecht, P., Lambert, M. and Christophe, J. (1979): FEBS Lett. <u>103</u>:229-233.
16. Rodbell, M. (1980): Nature <u>284</u>:17-22.
17. Said, S.I. and Mutt, V. (1979): Science <u>169</u>:1217-1218.
18. Staun-Olsen, P., Ottesen, B., Bartels, P.D., Nielsen, M.H., Gammeltoft, S. and Fahrenkrug, J. (1982): J. Neurochem. <u>39</u>: 1242-1251.
19. Watanabe, A.M., McConnaughey, M.M., Strawbridge, R.A., Flemming, J.W., Jones, L.R. and Besch, H.R. (1978): J. Biol. Chem. - <u>253</u>:4833-4836.
20. Westlind, A., Grynfarb, M., Hedlund, B., Bartfai, T. and Fuxe, K. (1981): Brain Res. <u>225</u>:131-141.

SEROTONIN AND A ROLE IN THE MODULATION OF CHOLINERGIC

TRANSMISSION IN THE MAMMALIAN PREVERTEBRAL GANGLIA

N.J. Dun, R.C. Ma, M. Kiraly and A.G. Karczmar

Department of Pharmacology
Loyola University
Stritch School of Medicine
2160 South First Avenue
Maywood, Illinois 60153 U.S.A.

INTRODUCTION

The view that sympathetic ganglia receive synaptic inputs exclusively from cholinergic preganglionic neurons situated in the
spinal cord is undergoing rapid revision as evidence accumulated
in the past few years clearly indicates that sympathetic ganglia,
particularly the abdominal prevertebral ganglia (celiac-superior
mesenteric plexus and inferior mesenteric ganglion) receive cholinergic as well as non-cholinergic inputs from the central and/or
peripheral nervous system (3, 4, 23).

A non-cholinergic input was first demonstrated in the bullfrog
sympathetic ganglia where appropriate stimulation elicited a long--
lasting depolarization that was resistant to cholinergic antagonists
(22). Subsequently, analogous slow non-cholinergic depolarizations
were reported in neurons of the guinea-pig inferior mesenteric (4,
5, 21) and myenteric (18) ganglia. The first suggestion that the
non-cholinergic input may arise from neurons other than the cholinergic
preganglionic neurons came from the study of guinea pig inferior
mesenteric ganglia where the non-cholinergic excitatory potential
(non-cholinergic epsp) is proposed to be generated by substance P
or a similar peptide released from collateral branches of peripheral
sensory fibers with their cell bodies in the dorsal root ganglia
(2, 4, 17, 20, 24).

In this study evidence is presented indicating that in a portion
of celiac neurons a slow depolarization mediated by the indoleamine,
serotonin (5-HT) may represent yet another example of non-cholinergic

inputs to the prevertebral ganglion cells and that its primary function is to provide a mechanism by which the cholinergic epsp is facilitated.

MATERIALS AND METHODS

Male guinea pigs weighing between 250-300 gm were used throughout this study. The celiac-superior mesenteric plexus together with its left greater splanchnic nerves was rapidly excised from the animal and transferred to the recording chamber. The plexus was continuously superfused with a modified Krebs solution equilibrated with 95% O_2 & 5% CO_2, and the temperature of the solution was maintained at about 34°C (4, 19).

Intracellular recordings were made from neurons of the isolated left celiac ganglia by means of fiber-containing glass microelectrodes filled with 3 M KCl, having a tip resistance of 30-60 MΩ. Appropriate compounds were dissolved in Krebs solution and applied in known concentrations to the ganglia by superfusion. Synaptic potentials were evoked mainly by stimulation of the greater splanchnic nerves by means of a suction electrode (19).

Immunohistofluorescent studies were carried out on post-fixed celiac-superior mesenteric plexus. The plexus was cut on a cryostat at 12 μm and the sections were processed for indirect immunohistofluorescence as described previously (19). Lastly, a modified HPLC-EDC system (15) was employed to measure quantitatively the ganglionic content of 5-HT.

RESULTS

Non-cholinergic epsp's. Single electrical stimulation applied to the greater splanchnic nerves elicited in the celiac neurons a short-lasting fast excitatory postsynaptic potential (f-epsp) which could be reversibly suppressed by nicotinic antagonists such as hexamethonium (0.1 mM) and d-tubocurarine (50 μM); this response is therefore the classical ganglionic response, mediated by a nicotinic action of ACh.

When the nerves were stimulated repetitively (10-20 Hz, 1-2 sec), the fast epsp was followed by a slowly rising and falling depolarization in about 80% of the celiac neurons tested. The slow depolarization was not blocked by cholinergic nicotinic or muscarinic antagonists but was reversibly eliminated in a low Ca/high Mg solution, indicating that the slow depolarization was initiated by a non-cholinergic transmitter(s) released synaptically. The slow depolarization will henceforth be termed as non-cholinergic epsp.

The duration of the non-cholinergic epsp elicited in the celiac neurons was long, ranging from 1 to 10 min, the average being about 3 min (7). The amplitude of the non-cholinergic epsp varied considera-

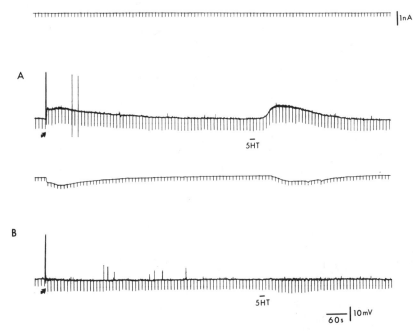

Figure 1.   Similar membrane resistance increase during the non-cholinergic epsp and 5-HT depolarization in a guinea pig celiac ganglion
cell.   A: repetitive stimulation of the left splanchnic nerves (20
Hz, 2 sec; indicated by a curved arrow) elicited a burst of action
potentials (dark vertical tracing) followed by a noncholinergic epsp.
The spikes in this and subsequent recordings are attenuated because
of the limitation of the frequency response of the pen recorder.
The preparation was superfused continuously with a Krebs solution
containing atropine (1 μM).   Top tracings of A and B represent the
amount of current (downward deflections) used to induce the hyperpolarizing electrotonic potentials (downward deflections of lower
tracings), and the lower tracings represent membrane potential change.
The amplitude of hyperpolarizing electrotonic potentials were used
to monitor the cell input resistance change.   5-HT (10 μM) was applied
to the ganglion for 15 sec as   indicated.   The slow depolarization
elicited by nerve stimulation and by 5-HT was accompanied by an
increase in membrane resistance (as shown by an increase of the
amplitude of the electrotonic potentials) of 57% and 50%, respectively;
B: non-cholinergic epsp and 5-HT depolarization were elicited in
the same manner as in A.   However, membrane depolarization was prevented by passage of hyperpolarizing currents to manually clamp
the membrane potential at the resting level.   Under these conditions,
the membrane resistance increase was 75% and 66%, respectively (From
Kiraly et al., 1983; 19).

bly from neuron to neuron, and the mean was about 6 mV (7). A representative recording of the fast epsp and non-cholinergic epsp elicited in a guinea pig celiac neuron is shown in Fig. 1.

5-HT Mimics the Non-cholinergic epsp. In an attempt to identify the transmitter(s) responsible for the generation of the non-cholinergic epsp, the effects on the celiac neurons of several putative transmitters including epinephrine, norepinephrine, dopamine, substance P and serotonin were investigated. In our preliminary study 5-HT was found to cause a slow depolarization similar to the non-cholinergic epsp in more than half of the cells examined, while the membrane effects of other transmitters were neither impressive nor consistent. Accordingly, 5-HT was chosen for more intensive study.

First, it could be demonstrated that a brief application (10-15 sec) of 5-HT (1-10 μM) produced a long lasting depolarization comparable to that induced by nerve stimulation (Fig. 1). More importantly, the slow depolarization induced by 5-HT in normal Krebs solution was quantitatively similar to that evoked by 5-HT in Ca-deficient (0.2 mM) Krebs solution, in all 12 cells tested. This result indicates that the slow depolarization was due to a direct action of 5-HT on postganglionic neurons from which the recordings were made, rather than being mediated via a 5-HT-induced, intraganglionic release of a depolarizing substance(s).

Second, the membrane resistance changes associated with the 5-HT depolarization and with the non-cholinergic epsp were found to be similar when evaluated in the same neurons. In the experiment shown in Fig. 1 there was a similar increase in membrane resistance during the course of membrane depolarization induced by either nerve stimulation or by 5-HT. A parallel change in membrane resistance was consistently observed in all 17 other cells investigated.

Lastly, the non-cholinergic epsp and 5-HT depolarization exhibited similar voltage dependence, i.e., membrane hyperpolarization increased the amplitude of the slow depolarizations whether due to presynaptic stimulation or to 5-HT, in the large majority of the cells tested. It should be noted that in a few cells the slow depolarizations were depressed by membrane hyperpolarization. Although the precise ionic mechanism underlying the non-cholinergic epsp and 5-HT depolarization remains to be determined (7), what needs to be stressed in the present context is that 5-HT and the transmitter(s) mediating the non-cholinergic epsp caused very similar electrophysiological changes of the celiac neurons.

Desensitization of Non-cholinergic epsp by 5-HT. That tachyphylaxis develops rapidly to 5-HT has been reported with respect to a number of tissues; this tachyphylaxis has been employed as one of the pharmacological criteria in establishing the action of 5-HT (14). In this study, the non-cholinergic epsp was found to be abolished after

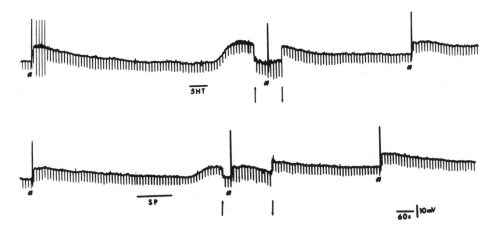

Figure 2.   Effects of 5-HT and substance P on membrane potential
and on non-cholinergic epsp of a celiac ganglion cell.   Upper tracing: repetitive stimulation (20 Hz, 2 sec; indicated by a curved
arrow) evoked a burst of action potentials followed by a large and
long lasting non-cholinergic epsp.   The ganglion was continuously
superfused with a Krebs solution containing atropine (1 μM).   Superfusing the ganglion with 5-HT (1 μM) for 1 min caused a large depolarization accompanied by a small decrease in membrane resistance
at the rising phase.   However, this decrease was probably due to
membrane rectification as there was a large increase in membrane
resistance when membrane potential was momentarily restored to the
resting level by hyperpolarizing currents (between the two arrows).
Under these conditions, nerve stimulation (curved arrow) elicited
only action potentials that were not followed by a detectable non-
cholinergic epsp, indicating that the failure to induce the latter
is likely to be due to receptor desensitization.   A noncholinergic
epsp was elicited a few min after washing.   Lower tracing: superfusing
substance P (1 μM) to the same neuron for 2 min also caused a membrane
depolarization.   At the peak of substance P-induced depolarization,
membrane potential was again temporarily returned to the resting
level as indicated between the two arrows. Nerve stimulation (curved
arrow) in this case elicited a non-cholinergic epsp comparable to
that before substance P superfusion, suggesting that substance P
and the transmitter mediating the non-cholinergic epsp probably acted
on different receptors (From Kiraly et al., 1983; 19).

superfusing the ganglia with excess 5-HT.   This suppression of the
non-cholinergic epsp occurred even when hyperpolarizing current
was employed to restore the membrane potential to the resting level,
following the application of 5-HT and prior to the presynaptic stimulation (Fig. 2, upper tracing).   After the 5-HT induced depolarization
had subsided, repetitive nerve stimulation could again evoke a large
non-cholinergic epsp (Fig. 2, upper tracing).

In view of our previous finding that substance P or a similar peptide may be the mediator of a non-cholinergic epsp elicited in a portion of inferior mesenteric ganglion cells (4, 5, 6, 17), it was important to know whether substance P can cause a slow depolarization in the celiac neurons and whether or not the non-cholinergic epsp can be desensitized by this peptide. The number of celiac neurons sensitive to substance P was relatively small as compared to the number of inferior mesenteric ganglion cells that were depolarized by substance P (8, 9). More importantly, the non-cholinergic epsp elicited in the majority of celiac neurons was not affected by prolonged application of substance P. In the experiment shown in Fig. 2, substance P caused a slow depolarization in a manner that is quite similar to that produced by 5-HT; however, the non-cholinergic epsp elicited in this neuron was not at all affected by substance P (lower tracing), whereas it was abolished by 5-HT (upper tracing). Thus, there was no cross-desensitization between substance P and the transmitter(s) mediating the non-cholinergic epsp.

Pharmacological Analysis. The serotoninergic nature of the non-cholinergic epsp elicited in celiac neurons was further evaluated by means of several pharmacological agents known to affect serotoninergic transmission.

Cyproheptadine (10-50 μM), a 5-HT receptor blocker effective at several sites (16) markedly suppressed the non-cholinergic epsp as well as the 5-HT depolarization (n=16); the cholinergic epsp's were also slightly to moderately depressed. D-tubocurarine (10-50 μM) which has been reported to block 5-HT mediated responses in molluscan neurons (13) was ineffective in any of the 6 neurons tested. Methysergide (1-50 μM) was more effective in blocking the cholinergic epsp than in blocking the non-cholinergic epsp and 5-HT-induced depolarization.

When applied to the ganglia for 10-15 min, fluoxetine (30-50 μM), a 5-HT reuptake blocker (12) significantly increased the amplitude as well as the duration of the depolarization evoked either synaptically or by 5-HT in all 7 cells investigated; the effect was fully reversible after washing with Krebs solution (19). L-tryptophan (10-50 μM), a precursor of 5-HT (11), enhanced selectively the non-cholinergic epsp, while the 5-HT induced depolarization was not appreciably increased in 8 cells tested (9).

Ganglionic Content of 5-HT. This collective electrophysiological and pharmacological analysis of the non-cholinergic epsp and response to 5-HT suggested that 5-HT is the mediator that generates the slow depolarization in a population of celiac neurons. Since the presence of 5-HT in the guinea pig celiac-superior mesenteric plexus has not been reported heretofore, a modified HPLC-EDC system was employed to ascertain that 5-HT is present in the ganglia. The 5-HT content in the ganglia freshly removed from animals was rather low, but it could be increased dramatically when the ganglia were incubated in vitro with Krebs solution containing L-tryptophan (50 μM) for 30 min (Fig. 3). Moreover, addition of fluoxetine to the incubation medium effectively prevented the increase of ganglionic content of 5-HT incuded by L-tryptophan.

Localization of 5-HT-Like Immunoreactivity. In the ganglia fixed immediately after removal from the animal or a few hours after electro-physiological study, 5-HT-like immunoreactivity was generally in-conspicuous. A number of small neurons comparable in their size to the small intensely fluorescent cells (10) was found to exhibit 5-HT-like immunoreactivity. However, when the ganglia were superfused in vitro for 30-60 min with L-tryptophan (50 μM) prior to fixation, dense but unevenly distributed networks of nerve fibers exhibiting 5-HT-like immunoreactivity could be observed surrounding many gangli-onic neurons (Fig. 4). Particularly strong immunofluorescence was visualized in nerve fibers at the areas near the entry of the greater

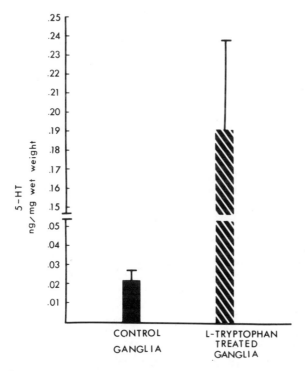

Figure 3. 5-HT content of guinea pig celiac-superior mesenteric plexus measured by a modified HPLC-EDC system. The 5-HT content was markedly increased after perfusing the ganglia with L-tryptophan (50 μM) for 60 min in vitro. The bars represent mean ± S.D. (n=6).

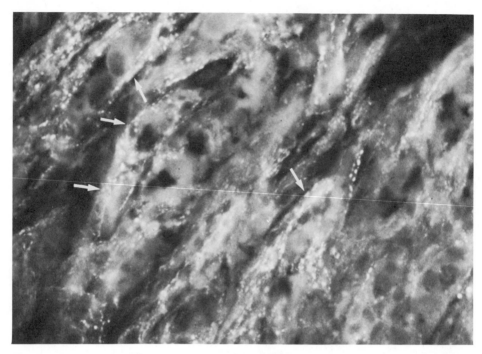

Figure 4. Immunofluorescence micrograph of a guinea-pig celiac-superior mesenteric plexus after incubation with 5-HT antiserum (x500). The plexus was superfused with tryptophan (50 μM) for 60 min prior to fixation. Networks of fluorescent varicose fibers are seen surrounding several ganglion cells (arrows) (From Kiraly et al., 1983; 19).

splanchnic nerves, and in some instances, bead-like fluorescent fibers could be traced along the splanchnic nerves. For the control study, little or no fluorescence was observed after incubation of the ganglia with antibodies pre-absorbed with excess 5-HT conjugated to bovine serum albumin (a gift from Dr. Mark Brownfield, University of Wisconsin).

Facilitation of Cholinergic epsp. A possible role of non-cholinergic epsp in the modulation of the primary transmission pathway of the ganglia, the cholinergic epsp, was evaluated in the next series of experiments.

Subthreshold f-epsp's were evoked at a low frequency (0.3 Hz) by stimulation of the splanchnic nerve, and a non-cholinergic epsp was induced by a train of supramaximal stimuli delivered to the celiac nerves. As shown in Fig. 5, the subthreshold f-epsp's were markedly augmented during the course of the slow depolarization, resulting in spike discharges (left upper tracing). After the amplitude of f-epsp's had returned to control level, the membrane potential

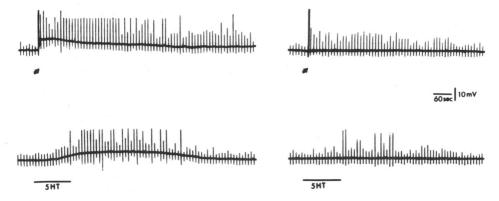

Figure 5.   Facilitation of cholinergic epsp's by non-cholinergic transmission and 5-HT.   Top tracings: subthreshold cholinergic epsp's were evoked by stimulation of left greater splanchnic nerves at 0.3 Hz (small vertical deflections), and a non-cholinergic epsp was induced by stimulation of the coeliac nerve (20 Hz, 2 sec, curved arrow).   The subthreshold cholinergic epsp's were effectively enhanced and reached the threshold; this effect persisted throughout the course of the non-cholinergic depolarization.   After a 10 min washing period, the non-cholinergic epsp was again evoked, except that this time the membrane potential was held at the resting level by passage of hyperpolarizing current (right top tracing).   Under these conditions, some of the cholinergic epsp's still evoked spikes.   **Lower tracings:** application of 5-HT caused a slow depolarization and facilitation of the amplitude of cholinergic epsp's (left lower tracing).   When the 5-HT depolarization was nullified by hyperpolarizing current, some of the subthreshold cholinergic epsp still evoked spikes (right lower tracing).

was manually clamped and the non-cholinergic epsp was again elicited; under these conditions, a potentiation of the f-epsp's could still be observed, although it was of a lesser magnitude (right upper tracing).   Furthermore, the f-epsp's could be also augmented by means of the application of 5-HT, in a manner similar to the augmentation produced by nerve stimulation (lower tracings).

DISCUSSION

On the basis of our electrophysiological, pharmacological, neurochemical and histofluorescent findings 5-HT appears to be the mediator responsible for the generation of a non-cholinergic epsp in a portion of celiac neurons.   Firstly, the electrophysiological action of 5-HT on celiac neurons is very similar to that of the transmitter released synaptically.   Secondly, perfusion with 5-HT reversible abolishes the non-cholinergic epsp; this observation is compatible with the phenomenon of receptor desensitization that can be readily obtained in tissues endowed with serotoninergic receptors (14). Thirdly, the slow depolarization, whether evoked presynaptically or by 5-HT, was affected by a number of pharmacological agents in a manner consistent with the hypothesis that 5-HT is the mediator of the non-cholinergic epsp. Finally, 5-HT is present in the ganglia, and 5-HT-like immunoreactivity is localized primarily in the nerve fibers which seem to come into close proximity with the ganglionic neurons.

The origins of 5-HT-containing fibers in the celiac ganglion is of interest. It was proposed that the substance P-containing fibers present in the guinea pig inferior mesenteric ganglion are collateral branches of peripheral sensory fibers originating from dorsal root ganglia; thus, these fibers may be involved in the transmission of sensory signals (2, 4, 17, 20, 24). Whether or not 5-HT fibers present in the celiac ganglia may also derive from dorsal root neurons, and whether or not they subserve a sensory transmission remains to be investigated. Alternatively, some enteric neurons which are known to contain 5-HT-like immunoreactivity (1) may project their fibers to the prevertebral ganglia. Whichever the case may be, the 5-HT mediated transmission demonstrated here may constitute yet another example of non-cholinergic afferent inputs to the sympathetic neurons from the central and/or peripheral nervous system.

What might be the cellular function of non-cholinergic transmission in celiac neurons and how does it affect cholinergic transmission? The most apparent effect of non-cholinergic transmission as demonstrated here was a facilitation of cholinergic epsp's. Two possible mechanisms should be considered in this context. First, the depolarization that is generated during the non-cholinergic transmission would bring the cell closer to the threshold level; second, the increased membrane resistance that is often associated with the slow depolarization may amplify the current flow produced by the cholinergic epsp's, resulting in a greater voltage deflection.

As most of the central cholinergic inputs to the prevertebral ganglion cells are subthreshold, and as summation of several inputs is required to discharge the neuron (23), the non-cholinergic transmission, which has a time course of minutes, would be effective in maximizing the likelihood of temporal summation of cholinergic epsp's impinging upon a ganglion cell. Furthermore, an increase in membrane resistance should increase space constant of the soma-dendritic membrane, thus increasing the likelihood of spatial summation of cholinergic epsp's. Taken together, our results suggest that the non-cholinergic mechanism provides an effective means of integrating input and output signals of the sympathetic neurons.

Acknowledgement. This study was supported in part by NS15848, BRSG RR05368 and U.S. Army Medical Research and Development Command Contract DAMD-17-83-6-3313.

REFERENCES

1. Costa, M., Furness, J.B., Cuello, A.C., Verhofstad, A.A.J., Stein-
    busch, H.W.J. and Elde, R.P. (1982): Neurosci. 7:351-363.
2. Dalsgaard, C.J. Hokfelt, T., Elfvin, L.G., Skirboll, L. and Emson,
    P. (1982): Neurosci. 7:647-654.
3. Dun, N.J. (1983): Autonomic Ganglia (ed) L.G. Elfvin, John Wiley
    & Sons, Chichester, pp. 345-366.
4. Dun, N.J. and Jiang, Z.G. (1982): J. Physiol. (London) 325:145-
    159.
5. Dun, N.J. and Karczmar, A.G. (1979): Neuropharmacol. 18:215-218.
6. Dun, N.J. and Kiraly, M. (1983): J. Physiol. (London) 340:107-120.
7. Dun, N.J. and Ma, R.C. (1984): J. Physiol. (London) 351:47-60.
8. Dun, N.J. and Minota, S. (1981): J. Physiol. (London) 321:259-271.
9. Dun, N.J., Kiraly, M. and Ma, R.C. (1984): J. Physiol. (London)
    351:61-76.
10. Elfvin, L.G., Hökfelt, T. and Goldstein, M. (1975):   J. Ultra-
    struc. Res. 51:377-396.
11. Fernstrom, J.D. and Wurtman, R.J. (1971): Science 173:149-152.
12. Fuller, R.W. and Wong, D.T. (1977): Fed. Proc. 36:2154-2158.
13. Gerschenfeld, H.M. and Paupardin-Tritsch, D. (1974): J. Physi-
    ol. (London) 243:427-456.
14. Gyermek, L. (1966): In 5-hydroxytryptamine and Related Indole-
    alkylamines (ed) V. Erspamer, Handbuch der Experimentellen Pharma-
    kologie, Springer-Verley, Berlin, pp. 471-528.
15. Hadjiconstantinou, M. Potter, P.E. and Neff, N.H. (1982):   J.
    Neurosci. 2:1836-1839.
16. Haigler, H.J. and Aghajanian, G.K. (1977): Fed. Proc. 36:2159-
    2164.
17. Jiang, Z.G., Dun, N.J. and Karczmar, A.G. (1982): Science 217:
    739-741.
18. Katayama, Y. and North, R.A. (1978): Nature 274:387-388.
19. Kiraly, M., Ma, R.C. and Dun, N.J. (1983): Brain Res. 275:378-
    383.
20. Matthews, M.R. and Cuello, A.C. (1982): Proc. Natl. Acad. Sci.
    U.S.A. 79:1668-1672.
21. Neild, T.O. (1978): Brain Res. 140:231-239.
22. Nishi, S. and Koketsu, K. (1968): J. Neurophysiol. 31:109-121.
23. Szurszewski, J.H. (1981): Ann. Rev. Physiol. 43:53-68.
24. Tsunoo, A., Konishi, S. and Otsuka, M. (1982): Neurosci. 7:
    2025-2037.

CHOLINERGIC AND PEPTIDERGIC REGULATION OF GANGLIONIC

TYROSINE HYDROXYLASE ACTIVITY

R.E. Zigmond and N.Y. Ip

Department of Pharmacology
Harvard Medical School
Boston, MA 02115 U.S.A.

INTRODUCTION

Acetylcholine has traditionally been thought to be the sole preganglionic neurotransmitter in sympathetic ganglia. However, this view was challenged 15 years ago by Nishi and Koketsu in their studies on a frog sympathetic ganglion (10). These workers found that certain of the postsynaptic electrophysiological consequences of preganglionic nerve stimulation could not be abolished by cholinergic antagonists. More recently, Jan, Jan and Kuffler (8, 9) identified the neurotransmitter responsible for these non-cholinergic effects as being a peptide that resembles, but is not identical to, luteinizing hormone-releasing hormone. We have recently obtained evidence that a non-cholinergic transmitter is released also by preganglionic neurons in a mammalian sympathetic ganglion, the rat superior cervical ganglion, and that this neurotransmitter is different from the non-cholinergic transmitter found in frog sympathetic ganglia. Our conclusions come from studies on a postsynaptic biochemical consequence of preganglionic nerve stimulation rather than on an electrophysiological consequence. The biochemical parameter we have been investigating is the activity of the enzyme tyrosine 3-monooxygenase (tyrosine hydroxylase; TH; EC 1.14.16.2), the enzyme which catalyzes the rate-limiting step in catecholamine biosynthesis.

METHODS

Superior cervical ganglia were removed from adult male Sprague-Dawley rats, desheathed, incubated in a physiological medium, and maintained at 37°C in an atmosphere of 95% O_2/5% CO_2 (5). Following a preincubation period, the measurement of TH activity was initiated by addition of the dihydroxyphenylalanine (DOPA) decarboxylase (EC

4.1.1.28) inhibitor brocresine (150 μM). At the end of the incubation period, the ganglia were homogenized, the homogenates combined with their respective incubation media, and the DOPA content in the combined sample measured by HPLC using electrochemical detection (3). Control experiments established that DOPA is not detectable in ganglia incubated in the absence of brocresine. Furthermore, the accumulation of DOPA in the presence of brocresine can be completely inhibited by addition of the TH inhibitor 3-iodo-tyrosine. Finally, once formed, DOPA is stable in the ganglion and medium for at least 30 min. Therefore, we concluded that the rate of DOPA accumulation is a valid measure of tyrosine hydroxylation (5).

For investigation of the effects of carbachol, ganglia were preincubated for 30 min in control, brocresine-free medium, and then incubated for 30 min in medium containing carbachol (0.1 mM) and brocresine (150 μM). When the effects of hexamethonium and atropine were examined, these cholinergic antagonists were included in both the preincubation and incubation media. A similar protocol was used to examine the effects of a variety of neuropeptides at a concentration of 10 μM. When lower concentrations of neuropeptides were used, ganglia were preincubated for 60 min in medium containing the peptide, after which brocresine was added and the incubation was continued for an additional 15 min. In most experiments with peptides, bovine serum albumin (1 mg/ml) was added to the incubation medium. In one experiment, groups of animals either had their cervical sympathetic trunks cut 4 days prior to the experiment or were sham-operated. To examine the effects of preganglionic nerve stimulation, ganglia were incubated in brocresine-containing medium, and the cervical sympathetic trunk was stimulated at 10 Hz for 30 min via a suction electrode. When the ability of hexamethonium and atropine to block the effects of nerve stimulation was examined, ganglia were preincubated with these antagonists for 10 min before the stimulation began.

The data are expressed as the mean rates of DOPA synthesis per ganglion per 15 min or per hour $\pm$ S.E.M. or as the mean percentage of the control rate. The significance of differences between groups was assessed using Student's t-test for two means (two-tailed).

RESULTS

Incubation of ganglia with carbachol (0.1 mM) increased the rate of DOPA synthesis 4.5-fold (Table 1). The muscarinic antagonist atropine (6 μM) produced a small decrease in the response to carbachol, but this effect was not statistically significant. The nicotinic antagonist hexamethonium (3 mM) produced a large (76%), though incomplete, inhibition of the response to carbachol. Addition of both hexamethonium and atropine completely blocked the response to carbachol. Thus, carbachol appears to increase TH activity by both a nicotinic and a muscarinic mechanism. This interpretation is

Table 1

Blockade of the effect of carbachol by cholinergic antagonists

	DOPA Synthesis pmol/ganglion/h
Control	78 ± 14
Carbachol	353 ± 38
Carbachol + Atropine	289 ± 16
Carbachol + Hexamethonium	143 ± 17
Carbachol + Hexamethonium + Atropine	67 ± 2

Ganglia were preincubated for 30 min and then incubated with carbachol (0.1 mM) and brocresine (150 μM) for 30 min. When atropine (6 μM) and/or hexamethonium (3 mM) were examined, they were included in both the preincubation and incubation media. The data represent the mean ± S.E.M. of 3-6 ganglia.

supported by the finding that both dimethylphenylpiperazinium, a nicotinic agonist, and bethanechol, a muscarinic agonist, produced concentration-dependent increases in DOPA synthesis (5). The maximum increases produced by these nicotinic and muscarinic agonists were 4-fold and 2-fold respectively.

Preganglionic nerve stimulation also elevated TH activity. Stimulation at 10 Hz for 30 min produced approximately a 4-fold increase in the rate of DOPA synthesis (Fig. 1). This effect of nerve stimulation was not significantly inhibited by atropine (6 μM) but was decreased by 43% by hexamethonium (3 mM). Addition of both hexamethonium and atropine produced no further decrease in the response to preganglionic nerve stimulation than did addition of hexamethonium alone. Thus, preganglionic nerve stimulation in the presence of both hexamethonium and atropine, produced about a 2.5-fold increase in the rate of DOPA synthesis (Fig. 1). Increasing the concentration of both cholinergic antagonists 10-fold did not further decrease this biochemical response to nerve stimulation (7). These data suggest that the increase in TH activity produced by preganglionic nerve stimulation is mediated in part by acetylcholine and in part by a second (non-cholinergic) transmitter.

As a first step in an attempt to identify this non-cholinergic transmitter, the ability of 14 neuropeptides to increase TH activity was examined. A standard protocol was used for this initial screening. Each peptide was tested at a concentration of 10 μM during a 30 min incubation. The peptides examined included four that have

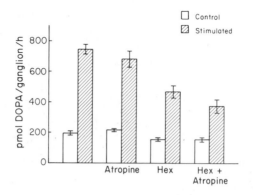

Figure 1. **Increased DOPA synthesis during preganglionic nerve stimulation.** The rate of DOPA synthesis was measured during stimulation of the preganglionic cervical sympathetic trunk at 10 Hz for 30 min. In certain ganglia, atropine (6 μM) and/or hexamethonium (Hex, 3 mM) were added 10 min prior to the beginning of stimulation and were included during the period of stimulation. The data for the atropine and the Hex + atropine groups represent the mean $\pm$ SEM of 3 ganglia. All other groups contained 7 ganglia. Reprinted from Ip et al. (7).

been detected in the rat superior cervical ganglion by immunocytochemistry, namely, enkephalin, somatostatin, substance P, and VIP. Secretin and VIP increased TH activity by 3- and 4-fold, respectively. Angiotensin II, bombesin, bradykinin, cholecystokinin octapeptide, glucagon, insulin, luteinizing hormone-releasing hormone, [D-Ala^2]-Met-enkephalinamide, motilin, neurotensin, somatostatin, and substance P produced no effects (6).

The effects of secretin and VIP were examined in more detail. Secretin produced a significant elevation of DOPA synthesis at 1 nM and caused a maximal effect at 100 nM. VIP significantly elevated DOPA synthesis at 100 nM and appeared to cause a near-maximal stimulation at 10 μM (Fig. 2). Though more potent than VIP, secretin consistently produced a smaller maximal increase in enzyme activity (6). Recently a third peptide, PHI, which is structurally related to both secretin and VIP, has been found to elevate TH activity (4).

Since our experiments with carbachol and with preganglionic nerve stimulation indicate that TH activity can be elevated by cholinergic stimulation, we sought to determine whether secretin and VIP produce their effects indirectly by causing the release of acetylcholine from preganglionic nerve terminals. For this purpose we examined the ability of these peptides to increase enzyme activity in ganglia taken from animals whose cervical sympathetic trunks had been cut 4 days previously. This 4-day interval allows time

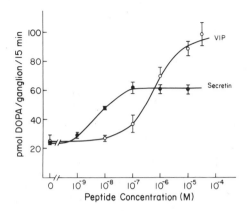

Figure 2. Dose-response curves for the effects of secretin and VIP on the rate of DOPA synthesis. Ganglia were preincubated for 60 min in the presence of various concentrations of secretin and VIP. Brocresine was then added, and the rate of DOPA synthesis was measured during the subsequent 15 min. Each point represents the mean ± SEM of 4 ganglia for VIP and of 8 ganglia for secretin. Reprinted from Ip et al. (6).

for the preganglionic nerve terminals to degenerate (11). In addition, we examined the abilities of these peptides to increase enzyme activity in intact ganglia incubated in the presence of the cholinergic antagonists hexamethonium and atropine. Under both conditions, the effectiveness of secretin (Fig. 3) and VIP (6) in elevating TH was unaltered. Thus, these data suggest that both secretin and VIP act directly on ganglionic neurons to stimulate TH activity.

To examine whether there is any interaction between the cholinergic and peptidergic regulation of TH activity, the effect of various concentrations of secretin from 1 nM to 1 μM were examined in the absence and in the presence of a low concentration (3 μM) of carbachol. Carbachol produced a potentiation of the enzyme response at all concentrations of secretin examined (Ip, Baldwin and Zigmond, submitted).

DISCUSSION

Previous work from this laboratory established that preganglionic nerve stimulation causes a delayed but sustained increase in TH activity in the rat superior cervical ganglion (2, 12, 13). Maximum enzyme activity was seen three days after a 30-90 minute period of nerve stimulation. The magnitude of the increase varied with the frequency and duration of stimulation. The largest increase seen with the stimulation conditions examined was 2-fold (13). This

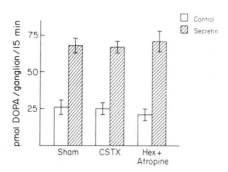

Figure 3. The effect of decentralization and cholinergic antagonists on the increase in DOPA synthesis produced by secretin. Ganglia were removed from animals that had had their preganglionic cervical sympathetic trunks cut 4 days previously (CSTX) or had been sham-operated (Sham). The ganglia were preincubated with secretin (1 μM) or control medium for 60 min prior to addition of brocresine. The rate of DOPA synthesis was then measured during the subsequent 15 min. Certain ganglia from sham-operated animals were preincubated with hexamethonium (Hex) (3 mM) and atropine (6 μM) with or without secretin (1 μM) for 60 min prior to addition of brocresine. Values represent the mean $^+_-$ SEM of 3 or 4 ganglia. Reprinted from Ip et al. (6).

increase in enzyme activity could be completely blocked by the nicotinic antagonists hexamethonium or chlorisondamine (1, 2). Immunoprecipitation studies using an antibody raised against TH established that the increase in enzyme activity could be completely accounted for by an increase in the amount of immunoprecipitable enzyme (14).

In this chapter, we have presented evidence that preganglionic nerve stimulation also produces a rapid elevation of TH activity. This increase in enzyme activity can be seen within 15 min of the onset of stimulation (7). While we have not yet established the mechanism of this effect, it is highly likely, due to its time course, that it involves activation of pre-existing TH molecules. While part of the increase in TH activity produced by preganglionic nerve stimulation could be blocked by hexamethonium, about half of the increase was resistant to both hexamethonium and atropine even at very high concentrations (30 mM and 60 μM respectively). In addition, this hexamethonium-and atropine-resistant effect was insensitive to the acetylcholinesterase inhibitor physostigmine (7). Based on these data, we conclude that the rapid increase in TH activity produced by nerve stimulation is mediated in part by acetylcholine and in part by a second (non-cholinergic) transmitter. Furthermore, the cholinergic mechanism appears to be mediated entirely via nicotinic receptors. No evidence was observed for a muscarinic component in

this increase. Nevertheless, it remains possible that an effect
of muscarinic transmission would be present under certain stimulation
conditions (e.g., stimulation at a higher frequency).

Secretin, VIP and PHI are also capable of producing a rapid
elevation of TH activity. Further investigations will be directed
towards determining whether one or more of these peptides is involved
in the non-cholinergic regulation of this enzyme activity during
preganglionic nerve stimulation. A low concentration of carbachol
potentiates the response of the superior cervical ganglion to secretin,
thus raising the possibility that acetylcholine and a peptide might
interact in the ganglion in the regulation of catecholamine biosynthesis.

Acknowledgements. This work was supported by NIH grant NS 12651.
REZ is the recipient of an NIMH Research Scientist Development Award
(MH 00162) and NYI is a postdoctoral trainee of the USPHS (training
grant NS 07009). We would like to acknowledge the collaboration
of Dr. Robert Perlman, Mr. Calvin Ho, and Ms Claire Baldwin in certain
of the experiments reported here.

REFERENCES

1. Chalazonitis, A. and Zigmond, R.E. (1980): J. Physiol. (Lond.)
    300:525-538.
2. Chalazonitis, A., Rice, P.J. and Zigmond, R.E. (1980): J. Phar-
    macol. Exp. Ther. 213:139-143.
3. Erny, R.E., Berezo, M.W. and Perlman, R.L. (1981): J. Biol.
    Chem. 256:1335-1339.
4. Ip, N.Y. and Zigmond, R.E. (1984): Peptides 5:309-312.
5. Ip, N.Y., Perlman, R.L. and Zigmond, R.E. (1982): J. Pharma-
    col. Exp. Ther. 223:280-283.
6. Ip, N.Y., Ho, C.K. and Zigmond, R.E. (1982): Proc. Natl. Acad.
    Sci. USA 79:7566-7569.
7. Ip, N.Y., Perlman, R.L. and Zigmond, R.E. (1983): Proc. Natl.
    Acad. Sci. USA 80:2081-2085.
8. Jan, Y.N., Jan, L.Y. and Kuffler, S.W. (1979): Proc. Natl.
    Acad. Sci. USA 76:1501-1505.
9. Jan, Y.N., Jan, L.Y. and Kuffler, S.W. (1980): Proc. Natl.
    Acad. Sci. USA 77:5008-5012.
10. Nishi, S. and Koketsu, K. (1968): J. Neurophysiol. 31:109-121.
11. Raisman, G., Field, P.M., Ostberg, A.J.C., Iversen, L.L. and
    Zigmond, R.E. (1974): Brain Res. 71:1-16.
12. Zigmond, R.E. and Ben-Ari, Y. (1977): Proc. Natl. Acad. Sci.
    USA 74:3078-3080.
13. Zigmond, R.E. and Chalazonitis, A. (1979): Brain Res. 164:
    137-152.
14. Zigmond, R.E., Chalazonitis, A. and Joh, T. (1980): Neurosci.
    Lett. 20:61-65.

REGULATION OF ACETYLCHOLINE RELEASE FROM RODENT CEREBRUM BY

PRESYNAPTIC RECEPTORS, METHIONINE ENKEPHALIN AND SUBSTANCE P

B.V. Rama Sastry, N. Jaiswal and O.S. Tayeb

Department of Pharmacology
Vanderbilt University School of Medicine
Nashville, Tennessee 37232 U.S.A.

INTRODUCTION

Acetylcholine (ACh) is released both spontaneously and upon electrical stimulation of the nerve (9). Both are dependent upon the influx of extracellular Ca^{++} ions. The occurrence of spontaneous release of ACh suggests that there is a steady state level of ACh in the biophase of synaptic gaps and that there may be equilibrium between the rates of influx and efflux for Ca^{++} ions. An ACh-amplification mechanism which suggests that the amount of ACh released by a nerve impulse is too little to effect synaptic transmission has been postulated by Koelle (8), but this ACh quickly releases further quantities of ACh to induce a postsynaptic response. A positive feedback mechanism is implicit in Koelle's postulate because when the released amount of ACh is small, it can flood the nerve terminal to activate presynaptic cholinergic receptors, which results in the stimulation of ACh release. These presynaptic cholinergic receptors seem to be of muscarinic type (Ms) because they are activated by classical muscarinic agonists. So far, no selective agonists or antagonists have been described for these cholinergic receptors.

Several studies have indicated that there are presynaptic muscarinic receptors (Mi) in myentric plexus of guinea pig ileum. Stimulation of these receptors by exogenous ACh and classical muscarinic agonists decreased the amount of ACh released, whereas blockade by antimuscarinic drugs increased the amount of ACh released (7, 23). When the guinea pig longitudinal ileal muscle was incubated for 5-10 min in Tyrode with 5-hydroxymethyl-furfurlytrimethylammonium (5-HMFT, 1-100 nM), a furan analog of muscarine, no response was observed. When the muscle was washed with Tyrode solution, a contraction of the muscle (after response, AR) was observed (1, 14-16).

AR to 5-HMFT has been shown to involve release of ACh and is antagonized by atropine or methionine enkephalin (MEK). 5-HMFT blocked
nicotine induced release of ACh in this tissue. Therefore, a negative
feedback system has been postulated which consists of presynaptic
muscarinic receptors, ACh and MEK for release of ACh.

Substance P (SP) and MEK have opposite actions on ACh release.
While SP exhibits a facilitatory effect on the responses of guinea
pig ileum to transmural nerve stimulation, MEK inhibits these responses
(4, 5). The distribution of ACh, MEK and SP in mammalian brain are
similar (3, 22). Further, the similar distribution of SP and MEK
in pain pathways and their opposite actions on pain perception suggest
that they be involved in the positive and negative feedback loops
of ACh release.

Many hormones and neurotransmitters bind to surface receptors
which transmit information to transducing elements responsible for
generating intracellular signals, cyclic AMP and Ca^{++} (2). The
generation of these messengers is mediated through two different
types of receptors in adrenergic, serotonergic, histaminergic and
even peptidergic systems. In the case of those receptors which
function through cyclic AMP, several steps linked to adenylate cyclase
have been delineated. There is considerable interest in the nature
of transducing mechanisms responsible for raising the intracellular
levels of Ca^{++}. One of the main biochemical events seems to be
the hydrolysis of phosphatidylinositol (PI) in coupling surface
receptors to Ca^{++} gates (11, 25). Classical muscarinic agonists
increase PI turnover generating phosphatidic acid (PA) which increases
Ca^{++} influx in several secretory systems (25). Therefore, questions
arise whether two types of muscarinic receptors (Ms and Mi) may be
involved in the generation of PA, Ca^{++} influx, and therefore regulation
of ACh release by positive and negative feedback systems.

Mouse or rat cerebral slices were used to delineate the steps
in these positive and negative feedback systems. The simultaneous
release of ACh, MEK and SP from the slices, the effects of long
acting methionine analog, D-ala-enkephalinamide (DALA) and SP on
ACh release and Ca^{++} uptake, and the effects of disteroylphosphatidic
acid (DSPA) on ACh release were studied. These studies suggest
that muscarinic receptors (Ms), release of SP, activation of PA
generation and Ca^{++} influx are components of the positive feedback
system; and muscarinic receptors (Mi), release of MEK, inhibition
of PA generation and Ca^{++} influx are components of the negative
feedback system.

MATERIALS AND METHODS

The details for the construction of the superfusion baths, chemicals required for superfusion, sources for [^{3}H]-choline chloride and
[^{125}I]-opioid peptides have been described elsewhere (17-19, 21).

        Cerebra from male Swiss mice weighing 20-30 g were used in all
studies.   These studies were also confirmed in the cerebra of male
Fischer 344 rats (21).

**Superfusion of Rodent Cerebral Slices and Measurement of [^{3}H]-ACh
Release.** In these studies, mouse or rat cerebral slices were incubated
with [^{3}H]-choline, which results in the formation of [^{3}H]-ACh in
the nerve terminals. The spontaneous and evoked release of [^{3}H]-ACh
were measured according to the procedures described elsewhere (18,
21).   Under the conditions used, some of the [^{3}H]-ACh which is hydrolyzed by cholinesterases is recovered as [^{3}H]-choline.   Choline by
itself is not released under these conditions.   Therefore, total
radioactivity in the superfusate gives an index of [^{3}H]-ACh released.
The release of [^{3}H]-ACh was expressed in terms of the percent of
the total labeled choline taken up by the brain slices or μmol/g
wet tissue, whichever is appropriate.   This method for measuring
[^{3}H]-ACh was validated by several criteria as described previously
(21).

Radioimmunoassays (RIA) for MEK and SP in the Superfusates. Aliquots
of superfusates (0.2 ml) were mixed with an equal volume of 0.1 N
HCl and stored at -20°C until the time of the assay.   Details of
the RIA for opioid peptides and the assay for [^{125}I] have been described in our previous publications (17, 19).

Miscellaneous Methods. The tissue viability during superfusion was
evaluated by following the loss of lactic dehydrogenase (LDH).   The
product, [^{3}H]-ACh, in the superfusion medium was identified by thin
layer chromatography.   Calcium uptake was determined using [^{45}Ca]$^{++}$
and lanthanium chloride wash to remove nonspecific binding of [^{45}Ca]$^{++}$
to the tissue (26).

RESULTS

Release of [^{3}H]-ACh. The first phase of release of [^{3}H]-ACh increased
with respect to time until it reached a peak at 5 min.   It then
declined exponentially with a half-time of 35 min (Fig. 1A).   Increase
in the potassium concentration in the superfusion fluid to 20 mM
increased the amount of [^{3}H]-ACh released.   Field electrical stimulation of the slices increased the amount of [^{3}H]-ACh released for
5 min (Fig. 1B).   Absence of calcium in the superfusion fluid depressed
the amount of the spontaneous as well as the electrically evoked
release of [^{3}H]-ACh.   Presence of hemicholinium-3 (HC-3, 10 μM) in
the bath caused a small but significant increase of the amount of
[^{3}H]-ACh released (Fig. 1C).   Therefore, HC-3 was included in the
superfusion medium unless otherwise mentioned.

        After the pattern of [^{3}H]-ACh was established, the effects of
pharmacological agents on the cumulative spontaneous release of

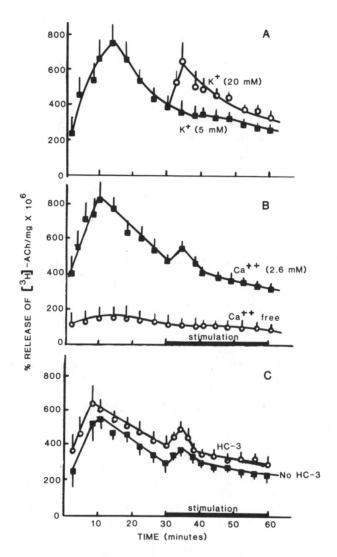

Figure 1.   Release of [³H]-ACh from slices of hemicerebrum of the
mouse.   Each point is a mean ± standard error from 6 experiments.
Stimulation represents period of electrical stimulation. The amount
of [³H]-ACh released for each period was calculated as a percentage
of total labeled choline uptake per mg of wet tissue.   The ordinate
represents these percentages multiplied by 10^{-6} (i.e., the figure,
800, on the ordinate, indicates that 800×10^{-6}% total labelled choline
taken up by the tissue was released as [³H]-ACh).

[^{3}H]-ACh during 0-30 min and evoked release by field electrical stimulation during 30-40 min were studied.

Simultaneous Release of [^{3}H]-ACh, MEK and SP. All three, [^{3}H]-ACh,
MEK and SP, were spontaneously released from the cerebral slices.
The details have been described elsewhere (18). The peak for SP
release preceded the peaks for the release of [^{3}H]-ACh and MEK.
The peak for SP was followed within 3 min by the peak release of
[^{3}H]-ACh. The peak release of [^{3}H]-ACh was followed by the release
of MEK with a broad peak which started within 3 min of the peak
release of [^{3}H]-ACh. The release of MEK was accompanied by decreases
in the amounts of [^{3}H]-ACh and SP released.

Effects of SP, DALA and Naloxone on [^{3}H] Release and [^{45}Ca]$^{++}$ Uptake.
The long acting enkephalin analog, DALA (17 nM) depressed spontaneous
[^{3}H]-ACh release during 0-30 min by about 35% (Fig. 2A). It did
not exhibit any effect in Ca^{++} free medium. Naloxone (55 nM) increased
[^{3}H]-ACh release by about 35%, indicating that endogenously released
enkephalins have inhibitory effects on ACh release. Substance P
(650 nM) increased [^{3}H]-ACh release by about 35%. The effects of
DALA, naloxone and SP on evoked release of [^{3}H]-ACh are similar to
those effects described above for spontaneous release.

DALA (17-350 nM) inhibited [^{45}Ca]$^{++}$ uptake in cerebral slices
by 25-55% whereas SP (70-650 nM) enhanced [^{45}Ca]$^{++}$ uptake by 105-165%.

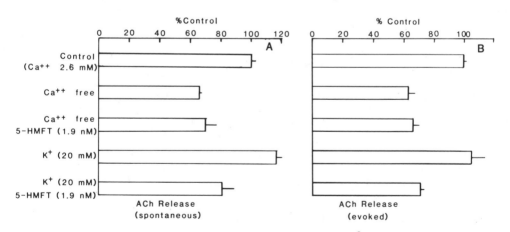

Figure 2. Effects of 5-HMFT on the released [^{3}H]-ACh from mouse
cerebral slices. **A:** Cumulative spontaneous release during 0-30 min.
B: Evoked release during 30-40 min by field electrical stimulation.
Each bar is a mean ± standard error from 6 values. The superfusion
medium contained HC-3 (10 μM).

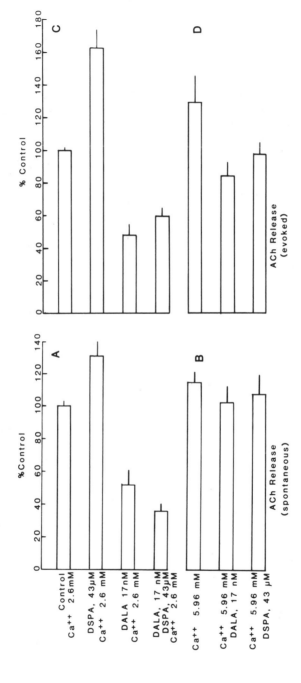

Figure 3. **Effects of disteroylphosphatidic acid (DSPA) and D-alaenkephalinamide (DALA) on the spontaneous (A and B) and evoked (C and D) release of [³H]-ACh from the mouse cerebral slices.** [³H]-ACh from the mouse cerebral slices at 2.6 mM CA⁺⁺ (A and C) and 5.96 mM Ca⁺⁺ (B and D). Each bar is a mean ± standard error from 6 values.

Effects of 5-HMFT on the Release of [^{3}H]-ACh, MEK and SP. 5-HMFT (1.9 nM) inhibited both spontaneous and evoked release of [^{3}H]-ACh (Fig. 2). It also depressed [^{3}H]-ACh release in K^+ (20 mM)-depolarized slices. It depressed MEK release and increased SP release. It did not have any effect in Ca^{++} free medium.

Effect of DSPA and DALA on [^{3}H]-ACh release. DSPA increased and DALA decreased [^{3}H]-ACh release at 2.6 mM Ca^{++} (Fig. 3). The effects of DSPA and DALA were partially nullified by increasing the level of Ca^{++} in the medium to 5.96 mM. These observations indicate that DSPA and DALA exert their effects by modifying Ca^{++} influx at the normal Ca^{++} level (2.6 mM). DSPA increases the uptake of [^{45}Ca]$^{++}$ by slices while DALA decreases it.

Effects of 5-HMFT on DSPA-induced release of [^{3}H]-ACh. DSPA (43 μM) significantly increased the rate of spontaneous as well as evoked release of [^{3}H]-ACh at normal concentration of Ca^{++} (2.6 mM) by about 35-70%. This effect was antagonized by 5-HMFT (1.9 nM). At a higher concentration of Ca^{++} (5.96 mM), DSPA did not exert a significant effect, but the effect of 5-HMFT was still exhibited.

DISCUSSION

There seems to be a homeostatic relationship among the rates of release of ACh, MEK and SP and Ca^{++} fluxes in the mouse cerebrum. This homeostatic relationship is of significance in the autoregulation of ACh release involving two types of presynaptic muscarinic receptors (Ms and Mi) and the level of ACh in the biophase of the synaptic gap. This autoregulation of ACh occurs through the operation of two feedback systems, one positive and one negative. The components of the positive feedback system include an Ms receptor, SP, and activation of Ca^{++} influx, while the negative feedback system includes an Mi receptor, MEK, and inhibition of Ca^{++} influx. Low levels of ACh in the biophase of the cholinergic synaptic gap may trigger the positive feedback system, and high levels of ACh may trigger the negative feedback system. In order to assess the significance of these feedback systems, the function of each component should be evaluated.

MEK decreases the rate of release of ACh from the mouse cerebral slices. This action is accompanied by a decrease in the uptake of Ca^{++} by the cerebral slices. This result is supported by the observation that morphine decreases both the release of ACh (27) and Ca^{++} uptake (12). The time course for the simultaneous release of MEK and ACh, the effect of exogenously applied MEK analog (DALA) on the pattern of ACh release, and the effect of naloxone in increasing the rate of ACh release suggest that the rate of MEK release is increased in response to the increased rate of release of ACh. These observations support the suggested role of enkephalin as a negative feedback regulator of the release of ACh (1, 16).

SP increases the rate of release of ACh by an action mediated through increased uptake of Ca^{++} by the cerebral slices. A similar action for SP on Ca^{++} uptake was found in the parotid gland (13). Further, the increased release of opioid peptides has been suggested in response to analgesic doses of SP (10). The time course of the simultaneous release of SP and ACh, and the effect of exogenous SP on the pattern of ACh release, suggest that ACh is released in response to the preceded peak release of SP (18). These observations indicate a role for SP as a positive feedback regulator of the release of ACh.

There are several well established findings which suggest that there may exist two types of muscarinic receptors: a) ACh depolarizes intestinal smooth muscle membrane and causes contraction, whereas it hyperpolarizes the sino-atrial membrane of the heart and decreases heart rate. b) The classical anticholinergic agent, atropine, causes central nervous system (CNS) excitation whereas another anticholinergic agent, scopolamine, causes CNS depression. c) Some cholinergic agents cause release of ACh by activating presynaptic muscarinic receptors (Ms), whereas some muscarinic agonists depress ACh release in nervous tissue (Mi). However, no clear cut evidence is available for existence of sub-types of muscarinic receptors due to the lack of specific agonists. 5-HMFT was found to be selective to activate muscarinic receptors of Mi type in the Auerbach plexus of the guinea pig ileum (1, 15, 16).

According to the present investigations, the negative feedback system seems to operate through a muscarinic receptor (Mi). Excess ACh in synaptic gaps activates the muscarinic receptor (Mi) and leads to a decrease in Ca^{++} influx and ACh release. In a medium containing Ca^{++} (2.6 mM), 5-HMFT (1.9 nM) decreased both spontaneous and electrically evoked release of ACh by 50%. Its effect was not significant in Ca^{++} free medium. It decreased K^+ (20 mM) induced ACh release. Scopolamine (10 nM) abolished the effect of 5-HMFT on ACh, but atropine (1 μM) did not (6). Classical muscarinic receptor agonists activate the hydrolysis of PI and the resulting PA increases the influx of Ca^{++}, which results in the increased rate of release of ACh. Therefore, the positive feedback system should operate through a muscarinic receptor of Ms type.

If the rate of release of ACh were to be regulated by the above feedback mechanisms, they should be affected in conditions producing ACh deficits (e.g., aging and senile dementia). The patterns of release of ACh, MEK and SP from the cerebral cortical slices of Fischer 344 rats, ages 3-33 months, support the above feedback systems. The rates of release of ACh and MEK decreased while the rate of release of SP increased as a function of age (20, 24). Alterations in the release of MEK may be a regulatory consequence of the decreased rate of ACh release as a function of age.

In summary, the findings of this study indicate the operation of a homeostatic relationship between the release of ACh, MEK and SP. This relationship may be of physiological importance in the regulation of the release of ACh. If the amount of ACh in the synaptic gap is low, a positive feedback loop is triggered causing the release of SP either directly or by a disinhibition phenomenon. Low ACh does not stimulate the negative feedback loop. Hence, no MEK is released, leading to a relief of an inhibitory action of MEK on the release of SP. The released SP would increase further release of ACh by increasing the uptake of Ca^{++}. This process would continue until a peak release of ACh was reached. At this point, the high amount of ACh in the synaptic gap would trigger a negative feedback loop, inducing the release of MEK, which in turn would limit any further release of ACh and SP by decreasing the uptake of Ca^{++}. These feedback systems seem to operate through two different types of muscarinic receptors.

Acknowledgements. This study was supported by US PHS-NIH grants AG-02077, ES-03172, HD-10607 and The Council for Tobacco Research, U.S.A., Inc.

REFERENCES

1. Bass, A.D., Sastry, B.V.R. and Owens, L.K. (1979): Fed. Proc. 38:273.
2. Berridge, M.J. (1981): In Drug Receptors and Their Effectors (ed) N.J.M. Birdsall, MacMillan, London, pp. 75-86.
3. Brownstein, M.J., Mroz, E.A., Kizer, J.S., Palkovits, M. and Leeman, S.E. (1976): Brain Res. 116:299-305.
4. Goldstein, A. (1976): Science 193:1081-1088.
5. Hedqvist, P. and von Euler, U.S. (1975): Acta Physiol. Scand. 95:341-343.
6. Jaiswal, N., Tayeb, O.S. and Sastry, B.V.R. (1982): Fed. Proc. 41:1475.
7. Kilbinger, H. and Wessler, I. (1980): Neurosci. 5:1331-1340.
8. Koelle, G.B. (1971): Ann. N.Y. Acad. Sci. 183:5-20.
9. MacIntosh, F.C., Collier, B. (1976): In Neuromuscular Junction (ed) E. Zaimis, Springer-Verlag, New York, pp. 99-228.
10. Malik, J.B. and Goldstein, J.M. (1978): Life Sci. 23:835-844.
11. Michell, R.H. (1975): Biochim. Biophys. Acta 415:81-147.
12. Munoz, F.G., Guerrero, M.D.L. and Way, E.L. (1979): Science 206:89-91.
13. Putney, J.W., Van de Walle, C.M. and Leslie, B.A. (1978): Mol. Pharmacol. 14:1046-1053.
14. Sastry, B.V.R., Rowell, P.P., Ochillo, R.F. and Chaturvedi, A. K. (1977): The Pharmacologist 19:220.
15. Sastry, B.V.R., Owens, L.K. and Chaturvedi, A.K. (1978): Fed. Proc. 37:609.
16. Sastry, B.V.R., Owens, L.K. and Chaturvedi, A.K. (1978): Absts. 7th Int. Cong. Pharmacol., Paris, p. 173.

17. Sastry, B.V.R., Tayeb, O.S., Janson, V.E. and Owens, L.K. (1981): Placenta, Suppl. 3:327-337.
18. Sastry, B.V.R. and Tayeb, O.S. (1982): Adv. Biosci. 38:165-172.
19. Sastry, B.V.R., Janson, V.E., Owens, L.K. and Tayeb, O.S. (1982): Biochem. Pharmacol. 31:3519-3522.
20. Sastry, B.V.R., Jaiswal, N. and Tayeb, O.S. (1983): Irish J. Med. Sci. 152(Suppl. 1, Abstracts):55.
21. Sastry, B.V.R., Janson, V.E., Jaiswal, N. and Tayeb, O.S. (1983): Pharmacology 26:61-72.
22. Smith, T.W., Hughes, J., Kosterlitz, H.W. and Sosa, R.P. (1976): In Opiate and Endogenous Opioid Peptides (ed) H.W. Kosterlitz, North Holland, Amsterdam, pp. 57-67.
23. Szerb, J.C. (1980): Naunyn-Schiedeberg's Arch. Pharmacol. 311: 119-127.
24. Tayeb, O.S. and Sastry, B.V.R. (1981): Gerontologist 21:151 (Special issue of abstracts - October).
25. Thompson, Jr. (1980): In The Regulation of Membrane Lipid Metabolism CRC Press, Boca Raton, Florida, pp. 188-189.
26. Weiss, G.B. (1981): In New Perspectives of Calcium Antagonists (ed) G.B. Weiss, Amer. Physiol. Soc., Bethesda, Maryland, pp. 83-94.
27. Yaksh, T.L. and Yamamura, H.I. (1977): Neuropharmacol. 16:227-233.

SELECTIVE FACILITATORY EFFECTS OF VASOACTIVE INTESTINAL POLYPEPTIDE ON MUSCARINIC MECHANISMS IN SYMPATHETIC AND PARASYMPATHETIC GANGLIA OF THE CAT

M. Kawatani, M. Rutigliano and W.C. deGroat

Department of Pharmacology
Medical School
University of Pittsburgh
Pittsburgh, Pennsylvania 15261   U.S.A.

INTRODUCTION

Vasoactive intestinal polypeptide (VIP) is widely distributed in the mammalian peripheral nervous system (11, 13, 15, 16, 17, 18) and is frequently localized at sites of cholinergic transmission (17, 18, 26, 27, 28). At some cholinergic synapses VIP has been identified as a cotransmitter released with acetylcholine (ACh) (5, 17, 26, 28, 29). The functions and interactions of these co-transmitters have been studied extensively in the cholinergic innervation of the salivary and sweat glands of the cat (2, 26, 27, 28, 29). At both sites it has been demonstrated that neurally released VIP and ACh can produce independent responses, i.e., vasodilation by VIP and secretion by ACh. In addition, in the salivary gland, VIP and ACh have synergistic effects on secretory cells. This synergistic interaction is unusual since VIP does not directly stimulate glandular secretion but rather enhances the secretory effect of ACh, possibly by increasing the affinity for ACh of the muscarinic receptors on the gland cells (30). Thus, VIP appears to be a neuromodulator as well as neurotransmitter at cholinergic synapses in salivary glands.

VIP has been identified with immunohistochemical techniques in neurons and nerve terminals in autonomic ganglia (13, 15, 16, 19, 20, 25), where ACh is known to be the principal excitatory neurotransmitter. However, the function of VIP in ganglia has not been determined. In the present experiments we have examined the effects of exogenous VIP on transmission and on the responses to cholinergic agonists in autonomic ganglia of the cat. Two ganglia were studied: (1) parasympathetic ganglia of the urinary bladder, where VIP is

1057

localized in neurons and in dense collections of terminals surrounding the ganglion cells (Fig. 1) (13, 15, 16, 19, 20, 25), and (2) the superior cervical ganglion, a sympathetic ganglion exhibiting relatively modest numbers of VIP terminals (16, 25).

MATERIALS AND METHODS

Experiments were conducted in situ in chloralose or dial urethane anesthetized cats. Bladder ganglia (parasympathetic) or superior cervical ganglia (sympathetic) were decentralized and postganglionic nerves were isolated for monophasic recording. Drugs were administered to the ganglia by close intra-arterial injection. Postganglionic discharges were elicited by electrical stimulation of preganglionic nerves or by injection of nicotinic or muscarinic ganglionic stimulating agents (see Fig. 1). More detailed descriptions of the methodology can be obtained in previous papers (7, 9, 10).

In some experiments, vesical ganglia were removed for immunohistochemical studies. In these experiments the animals were perfused with Krebs-Ringer solution and then with fixative (4% paraformaldehyde in 0.1 M phosphate buffer, pH 7.2). Tissue saturated with 30% sucrose solution was then sectioned in a cryostat (28 μm) and processed for VIP by indirect immunohistochemical methods using the Sternberger technique (36). Some sections were also processed for acetylcholinesterase using the modified thiocholine method (22, 25).

RESULTS

An example of VIP-immunoreactivity in a parasympathetic ganglion on the surface of the urinary bladder is shown in Figure 1. Axons and varicosities containing VIP surround the ganglion cells. VIP was also detected in ganglion cells which stained for acetylcholinesterase and therefore were presumably cholinergic. VIP-immunoreactivity was not reduced 4-8 weeks following transection of the ipsilateral pelvic and hypogastric nerves. Since these nerves carry efferent and afferent inputs to the bladder from the spinal cord and dorsal root ganglia and also from the paravertebral and prevertebral sympathetic ganglia, it must be concluded that VIP-containing varicosities arise from neurons in the bladder ganglia, pelvic plexus or the bladder wall and not from extrinsic sources.

As indicated in Figure 1, experiments were conducted to examine the effects of VIP on nicotinic and muscarinic mechanisms in both sympathetic and parasympathetic ganglia. VIP was tested on ganglionic responses elicited by endogenously released ACh as well as on the responses elicited by the administration of exogenous ACh and other cholinomimetic agents.

The administration of VIP in varying doses intra- arterially (0.01-30 μg) to the superior cervical ganglia (Fig. 2A) or to the

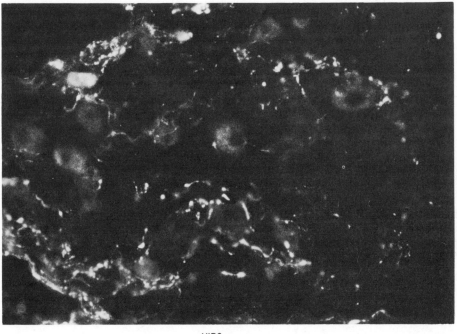

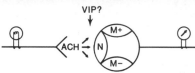

Figure 1. (A) Photomicrograph showing VIP-containing axons and varicosities in bladder ganglia of the cat; (B) Diagram summarizing
cholinergic mechanisms in autonomic ganglia and indicating the design
of the present experiments in which the effects of VIP were examined
on nicotinic (N), muscarinic excitatory (M+) and muscarinic inhibitory
(M-) responses in ganglia. Neural activity recorded on postganglionic
nerves was elicited by electrical stimulation of preganglionic nerves
or by the administration of exogenous acetylcholine or other cholinergic agonists.

bladder ganglia did not alter the amplitude of the postganglionic
action potentials elicited by electrical stimulation of the preganglionic nerves. However, in the same experiments, other agents
which are known to have ganglionic inhibitory and facilitatory actions
(eg., hexamethonium, leucine-enkephalin, norepinephrine, tetramethylammonium, and acetyl-β-methylcholine) produced the expected changes
in the evoked postganglionic potentials.

Although VIP did not alter nicotinic transmission it did affect
muscarinic ganglionic transmission which could be unmasked by treating

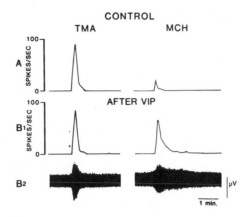

Figure 2. Effect of VIP on nicotinic and muscarinic transmission
in the superior cervical ganglion of the cat. (A) Computer averages
of five action potentials show that VIP (20 μg) had no effect on
nicotinic transmission elicited by the stimulation of the preganglionic fibers in the cervical sympathetic trunk. (B) VIP (20
μg) produced a large facilitation of muscarinic firing which was
initiated by treatment with 217AO, an anticholinesterase agent.
(C) Graph comparing the effect of VIP on nicotinic transmission
and muscarinic transmission. Abscissa represents time (minutes).
Ordinate represents percent of control: □ -- □ , area of computer
averaged action potentials for nicotinic transmission and ●—● ,
rate meter records of multiunit firing for muscarinic transmission.
The graph shows that the facilitation of muscarinic firing persisted
for more than five minutes and that nicotinic transmission was not
changed by VIP. The arrows in A correspond to stimulation artifacts,
and in B and C to the injections of VIP.

the ganglia with an irreversible anticholinesterase drug (217AO,
10-50 μg, i.a.) (7, 8, 10). The muscarinic postganglionic discharge
induced by 217AO was apparent as low amplitude asynchronous postganglionic firing. Previous studies (7, 10) have established that
this firing is dependent upon the activation of muscarinic receptors
by endogenously released ACh. The firing is markedly enhanced by
repetitive stimulation (30 Hz for 30 sec) of the preganglionic nerves,
is resistant to blockade by hexamethonium, but is completely abolished
by small doses of atropine (7, 10). VIP (5-20 μg, i.a.) markedly
enhanced the postganglionic firing in 217AO-treated ganglia (Fig. 2).
The VIP-induced discharge was not affected by hexamethonium in doses
that blocked nicotinic transmission but was blocked by atropine
(1-5 μg, i.a.) or by the selective muscarinic antagonist, pirenzepine
(1-5 μg, i.a.) (4, 14). The facilitatory effect of VIP persisted
for 5-10 min depending upon the dose administered. VIP elicited

reproducible effects when administered at 20-30 min intervals. VIP did not elicit postganglionic firing in untreated ganglia.

The differential effects of VIP on nicotinic and muscarinic mechanisms was also evident when VIP was tested against drugs that directly stimulate postsynaptic cholinergic receptors. As shown in Figure 3, VIP administered 10-50 seconds prior to cholinergic agonists did not alter the postganglionic discharge elicited by a nicotinic stimulant, tetramethylammonium (TMA, 1-5 µg, i.a.) but markedly enhanced the discharge elicited by the muscarinic agent, acetyl-β-methylcholine (MCH, 10-30 µg, i.a.). The facilitatory effect of VIP occurred in both the superior cervical ganglia and bladder ganglia with doses ranging from 5 to 30 µg, i.a. (Fig. 3B). The facilitatory effect of VIP persisted for 10-20 min. The effect of VIP was also examined on peptidergic excitation of bladder ganglion cells. In these experiments VIP (0.5-30 µg, i.a.) did not alter the response to substance P (0.5-20 µg, i.a.).

A similar selective facilitatory response was also observed in the superior cervical ganglion when VIP was administered prior to an injection of ACh which stimulates both nicotinic and muscarinic receptors. As illustrated in Figure 4, the postganglionic discharge evoked by ACh can consist of an early nicotinic discharge (ED) and a late muscarinic discharge (LD) (37). VIP markedly enhanced and prolonged the late discharge without altering the early discharge.

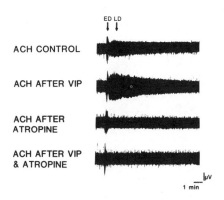

Figure 3. The effect of VIP on nicotinic and muscarinic discharges in vesical ganglia of the cat. (A) Rate meter records of multiunit discharges elicited by both a nicotinic agonist, tetramethylammonium (TMA, 5 µg) and a muscarinic agonist acetyl-β-methylcholine, (MCH, 25 µg). (B₁) Rate meter records show that after the administration of VIP (40 µg) the TMA discharge was unchanged while the discharge elicited by MCH was markedly enhanced. (B₂) Oscilloscope records of the discharges from which the rate meter records in B₁ were obtained.

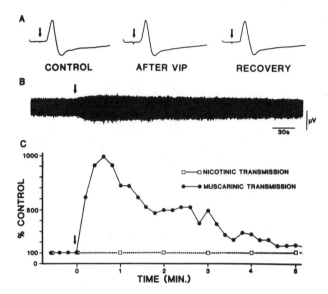

Figure 4.   Effect of VIP (0.5 μg) on the acetylcholine (ACh, 10
μg) evoked discharge in the superior cervical ganglion of the cat.
Top discharge is a control showing the early (nicotinic, ED) and
late (muscarinic, LD) discharges.   The second record shows that
the early discharge is unchanged while the late discharge is markedly
enhanced following the administration of VIP.   The third record
shows the abolition of the late discharge after treatment with atropine
(5 μg).   The final record shows that VIP does not alter the early
response which remains after treatment with atropine.

This facilitatory response persisted for 10-20 min.   Following the
administration of atropine (1-5 μg, i.a.), which blocked the late
ACh-discharge (Fig. 4), VIP did not change the response to injected
ACh (Fig. 4).

        In contrast to the facilitatory effect of VIP on muscarinic
excitation in ganglia, VIP had the opposite effect on muscarinic
depression of ganglionic transmission.   It is known from previous
studies (35, 37) that the administration of muscarinic agonists,
such as acetyl-β-methylcholine or bethanechol can depress transmission
in both sympathetic and parasympathetic ganglia.   In the present
experiments on bladder parasympathetic ganglia it was shown that
VIP (10-30 μg, i.a.) administered 10-30 sec prior to an injection
of acetyl-β-methylcholine antagonized or completely blocked the muscarinic depression of transmission.   However, VIP did not alter
the depression of transmission elicited by leucine-enkephalin (1-10
μg/kg, i.a.), GABA (1-10 μg/kg, i.a.), norepinephrine (1-10 μg/kg,

i.a.) or stimulation of sympathetic inhibitory nerves (hypogastric nerves) (9).

DISCUSSION

The present results indicate that exogenous VIP has very selective actions on cholinergic mechanisms in autonomic ganglia. VIP enhanced muscarinic transmission and the ganglionic excitatory responses to muscarinic agonists, suppressed muscarinic inhibitory responses, but did not alter nicotinic transmission or the responses to nicotinic agonists. These observations suggest that VIP must have a very specific postsynaptic action. It is also possible that VIP enhances the interaction of acetylcholine with the muscarinic excitatory receptors or enhances the transduction mechanism leading to muscarinic depolarization and ganglion cell firing. The failure of VIP to enhance nicotinic mechanisms or to directly excite the untreated ganglion would suggest that VIP does not act presynaptically to enhance transmitter release or postsynaptically to depolarize the ganglion cells.

The facilitatory interaction between VIP and muscarinic agonists in ganglia is similar to the VIP-induced facilitation of muscarinic transmission in the salivary glands, where VIP enhances muscarinic excitation of the gland cells but has no direct excitatory effect on the gland (26, 27, 28, 29). Radioligand-receptor binding studies suggest that the facilitatory effect of VIP may be related to an enhancement of the affinity of the muscarinic receptors for muscarinic agonists (30). A similar mechanism could contribute to the effects of VIP in autonomic ganglia.

An increase in cyclic AMP levels might also be involved in the selective ganglionic facilitatory effects of VIP. VIP is known to increase cyclic AMP concentrations in various tissues including autonomic ganglia (38); and cyclic AMP has been implicated (21, 24) in the facilitatory effect of other substances (eg., dopamine) on muscarinic responses in ganglia. In addition, exogenous cyclic AMP applied extracellularly or by intracellular injection enhances the ganglionic depolarizing response to muscarinic agonists (21, 24).

It is interesting that isoproterenol, a beta adrenergic receptor agonist which also increased cyclic AMP levels in ganglia (3, 23, 34, 39) duplicates the effects of VIP on cholinergic induced discharges in the ganglion (i.e., enhancement of muscarinic but no effect on nicotinic firing; 7, 8). However, isoproterenol also enhances nicotinic transmission, which may be related to a presynaptic action of the drug (23, 32).

The present results also indicate that VIP can exert differential effects on muscarinic ganglionic mechanisms; i.e., enhancement of muscarinic facilitation and suppression of muscarinic inhibition.

These findings suggest: 1) that VIP might have different effects on the muscarinic receptors which mediate facilitation and inhibition or 2) that VIP might selectively affect the transduction mechanisms underlying facilitation and inhibition. In this regard, it is noteworthy that both sites in the peripheral system where VIP facilitates muscarinic responses (salivary gland and ganglia) the receptors have been identified as the M_1 type (1). Unfortunately, the receptor type mediating muscarinic inhibition in ganglia has not been determined.

The presence of VIP in ganglia and the unusual effects of exogenous VIP on ganglionic transmission raise questions about the possible physiological role of VIP at ganglionic synapses. In bladder parasympathetic ganglia VIP is contained in cholinergic ganglion cells and in axons and varicosities surrounding the ganglion cells (13, 15, 16, 18, 19, 20, 25). Since VIP varicosities did not disappear following chronic decentralization (6, 20) they must originate from neurons intrinsic to the ganglia or the peripheral plexus; possibly as axon collaterals or dendrites of the principal cholinergic ganglion cells. Thus VIP may be a local modulator of cholinergic transmission. However, since exogenous VIP only affects muscarinic mechanisms in the ganglia and since these mechanism are considered to be primarily modulatory, the VIP system may represent a second order modulator by which the peptide only indirectly alters ganglionic transmission.

The present results may also provide some insight into the interactions of VIP and ACh at cholinergic postganglionic nerve terminals where the two substances can be released as cotransmitters. It has been speculated that muscarinic inhibitory receptors on adrenergic and cholinergic terminals (5, 12, 28, 31, 33) may be similar to muscarinic inhibitory receptors in ganglia. Thus it is tempting to speculate that VIP released as a cotransmitter may block muscarinic presynaptic inhibition at the cholinergic nerve terminals in the same way that it blocks muscarinic inhibition in ganglia. Therefore, VIP may enhance ACh release (5); and at certain synapses, such as those in the salivary gland, may have both a presynaptic and postsynaptic action to facilitate cholinergic transmission.

Acknowledgements. We thank Susan Erdman and Mary Backes for excellent technical assistance. Supported by NSF grant BNS 82-08348, NIH grants AM 31788, and NS 18075, and clinical research grant MH30915.

REFERENCES

1. Birdsall, N.J.M., Burgen, A.S.V., Hammer, R., Hulme, E.C. and Stockton, J. (1980): Scand. J. Gastroent. 15:Suppl. 66:1-4.
2. Bloom, S.R. and Edwards, A.V. (1980): J. Physiol. 300:41-53.
3. Brown, D.A., Caulfield, M.P. and Kirby, P.J. (1979): J. Physiol. 290:441-451.

4. Brown, D.A., Forward, A. and Marsh, S. (1980): Br. J. Pharmac. 71:362-364.
5. Burnstock, G. (1983): In: Dale's Principle and Communication Between Neurons (ed) N.N. Osborne, Pergamon Press, Oxford and New York, pp. 7-35.
6. Dail, W.G., Moll, M.A., and Dziurzynski, R.A. (1983): Neurosci. Abst. 9:292.
7. deGroat, W.C. and Volle, R.L. (1966): J. Pharmac. Exp. Ther. 154:200-215.
8. deGroat, W.C. (1967): Circulation Research 20-21: Suppl.3: 135-145.
9. deGroat, W.C. and Saum, W.R. (1972): J. Physiol. 220: 297-314.
10. deGroat, W.C. and Saum, W.R. (1976): J. Physiol. 256: 137-158.
11. Fahrenkrug, J. (1979): Digestion 19:149-169.
12. Fozard, J.R. and Muscholl, E. (1972): Br. J. Pharmac. 45:616-629.
13. Håkanson, R., Sundler, F. and Uddman, R. (1982): In: Vasoactive Intestinal Peptide (ed) S. Said, Raven Press, New York, pp. 121-144.
14. Hammer, R., Berrie, C.P., Birdsall, N.J.M., Burgen, A.S.V. and Hulme, E.C. (1980): Nature 283:90-92.
15. Hökfelt, T., Schultzberg, M., Elde, R., Nilsson, G., Terenius, L., Said, S. and Goldstein, M. (1978): Acta Pharmacol. et Toxicol. 43:79-89.
16. Hökfelt, T. (1979): Neurosci. Res. Prog. Bull. 17: 424-443.
17. Hökfelt, T., Johansson, O., Ljungdahl, A., Lundberg, J.M. and Schultzberg, M. (1980): Nature 284:515-521.
18. Hökfelt, T., Schultzberg, M., Lundberg, J.M., Fuxe, K., Mutt, V., Fahrenkrug, J. and Said, S. (1982): In: Vasoactive Intestinal Peptide (ed) S. Said, Raven Press, New York, pp. 65-90.
19. Kawatani, M. and deGroat, W.C. (1983): The Pharmacologist 25:157.
20. Kawatani, M., Rutigliano, M. and deGroat, W.C.: Brain Research (In press).
21. Kobayashi, H., Hashiguchi, T. and Ushiyama, N.S. (1978): Nature 271:268-270.
22. Koelle, G.B. (1951): J. Pharmac. Exp. Ther. 103: 153-171.
23. Kuba, K., Kato, E., Kumamoto, E., Koketsu, K. and Hirai, K. (1981): Nature 291:654-656.
24. Libet, B. (1970): Fed. Proc. 29:1945-1956.
25. Lundberg, J.M., Hökfelt, T., Schultzberg, M., Uvnas-Wallensten, K., Kohler, C. and Said, S. (1979): Neuroscience 4:1539-1559.
26. Lundberg, J.M. (1981): Acta Physiol. Scand. Suppl. 496:1-57.
27. Lundberg, J.M., Änggård, A. and Fahrenkrug, J. (1981): Acta Physiol. Scand. 113:317-327.
28. Lundberg, J.M., Änggård, A. and Fahrenkrug, J. (1981): Acta Physiol. Scand. 113:328-336.
29. Lundberg, J.M., Änggård, A. and Fahrenkrug, J. (1982): Acta Physiol. Scand. 114:329-337.
30. Lundberg, J.M., Hedlund, B. and Bartfai, T. (1982): Nature 295: 147-149.
31. Marchi, M., Paudice, P., Caviglia, A. and Raiteri, M. (1983): Eur. J. Pharmac. 91:63-68.

32. Medgett, I.C. (1983): Br. J. Pharmac. $\underline{78}$:17-27.
33. Norström, Ö. and Bartfai, T. (1980): Acta Physiol. Scand. $\underline{108}$: 347-353.
34. Quenzer, L., Yahn, D., Alkadhi, K. and Volle, R.L. (1979): J. Pharmac. Exp. Ther. $\underline{208}$:31-36.
35. Saum, W.R. and deGroat, W.C. (1972): Science $\underline{175}$:659-661.
36. Sternberger, L.A. (1979): Immunocytochemistry: John Wiley and Sons, New York.
37. Takeshige, C., Pappano, A.J., deGroat, W.C. and Volle, R.L. (1963): J. Pharmac. Exp. Ther. $\underline{141}$:333-342.
38. Volle, R.L. and Patterson, B.A. (1982): J. Neurochem. $\underline{39}$:1195-1197.
39. Wamsley, J.M., Black, A.C., Redick, J.A., West, J.R. and Williams, T.H. (1978): Brain Res. $\underline{156}$:75-82.

INTERACTION OF CYCLIC AMP WITH MUSCARINIC AUTOINHIBITION OF [3H]-ACETYLCHOLINE SECRETION IN GUINEA-PIG ILEUM MYENTERIC PLEXUS

P. Alberts, V. Ögren and Å. Sellström

Division of Experimental Medicine
National Defense Research Institute
FOA-4
S-901 82 Umeå, SWEDEN

INTRODUCTION

Acetylcholine (ACh) secretion from guinea-pig ileum myenteric plexus is affected by muscarinic agents (1, 4, 5, 6, 9, 11). The secretion is also influenced by inhibitors of phosphodiesterase (3, 10, 12). Addition of 8-Br cyclic AMP enhanced the secretion of [3H]-ACh from guinea-pig ileum myenteric plexus, while the 8-Br analog of cyclic GMP was without effect (2).

We have therefore studied the role of endogenous cyclic nucleotides in muscarinic receptor mediated control of ACh secretion in the guinea-pig ileum myenteric plexus, using two inhibitors of phosphodiesterase.

MATERIALS AND METHODS

The longitudinal muscle with the myenteric plexus attached was prepared as described by Paton and Zar (8, cf. 1). The transmitter stores of the cholinergic nerves were labelled by preincubation with [3H]-choline (10^{-6} M, New England Nuclear) for 60 min at 37°C. During the preincubation the preparation was stimulated electrically at 0.2 Hz. After incubation hemicholinium-3 (10^{-5} M) and eserine (10^{-5} M) were added to the Tyrode solution.

The preparation was mounted under a passive load of 10 mN and superfused with the Tyrode solution at a rate of 1 ml/min. Stimulation was with trains of 150 shocks (0.5 ms, 25 V/cm) at 0.5 Hz with a Grass S4C stimulator. The electrically evoked overflow of [3H] collected in the superfusing medium was calcium dependent and tetrodotoxin sensitive. It has previously been shown that under these experimental conditions the evoked secretion mainly consists of $[^3H]$-ACh (1). Therefore the evoked efflux of total $[^3H]$ is used as a measure of $[^3H]$-ACh secretion. After the experiments the tissue content of $[^3H]$ was extracted in 0.4 M perchloric acid. The results are expressed as the evoked fractional secretion of total $[^3H]$ contained in the preparation at the time of stimulation (cf. 1).

RESULTS AND DISCUSSION

Addition of 3-isobutyl-1-methylxanthine (IBMX, 3 mM) to inhibit neuronal phosphodiesterase enhanced the secretion of $[^3H]$-ACh by 113 $\pm$ 33% (n = 5, p < 0.01). A structurally different inhibitor of phosphodiesterase, SQ 20,006 (0.3-0.5 mM), enhanced the secretion by 40-120%. Higher concentrations of SQ 20,006 caused further enhancement that deviated from a simple "activity" isotherm.

The secretion of $[^3H]$-ACh is enhanced by blockade of muscarinic receptors by atropine. To study the effect of muscarinic receptor blockade on the inhibition of cyclic nucleotide phosphodiesterase the effect of IBMX on the secretory response was tested in the presence of atropine.

The effect of IBMX on the secretion of $[^3H]$-ACh was diminished by the addition of atropine. The enhancement caused by IBMX (3 mM) in the presence of 30 nM atropine was 36 $\pm$ 18% (n = 6, p < 0.05) and only 8 $\pm$ 8% (n = 12) in the presence of 10^{-6} M atropine.

Since the enhancing effects of atropine and IBMX were not additive the results suggest that endogenous cyclic nucleotides may be involved in muscarinic autoinhibition of $[^3H]$-ACh secretion in guinea-pig ileum myenteric plexus. Inhibition of adenylate cyclase by muscarinic receptor activation has been shown in rat striatum (7). The present results are compatible with the existence of such a mechanism in the myenteric cholinergic terminals.

REFERENCES

1. Alberts, P., Bartfai, T. and Stjärne, L. (1982): J. Physiol. (Lond.) 329:93-112.
2. Alberts, P. and Stjärne, L. (1982): Acta Physiol. Scand. 115: 269-272.
3. Cook, M.A., Hamilton, J.T. and Okwuasaba, F.K. (1978): Nature 271:768-771.
4. Gustafsson, L., Hedqvist, P. and Lundgren, G. (1980): Acta Physiol. Scand. 110:401-411.
5. Kilbinger, H. (1977): Naunyn-Schmiedeberg's Arch. Pharmacol. 300:145-151.
6. Nordström, Ö., Alberts, P., Westlind, A., Unden, A. and Bartfai, T. (1983): Mol. Pharmacol. 24:1-5

7. Olianas, M.C., Onali, P., Neff, N.H. and Costa, E. (1983):
    Mol. Pharmacol. 23:393-398.

8. Paton, W.D.M. and Zar, M.A. (1968): J. Physiol. (Lond.) 194:13-33.

9. Sawynok, J. and Jhamandas, K. (1977): Can. J. Physiol. Pharma-
    col. 55:909-916.

10. Sawynok, J. and Jhamandas, K. (1979): Can. J. Physiol. Pharma-
    col. 57:853-859.

11. Szerb, J.C. (1980): Naunyn-Schmiedeberg's Arch. Pharmacol.
    311:119-127.

12. Vizi, E.S. and Knoll, J. (1976): Neuroscience 1:391-398.

PARASYMPATHETIC INNERVATION OF THE HEART:

ACETYLCHOLINE TURNOVER IN VIVO

O.M. Brown, J.J. Salata[*] and L.A. Graziani

Department of Pharmacology
State University of New York College of Medicine
Upstate Medical Center
Syracuse, New York 13210 U.S.A.

[*]Current Address:
Krannert Institute of Cardiology
Department of Medicine
Indiana University School of Medicine
1001 W. 10th Street
Indianapolis, Indiana

INTRODUCTION

Parasympathetic innervation of the heart plays an important role in the dynamic control of cardiac function. Biochemical approaches used to study cardiac vagal innervation have included measurement of levels of acetylcholine (ACh) (1, 2, 22, 33), and the activities of both choline acetyltransferase (8, 28, 30) and cholinesterase (32). These methods have been useful in characterizing the normal distribution of cardiac vagal innervation and its alterations following certain perturbations such as bilateral or unilateral vagotomy and transplantation.

In an attempt to better describe the parasympathetic innervation of the heart we have employed a direct chemical approach which allows for the characterization of the dynamics of cardiac ACh. We report here on a method and a description of choline (Ch) uptake and ACh synthesis in rat heart in vivo.

METHODS

Non-deuterated (d_0) Ch and Ch esters were obtained from Sigma Chemical Co. (St. Louis, MO). Choline-1,1,2,2-d_4 bromide (d_4-ChBr), choline-N,N,N-trimethyl-d_9 bromide (d_9-ChBr), acetylcholine-1,1,2,2-d_4 bromide (d_4-AChBr) and acetylcholine-N,N,N-trimethyl-d_9 bromide (d_9-AChBr) were purchased from Merck and Co., Inc. (Dorval, Quebec). All other chemicals were obtained from commercial sources and were of analytical reagent grade. Male Sprague-Dawley rats from Taconic Farms (Germantown, NY) were used in these experiments; weighing 250-400g for the time course study, and 175-250g for the vagus stimulation studies.

Choline uptake. To characterize Ch uptake and ACh synthesis, the time course of deuterium labeled Ch and ACh were followed in the rat heart. Rats were anesthetized with a combination of chloral hydrate (300 mg/kg) and sodium pentobarbital (30 mg/kg). At time zero a bolus injection of d_4-Ch (20 μmol/kg) was administered into the external jugular vein. At various time points thereafter the heart was quickly removed and homogenized for analysis.

Vagus stimulation. To determine the effects of vagal activity on heart rate, Ch uptake and ACh synthesis, another experiment was performed which employed bilateral vagus stimulation. Rats were anesthetized and, following a midline cervical incision, the trachea was intubated and both cervical vagus nerve trunks were exposed and placed over wire electrodes. The vagi were given electrical pulse train stimulation from a Grass S88 Stimulator: twin 5 volt pulses 25 msec apart and 0.5 msec in duration were administered at a pulse rate (4-5/sec) which entrained the heart to a rate that was approximately 70% of the resting (pre-stimulation) rate. Heart rate (EKG) was monitored with a Grass Polygraph. After a brief initial test stimulation, d_4-Ch was injected at time zero, pulse train stimulation was applied at 2 min for three minutes and the animals were sacrificed at 5 min (Table 1) (control rats received the same treatment but no stimulation). The rat hearts were dissected (in situ) into two areas for analysis: a) right atrium + S.A. node (the area richest in ACh); and b) the remainder of the heart.

Sample Preparation. Rat heart samples were prepared for Ch and ACh analysis using methods similar to those in our previous reports (1-4). Each sample was rapidly blotted and weighed, and placed into a tube containing 5 ml acetonitrile (CH_3CN), 1 ml 10^{-4}N HCl, 250 nmol tetramethylammonium iodide (TMAI) (as coprecipitant), and 4 nmol d_9-ACh and 30 nmol d_9-Ch as internal standards. The samples were homogenized and allowed to stand on ice for 20 min. The homogenate was centrifuged and the resulting supernatant was extracted with toluene followed by a further extraction with hexane. The resulting aqueous solution was frozen and lyophilized.

All Ch (d_0, d_4 and d_9) was quantitatively esterified to propionylcholine (PCh) by using a method similar to that of Stavinoha and Weintraub (31). One hundred microliters CH_3CN and 100 μl redistilled propionyl chloride were added to the sample and it was

Table 1

Effects of Pulse Train Bilateral Vagus Stimulation on
Heart Rate: Time Line Protocol and Results from 16 Rats
Sacrificed 5 Minutes After d_4-Ch Injection[a]

The Times of Experimental Perturbations and Heart
Rate Readings (letters A-F) are Indicated by Arrows

Inject d_4-Ch   Stimulate   Sample

```
      ↓                       ↓                          ↓
──────┼───────────────────────┼──────────────────────────┼────
      0                       2       Minutes             5

  ↑      ↑      ↑             ↑         ↑                 ↑
  A      B      C             D         E                 F
```

Reading	A	B	C	D	E	F
Time (min)	-0.5	0	0.5	1.5	2.2	4.7
Percent of Resting Rate $\pm$ SEM	100[b]	82.3 $\pm$2.8	97.3 $\pm$1.6	100.9 $\pm$1.1	67.6 $\pm$2.6	71.0 $\pm$2.8

[a]Electrical stimulation parameters are described in Methods
[b]Resting rate = 373 $\pm$ 9 beats/min

capped and placed in a 60°C oven for 30 min. The sample was dried under a stream of nitrogen and solubilized in 2 ml 10^{-4}N HCl.

Choline esters were precipitated by the addition of KI I_2. The precipitate was dissolved and extracted twice with ether to remove I_2. One-tenth milliliter 0.1 N $Na_2S_2O_3$ was added to reduce any remaining I_2 to I^-. This solution was lyophilized and the residue was taken up in CH_3CN. A suitable aliquot was pipetted onto a platinum pyrolysis ribbon for analysis by pyrolysis mass fragmentography (PMF).

Pyrolysis Mass Fragmentography (PMF). All heart samples and standards were analyzed for their content of deuterated and non-deuterated Ch (PCh) and ACh by PMF. A Finnigan gas chromatograph-mass spectrometer (GC/MS) fitted with a pyrolyzer, data system and multiple ion detector was used to perform these analyses. The PMF conditions have been described in detail earlier (4).

It has been well established that the most abundant ion (Base Peak) resulting from the electron ionization PMF of ACh (d_0-ACh) and other Ch esters is the dimethylmethyleneimmonium (DMMI) ion, $[(CH_3)_2\text{-}N=CH_2]^+$ (m/e 58) (2). We therefore programmed the GC/MS to monitor m/e 58 in analyzing samples for ACh and PCh (from Ch). When subjected to PMF, deuterated variants of ACh and PCh generate the corresponding analogs of DMMI with larger m/e ratios (4, 20, 29). Thus d_4-ACh and d_4-PCh (from d_4-Ch) produce the fragment, $[(CH_3)_2\text{-}N=CD_2]^+$, which is monitored at m/e 60. Likewise, d_9-ACh and d_9-PCh produce the fragment, $[(CD_3)_2\text{-}N=CH_2]^+$, monitored at m/e 64.

In the analysis of heart samples, unlabeled (d_0-) ACh and PCh (m/e 58) represented endogenous ACh and Ch from the heart; d_4-Ch (d_4-PCh) (m/e 60) represented the amount of labeled Ch taken up by the heart samples following i.v. injection, which was then partially converted to d_4-ACh (m/e 60); and d_9-ACh and d_9-Ch (d_9-PCh) (m/e 64) were added as internal standards for quantitative analysis.

Quantitation and Turnover Calculations. Standard curves were prepared by adding increasing amounts of d_0-AChI, d_0-ChI, d_4-AChBr and d_4-ChBr to tubes containing 4 nmol d_9-AChBr and 30 nmol d_9-ChBr as internal standards and subjecting them to the extraction and PMF methods. The PMF peak for each variant of Ch or ACh contains a small contribution from their other variants. Mathematical correction for this phenomenon was achieved by using the procedure of Jenden and coworkers to solve three simultaneous equations for each sample (and standard) peak (20).

The following PMF peak area ratios were determined (for all samples and standards): d_0-ACh/d_9-ACh; d_4-ACh/d_9-ACh; d_0-PCh/d_9-PCh and d_4-PCh/d_9-PCh. Appropriate standard curves were produced by plotting these ratios for the standards as a function of the concentrations of added Ch or ACh. Specific activity values for each sample were calculated as the mole ratio of the d_4-variant to the total: nmol d_4/(nmol d_4 + nmol d_0).

Turnover rates for ACh were calculated from a kinetic model with steady-state assumptions as described by Zilversmit (34). We have previously concluded that the product-precursor relationship in this preparation is an open system and that the principles necessary for steady-state calculations of turnover apply (4). Jenden and coworkers (18, 19) and Costa and his coworkers (10, 26, 27) have modified the Zilversmit model for ACh turnover. They have developed the following precursor-product relationship

$$\frac{d\ SA\ ACh}{dt} = k\ (SA\ Ch - SA\ ACh)$$

where SA ACh is the specific activity of ACh (product), SA Ch is the specific activity of Ch (precursor) and k is the turnover rate

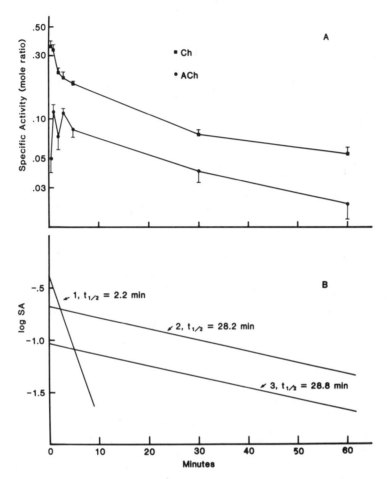

Figure 1. (A) Semilog plots of the specific activity (SA) of Ch (squares) and ACh (circles) in rat heart as a function of time following pulse i.v. injection of d_4-Ch. Lines are plotted from data in reference (4) and are expressed as mean ± SEM. Each point represents between 8 and 21 animals. (B) Regression lines for the above SA data: 1) SA Ch from 0.5 to 2 min, log y = -0.139 x -0.362, r^2 = 0.96; 2) SA Ch from 2 to 60 min, log y = -0.0107 x -0.677, r^2 = 0.94; 3) SA ACh from 2 to 60 min, log y = -0.0106 x -1.030, r^2 = 0.94.

constant of the product. This constant (k) was calculated by the method of finite differences from SA ACh and SA Ch values obtained at two different times (4, 18, 26, 27). The turnover rate of ACh (TR ACh) was determined by multiplying the total concentration of ACh by the rate constant, k.

RESULTS

We have employed stable isotope labeling techniques in an attempt to study ACh synthesis in heart. Rats were injected with a pulse of d_4-Ch (20 μmol/kg, i.v.) and at various times thereafter their hearts were removed and analyzed by PMF for both Ch and ACh. Following analysis, the SA of Ch and ACh were calculated and plotted as a function of time after injection of d_4-Ch (Fig. 1A). Rat hearts accumulated d_4-Ch rapidly and efficiently such that approximately 35% of total heart Ch was labeled at 0.5 min. From this maximum, the SA of Ch appears to decay in two phases, i.e., with two different rates. There is an initial rapid fall from 0.5 to 2 min with a half-time of 2.2 min (Fig. 1B). A more gradual decline in choline SA with a half-time of 28.2 min was found for the time period between 2 and 60 min.

Figure 1A also shows a plot of the changes in specific activity of ACh following injection of d_4-Ch. The SA of ACh rose to a maximum of 0.11 at 1 min and declined thereafter, with a slight dip at 2 min. The decline from 2 to 60 min is very similar to that for Ch, with a half-time of 28.8 min and a near identical regression slope (Fig. 1B). It is apparent from the SA curves that once equilibrium between the labeled pools is reached (at about 2 min) the rates of decay in the SA of Ch and ACh are parallel from 2 to 60 min (Fig. 1B).

For the time intervals studied, the turnover rate constant, k, of ACh was calculated by the method of finite differences (4, 18, 27). Using these values and the steady-state ACh concentration, the acetylcholine turnover rate was calculated from the SA regression lines in Figure 1B (4). A value of 0.144 nmol/g/min for TR ACh was obtained for the period from 2 to 60 min.

Vagus stimulation slowed the heart to a rate approximately 70% of resting rate (373 beats/min). This decreased rate was main-tained for the duration of the 3 min stimulation period (Table 1). Both the right and the left vagus nerves were apparently effectively stimulated as we observed not only slowing of sinus rhythm, but also many instances of atrioventricular block (2:1, 3:2 and 3:1). We also noted a short-lived (30 sec or less) bradycardia caused by the injection of d_4-Ch (Table 1).

The specific activity of ACh was higher in the hearts of stimu-lated animals as compared to non-stimulated controls (Table 2). Especially affected was the SA of ACh in the right atrium + S.A. node area: 0.244 $\pm$ 0.020 stimulated, 0.174 $\pm$ 0.025 controls. Vagus stimulation resulted in a decrease in the SA of Ch in right atrium + S.A. node but an increase in the remainder of the heart (Table 2).

Table 2

Effects of vagus stimulation on the specific
activity of choline and acetylcholine in rat heart[a]

| | Specific Activity (mole ratio ± SEM)[b] | | | |
| | Choline | | Acetylcholine | |
Sample	Stimulated	Control	Stimulated	Control
S.A. Node + Right Atrium	0.204 ±.030	0.196 ±.040	0.244[*] ±.020	0.174 ±.025
Remainder of Heart	0.145 ±.015	0.113 ±.014	0.068 ±.012	0.050 ±.020
Whole Heart[c]	0.147 ±.014	0.118 ±.013	0.095 ±.014	0.067 ±.019
(N)	(15)	(10)	(15)	(10)

[a]Injection and stimulation protocol is described in Methods and in Table 1. Same animals as in Table 1. [b]Calculated as described in Methods and Reference 4. [c]Mathematically reconstructed. [*]$p <$ 0.05 when Stimulated compared to Control with Student's t-test. (This test was not performed on reconstructed whole heart data.)

DISCUSSION

In an effort to characterize in vivo Ch uptake and ACh synthesis in the heart we monitored labeled Ch and ACh following pulse injections of deuterium-labeled choline (d_4-Ch) into anesthetized rats. The changes found in Ch and ACh specific activities with time (Fig. 1A) were somewhat similar to those seen in brain (17, 19). However, the maximum values achieved in heart (SA Ch = 35% and SA ACh = 11%) (4) are much greater than those reported for brain (SA Ch - 9%, SA ACh ≈ 3%) (17, 19). This difference is in part due to the concentration gradient for Ch being from the brain to the blood (6, 7, 9). The heart appears not to have such a gradient, as injections of Ch in amounts which cause no increase in guinea pig brain Ch content increase heart Ch content by over 400% (15).

We found that after a short period of equilibration a steady-state for ACh synthesis was achieved, as the SA of Ch and ACh declined at near identical rates (Fig. 1B). ACh released from neurons is ordinarily hydroylzed to form acetate and Ch. Much of this Ch is taken back up to again form ACh. Such recycling of deuterated Ch would not adhere to steady-state kinetic principles which are necessary

for turnover calculations (5, 17, 27). However, this reutilization
of Ch may not be as important a mechanism in heart as it is in most
other organs (25). Lindmar, et al. (24) have recently concluded
that the rate of ACh hydrolysis in the interstitial spaces of the
heart is low and that diffusion and washout are important factors
for synaptic removal of ACh. With the above in mind, we made the
assumption that ACh synthesis in the heart is an open system (i.e.,
none of the product returns to precursor) and that the principles
necessary for steady-state calculations of turnover apply (4, 34).
From our regression lines of the decay in specific activity of Ch
and ACh (Fig. 1B) we calculated a half-time for Ch of 28.2 min, a
half-time for ACh of 28.8 min and a value of 0.144 nmol/g/min for
the turnover rate of ACh in rat heart. With a steady-state level
of 8.6 nmol ACh/g (4), these results indicate that the total pool
of ACh in rat heart turns over approximately once every hour.

We are not certain as to why the specific activity curves for
precursor and product do not cross (Fig. 1A) as expected according
to the mathematical model described by Zilversmit (34). A possible
explanation for this inconsistency is that more than one functional
compartment exists for ACh in the heart; this is certainly the case
for brain ACh (18). However, when the parasympathetic neurons in
the right atrium + S.A. node are activated by electrical stimulation
of the vagus nerves, the specific activity for Ch and ACh apparently
do cross one another at some point in time, since the SA of ACh in
this area from vagus stimulated rats is much higher than the corresponding SA of Ch at 5 min (Table 2).

Although few studies have been reported on in vivo Ch uptake
and ACh synthesis in peripheral organs, some of the present results
can be compared to other work. Haubrich et al. (14) followed the
fate of injected (tritiated) Ch in anesthetized guinea pigs. They
did not indicate complete time course data, but did report a figure
of 20 min for the half-life of Ch in guinea pig heart. This is
similar to the value of 28.2 min which we found for the half-life
of Ch in rat heart. Haubrich et al. (14) also reported a "synthesis
rate" of 0.21 nmol/g/min (calculated from one time point only) for
ACh in guinea pig heart. Hanin and coworkers have studied the time
course of Ch uptake and ACh synthesis in rat salivary gland (10-12).
They reported a Ch half-life of "longer than 20 min". They also
calculated a value of 0.175 nmol/g/min for the ACh turnover rate
in rat salivary gland based on two time points, 32 and 64 min post-
injection (10).

To help characterize the effects of vagus nerve activity on
Ch uptake and ACh synthesis in vivo, we subjected some rats to bilateral electrical stimulation of the cervical vagus nerves. The
pulse train stimulation used was effective in entraining the heart
rhythm (21, 22) to the predetermined rate of 70% of resting (Table
1). This vagal inhibition of the heart was accompanied by an enhance-

ment in the ACh synthesis rate, as indicated by increased ACh specific activity (Table 2). This effect was most dramatic in the right atrium + S.A. node area, which is the area richest in ACh and other indicators of cholinergic innervation (1).

The brief period of bradycardia following the injection of d_4-Ch (Table 1) may be explained by an agonist action of Ch on pre- or post-ganglionic parasympathetic ACh receptors. There is evidence for weak activity of Ch at both nicotinic and muscarinic ACh receptors (13, 16).

We have thus invoked deuterated isotopes and PMF techniques to describe the dynamics of Ch and ACh in the rat heart *in vivo*. The present model is sensitive to conditions expected to alter the rate of ACh synthesis (resting vs. vagus stimulation). The cardiac inhibition resulting from vagus stimulation was accompanied by an increase in the *in vivo* synthesis rate of heart ACh.

Acknowledgements. This work was supported by a grant-in-aid from the American Heart Association with funds contributed in part by the Broome County, N.Y. Chapter of the American Heart Association. We are grateful to Mary Anne Carroll and Patricia (Woyciesjes) Relyea for providing excellent assistance and valuable advice.

REFERENCES

1. Brown, O.M. (1976): Am. J. Physiol. 231:781-785.
2. Brown, O.M. (1981): Life Sci. 28:819-825.
3. Brown, O.M., Post, M.E. and Mallov, S. (1977): J. Stud. Alcohol 38:603-617.
4. Brown, O.M. and Salata, J.S. (1983): Life Sci. 33:213-224.
5. Cheney, D.L. and Costa, E. (1977): Ann. Rev. Pharmacol. Toxicol. 17:369-386.
6. Choi, R.L., Freeman, J.J. and Jenden, D.J. (1975): J. Neurochem. 24:735-741.
7. Dross, K. and Kewitz, H. (1972): N.S. Arch. Pharmacol. 274: 91-106.
8. Eskrom, J. (1978): Acta Physiol. Scand. 102:116-119.
9. Freeman, J.J., Choi, R.L. and Jenden, D.J. (1975): J. Neurochem. 24:729-734.
10. Hanin, I., Cheney, D.L., Trabucchi, M., Massarelli, R., Wang, C.T. and Costa, E. (1973): J. Pharmacol. Exp. Ther. 187:68-77.
11. Hanin, I., Massarelli, R. and Costa, E. (1972): Adv. Biochem. Psychopharm. 6:181-202.
12. Hanin, I., Massarelli, R. and Costa, E. (1972): J. Pharmacol. Exp. Ther. 181:10-18.
13. Haubrich, D.R., Risley, E.A. and Williams, M. (1979): Biochem. Pharmacol. 28:3673-3674.
14. Haubrich, D.R., Wang, P.F.L. and Wedeking, P.W. (1975): J. Pharmacol. Exp. Ther. 193:246-255.

15. Haubrich, D.R., Wedeking, P.W. and Wang, P.F.L. (1974): Life Sci. 14:921-927.
16. Holz, R.W. and Senter, R.A. (1981): Science 214:466-468.
17. Jenden, D.J. (1974): In The Phenothiazines and Structurally Related Drugs (eds) L.S. Forrest, C.J. Carr and E. Usdin, Raven Press, New York, pp. 769-777.
18. Jenden, D.J. (1978): Adv. Behav. Biol. 24:139-162.
19. Jenden, D.J., Choi, L. Silverman, R.W., Steinborn, J.A., Roch, M. and Booth, R.A. (1974): Life Sci. 14:55-63.
20. Jenden, D.J., Roch, M. and Booth, R.A. (1973): Anal. Biochem. 55:438-448.
21. Levy, M.N., Iano, T. and Zieske, H. (1972): Circ. Res. 30: 186-195.
22. Levy, M.N., Martin, P.J., Iano, T. and Zieske, H. (1969): Circ. Res. 25:303-314.
23. Lewartowski, B. and Bielecki, K. (1963): J. Pharmacol. Exp. Ther. 142:24-30.
24. Lindmar, R., Loffelholz, K. and Weide, W. (1982): N.S. Arch. Pharmacol. 318:295-300.
25. Löffelholz, K. (1981): Am. J. Physiol. 240:H431-H440.
26. Neff, N.M., Spano, P.F., Gropetti, A., Wang, C.T. and Costa, E. (1971): J. Pharmacol. Exp. Ther. 176:701-710.
27. Racagni, G., Cheney, D.L., Trabucchi, M., Wang, C. and Costa, E. (1975): Life Sci. 15:1961-1975.
28. Roskoski, Jr., R., Schmid, P.G., Mayer, H.E. and Abboud, F.M., (1975): Circ. Res. 36:547-552.
29. Salata, J.J. and Brown, O.M. (1980): Fed. Proc. 39:1832.
30. Slavikova, J. and Tuček, S. (1982): Pflugers Arch. Physiol. 392:225-229.
31. Stavinoha, W.B. and Weintraub, S.T. (1974): Science 183:964-965.
32. Tuček, S. and Vlk, J. (1962): Physiol. Bohemoslov. 11:319-327.
33. Vlk, J. and Tuček, S. (1961): Physiol. Bohemoslov. 10:65-71.
34. Zilversmit, D.B. (1960): Am. J. Med. 29:832-848.

TENDENCY FOR REPETITIVE VAGAL ACTIVITY TO

SYNCHRONIZE CARDIAC PACEMAKER CELLS

M.N. Levy and T. Yang

The Mt. Sinai Medical Center, and
Case Western Reserve University
Cleveland, Ohio   44118   U.S.A.

INTRODUCTION

There are thousands of cells in the sinoatrial (S-A) node of the mammalian heart that possess the property of automaticity; i.e., each is able to generate impulses independently. The cell-to-cell communication among these automatic cells is notoriously poor. Very little information is available about the natural mechanisms that serve to coordinate the activities of the cells in the S-A node. Recent evidence (4, 7, 8, 18, 19) indicates that the "phase-dependency" of the cardiac response to vagal stimulation is one mechanism that provides some coordination of automatic cell firing; "phase-dependency" refers to the tendency for the cardiac response to a vagal action potential to depend on the timing of that action potential within the cardiac cycle.

Brown and Eccles (1) demonstrated in 1934 that the chronotropic response to a single vagal stimulus depended on the phase of the cardiac cycle at which the stimulus was given. More recent investigations have shown that the responses to repetitive vagal stimulation are also phase-dependent. If the vagus nerves are stimulated once each cardiac cycle at a predetermined phase of the cycle (so-called phase-coupled stimulation), the magnitude of the chronotropic response will depend on the timing of the stimuli within the cardiac cycle (2, 8, 12, 13). In the experiment shown in Figure 1, the cardiac segments of the transected cervical vagi of an anesthetized dog received one supramaximal pulse each cardiac cycle (12). When the stimuli (St) were delivered at the beginning of atrial depolarization (i.e., at a P-St interval of zero), the cardiac cycle length (P-P interval) increased to a value of 700 msec. The control value prior to vagal stimulation was 390 msec. As the stimuli were given at

1081

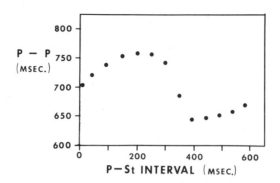

Figure 1. The effects of phase-coupled vagal stimuli on the cardiac cycle length (P-P interval) in an anesthetized dog. The chronotropic effect varied with the P-St interval, which is the time from the onset of atrial depolarization (P wave) till the beginning of the stimulus (St). (From Levy et al., 13, with permission of the American Heart Association.)

progressively later phases of the cardiac cycle, the P-P interval continued to increase. The maximum value of 760 msec was reached when the P-St interval was about 225 msec.

As the P-St intervals were increased above 225 msec (Fig. 1), the P-P intervals then progressively diminished until a minimum chronotropic response was obtained at a P-St interval of about 390 msec. When the nerves were stimulated at still later phases of the cycle, the chronotropic responses progressively increased again. Such a curve of the P-P intervals, plotted as a function of the P-St interval, constitutes a "phase-response curve" for the automatic cells in the S-A node.

The phase-dependency of the chronotropic response to repetitive vagal stimulation confers a peculiar characteristic to such stimulation; i.e., such repetitive stimuli tend to entrain the S-A nodal pacemaker cells (2, 8, 12, 13). The mechanism responsible for synchronization can be understood by analyzing the schematic phase-response curve in Figure 2. Let stimuli (St) be given to the vagus nerves at an absolutely constant frequency. The stimulator in this hypothetical experiment is independent of the cardiac activity; i.e., the stimulator is not triggered by any event in the cardiac cycle. Let the frequency of that stimulation be such that, temporarily, the prevailing frequency of the cardiac pacemaker precisely equals the stimulation frequency. Under these conditions, the P-St intervals will temporarily be constant from cycle to cycle; i.e., in Figure 2, $P_1-St_1 = P_2-St_2 = P_3-St_3$, etc. Let such a repetitive stimulus occur at point St_s (representing a stable equilibrium) in the phase-response curve in the upper part of the figure. Note that point

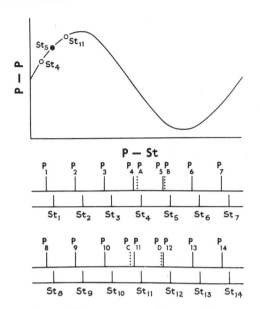

Figure 2. Schema that illustrates the synchronizing tendency of evenly spaced vagal stimuli. St_S represents a stable equilibrium point. P_A and P_C represent momentary perturbations of the otherwise constant rhythm. P_B and P_D indicate the partial restoration of the regular rhythm produced by the synchronizing effect. (From Levy et al., 13, with permission of the American Heart Association.)

St_S lies on a region of the phase-response curve that has a positive slope.

Now let a momentary perturbation occur during the fourth beat, such that the P wave begins later than expected (at P_A rather than P_4 in the lower half of Fig. 2). The next stimulus (St_4) will then occur at an earlier phase of that cycle; i.e., the P_A-St_4 interval will be reduced. St_4 will then fall on a portion of the phase-response curve at which the response will be attenuated, as may be seen in the upper half of Figure 2. Therefore, the next cardiac cycle will not be as long as the preceding cycles, i.e., P_A-P_B < P_1-P_2. If the correction is only partial in one beat, then P_B will occur later in time than would the normal P wave (P_5), had there been no perturbation. Hence, St_5 would occur somewhere between St_4 and St_S, and the next cardiac cycle would be slightly shorter than normal, but longer than P_A-P_B. Subsequent stimuli would approach the position of St_S in the phase-response curve. By a similar method of analysis, it can easily be seen that if the brief perturbation consisted of a shortening rather than a lengthening of the P-P interval (e.g., P_{10}-P_C near the bottom of the figure), the same negative feedback mechanism would operate to maintain a stable synchronization.

As a consequence of this synchronizing tendency, the heart rate tends to assume the same frequency as the vagal stimuli when the frequency of vagal stimulation is changed within a critical range. Within such a limited range of frequencies, an increase in the frequency of vagal stimulation from f_1 to f_2 will cause an equal increment in heart rate, from f_1 to f_2. This is a paradoxical response. In view of the well known fact that the vagus nerves are inhibitory to the heart, an increase in stimulation frequency would be expected to induce a reduction in heart rate!

When the vagal stimulation consists of repetitive, brief bursts of pulses, rather than single, equally spaced pulses, the tendency for the S-A nodal cells to be entrained by the vagal activity is considerably enhanced (2, 12). The spontaneous activity recorded from efferent vagal fibers to the heart consists of such repetitive bursts of action potentials, which are clustered within a discrete portion of each cardiac cycle (10, 11). Hence, the naturally occurring efferent vagal activity probably possesses this potent synchronizing effect.

Figure 3 illustrates the changes in cardiac cycle length (P-P interval) that occur when the vagi are stimulated by such repetitive bursts of pulses, and the frequency of those bursts is progressively increased (12). It is evident that the change in cycle length evoked by the vagal stimulation is not a continuous, monotonic function of the stimulation frequency. Instead, the P-P interval tracing discloses several discontinuities, such that at certain frequencies

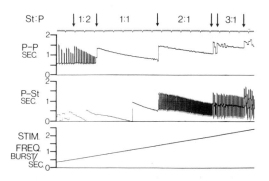

Figure 3. Changes in cardiac cycle length (P-P interval) as the frequency of vagal stimulus bursts was progressively increased in an anesthetized dog. The P-St interval denotes the time from the beginning of atrial depolarization (P wave) till the beginning of the vagal stimulus burst (St). The numbers between the arrows denote the ratio of vagal stimulus bursts to heart beats. (Modified from Levy et al., 12, with permission of the American Heart Association.)

a minute change in stimulation frequency evoked a pronounced change in cardiac cycle length. Between each of the discontinuities (indicated by the arrows), the cardiac rate displayed a paradoxical response; i.e., the cardiac cycle length decreased (rather than increased) as the frequency of stimulation was raised. The ratio (St:P) inserted between the arrows denote the ratios of the stimulus bursts (St) to the number of atrial depolarizations (P waves). As the burst frequency was increased from 0.76 to 1.33 bursts/sec. (between the arrows that signify an St:P ratio of 1:1), the P-P interval decreased from 1.32 to 0.75 sec; that is, the heart rate had increased from 0.76 to 1.33 beats/sec. This frequency was identical to the stimulation frequency, which indicates that the heart had been beating once for each vagal stimulus burst over the entire frequency range from 0.76 to 1.33 Hz. Over other frequency ranges, synchronization had also prevailed, but in other St:P ratios (Fig. 3). Note that, within a given range of synchronization, the changes in cardiac cycle length were accompanied by changes in the P-St interval. Other experiments (12), in which a phase-coupled stimulation protocol was used, have established that the changes in cycle length were indeed causally related to the associated changes in the P-St interval; that is, the chronotropic response was phase-dependent (e.g., Fig. 1).

The same type of phase-response behavior that characterizes the S-A nodal pacemaker cells is equally prominent for automatic cells in the A-V junction (18). Furthermore, the phasic effects of vagal activity on the cardiac pacemaker regions can be demonstrated reflexly. Stimulation of afferent fibers in the carotid sinus or aortic depressor nerves in dogs will evoke reflex effects on heart rate that depend on the time within the cardiac cycle that the stimuli are delivered (14, 15). Reflex vagal excitation by a neck-suction device in human subjects has also disclosed that an analogous phase-response relationship applies in man (3).

In the S-A node, the clusters of pacemaker cells are in poor communication with each other; conduction velocity is very slow from one cluster to th next (5, 9). The pronounced synchronizing tendency of repetitive vagal activity probably serves to coordinate the automaticity of such diverse cell clusters within the S-A node (9).

When bursts of pulses are delivered to the vagus nerves once each cardiac cycle, and the timing of the bursts is gradually changed on successive cardiac cycles until the entire cycle is scanned (phase-coupled stimulation), chronotropic responses similar to that shown in Figure 4 are often obtained (12, 19). Note that at a certain critical time (t_c) in the cardiac cycle, a very small change in the phase of stimulation evoked a large change in the chronotropic response (from cycle length L_1 to cycle length L_2). When such large changes in cycle length can be evoked by such a small change in the phase

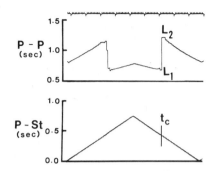

Figure 4. The effect of the timing of one burst of stimuli to the right vagus nerve each heart beat on the cardiac cycle length (P-P interval) in an anesthetized dog. The time (P-St interval) from the beginning of each atrial depolarization (P wave) to the beginning of the next vagal stimulus burst (St) was varied as an ascending and descending ramp-like function of time (lower channel). Note that at a critical time (t_c) in the cardiac cycle, a small decrement in the phase of the stimulus causes the cycle length to increase suddenly from the minimum (L_1) to the maximum (L_2) value. Downward deflections of the time marker, 10-second intervals. (Modified from Yang et al., 19.)

of the stimuli, the individual automatic cells in the S-A node must be tightly synchronized with one another, as explained below.

The marked disparity in the chronotropic responses at the times that correspond to the maximum and minimum values on the phase-response curve undoubtedly reflects pronounced phase-dependent variations in responsiveness of the pacemaker cells. The negative chronotropic effect of acetylcholine (ACh) is exerted on a pacemaker cell during its phase of slow diastolic depolarization (4, 7, 8, 17). ACh elicits a hyperpolarization and a change in the slope of the pacemaker potential. During the time that a pacemaker cell is undergoing an action potential, the chronotropic effect of ACh would be minimal. The ACh would merely tend to diminish the duration of that action potential (4), which effect would actually tend to shorten rather than lengthen the cardiac cycle.

The abrupt change in responsiveness of a pacemaker cell to ACh probably occurs appreciably earlier than the upstroke of the action potential (4, 7, 8), principally because of the substantial latent period required for ACh to alter the potassium conductance of the pacemaker cell membrane (6). The latent period for iono-phoretically applied ACh to induce hyperpolarization varied from about 50 to 150 ms in isolated S-A node preparations from rabbits (16) and kittens (7).

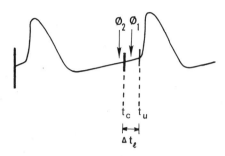

Figure 5. The postulated changes in transmembrane potential in a hypothetical pacemaker cell in the S-A node. Time t_u is the beginning of the upstroke of the action potential. Time t_c represents a critical point in the cycle at which the chronotropic responsiveness to ACh suddenly changes. The time difference between t_c and t_u is equal to the latent period (Δt_1) for ACh to cause the cell membrane to become hyperpolarized. A stimulus given at a time ($\emptyset_2$) just before t_c will evoke a maximum negative chronotropic response, whereas a stimulus given at a time ($\emptyset_1$) just after t_c will be minimally effective. (Modified from Yang et al., 19.)

Thus, the abrupt change in responsiveness of a pacemaker cell to ACh occurs during the period of phase 4 depolarization. It precedes the next expected action potential by a time interval equal to the latent period for the ACh effect on the cell membrane. If t_c represents this critical time during phase 4 depolarization, t_u represents the beginning of the upstroke of the next expected action potential, and Δt_1 is the latent period for the ACh effect, then $t_c = t_0 - \Delta t_1$ (Fig. 5). If ACh is released from the vagal nerve endings shortly before time t_c ($\emptyset_2$, Fig. 5), it will be able to hyperpolarize the cell membrane prior to the upstroke of the next expected pacemaker action potential. Hence, it will extend maximally that same pacemaker cycle (8).

If the ACh is released shortly after time t_c (point $\emptyset_1$, Fig. 5), however, it will not be able to postpone the next action potential, because the upstroke of the next action potential will be generated before the latent period for the ACh effect is terminated. The remaining ACh will hyperpolarize the cell membrane during the subsequent pacemaker cell cycle. However, the effect will be less pronounced (8), because much of the ACh will have been hydrolyzed by acetylcholinesterase during the course of the intervening action potential. Hence, the negative chronotropic effect of ACh released at time $\emptyset_1$ will be less pronounced than the influence of an equivalent amount of ACh released at time $\emptyset_2$. Times $\emptyset_1$ and $\emptyset_2$ are analogous to the phases of the cardiac cycle at which the minimum and maximum chronotropic responses are evoked (e.g., L_1 and L_2 in Fig. 4).

As stated above, when the efferent vagal activity occurs in repetitive bursts, the pacemaker cells in the S-A node tend to become synchronized with those bursts of neural activity (Fig. 3). It follows, therefore, that the pacemaker cells in the node must also be synchronized with one another. If the response of the S-A node as a whole is similar to that of a single cell (Fig. 5), it may be concluded that all of the individual pacemaker cells in the structure must be behaving almost identically.

The extent of entrainment of pacemaker cells in the S-A node is reflected by the phase-difference between the minimum and maximum responses of the phase-response curve, as illustrated in Figure 6. Panel A represents the transmembrane potentials of three hypo-thetical pacemaker cells that are entrained very tightly, whereas panel B depicts the potentials of these cells when they are more loosely synchronized. In accordance with the symbols used in Figure 5, the heavy vertical line that appears late during each phase 4 depolarization represents the critical time (t_c) at which there is a sudden change in the responsiveness of the cell membrane to ACh.

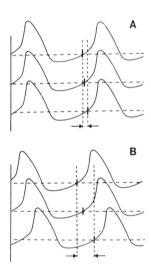

Figure 6. A diagram of two groups of action potentials in hypothetical pacemaker cells in the S-A node. In panel A, the cells are tightly synchronized, whereas in panel B, the synchronization is looser. The heavy vertical line before the upstroke of each action potential represents the critical time (t_c in Fig. 5) at which the responsiveness of the pacemaker cells to ACh suddenly changes. The dashed vertical lines mark the earliest and latest critical times, respectively, of the two groups of pacemaker cells in the S-A node. (Modified from Yang et al., 19.)

If the minimum-to-maximum phase-difference is relatively small, such as in the experiment shown in Figure 4, it may be concluded that all of the pacemaker cells must be aligned temporally such that the critical times for each cell occur within a few milliseconds of one another. This tight synchronization is represented in Figure 6A by the closeness of the dashed vertical lines, which represent the temporal limits of times t_c. Conversely, when the minimum-to-maximum phase-difference is relatively great, such as in the experiment shown in Figure 1, the temporal alignment of critical times is probably not as tight. This looser entrainment is represented by the greater horizontal distance between the vertical dashed lines in Figure 6B. The minimum-to-maximum phase-difference tends to decrease as the number of pulses in each burst of vagal stimuli is increased (19). This suggests that the greater the intensity of the repetitive bursts of vagal impulses, the greater is its tendency to synchronize the pacemaker cells within the S-A node.

Thus, the pumping action of the heart tends to synchronize the activity of the automatic cells that initiate each heart beat. As the heart ejects blood each cardiac cycle, the rise in arterial blood pressure stimulates the arterial baroreceptors. This induces a reflex burst of action potentials down the efferent vagal fibers to the heart each cardiac cycle. These pulse-synchronous bursts of action potentials, in turn, serve to synchronize the activity of the multitude of automatic cells that comprise the S-A node, the natural pacemaker site in the heart.

Acknowledgement. This work was supported by U.S.P.H.S. Grant HL 10951.

REFERENCES

1. Brown, G.L. and Eccles, J.C. (1934): J. Physiol. (London) 82:211-240.
2. Dong, E., Jr. and Reitz, B.A. (1970): Circ. Res. 27:635-646.
3. Eckberg, D.L. (1980): Circ. Res. 47:208-216.
4. Goto, J., Toyama, J. and Yamada, K. (1983): J. Electrocardiol. 16:45-52.
5. Hariman, R.J., Hoffman, R.B. and Naylor, R.E. (1980): Circ. Res. 47:775-791.
6. Hill-Smith, I. and Purves, R.D. (1978): J. Physiol. (London) 279:31-54.
7. Jalife, J. and Moe, G.K. (1979): Circ. Res. 45:595-607.
8. Jalife, J., Slenter, V.A.J., Salata, J.J. and Michaels, D.C. (1983): Circ. Res. 52:642-656.
9. James, T.N. (1973): Circ. Res. 32:307-313.
10. Katona, P.G., Poitras, J.W., Barnett, G.O. and Terry, B.S. (1970): Am. J. Physiol. 218:1030-1037.
11. Kunze, D.L. (1972): J. Physiol. (London) 222:1-15.
12. Levy, M.N., Iano, T. and Zieske, H. (1972): Circ. Res.

　　　　30:286-295.

13.　Levy, M.N., Martin, P.J., Iano, T. and Zieske, H. (1969):
　　　Circ. Res. 25:303-314.

14.　Levy, M.N. and Zieske, H. (1969):　Circ. Res. 24:303-311.

15.　Levy, M.N. and Zieske, H. (1972):　Circ. Res. 30:634-641.

16.　Osterrieder, W., Yang, Q.F. and Trautwein, W. (1981):
　　　Pflugers Arch. 389:283-291.

17.　Spear, J.F., Kronhaus, K.D., Moore, E.N. and Kline, R.P. (1979):
　　　Circ. Res. 44:75-88.

18.　Wallick, D.W., Levy, M.N., Felder, D.S. and Zieske, H. (1979):
　　　Am. J. Physiol. 237:275-281.

19.　Yang, T., Jacobstein, M.D. and Levy, M.N. (1984):　Am. J.
　　　Physiol. 246:H585-H591.

TRANSIENT INOTROPIC AND DROMOTROPIC RESPONSES

TO BRIEF VAGAL STIMULI

P. Martin

Division of Investigative Medicine
Mt. Sinai Medical Center, University Circle
Cleveland, Ohio 44106

INTRODUCTION

It is well known that vagal activity depresses cardiac chrono-tropic, dromotropic, and inotropic functions. However, much of the previous work in this area has concentrated on steady-state effects of continuous trains of vagal stimuli. This report is a brief synthesis of some of our work in this area, and will be concerned primarily with dromotropic and atrial inotropic effects.

METHODS

Surgical Procedures. Mongrel dogs (12.0 to 25.2 kg, mean = 18.8 $\pm$ 3.3) of either sex were anesthetized with pentobarbital sodium (30 mg/Kg IV) and given propranolol (1 mg/Kg). The chest was opened by transsternal incision at the 4th intercostal level, and intermittent positive-pressure ventilation was immediately instituted via a tracheal cannula. To evaluate atrial contractility, a small balloon was inserted into the right atrial appendage and connected to a strain gauge by a 1 to 2 cm length of rigid tubing (10). The balloon had an unstretched capacity of about 2 ml, and it was filled with 1 to 2 ml of saline. The balloon did not interfere with blood flow through the atrium. Bipolar electrode catheters were inserted into the left atrium to record the left atrial electrogram (P-wave), and into the right ventricle to measure the ventricular electrogram (R-wave). Both cervical vagi were ligated, and bipolar electrodes were attached to the distal ends of both nerves.

Measurement of Variables. The basic computer, instrumentation, and recording techniques have been described previously (8, 10). Briefly, the electrograms (P-wave and R-wave) and pressures were recorded on an oscillograph and also coupled to a parallel logic analog computer (EAI-580) that served to generate a vagal stimulus burst at a precise time in the cardiac cycle. The delay from the beginning of the stimulus burst to the beginning of the atrial depolarization was denoted as the P-St interval. The atrioventricular conduction time (P-R interval) and cardiac cycle duration (P-P interval) were computed on a beat-by-beat basis, and these derived variables and the P-St interval were also recorded on the oscillograph. The voltage analog of the pressure generated in the balloon in the atrial appendage was input to the analog computer. The absolute systolic and diastolic levels of each contraction were measured and subtracted on a beat-by-beat basis, and also recorded on the oscillograph. This difference, the atrial contractile amplitude (denoted as ACA), was taken as the index of atrial myocardial contractile force. The P-R, P-P, and P-St intervals and the ACA measurements were used to construct "vagal effect curves." These curves show the time course of the vagal effect on P-R conduction time and atrial contractility (8, 10). Briefly, the contractile responses to a sequence of vagal stimulus bursts, given about 2 minutes apart and over a range of P-St intervals that encompass the entire cardiac cycle, are combined on one curve. The abscissae of these curves are the times that have elapsed from the onset of the stimulus, and the ordinates are the percentage changes in contractile amplitude that occurred on each of several subsequent cardiac cycles after the stimulus.

Experiment Protocol. Brief vagal stimulus bursts were given about 2 minutes apart when the heart was in spontaneous sinus rhythm, and during right atrial pacing. The pulse width of the stimulus impulses was 1 ms, the interpulse interval was 5 ms, and the amplitude was supramaximal (10-15 v). We varied two characteristics of the vagal stimulus bursts: the number of pulses per stimulus burst (N), and the P-St interval. The right atrium was paced in some experiments by an isolated Grass S4 stimulator through two clip electrodes, one attached near the sinus node, and the other at the base of the atrial appendage. The paced heart period was set just shorter than the spontaneous P-P interval. Additional vagal effect curves were then generated with the identical stimulus protocols described above. Statistical analyses were performed by appropriate analyses of variance (ANOVA) and student's t-tests. The results were expressed as means $\pm$ S.D., and a difference between means was considered significant for $p < 0.05$.

RESULTS

Figure 1 (panel A) illustrates a typical vagal effect curve for changes of P-R conduction while the heart was paced at each of two different levels. It is seen that there is a latency of about 200 ms after the stimulus before any effect appears. The response then rapidly rises to a peak at about 450 ms, and then gradually returns toward control over the next several seconds. The sharp

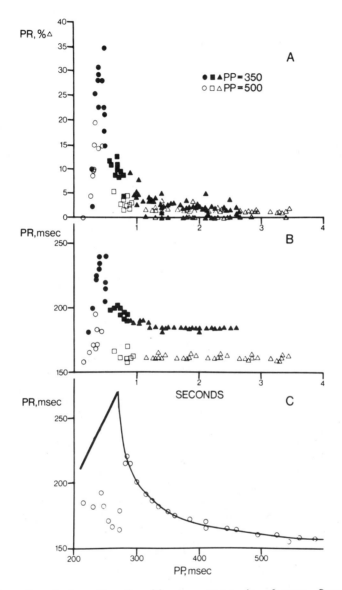

Figure 1. Panel A: Vagal effect curves (% change from control) for the same animal and stimulus parameters, but at different cardiac pacing intervals. Panel B: Same data as in panel A, but plotted as absolute values of the P-R interval. Panel C: Plot of P-R interval versus (steady-state) pacing interval in the same animal.

rise in these curves indicates that the P-R conduction response is quite sensitive to the P-St interval; e.g., one P-St may induce a peak response of 50% of maximal, but a P-St that occurs 5-10 ms later in the cardiac cycle will often induce no response at all in that next cardiac cycle, and only a low-to-moderate response on the very next cycle.

The closed symbols of Fig. 1 illustrate that the amplitude of the P-R interval response increases with paced heart rate. This is partially explained by replotting the data of panel A on an absolute scale, as shown in panel B of Fig. 1. This shows that the basal control value of the P-R interval is greater at the faster heart rate (shorter P-P interval). This is a well-described phenomenon, and the mechanism for it is based, in part, on the relative refractory properties of A-V nodal cells. That is, at shorter heart periods, there is less time for nodal cells to fully recover their excitable state; thus, a given incident excitation finds the cells relatively more depressed and conduction time is prolonged. Similarly, the A-V response to a given vagal stimulus will depend on the background (control) refractory state of the nodal cells (14). This is because the incremental delay in A-V conduction is not uniform as the membrane state becomes increasingly more refractory; rather, the incremental delay is a power function, as illustrated in panel C of Fig. 1.

For the response of A-V conduction time, the effects of a change in heart period and the simultaneous application of brief vagal stimulus do not summate linearly (there is a nonlinear interaction between these variables). That is, a sudden increase in heart period alone would be expected to decrease A-V conduction time, as shown by the steady-state results of Fig. 1-C; a brief vagal burst alone would be expected to increase the P-R interval, as shown in Fig. 1A,B. But this same vagal burst is also well known to suddenly increase the P-P interval in the unpaced heart. Thus, it may have been predicted that in the unpaced heart the P-R interval would have evoked little change in response to a brief vagal burst (the oppositely directed effects of heart period and ACh would approximately cancel). We rather found that the P-R interval was actually significantly shorter under these combined influences than when a sudden increase in heart period occurred alone. This is illustrated in Fig. 2 for one animal; the composite results were consistent with the findings in this dog. This was termed a paradoxical response in that A-V conduction in the unpaced heart was faster in the presence than in the absence of vagal activity for the transient responses (8). The reasons for this paradoxical response are complex, but two important mechanisms appear to be the shifts of pacemaker site and the alterations of atrial activation patterns. These mechanisms have been amply demonstrated, subsequently, to accompany the chronotropic responses to brief vagal activity (14).

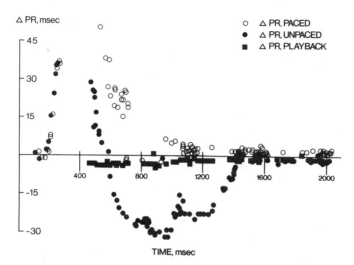

Figure 2. Vagal effect curves for P-R interval changes (PR) for paced (open symbols) and unpaced (closed symbols) heart preparations. The curve comprised of solid squares represents the AV conduction responses obtained with no vagal stimulation but by pacing the heart with the variable sequence of indentical P-P intervals (under computer control) recorded with vagal stimulation (adapted from 8).

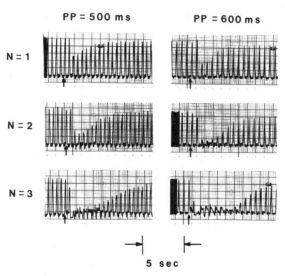

Figure 3. Oscillographic tracings of atrial balloon pressure in response to 1, 2, or 3 stimuli per vagal burst (N), given at the time marked by the arrow in each panel, and with each set of stimuli at 2 different values of cardiac pacing interval (PP) (adapted from 10).

The atrial inotropic response to brief vagal bursts is illustrated
in Fig. 3. When the heart was paced at 500 ms, a stimulus burst
of a single impulse reduced the atrial contraction amplitude (ACA)
by about half, and a burst of three impulses reduced ACA to negligible
values for 3-4 sec. An even greater effect was produced when the
atrial pacing interval was increased to 600 ms (Fig. 3). This dependence of the response amplitude on the atrial pacing interval
was consistent across all animals (10). Additional experiments
have indicated that the atrial contractile response was also dependent
on the time in the cardiac cycle (P-St interval) at which the stimulus
burst was given. When the P-St was changed by 100 ms, the inotropic
response on the next contraction changed by as much as 40% (10).

The time course of the response is illustrated by Fig. 4. As
opposed to the A-V conduction response (Fig. 2), the time course
of the inotropic response is very similar for paced- and unpaced-heart
preparations (not shown). This time course of the inotropic response
typically has a latent period of about 0.5 sec, reaches a nadir at
about 2 sec after the stimulus, and then gradually returns to control
in about 15 sec. A typical chronotropic response to the identical
vagal stimuli is shown in the upper panel of Fig. 4. Of particular
interest is that the time course of the inotropic response is nearly
identical to the time course of the slow secondary phase of the
chronotropic response. This suggests that the rate limiting steps
in these two divergent cardiac responses may be related.

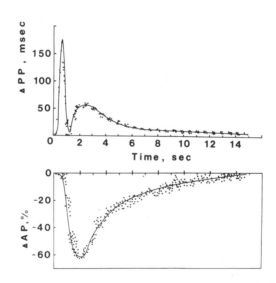

Figure 4. Vagal effect curves for cardiac interval (P-P, top panel)
in unpaced-heart preparation and atrial balloon pressure (AP, lower
panel) in the paced-heart preparation from the same animal (adapted
from 10).

The residual effects of the cardiac responses (after termination of the responses) to brief vagal stimuli are also important. Fig. 5 illustrates the results of testing the A-V conduction response to two successive, identical, brief bursts of vagal stimuli. The only variable was the time period (T) between the bursts (the P-St and P-P intervals were held constant). The amplitude of the response to the second burst (P2) fell to 37% of the amplitude of the first (P1) at a T of 3.4 sec. The amplitude of the second response then gradually increased, but did not return to the level of the first until about a minute had elapsed between the stimulus bursts. The

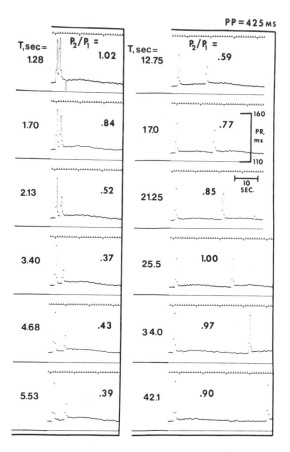

Figure 5. The A-V conduction response to pairs of brief vagal bursts, where the members of each pair were separated by a continuously increasing time interval (T). The peak response to the first burst of each pair is P1, the peak response to the second is P2, and the ratio of these (P2/P1) is indicated for each panel of the figure (adapted from 7).

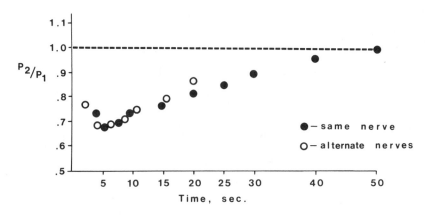

Figure 6. Composite data related to Fig. 5 where P2/P1 is plotted versus T (solid symbols). The open symbols represent a variation on the experiment where the first burst of each stimulus pair (evoking P1) was delivered to one vagus nerve, but the second stimulus burst (evoking P2) was delivered to the opposite vagus nerve (adapted from 7).

composite data (Figure 6) illustrate that a decrease of acetylcholine release with the second stimulus burst is not a likely mechanism for the phenomenon. In this experiment, one vagus nerve was used for the first response, and the other vagus was used for the second of each response pair. The same depression of the second response amplitude relative to the first (P2/P1) was found in this experiment (open symbols) as when both vagal bursts of each pair were delivered to the same nerve (closed symbols; 7). This same dependence of the response to brief vagal bursts upon the recent history of vagal activity was also found for other cardiac variables. The composite data of Fig. 7 illustrates the phenomena for atrial contractile force and heart period as well as for A-V conduction (11).

DISCUSSION

The effects of vagal stimulation on cardiac function have provided a fertile experimental area for cardiovascular physiologists for many years. However, most such studies have concentrated on the effects of continuous trains of vagal stimuli. Although transient effects were first described some time ago (1), it has been widely recognized only in more recent years that the transient effects of brief vagal stimuli may also have an important role in cardiovascular control. This recognition was fostered by at least two factors: a) the demonstration that efferent vagal activity naturally occurs in discrete bursts (4, 5), usually related to the arterial systolic pressure pulse; and b) the demonstration that brief vagal bursts

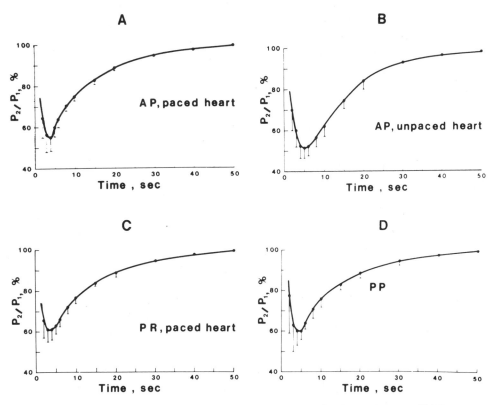

Figure 7. Composite data similar to Fig. 6, but for a different experimental series where chronotropic and inotropic responses were measured as well as the dromotropic responses (adapted from 11).

can produce profound transient alterations of cardiac function, as discussed above.

Thus, the potential exists for the autonomic nervous system to exercise cardiac feedback control on a beat-to-beat basis. For example, the arterial systolic pressure pulse, via the baroreceptor reflex, can induce a synchronous burst of vagal activity. This vagal burst, depending on the instantaneous heart period and consequent P-St interval, could reflexely alter heart period, A-V conduction, and atrial contractility over a wide range of values. This can serve as an important basis for several clinically observed arrhythmias (15). When the atrial contribution to ventricular filling is important, the transient effects of brief vagal bursts on atrial contractile function could also affect the dynamics of ventricular filling and pumping.

The transient cardiac effects of brief vagal stimuli are now realized to be much more complex than understood earlier. A prime example of this is the interaction of the direct vagal effects with other cardiac variables, e.g., simultaneous changes in heart period, as noted above. Another factor of unknown physiologic significance is the time-history dependency that is described in detail above (Figs. 5,6). Since this effect exhibits a virtually identical time course for all cardiac variables, it is tempting to ascribe the effect to a mechanism, such as receptor function, that is common to all variables. Cardiac responses gradually return (or "fade") back towards control levels during the application of continuous trains of vagal stimuli (2,3,6,12,13). However, the "fade" phenomenon does not have a similar time course for all cardiac variables, and may thus represent dissimilar (but possibly overlapping) mechanisms. Nevertheless, the fade of chronotropic and dromotropic responses do have a time course very similar to the time course of the transient time history dependency shown in Fig. 7 (3,6,12,13), and this suggests that these mechanisms may be related. It is clear however, that the interactions between various transient cardiac events can be quite complex indeed, and much work is needed to clarify these phenomena.

Acknowledgement. This work was supported by U.S.P.H.S. grants HL 22484 and HL 10951.

REFERENCES

1. Brown, G.L. and Eccles, J.C. (1934): J. Physiol. (London) 82:242-257.
2. Gertjegerdes, W., Ravens, U. and Ziegler, A. (1979): J. Cardiovasc. Pharmacol. 1:235-243.
3. Jalife, J., Hamilton, A.J. and Moe, G.K. (1980): Am. J. Physiol. 238:H439-H448.
4. Jewett, D.L. (1964): J. Physiol. (London) 175:321-357.
5. Katona, P.G., Poitras, J.W., Barnett, G.O. and Terry, B.S. (1970): Am. J. Physiol. 218:1030-1037.
6. Loeb, J.M., Dalton, D.P. and Moran, J.M. (1981): Am. J. Physiol. 241:H684-H690.
7. Martin, P.J. (1976): Circ. Res. 38:448-453.
8. Martin, P. (1977): Circ. Res. 40:81-89.
9. Martin, P. (1977): Circ. Res. 41:593-599.
10. Martin, P. (1980): Am. J. Physiol. 239:H333-H341.
11. Martin, P. (1980): Am. J. Physiol. 239:H494-H500.
12. Martin, P., Levy, M.N. and Matsuda, Y. (1982): Am. J. Physiol. 243:H219-H225.
13. Martin, P. (1983): Am. J. Physiol. 245:H584-H591.
14. Randall, W.C., Rinkema, L.E., Jones, S.B., Moran, J.F. and Brynjolfsson, G. (1982): Am. J. Physiol. 242:H98-H106.
15. Spear, J.F., Kronhaus, K.D., Moore, E.N. and Kline, R.P. (1979): Circ. Res. 44:75-88.

BRAIN ACETYLCHOLINE IN HYPERTENSION AND BEHAVIOR: STUDIES USING N-(4-DIETHYLAMINO-2-BUTYNYL)-SUCCINIMIDE

H.E. Brezenoff, N. Hymowitz*, W.M. Coram and R. Giuliano

Department of Pharmacology
and
*Department of Psychiatry
University of Medicine and Dentistry of New Jersey
100 Bergen Street
Newark, NJ 07103 U.S.A.

INTRODUCTION

Acetylcholine (ACh) in brain is thought to play a role in a large array of centrally mediated physiological functions. We have been examining central cholinergic involvement in two processes: regulation of blood pressure and schedule-controlled behavior.

Blood pressure. Administration of acetylcholinesterase (AChE) inhibitors in the rat evokes a centrally mediated increase in blood pressure. The pressor response is accompanied by an increase in sympathetic nerve activity and is abolished by atropine, but not by methylatropine; by transection of the spinal cord; by systemic administration of alpha-adrenergic blocking agents; and by depletion of brain ACh following injection of hemicholinium-3 (HC-3) into the cerebral ventricles [see (3) for review]. An increase in blood pressure also occurs in humans following systemic injection of physostigmine (1, 10, 12, 14) or arecoline (13).

Blockade of central muscarinic receptors with atropine, or depletion of brain ACh, does not reduce blood pressure in normotensive rats. This led us to speculate that the central cholinergic system mediating a rise in blood pressure normally was quiescent, causing a hypertensive effect only when specifically activated. This hypothesis is supported by the finding that both atropine (5) and centrally administered HC-3 (2, 4) reduce blood pressure in the spontaneously hypertensive rat.

1101

Schedule-controlled behavior. In addition to their effects on blood pressure, AChE inhibitors consistently suppress rates of schedule-controlled behavior in a variety of animal species (6, 16-18). Although one would expect cholinergic antagonists to exert opposing effects, both atropine and scopolamine also reduce response rates (6, 11, 16-18). In spite of its own inhibitory effects, atropine has been reported to antagonize to varying degrees the suppressant effects of physostigmine.

N-(4-diethylamino-2-butynyl)-succinimide, or DKJ-21, is reported to be a selective antagonist of central, as opposed to peripheral, muscarinic receptors (7, 8). Thus, we have available antimuscarinic drugs with actions localized to the periphery (methylatropine), to the CNS (DKJ-21) or to both sites (atropine). By comparing the effects of these agents it should be possible to distinguish between central and peripheral antimuscarinic effects. The studies to be reported herein describe the cardiovascular and behavioral effects of DKJ-21, and its antagonism of the effects of physostigmine on these systems.

METHODS

Behavior. Six experimentally naive male Sprague-Dawley rats initially weighing 325-375 g were maintained at 80% of initial free-feeding weight. They were housed individually with water freely available in the home cage unless specified otherwise.

The rats were trained to press a lever under a fixed-interval 50-second schedule of food reinforcement. Under this paradigm, a food pellet does not become available until 50 seconds after the previous food pellet has been obtained. After that time, a single press of the lever will release the pellet into the food tray. This results in a "scalloped" pattern of responding, with low rates of responding during the early portion of the interval and higher rates toward the end. Experiments were performed in a Grason-Stadler sound attenuating operant rat chamber. Programming was accomplished by electromechanical equipment, and responses were monitored by counters and a cumulative recorder. Noyes food pellets (0.045 g) served as reinforcers. Each session lasted 50 min.

Physostigmine was injected 5 minutes prior to the beginning of the experiment. The antagonists were injected 15 minutes prior to placing the animals in the chambers (10 minutes prior to injection of physostigmine when antagonism was being studied).

Drugs were dissolved in saline and were administered i.p. in a volume of 0.1 ml/100 g body weight. All doses refer to the salt.

Hypertension. Spontaneously hypertensive rats (SHR), 5 to 36 weeks of age were used in this study. For direct recording of blood pressure in the unanesthetized rat, the animals were first anesthetized with ether, and the carotid artery or the caudal vein was cannulated at the same time, and the rats were allowed 24 hours to recover prior to the experiment.

In several experiments, blood pressure was recorded noninvasively via tail-cuff occlusion (Narco Biosystems). The animals were placed in a warm environment (32°C) for one hour prior to recording. The first one or two days of recording were not counted, to allow acclimation to the experimental conditions.

RESULTS

Behavior. All of the animals acquired stable response rates which were maintained throughout the study. The animals revealed characteristic "scalloped" response rates, with low rates in the initial portion of the interval and high rates towards the end.

Table 1

Effect of Muscarinic Blockers Alone and in Combination
with Physostigmine on Operant Responding

ATROPINE		METHYLATROPINE		DKJ-21	
Dose[1]	% of baseline	Dose	% of baseline	Dose	% of baseline
		- Effect of Blockers Alone -			
0.1	92	0.05	66	5	103
0.5	68	0.1	45	10	104
1.0	48	0.5	24	20	97
2.0	39			40	94
4.0	33			100	90
		- Effect of Physostigmine Alone (mean $\pm$ SEM)[2] -			
44 $\pm$ 7		34 $\pm$ 11		48 $\pm$ 7	
		- Effect of Physostigmine After Blockers -			
0.1	51			40	84
0.5	53	0.5	21	80-100	100

[1]Doses are given as mg/kg, i.p. [2]Reproducible responses to physostigmine were obtained at weekly intervals, in individual animals, with doses of 250-600 μg/kg.

Table 1 shows the effects of the antimuscarinic drugs as a percent of the previous predrug session response rate. Atropine (0.5-4.0 mg/kg) and methylatropine (0.05-0.5 mg/kg) decreased responding in a dose-dependent manner. Rate-dependent effects generally were not observed; response rates in each portion of the interval decreasing to a similar extent.

Unchewed and partly chewed food pellets were found in the excreta tray following administration of atropine and methylatropine. Since both drugs produce xerostomia, we speculated that dry mouth may have caused difficulty in eating. This in turn might result in a motivational decrease in lever-pressing. To test this hypothesis, we examined the effects on responding produced by 24-hour water deprivation. This procedure also reduced response rates and caused numerous food pellets to remain uneaten.

Physostigmine also reduced response rates in each portion of the 50-sec interval (Table 1). Threshold effects initially were observed at doses of 50-100 µg/kg. Considerable variability was noted and tolerance generally developed to repeated administration, especially at the lower doses. The dose and/or the interval between doses was increased until reproducible effects were obtained.

To test antagonism of the rate-suppressing effects, doses of physostigmine were selected which, upon weekly injection, consistently produced submaximal inhibition. These doses ranged between 250 and 600 µg/kg in the different subjects. The degree of suppression produced by physostigmine did not decrease in the presence of atropine or methylatropine (Table 1). The combination of methylatropine and physostigmine generally reduced response rates more than either compound alone.

In contrast to atropine and methylatropine, DKJ-21 (40-100 mg/kg) antagonized the rate-reducing effects of physostigmine in each animal (Table 1). Antagonism of physostigmine was achieved with doses of DKJ-21 which had little or not effect on response rates when administered alone.

Hypertension. Control systolic blood pressure in normotensive WKY controls and in SHR animals averaged 119 ± 4 and 201 ± 2 mm Hg, respectively. Intravenous injection of DKJ-21, in doses up to 100 mg/kg, exerted no effect on blood pressure in WKY control rats. In contrast, doses between 6.25 and 25 mg/kg produced variable, but dose-related, decreases in SHR animals (Table 2). The variability was due primarily to two factors: First, about one-third of the animals either failed to respond, or responded with less than a 20 mm Hg fall in pressure, at any dose up to 100 mg/kg. In contrast, other SHR animals showed up to a 75 mm Hg fall at these same doses (Table 2). A second cause of the variability was the presence of undulations in blood pressure, which occurred in many of the animals and which could be seen under continuous direct recording. When monitored by tail-cuff occlusion, blood pressure at the 15 minute recordings varied by as much as 20 mm Hg. In contrast, the pressures following administration of saline were stable within a few mm Hg.

An unexpected finding was that the pressures still were reduced 24 hr after injection of DKJ-21. Indeed, in some rats, DKJ-21 exerted a greater hypotensive effect the day following the injection. Blood pressure returned to predrug levels by the third day.

To test whether the antihypertensive doses of DKJ-21 were acting at peripheral sites to reduce blood pressure, the following experiments were performed. Blood pressure changes in SHR were monitored in response to control i.v. injections of noradrenaline (0.1 μg/kg), angiotensin I (0.1 μg/kg), dimethylphenylpiperazinium (DMPP) (200 μg/kg), and ACh (0.2 μg/kg). These injections were followed by 25 mg/kg of DKJ-21. Thirty min later, the autonomic challenges were repeated. In no instance did DKJ-21 affect the magnitude or the duration of the vascular responses to the test compounds.

Table 2

Maximum Decreases in Systolic Blood Pressure
in SHR Animals After Injection of DKJ-21

	6.25 mg/kg[1]		12.5 mg/kg		25.0 mg/kg	
	Peak[2]	Time[3]	Peak	Time	Peak	Time
	37	90	52	90	75	90
	27	135	50	45	50	15
	27	135	30	135	50	90
					45	150
	20	45	22	60	30	75
	10	30	20	75		
	5	30	18	90	10	150
	----	----	----	----	----	----
$\overline{X}$	21	78	32	83	43	88
SEM	5	20	6	13	9	22

[1]Doses are given in mg/kg, i.v. [2]Maximum fall in mm Hg. [3]Time, in minutes, of maximum fall

To determine whether DKJ-21 blocked central muscarinic receptors involved in cardiovascular function, the following experiment was performed. Physostigmine was injected i.v. in a dose of 75 μg/kg, which caused a rise in blood pressure averaging 60 $\pm$ 5 mm Hg. Prior administration of DKJ-21 (25 mg/kg) resulted in a 60% reduction in the pressor response to physostigmine. The pressor response still was reduced by 54% when tested 24 hr after DKJ-21 injection.

DISCUSSION

The role of brain ACh in schedule-controlled behavior is confounded by the fact that muscarinic agonists, cholinesterase inhibitors, and muscarinic antagonists, all suppress response rates (6, 11, 16-18). Although low doses of atropine have been reported to increase responding, the antagonism of the rate-suppressant effect of physostigmine is inconsistent and incomplete.

In addition to its central effects, atropine is a strong antagonist at peripheral muscarinic receptors. In fact, it is a more effective blocker at peripheral than at central sites (7, 8, 15). Therefore, it is possible that the multiplicity of active sites may interact to alter an apparently unrelated response.

DKJ-21 is one of a series of compounds designed to selectively block central muscarinic receptors (8, 14). Although not very potent, the ratio of central to peripheral receptor blockade is quite high. DKJ-21 can inhibit completely the tremors induced by oxotremorine at doses which cause neither mydriasis nor salivation (15). It also is an effective antagonist at ganglionic muscarinic receptors (7), which may resemble muscarinic receptors in the central nervous system (9).

The importance of the distinction between central and peripheral receptors is striking in our behavioral experiments. It is apparent that the suppression of lever-pressing behavior caused by atropine is due, at lease in large measure, to the production of dry mouth, making it difficult for the animals to eat. The similar suppression of response-rate and appearance of uneaten food pellets following water deprivation supports that conclusion.

In contrast to atropine, DKJ-21 does not inhibit salivation or block contractions of the ileum (15). Likewise, in opposition to atropine, it did not interfere with lever-pressing behavior, and it completely prevented the rate-suppressant effects of physostigmine. These observations suggest that physostigmine acts centrally to suppress schedule-controlled behavior, and that blockade of central muscarinic receptors can prevent that action. Unfortunately, the peripheral effects of atropine may be sufficient to mask the central antagonistic effect.

The effect of DKJ-21 on blood pressure in the SHR is consistent with previous reports from this laboratory. That is, blockade of central muscarinic receptors with atropine (5), and depletion of brain ACh with HC-3 (2), both reduce blood pressure in this model of hypertension. DKJ-21 does not appear to exert significant peripheral vascular effects. At antihypertensive doses, it did not block the vasodepressor effect of acetylcholine, or the pressor responses to noradrenaline, angiotensin I or DMPP.

Although DKJ-21 did not prevent the vasodepressor effect of ACh, it did inhibit the pressor response to physostigmine. This suggests that DKJ-21 blocks those central muscarinic receptors mediating a rise in blood pressure. The variability in the antihypertensive response to DKJ-21 can not be explained at this time.

In conclusion, these experiments indicate that DKJ-21 selectively inhibits central muscarinic mechanisms involved in both behavior and cardiovascular control. This compound could prove to be an important adjunct in studying central muscarinic mechanisms where the peripheral effects of atropine may be an interfering factor. In addition, the antihypertensive effect of DKJ-21 in SHR suggests a potential new direction for antihypertensive therapy.

Acknowledgement. This work was supported in part by USAMRDC contract No. DAMD17-82-C-2172.

REFERENCES

1. Aquilonius, S.M. and Sjostrom, R. (1971):   Life Sci. $\underline{10}$:405-414.
2. Brezenoff, H.E. and Caputi, A.P. (1980):   Life Sci. $\underline{26}$:1037-1045.
3. Brezenoff, H.E. and Giuliano, R. (1982):   Ann. Rev. Pharmacol. Toxicol. $\underline{22}$:341-381.
4. Bucccafusco, J.J. and Spector, A. (1980):   J. Cardiovasc. Pharmacol. $\underline{2}$:347-355.
5. Caputi, A.P., Camilleri, B.H. and Brezenoff, H.E. (1980):   Eur. J. Pharmacol. $\underline{66}$:103-109.
6. Chait, L.D. and Balster, R.L. (1979):   Pharmacol. Biochem. Behav. $\underline{11}$:37-42.
7. Dahlbom, R., Karlén, B., George, R. and Jenden, D.J. (1966):   Life Sci. $\underline{4}$:431-442.
8. Dahlbom, R., Karlén, B., George, R. and Jenden, D.J. (1966):   J. Med. Chem. $\underline{9}$:843-846.
9. Fisher, A., Weinstock, S., Gitter, S. and Cohen, S. (1976):   Eur. J. Pharmacol. $\underline{37}$:329-338.
10. Janowski, D.S., Risch, C., Huey, L., Judd, L.L. and Rausch, J. (1983):   Psychopharmacol. Bul. $\underline{19}$:675-681.
11. Longo, V.G. (1966):   Pharmacol. Rev. $\underline{18}$:965-996.

12. Nattel, S., Bayne, L. and Ruedy, J. (1979): Clin. Pharmacol. Ther. 25:96-102.
13. Nutt, J., Rosin, A. and Chase, T.A. (1978): Neurology 28:1061-1064.
14. Pandit, U.A., Kothary, S.P., Samara, S.K., Domino, E.F. and Pandit, S.K. (1983): Anesth. Analg. 62:679-685.
15. Ringdahl, B. and Jenden, D.J. (1983): Life Sci. 32:2401-2413.
16. Sanger, D.J. and Blackman, D.E. (1976): Pharmacol. Biochem. Behav. 4:73-83.
17. Vaillant, G.E. (1967): J. Pharmacol. Exp. Ther. 157:636-648.
18. Wenger, G.R. (1979): J. Pharmacol. Exp. Ther. 209:137-143.

INHIBITION OF BRAIN ACETYLCHOLINE BIOSYNTHESIS BY

CLONIDINE AND METHYLDOPA: RELEVANCE TO HYPERTENSIVE DISEASE

J. Buccafusco

Department of Pharmacology and Toxicology
and
Department of Psychiatry
Medical College of Georgia
and
Veterans Administration Medical Center
Augusta, Georgia U.S.A.

INTRODUCTION

Clonidine and methyldopa are antihypertensive drugs employed widely throughout the world. It is generally agreed that these agents owe their antihypertensive properties to an action within the central nervous system which results in a reduction in central sympathetic drive and a fall in arterial pressure. In this regard, clonidine is approximately 1000 fold more potent than methyldopa. Both drugs, however, can stimulate central alpha-adrenergic receptors, clonidine directly, and methyldopa, through an active metabolite (methylnorepinephrine). It is precisely this property, central alpha-adrenergic receptor stimulation, which accounts for their ability to lower arterial pressure in various animal models.

The preponderance of recent evidence, however, indicates that these alpha receptors are not located on catecholaminergic neurons. For example, the antihypertensive action of clonidine is not affected by drug treatments which cause severe depletion of brain catecholamines or which destroy adrenergic nerve endings (16, 31, 34). Clonidine also is known to produce many other centrally mediated pharmacological effects. As with its antihypertensive action, the ability of clonidine to: inhibit efferent sympathetic nerve activity (18, 25); enhance reflex vagal bradycardia (24); produce antinociception (30); evoke a centrally mediated mydriatic effect (26) and inhibit efferent

1109

phrenic nerve activity (33); have all been reported to occur independently of central noradrenergic neurons. Furthermore, it has been known for several years that while methyldopa requires functioning brain catecholaminergic neurons to exert its antihypertensive effects (presumably the support is biosynthesis to an active metabolite), marked depletion of brain catecholamines does not alter this action (20).

These findings imply that the central adrenergic receptors which mediate the antihypertensive actions of clonidine and methyldopa are located on non-catecholaminergic neurons. One alternate hypothesis would be that clonidine's antihypertensive action is related to its ability to inhibit the synthesis or release of brain acetylcholine (ACh). It has been recognized for several years that clonidine can produce a marked inhibition of ACh release from parasympathetic nerves in several tissues (14, 15, 17, 23) and from preganglionic sympathetic neurons (3, 7). Direct evidence for such an inhibitory action of clonidine on central cholinergic neurons was first provided by this laboratory in 1980 (9).

The purpose of this study was to determine whether clonidine and methyldopa, two centrally-acting antihypertensive drugs with different chemical structures and neuronal metabolism, could inhibit the biosynthesis of regional brain ACh, at doses which elicit a significant antihypertensive response in an animal model of human essential hypertension, the spontaneously hypertensive rat (SHR).

MATERIALS AND METHODS

Surgical procedures and drug administration. Male, SHR 15-18 weeks of age (obtained from Taconic Farms, Inc.) were anesthetized with methohexital, for implanting chronic intracerebroventricular (i.c.v.) cannula guides (10), and for implanting catheters in the lower abdominal aorta and jugular vein (12). At the time of the experiment (allowing one week recovery from the brain surgery) arterial blood pressure was recorded directly from the exteriorized arterial line in rats freely-moving in their home cage. SHR were infused with clonidine, methyldopa or saline over 3-5 minutes through the exteriorized venous line. Before drug administration or at the time of maximal arterial pressure decrease, 20 μCi of [^{3}H]-choline (80 Ci/ mmole) were injected $\underline{via}$ the i.c.v. route to label endogenous stores of brain choline metabolites (10). Rats were killed at several time intervals after isotope injection by a beam of microwave irradiation (2450 MHz, 3.8 kW for 2.2-2.3 seconds) focused on the skull.

Estimation of brain ACh turnover rates. Choline metabolites were extracted from regionally dissected brain tissue using formic acid: acetone. [^{3}H]-ACh was isolated by thin layer chromatography, and endogenous ACh levels measured by radioimmunoassay (32). [^{3}H]-ACh levels are expressed as a percent of the total dpm of [^{3}H] extracted.

Endogenous ACh levels are expressed as nmoles per gram tissue, wet weight. ACh specific activity is the ratio of [^{3}H]-ACh:endogenous ACh levels. In the methyldopa experiments the fractional rate constant (k) of ACh (fractional turnover rate) was obtained by decomposing the curve of the semilog plot of specific activity as a function of time into two single exponentials. A tangent to the descending portion of the curve was drawn and the values for the early ascending portion of the curve subtracted. These differences were replotted, and a new curve was obtained by linear regression (r > 0.90 in all cases). The resultant T 1/2 gave k (min^{-1}) by the equation 1/k = 1.44 T 1/2. The turnover rate (nmoles/g/min) of ACh is the product of k and the ACh content (10).

Values are presented as means $^+$ S.E.M. The difference between the means of two groups was tested using Student's t-test for unpaired data, and was considered significant when p < 0.05 (two tailed analysis).

RESULTS

Pulse i.c.v. injection of [^{3}H]-choline rapidly labelled stores of endogenous choline. Also, measurement of [^{3}H]-ACh formation at an early time interval (2 min) after injection yielded a good estimate of relative cholinergic activity, providing steady-state levels of endogenous ACh were maintained throughout the experiment (10).

Clonidine does not produce changes in regional brain ACh levels in single doses up to 300 μg/kg (9, 11). Intravenous injection of clonidine (30 μg/kg) did however produce a significant reduction in arterial pressure and heart rate by 100 min in unanesthetized SHR (Table 1). This proved to be the time of maximal blood pressure reduction for this dose of clonidine. Figure 1 illustrates the effect of clonidine on the formation of [^{3}H]-ACh following i.c.v. injection of [^{3}H]-choline in SHR. Clonidine significantly inhibited the formation of [^{3}H]-ACh (up to 45%) in all brain regions (excepting striatum) with the greatest effect occurring in the hypothalamus and lower brainstem.

Like clonidine, intravenous injection of methyldopa produced a significant reduction in mean arterial pressure (Fig. 2). In this case the maximal antihypertensive response to methyldopa (200 mg/kg) was observed at about 2 hr following injection. In view of the marked effect of clonidine in hypothalamic and lower brainstem regions, these analogous areas also were investigated in methyldopa experiments. Since knowledge of the effects of methyldopa on ACh levels was not available, a more complete turnover analysis was performed.

Figure 3 illustrates the profile of ACh specific activity time curves for rostral and caudal hypothalamic segments. A similar

Table 1

Change in Blood Pressure (BP) and Heart Rate (HR) in SHR 100 min
Following Intravenous Injection of Clonidine (30 µg/kg)

Preclonidine BP (mmHg)		Change in BP (mmHg)
Systolic	209 ± 11	-27 ± 4
Diastolic	132 ± 13	-22 ± 7

Preclonidine HR		Change in HR
Beats/Min	387 ± 15	-32 ± 11

Each value represents the mean ± S.E.M. of 6 experiments.

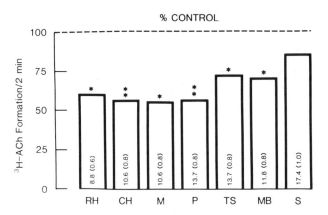

Figure 1. [³H]-ACh formation expressed as a percent of total extractable [³H] from several brain regions of SHR at 2 min following
pulse tracer injection of [³H]choline (20 µCi, i.c.v.) in control
animals, and in rats pretreated 100 min earlier with clonidine (30
µg/kg, i.v.). Clonidine data are expressed as percent of control
[³H]-ACh formation with the actual mean control values, % dpm [³H]-ACh
± (S.E.M.), shown within the bars. RH = rostral hypothalamus, CH
= caudal hypothalamus, M = medulla oblongata, P = pons, TS = thalamus-
septum, MB = midbrain, S = striatum, N = 5-8, * = p < 0.05, ** = p
< 0.01 compared to control means.

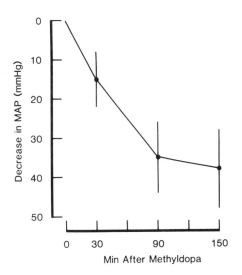

Figure 2.   Effect of i.v. injection of methyldopa (200 mg/kg) on
mean arterial pressure (MAP) of freely-moving SHR. Each point represents the mean ± S.E.M. of 5 experiments.

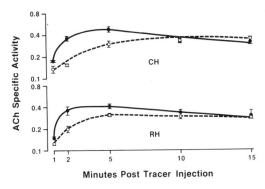

Figure 3.   Specific activity of brain ACh as the ratio of % [³H]-ACh
formed: endogenous ACh (nmoles/g) in control (filled circles) and
in SHR treated 2 hr earlier with methyldopa (200 mg/kg, i.v., open
circles).   Each point represents the mean ± S.E.M. (vertical lines)
of 4-8 experiments.   RH = rostral hypothalamus, CH = caudal hypothalamus.   The brain was labeled by a 10 sec injection of 20 μCi
of [³H]-choline, i.c.v.

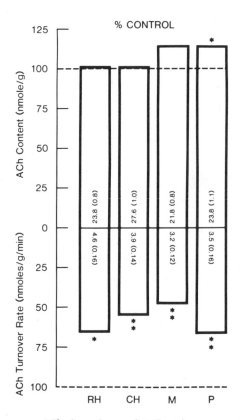

Figure 4. Endogenous ACh levels and ACh turnover rates from 4 brain regions of control SHR and from SHR pretreated 2 hr earlier with 200 mg/kg of methyldopa, i.v. Methyldopa data is expressed as percent of control ACh content or ACh turnover rate with the actual mean control values ± (S.E.M.) shown within the bars. Each value represents the mean of 8 experiments. * = p < 0.05, ** = p < 0.01 compared to control means. See Figure 1 for abbreviations.

analysis was employed for pons and medulla oblongata. Inspection of the curves reveals a marked inhibition of the initial slopes in methyldopa treated SHR. Calculation of relative turnover rates revealed a 34-54% reduction produced by methyldopa, with the greatest effect occurring in the medulla oblongata and the caudal hypothalamus (Fig. 4).

Since both clonidine and methyldopa reduced brain ACh biosynthesis in SHR, experiments were performed to determine whether ACh biosynthesis was enhanced in SHR as compared with normotensive Wistar Kyoto (WKY) control animals. Preliminary results for 4 brain regions are illustrated in Table 2. Although ACh levels were not significantly

Table 2

Regional Brain Acetylcholine Specific Activity Following
Central Injection of [^{3}H]-Choline in SHR and WKY

Brain Region	Strain	Acetylcholine Specific Activity (% DPM [^{3}H]-Acetylcholine Formed/2 Min/nmole/G)
Cerebral Cortex	WKY	0.42 ± 0.03
	SHR	0.32 ± 0.05
Hippocampus	WKY	0.35 ± 0.04
	SHR	0.32 ± 0.03
Rostral Hypothalamus	WKY	0.16 ± 0.02
	SHR	0.25 ± 0.02*
Caudal Hypothalamus	WKY	0.16 ± 0.01
	SHR	0.24 ± 0.01*

20 μCi of [^{3}H]-choline (80 Ci/mmole) were injected i.c.v., and animals subjected to focused microwave irradiation to the head 2 min later. *p < 0.01; each value is mean ± S.E.M. of 7 experiments.

altered between the two strains, SHR revealed a higher (up to 56%) synthesis rate for [^{3}H]-ACh compared to WKY in the hypothalamus, but not in cerebral cortex or hippocampus.

DISCUSSION

It is clear from the many studies from several laboratories, as recently reviewed (5), that brain ACh plays a prominent role in the central control of arterial pressure. Pharmacologic activation of central cholinergic neurons in many vertebrate species (5) and in man (1, 22, 28, 29) produces a hypertensive response. More direct evidence for the role of central cholinergic neurons in hypertension has been obtained for experimental hypertension, particularly genetic hypertension. For example, interference with central cholinergic transmission elicited through depletion of endogenous stores of the neurotransmitter (4, 8) or by blocking central muscarinic receptors (13) in SHR, leads to a marked reduction in blood pressure. Furthermore indirect evidence of enhanced brain cholinergic neuronal activity in SHR as compared to normotensive control animals has been obtained (8, 27). Also, enzyme markers for brain cholinergic neurons, acetylcholinesterase and choline acetyltransferase, and brain muscarinic receptor binding have been reported to be altered in SHR (2, 19, 21, 35).

The findings of our preliminary studies of enhanced hypothalamic cholinergic activity in hypertensive, compared with normotensive rats is consistent with the hypothesis of a central cholinergic neuronal involvement in at least some forms of hypertension. Cholinergic neurons within the hypothalamus are known to participate in cardiovascular regulation (5, 6), and a powerful cholinergic pressor system is known to exist within the posterior hypothalamic nucleus (6).

Clonidine and methyldopa are centrally-acting antihypertensive drugs with similar mechanisms of action. However, there is a 1000 fold difference between their clinically effective antihypertensive doses. The present study indicates that a similar degree of central cholinergic inhibition is produced by clinically relevant doses of both agents. It should be pointed out that the inhibitory effects of clonidine on ACh biosynthesis is mediated through stimulation of central α-adrenergic receptors (9, 11). This would be a necessary requirement for any model of the mechanism of clonidine's antihypertensive response. In fact, it has been recognized for several years that clonidine can produce marked inhibition of ACh release from parasympathetic nerves in several peripheral tissues (14, 15, 17, 23) and from preganglionic sympathetic neurons (3, 7). The possibility that this interaction also occurs within the CNS has been suggested by this study. The fact that methyldopa must first undergo metabolism within brain catecholaminergic neurons to an active metabolite, methylnorepinephrine, suggests that these neurons directly innervate or are closely juxtaposed to cholinergic neurons involved in cardiovascular regulation.

These findings indicate that experimental hypertension is associated with enhanced cholinergic activity in certain brain regions, perhaps localized to the hypothalamus or lower brainstem. The ability of clonidine and methyldopa to reduce this exaggerated cholinergic activity in SHR to levels observed for the normotensive animal suggests that these agents, at clinically relevant doses, produce their antihypertensive effects through a central "anticholinergic" action. The fact that hypertension is produced in man following chemical stimulation of brain cholinergic receptors may point to similar conclusions concerning some forms of hypertensive disease.

Acknowledgement. The author would like to acknowledge the excellent technical assistance provided by Ms. Laura F. Crouch. This work was supported by NIH grant #HL30046 and the Veterans Administration.

REFERENCES

1. Aquilonius, S.M. and Sjöstrom, R. (1971): Life Sci. 10:405-414.

2. Bagjar, J., Hrdina, V. and Golda, V. (1979): In Progress in Brain Research, (ed) S. Tuček, Vol. 49, Elsevier Pub. Co., New York, p. 471.
3. Bently, G.A. and Li, D.M.F. (1968): Eur. J. Pharmacol. 4: 124-134.
4. Brezenoff, H.E. and Caputi, A.P. (1980): Life Sci. 26:1037-1045.
5. Brezenoff, H.E. and Giuliano, R. (1982): Ann. Rev. Pharmacol. Toxicol. 22:341-381.
6. Buccafusco, J.J. and Brezenoff, H.E. (1979): Brain Res. 165:295-310.
7. Buccafusco, J.J. and Spector, S. (1980): Experientia 36:671-672.
8. Buccafusco, J.J. and Spector, S. (1980): J. Cardiovascular Pharmacol. 2:347-355.
9. Buccafusco, J.J., Finberg, J.P.M. and Spector, S. (1980): J. Pharmacol. Exp. Ther. 212:58-65.
10. Buccafusco, J.J. (1982): Biochem. Pharmacol. 31:1599-1607.
11. Buccafusco, J.J. (1982): J. Pharmacol. Exp. Ther. 222:595-599.
12. Buccafusco, J.J. (1983): Pharmacol. Biochem. Behav. 18:209-215.
13. Caputi, A.P., Camilliere, B.H. and Brezenoff, H.E. (1980): Eur. J. Pharmacol. 66:103-109.
14. Deck, R., Oberdorf, A. and Kroneberg, G. (1971): Arzneim. Forsch. 21:1580-1584.
15. Del Tacca, M., Soldani, G., Bernardini, C., Martinotti, E. and Impicciatore, M. (1982): Eur. J. Pharmacol. 81:255-261.
16. Finch, L., Buckingham, R.E., Moore, R.A. and Bucher, T.J. (1975): J. Pharm. Pharmacol. 27:181-186.
17. Green, J.G., Wilson, H. and Yates, M.S. (1979): Eur. J. Pharmacol. 53:297-300.
18. Haeusler, G. (1974): Naunyn Schmeidebergs Arch. Pharmacol. 286:97-111.
19. Helke, C., Muth, E.A. and Jacobowitz, D.M. (1980): Brain Res. 183:425-436.
20. Henning, M. and Rubenson, A. (1971): J. Pharm. Pharmacol. 23:407-411.
21. Hershkowitz, M., Eliash, S. and Cohen, S. (1983): Eur. J. Pharmacol. 86:229-236.
22. Janowsky, D.S., Risch, S.C., Judd, L.L., Huey, L.Y. and Parker, D.C. (1981): Psychopharmacol. Bull. 3:129-132.
23. Kaess, H. and von Mikulicz-Radecki, J. (1971): Eur. J. Clin. Pharmacol. 3:97-101
24. Kobinger, W. and Pichler, L. (1975): Eur. J. Pharmacol. 30: 56-62.
25. Kobinger, W. and Pichler, L. (1976): Eur. J. Pharmacol. 40: 311-320.
26. Koss, M.C. and Christensen, H.D. (1979): Naunyn Schmeidebergs Arch. Pharmacol. 307:45-50.
27. Kubo, T. and Tatsumi, M. (1979): Naunyn Schmeidebergs Arch. Pharmacol. 306:81-83.

28. Nattel, S., Bayne, L. and Ruedy, J. (1979): Clin. Pharmacol. Ther. 25:96-102.

29. Nutt, J.G., Rosin, A. and Chase, T.N. (1978): Neurology 28:1061-1064.

30. Paalzow, G. and Paalzow, L. (1976): Naunyn Schmiedebergs Arch. Pharmacol. 292:119-126.

31. Reynoldson, J.A., Head, G.A. and Korner, P.I. (1979): Eur. J. Pharmacol. 55:257-654.

32. Spector, S., Felix, A., Semenuk, G. and Finberg, J.P.M. (1978): J. Neurochem. 30:685-689.

33. Von Tauberger, G., Thoneick, H.-U. and Dulme, H.-J. (1978): Arzneim. Forsch. 28:651-654.

34. Warnke, E. and Hoefke, W. (1977): Arnzeim. Forsch. 27(12):2311-2313.

35. Yamori, Y. (1975): In Regulation of Blood Pressure by the Centra Nervous System, (eds) G. Onesti, M. Fernandes, K.E. Kim, Grun & Stratton, New York, pp. 77-86.

ACETYLCHOLINE - PROSTANOID INTERACTION IN

THE PULMONARY CIRCULATION

J.D. Catravas, J.J. Buccafusco and H. El-Kashef

Department of Pharmacology
Medical College of Georgia, Augusta
Georgia 30912 USA

INTRODUCTION

Parasympathetic innervation of pulmonary arteries and/or veins has been identified histologically in many species and is achieved via the tenth cranial nerve (vagus; 5, 6). The pulmon- ary hemodynamic alterations resulting from vagal stimulation or administration of muscarinic agonists, however, remain controversial. Over the last fifty years, sporadic investigations have proposed that acetylcholine (ACh) causes vasoconstriction in the pulmonary circulation of the dog, monkey, rabbit, rat, guinea pig, cat, frog and sheep; however others have observed vasodilation in the pulmonary circulation of dog, cat and pig (for review, see ref. 1). Some of the factors that could be responsible for the apparent discrepancy in these findings include adrenergic discharge or concomitant bronchocon- striction contributing to pulmonary vasoconstriction, and choli- nergic - induced bradycardia, systemic hypotension and decreased cardiac output causing secondary pulmonary vasodilation.

The nervous and humoral system(s) primarily responsible for the control of pulmonary vascular tone under normal or abnormal conditions remain elusive. Because of the ubiquitous presence of ACh in all mammalina organisms, the existence of parasympathetic innervation in the pulmonary vasculature of most species and the confusing in- formation available on the actions of cholinergic agonists in the pulmonary circulation, we have decided to undertake a systematic investigation of the role of the parasympathetic nervous system in mammalian lung vasculature, utilizing appropriate experimental designs.

The present study examines the effects of ACh on pulmonary vascular resistance of the rabbit lung perfused in situ at constant flow; it also begins to investigate the effects of ACh on the pulmonary circulation of the anesthetized rabbit, in vivo, vis-a-vis changes in airway smooth muscle and systemic vascular resistance.

MATERIALS AND METHODS

The animals used in this study were male New Zealand albino rabbits weighing between 2.5 and 3.5 kg. They were individually housed, and were given free access to food and water.

In Situ Preparation. The animals were anesthetized with an infusion of a mixture of urethane (200 mg/ml) and allobarbital (50 mg/ml; a gift of Ciba-Geigy Corp.) into a marginal ear vein at volumes individually adjusted to produce surgical anesthesia. The neck and chest areas were shaved and a cannula was introduced into the trachea and connected to a Harvard Intermediate respirator. Artificial respiration was begun with oxygen-enriched air (40% O_2) at tidal volumes and rates necessary to maintain normal blood pH and pCO_2 levels and pO_2 > 100 mmHg. A catheter was introduced into the left carotid artery, 5,000 IV heparin were administered and the animal was exsanguinated. The chest was opened via a mid-sternal incision and catheters were placed into the main pulmonary artery and left atrium. Perfusion through the pulmonary artery was begun immediately by means of an adjustable peristaltic pump, utilizing a physiologic salt solution (4) containing 5% bovine serum albumin and maintained at 37°C in a water bath. Inspiratory gases were switched to a mixture of room air and 95% O_2 - 5% CO_2 at rates and volumes that maintained the pH, pCO_2 and pO_2 of the arterial outflow perfusate at 7.3-7.4, 30-35 mmHg and >100 mmHg, respectively. The initial 30-40 ml of the perfusion solution were discarded in order to clean the vasculature of red cells; at that time, the left atrial outflow tubing was returned to the perfusion reservoir (maintained at 200 ml) to establish a recirculating system. Left atrial pressure, pulmonary arterial pressure and airway pressure were continuously monitored by means of Statham transducers connected to a NARCO physiologic recorder. The transducers were calibrated at mid-chest level. Left atrial pressure was maintained at 0 mmHg by means of a needle valve positioned in series to the outflow tubing. Airway pressure was maintained at 15 cm water by changing minute ventilation, while inspiratory gas concentration was altered to maintain control pH, pCO_2 and pO_2 values. Drugs were added directly to the perfusion reservoir at doses necessary to produce the desired concentrations. At the end of the experiment, dry/wet lung weight was determined to assure that the data were not collected from an edematous preparation. The experiments lasted up to 2.5 hours.

In Vivo Preparation. The rabbits were anesthetized and the trachea intubated as described above. Catheters were placed in both jugular veins and the left carotid artery, and the chest was opened by a mid-sternal incision. The right jugular catheter was advanced to the main pulmonary artery, and another catheter was placed directly into the left atrium. Flow was measured via an electromagnetic flow meter with a probe placed around the ascending aorta. Pulmonary arterial, systemic arterial, left atrial and airway pressures were continuously monitored. Drugs were administered into the left jugular catheter.

Statistical comparisons of means utilized Student's test or the analysis of variance (ANOVA), when appropriate (11). Differences were considered significant when $P<0.05$. All data processing and analyses were performed with the aid of a DEC Rainbow computer (Digital Equipment Corp., Maynard, MA).

RESULTS

The effects of cumulative doses of ACh on pulmonary vascular resistance (PVR) in the isolated rabbit lung perfused in situ is shown in Figure 1. ACh produced a powerful, dose-dependent increase in PVR in all animals tested. Because of significant differences in the starting tone of a few of the animals, they were divided into "low pressure" and "high pressure" groups to examine any differences in their response to ACh. As seen in Figure 1, the main difference between the two groups is their sensitivity to the pressor actions of ACh, with the "high pressure" group exhibiting a steeper log dose-response curve. In all animals, ACh concentrations above 10^{-6} M created rapid, irreversible edema that could be prevented with the addition of atropine in the perfusate reservoir. However, in contrast to the picture seen with the rapid reversal by atropine of ACh-induced systemic hypotension, in vivo, in this preparation up to twenty min were required for the pressure to return to baseline after the addition of atropine in the reservoir. This phenomenon suggested the existence of a second mediator of the pulmonary vasoconstrictor actions of ACh, and this was investigated in experiments where the actions of several agents known to increase PVR were inhibited.

Table 1 shows that blockade of histamine H-1 and H-2 receptors or angiotensin II receptors was ineffective in altering the vasoconstrictor actions on ACh. However, when prostaglandin synthesis was inhibited by adding indomethacin (10^{-5} M) or aspirin (10^{-3} M) into the reservoir twenty min before ACh, complete inhibition of the effects of ACh was observed up to 10^{-4} M (highest concentration of ACh tested).

In all experiments, ACh at concentrations between 10^{-10} and 10^{-8} M produced no effect on PVR. To examine the possibility that this, as well as the vasoconstrictor actions of ACh observed at higher doses, was due to a low initial vascular tone, a number of experiments were performed where starting vessel tone was increased (80-100%) by pretreatment with KCl or phenylephrine. As can be

Table 1

Effect of Pretreatment with Pharmacologic Modulators on
ACh-Induced Pulmonary Vasoconstriction and Bronchial
Constriction in the Isolated Rabbit Lung Perfused In Situ

TREATMENT	CHANGE FROM CONTROL RESPONSE	
	PVR	Airway Pressure
H1 + H2 Receptor Blockade	-	-
A II Receptor Blockade	-	-
KCl	-	-
Phenylephrine	-	-
Muscarinic Receptor Blockade	+	+
Inhibition of PG synthesis	+	-

- : no inhibition of control response to ACh; + : complete inhibition
of control response to ACh; A II : angiotensin II; PG: prostaglandins. H1 + H2 blockade was achieved by pretreatment with diphen-
hydramine (10^{-6} M) and cimetidine (10^{-5} M). A II receptor and muscarinic receptor blockade was achieved by pretreatment with saralasin
(2.5×10^{-6} M) and atropine (10^{-5} M), respectively. PG synthesis
was inhibited by 20 min pretreatment with aspirin (10^{-3} M) or indomethacin (10^{-5} M). KCl and phenylephrine concentrations (used to
raise pulmonary vascular tone) were 10^{-2} M and 10^{-3} M, respectively.

seen in Table 1, altered vascular tone did not affect the actions
of ACh on PVR at any dose tested (10^{-10} M - 10^{-6} M).

Table 1 also shows the actions of ACh on airway pressure. In
most experiments airway pressure was maintained at 15 cm water to
prevent any influences of bronchial constriction on PVR. In four
animals, however, airway pressure was allowed to rise and it did,
in response to ACh, in a dose dependent fashion. As with PVR, airway
pressure changes were not affected by histamine or angiotensin receptor
antagonism, and were blocked by atropine. However, in contrast to
the pulmonary vasculature, ACh-induced increase in airway pressure
was not affected by inhibition of prostaglandin synthesis, indicating
a difference in the mechanism responsible for ACh-induced constriction
of pulmonary vascular and airway smooth muscle.

The effects of ACh infusion in the anesthetized rabbit, in vivo,
are shown in Figure 2. As with the in situ preparation, ACh caused
a pronounced increase in pulmonary arterial pressure (PAP) despite
a concurrent decrease in systemic arterial pressure (SAP). However,
as with airway pressure, indomethacin did not affect the ACh-induced
systemic hypotension, while it completely obliterated the increase

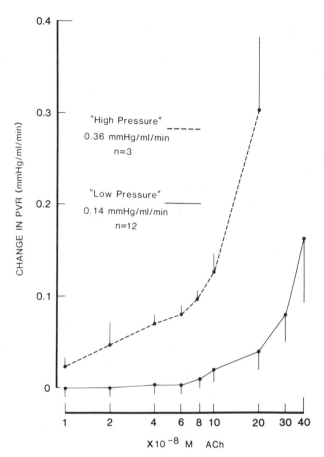

Figure 1. **Effects of cumulative doses of ACh on pulmonary vascular resistance (PVR) in the rabbit lung perfused _in situ_.** Rabbits were perfused at constant flow (92 $\pm$ 3 ml/min) and were divided into "high pressure" or "low pressure groups" depending on the starting pulmonary arterial pressure values. Each dose of ACh was added at the peak of the response from the previous dose. Edema occurred above 2 x 10^{-7} M ACh ("high pressure" group) or 4×10^{-7} M ACh ("low pressure" group).

in PAP. ACh-induced changes in both PAP and SAP were inhibited by atropine, and subsequent administration of a bolus of epinephrine showed reactive pulmonary and systemic vascular beds.

DISCUSSION

Utilizing carefully controlled experimental models, we have demonstrated that ACh produces a potent constrictor effect in the

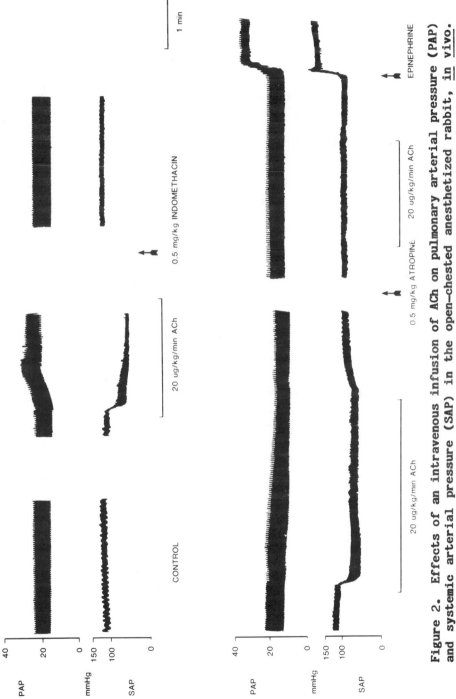

Figure 2. Effects of an intravenous infusion of ACh on pulmonary arterial pressure (PAP) and systemic arterial pressure (SAP) in the open-chested anesthetized rabbit, in vivo. Twenty min elapsed between idomethacin administration and the next ACh infusion.

pulmonary circulation of rabbits, and that this vasoconstriction is dependent on the release of a prostanoid.

While an ACh-prostanoid interaction has not been previously observed in the lung, ACh has been shown to release prostaglandins or prostaglandin-like material from various tissues such as adrenal glands (10), vas deferens (3) and heart (9). The mammalian lung is a rich source of prostaglandins (8) and their rapid turnover could easily accommodate the pattern of ACh changes observed here.

One interesting result from our experiments is that prostanoids appear to mediate the pulmonary vascular but not the systemic vascular or airway smooth muscle effects of ACh. This distinction could prove essential in the ability of ACh to control pulmonary vascular tone. The source and nature of the prostanoid(s) mediating the pulmonary actions of ACh remain unknown. Experiments are under way in our laboratory to determine that, as well as the involvement of pulmonary endothelium. Vascular endothelium from the lung, in particular, but from other beds as well, is rich in metabolic activities, including synthesis and release of prostaglandins (2, 8). Furthermore, systemic vascular endothelium has been shown to be a necessary component for the action of many vasoactive agents, including ACh (7). Thus, the ubiquitous nature of vascular endothelium, and the differences in metabolic activities among endothelia from various vascular beds could play an important role in the opposing actions of ACh in the rabbit systemic and pulmonary circulations.

Acknowledgements. The authors are pleased to acknowledge the expert technical assistance of Mr. Brian Meister, Ms. Tracy Dendle and Ms. Nancy Quinn. This work was supported in part by the American Lung Association, Georgia Affiliate.

REFERENCES

1.   Aviado, D.M. (1965): ed. The Lung Circulation, Pergamon Press, New York.

2.   Bakhle, Y.S. (1981): Bull, Europ. Physiopath. Resp. 17: 491-508.

3.   Borda, E., Agostini, M.D.C., Peredo, H., Gimeno, M.F. and Gimeno, A.L. (1983): Arch. Int. Pharmacodyn 263:245-253.

4.   Effros, R., Shapiro, L. and Silverman, P. (1980): J. Appl. Physiol. 49:589-600.

5.   El-Bermani, A.W. Bloomquist, E.I. and Montvilo, J.A. (1982): Thorax 37:703-710.

6.   Fisher, A. (1965): Acta Anat (Basel) 60:481-496.

7.   Furchgott, R.F. and Zawadzki, J.V. (1980): Nature 288: 373-376.

8.   Johnson, A.R., Callahan, K.S., Tsai, S.C. and Campbell, W.B. (1981): Bull. Europ. Physiopath. Resp. 17:531-551.

9.   Junstad, M. and Wennmalm, A. (1974): Br. J. Pharmacol. 52:

      375-379.
10.   Ramwell, P.W., Shaw, J.E., Douglas, W.W. and Poisner, A.M.
      (1966): Nature 210:273-274.
11.   Snedecor, G.W. and Cochran, W.G. (1980): Statistical Methods,
      7th ed., Iowa State Press, Ames.

LONG-TERM EFFECTS OF AF64A ON LEARNING AND

MEMORY PROCESSES IN THE RAT

T.J. Walsh[*], D.L. DeHaven[**], A. Russell[+] and I. Hanin[+]

[*]Laboratory of Behavioral and Neurological Toxicology
NIEHS
Research Triangle Park, North Carolina   27709 U.S.A.

[**]Biological Sciences Research Center
University of North Carolina School of Medicine
Chapel Hill, North Carolina   27514 U.S.A.

[+]Western Psychiatric Institute and Clinic
University of Pittsburgh School of Medicine
Pittsburgh, Pennsylvania   15213 U.S.A.

INTRODUCTION

Acetylcholine (ACh) and the cholinergic system have long been implicated in cognitive processes. Studies of the molecular events which contribute to learning and memory consolidation suggest an important, if not essential, role for this transmitter system in associative mechanisms. For example, cholinergic activity indexed by ACh concentrations and high affinity choline transport (HAChT) is enhanced in a region-dependent manner subsequent to a learning experience (18, 25). Furthermore, drugs that inhibit cholinergic tone, and lesions of either the nucleus basalis of Meynert or the medial septum, structures providing the cholinergic input to the neocortex and the limbic system, also impair cognitive processes (1, 2, 6, 9, 10, 15, 21, 29). In contrast, cholinomimetics, like physostigmine and arecoline, typically improve these processes (1, 3).

Insights into the function of the cholinergic system have also been provided by studies of neuropsychiatric disorders which are characterized by cognitive decline or mental retardation. For example, the earliest clinical signs of senile dementia of the Alzheimer type (SDAT) are deterioration of cognitive and mnemomic ability

(27). The morphological, neurochemical, and behavioral abnormalities associated with SDAT indicate a primary involvement of the cholinergic system in this disease. Histological analysis of post-mortem tissue has demonstrated a significant loss of neurons in the nucleus basalis of Meynert in SDAT patients (4, 24, 34). This degeneration of cholinergic neurons results in a decrease of presynaptic cholinergic markers in the neocortex and limbic system. For example, the neurochemical deficits observed in SDAT include decreases in HAChT, choline acetyltransferase (ChAT) activity, ACh synthesis and ACh concentrations in neocortex, hippocampus and amygdala (26, 27). There is also evidence that the learning and memory impairments associated with SDAT are related to the concomitant loss of cholinergic function. For example, there is a significant correlation between cognitive impairment and decreased ChAT activity in the cerebral cortex observed in SDAT (23). In essence, the degree of dementia is proportional to the loss of cholinergic tone. Furthermore, the constellation of cognitive deficits observed in SDAT can be reproduced in healthy young adults by the administration of scopolamine, a muscarinic antagonist (5). The neurochmeical specificity of these behavioral impairments is highlighted by the observation that physostigmine, but not amphetamine, can attenuate these memory deficits. These studies indicate that the learning and memory impairments associated with SDAT are probably dependent upon the chronic disability of cholinergic processes subserving cognition. Morphological evidence indicates that the dementias characteristic of Parkinson's Disease and Down's Syndrome might also be related to the degeneration of cholinergic neurons in the basal forebrain (19, 24).

Recent data indicate that neurotoxic analogs of choline might be used to develop animal models of chronic cholinergic hypofunction (17). Ethylcholine mustard aziridinium ion (AF64A) is similar to choline, but in addition to its choline-like structure it contains a reactive aziridinium moiety which irreversibly inhibits HAChT (8), the rate limiting step in ACh synthesis. Administration of AF64A into either the cerebroventricles or the hippocampus of rodents produces a persistent reduction of various presynaptic cholinergic markers including HAChT, ChAT activity, and ACh concentrations (17). Thus, AF64A might be a useful tool for investigating the neurobiology of the cholinergic system and also for developing animal models of SDAT.

The experiments presented here have examined the behavioral and biochemical effects of AF64A administered into the cerebroventricles of rats. The long-term effects of this cholinotoxin on retention of a one-trial passive avoidance response and performance in a radial-arm maze were also examined. Some of the work reported in this chapter has also recently been published elsewhere (30).

MATERIALS AND METHODS

Male Fischer-344 rats were used in the following experiments. All animals were 90-120 days of age and weighed between 250-300 grams at the beginning of the study.

Immediately prior to surgery rats were anesthetized with sodium pentobarbital (45 mg/kg) and positioned in a stereotaxic instrument. A midline incision was made in the scalp and two holes were drilled in the skull for placement of the injection cannula. They were then infused, bilaterally, with either 7.5 or 15 nmol of AF64A or artificial CSF using a 34-ga. injection cannula interfaced with a Sage Instruments Infusion Pump. A total volume of 2.5 μl of solution was delivered into each lateral ventricle (-0.5 mm posterior to bregma, 1.5 mm lateral to the sagittal suture and 2.5 mm vertical from dura) at a rate of 0.5 μl per min.

Solutions of AF64A were prepared from 10 mM acetylethylcholine mustard HCl according to methods described by Fisher and colleagues (8).

Passive Avoidance. Retention of a one-trial passive avoidance response was assessed 35 days after dosing. During training each rat was placed into the smaller lighted half of a two-compartment shuttle box. Sixty sec later the door separating the compartments was raised and the rat's latency to enter the dark compartment was recorded. Following the rat's entry into the dark compartment, the door was closed and a 0.8 mA footshock was delivered for 1 sec through the grid floor. Twenty-four hours later retention was assessed. During the retention test each rat was again placed into the lighted compartment for 60 sec, the door was then opened and the number of partial entries into the dark compartment was recorded, as was the latency to completely enter the dark compartment.

Radial-Arm Maze. Short-term spatial memory in an eight arm radial maze was assessed 60-80 days after dosing. The maze was modified from Olton and Samuleson (20) and consisted of an octagonal central arena (50 cm across and 31 cm high) from which eight equally spaced arms (59 cm long, 10 cm wide, 12.5 cm high) radiated.

Rats were reduced to 85% of their free-feeding weight by limiting their daily ration of food. Prior to a test session each arm was baited with one 45 mg food pellet. During a trial each rat was placed into the central arena and removed after it either obtained all 8 pellets or 10 min elapsed, whichever came first. The most efficient strategy for retrieving all 8 pellets is for the rat to enter each of the arms only once (i.e., 8 pellets in 8 arm choices). During a trial the animals must "remember" which arms have and have not already been visited. Further discussion of the characteristics of this task can be found in Olton et al. (21). Animals were tested for a total of 15 sessions. The following indices of maze performance were automatically recorded by the output of photocell circuits located throughout the maze and interfaced with a PDP 8 computer: 1) the number of different arms entered in the first eight choices; 2) the total number of arms entered to obtain all 8 pellets during a single trial; and 3) the pattern of spatial responding (i.e., the sequence of arm entries) in the maze.

Neurochemical Analysis. Animals were killed 120 days after dosing by microwave irradiation focused to the skull. Rats were subsequently decapitated and their brains were removed and dissected into frontal cortex, hippocampus and corpus striatum.

Regional concentrations of ACh and choline were determined by a combined gas chromatographic/mass spectrometric method described by Hanin and Skinner (11). Dopamine (DA), dihydroxy-phenylacetic acid (DOPAC), homovanillic acid (HVA), serotonin (5-HT) and 5-hydroxy-indoleacetic acid (5-HIAA) were determined by reverse phase HPLC with electrochemical detection using 5-hydroxy-N-methyltryptamine as the internal standard (13). Norepinephrine (NE) was measured using the HPLC-EC procedure of Keller et al. (12) modified by Mailman and colleagues (16), using dihydroxybenzylamine as the internal standard.

RESULTS

Following surgery the rats treated with AF64A were hypokinetic and unreactive to environmental stimuli in their home cages. Within 7 days of dosing 41% of the 30 nmole group (15 nmoles in each ventricle) and 17% of the 15 nmole group (7.5 nmoles in each ventricle) died of unknown causes. Signs of systemic toxicity such as aphagia, adipsia, diarrhea, ataxia or convulsions were not observed. In the surviving animals AF64A administration resulted in a transient loss of weight (8%) which was evident at 1 and 7 but not 14 days after dosing. The time-dependent effects of this dosing regimen on locomotor activity and hot-place latencies have been reported elsewhere (30).

Passive Avoidance: 35 Days After Dosing. Both doses of AF64A impaired retention of the passive avoidance response (see Table 1). During the retention test none of the control animals entered the chamber in which they had been shocked, for the duration of 600 sec. In contrast to the control performance, 36% of the 15 nmol and 50% of the 30 nmol AF64A-treated rats entered the dark compartment in less than 300 sec.

The initial step-through latencies of the three groups did not significantly differ during the training trial [$F(2,33) = 1.21$, $p > 0.10$]. All animals were oriented to the apparatus and entered

Table 1

Effects of AF64A on Retention of a Step-Through
Passive Avoidance Response

Group	Training Step-Through Latencies	No. Partial Entries	Retention Latencies
CSF	18 ± 6	1.75 ± 0.45	600
AF64A 7.5 nmoles per ventricle	27 ± 11	14.00 ± 2.85*	410 ± 60*
AF64A 15 nmoles per ventricle	12 ± 10	11.80± 4.81*	329 ± 72**

Data represent means ± S.E.M. $p < 0.05$; **$p < 0.01$ vs. CSF control; Fisher's Least Significance Difference test.

the dark compartment within 35 sec following the opening of the guillotine door. All rats also appeared reactive to the level of footshock used during training.

Retention of the passive avoidance response, indexed by the number of partial entries [$F(2,33) = 3.95$, $p < 0.05$] and the step-through latencies [$F(2,33) = 4.06$, $p < 0.05$] during the retention text, was significantly different between groups. The retention latencies of both the 15 and 30 nmol groups were shorter (27-42%, $p < 0.01$) during the 24 hour retention test while the number of partial entries was increased (765%, $p < 0.05$) in both groups relative to the controls. Post-hoc analyses demonstrated that both doses of AF64A had comparable effects on retention of the passive avoidance response.

Radial-Arm Maze: 60-80 Days After Dosing. Radial-arm maze performance, assessed 60-80 days after dosing, was also disrupted by both doses of AF64A (see Table 2). By the end of the 15 sessions the control group was entering approximately 7.5 new arms during their first eight choices. In contrast, the 15 and 30 nmol AF64A groups were entering 5-6 new arms in their first eight selections. A 3 x 5 repeated ANOVA on the mean number of correct responses during each block of 3 trials demonstrated a significant treatment [$F(2,32) = 3.50$, $p < 0.05$] and day [$F(4,128 = 8.76$, $p < 0.001$) effect. Univariate analysis of each individual block of trials revealed a significant ($p < 0.05$) between group difference during all blocks except the first. Therefore, the deficits in maze performance observed in the AF64A groups were evident throughout the period of testing.

Table 2

Effects of AF64A on Radial-Arm Maze Performance.   Data Represent
the Mean $\pm$ S.E.M. Correct Responses in the First Eight
Arm Selections

Group	Trials: 1-3	4-6	7-9	10-12	13-15
CSF	5.0 ± 0.3	5.7 ± 0.3	6.8 ± 0.1	6.9 ± 0.1	7.33 ± 0.1
AF64A 7.5 nmoles per ventricle	3.6 ± 0.3	3.8 ± 0.2	4.2 ± 0.2	4.1 ± 0.2	5.20 ± 0.2
AF64A 15 nmoles per ventricle	3.4 ± 0.4	4.0 ± 0.5	4.9 ± 0.6	4.7 ± 0.3	5.70 ± 0.2

The AF64A-treated rats made significantly ($p < 0.05$) fewer correct
responses in their first eight choices during each block of trials,
except the first.

The total number of arm entries required to obtain all 8 pellets
was also increased by AF64A administration (data not presented).
During the last block of 3 trials the controls required 8.5 arm
entries on the average to retrieve all 8 pellets in a single session.
The groups treated with 15 and 30 nmol of AF64A, however, required
approximately 12.5 choices to complete a trial. A 3 x 5 ANOVA revealed
a significant treatment [$F(2,32)= 4.41$, $p < 0.05$] and day [$F(4,128)$
$= 10.23$, $p < 0.001$] effect but an insignificant treatment x day
interaction [$F(8,128)=0.59$, $p > 0.10$].

The spatial pattern of responding in the maze was also affected
by AF64A administration. While the CSF group entered the adjacent
arm on approximately 65% of their choices the treated rats exhibited
little preference in their choice of arms 1 (32%), 2 (26%) or 3
(25%) distant to the one just exited. All of the animals refrained
from re-entering an arm just visited.

Neurochemical Measures: 120 Days After Dosing. AF64A had no long-
term effects on the concentrations of catecholamines, indoleamines
or their metabolites in hippocampus, striatum or frontal cortex
(30). This cholinotoxin did, however, decrease the concentrations
of ACh in the hippocampus and frontal cortex (see Table 3).

Independent two-way treatment x region ANOVAs on each of these neurochemical parameters revealed that AF64A produced no significant treatment or treatment x region effects on NE, DA, DOPAC, HVA, 5-HT or
5-HIAA (all p's > 0.10). AF64A did, however, have a significant

Table 3

Concentrations of Acetylcholine and Choline in Brain Regions
of AF64A-Treated Rats

	Striatum	Hippocampus	Frontal Cortex
Acetylcholine (nmoles/gram tissue)			
CSF	105.07 ± 11.73	39.36 ± 4.53	53.00 ± 3.39
AF64A 7.5 nmoles per ventricle	138.11 ± 34.66	$15.09 \pm 3.46^{**}$	53.30 ± 9.50
AF64A 15 nmoles per ventricle	77.57 ± 21.00	$22.37 \pm 6.01^{*}$	$20.00 \pm 7.79^{**}$
Choline (nmoles/gram tissue)			
CSF	33.43 ± 5.57	46.30 ± 11.14	41.71 ± 7.21
AF64A 7.5 nmoles per ventricle	46.00 ± 15.68	40.06 ± 15.86	30.80 ± 8.07
AF64A 15 nmoles per ventricle	26.71 ± 9.87	49.96 ± 15.96	27.31 ± 9.67

Values are means $\pm$ S.E.M. $^*p < 0.05$ vs. CSF control; $^{**}p < 0.01$
vs. CSF control; Fisher's Least Significant Difference Test.

treatment effect on regional ACh content ($p < 0.05$). Univariate
ANOVAs followed by group by group comparisons demonstrated that
while both doses of AF64A produced a significant 44-62% decrease
of ACh in the hippocampus relative to the controls ($p < 0.05$ for
both doses), only the higher dose (30 nmol) caused a significant
($p < 0.05$) (62%) depletion of ACh in the frontal cortex. Neither
dose affected the concentration of ACh in the striatum or the content
of choline in any of the three regions examined.

DISCUSSION

Intraventricular, bilateral administration of either 7.5 or 15
nmol of AF64A, produced persistent cognitive impairments in rats.
Retention of a step-through passive avoidance response was disrupted
even 35 days after AF64A administration. Performance in an 8 arm
radial maze was impaired throughout the period of testing (i.e., up
to 90 days following treatment). The neurochemical effects of this

dosing regimen, assessed 120 days after dosing, involved only the cholinergic system, which supports the contention that AF64A is a specific cholinotoxin. The concentration of ACh in the hippocampus was significantly reduced (44-62%) in both AF64A groups, and the content of ACh in the frontal cortex was also decreased (62%) in the 30 nmol group. There was no effect of AF64A on levels of ACh in the striatum. Furthermore, the regional concentrations of catecholamines, indoleamines, their metabolites, or choline were not affected.

Despite the apparent specificity of AF64A's cytotoxic effects, adaptive alterations in other transmitter systems are likely to occur following administration of any neurotoxin. Therefore, it can be difficult to correlate behavioral changes occurring shortly after treatment with the long-lasting neurochemical effects of a compound. Nevertheless, the specificity and persistence of the neurochemical changes induced by AF64A, as well as the similarity of the behavioral effects to those resulting from experimental manipulation of the cholinergic system, argue that the cognitive deficits observed in this study resulted from a specific cholinergic lesion.

Radial-arm maze performace was disrupted by both AF64A doses. The impairment was characterized by fewer correct arm entries in the first eight choices and a greater number of entries required to obtain all 8 pellets during a session. The AF64A-treated rats readily explored the maze, and performed the motor responses to obtain and consume the pellets following a correct arm entry. The deficits in maze performance were not due to changes in motivation or locomotor behavior. Rather, the data suggest that the AF64A groups were unable to inhibit responding to previously entered arms. One hypothesis which might account for their impaired performance is that their "working memory" of which arms had or had not been chosen during a test session was impaired. Further experiments are addressing the nature of the cognitive deficits observed in the present study.

The results of this study support the contention that the cholinergic input to the hippocampus is required for efficient performance of a radial-arm maze task. Lesions of the hippocampus, its associated projections or of the medial septum, which provides cholinergic input to this structure, impair performance of this task (2, 21, 29). Furthermore, systemically administered scopolamine impairs, while physostigmine improves maze performance (3, 6). Since both AF64A doses impaired maze performance but only the higher dose (30 nmol) reduced the concentration of ACh in the cortex, it appears that the depletion of ACh in the hippocampus is sufficient to account for these long-term deficits. The neocortex has, however, been shown to play some role in spatial learning (14), and we are investigating the contribution of the cortical cholinergic innervation to radial-arm maze performance.

The cholinergic input to the hippocampus is an important neural substrate of maze behavior in the rat. Performance of this task is dependent upon neural circuitry and neurochemical systems which are known to be affected in disorders such as SDAT (27). Spatial learning and memory is also disrupted in aged humans and rodents (22, 28). Therefore, this task might be used to further examine the neurobiological and cognitive mechanisms which are disrupted by the aging process. Potentially useful dietary, environmental and pharmacological strategies designed to enhance cholinergic tone and attenuate age-related memory disorders might also be assessed by this task.

In conclusion, AF64A, a neurotoxic analog of choline, produced specific biochemical and behavioral changes in the rat. Intraventricular administration of this compound produced a long-term decrease of cholinergic function together with cognitive impairments. These data suggest that AF64A is a select cholinotoxin which can be used to investigate the neurobiology of the central cholinergic system. Furthermore, since AF64A produces some of the characteristic neurochemical and cognitive deficits observed in SDAT it should be an important tool for elucidating the underlying mechanisms of this disorder.

NOTE ADDED AFTER COMPLETION OF MANUSCRIPT

Since these studies were conducted, these experiments have been repeated successfully by us, with comparable results, using 3 nmoles AF64A, administered bilaterally, into each lateral ventricle.

REFERENCES

1. Bartus, R.T., Dean, R.L., Beer, B. and Lippa, A.S. (1982): Science 217:408-417.
2. Beatty, W.W. and Carbone, C.P. (1980): Physiol. Behav. 24:675-678.
3. Buresova, O. and Bures, J. (1982): Psychopharmacology 77:268-271.
4. Coyle, J.T., Price, D.L. and DeLong, M.R. (1983): Science 219: 1184-1190.
5. Drachman, D.A. and Sahalian, B.J. (1980): In The Psychobiology of Aging: Problems and Perspectives (ed) D.G.Stein, Elsevier, North Holland, pp. 348-368.
6. Eckerman, D.A., Gordon, W.A., Edwards, J.D., MacPhail, R.C. and Gage, M.I. (1980): Pharmacol. Biochem. Behav. 12:595-602.
7. Fisher, A. and Hanin, I. (1980): Life Sci. 27:1615-1643.
8. Fisher, A., Mantione, C.R., Abraham, D.J. and Hanin, I. (1982): J. Pharmacol. Exp. Ther. 222:140-145.
9. Flicker, C., Dean, R.L., Watkins, D.L., Fisher, S.K. and Bartus, R.T. (1983): Pharmacol. Biochem. Behav. 18:973-981.
10. Gray, J.A. and McNaughton, J. (1983): Neurosci. Biobehav. Rev. 7:119-188.
11. Hanin, I. and Skinner, R.F. (1975): Anal. Biochem. 66:568-583.

12. Keller, R., Oke, A., Mefford, I. and Adams, R.N. (1976): Life Sci. 19:995-1004.

13. Kilts, C.D., Breese, G.R. and Mailman, R.B. (1981): J. Chromatog. 225:347-357.

14. Kolb, B., Sutherland, R.J. and Whishaw, I.Q. (1983): Behav. Neurosci. 907:13-27.

15. Lo Conte, G., Bartolini, L., Casamenti, F., Marconi-Pepeu, I. and Pepeu, G. (1982): Pharmacol. Biochem. Behav. 17:933-937.

16. Mailman, R.B., Krigman, M.R., Frye, G.D. and Hanin, I. (1983): J. Neurochem. 40:1423-1429.

17. Mantione, C.R., Fisher, A. and Hanin, I. (1984): Life Sci. 35:33-41.

18. Matthies, H., Rauca, C.H. and Liebman, H. (1974): J. Neurochem. 23:1109-1113.

19. Nakano, I. and Hirano, A. (1983): Ann. Neurol. 13:87-91.

20. Olton, D.S. and Samuelson, R.J. (1976): J. Exp. Psychol. 2:97-116.

21. Olton, D.S., Becker, J.T. and Handelman, G.E. (1979): Behav. Brain Sci. 2:313-365.

22. Perlmutter, M., Metzger, R., Nezworski, T. and Miller, K. (1981): J. Gerontol. 36;59-65.

23. Perry, E.K., Tomlinson, B.E., Blessed, G., Bergman, K., Gibson, P.H. and Perry, R.H. (1978): Brit. Med. J. 2:1457-1459.

24. Price, D.L., Whitehouse, P.J., Struble, R.G., Clark, A.W., Coyle, J.T., DeLong, M.R. and Hedreen, J.C. (1982): Neurosci. Comment. 1:84-92.

25. Raaijmakers, W.G.M. (1982): In Neuronal Plasticity and Memory Formation (eds) C. Marsan and H. Matthies, Raven Press, New York, pp. 373-385.

26. Sims, N.R., Bowen, D.M., Allen, S.J., Smith, C.C.T., Neary, D., Thomas, D.T. and Davison, A.N. (1983): J. Neurochem. 40: 503-509.

27. Terry, R.D. and Davies, P. (1980): Ann. Rev. Neurosci. 3:77-95.

28. Wallace, J.E., Drauter, E.E. and Campbell, B.A. (1980): J. Gerontol. 35:355-363.

29. Walsh, T.J., Miller, D.B. and Dyer, R.A. (1982): Neurobehav. Toxicol. Teratol. 4:177-183.

30. Walsh, T.J., Tilson, H.A., DeHaven, D.L., Mailman, R.B., Fisher, A. and Hanin, I. (1984): Brain Research 321:91-102.

31. Whitehouse, P.J., Price, D.L., Struble, R.G., Clark, A.W., Coyle, J.T. and DeLong, M.R. (1982): Science 215:1237-1239.

BIOCHEMICAL AND BEHAVIORAL EFFECTS

OF ETHYLCHOLINE MUSTARD AZIRIDINIUM ION

F. Casamenti, L. Bracco[*], F. Pedata and G. Pepeu

Department of Pharmacology
University of Florence
Viale G.B. Morgagni 65
50134 Florence, Italy

[*]Neurological Clinic
University of Florence
Careggi
50134 Florence, Italy

INTRODUCTION

The most important biochemical change detected in the brain of patients affected by senile dementia of Alzheimer type is a decrease in choline acetyltransferase (ChAT) activity in the cerebral cortex and hippocampus (17). The decrease is caused by a loss of cholinergic fibers originating from the magnocellular forebrain nuclei and the septum (2).

An animal model showing a similar loss of cholinergic fibers in the cerebral cortex and hippocampus could be very useful for understanding the pathogenesis of senile dementia and detecting potentially beneficial drugs.

Ethylcholine mustard aziridinium ion (ECMA; AF64A) has recently been proposed as a specific neurotoxin for the cholinergic neurons (12). Injected intracerebroventricularly (i.c.v.), ECMA brought about a long-lasting decrease in ChAT activity, and high affinity choline uptake (HAChT) and acetylcholine (ACh) levels in mouse cerebral cortex and hippocampus (3). In the latter a decrease of the pre-synaptic cholinergic markers was also found after intrahippocampal injection in the rat (13). Animals thus treated could represent a potential model of Alzheimer disease (8).

In order to ascertain the similarities between the changes induced by ECMA in cholinergic mechanisms and those found in senile dementia, we studied the magnitude and time course of the changes in ChAT activity and HAChT in different brain areas and the cortical ACh output after i.c.v. injection in the rat. The effect of ECMA administration on the acquisition of a passive avoidance conditioned responses was also investigated.

MATERIALS AND METHODS

The experiments were carried out on male Wistar rats of 140-150g body weight. Under ether anaesthesia a hole was drilled in the skull and ECMA or its vehicle were injected unilaterally in the lateral ventricles using a Hamilton microsyringe.

Seven or twenty days after the injection, the rats were killed by decapitation, the brain quickly removed and dissected. To increase reproducibility the samples for ChAT determinations were prepared with the help of a LKB cryomicrotome. For ChAT determination the samples were homogenized in 20 vols of 10 mM EDTA sodium salt and 0.2% (v/v) Triton X 100. For HAChT determination the samples were homogenized in 20 vols of ice cold 0.32 M sucrose. ChAT activity was assayed by the micromethod of Fonnum (5). HAChT activity determinations were performed on a crude mitochondrial preparation according to the method of Simon et al. (18). Protein content of the homogenates was measured using the method of Lowry (11).

ACh output from the cerebral cortex was investigated 20 days after ECMA, following amphetamine, or its vehicle administration (1g/Kg i.p.), by the cortical cup technique in urethane anesthetized rats (14). ACh diffusing into the Perspex cylinder placed on the exposed cortical surface was quantified by bioassay on the dorsal muscle of the leech (15).

Twenty days after the injection, the acquisition of a passive avoidance conditioned response was studied, shortly before decapitation or the cortical cup application, by means of an apparatus derived from that described by Jarvik and Kopp (9). The rats were trained in a one-trial passive avoidance task. The retest was carried out 30 min after training. The rats were placed in an illuminated Plexiglas box. After 1 min a guillotine door was opened and the latency between the opening and the entrance into a dark box was measured. When the rat walked into the dark box it received a 1.5 mA foot shock through an electrified grid floor. The trial was terminated when the rat ran back into the illuminated compartment from which it was then removed. The rat was allowed to remain in the illuminated compartment up to 120 sec without walking into the dark compartment before being removed. Better performance was indicated by longer retest latencies.

ECMA was a gift of ISF, Trezzano sul Naviglio, Milano, Italy. The identity of the substance was checked by IR, NMR and MS analysis and its purity assessed by chromatographic analysis was 98%. One hr before administration the compound was dissolved in concentrated NaOH at 11.5. After 20 min the pH was adjusted to 7 with concentrated HCl (13).

Statistical analysis was carried out using Student's two tailed t-test.

RESULTS

All rats injected i.c.v. with ECMA 32 nmol/10 μl showed abnormal posture and hyperexcitability. Within a week they lost 20-30% of their body weight and their mortality ranged between 30 and 50%. Macroscopic brain examination of the rats dying within the first week revealed an area of necrosis around the injected ventricle. In the surviving rats the gross behavior returned to normal and the brain showed no macroscopic damage. Higher doses of ECMA caused 100% mortality within 3-4 days.

No behavioral changes, weight loss and mortality were found in vehicle-injected rats.

Table 1 shows that 7 days after ECMA administration there was a significant decrease in ChAT activity in all brain regions investigated. The larger decrease occurred in the hippocampus where it persisted 20 days post injection. Conversely, by this time ChAT activity had returned to normal levels in the cerebral cortex and striatum.

Seven days after ECMA injection, there was a slight but not significant decrease in HAChT activity in the cerebral cortex and hippocampus, which became more pronounced and statistically significant 20 days post injection as shown in Table 2. However no difference in spontaneous ACh release from the cerebral cortex was found, 20 days after injection, between vehicle-(1.12 ± 0.11 ng/min/cm^2, n=7) and ECMA-(1.15 ± 0.2 ng/min/cm^2, n=4) treated rats. Amphetamine (1 mg/Kg i.p.) was able to induce a statistically significant increase in cortical output in both groups of rats as illustrated in Fig. 1. The increase seems to be somewhat delayed in the ECMA-treated rats.

Table 3 shows the retest latencies after one trial training session for the acquisition of a passive avoidance conditioned response. It can be seen that 20 days after injection the rats injected with ECMA showed a statistically significant decrease in the retest latencies. This indicates an impairment in the acquisition of the conditioned response.

Table 1

Effect of ECMA (32 nmol/10 μl) injected i.c.v.
on ChAT activity in different brain regions

Treatment	Days post injection	N. rats	Brain regions	ChAT activity μmol/h/100 mg prot. $\pm$ S.E.	% change
Vehicle	7	4	Frontal cortex	4.39 $\pm$ 0.27	
			Striatum	22.50 $\pm$ 1.25	
			Hippocampus	6.40 $\pm$ 0.6	
ECMA	7	8	Frontal cortex	3.28 $\pm$ 0.28*	-26
			Striatum	15.41 $\pm$ 0.61**	-32
			Hippocampus	3.49 $\pm$ 0.46***	-48
ECMA	20	3	Frontal Cortex	4.13 $\pm$ 0.07	-07
			Striatum	23.06 $\pm$ 1.4	+02
			Hippocampus	3.84 $\pm$ 0.38**	-41

Values significantly different from vehicle-injected. *$p<0.05$; **$p<0.01$; ***$p < 0.001$.

Table 2

Effect of ECMA (32 nmol/10 μl i.c.v.) on high
affinity choline uptake (pm/4 min/mg protein)

Treatment	Days after injection	CORTEX	% change	HIPPOCAMPUS	% change
Vehicle		2.74 $\pm$ 0.30		3.39 $\pm$ 0.31	
ECMA	7	2.25 $\pm$ 0.22	-18	2.66 $\pm$ 0.40	-22
ECMA	20	2.03 $\pm$ 0.06*	-26	2.52 $\pm$ 0.16*	-26

Each point is the mean $\pm$ S.E. of 4 rats. *$p < 0.05$.

Table 3

Effect of ECMA on Passive Avoidance
Conditioned Response

Treatment	Dose nmol i.c.v.	N. rats	Retest latencies (sec) mean ± S.E.
Vehicle	-	21	93.5 ± 8.5
ECMA	32	40	56.4 ± 6.8*

*Significantly different from vehicle: p < 0.001.

DISCUSSION

In our experiments, 20 days after i.c.v. ECMA administration, the rats showed a decrease in ChAT and HAChT activity in the hippocampus associated with an impairment in the acquisition of a passive avoidance conditioned response. Conversely in the cerebral cortex there was a complete recovery in ChAT activity and a normal spontaneous and amphetamine-evoked ACh release. it is likely that these differences are due to a higher concentration of ECMA reaching the hippocampus than other brain regions because of its contiguity with the lateral ventricles.

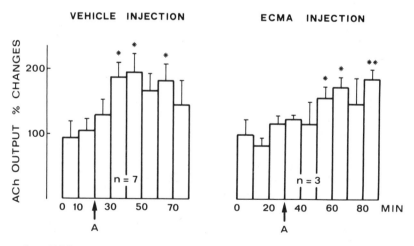

Figure 1. Effect of amphetamine (A) 1 mg/kg i.p. on ACh release from the cerebral cortex in urethane anesthetized rats 20 days after i.c.v. injection of ECMA 32 nmol/10 μl. The numbers indicate the number of experiments. *p < 0.01, **p < 0.001

A complete recovery in cortical and striatal ACh levels was observed (3) in mice 20 days after the i.c.v. injection of 65 nmol/ mouse ECMA while the level remained low in the hippocampus.

In our experiments only HAChT was reduced in the cerebral cortex 20 days after injection. However HAChT activity can be considered an indication of the presence and activity of the cholinergic neurons (1). Therefore the decrease in HAChT activity may result from the impairment of the uptake mechanism in some cholinergic neurons and from a compensatory increase in activity in others. In this regard Mantione et al. (13) found no correlation between the decrease in ChAT and HAChT activity in the hippocampus of rats 5 days after local injection of ECMA. Furthermore, it has been shown that electrolytic lesions of the nucleus basalis are followed by a persistent decrease in ChAT activity and a recovery in HAChT activity (16).

In conclusion our results confirm earlier reports showing that i.c.v. administration of ECMA can be used to induce a long lasting impairment of the hippocampal cholinergic system. It is pertinent to mention that the selective inhibition of ChAT activity in the hippocampus is followed by an impairment in the acquisition of a passive avoidance conditioned response (7) similar to that observed in our experiments.

The use of electrolytic lesions (10) or of the local injection (4, 6) of the excitotoxic aminoacids in the basal forebrain nuclei seems therefore to be the most reliable procedure for obtaining a diffuse impairment of the cortical cholinergic network reminiscent of that occurring in Alzheimer's disease.

Acknowledgement. This work was supported by CNR grant n. 8202043.04.

REFERENCES

1. Antonelli, T., Beani, L., Bianchi, C., Pedata, F. and Pepeu, G. (1981): Br. J. Pharmac. 74:525-531.
2. Fibiger, H.C. (1982): Br. Res. Rev. 4:327-388.
3. Fisher, A., Mantione, C.R., Abraham, D.J. and Hanin, I. (1982): J. Pharmacol. Exp. Ther. 222:140-145.
4. Flicker, C., Dean, R.L., Watkins, D.L., Fisher, S.K. and Bartus, R.T. (1983): Pharmacol. Biochem. Behav. 18:973-981.
5. Fonnum, F. (1975): J. Neurochem. 24:407-409.
6. Friedman, E., Lerer, B. and Kuster, J. (1983): Pharmacol. Biochem. Behav. 19:309-312.
7. Glick, S.D., Mittag, T.W. and Green, J.P. (1973): Neuropharmacology 12:291-296.
8. Hanin, I., Mantione, C.R. and Fisher, A. (1982): In Alzheimer's Disease: A Report of Progress (Aging vol. 19) (eds.) S. Corkin, K.L. Davis, J.H. Growdon, E. Usdin and R.J. Wurtman, Raven Press, New York 267-270.

9.   Jarvik, M.E. and Kopp, R. (1967):   Psychol. Rep. 21:221-224.
10. Lo Conte, G., Bartolini, L., Casamenti, F., Marconcini-Pepeu,
     I. and Pepeu, G. (1982): Pharmacol. Biochem. Behav. 17:933-937.
11. Lowry, O.M., Rosebrough, N.J., Farr, A.L. and Randall, R.J.
     (1951): J. Biol. Chem. 193:265-275.
12. Mantione, C.R., Fisher, A. and Hanin, I. (1981):   Science 213:
     579-580.
13. Mantione, C.R., Zigmond, M.J., Fisher, A. and Hanin, I. (1983):
     J. Neurochem. 41:251-255.
14. Mulas, A., Mulas, M.L. and Pepeu, G. (1974): Psychopharmacology
     (Berl.) 30:223-230.
15. Murnaghan, M.F. (1958):   Nature (London) 182:317.
16. Pedata, F., Lo Conte, G., Sorbi, S., Marconcini-Pepeu, I. and
     Pepeu, G. (1982):   Brain Res. 233:359-367.
17. Price, D.L., Whitehouse, P.J., Struble, R.G., Clark, A.W., Coyle,
     J.T., De Long, M. and Hedreen, J.C. (1982): Neurosci. Comment.
     1:84-92.
18. Simon, J.R., Atweh, S. and Kuhar, M.J. (1976):   J. Neurochem.
     26:909-922.

HISTOCHEMICAL AND BIOCHEMICAL EFFECTS OF THE INJECTION OF AF64A INTO THE NUCLEUS BASALIS OF MEYNERT:RELEVANCE TO ANIMAL MODELS OF SENILE DEMENTIA OF THE ALZHEIMER TYPE

M.R. Kozlowski and R.E. Arbogast

Department of Medicinal Sciences
Pfizer Central Research
Groton, Connecticut 06340 U.S.A.

INTRODUCTION

The fourth largest cause of death in the United States is senile dementia of the Alzheimer type (SDAT; 9). The initial symptom of this disorder is loss of short term memory. This is followed over a period of years by progressive dementia, complete loss of ability to care for oneself and finally, death (16). The most universal neuropathological change seen in victims of SDAT is a loss of acetylcholine in the cerebral cortex (3, 13). This loss appears due to the death of neurons in the nucleus basalis of Meynert (nbM), an area which supplies the cortex with a substantial proportion of its cholinergic innervation (8, 17). Thus animals with selective lesions of the nbM may be useful models of SDAT in humans.

Recently a compound has been developed which appears well suited to creating such lesions. This compound, ethylcholine aziridinium ion (AF64A), has been demonstrated to cause selective toxic effects on cholinergic neurons (5, 10). The primary basis of these toxic effects appears to be the irreversible binding of AF64A to the high affinity choline uptake sites present on cholinergic neurons (4). AF64A causes a chronic decrease in functioning of forebrain cholinergic systems when injected into the cerebral ventricles. A similar decrease restricted to the hippocampus and striatum can be produced when AF64A is injected directly into these structures (10 and 14, respectively).

The present study examines the effects of AF64A when injected into the nbM. This study addresses the question of whether such injection, by selectively damaging cholinergic nbM neurons, can

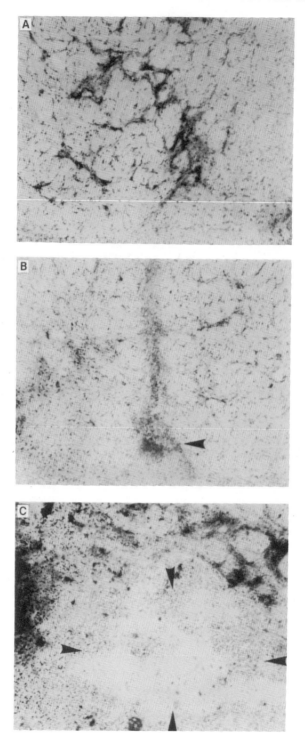

produce a deficit in cortical cholinergic functioning which mimics that seen in SDAT.

METHODS

A. **Preparation of AF64A.** AF64A was obtained as the nitrogen mustard of acetylethylcholine from Research Biochemicals Inc. The aziridinium ion was prepared as previously described (10). Briefly, AF64A was dissolved in water at 10 nmoles/μl, and the pH was raised to 11.5-11.7 with concentrated sodium hydroxide and maintained at this level for 20 min. The pH was then adjusted to 7.0 with concentrated hydrochloric acid and finally titrated to 7.4 with solid sodium bicarbonate. Solutions for injection were prepared by diluting this stock solution at least 10 fold with isotonic saline adjusted to pH 7.4 with solid sodium bicarbonate. The titrated saline solution was used alone for vehicle injections.

B. **Intracerebral Injection Procedure.** The nbM was visualized in rat brain sections by acetylcholinesterase (AChE) histochemistry (6) combined with Nissl staining. The exact stereotaxic coordinates for this structure were interpolated from those of radiofrequency lesions made at known sites around the nbM. The AF64A injections were made in male, Sprague-Dawley rats (Charles River; 270-300 g) anesthetized with equithesin and positioned in a stereotaxic apparatus. AF64A (0.01 to 0.2 nmole in 1 μl) or its vehicle (1 μl saline) was injected unilaterally at two sites along the rostrocaudal extent of the nbM in the right hemisphere. The solution was injected at a rate of 0.4 μl/min and the cannula was allowed to remain in place for 5 min after the termination of the injection. The locations of the cannula tips were examined in representative animals whose brains were sectioned and stained for AChE and Nissl substance. In all cases, the cannula tips were located either within or immediately adjacent to the area of the nbM.

C. **Biochemical and Histochemical Procedures.** The animals were sacrificed 7 to 14 days after surgery and their brains taken for either histochemical examination of AChE activity in the area of the nbM or measurement of choline acetyltransferase (ChAT) levels

---

Figure 1. **The effects of AF64A on AChE and Nissl staining in the nbM.** (A) The area of the nbM from the intact hemisphere of a rat given contralateral injections of 0.02 nM AF64A. Cholinergic cell bodies stained for AChE activity appear as dark dots. (B) nbM from the lesioned hemisphere of the same rat. Note the small area of non-specific damage surrounding the cannula tip (arrow) and the reduced number of AChE-stained cell bodies. (C) Massive, non-specific damage produced by injections of 0.2 nM AF64A into the nbM (bounded by arrows).

in cortex. AChE activity was visualized by the procedure of Hardy et al. (6) as previously described for brain sections (11). AChE staining in the area of the nbM in the right (lesioned) hemisphere was compared to that in the left (intact) hemisphere as well as to that in brains of naive animals. For measurement of ChAT activity, the cortex of each hemisphere was divided into three parts: anterior, central and posterior. Each area was assayed for ChAT activity by the procedure of Schrier and Shuster (15).

RESULTS

A. AChE Histochemistry. The ventral and ventromedial globus pallidus stained diffusely for AChE activity. Interspersed within the diffusely stained regions were many small, heavily stained areas. Counterstaining for Nissl substance revealed that these intensely stained areas corresponded to medium-sized cell bodies. Their location, size and density of AChE activity suggested that these cells belonged to the nbM (7, 12). Furthermore, the area of diffuse AChE activity encompassed the anatomically described location of the nbM (7, 12).

Injections of vehicle or of 0.01 nmole (per site) of AF64A into the area of the nbM had no visible effect on AChE staining. However, administration of 0.02 and 0.05 nmoles produced an obvious loss of both the diffuse AChE staining as well as the number of darkly labeled cell bodies in the area of the nbM (Fig. 1A and 1B). These doses produced only a small area of non-specific damage around the cannula tip. The injection of 0.1 and 0.2 nmoles of AF64A produced substantial non-specific damage at the injection site (Fig. 1C). Because of the extent of this non-specific damage, no biochemical assays were performed on animals given these doses.

B. Cortical ChAT Activity. In agreement with the results of the AChE histochemistry, injections of either vehicle or 0.01 nmoles of AF64A into the region of the nbM produced no decrease in cortical ChAT activity in any region of the injected hemisphere when compared to the intact hemisphere (Table 1). However, injections of 0.02 and 0.05 nmole significantly reduced ChAT activity in the central cortex of the injected hemisphere by 8% and 14% respectively, when compared to the intact hemisphere (Table 1). No asymmetries in ChAT activity were found in either the anterior or posterior cortical regions at these doses. Cortical ChAT levels in the intact hemispheres of AF64A-treated animals did not differ from those in vehicle injected animals, suggesting that AF64A produced no contralateral effects on this enzyme activity.

DISCUSSION

The discovery that the loss of cortical ChAT activity observed in SDAT parallels the death or dysfunction of cholinergic neurons in the nbM (17) suggests that specific damage to this nucleus may

Table 1

The Effect of AF64A Injections into the nbM on Cortical ChAT
Activity.   The Average Value for ChAT Activity in the
Non-Lesioned Hemisphere of the Vehicle Group
(Central Cortex) was 14.57 $\pm$ 0.91 nmole ACh/mg protein/h.

Dose of AF64A (nM)	n	ChAT activity in lesioned hemisphere as a percentage of intact hemisphere		
		Anterior Cortex	Central Cortex	Posterior Cortex
0 (vehicle)	5	111 $\pm$ 9	106 $\pm$ 7	87 $\pm$ 6
0.01	5	100 $\pm$ 4	97 $\pm$ 8	121 $\pm$ 13
0.02	5	92 $\pm$ 3	92 $\pm$ 2*	90 $\pm$ 6
0.05	5	117 $\pm$ 4	86 $\pm$ 7*	100 $\pm$ 9

*Significant decrease in ChAT activity in the lesioned hemisphere
compared to the intact hemisphere (t-test for paired comparisons,
$p < .05$).

be useful in producing animal models of SDAT. An obvious candidate
for producing such damage is AF64A, on the basis of its demonstrated
specific toxic effects on cholinergic neurons (5, 10 however, see
also 1, 2).   The results of the present study suggest that local
injections of AF64A in the vicinity of the nbM can both destroy
neurons tentatively identified as belonging to the nbM and lower
cortical ChAT levels.   However, the small decrease in ChAT levels
and the restriction of this decrease to the central area of cortex
raises at least two questions.   First, could the decrease in cortical
ChAT levels be due to a small amount of non-specific damage to the
nbM that escaped notice by the qualitative examination of the lesion
site performed in this study?   Second, can AF64A be used to produce
a decrease in ChAT levels large enough to compare with that which
occurs in SDAT without also producing extensive non-specific damage?

In response to the first question, a non-specific lesion of
the nbM of sufficient magnitude to produce a cortical ChAT deficit
should have been detectable by the methods employed in this study,
since the small amount of non-specific damage resulting from vehicle
injections, which did not alter cortical ChAT, was visible.   Instead
of being due to non-specific damage, the small magnitude of the
ChAT loss may be the logical correlate of the small percentage of
the area of the nbM depleted of AChE activity by the AF64A.   Furthermore, the localized nature of the decrease in ChAT activity may
reflect the topography of nbM projections to cortex (12).   The question

of whether AF64A can produce large ChAT depletions in cortex without extensive non-specific damage to the nbM cannot be answered at this time. Clearly, the production of larger ChAT depletions cannot be achieved by increasing the concentration of AF64A injected into the nbM without producing substantial non-specific damage (present results).

In conclusion, while much additional study is necessary, the present results show that injections of low concentrations of AF64A into the nbM can specifically damage this nucleus and decrease cortical ChAT activity. Thus, such injections may be useful in creating animal models of SDAT.

REFERENCES

1. Asante, J.W., Cross, A.J., Deakin, J.F.W., Johnson, J.A. and Slater, H.R. (1983): Br. J. Pharmacol. 80:573P.
2. Caulfield, M.P., May, P.J., Peddler, E.K. and Prince, A.K. (1983): Br. J. Pharmacol. 79:287P.
3. Davies, P. (1979): Brain Res. 171:319-327.
4. Fisher, A. and Hanin, I. (1980): Life Sci. 27:1615- 1634.
5. Fisher, A., Mantione, C.R., Abraham, D.J. and Hanin, I. (1982): J. Pharmacol. Exp. Ther. 222: 140-145.
6. Hardy, H., Heimer, L., Switzer, R. and Watkins, D. (1976): Neurosci. Lett. 3:1-5.
7. Houser, C.R., Crawford, G.D., Barber, R.P., Salvaterra, P.M. and Vaughn, J.E. (1983): Brain Res. 266:97-119.
8. Johnston, M.V., McKinney, M. and Coyle, J.T. (1979): Proc. Natl. Acad. Sci. 76:5392-5396.
9. Katzman, R. (1976): Arch. Neurol. 33:217-218.
10. Mantione, C.R., Zigmond, M.J., Fisher, A. and Hanin, I. (1983): J. Neurochem. 41:251-255.
11. Marshall, J.F., Van Oordt, K. and Kozlowski, M.R. (1983): Brain Res. 274:283-289.
12. Mesulam, M.-M., Mufson, E.J., Levey, A.I. and Wainer, B.H. (1983): J. Comp. Neurol. 214:170-197.
13. Perry, E.K., Gibson, P.H., Blessed, G., Perry, R.H. and Tomlinson, B.E. (1977): J. Neurol. Sci. 34:274-265.
14. Sandberg, K., Sanberg, P.R., Hanin, I., Fisher, A. and Coyle, J.T. (1984): Behav. Neurosci. 98:162-165.
15. Schrier, B.K. and Shuster, L. (1967): J. Neurochem. 14:977-985.
16. Sinex, F.M. and Myers, R.H. (1982): In: Alzheimer's Disease, Down's Syndrome and Aging (eds) F.M. Sinex and C.R. Merril, The New York Academy of Sciences, New York, pp. 3-13.
17. Whitehouse, P.J., Price, D.L., Struble, R.G., Clark, A.W., Coyle, J.T. and DeLong, M.R. (1982): Science 215:1237-1239.

NEUROTOXIC EFFECTS OF NITROGEN MUSTARD ANALOGUES OF CHOLINE

E.H. Colhoun

Department of Pharmacology and Toxicology
The University of Western Ontario
London, Ontario, Canada   N6A 5C1

INTRODUCTION

The first synthesis of acetylcholine mustard (AChM) can be attributed to Hanby and Rydon (11), and the toxicology of it and other nitrogen mustard analogues of choline to Anslow et al. (1). Jackson and Hirst (15) modified the synthesis procedure to give higher yield, and showed clearly that the biological activity of AChM was due to formation of the aziridinium ion (Az) isomer in a polar solvent. The framework molecular model of the parent mustard compound, its Az ion isomer and acetylcholine (ACh) is illustrated in Fig. 1. Note the similarity between ACh and the Az ion isomer and in particular the ammonium head. In our laboratory the synthesis technique for the production of AChMAz has, with modifications, been utilized for the synthesis of other mustard analogues (Fig. 2). These compounds are acetylmonoethycholine mustard (AMEChM) and ethoxycholine mustard (EChM). Choline mustard (ChM) and monoethylcholine mustard (MEChM) Az were obtained by alkaline hydrolysis of the acetylated analogues. The aziridinium ion isomers are denoted by the Az suffix.

Until our work on the cholinergic nerve terminal (24) most of the experiments with AChMAz were carried out on various muscle preparations (5, 13, 15, 20, 22). These experiments showed that AChMAz was a potent cholinergic agonist with possible alkylating properties. Clement (5) was the first to recognize that the toxicity of AChMAz was due to its _in vivo_ hydrolysis to ChMAz, and in pursuit of the mechanism of action showed that transport of choline in human erythrocytes was antagonized by ChMAz (6). According to Hanin et al. (12) ChMAz, MEChMAz and a mustard analogue of hexamethonium (3) are candidates for the role of cholinotoxins, with the mustard analogues of choline being the most likely agents as selective toxins in the

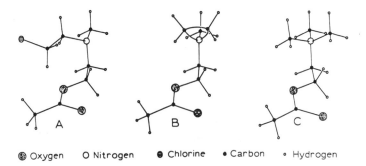

⊛ Oxygen   O Nitrogen   ● Chlorine   • Carbon   ∘ Hydrogen

Figure 1.   Framework molecular models of acetylcholine mustard (A),
and acetylcholine mustard aziridinium ion (B), illustrating the
latter's structural relationship to acetylcholine (C).   Jackson (14).

$$
\text{General Formula} = \quad
\begin{array}{ccccc}
H & H & & R_2 & \\
| & | & & \diagup & \\
R_1 - C & - C & - N & - R_3 \\
| & | & & \diagdown & \\
H & H & & R_4 &
\end{array}
$$

Compound	R_1	R_2	R_3	R_4
Choline	OH	CH_3	CH_3	CH_3
ChM Az	OH	CH_3	$CH_2 \diagup\!\!\!\!-\!\!\!\!\diagdown CH_2$	
MECh	OH	CH_2CH_3	CH_3	CH_3
MEChM Az	OH	CH_2CH_3	$CH_2 \diagup\!\!\!\!-\!\!\!\!\diagdown CH_2$	
AMEChM Az	CH_3COO	CH_2CH_3	$CH_2 \diagup\!\!\!\!-\!\!\!\!\diagdown CH_2$	
ACh	CH_3COO	CH_3	CH_3	CH_3
AChM Az	CH_3COO	CH_3	$CH_2 \diagup\!\!\!\!-\!\!\!\!\diagdown CH_2$	
ECh	CH_3CH_2O	CH_3	CH_3	CH_3
EChM Az	CH_3CH_2O	CH_3	$CH_2 \diagup\!\!\!\!-\!\!\!\!\diagdown CH_2$	
HN2	Cl	CH_3	$CH_2 \diagup\!\!\!\!-\!\!\!\!\diagdown CH_2$	

Figure 2.   Molecular structures of nitrogen mustard analogs of choline
in relation to acetylcholine and choline.

cholinergic nervous system. In vivo evidence for this hypothesis is, as yet, contradictory.

CHEMISTRY

In our laboratory, ChMAz and MEChMAz are derived by the alkaline hydrolysis of the parent esters, AChMAz and AMEChMAz. It is important to note that the Az ion isomers are first generated in a polar solvent and that the maximum concentration of ion is about 70% (Fig. 3) at pH 7.5 at 22°C. More ion can be generated at 37°C but the solvated ion decomposes more rapdily to the hydroxyethylamine, by dimerization to the piperanzinium salt (polymer) or reversion to the starting material. The major decay product appears to be hydrolysis to the alcohol (14) when ion formation is carried out at 23°C. At ambient temperature the decline in the Az ion is about 5% in 90 min. The decay products appear to have no biological activity.

We have found that 10 mg/ml of neat AChMAz (prepared as the oil) is optimum for solvating the Az ion. Concentrations above this result in lower ion production (Fig. 4), due to increased formation of the dimeric piperazinium salt (14). Although we have prepared ChM as the hydrochloride we routinely prepare it by hydrolysis of AChMAz at pH 11.0 for 10 min. The amount of ChMAz is about 40% when titrated against thiosulphate. Some of the differences in biological activity of the nitrogen mustard analogues of choline may in fact be due to either differences in ion concentration or

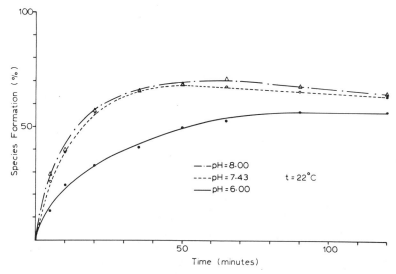

Figure 3. Effect of pH on the formation of acetylcholine mustard aziridinium ion. Jackson (14).

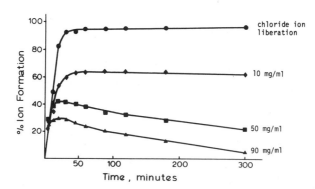

Figure 4. The formation and stability of the aziridinium ion of acetylcholine mustard in aqueous solution is a function of acetylcholine mustard concentration and time. Aziridinium ion formation was inversely related to the concentration of acetylcholine mustard, and stability of the ion was greater at lower acetylcholine mustard concentrations.

corrections to 100% of the actual Az ion itself. Reports in the literature about the biological effects of these mustard analogues of choline should be standardized to a common Az ion concentration. In our laboratory MEChMAz was prepared in a similar manner to ChMAz. Dr. M. Hirst synthesized EChM and its biological activity was first reported by Pugh et al. (21). In 1977, Rylett and Shazani in our laboratory synthesized AMEChMAz and MEChMAz by an adaption of the procedure for AChM (see 23).

TOXICOLOGY

All of the nitrogen mustard analogues of choline synthesized in our laboratory have been tested for in vivo toxicity in mice. Also the effect of AChMAz and ChMAz has been investigated in the anaesthetized rat (16). The data outlined in Table 1 summarize the toxicity of the various compounds. The acetylated compounds and EChM are cholinergic agonists with acute toxicity similar to ACh when given intravenously. The toxicity of these compounds relative to that found for ACh is given and it is evident that in most instances they are more toxic than ACh. It should be noted, however, that the effect of ACh in vivo is evanescent (8), and its rapid hydrolysis to choline is insidious in determining toxicity. ChMAz and MEChMAz had effects similar to hemicholinium-3 (HC-3), and the comparative effects of AChMAz, ChMAz and HC-3 were recorded as kinetograms obtained by placing mice in an activity chamber (kinetograph), and recording their spontaneous activity. Although HC-3 produced toxicity faster than either ChMAz or MEChMAz the signs of toxicity were strikingly similar for all three compounds, and it was concluded that the mustard

Table 1

The Acute Toxicity (24 hr) of Nitrogen Mustard Analogs of
Choline Given Intravenously to Mice

Compound	LD_{50} mg/kg*	Potency relative to ACh	Signs of Toxicity
ACh	24.6	1.0	Agonist-piloerection, lacrimation, salivation, urination, defecation, heart block, death, 0-1 min
AChMAz	3.1	0.13	Similar to ACh-survivors at the minimum lethal dose exhibit toxicity signs of ChMAz
ECMAz	7.0	0.28	Similar to ACh and AChMAz death 1-3 min
ECMAz (i.p.)	89.6	3.6	Similar to ACh but signs of toxicity delayed
ECMAz	56.0	2.2	Cytotoxicity 36 hr-7 days noncholinergic
Choline (Ch)	---	est. 20-27	Weak agonist
ChMAz	2.8	0.11	1-3 min. no effect; 3-5 min. quiescence; 5-10 min terminal respiratory convulsions duration 30-60 sec
MEChMAz (AF64A)	4.6	0.19	Similar to ChMAz
MECh	---	est. 35	Weak agonist
ECh (ethoxycholine)	---	---	Agonist; similar to ECMAz
AMECh (acetylmonoethylcholine)	---	---	Agonist
HC-3	0.089	0.004	Similar to ChMAz or MEChMAz; (hemicholinium) time span to death shorter

*The LD_{50} values for nitrogen mustard analogs of choline have been
corrected to 100% Az ion.

analogues of choline were acting presynaptically in the peripheral cholinergic system and causing death through respiratory paralysis, with the diaphragm being the target site. At lower than the acute toxic dose, AChMAz could produce toxicity through hydrolysis to ChMAz; similarily for AMEChMAz. The effects of ChMAz could be prevented by pretreating mice with choline or pentobarbital sodium; DL-amphetamine and other central stimulants potentiated the toxic effect of ChMAz.

Studies with the pentobarbital-anesthetized rat showed, by measuring respiration, phrenic nerve discharge and blood pressure, that ChMAz had effects similar to HC-3 given intravenously. In contrast to HC-3 in the rat, however, ChMAz appeared to cross the blood-brain barrier and depress phrenic nerve activity as well as inhibiting contractions of the diaphragm. It has been suggested that MEChMAz might be given peripherally to achieve an effect in the central nervous system (12). Our results with ChMAz in the rat would support this idea, but the relative amount of compound transported into the brain might be negligible compared to its concentration and action in the peripheral nervous system.

CELLULAR MECHANISM OF ACTION OF ChMAz AND MEChMAz

The mechanism of action of some of the nitrogen mustard analogues of choline at presynaptic and postsynaptic sites in the cholinergic nervous system is shown in Fig. 5. The actions of AChMAz and ChMAz are compared with ACh and choline. Extraneuronal choline either from the hydrolysis of ACh, synthesis in the liver or from a dietary source, is thought to be transported via the sodium-dependent high-affinity choline carriers intraneuronally, and there synthesized by choline acetyltransferase (ChAT) (with acetyl Coenzyme-A as the cosubstrate) to ACh. The source of acetyl groups for Coenzyme-A in the cytosol is generated by mitochondria in the form of citrate. In the rat, a coupling may exist between the high-affinity choline carrier and ChAT, with the possibility that an isoenzyme of ChAT may be membrane-bound and thus in intimate proximity to the site of the choline carriers. Through transport on the low affinity carriers choline is utilized for the production of phosphorylcholine and incorporated into structural components in the nerve ending. The schematic shown (Fig. 5) indicates clearly that AChMAz and ChMAz can interact in the nerve terminal and junction similar to choline or ACh. MEChMAz would act like ChMAz, but with a lower order of potency (26). The results of Rylett and Colhoun (24, 25, 27, 28) describe the postulated mechanism of action of ChMAz with high-affinity choline carriers and inhibition of ChAT.

Our most recent results suggest that ChMAz alkylates synaptosomal choline carriers. Moreover, because of inhibition of synaptosomal ChAT at low ChMAz concentrations, we feel that in rat brain synaptosomes a coupling may exist between ChAT and sodium-dependent,

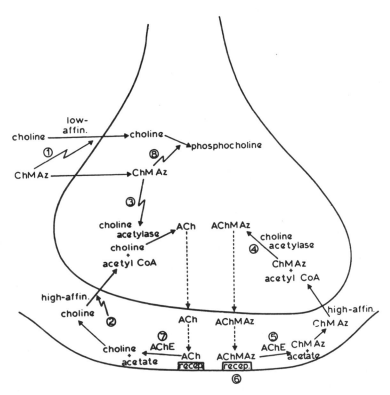

Figure 5. A schematic of a cholinergic synapse showing the proposed sites of action of acetylcholine mustard aziridinium ion and choline mustard aziridinium ion. Rylett (23).

high-affinity choline carriers. The nature of the coupling has yet to be determined, and a difference may exist for guinea-pig and rat brain synaptosomes (28).

ACTIONS OF CHOLINE MUSTARD Az IN THE CENTRAL NERVOUS SYSTEM

Fisher and Hanin (9) have suggested that analogues of choline, particularly as the nitrogen mustard analogues, were potential tools in developing animal models of cholinergic hypofunction. Using MEChMAz (AF64A) as a cholinotoxin, Hanin and colleagues (10, 12, 17-19) have demonstrated that this compound may have selective cholinergic actions in the central nervous system of mice and rats. The result of Caulfield et al. (4) and Asante et al. (2) for MEChMAz appear, however, to be different, and they indicate that it may lack specificity and be unable to produce an animal model for behavioural disorders.

To date, similar experiments have not been carried out with ChMAz, which is interesting in view of the published data on the mechanism of action of this nitrogen mustard analogue of choline, and its close similarity in chemical structure to choline. Preliminary experiments recently performed in our laboratory and summarized in Table 2 show some of the biochemical effects of ChMAz injected into areas of rat brain. As expected, we obtained evidence of inhibition of both ChAT and the high-affinity choline carriers when 20 nmoles of ChMAz were injected bilaterally into the lateral ventricles or medial septum. We examined the activity of ChAT and high-affinity transport of choline in hippocampus, striatum and cortex. Depression of ChAT and transport of choline was found only in the hippocampus following injection of ChMAz into the medial septum, suggesting selective lesions (7). Preliminary studies carried out in collaboration with Dr. E. McGeer and Dr. P. McGeer, Kinsman Laboratory, The University of British Columbia, Vancouver, have revealed necrosis

Table 2

Effect of ChMAz Injected Into Regions of Rat Brain On Choline Acetyltransferase (ChAT) Activity and High-Affinity Transport of Choline (HAChT)

(a): Lateral Ventricles (7 days after injection)

Dose ChMAz nmoles	Biochemical Parameter	% Decrease Relative to Control		
		Cortex	Hippocampus	Striatum
20	ChAT	20	32	0
	HAChT	25	35	0
40	ChAT	30	45	13
	HAChT	30	45	15

(b): Medial Septum (7 days after injection)

20	ChAT	15	60	0
	HAChT	15	80	0
8	ChAT	0	30	0
	HAChT	0	33	0
1	ChAT	0	10	0
	HAChT	0	15	0

at the site of injection, which may be a factor in depression of ChAT activity in rat brain. The data tended to show less necrosis when the dose of ChMAz was reduced, but with a concomitant lack of effect against ChAT.

In experiments carried out in our laboratory with ChMAz we have thus found evidence of lesions at the site of injection. Some years ago we found that AChMAz produced lesions when injected into the body cavity of an invertebrate (5). The apparent ability of ChMAz, MEChMAz and AChMAz to produce tissue necrosis at the site of injection suggests a common action of the beta-haloethylamines, which must be taken into consideration when measuring effects following injection of any one of them into the central nervous system.

Acknowledgements. In compiling this manuscript on the neurotoxic effects of nitrogen mustard analogues of choline, I express my indebtedness to my colleagues in the Department of Pharmacology and Toxicology who assisted with the many experiments over the past 10 years. In particular I wish to thank Dr. M. Hirst, Dr. J. Clement, Ms. Sharon Pugh, Mr. E. Kurek and Dr. R.J. Rylett, all of whom in their own way have contributed to my understanding of the mechanism of action of the aziridinium ion isomers of choline.

REFERENCES

1. Anslow, W.P., Karnovsky, D.A., Val Jager, B. and Smith, H.W. (1947): J. Pharmacol. Exp. Therap. 91:224-235.
2. Asante, J.W., Cross, A.J., Deakin, J.F.W., Johnson, J.A. and Slater, H.R. (1983): Proc. Br. Pharmacol. Soc. P. 82.
3. Barker, L.A. (1979): In Brain Acetylcholine and Neuropsychiatric Disease (eds) K.L. Davis and P.A. Berger, Plenum Press, New York, pp. 515-532.
4. Caulfield, M.P., May, P.J., Pedder, E.K. and Prince, A.K. (1983): Br. J. Pharmacol. Proc. Suppl. 79:287P.
5. Clement, J.G. (1974): Ph.D. Thesis, University of Western Ontario, London, Ontario, Canada. Toxicity and RelatedPharmacological Actions of Acetylcholine Mustard.
6. Clement, J.G. and Colhoun, E.H. (1974): Pharmacologist 10:273.
7. Colhoun, E.H., Brajac, D. and Rylett, R.J. (1983): Soc. for Neurosciences 9:266.
8. Dale, H.H. (1914): J. Pharmacol. Exp. Ther. 6:147-190.
9. Fisher, A. and Hanin, I. (1980): Life Sci. 27:1615-1634.
10. Fisher, A., Mantione, C.R., Grauer, E., Levy, A. and Hanin, I. (1983): In Behavioral Models and the Analysis of Drug Action (eds) M.Y. Speigelstein and A. Levy, Elsevier Scientific Publishing Company, Amsterdam, pp. 333-342.
11. Hanby, W.E. and Rydon, N.H. (1947): J. Chem. Soc. 513-519.
12. Hanin, I., DeGroat, W.C., Mantione, C.R., Coyle, J.T. and Fisher, A. (1983): Banbury Report 15:243-253.

13. Hudgins, P.M. and Stubbins, J.F. (1972): J. Pharmacol. Exp.Ther. 182:303-311.
14. Jackson, C.H. (1972): Ph.D. Thesis, University of Western Ontario, London, Ontario, Canada. Pharmacological Actions of a Mustard Derivative of Acetylcholine.
15. Jackson, C.H. and Hirst, M. (1972): J. Med. Chem. 15:1183-1184.
16. Kurek, E.J.C. (1978): M.Sc. Thesis, University of Western Ontario, London, Ontario, Canada. Further Investigation ofthe Toxicity of Acetylcholine Mustard Aziridinium Ion and Analogues.
17. Mantione, C.R., Fisher, A. and Hanin, I. (1981): Science 213: 579-580.
18. Mantione, C.R., DeGroat, W.C., Fisher, A. and Hanin, I. (1981): Pharmacologist 23:224.
19. Mantione, C.R., Zigmond, M.J., Fisher, A. and Hanin, I. (1982): Trans. Am. Soc. Neurochem. 13:387.
20. Parker, T.S., Macri, J.R. and Jenden, D.J. (1976): Proc. West. Pharmacol. Soc. 19:79.
21. Pugh, S.C., Hirst, M. and Colhoun, E.H. (1974): Can. Fed. Biol. 17:56.
22. Robinson, D.A., Taylor, D.G. and Young, J.M. (1975): Br. J. Pharmacol. 53:363-370.
23. Rylett, R.J. (1980): Ph.D. Thesis, University of Western Ontario, London, Ontario, Canada. Actions of Choline Mustard Aziridinium Ion at the Cholinergic Nerve Terminal.
24. Rylett, R.H. and Colhoun, E.H. (1977): Can. J. Physiol.Pharmacol. 55:769-772.
25. Rylett, R.J. and Colhoun, E.H. (1979): J. Neurochem. 32:553-558.
26. Rylett, R.H. and Colhoun, E.H. (1980): J. Neurochem. 34:713-719.
27. Rylett, R.J. and Colhoun, E.H. (1980): Life Sci. 26:909-914.
28. Rylett, R.J. and Colhoun, E.H. (1985): J. Neurochem. 43:787-794.

SYNTHESIS, STORAGE AND RELEASE OF CHOLINE ANALOG ESTERS

B. Collier and S.A. Welner

Department of Pharmacology
McGill University
McIntyre Building, 3655 Drummond Street
Montreal H3G 1Y6, Canada

INTRODUCTION

It is well accepted that the choline for acetylcholine (ACh) synthesis is delivered to choline acetyltransferase by a process of transport that carries choline from the extracellular space to the cytosol of the cholinergic nerve terminal (see reviews by 11, 12). The acetyl group for ACh synthesis clearly derives from acetyl-- CoA, although the mechanism by which acetyl-CoA is delivered to choline acetyltransferase remains unclear (see 5, 19). The rate of ACh synthesis is known to be regulated according to need so that activity in a cholinergic neurone that releases transmitter turns on ACh synthesis (see 15, 19). The question is: how is this control achieved?

Over the last few years, we have used analogs of choline in an attempt to provide information about ACh synthesis, storage and release mechanisms. In the last two meetings of this series of International Cholinergic meetings, we presented data dealing with the regulation of the choline uptake process in preganglionic sympathetic nerve terminals, and with the relationship between stimulation-induced increased choline uptake and stimulus-induced increased ACh synthesis (8, 9). The present chapter is about experiments that are a continuation of those mentioned above.

PRECURSOR DELIVERY AND SYNTHESIS REGULATION

It is normally considered that ACh synthesis is not controlled by inhibition of the reaction by product, nor by changes of enzyme activity per se (e.g., 15, 19). So, the focus has been toward regulation by precursor delivery or by mass action (e.g., 10).

If precursor delivery regulates ACh synthesis, then it seems reasonable to postulate that neuronal activity that releases transmitter and activates synthesis should activate precursor delivery. For choline, we tested this premise for the cholinergic nerve terminals of sympathetic ganglion, with choline analogs like homocholine or triethylcholine (7, 17); these compounds are reasonable substrates for the choline transport process, but are not very good substrates for choline acetyltransferase. Preganglionic nerve activity enhanced choline analog accumulation into cholinergic terminals, indicating that the choline uptake mechanism is regulated in a way that could be functional in driving increased ACh synthesis. For acetyl-CoA, we obtained similar evidence for increased availability to choline acetyltransferase during neuronal activity, by inference from studies of ACh synthesis from acetate (13). Glucose-derived acetyl-CoA diluted acetate-derived acetyl-CoA during neuronal activity, which was interpreted to indicate that delivery of mitochondrial acetyl-CoA for ACh synthesis was enhanced during nerve stimulation.

Thus, there is evidence for the increased delivery of both ACh precursors during synaptic activity. But the two processes appeared evident with equal magnitude in the presence of enough Mg^{++} to block transmitter release, yet ACh content did not increase as it might be expected to do if the increased delivery of precursors increased synthesis. The conclusion from these data that seemed most attractive was that increased precursor delivery during activity was facilitatory to increased ACh synthesis, but alone, not enough to explain the phenomenon.

MASS ACTION AND SYNTHESIS REGULATION

The control of ACh synthesis as the result of mass action implies that the choline acetyltransferase reaction is at equilibrium in a resting nerve terminal and that the equilibrium is displaced to the right by transmitter release (18, 19). A reasonable corollary of this idea is that ACh uptake by synaptic vesicles is the process that is limiting to ACh synthesis. This concept is one I can't find written down explicitly, although it is one I know has arisen in discussions with, for example, both F.C. McIntosh and V.P. Whittaker over the last few years (thus, the idea is likely theirs, not mine!). Anyway, the hypothesis (see Fig. 1) is based upon the well-known assumption that ACh release occurs from synaptic vesicles, that following exocytosis there is vesicle re-cycling, and that empty vesicles are charged by ACh taken up from the cytoplasmic pool. Thus, it is vesicle uptake of ACh that displaces the equilibrium of the choline acetyltransferase reaction, if that equilibrium were to exist.

To test this idea, we have sought for a choline analog that would be taken up by cholinergic nerve terminals, that would be acetylated to generate a false ester within the nerve terminal;

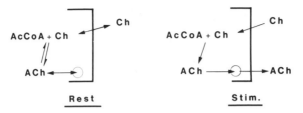

Figure 1. Schematic representation of the idea that the choline acetyltransferase reaction is in equilibrium and that ACh synthesis activation, following release, is the consequence of ACh uptake into synaptic vesicles.

but we also wanted that false ester to be a poor substrate for the ACh uptake mechanism of vesicles.

SUBCELLULAR DISTRIBUTION AND RELEASE OF FALSE TRANSMITTERS

Previous studies with false transmitters provided clues about what choline analogs to make with the objective stated above. When rat brain slices are incubated with a normal Krebs medium containing choline analogs, the acetylated products can be measured in subcellular fractions prepared from those slices (see 2, 3). Neither ethylcholine esters, nor the homocholine ester, distinguish themselves readily from ACh; but the acetyl esters of triethylcholine and of homocholine most nearly do so (Fig. 2). These two compounds are also the false transmitters in Torpedo nerve terminals that most clearly differ from ACh in their subcellular distribution (14, 20). Thus, on the basis of this evidence, we made the ethyl derivatives of homocholine.

THE ETHYL HOMOCHOLINES

We synthesized the three ethyl analogs of homocholine:

$$\begin{matrix} CH_3 \\ CH_3 \\ C_2H_5 \end{matrix} \!\!\!\searrow\!\! \overset{+}{N} - CH_2 - CH_2 - CH_2 - OH \qquad \text{monoethylhomocholine}$$

$$\begin{matrix} CH_3 \\ C_2H_5 \\ C_2H_5 \end{matrix} \!\!\!\searrow\!\! \overset{+}{N} - CH_2 - CH_2 - CH_2 - OH \qquad \text{diethylhomocholine}$$

$$\begin{matrix} C_2H_5 \\ C_2H_5 \\ C_2H_5 \end{matrix} \!\!\!\searrow\!\! \overset{+}{N} - CH_2 - CH_2 - CH_2 - OH \qquad \text{triethylhomocholine}$$

The three analogues were tested as substrates for the choline uptake process by incubating [14C]analogs with synaptosomes prepared from

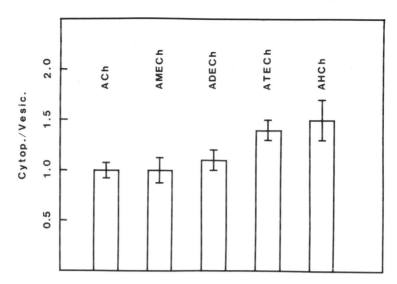

Figure 2. The subcellular distribution of ACh and four false transmitters. The ratio of ester in cytoplasmic fraction to that in vesicle-bound fraction is shown. The false transmitters are acetylmono-
ethylcholine (AMECh), acetyldiethylcholine (ADECh), acetyltriethyl-
choline (ATECh) and acetylhomocholine (AHCh).

rat forebrain. The mono- and the di-ethyl homocholines were transported, but the triethyl compound was not.

After their uptake into nerve terminals, the mono- and di-ethyl
homocholines were acetylated. This was shown by incubating slices
of cerebral cortex with $[^{14}C]$ analogs and measuring acetyl products
in tissue extracts. The subcellular distribution of acetyldiethyl-

Table 1

Subcellular Distribution of ACh and of
Acetyldiethylhomocholine

	cytoplasmic/vesicular
Acetylcholine	1.3
Acetyldiethylhomocholine	16.9

n = 4 in each case

Table 2

Spontaneous and Evoked Release of ACh and
Acetyldiethylhomocholine From Cortex Slices

Medium	Acetylcholine	Acetyldiethylhomocholine
	(dpm/mg wet weight/5 min)	
Normal	15.6 $\pm$ 2.4	24.6 $\pm$ 4.5
Normal	14.6 $\pm$ 2.1	23.2 $\pm$ 3.7
High K$^+$	53.0 $\pm$ 7.2	16.8 $\pm$ 3.1

n = 4 in each case

homocholine was compared to that of ACh by incubating slices with [^{14}C] diethylhomocholine and [^{3}H] choline, homogenizing the slices, preparing synaptosomes, lysing the synaptosomes and measuring esters in vesicle-rich and cytosolic fractions. The acetyldiethylhomocholine was less well accumulated by synaptic vesicles than was ACh (Table 1).

The release of acetyldiethylhomocholine was compared to that of ACh by incubating slices with the precursors and subsequently collecting transmitters released into normal medium and into K$^+$-rich medium. Both esters effluxed into normal medium, but K$^+$ released only ACh, not acetyldiethylhomocholine (Table 2).

Thus, acetyldiethylhomocholine differs in two ways from the false esters previously studied: it accumulates poorly into synaptic vesicles, and it appears not to be released by stimuli that release ACh; the false transmitters previously studied are releasable (see 6). In other words, the transmitter release mechanism has the same structural specificity as has the process responsible for ACh seques-tration by vesicles, as if a vesicle releases its transmitter in response to stimuli and the cytosol does not.

ACETYLDIETHYLHOMOCHOLINE AND ACh SYNTHESIS REGULATION

As mentioned earlier, if vesicle uptake of ACh plays a key role in ACh synthesis regulation, a false ester that is largely excluded from vesicles should provide information about the phenom-enon. We have studied this with superior cervical ganglia. The activation of synthesis by activity was measured as increased in-corporation of choline or choline analog into its acetylester induced by preganglionic nerve stimulation. For ACh and the other releasable false transmitters, stimulation activates their synthesis; but it was so too (by 7-fold) for acetyldiethylhomocholine (Table 3). Compatible with the study on brain, acetyldiethylhomocholine was not releasable from ganglia by nerve impulses.

The activation of ACh synthesis in ganglia during preganglionic stimulation is reduced when transmitter release is reduced by the presence of increased Mg^{++}. A similar result pertains to the synthesis of acetyldiethylhomocholine (Table 3), even though the release of this compound is not, of course, affected. This result might suggest that, under normal conditions, when release of ACh occurs, the uptake into vesicles of endogenous ACh relieves the restraint of cytoplasmic ACh to the choline acetyltransferase reaction, and that the increased acetylation of cytoplasmic analog reflects this. This idea was tested using the compound 2-(4-phenylpiperidino)cyclohexanol or AH5183 (4, 16) which is a potent inhibitor of the process responsible for ACh uptake by vesicles (1, see also chapter by Parsons in this book).

The compound AH5183, if it blocks the uptake of endogenous ACh by vesicles, should prevent any stimulus-induced loss of cytoplasmic ACh and, thereby, prevent any disturbance of the choline acetyltransferase reaction's equilibrium. Thus, if mass action regulates ACh synthesis, AH5183 should prevent the stimulus-induced activation of the conversion of diethylhomocholine to its acetyl ester; it did not (Table 3). Therefore, either the hypothesis is untenable or the AH5183 is ineffective. Toward characterizing the compound, we have measured its effect on ACh release from ganglia during preganglionic stimulation: initial release was in normal amount, but was sustained less-well than in the drug's absence.

Table 3

The Change in the Synthesis of Acetyldiethylhomocholine by Ganglia During Preganglionic Nerve Stimulation

Condition	Increased Ester formed during stim. (d.p.m.)
Normal	890 ± 284
Increased Mg^{++}	348 ± 129
AH5183	1174 ± 258

n = 4 in each case

The evoked release of ACh in the presence of the compound fell to an immeasurable level at a time when total tissue stores of ACh were far from zero, a result compatible with its presumed major action.

Thus it appears that neither mass action nor precursor delivery per se is the key factor that regulates the synthesis of ACh. The present inference from the activation of acetyldiethylhomocholine synthesis during activity suggests that ACh synthesis is normal during activity when transmitter release is blocked by AH5183, but that synthesis is reduced when transmitter release is blocked by Mg^{++}. As Mg^{++} blocks Ca^{++} influx but AH5183 does not, it is suggested that the increased influx of Ca^{++} associated with nerve terminal impulses might be involved in the process of ACh synthesis activation. This hypothesis will be tested in time for the next meeting of this series.

Acknowledgements. This work was supported by the Medical Research Council of Canada with operating grants and a Studentship (to S.A.W.). Dr. S.M. Parsons generously provided us with the compound AH5183, as well as some solvent.

REFERENCES

1. Anderson, D.C., King, S.C. and Parsons, S.M. (1983): Molec. Pharmacol. 24:48-54.
2. Boksa, P. and Collier, B. (1980): J. Neurochem. 34:1470-1482.
3. Boksa, P. and Collier, B. (1980): J. Neurochem. 35:1099-1104.
4. Brittain, R.T., Levy, G.P. and Tyers, M.B. (1969): Brit. J. Pharmacol. 36:173-174.
5. Clark, J.B., Booth, R.F.G., Harvey, S.A.K., Leong, S.F. and Patel, T.B. (1982): In Neurotransmitter Interaction and Compartmentation (ed) H.F. Bradford, Plenum Press, New York, pp. 431-460.
6. Collier, B., Boksa, P. and Lovat, S. (1979): Progr. Brain. Res. 49:107-121.
7. Collier, B. and Ilson, D. (1977): J. Physiol. 264:489-509.
8. Collier, B., Ilson, D. and Lovat, S. (1977): In Cholinergic Mechanisms and Psychopharmacology (ed) D.J. Jenden, Plenum Press, New York, pp. 457-464.
9. Collier, B. and O'Regan, S. (1981): In Cholinergic Mechanisms (eds) G. Pepeu and H. Ladinsky, Plenum Press, New York, pp. 97-107.
10. Jenden, D.J. (1979): In Nutrition and the Brain (eds) A. Barbeau, J.H. Growdon and R.J. Wurtman, Raven Press, New York, pp. 13-24.
11. Jope, R.S. (1979): Brain Res. Rev. 1:313-344.
12. Kuhar, M.J. and Murrin, L.C. (1978): J. Neurochem. 30:15-21.
13. Kwok, Y.N. and Collier, B. (1982): J. Neurochem. 39:16-26.
14. Luqmani, Y.A., Sudlow, G. and Whittaker, V.P. (1980): Neuroscience 5:153-160.
15. MacIntosh, F.C. and Collier, B. (1976): In Neuromuscular Junction (ed) E. Zaimis, Springer-Verlag, Berlin, pp. 99-228.
16. Marshall, I.G. (1970); Brit. J. Pharmacol. 38:503-516.

17. O'Regan, S. and Collier, B. (1981): Neuroscience 6:511-520.
18. Potter, L.T., Glover, V.A.S. and Saelens, J.K. (1968): J. Biol. Chem. 243:3864-3870.
19. Tuček, S. (1978): In Acetylcholine Synthesis in Neurons (eds) Chapman and Hall, London, pp. 1-259.
20. Whittaker, V.P. and Luqmani, Y.A. (1980): Gen. Pharmacol. 11:7-14.

A NEW PHARMACOLOGICAL TOOL TO STUDY ACETYLCHOLINE

STORAGE IN NERVE TERMINALS

S.M. Parsons, D.C. Anderson, B.A. Bahr and G.A. Rogers

Department of Chemistry
University of California
Santa Barbara, California 93106 U.S.A.

INTRODUCTION

Synaptic vesicles isolated from the electric organ of Torpedo californica exhibit MgATP stimulated uptake of tritium labelled acetylcholine, [^{3}H]ACh. The process probably is mediated by a proton pumping ATPase which acidifies the interior of the vesicles (7, 8, 17, 18). A hypothesized separate ACh transporter protein in the vesicle membrane appears to utilize the vesicle membrane pH gradient to drive concentrative uptake of exogenous [^{3}H]ACh (1, 15, 16).

Recently we have studied the pharmacological characteristics of the ACh transport system (2, 3). A wide range of drugs was screened for effects on the uptake system, which was shown to be distinguishable from all other known cholinergic proteins. From among greater than 80 tested drugs we found two to be particularly potent inhibitors of ACh transport. These are dl-trans-2-(4-phenylpiperidino)cyclo-hexanol (called AH5183) and the tetraphenylborate anion, which exhibited IC$_{50}$ values of about 40 and 300 nM, respectively. These drugs had been hypothesized by I.G. Marshall and colleagues to block ACh storage by synaptic vesicles in vivo (6,10-12). The study described in this paper examined the mechanism(s) by which AH5183 acts to inhibit ACh uptake. Also, resolution of AH5183 into its optical isomers has been accomplished so that stereoselectivity of the drug action could be examined.

MATERIALS AND METHODS

Torpedo californica electric organ synaptic vesicles were isolated as described (1). Experiments were conducted at about 0.5 mg vesicle protein/ml unless otherwise stated. The optical isomers of AH5183

were resolved by fractional crystallization of the (-)- or (+)-di-
p-toluoyltartaric acid (Chemical Dynamics Corp.) salts followed by
conversion to the hydrochloride salts.

For determination of the rates of leakage of endogenous ACh from
vesicles, external choline (arising from leakage of ACh during vesicle
purification) first was removed by centrifugation-gel filtration
of twenty 200 μl portions of purified vesicles through 2 ml Sephadex
G50 columns at 4°C, equilibrated in 0.6 M glycine, 0.2 M Hepes, 1
mM EDTA, 1 mM EGTA, 0.02% (w/v) KN$_3$, adjusted to pH 7.4 with 0.8 N
KOH (buffer A), followed by pooling. Ten μl portions were assayed
periodically for external choline by the chemiluminescent method
of Israël and Lesbats (6) in 0.2 M sodium phosphate at pH 8.6. A
control experiment utilizing vesicles treated with 0.4 mM paraoxon
demonstrated that the assayed Ch was originally ACh.

[^{3}H]ACh uptake ratios, which give the concentration of [^{3}H]ACh
inside the vesicles compared to the concentration outside, were
determined as described (1) by centrifugation-gel filtration after
uptake in buffer A containing 5 mM MgATP, 40 mM KHCO$_3$ and an ATP
regeneration system. In later experiments [^{3}H]ACh taken up by vesicles
under the same active transport conditions was determined by filtration
of 50 μl of vesicles through Millipore filters (type HAWP 013 00)
followed by washing with buffer A containing 0.3 mM AH5183.

RESULTS

One mechanism by which a tertiary amine like AH5183 might inhibit
the ACh active transport system is to act as a protonophoric uncoupler
or an internal base which quenches the putative proton gradient
generated by the ATPase. This possibility can be tested by examining
the effect of the drug on endogenous stores of ACh, since uncouplers
induce rapid efflux of endogenous ACh from isolated Torpedo electric
organ synaptic vesicles (14). Figure 1A shows that in the absence
of MgATP purified vesicles spontaneously leak nearly all of their
endogenous ACh over a 6 hr. period at 23°C. This corresponds to
leakage of a very large number of ACh molecules per synaptic vesicle
since a fully loaded vesicle contains about 40,000 ACh molecules.
In the presence of exogenous MgATP loss of endogenous ACh was much
slower (Fig. 1B). This occurred even though the supplied ATP would
have been depleted in an estimated 2 hr. by the vesicular ATPase
activity. A relatively very high concentration of AH5183 had no
significant effect on either leakage process. This result is inconsistent with an uncoupler-like activity for AH5183.

A different type of behavior was obtained when active uptake
of exogenous [^{3}H]ACh was allowed to proceed for a while before AH5183
was added to the vesicle suspension. As shown in Figure 2A, addition
of AH5183 after 30 min of uptake led to rapid loss of about one-fourth
of the accumulated [^{3}H]ACh. This can be compared to loss of about

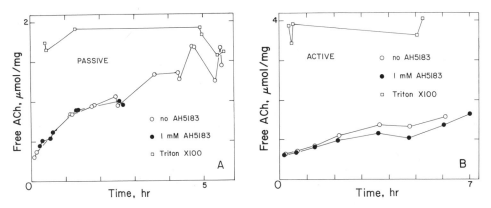

Figure 1.   Effect of AH5183 on the spontaneous leakage of endogenous ACh.   Purified vesicles were incubated at 23°C under isosmotic conditions and external ACh per mg vesicle protein was determined periodically.   Endogenous ACh which leaked from the vesicles was converted rapidly during the incubation to Ch by contaminating esterase, thus insuring that leakage was irreversible, and it was this which actually was determined by the assay.   ACh remaining inside the vesicles was not detected except in the presence of Triton X100.   Leakage was studied in the absence of AH5183, ○; in the presence of 1 mM AH5183, ●; or in the presence of 1% Triton X100, □ , which gives a measure of the total ACh present inside and outside the vesicles.   In A, vesicles were incubated in the absence of an energy source while in B they were incubated with 10 mM ATP, 10 mM potassium phosphoenolpyruvate, 40 mM $KHCO_3$, and 85 μg pyruvate kinase per ml.   Adapted from Anderson et al. (4).

one half of the accumulated [^{3}H]ACh which was caused by the uncoupler FCCP.   In the absence of drugs uptake continued to occur up to 60 min of incubation.   When the concentration dependence for the AH5183-mediated loss was determined, 4 x 10^{-8} M drug was required for the half effect, which is similar to the IC_{50} value obtained when the drug is present at the beginning of [^{3}H]ACh active uptake (Fig. 2B).   Also, shown in Figure 2B is the observation that AH5183 has no effect on the larger efflux of accumulated [^{3}H]ACh which is induced by the uncoupler CCCP, in agreement with Figure 1.   The actual concentration of [^{3}H]ACh released by AH5183 was about 4 x 10^{-7} M in Figure 2B.

A potential dependence of the inhibitory potency of AH5183 on the synaptic vesicle concentration was studied and the results are shown in Figure 3.   The more potent 1 optical isomer of AH5183 (see below) was present at the beginning of [^{3}H]ACh active uptake in this study.   Several anomalies were observed.   First, in the absence of the drug, less than 2-fold greater uptake of [^{3}H]ACh occurred

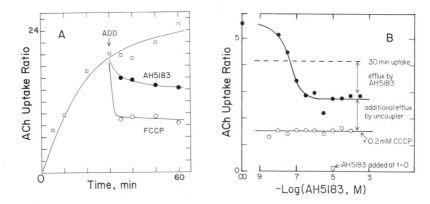

Figure 2. Displacement of exogenous [³H]ACh by AH5183. A: [³H]ACh at 50 μM was actively transported for 30 minutes at 23°C after which either nothing was added to the uptake solution, □ ; 3 μM AH5183 was added, ● ; or 10 μM FCCP was added, ○. Two-hundred-fifty μl portions of vesicle suspensions were assayed periodically for [³H]ACh content. B: [³H]ACh (90 mCi/mmol) at 200 μM was taken up by vesicles (0.30 mg protein/ml) as above for 30 minutes after which AH5183 at the indicated concentration was added, ● ; or 0.2 mM CCCP plus AH5183 at the indicated concentration was added, ○. Incubation was continued 30 minutes more after which 125 μl of vesicles were sampled for [³H]ACh content. Adapted from Anderson et al. (4).

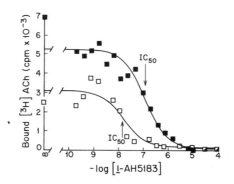

Figure 3. Effect of vesicle concentration on inhibition of ACh transport by 1-AH5183. Purified vesicles at 2.4 mg protein/ml, which is an estimated 120 nM vesicles, ■; and at 0.24 mg protein/ml, or 12 nM vesicles, □; actively transported [³H]ACh at 50 μM concentration and 23°C for 30 minutes in the presence of the indicated concentration of 1-AH5183, after which uptake was assayed by filtration. Hyperbolic inhibition curves were fit to the data. The IC_{50} value at the high vesicle concentration was 115 ± 34 nM whereas it was 16 ± 7 nM at the lower concentration. Adapted from Anderson et al. (4).

at a 10-fold greater vesicle concentration. Second, at the higher
vesicle concentration the IC_{50} value for the best fit to a hyperbolic
titration curve was eight times higher than for the low vesicle
concentration. Thus, the [^{3}H]ACh transport specific activity and
the IC_{50} value for drug inhibition depend on the vesicle concentration.

Inhibition by AH5183 of the steady state initial velocity active
transport of [^{3}H]ACh by vesicles was studied. Inhibition was non-
competitive with respect to [^{3}H]ACh with a K_i of 41 $\pm$ 7 nM. This is
not consistent with direct competition between [^{3}H]ACh and AH5183
for a transporter active site on the outside of the synaptic vesicle
(Fig. 4).

The separated optical isomers of AH5183 differ in potency by
55-fold, with the $\underline{l}$ isomer having an approximate IC_{50} of 20 nM
and the $\underline{d}$ isomer an IC_{50} of about 1100 nM. Thus, binding is
stereoselective, as is expected for binding to a protein.

DISCUSSION

The results presented here eliminate the possibility that AH5183
inhibits active uptake of [^{3}H]ACh by acting as a nonspecific uncoupler
of synaptic vesicles. Its effects on vesicular ACh are either non-
existent or of limited impact as compared with authentic uncouplers,

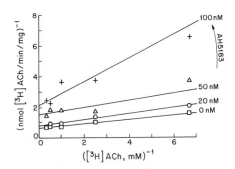

Figure 4. Reciprocal initial velocity dependence __versus__ reciprocal
[^{3}H]ACh, in the presence of changing fixed concentrations of AH5183
at 0, 20, 50, and 100 nM. Vesicles at 23°C were allowed to actively
transport ACh at the indicated concentrations of ACh and AH5183 for
2 minutes before addition of [^{3}H]ACh, after which net uptake of [^{3}H]ACh
was determined 2, 4, and 6 minutes later by filtration. The rate
of uptake was determined by linear regression analysis of cpm __versus__
time data followed by conversion to nmol ACh/min/mg vesicle protein
which is graphed in the reciprocal. In the absence of AH5183 the
maximal velocity of uptake was 1.6 nmol/min/mg vesicle protein and
the apparent transport dissociation constant for ACh was about 0.22
mM. The noncompetitive inhibition constant for AH5183 was 41 $\pm$ 7
nM. Adapted from Bahr and Parsons (5).

depending on the type of experiment conducted. Authentic uncouplers lead to a large loss of internal ACh, apparently as a result of dissipation of the stabilizing electrochemical proton gradient (14). Moreover, the great sensitivity of the potency of the drug to its stereochemical structure eliminates nonspecific membrane effects as the origin of transport inhibition.

It is important to recognize that the experiments reported in Figures 1 and 2 were quite different from each other. In the experiment reported in Figure 1 the drug had no apparent effect on the efflux from the inside of the vesicle of a very large number of endogenous ACh molecules, whereas in Figure 2, the drug appeared to displace a fraction of the much smaller quantity of exogenous radioactive ACh which had been newly taken up by the vesicles in vitro. The former experiment would not be sensitive to drug effects which have stoichiometries comparable to the synaptic vesicle concentration, whereas the latter experiment could be.

This difference is important because of the following considerations. The synaptic vesicle concentration of Figure 2B can be estimated to be 1.5×10^{-8} M on the basis of the protein concentration (1). The amount of ACh which was displaced from the vesicles by AH5183 was 4×10^{-7} M, or about 26 molecules per vesicle. It is probable that this represents a more loosely bound population of [^{3}H]ACh than the remaining [^{3}H]ACh which is resistant to displacement by AH5183. One possibility is that the displaced [^{3}H]ACh was bound reversibly to ACh transporters in the vesicle membrane, whereas the resistant [^{3}H]ACh was fully inside of the vesicle. However, the IC_{50} value of racemic AH5183 which accomplished displacement of [^{3}H]ACh was only about 4×10^{-8} M. Since it actually is the $\underline{1}$ optical isomer which is active at low concentrations, the corrected IC_{50} value was about 2×10^{-8} M. Since half of the AH5183 binding sites in the vesicles presumably would be occupied in a simple competitive situation by the drug at the IC_{50} concentration, there was no more than 4×10^{-8} M of binding site for the drug. In other words, at least ten molecules of [^{3}H]ACh were displaced from the vesicles for each binding site occupied by AH5183. Note that this last ratio calculation does not depend on knowledge of the concentration of synaptic vesicles, which is not known with as great a certainty as the drug and [^{3}H]ACh concentrations.

Because the concentration of AH5183 required for effective inhibition is comparable to the estimated synaptic vesicle concentration the possibility arises that the AH5183 binding site concentration is comparable to the IC_{50} value. Certainly much lower binding site concentrations would exacerbate the apparent substoichiometric relationship discussed above. If the binding site concentration were comparable to the IC_{50} value, the IC_{50} value should increase at higher synaptic vesicle concentration. Such was found to be the case. An alternative explanation for the strong vesicle concentration dependence is that severe nonspecific binding of the drug to the vesicle membrane occurs. This would require, however, that the specific binding site concentration be significantly less than the IC_{50} value since most of the drug would be not available for receptor binding. These possibilities theoretically can be distinguished by careful analysis of the shape of the inhibition curve, but this was not possible here because of the data scatter. This experiment also indicated that the amount of uptake of [^{3}H]ACh was less than expected at high synaptic vesicle concentration. Such behavior is consistent with involvement of an endogenous dissociable factor in [^{3}H]ACh uptake or with transport-inhibiting aggregation of vesicles.

Thus, it seems likely that a stereoselective receptor for AH5183 is present in isolated vesicles at concentrations comparable to, or less than the vesicle concentration. This might indicate that the receptor plays a role in ACh storage which is indirect and which involves an amplification mechanism or that only a small fraction of the vesicles is competent to transport ACh. Much additional work is required to clarify the alternatives.

Acknowledgement. This work was supported by a grant from the Muscular Dystrophy Association and by grant NS 15074 from the National Institute of Neurological and Communicative Disorders and Stroke. We thank Marianne Henry for preparation of synaptic vesicles.

REFERENCES

1. Anderson, D.C., King, S.C. and Parsons, S.M. (1982): Biochemistry 21:3037-3043.
2. Anderson, D.C., King, S.C. and Parsons, S.M. (1983): Molec. Pharmacol. 24:55-59.
3. Anderson, D.C., King, S.C. and Parsons, S.M. (1983): Molec. Pharmacol. 24:48-54.
4, Anderson, D.C., Bahr, B.A. and Parsons, S.M. (1985): J. Neurochem., submitted.
5. Bahr, B.A. and Parsons, S.M. (1985): J. Neurochem., submitted.
6. Bowman, W.C. and Marshall, I.G. (1972): Internat. Encyclo. Pharmacol. therp., Sec. 14 1:377-378.
7. Breer, H., Morris, S.J. and Whittaker, V.P. (1977): Eur. J. Biochem. 80:313-318.
8. Fuldner, H.H. and Stadler, H. (1982): Eur. J. Biochem. 121:519-524.
9. Israël, M. and Lesbats, B. (1981): J. Neurochem. 37:1475-1483.
10. Marshall, I.G. (1970): Br. J. Pharmacol. 38:503-516.
11. Marshall, I.G. (1970): Br. J. Pharmacol. 40:68-77.
12. Marshall, I.G. and Parsons, R.L. (1975): Br. J. Pharmacol. 54:333-338.
13. Michaelson, D.M. and Ophir, I. (1980): J. Neurochem. 34:1483-1490.

14. Michaelson, D.M., Pinchasi, I., Angel, I., Ophir, I., Sokolov-sky, M. and Rudnick, G. (1979): In Molecular Mechanisms of Biological Recognition (ed) M. Balaban, Elsevier/Holland, pp. 361-372.

15. Parsons, S.M., Carpenter, R.S., Koenigsberger, R. and Rothlein, J.E. (1982): Fed. Proc. 41:2765-2768.

16. Parsons, S.M. and Koenigsberger, R. (1980): Proc. Nat'l. Acad. Sci. U.S.A. 77:6234-6238.

17. Rothlein, J.E. and Parsons, S.M. (1979): Biochem. Biophys. Res. Comm. 88:1069-1076.

18. Rothlein, J.E. and Parsons, S.M. (1982): J. Neurochem. 39:1660-1668.

ALKYLATING DERIVATIVES OF OXOTREMORINE HAVE

IRREVERSIBLE ACTIONS ON MUSCARINIC RECEPTORS

F. J. Ehlert and D.J. Jenden

Department of Pharmacology, School of Medicine and
Brain Research Institute
University of California
Los Angeles, California 90024

INTRODUCTION

The introduction of a β-chloroethylamino group into the structure of a drug has proven to be a successful means of developing irreversible ligands for α-adrenergic (10), muscarinic (6) and opiate receptors (2). Muscarinic agents would appear to be ideally suited for this type of chemical modification since the cationic aziridinium ring derived from a β-chloroethylmethylamino compound is structurally very similar to the trimethyl ammonium group of acetylcholine. This has been born out by the irreversible antagonist benzilylcholine mustard (6) which has proven to be a useful investigational tool in muscarinic receptor pharmacology. Recently, we and our colleagues have described some β-chloroethylamine derivatives of oxotremorine that are potent muscarinic agonists that bind irreversibly to the muscarinic receptor (3, 13, 14). The compounds, N-[4-(2-chloroethyl-methylamino)-2-butynyl]-2-pyrrolidone (BM 123) and N-[4-(2-chloro-methylpyrrolidino)-2-butynyl]-2-pyrrolidone (BM 130) (see Figure 1), cyclize spontaneously in neutral aqueous solutions to form aziridinium ions, which are responsible for their pharmacological effects (3, 13, 14). BM 123 and BM 130 stimulate contractions of the guinea pig ileum, and these effects are blocked by atropine, but not by hexamethonium (3, 13, 14). When injected into rats, BM 123 and BM 130 produce typical muscarinic effects including chromodacryorrhea, diarrhea, hypothermia, lacrimation, salivation and tremor. In this report, we describe some irreversible muscarinic receptor binding characteristics of BM 123 and BM 130, as well as the persistent inhibitory effect of BM 123 on acetylcholine release from the myenteric plexus of the guinea pig ileum.

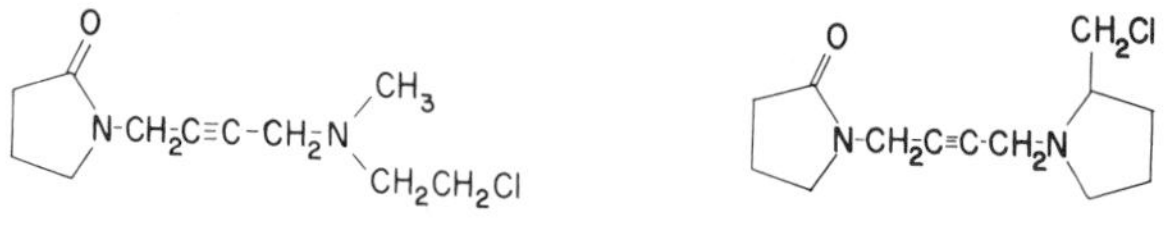

Figure 1. Structures of BM 123 and BM 130.

METHODS

Muscarinic Receptor Binding Assay. Binding assays were run on homogenates of the cerebral cortex and heart from male Sprague Dawley rats (200 - 250 g) and the longitudinal muscle of the ileum from male guinea pigs (English short hair, 350 - 400 g). The cerebral cortex was homogenized with a Potter Elvehjem glass homogenizer in 50 volumes of 0.05 M phosphate buffer (81 mM Na^+, 9.1 mM K^+, 50 mM PO_4, pH 7.4). The freshly excised heart was perfused through the aorta with ice cold saline and minced with scissors. The whole ileum was excised and mounted on a 1 ml pipet, and the outer longitudinal muscle layer was rubbed off with a cotton swab and minced with scissors. Both the minced heart and ileum were homogenized with a Polytron in 20 volumes of 0.05 M phosphate buffer, and poured through 3 layers of cheese cloth. Solutions of BM 123 and BM 130 were made up in 0.05 M phosphate buffer and incubated at room temperature for 60 and 15 min, respectively, to allow formation of the aziridinium ions. The drug solutions were then put on ice, and used as soon as possible. The various tissue homogenates were preincubated at 37°C for 10 min; an aliquot of BM 123 or BM 130 was added, and the tissue was incubated further at 37°C for the times indicated below. The incubation was stopped by addition of thiosulfate (1.0 mM) and by washing the homogenates four times by centrifugation at 30,000 x g for 10 min, followed by resuspension in fresh phosphate buffered saline containing GTP (0.1 mM) and $MgSO_4$ (10 mM). The final pellets were frozen at -20°C and thawed the next day for measurement of specific [3H](-)3-quinuclidinyl benzilate ([3H](-)QNB) binding. The pellets from cerebral cortex, ileum, and heart were resuspended to a concentration representing 10, 20, and 100 mg original wet tissue weight per ml 0.05 M phosphate buffer, respectively.

In other experiments, mice were injected intravenously with BM 123 in the parent mustard form by diluting an ethanolic solution of BM 123 twenty fold with ice cold saline immediately before injection. The mice were killed twenty-four hours later, and their forebrains were homogenized with a Potter Elvehjem homogenizer in 10 volumes of 0.05 M phosphate buffer. The homogenates were washed twice by centrifugation at 30,000 x g for 10 min followed by resuspension in fresh phosphate buffer. The final pellets were resuspended to a concentration representing 10 mg original wet tissue weight per ml 0.05 M phosphate buffer.

The specific binding of $[^3H](-)QNB$ (33.2 Ci/mmol; New England Nuclear, Boston, Massachusetts) was measured by the filtration assay of Yamamura and Snyder (15) with the following modifications. Aliquots (0.1 ml) of tissue homogenate were incubated with $[^3H](-)QNB$ in a final volume of 2 ml containing 0.05 M phosphate buffer. All assays were run in triplicate. Non-specific binding was defined as that binding occurring in the presence of 10 μM atropine. Protein was measured by the method of Lowry et al. (9), using bovine serum albumin as the standard.

Assay of Acetylcholine Release from Guinea Pig Meyenteric Plexus. Strips of ileum 4 cm in length were prepared from guinea pigs of either sex (300 - 500 g) and placed on glass pipettes to allow removal of the longitudinal muscle layer and attached myenteric plexus, as described by Paton and Vizi (12). The strips were mounted on platinum electrodes embedded 12 mm apart in a Perspex rod, and arranged in a tube containing 3 ml Krebs solution at 37°C (NaCl: 138 mM; KCl: 5 mM; $CaCl_2$: 2 mM; $MgCl_2$: 1 mM; KH_2PO_4: 0.4 mM; $NaHCO_3$: 12 mM; glucose: 11 mM), gassed with 95% O_2, 5% CO_2, and containing physostigmine (30 μM). The strips were equilibrated for 60 min, and thereafter the bath fluid was changed every 10 min. The experiment was divided into groups of three successive 10 min epochs. The first was used to measure resting release; during the second the strip was stimulated (0.1 Hz, 1 msec, 200 mA) to measure evoked release; while the third epoch provided for recovery, the fluid being discarded. Under these conditions the strips function for several hours without apparent fatigue. After 3 such cycles (Pretreatment phase), the strips were immersed in a test solution without stimulation for 30 minutes (Treatment phase), and thereafter the design was resumed (Test phase). N-Methylatropine (10^{-7} M) was always present during the Test phase. Acetylcholine was measured in the incubation medium by gas chromatography/mass spectrometry (GCMS), using a deuterium labelled internal standard as previously described (5, 7).

RESULTS

When homogenates of various tissues of the rat were incubated with BM 123 and BM 130 for 20 min at 37°C, washed extensively, and then assayed for $[^3H](-)QNB$ binding, an inhibition of binding was observed. Figure 2 shows the results of $[^3H](-)QNB$ binding measurements made at a single $[^3H]$ligand concentration of 0.4 nM in homogenates of the rat cerebral cortex and heart and the longitudinal muscle of the guinea pig ileum that had been exposed to the indicated concentrations of BM 123 and BM 130 for 20 min. It can be seen that BM 123 and BM 130 caused a concentration dependent reduction in $[^3H](-)QNB$ binding in the various tissues, that persisted after extensive washing.

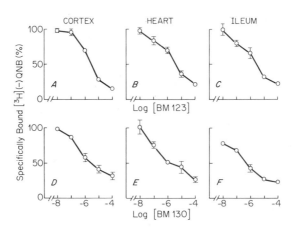

Figure 2. Effect of BM 123 and BM 130 on [^{3}H](-)QNB binding in rat cerebral cortex (A,D) and heart (B,E) and the longitudinal muscle of the guinea pig ileum (C,F). Tissue homogenates were incubated with the indicated concentrations of BM 123 and BM 130 for 20 min at 37°C. The homogenates were washed extensively and then assayed for specific [^{3}H](-)QNB binding at a [^{3}H]ligand concentration of 0.4 nM. The data represent the means $\pm$ S.E.M. of 3 - 5 binding measurements.

To determine the effect of BM 123 and BM 130 on the equilibrium binding parameters of [^{3}H](-)QNB, homogenates of the cerebral cortex were incubated with BM 123 (10 μM) and BM 130 (10 μM) for 20 min and then washed extensively. Specific [^{3}H](-)QNB binding was assayed at various [^{3}H]ligand concentrations, and these results are shown in Figure 3. Nonlinear regression analysis of the data showed that the predominant effect of preincubation with BM 123 and BM 130 was to reduce the binding capacity of [^{3}H](-)QNB from 0.85 pmol/mg protein in controls to 0.16 and 0.28 pmol/mg protein in BM 123 and BM 130 treated homogenates, respectively. No significant effect of BM 123 or BM 130 on the apparent affinity of [^{3}H](-)QNB was detected (control K_D = 0.040 nM). The reduction in binding capacity caused by preincubation with BM 123 and BM 130 was completely prevented by atropine (1.0 μM). Measurements of [^{3}H](-)QNB binding in cortical homogenates that had been preincubated with both atropine and BM 123 or atropine and BM 130 were not significantly different from control (see Fig. 3).

An irreversible inhibition of muscarinic receptor binding capacity was also shown following in vivo administration of BM 123. Mice were injected intravenously with BM 123 (5.0 μmole/kg) and killed 24 hours later. Specific [^{3}H](-)QNB binding was measured at several [^{3}H]ligand concentrations as shown in Figure 4. Nonlinear regression analysis of the binding isotherms showed that BM 123 treatment caused

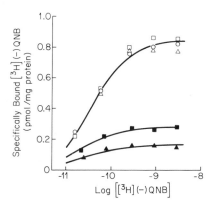

Figure 3. Effect of BM 123 and BM 130 on [^{3}H](-)QNB binding to the rat cerebral cortex. Specific [^{3}H](-)QNB binding was measured in control (o), BM 123 treated (▲), BM 130 treated (■), BM 123 plus atropine treated (△), and BM 130 plus atropine treated (□) homogenates of the rat cerebral cortex. The curves represent the least squares fit to the data from control, BM 123 treated, and BM 130 treated homogenates. Each point represents the mean binding value of three measurements.

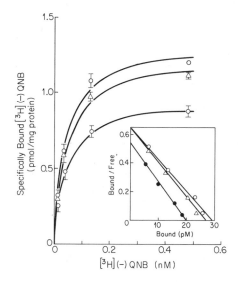

Figure 4. Effect of in vivo administration of BM 123 on [^{3}H](-)QNB binding to the mouse forebrain. Specific [^{3}H](-)QNB binding was measured in homogenates of the forebrains from control (o), BM 123 treated (●), and BM 123 plus atropine treated (△) mice. The data points represent the mean binding values $\pm$ S.E.M. of five measurements. Inset: Scatchard analysis of the mean binding values.

Table 1

| Drug Present During | | ACh Release Rate |
Pretreatment Phase	Treatment Phase	fmol mg^{-1} min^{-1}
None	None	478$\pm$64
None	BM 123 (2 x 10^{-5} M)	186$\pm$26
Methylatropine (10^{-7} M)	Methylatropine (10^{-7} M)	429$\pm$85
Methylatropine (10^{-7} M)	Methylatropine (10^{-7} M) + BM 123 (2 x 10^{-5} M)	353$\pm$55
None	Oxotremorine (10^{-5} M)	526$\pm$74

Effects of oxotremorine and BM 123 on evoked release of acetylcholine from guinea pig myenteric plexus. Physostigmine (3 x 10^{-5} M) was present throughout, and N-methylatropine (10^{-7} M) was always present during the Test phase. Figures given for release are averages of stimulated (0.1 Hz) less resting release averaged over 150 min period (Test phase) following Treatment phase. Means and standard errors of data from six preparations are presented.

a significant reduction in the binding capacity of [^{3}H](-)QNB, from 1.35 pmol/mg protein in controls, to 0.97 pmol/mg protein in BM 123 treated mice. Premedication of the mice with atropine (5.0 μmole/kg) 15 min before injection with BM 123 prevented the loss in receptor capacity. BM 123 treatment had no significant effect on the dissociation constant of [^{3}H](-)QNB (control K_D = 0.039 nM).

Table 1 summarizes the results of some experiments on the effects of BM 123 on evoked acetylcholine release. None of the conditions used in these experiments altered resting release, and the data presented reflect the incremental release induced by 0.1 Hz stimulation. BM 123 (2 x 10^{-5} M) produced a highly significant inhibition of release (P = 1.5 x 10^{-3}), despite the fact that it was washed out before the Test phase, and N-methyl atropine (10^{-7} M) was present during the Test phase. In contrast, the effect of oxotremorine (10^{-5} M), which is a potent inhibitor of acetylcholine release (8), did not outlast its presence in the solution under these conditions and its effect was blocked by N-methylatropine even if it was present. Although N-methylatropine (10^{-7} M) did not reverse the effect of BM 123, the same concentration almost completely prevented the effect of BM 123 when both were present together.

We conclude from these results that, like its structural congener oxotremorine, BM 123 inhibits evoked acetylcholine release by an action at presynaptic autoreceptors, and that like oxotremorine, its effect is prevented competitively by N-methylatropine. In contrast

to that of oxotremorine, the inhibitory effect of BM 123 is sustained
and cannot be reversed by subsequent treatment with N-methylatropine.
These data are consistent with a sustained and possibly irreversible
agonist effect of BM 123.

DISCUSSION

The results presented here strongly suggest that BM 123 and
BM 130 bind irreversibly to muscarinic receptors from various tissues.
The irreversible nature of the interaction of these compounds with
muscarinic receptors is readily illustrated by the results of our
binding experiments with the specific muscarinic antagonist $[^3H](-)$
QNB. An inhibition of $[^3H](-)$QNB binding was measured in the cerebral
cortex, heart and ileum after homogenates of these tissues had been
incubated with BM 123 and BM 130 and washed extensively with a buffer
containing GTP and excess NaCl, both of which reduce the affinity
of agonists for muscarinic receptors (1, 4). Thus, it is unlikely
that the aziridinium ions still could have been reversibly bound
to the receptor after such washing. In cerebral cortex, this inhibition of $[^3H](-)$QNB binding was prevented by atropine and was shown
to be the result of a reduction in binding capacity without an accompanying change in apparent affinity. Similar results have been
observed in studies of the binding of $[^3H]$N-methylscopolamine to
the rat cerebral cortex and guinea pig ileum (3, 13, 14). Collectively, the results described above are consistent with the postulate
that the aziridinium ions of BM 123 and BM 130 bind covalently to
the recognition site of the muscarinic receptor.

It is clear from a number of previous observations that the
aziridinium ions of BM 123 and BM 130 are potent muscarinic agonists
which induce a variety of muscarinic effects (3, 13, 14). Both
the parent 2-chloroethylamines and their alcoholic hydrolysis products
are largely inactive and devoid of irreversible binding properties
(3, 13, 14).

It remains to be established, however, whether the alkylated
receptor complex is an active or inactive one. If we consider the
guinea pig myenteric plexus preparation, BM 123 caused inhibition
of the evoked release of ACh that persisted after extensive washing
with atropine. Thus, in this model system, BM 123 behaves as an
irreversible agonist, presumably because the covalent binding of
BM 123 to the muscarinic receptor generates an active conformation.

The consequences of muscarinic receptor alkylation by BM 123
and BM 130 in the guinea pig ileum are qualitatively different with
regard to their effects on contractility. Initially, the aziridinium
ions of BM 123 and BM 130 stimulate contraction with potencies similar
to that of oxotremorine. After the initial stimulation, however,
there is a long lasting period of cholinergic blockade. It was
shown that the ED_{50} values of oxotremorine-M increased several fold

after exposing the ileum to BM 123 for 30 min (3, 13, 14). This shift in the agonist dose-response curve can be likened to that which occurs following inactivation of muscarinic receptors with an irreversible antagonist. Provided that the agonist used is highly efficacious and only requires a small percentage of the receptors to be occupied for maximum response, the shift in the agonist dose response curve caused by alkylation of a portion of the receptors by an antagonist is related quantitatively to percent receptor occupancy by the antagonist according to the following relationship (11):

$$Y = (R - 1)/R \times 100$$

where Y equals the percent receptors alkylated by the antagonist and R is equal to the ratio of ED_{50} values of the agonist measured after inactivation of receptors divided by that measured before inactivation.

In the guinea pig ileum, we have found that the loss in $[^3H](-)QNB$ binding sites following exposure to BM 123 and BM 130 agrees reasonably well with the estimate of receptor occupancy calculated pharmacologically by application of the equation described above (14). Thus, with regard to those receptors mediating contractility, the persistent lack of sensitivity of the guinea pig ileum is not inconsistent with the alkylated receptor complex being an antagonistic one. It has been suggested, however, that continuous receptor stimulation by an irreversibly bound agonist could result in desensitization and a consequent lack of sensitivity quantitatively similar to that produced by an irreversibly bound antagonist (3). Thus, on the basis of whole tissue responses alone, it might not be possible to distinguish between an irreversible desensitization and irreversible antagonism.

Studies of muscarinic responses that are more closely coupled to the receptor or that are more resistant to desensitization may help to settle this question. In this connection, inhibitory muscarinic responses, such as presynaptic inhibition of acetylcholine release, may be inherently less likely to be desensitized. It is hoped that the irreversible oxotremorine analogues described in this study will prove to be useful tools for studying mechanisms of receptor activation, coupling and desensitization.

Acknowledgements. We thank Ruth A. Booth, Kathleen M. Rice and Margareth Roch for their excellent technical assistance, and Nelly Canaan for her expert editorial assistance. This work was supported by U.S. Public Health Service Grant MH 17691 and U.S. Army Medical Research and Development Command Contract DAMD17-83-C-3073.

REFERENCES

1. Birdsall, N.J.M., Berrie, C.P., Burgen, A.S.V. and Hulme, E.
    C. (1980): In Receptors for Neurotransmitters and Peptide Hormones
    (eds) G. Pepeu, M.J. Kuhar and S.J. Enna, Raven Press, New York,
    pp. 107-116.
2. Caruso, T.P., Larson, D.L., Portoghese, P.S. and Takemori, A.
    E. (1980): J. Pharmacol. Exp. Ther. 213:539-544.
3. Ehlert, F.J., Jenden, D.J. and Ringdahl, B. (1984): Life Sci.
    34:985-991.
4. Ehlert, F.J., Roeske, W.R., Rosenberger, L.B. and Yamamura, H.
    I. (1980): Life Sci. 26:245-252.
5. Freeman, J.J., Choi, R.L. and Jenden, D.J. (1975): J. Neuro-
    chem. 24:729-734.
6. Gill, E.W. and Rang, H.P. (1966): Mol. Pharmacol. 2:284-297.
7. Jenden, D.J., Roch, M. and Booth, R.A. (1973): Analyt. Biochem.
    55:438-448.
8. Kilbinger, H. (1978): In Cholinergic Mechanisms and Psychopharma-
    cology (ed) D.J. Jenden, Plenum Press, New York, pp. 401-410.
9. Lowry, O.H., Rosebrough, N.J., Farr, A.L. and Randall, R.J.
    (1951): J. Biol. Chem. 193:265-275.
10. Nickerson, M. and Gump, W.S. (1949): J. Pharmacol. Exp. Ther.
    97:25-47.
11. Paton, W.D.M. (1961): Proc. Roy. Soc. Lond. B. 154:21-69.
12. Paton, W.D.M. and Vizi, E.S. (1969): Brit. J. Pharmacol. 35:
    10-28.
13. Ringdahl, B., Ehlert, F.J. and Jenden, D.J. (1983): Fed. Proc.
    42:1148.
14. Ringdahl, B., Resul, B., Ehlert, F.J., Jenden, D.J. and Dahlbom,
    R. (1984): Mol. Pharmacol. 26:170-179.
15. Yamamura, H.I. and Snyder, S.H. (1974): Proc. Natl. Acad. Sci. USA
    71:1725-1729.

ACTION OF PHOSPHOLIPASE NEUROTOXINS ON <u>TORPEDO</u> SYNAPTOSOMES:

CHANGES IN MEMBRANE POTENTIAL AND PHOSPHOGLYCERIDE COMPOSITION

M.J. Dowdall, P. Fretten and P.G. Culliford

Department of Biochemistry
University of Nottingham Medical School
Queen's Medical Centre
Nottingham NG7 2UH   United Kingdom

INTRODUCTION

A number of snake neurotoxins with intrinsic phospholipase A_2 activity are potent neuromuscular blocking agents which inhibit transmitter release by a complex, but as yet poorly characterized, sequence of events (14, 15). With β-bungarotoxin (β-BtX), the best-known toxin of this group, there is a well documented triphasic effect on evoked transmitter release from motor nerve terminals (see 14 for literature citations) in which both phospholipase-independent and phospholipase-dependent actions are involved (3, 14). Following an early rapid inhibitory phase which is independent of phospholipase activity, transmitter release from β-BtX-poisoned junctions transiently recovers before entering a third and final stage in which release gradually decreases to zero. Since the last two phases involve phospholipase activity they presumably reflect changes in the phospholipid composition of presynaptic membrane systems.

It is generally believed that the initial phase reflects a specific binding of β-BtX to the presynaptic plasma membrane thereby initiating the neurotoxic sequence. A very similar pharmacological pattern has been observed with other toxins in this group [notexin (NtX), crotoxin and taipoxin (TpX)] and it seems reasonable to assume that in each case binding to the preterminal membrane precedes subsequent catalytic events ('<u>Corpora non agunt nisi fixata</u>').

<u>Torpedo</u> synaptosomes provide a particularly appropriate experimental preparation for examining the biochemical changes which underly these observed pathophysiological effects (see 8). Previous studies have shown their potent inhibitory action on membrane uptake systems

1187

in isolated <u>Torpedo</u> synaptosomes (6, 7) and, at relatively high concentrations, their lytic effects as well (12). In the present study these observations are extended to cover the effect of neurotoxins on synaptosomal membrane potential and phosphoglyceride composition. Experimental conditions were chosen to match those used in previous studies with <u>Torpedo</u> synaptosomes. In this way valid comparisons could be made between phospholipase activity <u>at the site of action</u> and other effects. A preliminary account of some of this work has been given earlier (9).

MATERIALS AND METHODS

Synaptosomes were prepared from <u>Torpedo marmorata</u> electric organ using the method of Morel et al. (17) as described previously (6). In some experiments where membrane potential was estimated the <u>Torpedo</u> Ringer solution used for tissue comminution (280 mM NaCl, 3 mM KCl, 1.8 mM $CaCl_2$, 3.4 mM $MgCl_2$, 0.1 M sucrose, 0.3 M urea, 5.5 mM D-glucose, 10 mM Tris pH 7.4) was modified by reducing NaCl to 50 mM and maintaining isotonicity by the addition of 0.51 M L-glycine. In these experiments the low Na^+ environment was also maintained throughout the fractionation procedure by the substitution of L-glycine for the "missing" NaCl (see also 21).

Synaptosomal membrane potential was estimated using two different methods: carbocyanine fluorescence and $[^3H]$-lipophilic cation distribution. The fluorescence procedure, which was used both qualitatively and quantitatively, employed the dye 3,3'-dipropylthiacarbocyanine $[DiS-C_3-(5)]$ as a reporter of membrane potential (24). Fluorescence changes were measured, using excitation and emission wavelengths of 600 and 660 nm, respectively, after addition of synaptosomes derived from 150 mg tissue to 2 ml of <u>Torpedo</u> Ringer solution containing 2.1 μM dye at room temperature.

Qualitative changes in membrane potential were assessed by recording changes in fluorescence for 10 min. following addition of toxins. Increased fluorescence was interpreted as depolarization, and decreased fluorescence as hyperpolarization.

Quantitative determinations using the dye were more complicated and involved titration of the fluorescence signal as a function of varying external KCl concentration and independent measurements of intrasynaptosomal space and synaptosomal K^+ content. In these assays valinomycin (final concentration 1 μM) was added to synaptosome suspensions in dye after an initial 5 min. equilibration. The difference in fluorescence (ΔF) before and after valinomycin was recorded in arbitary units for <u>Torpedo</u> Ringer solution at increasing KCl concentrations (3-15 mM).

From a plot of ΔF against $Log_{10}[K^+]_o$ the concentration of $[K^+]_o$ at which valinomycin caused no change in fluorescence (i.e. the

null-point) was determined and used together with measurements of
$[K^+]_i$ (see below) to calculate the membrane potential using the
Nernst equation: $\Psi = -59 \, \text{Log}[K^+]_i/[K^+]_o$ m.V. For synaptosomes treated
with toxins, addition was 8 min. prior to valinomycin.

For the measurement of lipophilic cation distribution synaptosomes
were incubated with 1 μM $[^3H]$-triphenylmethyl phosphonium ($[^3H]$-TPMP)
for 45 min at 20-25°C in normal _Torpedo_ Ringer solution (3 mM K^+)
or high K Ringer solution (125 mM K^+ and correspondingly less Na^+),
and then sedimented by centrifugation in a microhaematocrit centrifuge
at 12,000$_{gav}$ for 10 min. Measurement of $[^3H]$ in pellets (corrected
for non-potential related accumulation by subtration of high K condition from normal K condition) and supernatants together with parallel
estimates of intrasynaptosomal space (see below) were used to estimate
the ratio of $[TPMP]_i:[TPMP]_o$ from which membrane potential was computed
using the Nernst equation.

Intrasynaptosomal space was determined by subtraction of extra-
synaptosomal space (measured using $[^{14}C]$sorbitol) from total space
(measured with $[^3H]H_2O$) in synaptosomes pelleted by centrifugation
at 100,000$_{gav}$ for 60 min from synaptosomes (2.5 g original tissue)
preincubated at 0°C for 1 h in 8 ml _Torpedo_ Ringer solution containing
$[^{14}C]$sorbitol (0.4 μCi/ml) and $[^3H]H_2O$ (2 μCi/ml). Replicate synaptosomal pellets produced in parallel were lysed in 0.05% Triton-X-100
prior to K^+ determination by atomic absorption spectrophotometry.

Synaptosomal phospholipids were extracted using chloroform:
methanol 1:2 (25) and resolved by 2D thin layer chromatography
(T.L.C.) on silica-gel plates using double development in chloro-
form:methanol:18 M ammonium hydroxide (13:5:1, by vol.) in the first
dimension followed by a single run in chloroform:methanol:acetone:
glacial acetic acid:water (6:2:8:2:1, by vol.) in the second. Separated phosphoglycerides and lysophosphoglycerides were visualized
using I_2 vapor, removed together with associated silica and used
for measurement of phosphate content (2). For determining the fatty
acid compositions of individual phosphoglycerides α-tocopherol was
used as an antioxidant and applied to T.L.C. plates as a spray,
before and after development in the second solvent system. Visualization after separation was achieved using rhodamine 6G prior to
conversion of esterified fatty acids to their methyl esters, using
sodium methoxide in benzene, according to the method of Glass (13).
Fatty acid methyl esters were separated using gas chromatography
and detected by flame ionization. Identification was made by comparison of retention times with standards.

Phospholipase A_2 activity of toxins towards _Torpedo_ synaptosomes
was measured after prelabelling of synaptosomal phospholipids with
9,10(n)-$[^3H]$palmitic acid by an 18 h room temperature (20-25°C)
incubation of finely sliced electric organ in _Torpedo_ Ringer solution
containing $[^3H]$palmitate (SA. 0.5 Ci/mmole) and anti-bacterial and

anti-fungal agents. Tissue was washed free of incubation medium prior to synaptosome isolation by the standard technique. Further unmetabolized [^{3}H]palmitate was removed by two successive 30 min. incubations at 0°C in 0.4 M NaCl, 10 mM phosphate buffer pH 7.4 containing 1 mM K$^+$-palmitate and 10 mg/ml BSA (fatty acid free), each followed by a 30 min. centrifugation at 12,000$_{gav}$ to separate medium from synaptosomes. They were either used immediately or resuspended in 0.2 M sucrose/ 0.3 M NaCl and stored in aliquots at liquid N$_2$ temperature until required for assay. Release of [^{3}H]-palmitate from phosphoglycerides was measured following separation of fatty acids from other lipids by 1D T.L.C. on plastic-backed silica sheets using two solvent systems: (1) Chloroform:methanol: acetone:glacial acetic acid:water (6:2:8:2:1, by vol.) followed by (2) hexane:diethylether: glacial acetic acid (60:40:1). The [^{3}H] content of separated fatty acids was determined by liquid scintillation photometry. Full details of this radiometric assay will be published elsewhere (Culliford and Dowdall, manuscript in preparation).

Toxins and their derivatives used in this study were kindly provided by Dr. D. Eaker, Institute of Biochemistry, Uppsala, Sweden. Abbreviations are as used in previous publications (6-9, 11, 12).

RESULTS

Toxin-Induced Changes in Transmembrane Potential. Representative records showing the effect of NtX and TpX on fluorescence output from synaptosomes treated with DiS-C$_3$-(5) are given in Fig. 1. The concentration-dependent decrease in fluorescence suggests that these toxins hyperpolarize the synaptosomal membrane. By contrast NtX-II-5 (0.01 μg/ml) and α-BtX (2.5 μg/ml) had no effect on the fluorescence signal. In these preliminary studies toxin concentrations were chosen to be below the threshold for lysis (12). This was of critical importance since any release of dye accompanying lysis would increase fluorescence and thus mask membrane potential-dependent dye movements. Modification of TpX with p-bromophenylacyl bromide produces a derivative (PBP)$_2$TpX, with no detectable phospholipase activity using conventional assays (11) but one which retains some hyperpolarizing action on Torpedo synaptosomes, albeit reduced (Fig. 1). In all of these qualitative experiments addition of valinomycin alone to synaptosomes also caused a reduction in fluorescence in accordance with its anticipated hyperpolarizing action.

Taken together therefore, these results suggest that some, but perhaps not all, toxins exert a hyperpolarizing action on the preterminal membrane. By contrast previous reports of others (20, 22, 23) have claimed a depolarization of rat brain synaptosomes with β-BtX and NtX-II-5. Confirmatory evidence for hyperpolarization was therefore sought using two independent methods, both of which would also provide quantitative estimates of transmembrane potential.

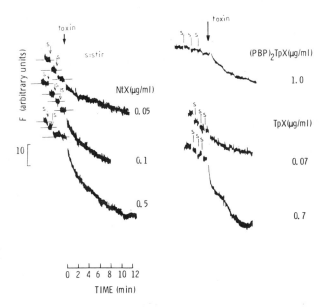

Figure 1. **Decreased carbocyanine dye fluorescence in synaptosomes treated with neurotoxins.** <u>Torpedo</u> synaptosomes pre-equilibrated with 2 μM Di-S-C$_3$-(5) in <u>Torpedo</u> Ringer solution were treated with the toxins indicated, and fluorescence (F) changes recorded. Note that toxin-induced changes were significantly greater than those produced by stirring (S) alone, and were also concentration-dependent. Changes in fluorescence due to synaptosomal lysis were minimized by using non-lytic toxin concentrations. The results are consistent with hyperpolarization.

Null-point titrations of fluorescence against $[K^+]_o$ are shown in Fig. 2 for individual synaptosome preparations, in the presence and absence of various toxins or their PBP-derivatives. In each case the control titration curves were shifted to the left in the presence of toxins giving a reduced value for null-point $[K^+]_o$. Tabulated estimates of synaptosomal membrane potentials derived from such measurements are given in Table 1. Although these results clearly indicate toxin-induced hyperpolarization, an element of uncertainty exists since, for purposes of comparison, the assumption has been made that toxins did not affect $[K^+]_i$. However, in one preparation where this was checked for three of the toxins $[K^+]_i$ was only marginally affected, and the corrected estimates of membrane potential are therefore not substantially different from the originals (see bracketed values in Table 1).

Estimates of the absolute value of synaptosomal membrane potentials using Di-S-C$_3$-(5) and the null-point titration procedure described above, although always negative, were frequently well below

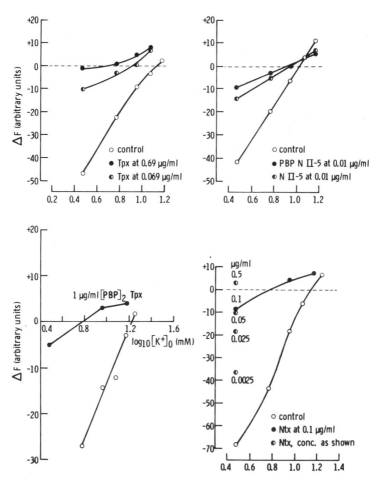

Figure 2.   Carbocyanine dye fluorescence as a function of $[K^+]_o$ in synaptosome preparations and the effect of neurotoxins.   Torpedo synaptosomes were pre-equilibrated with dye (as described in legend to Fig. 1) in Torpedo Ringer solutions of varying KCl concentrations (3-15 mM).   The change in fluorescence (ΔF) following addition of valinomycin is plotted as a function of $[K^+]_o$ to determine the null-points for control and toxin-treated preparations.   Note the approximate linearity of the plots and the lowered null-points for toxin-treated synaptosomes.   Membrane potentials, calculated as described in the text, are given in Table 1.

the expected physiological range.   This variation is reflected in the mean estimate for controls and its wide S.E.M. value (legend to Table 1).   Since the null-point titration procedure almost certainly underestimates membrane potential due to the depolarizing action

Table 1

Effect of Neurotoxins on Synaptosomal Membrane Potential

Toxin	Concn (μg/ml)	Preparation	Calculated Membrane Potential (mV)		
			Control	+Toxin	
NtX	0.01	1	- 5.6	-10.4	
NtX	0.10	2	-12.2	-33.5	
β-BtX	2.50	3	-51.8	-63.2	
β-BtX	2.50	4	-11.9	-18.1	(-13.9)
TpX	0.07	4	-11.9	-20.7	(-23.3)
TpX	0.07	5	- 5.9	-13.0	
TpX	0.07	3	-51.8	-66.3	
TpX	0.70	3	-51.8	-79.6	
(PBP)$_2$ TpX	1.00	4	-11.9	-38.5	(-34.4)

Membrane potentials calculated from estimates of null-point $[K]_o$ (see Fig. 2) and $[K]_i$ (see Materials and Methods section) are given for five independent synaptosome preparations. For toxin-treated synaptosomes it was assumed that $[K]_i$ was unaltered although with preparation 4 this was corrected for; see bracketed values in right hand column. Using this technique the mean membrane potential for 8 independent preparations (those shown plus three others) was -24.3 ± S.E.M. 6.7 mV.

Table 2

Effect of PBP-Toxins on Synaptosomal Membrane Potential

Toxin Derivative	Concn (μg/ml)	Calculated Membrane Potential (mV)	
		Control	+Toxin Derivative
PBP-NtX-II-5	0.01	-50.1	-56.6
	0.10	-50.1	-55.9
PBP-α chain TpX	1.00	-64.5	-71.4

Membrane potentials were estimated from the distribution of $[^3H]$-TPMP as described in Materials and Methods section, using synaptosomes isolated in a low Na^+ environment. Mean potential for controls in independent synaptosome preparations was 43.0 ± S.E.M. 5.9 mV (n = 8) for low Na^+ and 28.9 ± S.E.M. 3.5 mV (n = 14) for normal Na^+.

of elevated $[K^+]_o$, a second quantitative procedure using $[^3H]TPMP$ was also used. Mean estimates were marginally greater using this method, but again, considerable variation was seen between different preparations (legend to Table 2).

Finally in an attempt to increase resting potentials to a more physiological range, a low Na^+ medium was used for synaptosome preparation (see Materials and Methods section). In paired experiments in which synaptosomes isolated from the same tissue but in either low Na^+ or the standard high Na^+ were compared using $[^3H]TPMP$ uptake, a statistically significant elevation of membrane potential ($\simeq$ 8mv) was seen for the low Na^+ preparations (see legend to Table 2).

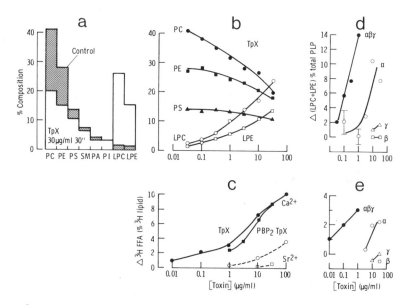

Figure 3. Phospholipase activity of TpX, its subunits and $(PBP)_2TpX$ towards <u>Torpedo</u> synaptosomes. Phospholipolysis in <u>Torpedo</u> synaptosomes was evaluated by measuring compositional changes (a), (b) and (d) and the release of $[^3H]$palmitic acid from prelabelled phospholipids [(c) and (e)] as described in the text. Major compositional changes caused by relatively high TpX concentrations are shown in (a) and at lower concentrations in (b). Phospholipase activity, expressed as the change in lyso-lipid as a % of total phospholipid phosphate (PLP), is indicated in (c), and can be compared with activity computed from release of $[^3H]$-free fatty acid (FFA) shown in (c) and (e). Note that in (d) and (e) the activities of individual subunits (α, β, and γ) are markedly less than the ternary complex ($\alpha\beta\gamma$). In (c) the activities of TpX and $(PBP)_2TpX$ were greatly reduced in <u>Torpedo</u> Ringer solution when Ca^{2+} was replaced by Sr^{2+}.

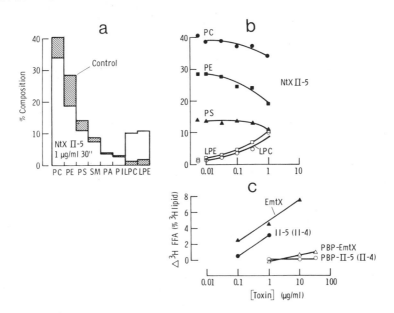

Figure 4.   Phospholipase activity of NtX-II-5 and related toxins towards <u>Torpedo</u> synaptosomes and the effect of histidine modification with p-bromophenylacyl bromide.   Compositional changes to <u>Torpedo</u> synaptosomes exposed to NtX-II-5 are shown in (a) and (b).   Release of [³H]palmitate from prelabelled synaptosomes is shown in (c) for NtX-II-5, and two related toxins NtX and EmtX.   Note that chemical modification of all three toxins with p-bromophenylacyl bromide elimi- nates phospholipase activity.

Using these preparations two derivatized toxins, PBP NtX-II-5 and PBP   TpX, were shown to produce slight hyperpolarization of the synaptosomal membrane, as measured using [³H]TPMP (see Table 2).

Effect of Toxins on Synaptosomal Phosphoglycerides.   Changes in the phosphoglyceride composition of synaptosomes after exposure to toxins are illustrated in Figs. 3 and 4.   The composition of control preparations was remarkably similar to those previously reported for purified <u>Torpedo</u> synaptic vesicles (1) and mammalian brain synap- tosomes (10), with PC and PE respectively being the most abundant species.   After 30 min. exposure to relatively high concentrations of TpX or NtX-II-5 the phosphoglyceride compositon was markedly altered, with the most obvious change being loss of PC and PE, and a concomitant increase in the corresponding lyso-phosphoglycerides. Under the same conditions β-BtX (10 μg/ml) had no detectable effect on composition.   Lower concentrations of TpX and NtX-II-5 produced smaller but measurable changes in PC and PE, as can be seen in Figs. 3 and 4, where changes in the major phospholipids are plotted as a function of toxin concentration.

The thresholds for detectable changes were 0.03 and 0.1 μg/ml for NtX-II-5 and TpX, respectively, which are comparable to thresholds for uptake inhibition (6), but slightly above those required to see changes in membrane potential (Fig. 2). When tested individually all three subunits of TpX were markedly less effective than the ternary complex in producing synaptosomal lyso-PC and lyso-PE (Fig. 3); only the α-subunit had sufficient activity to be reliably detected, but even here this was $\simeq$ 300X lower on a molar basis than TpX. These results provide considerable support for the idea of a synergistic action between subunits at the toxin target site (8, 11).

Confirmatory evidence for phospholipase A_2 activity and its further characterization was sought using a different analytical approach, namely release of free fatty acids. Release of endogenous fatty acids was tried initially but proved to be troublesome for quantitative studies since toxin induced release was spread over the complete spectrum of fatty acids and in addition was small compared to baseline levels. However, this approach provided invaluable information on the fatty acid composition of the component phosphoglycerides, and was particularly useful in showing palmitate to be the most abundant species of fatty acid lost from phosphoglycerides during exposure to TpX (Table 3). This was unexpected since, conventionally, it is generally believed that the 2-position of phosphoglycerides is occupied predominantly by unsaturated fatty acids. Since palmitate was by far the most abundant species at C-2 of PC and the second most abundant species at C-2 of PE, it was chosen as the precursor for radiolabelling nerve terminal phosphoglycerides. Such in vitro incorporation of [^{3}H]-palmitate formed the basis of a radiometric phospholipase A_2 assay in which release of free [^{3}H]-palmitate from prelabelled synaptosomal lipids was easily measured (see Materials and Methods section).

Results using this assay procedure are shown in Figs. 3 and 4 where they can be compared with those using change in phospholipid composition. In the case of TpX and its subunits (Fig. 3) it is clear that the two different assays gave essentially the same pattern, although the radiometric assay was marginally more sensitive. Thus the relative activities of individual subunits as compared to the TpX complex were remarkably similar by both techniques.

Replacement of Ca^{2+} by Sr^{2+} reduced the phospholipase activity of TpX by $\simeq$ 100X in accordance with its known requirement for Ca^{2+}. Contrary to expectation the activity of (PBP)$_2$TpX was only slightly less than the native toxin. Previous measurements of the phospholipase activity of TpX before and after modification, using egg yolk PC dispersion, showed a >40X reduction in activity (11). The present results suggest that the surviving unmodified γ-subunit is largely responsible for the "residual" activity in (PBP)$_2$TpX towards Torpedo

Table 3

Fatty Acid Composition of Phospholipids in _Torpedo_ Synaptosomes

Fatty Acid	Total Phospholipid Extract	PC Overall	PC C-2	PE Overall	PE C-2
14:0	0.4	1.4	2.4	0.5	0.1
16:0	29.4	32.2	56.0	10.2	15.8
16:1	4.2	5.2	5.4	5.3	6.2
18:0	9.8	0.7	0.7	3.9	5.8
18:1	10.3	9.3	7.8	10.0	10.1
18:3	1.8	8.5	1.6	10.0	6.6
20:4	1.9	7.5	4.5	9.5	8.8
20:5	4.4	7.3	6.8	8.5	3.4
22:6	31.7	5.7	5.3	13.8	20.7
24:1	1.7	4.8	2.0	6.1	4.5

Composition (mole %)

Fatty acid analysis was by gas chromatography and is shown for the two major phospholipids separated by 2D T.L.C. Minor fatty acids not given. The compositions of PC + PE at C-2 were computed from their change in fatty acid content and that of their lyso derivatives, on treatment of synaptosomes with 30 μg/ml TpX for 30 min. at 20-25°C. Similar results were obtained after less extensive lipolysis (1 and 10 μg/ml TpX).

synaptosomes. The implications of this interpretation are considerable; the hitherto assumed negligible phospholipase activity of (PBP)$_2$TpX at its site of action has been used as evidence that binding alone might result in changes in membrane-associated processes such as transport (6, 8, 11). For further comment on this see Discussion section.

That modified single chain toxins have greatly reduced phospholipase activity can be seen in Fig. 4, where all three toxins tested had <0.3% of the activity of that measured for their native forms. The relative activities of all toxins assayed radiometrically were 3.38:1.07:1.07:1.0:<0.07 for EmtX, NtX, NtX-II-5, TpX and β-BtX, respectively. Where comparison with results obtained from phospholipid composition was possible, these results were similar. The radiometric procedure thus provides a convenient and reliable method for measuring phospholipase activity in this system.

DISCUSSION

Previous studies (20, 22) using rat brain synaptosomes and fluorescent dyes as reporter molecules of membrane potential have shown that β-BtX causes depolarization. By contrast the present investigation in which the membrane potential of <u>Torpedo</u> synaptosomes has been monitored using essentially the same technique provides no evidence for β-BtX-induced depolarization under conditions where it is known to inhibit synaptosomal transport systems (6). A more quantitative method, again using fluorescent dye (24), also failed to provide any evidence for depolarization; if anything, a slight hyperpolarizing action was indicated.

It therefore seems that the mechanism of action of β-BtX depends on the type of nerve terminal under study, and this possibly reflects differences in the nature and distribution of the toxin binding sites (see also 16). Since β-BtX is composed of two unrelated but covalently-linked subunits, it is conceivable that two different classes of binding site exist on nerve terminal membranes. Some support for this idea comes from recent studies on the binding of [3H]β-BtX to rat brain synaptic membranes, where the evidence clearly points to an interaction via the smaller (non-phospholipase A_2) subunit (18, 19). The action of β-BtX on nerve terminals may thus be multifarious, particularly if two classes of binding site are independently distributed amongst different types of nerve terminal.

Depolarization of rat brain synaptosomes has also been reported for NtX-II-5, on the basis of increased carbocyanine dye fluorescence (23). This was not seen in the present study with <u>Torpedo</u> synaptosomes. The difference in this case might well be due to lytic effects; these being rigorously excluded in the present study but not in the earlier one with rat brain synaptosomes. With <u>Torpedo</u> synaptosomes, exposed to relatively low NtX-II-5 concentrations, there was, as with β-BtX, some evidence for a slight hyperpolarization.

The other native toxins used, NtX and TpX, had a clearcut hyperpolarizing effect on <u>Torpedo</u> synaptosomes. The question of the role of phospholipase activity in this phenomenon was examined by testing the effect of PBP-toxin derivatives on synaptosomal membrane potential, and by measuring the phospholipase activity of the toxins and PBP-derivatives, with <u>Torpedo</u> synaptosomes as substrate, under identical conditions to those employed in evaluating membrane potential (i.e. physiological). This produced some surprising results.

All toxins except β-BtX were found to have substantial phospholipase activity under these conditions. The results with TpX showed activities $\simeq$ 30X greater than measured using crude egg yolk phosphatidylcholine (PC) dispersions (11), but comparable to measurements where purified PC was used (5). Clearly the absolute activities of phospholipase neurotoxins are critically dependent on the substrate used for assay. This therefore vindicates our approach for measuring toxin phospholipase activity.

The effect of toxin modification with p-bromophenylacylbromide on phospholipase activity was as expected for the single chain toxins studied. Thus NtX, NtX-II-5 and EMtX were all inactivated by this treatment. For TpX however, this was not the case and it is suggested that the unmodified γ-subunit is responsible for the remaining ($\approx 50\%$) activity. From this it follows that the contribution of the γ-subunit to phospholipase activity of the native complex is about 50%. This is some 2240X greater than its activity alone and should be compared to a calculated 133-fold activation of the α-subunit (β chain assumed to have no activity).

Clearly the subunits of TpX can interact synergistically to produce a very potent phospholipase A_2. The Ca^{2+}-dependence of phospholipase activity in $(PBP)_2TpX$ (Fig. 3) also now provides an explanation for its Ca^{2+} requirement for inhibiting choline transport into T-sacs (8). Explanations based on change in binding affinity previously proffered to account for this are now redundant. In fact, the finding that $(PBP)_2TpX$ is almost as catalytically active as TpX has forced a complete re-evaluation of existing models for the mechanism of action of TpX.

Since phospholipase activity and inhibition of transport processes are both only marginally reduced by covalent modification, the latter probably depends upon the former. By contrast neurotoxicity is greatly reduced (11). Conceivably $\alpha-$ and γ-subunits interact at separate sites in the synaptic plasma membrane, but collectively contribute to phospholipolysis leading to changes in membrane structure and inhibiton of transport processes. The complete neurotoxic response would then require that the phospholipase action of the α chain alone be sufficient for further irreversible changes with the activity of the γ chain potentiating this action.

Whether or not toxin phospholipase activity is necessary for membrane hyperpolarization is difficult to answer with certainty, although the somewhat patchy information tends to suggest that it is not. The best evidence for this comes from the two toxins, β-BtX and NtX-II-5, with weak hyperpolarizing activities. Under conditions where phospholipase activity was below detectability (native β-BtX and PBP-NtX-II-5) Torpedo synaptosomes behaved as if hyperpolarized. Moreover for NtX-II-5 elimination of phospholipase activity (Fig. 4) by covalent modification barely altered its apparent hyperpolarizing activity (Fig. 2). Additional evidence comes from the toxins, NtX and TpX, producing larger hyperpolarizations. For NtX, hyperpolarization could be detected at 0.01 μg/ml, the threshold concentration for detecting phospholipase activity (cf. Table 1 and Fig. 4). The modified α-subunit of TpX, with <10% of native phospholipase activity, was also found to induce a slight hyperpolarization at a concentration (1 μg/ml) where lipolysis could not be detected (cf. Table 2 and Fig. 3).

The soundest conclusion to be drawn from these observations is that toxin-induced alterations in membrane potential are independent of gross phospholipase activity, and may be due to binding alone or binding and very limited catalysis.

CONCLUDING REMARKS

Presynaptic phospholipase A_2 neurotoxins have a variety of effects on Torpedo synaptosomes which can be detected using biochemical and biophysical techniques. These include membrane hyperpolarization, structural changes in the synaptic membrane, inhibition of membrane transport systems and release of cytoplasmic enzymes. By measuring phospholipolysis in synaptosomes under conditions identical to those where these effects have been observed, it has been possible to partially separate phospholipase-dependent effects from those which are independent of phospholipase activity. All these changes, except membrane hyperpolarization, appear to be directly attributable to enzyme activity, whereas binding alone might be sufficient to cause hyperpolarization. This would provide an explanation for the initial inhibitory action of these toxins on transmitter output from the neuromuscular junction, which is also phospholipase-independent (3).

The fact that all toxins in this group produce similar pharmacological and biochemical responses argues for a common binding site for these somewhat structurally diverse toxin molecules. This is possibly close to, or part of, an ion channel in the preterminal membrane. Arguments against a common molecular site of action have been put forward on pharmacological (4) and biochemical (18) grounds. Torpedo synaptosomes provide a system where further investigations of molecular specificity might prove profitable. In particular it is suggested that these sites might be mapped by exploiting the phospholipase activity of the toxins. Thus it would be interesting to know if toxins with low or blocked phospholipase activity antagonize the phospholipolytic actions of others in this group. These studies are currently underway.

Acknowledgements. Support from the Wellcome Trust (MJD and P.F.) and the MRC (PGC) is gratefully acknowledged. We also thank Dr. D. Eaker, Institute of Biochemistry, University of Uppsala, for supplying toxins, and Prof. J.N. Hawthorne for his interest.

REFERENCES

1. Baker, R.R., Dowdall, M.J. and Whittaker, V.P. (1975): Brain. Res. 100:629-644.
2. Bartlett, G.R. (1959): J. Biol. Chem. 234:466-468.

3. Caratsch, C.G., Maranda, B., Miledi, R. and Strong, P.N. (1981): J. Physiol. 319:179-191.

4. Chang, C.C. and Su, M.J. (1980): Toxicon 18:641-648.

5. Chang, C.C., Su, M.J. and Eaker, D. (1977): Naunyn-Schmeideberg's Arch. Pharmacol. 299:166-161.

6. Dowdall, M.J. and Fretten, P. (1981): In Cholinergic Mechanisms: Phylogenetic Aspects, Central and Peripheral Synapses, and Clinical Significance (eds) G. Pepeu and H. Ladinsky, New York, Plenum Press, pp. 249-259.

7. Dowdall, M.J., Fohlman, J.P. and Eaker, D. (1977): Nature 269: 700-702.

8. Dowdall, M.J., Fohlman, J.P. and Watts, A. (1979): In Advances in Cytopharmacology, Vol. 3 (eds) B. Ceccarelli and F. Clementi, New York, Raven Press, pp. 63-76.

9. Dowdall, M.J., Fretten, P. and Culliford, P.G. (1982): In Presynaptic Receptors (ed) J. deBelleroche, United Kingdom, Ellis Horwood, pp. 195-206.

10. Eichberg, J., Whittaker, V.P. and Dawson, R.M.C. (1964): Biochem. J. 92:91-100.

11. Fohlman, J., Eaker, D., Dowdall, M.J., Lullmann-Rauch, R., Sjodin, T. and Leaners, S. (1979): Eur. J. Biochem. 94:531-540.

12. Fretten, P., Dowdall, M.J. and Culliford, P.G. (1981): Biochem. Soc. Trans. 9:411-412.

13. Glass, R.L. (1971): Lipids 6:919-925.

14. Howard, B.P. and Gundersen, C. (1980): Ann. Rev. Pharmacol. Toxicol. 20:307-336.

15. Lee, C.Y. (1979): In Advances in Cytopharmacology, Vol. 3 (eds) B. Ceccarelli and F. Clementi, New York, Raven Press, pp. 1-16.

16. Miura, A., Muramatsu, I., Fujiwara, M., Hayashi, K. and Lee, C.Y. (1981): J. Pharmacol. Exp. Ther. 217:505-509.

17. Morel, N., Israel, M., Manaranche, R. and Mastour-Frachon, P. (1977): J. Cell. Biol. 75:43-55.

18. Othman, I.B., Spokes, J.W. and Dolly, J.O. (1982): Eur. J. Biochem. 128:267-276.

19. Othman, I.B., Wilkin, G.P. and Dolly, J.O. (1983): Neurochem. Int. 5:487-496.

20. Ng., R. and Howard, B.D. (1978): Biochemistry 17:4978-4986.

21. Richardson, P.J. and Whittaker, V.P. (1981): J. Neurochem. 36: 1536-1542.

22. Sen, I. and Cooper, J.R. (1978): J. Neurochem. 30:1369-1375.

23. Sen, I., Baba, A., Schulz, R.A. and Cooper, J.R. (1978): J. Neurochem. 31:969-976.

24. Tsien, R.Y. and Hiady, S.B. (1978): J. Membr. Biol. 38:73-97.

25. Yagihara, Y., Bleasdale, J.E. and Hawthorne, J.N. (1973): J. Neurochem. 21:173-190.

STUDIES OF TWO NOVEL PRESYNAPTIC TOXINS

W.O. McClure, D.E. Baxter, R. Brusca, R.D. Crosland[*],
T.H. Hsiao[**], M.L. Koenig, J.V. Martin[***]
and J.E. Yoshino[****]

Section of Neurobiology
University of Southern California
University Park
Los Angeles, California 90089-0371 U.S.A.

INTRODUCTION

The use of toxins has been widespread in examining a number of biological phenomena. Every undergraduate in an introductory course in biochemistry is aware of the toxins which inhibit the electron transport system and oxidative phosphorylation, and appreciates the value of these molecules in the understanding of complex biological processes. Our laboratory has long agreed with others in our field that toxins would be useful - in fact, would probably be essential - in unravelling the molecular mechanisms which control release of neurotransmitters at synaptic terminals. We have limited our work to toxins which are active at the presynaptic terminal, for we wish to consider only the relese of neurotransmitter. We have further restricted our attention to toxins which stimulate

---

[*]Present Address: Department of Pharmacology, University of California at Los Angeles, Los Angeles, CA 90025

[**]Present Address: Department of Biology, Utah State University, Logan, UT 84332

[***]Present Address: Building 10, National Institute of Mental Health, Bethesda, MD 20205

[****]Present Address: Department of Biochemistry, Virginia Commonwealth University, Richmond, VA 23298

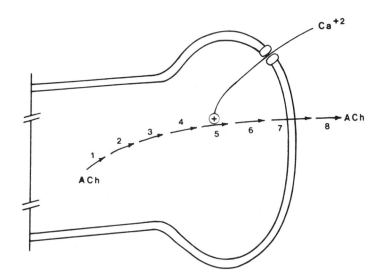

Figure 1. Simplified scheme for the reactions involved in the release of acetylcholine (ACh) at a nerve terminal. ACh is indicated moving from an initial internal state through a series of transitions which finally cross the membrane (Step 7) and lead to ACh in the cleft, free of association with presynaptic elements. Each step refers to a change in state of the ACh, but does not imply a change in the covalent bond structure of the neurotransmitter. Synaptic vesicles are involved in some of the defined states, but we do not now know which ones. For the sake of description, eight reactions are considered, of which six are internal and one external to the membrane. The actual number of steps is unknown.

release, and have avoided working with those which inhibit release. It is worthwhile to examine the logic behind this decision. In Fig. 1 is presented a sketch of the biochemical steps involved in a mechanism of release of the neurotransmitter, acetylcholine (ACh).

In this model the numbered steps indicate any change in the chemical state of a molecule of ACh. A change in state could involve a movement from one intracellular site to another, such as being introduced into or expelled from a vesicle. It could also involve simply a change in binding, as a transfer of ACh from one protein to another, or from a bound to an unbound state. The model presented here is greatly simplified, but not to the point of being unrealistic. Almost any mechanism of release can be reduced to a description similar to that of Fig. 1.

The choice of which of several toxins to study can be clarified to some extent by considering the logic associated with Fig. 1. It is clear that a toxin which blocks any of the steps of Fig. 1

will inhibit release of neurotransmitter. Conversely, a toxin which inhibits release could act at any of the steps involved in release. Because of this inherent uncertainty in the site of action of inhibitory toxins, we have chosen not to examine them. The situation with an excitatory toxin is much more clear. It is known that release of ACh occurs continually at a low level, but is stimulated by several orders of magnitude over a period of a few microseconds by the introduction of Ca^{+2} ions into the cytoplasm of the terminal (4, 5, 6, 7). The Ca^{+2} must act on some step in the pathway to increase the rate of release. In Fig. 1, this step is arbitrarily assumed to be 5. Since the overall rate of release is increased when the rate of step 5 is increased, we must conclude that step 5 is rate-limiting. Following the same logic, an excitatory toxin must act also at the rate-limiting step, for action at any other step would result in only a minimal increase in rate. It follows that the action of an excitatory toxin is probably directed toward the usual controlling step, and that such a toxin might be usefully employed as a tool with which to study the reactions controlling release. As a corollary, it follows that a stimulating toxin probably interacts with the same step at which Ca^{+2} acts to stimulate release.

It is important to realize that nothing in these arguments requires that the controlling step be physically located at the membrane. It is certainly attractive to assume that this is so; nonetheless, a cytoplasmic rate-limiting step would be quite as effective as one localized on or in the membrane.

Over the past few years we have searched for, and have found, a number of toxins which stimulate release. In this report, we shall review briefly the major points of interest of two of these: leptinotarsin (LPT), a toxin from the hemolymph of the leptinotarsid beetles; and iotrachotin (IOT), a toxin from the wound exudate of the purple bleeder sponge of the Western Caribbean, Iotrachota biratulata.

MATERIALS AND METHODS

The methods used in these studies have been reported elsewhere (1, 10). A brief recapitulation of the major methods will be given.

Sources of Material. β-leptinotarsin-d and β-leptinotarsin-h were isolated from lyophilized hemolymph of fourth instar larvae of beetles of the species Leptinotarsa decemlineata (10) and L. haldemani (1), respectively. Both toxins represent the major activity of the respective sample, and have been purified by conventional techniques to about 50% homogeneity. Iotrochotin was isolated, also by conventional techniques, from the fluid released from I. biratulata when the organism was squeezed. Iotrochotin is homogeneous by several criteria, including polyacrylamide gel electrophoresis, both in the presence and absence of ionic detergents.

Assay. Synaptosomes were prepared by the method of Gray and Whittaker (2) from freshly dissected rat brain. For most of the work, the P2 pellet was resuspended in physiological saline and used. This preparation consists of about equal amounts of synaptosomes, mito-chondria, and myelin. When the contamination by mitochondria or myelin might confuse the interpretation of an experiment, these three fractions were separated by centrifugation on sucrose density gradients.

Synaptosomes were loaded with tritiated choline with a 30 min preincubation at 37°C, and exposed to toxin for a period of time. The reaction was terminated by sedimenting the synaptosomes in a Microfuge (Beckman Instruments, Fullerton, CA), after which an aliquot of the supernatant was removed for analysis. Radioactivity was measured using liquid scintillation spectrometry. Separation of ACh from Ch was carried out, when desired, by a modification of the method of Goldberg and McCaman (3).

RESULTS

Leptinotarsin. The β-leptinotarsins are a set of proteins with mo-lecular weights in the range of 40-50,000 (1, 8, 10). They appear to be very similar, but not identical, when isolated from different species of the genus <u>Leptinotarsa</u>. Because of their similarity, both will be considered together in the following text. Where rele-vant, differences will be noted.

Leptinotarsin stimulates release of ACh from rat brain synap-tosomes. The toxin stimulates release in a dose-dependent manner, with increasing concentrations of toxin producing increasing amounts of release. The variation of release with the concentration of toxin is well described by Hill's equation, but the degree of co-operativity, as measured by the Hill coefficient (n_H) is different for the toxins from the two species (Fig. 2). Leptinotarsin from <u>L. haldemani</u> shows no cooperativity ($n_H = 1.0$), while the toxin from <u>L. decemlineata</u> has some cooperativity ($n_H = 2.0$).

That the toxin from <u>L. decemlineata</u> has more complex kinetics than does that from <u>L. haldemani</u> is indicated also from the time dependence of release. Toxin from <u>L. haldemani</u> yields release which has a simple exponential dependence upon the time of measurement. In contrast, release stimulated by toxin from <u>L. decemlineata</u> is characterized by more complex kinetics.

The mechanism by which LPT stimulates release has been considered in some detail. To examine the role of membrane channels, we have altered the ionic composition of the solution bathing the synaptosomes, and have examined the action of other toxins which are known to interact with various channels. Replacement of external Na^+ ions by glucosammonium ions has no effect upon the action of LPT. In a

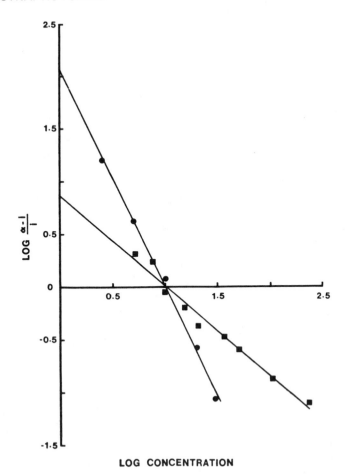

Figure 2. Hill plot of the variation of release with the concentration
of toxin for β-leptinotarsin-h (■) and β-leptinotarsin-d (●).   alpha:
maximal release, at saturating concentration of toxin. Concentrations
of toxin are normalized to the concentration which yields half-maximal
release for the toxin in question. The half-saturating concentrations
are: β-leptinotarsin-d, 5.1 nM; β-leptinotarsin-h, 0.88 nM. Measurements were conducted using rat brain synaptosomes labelled with tritiated choline, as given in the Methods.

similar way, tetrodotoxin, which blocks the Na^+ channel, is without
affect on release stimulated by LPT. These data make an action of
LPT on the Na^+ channel quite unlikely.   It is also unlikely that
LPT acts on the K^+ channel, for altering external levels of K^+ does
not affect the action of the toxin. Addition of tetraethylammonium

ions to the external solution is also ineffective in altering the
action of LPT (Table 1).

The Ca^{+2} channel is probably involved in release stimulated by
LPT. Several lines of evidence support this assertion. If Ca^{+2}
ions are removed from the external solution, LPT can no longer cause
release (Table 2). Release can be re-established in Ca^{+2}-free solutions by supplying either Ba^{+2} or Sr^{+2} ions, both of which are known
to be able to replace Ca^{+2} ions in release. Finally, the action
of LPT can be inhibited by ions which are known to act as inhibitors
of release stimulated by Ca^{+2}. Both Co^{+2} and Cd^{+2} can completely
inhibit release stimulated by LPT, even in solutions containing
physiological levels of Ca^{+2} ions (Table 2).

It is possible that LPT interacts with synaptosomes by acting
as an ionophore, in a manner analogous to the organic Ca^{+2} ionophore,
A23187. We feel that this explanation is unlikely. Liposomes can
be made by sonicating pure lipids suspended in solutions containing
$[^{45}Ca]^{+2}$, and then rinsed to remove external radioactivity. LPT does
not cause liposomes prepared in this way to release Ca^{+2}, which
suggests that the toxin does not interact with these vesicles in a
way characteristic of a Ca^{+2} ionophore. Furthermore, LPT does not
depolarize the rat neuromuscular junction, which it should do if
it were allowing the influx of Ca^{+2} ions.

Evidence links LPT to an action on the voltage-sensitive Ca^{+2}
channel of the presynaptic membrane. Changes in the membrane potential
of synaptosomes can be measured using the carbocyanine dyes which

Table 1

Effect of Various Agents Upon Release Stimulated By
β-Leptinotarsin

Agent	Release (% of control)
control	100 ± 7
+ tetrodotoxin, $1\mu M$	126 ± 21
- Na^+ + glucosammonium	106 ± 10
- Na^+ + sucrose	120 ± 17

Release of $[^3H]ACh$ from rat brain synaptosomes stimulated with 4.2
μM β-leptinotarsin was studied. Solutions of the suspending medium
were modified as indicated, using the following concentrations:
TTX, 1 μM; glucosammonium chloride, 128 mM to replace an equal reduction of NaCl; sucrose, 256 mM to replace a reduction of 128 mM
NaCl. For details see Methods.

Table 2

Effect of Agents Which Modify Calcium Responses Upon
Release Stimulated by β-Leptinotarsin

Agent	Release (% of control)
control, + Ca^{+2}	100 $\pm$ 12
- Ca^{+2} + EGTA	0^a
- Ca^{+2} + Sr^{+2}	109 $\pm$ 13
- Ca^{+2} + Ba^{+2}	117 $\pm$ 14
+ Ca^{+2} + Cd^{+2}	0^a
+ Ca^{+2} + Co^{+2}	0.7 $\pm$ 3.6

Concentrations used were: β-leptinotarsin-h, 9.5 nM; Ca^{+2}, in control experiments with EGTA, 1.2 mM; EGTA, 1 mM; Sr^{+2} and Ba^{+2}, 1.2 mM. In experiments involving Cd^{+2} and Co^{+2}, the concentration of Ca^{+2} was 2 mM and both Cd^{+2} and Co^{+2} were 50 mM. ^{a}no detectable release

bind to the membrane and alter their fluorescence in a manner which is a function of the membrane potential (9). Measurements using one of these dyes (di-O-C_5) indicate that LPT depolarizes synaptosomes, with a time course which closely parallels release. The depolarization caused by LPT depends upon neither Na^+ nor K^+ in the external solution, but is abolished when Ca^{+2} ions are removed from the bathing solution. These data suggest that LPT stimulates uptake of Ca^{+2} into synaptosomes.

It should be possible to measure uptake of Ca^{+2} into synaptosomes stimulated by LPT. In suspensions of synaptosomes incubated with $[^{45}Ca]^{+2}$, treatment with LPT causes an influx of radioactivity. Furthermore, the increase in $[^{45}Ca]^{+2}$ inside synaptosomes closely parallels the release of ACh into the outside solution (Fig. 3).

Ca^{+2} ions are known to activate release in a highly cooperative manner. If Co^{+2} and Cd^{+2} inhibit by competing with Ca^{+2} for a site in the release process, these ions might also exert a high level of cooperativity in their inhibition of release stimulated by leptinotarsin in the presence of a fixed concentration of Ca^{+2}. Cd^{+2} does inhibit in this manner. Inhibition of release by varying concentrations of Cd^{+2} is well satisfied by the Hill equation (Fig. 4). The calculated value of the Hill coefficient in this experiment was 4.7, which suggests that between 4 and 5 ions of Cd^{+2} are required to inhibit completely the release stimulated by leptinotarsin.

Iotrochotin. Studies with IOT are not as far advanced as those with LPT. Iotrochotin has been purified to homogeneity as judged

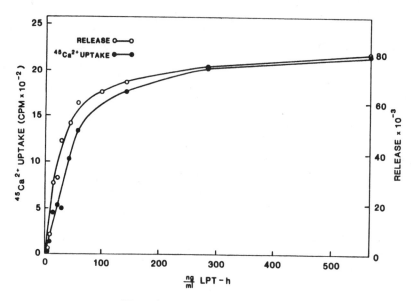

Figure 3. Uptake of $[^{45}Ca]^{2+}$ into synaptosomes stimulated by β-leptinotarsin-h. Purified synaptosomes were suspended in solutions containing 7 μM choline chloride and incubated for 30 min at 37°C in
a shaking water bath. After washing, aliquots of synaptosomes were
added to solutions containing 0.2 μCi of $[^{45}Ca]Cl_2$ and various amounts
of β-leptinotarsin-h and incubated for 10 min at 37°C. The synaptosomes were pelleted, washed twice, resuspended, and counted. Uptake
of $[^{45}Ca]^{2+}$ (●) was calculated as the difference between the radioactivity in suspensions originally containing β-leptinotarsin-h and
the radioactivity in equivalent suspensions lacking toxin. For comparison, data describing release of tritiated materials from aliquots
loaded with $[^3H]$choline are also given (o). From Crosland et al. (1).

by its behavior on chromatography and electrophoresis. The toxin
is a protein with an apparent molecular weight of 18,000. A similar
value of the molecular weight is obtained using either gel exclusion
chromatography or electrophoresis on polyacrylamide gels in the
presence of sodium dodecyl sulfate and reducing agents, which suggests
that the toxin is a single polypeptide chain. Iotrochotin releases
radioactivity from synaptosomes pre-labelled with tritiated choline.
Fractionation of the released radioactivity indicates that about
60% of it is ACh, while the remainder is choline. A similar ratio
of ACh:Ch is seen in synaptosomes which had not been exposed to
toxin, but had been lysed in hypotonic media. The fact that the
composition of released radioactivity is the same as that in the
terminal suggests that IOT may not act to release ACh in a preferential
manner. To control for the possibility that IOT may simply poke
holes in the membrane, we examined the release of two proteins,

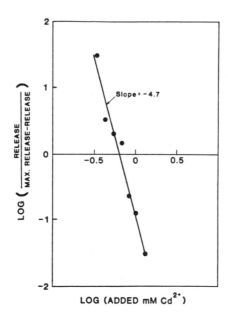

Figure 4. Hill plot of the effect of varying concentrations of Cd^{2+} on release from synaptosomes stimulated by β-leptinotarsin-h. Synaptosomes previously incubated with [^{3}H]choline were pelleted, washed, and added to solutions containing β-leptinotarsin-h (0.9 μM) and various concentrations of $CdCl_2$. Maximal release = 58,000 cpm. From Crosland et al. (1).

lactate dehydrogenase and choline acetyltransferase, from synaptosomes in which release of radioactivity was being simultaneously measured. The release of the two proteins was not stimulated by IOT under conditions causing release of large amounts of radioactivity, suggesting that this toxin does not grossly disrupt the membrane.

Release stimulated by IOT has an interesting dependence upon the time of observation. Release observed at a given concentration of toxin is complete, within experimental error, at the earliest times measured. If the concentration of toxin is increased, a time course of similar shape is seen, but the amount of ACh released becomes greater. If the concentration of IOT is increased even further, saturation is eventually reached at the point when all the synaptosomal store of radioactivity is released before the first time point. The concentration of toxin necessary to achieve half the saturating (maximal) release is about 0.5 nM. The time dependence of this system is that expected for a simple titration of releasing sites with toxin.

The action of IOT has been examined using experiments similar to those considered for β-LPT. Release stimulated by IOT does not seem to involve either the Na^+ or K^+ channels in synaptosomes. In contrast to β-LPT, removing Ca^{+2} from the bathing solution has no affect upon release stimulated by IOT. These data suggest that release stimulated by this toxin does not involve any of the usual presynaptic channels for ions. It is possible, although not proven by these data, that IOT acts at the synaptic site which regulates release. If so, IOT may be of value as a tool in the study of the molecular aspects of release.

DISCUSSION

Two new proteinaceous toxins which act at the presynaptic terminal to stimulate release of ACh have been found, purified, and partially characterized. One of these, IOT, is only in the initial stages of our study. The early data appear promising, and are consistent with the possibility that the toxin acts directly on the molecular system controlling the rate of release. The second toxin, β-LPT, probably acts at the presynaptic voltage-sensitive Ca^{+2} channel, where it forces the channel open and allows an influx of ions.

The goal of these studies is to use either or both of the toxins as tools in the study of their respective receptors, planning thereby to begin to attack both the Ca^{+2} channel and the release mechanism on a molecular level. Several experimental results suggest that either or both toxins may be useful in this regard. First, both interact with their receptors at low concentrations. Since the true dissociation constant for the toxin:receptor interaction is equal to or less than the concentration which half-stimulates release, the affinity of these toxins for their receptors is probably in the range of $1-5 \times 10^{-10}$ M. This represents sufficiently firm binding to be useful as an affinity ligand. Second, the toxins stimulate release of not only ACh, but also two other neurotransmitters, norepinephrine and 4-aminobutyric acid. The fundamental mechanisms controlling release are likely to be very similar for all neurotransmitters, so a toxin which acts on several neurotransmitters is likely to be interacting with a point of central importance in release. Third, the fact that both toxins are proteins suggest that both act externally, and probably interact with receptors in the outer face of the membrane. This is not meant to suggest that all proteinaceous toxins act outside the cell, for several are known which can gain access to the cytoplasmic space. Nonetheless, most toxic proteins act externally, and it seems most reasonable to suppose that these do also, until proven otherwise.

In order to function as an affinity ligand, these two toxins must be labelled in some way. Studies aimed at this goal are currently underway. When labelled homogeneous derivatives are available, they should be quite useful in further understanding of the mechanism of release of acetylcholine.

Acknowledgements. This work has been supported by funds from the Dreyfus Foundation, the National Science Foundation, and Nelson Research, Irvine, California. We thank Ms. D.L. Lindel-Coleman for extraordinarily competent technical assistance.

REFERENCES

1. Crosland, R.D., Hsiao, T.H. and McClure, W.O. (1984): Biochemistry 23:734-741.
2. Gray, E.G. and Whittaker, V.P. (1962): J. Anat. 96:79-87.
3. Goldberg, A.M. and McCaman, R.E. (1973): J. Neurochem. 31:1005-1013.
4. Katz, B. and Miledi, R. (1967): J. Physiol. (London) 189:535-544.
5. Katz, B. and Miledi, R. (1967): J. Physiol. (London) 192:407-436.
6. Llinas, R., Blinks, J.R. and Nicholson, C. (1972): Science 176:1127-1128.
7. Llinas, R., Steinberg, I.Z. and Walton, K. (1976): Proc. Nat. Acad. Sci. USA 73:2918-2922.
8. McClure, W.O., Abbott, B.C., Baxter, D.E., Hsiao, T.H., Satin, L.S., Siger, A. and Yoshino, J.E. (1980): Proc. Nat.Acad. Sci. USA 77:1219-1223.
9. Sims, P.J., Waggoner, A.S., Wang, C.H. and Hoffman, J.F. (1974): Biochemistry 13:3315-3330.
10. Yoshino, J.E., Baxter, D.E., Hsiao, T.H. and McClure, W.O. (1980): J. Neurochem. 34:635-642.

CONFERENCE ON DYNAMICS OF CHOLINERGIC FUNCTION:

OVERVIEW AND COMMENTS

A.G. Karczmar

Loyola University Medical Center
Department of Pharmacology and Therapeutics
2160 South First Avenue
Maywood, Illinois 60153 U.S.A.

TRIBUTE

*"Hoc enim per rationes logicas plenum sciri non poterit,
sed metaphysica determinanda relinquntur." Albertus
Magnus (Albert von Bollstadt), 1193(?) to 1280:
Liber de Praedicabilibus. Opera Omnia, Vol. 1,
A. Borguet, Ed., L. Vives, Paris, 1980.*

With each successive International Cholinergic Meeting its
program increases and so do the logistic and fiscal problems. For
this, the Fifth Meeting, these problems were particularly difficult
to resolve, as the interest in the cholinergic system and the desire
to participate were never so high, yet the state of the world's
economy made the financing of the Meeting particularly difficult.

Nevertheless, Israel Hanin - with a minor assist from his Committee - managed in the face of these difficulties to organize a
Meeting that was major in size and scope; significant in its contributions; and one that was run smoothly, effectively, and on time.
He worked hard and successfully to gather sizable contributions
from the private sector in the U.S.A. and abroad, and from the federal
U.S.A. sources; he made a special effort to find a suitable congenial
and charming site for the Meeting; and especially, he put forward
an interesting and coherent program. He well deserves an enthusiastic
vote of thanks.

No less gratefully we should acknowledge the help of the Staff
for this Meeting. Their tasks were innumerable as they included

1215

correspondence - and stuffing the envelopes, preparation of kits and name tags for the Meeting, airport duties, care of lights, projections and sound, cafeteria and snack services, information and guidance activities, etc., etc... This staff consisted of Denise Arst, Linda Barilaro, Matthew Clancy, Dr. Gary Hasey, Leda Hanin, Dahlia Hanin, Ursula Kopp, Dr. Steven Leventer, Tony Lieteau, Joyce O'Leary, Denise Sorisio and Dr. Ellen Witt. Without them there would have been no Meeting!

INTRODUCTION

The symposium on the Dynamics of Cholinergic Function was the Fifth Meeting of the International Cholinergic Group. It followed the inaugural Skokloster Symposium (41) and the Meetings in Boldern, Switzerland (112), La Jolla (50) and Florence (93). It pleases me that most of the organizers and chairpersons of the early Meetings such as Edith Heilbronn, Peter Waser, Don Jenden, Giancarlo Pepeu and Herbert Ladinsky were present at, and contributed papers for, this particular Symposium.

I would like to make a few initial remarks.

First of all, the number of participants and papers has increased from meeting to meeting; hopefully, the quality and sophistication of the contents has improved as well. Thus, sixty investigators participated in the Skokloster Meeting and 115 in the Florence Meeting, while there were 165 participants at the Oglebay Park Meeting; these 165 participants delivered 43 presentations and presented 87 posters. The query that could be raised, therefore, is why do we need so many meetings in the cholinergic area, and why must we increase the number of the presentations from one meeting to another?

There may be several answers to this question. One answer is given by Murphy's law: once an organization is started it has a growth of its own, whether justified or not. It may also be claimed that we are never satisfied with what we have done at any particular meeting, so we continue to have more meetings. This can be illustrated by the story of an Italian tenor who was recalled back by his audience for more and more encores of an aria; as he thanked them and asked what he could do to repay them, a member of the audience told him: "We want you to do it again till you get the tune right." Personally, I would be gratified if the reason for these repeated performances were the rapid growth of the cholinergic field. It is sufficient to think of what such immortal cholinergikers as Otto Loewi, Henry Dale, or William Feldberg, would have said if we confronted them with such subjects of our discussions at this Meeting as channel-receptors macromolecules, second messengers, autoreceptors and the use of autoimmune techniques for the tracing of cholinergic pathways. In fact, if we just look back to the 1970 Skokloster meeting and compare its subject index with the index for the present Meeting we come up with the same idea, and that is that the cholinergic field, which provided in the past - and some of this past was reviewed at this Meeting with intimate insight by Frank MacIntosh and Bo Holmstedt[*] - such tremendous advances as the discovery of the chemical transmission, miniature endplate potentials, elementary events, etc., continue to be in the forefront of the Neurosciences.

It is true that at the Skokloster Meeting we were already introduced to the sensitive gas chromatography method for the determination of acetylcholine (ACh), but there was only brief and tentative mention of mass spectrometry. At that Meeting the cholinergic receptor was barely alluded to, while several nicotinic and muscarinic receptors and their presynaptic and postsynaptic roles were emphasized at the present Meeting. Similarly, a notable advance was made in the area of cholinergicity of a number of behaviors: while in Skokloster the only presentation concerning behavioral correlates of the cholinergic system was that of Crossland on the abstinence syndrome, at this as well as at the Boldern, La Jolla and Florence Meetings, a number of behavioral states and definitive information concerning the change in cholinergic synapses during aging and certain neurological states were presented. And as almost absolute novelties, several types of second messengers, receptors versus channels, and interactions between cholinergic and other neurotransmitters were discussed. Yet, as the participants of the Skokloster meeting, and even more so, Dale and his contemporaries would have been astounded if they could listen to us, we certainly would be just as shocked if we could - and to some of us this would be a miracle - participate in the 10th Meeting of this Group.

The final comment that I would like to make is that I stand humble presenting this summary, as I know that the prior commentators for the earlier Meetings of this group were such individuals as George Koelle.

TOPICS OF THE PRESENT MEETING

The topics covered at this Meeting concern 10 important areas in the cholinergic field (Table 1). Let me summarize and comment on the information presented with respect to these topics.

---

[*]Un-dated name quotations refer to authors who contributed chapters to this book; dated references (such as, eg., Shute and Lewis, 1967, see below) are listed in the bibliography for this chapter.

Table 1

I. Anatomy of the Central Cholinergic System

II. Aging, SDAT and Other Clinical Conditions

III. Cholinergic Pre- and Post-Synaptic Receptors

IV. Acetylcholine Release

V. Cholinesterases, Anticholinesterases and Reactivators

VI. Acetylcholine Synthesis, Metabolism and Precursors

VII. Interactions of Acetylcholine With Other Neurotransmitter Systems

IX. Cholinergic Mechanisms in Physiological Function Including Cardiovascular Events

X. Neurotoxic Agents and False Transmitters

I. CHOLINERGIC PATHWAYS: ANATOMY OF THE CENTRAL CHOLINERGIC SYSTEM

Classically, marked advances were made in delineating the cholinergic pathways by using histochemical assays for cholinesterases, developed originally by Friedenwald and Koelle, and subsequently employed so effectively by Koelle (cf. 68) and by Shute and Lewis (101). This development illustrates the advantage of identifying a neurotransmitter or the enzymes that are crucial for its action. As emphasized at this Meeting by Butcher and Woolf, the current methods include retrogradely and anterogradely transported neuronal labels; immunohistochemical and cytochemical tests for acetylcholinesterase (AChE) and choline-o-acetyltransferase (ChAT); methods concerning the identification of the axons and the somata; procedures concerning the localization of nicotinic and muscarinic receptors, whether by binding or autoradiographic methods as illustrated in this book by Schwartz and Kellar and by Yamamura and his group; as well as methods for measuring ACh release from strategic brain sites, developed by the late Nick Giarman with Pepeu, which led to the results described at this Meeting by Pepeu and his associates.

These various methods yield consistent data, as it appears from the presentations of Butcher and Woolf, the McGeers with Kimura and Peng, Watson et al., Struble and his associates, Pepeu, Casamenti and their associates, Salvaterra et al., Wamsley and Gehlert, and

Schwartz and Kellar.   Besides the sites known for many years as cholinergic such as somatic and parasympathetic cell bodies of cranial nerves and the preganglionic sympathetic neurons as well as alpha and gamma motoneurons, we now are aware of the pathway which projects from the medial basal forebrain cholinergic system, and which inner- vates the septal nuclei as well as the limbic structures, the cortex and the brain stem.   There are also projections from pedunculo-pontine and parabrachial nuclear complex and from the giganto- and magno- cellular elements of tegmental nuclei to the thalamus, habenula and interpeduncular nucleus.   Finally, less well ascertained pathways are the ascending and descending pathways originating from the brain stem reticular formation.   Additional, localized cholinergic networks or neuronal fields are present in the retina and the basal ganglia; these were recently defined at length by McGeer's group (80) and earlier, by Fibiger (28).

The relative ubiquity of the cholinergic nervous system should be stressed at this time.   There is no wonder that the cholinergic nervous system is implicated in so many behaviors and that the choli- nergic agonists and antagonists have so many behavioral effects (see below, Sections II and VIII).   An interesting point in this context is that this ubiquitous central cholinergic system responds preponderantly but not solely, muscarinically (cf. eg., 69).   The new wrinkle presented at this Meeting by Watson et al. is that there may be several subpopulations of central muscarinic receptors.

Another point should be made with respect to receptor research presented in this Section of the Meeting by Annica Dahlström and her associates.   Dahlström identified presynaptic cholinergic receptors which are transported via axonal transport from the neurons to the nerve terminals.   This work should be related to the studies also presented at this Meeting by Bierkamper, and by Aizenman and their associates, as it appears that there may be muscarinic and nicotinic nerve terminal receptors both acting in a negative feed-back manner. This point will be discussed subsequently (see below, Sections III and IV).

Still another exciting finding was presented within the context of the anatomy of the cholinergic system by Sorbi and his associates. It appears that there is a great individual variability with regard to the lateralization of the cholinergic enzymes in the human cortex; furthermore, there is a correlation between this cholinergic asymmetry and the asymmetry of the energy-related enzymes.

II.   AGING, SDAT AND OTHER CLINICAL CONDITIONS

Aging and SDAT.   The role of the cholinergic system in aging and senile dementia of Alzheimer's type (SDAT) - perhaps a premature form of aging - was proposed in the late 70's by Bowen and by Perry (see Bowen in this book and 94).   Some recent corroborative evidence concerns the possible cholinergic basis of memory impairment in aging (see for instance, 21). It should be pointed out in this context that lesions of the hippocampus, a target structure for cholinergic neurons (see above), notoriously cause amnesia and learning deficits, whether in man or animals (for references see 21 and 75). At this Meeting a corroborative finding was made by Lerer and her coinvestigators, as they demonstrated that, in rats, lesions of the magnocellular nuclei depleted ChAT of frontal and parietal cortex and caused impairment of some - but not all - types of learning.

Improvement of this age-related learning deficit, in animals, by the administration of either ACh precursors or anticholinesterases (antiChE's) is also consistent with the concept that a cholinergic deficit underlies this malfunction of learning (10). It should be pointed out, however, that cholinergic rectification of learning deficits in aged animals or in man may be a matter of controversy. In fact, presentations by Mervis et al. and of K.L. Davis et al. at this Meeting did not yield any clear cut data on this matter. For example, K.L. Davis et al. confirmed their earlier findings that physostigmine causes only a slight improvement in memory and overall status of dementia; furthermore, that effect was not obtained in a significant fraction of patients. Neither is it clear at this time, what is the effect of precursors such as choline in the improvement of memory (see Sections IV and VI). It should be remembered in this context that choline and the components of the phosphatidyl inositol cycle may have effects - including those on aging - that may be independent of their contribution, if any, to the synthesis of ACh. These effects may involve transmission (92) or, perhaps, the membrane itself. Thus, at this Meeting Mykita and his associates demonstrated that cultured neurons that showed pathology under the conditions of reduced atmosphere of CO_2 may be protected from the damage, or their pathology reversed, by the treatment with cytidine diphosphocholine. They proposed, therefore, the use of cytidine diphosphocholine in the treatment of cerebral vascular disease.

It is pertinent for this topic that the cholinergic system represents an age-related continuum. Indeed, the number of both muscarinic and nicotinic receptors increases in several brain parts- such as hippocampus - during early postnatal phases of mouse growth, as shown in this book by Nordberg. Similarly, a cholinergic maturation characterizes chick ganglia and iris, as presented at this and the previous Meeting, by Giacobini (see 32).

Altogether, past investigations indicate that following a period of adulthood, there may be cholinergic senescence. This point of view was presented at this Meeting by a number of investigators, particularly with respect to the CNS. For instance, Gibson and Peterson found that the release of ACh from brain slices of mice decreases with age, and they related this phenomenon to an age-related decrease in calcium uptake as well as mouse performance. What is

consistent with their hypothesis is that they could compensate for
this behavioral deficit by means of 3-4 diaminopyridine, a releaser
of ACh.   Similarly, Pontecorvo and Bartus confirmed their earlier
data indicating, on biochemical and electrophysiological grounds,
that aging in rats is a cholinergic process, and found additional
evidence for this concept as their aged rats behaved similarly to
young rats with nucleus basalis lesions induced by means of the
neurotoxin, ibotenic acid.

Furthermore, several presentors in this Section of the Meeting
described data indicating that cholinergic deficits may underlie
both aging and SDAT.   Thus, London and Waller found a decrease in
ChAT, although not in muscarinic binding, in the neocortical and
hippocampal samples obtained from Alzheimer's dementia patients.
These authors stressed that there may be a complex relationship
between ChAT and muscarinic binding during rodent aging, and that
this relationship may depend on species.   Again, Bigl et al. found
that the distribution of the neuritic plaques observed in autopsied
brains of SDAT patients matches that of cholinergic neurons, such
as those of the nucleus basalis, that appear to be reduced in number
in the course of Alzheimer's dementia.   Also, Price et al. stressed
that, in primates, basal forebrain cholinergic projections to medial
septum, diagonal band and nucleus basalis exhibit high incidence
of plaques during aging; these are the sites that exhibit cholinergic
deficits in SDAT patients.

In a similar vein, severity of the dementia of SDAT patients
was related, by Bowen and his associates, to decrements in ACh synthesis in the middle temporal gyrus samples obtained with diagnostic
craniotomy; and to lowering of ACh levels in the cerebrospinal fluid
(CSF), by K.L. Davis and his colleagues.   Furthermore, Etienne and
his associates found a striking loss of large neurons of the nucleus
basalis in brains from SDAT patients; this loss involving large
neurons exhibiting normally high AChE and, as shown by the McGeers,
also exhibiting high ChAT activity.   Interestingly, there was no
apparent loss of striatal neurons in the same tissues studied.   It
must be, however, stressed that Etienne and his associates did not
find any progress in cell loss when they compared early and late
SDAT patients.

If indeed there is a lesion or a loss of cholinergic neurons
in aging or SDAT, the interesting point raised by J.N. Davis and
his associates should be considered. These investigators demonstrated
that in rodents (and possibly in primates) the lesions to a number
of basal forebrain cholinergic neurons that caused neuronal losses
in the septo-hippocampal pathway, the neocortex, or habenula, resulted
in the replacement of the neurons in question by nerve fibers normally
innervating blood vessels ("sympathetic ingrowth").   They suggested
that a trophic factor generated by the septo-hippocampal pathway
may be the cause of the ingrowth.   This finding is consistent with

that of Morrow et al. (85) and Loy et al (74), who have demonstrated that sectioning of the medial septum and diagonal band of the rat increases the number of alpha-adrenergic receptor sites, and causes ingrowth of sympathetic axons in the hippocampus.

Altogether, the data presented by various investigators at this Meeting as to the cholinergic deficit during animal and human aging and in SDAT are consistent. On the other hand, cholinergic senescence of the periphery is not as clear. Specifically, at this Meeting Giacobini stressed that one of the markers of human senescence is the age-related senile miosis. Similarly, Smith found that ACh content per endplate is markedly decreased in aged rats (see also 104). However, he found an increase in the spontaneous release of ACh, which he related on the basis of his morphometric findings, to the increased number with age in the nerve terminals - possibly a compensation for the decrease in ACh content or synthesis. On the other hand, other investigators found a decrease in spontaneous release of ACh with age in mice; they found also that ACh stores of the nerve terminals are depleted more readily upon prolonged activity, in the old, rather than in the young mice (40, 89).

In view of these findings it would be useful to have a readily accessible marker that would reflect the central cholinergic loss. For several years red blood cells and platelets were studied as possible models of aging. At this Meeting, somewhat inconsistent data were presented by several investigators. Blass and Hanin and their associates found that Alzheimer's patients exhibit a high ratio of red cell choline to plasma choline as compared to controls. However, Domino and his associates stressed that while the plasma choline increases, and its turnover decreases with age, in monkeys and humans, the choline concentration of red blood cells does not; what is more important, they could not find any change in plasma or red blood cell choline levels in SDAT or tardive dyskinesia pa-tients. Finally, Sherman and Friedman found that both plasma and red blood cell choline increase with age, and that this increase was similar in elderly controls versus SDAT patients; on the other hand, they could correlate choline red blood cell levels with precursor (and piracetam) therapy and, in fact, the small percentage of patients that responded in their hands to precursors, showed particularly high red blood cell choline levels.

Two other markers were suggested by K.L. Davis and his group, namely the reduction in SDAT patients in the mean plasma cortisol concentration, which was correlated inversely with the CSF concen-tration of 3-methoxy, 4-hydroxy phenylglycol (MHPG). Furthermore, CSF concentrations of ACh correlated inversely with the degree of cognitive impairment in SDAT patients. It should be emphasized that as it is well known (cf. 18) that cholinergic stimulants and physostigmine release corticotropin releasing factors (CRF; this effect being antagonizable by perhaps both atropine and hexamethonium),

as well as increase plasma levels of cortisol. Thus, the finding
of an increase of cortisol concentration in SDAT patients is unexpected
and perhaps inconsistent with the conceptualization of the cholinergic
deficit in these patients.

Additional diagnostic measures based on sophisticated procedures
may be available in the future. For instance, Eckernas, Aquilonius
and their associates, employed radioactive choline and positron
emission tomography in monkeys, to study the regional kinetics of
choline-related radioactivity. They suggested that certain components
of these kinetics may be employed to measure choline metabolism
and ACh formation.

In this context, Janowsky, Risch and Gillin emphasized that
physostigmine causes a change in a number of hormones and neurotransmitters of the serum, and that these effects, at least in psychiatric patients, are coupled with changes in mood. Physostigmine
also had significant effects on diastolic and systolic blood pressure.
While this may suggest that physostigmine and its responses may be
used diagnostically by monitoring blood levels of certain substances
in several conditions, Janowsky and his colleagues also stressed
that these findings illustrate that ACh may be a major moderator
of stress, since it increases the sympathetic outflow and releases
opioids, which again may have a diagnostic as well as therapeutic
significance.

Another important consideration that emerges from the concept
that the cholinergic deficit underlies aging and SDAT is whether
or not dementia and aging may be treated by means of ACh precursors.
It should be pointed out that precursors such as choline and lecithin
may not generate ACh preferentially, since they also participate
in phospholipid metabolism, including that of the liver and cell
membranes (see also Sections IV and VI). Pertinent data on this
matter were presented at our earlier Meetings by Ansell and his
associates (8, 9) and also by Blusztajn and Wurtman (12). The matter
revolves around the question as to whether or not precursor dosing
leads to increased synthesis of ACh. This matter is still controversial, as illustrated in this Meeting by K.L. Davis and his associates and by Mervis et al. In our own laboratories, even repeated
administration of maximally tolerated doses of choline to either
young or old mice led to a very marked increase of brain choline
but not ACh levels; furthermore, ACh turnover generally decreased
(as could be expected), due either to precursor-generated negative
feedback, or to the agonist effect of high doses of choline (67).
This controversial matter was discussed elsewhere at this Meeting
(see Sections IV and VI, below). The interesting point made in
this Session of the Meeting was that the use of such precursors is
not the only possible way of antagonizing cholinergic deficits seen
either in aging or SDAT.

Piracetam, a nootropic drug, that is an agent that improves cognitive function (34), may improve learning and antagonize scopolamine-induced amnesia by acting on cholinergic mechanisms. At this Meeting, the Florentine investigators Pedata, Pepeu and their associates confirmed that piracetam and its analogs facilitate learning in mature rats, and antagonize its impairment by hemicholinium and atropine. They suggested that piracetam may act by stimulating ACh utilization, as indicated by the interaction between the ACh depleting effect of hemicholinium, and piracetam or its analogs.

Other Clinical Conditions. It was also attempted at this Meeting to relate other than aging disease entities to changes in the cholinergic system and its components. Affective disorders may show such a link. Sitaram and coinvestigators had demonstrated in the past that patients with major affective disorders showed increased sensitivity to the induction of rapid eye movement (REM) sleep by muscarinic agonists; at this Meeting they reported that endogenously depressed patients show a shortened REM sleep latency compared to normals and patients suffering from other types of depression, and that these patients exhibited a supersensitive response of several parameters of REM sleep to arecoline. The anterior pituitary is also supersensitive to cholinergic agonists in depressive illness. Risch et al. showed that depressives responded with a marked increase in plasma concentrations of ACTH and endorphins to physostigmine treatment, as compared to controls or psychiatric cases not presenting with major depressive illness.

Sitaram et al. had suggested in 1982 that the muscarinic hypersensitivity of patients with major affective disorders may represent a genetic alteration at the receptor level, and this is consistent with the findings of supersensitivity concerning such different phenomena as REM sleep and pituitary sensitivity. Furthermore, at this meeting Gershon and Nadi reported that muscarinic receptors of the fibroblasts of patients suffering from affective disorders showed increased muscarinic receptor density as compared to the controls. However, Gershon and Nadi stressed also that this characteristic is heterogeneous, since certain pedigrees exhibited no change in the muscarinic receptor density. Furthermore, the generalization that muscarinic cholinergic supersensitivity is a genetically controlled trait marker for major affective illnesses was not supported at this meeting by the data presented by Kaufman, Gillin and their associates, and by Stanley and his coinvestigators. These two groups of scientists did not find changes in muscarinic binding either with respect to the number of binding sites, or their affinity in the frontal cortex, hypothalamus and pons of suicide cases, as compared with controls. However, not all suicides meet the diagnostic criteria for major affective or bipolar illness.

The red cell choline biokinetics have already been referred to above as possible markers for SDAT; they may serve as such also in affective disorders. This possibility exploits the finding of Martin of an active accumulation of choline in the human red blood cells, an inhibition by lithium treatment of red blood cell choline transport, and the subsequent indication that choline transport and Li action may be changed in affective disorders. At this Meeting, Mallinger, Hanin and their associates continued the earlier work of Hanin and his group (see 38) as they demonstrated that there is an inverse correlation between choline transport into, and out of the cells, and red blood cell choline content. They confirmed also that depressed patients show a high choline content in their red blood cells, and that lithium antagonizes irreversibly the choline transport and elevates the choline levels of red blood cells; thus, lithium-treated depressive patients show red blood cell choline that are higher than those of untreated depressives, whose choline levels are, in turn, higher than those of normal subjects. At this Meeting, K. Martin emphasized that the characteristics of choline influx and efflux and of the lithium action in the human red blood cells are not shared by other species; in fact, only certain genetic lines of sheep have a choline transport system similar to that found in human cells, but even in this case, this system differs from that of man, since lithium does not inhibit the sheep choline transport irreversibly.

Finally, Aquilonius and his associates demonstrated that cholinergic markers may be diagnostic in diseases other than depressive disease and SDAT. Employing cryosections of human spinal cord, they found that ChAT was reduced in motor neurons, and in the dorsal horn areas of patients with amyotropic lateral sclerosis.

Certain comments are appropriate at this time. First of all, as is well known today, the cholinergic system is not the only system that may be involved in aging or, for that matter, in affective disease. Thus, the involvement of other neurotransmitters in aging was stressed in recent reviews of (33, 43, 47). Furthermore, besides changes in the neurotransmitters other pertinent changes obviously occur in aging, eg., in brain blood flow, energy metabolism, etc. Another important point to be raised is that, if indeed cholinergic and other neurons are lost in Alzheimer's disease or aging, the precursors ultimately will not be able to do their job, since there will be no cholinergic neurons to take them up; this situation is analogous to that which may exist with respect to dopaminergic neurons in late stages of Parkinsonian disease. Under these circumstances, cholinergic agonists may be more useful than precursors, analogously again to the use of dopaminergic agonists in late Parkinsonian disease. This possibility, however, would be limited if the loss of cholinergic neurons led to a parallel change in cholinergic receptors, eg., down-scaling of the receptors. It happens that there is a controversy as to the state of cholinergic receptors in SDAT (compare, eg., 107 with 90). Another pertinent point is that at this time the knowledge of choline and ACh metabolism and of the relationship of this metabolism to the generation of phospholipids versus that of ACh is insufficient; thus, we may not know at this time how to exploit - precursor-wise or by other means - phospholipid and choline metabolism. This problem is taken up at another point of this Meeting (see below, Section IV).

III.   CHOLINERGIC PRE- AND POST-SYNAPTIC RECEPTORS

This series of presentations deals with muscarinic and nicotinic receptors, whether located presynaptically or postsynaptically, and their role in the regulation of the release of ACh. These studies expand on the earlier findings concerning the differentiation between receptor sub-populations of these two receptor types (cf., eg., 11).

Muscarinic Receptors. Dahlbom, Resul and associates demonstrated that oxotremorine-like agonists and antagonists act on a specific receptor site, while classical muscarinic antagonists bind also to accessory areas located near receptor sites. Furthermore, Ringdahl claimed that the muscarinic affinity and efficacy of different oxotremorine analogues were not related, which indicates that structural requirements for occupation and activation of muscarinic receptors, at least in the guinea pig ileum, are different. Similarly, Nordström, Danielsson and their associates demonstrated that methyl-pyrrolidino-2-butynyl acetamide (BM-5) exhibits, depending on the site of action, either agonist or antagonist properties, indicating differences between central and peripheral muscarinic receptors. In fact, there may be muscarinic receptors that cannot be easily classified at this time, since the Arrhenius Laboratory investigators demonstrated that, while alaproclate synergizes with oxotremorine and physostigmine as a tremorigen as well as antagonizes in vitro high affinity muscarinic agonist receptor binding, it exhibits limited effects on the muscarinic receptor, and is a specific serotonin uptake blocker.

Muscarinic receptors may also differ in their susceptibility to sensitivity change due to chronic treatment with agonists or antagonists. Marchi and Raiteri were concerned with muscarinic presynaptic autoreceptors regulating the release of ACh at the cortical and hippocampal sites, and the muscarinic presynaptic heteroreceptors mediating striatal dopamine release. They demonstrated that the autoreceptors are more susceptible to the development of tolerance due to repeated treatment with agonists, and the development of hypersensitivity due to chronic treatment with antagonists, than the heteroreceptors. It is of interest in this context that, in the sixties, Russell and his associates had first demonstrated that central, behavioral actions of organophosphorus drugs are tolerance-liable; he presented these data at one of our earlier meetings (100). It should also be pointed out that the heteroreceptors involved in the release of dopamine are analogous to those that were classically studied by Burn and Rand (13), since they proposed the presence of a "cholinergic link" in catecholaminergic transmission, a concept widely criticized in the 50's and 60's (see below, and 54).

Nicotinic Receptors. There may also be several sub-categories of nicotinic receptors. Larsson and Nordberg demonstrated that in the hippocampus, nicotine and d-tubocurarine (dTC) exhibit two binding sites, while bungarotoxin exhibits only one binding site; nicotine and dTC binding sites included a high and low affinity site.

An interesting finding concerning the nicotinic receptor was described by Mautner as he showed that the crustacean axonal membrane includes a receptor bound by bungarotoxin, this binding being prevented by dTC. Mautner believes that the receptor in question is a peptide with a molecular weight of 50,000D, which may constitute a single channel receptor for potassium, since the channel opening was blocked with tetraethylammonium (TEA). It is of nostalgic interest that this receptor of the axonal membrane, which appears to be cholinergic in nature, seems to be analogous with the cholinergic receptor which the late David Nachmansohn[**], a past colleague of Henry Mautner's, had studied for many years, and which Nachmansohn felt underlies axonal conduction (87). Going back even further in history, this receptor that concerns conduction rather than transmission was studied initially by von Muralt (cf. 86).

The possible existence of sub-populations of muscarinic and nicotinic receptors is of both basic and implied importance. From a basic point of view, the diversity of pre- and post-synaptic receptors suggests that a single neurotransmitter, in this case ACh, may exert site specific effects and provide moment-to-moment transmission modulation. If the muscarinic or nicotinic presynaptic receptors differ in their affinity or efficacy from the postsynaptic muscarinic and nicotinic receptors, transmission will be subject to computer-like modulations, depending on the balance between the characteristics of the pre- and post-synaptic receptors. On a clinical level, the concept of the differences between characteristics of receptors of various brain regions and various peripheral sites offers the possibility of development of drugs that may have region-specific differential action. This point will be returned to at the end of these comments.

A major question with respect to, particularly, the presynaptic receptors is their physiological role. This question was addressed during the next session.

[**]Dr. Nachmansohn's death coincided with the last day of this Conference.

IV. ACETYLCHOLINE RELEASE

This section concerns several important aspects of the release
of ACh from nerve terminals, such as the role of vesicular and non-
vesicular pools of ACh; the source of choline and its role in ACh
release; the release function of specific proteins of the nerve
terminal membrane; and the role of nerve terminal receptors in presynaptic regulation of ACh release.

Cytoplasmic and Vesicular ACh. The classical hypothesis of the
quantal release of ACh which is related to the confinement of ACh
quanta in the vesicles was first challenged at an earlier one of
these meetings by Marchbanks (see for instance, 78). He proposed
that cytoplasmic ACh participates in a non-vesicular, but quantal
in character, release of ACh, and that cytoplasmic ACh and the current
influx carrier are linked (see also 76). At this Meeting Molenaar
and Polak related in the electric organ of the Torpedo the total
ACh pool, the vesicular pool, and the amount of ACh in vesicles,
to ACh release. They felt that their results favor the hypothesis
of a vesicular release of ACh. While they stated also that the
resting or spontaneous release of ACh may be non-quantal in nature,
and that potassium depolarization causes an increase in this non-
quantal ACh release, yet such an increase was negligible when compared
with the depolarization induced quantal release. Consequently,
they suggested that "a significant non-quantal contribution to quantal
acetylcholine release induced by the action potential is improbable."
The findings of Carroll may also be pertinent in this context. He
reported that veratradine or potassium depolarization of minces of
mouse forebrain led to the hydrolysis of cytoplasmic ACh rather
than to its release. He hypothesized that this hydrolysis may facilitate the release of ACh from vesicles by providing substrate
for ACh formation. He also found that the inhibition of intraterminal
AChE stimulated calcium-independent spontaneous release of ACh;
thus, ACh may play a role in regulating its own spontaneous release.

The question of pools may be further complicated by the findings
from Victor Whittaker's laboratory (for the earlier work, see 114).
At this Meeting Agoston, Whittaker and their associates demonstrated
that there are several pools of vesicles in the myenteric plexus
of the guinea pig ileum, which is consistent with the earlier data
from the Gottingen Laboratory concerning Torpedo. These pools represent newly synthesized versus stored ACh, and the ratio between
these two pools and the two vesicular populations depends on the
species. Another population of vesicles was studied by Boyne and
Phillips who, using an elegant quick-freeze technique, demonstrated
that "the membrane attached vesicles are the immediate source of
the metabolically active vesicle population" of the stimulated electric
organ of Narcine brasiliensis; furthermore, these vesicles selectively
took up calcium when exposed to maximal concentrations of calcium
entering the nerve terminal during its depolarization. This could

activate selective ACh turnover in this specific population, "thereby assuring the preferential release of newly synthesized neurotransmitter."

A somewhat different conclusion with respect to the significance of vesicular versus cytoplasmic ACh was reached by Dunant and his associates. They employed a sophisticated technique that included a freeze-fracture procedure which allowed them to identify morphological changes that occur within one msec following the impulse. They employed an amino-pyridine. This ACh releaser generated a long-lasting giant electroplaque potential of the Torpedo which could be related to cytoplasmic ACh which had been utilized and renewed, rather than to vesicular ACh. This conclusion was reached based on the observation that the vesicular population remained stable, and no transfer of ACh appeared to occur between the cytoplasm and the vesicles. On the other hand, at the time of transmission there was a dramatic increase in the number of large intra-membrane particles.

Nerve Terminal Proteins. Several papers focused on the role of the proteins of the nerve terminal or the synaptosomes. Thus, Ducis working with the vesiculated synaptic plasma membranes of the Torpedo used fractionation and related methods to identify the proteins of the lysed membranes under conditions of optimal transport of choline. Six membrane fractions were tentatively identified. These fractions included myosine, a nucleotidase, AChE and ChAT; the two remaining components with M_r of 57 and 28 appeared to be the candidates for choline carriers. Morel, Israël and their associates, also worked with cholinergic synaptosomes prepared from Torpedo electric organ, and identified the proteins by means of appropriate antibodies; these proteins, included ectocellularly, oriented presynaptic AChE as well as a 67,000 dalton protein. When AChE and the 67,000 dalton protein were removed by selective proteolysis, ACh release induced by potassium was not affected. Thus, these two glycoproteins are not involved in ACh release. Furthermore, these investigators managed to reconstitute presynaptic membranes employing lyophilized synaptosomal membranes; these reconstituted proteo-liposomes can be filled with ACh, which they release upon activation with neurotoxins or a calcium ionophore.

Still another aspect of nerve terminal proteins was investigated by Howard and Mellega. They used in their investigation a clonal line of rat pheochromocytoma cells which release both dopamine and ACh. While this release, contrary to that from nerve terminals, does not require ATP, the role of ATP is taken over presumably by a modification of protein 1_B, which seemed to be an equivalent of the mammalian brain nerve terminal proteins that undergo increased phosphorylation during the process of the neurotransmitter release. However, the 1_B-like 80,000D protein of the pheochromocytoma clonal

line studied by Howard and Melega did not exhibit detectable phosphorylation.

Source of Choline for Release of Acetylcholine. Another aspect of
the release of ACh was taken up by Maire, Blusztajn and Wurtman.
Evaluating the release of ACh from the rat's striatum these investigators could not account for choline and ACh of the effluent in
terms of the initial and final levels of choline and ACh. This
finding, as well as the data obtained with hemicholinium, suggested
to these investigators "that the choline originating from choline
phospholipids is liberated into the extra-cellular space, and it
then can be used for ACh synthesis only if the high affinity uptake
of choline is functioning." This finding is of importance with respect
to the use of precursors in SDAT (see above) and to choline metabolism
(see Section VI).

It is of interest in this context that Consolo and her associates
demonstrated a commonality of uptake mechanisms between the nerve
terminal and the muscle. They showed that denervation, as well as
hemicholinium, decrease ACh content of three different types of
rat muscles; this finding does not, of course, explain the enigmatic
presence of ACh and choline in the muscle, and their relationship
to cholinergic activity.

Nerve Terminal Receptors. In the 50's and 60's, Burn (cf. 13) and
Koelle (cf. 68) proposed that nerve terminals are endowed with cholinergic receptors involved in the release of transmitters. Burn
referred to what would be termed today as heteroreceptors. He described cholinergic receptors situated on adrenergic nerve terminals,
suggesting that these receptors may be involved in the release of
catecholamines, thus constituting a "cholinergic link" (see above).
Koelle described cholinergic autoreceptors, as he presented evidence
suggesting that the receptors in question respond in a positive
feedback fashion ("percussively") to ACh released from these very
receptors. It should be pointed out that Burn's heteroreceptors
are presumably muscarinic in nature, while Koelle's autoreceptors
appeared to be nicotinic in nature.

At our Meeting of 1977, Szerb presented evidence as to muscarinic
nerve terminal feedback regulating ACh release in both the peripheral
and central nervous system (106). At this Meeting, additional findings
concerning both the muscarinic and nicotinic presynaptic receptors
were presented. The findings of Kilbinger, Cooper, and their associates, were consistent with those of Szerb. They showed that the
release of ACh, evoked from the myenteric plexus of the guinea pig
ileum, was decreased in a dose dependent manner by muscarinic agonists;
these effects were antagonized by muscarinic antagonists. Similar
findings were presented by Adler et al. who found that DFP decreases
the frequency of miniature endplate potentials in cultures in which
myotubes and neuroplastoma cells are made to generate functional

synapses; and by Mattio, Richardson and Giacobini, who demonstrated that DFP decreases, in a muscarinic manner, ACh release at the pupil. In addition, Kilbinger described related muscarinic nerve terminal receptors involved in a transient increase of spontaneous outflow of ACh, while Cooper and his associates presented evidence indicating that there is a nicotinic nerve terminal receptor which is involved in a positive feedback effect on ACh release.

On the other hand, working with the nerve terminal of the phrenic nerve of the rat, Bierkamper and Aizenman, as well as Häggblad and Heilbronn, described a nicotinic receptor involved in the inhibition of ACh release. In agreement with this concept, these two groups of investigators found that alpha-bungarotoxin increased the evoked ACh release; however, in the hands of Bierkamper and Aizenman dTC unexpectedly decreased ACh release. Nicotinic receptors seem also to be present in the brain 69), and Schwartz and Kellar demonstrated their presence in synaptosomal fractions of several parts of rat brain. They also presented evidence that these include presynaptic heteroreceptors concerned with release of serotonin and catecholamines in the case of striatum and hypothalamus, and postsynaptic receptors in the case of cortex and thalamus.

It should be added that several groups of investigators were concerned in this Meeting with non-cholinergic nerve terminal receptors that also regulate ACh release. For instance, Vizi and Somogyi expanded on their earlier work concerning the inhibitory action of alpha adrenoceptors on ACh release, and added adenosine and opioid μ receptors to the list of non-cholinergic nerve terminal receptors regulating ACh release. In addition, they presented findings indicating that each neuron may exhibit a number of different receptors rather than there being subpopulations of neurons endowed with only one, whether muscarinic or non-cholinergic receptor; and that these several presynaptic recognition sites characterize both central and peripheral neurons. Similarly, Cooper and his associates identified a number of non-cholinergic receptors, including an inhibitory purinergic receptor at the nerve endings of the guinea pig ileum myenteric plexus. They pointed out, however, that the release characteristics of the synaptosomes may differ from those of the myenteric plexus since morphine, enkephalins, somatostatin, and a number of peptides did not affect synaptosomal release of ACh.

There may be still another nerve terminal channel which is independent of those already mentioned. This is a potassium channel in which Thesleff (109) proposed that its blockade by aminopyridines increases transmitter release. Albuquerque (5) also has presented evidence indicating a similar effect of phencyclidine. At this Meeting, Schwarz et al. demonstrated the facilitation by mono- and di-aminopyridines of ACh release in the rat hippocampus, and suggested that aminopyridines act on fast potassium channels linked with calcium influx, while phencyclidine may act on a potassium channel not linked with calcium.

Released ACh may act also on autoreceptors concerned with mobilization of the choline of the phospholipids of the nerve terminal membrane. Löffelholz and his associates demonstrated that the muscarinic receptor activitation facilitated the release of choline from rat cortex and the chicken heart. They also presented data showing that this efflux of choline depends on the cholinergic, muscarinic activation of phospholipid metabolism. It is of interest that cyclic AMP, which is known to facilitate the release of ACh, also facilitates choline release; Löffelholz and his associates presented data suggesting that the cyclic AMP-dependent facilitation of choline release is blocked by cholinergic muscarinic activitation. Thus, cyclic AMP-mediated and cholinergically mediated choline release form a dipole.

Of great significance is the question of whether the auto-and heterocholinergic receptors serve as physiological regulators of ACh release. At this Meeting, Briggs and his associates presented data indicating that the tetanic presynaptic stimulation of the rat superior cervical ganglion leads to a long-term increase of ACh release, confirming the hypothesis enunciated much earlier. Furthermore, Briggs and his associates found that only evoked, rather than spontaneous release, is thus affected.

V. CHOLINESTERASES, ANTICHOLINESTERASES AND REACTIVATORS

This session of the Meeting revealed the profound revolution that has occurred in the last few years with regard to cholinesterases (ChEs) and antiChE drugs.

One aspect of this revolution concerns synaptic effects that do not arise from accumulation of ACh; this accumulation constituting the classical effect of antiChEs. Thus, Albuquerque and his collaborators continued the work initiated in 1974 by Kuba et al. (70), that demonstrated direct receptor and/or channel effects of organophosphorus (OP) ChEs such as DFP. At the present Meeting Albuquerque and his associates showed that OP drugs as well as tertiary and quaternary carbamates exhibit direct agonistic properties, since they activate ACh receptors. These receptors may not be coupled with activation of a homogeneous populations of channels, as there may be sub-population of channels that differ with regard to their opening frequency and/or life-time. Furthermore, antiChEs may act as channel activators or channel blockers, and they may block the channel either in its open or closed conformation; the complexity of these effects being illustrated by the double exponential effects of the membrane potential on the decay of the EPC current and/or complicated membrane voltage - membrane current relationships that deviate from linear relationships. Finally, the data confirmed earlier publications from Dr. Albuquerque's laboratory (3, 6) concerning the desensitization that may depend on action on the receptor-channel macromolecule that leads first to burst-like activity of the elementary events representing the channel activation, and then to the cessation of receptor activation.

A related effect, presumably also not resulting from ACh accumulation, was observed in Aplysia by Tauc and his associates. The effect of a non-hydrolyzable cholinomimetic, carbachol, on the H-type but not its effect on D-type cells, was augmented by OP and carbamate antiChEs. As is well known, the hyperpolarizing H-type synapse functions via the chloride channels, while the D-type cells function in terms of cationic channels. Since Tauc and his associates did not observe a change in the conductance or life-time of chloride channels opened by the carbachol-activated H-type synapses, they proposed that the inhibition of AChE increases the number of ACh receptors that can be activated by carbachol. In support of their contention, they showed that similar effects could be obtained with detergents that may act by removing the molecular action of AChE on the ACh receptor.

A non-classical, morphologic effect of antiChEs was described by Dettbarn and Misulis, although in this case ACh accumulation may have been involved. Dettbarn continued the study of the morphological effects of neostigmine and OP drugs on the endplate and the muscle, a phenomenon that he described for the first time several years ago with Laskowski (19, 71, 72). At this time, Dettbarn and Misulis demonstrated that even a single acute administration of AChE inhibitors may produce a myopathy and complete, although localized necrosis. Furthermore, these effects depend on inactivation of AChE since they are preventable by 2-PAM as well as, interestingly enough, reversible ChE inhibitors. Dettbarn believes that the events that contribute to necrosis may involve increased uptake of calcium, leading to calcium overload and energy depletion and/or calcium activation of proteolytic enzymes.

Still another interesting phenomenon concerning OP drugs is the development of tolerance of their effects on behavior (see also above, Section II, and 100). Ho and his associates found that certain behaviors are not subject to tolerance upon repeated administration of DFP. Furthermore, while several behavioral parameters such as tremor, hind limb abduction and hypothermia, show a biphasic effect upon chronic administration of DFP - an early increase followed by tolerance - yet the rate of the development of tolerance differs from one behavior to another. Ho and his associates further demonstrated that DFP increases GABA and glutamate brain levels and the number of GABA and dopamine receptors, while reducing the number of muscarinic receptors (see Section VII of this Summary). They therefore proposed that these effects of DFP may constitute a compensation for excessive cholinergic activity produced by DFP, thus leading to tolerance. This is a novel way of looking at tolerance since, heretofore, tolerance was understood as being due to changes in the cholinergic receptor (such as desensitization) or in metabolism of antiChEs.

Altogether, the phenomena described by Tauc, Albuquerque, and Dettbarn and their coinvestigators, appear to deal with non-classical, novel effects of ChE inhibitors. This naturally increases the range and the versatility of synaptic cholinergic responses. An additional basis for this versatility may be due to certain characteristics of ChEs. Polymorphism of these enzymes is pertinent in this context. It is well known that ChEs represent a family of enzymes and that besides such well known types of ChEs as butyrylcholinesterases (BuChEs) and AChEs, there are also aryl cholinesterases, cholinesterases (β-esterases) studied by Aldridge (7), as well as isozymes of both AChE and BuChE. It is of interest that the role of some of these enzymes has not been classified as yet.

At this Meeting Massoulie classified the various polymorphic forms of AChE and BuChE as either globular or isometric, depending on the presence or absence of a collagen-like tail. He sub-divided, furthermore, the globular forms into detergent insensitive and amphipatic molecules, depending on their capacity to interact with non-denaturing detergents. He suggested that each ChE may have a specific precursor, a concept similar to that proposed earlier for the BuChE-AChE relation by Koelle (see, however, below). Furthermore, every tissue may possess a characteristic spectrum of molecular forms of ChEs; this suggests that the various forms of ChE may exhibit special functions related to their strategic sites as well as differential sensitivities to inhibitors.

A controlling factor with respect to ChEs may be trophic in nature, and very pertinent findings were presented at this Meeting by Koelle and his associates. For several years now, Koelle has been concerned with the relationship between AChE and BuChE, and between postsynaptic ChEs and cholinergic innervation, as the postsynaptic AChE and BuChE activity decreases conspicuously but not to the same extent following ganglionic denervation. After several years of studies Koelle presented evidence that the nerve tissue contains a neurotrophic factor - not identified as yet - which maintains postsynaptic AChE and BuChE. Earlier, he hypothesized that BuChE may serve as a precursor of AChE; as he pointed out, this speculation may be inconsistent with the finding that different genes appear to control synthesis of these two enzymes (105).

Other factors influencing antiChE action include the distribution and biokinetics of OP drugs, which was studied so far to a relatively limited extent. At this particular Meeting, Waser showed a novel phenomenon of re-distribution of Sarin following oxime treatment. This relates to the findings presented elsewhere concerning the storage of active OP drugs in the organism (see, for instance, 30). The investigation of the distribution of OP drugs and its changes was made possible by the development of sensitive detection methods such as those presented at this Meeting by Waser, and by Nordgren and Holmstedt. These methods, based on gas-chromatography and mass-spectrometry, lend themselves also to the evaluation of other pertinent agents, such as succinylcholine, as presented by Nordgren and Holmstedt at this Meeting.

Protection against OP drugs was also discussed in this Section. Shih and his colleagues described the antagonism of OP toxicity by atropine sulfate and oximes, and emphasized that oxime-induced re-activation of peripheral and, on intracerebral administration, central phosphorylated ChEs, may not correlate with oxime antagonism of OP drugs. Nor was this antagonism related to lowering of brain levels of ACh by atropine-oxime combinations. For example, the oxime HI6 produced a high degree of protection that was not related to its reactivating capacity.

Another well known concern is that of the relationship between OP action on ChEs, OP-induced ACh accumulation and ACh turnover. Karlén, Lundgren, Lundin and Holmstedt showed in this meeting that the increase in ACh levels produced by the OP drugs caused a decrease in the turnover of ACh; however, this relationship was not always linear and the rate of ACh accumulation differed from one OP agent to another, and from one tissue to another.

This interesting session also concerned a number of classical aspects of the action of antiChEs, such as the reactivation of the phosphorylated enzymes and/or effect of ACh accumulation on ACh turnover. More novel is the notion of myopathic effects of antiChEs shown at this Meeting and elsewhere by Dettbarn as well as by other investigators (cf., eg., 48). In fact, many years ago we stressed the effect of antiChEs on early ontogenesis and morphogenesis (52, 53); it was actually emphasized (o.c.) that some of these effects occur before ontogenetic emergence of ChEs and/or ACh. Axonal and neuronal neurotoxicity constitute other morphopathological effects of these drugs, a matter that is a subject of past and present in-vestigations (2, 51).

Another novel action is that which may be due to direct membrane action of antiChEs. It includes channel and/or receptor agonist action of various types of antiChEs including OP drugs and carbamates, as demonstrated at this Meeting by Tauc, and by Albuquerque. An important aspect of this action of antiChEs is their desensitizing effect described here by Albuquerque; that this effect is independent of ACh accumulation was earlier shown by Karczmar and Ohta (64). This effect is important clinically, as it may underlie refractoriness to anticholinesterases, which develops readily in myasthenia gravis patients treated with these drugs (64, 108).

All these aspects of the functions of ChEs and of the effects of antiChEs raised in this session reveal the great variety of such functions and effects.

VI.   ACETYLCHOLINE SYNTHESIS, METABOLISM AND PRECURSORS

Contrary to the classical views, Ansell (cf. 8), Tuček (110) and Blusztajn and Wurtman (12) proposed that the synthesis of releasable ACh may depend only to some limited extent on recycling and nerve terminal uptake of choline resulting from ACh hydrolysis. Tuček (110) embraces the view that free extracellular choline is the principal source of choline precursor for the synthesis of ACh in the neurons; this choline may result from several metabolic pools, including liver and neuronal phospholipid metabolism. Choline pathways are important determinants of the capacity of the precursor treatment, whether with phosphatidyl choline (lecithin) or choline to increase concentrations, turnover and/or release of ACh in the brain.

Several papers presented at this Meeting have a bearing on this matter. Corradetti and his associates from the University of Florence studied the source of choline released into a cup applied to the denuded rat cerebral cortex, particularly when the release was augmented by means of cholinomimetics. They ruled out the possibility that this increase is due to the augmentation by muscarinic stimulation of blood vessel permeability; their results indicated that a relatively small percentage of choline efflux induced by muscarinic stimulation of the brain is generated by the plasma, while most of it results from the mobilization of free choline from the cortical cell. This is consistent with Löffelholz's results, presented earlier (see above). Furthermore, Jope found that chronic administration of lecithin to rats led to high levels of choline in the blood just as it induced high blood levels in man and in the animal brain (39, 67), but did not affect ACh brain levels; what is important is that there was also an increase in the fatty acids, cholesterol, triglycerides, and phosphocholine, serving as a "sink" for lecithin (see also 83). Thus, phosphatidyl choline or choline participate in phospholipid metabolism. Conversely, phospholipid metabolism contributes to choline generation, as was also shown at this Meeting, by Zeisel. His data indicated that membrane-containing brain fractions, and to a lesser extent phosphorylcholine and glycerophosphorylcholine, contribute to the formation to choline. ATP may also regulate the formation of choline, since in the presence of adequate ATP concentration most of the lysophopatidyl choline was reacetylated, rather than degraded to free choline. These findings are consistent with the concept of Blusztajn and Wurtman (o.c.) that brain can synthesize choline via methylation of phosphatidyl ethanolamine and the formation of phosphatidyl choline.

It may be suggested, on the basis of Wecker's data presented at our previous Meeting (113), that whether the precursor such as

choline is routed to phospholipids or to ACh may depend on ACh turnover. She has shown in an interesting paradigm that depletion of
choline and/or ACh induced by atropine or fluphenazine, can be antagonized by the precursor, by some mechanism which may have no
effect on resting levels of ACh. At this Meeting, she demonstrated
that while dietary supplementation with choline did not affect brain
concentrations of ACh or choline, or the activity of ChAT and AChE,
or finally, the density and the characteristics of cholinergic receptors, it antagonized pentylenetetrazol-induced depletion of brain
ACh. Of further interest is that Wecker could also increase with
a choline-supplemented diet the phospholipid-protein ratio of the
microsomes, but not of the synaptosomes of the rat striata.

The precursor protection from ACh depletion may not however
necessarily involve the metabolic activity of the precursor. Flentge
et al. found that choline antagonism of the fluphenazine-induced
decline in ACh was combined with an increase in dihydroxyphenylacetic
acid (DOPAC) levels; in fact, following fluphenazine administration
ACh and DOPAC levels were strongly correlated. Flentge and his
associates concluded from these findings that "the primary action
of choline might be the prevention of dopaminergic receptor blockade
by fluphenazine; as a consequence, normal acetylcholine release
and acetylcholine levels may be restored." A particular finding
by Wecker should be considered in this context; she demonstrated
that choline protects from nicotine toxicity--this suggests that
the choline-supplemented diet is pharmacologically active and, specu-
latively, that this activity of choline underlies its protection
against ACh depletion.

The notion that the contribution of the precursor to phospholipid
versus ACh metabolism may depend on the brain activity and on the
stimulation of the cholinergic neurons is consistent with the data
presented by Rylett. She demonstrated that in some species such
as the guinea pig, the synaptosomal high affinity choline transport
stimulation by depolarization is coupled with acetylation of choline,
and that the increase in choline uptake is used exclusively in the
guinea pig for the manufacture of ACh, rather than for phospholipid
metabolism. Interestingly, choline mustard aziridinium, an inhibitor
of active sodium dependent choline transport in brain synaptosomes,
produced an irreversible decrease in synaptosomal content of ChAT,
in parallel with alkylation of the high affinity choline carrier
(see below, Section IX). As Rylett showed that these interesting
phenomena cannot be generalized from one species to another, more
work is needed to assess the significance of the coupling between
depolarization-stimulated choline uptake and ACh versus phospholipid
synthesis.

A related aspect of the relationship between neuronal activity
and ACh synthesis concerns the feedback form the target tissue.
According to Tuttle and Pilar, effective neuromuscular transmission

capable of sustaining of the contracture arises ontogenetically in the avian ciliary ganglion-iris system in parallel with the development of responsiveness of ACh synthesis to demand subsequent to release. On the other hand, when the ciliary ganglion neurons are removed from the embryo and cultured in the absence of the target tissue, ACh synthesis remains unresponsive to demand, in spite of the presence of high levels of ChAT and capacity to release ACh. However, a release demand is accrued when the neurons are plated onto pectoral myotubes in culture.

Coupling between ChAT activity, choline uptake and precursor availability was studied by O'Neill and his associates, using compounds that either inhibit ChAT activity, choline uptake, or both. Among these compounds, Benzylidene-3-quinuclidinones may react particularly readily with nucleophilic groupings of the ChAT molecule, and may help in resolving the significance of this relationship for ACh synthesis.

Still another question is that of the role of glial cells in ACh synthesis. Culturing the chick brain neurons with and without their glial cells, Kasa demonstrated that choline-superfused glial cells released choline which could be utilized in the synthesis of neuronal ACh and phospholipids, provided that the neurons had established contact with glial cells. On the other hand, the steady-state of ACh was higher in neurons cultured in the absence of glial cells rather than in mixed cultures. This steady state, the choline efflux, and the relationship between glia and neurons with regard to this efflux, may depend on a Ba^{2+} sensitive K^{+} channel, as indicated at his Meeting by Hoffman and coinvestigators. Altogether, these findings suggest an interaction between synthesis, release and utilization of ACh.

Another question was that of the relation between the effect of precursors on choline or ACh metabolism on the one hand and pharmacological or behavioral effects on the other. Thus, Richter showed that the blocking action of intraventricularly administered barbiturates on choline uptake was not correlated, dose-effect wise, with the loss of the righting reflex (see also 98, 102); parenthetically, additional data led Richter to believe that barbiturate block of choline uptake depends on their GABA-mimetic action on the hypothalamus. In a similar vein, Hanin et al. have shown that choline and ACh levels increased in the rat cortex, plasma and red blood cells, but not in the hippocampus or in the striatum after 21 days of feeding with choline enriched diet; yet there was no effect of a choline-poor diet. Furthermore, the choline enriched diet caused hyperactivity, although it did not affect sleep EEG parameters.

This series of papers consitituted an important contribution to the present status of our knowledge of the relationship between ACh synthesis, ACh precursors and phospholipid metabolism. It appears that in the resting state of the animal precursor rich diet has little effect on ACh levels, synthesis and turnover (see above, Sections II and IV and 66, 67); such an effect may, however, occur in the case of active neurons. Even in the latter case, the effects of the precursors on ACh metabolism may also reflect their effects on non-cholinergic neurons and/or their pharmacological actions. Furthermore, some of the data presented indicate a dichotomy between the effects of precursor treatment on choline and ACh levels on the one hand and their behavioral effectiveness on the other. In this context, the contribution to and involvement of the precursors in, phospholipid metabolism and in membrane formation must be stressed.

VII. SECOND MESSENGER MECHANISMS

The concept that ACh as well as other transmitters, produce membrane effects via the activation of second messengers such as cyclic nucleotides or the components of the phosphatidyl inositol cycle, dates back from the ideas of Sutherland; this concept was proposed recently in an elegant and complete form by Greengard (37). Greengard proposed that ACh activates cyclic AMP and cyclic GMP that mediate phosphorylation of the neuronal membrane; these cyclic nucleotide phenomena underlie cholinergically mediated conductance changes that result in transmission, in, e.g., the cerebellum.

Earlier, Hokin and Hokin (46) proposed that muscarinic activation of phosphatidyl inositol metabolism underlies transport phenomena in exocrine glands and adrenal medulla (see also 81). At this Meeting, Heilbronn and her associates confirmed that a cyclic AMP dependent protein kinase phosphorylates the nicotinic receptor, and they identified two subunits of the receptor as the sites for phosphorylation. Further identification of protein kinase involved in the cholinergically mediated phosphorylation and activation of tyrosine hydroxylase of adrenal chromaffin cells was carried out by Waymire and his associates. Their data discriminated between cyclic AMP dependent protein kinase and calcium dependent lipid sensitive protein kinase (kinase A and C, respectively); it appeared that kinase C may be the main mediator of cholinergic activation of tyrosine hydroxylase.

An interesting aspect regarding the second messengers is the possibility that the phosphoinositide and nucleotide phenomena may be linked. For example, in the heart, muscarinic agonists inhibit catecholamine-stimulated cyclic AMP response and activate phosphoinositide metabolism. At this Meeting, J.H. Brown and S.L. Brown were concerned with the interesting question as to whether the same receptor is concerned with these two effects. Their data indicate that the phosphoinositide versus the cyclic AMP response depend on low and high affinity muscarinic receptor states, respectively; carbachol being capable of activating the receptor in either case, while oxotremorine induces the nucleotide response and antagonizes carbachol induction or recognition of the low affinity receptor.

The adenylate cyclase response differs, however, from the phospho-inositide response in its sensitivity to the agonists and responsive-ness to oxotremorine.

Ca^{2+} gating may be an important factor in the second messenger role of phosphoinositide cascade (81). However, Hawthorne and Swilem pointed out that muscarinic activation of the phosphoinositide cascade which depresses catecholamine secretion from the chromaffin cells of the adrenal medulla does not depend on the influx of the calcium ion, although it may be concerned with the activation of protein C kinase; on the other hand, the calcium influx caused by nicotinic drugs induces catecholamine secretion of the chromaffin cells.

Papers presented at this Meeting expand our understanding of the mechanisms underlying the action of cholinergically activated second messengers. Thus, Waymire et al. provided at this Meeting further identification of the protein kinase involved in the choli-nergic activation of tyrosine hydroxylase that has been studied for many years now by Costa and his associates (e.g., 17); interest-ingly, cyclic AMP dependent kinase may not be the mediator in this phenomenon. It may be added in this context that while cyclic AMP and cyclic GMP are, most probably, mediators of action of ACh at certain sites, they may not be thus involved at other sites, as for instance at the autonomic ganglia (22, 23).

In the early days of the demonstration of cholinergic transmission and of ACh-dependent conductance change, whether at muscarinic or nicotinic receptor sites, it was not felt necessary to think of a special explanation for the ACh generated conductance change. It was assumed that this immensely bioactive compound, which in fact can burn skin when applied to it in high concentration, does not need an assist in order to produce the permeability change. Yet, obviously, this was simply a case of the investigators not only missing a link, but missing the need for the link. Phosphorylations depending on cyclic nucleotides, and the phosphatidyl inositol cycle and related phenomena such as calcium gating and/or formation of a calcium ionophore (92, 96), provide such links and a mechanism for the modulation of membrane permeability. It should be added that at this Meeting the classical muscarinic activation of phosphatidyl inositol cascade was emphasized, while its nicotinic activation is a matter of controversy (see, for instance, 20, 81 and 99 versus 1).

VIII. INTERACTION OF ACETYLCHOLINE WITH OTHER NEUROTRANSMITTER SYSTEMS

When the ubiquity of cholinergic pathways in the CNS became known, it should have been taken as a truism that any effects on the cholinergic system will inevitably involve changes in other neuro-transmitter systems. Yet it appears historically that more purist ideas emerged initially, perhaps related to Dale's principle that only one transmitter can exist at various endings of a neuron, although

Dale's dictum does not at all imply the independence of cholinergic and other neurotransmitter systems. Perhaps this author's laboratories were among the first stressing the obvious, which is that activation of the central cholinergic system involves changes in other neurotransmitter systems as well (35).

A more surprising finding was that cholinergic synapses, considered historically as pure and monotransmitter in nature such as those of the ganglia or the neuromyal junction, are also multitransmitter in function.

The present Meeting included papers dealing with interaction between the cholinergic and other neurotransmitter systems both in the CNS and at the periphery. Furthermore, the converse phenomenon, that is the effect of other than cholinergic transmitters on cholinergic transmission, was also stressed. For instance, Blaker, Cheney and Costa expanded their earlier demonstration that muscimol, a GABA A receptor agonist, when given intraseptally, affects the turnover rate of ACh in the hippocampus. They found a decrease in ACh turnover which was correlated, upon septal injection of muscimol, with increased lever response rate during extinction of conditioned behavior that arises normally upon cessation of reinforcement. On the other hand, bicuculline, a GABA A antagonist, induced a decrease in responding. These findings are consistent with the notion that the activation of the central cholinergic system causes an increase in habituation rate (15). Higher doses of muscimol decreased ACh turnover in brain parts other than the hippocampus, such as the cortex, and produced sedation. Furthermore, intraseptally administered beta endorphin which, as found earlier by the Saint Elizabeths Hospital investigators, decreases hippocampal ACh turnover via activation of GABA interneurons, exerted a behavioral effect similar to that of muscimol, while substance P, which decreases the hippocampal turnover of ACh via a non-GABAergic mechanism, did not affect the paradigm in question. As baclofen, a GABA B agonist, exerted behavioral effects similar to, but 5 times more potent than those of muscimol, but did not change hippocampal ACh turnover, the authors hypothesized that the behavioral effect of muscimol depends on the inhibition of stimulatory cholinergic input to the hippocampus while baclofen acts as a presynaptic inhibitor of hippocampal efferents to the septum. Furthermore, Cheney and Panula pointed out that this interpretation is in agreement with the notion that the septum provides cholinergic interneurons which project to area CA3 of the hippocampus, where they impinge on glutaminergic pyramidal cells and complete the septohippocampal loop by innervating GABAergic neurons. Finally, Cheney's data indicate that there may be two types of GABAergic neurons in the septal nuclei, the small cells in the lateral septal nucleus which are associated with the glutaminergic feedback from the hippocampus, while the large cells in the medial septum and diagonal band are associated with the cell bodies and dendrites of the cholinergic neurons projecting to the hippocampus.

A similar complexity and interaction between the cholinergic and other systems exists in the striatum and the cortico-striatal pathways. In fact, the complexity of the striatal input via the substantia nigra and of the striatal-globus pallidus system is well known, since these systems include GABAergic, peptidergic, endorphinergic as well as catecholaminergic inputs. At the present Meeting, Ladinsky and his associates reported on a corticostriatal pathway that may act in an excitatory manner on the striatal cholinergic neurons or interneurons. Undercutting of the cortex depressed striatal ACh synthesis as well as glutamate uptake without affecting ACh and choline contents, ChAT, AChE, catecholamines or serotonin. Furthermore, these investigators showed that decortication prevented the increase in striatal ACh content induced by cholinergic agonists or ACh releasers and by GABAergic and catecholaminergic or dopaminergic agonists. On the other hand, the decrease in ACh content induced by neuroleptics was not affected by decortication.

Other studies presented at this Meeting should help in identifying the transmitter of the corticostriatal excitatory pathway abutting upon the cholinergic neurons of the striatum. Scatton and Fage provided evidence indicating that excitatory input from the cortex to the striatal cholinergic interneurons may depend on three types of amino acid receptors: results obtained with aminoergic neurotoxins, agonists and antagonists led to the conclusion that the effective excitatory amino acids receptors which are located on the dendrites of striatal cholinergic interneurons are of the N-methyl D-aspartate type; it appeared also that the actions of these receptors are independent of the nigrostriatal pathways. Furthermore, N-methyl D-aspartate failed to evoke ACh release from sites where cholinergic afferents rather than interneurons are found such as in hippocampus, interpeduncular nuclei and olfactory bulb. Excitatory amino acids acting at these sites give rise to action potentials in striatal cholinergic interneurons, presumably resulting in the release of ACh from cholinergic nerve terminals. Preliminary data also suggested that, in rats, the release of ACh and methionine enkephalin decreases while that of substance P increases, with age.

Noradrenergic mechanisms also interact with ACh turnover. The studies of Robinson suggested that the dorsal noradrenergic bundle originating in the locus coeruleus contributes to the activation by amphetamine of ACh turnover in several brain parts (including cortex and hippocampus). This effect of amphetamine was reduced by 6-OHDA induced lesions of the dorsal noradrenergic bundle.

Other pathways also appear to be involved in this phenomenon. ACh turnover of the cerebral cortex was studied by Wood, who found a number of regulatory inputs into the substantia innominata-cortical cholinergic pathway. These inputs include an inhibitory GABAergic pathway from nucleus accumbens, and inhibitory μ and δ opioid receptors, as well as possibly an alpha 1 excitatory component which depress and augment, respectively, ACh turnover. Serotonergic and adrenergic system did not seem to alter cortical ACh turnover generated by input from the substantia innominata.

Other studies reported in this session concerned the multiple transmitter modulation of ganglionic transmission. Thus, deGroat and his associates indicated that leucine-enkephalin acts as an inhibitory transmitter at cholinergic synapses in the sacral para-sympathetic outflow to the urinary bladder of the cat. This inhibition seems to depend on the S opioid receptors in the cholinergic nerve terminals. Additional supportive evidence came from the evaluation of the effects of enkephalins and naloxone, in vitro studies of the effect of leucine-enkephalin on the EPSPs of the parasympathetic ganglion of the bladder and the corresponding in situ studies, and, finally, the earlier demonstration from deGroat's laboratory of the presence of leucine-enkephalin in the preganglionic neurons of the lateral band of the sacral parasympathetic nucleus and of its transport in the preganglionic axons innervating the urinary bladder. DeGroat and his associates demonstrated also that vasoactive intestinal polypeptide (VIP) facilitates the muscarinic excitatory response in the sympathetic and parasympathetic ganglia of the cat; they correlated this finding with the indication that in the peripheral system VIP and ACh may be released by the same neurons. On the other hand, VIP did not alter nicotinic firing of the bladder ganglia or of the superior cervical ganglia, and did not affect the inhibition of transmission of bladder ganglia due to sympathetic stimulation or the administration of norepinephrine, GABA, and leucine-enkephalin.

A similar interaction between VIP and muscarinic receptors was described at this Meeting by Hedlund and her associates, following their demonstration of the coexistence in the cat and rat sub-man-dibular salivary gland of ACh and VIP. For instance, occupancy of muscarinic receptors by either agonists or, interestingly enough, antagonists, enhanced the VIP-mediated stimulation of adenylate cyclase in a GTP-dependent manner, while VIP attenuated the muscarinic receptor mediated increase in cyclic GMP. Finally, the chronic treatment with atropine resulted not only in muscarinic supersensi-tivity but also in the depletion of VIP of the salivary gland, and a marked increase in VIP binding. Another peptide effective at the sympathetic ganglia of some species is substance P, as was shown in the past for the guinea-pig inferior mesenteric ganglion (cf. 24), and for the amphibian sympathetic ganglia (49).

At this Meeting, Dun and his associates presented the continuation of their studies of multiple neurotransmitter interactions that take place in the autonomic ganglia as they addressed themselves to the role of serotonin in the modulation of cholinergic transmission of prevertebral (celiac) guinea-pig ganglia. In a portion of these ganglia presynaptically elicited slow depolarization may be mediated by serotonin, since this depolarization did not partake of the charac-

teristics of muscarinic or nicotinic cholinergic transmission, but
could be simulated as well as desensitized by, serotonin.   These
findings were consistent with the pharmacological analysis of the
effects of serotonin and of preganglionic stimulation by means of
appropriate serotonin antagonists and serotonin precursor.   These
data furthermore were consistent with immunohistochemical and HPLC
data indicating the presence of serotonin in the celiac ganglia,
and with the demonstration that serotonin is released by depolarization
in a calcium dependent manner.

The source of the serotonergic fibers is not entirely clear
at this time. The physiological role of this non-cholinergic potential
may be related to its facilitatory effect on the cholinergic (nicotinic) fast excitatory postsynaptic potential of the celiac ganglia
of the guinea pig.

An interesting question was addressed by Zigmond and Ip. They
expanded on their earlier findings indicating that tyrosine hydroxylase, the rate limiting enzyme in catecholamine biosynthesis, is
activated nicotinically in the rat sympathetic ganglion by carbachol;
they demonstrated that a similar effect on tyrosine hydroxylase
may be obtained by preganglionic stimulation (see also above). As
this effect was incompletely blocked by hexamethonium and not at
all by atropine, a non-cholinergic transmitter may be involved.
Further investigations identified VIP and secretin as mediators, a
number of other peptides and bioactive substances not being effective;
the effect of the peptides was not mediated by their capacity to
release ACh from preganglionic nerve terminals.

The findings presented in this session clearly demonstrate
the physiological and behavioral significance, as well as the mutual
interaction between cholinergic and other neurotransmitters.   In
fact, there is no behavior or function that is regulated by only
one neurotransmitter.   Indeed, the sites studied more extensively
at this Meeting such as the striatum or the hippocampus are affected
by a large variety of neurotransmitters, and the circuitry that is
involved is most complex in that it concerns both neuronal soma
and nerve terminal. It is also of interest that as many as perhaps
a dozen of peptides may be involved in the regulation of both peripheral and central transmission (see also 25, 45).

The most interesting concept that may emerge from the studies
presented here is that a neuron or an interneuron may release more
than one neurotransmitter; this concept may contravene the classical
hypothesis of Dale. It is, however, important to mention, in this
context, a clarification offered on this concept by J.C. Eccles.
One of the possible readings of Dale's concept is that whatever a
neuron releases at one nerve ending it releases at all of them.
While it was thought originally that Dale wished to predict that
whenever ACh release is demonstrated at a nerve terminal of the

neuron its release will be demonstrable at its other terminals, and while this principle allowed Eccles to devise a demonstration of the central cholinergic transmission at the spinal Renshaw cell (26), Dale's concept may be still in full force today if construed to refer to more than one transmitter. It should be stressed that the best indication that more than one neurotransmitter may be released from a nerve ending emerges from the studies of autonomic ganglia such as presented at this Meeting by Dun, by deGroat, and elsewhere by Hokfelt (44; see also 25, 49).

IX. CHOLINERGIC MECHANISMS IN PHYSIOLOGICAL FUNCTION, INCLUDING CARDIOVASCULAR EVENTS

It is well known that the cholinergic system contributes to a large number of physiological functions and behaviors; the former extend from thermocontrol and the control of feeding and thirst, to regulation of sexual function (see for instance, 56, 57, 59, 82). Certain functions and behaviors with neuropsychiatric connotations discussed at this Meeting have already been described (see Section II); additional findings concerning classical cholinergically-related physiological phenomena such as cardiac function, respiration, circulation and blood pressure, were also presented.

Several papers concerned the vagal activity as related to ACh, cardiac automaticity and conduction. O. Brown and his associates related ACh synthesis to vagal activity; they exploited a sensitive chemical method for the determination of choline uptake and ACh turnover in the rat heart to demonstrate that stimulation of the vagus markedly increased the turnover of ACh in right atrium and the SA node, the areas richest in ACh, as well as in the remainder of the heart. Of additional interest was their finding that only a fraction of injected choline was utilized for the formation of ACh and that the decay of choline taken up followed a biphasic relationship, indicating a complexity of the pools contributing to choline uptake. Levy and Young demonstrated that the right and left vagi differ in their ability to affect the cardiac cycle length (chronotropic effect) and with respect to the curves that relate the moment (within the cardiac cycle) of the application of the vagal stimulus to the chronotropic effectiveness of the stimulus (phase-response curve). The phase-response curve may reflect the extent to which the individual SA node cells are synchronized with one another during phase-coupled vagal stimulation. Levy and Young felt that the disparities between the effects of the right and left vagi are not due to the differences in the cardiac innervation by the two vagi but to innate differences in the magnitude of the chronotropic response of these two nerves. Similarly, P. Martin reported that the AV conduction response and the contractile (inotropic) response to vagal stimulation is dependent on the time in the cardiac cycle, the pacing rate, and such variables as sympathetic activity and the hysteresis, i.e., the elapsed time between the two consecutive vagal stimuli.

With respect to cholinergic control of blood pressure, Brezenoff et al. found that high blood pressure of spontaneously hypertensive rats was specifically blocked by atropinic antagonists and intraventricularly administered hemicholinium, which also blocked centrally mediated pressor response to physostigmine. Parenthetically, it is well known that certain antiChEs induce both a peripherally and centrally mediated pressor response (cf., 111). It was of interest that atropinic agents did not block the pressor responses to noradrenaline or angiotensin, nor did they cause a fall of blood pressure in normotensive rats. These data, suggesting that cholinergic mechanisms contribute to spontaneous hypertension, seem to be consistent with those of Buccafusco. He reported that clonidine and methyldopa, which are capable of lowering the blood pressure of spontaneously hypertensive rats, also lower ACh turnover in the medulla, thalamus, midbrain and hypothalamus, but not in the striatum. Furthermore, Buccafusco found that intraventricularly administered choline increased ACh synthesis in both spontaneously hypertensive and normotensive rats. Parenthetically, this last set of data may be inconsistent with some of the results presented in another session of this Meeting (see above, Sections II and VI). Buccafusco hypothesized that the antihypertensive effects of clonidine and methyldopa may relate to their blockade of ACh synthesis.

Catravas and his associates studied the cholinergic contribution to pulmonary circulation. They demonstrated that ACh increases pulmonary arterial pressure and pulmonary vascular resistance; these effects were augmented by physostigmine and blocked, after a delay, by atropine. Histamine and angiotensin receptor blockers did not affect pulmonary hypertension induced by ACh which was blocked by indomethacin; however, indomethacin did not affect the ACh-induced fall in systemic arterial pressure, and the increase in airway pressure, which reflects bronchogenic effect of ACh. Catravas and his associates concluded that muscarinic receptor-mediated pulmonary arterial pressure response but not the airway or systemic arterial response requires the release of a prostanoid.

These papers, that concerned vagal regulation of cardiac functions, illustrate the relationship between ACh synthesis and the parasympathetic vagal activity on the one hand, and the capacity of the vagus to provide a moment-to-moment control of the cardiac beat on the other. Interesting but not definitive data were provided in this session as to the contribution of the cholinergic system to hypertension and to the control of pulmonary circulation. It should be stressed in this context that there is relatively little evidence that atropinic drugs or cholinergic agonists exert, in man, any significant control of blood pressure.

X. NEUROTOXIC AGENTS AND FALSE TRANSMITTERS

Since the advent of the neurotoxins that affect either the nerve terminal such as black widow venom, or axonal conductance such as tetrodotoxin, and particularly since the advent of irreversible cholinergic nicotinic ligands such as bungarotoxin, neurotoxins have become useful tools for the study of mechanisms and sites of cholinergic transmission (88, 115). A number of presentations at this Meeting dealt with neurotoxins and related agents agents such as ethylcholine aziridinium, AF64A, which is an alkylating analog of choline, and a depletor of ACh in the brain in vivo (29, 77).

Hanin's group has reported that AF64A inhibits the activity of ChAT in vivo, but not in vitro, and explained these effects of AF64A as possibly being due to its being acetylated to a false trans- mitter; they demonstrated the formation of acetylethylcholine azi- ridinium by rat brain synaptosomes in the presence of acetylCoA, this compound being subject to AChE hydrolysis. In addition, Walsh, Hanin and their associates demonstrated at this Meeting that, when injected into the lateral ventricles of rats, AF64A depleted ACh of the hippocampus and cortex without affecting levels of catechola- mines, indoleamines and their metabolites, or choline. These effects were well correlated with transient effects on sensory motor function, as well as with long term deficits in cognitive behavior, impairment in maze performance, and hyperalgesia. Altogether, Mantione, Hanin, Fisher, Walsh et al. have suggested that AF64A disrupts transmission and causes learning deficits by a possible combination of effects, including an initial inhibition of choline uptake, false transmitter effect, irreversible inhibition of ChAT, and ultimately, disruption of the cholinergic nerve terminal.

Casamenti, Pepeu and their associates also studied the effects of this compound, and found that injection of AF64A into rat ventricles caused a significant decrease of ChAT activity in the hippocampus, but lesser decreases in the striatum and in the cortex. There was also an insignificant decrease in high affinity choline uptake in these brain areas. Furthermore, AF64A did not affect ACh release, whether spontaneous or induced by amphetamine. Casamenti et al. found a correlation between inhibition by AF64A of hippocampal ChAT and the impairment of passive avoidance conditioning response; hyper- excitability developed subsequently, followed by motor deficit and death. Kozlowski and his associates introduced still another element into this area of investigation. They found that AF64A, injected into the nucleus basalis of rats, produced a dose-dependent decrease in AChE staining of the nucleus, with high doses causing in addition a decrease in cortical ChAT activity; the latter event may have been caused by damage to cholinergic neurons projecting from the nucleus basalis to the cortex.

Aziridinium ion isomers of chloroethylamine are ACh mustard analogs of AF64A, some of which have been shown to react with cholinergic receptors as cholinergic agonists. As shown by Colhoun, some of these compounds, such as choline mustard aziridinium and monoethyl mustard aziridinium ions, produced hemicholinium like toxicity in mice and hemicholinium like effects on the rat phrenic nerve diaphragm preparation, and inhibited high affinity choline transport in rat synaptosomes. Moreover, consistent with the data presented by Mantione and coinvestigators, these neurotoxins inhibited ChAT and served as a precursor of a false transmitter. Of particular interest is the finding by Colhoun that ethylcholine mustard aziridinium (AF64A) and choline mustard aziridinium ions, injected in the medial septum, resulted in selective hippocampal lesions. These data are in concordance with Hanin et al., who have suggested earlier that these compounds may serve as selective "cholinotoxins" for animal model studies of cholinergic hypofunction.

Other choline analogs, choline esters and homocholine ester derivatives were studied by Collier. Similar to the aziridinium ion compounds, these compounds appear to enter cholinergic nerve endings of the cortex via the choline uptake system, and to become precursors of releasable false transmitters. Similar phenomena may occur in perfused ganglia. The synthesis of these false transmitters was stimulation-dependent, as is the synthesis of ACh.

AH5183 (a piperidino cyclo-hexanol), as shown by Parsons and his associates, was still another inhibitor of the choline transport system of the nerve terminal, and of the ACh transport system of synaptic vesicles. This compound seems to occupy ACh sites in the synaptic vesicles of the electric organ, causing the release of ACh from the vesicles; this activity residing in the L-isomer. Several analogs of AH5183 were synthesized, and their effectiveness in causing the loss of ACh in vitro was shown to be well correlated with the LD_{50} values obtained in mice. Thus, inhibition of ACh storage is an important component of the toxicity of AH5183 and related compounds.

Alkylated oxotremorine analogs constitute yet another neurotoxin type; Jenden and his coinvestigators reported that these compounds demonstrate either a high or a low affinity to muscarinic mouse receptors, their binding being antagonizable by atropine. Furthermore, the compounds in question induced a long-lasting blockade of the action of muscarinic agonists, which could be due either to desensitization, or to the loss of these receptors.

Another type of neurotoxin, the snake neurotoxins such as taipoxin, bungarotoxin, notexin and other compounds which exhibited phospholipase A_2 activity, was studied by Dowdall and his associates. These compounds are potent neuromuscular blocking agents that seem to target on the presynaptic membrane and to affect transmitter release. This action exhibited several phases; toxicity could be related to the final phase of progressive decrease of ACh release. The compounds in question induced hypopolarization of Torpedo synaptosomal membranes. However, among these were toxins that did not show phospholipase activity; furthermore, different toxins seemed to have distinct action sites at the synaptosomal membrane.

An opposite action was exhibited by the annelid neurotoxin, Glycera convoluta venom. According to Manaranche and his associates, this compound increases the frequency of miniature endplate potentials, depletes ACh of neuromuscular junctions and electric organs, and increases calcium dependent ACh release. These activities seem to be related to the binding of the neurotoxin to the ectocellularly exposed protein of the presynaptic plasma membrane. Manaranche and his associates suggested that these effects may depend on the rearrangement of intramembrane particles, and an effect on calcium uptake mechanisms.

Yet another interesting group of toxins described at this Meeting exhibits phospholipase activity. Their effect on membrane phospholipids may regulate transmitter release as described by Dowdall; this effect may be also mediated by their generation of phospholipid metabolites such as phosphatidic acid, which may act as Ca^{2+} carriers (91, 92, 96).

As already indicated, neurotoxins constitute important tools for the identification of several pre- and post-synaptic sites involved in cholinergic transmission. Some of the pertinent data are controversial, as for instance, earlier findings of Albuquerque, Eldefrawi and their associates (cf. 27) concerning a differential effect of α-bungarotoxin and histrionicotoxin on what they considered the channel protein, versus the receptor protein. The data presented at this Meeting exemplify the versatility of neurotoxin action: neurotoxins may act as false transmitters, inhibitors of choline uptake in the nerve terminal and/or ACh uptake into the synaptic vesicles, facilitators or inhibitors of ACh release, or finally, inhibitors of ChAT. At this time, the main mechanism of actions of such interesting neurotoxins as AF64A is not clear as yet; a combined effect is possible which may also explain the present controversy with regard to the action of these drugs. Besides being convenient tools for the study of cholinergic phenomena, the neurotoxins provide us with important models. For instance, the nitrogen mustard analogs of choline studied by Hanin, Colhoun, Rylett, and their coinvestigators, may provide us with a model of Alzheimer's disease.

FINAL COMMENTS

I would like now to emphasize what was achieved at this Meeting and what conclusions were reached. I would like also to refer to problems that remain to be resolved, as well as to topics that were covered very sparingly or not at all; obviously, these unresolved questions should be considered for the agenda of our future Meetings. This matter of unresolved problems as well as certain other matters that loom large in my mind will lead me finally to my dreams as to what might be presented either at our next Meeting in England or in a more distant future, and to my analysis of what are the ultimate goals that we should focus on.

Table I summarizes the topics that were covered or at least referred to in the Meeting. It should be emphasized that while the problems inherent in some of these topics were resolved, in most cases the answers are still in abeyance. For example, the major problems that concern the cholinergic pathways can be considered as resolved since, on the basis of results obtained with several efficient markers of the cholinergic synapses we delineated at this Meeting, with a certain degree of finality, the important central cholinergic pathways, particularly those emanating from the medial basal forebrain cholinergic system. This is not to mean that the complete map of the central cholinergic pathways is available at this time, but it does mean that only confirmatory information and final touches are needed.

Similarly, it appears that the significance of the cholinergic system with regard to aging and particularly to SDAT, was fully demonstrated. Yet, this does not mean that transmitters other than ACh, and pathologies not directly concerned with the transmitters are not involved in either aging or SDAT. Furthermore, it must be stated that aging and SDAT constitute an on-going problem, many questions remaining unanswered at this time. Questions related to this issue that should be addressed at the next Meeting, and that hopefully may be reasonably expected to be answered at that time, are as follows: Is SDAT a speeded up process of aging, or is it a specific entity? If cholinergic neurons are particularly sensitive to aging, or SDAT, or both, what makes these neurons so sensitive as compared to the other neurons? What are the mechanisms involved in the loss of the cholinergic neurons whether in SDAT or aging? Are cholinergic neurons vulnerable in certain individuals who are then prone to SDAT while catecholaminergic neurons may be vulnerable in other individuals who are then prone to Parkinson's Disease, and what are the mechanisms that underlie this hypothetical differ- ential sensitivity? Is there any relationship between SDAT and Huntington's chorea or tardive dyskinesia, as it appears that choli- nergic agonists may be useful in either? Finally, there is the question of the cholinergic treatment of SDAT and aging, and this point will be returned to subsequently.

Similarly, while much was done with respect to cholinergic, both pre- and post-synaptic receptors, yet, this area is in consider- able flux and many points have to be clarified. At this meeting, it became manifest that there are subpopulations of muscarinic and nicotinic receptors, whether at the post-or pre-synaptic sites. A question which is pertinent in this context is whether or not the sensitivity of postsynaptic and presynaptic receptors to specific antagonists is sufficiently differentiated to allow us to exploit this specificity for the design of drugs; this point will be returned to later. Also, evidence is forthcoming indicating that the pre-synaptic receptors of several types regulate transmitter release; feedback regulation may be involved, and it may constitute the physiological role of presynaptic receptors, although the precise delineation of this role is not available at this time. Similarly, the role of the postsynaptic receptors is not too clearly understood. The propositions here range from the concept of the cooperation of postsynaptic receptors in maintaining effective transmission, to the concept that long latency processes of facilitation and inhibition result from the interplay between these receptors, these processes possibly underlying such functions as, e.g., memory formation (73).

Still another area that is in flux is that of ChEs and of antiChE drugs. Following the discovery of carbamate and OP antiChEs, it was thought that these compounds represent the rare category of drugs that have a single mechanism of action, and that they work solely by inhibition of ChEs and accumulation of ACh. Gradually, it became clear that it is not so, and much pertinent evidence was presented at this Meeting. Thus, there is no doubt today that antiChEs specifically, and cholinergic agonists generally, exhibit morphologic and direct receptor and channel effect that do not depend on the classical antiChE action; these phenomena were discussed at this Meeting and elsewhere (62). While certain myopathic actions of these drugs may depend on ACh accumulation as described at this Meeting by Dettbarn, yet there are morphologic including ontogenetic and neurotoxic effects of these drugs that admittedly do not (2, 51-53). Mechanisms for these non-classical effects elude us - do they depend on the phosphorylation by the organophosphorus drugs of non-ChE proteins? Do they depend on enzymes other than ChEs (51)?

Another unsolved problem is that of the relationship of BuChE to AChE. In fact, the physiological significance of BuChE is still in abeyance. Furthermore, trophic (cf. Koelle, this Meeting) and genetic (105) control of ChE synthesis are areas awaiting final solution.

Yet another problem that pertains to this area is that of the treatment of antiChE poisoning. If we were unfortunate enough to be exposed accidentally or in the course of chemical warfare to OP drugs, how could we antagonize effectively their poisonous effects? It should be emphasized that oximes, atropine, central anticonvulsants and reversible antiChEs capable of protecting AChE from irreversible antiChEs are not very effective antidotes, and offer only partial protection, particularly with regard to OP antiChEs capable of producing rapid aging of phosphorylated ChE.

Another on-going topic of investigation is that of transmitter interaction. This problem has been with us ever since the proposal of the existence of a link between cholinergic and adrenergic transmission by Burn and Rand (13), and since the demonstration of the effect of accumulation of ACh on levels and synthesis of catecholamines (35). At this Meeting, further evidence for interaction of various neurotransmitters with ACh was demonstrated both for the CNS and the autonomic ganglia; similar evidence is available even for such a classical cholinergic site as the neuromyal junction (cf., e.g., 4, 31). Much more work has to be conducted in this area; for instance, the knowledge of the anatomical circuitry that must underlie neurotransmitter interaction in the ganglia and the CNS is limited, although exciting information was presented at this Meeting and elsewhere with regard to the basal ganglia and septo-hippocampal and ponto-cortical pathways.

Another problem which is in the flux is that of neurotoxins. Traditionally, this field is of great importance for delineation of the sites and mechanisms of cholinergic transmission; this appears clearly from the 1974 review of Narahashi (88). It suffices to adduce the importance of muscarinic and nicotinic ligands for the mapping of the central muscarinic and nicotinic receptors and for establishing their molecular structure (97). At this Meeting it was stressed that neurotoxins may exhibit mixed effects that include both pre- and post-synaptic sites; it was of particular interest that some of them - as for instance AF64A - may work via false transmitter mechanisms. An additional significance of neurotoxins is that they may serve as useful models for the study of such disease entities as multiple sclerosis or SDAT.

ACh metabolism and its relationship to choline continues to attract our attention. The question of the significance of exogenous choline for the formation of ACh was raised at this and our earlier Meetings, and it apears that choline and lecithin may not readily participate in formation of ACh under resting conditions, while they may do so under the conditions of stress. In the same vein, it was shown at this Meeting that there may be an interchange between neuronal membrane phospholipids and the choline pool(s) that may subserve ACh formation. In this context, perhaps we should begin to employ consistently the terms "metabotropic" and "synaptotropic" ACh that were devised by the McGeers and Eccles (79), to indicate that ACh participates in non-transmissive and transmissive phenomena, respectively, and we should remember in this context the presence of ACh in non-neuronal tissues including pre-neural embryonic layers (52). The terms "metabotropic" and "synaptotropic" choline may be added at this time to our cholinergic vocabulary. It should be pointed out that the concept of metabotropic and synaptotropic choline and ACh is important in its own right, and also with respect to our capacity to compensate via precursors for the cholinergic deficit of SDAT or tardive dyskinesia.

Still another topic that was brought up but not explored at sufficient length at this Meeting was that of second messengers. Cyclic nucleotides and phosphorylating protein kinases were touched upon, but the question as to the sites where cyclic nucleotides indeed underlie the action of cholinergic agonists remains unresolved to date (see for instance, 37 versus 63). Furthermore, phosphatidyl inositol cascade, classically important for the functioning of exocrine glands and muscarinic responses (46, 81) can be activated also by nicotinic stimulation, as indicated at this Meeting by Hawthorne and Swilem. It must be added that phosphatidyl inositol metabolism generates compounds such as phosphatidic acid, that serve as calcium carriers, thus contributing to the pre- and post-synaptic regulation of cholinergic transmission. Finally, the interaction between these two second messenger systems, raised at this Meeting, deserves futher attention.

While so far I have stressed problems that were addressed but not necessarily resolved at this Meeting, let me now enumerate topics that were barely alluded to. Affective conditions in man or control of circulation were briefly discussed at this Meeting, yet, there was no thorough examination of several behaviors notoriously correlated with cholinergic system such as for instance REM sleep or aggression, although these two and additional 20 behaviors have cholinergic correlates (60, 62, 82), and cholinergic components of schizophrenia were emphasized recently by several investigators (65, 103). Two additional aspects of this particular matter should be emphasized at this time. The EEG activity as well as other evoked potentials constitute a reflection of these behaviors. Yet, at this Meeting very little attention was paid whether to many behaviors or to their electrophysiological counterparts and monitors. Also, multitransmitter interactions were emphasized at this Meeting, and the behaviors in question, while exhibiting cholinergic correlates, are all multi-transmitter in nature; yet none of these behaviors was discussed at this Meeting, with the full implications of their multitransmitter nature, circuitry involved and the electrophysiological and evoked potential correlates.

In my concluding remarks I would like to speak of the dreams for the future--some of these dreams may not come to fruition at our meeting in England but may become a reality in 25 or 50 years. First, there is a clinical dream. For those of us that have investigated the cholinergic system for decades it is disappointing that this system, so important for basic pharmacology and neuropharmacology, is exploited in the clinic to a limited extent only (cf. 58, 61). Certainly, we use cholinergic and anticholinergic drugs in anesthesia, sometimes in Parkinson's Disease and Myasthenia Gravis, sometimes to relax the gut or dry up secretions, and the old-timers are thrilled that finally, cholinergic drugs have begun to be used, at least experimentally, in manias, SDAT and in the control of sleep. However, the results presented at this Meeting suggest that many of these experimental approaches are doomed to early failure. The use of the cholinergic drugs will nevertheless become more effective, if research such as that exemplified at this meeting by Dahlbom is continued, to yield better knowledge of specificity of cholinergic receptors and differences between these receptors, particularly at a regional level. If this specificity is established with respect to, say, muscarinic receptors of the myenteric plexus versus muscarinic receptors of the CNS, and with respect to other subpopulations of regional muscarinic and nicotinic receptors, dramatic advances may be made in the use of cholinergic drugs whether in ulcer, familial dysautonomias, or in manias.

Still in the area of clinical dreams, further development concerning our understanding of the phospholipids of the neuronal membrane, choline metabolism and choline-phospholipid exchange may teach us how to ultimately combat aging and SDAT. Better knowledge of choline and phospholipid pools and metabolism should lead to the development of more efficacious ACh precursors, i.e., compounds used preferentially in ACh rather than phospholipid biosynthesis. Yet, ultimately, even the best precursors cannot compensate for the loss of cholinergic neurons and cholinoceptive receptors during aging or SDAT. However, if we ever learn how to increase the plasticity of the brain and of its cholinergic receptors, or how to render the cholinergic receptor membranes less vulnerable to aging and disease, including a possibly immunoreactive damage, then indeed we may have made significant strides in the battle against aging as well as in Myasthenia Gravis.

My final dream is of a basic and philosophical nature. At the next Meeting and at subsequent meetings there may be sessions devoted to improving our understanding of the circuitry and multitransmitters involved in such behaviors as aggression or schizoid malfunction. But, even if such sessions are markedly successful, will these sessions result in bridging the gap between organic knowledge and behavior? Will we ever understand how cognitive behavior works in organic terms and how we can describe and present a model of consciousness? Is any hope in this respect unduly reductionistic (95)? Is this a problem that belongs to Carnap's (16) category of nonsensical paradigms and open-ended systems (36) or is it a component of Turing's enigma (42); that is, is this a question for which there cannot ever be either a logical or experimental answer (55)? In this context, please look up the quotation from Albertus Magnus which is adduced as a motto for this presentation.

As I comment on what was accomplished at this Meeting, and as I dream on what may be accomplished at the next and subsequent Meetings of this group, my appetite is whetted and I am reminded of an Oscar Wilde story. As Oscar was examined at Cambridge for a first in Greek, he was asked to translate from the Greek version of the Gospel according to St. Matthew. After he translated a few paragraphs, on of the examiners stopped him and said, "Thank you Mr. Wilde, this will do." Oscar responded, "Oh please, let me continue, I can't wait to see what happens!"

Acknowledgement. Published and unpublished investigations of these laboratories referred to in this paper were supported in part by National Institutes of Health Grants NS06455, GM77 and BRSG 447-22, the Bellweber Foundation and the Potts Foundation.

REFERENCES

1. Abdel-Latif, A.A., Akhtar, R.A. and Hawthorne, J.W. (1977): Biochem. J. 162:61-73.
2. Abou-Donia, M.D., Graham, D.G. and Komcil, A.A. (1979): Toxicol. Appl. Pharmacol. 49:293-299.
3. Akaike, A., Ikeda, S.R., Brookes, N., Pascuzzo, G., Rickett, D.L. and Albuquerque, E.X. (1984): Molec. Pharmacol. 25:102-112.
4. Akasu, T., Karczmar, A.G. and Koketsu, K. (1983): Eur. J. Pharmacol. 88:63-70.
5. Albuquerque, E.X., Aquayo, L.G., Warnick, J.E., Weinstein, H., Glick, S.D., Maayani, S., Ickowicz, R.K. and Blanstein, M.D. (1981): Proc. U.S.A. Natl. Acad. Sci. 78:7792-7796.
6. Albuquerque, E.X., Akaike, A., Shaw, K.-P. and Rickett, D.L. (1984): Fund. Appl. Toxicol. 4:S27-S33.
7. Aldridge, W.N. (1954): Chem. and Ind. 473-476.
8. Ansell, G.B. and Spanner, S. (1981): In: Cholinergic Mechanisms (eds) G. Pepeu and H. Ladinsky, Plenum Press, New York, pp. 393-403.
9. Ansell, G.B. and Spanner, S. (1982): In: Phospholipids in the Nervous System (eds) L.A. Horrocks, G.B. Ansell and G. Porcellatti, Raven Press, New York, 1:137-144.
10. Bartus, R.T., Dean, R.L., Goas, J.A. and Lippa, S.A. (1980): Science 209:301-309.
11. Birdsall, M.J.M. and Hulme, E.L. (1976): J. Neurochem. 27:7-16.
12. Blusztajn, J.K. and Wurtman, R.J. (1983): Science 221:614-620.
13. Burn, J.H. and Rand, M.J. (1959): Nature 184:163-165.
14. Carlton, P.L. (1968): In: Psychopharmacology - A Review of Progress 1957-1967 (ed) D.H. Efron, U.S. Govt. Printing Office, PHS Publ. No. 836, Washington, D.C., pp. 125-135.
15. Carlton, P.L. (1968): Prog. Brain Res. 24:48-60.
16. Carnap, R. (1937): The Logical Syntax of Language, Harcourt, New York.
17. Costa, E., Chuang, D.M., Guidotti, A. and Hollenbeck, R. (1978): In: Cholinergic Mechanisms and Psychopharmacology (ed) D.J. Jenden, Plenum Press, New York, pp. 267-284.
18. Davis, B.M. and Davis, K.L. (1979): In: Brain Acetylcholine and Neuropsychiatric Disease (eds) K.L. Davis and P.A. Berger, Plenum Press, New York, pp. 445-458.
19. Dettbarn, W-D. (1984): Fund. Appl. Toxicol. 4:S18-S26.
20. DeRobertis, E. (1967): Science 156:907-914.

21. Drachman, D.A. (1978): In: Psychopharmacology: A Generation of Progress (eds) M.A. Lipton, A. DiMascio and K.F. Killam, Raven Press, New York, pp. 651-662.

22. Dun, N.J., Kaibara, K. and Karczmar, A.G. (1978): Brain Res. 150:658-661.

23. Dun, N.J. and Karczmar, A.G. (1978): Proc. Natl. Sci. U.S.A. 75:4029-4032.

24. Dun, N.J. and Minota, S. (1981): J. Physiol. (Lond.) 321: 259-271.

25. Dun, N.J. and Karczmar, A.G. (1985): In: Vertebrate Autonomic and Enteric Ganglia: Anatomy, Electrophysiology and Neuropharmacology (eds) A.G. Karczmar, K. Koketsu and S. Nishi, Plenum Press, New York (in press).

26. Eccles, J.C., Fatt, P. and Koketsu, K. (1954): J. Physiol. (Lond.) 126:524-562.

27. Eldefrawi, M.E., Eldefrawi, A.T., Mansour, N.A., Daly, J.W., Witkop, B. and Albuquerque, E.X. (1978): Biochemistry 17:5474-5484.

28. Fibiger, H.C. (1982): Brain Res. Rev. 4:327-388.

29. Fisher, A. and Hanin, I. (1980): Life Sci. 27:1615-1634.

30. Fonnum, F. and Sterri, S.H. (1981): Fund. Appl. Toxicol. 1:143-147.

31. Gallagher, J.P. and Karczmar, A.G. (1973): Neuropharmacol. 12:783-792.

32. Giacobini, E. and Marchi, M. (1981): In: Cholinergic Mechanisms (eds) G. Pepeu and H. Ladinsky, Plenum Press, New York, pp. 1-23.

33. Giacobini, E., Filogamo, G., Giacobini, G. and Vernadakis, A. (eds) (1982): The Aging Brain: Cellular and Molecular Mechanisms of Aging in the Nervous System, Raven Press, New York.

34. Giurgea, C. (1972): Actualites Pharmacol. 25:115-156.

35. Glisson, S.N., Karczmar, A.G. and Barnes, L. (1974): Neuropharmacology 13:623-632.

36. Godel, K. (1931): Monatshefte f. Math. U. Phys. 38:173-198.

37. Greengard, P. (ed) (1978): Cyclic Nucleotides, Phosphorylated Proteins, and Neuronal Function, Raven Press, New York.

38. Hanin, I., Spiker, D.G., Mallinger, A.G., Kopp, U., Himmelhoch, J.M., Neil, J.F. and Kupfer, D.J. (1981): In: Cholinergic Mechanisms, (eds) G. Pepeu and H. Ladinsky, Plenum Press, New York, pp. 901-920.

39. Harris, C.M. (1981): Test of memory-enhancing effect of phosphatidyl choline, Ph.D. Thesis, Graduate College of the University of Illinois at the Medical Center, Chicago, Illinois.

40. Hasuo, H., Karczmar, A.G. and Nishimura, T. (1983): Proc. Soc. Neurosci. 9:925.

41. Heilbronn, E. and Winter, A. (eds) (1970): Drugs and Cholinergic Mechanisms in the CNS, Forsvarfets Forkskningsanstalt, Stockholm, Sweden.

42. Hodges, A. (1983): Alan Turning: The Enigma, Touchstone Brooks, New York.

43. Hoffmeister, F. and Mueller, C. (eds) (1979): Brain Function in Old Age, Springer-Verlag, Berlin.

44. Hökfelt, T., Elfvin, L.G., Schultzber, M., Goldstein, M. and Nilsson, G. (1977): Brain Res. $\underline{132}$:29-41.

45. Hökfelt, T., Johansson, O., Ljungdahl, A., Lundberg, J., Schultz-berg, M., Terenius, L., Goldstein, M., Elde, R., Steinbusch, H. and Verhofstad, A. (1979): In: Neurotransmitters (ed) P. Simon, Adv. Pharmacol. Therap., Pergamon Press, Oxford, $\underline{2}$:131-143.

46. Hokin, M.R. and Hokin, L.E. (1960): Int. Rev. Neurobiol. $\underline{2}$: 99-136.

47. Hoyer, S. (ed) (1982): The Aging Brain, Springer-Verlag, Berlin.

48. Hudson, C.S., Rash, J.E., Tiedt, T.N. and Albuquerque, E.X. (1978): J. Pharmacol. Exp. Therap. $\underline{205}$:340-356.

49. Jan, Y.N., Jan, L.Y. and Kuffler, S.W. (1979): Proc. Natl. Acad. Sci. U.S.A. $\underline{76}$:1501-1505.

50. Jenden, D.J. (ed) (1977): Cholinergic Mechanisms and Psycho-pharmacology, Plenum Press, New York.

51. Johnson, M.K. (1980): Dev. Toxicol. Environ. Sci. $\underline{8}$:27-38.

52. Karczmar, A.G. (1963a): In: Cholinesterases and Anticholines-terase Agents (ed) G.B. Koelle, Handbch. d. Exper. Pharmakol., Erganzungswk. $\underline{15}$, Springer-Verlag, Berlin, pp. 799-832.

53. Karczmar, A.G. (1963b): In: Cholinesterases and Anticholines-terase Agents (ed) G.B. Koelle, Handbch, d. Exper. Pharmakol., Erganzungswk. $\underline{15}$, Springer-Verlag, Berlin, pp. 179-186.

54. Karczmar, A.G. (1967): In: Physiological Pharmacology (eds) W.S. Root and F.G. Hofman, Academic Press, New York, $\underline{3}$:163-322.

55. Karczmar, A.G. (1972): In: Brain and Human Behavior (eds) A.G. Karczmar and J.C. Eccles, Springer-Verlag, Berlin, pp. 1-20.

56. Karczmar, A.G. (1978): In: Neuropsychopharmacology, Proceedings of the 10th Congress, CINP (eds) P. Deniker, C. Radouco-Thomas and A. Villeneuve, Pergamon, Oxford, $\underline{1}$:581-608.

57. Karczmar, A.G. (1979): In: Nutrition and the Brain (eds) A. Barbeau, J.H. Growdon and R.J. Wurtman, Raven Press, New York, $\underline{5}$:141-175.

58. Karczmar, A.G. (1979): Drug Therapy $\underline{4}$:31-42.

59. Karczmar, A.G. (1980): In: Modern Problems in Pharmacopsy-chiatry (ed) T. Ban, Karger, Basel, pp. 1-76.

60. Karczmar, A.G. (1981a): Psychopharmacol. Bull. $\underline{17}$:68-70.

61. Karczmar, A.G. (1981b): In: Cholinergic Mechanisms (eds) G. Pepeu and H. Ladinsky, Plenum Press, New York, pp. 853-869.

62. Karczmar, A.G. (1984): Fund. Appl. Toxicol. $\underline{4}$:1-17.

63. Karczmar, A.G. and Dun, N.J. (1978): In: Psychopharmacology: A Generation of Progress (eds) M.A. Lipton, A. DiMascio and K.F. Killam, Raven Press, New York, pp. 293-305.

64. Karczmar, A.G. and Ohta, Y. (1981): Fund. Appl. Toxicol. $\underline{1}$: 135-142.

65. Karczmar, A.G. and Richardson, D.L. (1985): In: Central Choli-nergic Mechanisms and Adaptive Disfunctions (eds) M.M. Singh, D.M. Warburton and H. Lal, Plenum Press, New York, pp. 193-221.

66. Kindel, G. and Karczmar, A.G. (1981): Fed. Proc. $\underline{40}$:269.

67. Kindel, G. and Karczmar, A.G. (1982): Fed. Proc. <u>41</u>:1232.
68. Koelle, G.B. (1963): In: <u>Cholinesterases and Anticholinesterase Agents</u> (ed) G.B. Koelle, Handbch. d. exper. Pharmakol., Erganzungswk <u>15</u>, Springer-Verlag, Berlin, pp. 187-298.
69. Krnjević, K. (1974): Physiol. Rev. <u>54</u>:418-540.
70. Kuba, K., Albuquerque, E.X., Daly, J. and Barnard, E.A. (1974): J. Pharmacol. Exp. Ther. <u>189</u>:499-512.
71. Laskowski, M.B. and Dettbarn, W-D. (1975): J. Pharmacol. Exp. Ther. <u>194</u>:351-361.
72. Laskowski, M.B. and Dettbarn, W-D. (1977): Ann. Rev. Pharmacol. Toxicol. <u>17</u>:387-409.
73. Libet, B., Kobayashi, H. and Tanaka, T. (1975): Nature <u>258</u>: 155-157.
74. Loy, R., Milner, T.A. and Moore, T.Y. (1980): Exper. Neurol. <u>67</u>:339-411.
75. Luria, A.R. (1973): <u>The Working Brain</u>, Basic Books, Inc., Publs., New York.
76. MacIntosh, F.C. (1977): In: <u>Cholinergic Mechanisms and Psychopharmacology</u> (ed) D.J. Jenden, Plenum Press, New York, pp. 297-323.
77. Mantione, C.R., Fisher, A. and Hanin, I. (1981): Science <u>213</u>: 579-580.
78. Marchbanks, R.M. (1981): In: <u>Cholinergic Mechanisms</u> (eds) G. Pepeu and H. Ladinsky, Plenum Press, New York, pp. 489-496.
79. McGeer, P.L., Eccles, J.C. and McGeer, E.G. (1978): <u>Molecular Neurobiology of the Mammalian Brain</u>, Plenum Press, New York.
80. McGeer, P.L., McGeer, E.G. and Peng, J.H. (1984): Life Sciences <u>34</u>:2319-2338.
81. Michell, R.H. (1975): Biochim. Biophys. Acta <u>A15</u>:81-147.
82. Myers, R.D. (1974): <u>Handbook of Drug and Chemical Stimulation of the Brain</u>, Reinhold, New York.
83. Millington, W.R. and Wurtman, R.J. (1982): J. Neurochem. <u>38</u>: 1748-1752.
84. Molgo, J., Lemeignan, M. and Lechat, P. (1975): C.R. hebd. Seanc. Acad. Sci. Paris <u>281</u>:1637-1639.
85. Morrow, A.L., Loy, R. and Creese, I. (1980): Proc. Natl. Acad. Sci. U.S.A. <u>80</u>:6718-6722.
86. Muralt, A.V. (1946): <u>Die Signalubermittlung im Nerven</u>, Verlag Birkhouser, Basel.
87. Nachmansohn, D. (1959): <u>Chemical and Molecular Basis of Nerve Activity</u>, Academic Press, New York.
88. Narahashi, T. (1974): Physiol. Rev. <u>54</u>:813-889.
89. Nishimura, T., Hasuo, H. and Karczmar, A.G. (1984): Fed. Proc. <u>43</u>:372.
90. Nordberg, A., Adolfsson, R., Marcusson, J. and Winblad, B. (1982): In: <u>The Aging Brain: Cellular and Molecular Mechanisms of Aging in the Nervous System</u>, (eds) E. Giacobini, G. Giacobini, G. Filogamo and A. Vernadakis, Raven Press, New York, pp. 231-245.
91. Ohta, Y. and Karczmar, A.G. (1980): The Pharmacologist <u>22</u>:181.

92. Ohta, Y. Karczmar, G.S. and Karczmar, A.G. (1981): Abstracts, Eighth Internl. Congr. Pharmacol., p. 646.

93. Pepeu, G. and Ladinsky, H. (eds) (1981): Cholinergic Mechanisms, Plenum Press, New York.

94. Perry E.K. and Perry, R.H. (1982): In: The Aging Brain, (ed) S. Hoyer, Springer-Verlag, Berlin, pp. 140-145.

95. Popper, K.R. and Eccles, J.C. (1977): The Self and Its Brain, Springer International, Berlin.

96. Putney, J.W., Jr., Weiss, S.J., Van de Valle, C.M. and Haddas, R.A. (1980): Nature 28:345-347.

97. Raftery, M.A., Bianca, M., Conti-Tronconi, S., Dunn, M.J., Crawford, R.D. and Middlemas, D. (1984): Fund. Appl. Toxicol. 4: S34-S51.

98. Richter, J.A. and Gormley, A. (1982): J. Pharmacol. Exper. Therap. 222:778-785.

99. Rosenberg, P. (1973): J. Pharm. Sci. 62:1552-1554.

100. Russell, R.W. (1977): In: Cholinergic Mechanisms and Psychopharmacology, (ed) D.J. Jenden, Plenum Press, New York, pp. 709-732.

101. Shute, L.L.D. and Lewis, P.R. (1967): Brain 90:497-520.

102. Simon, J.R., Atweh, S. and Kuhar, M.J. (1976): J. Neurochem. 26:909-922.

103. Singh, M.M. and Kay, S.R. (1976): World J. Psychosynth. 8:34-41.

104. Smith, D.O. (1984): J. Physiol. (Lond.) 347:161-176.

105. Soreq, H., Levin-Sonkin, D. and Razon, N. (in press): The EMBO J.

106. Szerb, J.C. (1977): In: Cholinergic Mechanisms and Psychopharmacology (ed) D.J. Jenden, Plenum Press, New York, pp. 49-60.

107. Terry, R.D. and Davies, P. (1980): Ann. Rev. Neurosci. 3:77-95.

108. Thesleff, S. and Quastel, D.M.J. (1965): Ann. Rev. Pharmacol. 5:263-284.

109. Thesleff, S. (1980): Neurosci. 5:1413-1419.

110. Tuček, S. (1983): In: Enzymes in the Nervous System, (ed) A. Lajtha, Plenum Press, New York, 4:219-248.

111. VanMeter, W.G., Karczmar, A.G. and Fiscus, R.R. (1978): Arch. Int. Pharmacodyn. 23:249-260.

112. Waser, P.G. (ed) (1975): Cholinergic Mechanisms, Raven Press, New York.

113. Wecker, L. and Goldberg, A.M. (1981): In: Cholinergic Mechanisms, (eds) G. Pepeu and H. Ladinsky, Plenum Press, New York, pp. 451-461.

114. Whittaker, V.P. (1977): In: Cholinergic Mechanisms and Psychopharmacology, (ed) D.J. Jenden, Plenum Press, New York, pp. 323-345.

115. Witkop, B. and Brossi, A. (1984): In: Natural Products and Drug Development, (eds) P. Krugsgaarch-Larsen, S. Brogger Christensen and H. Kofod, Munksgaard Publs., Copenhagen, pp. 283-300.

(Left to Right)

ROW 1: (Seated) John Blass, George B. Koelle, Peter G. Waser, Donald J. Jenden, Israel Hanin, Frank C. MacIntosh, Alexander G. Karczmar, Edith Heilbronn, Giancarlo Pepeu, Alan M. Goldberg

ROW 2: Victor J. Nickolson, Nae J. Dun, Stanley M. Parsons, Agneta Nordberg, Ezio Giacobini, B.V. Rama Sastry, Kathleen A. Sherman, Mario Marchi, Michael Stanley, Larry L. Butcher, Fiorella Casamenti, Tsung-Ming Shih, Herbert Ladinsky, Silvano Consolo, Kenneth L. Davis, Darwin L. Cheney, Janusz B. Suszkiw, Michael R. Kozlowski

ROW 3: Dean O. Smith, Steven H. Zeisel, Susan E. Robinson, Barbara Lerer, R. Jane Rylett, Rochelle D. Schwartz, Joan Heller Brown, Marie-Louise Tjörnhammar, Britta Hedlund, David S. Janowsky, Natraj Sitaram, Linda M. Barilaro, Paul M. Salvaterra, Denise Sorisio, Elias Aizenman, Ileana Pepeu, Aurora V. Revuelta, Felicita Pedata, Clementina Bianchi, Lorenzo Beani, Henry G. Mautner

ROW 4: S. Craig Risch, Guillermo Pilar, E. Sylvester Vizi, Thomas J. Walsh, Sikander L. Katyal, Rob L. Polak, Roni E. Arbogast, Jean Massoulié, Denes Agoston, Brian Collier, Lynn Wecker, Bruce Howard, Richard S. Jope, Bernard Scatton, Matthew Clancy, Paul T. Carroll

ROW 5: William G. VanMeter, Michael Adler, Peter Kasa, Annica B. Dahlström, Gary E. Gibson, Peter C. Molenaar, Ingrid Nordgren, John D. Catravas, Judith Richter, David M. Bowen, Mark Watson, Renato Corradetti, Lorenza Eder-Colli, Marvin Lawson, Ing K. Ho, Jack C. Waymire

ROW 6: Paul L. Wood, Matthew N. Levy, Jean-Claude Maire, Frans Flentge, Richard Dahlbom, Pierre Etienne, George G. Bierkamper, Robert G. Struble, A.J. Vergroessen, Seana O'Regan, Robert Manaranche, Maurice Israel, Yacov Ashani, Abraham Fisher, Steven Leventer, Alan G. Mallinger

ROW 7: Anders Undén, Edward F. Domino, William D. Blaker, Peteris Alberts, Johann Häggblad, Daniel L. Rickett, Sten-Magnus Aquilonius, Serge Mykita, Hans Selander, Oliver Brown, Henry Brezenoff, Sven-Åke Eckernäs, Frederick J. Ehlert, Björn Ringdahl, Volker Bigl, Duane Hilmas, Clark A. Briggs, Nicolas Morel

ROW 8: Bo Karlén, Michael J. Dowdall, John J. O'Neill, Heinz Kilbinger, Wolf-D. Dettbarn, Konrad J. Martin, Konrad Löffelholz, Roy D. Schwarz, Jerry J. Buccafusco, Ernst Wulfert, Howard J. Colhoun, Paul Martin, Jack R. Cooper, Christer Larsson, Harry M. Geyer, Michael J. Pontecorvo, William E. Houston, Jurgen von Bredow, Yves Dunant